REVIEWS IN MINERALOGY AND GEOCHEMISTRY

Volur

Carbo

EDITORS

Robert M. Hazen
Geophysical Laboratory, Carnegie Institution of Washington
Washington, DC 20015, U.S.A.

Adrian P. Jones
Earth Sciences, University College London
Gower Street, London WC1E 6BT, United Kingdom

John A. Baross
School of Oceanography and Astrobiology Program
University of Washington
Seattle, Washington 98195, U.S.A.

ON THE FRONT COVER: Earth as seen from space. Elektro-L weather satellite image collected May 2011 by the Russian Federal Space Agency Research Center for Earth Operative Monitoring (NTS OMZ). Image was processed by James Drake (*http://infinity-imagined.tumblr.com/*) and can be found at *http://planet--earth.ca.*

Inset images (top to bottom): Cartoon illustrating the range of deep-Earth processes that likely influence the long-term carbon cycle. See Figure 1 in Chapter 7 by Dasgupta. Predicted high-pressure orthorhombic structure of $CaCO_3$. See Figure 14b in Chapter 3 by Oganov et al. The biomineralized calcium carbonate eye lenses of the trilobite Erbenochile erbeni. See Figure 5 in Chapter 4 by Hazen et al. Carbonatite lava flow from an eruption of Oldoinyo Lengai, Tanzania. Photo courtesy of Tobias Fischer.

Series Editor: ***Jodi J. Rosso***

MINERALOGICAL SOCIETY OF AMERICA
GEOCHEMICAL SOCIETY

Reviews in Mineralogy and Geochemistry, Volume 75

Carbon in Earth

ISSN 1529-6466
ISBN 978-0-939950-90-4

THE MINERALOGICAL SOCIETY OF AMERICA
3635 CONCORDE PARKWAY, SUITE 500
CHANTILLY, VIRGINIA, 20151-1125, U.S.A.
WWW.MINSOCAM.ORG

Carbon in Earth

75 *Reviews in Mineralogy and Geochemistry* **75**

FROM THE SERIES EDITOR

Every RiMG volume is unique and special and, over the years, RiMG volumes have cumulatively covered the atomic to global scales for a variety of topics related to mineralogy and geochemistry. *Carbon in Earth* continues in this tradition. In *Carbon in Earth*, the reader will learn about the intricate aspects of atomic mineral crystal structures, models of carbon emissions from volcanoes, and how to probe the hidden secrets of carbon in the Earth's mantle and core, in addition to much, much more. It is an enjoyable read even for those not actively involved in this area of research. *Carbon in Earth* is the result of the monumental efforts by volume editors Bob Hazen, Adrian Jones, and John Baross, who deftly organized and wove together the diverse subjects covered in this volume. Gentlemen, a job well done.

The RiMG series has spanned several generations of publishing technologies since its inception almost 40 years ago. In its earliest days, handwritten manuscripts were scrupulously typed and typewriters were the cutting edge. For years, Paul Ribbe painstakingly cut, positioned, and pasted figures and tables onto paper. Thankfully, the late 80's and 90's brought the advent of personal computers and desktop publishing. The publishing world went digital and so did RiMG. Since 2003, RiMG has been produced electronically and volumes are now readily available in both print and electronic formats. This volume has brought the RiMG series into yet another new technological direction through incorporation of *interactive* digital elements. I hope you take the time to check it out! As an Open Access volume, any interested reader can easily find and access *Carbon in Earth* through the MSA website (*www.minsocam.org/MSA/RIM*).

All supplemental materials associated with this volume can be found at the MSA website. Errata will be posted there as well.

Jodi J. Rosso, Series Editor
West Richland, Washington
January 2013

PREFACE

Carbon in Earth is an outgrowth of the Deep Carbon Observatory (DCO), a 10-year international research effort dedicated to achieving transformational understanding of the chemical and biological roles of carbon in Earth (*http://dco.ciw.edu*). Hundreds of researchers from 6 continents, including all 51 coauthors of this volume, are now engaged in the DCO effort. We proposed this volume of the Reviews in Mineralogy and Geochemistry (RiMG) series as a benchmark for our present understanding of Earth's carbon—both what we know and what we have yet to learn. Our ambition is to produce a second, companion volume to mark the progress of this decadal initiative.

This volume addresses a range of questions that were articulated in May 2008 at the First Deep Carbon Cycle Workshop in Washington, DC. At that meeting 110 scientists from a

1529-6466/13/0075-0000$00.00 DOI: 10.2138/rmg.2013.75.0

dozen countries set forth the state of knowledge about Earth's carbon. They also debated the key opportunities and top objectives facing the community. Subsequent deep carbon meetings in Bejing, China (2010), Novosibirsk, Russia (2011), and Washington, DC (2012), as well as more than a dozen smaller workshops, expanded and refined the DCO's decadal goals. The 20 chapters that follow elaborate on those opportunities and objectives.

A striking characteristic of *Carbon in Earth* is the multidisciplinary scientific approach necessary to encompass this topic. The following chapters address such diverse aspects as the fundamental physics and chemistry of carbon at extreme conditions, the possible character of deep-Earth carbon-bearing minerals, the geodynamics of Earth's large-scale fluid fluxes, tectonic implications of diamond inclusions, geosynthesis of organic molecules and the origins of life, the changing carbon cycle through deep time, and the vast subsurface microbial biosphere (including the hidden deep viriosphere). Accordingly, the collective authorship of *Carbon in Earth* represents laboratory, field, and theoretical researchers from the full range of physical and biological sciences.

A hallmark of the DCO is our desire to implement advanced strategies in communications, data management, engagement, and visualization. Accordingly, this volume incorporates some novel aspects. Thanks to sponsorship by the Alfred P. Sloan Foundation, which continues to provide significant support for the DCO, this is the first of the RiMG series to be published as an open access volume. We have thus been able to focus on the electronic publication to incorporate a number of novel features, including hyperlinks to websites and databases, video and animations, and direct links to many references.

This effort has benefitted immeasurably from the contributions of numerous individuals. In particular we are grateful to Jesse Ausubel and his colleagues at the Alfred P. Sloan Foundation for generous support in the production and publication of this open access volume. We thank the many scientists who contributed thoughtful and constructive reviews of chapters, including J. Ausubel, J. Brodholt, G. Bulanova, J. Deming, D. Dobson, H. Downes, M. Guthrie, R. J. Hemley, O. Lord, A. Mangum, R. McDuff, C. Moyer, T. Phelps, S. Russell, C. M. Schiffries, M. Schulte, C. B. Smith, K. Stedman, D. A. Sverjensky, A. Templeton, and A. Woolley, as well as several anonymous reviewers. We are indebted to Lauren Cryan and Morgan Phillips, who provided critical technical assistance. J. Alex Speer of the Mineralogical Society of America offered key advice on the production of this open access volume. Finally, we are extremely indebted to RiMG Series Editor Jodi Rosso, who has worked tirelessly, with exceptional skill and thoughtful creativity, in producing this volume. Her continuing efforts to foster this extraordinary series are of immeasurable benefit to the international scientific community.

Robert M. Hazen
Geophysical Laboratory
Carnegie Institution of Washington

Adrian P. Jones
University College London

John A. Baross
University of Washington

Carbon in Earth

75 *Reviews in Mineralogy and Geochemistry* **75**

TABLE OF CONTENTS

1 Why Deep Carbon?

Robert M. Hazen, Craig M. Schiffries

2 Carbon Mineralogy and Crystal Chemistry

Robert M. Hazen, Robert T. Downs
Adrian P. Jones, Linda Kah

3 Structure, Bonding, and Mineralogy of Carbon at Extreme Conditions

Artem R. Oganov, Russell J. Hemley,
Robert M. Hazen, Adrian P. Jones

4 Carbon Mineral Evolution

Robert M. Hazen, Robert T. Downs, Linda Kah, Dimitri Sverjensky

5 The Chemistry of Carbon in Aqueous Fluids at Crustal and Upper-Mantle Conditions: Experimental and Theoretical Constraints

Craig E. Manning, Everett L. Shock, Dimitri A. Sverjensky

6 Primordial Origins of Earth's Carbon

Bernard Marty, Conel M. O'D. Alexander, Sean N. Raymond

7 Ingassing, Storage, and Outgassing of Terrestrial Carbon through Geologic Time

Rajdeep Dasgupta

8

Carbon in the Core: Its Influence on the Properties of Core and Mantle

Bernard J. Wood, Jie Li, Anat Shahar

9

Carbon in Silicate Melts

Huaiwei Ni, Hans Keppler

10 Carbonate Melts and Carbonatites

Adrian P. Jones, Matthew Genge
Laura Carmody

11 Deep Carbon Emissions from Volcanoes

Michael R. Burton,
Georgina M. Sawyer,
Domenico Granieri

12 Diamonds and the Geology of Mantle Carbon

Steven B. Shirey, Pierre Cartigny, Daniel J. Frost, Shantanu Keshav, Fabrizio Nestola, Paolo Nimis, D. Graham Pearson, Nikolai V. Sobolev, Michael J. Walter

13 Nanoprobes for Deep Carbon

Wendy L. Mao, Eglantine Boulard

14 On the Origins of Deep Hydrocarbons

Mark A. Sephton, Robert M. Hazen

15 Laboratory Simulations of Abiotic Hydrocarbon Formation in Earth's Deep Subsurface

Thomas M. McCollom

16 Hydrocarbon Behavior at Nanoscale Interfaces

David R. Cole, Salim Ok, Alberto Striolo, Anh Phan

17 Nature and Extent of the Deep Biosphere

Frederick S. Colwell, Steven D'Hondt

18 Serpentinization, Carbon, and Deep Life

Matthew O. Schrenk, William J. Brazelton, Susan Q. Lang

19 High-Pressure Biochemistry and Biophysics

Filip Meersman, Isabelle Daniel,
Douglas H. Bartlett, Roland Winter
Rachael Hazael, Paul F. McMillan

20 The Deep Viriosphere: Assessing the Viral Impact on Microbial Community Dynamics in the Deep Subsurface

Rika E. Anderson, William J. Brazelton,
John A. Baross

Reviews in Mineralogy & Geochemistry
Vol. 75 pp. 1-6, 2013

Why Deep Carbon?

Robert M. Hazen and Craig M. Schiffries

Geophysical Laboratory, Carnegie Institution of Washington
5251 Broad Branch Road NW
Washington, DC 20015, U.S.A.

rhazen@ciw.edu *cschiffries@ciw.edu*

All chemical elements are special, but some are more special than others. Of the 88 naturally occurring, long-lived elements on Earth, carbon stands alone. As the basis of all biomolecules, no other element contributes so centrally to the wellbeing and sustainability of life on Earth, including our human species. The near-surface carbon cycle profoundly affects Earth's changeable climate, the health of ecosystems, the availability of inexpensive energy, and the resilience of the environment. No other element plays a role in so diverse an array of useful solid, liquid, and gaseous materials: food and fuels; paints and dyes; paper and plastics; abrasives and lubricants; electrical conductors and insulators; thermal conductors and insulators; ultra-strong structural materials and ultra-soft textiles; and precious stones of unmatched beauty. No other element engages in such an extraordinary range of chemical bonding environments: with oxidation states ranging from −4 to +4, carbon bonds to itself and more than 80 other elements. Carbon's chemical behavior in Earth's hidden deep interior epitomizes the dynamic processes that set apart our planet from all other known worlds.

Past consideration of the global carbon cycle has focused primarily on the atmosphere, oceans, and shallow crustal environments. A tremendous amount is known about these parts of Earth's carbon cycle. By contrast, relatively little is known about the deep carbon cycle (Fig. 1). Knowledge of the deep interior, which may contain more than 90% of Earth's carbon (Javoy 1997), is limited (Table 1). Basic questions about deep carbon are poorly constrained:

- How much carbon is stored in Earth's deep interior?
- What are the reservoirs of that carbon?
- How does carbon move among reservoirs?
- Are there significant carbon fluxes between Earth's deep interior and the surface?
- What is the nature and extent of deep microbial life?
- Are there deep abiotic sources of methane and other hydrocarbons?
- Did deep organic chemistry play a role in life's origins?

Key unanswered questions guide research on carbon in Earth. Perhaps most fundamental are questions related to understanding the physical and chemical behavior of carbon-bearing phases at extreme conditions representative of the deep interiors of Earth and other planets. We need to make an inventory of possible C-bearing minerals at pressures and temperatures to hundreds of gigapascals and thousands of degrees and we must characterize the physical and thermochemical properties of those phases at relevant pressure-temperature conditions both experimentally and theoretically (Hazen et al. 2013a, Chapter 2; Oganov et al. 2013, Chapter 3). We need to investigate the diversity of carbon minerals through more than 4.5 billion years of Earth history (Hazen et al. 2013b, Chapter 4). We must also understand the behavior of

1529-6466/13/0075-0001$00.00 DOI: 10.2138/rmg.2013.75.1

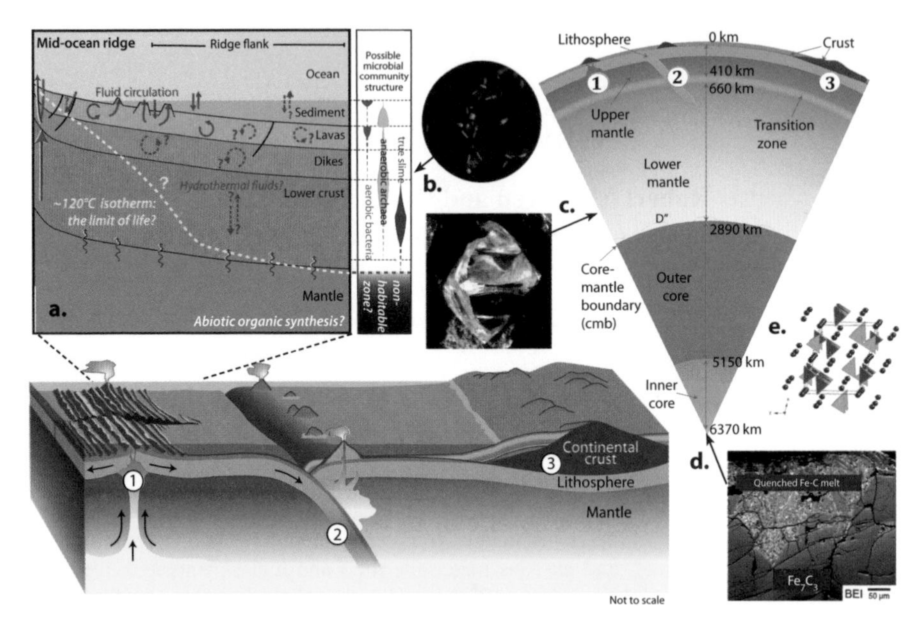

Figure 1. Earth's Deep Carbon Cycle. There is still much to learn about the cycling, behavior, and storage of Earth's deep carbon, from crust to core. Poorly quantified reservoirs of carbon include: (a) Microbial community structures on mid-ocean ridge flanks (adapted from images by Rosalind Coggon and Benoit Idlefonse). Other deep biosphere habitats affecting subsurface carbon cycling also exist. (b) An epifluorescence micrograph shows an iron-reducing enrichment culture from a serpentinite-hosted habitat. (image: Matt Schrenk). (c) Diamonds and their inclusions provide glimpses into Earth's deep interior (photo: U.S. Geological Survey). Theoretical and experimental studies allow us to speculate on carbon's role even deeper in Earth. (d) Theoretically, iron carbide (F_7C_3) is a potential constituent of Earth's solid inner core (image: Yoichi Nakajima). (e) Experimental studies suggest the existence of magnesium-iron carbon-bearing structures similar to phase II of magnesite at high pressures corresponding to depths greater than 1800 kilometers (image: Eglantine Boulard).

carbon-bearing fluids, and compile a comprehensive database of thermochemical properties and speciation of C-O-H-N fluids extending to upper mantle pressure and temperature conditions (Manning et al. 2013, Chapter 5).

The nature and extent of carbon reservoirs in Earth's deep interior is fundamental to understanding carbon in Earth. Primitive carbonaceous chondritic meteorites provide one proxy for the composition of the earliest Solar System and, by extension, Earth at the time of its formation (Marty et al. 2013, Chapter 6). However, the several weight percent carbon in those meteorites is one to two orders of magnitude greater than Earth's confirmed carbon reservoirs. What happened to the missing carbon? One key to answering this question is a comprehensive characterization of mantle carbon reservoirs and quantification of deep fluxes of carbon-bearing fluids to and from the mantle through Earth history—the essence of the deep carbon cycle (Dasgupta, 2013, Chapter 7). Current estimates of the carbon flux into Earth's mantle through tectonic subduction exceed by an order of magnitude the known carbon flux emitted by volcanoes—an untenable balance that would deplete surface reservoirs of carbon in significantly less than a billion years. Added to these uncertainties, we do not know the concentration, chemical bonding, or mineralogy of carbon in Earth's core (Wood et al. 2013, Chapter 8). Could trace amounts of carbon (a few parts per million) enter mantle silicate minerals or melts (Ni and Keppler 2013, Chapter 9)? These potential volumetrically large but diffuse carbon reservoirs contrast with such concentrated deep carbon phases as carbonate

Table 1. Possible deep carbon reservoirs

Reservoir	Composition	Structure	[C] (mole %)	Depth (km)	Abundance (wt %)
Diamond	C	diamond	100	>150	<< 1
Graphite	C	graphite	100	<150	<< 1
Carbides	SiC, FeC, Fe_3C	moissanite, cohenite	25-50	?	?
Carbonates	$(Ca,Mg,Fe)CO_3$	unknown	20	0 to ?	?
Metal	Fe,Ni	kamecite/awaurite	minor?	?	?
Silicates	Mg-Si-O	various	trace?	?	?
Oxides	Mg-Fe-O	various	trace?	?	?
Sulfides	Fe-S	various	trace?	?	?
Silicate melts	Mg-Si-O		trace?	?	?
CHON fluids	C-H-O-N		variable	?	?
Methane	CH_4		20	?	?
Methane clathrate	$[H_2O+CH_4]$	clathrate	variable	?	?
Hydrocarbons	C_nH_{2n+2}		variable	?	?
Organic species	C-H-O-N		variable	?	?
Deep life	C-H-O-N-P-S		variable	<15	?

Modified after Hazen et al. 2012.

magmas (Jones et al. 2013, Chapter 10) and diamonds (Shirey et al. 2013, Chapter 12), as well as volcanic emissions of carbon species (Burton et al. 2013, Chapter 11), which provide special insights to Earth's dynamic state and complex history.

Progress in all of these areas depends on the development of new instrumentation, including environmental chambers to access C-bearing samples in new regimes of *P-T* under controlled conditions (e.g., pH, f_{O_2}) and with increased sample volumes, as well as enhanced analytical facilities for investigating carbon-bearing samples at the nanometer scale (Mao and Boulard 2013, Chapter 13).

A central aspect of Earth's carbon cycle—one fundamentally tied to economic and environmental concerns—is the nature, sources, and evolution of subsurface organic molecules, including hydrocarbons and biomolecules. While a biologically mediated origin is postulated for most of Earth's so-called "fossil fuels," debates continue on the genesis of some deep hydrocarbons (Sephton and Hazen 2013, Chapter 14). To understand possible deep abiotic organic synthesis, including prebiotic processes that set the stage for life, we need to exploit experimental procedures to mimic deep hydrothermal geochemical environments (McCollom, Chapter 15). In this regard, a comprehensive understanding of diverse mineral-molecule nano-scale interactions at the fluid-rock interface is required (Cole et al. 2013, Chapter 16).

Any overview of carbon in Earth must assess the fascinating role of deep subsurface microbial life, which, though sparse, is widespread and thus may represent a significant fraction of Earth's total biomass. In our efforts to assess the nature and extent of the deep microbial biosphere, we are conducting a global 3-D census of deep microbial life, including both terrestrial and marine ecosystems (Colwell and D'Hondt 2013, Chapter 17). Central to any discussion of deep biology is energy flow, as exemplified by zones of serpentinization—zones of intense hydrogen production that possibly represent the oldest ecosystem on Earth (Schrenk et al. 2013, Chapter 18). Any understanding of the nature and survival of deep life entails investigations of biomolecular adaptations under extreme conditions, which in turn necessitates a new laboratory-based approach to studying life at extreme pressures and temperatures (Meersman et al. 2013, Chapter 19). Most enigmatic of all are deep viruses, which promote rapid microbial turnover rates and lateral gene transfer and thus may play the dominant role in evolution of subsurface ecosystems (Anderson et al. 2013, Chapter 20).

FRONTIERS OF DEEP CARBON RESEARCH

Carbon in Earth represents a synthesis of a diverse body of research in physics, chemistry, biology, and the Earth and space sciences. The richness of this collection points to the potential for new discoveries, as the findings in one scientific domain often inform those in another seemingly unrelated field. (Indeed, one colleague has suggested that we should have the authors read the entire volume and then start writing their chapters over again!)

One example illustrates this kind of potential cross-fertilization. In Chapter 4, Hazen et al. (2013) point to dramatic changes in Earth's near-surface carbon mineralogy through more than 4 billion years of Earth history. The innovation of carbonate biomineralization and the rise of the terrestrial biosphere, in particular, have increased the crustal diversity, volume, and distribution of carbon-bearing minerals and other substances. In Chapter 3, Oganov et al (2013) consider high-pressure carbon-bearing minerals, including those in Earth's mantle and core. In Chapter 7, Dasgupta (2013) argues that the upper mantle geotherm of the past billion years is sufficiently cool for subducted carbonates to remain largely sequestered in the mantle—a situation that contrasts with prior eons of Earth history. And in Chapter 11, Burton et al. (2013) catalog all known volcanic emissions of carbon and conclude that these varied sources collectively represent only a small fraction of the carbon that is being subducted. Taken together, these observations suggest a possibly dramatic Phanerozoic increase in the amount of subducted carbon that remains sequestered in the mantle—a trend that could lead to significant depletion of crustal carbon, and thus adverse biological consequences, within a few hundred million years. Further research is needed to evaluate the extent to which life might contribute to its own demise by contributing to a net transfer of carbon from the crust to the mantle. Similar insights await the thoughtful and diligent reader of this volume.

Carbon in Earth, though extensive, is by no means encyclopedic and significant swaths of Earth's carbon story are missing from these pages. For example, we all but neglect the nature and origins of coal—one of the crust's most concentrated and extensive carbon reservoirs (see, however, Manning et al. 2013). In spite of a once-thriving research community (Van Krevelen 1993; Davidson 2004; and references therein), and coal's continuing economic importance and environmental implications, research on this fascinating substance has all but ceased. Details of coal's structural chemistry and complex maturation processes remain obscure, and thus represent a promising potential direction for future research on carbon in Earth.

Similarly missing is a comprehensive treatment of non-volcanic fluxes of deep carbon, including deep crustal sources and sinks. On the one hand, significant CO_2 is released through regional metamorphism of carbonate-rich sedimentary sequences (Bowen 1940; Ferry 1992; Kerrick and Caldeira 1993, 1998; Bickle 1996; Ague 2000, 2012). On the other hand, retrograde metamorphism may be a sink for CO_2. Research on retrograde metamorphism is key for understanding whether or not fluids transport significant CO_2 from the deep crust to the shallow hydrosphere and atmosphere. In some deep crustal zones CO_2-rich fluids may drive dehydration reactions—the process of charnockitization—and contribute to the stabilization of continental cratons (Janardhanan et al. 1979; Newton et al. 1980). The net effects of these metamorphic processes on the carbon cycle today, not to mention their variations through Earth history, are not well understood.

Several chapters in this volume allude to the central roles played by plate tectonics in Earth's carbon cycle, notably in subduction zones (e.g., Daspgupta 2013) and back-arc and ridge volcanism (Burton et al. 2013; Jones et al. 2013). Nevertheless, important aspects of large-scale geodynamics and geophysical modeling are lacking from *Carbon in Earth*. For example, although Shirey et al. (2013) present remarkable data on diamond inclusions that may point to the start of modern-style lateral tectonics and the Wilson cycle approximately 3 billion years ago, subsequent effects of the "supercontinent cycle" on the carbon cycle through deep time are not well understood. Nor do our models of carbon cycling yet incorporate the impacts of episodic megavolcanism associated with massive flood basalts—events that correlate with several intervals of Phanerozoic mass extinctions, and thus must have affected the distribution of crustal carbon. And we have yet to bring to bear the full geophysical arsenal of techniques to probe the nature and distribution of carbon-bearing solid and fluids in Earth's deep interior.

Other telling gaps in our knowledge of deep carbon relate to the fascinating deep biosphere. Remarkable discoveries of subsurface microbial life (Colwell and D'Hondt 2013) and associated viruses (Anderson et al. 2013) hint at a surprising hidden diversity, primarily within the microbial domains of archaea and bacteria. However, as Colwell and D'Hondt (2013) point out, surprising discoveries of a rich subsurface community of eukayotes promises an even richer deep taxonomy (Monastersky 2012; Orsi et al. 2012). These findings, coupled with advances in single-cell genomics (Stepanauskas and Sieracki 2007; Stepanauskas et al. 2012), predict a coming decade of extraordinary discovery.

At this stage in our pursuit of a fundamental understanding of carbon in Earth, our knowledge baseline as set forth in the following chapters is significant, but the unanswered questions—what we know we don't know—far outweigh what we know. That's an energizing state for any scientific pursuit, for the unknown is what beckons us to the laboratory, the field, and the computer.

And yet, as we embark on this decadal quest, the greatest lure is not so much the assurance that we will fill in many gaps in what we know we don't know, but rather that we will discover entirely new, unanticipated phenomena. Even though we are unable to articulate what those discoveries might be, we can be assured that such adventures lie in wait for the curious and prepared mind.

ACKNOWLEDGMENTS

We thank members of the Deep Carbon Observatory (DCO) Executive Committee, International Science Advisory Committee, and Secretariat for contributions to this manuscript. We are grateful to Andrea Mangum, who designed and executed the figure. The authors acknowledge generous support from the Deep Carbon Observatory, the Alfred P. Sloan Foundation, and the Carnegie Institution of Washington.

REFERENCES

Ague JJ (2000) Release of CO_2 from carbonate rocks during regional metamorphism of lithologically heterogeneous crust. Geology 28:1123-26

Ague JJ (2012) Deep crustal metamorphic carbon cycling in collisional orogens. What do we really know? Am Geophys Union Annual Meeting, abstract V41C-01

Anderson RE, Brazelton WJ, Baross JA (2013) The deep viriosphere: assessing the viral impact on microbial community dynamics in the deep subsurface. Rev Mineral Geochem 75:649-675

Bickle MJ (1996) Metamorphic decarbonation, silicate weathering and the long-term carbon cycle. Terra Nova 8:270-276

Bowen NL (1940) Progressive metamorphism of siliceous limestone and dolomite. J Geol 48:225-274

Burton MR, Sawyer GM, Granieri D (2013) Deep carbon emissions from volcanoes. Rev Mineral Geochem 75:323-354

Cole DR, Ok S, Striolo A, Phan A (2013) Hydrocarbon behavior at nanoscale interfaces. Rev Mineral Geochem 75:495-545

Colwell FS, D'Hondt S (2013) Nature and extent of the deep biosphere. Rev Mineral Geochem 75:547-574

Dasgupta R (2013) Ingassing, storage, and outgassing of terrestrial carbon through geologic time. Rev Mineral Geochem 75:183-229

Davidson RM (2004) Studying the Structural Chemistry of Coal. IEA Coal Research, London, UK

Ferry JM (1992) Regional metamorphism of the Waits River Formation, eastern Vermont: delineation of a new type of giant metamorphic hydrothermal system: J Petrol 33:45-94

Hazen RM, Downs RT, Jones AP, Kah L (2013a) Carbon mineralogy and crystal chemistry. Rev Mineral Geochem 75:7-46

Hazen RM, Downs RT, Kah, L, Sverjensky D (2013b) Carbon mineral evolution. Rev Mineral Geochem 75:79-107

Hazen RM, Hemley RJ, Mangum AJ (2012) Carbon in Earth's interior: Storage, cycling, and life. Eos Trans Am Geophys Union 93:17-18

Janardhanan S, Newton RC, Smith JV (1979) Ancient crustal metamorphism at low P_{H_2O}: Charnockite formation from Kabbaldurga, south India. Nature 278:511-514

Javoy M (1997) The major volatile elements of the Earth: Their origin, behavior, and fate, Geophys Res Lett 24:177-180. DOI: 10.1029/96GL03931

Jones AP, Genge M, Carmody L (2013) Carbonate melts and carbonatites. Rev Mineral Geochem 75:289-322

Kerrick DM, Caldeira K (1993) Paleoatmospheric consequences of CO_2 released during early Cenozoic regional metamorphism in the tethyan orogen. Chem Geol 108:201-230

Kerrick DM, Caldeira K (1998) Metamorphic CO_2 degassing from orogenic belts. Chem Geol 145:213-232

Manning CE, Shock EL, Sverjensky D (2013) The chemistry of carbon in aqueous fluids at crustal and upper-mantle conditions: experimental and theoretical constraints. Rev Mineral Geochem 75:109-148

Mao WL, Boulard E (2013) Nanoprobes for deep carbon. Rev Mineral Geochem 75:423-448

Marty B, Alexander CMO'D, Raymond SN (2013) Primordial origins of Earth's carbon. Rev Mineral Geochem 75:149-181

McCollom TM (2013) Laboratory simulations of abiotic hydrocarbon formation in Earth's deep subsurface. Rev Mineral Geochem 75:467-494

Meersman F, Daniel I, Bartlett DH, Winter R, Hazael R, McMillain PF (2013) High-pressure biochemistry and biophysics. Rev Mineral Geochem 75:607-648

Monastersky R (2012) Ancient fungi found in deep-sea mud. Nature 492:163

Newton RC, Smith JV, Windley BF (1980) Carbonic metamorphism, granulites and crustal growth. Nature 288:45-50

Ni H, Keppler H (2013) Carbon in silicate melts. Rev Mineral Geochem 75:251-287

Oganov AR, Hemley RJ, Hazen RM, Jones AP (2013) Structure, bonding, and mineralogy of carbon at extreme conditions. Rev Mineral Geochem 75:47-77

Orsi W, Biddle J, Edgcomb V (2012) Active fungi amidst a marine subsurface RNA paleome. Am Geophys Union Annual Meeting, abstract B41F-02

Schrenk MO, Brazelton WJ, Lang SQ (2013) Serpentinization, carbon, and deep life. Rev Mineral Geochem 75:575-606

Sephton MA, Hazen RM (2013) On the origins of deep hydrocarbons. Rev Mineral Geochem 75:449-465

Shirey SB, Cartigny P, Frost DJ, Keshav S, Nestola F, Nimis P, Pearson DG, Sobolev NV, Walter MJ (2013) Diamonds and the geology of mantle carbon. Rev Mineral Geochem 75:355-421

Stepanauskas R, Onstott TC, Lau CY, Kieft TL, Woyke T, Rinke C, Sczyrba A; van Heerden E (2012) Single cell genomics of subsurface microorganisms. Am Geophys Union Annual Meeting, abstract B41F-01

Stepanauskas R, Sieracki ME (2007) Matching phylogeny and metabolism in the uncultured marine bacteria, one cell at a time. Proc Natl Acad Sci USA 104:9052-9057

Van Krevelen DW (1993) Coal: Typology, Physics, Chemistry, Constitution. 3rd edition. Elsevier, Amsterdam

Wood BJ, Li J, Shahar A (2013) Carbon in the core: its influence on the properties of core and mantle. Rev Mineral Geochem 75:231-250

Reviews in Mineralogy & Geochemistry
Vol. 75 pp. 7-46, 2013

Carbon Mineralogy and Crystal Chemistry

Robert M. Hazen

Geophysical Laboratory, Carnegie Institution of Washington
5251 Broad Branch Road NW
Washington, DC 20015, U.S.A.

rhazen@ciw.edu

Robert T. Downs

Department of Geosciences, University of Arizona
1040 East 4th Street
Tucson, Arizona 85721-0077, U.S.A.

rdowns@u.arizona.edu

Adrian P. Jones

Earth Sciences, University College London
Gower Street
London WC1E 6BT, United Kingdom

adrian.jones@ucl.ac.uk

Linda Kah

Department of Earth & Planetary Sciences
University of Tennessee
Knoxville, Tennessee 37996-4503, U.S.A.

lckah@utk.edu

INTRODUCTION

Carbon, element 6, displays remarkable chemical flexibility and thus is unique in the diversity of its mineralogical roles. Carbon has the ability to bond to itself and to more than 80 other elements in a variety of bonding topologies, most commonly in 2-, 3-, and 4-coordination. With oxidation numbers ranging from −4 to +4, carbon is observed to behave as a cation, as an anion, and as a neutral species in phases with an astonishing range of crystal structures, chemical bonding, and physical and chemical properties. This versatile element concentrates in dozens of different Earth repositories, from the atmosphere and oceans to the crust, mantle, and core, including solids, liquids, and gases as both a major and trace element (Holland 1984; Berner 2004; Hazen et al. 2012). Therefore, any comprehensive survey of carbon in Earth must consider the broad range of carbon-bearing phases.

The objective of this chapter is to review the mineralogy and crystal chemistry of carbon, with a focus primarily on phases in which carbon is an essential element: most notably the polymorphs of carbon, the carbides, and the carbonates. The possible role of trace carbon in nominally acarbonaceous silicates and oxides, though potentially a large and undocumented reservoir of the mantle and core (Wood 1993; Jana and Walker 1997; Freund et al. 2001; McDonough 2003; Keppler et al. 2003; Shcheka et al. 2006; Dasgupta 2013; Ni and Keppler 2013; Wood et al. 2013), is not considered here. Non-mineralogical carbon-bearing phases

1529-6466/13/0075-0002$00.00 DOI: 10.2138/rmg.2013.75.2

treated elsewhere, including in this volume, include C-O-H-N aqueous fluids (Javoy 1997; Zhang and Duan 2009; Jones et al. 2013; Manning et al. 2013); silicate melts (Dasgupta et al. 2007; Dasgupta 2013; Manning et al. 2013); carbonate melts (Cox 1980; Kramers et al. 1981; Wilson and Head 2007; Walter et al. 2008; Jones et al. 2013); a rich variety of organic molecules, including methane and higher hydrocarbons (McCollom and Simoneit 1999; Kenney et al. 2001; Kutcherov et al. 2002; Sherwood-Lollar et al. 2002; Scott et al. 2004; Helgeson et al. 2009; McCollom 2013; Sephton and Hazen 2013); and subsurface microbial life (Parkes et al. 1993; Gold 1999; Chapelle et al. 2002; D'Hondt et al. 2004; Roussel et al. 2008; Colwell and D'Hondt 2013; Schrenk et al. 2013; Meersman et al. 2013; Anderson et al. 2013).

The International Mineralogical Association (IMA) recognizes more than 380 carbon-bearing minerals (*http://rruff.info/ima/*), including carbon polymorphs, carbides, carbonates, and a variety of minerals that incorporate organic carbon in the form of molecular crystals, organic anions, or clathrates. This chapter reviews systematically carbon mineralogy and crystal chemistry, with a focus on those phases most likely to play a role in the crust. Additional high-temperature and high-pressure carbon-bearing minerals that may play a role in the mantle and core are considered in the next chapter on deep carbon mineralogy (Oganov et al. 2013).

SYSTEMATIC CARBON MINERALOGY

Carbon, a non-metal that typically forms covalent bonds with a variety of other elements, is the most chemically adaptable element of the periodic table. In an ionic sense, element 6 can act as a cation with oxidation number +4, as in carbon dioxide (CO_2) or in the carbonate anion (CO_3^{-2}). Alternatively, carbon can act as an anion with oxidation number as low as −4, as in methane (CH_4) and in other alkanes. Carbon also frequently displays a range of intermediate oxidation number states from +2 in carbon monoxide (CO) or −2 in methanol (CH_3OH), as well as occurring in its neutral state (C) in a variety of carbon allotropes.

Carbon chemistry is also enriched by the ability of C to form single, double, or triple bonds, both with itself and with a wide range of other chemical elements. In many carbon compounds each C atom bonds to 4 other atoms by single bonds, as in methane and carbon tetrafluoride (CF_4). But carbon commonly forms double bonds with itself, for example in ethene ($H_2C{=}CH_2$), oxygen in carbon dioxide (O=C=O), or sulfur in carbon disulfide (S=C=S), and it can form triple bonds with itself, as in ethylene (commonly known as acetylene; HC≡CH), or with nitrogen, as in hydrogen cyanide (HC≡N). Carbon's remarkably diverse mineralogy arises in part from this unmatched range of valence states and bond types.

In the following sections we review the systematic mineralogy of carbon, including carbon allotropes, carbides, carbonates, and minerals that incorporate organic carbon molecules.

Carbon allotropes

The element carbon occurs in several allotropes, including graphite, diamond, lonsdaleite, fullerenes (including buckyballs and carbon nanotubes), graphene, and several non-crystalline forms (Table 1). These varied carbon allotropes exhibit extremes in physical and chemical properties—variations that reflect differences in their atomic structures. Diamond and lonsdaleite are the hardest known substances, whereas graphite is among the softest. Transparent diamond is an exceptional electrical insulator while possessing the highest known thermal conductivity at room temperature, whereas opaque graphite is an electrical conductor and thermal insulator. Isotropic diamond is widely used as a tough abrasive, whereas anisotropic graphite is employed as a lubricant. Diamond possesses the highest nuclear density of any known condensed phase at ambient conditions (Zhu et al. 2011) and is predicted to have high-pressure phases of even greater density (Oganov et al. 2013), whereas some carbon nanogels feature among the lowest known nuclear densities.

Table 1. Allotropes of native carbon.

Name	Structure Description	Reference
Graphite	Hexagonal; stacked flat layers of 3-coordinated sp^2 C	Klein and Hurlbut (1993)
Diamond	Cubic; framework of 4-coordinated sp^3 C	Harlow (1998)
Lonsdaleite	Hexagonal; framework of 4-coordinated sp^3 C	Frondel and Marvin (1967a); Bundy (1967)
Chaoite (also "Ceraphite")	A disputed form of shocked graphitic C from impact sites	El Goresy and Donnay (1968); McKee (1973)
Fullerenes		
Buckyballs	Closed cage molecules sp^2 C: C_{60}, C_{70}, C_{76}, etc.	Kroto et al. (1985)
Nano onions	Multiple nested buckyballs with closed cages	Sano et al. (2001)
Nanotubes	Cylindrical fibers of sp^2 C, single tubes or nested	Belluci (2005)
Nanobuds	Buckyballs covalently bonded to the exteriors of nanotubes	Nasibulin et al. (2007)
Buckyball and chain	Two buckyballs linked by a carbon chain	Shvartsburg et al. (1999)
Graphene	One-atom-thick graphitic layers with sp^2 bonding	Geim and Novoselov (2007)
Carbon Nanofibers (also known as "Graphite Whiskers")	Graphene layers arranged as cones, cups, or plates	Hatano et al. (1985); Morgan (2005)
Amorphous Carbon	Non-crystalline carbon, with disordered sp^2 and sp^3 bonds. Sometimes inaccurately applied to soot and coal.	Rundel (2001)
Glassy Carbon (also "Vitreous Carbon")	Dense, hard, non-crystalline 3-dimensional arrangement of sp^3 bonds, possibly related to a fullerene structure	Cowlard and Lewis (1967)
Carbon Nanofoam (also "Carbon Aerogel")	Nano-clusters of carbon in 6- and 7-rings, forming a low-density three-dimensional web.	Rode et al. (2000)
Linear Acetylenic Carbon (also known as Carbyne)	Linear molecule of acetylenic C (…C=C=C=C…)	Heimann et al. (1999)

Here we focus on the essential characteristics of the three most common minerals of native carbon—graphite, diamond, and lonsdaleite—all of which play roles in Earth's subsurface carbon cycle. For more comprehensive reviews of the chemical and physical properties of these carbon polymorphs see Bragg et al. (1965), Deer et al. (1966), Field (1979), Davies (1984), Klein and Hurlbut (1993), Harlow (1998), and Zaitsev (2001).

Graphite. Carbon forms covalent bonds with itself in all of the carbon polymorphs, in which C adopts one of two coordination environments (Table 1). In graphite, graphene, buckyballs, nanotubes, and several types of amorphous and glassy carbon, the carbon atoms are in planar three-coordination. This trigonal coordination, with typical C-C distances of ~1.42 Å and C-C-C angles close to 120°, results from the hybridization of carbon's electrons into three orbitals, known as the sp^2 bonding configuration because the $2s$ orbital mixes (or hybridizes) with two $2p$ orbitals to form three sp^2 orbitals.

The hexagonal layered structure of graphite (Fig. 1a; Animation 1) features electrically neutral, monoatomic, flat carbon layers, which bond to each other through van der Waals attractions and which are separated from each other by 3.41 Å. This layered structure leads to the distinctive, highly anisotropic properties of graphite (Fig. 1b). Individual carbon layers, which are known as graphene when meticulously separated from a graphite crystal or vapor deposited, are extremely strong and display unique electronic and mechanical properties (Geim and Novoselov 2007; Geim 2009). Weak van der Waal's forces between these layers lead to graphite's applications in lubricants and as pencil "lead."

Figure 1. Graphite. (a) The graphite (native C) crystal structure (hexagonal, space group $P6_3/mmc$; a = 2.464 Å; c = 6.736 Å; Z = 4) incorporates monoatomic layers of 3-coordinated carbon, linked by van der Waals interactions. (b) Natural graphite crystals reflect the hexagonal crystal structure. This 1 mm diameter crystal displays basal pinacoid and prism faces. The crystal, associated with calcite, is from the Crestmore quarries, Riverside Co., California. Photo courtesy of John A. Jaszczak. [Animation 1: *For readers of the electronic version, click the image for an animation of the graphite crystal structure.*]

Natural graphite is a common crustal mineral that occurs most abundantly in metamorphic rocks in pods and veins as a consequence of the reduction and dehydration of sediments rich in carbon (e.g., Klein and Hurlbut 1993). Commercial metamorphic graphite deposits, notably from China, India, and Brazil, are exploited in the manufacture of pencils, lubricants, steel, and brake linings. Graphite is also found as an accessory mineral in some igneous rocks and as a micro- or nano-phase in a variety of meteorites. In these diverse lithologies graphite is commonly precipitated in veins from reduced C-O-H fluids (Rumble and Hoering 1986; Rumble et al. 1986). In addition, graphite can crystallize from the vapor state in the carbon-rich, expanding hot envelopes of energetic late-stage stars, including supernovas.

Diamond and lonsdaleite. In contrast to the sp^2 bonding environment of carbon in graphite, each carbon atom can bind to 4 adjacent C atoms in tetrahedral coordination, as exemplified by the diamond and lonsdaleite polymorphs (Figs. 2 and 3; Animations 2 and 3). The C-C distance in these minerals are ~1.54 Å, while C-C-C angles are close to the ideal tetrahedral value of 109.5°. This tetrahedral bonding configuration reflects the hybridization of one 2*s* and three 2*p* orbitals from each carbon atom to form four sp^3 orbitals.

The structures of cubic diamond and hexagonal lonsdaleite are similar: both forms of carbon feature tetrahedral coordination of C in a three-dimensional framework. However, given a flat layer of linked carbon tetrahedra with all vertices pointed in the same direction (layer A), there exist two ways to stack subsequent layers, in orientations described as B or C. The diamond structure represents a three-layer stacking sequence of [...ABCABC...] along the (111) cubic direction, whereas lonsdaleite stacking is two-layer [...ABAB...] along the (001) hexagonal direction. A similar stacking difference distinguishes the cubic and hexagonal forms of a number of topologically similar mineral pairs, such as the polymorphs of zinc sulfide (ZnS), sphalerite and wurtzite, respectively. Note also that the stacking sequence of these carbon polymorphs can incorporate errors, as is often the case with lonsdaleite that has formed by impact shock of graphite (Frondel and Marvin 1967).

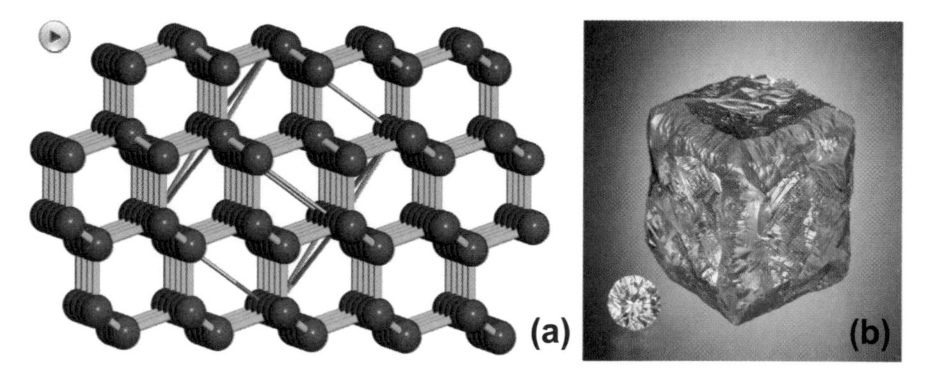

Figure 2. (a) The diamond (native C) crystal structure (cubic, space group $Fd\bar{3}m$; a = 3.560 Å; Z = 8) features a framework of tetrahedrally coordinated carbon atoms. (b) A natural diamond crystal reflects the cubic crystal structure. The semi-translucent diamond cube (ref. no. S014632) is 24.3 × 21.8 × 21.7 mm in size and weighs 156.381 carats (31.3 gm) and is shown with a 1-carat diamond for scale. It is represented to be from Ghana. Photo courtesy of Harold and Erica Van Pelt. [Animation 2: *For readers of the electronic version, click the image for an animation of the diamond crystal structure.*]

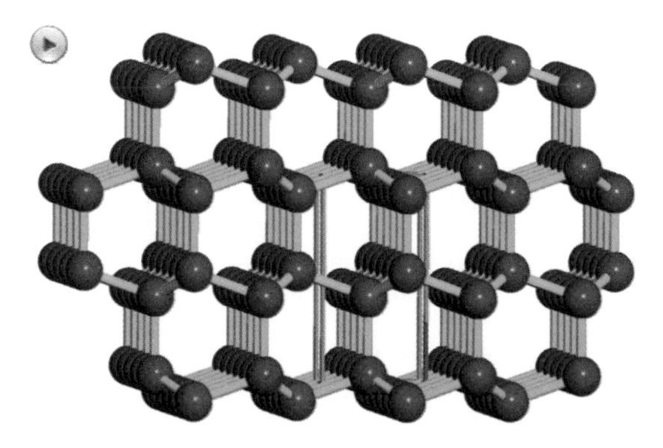

Figure 3. The lonsdaleite (native C) crystal structure (hexagonal, space group $P6_3/mmc$; a = 2.52 Å; c = 4.12 Å; Z = 4). [Animation 3: *For readers of the electronic version, click the image for an animation of the lonsdaleite crystal structure.*].

The framework structures of diamond and lonsdaleite lead to their superlative physical and chemical properties (Davies 1984). Notably, the exceptional hardness and strength of these phases arises from the strong three-dimensional network of C-C bonds. Recent theoretical studies on diamond and lonsdaleite suggest that diamond possesses the greater strength (Pan et al. 2009; Lyakhov and Oganov 2011).

The key to understanding the contrasting properties of the natural carbon allotropes is their different pressure-temperature stability fields. The deep origin of diamond was first recognized in the years following the widely publicized discovery of the 20-carat Eureka diamond by children playing in a dry central South African streambed in 1866, and the even more dramatic 83.5-carat Star of South Africa diamond two years later. A subsequent diamond rush brought more than 10,000 prospectors to the semi-desert region, and inevitably led to the recognition of diamond in their volcanic host rock, dubbed kimberlite (Davies 1984; Hazen 1999). The South African kimberlites' cone-shaped deposits, which cut vertically through shattered country rock, spoke of violent explosive eruptions from great depth (Lewis 1887; Bergman 1987; Mitchell 1995).

The hypothesis that diamond comes from depth was reinforced by determination of the crystal structures of diamond versus graphite (Bragg and Bragg 1913; Hull 1917; Hassel and Mark 1924; Bernal 1924). The higher coordination number of carbon in diamond (4-fold) compared to graphite (3-fold), coupled with its much greater density (3.51 vs. 2.23 g/cm^3), provided physical proof that diamond was the higher-pressure polymorph. Early determinations of the carbon phase diagram (Rossini and Jessup 1938), though refined in subsequent years (Bundy et al. 1961; Kennedy and Kennedy 1976; Day 2012; Fig. 4), revealed that diamond is the higher-pressure, lower-temperature form and that, under a normal continental geotherm, diamond likely forms at depths greater than 100 kilometers. By contrast, the steeper geothermal gradients of the oceanic crust and mantle preclude diamond formation, at least within the sub-oceanic upper mantle and transition zone.

Efforts to synthesize diamond under extreme laboratory conditions extend back to the early 1800s, long before the crucial role of high-pressure was recognized (Mellor 1924;

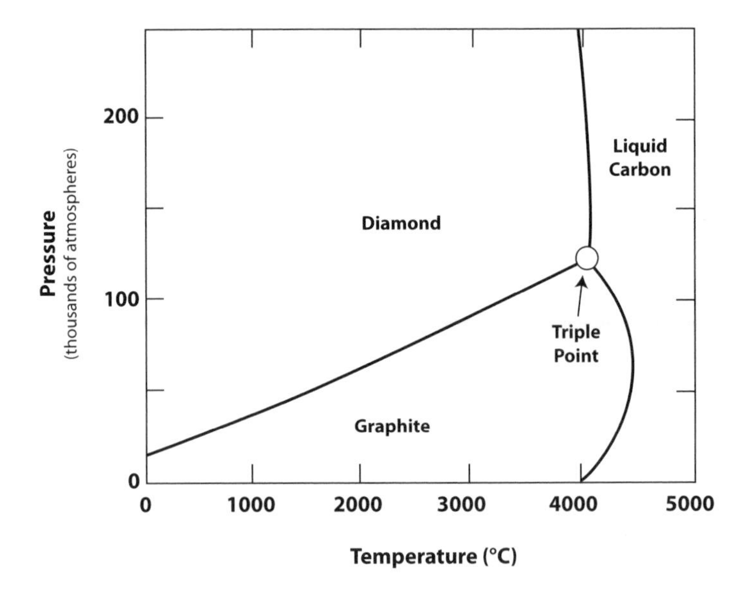

Figure 4. The carbon phase diagram (after Hazen 1999).

Hazen 1999). Among the most renowned 19th-century chemists to try his hand at diamond making was Frederick-Henri Moissan, who won the Nobel Prize for his risky isolation of the dangerous element fluorine. Moissan employed a novel electric arc furnace to generate record temperatures ~3000 °C. Moissan initially thought he was successful in synthesizing diamond but is now known to have formed hard, transparent crystals of silicon carbide—a compound that he also discovered in nature, and what is now known as the mineral moissanite (Moissan 1904a; Hazen 1999; see below). Numerous other heroic efforts prior to 1950 also failed (Bridgman 1931, 1946; von Platen 1962; Coes 1962). In spite of the relatively accessible pressure-temperature regime of diamond stability > 4 GPa (approximately 40,000 atm or 40 kbar), the transition from graphite to diamond is kinetically inhibited. The keys to facile diamond synthesis—employing a liquid metal flux coupled with sustained pressures above 5 GPa and temperatures above 1200 °C—were not achieved until the post-World War II efforts by scientists at the General Electric Research Laboratory in Schenectady, New York (Bundy et al. 1955; Suits 1960, 1965; Wentorf 1962; Hall 1970; Strong 1989; Hazen 1999). The breakthrough of high-pressure diamond synthesis by Francis Bundy, Tracy Hall, Herbert Strong, and Robert Wentorf in December of 1954 led to what is today a multi-billion dollar industry that supplies annually hundreds of tons of industrial diamond abrasives.

Two other synthesis techniques have expanded the varieties of diamond available for commercial exploitation. Of special interest are efforts to generate diamond and lonsdaleite under shock conditions that mimic bolide impact events (DeCarli and Jamieson 1961; Beard 1988). In some meteorites lonsdaleite and/or diamond is found to replace graphite in crystallites that retain a hexagonal shape (Langenhorst et al. 1999; El Goresy et al. 2001; Langenhorst and Deutsch 2012). In this rapid, solid-state martensitic transition the flat, graphitic sp^2 planes of carbon atoms shift relative to each other and buckle to produce a sp^3 array of C layers that are rather disordered in their stacking arrangement.

The subsequent discovery of techniques for diamond and lonsdaleite synthesis by vapor deposition at low-pressure conditions that mimic diamond formation in expanding stellar envelopes has greatly increased the potential for diamond use in science and industry (Angus et al. 1968; Derjaguin and Fedoseev 1968; Angus and Hayman 1988; Spear and Dismukes 1994; Irifune and Hemley 2012). These varied efforts in diamond synthesis have produced exceptional new materials, including isotopically pure diamonds with the highest recorded thermal conductivity, semiconducting diamonds, nano-crystalline polishing powders, lonsdaleite crystals that are harder than many natural diamond, and a range of deeply colored flawless synthetic gemstones up to 10 carats (Liang et al. 2009; Meng et al. 2012).

Intense research on natural diamond is also providing important insights regarding Earth's geochemical and tectonic evolution. A number of recent studies focus on diamond's rich and revealing suites of oxide, silicate, carbide, and sulfide inclusions from depths of up to perhaps 850 kilometers—mantle samples that provide evidence for aspects of geochemical and tectonic evolution over more than 3 billion years of Earth history (Shirey et al. 2002; McCammon et al. 2004; Sommer et al. 2007; Pearson et al. 2007; Gübelin and Koivula 2008; Shirey and Richardson 2011; Walter et al. 2011; Shirey et al. 2013).

Carbides

Carbides, which form when carbon bonds to a less electronegative element, are represented by dozens of synthetic compounds that have a range of industrial applications (Ettmayer and Lengauer 1994). The International Mineralogical Association has recognized 10 different naturally occurring carbide minerals (Tables 2 and 3; *http://rruff.info/ima/*). Although rare and volumetrically trivial as reservoirs of carbon in the crust, they may represent a significant volume of carbon in Earth's deep interior, and thus may provide insight to the deep carbon cycle (Dasgupta 2013; Wood et al. 2013).

Table 2. Natural carbide minerals (data compiled from *http://MinDat.Org* locality and species databases).

Mineral (Key)	Formula	Paragenesis*	Locality register**
Cohenite (C)	$(Fe,Ni,Co)_3C$	1,2,4	G1, G2, M1, R1, R2
Haxonite (H)	$(Fe,Ni)_{23}C_6$	2	M1
Isovite (I)	$(Cr,Fe)_{23}C_6$	4	R3, R4
Khamrabaevite (K)	(Ti,V,Fe)C	2,4?	M1, T1, U1, U2
Moissanite (M)	SiC	2,3,4	C1-C6, M1, R5, R6, S1, T2, U2
Niobocarbide (N)	(Nb,Ta)C	4	R7
Qusongite (Q)	WC	1,3,4	C3, C4, C7
Tantalcarbide (Ta)	(Ta,Nb)C	4	R7
Tongbaite (To)	Cr_3C_2	4	C8, R3
Yarlongite (Y)	$(Cr_4Fe_4Ni)C_4$	4	C4

*1 = coal fire or intrusion in coal/graphite; 2 = meteorite; 3 = kimberlite; 4 = other ultramafic
**See Table 3 for locality key

All metal carbides are refractory minerals; they have relatively high solidus and liquidus temperatures, with melting points typically above 2400 °C (Nadler and Kempter 1960; Lattimer and Grossman 1978). A number of these minerals are found in association with diamond and other high-pressure phases, as well as with assemblages of unusual reduced minerals, including native Al, Fe, Si, Sn, W, and more than a dozen other native metallic elements, as well as exotic sulfides, phosphides, and silicides (Table 4). Carbides, along with diamond, may thus represent Earth's deepest surviving minerals, and may prove a relatively unexplored window on the nature of the deep mantle environment and the deep carbon cycle.

Moissanite. Moissanite (α-SiC), also known commercially as carborundum when sold as an abrasive (hardness 9.5), is the most common of the natural carbides. Produced synthetically for more than a century (Acheson 1893), moissanite is used in numerous applications, including automobile parts (e.g., brakes and clutches), bulletproof vests, light-emitting diodes, semiconductor components, anvils for high-pressure research and, since 1998, it has been marketed as inexpensive diamond-like artificial gemstones (Bhatnagar and Baliga 1993; Xu and Mao 2000; Madar 2004; Saddow and Agarwal 2004).

Since its discovery in 1893 by Henri Moissan in mineral residues from the Canyon Diablo meteor crater in Arizona (Moissan 1904b), natural moissanite has been found in dozens of localities, including meteorites, serpentinites, chromitites, ophiolite complexes, and in close association with diamond in kimberlites and eclogites (Lyakhovich 1980; Leung et al. 1990; Alexander 1990, 1993; Di Pierro et al. 2003; Lee et al. 2006; Qi et al. 2007; Xu et al. 2008; Trumbull et al. 2009; Shiryaev et al 2011; see also *http://MinDat.org*). While some occurrences of moissanite are apparently of near-surface origin, including sites of forest fires and contact metamorphism of silicate magmas with coal beds (Sameshima and Rodgers 1990), silicon carbide also represents one of the deepest mantle minerals known to reach the surface. The discovery of moissanite inclusions in diamond (Moore and Gurney 1989; Otter and Gurney 1989; Leung 1990; Gorshkov et al. 1997), combined with observations of native silicon and Fe-Si alloy inclusions in moissanite (Trumbull et al. 2009), may point to an origin in reduced mantle microenvironments. However, further research is required to determine the range of oxygen fugacities under which moissanite is stable at mantle pressures and temperatures.

Occurrences of silicon carbide in mantle-derived kimberlites and several ophiolite complexes reflect its stability at high pressure and very low oxygen fugacity (Mathez et al. 1995). Secondary ion mass spectrometric (SIMS) analysis shows that ophiolite-hosted

Table 3. Natural carbide locality register, with associated reduced and/or high-*P* mineral species (data compiled from *http://MinDat.Org* locality and species databases). Several moissanite localities of poorly-defined paragenesis are not included.

Code	Locality	Carbides*	Type**	Associated Phases***
Canada				
C1	Jeffrey Mine, Quebec	M	Ser	Gra, Cu, Awa, IMA, Hea, Pyr, Sha
China				
C2	Fuxian kimberlite field, Liaoning Prov.	M	Kim	Dia, Gra, Lon, Fe
C3	Mengyin kimberlite field, Linyi Prefecture	M, Q	Kim	Dia, Cr, Fe, Pb, W, Pen, Pol, Coe
C4	Luobusa ophiolite, Tibet	M, Q, Y	Oph	Dia, Gra, Al, Cr, Cu, In, Fe, Ni, Os, Pd, Rh, Ru, Si, Ag, Sn, Ti, W, Zn, Zan, Ala, Wus, Luo
C5	Picrite lava outcrop, Yunnan Prov.	M	Ser	Cu, Zn
C6	Dongjiashan Hill, Dabie Mountains	M	Ser	Dia, Si
C7	Yangliuping N-Cu-PGE Deposit, Sichuan Prov.	Q	UGr	Cr, Pt, Dan, Cub, Lau, Lin, Vio
C8	Liu Village, Henan Prov.	To	Ult	Cr, Cu, Fe, Pb, Alt, Awa, Bi2, Cub, Pen, Pyr, Vio,
Germany				
G1	Bühl, Weimar, Kassel, Hesse	C	MCo	Fe
Greenland				
G2	Disko Island, Kitaa Prov.	C	Ult	Gra, Fe, Pb, Arm, Sha, Tro, Sch
Meteorites				
M1	Various localities	C, H, K, M		
Russia				
R1	Khungtukun Massif, Khatanga, Siberia	C	MCo	Gra, Cu, Fe, Tro, Wus
R2	Coal Mine 45, Kopeisk, Urals	C	CoF	Gra, Tro
R3	Is River, Isovsky District, Urals	I, To	Pla	Os, Pt, Iso, Tet, Cup, Mal, Fer, Gup, Xif
R4	Verkhneivinsk, Neiva River, Urals	I	Pla	Os, Ru, Any
R5	Billeekh intrusion, Saha Rep., Siberia	M	Gab	Al, Sb, Cd, Cu, Fe, Pb, Sn, Zn
R6	Avacha volcano, Kamchatka	M	Pic	Dia
R7	Avorinskii Placer, Baranchinsky Massif, Urals	N, Ta	Pla	Jed, Ru
South Africa				
S1	Monastery Mine, Free State Prov.	M	Kim	Dia, Maj
Tajikistan				
T1	Chinorsai intrusion, Viloyati Sogd	K		Gra, Bi, Fe, Bis
Uzbekistan				
U1	Ir-Tash Stream Basin, Tashkent Viloyati	K	???	Gra, Gue, Sue
U2	Koshmansay River, Tashkent Viloyati	K, M	Lam	Ala, Mav

*See Table 2 for key to carbides

**CoF = coal fire; GCo = gabbro intrusion in coal; Gra = granodiorite; Kim = kimberlite; Lam = lamproite; Oph = ophiolite; Pic = picrite; Pla = placer; Ser = serpentinite; UGr = ultramafic intrusion in graphite schist; Ult = unspecified ultramafic

***See Table 4 for key to associated phases

Table 4. Select reduced and/or high-*P* mineral species associated with natural carbides (data compiled from *http://MinDat.Org* locality and species databases). For locality key see Table 3.

Mineral (Key)	Formula	Localities
Carbon Allotropes		
Diamond (Dia)	C	C2,C3,C4,C6,R6,S1
Lonsdaleite (Lon)	C	C2
Graphite (Gra)	C	C1,C2,C4,G2,R1,R2,T1,U1
Native Metals		
Aluminum (Al)	Al	C4,R5
Antimony (Sb)	Sb	R5
Bismuth (Bi)	Bi	T1
Cadmium (Cd)	Cd	R5
Chromium (Cr)	Cr	C3,C4,C7,C8
Copper (Cu)	Cu	C1,C4,C5,C8,R1,R5
Indium (In)	In	C4
Iron (Fe)	Fe	C3,C4,C8,G1,G2,R1,R5,T1
Lead (Pb)	Pb	C3,C8,G2,R5
Nickel (Ni)	Ni	C4
Osmium (Os)	Os	C4,R3,R4
Palladium (Pd)	Pd	C4
Rhodium (Rh)	Rh	C4
Ruthenium (Ru)	Ru	C4,R4,R7
Silicon (Si)	Si	C4,C6
Tin (Sn)	Sn	C4,R5
Titanium (Ti)	Ti	C4
Tungsten (W)	W	C4
Zinc (Zn)	Zn	C4,C5,R5
Metal Alloys		
Altaite (Alt)	PbTe	C8,G2
Anyuite (Any)	$Au(Pb,Sb)_2$	R4
Awaruite (Awa)	Ni_3Fe	C1,C4,C8
Danbaite (Dan)	$CuZn_2$	C7
IMA2009-083 (IMA)	Ni_3Sn	C1
Isoferroplatinum (Iso)	Pt_3Fe	R3
Jedwabite (Jed)	$Fe_7(Ta,Nb)_3$	R7
Tetraferroplatinum (Tet)	PtFe	R3
Phosphides		
Schreibersite (Sch)	$(Fe,Ni)_3P$	G2

Mineral (Key)	Formula	Localities
Transition Metal Sulfides		
Alabandite (Ala)	MnS	C4,U2
Bismuthinite (Bi1)	Bi_2S_3	T1
Bismutohauchecornite (Bi2)	$Ni_9Bi_2S_8$	C8
Cubanite (Cub)	$CuFe_2S_3$	C1,C7,C8
Cuproiridsite (Cup)	$(Cu,Fe)Ir_2S_4$	R3
Heazlewoodite (Hea)	Ni_3S_2	C1
Laurite (Lau)	RuS_2	C7
Linnaeite (Lin)	Co_3S_4	C7
Malanite (Mal)	$Cu(Pt,Ir)_2S_4$	R3
Pentlandite (Pen)	$(Fe,Ni)_9S_8$	C4,C7,C8
Polydymite (Pol)	Ni_3S_4	C3
Pyrrhotite (Pyr)	Fe_7S_8	C1,C4,C7,C8
Shandite (Sha)	$Ni_3Pb_2S_2$	C1,G2
Troilite (Tro)	FeS	G2,R1,R2
Violarite (Vio)	$FeNi_2S_4$	C7,C8
Shandite (Sha)	$Ni_3Pb_2S_2$	C1,G2
Transition Metal Oxides		
Armalcolite (Arm)	$(Mg,Fe)Ti_2O_5$	G2,R1
Cuprite (Cup)	Cu_2O	C1
Wustite (Wus)	$Fe_{1-x}O$	C4,R1
Cuprite (Cup)	Cu_2O	C1
High-Pressure Oxides and Silicates		
Coesite (Coe)	SiO_2	C3,C4
Kyanite (Kya)	Al_2SiO_5	C4
Majorite (Maj)	$MgSiO_3$	S1
Silicides		
Fersilicite (Fer)	FeSi	R3
Gupeiite (Gup)	Fe_3Si	R3,U1
Luobusaite (Luo)	$Fe_{0.84}Si_2$	C4
Mavlyanovite (Mav)	Mn_5Si_3	U2
Suessite (Sue)	$(Fe,Ni)_3Si$	U1
Xifengite (Xif)	Fe_5Si_3	R3
Zangboite (Zan)	$TiFeSi_2$	C4

moissanite has a distinctive ^{13}C-depleted isotopic composition ($\delta^{13}C$ from −18 to −35‰, $n = 36$), which is significantly lighter than the main carbon reservoir in the upper mantle ($\delta^{13}C$ near −5‰). Alternatively, significant isotope fractionation between carbide and diamond has been observed in high-pressure experiments (−7 per mil at 5 GPa; Mikhail et al. 2010; Mikhail 2011), greatly complicating the potential for identification of carbon reservoirs through carbon isotope systematics of mantle-derived samples (Mikhail et al. 2011). It has been suggested that moissanite may also occur in the lower mantle, where the existence of ^{13}C-depleted carbon is strongly supported by studies of extraterrestrial carbon (Trumbull et al. 2009).

Moissanite, like diamond, also forms by vapor deposition (Hough et al. 1997) and it is relatively common in space in the envelopes of carbon-rich AGB stars. It is subsequently carried as a pre-solar guest in carbonaceous chondrites. The origins of moissanite and other carbides have been inferred from unusual variations in both C-isotopes and N-isotopes (Daulton et al. 2003). Meteoritic carbides commonly contain dissolved nitrogen, while the comparable family of nitride minerals [e.g., osbornite (TiN)] contains some dissolved carbon. This mutual limited solubility of N-C in minerals persists through natural diamond (C), which also contains minor N, and in the future may be useful for understanding crystallization histories and source reservoirs.

Moissanite is also relatively widely distributed as micro-crystals in the ejecta from some meteorite impact craters formed in continental crust, and it may be associated with impact diamond and a variety of iron silicides like suessite (Ernston et al. 2010). Low-pressure SiC is also found in the KT impact layer (Hough et al. 1995; Langenhorst and Deutsch 2012), though unlike impact diamond, SiC is not ubiquitous in crustal impact deposits (Gilmour et al. 2003). Indeed, another possibility for deep SiC formation, given the antiquity of some kimberlite/diamond hosted SiC, might be residues from giant impact processes during formation of Earth's Moon, since at that time materials from a cross-section through the upper mantle were violently exposed to the vacuum of space.

Ideal moissanite has a hexagonal structure closely related to that of lonsdaleite (and identical to wurtzite), in which every atom is tetrahedrally coordinated and corner-linked tetrahedral layers are stacked ideally in a two-layer [...ABAB...] configuration (Fig. 5). The Si-C distance of 1.86 Å is appreciably longer than that of the carbon polymorphs because of the greater size of Si compared to C. Moissanite is known to recrystallize at temperatures between 1400 and 1600 °C to β-SiC, which is cubic and similar to the diamond structure.

More than 250 stacking polytypes of SiC have been documented, notably hexagonal forms with 4-layer (4H) and 6-layer (6H—the most common terrestrial polytype; Capitani et al. 2007) repeats {[...ABAC...] and [...ABCACB...], respectively}, and a rhombohedral form with a 15-layer (15R) sequence [...ABCBACABACBCACB...]. Hundreds of other polytypes with repeat sequences from dozens to hundreds of layers (i.e., 141R and 393R) have also

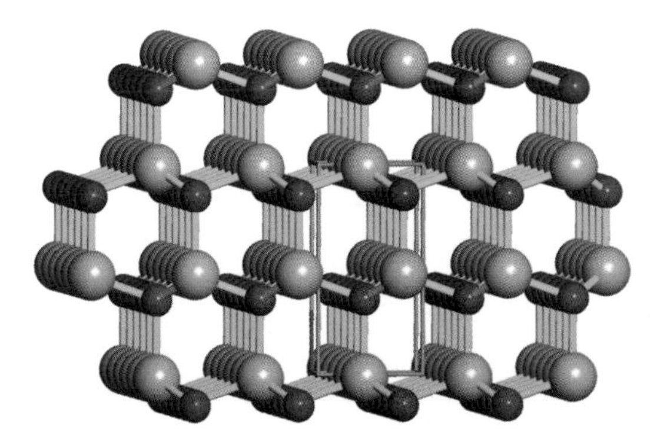

Figure 5. The moissanite (SiC) crystal structure (hexagonal, space group $P6_3mc$; $a = 3.081$ Å; $c = 5.031$ Å; $Z = 2$). Carbon and silicon atoms appear in blue and grey, respectively.

been characterized (Krishna and Verma 1965; Lee et al. 2006; Capitani et al. 2007; Shiryaev et al. 2011; see also *http://img.chem.ucl.ac.uk/www/kelly/LITERATURESICWEB.HTM#4* and *http://MinDat.org*). Such complex polytypes have inspired a variety of models related to possible growth models, including screw dislocation or spiral growth mechanism (e.g., Frank 1949, 1951; Verma and Krishna 1966), the faulted matrix model (Pandey and Krishna 1975a, 1975b, 1978), one-dimensional disorder theory (Jagodzinski 1954a, 1954b), and the axial nearest neighbor Ising model (Price and Yeomans 1984).

A curious crystal chemical aspect related to silicon carbide is the apparent absence of Si-C bonding in other naturally occurring compounds (Nawrocki 1997; Franz 2007; Tran et al. 2011). Chemists have explored a rich landscape of synthetic organic silanes, silanols, and silicones, but natural examples of these potentially crystal-forming compounds have not yet been described.

Cohenite. The iron carbide cohenite [$(Fe,Ni,Co)_3C$], also called cementite when it occurs as a binding agent in steel, is second in abundance as a natural carbide only to moissanite. In nature cohenite is known primarily as an accessory mineral from more than a dozen iron meteorites (Brett 1967), but it also occurs occasionally with native iron in the crust, for example at Disko Island in central west Greenland, and the Urals in Russia. Though iron carbides are rare in nature, the low-pressure phase behavior of carbon in iron has been studied extensively by the steel industry (e.g., Brooks 1996).

Iron carbide occurs occasionally with native iron in the crust. For example, local occurrences of metallic iron with iron carbide ("cohenite") may result from thermal interaction and reduction of basalt with coal or other carbon-rich sediments (Melson and Switzer 1966; Pederson 1979, 1981; Cesnokov et al. 1998). Most famously, Fe-Ni carbide (Ni-poor cohenite) occurs in massive native iron, with schreibersite, sulfides and a variety of minerals in graphite-bearing glassy Tertiary basalts on and around Disko Island, Greenland (individual iron masses > 20 tons; Nordenskiöld 1872; Pauly 1969; Bird and Weathers 1977; Goodrich 1984; Goodrich and Bird 1985). A further 10 ton mass was discovered as recently as 1985, 70 km away from the original Disko iron location (Ulff-Moller 1986). Comparable massive and dispersed native iron containing not only iron carbide but also silicon carbide (H-6, moissanite), occurs together with a rich variety of more than 40 minerals including native metals (Al, Cu), in glass-bearing Permian doleritic sills on the Putorana Plateau, Siberia (Oleynikov et al. 1985). These occurrences are partly brecciated and, although superficially resembling meteoritic textures, they are considered to be terrestrial (Treiman et al. 2002). The origin of the Disko iron is still debated: detailed mapping of dispersed iron in regional basalts strongly favors large-scale interaction of carbon-rich sediments with volcanic lavas (Larsen and Pedersen 2009). However, their correlation with basal stratigraphic units on the Nussussuaq peninsula that preserve unambiguous Ir-bearing impact spherules has reintroduced the prospect for involvement of a meteorite impact in the origin of carbon-rich Disko iron (Jones et al. 2005) as originally invoked by its discoverer (Nordenskiöld 1872).

The structure of cohenite, (orthorhombic; space group *Pbnm*, a = 4.518 Å; b = 5.069 Å; c = 6.736 Å; Hendricks 1930), has been extensively studied. Many samples display different cell parameters, potentially related to a variety of causes such as quenching rates. The structure is composed of regular trigonal prisms of iron atoms with carbon at the center (Fig. 6; Animation 4). (Note that in the first structure experiments it was reported that cohenite features a framework of near regular CFe_6 octahedra, each with 6 iron atoms surrounding a central carbon atom. However, the positions of the carbon atoms were not determined and the assumption of octahedral coordination was incorrect.)

New carbides from Chinese ultramafic rocks. Two natural occurrences of exotic carbides deserve special note, because they point to a possibly rich and as yet largely unexplored deep

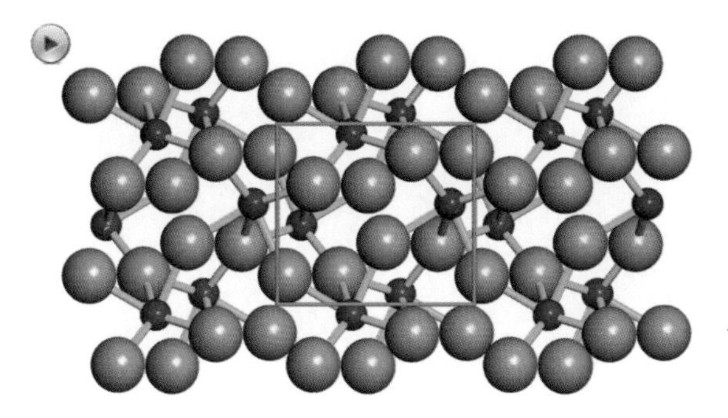

Figure 6. The structure of the iron carbide cohenite [$(Fe,Ni,Co)_3C$] (orthorhombic, space group *Pbnm*; a = 4.518 Å, b = 5.069 Å; c = 6.736 Å; Z = 4). Carbon and iron atoms appear in blue and gold, respectively. [Animation 4: *For readers of the electronic version, click the image for an animation of the iron carbide cohenite crystal structure.*]

carbide mineralogy. The first localities are associated with ultramafic rocks in Central China: (1) in podiform chromitites of the Luobusha ophiolite complex, Autonomous Tibetan Region; (2) within the Mengyin kimberlite field, Linyi Prefecture; (3) within the Yangliuping N-Cu-PGE Deposit, Sichuan Province; and (4) at Liu Village, Henan Province. These deposits incorporate minor amounts of three carbides that are unknown from any other region: qusongite (WC; Fang et al. 2009; Shi et al. 2009), yarlongite [$(Fe_4Cr_4Ni)C_4$; Nicheng et al. 2005, 2008], and tongbaite (Cr_3C_2; Tian et al. 1983; Dai et al. 2004). The association of these carbides with moissanite, cohenite, and khamrabaevite [(Ti,V,Fe)C], as well as other dense, high-temperature phases such as diamond, coesite (a high-pressure form of SiO_2), and varied native metals, including Fe, Ni, W, Cr, Pb, and W, points to a high-temperature, high-pressure origin in a reduced mantle environment (Robinson et al. 2001; Shi et al. 2009).

New carbides from placer deposits of the Urals. A second enigmatic carbide region is found in the Ural Mountains of Russia, within both the Avorinskii Placer, Baranchinsky Massif, and in sediments of the Neiva River near Verkhneivinsk. Placer deposits have yielded <0.3 mm-diameter grains of isovite [$(Cr,Fe)_{23}C_6$; Generalov et al. 1998], as well as euhedral crystals from the complex nonstoichiometric solid solution between niobocarbide and tantalocarbide [$(Nb,Ta)C_{1-x}$; Gusev et al. 1996; Novgorodova et al. 1997]. All known specimens of the latter two minerals, however, were collected early in the 20th century and the exact location of the placer deposit is currently unknown. These two possibly related placer deposits produce an enigmatic suite of other unusual minerals, including native rhenium and osmium (both with melting temperatures > 3000 °C), anyuite [$Au(Pb,Sb)_2$], and jedwabite [$Fe_7(Ta,Nb)_3$]. The source lithologies and paragenesis of these minerals are not known, though the concentration of Nb and Ta suggests a possible association with carbonatitic magmas. In any case, they point to the potential diversity of rare carbides in unusual geochemical environments.

Rhombohedral carbonates

By far the most abundant carbon-bearing minerals, both in the number of different species and in their total crustal volume, are the carbonates, of which more than 300 have received IMA approval (*http://rruff.info/ima/*; Fig. 7). Several previous compilations have reviewed carbonate minerals in detail (Reeder 1983a; Klein and Hurlbut 1993; Chang et al. 1997). Here we summarize key aspects of the mineralogy and crystal chemistry of select carbonates (Tables 5 and 6). Almost all of these minerals incorporate near-planar $(CO_3)^{2-}$ anions, with an equilateral triangle of oxygen atoms around the central carbon atom. Most C-O bond distances are between 1.25 and 1.31 Å and O-C-O angles are close to 120°. Rigid-body libration of these molecular anions contributes to distinctive characteristics of carbonate vibrational spectra, notably three prominent infrared absorption features at ~690-750, 840-900, and 1400-1490 cm^{-1} (Adler and Kerr 1963; White 1974; Chang et al. 1997) and the strong symmetric stretching modes found near 1100 cm^{-1} in Raman spectra (Rutt and Nicola 1974).

Figure 7. (*caption on facing page*)

Two types of rhombohedral carbonates—the calcite and dolomite groups—collectively represent by far the most abundant carbonate minerals in Earth's crust, with calcite ($CaCO_3$) and dolomite [$CaMg(CO_3)_2$] in massive sedimentary and metamorphic formations accounting for at least 90% of crustal carbon (Reeder 1983b). Orthorhombic carbonates in the aragonite group also play a significant role in Earth surface processes (Speer 1983), particularly through biomineralization (Stanley and Hardie 1998; Dove et al. 2003; Knoll 2003). These three groups are surveyed below.

Calcite and the calcite group. The most important carbonate minerals belonging to the calcite group (Table 5; Reeder 1983b) include calcite ($CaCO_3$), magnesite ($MgCO_3$), rhodocrosite ($MnCO_3$), siderite ($FeCO_3$), and smithsonite ($ZnCO_3$). It should be emphasized that these minerals seldom occur as pure end-members, but instead commonly form solid solutions with many divalent cations. The calcite structure (space group $R\bar{3}c$; e.g., Bragg 1914; Effenberger et al. 1981; Chang et al. 1997) has a topology similar to that of NaCl, with each Ca^{2+} coordinated to 6 $(CO_3)^{2-}$ groups, and each $(CO_3)^{2-}$ group in turn coordinated to 6 Ca^{2+} cations. However, the orientations of the $(CO_3)^{2-}$ groups, while the same within each layer, are 180° out of phase in successive layers, thus doubling the repeat distance in the *c* axial direction relative to a sodium chloride analog. Note that this layer-by-layer alternation of $(CO_3)^{2-}$ group orientations results in an oxygen atom distribution that approximates hexagonal close packing, with C and Ca occupying 3- and 6-coordinated interstices, respectively (Megaw 1973). The flattened shape of this carbonate anion results in an obtuse rhombohedral angle of 101°55′ (Fig. 8; Animation 5).

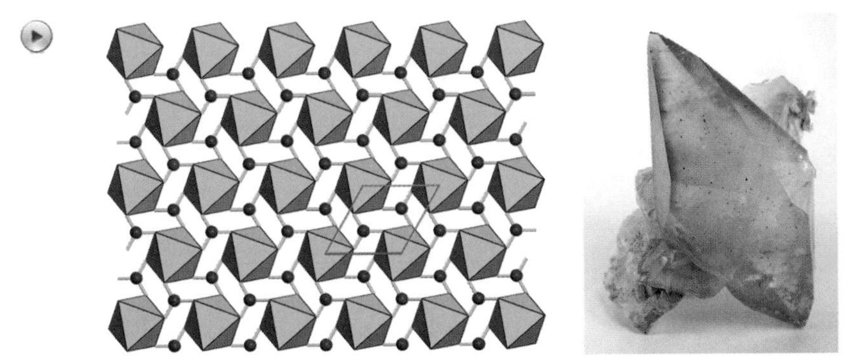

Figure 8. The structure of calcite ($CaCO_3$) (rhombohedral; space group $R\bar{3}c$; hexagonal setting a = 4.989 Å; c = 17.061 Å; Z = 6; rhombohedral setting a = 6.375; α = 46.1°; Z = 2). Calcite crystal from Elmwood Mine, Carthage, Tennessee. Photo courtesy of Rob Lavinsky. [Animation 5: *For readers of the electronic version, click the image for an animation of the calcite crystal structure.*]

Figure 7. (*figure on facing page*) The diversity of carbonate minerals. (a) Rhodochrosite ($MnCO_3$), rhombohedral crystals from the Home Sweet Home Mine, Mount Bross, Alma District, Park County, Colorado, USA. (b) Ankerite [$CaFe^{2+}(CO_3)_2$] from Brownley Hill mine, Nenthead, Cumbria, England showing the typical curved saddle-shaped rhombohedral crystals. (c) Artinite [$Mg_2CO_3(OH)_2{\cdot}3H_2O$], divergent sprays of clear colorless acicular crystals from San Benito County, California, USA. (d) Aurichalcite [$Zn_5(CO_3)_2(OH)_6$], divergent sprays of light blue lathlike crystals from Bisbee, Cochise County, Arizona, USA. (e) Hydromagnesite [$Mg_5(CO_3)_4(OH)_2{\cdot}4H_2O$], colorless bladed crystals from Paradise Range, Nye County, Nevada, USA. (f) Jouravskite [$Ca_3Mn^{4+}(SO_4)(CO_3)(OH)_6{\cdot}12H_2O$], yellow hexagonal prism associated with calcite from the Wessels mine, Kurumen, Kalahari Manganese fields, Cape Province, South Africa. (g) Stichtite [$Mg_6Cr_2CO_3(OH)_{16}{\cdot}4H_2O$], an aggregate of contorted purple plates from Dundas, Tasmania, Australia, (h) Zaratite [$Ni_3CO_3(OH)_4{\cdot}4H_2O$], green amorphous crust, intimately associated with népouite from Lord Brassy mine, Tasmania, Australia. All photos courtesy of the RRUFF project and irocks.com.

Table 5. Selected anhydrous carbonates, primarily species with more than 10 known localities (data compiled from *http://MinDat.Org* databases).

Name	Formula	Name	Formula	Name	Formula
Calcite Group		*Other Anhydrous Double and Triple Carbonates*		*Anhydrous Carbonate Oxides*	
Calcite	$CaCO_3$	Eitelite	$Na_2Mg(CO_3)_2$	Shannonite	$Pb_2O(CO_3)$
Gaspeite	$NiCO_3$	Nyerereite	$Na_2Ca(CO_3)_2$	Bismutite	$Bi_2O_2(CO_3)$
Magnesite	$MgCO_3$	Zemkorite	$Na_2Ca(CO_3)_2$	Beyerite	$CaBi_2O_2(CO_3)_2$
Otavite	$CdCO_3$	Bütschilite	$K_2Ca(CO_3)_2$	Rutherfordine	$(UO_2)(CO_3)$
Rhodocrosite	$MnCO_3$	Fairchildite	$K_2Ca(CO_3)_2$	Čejkaite	$Na_4UO_2(CO_3)_3$
Siderite	$FeCO_3$	Shortite	$Na_2Ca_2(CO_3)_3$		
Smithsonite	$ZnCO_3$	Paralstonite	$BaCa(CO_3)_2$	*Anhydrous Carbonate Sulfates*	
Sphaerocobaltite	$CoCO_3$	Barytocalcite	$BaCa(CO_3)_2$	Burkeite	$Na_4(SO_4)(CO_3)$
		Huntite	$CaMg_3(CO_3)_4$	Davyne	$(Na,Ca,K)_8(Si,Al)_{12}O_{24}(Cl,SO_4,CO_3)_{2-3}$
Dolomite Group		Alstonite	$BaCa(CO_3)_2$	Hanksite	$KNa_{22}(SO_4)_9(CO_3)_2Cl$
Dolomite	$CaMg(CO_3)_2$	Norsethite	$BaMg(CO_3)_2$		
Ankerite	$FeMg(CO_3)_2$	Sahamalite-(Ce)	$Ce_2Mg(CO_3)_4$	*Anhydrous Carbonate Silicates*	
Kutnohorite	$CaMn(CO_3)_2$	Burbankite	$(Na,Ca)_3(Sr,Ba,Ce)_3(CO_3)_5$	Spurrite	$Ca_5(SiO_4)_2(CO_3)$
Minrecordite	$CaZn(CO_3)_2$	Carbocernaite	$(Ca,Na)(Sr,Ce,La)(CO_3)_2$	Tilleyite	$Ca_5(Si_2O_7)(CO_3)_2$
		Benstonite	$Ba_6Ca_6Mg(CO_3)_{13}$		
Aragonite Group					
Aragonite	$CaCO_3$	*Anhydrous Carbonate Halides*			
Cerussite	$PbCO_3$	Bastnäsite Group	$(REE)(CO_3)F$		
Strontianite	$SrCO_3$	Brenkite	$Ca_2(CO_3)F_2$		
Witherite	$BaCO_3$	Phosgenite	$Pb_2(CO_3)Cl_2$		
		Kettnerite	$CaBiO(CO_3)F$		
Other Anhydrous Single Carbonates		Synchysite Group	$Ca(REE)(CO_3)_2F$		
Vaterite	$CaCO_3$	Huanghoite-(Ce)	$Ba(Ce,REE)(CO_3)_2F$		
Zabuyelite	$Li_2(CO_3)$	Horváthite-(Y)	$NaY(CO_3)F_2$		
Gregoryite	$Na_2(CO_3)$	Parisite Group	$Ca(REE)_2(CO_3)_3F_2$		
Natrite	$Na_2(CO_3)$	Röntgenite-(Ce)	$Ca_2Ce_3(CO_3)_5F_3$		
Olekminskite	$Sr_2(CO_3)_2$	Lukechangite-(Ce)	$Na_3Ce_2(CO_3)_4F$		
		Cordylite Group	$NaBa(REE)_2(CO_3)_4F$		

Table 6. Selected hydrous carbonates.

Name	Formula	Name	Formula
Aluminohydrocalcite	$CaAl_2(CO_3)_2(OH)_4 \cdot 3H_2O$	Leadhillite	$Pb_4(SO_4)(CO_3)_2(OH)_2$
Ancylite Group	$(Ca,Sr,Pb)(REE)(CO_3)_2(OH) \cdot H_2O$	Liebigite	$Ca_2(UO_2)(CO_3)_3 \cdot 11H_2O$
Artinite	$Mg_2CO_3(OH)_2 \cdot 3H_2O$	Macphersonite	$Pb_4(SO_4)(CO_3)_2(OH)_2$
Aurichalcite	$Zn_5(CO_3)_2(OH)_6$	Malachite	$Cu_2CO_3(OH)_2$
Azurite	$Cu_3(CO_3)_2(OH)_2$	Manasseite	$Mg_6Al_2CO_3(OH)_{16} \cdot 4H_2O$
Brianyoungite	$Zn_3CO_3(OH)_4$	Mcguinnessite	$CuMgCO_3(OH)_2$
Brugnatellite	$Mg_6Fe^{3+}CO_3(OH)_{13} \cdot 4H_2O$	Monohydrocalcite	$CaCO_3 \cdot H_2O$
Caledonite	$Cu_2Pb_5(SO_4)_3(CO_3)(OH)_6$	Nahcolite	$NaHCO_3$
Calkinsite-(Ce)	$(Ce,REE)_2(CO_3)_2 \cdot 4H_2O$	Nakauriite	$Cu_8(SO_4)_4(CO_3)(OH)_6 \cdot 48H_2O$
Callaghanite	$Cu_2Mg_2CO_3(OH)_6 \cdot 2H_2O$	Natron	$Na_2CO_3 \cdot 10H_2O$
Carbonatecyanotrichite	$Cu_4Al_2CO_3(OH)_{12} \cdot 2H_2O$	Nesquehonite	$MgCO_3 \cdot 3H_2O$
Chalconatronite	$Na_2Cu(CO_3)_2 \cdot 3H_2O$	Niveolanite	$NaBeCO_3(OH) \cdot 2H_2O$
Claraite	$Cu^{2+}{}_3CO_3(OH)_4 \cdot 4H_2O$	Pirssonite	$Na_2Ca(CO_3)_2 \cdot 2H_2O$
Coalingite	$Mg_{10}Fe^{3+}{}_2CO_3(OH)_{24} \cdot 2H_2O$	Pokrovite	$Mg_2CO_3(OH)_2$
Dawsonite	$NaAlCO_3(OH)_2$	Pyroaurite	$Mg_6Fe^{3+}{}_2CO_3(OH)_{16} \cdot 4H_2O$
Dypingite	$Mg_5(CO_3)_4(OH)_2 \cdot 5H_2O$	Reevesite	$Ni_6Fe^{3+}{}_2CO_3(OH)_{16} \cdot 4H_2O$
Fukalite	$Ca_4Si_2O_6(CO_3)(OH)_2$	Rosasite	$CuZnCO_3(OH)_2$
Gaylussite	$Na_2Ca(CO_3)_2 \cdot 5H_2O$	Scawtite	$Ca_7(Si_3O_9)_2(CO_3) \cdot 2H_2O$
Glaukospaerite	$CuNiCO_3(OH)_2$	Schröckingerite	$NaCa_3(UO_2)(SO_4)(CO_3)_3F \cdot 10H_2O$
Harkerite	$Ca_{12}Mg_4Al(SiO_4)_4(BO_3)_3(CO_3)_5 \cdot H_2O$	Sjögrenite	$Mg_6Fe^{3+}{}_2(CO_3)(OH)_{16} \cdot 4H_2O$
Hellyerite	$NiCO_3 \cdot 6H_2O$	Stichtite	$Mg_6Cr_2CO_3(OH)_{16} \cdot 4H_2O$
Hydrocerussite	$Pb_3(CO_3)_2(OH)_2$	Susannite	$Pb_4(SO_4)(CO_3)_2(OH)_2$
Hydromagnesite	$Mg_5(CO_3)_4(OH)_2 \cdot 4H_2O$	Takovite	$Ni_6Al_2CO_3(OH)_{16} \cdot 4H_2O$
Hydrotalcite	$Mg_6Al_2CO_3(OH)_{16} \cdot 4H_2O$	Tengerite-(Y)	$Y_2(CO_3)_3 \cdot 2\text{-}3H_2O$
Hydroxylbastnäsite-(Ce)	$CeCO_3(OH)$	Teschemacherite	$(NH_4)HCO_3$
Hydrozincite	$Zn_5(CO_3)_2(OH)_6$	Thaumasite	$Ca_3Si(OH)_6(SO_4)(CO_3) \cdot 12H_2O$
Ikaite	$CaCO_3 \cdot 6H_2O$	Thermonatrite	$Na_2(CO_3) \cdot H_2O$
Kalicinite	$KHCO_3$	Trona	$Na_3(HCO_3)(CO_3) \cdot 2H_2O$
Landsfordite	$MgCO_3 \cdot 5H_2O$	Tyrolite	$Ca_2Cu_9(AsO_4)_4(CO_3)(OH)_8 \cdot 11H_2O$
Lanthanite Group	$(REE)_2(CO_3)_3 \cdot 8H_2O$		

Calcite is widely distributed in Earth's crust; it appears most commonly within sedimentary rocks, where it occurs as the principal mineral of limestone, and as a natural cementing agent in many siliceous sandstone and shale units that were deposited under marine conditions. Calcite also dominates some metamorphic rocks such as marble and calcareous gneiss; occurs widely in hydrothermal systems, where it forms extensive vein networks; and is common in some unusual carbonate-rich igneous rocks such as carbonatites (Jones et al. 2013).

Although widely formed under Earth's near-surface conditions, distributions of calcite and other rhombohedral carbonate minerals have varied significantly through Earth history, principally as a consequence of feedbacks between the geosphere and biosphere (Knoll 2003; Hazen et al. 2008; Hazen et al. 2013). At present, the majority of calcium carbonate deposition occurs as calcite precipitated within shallow marine settings. In these settings, magnesium—whose concentration is nearly 4× that of calcium in normal marine fluids—is readily incorporated into the calcite crystal lattice, with Mg concentrations of marine calcites equal to a few to nearly 20 mol% $MgCO_3$ (MacKenzie et al. 1983; Morse and MacKenzie 1990; Morse et al. 2006; Berner and Berner 2012). Such Mg-bearing calcites are commonly referred to as magnesian calcite or "high Mg calcite" (HMC) and are distinguished from calcites with low Mg concentrations ("low Mg calcite" or LMC). Magnesium is one of a suite of divalent ions that readily co-precipitate with calcium in the calcite lattice. Because ionic co-precipitation reflects a combination of the ionic availability, temperature, and lattice structure, and because it can substantially affect the solubility and rate of dissolution of the resultant calcite, differential co-precipitation of ions within calcite has been, and continues to be, a subject of intense investigation as a means of unraveling the geologic history of the oceanic system (Morse and Mackenzie 1990).

One of calcite's remarkable and as yet largely unexplored features is its extraordinary range of crystal forms (e.g., Dana 1958). Habits range from the more common rhombohedral and scalenohedral crystal forms, to needle-like, platy, and equant shapes with expression of at least 300 different documented crystal forms (Fig. 9; see also specimen photographs on *http://MinDat.org*). Some of the most varied crystal habits are widely distributed in association with both biological skeletalization (Fallini et al. 1996; Dove et al. 2003) and speleogenesis (Frisia et al. 2000) and reflect a complex array of physical, chemical, and biological influences during crystallization. Because calcite crystal morphology is strongly affected by the kinetics of crystal growth, unraveling the differential effects of fluid saturation state, carbonate ion availability, ionic activity, the presence or absence of ionic inhibitors to nucleation and growth, and even the presence or absence of mineral catalyzing organic molecules (e.g., Cody and Cody 1991; Teng and Dove 1997; Teng et al. 1998; Orme et al. 2001), is critical to reveal as of yet untapped insights to Earth's crustal evolution. This need to document connections between environment and crystal form may be true, in particular, for our understanding of distinct carbonate morphologies such as "herringbone" calcite (Sumner and Grotzinger 1996; Kah et al. 1999) and "molar-tooth" calcite (Pollock et al. 2006) that show distinct environmental distributions through Earth history (see Hazen et al. 2013).

Other calcite group minerals. The magnesium carbonate magnesite ($MgCO_3$) forms primarily through alteration of Mg-rich igneous and metamorphic rocks, commonly in association with serpentine, as well as by direct precipitation from Mg-rich solutions and as a primary phase in mantle-derived carbonatites. Anhydrous magnesium carbonate commonly hydrates to form one of several secondary minerals (Table 6), including hydromagnesite [$Mg_4(CO_3)_3(OH)_2 \cdot 3H_2O$], artinite [$Mg_2CO_3(OH)_2 \cdot 3H_2O$], dypingite [$Mg_5(CO_3)_4(OH)_2 \cdot 5H_2O$], pokrovite [$Mg_2CO_3(OH)_2$], nesquehonite ($MgCO_3 \cdot 3H_2O$), and landsfordite ($MgCO_3 \cdot 5H_2O$).

Rhodochrosite, the manganese carbonate ($MnCO_3$), most commonly occurs as a vein-filling phase in hydrothermal ore districts. Limited solid solutions with Ca, Fe, and Mg end-members, as well as Zn, Ba, and Pb, are typical, as is partial alteration to manganese oxide

Figure 9. Crystal forms of natural calcite. (a) Rhombohedral cleavage fragment of optical grade material from near Presidio, Texas, USA, University of Arizona Mineral Museum 16674 (b) Scalenohedrons collected in the early 1800's from the Bigrigg mine, Cumbria, England. Bob Downs specimen; (c) Hexagonal prisms with rhombohedral terminations from Joplin, Missouri, USA UAMM 16545; (d) Herringbone growths of acicular crystals from the Southwest mine, Bisbee, Arizona, USA, UAMM 9499; (e) Hexagonal prisms with pyrite centers from Charcas, San Luis Potosi, Mexico, UAMM 1214; (f) Stalactite globules from Southwest mine, Bisbee, Arizona, USA, UAMM 9499; (g) Aggregate of bladed crystals from the Onyx cave, Santa Rita Mts, Arizona, USA, UAMM 5503; (h) Hexagonal prisms with pyramidal terminations from thje Camp Bird mine, Imogene Basin, Ouray County, Colorado, USA, UAMM 6703. All photographs by Alesha Siegal, University of Arizona.

hydroxides. Rhodochrosite is typically pale pink in color, though relatively rare deep rose pink specimens occur occasionally and are highly prized as semi-precious gemstones.

The iron carbonate siderite ($FeCO_3$) occurs in massive beds as an important component of some Precambrian banded iron sedimentary formations (Klein 2005), as well as in hydrothermal veins associated with ferrous metal sulfides. Siderite commonly incorporates Ca, Mn, and Co, and it forms a complete solid solution with magnesite in a variety of lithological settings, as well as with smithsonite ($ZnCO_3$) in hydrothermal lead-zinc ore deposits. Siderite is only stable under conditions of relatively low f_{O_2} (Hazen et al. 2013). It is metastable under ambient oxic conditions and typically decomposes to a suite of iron oxide-hydroxides such as goethite [FeO(OH)], and related hydrous phases [$FeO(OH)\cdot nH_2O$]—reactions that are accelerated by chemolithoautotrophic microbial activity.

Dolomite group. The dolomite group (space group $R\bar{3}$) is topologically identical to calcite, but in these double carbonate minerals two or more different cations occupy alternate layers perpendicular to the *c* axis (Wasastjerna 1924; Wyckoff and Merwin 1924; Reeder 1983b; Chang et al. 1997; Fig. 10; Animation 6). Important end-member minerals in this group (Table 5) include dolomite [$CaMg(CO_3)_2$], ankerite [$FeMg(CO_3)_2$], kutnohorite [$CaMn(CO_3)_2$], and minrecordite [$CaZn(CO_3)_2$]. Note that in both the calcite and dolomite mineral groups the $(CO_3)^{2-}$ anions lie perpendicular to the rhombohedral *c* axis and they librate with a helical motion along this axis (Gunasekaran et al. 2006; Fig. 11; Animation 7]).

The planar orientation of the CO_3^{2-} anions in the calcite and dolomite group minerals results in many of their distinctive properties, for example in their extreme optical anisotropy (maximum and minimum refractive indices differ by ~0.2 in these minerals), which causes the familiar double refraction seen through calcite cleavage rhombohedra. The near-perfect [104] cleavage of calcite and dolomite group minerals also arises from anisotropies in bonding; this cleavage plane results in the minimum number of broken Ca-O bonds and no broken C-O bonds. Strong bonding in the plane parallel to the CO_3^{2-} anions (i.e., the *a* axis of the hexagonal setting) compared to weaker Ca-O bonds in the perpendicular direction (the *c* axis of the hexagonal setting) leads to extreme anisotropy in calcite's thermal expansion, as well. The *c*-axis thermal expansion is positive ~ $+3.2 \times 10^{-5}$ °C^{-1}, whereas *a*-axis thermal expansion

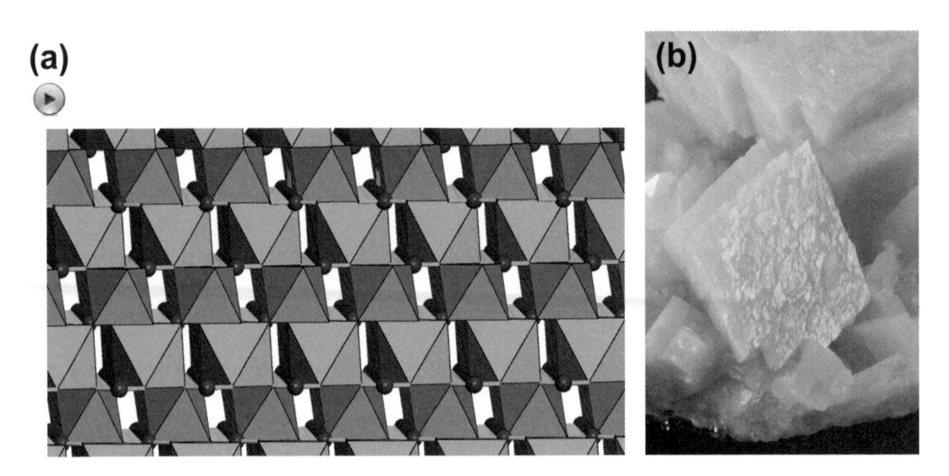

Figure 10. Dolomite. (a) The structure of dolomite [$MgCa(CO_3)_2$] (rhombohedral; space group $R\bar{3}$; hexagonal setting $a = 4.807$ Å; $c = 16.00$ Å; $Z = 3$; rhombohedral setting $a = 6.015$, $\alpha = 47.1°$; $Z = 1$). Blue spheres are carbon atoms, with light green CaO_6 octahedra and dark green MgO_6 octahedra. (b) Dolomite crystals reflect their rhombohedral crystal structure. [Animation 6: *For readers of the electronic version, click the image for an animation of the dolomite crystal structure.*]

Figure 11. The rigid body librating CO_3^{2-} unit of the calcite and dolomite mineral groups. Blue and red ellipsoids represent carbon and oxygen atoms, respectively. [Animation 7: *For readers of the electronic version, click the image for an animation with librating* CO_3^{2-}.]

is negative ~ -0.3×10^{-5} °C^{-1} (Markgraf and Reeder 1985). Note, however, that in magnesite and dolomite the shorter, stronger Mg-O bonds result in thermal expansion that is positive in both *a*- and *c*-axial directions.

Dolomite is by far the most abundant species among the dolomite group minerals. It forms primarily within both sedimentary and metamorphic deposits through the diagenetic replacement of calcite during interaction with Mg-rich fluids. Phase relations in the $CaCO_3$-$MgCO_3$-$FeCO_3$ system, including the phases calcite, magnesite, siderite, and dolomite, as well as magnesian calcite, as summarized by Chang et al. (1997), reveal extensive regions in pressure-temperature-composition space of coexisting calcite group and dolomite group minerals—a topology that is borne out by the common association of calcite and dolomite in sedimentary rocks.

Ankerite and kutnahorite are Fe^{2+}- and Mn^{2+}-bearing dolomites, respectively, with near-continuous solid solutions observed among the Mg, Fe, and Mn end-members (Essene 1983). These phases occur most commonly as a result of hydrothermal alteration of calcite by reduced fluids rich in Fe^{2+} and Mn^{2+}.

The aragonite group

A large number of $CaCO_3$ polymorphs enrich carbonate mineralogy (Carlson 1983; Chang et al. 1997). The aragonite group, which includes the aragonite form of $CaCO_3$ plus cerussite ($PbCO_3$), strontianite ($SrCO_3$), and witherite ($BaCO_3$), prevail in carbonates that contain cations with ionic radii as large or larger than calcium. The crystal structure of aragonite (Speer 1983; Fig. 12; Animation 8), first determined by Bragg (1924), is orthorhombic with the standard space group *Pnam* (with $c < a < b$). However, the structure is more conveniently described in a non-standard orientation with $a < c < b$, resulting in space group *Pmcn*. The structure in this orientation possesses alternating (001) layers of divalent metal cations and $(CO_3)^{2-}$ anions. Two types of $(CO_3)^{2-}$ layers (C_1 and C_2) alternate with two orientations of metal cations (A and B) in a stacking sequence [...AC_1BC_2...] (Speer 1983). Each $(CO_3)^{2-}$ anion is coordinated to 6 divalent metal atoms, and each metal atom is coordinated to 9 oxygen atoms (Fig. 12a).

Aragonite is a denser polymorph of $CaCO_3$ than calcite (2.95 vs. 2.71 g/cm^3), and has been recognized as a mineral characteristic of relatively high-pressure, low-temperature metamorphic environments (McKee 1962; Coleman and Lee 1962; Ernst 1965). Aragonite, however, is also one of the predominant carbonate minerals—along with calcite and dolomite—that comprise the vast amount of carbon mineralization in Earth's surface environments. Although aragonite is the least stable of these minerals, and undergoes rapid recrystallization to thermodynamically more stable forms, it has been recognized as a fundamental constituent of marine depositional environments since at least the Archean (Sumner and Grotzinger 2004), and represents a primary shallow-marine depositional facies throughout the Proterozoic (Grotzinger and Read 1983; Bartley and Kah 2004; Kah and Bartley 2011). Furthermore, since the onset of enzymatic

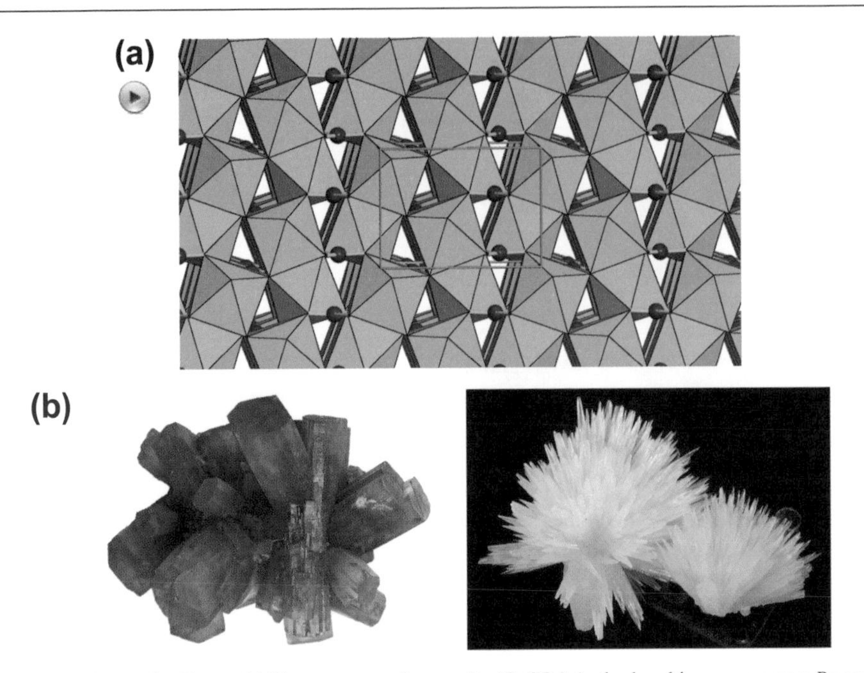

Figure 12. Aragonite Group. (a) The structure of aragonite ($CaCO_3$) (orthorhombic; space group *Pmcn*; a = 4.960 Å; b = 7.964 Å; c = 5.738 Å; Z = 4). Blue spheres are carbon atoms, whereas green polyhedral are CaO_9 groups. (b) The orthorhombic structure of aragonite is manifest in its common crystal forms, including tabular crystals from Tazouta Mine, Sefrou, Morocco (left) and acicular crystals from Transvall, South Africa. Photos courtsey of Rob Lavinski. [Animation 8: *For readers of the electronic version, click the image for an animation of the aragonite crystal structure.*]

biomineralization, more than 500 million years ago, aragonite has been a primary constituent of the fossil record, forming metastable skeletons of some calcareous algae (Bathurst 1976) as well as the skeletal components of a variety of invertebrates, including a wide variety of molluscs, scleractinian corals, and some bryozoans (Cloud 1962; Rucker and Carver 1969; Knoll 2003; Dove 2010). Three other aragonite group minerals, the Sr, Ba, and Pb carbonates strontianite, witherite, and cerussite, respectively, are all found primarily in relatively low-temperature hydrothermal or supergene environments, commonly associated with sulfates and metal sulfide ores (Smith 1926; Mitchell and Pharr 1961; Mamedov 1963; Speer 1977; Dunham and Wilson 1985; Wang and Li 1991). Strontianite, in particular, has also been found in association with enzymatic biomineralization, with strontianite comprising nearly 40% of the skeletal carbonate in some scleractinian corals (McGreegor et al. 1997), although it is uncertain whether strontianite in this case is primary or an early diagenetic phase resulting from recrystallization along the metastable strontianite-aragonite solid solution (Plummer et al. 1992).

Phase diagrams for the aragonite polymorphs of $CaCO_3$ and other calcite group minerals underscore the effects of divalent cation coordination number on carbonate structure type. Several calcite group minerals transform to the aragonite structure at high pressure (Carlson 1983; Yoshioka and Kitano 2011); conversely, the aragonite group minerals strontianite and witherite transform to the calcite structure at high temperature (Chang 1965).

Other anhydrous carbonates

More than 90 anhydrous carbonates other than the above mentioned rhombohedral and orthorhombic species have been described, though only a handful of these diverse phases are common (Table 5; *http://rruff.info/ima/*). Of these minerals, vaterite, huntite, and several minerals associated with carbonatites deserve special note.

Vaterite. Vaterite is a polymorph of $CaCO_3$ that is unstable under ambient conditions, yet plays an important role in biomineralization. Nanocrystalline vaterite rapidly converts to calcite, but can be stabilized indefinitely by a variety of hydrophilic organic molecules. Vaterite's critical role in biomineralization arises from the low energy required for its conversion to other crystalline forms (Xu et al. 2006; Soldati et al. 2008; Wehrmeister et al. 2012). Therefore, it is the preferred mineral phase (along with amorphous calcium carbonate, or ACC) for storing material critical to skeletal growth. Typically, metazoans process nan-aggregates of vaterite (or ACC; Addadi et al. 2003) to form mesoscale syntaxial structures, which may provide an energetically favorable pathway to the construction of larger skeletal elements (e.g., single-crystal echinoderm plates).

Details of the vaterite crystal structure, which must be analyzed using nano-scale powders by X-ray or electron diffraction methods, remains in doubt. It was once thought to be hexagonal (space group $P6_3/mmc$; a = 7.135 Å; c = 16.98 Å; Kahmi 1963). However, Le Bail et al. (2011) found evidence for an orthorhombic structure with 3-fold cyclic twinning (space group $Ama2$; a = 8.472 Å; b = 7.158 Å; c = 4.127 Å; Z = 4). By contrast, Mugnaioli et al. (2012) propose a monoclinic unit cell (a = 12.17 Å, b = 7.12 Å, c = 9.47 Å, β = 118.94°), which is a geometric transformation of the smallest hexagonal cell proposed by Kamhi (1963). All studies agree that vaterite features Ca in distorted 7 or 8 coordination, with an octahedron of 6 Ca-O bonds at ~2.4 Å, and two longer Ca-O bonds at >2.9 Å (Fig. 13; Animation 9). Thus, calcium coordination in vaterite is intermediate between that of calcite (6) and aragonite (9). Wang and Becker (2009) employed first-principles calculations and molecular dynamics simulations to elucidate details of the vaterite structure and CO_3^{2-} group orientations. Previous studies had suggested rotational disorder among carbonate groups, but Wang and Becker (2009) demonstrate a more stable configuration with an ordered CO_3^{2-} superstructure. Vaterite is commonly rotationally disordered when first crystallized, but it can achieve carbonate orientational order through annealing.

Huntite. Huntite [$CaMg_3(CO_3)_4$] occurs as a low-temperature mineral, both by direct precipitation from aqueous solutions enriched in Mg and as an alteration product of dolomite or magnesite (Kinsman 1967). Its hexagonal structure (space group $R32$; a = 9.503 Å; c = 7.821 Å) bears some similarities to rhombohedral carbonates. Magnesium is in octahedral

Figure 13. The crystal structure of vaterite ($CaCO_3$). (hexagonal, space group $P6_3/mmc$; a = 7.135 Å; c = 16.98 Å; Z = 16). Blue spheres and green polyhedral represent carbon and calcium-oxygen polyhedral, respectively. [Animation 9: *For readers of the electronic version, click the image for an animation of the vaterite crystal structure.*]

coordination in all of these carbonates, but calcium in huntite is in trigonal prismatic coordination, as opposed to octahedral coordination in calcite, dolomite, and other species (Dollase and Reeder 1986; Fig. 14 [Animation 10]).

Carbonatite carbonate mineralogy. Finally, a number of exotic carbonate minerals are found associated with carbonatites, which are defined as igneous rocks (extrusive or intrusive) with greater than 50% carbonate minerals (Tuttle and Gittins 1966; Bell 1989; Jones et al. 2013). These rare mantle-derived magmas are stable over a wide temperature range and can erupt as remarkably cool surface lavas at ~500 to 600 °C (Dawson et al. 1990; Church and Jones 1995). Carbonatites encompass a range of compositions, including those dominated by calcite (Ca), dolomite (Ca-Mg), ankerite (Ca-Fe), and alkali (Na-K+Ca) carbonate minerals. Furthermore, these magmas are often enriched in an unusual suite of elements in addition to carbon: alkalis, alkaline earths, fluorine, phosphorus, rare earth elements, and niobium (Deans 1966). As a result of these chemical complexities, approximately 30 different carbonate minerals have been identified from these varied sources (Kapustin 1980; Chang et al. 1997).

Several of the rhombohedral carbonates are found commonly in carbonatites, including calcite, magnesite, siderite, rhodochrosite, and dolomite, as well as orthorhombic aragonite and strontianite (Garson and Morgan 1978; Kapustin 1980; Dziedzic and Ryka 1983). These familiar minerals are accompanied by a number of exotic double carbonates include alstonite and barytocalcite [both polymorphs of $CaBa(CO_3)_2$], norsethite [$BaMg(CO_3)_2$], and the rare earth carbonates burbankite [$(Na,Ca)_3(Sr,Ba,Ce)_3(CO_3)_5$], sahamalite-(Ce) [$Ce_2Mg(CO_3)_4$] and carbocernaite [$(Ca,Na)(Sr,Ce,La)(CO_3)_2$]. Note that reference is sometimes made to "breunnerite" [$(Mg,Fe)CO_3$], but this Fe-rich magnesite is not a valid mineral species.

The alkali carbonatites are exemplified by Oldoinyo Lengai in Tanzania (the only currently active carbonatite volcano), and feature several minerals with the general formula [$(Na,K)_2Ca(CO_3)_2$] (Dawson 1962, 1966; Chang et al. 1997). The two sodium-rich end-members are nyerereite (typically with Na/K ~4.5, though the official IMA chemical formula lacks K) and a possible higher-temperature polymorph zemkorite (with Na/K ~6.2; again, the official IMA-approved formula lacks K); potassium end-member species are fairchildite and a lower-temperature form, bütschliite (Dawson 1962; Mrose et al. 1966; McKie and Frankis 1977; Yergorov et al. 1988; Chang et al. 1997). These phases are also accompanied by the rare sodium carbonate gregoryite [$Na_2(CO_3)$], which can only form in alkali-rich, alkaline earth-poor systems.

Figure 14. The structure of huntite [$CaMg_3(CO_3)_4$]. (hexagonal; space group *R*32; *a* = 9.503 Å, *c* = 7.821 Å; *Z* = 3). The huntite structure is unusual in the trigonal prism coordination of calcium, coupled with octahedral coordination of magnesium. [Animation 10: *For readers of the electronic version, click the image for an animation of the huntite crystal structure.*]

Carbonatites can also develop suites of rare earth element (REE) carbonate or carbonate-fluoride minerals, including the bastnäsite group [REE(CO_3)F, found with dominant REE = Ce, La, Nd, or Y], synchysite and parasite [both CaREE(CO_3)$_2$F, found with dominant REE = Ce, Nd, or Y]], huanghoite-(Ce) [Ba(Ce,REE)(CO_3)$_2$F], and the cordylite group [NaBaREE$_2$(CO_3)$_4$F, found with dominant REE = Ce or La]. REE-carbonatites are often geologically associated with U and Th minerals (Ruberti et al 2008; also see section on "Uranyl carbonates" below). Two of the world's largest economic ore bodies for REE are carbonatites: Mountain Pass California (age 1.37 Ga; Olson et al. 1954; Jones and Wyllie 1986; Castor 2008) and Bayan Obo, China (age 1.35 Ga: Le Bas et al. 1992; Yang et al. 2011). Finally, several hydrous carbonates, including the uncommon REE phases of the ancylite group [(Ca,Sr,Pb)(REE)(CO_3)$_2$(OH)·H_2O, found with dominant REE = Ce, La, or Nd], calkinsite-(Ce) [(Ce,REE)$_2$(CO_3)$_2$·4H_2O], and the lanthanite group [REE$_2$(CO_3)$_3$·8H_2O, found with dominant REE = Ce, La, or Nd], along with two closely associated hydrated Mg-Al carbonates, manasseite [$Mg_6Al_2(CO_3)(OH)_{16}\cdot 4H_2O$] and hydrotalcite [$Mg_6Al_2(CO_3)(OH)_{16}\cdot 4H_2O$], have been reported from some carbonatites (Kapustin 1980).

Hydrous carbonates

Most carbon-bearing minerals—more than 210 of the approximately 320 IMA approved carbonate species—are hydrated or hydrous (Table 6; *http://rruff.info/ima/*). These diverse species, whose classification has been systematized by Mills et al. (2009), occupy numerous specialized near-surface niches but are for the most part volumetrically minor. Most of these diverse minerals have mixed anionic groups, including more than 30 carbonate-sulfates, more than 40 carbonate-silicates, and more than 20 carbonate-phosphates, plus carbonates with uranyl, arsenate, borate, and other ionic groups, which represent near-surface alteration products of other minerals (see Tables 5 and 6 for some of the more common representative examples).

Malachite and azurite. The common hydrous copper carbonates, malachite [$Cu_2(OH)_2CO_3$] and azurite [$Cu_3(OH)_2(CO_3)_2$], which form in the oxidized zone of copper deposits, are typical of a large number of near-surface carbonate phases, including more than 20 other hydrous copper carbonates. These colorful minerals are among the thousands of mineral species that Hazen et al. (2008) identified as potentially biologically mediated (Hazen et al. 2013).

Uranyl carbonates. At least 30 carbonates incorporate the $(UO_2)^{2+}$ uranyl cation group. These fascinating and colorful phases, which commonly form as alteration products of near-surface uranium ore bodies, include 4 anhydrous carbonates: rutherfordine [$(UO_2)(CO_3)$], the isomorphous alkali uranyl carbonates agricolite and čejkaite [$K_4(UO_2)(CO_3)_3$ and $Na_4(UO_2)(CO_3)_3$, respectively], and widenmannite [$Pb_2(UO_2)(CO_3)_3$]. Most of the hydrous and hydrated uranyl carbonates are rare; 20 of these species are documented from five or fewer localities (*http://MinDat.org*), while only four of these hydrated uranium carbonate species are known from more than 30 localities: andersonite [$Na_2Ca(UO_2)(CO_3)_3\cdot 6H_2O$], bayleyite [$Mg_2(UO_2)(CO_3)_3(H_2O)_{12}\cdot 6H_2O$], liebigite [$Ca_2(UO_2)(CO_3)_3\cdot 11H_2O$], and schrockingerite [$NaCa_3(UO_2)(SO_4)(CO_3)_3F\cdot 10H_2O$]. These minerals are all Ca or Mg carbonates that presumably formed by alteration of calcite or dolomite in a relatively oxidized subsurface environment—conditions that must significantly postdate the Great Oxidation Event (Hazen et al. 2008, 2009; Sverjensky and Lee 2010; Hazen et al. 2013).

Hydrated calcium carbonates. In addition to the mineralogically complex hydrated carbonates outlined above, hydrated phases of calcium carbonate—i.e., ikaite [$CaCO_3\cdot 6H_2O$] and its pseudomorphs [which have been given a number of unofficial varietal names such as thinolite, glendonite, jarrowite, and fundylite (Browell 1860; Dana 1884; Shearman and Smith 1985; Ito 1996; Swainson and Hammond 2001)], plus monohydrocalcite ($CaCO_3\cdot H_2O$) and hydromagnesite [$Mg_5(CO_3)_4(OH)_2\cdot 4H_2O$]—play critical roles in the near-surface co-evolution of the geosphere and biosphere. Hydrated calcite is common in modern microbial systems and commonly represents an initial precipitate phase within microbially mediated carbonate deposits. Similarly, pseudomorphs after ikaite provide thermodynamic indicators of cold marine temperatures, and can potentially be used as a unique indicator of glaciated conditions in the geologic past, for example, as pseudomorphs after ikaite in the Mississippian of Alberta, Canada (Brandley and Krause 1997) and in glendonites in Neoproterozoic shallow shelf environments (James et al. 1998).

Minerals incorporating organic molecules

The IMA has recognized approximately 50 organic minerals, and it is likely that many more species remain to be identified and described (Table 7; *http://rruff.info/ima*). These diverse phases fall into three main categories—organic molecular crystals, minerals with organic molecular anions, and clathrates. Note that while many of these minerals are a direct or indirect consequence of biological activity, some minerals with organic molecules may have a nonbiological origin. Consequently, Perry et al. (2007) have also introduced the term "organomineral" to designate "mineral products containing organic carbon," but "not directly produced by living cells." As examples they cite carbon-bearing siliceous hot-spring deposits, desert varnish, stromatolites, and a variety of trace fossils.

Organic molecular crystals. Organic molecular crystals encompass those carbon-bearing minerals in which electrically neutral organic molecules crystalize into a periodic arrangement, principally through van der Waals interactions (Table 7). This diverse group includes a number of species known only as accessory minerals associated with coal; for example, 9 different hydrocarbon minerals, such as kratochvilite ($C_{13}H_{10}$), fichtelite ($C_{19}H_{34}$), dinite ($C_{20}H_{36}$), and evenkite ($C_{24}H_{48}$); acetamide (CH_3CONH_2); and the ring-shaped nickel porphyrin mineral abelsonite, $Ni(C_{31}H_{32}N_4)$. Burning coal mines also produce molecular crystals through sublimation, including kladnoite [$C_6H_4(CO)_2NH$] and hoelite ($C_{14}H_8O_2$; Jehlička et al. 2007). Other molecular crystals are unique to fossilized wood [flagstaffite ($C_{10}H_{22}O_3$)]; roots [refikite ($C_{20}H_{32}O_2$)]; or bat guano deposits in caves, for example the purines uricite ($C_5H_4N_4O_3$) and guanine ($C_5H_5N_5O$), and urea [$CO(NH_2)_2$]. It is intriguing to note that average temperatures on Earth are too high for many crystals of small organic molecules to form, including crystalline forms of carbon dioxide (which has been observed near the poles of Mars; Byrne and Ingersoll 2003), methane, ethane (C_2H_6), ethylene (C_2H_2), and propane (C_3H_8). One can anticipate that these and many other such minerals await discovery in the cold, hydrocarbon-rich near-surface environment of Saturn's moon Titan, as well as in the form of condensed phases in dense molecular clouds (Glein 2012).

Minerals with organic anions. Approximately 25 mineral species incorporate organic anions bonded to Ca, Mg, Cu, Na, and other metal cations (Table 7). Predominant among these organic salt minerals are more than a dozen oxalates with the $(C_2O_4)^{2-}$ anion. The most common oxalate is weddellite [$Ca(C_2O_4)\cdot 2H_2O$], which is found in such varied environments as bat guano, sediments derived from lichens, human kidney stones, cactus (saguaro, *Carnegiea gigantea*), and the depths of the Weddell Sea. Other oxalates include caoxite [$Ca(C_2O_4)\ 3H_2O$], glushinskite [$Mg(C_2O_4)\cdot 2H_2O$], humboldtine [$Fe(C_2O_4)\cdot 2H_2O$], lindbergite [$Mn(C_2O_4)\cdot 2H_2O$], natroxalate [$Na_2(C_2O_4)$], and oxammite [$(NH_4)_2(C_2O_4)\cdot H_2O$] (the latter a biomineral known exclusively from guano deposits).

Other organic anions in minerals include formate $(HCOO)^{-1}$, for example in formicate [$Ca(CHOO)_2$] and dashkovite [$Mg(HCOO)_2\cdot 2H_2O$]; acetyl $(CH_3COO)^{-1}$ in hoganite [$Cu(CH_3COO)_2\cdot H_2O$] and calclacite [$Ca(CH_3COO)Cl\cdot 5H_2O$] (the latter mineral known only from specimens of limestone stored in wooden drawers); methyl sulfonate $(CH_3SO_3)^{-1}$ in ernstburkite [$Mg(CH_3SO_3)_2\cdot 12H_2O$]; and thiocyanate $(SCN)^{-1}$ in julienite [$Na_2Co(SCN)_4\cdot 8H_2O$]. Finally, Rastsvetaeva et al. (1996) describe a Cu^{2+} succinate monohydrate phase that occurs as a consequence of washing copper mineral specimens with detergents.

Table 7. Representative minerals incorporating organic molecules.

Name	Formula
Molecular Crystals: Hydrocarbons	
Kratochvilite	$C_{13}H_{10}$
Fichtelite	$C_{19}H_{34}$
Dinite	$C_{20}H_{36}$
Evenkite	$C_{24}H_{48}$
Other Organic Molecular Crystals	
Acetamide	CH_3CONH_2
Abelsonite	$Ni(C_{31}H_{32}N_4)$
Kladnoite	$C_6H_4(CO)_2NH$
Hoelite	$C_{14}H_8O_2$
Flagstaffite	$C_{10}H_{22}O_3)$
Refikite	$C_{20}H_{32}O_2)$
Uricite	$C_5H_4N_4O_3$
Guanine	$C_5H_5N_5O$
Urea	$CO(NH_2)_2$
Minerals with Organic Anions: Oxalates	
Weddellite	$Ca(C_2O_4)\cdot 2H_2O$
Caoxite	$Ca(C_2O_4)\cdot 3H_2O$
Glushinskite	$Mg(C_2O_4)\cdot 2H_2O$
Humboldtine	$Fe(C_2O_4)\cdot 2H_2O$
Lindbergite	$Mn(C_2O_4)\cdot 2H_2O$
Natroxalate	$Na_2(C_2O_4)$
Oxammite	$(NH_4)_2(C_2O_4)\cdot H_2O$
Other Minerals with Organic Anions	
Formicate	$Ca(CHOO)_2$
Dashkovaite	$Mg(HCOO)_2\cdot 2H_2O$
Hoganite	$Cu(CH_3COO)_2\cdot H_2O$
Calclacite	$Ca(CH_3COO)Cl\cdot 5H_2O$
Ernstburkite	$Mg(CH_3SO_3)_2\cdot 12H_2O$
Julienite	$Na_2Co(SCN)_4\cdot 8H_2O$
Clathrates	
Chibaite	$SiO_2\cdot n(CH_4,C_2H_6,C_3H_8,C_4H_{10})$; ($n_{max} = 3/17$)
Melanophlogite	SiO_2
Methane hydrate*	H_2O [CH_4]

* Not yet an IMA approved mineral species.

Clathrate silicates. Clathrates comprise a third as yet poorly described group of minerals containing organic molecules. These minerals feature open three-dimensional framework structures that incorporate molecules in cage-like cavities. Chibaite and melanophlogite are silica clathrates with open zeolite-like SiO_2 frameworks that may contain hydrocarbons from methane to butane (C_4H_{10}) as the guest molecule (Skinner and Appleman 1963; Momma et al. 2011). Melanophlogite with CO_2 and/or methane as the guest molecule has been found in natrocarbonatite lavas at Oldoinyo Lengai (Carmody 2012).

Clathrate hydrates. Water-based clathrates, also known as gas hydrates or clathrate hydrates, are remarkable crystalline compounds that form at low temperatures (typically < 0 °C) and elevated pressures (> 6 MPa). These materials have attracted considerable attention because of their potential for applications to energy storage and recovery applications (Max

2003; Boswell 2009; Koh et al. 2009, 2011). Different gas clathrate hydrate structure types have a variety of cage sizes and shapes, which depend primarily on the size and character of the gas molecule.

The principal documented natural clathrate hydrate mineral is an as yet unnamed methane hydrate commonly known as "methane ice, which crystallizes in marine sediments of the continental shelves or in permafrost zones below a depth of ~130 (Hyndman and Davis 1992). Methane hydrate is structurally similar to the silica clathrates. Its cubic structure (space group *Pm3m*, a = 12 Å) features a three-dimensional H_2O framework with two types of cages partially filled with CH_4 molecules: a pentagonal dodecahedron (designated 5^{12}, or a cage formed by 12 interconnected 5-member rings of H_2O) and a tetrakaidecahedron (designated $5^{12}6^2$), each with average radii ~4 Å, each holding one CH_4 molecule (Fig. 15; Animation 11). Although the methane content is variable, methane ice holds on average ~0.17 mole of methane per mole of water, corresponding to a density of ~0.9 g/cm^3 (Max 2003).

The geographic distribution of methane hydrate is extensive, with hundreds of confirmed deposits (Hyndman and Spence 1992; Kvenvolden 1995; Milkov 2004). The total methane storage in clathrates was estimated by Allison and Boswell (2007) as 2×10^{16} m^3—a quantity orders of magnitude greater than that represented by all other natural gas reserves. Annual natural gas consumption in North America, by comparison, is ~6×10^{11} m^3 (Boswell 2009). In fact, the methane stored in clathrate hydrates may exceed the energy represented by known reserves of all other fossil fuels combined (Kvenvolden 1995; Grace et al. 2008).

In addition to this common phase, Guggenheim and Koster van Groos (2003) and Koster van Groos and Guggenheim (2009) have reported the synthesis of a possible new gas-hydrate phase that consists of a clay-methane hydrate intercalate. In addition, Chou et al. (2000) have reported other methane clathrate hydrate phases that occur exclusively at high pressures.

Mineral-molecule interactions

Finally, it should be noted that organic molecules often interact strongly with mineral surfaces, especially in aqueous environments (Hazen 2006; Jonsson et al. 2009; Bahri et al. 2011; Cleaves et al. 2011). Such interactions of organic species with mineral surfaces have received special attention for at least two reasons related to the biosphere. First, a number of

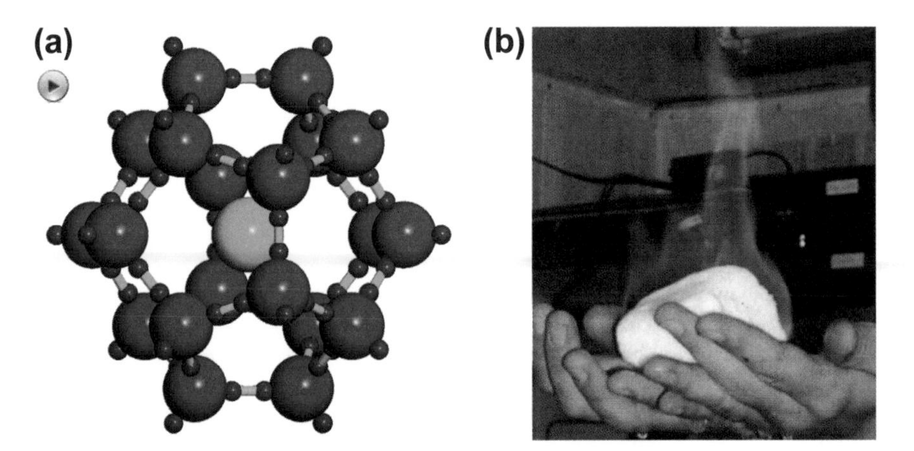

Figure 15. Methane hydrate. (a) The structure of methane hydrate [$H_2O{\cdot}CH_4$] (cubic; space group *Pm3m*; a = 12 Å). (b) Methane hydrate burns at room conditions, as the hydrate melts and methane is released. Photo courtesy of Jan Steffan, Expedition SO148, GEOMAR. [Animation 11: *For readers of the electronic version, click the image for an animation of the methane hydrate crystal structure.*]

minerals have been invoked as possibly playing key roles in the origins of life through the selection, concentration, protection, and templating of biomolecules, as well as the possible catalysis of biomolecules (Bernal 1951; Goldschmidt 1952; Orgel 1998; Lahav 1999; Schoonen et al. 2004; Hazen 2005, 2006). Specific hypotheses focus on the roles of hydroxides (Holm et al. 1993; Pitsch et al. 1995; Hill et al. 1998), quartz (Bonner et al. 1974, 1975; Evgenii and Wolfram 2000), feldspars and zeolites (Smith 1998; Parsons et al. 1998; Smith et al. 1999), carbonates (Hazen et al. 2001), phosphates (Weber 1982, 1995; Acevedo and Orgel 1986), borates (Ricardo et al. 2004; Grew et al. 2011), phosphides (Pasek et al. 2007), and sulfides (Wächtershäuser 1988, 1990, 1993; Russell et al. 1994; Russell and Hall 1997; Huber and Wächtershäuser 1998; Bebié and Schoonen 2000; Cody et al. 2000, 2001, 2004; Cody 2004). In this regard, clay minerals have received special attention for their potential ability to template and catalyze the polymerization of amino acids and nucleotides as steps in the origins of life (Cairns-Smith 1982, 2005; Cairns-Smith and Hartman 1986; Ferris et al. 1996; Ertem and Ferris 1996; Orgel 1998; Ferris 2005).

A second possible influence of mineral-molecule interactions on the biosphere also invokes clay minerals, which might have contributed significantly to the rise of atmospheric oxygen through the adsorption, concentration, and subsequent burial of significant amounts of organic matter from the terrestrial environment (Kennedy et al. 2006). This process of organic burial is one of the most efficient mechanisms for atmospheric oxidation (Berner et al 2000; Hayes and Waldbauer 2006; Hazen et al. 2013).

CONCLUSIONS: UNRESOLVED QUESTIONS IN CARBON MINERALOGY

More than two centuries of mineralogical research have revealed much regarding the varied C-bearing mineral phases in Earth's near-surface environment. Nevertheless, much remains to be learned. Among the most fundamental mineralogical questions—one evident from any museum display of carbonate minerals—is what physical, chemical, and biological processes lead to the remarkable range of calcite crystal forms? No other crystalline phase exhibits such a wide range of morphologies. What environmental factors influence calcite crystal forms? And why don't other rhombohedral carbonates display a similar variety?

Studies of minerals that incorporate organic molecules are in their infancy, and numerous other phases at the interface between the crystalline and biological worlds are likely awaiting discovery. Characterization of such phases may prove especially important in resolving debates regarding abiotic versus biotic origins of some deep organic molecules (Sephton and Hazen 2013; McCollom 2013). Furthermore, organic mineralogy may play a dominant role on Titan, as well as planets and moons in other C-rich star systems. Indeed, carbon mineralogy may provide one of the most sensitive geological indicators of the evolution and present state of other worlds.

ACKNOWLEDGMENTS

We thank John Armstrong, Russell Hemley, Andrea Mangum, Craig Schiffries, and Dimitri Sverjensky for invaluable discussions and suggestions during the preparation of this manuscript. We thank Lauren Cryan, Shaun Hardy, and Merri Wolf, who provided critical technical support. The authors gratefully acknowledge support from the Deep Carbon Observatory, the Alfred P. Sloan Foundation, the Carnegie Institution of Washington, the National Science Foundation, and NASA's Astrobiology Institute for support of this study.

REFERENCES

Acevedo OL, Orgel LE (1986) Template-directed oligonucleotide ligation on hydroxylapatite. Nature 321:790-792

Acheson G (1893) U.S. Patent 492,767. Production of Artificial Crystalline Carbonaceous Material.

Addadi L, Raz S, Weiner S (2003) Taking advantage of disorder: amorphous calcium carbonate and its roles in biomineralization. Adv Mater 15:959-970

Adler HH, Kerr PE (1963) Infrared absorption frequency trends for anhydrous normal carbonates. Am Mineral 48:124-137

Alexander CMO'D (1990) In situ measurement of interstellar silicon carbide in two CM chondrite meteorites. Nature 348:715-717, doi:10.1038/348715a0

Alexander CMO'D (1993) Presolar SiC in chondrites: how variable and how many sources? Geochim Cosmochim Acta 57:2869-2888

Allison E, Boswell R (2007) Methane hydrate, future energy within our grasp, an overview. DOE Overview Document, *http://www.fossil.energy.gov/programs/oilgas/hydrates/*

Anderson RE, Brazelton WJ, Baross JA (2013) The deep viriosphere: assessing the viral impact on microbial community dynamics in the deep subsurface. Rev Mineral Geochem 75:649-675

Angus JC, Hayman CC (1988) Low-pressure metastable growth of diamond and diamondlike phases. Science 241:913-921

Angus JC, Will HA, Stanko WS (1968) Growth of diamond seed crystals by vapor deposition. J Appl Phys 39:2915-2922

Bahri S, Jonsson CM, Jonsson CL, Azzolini D, Sverjensky DA, Hazen RM (2011) Adsorption and surface complexation study of L-DOPA on rutile (TiO_2) in NaCl solutions. Environ Sci Technol 45:3959-3966

Bartley JK, Kah LC (2004) Marine carbon reservoir, C_{org}-C_{carb} coupling, and the evolution of the Proterozoic carbon cycle. Geology 32:129-132

Bathurst RGC (1976) Carbonate Sediments and their Diagenesis (Developments in Sedimentology 12). Elsevier, New York

Beard J (1988) Explosive mixtures. New Scientist 5 November 1988:43-47

Bebié J, Schoonen MAA (2000) Pyrite surface interaction with selected organic aqueous species under anoxic conditions. Geochem Trans 1:47, doi: 10.1186/1467-4866-1-47

Bell K (ed) (1989) Carbonatites: Genesis and Evolution. Unwin Hyman, London

Belluci S (2005) Carbon nanotubes: physics and applications. Phys Status Solidi C 2:34-47

Bergman SC (1987) Lamproites and other potassium-rich igneous rocks: a review of their occurrences, mineralogy, and geochemistry. *In:* Alkaline Igneous Rocks. Fitton JG, Upton BJG (eds) Geol Soc London Spec Publ 30:103-119

Bernal JD (1924) The structure of graphite. Proc R Soc London A 106:749-777

Bernal JD (1951) The Physical Basis of Life. Routledge and Kegan Paul, London

Berner EK, Berner RA (2012) Global Environment: Water, Air and Geochemical Cycles. Princeton University Press, Princeton, New Jersey

Berner RA (2004) The Phanerozoic Carbon Cycle. Oxford University Press, Oxford, UK

Berner RA, Petsch SA, Lake JA, Beerling DJ, Popp BN, Lane RS, Laws EA, Westley MB, Cassar N, Woodward FI, Quick WP (2000) Isotope fractionation and atmospheric oxygen: implications for Phanerozoic O_2 evolution. Science 287:1630-1633

Bhatnagar M, Baliga BJ (1993) Comparison of 6H-SiC, 3C-SiC, and Si for power devices. IEEE Trans Electron Devices 40:645-655, doi:10.1109/16.199372

Bird JM, Weathers MS (1977) Native iron occurrences of Disko Island, Greenland. J Geol 85:359-371

Bonner WA, Kavasmaneck PR, Martin FS, Flores JJ (1974) Asymmetric adsorption of alanine by quartz. Science 186:143-144

Bonner WA, Kavasmaneck PR, Martin FS, Flores JJ (1975) Asymmetric adsorption by quartz: a model for the prebiotic origin of optical activity. Origins Life 6:367-376

Boswell R (2009) Is gas hydrate energy within reach? Science 325:957-958, doi: 10.1126/science.1175074

Bragg WH, Bragg WL (1913) The structure of diamond. Proc Roy Soc London A 89:277-291

Bragg WL (1914) The analysis of crystals by the x-ray spectrometer. Proc Roy Soc London A 89:468-489

Bragg WL (1924) The structure of aragonite. Proc Royal Soc London A 105:16-39

Bragg WL, Claringbull GF, Taylor WH (1965) Crystal Structures of Minerals. Cornell University Press, Ithaca, New York

Brandley RT, Krause FF (1997) Upwelling, thermoclines and wave-sweeping on an equatorial carbonate ramp. SEPM Spec Pub 56:365-390

Brett R (1967) Cohenite: its occurrence and a proposed origin. Geochim Cosmochim Acta 31:143-159

Bridgman PW (1931) The Physics of High Pressure. MacMillan, New York

Bridgman PW (1946) An experimental contribution to the problem of diamond synthesis. J Chem Phys 15:92-98

Brooks CR (1996). Principles of the Heat Treatment of Plain Carbon and Low Alloy Steels. ASM International, Materials Park, Ohio

Browell EJJ (1860) Description and analysis of an undescribed mineral from Jarrow Slake. Tyneside Naturalists Field Club 5:103-104

Bundy FP (1967) Hexagonal diamond—a new form of carbon. J Chem Phys 46:3437-3446, doi: 10.1063/1.1841236

Bundy FP, Bovenkerk HP, Strong HM, Wentorf RH Jr. (1961) Diamond-graphite equilibrium line from growth and graphitization of diamond. J Chem Phys 35:383-391

Bundy FP, Hall HT, Strong HM, Wentorf RH Jr (1955) Man-made diamonds. Nature 176:51-55

Byrne S, Ingersoll AP (2003) A sublimation model for Martian south polar ice features. Science 299:1051-1053, doi: 10.1126/science.1080148

Cairns-Smith AG (1982) Genetic Takeover and the Mineral Origins of Life. Cambridge University Press, Cambridge, UK

Cairns-Smith AG (2005) Sketches for a mineral genetic system. Elements 1:157-161

Cairns-Smith AG, Hartman H (1986) Clay Minerals and the Origin of Life. Cambridge University Press, Cambridge, UK

Capitani GC, Di Pierro S, Tempesta G (2007) The 6H-SiC structure model: further refinement from SCXRD data from a terrestrial moissanite. Am Mineral 92:403-407

Carlson WD (1983) The polymorphs of $CaCO_3$ and the aragonite-calcite transformation. Rev Mineral 11:191-225

Carmody, L. (2012) Geochemical characteristics of carbonatite-related volcanism and sub-volcanic metasomatism at Oldoinyo Lengai, Tanzania. PhD Dissertation, University College London, UK

Castor SB (2008) The Mountain Pass rare-earth carbonatite and associated ultrapotassic rocks, California. Can Mineral 46:779-806

Cesnokov B, Kotrly M, Nisanbajev T (1998) Brennende Abraumhalden und Aufschlüsse im Tscheljabinsker Kohlenbecken - eine reiche Mineralienküche. Mineralien-Welt 9:54-63

Chang LLY (1965) Subsolidus phase relations in the systems $BaCO_3$-$SrCO_3$, $SrCO_3$-$CaCO_3$, and $BaCO_3$-$CaCO_3$. J Geol 73:346-368

Chang LLY, Howie RA, Zussman J (1997) Rock-Forming Minerals. Volume 5B, 2nd Edition. Non-Silicates: Sulphates, Carbonates, Phosphates, Halides. Longman Group, Essex, UK

Chapelle FH, O'Neill K, Bradley PM, Methé BA, Ciufo SA, Knobel LL, Lovley DR (2002) A hydrogen-based subsurface microbial community dominated by methanogens. Nature 415:312-314

Chou I, Sharma A, Burruss RC, Shu J, Mao HK, Hemley RJ, Goncharov AF, Stern LA, Kirby SH (2000) Transformations in methane hydrates. Proc Natl Acad Sci USA 97:13484-13487

Church AA, Jones AP (1995) Silicate-carbonate immiscibility at Oldoinyo-Lengai. J Petrol 36:869-889

Cleaves HJ II, Crapster-Pregont E, Jonsson CM, Jonsson CL, Sverjensky DA, Hazen RM (2011) The adsorption of short single-stranded DNA oligomers to mineral surfaces. Chemosphere 83:1560-1567

Cloud PE (1962) Environment of calcium carbonate deposition west of Andros Island, Bahamas. US Geol Surv Prof Paper 350:1-138

Cody AM, Cody RD (1991) Chiral habit modifications of gypsum from epitaxial-like adsorption of stereospecific growth inhibitors. J Cryst Growth 113:508-529

Cody GD (2004) Transition metal sulfides and the origins of metabolism. Ann Rev Earth Planet Sci 32:569-599

Cody GD, Boctor NZ, Brandes JA, Filley TL, Hazen RM, Yoder HS Jr (2004) Assaying the catalytic potential of transition metal sulfides for abiotic carbon fixation. Geochim Cosmochim Acta 68:2185-2196

Cody GD, Boctor NZ, Filley TR, Hazen RM, Scott JH, Yoder HS Jr (2000) The primordial synthesis of carbonylated iron-sulfur clusters and the synthesis of pyruvate. Science 289:1339-1342

Cody GD, Boctor NZ, Hazen RM, Brandes JA, Morowitz HJ, Yoder HS Jr (2001) Geochemical roots of autotrophic carbon fixation: Hydrothermal experiments in the system citric acid, H_2O-(±FeS)-(±NiS). Geochim Cosmochim Acta 65:3557-3576

Coes L Jr (1962) Synthesis of minerals at high pressures. *In:* Modern Very High Pressure Techniques. Wentorf RH (ed) Butterworths, Washington, p 137-150

Coleman RG, Lee DE (1962) Metamorphic aragonite in the glaucophane schists of Cazadero, California. Am J Sci 260:577-595

Colwell FS, D'Hondt S (2013) Nature and extent of the deep biosphere. Rev Mineral Geochem 75:547-574

Cowlard FC, Lewis JC (1967) Vitreous carbon—a new form of carbon. J Mater Sci 2:507-512, doi: 10.1007/BF00752216

Cox KG (1980) Kimberlite and carbonatite magmas. Nature 283:716-717

D'Hondt S, Jørgensen BB, Miller DJ, Batzke A, Blake R, Cragg BA, Cypionka H, Dickens GR, Ferdelman T, Hinrichs K-U, Holm NG, Mitterer R, Spivack A, Wang G, Bekins B, Engelen B, Ford K, Gettemy G, Rutherford SD, Sass H, Skilbeck CG, Aiello IW, Guèrin G, House CH, Inagaki F, Meister P, Naehr T, Niitsuma S, Parkes RJ, Schippers A, Smith DC, Teske A, Wiegel J, Naranjo Padilla C, Solis Acosta JL (2004) Distributions of microbial activities in deep subseafloor sediments. Science 306:2216-2221, doi: 10.1126/ science.1101155

Dai M, Shi N, Ma Z, Xiong M, Bai W, Fang Q, Yan B, Yang J (2004) Crystal structure determination of tongbaite. Acta Mineral Sinica 24:1-6

Dana ES (1884) A crystallographic study of the thinolite of Lake Lahontan. US Geol Surv Bull 12:429-450

Dana ES (1958) A Textbook of Mineralogy, 4th edition. John Wiley & Sons, New York

Dasgupta R (2013) Ingassing, storage, and outgassing of terrestrial carbon through geologic time. Rev Mineral Geochem 75:183-229

Dasgupta R, Hirschmann MM, Smith ND (2007) Water follows carbon: CO_2 incites deep silicate melting and dehydration beneath midocean ridges. Geology 35:135-138, doi: 10.1130/G22856A.1

Daulton TL, Bernatowicz TJ, Lewis RS, Messenger S, Stadermann FJ, Amari S (2003) Polytype distribution of circumstellar silicon carbide: microstructural characterization by transmission electron microscopy. Geochim Cosmochim Acta 67:4743-4767

Davies G (1984) Diamond. Adam Hilger, Bristol, UK

Dawson JB (1962) Sodium carbonate lavas from Oldoinyo Lengai, Tanganyika. Nature 195:1075-1076

Dawson JB (1966) Oldoinyo Lengai—an active volcano with sodium carbonatite lava flows. *In:* Carbonatites. Tuttle OF, Gittins J (edd) Wiley Interscience, New York, p 155-168

Dawson JB, Pinkerton H, Norton GE, Pyle DM (1990) Physicochemical properties of alkali carbonatite lavas: data from the 1988 eruption of Oldoinyo Lengai, Tanzania. Geology 18:260-263.

Day HW (2012) A revised diamond-graphite transition curve. Am Mineral 97:52-62

Deans T (1966) Economic mineralogy of African carbonatites. *In:* Carbonatites. Tuttle OF, Gittins J (eds) Wiley Interscience, New York, p 385-413

DeCarli PS, Jamieson JC (1961) Formation of diamond by explosive shock. Science 133:1821-1823

Deer WA, Howie RA, Zussman J (1966) An Introduction to Rock-Forming Minerals. John Wiley & Sons, New York

Derjaguin BV, Fedoseev DV (1968) The synthesis of diamond at low pressures. Sci Am 233:102-109

Di Pierro S, Gnos E, Grobéty BH, Armbruster T, Bernasconi SM, Ulmer P (2003) Rock-forming moissanite (natural α-silicon carbide). Am Mineral 88:1817-1821

Dollase WA, Reeder RJ (1986) Crystal structure refinements of huntite, $CaMg_3(CO_3)_4$, with X-ray powder data. Am Mineral 71:163-166

Dove PM (2010) The rise of skeletal biomineralization. Elements 6:37-42

Dove PM, De Yoreo JJ, Weiner S (eds) (2003) Biomineralization. Reviews in Mineralogy and Geochemistry Volume 54. Mineralogical Society of America, Washington, DC

Dunham KC, Wilson AA (1985) Geology of the Northern Penine Orefield, Volume 2. Stainmore to Craven. Economic Memoir, British Geological Survey, London

Dziedzic A, Ryka W (1983) Carbonatites in the Tanjo intrusion (NE Poland). Arch Mineral 38:4-34

Effenberger H, Mereiter K, and Zemann J (1981) Crystal structure of magnesite, calcite, rhodochrosite, siderite, smithsonite, and dolomite, with discussion of some aspects of the stereochemistry of calcium type carbonates. Z Kristallogr 156:233-243

El Goresy A, Donnay G (1968) A new allotropic form of carbon from the Ries crater. Science 161:363-364, doi: 10.1126/science.161.3839.363

El Goresy A, Gillet P, Chen M, Künstler F, Graup G, Stähle V (2001) In situ discovery of shock-induced graphite-diamond phase transition in gneisses from the Ries crater, Germany. Am Mineral 86:611-621

Ernst WG (1965) Mineral paragenesis in Franciscan metamorphic rocks, Panoche Pass, California. Am Mineral 61:1005-1008

Ernston K, Mayer W, Neumair A, Rappenglück B, Rappenglück MA, Sudhaus D, Zeller KW (2010) The Cheimgau crater strewn field: evidence of a Holocene large impact event in southeastern Bavaria, Germany. J Siberian Fed Univ Eng Technol 1:72-103

Ertem G, Ferris JP (1996) Synthesis of RNA oligomers on heterogeneous templates. Nature 379:238-240

Essene EJ (1983) Solid solutions and solvi among metamorphic carbonates with applications to geological thermobarometry. Rev Mineral 11:77-96

Ettmayer P, Lengauer W (1994) Carbides: transition metal solid state chemistry. *In:* Encyclopedia of Inorganic Chemistry. King RB (ed) John Wiley & Sons, New York

Evgenii K, Wolfram T (2000) The role of quartz in the origin of optical activity on Earth. Origins Life Evol Biosphere 30:431-434

Falini G, Albeck S, Weiner S, Addadi L (1996) Control of aragonite or calcite polymorphism by mollusk shell macromolecules. Science 271:67-69

Fang Q, Bai W, Yang J, Xu X, Li G, Shi N, Xiong M, Rong H (2009) Qusongite (WC): a new mineral. Am Mineral 94:387-390

Ferris JP (2005) Mineral catalysis and prebiotic synthesis: montmorillonite-catalyzed formation of RNA. Elements 1:145-149

Ferris JP, Hill AR, Liu R, Orgel LE (1996) Synthesis of long prebiotic oligomers on mineral surfaces. Nature 381:59-61

Field JE (ed) (1979) The Properties of Diamond. Academic Press, New York

Frank FC (1949) The influence of dislocations on crystal growth. Discuss Faraday Soc No. 5:48-54

Frank FC (1951) The growth of carborundum: dislocations and polytypism. Philos Mag 42:1014-1021

Franz AK (2007) The synthesis of biologically active organosilicon small molecules. Curr Opin Drug Discovery Dev 10:654-671

Freund F, Staple A, Scoville J (2001) Organic protomolecule assembly in igneous minerals. Proc Natl Acad Sci USA 98:2142-2147

Frisia S, Borsato A, Fairchild AJ, McDermott F (2000) Calcite fabrics, growth mechanisms, and environments of formation in speleothems from the Italian Alps and SW Ireland. J Sediment Res 70:1183-1196

Frondel C, Marvin UB (1967) Lonsdaleite, a hexagonal polymorph of diamond. Nature 214:587-589, doi: 10.1038/214587a0

Garson MS, Morgan DJ (1978) Secondary strontianite at Kangankunde carbonatite complex, Malawi. Trans Inst Min Metall 87B:70-73

Geim AK (2009) Graphene: status and prospects. Science 234:1530-1534, doi: 10.1126/science.1158877

Geim AK, Novoselov KS (2007) The rise of graphene. Nature Mater 6:183-191

Generalov ME, Naumov VA, Mokhov AV, Trubkin NV (1998) Isovite $(Cr,Fe)_{23}C_6$—a new mineral from the gold-platinum bearing placers of the Urals. Zap Vseross Mineral O-va 127:26-37

Gilmour I, French BM, Franchi IA, Abbott JI, Hough RM, Newton J, Koeberl C (2003) Geochemistry of carbonaceous impactites from the Gardnos impact structure, Norway. Geochim Cosmochim Acta 67:3889-3903

Glein C (2012) Theoretical and experimental studies of cryogenic and hydrothermal organic geochemistry. PhD Dissertation, Arizona State University, Tempe, Arizona

Gold T (1999) The Deep Hot Biosphere. Copernicus, New York

Goldschmidt VM (1952) Geochemical aspects of the origin of complex organic molecules on the earth, as precursors to organic life. New Biol 12:97-105

Goodrich CA (1984) Phosphoran pyroxene and olivine in silicate inclusions in natural iron-carbon alloy, Disko Island, Greenland. Geochim Cosmochim Acta 48:1115-1126

Goodrich CA, Bird JM (1985) Formation of iron-carbon alloys in basaltic magma at Uivfaq, Disko Island: the role of carbon in mafic magmas. J Geol 93:475-492

Gorshkov AI, Bao YN, Bersho LN, Ryabchikov ID, Sivtsov AV, Lapina MI (1997) Inclusions of native metals and other minerals in diamond from kimberlite pipe 50, Liaoning, China. Int Geol Rev 8:794-804

Grace J, Collett T, Colwell F, Englezos P, Jones E, Mansell R, Meekison JP, Ommer R, Pooladi-Darvish M, Riedel M, Ripmeester JA, Shipp C, Willoughby E (2008) Energy from gas hydrates—assessing the opportunities and challenges for Canada. Report of the Expert Panel on Gas Hydrates, Council of Canadian Academies, September 2008

Grew ES, Bada JL, Hazen RM (2011) Borate minerals and origin of the RNA world. Origins Life Evol Biospheres 41:307-316, doi 10.1007/s11084-101-9233-y

Grotzinger JP, Read JF (1983) Evidence for primary aragonite precipitation, lower Proterozoic (1.9 Ga) dolomite, Wopmay orogen, northwest Canada. Geology 11:710-713

Gübelin EJ, Koivula JI (2008) Photoatlas of Inclusions in Gemstones. Opino, Basel, Switzerland

Guggenheim S, Koster van Groos AF (2003) New gas-hydrate phase: synthesis and stability of clay-methane hydrate intercalate. Geology 31:653-656, doi:10.1130/0091-7613(2003)031<0653:NGPSAS>2.0.CO;2

Gunasekaran S, Anbalagan G, Pandi S (2006) Raman and infrared spectra of carbonates of calcite structure. J Raman Spectrosc 37:892-899

Gusev AI, Rempel AA, Lipatnikov VN (1996) Incommensurate ordered phase in non-stoichiometric tantalum carbide. J Phys Condens Matter 8:8277-8293

Hall HT (1970) Personal experiences in high pressure. The Chemist July 1970:276-279

Harlow GE (ed) (1998) The Nature of Diamonds. Cambridge University Press, New York

Hassel O, Mark H (1924) Über die Kristallstruktur des Graphits. Z Phys 25:317-337

Hatano M, Ohsaki T, Arakawa K (1985) Graphite whiskers by new process and their composites, advancing technology in materials and processes. Science of Advanced Materials and Processes, National SAMPE Symposium 30:1467-1476

Hayes JM, Waldbauer JR (2006) The carbon cycle and associated redox processes through time. Philos Trans R Soc London 361:931-950

Hazen RM (1999) The Diamond Makers. Cambridge University Press, New York

Hazen RM (2005) Genesis: The Scientific Quest for Life's Origin. Joseph Henry Press, Washington, DC

Hazen RM (2006) Mineral surfaces and the prebiotic selection and organization of biomolecules. Am Mineral 91:1715-1729 Hazen RM, Ewing RC, Sverjensky DA (2009) Evolution of uranium and thorium minerals. Am Mineral 94:1293-1311

Hazen RM, Downs RT, Kah, L, Sverjensky D (2013) Carbon mineral evolution. Rev Mineral Geochem 75:79-107

Hazen RM, Filley TR, Goodfriend GA (2001) Selective adsorption of L- and D-amino acids on calcite: implications for biochemical homochirality. Proc Natl Acad Sci USA 98:5487-5490

Hazen RM, Hemley RJ, Mangum AJ (2012) Carbon in Earth's interior: storage, cycling, and life. Eos Trans Am Geophys Union 93:17-28
Hazen RM, Papineau D, Bleeker W, Downs RT, Ferry JM, McCoy TJ, Sverjensky DA, Yang H (2008) Mineral evolution. Am Mineral 93:1693-1720
Heimann RB, Evsyukov SE, Kavan L (eds) (1999) Carbyne and Carbynoid Structures. Physics and Chemistry of Materials with Low-Dimensional Structures, Volume 21. Kluwer Academic, Dordrecht, The Netherlands
Helgeson HC, Richard L, McKenzie WF, Norton DL, Schmitt A (2009) A chemical and thermodynamic model of oil generation in hydrocarbon source rocks. Geochim Cosmochim Acta 73:594-695
Hendricks BS (1930) The crystal structure of cementite. Z Kristallogr 74:534-545
Hill AR, Böhler C, Orgel LE (1998) Polymerization on the rocks: negatively-charged α-amino acids. Origins Life Evol Biosphere 28:235-243
Holland HD (1984) The Chemical Evolution of the Oceans and Atmosphere. Princeton University Press, Princeton, New Jersey
Holm NG, Ertem G, Ferris JP (1993) The binding and reactions of nucleotides and polynucleotides on iron oxide hydroxide polymorphs. Origins Life Evol Biosphere 23:195-215
Hough RM, Gilmour I, Pillinger CT, Arden JW, Gilkess KWR, Yuan J, Milledge HJ (1995) Diamond and silicon carbide in impact melt rock from the Ries impact crater. Nature 378:41-44
Hough RM, Gilmour I, Pillinger CT, Langenhorst F, Montanari A (1997) Diamonds from the iridium-rich K-T boundary layer at Arroyo el Mimbral, Tamaulipas, Mexico. Geology 25:1019-1022
Huber C, Wächtershäuser G (1998) Peptides by activation of amino acids with CO on (Ni,Fe)S surfaces: implications for the origin of life. Science 281:670-672
Hull AW (1917) A new method of x-ray crystal analysis. Phys Rev 10:661-696
Hyndman RD, Davis EE (1992) A mechanism for the formation of methane hydrate and sea-floor bottom-simulating reflectors by vertical fluid expulsion. J Geophys Res 97:7025-7041
Hyndman RD, Spence GD (1992) A seismic study of methane hydrate marine bottom simulating reflectors. J Geophys Res 97:6683-6698
Irifune T, Hemley RJ (2012) Synthetic diamond opens windows into the deep Earth. Eos Trans Am Geophys Union 93:65-66
Jagodzinski H (1954a) Fehlordnungserscheinungen und ihr Zusammenhang mit der Polytypie des SiC. Neues Jahrb Mineral Monatsh 1954:49-65
Jagodzinski H (1954b) Polytypism in SiC crystals. Acta Crystallogr 7:300
James NP, Narbonne GM, Sherman AB (1998) Molar-tooth carbonates: shallow subtidal facies of the Mid- to Late Proterozoic. J Sediment Res 68:716-722
Jana D, Walker D (1997) The impact of carbon on element distribution during core formation. Geochim Cosmochim Acta 61:2759-2763
Javoy M (1997) The major volatile elements of the Earth: their origin, behavior, and fate. Geophys Res Lett 24:177-180, doi: 10.1029/96GL03931
Jehlička J, Žáček V, Edwards HGM, Shcherbakova E, Moroz T (2007) Raman spectra of organic compounds kladnoite ($C_6H_4(CO)_2NH$) and hoelite ($C_{14}H_8O_2$): rare sublimation products crystallising on self-ignited coal heaps. Spectrochim Acta A68:1053-1057
Jones AP, Genge M, Carmody L (2013) Carbonate melts and carbonatites. Rev Mineral Geochem 75:289-322
Jones AP, Kearsley AT, Friend CRL, Robin E, Beard A, Tamura A, Trickett S, Claeys P (2005) Are there signs of a large Paleocene impact preserved around Disko Bay, West Greenland? Nuussuaq spherule beds origin by impact instead of volcanic eruption? *In:* Large Meteorite Impacts III. Special Paper 384. Kenkmann T, Horz F, Deutsch A (eds) Geological Society of America, Boulder, Colorado, p 281-298
Jones AP, Wyllie PJ (1986) Solubility of rare earth elements in carbonatite magmas, as indicated by the liquidus surface in $CaCO_3$-$Ca(OH)_2$-$La(OH)_3$. Appl Geochem 1:95-102
Jonsson CM, Jonsson CL, Sverjensky DA, Cleaves HJ, Hazen RM (2009) Attachment of L-glutamate to rutile (TiO_2): a potentiometric, adsorption, and surface complexation study. Langmuir 25:12127-12135
Kah LC, Bartley JK (2011) Protracted oxygenation of the Proterozoic biosphere. Int Geol Rev 53:1424-1442
Kah LC, Sherman AB, Narbonne GM, Kaufman AJ, Knoll AH, James NP (1999) $\delta^{13}C$ isotope stratigraphy of the Mesoproterozoic Bylot Supergroup, Northern Baffin Island: implications for regional lithostratigraphic correlations. Can J Earth Sci 36:313-332
Kamhi SR (1963) On the structure of vaterite, $CaCO_3$. Acta Crystallogr 16:770-772
Kapustin YL (1980) Mineralogy of Carbonatites. Smithsonian Institution, Washington, DC
Kennedy CS, Kennedy GC (1976) The equilibrium boundary between graphite and diamond. J Geophys Res 81:2467-2470
Kennedy MJ, Droser M, Mayer LM, Pevear D, Mrofka D (2006) Late Precambrian oxygenation; inception of the clay mineral factory. Science 311:1446-1449
Kenney JF, Shnyukov YF, Krayishkin VA, Tchebanenko II, Klochko VP (2001) Dismissal of claims of a biological connection for natural petroleum. Energia 22:26-34

Keppler H, Wiedenbeck M, Shcheka SS (2003) Carbon solubility in olivine and the mode of carbon storage in the Earth's mantle. Nature 424:414-416

Kinsman DJJ (1967) Huntite from an evaporate environment. Am Mineral 52:1332-1340

Klein C (2005) Some Precambrian banded iron-formations (BIFs) from around the world: their age, geologic setting, mineralogy, metamorphism, geochemistry, and origin. Am Mineral 90:1473-1499

Klein C, Hurlbut CS Jr (1993) Manual of Mineralogy, 21st edition. Wiley, New York

Knoll AH (2003) Biomineralization and evolutionary history. Rev Mineral Geochem 54:329-356

Koh CA, Sloan ED, Sum AK, Wu DT (2011) Fundamentals and applications of gas hydrates. Ann Rev Chem Biomol Eng 2:237-257

Koh CA, Sum AK, Sloan ED (2009) Gas hydrates: unlocking the energy from icy cages. J Appl Phys 106:061101, doi: 10.1063/1.3216463

Koster van Groos AF, Guggenheim S (2009) The stability of methane hydrate intercalates of montmorillonite and nontronite: implications for carbon storage in ocean-floor environments. Am Mineral 94:372-379

Kramers JD, Smith CB, Lock NP, Harmon RS, Boyd FR (1981) Can kimberlites be generated from an ordinary mantle? Nature 291:53-56

Krishna P, Verma AR (1965) On deduction of silicon-carbide polytypes from screw dislocations. Z Kristallogr 121:36-54

Kroto HW, Heath JR, O'Brien SC, Curl RF, Smalley RE (1985) C_{60}: Buckminsterfullerene. Nature 318:162-163

Kutcherov VG, Bendiliani NA, Alekseev VA, Kenney JF (2002) Synthesis of hydrocarbons from minerals at pressure up to 5 GPa. Proc Russian Acad Sci 387:789-792

Kvenvolden KA (1995) A review of the geochemistry of methane in natural gas hydrate. Org Geochem 23:997-1008

Lahav N (1999) Biogenesis: Theories of Life's Origin. Oxford University Press, New York

Langenhorst F, Deutsch A (2012) Shock metamorphism of minerals. Elements 8:31-36

Langenhorst F, Shafranovsky GI, Masaitis VL, Koivisto M (1999) Discovery of impact diamonds in a Fennoscandian crater and evidence for their genesis by solid-state transformation. Geology 27:747-750

Larsen LM, Pedersen AK (2009) Petrology of the Paleocene picrites and flood basalts in Disko and Nuussuaq, West Greenland. J Petrol 50:1667-1711

Lattimer JM, Grossman L (1978) Chemical condensation sequences in supernova ejecta. Moon Planets 19:169-184

Le Bail A, Ouhenia S, Chateigner D (2011) Microtwinning hypothesis for a more ordered vaterite model. Powder Diffr 26:16-21

Le Bas MJ, Keller J, Kejie T, Wall F, Williams CT, Peishan Z (1992) Carbonatite dykes at Bayan Obo, Inner Mongolia, China. Mineral Petrol 46:195-228

Lee J-S, Yu S-C, Bai W-J, Yang J-S, Fang Q-S, Zhang Z (2006) The crystal structure of natural 33R moissanite from Tibet. Z Kristallogr 221:213-217

Leung I (1990) Silicon carbide cluster entrapped in a diamond from Fuxian, China. Am Mineral 75:1110-1119

Leung I, Guo W, Friedman I, Gleason J (1990) Natural occurrence of silicon carbide in a diamondiferous kimberlite from Fuxian. Nature 346:352-354

Lewis HC (1887) On a diamantiferous peridotite, and the genesis of diamond. Geol Mag 4:22-24

Liang Q, Yan CS, Meng Y, Lai L, Krasnicki S, Mao H-K, Hemley RJ (2009) Recent advances in high-growth rate single crystal CVD diamond. Diamond Rel Mater 18:698-703

Lyakhov AO, Oganov AR (2011) Evolutionary search for superhard materials applied to forms of carbon and TiO_2. Phys Rev B 84:092103

Lyakhovich VV (1980) Origin of accessory moissanite. Int Geol Rev 22:961-970

Mackenzie FT, Bischoff WD, Bishop FC, Loijens M, Schoonmaker J, Wollast R (1983) Magnesian calcites: low-temperature occurrence, solubility and solid-solution behavior. Rev Mineral Geochem 11:97-144

Madar R (2004) Materials science: silicon carbide in contention. Nature 430:974-975 doi: 10.1038/430974a.

Mamedov KM (1963) Barite-witherite mineralization in mountain regions of the Turkmen, S.S.R. Ivz Akad Nauk Turkm SSR, Ser Fiz-Tekh, Khim Geol Nauk 1:78-82

Manning CE, Shock EL, Sverjensky D (2013) The chemistry of carbon in aqueous fluids at crustal and upper-mantle conditions: experimental and theoretical constraints. Rev Mineral Geochem 75:109-148

Markgraf SA, Reeder RJ (1985) High-temperature structure refinements of calcite and magnesite. Am Mineral 70:590-600

Mathez EA, Fogel RA, Hutcheon ID, Marshintsev VK (1995) Carbon isotopic composition and origin of SiC from kimberlites of Yakutia [Sakha], Russia. Geochim Cosmochim Acta 59:781-791

Max MD (2003) Natural Gas Hydrate in Oceanic and Permafrost Environments. Kluwer Academic Publishers, Dordrecht

McCammon CA, Satchel T, Harris JW (2004) Iron oxidation state in lower mantle assemblages II. Inclusions in diamonds from Kankan, Guinea. Earth Planet Sci Lett 222:423-434

McCollom TM (2013) Laboratory simulations of abiotic hydrocarbon formation in Earth's deep subsurface. Rev Mineral Geochem 75:467-494

McCollom TM, Simoneit BR (1999) Abiotic formation of hydrocarbons and oxygenated compounds during thermal decomposition of iron oxalate. Origins Life Evol Biosphere 29:167-186, doi: 10.1023/A:1006556315895

McDonough WF (2003) Compositional model for the Earth's core. *In:* Treatise on Geochemistry. Vol. 2. Mantle and Core. Carlson RW (ed) Elsevier, Oxford, UK, p 547-568

McGreegor RB, Pingitore NE, Lytle FW (1997) Strontianite in coral skeletal aragonite. Science 275:1452-1454

McKee B (1962) Aragonite in the Franciscan rocks of the Pacheo Pass area, California. Am Mineral 47:379-387

McKee DW (1973) Carbon and graphite science. Ann Rev Mater Sci 3:195-231

McKie D, Frankis EJ (1977) Nyererite: a new volcanic carbonate mineral from Oldoinyo Lengai, Tanzania. Z Kristallogr 145:73-95

Meersman F, Daniel I, Bartlett DH, Winter R, Hazael R, McMillain PF (2013) High-pressure biochemistry and biophysics. Rev Mineral Geochem 75:607-648

Megaw HD (1973) Crystal Structures: A Working Approach. Saunders, Philadelphia

Mellor JW (1924) Chapter 39: Carbon. *In:* A Comprehensive Treatise on Inorganic and Theoretical Chemistry. Longmans, London, p 710-771

Melson WG, Switzer G (1966) Plagioclase-spinel-graphite xenoliths in metallic iron-bearing basalts, Disko Island, Greenland. Am Mineral 51:664-676

Meng YF, Yan CS, Kransicki S, Liang Q, Lai J, Shu H, Yu T, Steele AS, Mao H-K, Hemley RJ (2012) High optical quality multicarat single crystal diamond produced by chemical vapor deposition. Phys Status Solidi A 209:101-104

Mikhail S (2011) Stable isotope fractionation during diamond growth and the Earth's deep carbon cycle. Ph.D thesis. University College London, UK

Mikhail S, Jones AP, Hunt SA, Guillermier C, Dobson DP, Tomlinson E, Dan H, Milledge H, Franchi I, Wood I, Beard A, Verchovsky S (2010). Carbon isotope fractionation between natural Fe-carbide and diamond; a light C isotope reservoir in the deep Earth and core? American Geophysical Union, Fall Meeting 2010, Abstract U21A-0001

Mikhail S, Shahar A, Hunt SA, Jones AP, Verchovsky AB (2011) An experimental investigation of the pressure effect on stable isotope fractionation at high temperature; implications for mantle processes and core formation in celestial bodies. Lunar Planet Sci Conf 42:1376

Milkov AV (2004) Global estimates of hydrate-bound gas in marine sediments: how much is really out there? Earth Sci Rev 66:183-197, doi:10.1016/j.earscirev.2003.11.002

Mills SJ, Hatert F, Nickel EH, Ferraris G (2009) The standardisation of mineral group hierarchies: application to recent nomenclature proposals. Eur J Mineral 21:1073-1080, doi: 10.1127/0935-1221/2009/0021-1994

Mitchell RH (1995) Kimberlites, Orangeites, and Related Rocks. Plenum, New York

Mitchell RS, Phaar RF (1961) Celestite and calciostrontianite from Wise County, Virginia. Am Mineral 46:189-195

Moissan H (1904a) The Electric Furnace. Edward Arnold, London

Moissan H (1904b) Nouvelles recherches sur la météorité de Cañon Diablo. Comptes Rendus 139:773-786

Momma K, Ikeda T, Nishikubo K, Takahashi N, Honma C, Takada M, Furukawa Y, Nagase T, Kudoh Y (2011) New silica clathrate minerals that are isostructural with natural gas hydrates. Nature Commun 2:196-197

Moore RO, Gurney JJ (1989) Mineral inclusions in diamond from the Monastery kimberlite, South Africa. Spec Publ Geol Surv Australia 14:1029-1041

Morgan P (2005) Carbon Fibers and their Composites. CRC Press, Boca Raton, Florida

Morse JW, Andersson AJ, Mackenzie FT (2006) Initial responses of carbonate-rich shelf sediments to rising atmospheric pCO_2 and "ocean acidification": role of high Mg-calcites. Geochim Cosmochim Acta 70:5814-5830

Morse JW, Mackenzie FT (1990) Geochemistry of Sedimentary Carbonates. Elsevier, Amsterdam

Mrose ME, Rose HJ, Marinenko JW (1966) Synthesis and properties of fairchildite and buetschliite: their relationship in wood-ash stone formation. Geol Soc Am Spec Paper 101:146

Mugnaioli E, Andrusenko I, Schüler T, Loges N, Dinnebier RE, Panthöfer M, Tremel W, Kolb U (2012) Ab-initio structure determination of vaterite by automated electron diffraction. Angew Chem 124:7148-7152, doi: 10.1002/anie.200123456

Nadler MR, Kempter CP (1960) Some solidus temperatures in several metal-carbon systems. J Phys Chem 64:1468-1471

Nasibulin AG, Pikhitsa PV, Jiang H, Brown DP, Krasheninnikov AV, Anisimov AS, Queipo P, Moisala A, Gonzalez D, Lientschnig G, Hassanien A, Shandakov SD, Lolli G, Resasco DE, Choi M, Tomanek D, Kauppinen EI (2007). A novel hybrid carbon material. Nature Nanotechnol 2:156-161.

Nawrocki, J. (1997) The silanol group and its role in liquid chromatography. J Chromatogr A779:29-71

Ni H, Keppler H (2013) Carbon in silicate melts. Rev Mineral Geochem 75:251-287

Nicheng S, Bai W, Li G, Xiong M, Fang Q, Yang J, Ma Z, Rong H (2008) Yarlongite: a new metallic carbide mineral. Acta Geol Sinica 83:52-56

Nicheng S, Ma Z, Xiong M, Dai M, Bai W, Fang Q, Yan B, Yang J (2005) The crystal structure of $(Fe_4Cr_4Ni)_9C_4$. Sci China Ser D Earth Sci 48:338-345

Nordenskiöld AE (1872) Account of an expedition to Greenland in the year 1870. Geol Mag 9:289-306, 355-368, 409-427, 449-463, 516-524

Novgorodova MI, Generalov ME, Trubkin NV (1997) The new TaC-NbC isomorphic row and niobocarbide—a new mineral from platinum placers of the Urals. Proc Russian Mineral Soc 126:76-95

Oganov AR, Hemley RJ, Hazen RM, Jones AP (2013) Structure, bonding, and mineralogy of carbon at extreme conditions. Rev Mineral Geochem 75:47-77

Oleynikov BV, Okrugin AV, Tomshin MD (1985) Native Metals Formation in Basic Rocks of the Siberian Platform. Yakutian Branch of the Soviet Academy of Science, Yakutsk

Olson JC, Shawe DR, Pray LC, Sharp WN (1954) Rare-earth mineral deposits of the Mountain Pass district, San Bernardino County, California. US Geol Surv Prof Paper 261

Orgel LE (1998) Polymerization on the rocks: theoretical introduction. Origins Life Evol Biosphere 28:227-234

Orme CA, Noy A, Wierzbicki A, McBride MT, Grantham M, Teng HH, Dove PM, DeYoreo JJ (2001) Formation of chiral morphologies through selective binding of amino acids to calcite surface steps. Nature 411:775-779

Otter ML, Gurney JJ (1989) Mineral inclusions in diamond from the Sloan diatreme, Colorado-Wyoming state line kimberlite district, North America. Spec Publ Geol Surv Australia 14:1042-1053

Pan Z, Sun H, Zhang Y, Chen C (2009) Harder than diamond: superior indentation strength of wurtzite BN and lonsdaleite. Phys Rev Lett 102:055503, doi: 10.1103/PhysRevLett.102.055503

Pandey D, Krishna P (1975a) A model for the growth of anomalous polytype structures in vapour grown SiC. J Cryst Growth 31:66-71

Pandey D, Krishna P (1975b) Influence of stacking faults on the growth of polytype structures II. Silicon carbide polytypes. Philos Mag 31:1133-1148

Pandey D, Krishna P (1978) Advances in Crystallography Oxford and IBH, New Delhi, India

Parkes RJ, Craig BA, Bale SJ, Getiff JM, Goodman K, Rochelle PA, Fry JC, Weightman AJ, Harvey SM (1993) Deep bacterial biosphere in Pacific Ocean sediments. Nature 371:410-413

Parsons I, Lee MR, Smith JV (1998) Biochemical evolution II: Origin of life in tubular microstructures in weathered feldspar surfaces. Proc Natl Acad Sci USA 95:15173-15176

Pasek MA, Dworkin JP, Lauretta DS (2007) A radical pathway for organic phosphorylation during schreibersite corrosion with implications for the origin of life. Geochim Cosmochim Acta 71:1721-1736

Pauly H (1969) White cast iron with cohenite, schreibersite, and sulphides from the Tertiary basalts on Disko, Greenland. Medd Dansk Geol Foren 19:8-30

Pearson DG, Canil D, Shirey SB (2007) Mantle samples included in volcanic rocks: xenoliths and diamonds. *In:* Treatise on Geochemistry. Vol. 2. Mantle and Core. Carlson RW (ed) Elsevier, Oxford, UK, p 171-275

Pedersen AK (1979) Basaltic glass with high-temperature equilibrated immiscible sulphide bodies with native iron from Disko, central West Greenland. Contrib Mineral Petrol 69:397-407

Pedersen AK (1981) Armalcolite-bearing Fe-Ti oxide assemblages in graphite-equilibrated salic volcanic rocks with native iron from Disko, central West Greenland. Contrib Mineral Petrol 77:307-324

Perry RS, Mcloughlin N, Lynne BY, Sephton MA, Oliver JD, Perry CC, Campbell K, Engel MH, Farmer JD, Brasier MD, Staley JT (2007) Defining biominerals and organominerals: direct and indirect indicators of life. Sediment Geol 201:157-179

Pitsch S, Eschenmoser A, Gedulin B, Hui S, Arrhenius G (1995) Mineral induced formation of sugar phosphates. Origins Life Evol Biosphere 25:297-334

Plummer LN, Busenberg E, Glynn PD, Blum AE (1992) Dissolution of strontiantire solid solutions in nonstoichiometric $SrCO_3$-$CaCO_3$-CO_2-H_2O solutions. Geochim Cosmochim Acta 56:3045-3072

Pollock MD, Kah LC, Bartley JK (2006) Morphology of molar-tooth structures in Precambrian carbonates: influence of substrate rheology and implications for genesis. J Sediment Res 76:310-323

Price GD, Yeomans JM (1984) The application of the ANNNI model to polytypic behaviour. Acta Crystallogr B40:448-454

Qi XX, Yang ZQ, Xu JS, Bai WJ, Zhang ZM, Fang QS (2007) Discovery of moissanitein retrogressive eclogite from the pre-pilot hole of the Chinese Continental Drilling Program (CCSD-PP2) and its geological implication. Acta Petrol Sinica 23:3207-3214

Rastsvetaeva RK, Yu D, Pushcharovsky DY, Furmanova NG, Sharp H (1996) Crystal and molecular structure of Cu(II) succinate monohydrate or "Never wash copper minerals with detergents". Z Kristallogr 211:808-811

Reeder RJ (1983b) Crystal chemistry of the rhombohedral carbonates. Rev Mineral 11:1-47
Reeder RJ (ed) (1983a) Carbonates: Mineralogy and Chemistry. Rev Mineral, Vol 11. Mineralogical Society of America, Washington, DC
Ricardo A, Carrigan MA, Olcott AN, Benner SA (2004) Borate minerals stabilize ribose. Science 303:196
Robinson PT, Malpas J, Cameron S, Zhou MF, Bai WJ (2001) An ultrahigh presure mineral assemblage from the Luobusa Ophiolite, Tibet. Eleventh Annual V. M. Goldschmidt Conference, Abstract 3138
Rode AV, Gamaly EG, Luther-Davies B (2000) Formation of cluster-assembled carbon nanofoam by high-repetition-rate laser ablation. Appl Phys A Mater Sci Process A70:135-144, doi: 10.1007/s003390050025
Rossini FD, Jessup RS (1938) Heat and free energy of formation of carbon dioxide, and of the transition between graphite and diamond. J Res Natl Bur Stand USA 21:491-513
Roussel EG, Cambon Bonavita M-A, Querellou J, Cragg BA, Webster G, Prieur D, Parkes RJ (2008) Extending the sub-sea-floor biosphere. Science 320:1046, doi: 10.1126/ science.1154545
Ruberti E, Gaston ER, Gomes CB, Comin-Chiaramonti P (2008) Hydrothermal REE fluorocarbonate mineralization at Barra do Itapirapua, a multiple stockwork carbonatite, southern Brazil. Can Mineral 46:901-914
Rucker JB, Carver RE (1969) A survey of the carbonate mineralogy of Cheilostome Bryozoa. J Paleontol 43:791-799
Rumble D III, Duke EF, Hoering TC (1986) Hydrothermal graphite in New Hampshire: evidence of carbon mobility during regional metamorphism. Geology 14:452-455
Rumble D III, Hoering TC (1986) Carbon isotope geochemistry of graphite vein deposits from New Hampshire, U. S. A. Geochim Cosmochim Acta 50:1239-1247
Rundel R (2001) Polycyclic aromatic hydrocarbons, phthalates, and phenols. *In:* Indoor Air Quality Handbook. Spengler JD, Samet JM, McCarthy JF (eds) McGraw-Hill, New York, p 34.1-34.2
Russell MJ, Daniel RM Hall AJ, Sherringham J (1994) A hydrothermally precipitated catalytic iron-sulphide membrane as a first step toward life. J Mol Evol 39:231-243
Russell MJ, Hall AJ (1997) The emergence of life from iron monosulphide bubbles at a submarine hydrothermal redox and pH front. J Geol Soc London 154:377-402
Rutt HN, Nicola JH (1974) Raman spectra of carbonates of calcite structure. J Phys C 7:4522-4528
Saddow SE, Agarwal A (2004) Advances in Silicon Carbide Processing and Applications. Artech House, Norwood, Massachusetts
Sameshima T, Rodgers KA (1990) Crystallography of 6H silicon carbide from Seddonville, New Zealand. Neues Jahrb Mineral Monatsh 1990:137-143
Sano N, Wang H, Chhowalla M, Alexandrou I, Amaratunga GAJ (2001) Synthesis of carbon 'onions' in water. Nature 414:506-507, doi: 10.1038/35107141
Schoonen MAA, Smirnov A, Cohn C (2004) A perspective on the role of minerals in prebiotic synthesis. AMBIO 33:539-551
Schrenk MO, Brazelton WJ, Lang SQ (2013) Serpentinization, carbon, and deep life. Rev Mineral Geochem 75:575-606
Scott HP, Hemley RJ, Mao HK, Hershbach DR, Fried LE, Howard WM, Bastea S (2004) Generation of methane in Earth's mantle: in situ high pressure-temperature measurements of carbon reduction. Proc Natl Acad Sci USA 101:14023-14026, doi: 10.1073/pnas.0405930101
Sephton MA, Hazen RM (2013) On the origins of deep hydrocarbons. Rev Mineral Geochem 75:449-465
Shcheka SS, Wiedenbeck M, Frost DJ, Keppler H (2006) Carbon solubility in mantle minerals. Earth Planet Sci Lett 245:730-742
Shearman DJ, Smith AJ (1985) Ikaite, the parent mineral of jarrowite-type pseudomorphs. Proc Geol Assoc 96:305-314
Sherwood-Lollar B, Westgate TD, Ward JA, Slater GF, Lacrampe-Couloume G (2002) Abiogenic formation of alkanes in the Earth's crust as a minor source for global hydrocarbon reservoirs. Nature 416:522-524
Shi N, Bai W, Li G, Xiong M, Fang Q, Yang J, Ma Z, Rong H (2009) Yarlongite: a new metallic carbide mineral. Acta Geol Sinica (Engl Ed) 83:52-56
Shirey SB, Cartigny P, Frost DJ, Keshav S, Nestola F, Nimis P, Pearson DG, Sobolev NV, Walter MJ (2013) Diamonds and the geology of mantle carbon. Rev Mineral Geochem 75:355-421
Shirey SB, Harris JW, Richardson SH, Fouch MJ, James DE, Cartigny P, Deines P, Viljoen F (2002) Diamond genesis, seismic structure, and evolution of the Kaapvaal-Zimbabwe craton. Science 297:1683-1686
Shirey SB, Richardson SH (2011) Start of the Wilson cycle at 3 Ga shown by diamonds from the subcontinental mantle. Science 333:434-436, doi: 10.1126/science.1206275
Shiryaev AA, Griffin WL, Stoyanov E (2011) Moissanite (SiC) from kimberlites: polytypes, trace elements, inclusions and speculations on origin. Lithos 122:152-164
Shvartsburg AA, Hudgins RR, Gutierrez R, Jungnickel G, Frauenheim T, Jackson KA, Jarrold MF (1999) Ball-and-chain dimers from a hot fullerene plasma. J Phys Chem 103:5275-5284
Skinner BJ, Appleman DE (1963) Melanophlogite, a cubic polymorph of silica. Am Mineral 48:854-867

Smith G (1926) A contribution to the mineralogy of New South Wales. Geol Surv New South Wales Mineral Res 34:1-145

Smith JV (1998) Biochemical evolution. I. Polymerization on internal, organophilic silica surfaces of dealuminated zeolites and feldspars. Proc Natl Acad Sci USA 95:3370-3375

Smith JV, Arnold FP Jr, Parsons I, Lee MR (1999) Biochemical evolution III: Polymerization on organophilic silica-rich surfaces, crystal-chemical modeling, formation of first cells, and geological clues. Proc Natl Acad Sci USA 96:3479-3485

Soldati AL, Jacob DE, Wehrmeister U (2008) Structural characterization and chemical composition of aragonite and vaterite in freshwater cultured pearls. Mineral Mag 72:579-592

Sommer H, Lieb KR, Hauzenberger C (2007) Diamonds, xenoliths and kimberlites: a window into the earth's mantle. UNESCO IGCP 557. Geochim Cosmochim Acta 71:A954

Spear KE, Dismukes JP (ed) (1994) Synthetic Diamonds: Emerging CVD Science and Technology. Wiley, New York

Speer JA (1977) The orthorhombic carbonate minerals of Virginia. Rocks Minerals 52:267-274

Speer JA (1983) Crystal chemistry and phase relationships of the orthorhombic carbonates. Rev Mineral 11:145-190

Stanley SM, Hardie LA (1998) Secular oscillations in the carbonate mineralogy of reef-building and sediment-producing organisms driven by tectonically forced shifts in seawater chemistry. Palaeogeogr Palaeoclimatol Palaeoecol 144:3-19

Strong H (1989) Early diamond making at General Electric. Am J Sci 57:794-802

Suits GG (1960) The Synthesis of Diamonds—A Case History in Modern Science. General Electric, Schenectady, New York

Suits GG (1965) Speaking of Research. Wiley, New York

Sumner DY, Grotzinger JP (1996) Herringbone calcite: petrography and environmental significance. J Sediment Res 66:419-429

Sumner DY, Grotzinger JP (2004) Implications for Neoarchaean ocean chemistry from primary carbonate mineralogy of the Campbellrand-Malmani Platform, South Africa. Sedimentology 51:1273-1299

Sverjensky DA, Lee N (2010) The Great Oxidation Event and mineral diversification. Elements 6:31-36

Swainson IP, Hammond RP (2001) Ikaite, $CaCO_3.6H_2O$: cold comfort for glendonites as palaeothermometers. Am Mineral 86:1530-1533

Teng HH, Dove PM (1997) Surface site-specific interactions of aspartate with calcite during dissolution: implications for biomineralization. Am Mineral 82:878-887

Teng HH, Dove PM, Orme C, DeYoreo JJ (1998) The thermodynamics of calcite growth: a baseline for understanding biomineral formation. Science 282:724-727

Tian P, Fang Q, Chen K, Peng Z (1983) A study on tongbaite—a new mineral. Acta Mineral Sinica 4:241-245

Tran NT, Min T, Franz AK (2011) Silanediol hydrogen bonding activation of carbonyl compounds. Chemistry Eur J 17:9897-9900, doi: 10.1002/chem.201101492

Treiman AH, Lindstrom DJ, Schwandt CS, Franchi IA, Morgan ML (2002) A "mesosiderite" rock from Northern Siberia, Russia: Not a meteorite. Meteorit Planet Sci 37:B13-B22

Trumbull RB, Yang J-S, Robinson PT, Di Pierro S, Vennemann T, Weidenbeck M (2009) The carbon isotope composition of natural SiC (moissanite) from Earth's mantle: new discoveries from ophiolites. Lithos 113:612-620

Tuttle OF, Gittins J (ed) (1966) Carbonatites. Wiley Interscience, New York

Ulff-Moller F (1986) A new 10 tons iron boulder from Disko, West Greenland. Meteorit Planet Sci 21:464

Verma AR, Krishna P (1966) Polymorphism and Polytypism in Crystals. Wiley, New York

von Platen B (1962) A multiple piston, high pressure, high temperature apparatus. *In:* Modern Very High Pressure Techniques. Wentorf RH (ed) Butterworths, Washington, p 118-136

Wächtershäuser G (1988) Before enzymes and templates: theory of surface metabolism. Microbiol Rev 52:452-484

Wächtershäuser G (1990) The case for the chemoautotrophic origin of life in an iron-sulfur world. Origins Life Evol Biosphere 20:173-176

Wächtershäuser G (1993) The cradle chemistry of life: on the origin of natural products in a pyrite-pulled chemoautotrophic origin of life. Pure Appl Chem 65:1343-1348

Walter MJ, Bulanova GP, Armstrong LS, Keshav S, Blundy JD, Gudfinnsson G, Lord OT, Lennie AR, Clark SM, Smith CB, Gobbo L (2008) Primary carbonatite melt from deeply subducted oceanic crust. Nature 454:622-625

Walter MJ, Kohn SC, Araujo D, Bulanova GP, Smith CB, Gaillou E, Wang J, Steele A, Shirey SB (2011) Deep mantle cycling of oceanic crust: evidence from diamonds and their mineral inclusions. Science 334:54-57, doi: 10.1126/science.1209300

Wang J, Becker U (2009) Structure and carbonate orientation of vaterite ($CaCO_3$). Am Mineral 94:380-386

Wang Z, Li G (1991) Barite and witherite in sedimentary rocks of the southeastern part of the Siberian platform. Tr Mineral Muz, Akad Nauk SSSR 22:207-210

Wasastjerna JA (1924) The crystal structure of dolomite. Soc Sci Fenn Commentat Phys-Math 2:1-14
Weber A (1982) Formation of pyrophosphate on hydroxyapatite with thioesters as condensing agents. BioSystems 15:183-189
Weber A (1995) Prebiotic polymerization: oxidative polymerization of 2,3-dimercapto-1-propanol on the surface of iron(III) hydroxide oxide. Origin Life Evol Biosphere 25:53-60
Wehrmeister U, Jacob DE, Soldati AL (2011) Amorphous, nanocrystalline, and crystalline calcium carbonates in biological materials. J Raman Spectrosc 42:926-935
Wentorf R (ed) (1962) Modern Very High Pressure Research. Butterworths, Washington
White WB (1974) The carbonate minerals. Mineral Soc Monograph 4:227-284
Wilson L, Head JW (2007) An integrated model of kimberlite ascent and eruption. Nature 447:53-57
Wood BJ (1993) Carbon in the core. Earth Planet Sci Lett 117:593-607, doi: 10.1016/ 0012-821X(93)90105-I
Wood BJ, Li J, Shahar A (2013) Carbon in the core: its influence on the properties of core and mantle. Rev Mineral Geochem 75:231-250
Wyckoff RWG, Merwin HE (1924) The crystal structure of dolomite. Am J Sci 8:447-461
Xu AW, Antonietti M, Colfen H, Fang Y-P (2006) Uniform hexagonal plates of vaterite $CaCO_3$ mesocrystals formed by biomimetic mineralization. Adv Funct Mater 16:903-908
Xu J, Mao H-K (2000) Moissanite: A window for high-pressure experiments. Science 290:783-787, doi: 10.1126/science.290.5492.783
Xu S-T, Wu W-P, Xiao W-S, Yang J-S, Chen J, Ji S-Y, Liu Y-C (2008) Moissanite in serpentinite from the Dabie Mountains in China. Mineral Mag 72:899-908
Yang K-F, Fan H-R, Santosh M, Hu F-F, Wang K-Y (2011) Mesoproterozoic carbonatitic magmatism in the Bayan Obo deposit, Inner Mongolia, north China—Constraints for the mechanism of super accumulation of rare earth elements. Ore Geol Rev 40:122-131
Yergorov NK, Ushchapovskaya ZF, Kashayev AA, Bogdanov GV, Sizykh YI (1988) Zemkorite, $Na_2Ca(CO_3)_2$, a new carbonate from Yakutian kimberlites. Dokl Akad Nauk SSSR 301:188-193
Yoshioka S, Kitano Y (2011) Transformation of aragonite to calcite through heating. Geochem J 19:245-249
Zaitsev AM (2001) Optical Properties of Diamond. Springer, Berlin
Zhang C, Duan ZH (2009) A model for C-O-H fluid in the Earth's mantle. Geochim Cosmochim Acta 73:2089-2102, doi: 10.1016/j.gca.2009.01.021
Zhu Q, Oganov AR, Salvado M, Pertierra P, Lyakhov AO (2011) Denser than diamond: ab initio search for superdense carbon allotropes. Phys Rev B 83:193410

Reviews in Mineralogy & Geochemistry
Vol. 75 pp. 47-77, 2013

3

Structure, Bonding, and Mineralogy of Carbon at Extreme Conditions

Artem R. Oganov

Department of Geosciences
State University of New York
Stony Brook, New York 11794-2100, U.S.A.

artem.oganov@stonybrook.edu

Russell J. Hemley, Robert M. Hazen

Geophysical Laboratory, Carnegie Institution of Washington,
5251 Broad Branch Road NW
Washington, DC 20015, U.S.A.

rhemley@ciw.edu rhazen@ciw.edu

Adrian P. Jones

Depatment of Earth Sciences, University College London
Gower Street, London WC1E 6BT United Kingdom

adrian.jones@ucl.ac.uk

INTRODUCTION

The nature and extent of Earth's deep carbon cycle remains uncertain. This chapter considers high-pressure carbon-bearing minerals, including those of Earth's mantle and core, as well as phases that might be found in the interiors of larger planets outside our solar system. These phases include both experimentally produced and theoretically predicted polymorphs of carbon dioxide, carbonates, carbides, silicate-carbonates, as well as very high-pressure phases of pure carbon. One theme in the search for possible high *P-T*, deep-Earth phases is the likely shift from sp^2 bonding (trigonal coordination) to sp^3 bonding (tetrahedral coordination) in carbon-bearing phases of the lower mantle and core, as exemplified by the graphite-to-diamond transition (Bundy et al. 1961; Davies 1984). A similar phenomenon has been documented in the preferred coordination spheres of many elements at high pressure. For example, silicon is ubiquitously found in tetrahedral coordination in crustal and upper mantle minerals, but adopts octahedral coordination in many high-pressure phases. Indeed, the boundary between Earth's transition zone and lower mantle may be described as a crystal chemical shift from 4-coordinated to 6-coordinated silicon (Hazen and Finger 1978; Finger and Hazen 1991). Similarly, magnesium and calcium commonly occur in octahedral 6-coordination in minerals at ambient conditions, but transform to 8- or greater coordination in high-pressure phases, as exemplified by the calcite-to-aragonite transformation of $CaCO_3$ and the pyroxene-to-perovskite and post-perovskite transformations of $MgSiO_3$ (Murakami et al. 2004; Oganov and Ono 2004). Consequently, a principal focus in any consideration of deep-Earth carbon minerals must include carbon in higher coordination, and even more complex bonding at more extreme conditions that characterize the interiors of larger planets.

1529-6466/13/0075-0003$00.00 DOI: 10.2138/rmg.2013.75.3

THEORETICAL CONSIDERATIONS

We briefly review theoretical methods used to examine dense carbon-bearing minerals, focusing on first-principles or *ab initio* approaches. To compute the energies, one can choose one among a hierarchy of theoretical approximations. The energetics of the phases considered in this chapter have also been studied using quantum chemistry methods. These approaches have been applied, for example, to pure carbon phases (Guth 1990; Che et al. 1999). A leading approach is density functional theory (Hohenberg and Kohn 1964; Kohn and Sham 1965), which in principle is an exact quantum-mechanical theory, but in practice requires approximations, such as the LDA (local density approximation: Perdew and Wang 1992). Early LDA calculations proved successful in predicting the high-pressure behavior of carbon (e.g., Fahy et al. 1986; Fahy and Louie 1977).

Recent extensions of the LDA include GGA (generalized gradient approximation: Perdew et al. 1996), meta-GGA (Tao et al. 2003), or higher-level approximations currently under development. The only approximate term in the equations is the exchange-correlation energy (the non-classical part of the electron-electron interaction energy), the most successful approximations of which are based on the properties of the electron gas, with more advanced approximations taking into account more non-local features, for example, gradient, Laplacian, or orbital kinetic energy density. The usual accuracy of such approximations as LDA and GGA is on the order of 1-2% for bond lengths and unit-cell parameters, where LDA usually underestimates and GGA overestimates bond lengths; ~15% for the elastic constants; and ~5% for vibrational frequencies. For phase transitions and chemical reactions, the GGA seems to perform much better than the LDA, with phase transition pressures accurate to within ~5 GPa (usually overestimated); however, for metal-insulator transitions errors of both approximations are typically much larger. For ionic and covalent materials and for normal metals (carbon allotropes and most carbonates and perhaps carbides belong to these classes) both LDA and GGA give good description of the structural properties and thermodynamics. Large errors in all compounds are documented for calculations of electronic excitation energies and band gaps (both LDA and GGA significantly underestimate band gaps); one must employ special methods, such as the *GW* method (Aryasetiawan and Gunnarsson 1998), to compute these parameters. Mott insulators represent a particular pathological case, where today's DFT too often gives unreasonable results. Until recently, DFT calculations could not adequately account for van der Waals interactions, but ways for incorporating these effects are now possible (Dion et al. 2004; Roman-Perez and Soler 2009).

Given the high degree of thermodynamic equilibrium reached in Earth's deep interior due to high temperatures and long geological timescales, it often suffices in studies of mantle and core mineralogy to consider free energies of relevant chemical equilibria without accounting for kinetics. The Gibbs free energy G of a particular phase can be calculated as:

$$G = E + PV - TS = F + PV = F - V\left(\frac{\partial F}{\partial V}\right)_T \qquad (1)$$

where P is the pressure, V the volume, T the temperature, S the entropy, F is the Helmholtz free energy, E_0 is the ground-state energy (at 0 K), and En are the energy levels of the system:

$$F = E_0 - k_B T \ln \sum_n e^{-E_a / k_B T} \qquad (2)$$

The sum within the logarithm in Equation (2) is called the partition function, and its rigorous calculation is complicated owing to the difficulty in obtaining energy levels experimentally or theoretically for their overwhelmingly large number in solids. In the simplest, harmonic approximation these energy levels are:

$$E_n = \hbar\left(\frac{1}{2}+n\right)\omega \tag{3}$$

where n is the quantum number, ω the frequency of the vibration, and $\hbar$ is the Planck constant. More sophisticated approximations exist, but the general point is that knowledge of the energy landscape allows one to compute all thermodynamic properties, including the entropy.

Predicting the most stable structure is a global optimization problem, and such problems are mathematically said to be of the NP-hard class and cannot be solved with a guarantee. To deal with such problems, one develops heuristic methods, the goal of which is to achieve high success rate and efficiency for a given class of problems. A variety of methods exists for predicting structures (e.g., Martoňák et al. 2003; Oganov and Glass 2006; Oganov et al. 2010; Pickard and Needs 2011; for an overview of different methods see Oganov 2010). A large number of results for carbon-bearing phases under pressure (Oganov et al. 2006, 2008; Li et al. 2009; Lyakhov and Oganov 2011; Wen et al. 2011; Zhu et al. 2011, 2012a; Bazhanova et al. 2012) have been obtained with the evolutionary algorithm USPEX (Oganov and Glass 2006; Oganov 2010; Lyakhov et al. 2010; Oganov et al. 2011).

A powerful computational method called evolutionary metadynamics has recently been developed by Zhu et al. (2012a). This method, which merges features of the USPEX method and metadynamics (Martoňák et al. 2003), is capable of rapidly finding the ground state structure and a large number of low-energy metastable structures, provided a reasonable initial structure. As in the original metadynamics method, the new method produces meaningful structural transformation mechanisms. However, equilibration is achieved not using molecular dynamics, which is prone to trapping in metastable states and often leads to amorphization during metadynamics simulations, but rather employs global optimization moves borrowed from the USPEX method. Unlike original metadynamics, this technique produces extremely rich sets of low-energy crystal structures, while using less computer time. Particularly attractive is a possibility to determine the most likely crystal structures accessible from the initial structure through metastable transitions.

ELEMENTAL CARBON

One of the remarkable features of carbon is existence of a broad range of metastable phases that can be formed near ambient conditions and their wide fields of kinetic stability. We consider both stable and metastable phases (Fig. 1) as well as both equilibrium and metastable transitions among the phases, and predictions of carbon allotropes at very high pressures beyond those of Earth. This section includes a discussion of metastable transitions of the lower density phase. Not only do these structures provide useful insight into carbon crystal chemistry and bonding, but transitions to these phases also may be relevant to natural impact phenomena involving these phases in nature.

Stable phases

The common naturally-occurring sp^2 and sp^3 allotropes of carbon occur in different crystallographic forms of graphite-like and diamond-like phases, respectively: hexagonal graphite and a rhombohedral stacking variant; as well as hexagonal lonsdaleite and cubic diamond (see Hazen et al. 2013). The large cohesive energy of diamond (717 kJ/mol), together with significant energy barriers, gives rise to its high degree of metastability and extremely high melting temperature (5000 K). The high activation barrier for a transformation between graphite and diamond explains why synthesis of diamond from graphite requires not only high pressures, but also high temperatures and often the presence of catalysts.

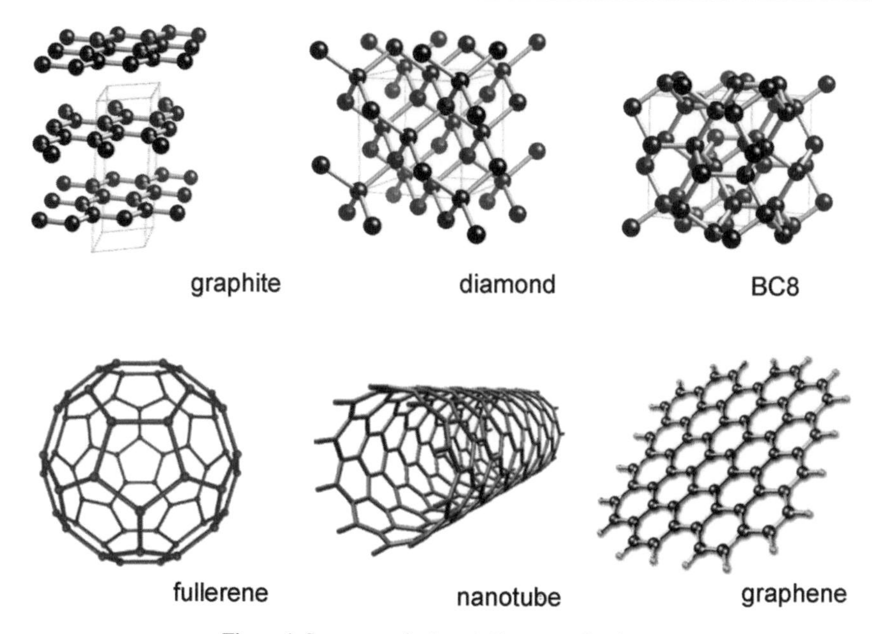

Figure 1. Structures of selected allotropes of carbon.

The structures and occurrences of naturally existing carbon allotropes, including graphite, diamond, and lonsdaleite, are reviewed elsewhere in this volume (Hazen et al. 2013). Here we focus on phase transitions in graphite. The graphite melting line at low pressures is now well established (Ludwig 1902; Basset 1939; Jones 1958; Gathers et al. 1974a,b; Kirillin et al. 1984; Savvatimskiy 2005; Scheindlin and Senchenko 1988; Fig. 2).

Since the original measurements of the compressibility (Adams 1921) and vibrational spectrum (Ramaswamy 1930; Raman 1961), the physical properties of carbon allotropes have been well determined over the range of conditions relevant to Earth's interior. For example, there is excellent agreement on the value of diamond's very high bulk modulus (440 GPa at 300 K) from a variety of experimental and theoretical techniques (Occelli et al. 2003). The singular strength of diamond correlates with its very high shear modulus (535 GPa at 300 K), which exceeds the bulk modulus (a phenomenon observed in only a few materials: Brazhkin et al. 2002). Theoretical methods now accurately reproduce the elastic properties of diamond and provide robust predictions of their variations over the entire *P-T* range of Earth's interior (Nuñez Valdez et al. 2012). The electronic properties of diamond are now well determined experimentally (Endo et al 2001), and there is excellent agreement with electronic structure calculations. On the other hand, the anharmonic properties of diamond are not fully understood (Gillet et al. 1998). The initial slope of the diamond melting line was for many years controversial, but now seems well established and has a positive value (Bundy et al. 1996). Recent very high *P-T* shock wave experiments indicate a maximum in the melting line near 1.5 TPa (Knudson et al. 2008; Eggert et al. 2010; Fig. 3). Early first-principles calculations (Galli et al. 1989) indicated that liquid carbon at low pressure and 5000 K, should have mainly two-fold and three-fold coordination. Theoretical studies of the carbon melting curve have continued to super-Earth conditions beyond the stability of diamond (see Correa et al 2006).

The types and concentrations of defects present in the material remain subjects of study (Gabrysch 2008). Recent advances in analytical methods confirm that most mantle diamonds

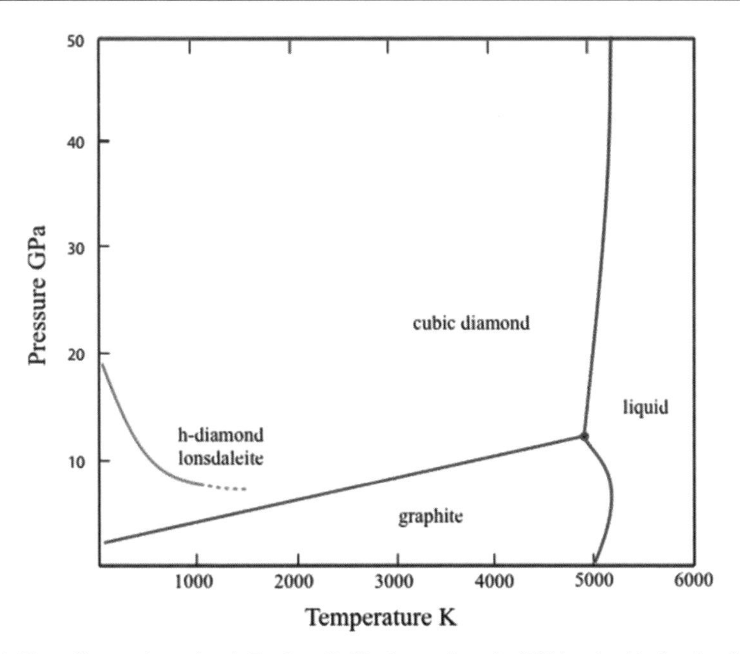

Figure 2. Phase diagram for carbon in Earth to 50 GPa (approximately 1500 km depth), showing the classic diamond-graphite line and field for liquid; note the liquidus for diamond has positive *dP*/*dT* consistent with calculations to higher pressure, and the only significant change to the early curve of Bundy et al. (1996). The red curve marks the metastable lonsdaleite or hexagonal (h-) diamond transition from Utsumi et al. (1996). At *P*-*T* conditions close to this red curve and above ~1070 K (800 °C), h-diamond from graphite is quenchable and recoverable, whereas below ~1070 K, h-diamond is unquenchable. The form of graphite used in the experiments has some influence, and additional metasatable complications for h-diamond are not yet fully understood (see Utsumi and Yagi 1991; Utsumi et al. 1994). Lonsdaleite is associated with formation in terrestrial impact craters and meteorites (see review in Ross et al. 2011) and has recently been discovered in ultrahigh pressure (UHP) metamorphic rocks (Godard et al. 2012).

are very pure except for variable low concentrations of nitrogen. McNeill et al. (2009) analyzed a suite of 10 diamonds from the Cullinan Mine (previously known as Premier), South Africa, along with other diamonds from Siberia (Mir and Udachnaya) and Venezuela. The concentrations of a wide range of elements for all the samples (expressed by weight in the solid) are very low, with rare earth elements along with Y, Nb, and Cs ranging from 0.01 to 2 ppb. Large ion lithophile elements (LILE) such as Rb and Ba vary from 1 to 30 ppb. Ti ranges from ppb levels up to 2 ppm.

The graphite-to-diamond and graphite-to-lonsdaleite transitions have been intensely investigated (see Hazen 1999; Hazen et al. 2013). The position of the diamond/graphite equilibrium line has been quite accurately established by thermodynamic calculations based upon the measured physical properties of graphite and diamond in the temperature range from 300 to about 1200 K (Rossini and Jessup 1938; Leipunskii 1939; Berman and Simon 1955; Day 2012) and by experiments on growing or graphitization of diamond (Bundy et al. 1961; Kennedy and Kennedy 1976). Shock compression experiments on hexagonal graphite particles yield mostly hexagonal diamond (e.g., Kudryumov et al. 2012), whereas in natural impact events, kinetics and grain size can influence the survival of lonsdaleite (DeCarli et al. 2002a).

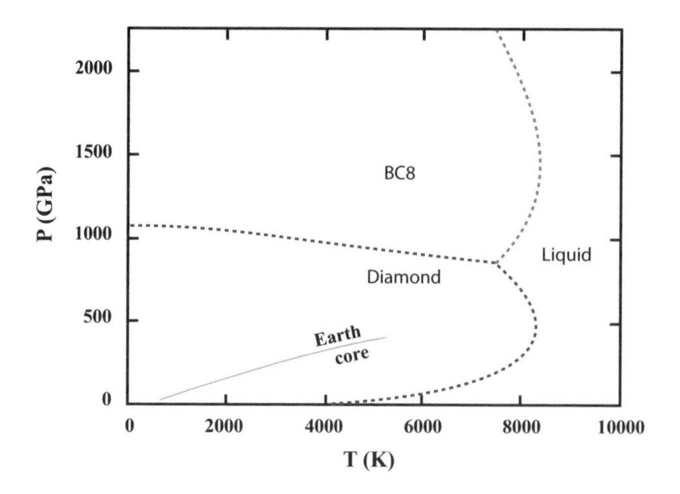

Figure 3. Calculated phase relations for diamond to super Earth conditions. Above 1000 GPa diamond is displaced by BC8. The green line marks the terrestrial geotherm (approximate), along which the maximum pressure at Earth's center corresponds to ~360 GPa in the inner core. Note that diamond is stable to approximately twice the pressure at Earth's center and does not melt below ~5000 K. Diamond is expected to be a planetary "survivor of catastrophe," such as the Moon-forming collision on Earth. Diagram based on Correa et al (2006). Note that on this diagram with extreme pressure range, the field for graphite is too small to depict (but tracks along the low-pressure axis).

Metastable phases

Because of the high cohesive and activation energy, carbon allotropes typically exist metastably well into a *P-T* region where a different solid phase is thermodynamically stable. For example, diamond survives indefinitely at room conditions, where graphite is the thermodynamically stable form. The metastability of diamond may also persist to very high pressures, well into the stability field of higher-pressure phases. We also mention that studies of metastable growth of diamond within Earth continue, though the mechanism is not well understood (Simakov 2012). It is also useful to point out that diamond can be easily produced outside of its stability field (e.g., by chemical vapor deposition), and that it can exhibit striking properties (Hazen 1999). For example, measurements show that the material has much higher toughness than natural diamond (Liang et al. 2009; Pan et al. 2009). Similarly, the natural hexagonal diamond or lonsdaleite (Bundy and Kasper 1967; Lonsdale 1971) is becoming increasingly recognized in both meteoritic and terrestrial materials (Ross et al. 2011; Shumilova et al. 2011), yet its occurrence cannot be interpreted in terms of origin, since fundamental understanding of its metastability remains unresolved (Godard et al. 2012; Kurdyumov et al. 2012).

Metastable allotropes of carbon include fullerenes, nanotubes, and graphene, as well as disordered or amorphous forms, such as glassy carbon and carbon black and possibly metastable solid forms referred to as carbynes (see Hazen et al. 2013, Table 1). The synthesis of carbynes (*sp* hybridized allotrope) has been reported (Kasatochkin et al. 1967; Chalifoux and Tykwinski 2010), but this finding remains controversial.

The proclivity of carbon to form well-defined kinetically stable structures persists at extreme conditions. For example, it has long been known that if graphite is compressed to ~15 GPa at low temperature (e.g., room temperature or below) it transforms to a different allotrope of carbon that is distinct from diamond (Aust and Drickamer 1963; Hanfland et al. 1989; Zhao and Spain 1989; Miller et al. 1997; Utsumi and Yagi 1991; Yagi et al. 1992; Mao et al. 2003). The phase was found to exhibit unusual properties. For example, Goncharov (1990) found that the transition pressure for single-crystal graphite can be lowered to about 14 GPa under uniaxial stress directed along the *c*-axis of the crystal. These Raman and optical observations were corroborated by Hanfland et al. (1989) and Utsumi et al. (1993). Hanfland et al. (1990) showed that the phase transition started at about 14 GPa using X-ray diffraction and Raman spectroscopy. The phase formed from single-crystal graphite was found to be transparent, and Miller et al. (1997) quenched the transparent form at low temperatures. X-ray Raman and diffraction measurements reported by Mao et al. (2003) showed mixed sp^2 and sp^3 bonding but the structure could not be determined. The phase was also found to be superhard (Fig. 4). Very recently, Wang et al. (2012b) used X-ray diffraction, Raman spectroscopy, and optical techniques to examine this phase.

Early calculations of the structure formed from compression of graphite were carried out by Kertesz and Hoffmann (1984) and Fahy et al. (1986). This metastable transition has been the subject of more recent theoretical studies. Using USPEX, Oganov and Glass (2006) found a number of interesting low-energy metastable carbon allotropes, among which was a monoclinic (space group *C2/m*) structure with tetrahedral coordination (sp^3 hybridization) of carbon atoms, and this structure contained 5- and 7-fold rings (Fig. 5). The structure

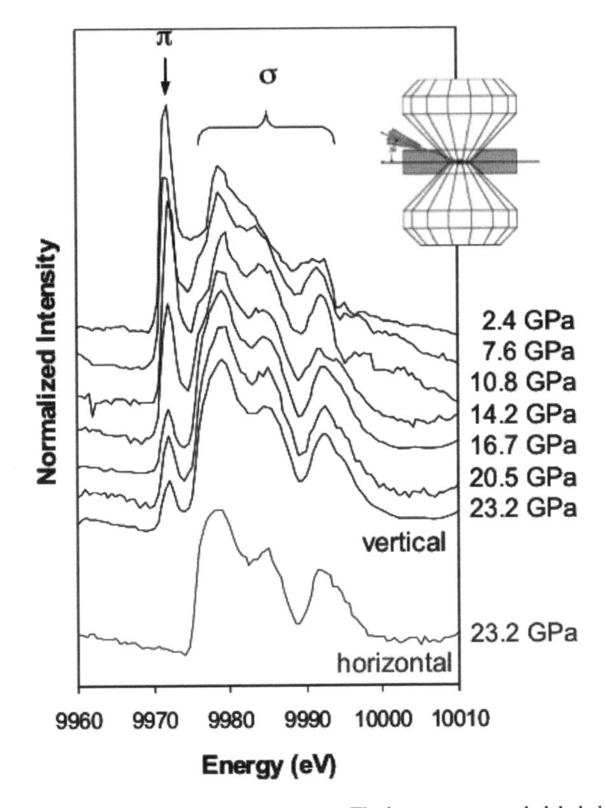

Figure 4. X-ray Raman spectra of graphite under pressure. The lower energy peak, labeled π, corresponds to 1s to π* transitions and the higher energy portion, labeled σ, corresponds to 1s to σ* transitions. The bottom spectrum, taken in the horizontal direction, probes bonds in the *a* plane, and the top spectra, taken in the vertical direction, probe the *c* plane. The inset shows the geometry of the diamond anvil cell experiment. The spectra reveal the development of mixed π and σ bonding in the high pressure phase formed by compression of graphite [Adapted with permission of The American Association for the Advancement of Science from Mao et al. (2003) *Science*, Vol. 302, p. 425-427, Fig. 1.]

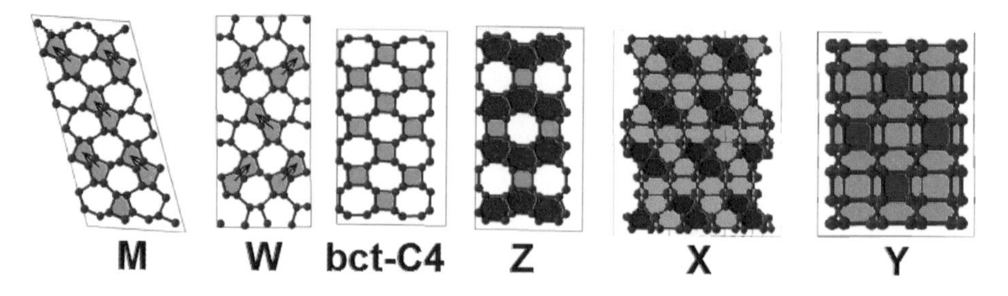

Figure 5. Calculated high-pressure metastable sp^3 structures of elemental carbon. [Used with permission of the American Physical Society from Zhu et al. (2012b) *Phys Rev B,* Vol. 85, 201407, Figs. 2c-f and 4a,b. *http://link.aps.org/doi/10.1103/PhysRevB.85.201407*]

has a 2D-analog: (2×1)-reconstructions of the (111) surfaces of diamond and silicon. Li et al. (2009) subsequently named this structure M-carbon (because of its monoclinic symmetry) and identified it as the "superhard graphite," based on the close match between theoretical and experimental diffraction data. The bct-C_4 structure (Baughman et al. 1997; Umemoto et al. 2010; Zhou et al. 2010) was proposed as an alternative, and that was shortly followed by descriptions of hypothetical W-carbon (Wang et al. 2011) and Z-carbon (Zhao et al. 2011; Amsler et al. 2012), as well as further structures (e.g., Selli et al. 2011; Niu et al. 2012). It is notable also that while Amsler et al. (2012) proposed Z-carbon to explain experimental measurements on "superhard graphite," Zhao et al. (2011) concluded that its X-ray diffraction is incompatible with experimental data on "superhard graphite," but can explain experimental data on over-compressed carbon nanotubes (Wang et al. 2004). With two evolutionary metadynamics simulations starting from graphite-2H and graphite-3R, Zhu et al. (2012b) found all of these structures and several new ones. All of these structures are metastable, superhard, and insulating; they all become more stable than graphite under pressure (7 to 27 GPa); they all contain carbon entirely in the tetrahedral coordination; and they all have buckled graphene sheets, suggesting possible mechanism of formation of these phases from graphite. Although more than a dozen different crystal structures have been suggested for the phase, the proposed low-symmetry, monclinic, M-carbon structure (Oganov and Glass 2006; Li et al. 2009) is the only one that fits the highest resolution high-pressure data, reported by Wang et al. (2012b).

Theory can discriminate between proposed high-pressure metastable structures and provide insights into the transition. The metastable high-pressure phase is formed solely because it has the lowest activation barrier at the indicated *P-T* conditions (e.g., 15 GPa and 300 K). Using transition path sampling (Dellago et al. 1998; Leoni and Boulfelfel 2010; Boulfelfel et al. 2012) found that the likeliest sp^3 structure of carbon to be formed by room-temperature compression of graphite-2H is indeed M-carbon. Boulfelfel et al. (2012) predict this transition has a lower energy barrier than transitions to other structures. The transition involves bending of graphene layers and formation of bonds between them—a mechanism that leads to the peculiar geometry of M-carbon with 5-and 7-fold rings. Shortly after this definitive theoretical confirmation of M-carbon, a definitive experimental confirmation followed (Wang et al. 2012b).

Fullerenes at pressure

We now consider transitions and structures formed from metastable low-pressure fullerene-related phases. The ambient phase of C_{60} has the high-symmetry face-centered cubic (*fcc*) structure. Reversible transitions are observed at low *P-T* conditions (see Hemley and Dera 2000). Under hydrostatic conditions the *fcc* structure of solid C_{60} remains kinetically stable to pressures of 20 GPa, whereas non-hydrostatic stress induces a phase transition to a new phase of lower symmetry (Duclos et al. 1991). Further studies revealed that non-hydrostatic compression of C_{60} induces the formation of diamond (Núñez-Regueiro et al. 1992) and other related structures above 15 GPa (Sundqvist 1999). Compression increases the possibility of interaction between the double bonds on nearest neighbor molecules, which results in the formation of polymers (Marques et al. 1996; Sundqvist 1999). He and Ne can be introduced into fullerene cages (see Saunders et al. 1996). The extent to which this could lead to the possibility of trapping gases under pressure in planetary interiors remains to be investigated.

Recently, Wang et al. (2012a) reported a long-range ordered material constructed from units of amorphous carbon clusters synthesized by compressing solvated fullerenes. Using X-ray diffraction, Raman spectroscopy, and quantum molecular dynamics simulation, they observed that, although C_{60} cages were crushed and became amorphous, the solvent molecules remained intact, playing a crucial role in maintaining the long-range periodicity. Once formed, the high-pressure phase is quenchable back to ambient conditions. The phase has very high strength, producing ring cracks in the diamond anvils used in the experiment (as does the phase produced from graphite; Mao et al. 2003).

Ultrahigh-pressure phases

A number of studies through the years have addressed the question of the upper limits of the diamond stability field and possible post-diamond phases. Experimental data in this *P-T* regime have been obtained by dynamic compression. Early LDA calculations by Fahy and Louie (1987) predicted that at pressures above 1 TPa diamond transforms to the so-called BC8 structure—an unusual tetrahedral structure that has been observed as a metastable state of silicon (Kasper and Richards 1964). Their total energy calculations (Fahy and Louie 1987) showed that diamond becomes less stable than the BC8 structure, but more stable structures could not be excluded. Evolutionary crystal structure prediction simulations (Oganov and Glass 2006) confirmed their conclusion that above 1 TPa the BC8 structure of carbon is thermodynamically stable. Other theoretical calculations, together with existing shock-wave data, are consistent with these predictions for BC8 (Knudson et al. 2008). Martinez-Canales et al. (2012) employed DFT methods to examine structures with up to 12 carbon atoms per unit cell at pressures between 1 TPa and 1 petapascal (1000 TPa). They propose a sequence of denser carbon allotropes: The BC8 phase is predicted to transform to the simple cubic structure at 2.9 TPa, followed by a soft-phonon transition to a simple hexagonal structure at 6.4 TPa, the electride *fcc* structure at 21 TPa, a double hexagonal close-packed (*hcp*) structure at 270 TPa, and ultimately a body-centered cubic (*bcc*) structure at 650 TPa. It is now possible to explore these ultrahigh-pressure regimes with dynamic compression techniques using megajoule class lasers such as the National Ignition Facility (NIF; e.g., Smith 2011).

CARBIDES

Carbides of iron have been considered as potentially significant components of Earth's core based on early comparisons with meteoritic iron (Ringwood 1960; Lovering 1964; Ahrens 1982; Hazen et al. 2013) and the requirement for a light alloying element to match observed core densities (Birch 1964; Wood et al. 2013). The stoichiometry of high-pressure Fe-carbide is uncertain; if Fe_7C_3 is more stable than Fe_3C, then up to 1.5% carbon could be present in the inner core (Mookherjee 2011; Mookherjee et al. 2011). The incorporation of carbon in the fluid outer core has been studied based on extrapolations from low *P-T* data as well as various theoretical techniques (Terasaki et al. 2010; Zhang and Yin 2012; and references therein). The widespread existence of carbides in Earth's silicate mantle (particularly Fe-carbide) has been predicted for depths >250 km based on experimentally determined f_{O_2} conditions (Frost et al. 2004; Rohrbach et al. 2007, 2010; Frost and McCammon 2008; Rohrbach and Schmidt 2011; Stagno and Frost 2011) and the reactivity of carbon and iron at high pressures and temperatures (Rouquette et al. 2008; Oganov et al 2008; Lord et al. 2009).

Mantle iron carbide stability is illustrated schematically in Figure 6 as a function of depth and oxygen fugacity. Despite the prediction for abundant Fe-carbides in the mantle, terrestrial occurrences are very rare, and are typically associated with diamond. Fe-carbides have been documented as inclusions within diamonds from Jagersfontein, South Africa (Jones at al. 2008); Juina, Brazil (Kaminsky and Wirth 2011); and the 23rd Party Congress Kimberlite, Russia (Bulanova and Zayakina 1991). Scarce occurrences of "native iron" inclusions in diamond (Stachel et al. 1998; Bulanova et al. 2010) likely include unrecognized Fe-carbide. Other fragmentary evidence to support carbide in the mantle includes unique spherical micro-inclusions of Fe-carbide ("cohenite") observed in garnet intergrowths from polycrystalline mantle diamond (Jacob et al. 2004). The mantle Fe-rich carbide compositions from Jagersfontein are the most numerous and show limited compositional variation of (Fe-Ni-Cr) with minor constituent silicon and trace Co (Table 1; Fig. 7; Hazen et al. 2013). Chemically these mantle carbides appear to be mixtures of various end members like haxonite [$(Fe,Ni)_{23}C_6$] and isovite [$(Cr,Fe)_{23}C_6$], but there is no experimental information on their phase relations at high-pressure mantle conditions.

The equation of state and elastic constants of cohenite have been studied to at least 187 GPa (Scott et al. 2001; Fiquet et al. 2009; Sata et al. 2010). Theoretical studies employing the random sampling approach by Weerasinghe et al. (2011) suggest that at inner core pressures (330 to 364 GPa) cohenite transforms to a denser orthorhombic *Cmcm* structure. Evolutionary

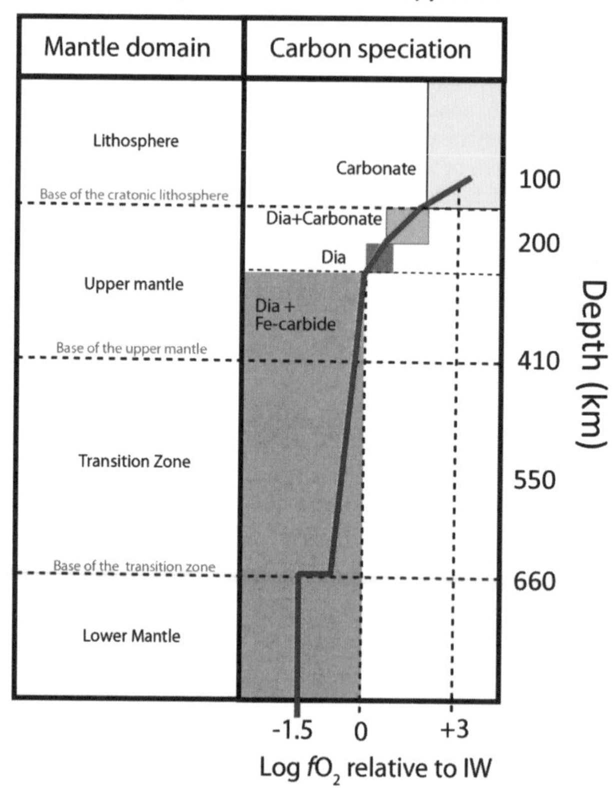

Figure 6. The speciation of carbon as a function of f_{O_2} with depth in an ambient pyrolitic mantle (Mikhail et al. 2011; Rohrbach and Schmidt 2011). Note the dominance in the terrestrial mantle for reduced carbon species relative to carbonate.

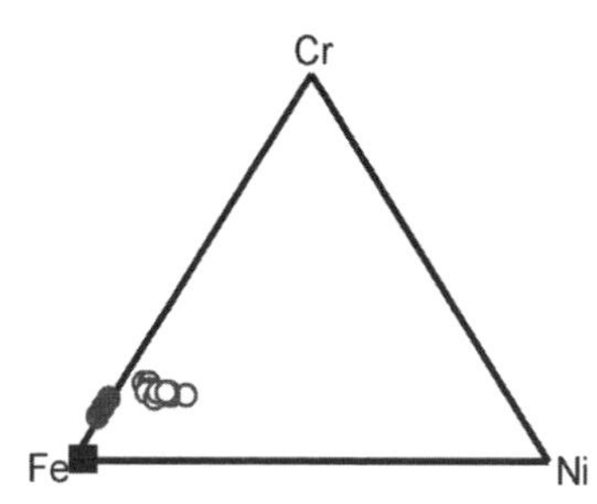

Figure 7. Plot of Fe-Ni-Cr compositions of mantle carbide inclusions within diamond at Jagersfontein, based on electron microprobe analyses (after Jones et al. 2008). Different diamonds can host different carbide compositions: solid red circles = Jag1 ($n = 9$), open blue circles = Jag2 ($n = 5$) and solid black square (Jag10 $n = 17$). These carbides also contain minor Co, Si and O (see Table 1).

Table 1. Mantle carbide compositions (weight percent, analyzed by WDS electron microprobe) of inclusions in diamond from Jagersfontein (Jones et al. 2008), from microspherules from garnet in polycrystalline diamond from Venetia (average Jacob et al. 2004), compared with theoretical compositions for cohenite, and cementite, and natural "cohenite" from enigmatic crustal basaltic rocks at Disko (Goodrich and Bird 1985; see also Bird et al. 1981 and Jones et al. 2005).

	Jag b20	Jag b26	Jag 5	Jag 10	Venetia (Average)	Cohenite	Cementite	Disko "cohenite"
Fe	65.43	80.92	88.42	90.28	90.41	54.92	94.73	90.7 – 93.4
Cr	15.07	8.99	2.04	0.02	n.a.	0	0	n.a.
Ni	9.32	0.4	0.33	0.1	1.61	28.86	0	0.7 – 3.1*
Co	0	0	0.21	0.18	0.37	9.66	0	0.4 – 0.7
Si	0.78	0.55	1.65	0.78	0.03	0	0	n.a.
S	0.14	0.06	0.02	0.07	0.21	0	0	n.a.
Al	0.43	0.02	0.15	0.07	n.a.	0	0	n.a.
O	2.08	1.55	0.97	3.99	n.a.	0	0	n.a.
C	5.32	6.21	5.86	4.51	6.53	6.56	5.27	5.2 – 7.2
Total	98.57	98.70	99.44	100		100	100	

structure prediction simulations (Bazhanova et al. 2012) found an even more stable structure with space group $I\bar{4}$, which at 0 K is stable in the pressure range 310 to 410 GPa. Both the random sampling study (Weerasinghe et al. 2011) and the evolutionary prediction (Bazhanova et al. 2012) agree that at inner core pressures Fe_3C is not thermodynamically stable. Rather, Fe_2C with the *Pnma* space group appears to be the most stable iron carbide at pressures of Earth's inner core. Bazhanova et al. (2012) found that incorporation of ~13 mol% carbon into the inner core leads to perfect matching of its density, which sets the upper limit for the carbon content of the core. More complex chemistries (e.g., ternary phase) remain to be studied using these techniques. In addition, it would be useful to examine higher *P-T* conditions relevant to super-Earths. We note that mixed metal carbides exhibit interesting electronic properties such as superconductivity [e.g., MgC_xNi_3 ($x \approx 1$), which has a perovskite structure (He et al. 2001)].

MOLECULAR FRAMEWORK STRUCTURES

Carbon dioxide

A number of molecular solids may play a role in carbon mineralogy, both at low temperatures and at high pressures (Hemley and Dera 2000). The crystalline phases of carbon dioxide exemplify the potential for varied crystal chemical contexts. The familiar low-temperature phase of carbon dioxide, also known as "dry ice," incorporates isolated linear CO_2

molecules in a cubic structure (space group *Pa*3, *a* ~ 5.5 Å), with 4 linear CO_2 molecules per unit cell (Keesom and Köhler 1934; Aoki et al. 1994). Structure refinement of this phase at 1.0 GPa gives a C-O bond length of 1.168(1) Å (Downs and Somayazulu 1998), very close to that of the free molecule. This phase, which is a mineral in the cold polar regions of Mars (Byrne and Ingersoll 2003), can also form at room temperature and pressures above ~0.3 GPa. Phase III crystallizes in a *Cmca* structure. The reported phase II (0.5-2.3 GPa) is less well characterized (Liu 1983), and Downs and Somayazulu (1998) find that phase I is stable in at least part of this pressure range. It has also been reported that a distorted phase IV forms between the stability field of I and III (Olijnyk and Jephcoat 1998). CO_2 III was identified by Raman spectroscopy (Hanson 1985; Olijnyk et al. 1988). At pressures >10 GPa carbon dioxide transforms to one of at least three other molecular solids, denoted phases II, III, and IV, as well as a possible phase VII, depending on the temperature (Aoki et al. 1994; Park et al. 2003; Sun et al. 2009; Datchi et al. 2009). However, these relatively low-density forms of crystalline CO_2 are not expected to persist at mantle conditions.

The existence of CO_2-based framework structures analogous to SiO_2 with the coordination of carbon by oxygen increased from three- to four-fold was first shown by Raman studies, which provided direct evidence for a polymeric structure following laser heating at pressures above ~40 GPa and above 2000 K (Iota et al. 1999). The phase was originally called "quartz-like" based on the close similarity to the vibrational spectra of quartz and named CO_2-V (Iota et al. 1999). First-principles theory has confirmed the stability of framework structures with tetrahedral carbon; i.e., relative both to the molecular solid and to decomposition to elemental carbon and hydrogen (Serra et al. 1999). Subsequent X-ray diffraction data appeared to be best fit with the tridymite structure (Yoo et al 1999). The stability of the phase has been examined with experiments (Tschauner et al. 2001; Santoro et al. 2006) and theoretical calculations (Holm et al. 2000; Dong, et al. 2000; Bonev et al. 2003; Oganov et al. 2008). An additional possible phase VI with octahedrally-coordinated carbon reported by Iota et al. (2008) has not been confirmed by other experiments or theory. The structures of these phases were until recently under debate. At the same time, it was not possible to rule out that there may be even more stable structures. Evolutionary crystal structure prediction calculations (Oganov et al. 2008) were shown to be consistent with β-cristobalite as the thermodynamically stable crystal structure (Fig. 8).

The structure of phase V has been recently determined by high-pressure X-ray diffraction measurements (Fig. 9; Datchi et al. 2012; Santoro et al. 2012). Using high-pressure diamond anvil cells combined with high-temperature laser heating, the authors of both studies formed CO_2-V and probed the crystal structure using synchrotron X-ray diffraction measurements. Through Rietveld analysis methods, both groups determined that CO_2-V does in fact have the β-cristobalite structure and is composed of CO_4 tetrahedral units. Interestingly, the CO_4 units possess O-C-O bond angles that average 109.5°, in stark contrast to molecular carbon dioxide. Structural determination as a function of pressure revealed the bulk modulus of 136 GPa, significantly less than the value reported previously (Yoo et al. 1999).

The melting curve of CO_2 has been determined to 15 GPa (Giordano et al. 2006; Giordano and Datchi 2007). Quenched samples after laser heating to 2000-3000 K at 28-75 GPa showed that CO_2 breaks down to form oxygen and diamond at lower temperature with increasing pressure (Tschauner et al. 2001). Recently, the solid–solid phase transitions, melting behavior, and chemical reactivity of CO_2 to these pressures and at temperatures up to 2500 K using *in situ* Raman spectroscopy in laser-heated diamond anvil cells were reported by Litasov et al. (2011). Molecular CO_2 melts to a molecular fluid up to 33 (±2) GPa and 1720 (±100) K, where it meets a solid-solid phase line to form a triple point (Fig. 10). At higher pressure, non-molecular phase V does not melt but instead dissociates to carbon and oxygen with a transition line having a negative *P-T* slope. A comparison with *P-T* profiles of Earth's mantle indicates that CO_2-V

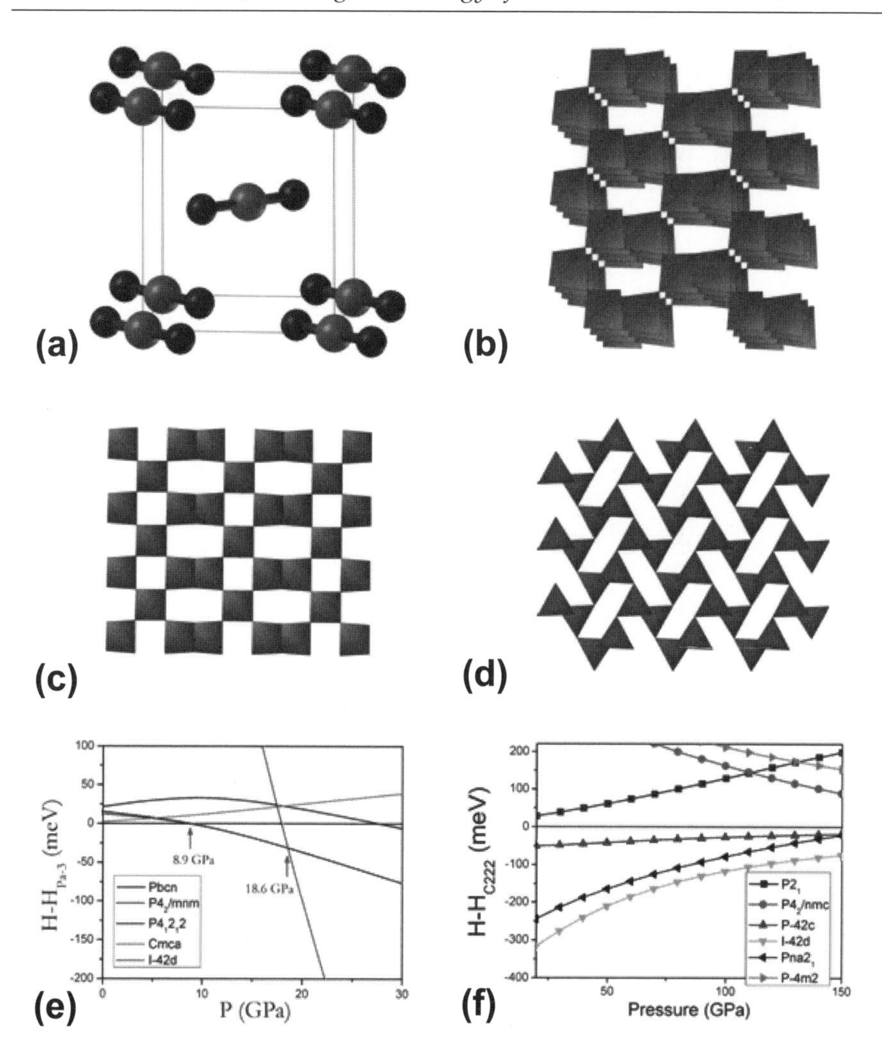

Figure 8. CO_2 structures: (a) molecular *P*4$_2$/*mnm* structure, stable at lower pressures than polymeric CO_2-V phase; (b) β-cristobalite-type structure of CO_2-V with tetrahedral coordination of carbon atoms; (c) polymeric *C*222$_1$ structure, (d) metastable polymeric *Pna*2$_1$ structure. At the bottom are the calculated enthalpies of candidate structures of CO_2: (e) in the low-pressure region, relative to the molecular *Pa*3 structure; (f) in the high-pressure region, relative to the non-molecular *C*222 structure. [Adapted with permission of Elsevier from Oganov et al. (2008) *Earth Planet Sci Lett*, Vol. 273, p. 38-47, Figs. 2 and 3.]

can be stable near the top of the lower mantle and dissociates at greater depths. Decarbonation reactions of subducted carbonates in the lower mantle would produce diamond and fluid oxygen, which in turn significantly affects redox state, increasing oxygen fugacity by several orders of magnitude. The reaction of free oxygen with lower mantle minerals such as Mg-perovskite could create significant conductivity anomalies (Litasov et al. 2011). Decomposition of CO_2 was found to be unlikely by first-principles calculations (Oganov et al. 2008; Boates et al 2012); as the computed enthalpy of decomposition of CO_2 is extremely high (3.3 eV at 50 GPa), discrepancy between experiment and theory that needs to be resolved. Recently, evidence for coesite-structured CO_2 has been reported (Sengupta and Yoo 2009). Still higher-pressure

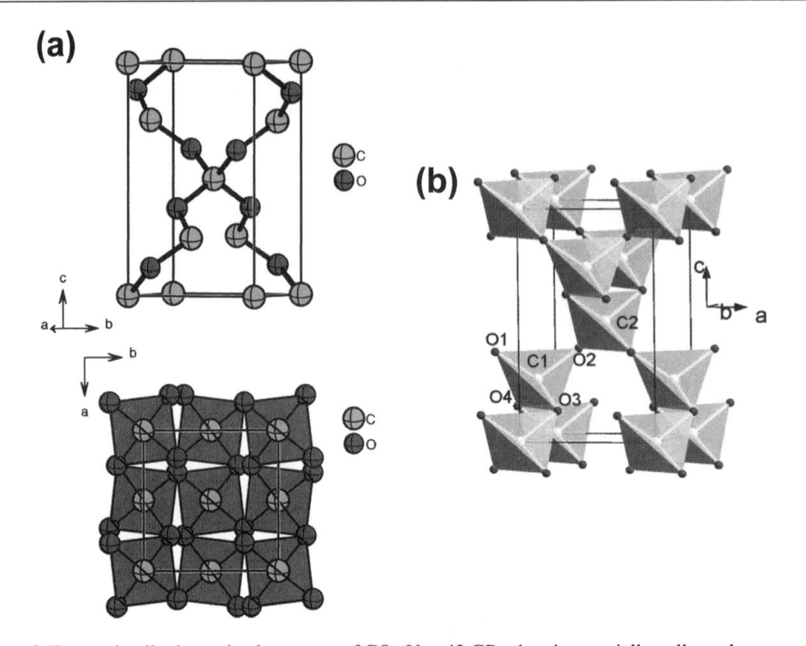

Figure 9. Expermintally determined structure of CO_2-V at 43 GPa showing partially collapsed arrangement of CO_4 tetrahedra. (a) Santoro et al. (2012) [Reproduced with permission of the National Academy of Sciences from *Proceedings of the National Academy of Sciences USA*, Vol. 109, p. 5176, Fig. 2.]; (b) Datchi et al. (2012) [Reproduced with permission of American Physical Society from *Phys Rev Lett*, Vol. 108, 125701, Fig. 2(a). *http://link.aps.org/doi/10.1103/PhysRevLett.108.125701*]

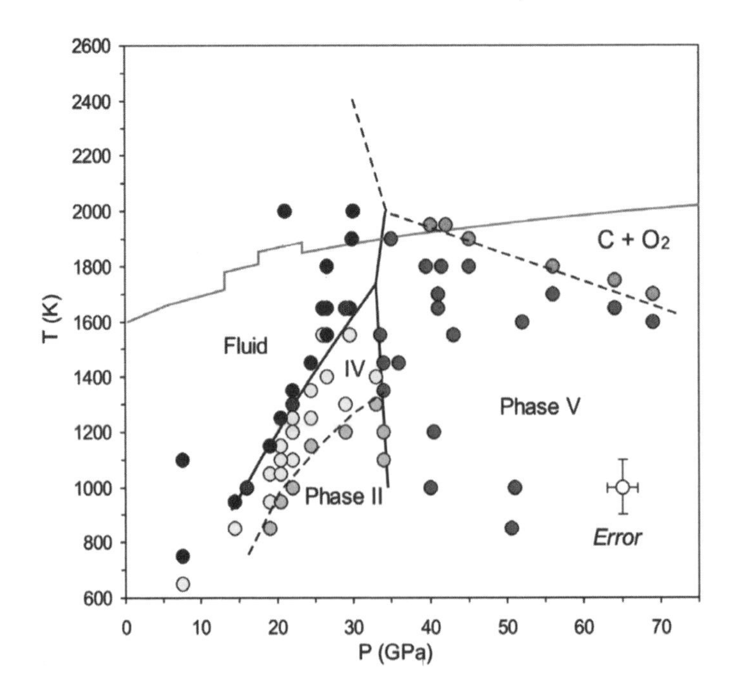

Figure 10. Phase diagram for CO_2. The light grey line is an estimated Earth geotherm. [Adapted with permission of Elsevier from Litasov et al. (2011), *Earth Planet Sci Lett*, Vol. 309, p. 318-323, Fig. 3.]

structures of CO_2 are predicted (Fig. 11), including dissociation at TPa pressures due to the predicted higher density of the elemental assemblage is predicted (Fig. 12).

Other compounds

Early work on the high-pressure behavior of other binary carbon compounds, in particular formed from simple molecular phases, was reviewed by Hemley and Dera (2000). Recently, structures of carbon monoxide have been investigated theoretically using DFT (Sun et al. 2011). Both three-dimensional frameworks and layered structures were found above 2 GPa, although these are metastable. Some of these new structures of CO (Fig. 13) could be metallic and others might be recoverable to ambient pressure.

The high-pressure behavior of clathrate hydrates, which exist in nature in several forms depending on pressure and temperature and gas composition, are of special interest. Recent high-pressure neutron scattering studies have provided definitive experimental results for testing and refining intermolecular potentials used to describe the hydrophobic interactions in these dense molecular phases (Tulk et al. 2012). These measurements yielded detailed data on the structure of a high-pressure form of methane clathrate hydrate and gave precise information on the number of methane molecules that could fit into the cages of the inclusion compound. The results showed that the correct occupancy for the largest cages in this beautiful structure is three methane molecules. The study provides tests of models for hydrophobic interactions and methane-methane interactions since the high-pressure structure H clathrate contains cages that are both singly and triply occupied.

Similar types of high-pressure transformations to denser forms can be anticipated for hydrocarbon phases. For example, liquid methane has been observed on Titan's surface and crystalline forms may occur at depth (e.g., Shin et al. 2012). At room temperature and 1.59 GPa methane crystallizes to the low-temperature (methane I) form, in which rotationally disordered CH_4 molecules occupy positions of a face-centered cubic (space group *Fm*3*m*) lattice (Hazen et al. 1980). The higher-pressure behavior of crystalline methane is complex, with several

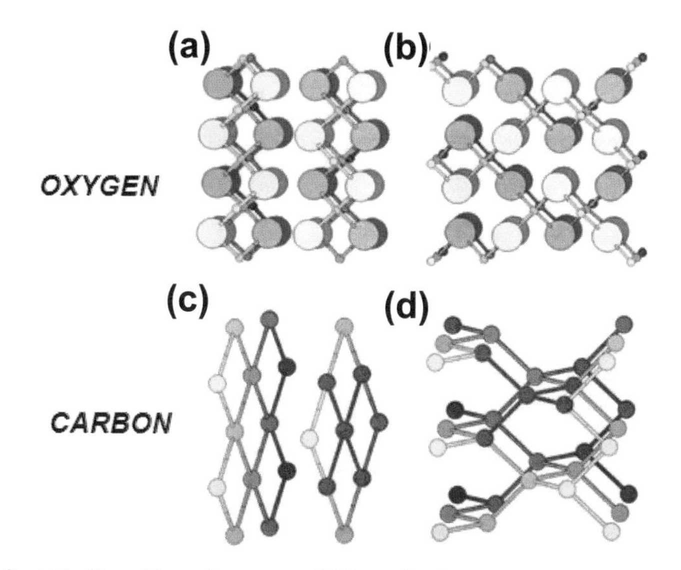

Figure 11. β-cristabolite and layered structures of CO_2 predicted at pressures above 200 GPa. (a) and (b) show the *fcc* oxygen sublatice and (c) and (d) show the carbon sublattice. [Adapted with permssion of the American Physical Society from Lee et al. (2009) *Phys Rev B*, Vol. 79, 144102, Fig. 4. *http://link.aps.org/doi/10.1103/PhysRevB.79.144102*.]

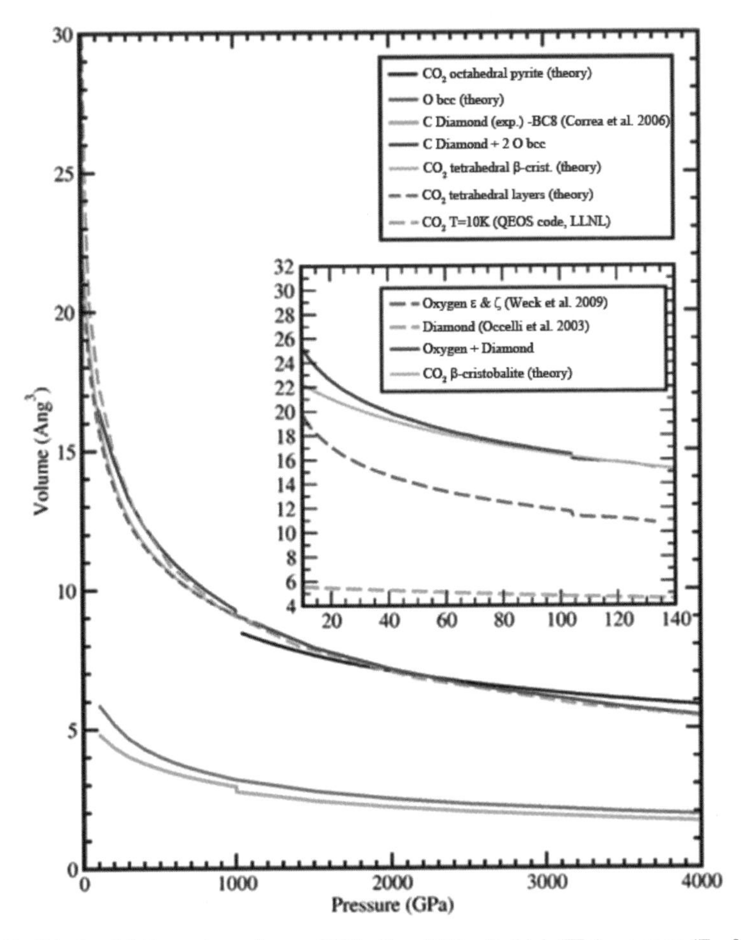

Figure 12. Calculated *P-V* equations of state of CO_2, C, and O to ultrahigh (TPa) pressures (T = 0 K). The volume change associated with the calculated breakdown of CO_2 to elemental C and O is indicated. The lower pressure range is shown in the inset (J. Montoya and R.J. Hemley, unpublished).

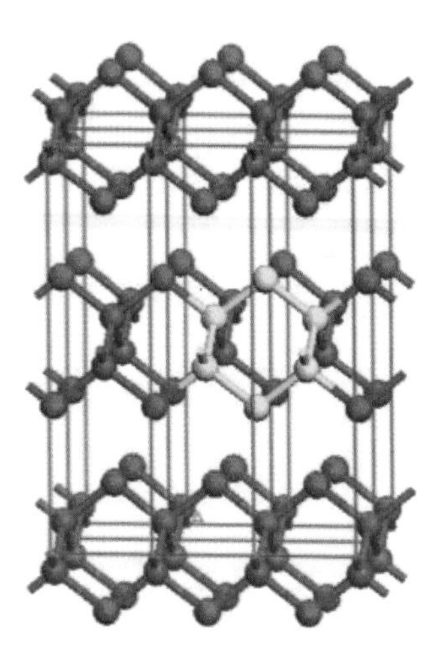

Figure 13. Predicted high-pressure structure of CO with the *Cmcm* layered structure. The grey and red spheres represent C and O atoms, respectively, and the yellow spheres identify the six-membered ring in the structure. This represents only one of several high-pressure structures predicted for CO under pressure (Used with permission of the American Physical Society from Sun et al. (2011) *Phys Rev Lett*, Vol. 106, 145502, Fig. 2d. *http://link.aps.org/doi/10.1103/PhysRevLett.106.145502*]

known phases. Above 5.25 GPa and room temperature solid methane transforms to structures in which the molecules are rotationally locked (Hirai et al. 2008). These studies point to the vast unexplored phase stability of high-pressure molecular hydrocarbons (see, Sephton and Hazen 2013).

CARBONATES

Behavior of *sp*^{2} carbonates

High-pressure phase transitions in rhombohedral carbonates epitomize the challenges of understanding deep carbon mineralogy, because a variety of reversible metastable transitions to slightly denser sp^2 forms are possible, and these forms are less kinetically inhibited than reconstructive transitions to sp^3 forms. Calcite provided one of the earliest experimental demonstrations of distinctive metastable high-pressure carbonate phases (Bridgman 1939; Jamieson 1957; Merrill and Bassett 1975). At ~1.5 GPa and 25 °C calcite transforms displacively to a monoclinic form known as calcite-II with space group $P2_1/c$, with a second reversible transition at ~2.2 GPa to calcite-III (Suito et al. 2001). Catalli and Williams (2005) report yet another transition to a presumably distorted modification of calcite-III at P ~25 GPa, which reverts to calcite-III with significant hysteresis on lowering pressure.

Less attention has been paid to the high-pressure polymorphs of other rhombohedral carbonates (Hazen et al. 2013). The high-pressure behavior of ferrous iron and magnesium carbonates, which are likely candidates for deep-Earth carbonates, are of special interest (Biellmann et al. 1993; Isshiki et al. 2004; Panero and Kabbes 2008; Seto et al. 2008). The high-spin (HS) to low-spin (LS) magnetic transition in siderite measured by means of X-ray emission spectroscopy (XES) has been reported to occur at 50 GPa (Mattila et al. 2007). The structure does not change through the spin transition, but there is a discontinuous volume change (Lavina et al. 2010). However, at 10 GPa and 1800 K, siderite breaks down to form a new iron oxide Fe_4O_5, isostructural with calcium ferrite, though the residual carbon phase was not identified (Lavina et al. 2011).

Dolomite [$CaMg(CO_3)_2$] is considered to be a major constituent of subducted carbonates (e.g., Zhu and Ogasawara 2002); therefore its phase stability and equation of state at high pressures and temperatures is important to understanding Earth's deep carbon cycle. At pressures below 7 GPa and temperatures between ~500 and 1000 °C dolomite is observed to decompose to aragonite plus magnesite (Martinez et al. 1996; Sato and Katsura 2001). As a result, dolomite is normally not considered a potential carrier to transport carbon to Earth's deep interior. However, two higher-pressure phases of dolomite have been documented experimentally. Dolomite-II, which has the calcite-III structure, occurs at pressure >20 GPa (Santillán et al. 2003). Mao et al. (2011) observed a transition to a significantly denser monoclinic phase called dolomite-III at 41 GPa and 1200 °C, with stability extending to 83 GPa. This study by Mao et al. (2011) provides evidence that the addition of minor amounts of iron can stabilize dolomite in polymorphs that exist at P-T conditions of subducting slabs, thereby providing a mechanism to carry carbonate into the deep mantle. X-ray diffraction/laser heating techniques were used to study high-pressure and temperature dolomite polymorphs to 83 GPa and 1700 K of a natural Fe-bearing dolomite from Windham, Vermont with a composition of $Ca_{0.988}Mg_{0.918}Fe_{0.078}Mn_{0.016}(CO_3)_2$. They observed two distinct phase transformations: 1) to dolomite-II at ~17 GPa and 300 K, and 2) to a new monoclinic phase (dolomite-III) between 36 and 83 GPa. Both high-pressure polymorphs were stable up to 1500 K, indicating that the addition of minor Fe stabilizes dolomite under the conditions akin to the deep mantle. Thus, Fe-dolomite may provide a means for delivering carbon to Earth's deep interior, though additional experiments are required to determine the structure of dolomite-III, examine the potential effects of Fe substitution on the phase transitions, and determine the redox properties of Fe-bearing carbonate in Earth's mantle. A recent experimental study examined the high-pressure vibrational and elastic properties of iron-bearing magnesite across the spin transition. Distinctive changes in the unit cell volumes and vibrational properties are indicative of a high-spin to low-spin transition at 40 GPa. The spin transition enhances the stability of the carbonate phase (Lin et al 2012).

The orthorhombic aragonite group minerals (Hazen et al. 2013) have received less study at high pressure. Holl et al. (2000) studied witherite to 8 GPa and found that it undergoes a displacive phase transition at approximately 7 GPa. The topology of the structure did not change, but the CO_3 groups shifted into the same plane as the Ba atoms, changing the symmetry to hexagonal.

High-pressure sp^3 carbonates

Given the behavior of other carbon-bearing systems at very high pressure, it seems likely that the sp^2 carbonates with trigonally coordinated carbon will transform to sp^3 polymorphs with tetrahedrally-coordinated carbon. At ~40 GPa, Ono et al. (2005) found a phase transition from aragonite to another phase, which was subsequently solved using the USPEX method and verified experimentally (Fig. 14; Oganov et al. 2006; Ono et al. 2007). The structure (orthorhombic, space group *Pmmn*) was called post-aragonite; it still has carbon in the sp^2-state (CO_3-anions), and is characterized by 12-fold coordination of Ca atoms and joint hexagonal

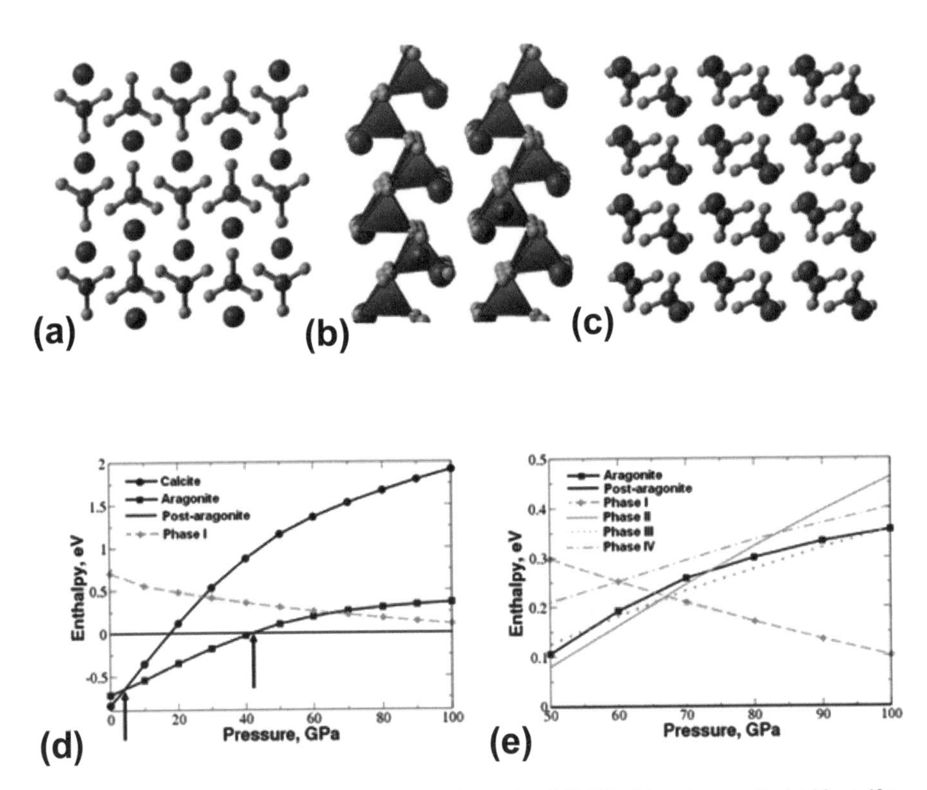

Figure 14. Predicted structures of high-pressure polymorphs of $CaCO_3$: (a) post-aragonite (stable at 42 to 137 GPa); (b) $C222_1$ phase (stable at > 137 GPa); (c) $P\bar{1}$ phase, probably stable in a narrow pressure range of a few GPa between aragonite and post-aragonite. (d) and (e) Enthalpies of candidate structures of $CaCO_3$ as a function of pressure, in eV per formula unit, relative to post-aragonite. [Adapted with permission of Elsevier from Oganov et al. (2006) *Earth Planet Sci Lett*, Vol. 241, p. 95-103, Figs. 1c-e, Fig. 3.]

close packing of oxygen and calcium atoms. At pressures above 137 GPa, a new structure was predicted using USPEX (Oganov et al. 2006), and once again the predicted structure matched the experimental X-ray diffraction patterns obtained at $P > 130$ GPa (Ono et al. 2007). The structure (space group $C222_1$) features chains of carbonate tetrahedra similar to silicate tetrahedral chains in pyroxenes. However, unlike pyroxenes, the $C222_1$ structure has large cations in one position (rather than two in pyroxenes). The presence of tetrahedral carbonate ions (with sp^3 hybridization) marks a dramatic shift in the chemistry of carbonates. Note that both post-aragonite and the $C222_1$ structures belong to new structure-types, which could not have been found by analogy with any known structures. Another new structure with $P\bar{1}$ symmetry was found by (Oganov et al. 2006) as metastable and energetically closely competitive with aragonite in a wide pressure range. It is predicted to be stable in a narrow (a few GPa) pressure interval around 42 GPa, between aragonite and post-aragonite. A phase with this structure was recently documented in experiments of Merlini et al. (2012).

Experiments demonstrate that magnesite is stable at least up to ~80 GPa (Fiquet et al. 2002) and transforms to a new phase above 100 GPa (Isshiki et al. 2004). Systematic search through databases of known crystal structures, combined with energy minimization, indicated that a pyroxene structure (space group $C2/c$) becomes energetically more favorable than magnesite above ~100 GPa (Skorodumova et al. 2005), but Oganov et al. (2006) found that the $C222_1$ structure (stable for $CaCO_3$) is preferable to the $C2/c$ pyroxene structure. USPEX simulations at 110 GPa and 150 GPa found a large number of low-enthalpy structures that are competitive in a wide pressure range and are more favorable than previously known structures (Oganov et al. 2008). Magnesite was found to be stable up to 82 GPa, whereas at higher pressures there are two phases with similar structures containing three-membered rings $(C_3O_9)^{6-}$ made of carbonate tetrahedra (Fig. 15) and Mg atoms in eight- and ten-fold coordination.

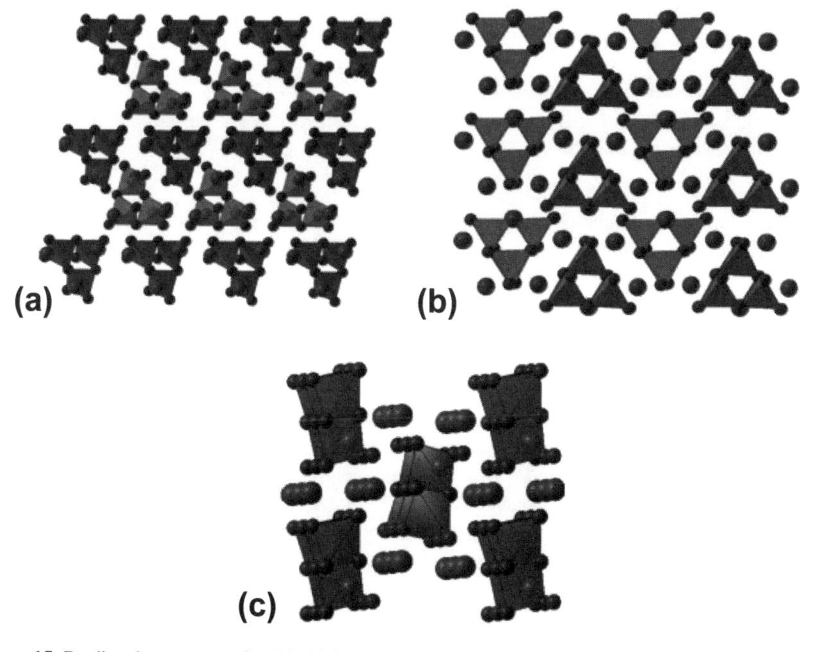

Figure 15. Predicted structures of stable high-pressure phases of $MgCO_3$: (a) post-magnesite phase II; (b) phase III; (c) $Pna2_1$-20 structure. [Adapted with permission of Elsevier from Oganov et al. (2008) *Earth Planet Sci Lett*, Vol. 273, p. 38-47, Fig. 5.]

These unusual structures were confirmed experimentally in a recent X-ray diffraction study by Boulard et al. (2011), who found that the Mg-Fe carbon-bearing compound can form at conditions corresponding to depths greater than 1,800 km (Fig. 16). Its structure, based on three-membered rings of corner-sharing $(CO_4)^{4-}$ tetrahedra, closely agrees with the earlier predictions (Oganov et al. 2008). This polymorph of carbonates concentrates $Fe^{(III)}$ as a result of intracrystalline reaction between $Fe^{(II)}$ and CO_3^{2-} groups schematically written as $4FeO + CO_2 \rightarrow 2Fe_2O_3 + C$. This oxidation-reduction reaction results in an assemblage consisting of the new high-pressure phase, magnetite, and diamond. Above ~160 GPa (Oganov et al. 2008), simulations predict a structure with space group $Pna2_1$ containing chains of corner-sharing carbonate tetrahedra to be stable.

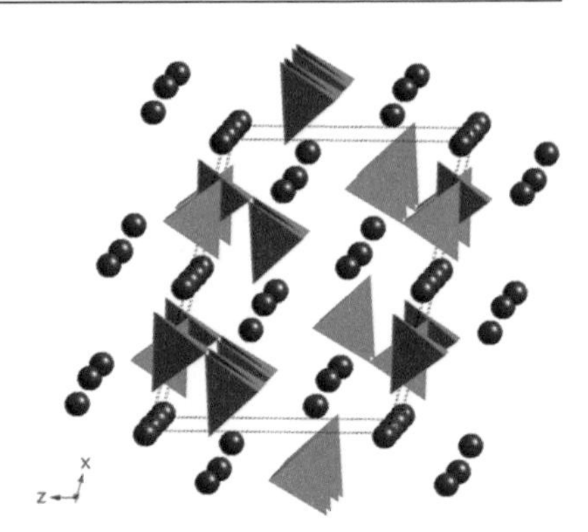

Figure 16. Experimentally determined structure of the high-pressure ferromagnesium carbonate phase with space group *P*21/*c*. $(CO_4)^{4-}$ tetrahedral appear in green and magnesium atoms are shown as violet spheres. [Reproduced with permission of the National Academy of Sciences from Boulard et al. (2011) *Proceedings of the National Academy of Sciences USA*, Vol. 108, p. 5184-5187, Fig. 4.]

The presence of tetrahedral carbonate-ions at very high pressures invites an analogy with silicates, but the analogy is limited. In silicates, the inter-tetrahedral angle Si-O-Si is extremely flexible (Belov 1961), which is one of the reasons for the enormous diversity of silicate structure types. Figure 17 shows the variation of the energy as a function of the Si-O-Si angle in the model $H_6Si_2O_7$ molecule (Lasaga and Gibbs 1987). One can see only a shallow minimum at Si-O-Si ≈ 135°, but a deep minimum at C-O-C ≈ 124° with steep energy variations for $H_6C_2O_6$. This difference suggests a much more limited structural variety of tetrahedral carbonates, compared to silicates. The C-O-C angles in the tetrahedral structures of CO_2, $CaCO_3$, and $MgCO_3$ are in the relatively narrow range from 115° to 125° at atmospheric pressure and decrease to 112° to 115° at 100 GPa. The differences between silicates and tetrahedral carbonates can also be seen from the large differences in the high-pressure structures of CO_2, $CaCO_3$, and $MgCO_3$ and the stable structures of SiO_2, $CaSiO_3$, and $MgSiO_3$.

The fundamental change of carbonate chemistry from sp^2 (triangular carbonate ions) to sp^3 (tetrahedral carbonate ions) has important consequences for the viscosity of carbonate melts, which at high pressures (presumably above ~80 to 130 GPa) is expected to be rather high (in contrast to very low viscosity of liquid carbonates at low pressures; Jones et al. 2013). This prediction remains to be confirmed experimentally.

Silicate carbonates

Under ambient conditions, CO_2 and SiO_2 are thermodynamically stable and do not react. Therefore, a natural question is whether or not compounds between CO_2 and SiO_2 are even possible under any conditions, and how they might relate to recently identified silica clathrate in carbonatite (melanophlogite: Beard et al. 2013). A recent study has reported the discovery that a silicon carbonate phase can form from carbon dioxide and silica under pressure (Santoro et al. 2011). In this work, dense carbon dioxide fills the micro-pores of silicalite, a pure SiO_2 zeolitic material, at high pressure. All the SiO_4 tetrahedra are on the surface of the micropores and consequently in contact with the CO_2. The large effective interaction area between the two materials greatly favors the chemical reaction between them. This silicon carbonate phase is obtained when the system is compressed to 18-20 GPa and heated at 600-700 K. The material can be temperature quenched and was characterized by optical spectroscopy and synchrotron

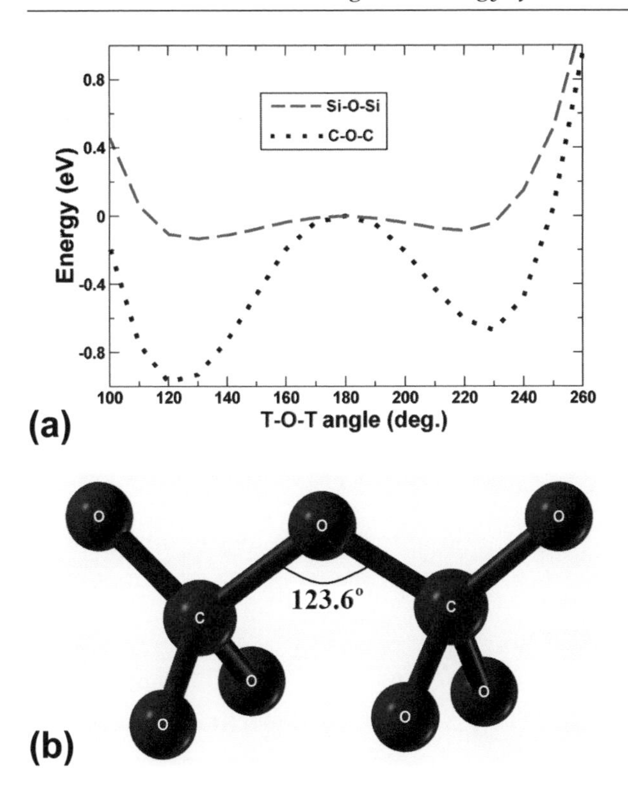

Figure 17. (a) Energy variation as a function of the T-O-T angle (red dashed line = Si; black dotted line = C). Calculations were performed on $H_6T_2O_7$ molecules; at each angle all T-O distances and O-T-O valence angles were optimized. (b) The minimum energy configuration of C_2O_7 has a 123.6° C-O-C angle. [Adapted with permission of Elsevier from Oganov et al. (2008) *Earth Planet Sci Lett*, Vol. 273, p. 38-47, Fig. 7.]

X-ray diffraction (Fig. 18). The structure of this new compound reflects that of the original silicalite crystal, although it is affected by stoichiometric and spatial disorder.

High-pressure silicate-carbonates with 4-coordinated carbon and 6-coordinated silicon represent an as yet unexplored crystal chemical possibility. More than 50 low-pressure silicate-carbonates with 3-coordinated carbon and 4-coordinated silicon are known, most notably the Ca-Si-carbonates spurrite [$Ca_5(SiO_4)_2(CO_3)$] and tilleyite [$Ca_5(Si_2O_7)(CO_3)_2$]. Some of these unusual crystal structures, which incorporate both planar CO_3 and tetrahedral SiO_4 units, may represent plausible structure topologies for high-pressure calcium carbonates with carbon in mixed 3- and 4-coordination.

Alternatively, polyhedral packing arrangements for tetravalent cations in both tetrahedral and octahedral coordination are well established. For example, (GeO_6) octahedra and (SiO_4) tetrahedra combine in the synthetic phases $Ca_2Ge_2Si_5O_{16}$ and $BaGeSi_3O_6$ (Robbins et al. 1968; Smolin 1969; Nevskii et al. 1979), while the rare mineral bartelkeite [$PbFeGe^{VI}(Ge^{IV}{}_2O_7)(OH)_2 \cdot H_2O$] incorporates both GeO_4 and GeO_6 polyhedra (Origlieri et al. 2012). These structure types, therefore, are possible isomorphs for silicate-carbonates at mantle pressures. In addition, Hazen et al. (1996) describe a related group of unusual high-pressure framework silicates with mixed 4- and 6-coordinated silicon (Fig. 19) with the general formula:

$$[(A^{1+}{}_{4-2x}B^{2+}{}_{x})Si^{VI}{}_{m}(Si^{IV}{}_{n}O_{2(m+n)+2}]$$

where A and B are monovalent and divalent cations, respectively, in 8 or greater coordination. All of these layered structures, including the benitoite form of $BaSi_4O_9$ and the wadeite form of $K_2Si_4O_9$ (Swanson and Prewitt 1983; Finger et al. 1995), possess densities intermediate between those of purely 4- or 6-coordinated phases, and thus must represent a transitional pressure

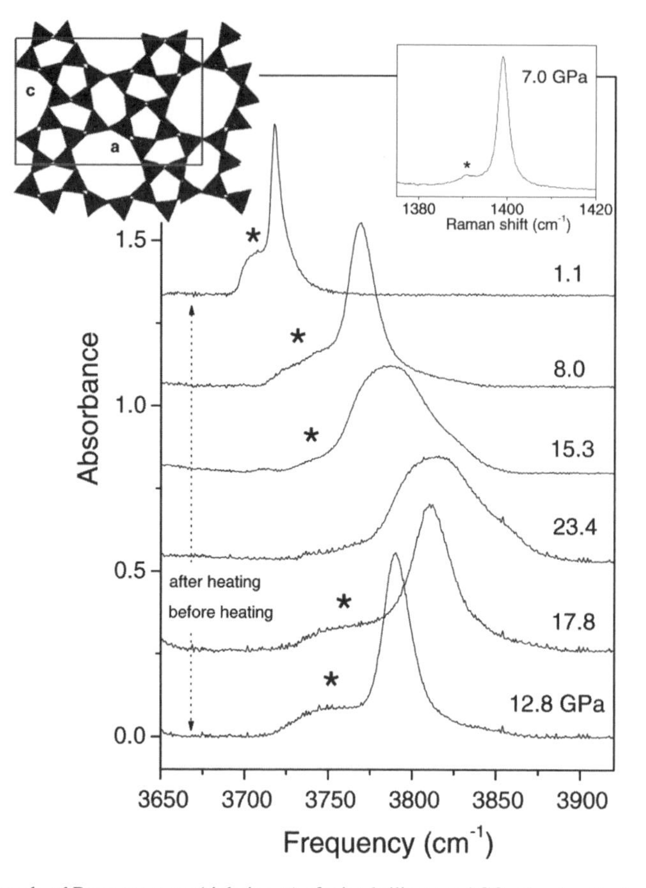

Figure 18. Infrared and Raman spectra (right insert) of mixed silicate and CO_2 showing the formation of the high pressure silicon carbonate phase. A schematic representation of the original silicate structure is shown in the left inset. [Reproduced with permission of the National Academy of Sciences from Santoro et al. (2011) *Proceedings of the National Academy of Sciences USA*, Vol. 108, p. 7689-7692, Fig. 1.]

range. Therefore, with the caveat that O-C-O angles are more restricted than O-Si-O angles (see above), a variety of plausible deep-Earth silicate-carbonate structures, including alkali and alkaline earth silicate-carbonate phases, should not be overlooked. At pressures significantly greater than those of Earth's lower mantle it is likely that silicon adopts a coordination greater than six; possible mixed carbon-silicate phases in this regime remains to be explored. These phases could be important components of carbon-rich super-Earths.

Finally, the nature of carbonates and silicate-carbonates at more extreme conditions, including those reached on dynamic compression, represents an unexplored area for future research. Transient shock pressures generated in bolide impacts exceed the duration of laboratory shock synthesis by orders of magnitude and, despite potential kinetic effects (DeCarli et al. 2002b), offer distinct potential for understanding the high-pressure behavior of carbon mineralogy. Combinations of natural observations, experiments, and modeling provide an important framework for understanding high-pressure phase equilibria in mixed carbon-silicate and simple carbonate systems over a very wide temperature interval (Martinez et al. 1998; Jones et al. 2000).

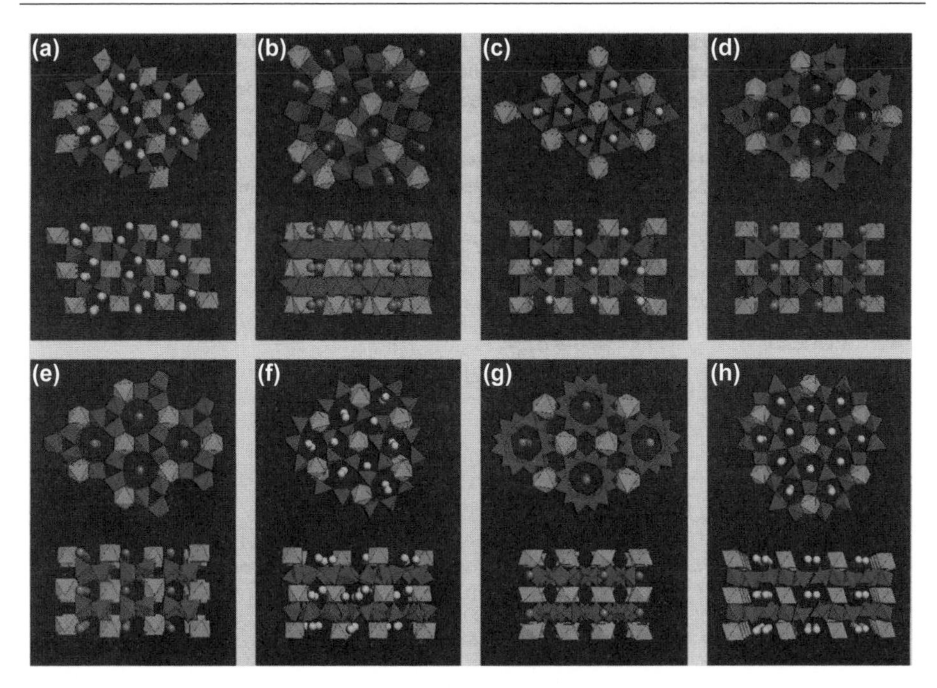

Figure 19. Observed and predicted framework silicates with SiO_4 tetrahedra (red), SiO_6 octahedra (blue), alkali cations (yellow spheres), and alkaline earth cations (green spheres) might represent carbonate-silicate structures at elevated temperature and pressure. Structures are shown in views parallel and perpendicular to layers of tetrahedral groups. (a) sodium silicate, $Na_4Si_2{}^{VI}Si_4{}^{IV}O_{14}$; (b) predicted calcium germanate-type, $Ca_2Si_2{}^{VI}Si_5{}^{IV}O_{16}$; (c) wadeite-type $K_4Si_2{}^{VI}Si_6{}^{IV}O_{18}$; (d) benitoite-type $Ba_2Si_2{}^{VI}Si_6{}^{IV}O_{18}$; (e) barium germanate-type $Ba_2Si_2{}^{VI}Si_6{}^{IV}O_{18}$; (f) sodium silicate, $Na_4Si_2{}^{VI}Si_6{}^{IV}O_{18}$; (g) predicted beryl-type $Ba_2Si_2{}^{VI}Si_6{}^{IV}O_{18}$; (h) sodium-calcium silicate, $Na_{1.8}Ca_{1.1}Si^{VI}Si_5{}^{IV}O_{14}$. [Reproduced with permission of The American Association for the Advancement of Science from Hazen et al. (1996) *Science*, Vol. 272, p. 1769-1770, Fig. 1.]

CONCLUSIONS

This survey of carbon mineralogy outlines much that is known regarding varied possible C-bearing mineral phases in deep planetary interiors. With the exception of the diamond polymorph of carbon and rare occurrences of Si and Fe carbides (Hazen et al. 2013), deep carbon phases have not been well documented, and with few notable exceptions, their high-pressure stability limits are still poorly understood. Furthermore, although the measured solubility of carbon in nominally acarbonaceous upper mantle silicates is extremely low (Keppler et al. 2003), the potential for carbon solid solution in lower mantle and core phases, including oxides, silicates, sulfides, and silicate melts, is not well constrained (Ni and Keppler 2013). Thus, while documented inventories of Earth's carbon represent only ~0.07 wt%, hidden reservoirs may represent in excess of 1.5% (Mason 1979; Kerridge 1985; Javoy 1997). This situation is analogous to our understanding of the deep hydrogen cycle 30 years ago (Smyth 2006; Smyth and Jacobsen 2006). Future high-pressure studies might consider the significant potential for storage and distribution of trace elements in C phases, including N, light elements, and especially noble gases, whose association with deep carbon is implied from volcanic degassing and from cosmochemistry (see Marty et al. 2013). We should also examine whether stability of new high-pressure C-bearing phases may influence isotopic fractionation at high-pressure conditions (Mikhail et al. 2011), which has enormous implications for reservoirs, fluxes, and geodynamic processes from the core to the atmosphere.

A great deal has been learned recently from both high-pressure experiments and computational theory. Recent advances point to the likelihood of novel *sp*^{3} phases with tetrahedrally-coordinated carbon, including topologies that differ from those of silicate minerals. High-pressure carbonates and carbonate-silicates will continue to be of special interest in modeling the extent and distribution of mantle carbon and elucidating the deep carbon cycle. High-pressure studies of carbon-bearing molecular phases also hold great promise for discovering novel phases with unusual properties. We thus anticipate that the next decade will see rapid and fundamental advances in the understanding of the mineralogy of carbon at extreme conditions.

ACKNOWLEDGMENTS

We thank J. Armstrong, R. E. Cohen, R. T. Downs, C. Schiffries, and D. A. Sverjensky for invaluable discussions and suggestions during the preparation of this manuscript. The authors gratefully acknowledge support from the Deep Carbon Observatory, the Alfred P. Sloan Foundation, the Carnegie Institution of Washington, the National Science Foundation, DARPA, and NASA's Astrobiology Institute for support of this study.

REFERENCES

Adams L (1921) The compressibility of diamond. J Wash Acad Sci 11:45-50
Ahrens TJ (1982) Constraints on core composition from shockwave data. Philos Trans R Soc London A306:37-47
Amsler M, Flores-Livas JA, Lehtovaara L, Balima F, Ghasemi SA, Machon D, Pailhes S, Willand A, Caliste D, Botti S, San Miguel A, Goedecker S, Marques MAL (2012) Crystal structure of cold compressed graphite. Phys Rev Lett 108:065501
Aoki K, Yamawaki H, Sakashita M, Gotoh Y, Takemura K (1994) Crystal structure of the high-pressure phase of solid CO_2. Science 263:356-358
Arapan S, Souza de Almeida J, Ahuja R (2007) Formation of sp^3 hybridized bonds and stability of $CaCO_3$ at very high pressure. Phys Rev Lett 98:268501
Aryasetiawan F, Gunnarson O (1998) The GW methods. Rep Prog Phys 61:237
Aust RB, Drickamer HG (1963) Carbon: A new crystalline phase. Science 140:817-819
Baughman R, Liu A, Cui C, Schields P (1997) A carbon phase that graphitizes at room temperature. Synth Met 86:2371-2374
Bazhanova ZG, Oganov AR, Gianola O (2012) Fe-C-H system at pressures of the Earth's inner core. Phys Usp 55:489-497
Beard AD, Howard K, Carmody L, Jones AP (2013) Melanophlogite pseudomorphing combeite: A new occurrence from the 2006 eruption of Oldoinyo Lengai, Tanzania. Submitted to American Mineralogist.
Belov NV (1961) Crystal Chemistry of Silicates with Large Cations. Russian Academy of Sciences Press, Moscow
Berman R, Simon F (1955) On the graphite-diamond equilibrium. Z Elektrochem 59:333
Biellmann C, Gillet P, Guyot F, Peyronneau J, Reynard B (1993) Experimental evidence or carbonate stability in Earth's lower mantle. Earth Planet Sci Lett 118:31-41
Birch F (1964) Density and composition of the mantle and the core. J Geophys Res 69:4377-4388
Boates B, Teweldeberhan AM, Bonev SA (2012) Stability of dense liquid carbon dioxide. Proc Natl Acad Sci 109:14808-14812
Bonev SA, Gygi F, Ogitsu T, Galli G (2003) High-pressure molecular phases of solid carbon dioxide. Phys Rev Lett 91:065501
Boulard E, Gloter A, Corgne A, Antonangeli D, Auzende A, Perrillat J, Fuyot F, Fiquet G (2011) New host for carbon in the deep Earth, Proc Natl Acad Sci USA 108:5184-5187
Boulfelfel SE, Oganov AR, Leoni S (2012) Understanding the nature of "superhard graphite". Sci Rep 2:471
Brazhkin VV, Lyapin AG, Hemley RJ (2002) Harder than diamond: dreams and reality. Philos Mag A 82:231-253
Bridgman PW (1939) The high-pressure behavior of miscellaneous minerals. Am J Sci 237:7-18
Bulanova G, Walter M, Smith CB, Kohn SC, Armstrong LS, Blundy J, Gobbo L (2010) Mineral inclusions in sublithospheric diamonds from Collier 4 kimberlite pipe, Juina, Brazil: subducted protoliths, carbonated melts and primary kimberlite magmatism. Contrib Mineral Petrol 160:489-510

Bulanova GP, Zayakina NV (1991) Graphite – iron – cohenite assemblage in the central zone of diamond from 23rd Party Congress kimberlite. Dokl Akad Nauk SSSR 317:706-709

Bundy FP, Bassett WA, Weathers MS, Hemley RJ, Mao HK, Goncharov AF (1996) The pressure-temperature phase and transformation diagram for carbon; updated through 1994. Carbon 34:141-153

Bundy FP, Bovenkerkm HP, Strong HM, Wentorf RH Jr (1961) Diamond-graphite equilibrium line from growth and graphitization of diamond. J Chem Phys 35:383-391

Bundy FP, Kasper JS (1967) Hexagonal diamond – A new form of carbon. J Chem Phys 46:3437

Byrne S, Ingersoll AP (2003) A sublimation model for Martian south polar ice features. Science 299:1051-1053

Catalli K, Williams Q (2005) A high-pressure phase transition of calcite-III. Am Mineral 90:1679-1682

Chalifoux WA, Tykwinski RR (2010) Synthesis of polyynes to model the sp-carbon allotrope carbyne. Nature Chem 2:967-971

Che JW, Cagin T, Goddard WA (1999) Generalized extended empirical bond-order dependent force fields including nonbonded interactions. Theor Chem Acc 102:346-354

Correa AA, Bonev SA, Galli G (2006) Carbon under extreme conditions: Phase boundaries and electronic properties from first-principles theory. Proc Natl Acad Sci 103:1204-1208

Datchi F, Giordano VM, Munsch P, Saitta AM (2009) Structure of carbon dioxide phase IV: Breakdown of the intermediate bonding state scenario. Phys Rev Lett 103:185701

Datchi F, Mallick B, Salamat A, Ninet S (2012) Structure of polymeric carbon dioxide CO_2-V. Phys Rev Lett 108:125701

Davies G (1984) Diamond. Adam Hilger, Bristol, UK

Day HW (2012) A revised diamond-graphite transition curve. Am Mineral 97:52-62

DeCarli PS, Bowden E, Jones AP, Price GD (2002a) Laboratory impact experiments versus natural impact events. GSA Special Paper 356:595-605

DeCarli PS, Bowden E, Sharp TG, Jones AP, Price GD (2002b) Evidence for kinetic effects on shock wave propagation in tectosilicates. Shock Compression of Condensed Matter-2001, Parts 1 and 2, Proceedings 620:1381-1384

Dellago C, Bolhuis PG, Csajka FS, Chandler D (1998) Transition path sampling and the calculation of rate constants. J Chem Phys 108:1964-1977

Dion M, Rydberg H, Schröder E, Langreth DC, Lundqvist BI (2004) Van der Waals density functional for general geometries. Phys Rev Lett 92:246401

Dong JJ, Tomfohr JK, Sankey OF, Leinenweber K, Somayazulu M, McMillan PF (2000) Investigation of hardness in tetrahedrally bonded nonmolecular CO_2 solids by density-functional theory. Phys Rev B 62:14685-14689

Downs RT, Somayazulu MS (1998) Carbon dioxide at 1.0 GPa. Acta Crystallogr C 54:897-898

Duclos S, Brister K, Haddon RC, Kortan AR, Thiel FA (1991) Effects of pressure and stress on C_{60} fullerite to 20 GPa. Nature 351:380-381

Eggert, JH. Hicks DG, Celliers PM, Bradley DK, McWilliams RS, Jeanloz R, Miller JE, Boehly TR, Collins GW (2010) Melting temperature of diamond at ultra-high pressure. Nature Phys 6:40-43

Endo K, Koizumi S, Otsuka T, Suhara M, Morohasi T, Kurmaev EZ, Chong DP (2001) Analysis of XPS and XES of diamond and graphite by DFT calculations using model molecules. J Comput Chem 22:102-108

Fahy S, Louie SG (1987) High-pressure structural and electronic properties of carbon. Phys Rev B 36:3373-3385

Fahy S, Louie SG, Cohen ML (1986) Pseudopotential total-energy study of the transition from rhombohedral graphite to diamond. Phys Rev B 34:1191-1199

Finger LW, Hazen RM (1991) Crystal chemistry of six-coordinated silicon: A key to understanding the Earth's deep interior. Acta Crystallogr B 47:561-580

Finger LW, Hazen RM, Fursenko BA (1995) Refinement of the crystal structure of $BaSi_4O_9$ in the benitoite form. J Phys Chem Solids 56:1389-1393

Fiquet G, Badro J, Gregoryanz E, Fei Y, Occelli F (2009) Sound velocity in iron carbide (Fe_3C) at high pressure: Implications for the carbon content of the Earth's inner core. Phys Earth Planet Inter 172:125–129

Fiquet G, Guyot F, Kunz M, Matas J, Andrault D, Hanfland M (2002) Structural refinements of magnesite at very high pressure. Am Mineral 87:1261-1265

Frost DJ, Liebske C, Langenhorst F, McCammon CA, Tronnes RG, Rubie DC (2004) Experimental evidence for the existence of iron-rich metal in the Earth's lower mantle. Nature 428:409-412

Frost DJ, McCammon CA (2008) The redox state of Earth's mantle. Annu Rev Earth Planet Sci 36:389-420

Gabrysch M (2008) Electronic properties of diamond. Ph.D. Thesis Uppsala University

Galli G, Martin R, Car R, Parrinello M (1989) Carbon: The nature of the liquid state. Phys Rev Lett 63:988-991

Gathers GR, Shaner JW, Young DY (1974a) High temperature carbon equation of state. UCRL Report 51644, doi: 10.2172/4265839

Gathers GR, Shaner JW, Young DY (1974b) Experimental, very high-temperature, liquid-uranium equation of state. Phys Res Lett 33:70-72

Gillet, P, Hemley RJ, McMillan PF (1998) Vibrational properties at high pressures and temperatures. Rev Mineral 37:525-590

Giordano VM, Datchi F (2007) Molecular carbon dioxide at high pressure and high temperature. Europhys Lett 77:46002

Giordano VM, Datchi F, Dewaele A (2006) Melting curve and fluid equation of state of carbon dioxide at high pressure and high temperature. J Chem Phys 125:054504

Godard G, Frezzotti ML, Palmeri R, Smith DC (2012) Origin of high-pressure disordered metastable phases (lonsdaleite and incipiently amorphized quartz) in metamorphic rocks: Geodynamic shock or crystal-scale overpressure? *In*: Ultrahigh Pressure Metamorphism: 25 Years After the Discovery of Coesite and Diamond. Dobrzhinetskaya LF, Faryad SW, Wallis S, Cuthbert S (eds) Elsevier, London, p 125-148

Goncharov AF (1990) Carbons at high pressure: Pseudomelting at 44 GPa. Zh Eksp Teor Fiz 98:1824

Goodrich CA, Bird JM (1985) Formation of iron-carbon alloys in basaltic magma at Uivfaq, Disko Island: The role of carbon in mafic magmas. J Geology 93:475-492

Guth JR, Hess AC, McMillan PF, Petuskey WT (1990) A valence force field for diamond from ab initio molecular orbital cluster calculations. J Phys Cond Matt 2:8007

Hanfland M, Hu JZ, Shu JF, Hemley RJ, Mao HK, Wu Y (1990) X-ray diffraction and Raman spectroscopy of graphite to 50 GPa. Bull Am Phys Soc 35:465-466

Hanfland M, Syassen K, Sonnenschein R (1989) Optical reflectivity of graphite under pressure. Phys Rev B 40:1951-1954

Hanson RC (1985) A new high-pressure phase of solid CO_2. J Phys Chem 89:4499-4501

Hazen RM (1999) The Diamond Makers. Cambridge University Press, New York

Hazen RM, Downs RT, Finger LW (1996) High-pressure framework silicates. Science 272:1769-1771

Hazen RM, Downs RT, Jones AP, Kah L (2013) Carbon mineralogy and crystal chemistry. Rev Mineral Geochem 75:7-46

Hazen RM, Finger LW (1978) Crystal chemistry of silicon-oxygen bonds at high pressure: Implications for the Earth's mantle mineralogy. Science 201:1122-1123

Hazen RM, Mao HK, Finger LW, Bell PM (1980) Structure and compression of crystalline methane at high pressure and room temperature. Appl Phys Lett 37:288-289

He T, Huang Q, Ramirez AP, Wang Y, Regan KA, Rogado N, Hayward MA, Hass MK, Slusky JS, Inumara K, Zandbergen HW, Ong NP, Cava RJ (2001) Superconductivity in the non-oxide perovskite $MgCNi_3$. Nature 411:54-56

Hemley RJ, Dera P (2000) Molecular crystals: high-temperature and high-pressure crystal chemistry. Rev Mineral Geochem 41:335-419

Hirai H, Konagai K, Kawamura T, Yamamoto Y, Yagi T (2008). Phase changes of solid methane under high pressure up to 86 GPa at room temperature. Chem Phys Lett 454:212-217

Hohenberg P, Kohn W (1964) Inhomogeneous electron gas. Phys Rev 136:B864-B871

Holl CM, Smyth JR, Laustsen HMS, Jacobsen SD, Downs RT (2000) Compression of witherite to 8 GPa and the crystal structure of $BaCO_3$ II. Phys Chem Miner 27:467-473

Holm B, Ahuja R, Belonoshko A, Johansson B (2000) Theoretical investigation of high pressure phases of carbon dioxide. Phys Rev Lett 85:1258-1261

Iota V, Yoo CS, Cynn H (1999) Quartzlike carbon dioxide: an optically nonlinear extended solid at high pressures and temperatures. Science 283:1510-1513

Iota V, Yoo CS, Klepeis J, Jenel Z, Evans W, Cynn H (2008) Six-fold coordinated carbon dioxide VI. Nature Mater 6:4-38

Isshiki M, Irifune T, Hirose K, Ono S, Ohishi Y, Watanuki T, Nishibori E, Takadda M, Sakata M (2004) Stability of magnesite and its high-pressure form in the lowermost mantle. Nature 427:60-63

Jacob DE, Kronz A, Viljoen KS (2004) Cohenite, native iron and troilite inclusions in garnets from polycrystalline diamond aggregates. Contrib Mineral Petrol 146:566-576

Jamieson JC (1957) Introductory studies of high-pressure polymorphism to 24,000 bars by x-ray diffraction with some comments on calcite II. J Geol 65:334-343

Javoy M (1997) The major volatile elements of the Earth: Their origin, behavior, and fate, Geophys Res Lett 24:177-180

Jones AP, Claeys P, Heuschkel S (2000) Impact melting of carbonates from the Chicxulub crater. Lecture Notes in Earth Sciences 91:343-361

Jones AP, Dobson D, Wood I, Beard AD, Verchovsky A, Milledge HJ (2008) Iron carbide and metallic inclusions in diamonds from Jagersfontein. 9th International Kimberlite Conference, 9IKC-A-00360

Jones AP, Genge M, Carmody L (2013) Carbonate melts and carbonatites. Rev Mineral Geochem 75:289-322

Jones AP, Kearsley AT, Friend CRL, Robin E, Beard A, Tamura A, Trickett S, Claeys P (2005) Are there signs of a large Paleocene impact, preserved around Disko Bay, West Greenland? Nuussuaq spherule beds origin by impact instead of volcanic eruption? Kenkmann T, Hörz F, Deutsch A (eds) Large Meteorite Impacts III. GSA Special Paper 384: 281-298

Kaminsky FV, Wirth R (2011) Iron carbide incluisons in lower-mantle diamond from Juina, Brazil. Can Mineral 49:555-572
Kasatochkin VI, Sladkov AM, Kudryavtsev YP, Popov NM, Korshak VV (1967) Crystalline forms of linear modification of carbon. Dokl Akad Nauk SSSR 117:358-360
Kasper JS, Richards SM (1964) The crystal structures of new forms of silicon and germanium. Acta Crystallogr 17:752-755
Keesom WH, Köhler JWL (1934) New determination of the lattice constant of carbon dioxide. Physica 1:167-174
Kennedy CS, Kennedy GC (1976) The equilibrium boundary between graphite and diamond. J Geophys Res 81:2467
Keppler H, Wiedenbeck M, Shcheka SS (2003) Carbon solubility in olivine and the mode of carbon storage in the Earth's mantle. Nature 424:414-416
Kerridge JF (1985) Carbon, hydrogen and nitrogen in carbonaceous chondrites. Geochim Cosmochim Acta 49:1707-1714
Kertesz M, Hoffman R (1984) The graphite-to-diamond transformation. J Sol State Chem 54:313-319
Kirillin AV, Kovalenko MD, Scheindlin MA (1984) Heating liquid carbon to 7000 K with a cw laser. J Exp Theor Phys Lett 40:781-783
Knudson MD, Desjarlais MP, Dolan DH (2008) Shock-wave exploration of the high-pressure phases of carbon. Science 322:1822-1825
Kohn W, Sham LJ (1965) Self-consistent equations including exchange and correlation effects. Phys Rev 140:A1133-A1138
Kurdyumov AV, Britun VF, Yarosh VV, Danlienko AI (2012) The influence of shock compression on the graphite transformations into lonsdaleite and diamond. J Superhard Mater 34:19-27
Lasaga AC, Gibbs GV (1987) Applications of quantum-mechanical potential surfaces to mineral physics calculations. Phys Chem Miner 14:107–117
Lavina B, Dera P, Downs RT, Yang W, Sinogeikin S, Meng Y, Shen G, Schiferl D (2010) Structure of siderite $FeCO_3$ to 56 GPa and hysteresis of its spin-pairing transition. Phys Rev B 82:064110-7
Lavina B, Dera P, Kim E, Meng Y, Downs RT, Weck PF, Sutton SR, Zhao Y (2011) Discovery of the recoverable high-pressure iron oxide Fe_4O_5. Proc Natl Acad Sci 108:17281-17284
Lee MS, Montoya JA, Scandolo S (2009) Thermodynamical stability of layered structures in compressed CO_2. Phys Rev B 79:144102
Leipunskii OI (1939) On synthetic diamonds. Usp Khim 8:1519
Leoni S, Boulfelfel SE (2010) Pathways of structural transformations in reconstructive phase pransitions: Insights from transition path sampling molecular dynamics. *In:* Modern Methods of Crystal Structure Prediction. Oganov AR (ed) Wiley-VCH, New York, p 181-221
Li Q, Ma Y, Oganov AR, Wang H, Wang H, Xu Y, Cui T, Mao H-K, Zou G (2009) Superhard monoclinic polymorph of carbon. Phys Rev Lett 102:175506
Liang Q, Yan CS, Meng YF, Lai J, Krasnicki S, Mao HK, Hemley RJ (2009) Enhancing the mechanical properties of CVD single-crystal diamond, J Phys Condens Matter 21:364215
Lin JF, Liu J, Jacobs C, Prakapenka VB (2012) Vibrational and elastic properties of ferromagnesite across the electronic spin-pairing transition of iron. Am Mineral 97:583-591
Litasov KD, Goncharov AF, Hemley RJ (2011) Crossover from melting to dissociation of CO_2 under pressure: Implications for the lower mantle. Earth Planet Sci Lett 309:318–323
Liu L (1983) Dry ice II, a new polymorph of CO_2. Nature 303:508-509
Lonsdale K (1971) Formation of lonsdaleite from single-crystal graphite. Am Mineral 56:333-336
Lord OT, Walter MJ, Dasgupta R, Clark SM (2009) Melting in the Fe-C system to 70 GPa. Earth Planet Sci Lett 284:157-167
Lovering JF (1964) Electron microprobe analysis of terrestrial and meteoritic cohenite. Geochim Cosmochim Acta 28:1745-1748
Ludwig A (1902) Die schmelzung der kohle. Zeit Elektrochem 8:273
Lyakhov AO, Oganov AR (2011) Evolutionary search for superhard materials applied to forms of carbon and TiO_2. Phys Rev B 84:092103
Lyakhov AO, Oganov AR, Valle M (2010) How to predict very large and complex crystal structures. Comput Phys Commun 181:1623-1632
Mao WL, Mao H-K, Eng PJ, Trainor TP, Newville M, Kao C, Heinz DL, Shu J, Meng Y, Hemley RJ (2003) Bonding changes in compressed superhard graphite. Science 302:425-427
Mao Z, Armentrout M, Rainey E, Manning CE, Dera P, Prakapenka VB, Kavner A (2011) Dolomite III: A new candidate lower mantle carbonate. Geophys Res Lett 38:L22303
Marques L, Hodeau JL, Núñez-Regueiro M, Perroux M (1996) Pressure and temperature diagram of polymerized fullerite. Phys Rev B 54:R12633-R12636

Martinez I, Perez EMC, Matas J, Gillet P, Vidal G (1998) Experimental investigation of silicate-carbonate system at high pressure and high temperature. J Geophys Res-Solid Earth 103:5143-5163

Martinez I, Zhang J, Reeder RJ (1996) In situ x-ray diffraction of aragonite and dolomite at high pressure and high temperature: Evidence for dolomite breakdown to aragonite and magnesite. Am Mineral 81:611-624

Martinez-Canales M, Pickard CJ, Needs RJ (2012) Thermodynamically stable phases of carbon at multiterapascal pressures. Phys Rev Lett 108:045704

Martoňák R, Laio A, Parrinello M (2003) Predicting crystal structures: The Parrinello-Rahman method revisited. Phys Rev Lett 90:075503

Marty B, Alexander CMO'D, Raymond SN (2013) Primordial origins of Earth's carbon. Rev Mineral Geochem 75:149-181

Mason B (1979) Data of Geochemistry, 6th edition. U.S.Geological Survey Prof Paper 440-B-1

Mattila T, Pylkkanen J-P, Rueff S, Huotari G, Vanko M, Hanfland M, Lehtinen M, Hamalainen K (2007) Pressure induced magnetic transition in siderite $FeCO_3$ studied by x-ray emission spectroscopy. J Phys Condens Matter 19:386206_2007

McNeill J, Pearson DG, Klein-David O, Nowell GM, Ottley CJ, Chinn I (2009) Quantitative analysis of trace element concentrations in some gem-quality diamonds. J Phys Condens Matter 21:364207

Merlini M, Crichton WA, Hanfland M, Gemmi M, Mueller H, Kupenko I, Dubrovinsky L (2012) The structures of dolomite at ultrahigh pressure and their influence on the deep carbon cycle. Proc Natl Acad Sci USA 109:13509-13514

Merrill L, Bassett WA (1975) The high-pressure structure of $CaCO_3$ (II), a high-pressure metastable phase of calcium carbonate. Acta Cryst B31:343-349

Mikhail S, Shahar A, Hunt SA, Jones AP, Verchovsky AB (2011) An experimental investigation of the pressure effect on stable isotope fractionation at high temperature; implications for mantle processes and core formation in celestial bodies. 42nd Lunar and Planetary Science Conference, March 7-11 (The Woodlands, TX), http://www.lpi.usra.edu/meetings/lpsc2011/pdf/1376.pdf

Miller ED, Nesting DC, Badding JV (1997) Quenchable transparent phase of carbon. Chem Mater 9:18-22

Mookherjee M (2011) Elasticity and anisotropy of Fe_3C at high pressures. Am Mineral 96:1530-1536

Mookherjee M, Nakajima Y, Steinle-Neumann G, Glazyrin K, Wu X, Dubrovinsky, L, McCammon C, Chumakov A (2011) High-pressure behavior of iron carbide (Fe_7C_3) at inner core conditions. J Geophys Res 116:B04201

Murakami M, Hirose K, Kawamura K, Sata N, Ohishi Y (2004) Post-perovskite phase transition in $MgSiO_3$. Science 304:855-858

Nevskii NN, Ilyukhin VV, Ivanova LI, Belov NV (1979) Crystal structure of calcium germinate. Sov Phys Dokl 24:135-136

Ni H, Keppler H (2013) Carbon in silicate melts. Rev Mineral Geochem 75:251-287

Niu H, Chen X-Q, Wang S, Li D, Mao WL, LiY (2012) Families of superhard crystalline carbon allotropes constructed via cold compression of graphite and nanotubes. Phys Rev Lett 108:135501

Nuñez Valdez MN, Umemoto K, Wentzkovitch RM (2012) Elasticity of diamond at high pressures and temperatures. Appl Phys Lett 100:171902

Núñez-Regueiro M, Monceau P, Hodeau JL (1992) Crushing C_{60} to diamond at room temperature. Nature 355:237-239

Occelli F, Loubeyre P, LeToullec R (2003) Properties of diamond under hydrostatic pressures up to 140 GPa. Nature Mater 2:151-154

Oganov AR (ed) (2010) Modern Methods of Crystal Structure Prediction. Wiley-VCH, Berlin

Oganov AR, Chen J, Gatti C, Ma Y-Z, Ma YM, Glass CW, Liu Z, Yu T, Kurakevych OO, Solozhenko VL (2009) Ionic high-pressure form of elemental boron. Nature 457:863-867

Oganov AR (ed) (2010). Modern Methods of Crystal Structure Prediction. Berlin: Wiley-VCH

Oganov AR, Glass CW (2006) Crystal structure prediction using ab initio evolutionary techniques: principles and applications. J Chem Phys 124:244704

Oganov AR, Glass CW, Ono S (2006) High-pressure phases of $CaCO_3$: crystal structure prediction and experiment. Earth Planet Sci Lett 241:95-103

Oganov AR, Lyakhov AO, Valle M (2011) How evolutionary crystal structure prediction works - and why. Acc Chem Res 44:227-237

Oganov AR, Ma Y, Lyakhov AO, Valle M, Gatti C (2010) Evolutionary crystal structure prediction as a method for the discovery of minerals and materials. Rev Mineral Geochem 71:271-298

Oganov AR, Ono S (2004) Theoretical and experimental evidence for a post-perovskite phase of $MgSiO_3$ in Earth's D" layer. Nature 430:445-448

Oganov AR, Ono S, Ma Y, Glass CW, Garcia A (2008) Novel high-pressure structures of $MgCO_3$, $CaCO_3$ and CO_2 and their role in the Earth's lower mantle. Earth Planet Sci Lett 273:38-47

Olijnyk H, Dauefer H, Jodl HJ, Hochheimer HD (1988) Effect of pressure and temperature on the Raman spectra of solid CO_2. J Chem Phys 88:4204-4212

Olijnyk H, Jephcoat AP (1998) Vibrational studies on CO_2 up to 40 GPa by Raman spectroscopy at room temperature. Phys Rev B 57:879-888

Ono S, Kikegawa T, Ohishi Y (2007) High-pressure phase transition of $CaCO_3$. Am Mineral 92:1246-1249

Ono S, Kikegawa T, Ohishi Y, Tsuchiya J (2005) Post-aragonite phase transformation in $CaCO_3$ at 40 GPa. Am Mineral 90:667-671

Origlieri MJ, Yang H, Downs RT, Posner ES, Domanik KJ, Pinch WJ (2012) The crystal structure of bartelkeite, with a revised chemical formula, $PbFeGe^{VI}(Ge^{IV}{}_2O_7)(OH)_2.H_2O$, isotypic wth high-pressure $P2_1/m$ lawsonite. Am Mineral 97:1812-1815

Pan Z, Sun H, Zhang Y, Chen C (2009) Harder than diamond: superior indentation strength of wurtzite BN and lonsdaleite. Phys Rev Lett 102:055503

Panero R, Kabbes JE (2008) Mantle-wide sequestration of carbon in silicates and the structure of magnesite II. Geophys Res Lett 35:L14307

Park JH, Yoo CS, Iota V, Cynn H, Nicol MF, Le Bihan T (2003) Crystal structure of bent carbon dioxide phase IV. Phys Rev B 68:014107

Perdew JP, Bruke K, Ernzerhof M (1996) Generalized gradient approximation made simple. Phys Rev Lett 78:3865-3868

Perdew JP, Wang Y (1992) Accurate and simple analytic representation of the electron-gas correlation energy. Phys Rev B 45:13244-13249

Pickard CJ, Needs RJ (2011) Ab initio random structure searching. J Phys Condens Matter 23:053201

Raman CV (1962) The infra-red absorption by diamond and its significance. Part I. Materials and Methods. Proc Indian Acad Sci A 55:1-4

Ramaswamy C (1930) Raman effect in diamond. Nature 125:704

Ringwood AE (1960) Cohenite as a pressure indicator in oron meteorites. Geochim Cosmochim Acta 20:155-158

Robbins C, Perloff A, Block S (1968) Crystal structure of BaGe $[Ge_3O_9]$ and its relation to benitoite. J Res Natl Bureau Standards A 70:385-391

Rohrbach A, Ballhaus C, Golla-Shindler U, Ulmer P, Kamenetsky VS, Kuzmin DV (2007) Metal saturation in the upper mantle. Nature 449:456-458

Rohrbach A, Ballhaus C, Ulmer P, Golla-Schindler U, Schonbohm D (2010) Experimental evidence for a reduced metal-saturated upper mantle. J Petrol 52:717-731

Rohrbach A, Schmidt MW (2011) Redox freezing and melting in the Earth's deep mantle resulting from carbon-iron redox coupling. Nature 472:209-212

Roman-Perez G, Soler JM (2009) Efficient implementation of a van der Waals density functional: Application to double-wall carbon nanotubes. Phys Rev Lett 103:096102

Ross A, Steele A, Fries MD, Kater L, Downes H, Jones A P, Smith CL, Jeniskens PM, Zolensky ME, Shaddad MH (2011) MicroRaman spectroscopy of diamond and graphite in Almahata Sitta and comparison with other ureilites. Meteor Planet Sci 46:364-378

Rossini FD, Jessup RS (1938) Heat and free energy formation of carbon dioxide and the transition between graphite and diamond. J Natl Bureau Standards C 21:491-513

Rouquette J, Dolejs D, Kantor IY, McCammon CA, Frost DJ, Prakapenka VB, Dubrovinsky LS (2008) Iron-carbon interactions at high temperatures and pressures. Appl Phys Lett 92:121912-121913

Santillán J, Williams Q, Knittle E (2003) Dolomite-II: A high-pressure polymorph of $CaMg(CO_3)_2$. Geophys Res Lett 30:1054-1059

Santoro M, Gorelli F, Bini R, Haines J, Cambon O, Levelut C, Montoya JA, Scandolo S (2012). Partially collapsed cristobalite structure in the non molecular phase V in CO_2. Proc Natl Acad Sci USA 109:5176-5179

Santoro M, Gorelli F, Haines J, Cambon O, Levelut C, Garbarino G (2011) Silicon carbonate phase formed from carbon dioxide and silica under pressure. Proc Nat Acad Sci 108:7689-7692

Santoro M, Gorelli FA, Bini R, Ruocco G, Scandolo S, Crichton WA (2006) Amorphous silica-like carbon dioxide. Nature 441:857-860

Sata N, Hirose K, Shen G, Nakajima Y, Ohishi Y, Hirao N (2010) Compression of FeSi, Fe_3C, $Fe_{0.95}O$, and FeS under the core pressure and implication for light element in the Earth's core, J Geophys Res 115:B09204

Sato K, Katsura T (2001) Experimental investigation on dolomite dissociation into aragonite + magnesite up to 8.5 GPa. Earth Planet Sci Lett 184:529-534

Saunders M, Cross RJ, Jiménez-Vázquez HA, Shimshi R, Khong A (1996) Noble gas atoms inside fullerenes. Science 271:1693-1697

Savvatimskiy AI (2005) Measurements of the melting point of graphite and the properties of liquid carbon (a review for 1963-2003). Carbon 43:1115-1142

Scott HP, Williams Q, Knittle E (2001) Stability and equation of state of Fe_3C to 73 GPa: Implications for carbon in the Earth's core. Geophys Res Lett 28:1875-1878

Selli D, Baburin IA, Martoňák R, Leoni S (2011) Superhard sp^3 carbon allotropes with odd and even ring topologies. Phys Rev B 84:161411

Sengupta A, Yoo CS (2009) Coesite-like CO_2: a missing analog to SiO_2. Phys Rev B 80:014118

Sephton MA, Hazen RM (2013) On the origins of deep hydrocarbons. Rev Mineral Geochem 75:449-465

Serra S, Cavazzoni C, Chiarotti GL, Scandolo S, Tosatii E (1999) Pressure-induced solid carbonates from molecular CO_2 by computer simulation. Science 284:788-790

Seto Y, Hamane D, Nagai T, Fujino F (2008) Fate of carbonates within oceanic plates subducted to the lower mantle, and a possible mechanism of diamond formation. Phys Chem Miner 35:223-229

Shin K, Kumar R, Udachin KA, Alavi S, Ripmeester JA (2012) Ammonia clathrate hydrates as new solid phases for Titan, Enceladus, and other planetary systems. Proc Natl Acad Sci 109:14785-14790

Shumilova TG, Mayer R, Isaenko SI (2011) Natural monocrystalline lonsdaleite. Dokl Earth Sci 441:1552-1554

Simakov SK (2010) Metastable nanosized diamond formation from a C-H-O fluid system. J Mater Res 25:2336-2340

Skorodumova NV, Belonoshko AB, Huang L, Ahuja R, Johansson B (2005) Stability of the $MgCO_3$ structures under lower mantle conditions. Am Mineral 90:1008–1011

Smith R (2011) Ramp compression of carbon above 50 Mbar on NIF. Bull Am Phys Soc 56:BAPS.2011.DPP.TI3.5, http://meetings.aps.org/link/BAPS.2011.DPP.TI3.5

Smolin Yu-I (1969) Crystal structure of barium tetragermanate Sov Phys Dokl 13:641

Smyth JR (2006) Hydrogen in high pressure silicate and oxide mineral structures. Rev Mineral Geochem 62:85-116

Smyth JR, Jacobsen SD (2006) Nominally anhydrous minerals and Earth's deep water cycle. *In:* Earth's Deep Water Cycle. van der Lee S, Jacobsen SD (ed) American Geophysical Union Monograph Series 168:1-11

Stachel T, Harris JW, Brey GP (1998) Rare and unusual mineral inclusions in diamonds from Mwadui, Tanzania. Contrib Mineral Petrol 132:34-47

Stagno V, Frost DJ (2011) Carbon speciation in the asthenosphere: Experimental measurements of the redox conditions at which carbonate-bearing melts coexist with graphite or diamond in peridotite assemblages. Earth Planet Sci Lett 300:72-84

Suito K, Namba J, Horikawa T, Taniguchi Y, Sakurai N, Kobayashi M, Onodera A, Shimomura O, Kikegawa T (2001) Phase relations of $CaCO_3$ at high pressure and high temperature. Am Mineral 86:997-1002

Sun J, Klug DD, Martoňák R, Montoya JA, Lee M-S, Scandola S, Tosatti E (2009) High-pressure polymeric phases of carbon dioxide. Proc Natl Acad Sci USA 106:6077-6081

Sun J, Klug DD, Pickard CJ, Needs RJ (2011) Controlling the bonding and band gaps of solid carbon monoxide with pressure. Phys Rev Lett 106:145502

Sundqvist B (1999) Fullerenes under high pressures. Adv Phys 48:1-134

Swanson DK, Prewitt CT (1983) The crystal structure of $K_2Si^{VI}Si_3^{IV}O_9$. Am Mineral 68:581-585

Tao JM, Perdew JP, Staroverov VN, Scuseria GE (2003) Climbing the density functional ladder: Nonempirical meta-generalized gradient approximation designed for molecules and solids. Phys Rev Lett 91:146401

Terasaki H, Nishida K, Shibazaki Y, Sakamaki T, Suzuki A, Ohtani E, Kikegawa T (2010) Density measurement of Fe_3C liquid using x-ray absorption image up to 10 GPa and effect of light elements on compressibility liquid iron. J Geophys Res 115:B06207

Tschauner O, Mao HK, Hemley RJ (2001) New transformations of CO_2 at high pressures and temperatures. Phys Rev Lett 87:075701

Tulk CA, Klug DD, dos Santos AM, Karotis G, Guthrie M, Molaison JJ, Pradhan N (2012) Cage occupancies in the high pressure structure H methane hydrate: A neutron diffraction study. J Chem Phys 136:054502

Umemoto K, Wentzcovitch RM, Saito S, Miyake T (2010) Body-centered tetragonal C_4: A viable sp^3 carbon allotrope. Phys Rev Lett 104:125504

Utsumi W, Yagi T (1991) Light-transparent phase formed by room-temperature compression of graphite. Science 252:1542-1544

Utsumi W, Yagi T, Taniguchi T, Shimomura O (1996) In situ x-ray observation of the graphite-diamond transition using synchrotron radiation. Proc 3d NIRIM International Symposium on Advanced Materials, 257-261

Utsumi W, Yamakata M, Yagi T, Shimomura O (1994) Diffraction study of the phase transition from graphite to hexagonal diamond under high pressures and high temperatures. *In:* Proceedings of the 14th AIRAPT Conference, Colorado Springs, CO. Schmidt SC, Shaner JW, Samara GA, Ross M (eds) Am Inst Phys Conf Proceedings 309:535-538

Utsumi W, Yamakata M, Yagi T, Shimomura O (1993) In situ x-ray diffraction study of the phase transition from graphite to hexagonal diamond under high pressures and high temperatures. *In:* High Pressure Science and Technology. AIP Conference Proceedings 309:535-538, doi: 10.1063/1.46474

Wang JT, Chen C, Kawazoe Y (2011). Low-temperature phase transformation from graphite to sp^3 orthrhombic carbon. Phys Rev Lett 106:075501

Wang L, Liu B, Li H, Yang W, Ding Y, Sinogeiken SV, Meng Y, Liu Z, Zeng XC, Mao WL (2012a) Long-range ordered carbon clusters: A crystalline material with amorphous building blocks. Science 337:825-828

Wang Y, Panzik JE, Kiefer B, Lee KKM (2012b) Crystal structure of graphite under room-temperature compression and decompression. Sci Reports 2:520

Wang ZW, Zhao YS, Tait K, Liao XZ, Schiferl D, Zha CS, Downs RT, Qian J, Zhu YT, Shen TD (2004) A quenchable superhard carbon phase synthesized by cold compression of carbon nanotubes. Proc Natl Acad Sci USA 101:13699-13702

Weck G, Desgreniers S, Loubeyre P, Mezouar M (2009) Single-crystal structural characterization of the metallic phase of oxygen. Phys Rev Lett 102:255503

Weerasinghe GL, Needs RJ, Pickard CJ (2011) Computational searches for iron carbide in the Earth's inner core. Phys Rev B 84:174110

Wen XD, Hand L, Labet V, Yang T, Hoffmann R, Ashcroft NW, Oganov AR, Lyakhov AO (2011) Graphane sheets and crystals under pressure. Proc Natl Acad Sci USA 108:6833-6837

Wood BJ, Li J, Shahar A (2013) Carbon in the core: its influence on the properties of core and mantle. Rev Mineral Geochem 75:231-250

Xie Y, Oganov AR, Ma Y (2010) Novel structures and high pressure superconductivity of $CaLi_2$. Phys Rev Lett 104:177005

Yagi T, Utsumi W, Yamakata M, Kikegawa T, Shimomura O (1992) High-pressure in situ x-ray diffraction study of the phase transformation from graphite to hexagonal diamond at room temperature. Phys Rev B 46:6031

Yoo CS, Cynn H, Gygi F, Galli G, Iota V, Nicol M, Carlson S, Hausermann D, Mailhiot C (1999). Crystal structure of carbon dioxide at high pressure: "Superhard" polymeric carbon dioxide. Phys Rev Lett 83:5527-5530

Zhang Y, Yin QZ (2012) Carbon and other light elements contents in the Earth's core based on first-principles molecular dynamics. Proc Natl Acad Sci 109:19579-19583

Zhao YX, Spain IL (1989). X-ray diffraction data for graphite to 20 GPa. Phys Rev B 40:993

Zhao Z, Bu X, Wang, LM, Wen B, He J, Liu Z, Wang HT, Tian Y (2011), Novel superhard carbon: C-centered orthorhombic C_8. Phys Rev Lett 107:215502

Zhou XF, Qian GR, Dong X, Zhang L, Tian Y, Wang HT (2010). Ab initio study of the formation of transparent carbon under pressure. Phys Rev B 82:134126

Zhu Q, Oganov AR, Lyakhov AO (2012a) Evolutionary metadynamics: a novel method to predict crystal structures. Cryst Eng Commun 14:3596-3601

Zhu Q, Oganov AR, Salvado M, Pertierra P, Lyakhov AO (2011) Denser than diamond: ab initio search for superdense carbon allotropes. Phys Rev B 83:193410

Zhu Q, Zeng Q, Oganov AR (2012b) Systematic search for low-enthalpy sp^3 carbon allotropes using evolutionary metadynamics. Phys Rev B 84:201407

Zhu Y, Ogasawara Y (2002) Carbon recycled into deep Earth: Evidence from dolomite dissociation in subduction-zone rocks. Geology 30: 947-950

Reviews in Mineralogy & Geochemistry
Vol. 75 pp. 79-107, 2013

Carbon Mineral Evolution

Robert M. Hazen

Geophysical Laboratory, Carnegie Institution of Washington
5251 Broad Branch Road NW, Washington, DC 20015, U.S.A.

rhazen@ciw.edu

Robert T. Downs

Department of Geosciences, University of Arizona
1040 East 4th Street, Tucson, Arizona 85721-0077, U.S.A.

rdowns@u.arizona.edu

Linda Kah

Department of Earth & Planetary Sciences, University of Tennessee
Knoxville, Tennessee 37996-4503, U.S.A.

lckah@utk.edu

Dimitri Sverjensky

Department of Earth & Planetary Sciences, Johns Hopkins University
Baltimore, Maryland 21218, U.S.A.

sver@jhu.edu

INTRODUCTION

The discovery of the extreme antiquity of specific minerals through radiometric dating (e.g., Strutt 1910), coupled with Norman L. Bowen's recognition of a deterministic evolutionary sequence of silicate minerals in igneous rocks (Bowen 1915, 1928), implies that Earth's crustal mineralogy has changed dramatically through more than 4.5 billion years of planetary history. Detailed examination of the mineralogical record has led to a growing realization that varied physical, chemical, and biological processes have resulted in a sequential increase in diversification of the mineral kingdom. This diversification has been accompanied by significant changes in the near-surface distribution, compositional range (including minor and trace elements), size, and morphology of minerals (Ronov et al. 1969; Nash et al. 1981; Zhabin 1981; Meyer 1985; Wenk and Bulakh 2004; Hazen et al. 2008, 2009, 2011). Variation in Earth's mineralogical character thus reflects the tectonic, geochemical, and biological evolution of Earth's near-surface environment (Bartley and Kah 2004; Hazen et al. 2009, 2012; Grew and Hazen 2010a,b; McMillan et al. 2010; Krivovichev 2010; Grew et al. 2011; Tkachev 2011).

The mineral kingdom's evolutionary narrative shares many features with the increased complexity inherent within other evolving systems, including the nucleosynthesis of elements and isotopes, the prebiotic synthesis of organic molecules, biological evolution through Darwinian natural selection, and the evolution of social and material culture (Hazen and Eldredge 2010). In particular, well-known biological phenomena such as diversification, punctuation, and extinction appear to be common traits within a wide range of complex, evolving systems.

1529-6466/13/0075-0004$00.00 DOI: 10.2138/rmg.2013.75.4

Perhaps more than any other element, carbon exemplifies these processes of "mineral evolution." Four episodes outline major events in the mineral evolution of carbon: (1) the synthesis of the first mineral, likely diamond, and perhaps a dozen other "ur-minerals" in the cooling envelopes of active aging stars; (2) the formation of carbon-bearing materials in the condensing solar nebula, as preserved in chondrite and achondrite meteorites; (3) the diversification of carbon mineralogy through physical and chemical processes in the dynamic crust and mantle; and (4) the profound influence of biospheric evolution on Earth's near-surface carbon mineralogy. Hazen et al. (2008) further divided the latter three eras into 10 stages of Earth's mineral evolution. Here we review each of these stages of carbon mineral evolution (Table 1).

STAGES OF CARBON MINERAL EVOLUTION

The story of carbon mineralogy began more than 13 billion years ago in the gaseous envelopes of the first generation of large stars. Prior to the first large energetic stars, perhaps at a time when the universe was a few million years old, there had never been a place that was both dense enough with mineral-forming elements and also cool enough to condense crystals. But when a dying star explodes into a supernova, or when an asymptotic giant branch star sheds its outer envelope in an intense stellar wind, the star's element-rich gaseous envelope expands and rapidly cools, the outer layers cooling first.

Hazen et al. (2008) speculate that the first mineral was diamond (possibly co-precipitated with lonsdaleite), which condensed in carbon-rich zones at temperatures below about 3700 °C. Next came graphite at a slightly cooler 3200 °C. Diamond and graphite were the first of the "ur-minerals"—the dozen or so mineral species that formed prior to the first stellar nebulae, planets, and moons. In addition to diamond and graphite there are at least two other carbon-bearing ur-minerals, the iron and silicon carbides cohenite and moissanite, as well as possible nano-particles of titanium, iron, molybdenum, and zirconium carbides. These carbides have crystallization temperatures that are almost as high as diamond, so in the zones of an exploding star where carbon mixed with silicon or iron, carbides would have formed readily. This relative abundance of carbon-bearing phases among the dozen or so ur-minerals thus reflects the carbon-rich composition of stellar envelopes.

Ur-minerals are concentrated in so-called "dense molecular clouds," which are vast, cold volumes of the galaxy where hydrogen, helium, and mineral-bearing dust achieve densities of 10^2 to 10^4 particles per cubic centimeter, compared to only 1 particle per cubic centimeter in the interstellar voids of the Milky Way galaxy (Williams et al. 2000). The ur-minerals are present as nanometer- to micron-sized particles in the dust grains, some of which are preserved in the fine-grained matrix of chondritic meteorites. These grains can be isolated and identified by their anomalous isotopic compositions (Alexander 1990, 1993; Nittler 2003; Messenger et al. 2003, 2006; Mostefaoui and Hoppe 2004; Stroud et al. 2004; Jones 2007; Vollmer et al. 2007). Second-generation stars and their associated planets arise through gravitational clumping of this primitive nebular material.

Diamond and other carbon-bearing ur-minerals may persist for billions of years, but they are not the dominant C-rich species in molecular clouds. Astronomical observations employing millimeter-wave spectroscopy reveal distinctive absorption spectra from more than 160 different small molecular species, primarily organic molecules (Allamandola et al. 1989; Ehrenfreund and Charnley 2000; Allamandola and Hudgins 2003; Kwok 2009; for a list see *http://en.wikipedia.org/wiki/List_of_interstellar_and_circumstellar_molecules*). Indeed, in the cryogenic temperature regime of the dense molecular cloud, several of the more abundant C-bearing molecules, including CO, CO_2, and CH_4, may condense onto dust particles to form their own ices, and thus may represent an additional, if ephemeral, source of pre-solar mineralogical diversity.

Table 1. Principal stages in the mineral evolution of carbon on Earth.*

Stage	Age (Ga)	Principal Carbon Minerals
The "Ur-minerals"	>4.6	diamond, lonsdaleite, graphite, moissanite, cohenite
The Era of Earth's Accretion		
1. Primary Chondrite Minerals	~4.56	as above, plus haxonite
2. Altered Meteorites	~4.56-4.55	as above, plus dolomite, calcite, magnesite, siderite, aragonite
The Era of Crust and Mantle Reworking		
3. Igneous Rock Evolution	4.55-4.0	as above; plus ankerite, rhodochrosite, kutnohorite
4. Granitization; pegmatites	4.0-3.5	as above; plus niveolanite, zabuyelite, other pegmatite carbonates
5. Plate Tectonics	> 3.0	as above; plus continental shelf deposits of methane hydrate
The Era of Biological Mineralization		
6. Anoxic Biosphere	3.9-2.5	as above; increased carbonate deposition, but no new carbon minerals
7. Great Oxidation Event	2.5-1.9	as above; plus azurite, malachite, >100 other transition metal carbonates
8. Intermediate Ocean	1.9-1.0	as above; no new carbon minerals, but new carbonate textures
9. Snowball Earth	1.0-0.55	as above; no new carbon minerals
10. Skeletal Biomineralization	0.55-present	as above; skeletal carbonate biomineralization; plus >50 organic minerals

*Adapted in part from Hazen et al. (2008) Table 1.

The era of Earth's accretion

The first stages of Earth's mineral evolution—diversification beyond the dozen ur-minerals (including the carbon-bearing phases diamond, lonsdaleite, graphite, moissanite, and cohenite; Hazen et al. 2013)—began approximately 4.57 billion years ago and are preserved in the rich record of meteorites. These diverse objects reveal mineral-forming processes that occurred in the solar nebula prior to the formation of Earth and other planets. The mineralogical evolution of meteorites can be divided into two distinct stages—first the primitive chondrites, with approximately 60 different primary condensate phases, and then the diverse achondrite meteorites with more than 250 different minerals, which reflect planetesimal differentiation and subsequent alteration by aqueous, thermal, and shock processes (Hazen et al. 2008; McCoy 2010).

Stage 1—Primary chondrite minerals. Unaltered type 3 chondrite meteorites, which represent the most primitive stage of planetary evolution (McCoy 2010), include a variety of carbon-rich stony meteorites that formed by agglomeration of fine-grained nebular material (Brearley and Jones 1998; Weisberg et al. 2006). MacPherson (2007) defined "primary" chondrite minerals as phases that formed directly through condensation, melt solidification, or solid-state recrystallization following the earliest stages of nebular heating by the Sun. These phases include all of the original ur-minerals, but the only new crystalline carbon-bearing phase reported among the approximately 60 known primary chondritic minerals is the iron-nickel carbide haxonite [$(Fe,Ni)_{23}C_6$]. This mineral has not yet been identified in terrestrial rocks and is known only from meteorites (Brearley and Jones 1998) but cohenite $(Fe,Ni,Co)_3C$ and other Fe-based carbides are also found in lunar soils and rocks (Goldstein et al. 1976).

Carbon-containing minerals constitute a volumetrically trivial fraction of most chondrites. Note, however, that type 3 carbonaceous chondrites incorporate all but the most volatile of the 83 stable geochemical elements in roughly their cosmochemical abundances. Carbon, as the fourth most abundant element in the cosmos (after hydrogen, helium, and oxygen), is thus present in significant concentrations, primarily as complex suites of condensed organic molecules that comprise as much as 5 weight percent of some meteorites (Grady et al. 2002; Sephton 2002; Martins et al. 2007; Herd et al. 2011) and is also particularly widespread as nanodiamond with a mean size ~2.6 nm and abundance reaching 0.15 weight percent (Grady et al. 2002). Schmitt-Kopplin et al. (2010) report extreme molecular diversity in these meteorites: "tens of thousands of different molecular compositions and likely millions of diverse structures." These varied molecular species presumably constitute the dominant primordial carbon source that was subsequently processed during the accretion and differentiation of Earth and other terrestrial planets and moons.

Stage 2—Altered chondrite and achondrite meteorites. The next stage of carbon mineral evolution arose as a consequence of the gravitational accumulation of chondrites into growing planetesimals, with associated alteration by aqueous, thermal, and shock processes. These modes of mineral alteration are reflected in approximately 150 different mineral species identified in altered chondrites (Rubin 1997a,b; Brearley and Jones 1998; Brearley 2006; MacPherson 2007). These events, which characterized perhaps the first 10 to 20 million years of the solar nebula, resulted in a number of new carbon-bearing minerals, notably the first carbonates (as well as hydroxide and sulfate minerals) as a consequence of low-temperature (typically < 100 °C) aqueous alteration. Carbonates are especially abundant as veins, isolated grains, and brecciated fragments in the rare CI chondrites, which typically contain ~5 weight percent rhombohedral carbonates, predominantly dolomite, but also calcite (including Mg-, Fe-, and Mn-rich varieties), magnesite, and siderite (Fredriksson and Kerridge 1988). Carbonates are also common in CM chondrites; calcite predominates in these materials, although dolomite and minor aragonite are also known (Kerridge and Bunch 1979; Barber 1981).

As planetesimals grew to diameters greater than 100 kilometers, they experienced compaction, melting, and differentiation into an iron metal-rich core and silicate mantle. Mineralogical diversity increased as a result of variable oxygen fugacity, volatile content, aqueous fluid pH and salinity, variable heat sources, and impact events of ever greater intensity (McCoy et al. 2006). Additional diversity was driven by the fractionation of silicate melts by partial melting, crystal settling, and potential liquid immiscibility (Shukolyukov and Lugmair 2002; Wadhwa et al. 2006). Diamond, lonsdaleite, and graphite, and the iron-nickel carbides cohenite and haxonite, represent important accessory phases in many iron meteorites, and reflect the fractionation of at least some carbon into reduced, refractory core minerals. Note, however, that the total carbon content of iron meteorites rarely exceeds 0.1 weight percent (e.g., Jarosewich 1990).

Impact shock metamorphism may have also acted to modify significantly some carbon minerals (Langenhorst and Deutsch 2012). For example, the shock transition of graphite to diamond or lonsdaleite had been inferred by a number of researchers (Hanneman et al. 1967; Masaitis et al. 1972; Carlisle and Braman 1991; Russell et al. 1992; Ross et al. 2011) prior to direct observation of the frozen solid-state transformation in samples from meteor craters (Langenhorst et al. 1999; El Goresy et al. 2001). In a number of these shocked samples, diamond/lonsdaleite-type phases with a disordered stacking sequence preserve the hexagonal crystal form of graphite. Carbonate minerals are also known to express a range of shock metamorphic textural changes and, ultimately, volatilization under the influence of hypervelocity impacts, notably reflected in the release of CO_2 (Martinez et al. 1998; Jones et al. 2000; Agrinier et al. 2001; Ivanov and Deutsch 2002; Deutsch and Langenhorst 2007). However, no novel carbonate minerals associated with shocked meteorites have been documented.

Of the total of approximately 250 mineral species that are known to occur in all types of meteorites (e.g., Mason 1967; Rubin 1997a,b; Brearley and Jones 1998; Gaffey et al. 2002; MacPherson 2007), only about a dozen are carbon bearing. Additional carbon mineral diversification had to await significant reworking of Earth's outer layers by a variety of physical, chemical, and biological processes.

The era of crust and mantle processing

The billion years following planetary accretion was a time of intense reworking of Earth's outer gaseous, liquid, and solids layers. A succession of giant impacts, punctuated by the Moon-forming collision of Mars-sized Theia at ~4.55 Ga, must have had a profound influence on Earth's near-surface mineralogy. The planet's volatile content, including carbon, may have been drastically reduced and, for a time, an ocean of magma encircled the globe (Tonks and Melosh 1993; Ruzicka et al. 1999; Touboul et al. 2007). The subsequent three stages of Earth's mineral evolution reflect the influences of a complex combination of igneous, metamorphic, and tectonic processes.

Stage 3—Igneous rock evolution. Following Earth's accretion, and again for perhaps half a billion years following the Moon-forming event, the mineralogical diversity of Earth evolved primarily by crystallization of igneous rocks (Bowen 1915, 1928), supplemented by a steady bombardment of asteroidal and cometary material. However, many details of those processes, especially the composition and differential role of volatiles that would contribute to or even dominate formation of the oceans, atmosphere, and near-surface fluids (the "exosphere"), remain uncertain. Carbon does not play a significant role in Bowen's classic reaction series, nor is it obvious how the carbon content of primitive volatiles affected the evolution of magmas through fractional crystallization, crystal-liquid separation, and other physical and chemical igneous processes (though see Manning et al. 2013; Ni and Keppler 2013; Jones et al. 2013; Sephton and Hazen 2013). We therefore have little understanding of the processes that first generated significant amounts of carbon minerals or carbonate melts at or near Earth's surface, and what carbon minerals may have been present.

Meteorites and comets would have continued to provide a small but steady supply of a few carbon minerals to Earth's surface: graphite, nano-diamonds, moissanite, iron carbides, and minor carbonates, along with much more significant quantities of carbon dioxide and varied organic species. The first carbon minerals to form at or near Earth's surface seem likely to have been restricted to graphite through subsurface reactions of reduced C-O-H fluids (Rumble and Hoering 1986; Rumble et al. 1986; Manning et al. 2013) and the rhombohedral carbonates, predominantly in the Mg-Ca-Fe-Mn system (i.e., magnesite, calcite, siderite, dolomite, ankerite, rhodochrosite, and kutnohorite; Hazen et al. 2013).

One possibility for Earth's Hadean and Archean Eon carbon mineralization is the generation of carbonate melts, perhaps through immiscibility in the upper mantle. However, no evidence for Paleoarchean (much less Hadean) carbonatites is known to have survived (Jones et al. 2013). Carbonatites appear to have increased steadily in volume through Earth history from the 2.609 Ga ± 6 Ma (Neoarchean) Siilinjärvi carbonatite to the present (Veizer et al. 1992; Jones et al. 2013), with the majority of older (Proterozoic and Phanerozoic Eon) carbonatites preserved as intrusive bodies. This temporal distribution may reflect a preservational bias resulting from the rapid chemical degradation of carbonatite lava flows. Only one carbonatite volcano, Oldoinyo Lengai in Tanzania, is active today, and its carbonate mineralogy alters rapidly when in contact with the atmosphere, because the alkali carbonates are water-soluble (Genge et al. 2001). Such rapid weathering, however, would source an array of variably alkaline fluids, such as those that drain into alkaline Lake Natron, which could result in precipitation of carbonate mineral salts.

Indeed, the major hurdle in documenting Earth's carbon mineral evolution prior to the Neoarchean (~2.8 Ga) is the extensive alteration, and general rarity, of Earth's early rock record (e.g., Papineau 2010). No outcrops with an unambiguous age older than about 4 billion years survive. Even if relatively unaltered early Hadean samples are found, for example in the guise of Earth meteorites on the Moon (Armstrong et al. 2002; Chapman 2002; Jakosky et al. 2004), it is unlikely that carbon minerals will be found. Consequently, models of the genesis of Earth's earliest carbonates must remain speculative.

Barring a significant role of igneous carbonate production, however, aqueous precipitation appears to be the most likely source of Earth's earliest carbonates. Evidence for early oceans may be found in the oldest known fragments of Earth, detrital zircon grains with ages of 4.4 to 4.0 Ga extracted from Archean quartzites in Western Australia. These zircon crystals preserve oxygen isotope ratios characteristic of relatively cool, wet host rocks—possibly evidence for early granitic continental crust (Cavosie et al. 2005; Harrison et al. 2005; Hopkins et al. 2008, 2010) or pre-continental mafic to ultramafic crust (Shirey et al. 2008). Wet crust also implies an early ocean, perhaps by 4.4 Ga.

Such an early appearance of the hydrosphere is consistent with rapid volcanic outgassing and fluid-rock interactions that would have released initial atmospheric components, predominately N_2, CO_2, and H_2O, with minor H_2S (e.g., Holland 1984). The immediate mineralogical consequence of these interactions would have been significant serpentinization in Earth's mafic crust, as well as the first significant production of clay minerals, with magnesite as a possible accessory mineral. An additional consequence of early crustal weathering may have been a rapid increase in the salinity of the warm, slightly acidic oceans (Hardie 1996, 2003; Lowenstein et al. 2001; Dickson 2002).

Under such conditions, the pH and ionic composition would have been the most important factors that might have controlled carbonate precipitation in the Hadean or Paleoarchean oceans. The pH of these most ancient environments would have been a critical constraint on carbonate formation, because the speciation of carbon dioxide in natural waters is strongly governed by pH. Under even mildly acidic conditions, carbon dioxide is present primarily as aqueous CO_2 and as carbonic acid, which would have inhibited precipitation of carbonate minerals. Another critical constraint on carbonate precipitation would have come from the ionic composition of Earth's surface fluids. One can speculate that strongly reducing conditions of the earliest Earth would have resulted in copious amounts of reduced iron, such that siderite may have been the primary carbonate mineral in both terrestrial and marine settings. Siderite formation points to at least four specialized chemical conditions: (1) the presence of aqueous Fe^{2+}; (2) anoxic ocean water with oxygen fugacity $< 10^{-68}$ to prevent iron removal by precipitation of ferric iron oxide-hydroxides; (3) low sulfide and sulfate ions, which would otherwise lead to pyrite precipitation; and (4) dissolved aqueous HCO_3^- (Moore et al. 1992; Ohmoto et al. 2004). Alternatively, under early Earth's substantially higher heat flows, both oceanic basalts and their hydrothermal weathering products would have contained substantially more magnesium, suggesting a potentially greater importance of magnesite. In fact, Mozley (1989) showed that Archean siderite compositions appear to reflect faithfully the aqueous environments in which they form. Fresh water precipitation yields relatively pure $FeCO_3$, whereas formation in a saline environment produces siderite with significant Mg and Ca substitution for Fe—compositional effects that reflect the differences in ionic composition of terrestrial and marine waters.

Stage 4—Granitization. Carbon mineral evolution must have been significantly influenced by the gradual formation of cratons, with their suites of relatively low-density siliceous rocks generated by eutectic melting of mafic and ultramafic lithologies, including tonalite-trondhjemite-granodiorite (TTG), granodiorite-granite-monzogranite (GGM), and high-K syenite-granite (SG) suites (Smithies and Champion 2000; Smithies et al. 2003; Zegers 2004). The nature of this earliest continental crust probably differed both petrologically and tectonically from that of today (see, however, Hopkins et al. 2008, 2010). Tectonic activity prior to 3.5 Ga may have been dominated by vertical processes through deep mantle plumes, rather than today's predominantly lateral tectonics (e.g., Van Kranendonk 2011; Shirey et al. 2013). Surface relief of the thin, hot, principally basaltic crust would have been correspondingly low, and the exposed surface was likely subject to both intense weathering and intense mechanical erosion (Dott 2003), notably from rapid cycling (< 8 hours) of tides generated by the much closer Moon (Lathe 2006; Varga et al. 2006). In such an environment, significant terrestrial deposition of carbonate minerals seems unlikely.

An important mineralogical consequence of granitization was the generation of eutectic fluids concentrated in rare pegmatophile elements and the eventual deposition of complex pegmatites. These processes of element selection and concentration are likely to have taken considerable time (e.g., Grew and Hazen 2009, 2010b); one of the oldest known complex pegmatite, the Tanco pegmatite in Manitoba, Canada, is 2.67 billion years old (London 2008). Carbonate minerals are not dominant species in complex pegmatites, but a few rare species including niveolanite [$NaBeCO_3(OH){\cdot}2H_2O$] and zabuyelite (Li_2CO_3) exemplify the striking mineral diversification that is possible with the formation of complex pegmatites.

Stage 5—Plate tectonics. By 3.0 Ga, or perhaps significantly earlier, lateral tectonics and continent formation had begun on Earth (e.g., Harrison et al. 2005; Workman and Hart 2005; Shirey et al. 2008; Silver and Behn 2008; Shirey and Richardson 2011; Van Kranendonk 2011). The initiation of plate tectonics, especially the beginning of large-scale subduction and the associated crustal reworking and arc volcanism, had significant mineralogical consequences. On the one hand, large-scale hydrothermal reworking of the crust and upper mantle associated with volcanogenic processes at subduction zones and ridges generated the first massive sulfide deposits and associated precious metal concentrations (Sangster 1972; Hutchinson 1973). These ore bodies arise as hydrothermal solutions interact with many cubic kilometers of rock (Barnes and Rose 1998)—fluids that selectively dissolve and concentrate incompatible elements and thus lead to novel mineralization. On the other hand, this modern plate tectonic cycle also generated, for the first time, buoyant continental crust that resulted in substantial amounts of subaerially exposed crust, as well as potentially extensive shallow marine continental shelves. The combination of these factors may have, in turn, fundamentally changed carbonate mineralization in the marine realm.

At atmospheric carbon dioxide concentrations potentially 100 times present atmospheric levels (Kasting 1987; Kasting et al. 1993) the acidity of terrestrial rainwater would have contrasted sharply with that of marine environments, whose pH would have been buffered by hydrothermal weathering reactions at mid-ocean ridges. Subaerially exposed continents would have experienced extensive chemical weathering and the transport of these weathering products to the shallow ocean. Because chemical weathering of silicate minerals produces bicarbonate (HCO_3^-) as the conjugate base to carbonic acid, extensive continental weathering would result in enhanced delivery of bicarbonate to the oceans. Enhanced delivery of bicarbonate would have come at the expense of the acidity generated by equilibration of natural waters with elevated atmospheric carbon dioxide. Neutralization of terrestrial waters through chemical weathering and the subsequent delivery of bicarbonate to the marine system would have acted as an effective buffer of marine pH, and would have allowed marine carbonate saturation to increase to the extent that it was balanced by marine alkalinity.

Along with increased carbonate saturation state, another consequence of cratonization was the generation, perhaps for the first time, of extensive shallow marine shelves. These shallow marine environments would have experienced systematically greater degrees of solar heating and wave agitation, both of which drive local carbonate saturation states higher via degassing of carbon dioxide. In epicratonic environments, which experience reduced advective interchange with the open ocean, evaporative concentration of shallow marine waters can further raise carbonate saturation states. Together, these processes would have yielded, perhaps for the first time, an abundance of environments conducive to carbonate mineral formation. Finally, it was within these shallow marine environments that light-dependent (i.e., photosynthetic-bearing) benthic microbial mats (Noffke 2011)—and their complex relationship with both carbonate mineralization and carbon species within the ocean-atmosphere system—began to flourish, forever changing Earth's carbon cycle (Allwood et al. 2009; Dupraz et al. 2009; Zerkle et al. 2012).

Another possible mineralogical innovation associated with the stabilization of continental shelves may have been the first extensive deposition of methane hydrates (Hazen et al. 2013). This as yet unnamed clathrate mineral is today one of the crust's most abundant and widely distributed carbon minerals (Kvenvolden 1995). Most researchers cite methanogenic microbes and the degradation of buried organic matter as the principal sources of methane, though some advocates postulate a significant abiotic mantle source for some crustal methane deposits (Gold 1999; Kenney et al. 2001; Kutcherov et al. 2002; McCollom 2013; Sephton and Hazen 2013). Consequently, methane hydrate may have first occurred early in Earth's history, perhaps as early as Stage 3.

The most important potential consequences of plate tectonics on carbon mineralogy relate to subduction of carbon, notably carbon-rich sediments. The sequestration of carbonates and organic matter has undoubtedly played, and continues to play, a major and evolving role in Earth's carbon cycle (Hayes and Waldbauer 2006; Kah and Bartley 2011; Dasgupta 2013). The Archean Eon saw a significantly smaller volume of carbonate mineralization than today, but what there was likely dissociated during subduction as a consequence of the steeper geothermal gradient (e.g., Zhu and Ogasawara 2002; Manning et al. 2013; Jones et al. 2013). However, the lower temperature gradient of today's subduction zones may preclude carbonate dissociation, with limestone and dolomite recycled into Earth's deep interior (Kraft et al. 1991; Gillet 1993; Seto et al. 2008).

The era of the evolving biosphere

The evolution of Earth's near-surface carbon mineralogy, notably the diversity and distribution of carbonate minerals, is intertwined with the origin and evolution of life. In this respect, the co-evolving geosphere and biosphere transformed Earth. Three principal sequential episodes frame this story. With the origins of life and the rise of chemolithoautotrophic microbial communities by ~3.5 Ga, Earth's surface environments saw an acceleration of certain mineral-forming processes, notably our earliest evidence for marine carbonate precipitation. Then, coincident with increased continental growth, the rise of photosynthetic organisms saw an increase in both organic and inorganic carbon formation, which culminated in the Great Oxidation Event (GOE) at ~2.4 Ga. The GOE represents a fundamental shift in the oxidation state of Earth's surface that dramatically altered the near-surface geochemical environment and ultimately led to thousands of new mineral species. The GOE also marked the onset of a nearly 1.5 billion year period of progressive oxygenation of Earth's surface that highlights the complexity of physical, chemical, and biological interactions during carbonate mineral formation. Finally, the innovation of skeletal biomineralization at ~0.6 Ga irreversibly altered the carbon cycle and, consequently, the nature and distribution of carbonate minerals.

Stage 6—The origins of life and the rise of chemolithoautotrophs. Life emerged on Earth by the Paleoarchean Era (> 3.5 Ga); however, the influence of life on Earth's mineral diversity appears to have been minimal prior to the Mesoproterozoic Era (~1.8 to 1.6 Ga). The earliest cellular consortia were chemolithoautotrophs that exploited redox couples already present in Earth's dynamic near-surface geochemical environment. Life appears to have survived by accelerating favorable oxidation-reduction reactions that were otherwise kinetically inhibited.

It is important to bear in mind that true catalysts allow reactions that are thermodynamically favorable to take place despite kinetic barriers to their occurrence. Catalysts (including chemolithoautotrophic microbes) thus speed the approach to thermodynamic equilibrium, but do not alter that final equilibrium.

Banded iron formations (BIFs) may represent the most dramatic of these microbial precipitates (e.g., LaBerge 1973; Anbar and Holland 1992; Konhauser et al. 2002, 2007; Kappler et al. 2005). BIFs, which constitute a major source of iron ore, are among the earliest Archean sedimentary formations (as old as 3.85 Ga) and they occur in pulses through the Archean-Proterozoic boundary at ~2.45 Ga (Isley and Abbott 1999; Klein 2005). It should be noted, however, that some BIFs may have an abiotic origin through hydrothermal processes (Jacobsen and Pimentel-Close 1988; Kimberley 1989; Bau and Möller 1993; Klein 2005).

In any event, the majority of BIFs appear to have precipitated from colloidal or gel SiO_2 that contained aqueous ferric and ferrous iron as well as dissolved Na^+, K^+, Mg^{2+}, Ca^{2+}, and HCO_3^- (Klein 1974). Redox and pH variations led to three contrasting mineralogies: a quartz-hematite-magnetite oxide facies, a pyrite-marcasite-pyrrhotite sulfide facies, and a siderite-ankerite-dolomite-calcite carbonate facies. Oxide facies dominate Archean BIFs, though sulfide and carbonate facies also occur (e.g., Dymek and Klein 1988; Dauphas et al. 2007).

A number of lines of evidence point to an anoxic surface environment prior to the Great Oxidation Event at ~2.4 Ga (Holland 1984; Farquhar et al. 2001; Hazen et al. 2008; Sverjensky and Lee 2010). Among these many lines of evidence for Archean anoxia are: (1) unweathered detrital grains of such redox-sensitive minerals as pyrite, siderite, and uraninite (UO_2) in South African and Canadian river deposits (Rasmussen and Buick 1999; England et al. 2002); (2) the absence of a cerium anomaly in the 2.5 Ga Pronto paleosols (Murakami et al. 2001); (3) the absence of Fe^{3+} hydroxides in paleosols older than 2.3 Ga (Rye et al. 1995; Holland and Rye 1997); and (4) mass-independent fractionation of sulfur isotopes (e.g., Farquhar et al. 2000, 2007; Ono et al. 2003; Papineau et al. 2007). As noted by Sverjensky and Lee (2010), significant near-surface Fe^{2+} provided an effective hematite-magnetite redox buffer for much of Earth's earliest history—a buffer that maintained the surface oxygen fugacity at $\sim 10^{-72}$ (Fig. 1).

The near-surface precipitation of ferrous iron carbonates, such as siderite-ankerite deposition exemplified by early Archean, highly metamorphosed calc-silicate deposits in >3.7 Ga rocks of Greenland (Dymek and Klein 1988; Rose et al. 1996) and northern Quebec (Dauphas et al. 2007), requires a reducing environment with an STP oxygen fugacity $< 10^{-68}$ at assumed values of CO_2 fugacity $< 10^{-2}$ (Fig. 2). Siderite and ankerite precipitation are thus consistent with other evidence for a highly reducing near-surface Archean environment. Note, however, that siderite is lacking in Archean paleosols that are contemporaneous with marine Fe^{2+} carbonate precipitation (Rye et al. 1995). This absence of iron carbonates may reflect relatively low concentrations of atmospheric CO_2. Note, however, that other researchers suggest that atmospheric CO_2 concentrations were significantly greater than today and experienced significant fluctuations prior to 2.5 Ga (e.g., Holland 1984; Rye et al. 1995; Sagan and Chyba 1997; Kaufman and Xiao 2003; Kah and Riding 2007). Alternatively, Ohmoto et al. (2004) cite the absence of paleosol siderite as possible evidence for trace levels of atmospheric O_2, which would have prevented the formation of siderite in soils, even under high CO_2 concentrations. Indeed, trace element data from black shale deposits suggests that photosynthesis in near-shore (and presumably terrestrial) environments may have acted as local sources of biospheric oxygen in an otherwise anoxic Archean world (Kendall et al. 2010).

It remains unresolved the extent to which biological activity influenced carbonate mineral formation in the Archean Eon. Oxygenic photoautotrophs, such as cyanobacteria, are recognized for their ability to raise local pH and carbonate saturation and thereby actively promote carbonate mineral precipitation (Thompson and Ferris 1990; Kranz et al. 2010). The evolution

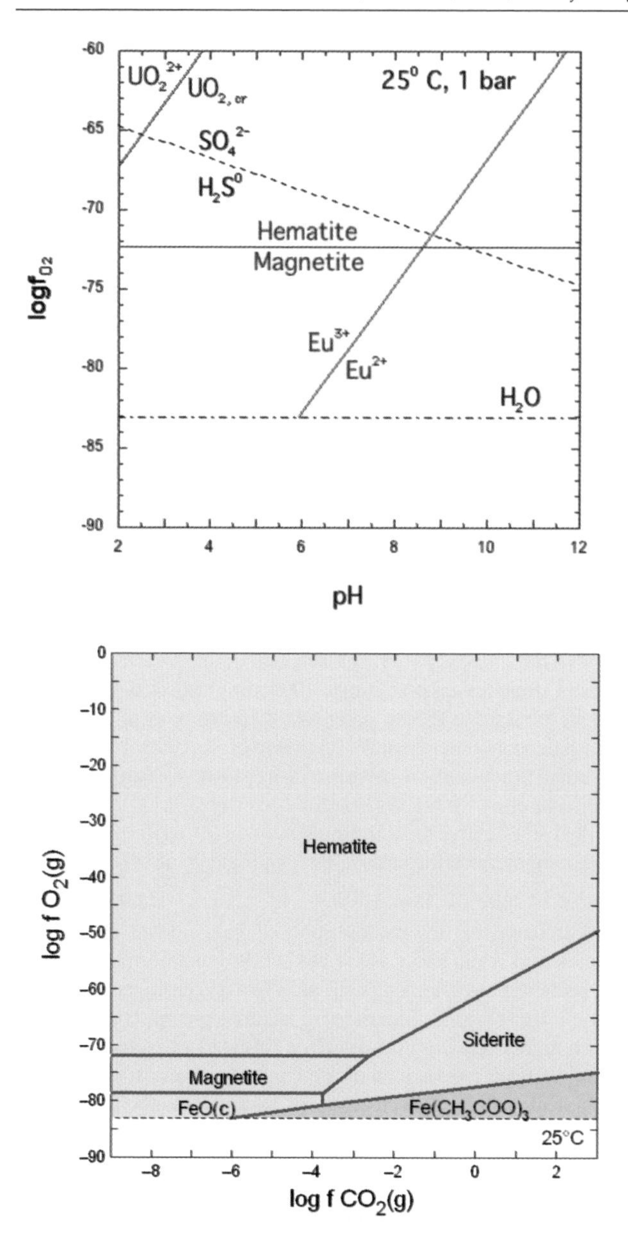

Figure 1. The presence of abundant Fe^{2+} in near-surface environments of the Archean Eon resulted in an effective hematite-magnetite buffer with f_{O_2} at STP constrained to lie near 10^{-72}. Observations of redox-sensitive uranium, sulfur, and europium species in Archean rocks are consistent with this value of f_{O_2} (Hazen et al. 2008; Sverjensky and Lee 2010). Thermodynamic data used to construct this diagram came from the following sources: U-species (Shock et al. 1997a); all other aqueous species (Shock et al. 1997b); minerals (Helgeson et al. 1978).

Figure 2. Ferrous iron carbonate stability requires log $f_{O_2} < -68$ at STP, assuming log $f_{CO_2} < -2$. This value is consistent with the STP hematite-magnetite buffer of log $f_{O_2} \sim -72$. The lower dashed line in the diagram at log $f_{O_2} = -83.1$ represents the lower stability of water at 1 atm H_2. Thermodynamic data used to construct this diagram came from the following sources: aqueous species (Shock et al. 1997b); minerals (Helgeson et al. 1978); the diagram was calculated with the aid of Geochemists Workbench (Bethke 1996).

of oxygenic photoautotrophs, however, appears not to have occurred until sometime after 2.7 Ga (Rasmussen et al. 2008). Similarly, changes in carbonate alkalinity associated with bacterial reduction of Mn and sulfate that promote carbonate mineral precipitation (Froelich et al. 1979; Canfield and Raiswell 1991) would not be expected until approximately 2.5 Ga, with the development of at least local oxygen oases and their associated redox gradients (Kaufman et al. 2007; Anbar et al. 2007; Kendall et al. 2010). Microbial influence on Archean carbonate mineralization is therefore more likely to reflect anoxygenic photoautotrophy (Olson and Blankenship 2004; Westall et al. 2006; Johnston et al. 2009) or methanogenesis (Battistuzzi et al. 2004; Ueno et al. 2006). Although either anoxygenic photoautotrophy or methanogenesis is capable of promoting increased carbonate saturation (Kenward et al. 2009; Bundeleva et al. 2012), experimental approaches suggest that active mineral precipitation within microbial mats requires not only metabolically increased carbonate saturation, but also the liberation of EPS-bound Ca during biofilm decomposition (Dupraz et al. 2009). Little evidence survives of significant non-ferroan carbonate precipitation prior to about 3.5 Ga, when a variety of stromatolitic forms provide unambiguous evidence for at least limited carbonate biomineralization (Walter et al. 1980; Lowe 1980; Buick et al. 1981; Byerly et al. 1986; Walter 1994; Grotzinger and Knoll 1999; Frankel and Bazylinski 2003; Allwood et al. 2006; Van Kranendonk 2006, 2007). However, these deposits are highly localized and extensive sedimentary carbonate deposits are rare until after 3.0 Ga. Even after the appearance of the earliest large-scale carbonate deposits in the ~2.95 Ga Steep Rock Group of northwestern Ontario (Jolliffe 1955; Wilks and Nisbet 1988; Tomlinson et al. 2003), nearly 250 million years passed before the first laterally extensive carbonate platforms at ~2.7 Ga (Sumner 1997). Increasing carbonate platform development in the late Archean Eon may reflect increased cratonization and the development of widespread, shallow marine environments with highly elevated carbonate saturation. Indeed it is these early epi- to pericratonal environments that preserve both neomorphic crystal textures characteristic of aragonite precipitation (Grotzinger 1989; Sumner and Grotzinger 2004) as well as the oldest known primary aragonite, which occurs as nanocrystals within dolomitic stromatolites of the 2.72 Ga Tumbiana Formation, Western Australia (Lepot et al. 2008). Carbonate mineral precipitation within these strongly supersaturated waters still may have been limited by regionally low oxygen and elevated concentrations of Fe^{2+}, which can effectively inhibit carbonate nucleation (Sumner and Grotzinger 1996b).

Whereas microbial consortia may have played a relatively minor role in the precipitation of carbonates in the Paleo- and Mesoarchean Eras, biological organisms appear to have played an increasing role in the Neoarchean Era, prior to the Paleoproterozoic Great Oxidation Event. The appearance of extensive Neoarchean carbonate platforms appears to have promoted marine fluids with elevated carbonate saturation. At this same time, the evolution of oxygenic photosynthesis would have led to considerable heterogeneity in benthic redox conditions (Kaufman et al. 2007; Anbar et al. 2007; Kendall et al. 2010). Critically, even small amounts of environmental oxygenation would have reduced the inhibitory effect of Fe^{2+} within the water column while driving increased microbial reduction of oxidized phases, which would have further enhanced local carbonate saturation. Evidence for a diverse set of microbe-mineral interactions at this time comes, in part, from petrographic observations in the ~2.52 Ga Cambellrand platform that record preferential nucleation of fibrous marine carbonate on specific microbial mat elements (Sumner 1997).

While microbial consortia may have played a localized role in the precipitation of carbonates, life played a relatively minor role in modifying Earth's Archean carbon mineralogy. That situation changed dramatically following global-scale biologically mediated changes in atmospheric chemistry of the Paleoproterozoic Great Oxidation Event.

Stage 7—Photosynthesis and the Great Oxidation Event. Numerous lines of evidence point to a dramatic rise in atmospheric oxygen concentrations from complete anoxia to perhaps a few percent of modern levels at approximately 2.4 to 2.25 Ga (Canfield et al. 2000; Kump et al. 2001; Kasting 2001; Kasting and Siefert 2002; Holland 2002; Towe 2002; Bekker et al. 2004; Barley et al. 2005; Catling and Claire 2005; Papineau et al. 2005, 2007; Kump and Barley 2007). This dramatic and relatively sudden change in atmospheric oxygenation marks the rise of oxygenic photosynthesis by cyanobacteria, an increased redox cycling within microbial mats, and is reflected in an unprecedented increase in the diversity of marine carbonate formation (Melezhik et al. 1997; Grotzinger and Knoll 1999). Oxygenation of Earth's surface environments is characterized by large-scale deposition of oxidized hematite-bearing BIFs as well as massive manganese oxide deposits in marine environments, some of which are accompanied by the Ca-Mn carbonate kutnohorite as found in quantity at the Mesoproterozoic Kalahari manganese fields, North Cape province, South Africa (Leclerc and Weber 1980; Dasgupta et al. 1992; Roy 2006).

The Paleoproterozoic rise in atmospheric oxygen (and concomitant gradual oxygenation of near-surface groundwater) likely represents the single most important event in the diversification of Earth's carbon mineralogy (Hazen et al. 2008; McMillan et al. 2010). Prior to 2.4 Ga, anoxic environments at Earth's surface precluded the formation of many carbonate species. Consider the familiar hydrated copper carbonates azurite [$Cu_3(OH)_2(CO_3)_2$] and malachite [$Cu_2(OH)_2CO_3$]. Calculations of their stability fields as a function of f_{O_2} and f_{CO_2} (Fig. 3) reveal that these (and many other) divalent copper minerals require $f_{O_2} > 10^{-43}$ at STP at presumed values of CO_2 fugacity $< 10^{-2}$ (e.g., Garrels and Christ 1965). However, prior to ~2.4 Ga the near-surface environment was constrained by the hematite-magnetite buffer with $f_{O_2} \sim 10^{-72}$ (Fig. 1). Consequently, neither azurite nor malachite was stable prior to the Great Oxidation Event. Similar arguments may be applied to the numerous carbonates of U^{6+}, Mo^{6+}, Hg^{2+}, and other redox sensitive cations (Hazen et al. 2009, 2012; McMillan et al. 2010).

It is interesting to note that most of the thousands of new mineral phases, including more than 100 new carbonate minerals that are associated with near-surface oxidation, did not first appear immediately following atmospheric oxidation at 2.4 Ga. In fact, recent surveys of the first appearances of the minerals of Be, B, and Hg point to a relatively protracted interval between 2.4 and 1.8 Ga with few new minerals, followed by a surprising pulse of mineral diversity between 2.0 and 1.8 Ga. Hazen et al. (2012) attributed this feature to mineralization associated with the assembly of the Columbia supercontinent, possibly coupled to a protracted interval of gradual subsurface oxidation (McMillan et al. 2010). Although oxygenation of surface environments remained limited at this time, as evidenced by biomarker evidence supporting photic zone anoxia (Brocks et al. 2005), this interval also corresponds to expansion of carbonate-hosted lead-zinc deposits, which reflect increased oxidative weathering, delivery of sulfate to, and bacterial reduction within, marine systems (Lyons et al. 2006).

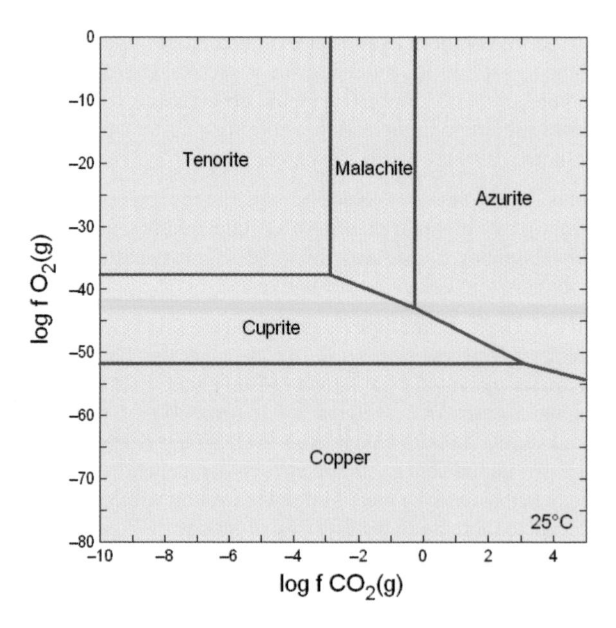

Figure 3. The stabilities of the copper carbonates azurite and malachite require log f_{O_2} > −43 at STP, assuming log f_{CO_2} < −2, which precludes their formation prior to the Great Oxidation Event. Thermodynamic data used to construct this diagram came from the following sources: aqueous species (Shock et al. 1997b); Cu carbonate minerals (Preis and Gamsjäger 2002); Cu oxide minerals (Helgeson et al. 1978); the diagram was calculated with the aid of Geochemists Workbench (Bethke 1996).

Stage 8—The intermediate ocean. The story of marine carbonate minerals in the aftermath of the GOE is complex and largely one of oceanic heterogeneity. The GOE represents a critical threshold beyond which Earth surface environments became oxidizing rather than reducing. Yet a second threshold of oxygenation—one necessary to bring a large portion of the oceanic substrate into the oxidative realm—may not have been passed until perhaps the final days of the Proterozoic Eon. In the intervening time, marine systems experienced a nearly 2 billion year interval in which Earth's surface experienced only gradual increases in oxygenation (Kah and Bartley 2011), potentially coupled with conjugate decreases in atmospheric CO_2 concentration (Bartley and Kah 2004). Within the marine realm, environmental evolution was reflected in (1) limited oxygenation of surface oceans (Brocks et al. 2005; Blumenberg et al. 2012); (2) increased riverine delivery of sulfate to surface oceans (Kah et al. 2004) and its subsequent microbial reduction, which formed broad regions of euxinia, particularly in near shore, shallow marine environments (Shen et al. 2002; Poulton et al. 2004; Lyons et al. 2009); and (3) the retention of perpetually anoxic, potentially ferruginous, deep oceans (Planavsky et al. 2011).

Together, restriction of oxygenated environments and development of strong redox gradients resulted in both spatial and temporal heterogeneity within the oceans. Spatial heterogeneity was also facilitated by warm, non-glacial conditions of the Mesoproterozoic Era (1.6-1.0 Ga) and the assembly of the supercontinent of Rodinia, both of which favored globally high sea levels that formed vast, shallow, epicratonic seaways. Reduction of advective mixing between these vast epicratonic seaways and open ocean environments, in combination with their low water volumes and a generally long residence time for fluids, produced localized environments that could show rapid chemical change in response to regional hydrodynamic conditions, thereby enhancing the heterogeneity of marine environments. Furthermore, within these varied environments, redox gradients were exploited by largely prokaryotic ecosystems that included both oxygenic and anoxygenic photosynthesizers (Dutkiewicz et al. 2003; Brocks et al. 2005; Blumenberg et al. 2012), a range of bacterial sulfate reducers and disproportionating sulfur oxidizers (Canfield and Teske 1996; Johnston et al. 2005), as well as an active methanogenic community (Guo et al. 2012). The Mesoproterozoic Era also saw the appearance and diversification in some marine environments of both unicellular and multicellular algae (Butterfield et al. 1990; Javaux et al. 2004; Knoll et al. 2006). Such biological diversification, of planktic algae in particular, appears to have played a critical role in the reorganization and potential stability of the marine carbon cycle (Ridgwell et al. 2003; Bartley and Kah 2004).

Ultimately, this confluence of physical, chemical, and biological conditions resulted in a global ocean in which a broad range of local parameters could act individually or in concert to overcome kinetic barriers to carbonate precipitation. As a result, the heterogeneity of Mesoproterozoic oceanic environments is marked by equally heterogeneous carbonate mineralization. Perhaps most striking—and in need of much further study—is the observation that biological and chemical heterogeneity of Mesoproterozoic oceans appear to have been reflected in a discrete partitioning of both carbonate crystal form and carbonate mineralogy. Current investigations suggest that seafloor precipitation of fibrous aragonite (Fig. 4a), which is common in Archean and Paleoproterozoic oceans, experienced a gradual restriction to peritidal, evaporative environments (Kah and Knoll 1996; Kah et al. 2012), where evaporative concentration may have greatly increased local saturation states.

Meanwhile, herringbone calcite (Fig. 4b) experienced a progressive restriction to deeper water environments. Herringbone calcite is an unusual carbonate morphology, believed to represent neomorphic replacement of high-magnesium calcite (Grotzinger and Kasting 1993), that consists of elongate crystals in which the *c*-axis rotates throughout growth, from parallel to perpendicular to crystal elongation. This unusual mode of crystal growth is most prevalent within Archean sea floor precipitates (Sumner and Grotzinger 1996a,b) and has been attributed both to inhibitory crystal growth effects of Fe^{2+} under conditions of regional anoxia (Sumner

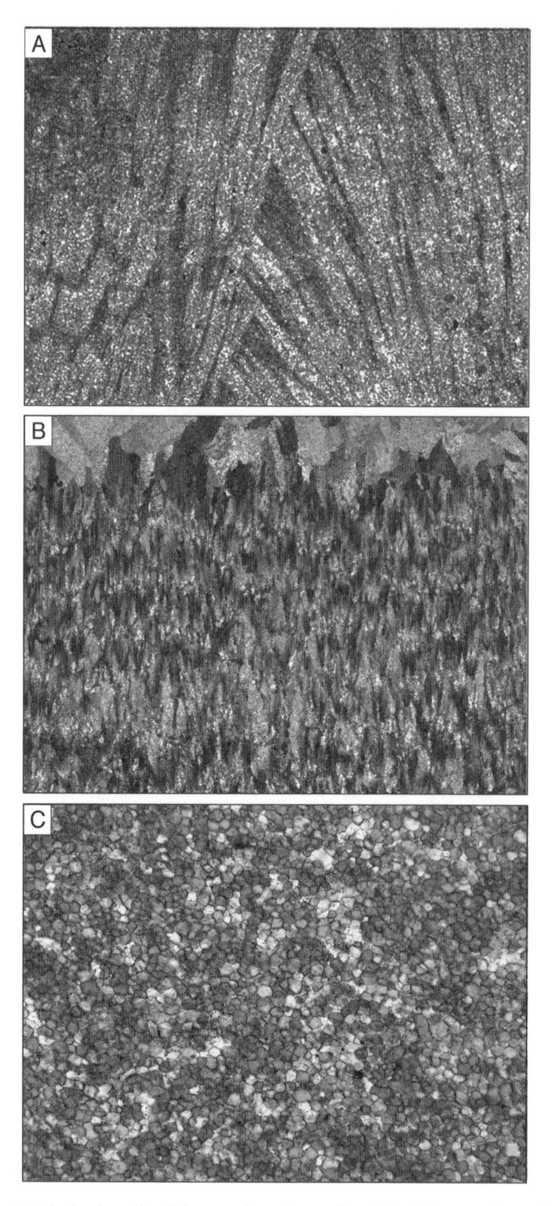

Figure 4. Differing morphologies of calcium carbonate are found in Mesoproterozoic strata and suggest that spatial variation in geochemical conditions may favor the abiotic precipitation of different carbonate minerals. A) Fans and splays of hexagonal, acicular crystals with blunt terminations suggest neomorphic replacement of original aragonite. B) Herringbone carbonate (here under crossed-polars) consists of elongate crystals, whose *c*-axis rotates from parallel to perpendicular to the elongated growth of the crystal. It has been interpreted to reflect neomorphic recrystallization of spherulitic crystal clusters, with microdolomite inclusions, which suggest that the original mineralogy may have been magnesian-calcite. C) Homogeneous crystal size distribution of molar-tooth calcite indicates precipitation through spontaneous nucleation and growth events, with limited Ostwald ripening. Laboratory experiments suggest original deposition as amorphous calcium carbonate. Long axis is 2.6 mm in A, 1.37 mm in B, and 6.9 mm in C.

and Grotzinger 1996b) and to increased crystal growth rates associated with the presence of locally elevated concentrations of dissolved inorganic carbon derived from re-mineralized organic matter (Tourre and Sumner 2000). In the Mesoproterozoic Era, extensive precipitation of herringbone carbonate is restricted to deeper water facies, with its presence in platform deposits closely associated with transgressive surfaces (Kah et al. 2006, 2009, 2012). Combined, these observations suggest that herringbone carbonate mineralization is strongly linked to oceanic redox gradients.

Between these shallow- and deep-water environments, carbonate mineralization appears to be more strongly linked to microbial activity, with finely crystalline calcite precipitates forming both within the water column and within benthic microbial mats. Microbial activity even appears to have fostered the formation of unique carbonate crystal morphologies that are preserved nearly exclusively within the Mesoproterozoic Era. "Molar-tooth" structure is an enigmatic, Precambrian carbonate fabric that consists of variously shaped voids and cracks that formed at or near the sediment-water interface and were filled with a uniform, equant microspar (Furniss et al. 1998; James et al. 1998; Bishop and Sumner 2006; Bishop et al. 2006; Pollock et al. 2006), referred to as molar-tooth microspar (Fig. 4c). Molar-tooth cracks likely formed via gas-sediment interaction during the microbial decomposition of sedimentary organic material (Furniss et al. 1998; Pollock et al. 2006). Void-filling microspar, which consists of uniformly sized spheroidal crystals and associated syntaxial overgrowths, is consistent with formation *via* spontaneous carbonate nucleation associated with the interaction of pore fluids enriched in dissolved organic matter with the infiltration of supersaturated seawater (Pollock et al. 2006). Although molar-tooth microspar has been suggested to have initially formed as either aragonite (Bishop and Sumner 2006) or vaterite (Furniss et al. 1998), experimental precipitation under Proterozoic-like conditions (elevated carbonate saturation and low sulfate) produces a mixture of amorphous calcium carbonate (ACC) with minor vaterite (Goodman and Kah 2007).

Stage 9—The snowball Earth. Characteristic heterogeneity in carbonate mineral formation continued into the Neoproterozoic, where marine ocean chemical environments experienced additional modification via continued biotic evolution and as a consequence of dramatic climate fluctuations between about 0.75 and 0.54 Ga. During the waning of the Proterozoic Eon, Earth environments are marked by at least three episodes of global glaciation (Hoffman et al. 1998; Kennedy et al. 1998). Evidence includes (1) low-latitude glacial deposits at or near sea level (Young 1995; Hoffman and Prave 1996; Halverson 2005; Evans 2006); (2) extreme carbon and sulfur isotope excursions (Hoffman et al. 1998; Gorjan et al. 2000; Hurtgen et al. 2005; Fike et al. 2006); (3) deposition of cap carbonates with abundant bottom-nucleated aragonite fans overlying glacial diamictites (Fairchild 1993; Kennedy 1996; Halverson et al. 2005); (4) distributions of Fe minerals (Young 1976; Canfield et al. 2007); and (5) Ir anomalies representing glacial cumulates (Bodiselitsch et al. 2005). According to the "snowball Earth" scenario (Kirschvink 1992; Hoffman and Schrag 2000), glacial cycles occurred when continents were clustered near the equator so that snow cover triggered runaway albedo feedback, possibly amplified by decreases in atmospheric CO_2 (Ridgwell et al. 2003; Donnadieu et al. 2004). During at least three snowball intervals of approximately 10 million years duration, surface-weathering processes decreased while atmospheric CO_2 concentrations from volcanoes increased significantly (Caldeira and Kasting 1992; Pierrehumbert 2004). Eventually, increased CO_2 led to rapid greenhouse warming, ice melting, and rapid deposition of thick aragonite crystal fans, suggesting highly supersaturated seawater. It is intriguing that melting of methane hydrates (Hazen et al. 2013) and the consequent release of methane may have provided a mineralogical accelerant to this period of positive climatic feedback (Jacobsen 2001).

Glacial melting in the Neoproterozoic Era was accompanied by a significant rise in atmospheric oxygen to perhaps 15 percent of today's value (Fike et al. 2006; Canfield et al. 2007). Resultant increases in clay mineral formation may have had an important influence on the carbon cycle and thus an indirect effect on carbon mineralogy. Atmospheric CO_2 enrichment, greenhouse warming, and oxygenation combined to enhance production of clay minerals in soils, for example by the bio-weathering of feldspar and mica (Schwartzman and Volk 1989; Paris et al. 1996; Barker et al. 1998; Ueshima and Tazaki 1998; Tazaki 2005). Accordingly, Kennedy et al. (2006) describe significant increases in Neoproterozoic clay mineral deposition. Increased clay mineral production may have enhanced sequestration of organic carbon, which readily adsorbs onto clay surfaces and can thus be buried in marine sediments (Hedges and Kiel 1995; Mayer et al. 2004). Because the oxidation of organic carbon provides a major sink for atmospheric oxygen, this clay-mediated burial of organic carbon may have indirectly contributed to a final Proterozoic rise in atmospheric oxygen, which heralded the evolution of skeletonizing metazoans (Knoll and Carroll 1999) and the most dramatic diversification of Earth's mineral inventory since the GOE.

Stage 10—The rise of skeletal biomineralization. By the beginning of the Phanerozoic Eon (0.542 Ga) biology dominated Earth's carbon mineralogy. The early Cambrian Period witnessed an abrupt increase in all major skeletal minerals (Runnegar 1987; Knoll 2003), including the carbonate minerals calcite, aragonite, and magnesian calcite, which are precipitated by numerous algal and animal phyla, and are volumetrically the most significant biominerals. In the early Paleozoic Era, however, non-enzymatic production of carbonate such as microbial carbonate (Riding 2002) continued to dominate over skeletal carbonate. In fact, enzymatic skeletal carbonate production in the Cambrian composes $< 5\%$ of the volume of marine carbonate (Pruss et al. 2010). It was not until after the Great Ordovician Biodiversification Event (GOBE; Harper 2006; Servais et al. 2009) that skeletal carbonate exceeded 15% of the volume of marine carbonate (Pruss et al. 2010).

Several lines of evidence suggest that changes in Earth's ocean chemistry may have played a fundamental role in the transition from non-enzymatic to enzymatic carbonate deposition. Microbial carbonate production, for instance, is controlled primarily by a combination of ambient carbonate saturation state, the availability of aqueous CO_2 and HCO_3, and pH changes within the microbial sheath in response to the photosynthetic uptake of CO_2 and the conversion of HCO_3 to CO_2 via biological carbon concentrating mechanisms (Merz 1992; Arp et al. 2001; Riding and Liang 2005). A dramatic decline in microbial carbonate production in the early Paleozoic Era (Riding 2006a,b) suggests a response in surface waters to a global decrease in pCO_2 and marine carbonate saturation.

Alternatively, it has been suggested that deep-ocean anoxia and the increased anaerobic cycling of organic carbon (Fischer et al. 2007; Higgins et al. 2009) may have reduced marine carbonate saturation in the early Paleozoic Era to a point at which enzymatic production of skeletal carbonate may have been prohibitively high (Pruss et al. 2010). In yet another model, the biological diversification of skeletonizing metazoans may have been inhibited prior to the late Middle Ordovician by deep-ocean anoxia that sequestered bioessential elements, thereby limiting nutrient sources for metazoans that rely on planktonic productivity (Servais et al. 2009). In this scenario, a sustained increase in skeletonizing metazoans could not have occurred prior to deep ocean ventilation and reduction of dysoxic to anoxic deep waters in the late Middle Ordovician (cf. Thompson and Kah 2012).

The complexity of a world that contains both enzymatic and non-enzymatic carbonate production is further illustrated by the biological and temporal variability in the mineralogy of calcium carbonate precipitated by corals, mollusks, and other invertebrates (Harper et al. 1997; Knoll 2003; Cohen and McConnaughey 2003; Palmer and Wilson 2004). Not only are specific carbonate minerals commonly associated with distinct groups of organisms (Lowenstam and Weiner 1989; Simkiss and Wilbur 1989), but organisms themselves can contain numerous different carbonate minerals, such as the remarkable example of the biological adaptation of magnesium carbonate in lenses of some phacopid trilobites with compound eyes (Fig. 5). Each

Figure 5. The compound eyes (with "eye shade") of the trilobite *Erbenochile erbeni*, as with other members of the order Phacopidae, incorporate single-crystal lenses of calcite with the hexagonal *c*-axes precisely oriented perpendicular to the lens. Furthermore, each lens is radially zoned with variable Mg/Ca to correct for chromatic aberration. Photograph courtesy of Adam Aronson.

lens is a single crystal of calcite with the hexagonal *c*-axes precisely oriented perpendicular to the lens. Furthermore, each lens is radially zoned with variable Mg/Ca to correct for chromatic aberration (Clarkson and Levi-Setti 1975; Lee et al. 2007).

The Phanerozoic rock record also preserved a series of temporal transitions in carbonate mineralogy. Calcite mineralization dominates the fossil record from the early Cambrian Period through the early Carboniferous Period (~540 to 300 Ma), but a dramatic and rather sudden shift to increased aragonite biomineralization is observed in fossils from the mid-Mississippian Period through the mid-Jurassic Period (~300 to 150 Ma). Calcite again became the dominant biocarbonate after the mid-Jurassic until approximately 35 million years ago, when aragonite biomineralization once again became common. The origin of these so-called "calcite-aragonite sea" intervals remains controversial. Because temporal variation in the abundance of calcite and aragonite mineralogies is observed in both skeletal and non-skeletal carbonates (Sandberg 1983, 1985; Wilkinson and Givens 1986), it is generally agreed that these temporal variations reflect global-scale changes in ocean chemistry. Additional correlation between "calcite-aragonite sea" intervals and the Mg/Ca concentration of marine minerals (Hardie 1996) suggests that marine Mg/Ca ratio (Mg/Ca < 2 favoring calcite; Mg/Ca > 2 favoring magnesian-calcite and aragonite) may play a primary role in this mineralogical transition.

Marine Mg/Ca ratio may also have substantial influence on the production of skeletal calcite versus aragonite. Stanley and Hardie (1998) showed that calcite-producing organisms are dominant components of the ecosystem during "calcite sea" intervals, and vice versa, which might reflect the differential stability of mineralogical phases, more robust mineralization of skeletons in thermodynamic stability with the ocean (Ries 2010), or even an evolutionary favorability within organisms whose skeletons are in equilibrium with the surrounding ocean (Porter 2007).

The interplay between ocean chemistry, biology, and carbonate mineralogy has continued through the Phanerozoic Eon. The evolution of planktonic calcifying organisms in the Mesozoic Era marked the first time in Earth's history when deep-ocean carbonate deposition was comparable to that of epicontinental environments (Ridgwell 2005). Widespread deposition of carbonate from planktic organisms not only buffers the isotopic composition of the marine carbonate system (Bartley and Kah 2004; Ridgwell and Zeebe 2005), but may also buffer the global exchange of CO_2 to an extent that future snowball events—such as that experienced by Earth in the late Neoproterozoic Era—would be impossible (Ridgwell et al. 2003).

In addition to the abundant and critically important carbonate minerals, Phanerozoic carbon mineralogy is enhanced by dozens of organic minerals (see Table 7 in Hazen et al. 2013), which have been identified from coal, black shales, oil shales, guano and other cave deposits, decayed wood, cacti, and other carbon-rich sources. Biologically-derived organic molecular minerals include the remarkable Ni-porphyrin abelsonite ($Ni^{2+}C_{31}H_{32}N_4$), the purine uricite ($C_5H_4N_4O_3$), and several hydrocarbons ranging in size from kratochvilite ($C_{13}H_{10}$) to evenkite ($C_{24}H_{48}$). More than two-dozen minerals with organic anions, notably oxalate $(C_2O_4)^{2-}$ are also known from Phanerozoic deposits.

CONCLUSIONS: UNRESOLVED QUESTIONS IN CARBON MINERAL EVOLUTION

Much remains to be learned about the mineral evolution of carbon minerals, particularly with respect to carbonate mineral evolution. Rhombohedral Mg-Ca-Fe-Mn carbonates are present in sedimentary, igneous, and metamorphic rocks throughout Earth history, but the paragenetic modes, production rates, and depositional environments of these carbonates have changed significantly. For example, biologically mediated Phanerozoic deposition of carbonates must play a significant, though as yet unresolved, role in the changing global carbon cycle through time. In particular, the onset of massive deposition of global platform carbonates from the Mesozoic Era, coupled with a gradually cooling geotherm and consequent stabilization of carbonates in subduction environments, must be altering the impact of subduction-mediated sequestration of carbon. The possibly significant long-term consequences to the global carbon cycle are not known.

Other aspects of carbon mineral evolution relate to (1) the incorporation of numerous trace and minor elements, including a variety of redox sensitive elements, into rhombohedral carbonates; (2) isotopic variation in carbon and other elements; and (3) temporal changes in the morphology of carbonates, especially the richly varied crystal forms of calcite (Hazen et al. 2013). Each of these aspects of carbon mineralogy hold the potential to reveal much about the co-evolution of the geosphere and biosphere.

It is also intriguing to speculate on the carbon mineral evolution of other planets and moons. In our solar system Titan presents an especially intriguing case. At surface temperatures < 100 K, and with a near-surface composition dominated by hydrocarbons, Titan features lakes of liquid methane and possibly ethane. In such an environment higher hydrocarbons become likely mineral phases. For example, Glein (2012) has proposed that acetylene could be a common evaporite mineral along the shores of Titan's hydrocarbon lakes.

Similar mineralogical oddities might dominate on planets in other star systems with very different metallicity, for example with significantly greater C/O (Bond et al. 2008, 2010; Carter-Bond et al. 2012; Michael Pagano, personal communication). In such a system hydrocarbons, carbides, and organo-silanes might dominate relative to carbonate minerals. In the extreme case, a carbon-rich planet might be dominated by graphite and diamond (Madhusudhan et al. 2012).

We conclude that the detailed study of carbon-bearing minerals through deep time can provide an unparalleled window into the tectonic, geochemical, and biological evolution of any terrestrial planet or moon.

ACKNOWLEDGMENTS

We thank John Armstrong, Joshua Golden, Edward Grew, Russell Hemley, Andrea Mangum, and Craig Schiffries for invaluable discussions and suggestions during the preparation of this manuscript. The authors gratefully acknowledge support from the Deep Carbon Observatory, the Alfred P. Sloan Foundation, the Carnegie Institution of Washington, the National Science Foundation, and NASA's Astrobiology Institute for support of this study.

REFERENCES

Agrinier P, Deutsch A, Schärer U, Martinez I (2001) Fast back-reaction of shock-released CO_2 from carbonates: an experimental approach. Geochim Cosmochim Acta 65:2615-2632

Alexander CMO'D (1990) In situ measurement of interstellar silicon carbide in two CM chondrite meteorites. Nature 348:715-717, doi:10.1038/348715a0

Alexander CMO'D (1993) Presolar SiC in chondrites: How variable and how many sources? Geochim Cosmochim Acta 57:2869-2888

Allamandola LJ, Hudgins DM (2003) From interstellar polycyclic aromatic hydrocarbons and ice to astrobiology. *In:* Solid State Astrobiology. Pirronello V et al. (ed), Kluwer, Dordrecht, p. 251-316

Allamandola LJ, Tielens AGGM, Barker JR (1989) Interstellar polycyclic aromatic hydrocarbons: the infrared emission bands, the excitation-emission mechanism and the astrophysical implications. Astrophys J Suppl Ser 71:733-775

Allwood AC, Walter MR, Kamber BS, Marshall CP, Burch IW (2006) Stromatolite reef from the Early Archaean era of Australia. Nature 441:714-718

Allwood AC, Grotzinger JP, Knoll AH, Burch IW, Anderson MS, Coleman ML, Kanik I (2009) Controls on development and diversity of Early Archean stromatolites. Proc Natl Acad Sci USA106:9548-9555

Anbar AD, Holland HD (1992) The photochemistry of manganese and the origin of banded iron formations. Geochim Cosmochim Acta 56:2595-2603

Anbar AD, Yun D, Lyons TW, Arnold GL, Kendall B, Creaser RA, Kaufman AJ, Gordon GW, Scott C, Garvin J, Buick R (2007) A whiff of oxygen before the Great Oxidation Event? Science 317:1903-1906

Armstrong JC, Wells LE, Gonzalez G (2002) Rummaging through Earth's attic for remains of ancient life. Icarus 160:183-196

Arp G, Reimer A, Reitner J (2001) Photosynthesis-induced biofilm calcification and calcium concentrations in Phanerozoic oceans. Science 292:1701-1704

Barber DJ (1981) Matrix phyllosilicates and associated minerals in C2M carbonaceous chondrites. Geochim Cosmochim Acta 45:945-970

Barker WW, Welch SA, Banfield JF (1998) Experimental observations of the effects of bacteria on aluminosilicate weathering. Am Mineral 83:1551-1563

Barley ME, Bekker A, Krapez B (2005) Late Archean to early Paleoproterozoic global tectonics, environmental change and the rise of atmospheric oxygen. Earth Planet Sci Lett 238:156-171

Barnes HL, Rose AW (1998) Origins of hydrothermal ores. Science 279:2064-2065

Bartley JK, Kah LC (2004) Marine carbon reservoir, C_{org}-C_{carb} coupling, and the evolution of the Proterozoic carbon cycle. Geology 32:129-132

Battistuzzi FU, Feijao A, Hedges SB (2004) A genomic timescale of prokaryotic evolution: insights into the origin of methanogenesis, phototrophy, and the colonization of land. BMC Evol Biol 4-44, doi:10.1186/1471-2148-4-44

Bau M, Möller P (1993) Rare earth element systematics of the chemically precipitated component in Early Precambrian iron-formations and the evolution of the terrestrial atmosphere-hydrosphere-lithosphere system. Geochim Cosmochim Acta 57:2239–2249

Bekker A, Holland HD, Wang P-L, Rumble D III, Stein HJ, Hannah JL, Coetzee LL, Beukes NL (2004) Dating the rise of atmospheric oxygen. Nature 427:117-120

Bethke CM (1996) Geochemical Reaction Modeling. Oxford University Press, New York

Bishop JW, Sumner DY (2006) Molar tooth structures of the Neoarchean Monteville Formation, Transvaal Supergroup, South Africa. I: Constraints on microcrystalline $CaCO_3$ precipitation. Sedimentology 53:1049-1068

Bishop JW, Sumner DY, Huerta NJ (2006) Molar tooth structures of the Neoarchean Monteville Formation, Transvaal Supergroup, South Africa. II: A wave-induced fluid flow model. Sedimentology 53:1069-1082
Blumenberg M, Thiel V, Riegel W, Kah LC, Reitner J (2012) Black shale formation by microbial mats lacking sterane-producing eukaryotes, late Mesoproterozoic (1.1 Ga) Taoudeni Basin, Mauritania. Precambrian Res 196-197:113-127
Bodiselitsch B, Koeberl C, Master S, Reimold WU (2005) Estimating duration and intensity of Neoproterozoic snowball glaciations from Ir anomalies. Science 308:239-242
Bond JC, Lauretta DS, Tinney CG, Butler RP, Marcy GW, Jones HRA, Carter BD, O'Toole SJ, Bailey J (2008) Beyond the iron peak: r- and s-process elemental abundances in stars with planets. Astrophys J 682:1234
Bond JC, O'Brien D, Lauretta D (2010) The compositional diversity of extrasolar planets. I. In situ simulations. Astrophys J 715:1050
Bowen NL (1915) The later stages of the evolution of the igneous rocks. J Geol 23, Supplement 1-91
Bowen NL (1928) The Evolution of the Igneous Rocks. Princeton University Press, Princeton, New Jersey
Brearley AJ (2006) The action of water. *In:* Meteorites and the Early Solar System II. D.S. Lauretta DS, McSween HY Jr (ed) University of Arizona Press, Tucson, p 587-624
Brearley AJ, Jones RH (1998) Chondritic meteorites. Rev Mineral Geochem 36:3.1-3.398
Brocks JJ, Love GD, Summons RE, Knoll AH, Logan GA, Bowden S. (2005) Biomarker evidence for green and purple sulfur bacteria in a stratified Palaeoproterozoic sea. Nature 437:866-870
Buick R, Dunlop JSR, Groves DI (1981) Stromatolite recognition in ancient rocks: an appraisal of irregularly laminated structures in an early Archean chert-barite unit from North-Pole, Western-Australia. Alcheringa 5:161-181
Bundeleva IA, Shirokova LS, Bénézeth P, Pokrovsky OS, Kompantseva EI, Balor S (2012) Calcium carbonate precipitation by anoxygenic phototrophic bacteria. Chem Geol 291:116-131
Butterfield NJ, Knoll AH, Swett K (1990) A bangiophyte red alga from the Proterozoic of arctic Canada. Science 250:104-107
Byerly GR, Lowe DR, Walsh MM (1986) Stromatolites from the 3,300–3,500-Myr Swaziland Supergroup, Barberton Mountain Land, South Africa. Nature 319:489-491
Caldeira K, Kasting JF (1992) Susceptibility of the early Earth to irreversible glaciation caused by carbon dioxide clouds. Nature 359:226-228
Canfield DE, Habicht KS, Thamdrup B (2000) The Archean sulfur cycle and the early history of atmospheric oxygen. Science 288:658-661
Canfield DE, Poulton SW, Narbonne GM (2007) Late-Neoproterozoic deep-ocean oxygenation and the rise of animal life. Science 315:92-95
Canfield DE, Raiswell R (1991) Precipitation and dissolution of carbonates: implications for fossil preservation. *In:* Taphonomy: Releasing the Data Locked in the Fossil Record. Allison PA, Briggs DEG (ed) Plenum Press, New York, p 411-453
Canfield DE, Teske A (1996) Late Proterozoic rise in atmospheric oxygen concentration inferred from phylogenetic and sulphur isotope studies. Nature 382:127-132
Carlisle DB, Braman DR (1991) Nanometre-size diamonds in the Cretaceous/Tertiary boundary clay of Alberta. Nature 352:708-709
Carter-Bond JC, Obrien DP, Raymond SN (2012) The compositional diversity of extrasolar terrestrial planets: II. Migration simulations. Astrophys J 760:44, doi: 10.1088/0004-637X/760/1/44
Catling DC, Claire MW (2005) How Earth's atmosphere evolved to an oxic state: A status report. Earth Planet Sci Lett 237:1-20
Cavosie AJ, Valley JW, Wilde SA (2005) Magmatic $\delta^{18}O$ in 4400-3900 Ma detrital zircons: A record of the alteration and recycling of crust in the Early Archean. Earth Planet Sci Lett 235:663-681
Chapman CR (2002) Planetary science: Earth's lunar attic. Nature 419:791
Clarkson ENK, Levi-Setti R (1975) Trilobite eyes and the optics of Des Cartes and Huygens. Nature 254:663-667
Cohen AL, McConnaughey TA (2003) Geochemical perspectives on coral mineralization Rev Mineral Geochem 54:151-187
Dasgupta R (2013) Ingassing, storage, and outgassing of terrestrial carbon through geologic time. Rev Mineral Geochem 75:183-229
Dasgupta S, Roy S, Fukuoka M (1992) Depositional models for manganese oxide and carbonate deposits of the Precambrian Sausar Group, India. Econ Geol 87:1412-1418
Dauphas N, Cates NL, Mojzsis SJ, Busigny V (2007) Recognition of sedimentary protoliths with iron isotopes in the >3750 Ma Nuvvuagituk supracrustal belt, Canada. Earth Planet Sci Lett 254:358-376
Deutsch A, Langenhorst F (2007) On the fate of carbonate and anhydrite in impact processes—evidence from the Chicxulub event. Geolog Föreningens I Stockholm Förhandlingar 129:155-160
Dickson JAD (2002) Fossil echinoderms as monitor of the Mg/Ca ratio of Phanerozoic oceans. Science 298:1222-1224

Donnadieu Y, Goddéris Y, Ramstein G, Nédélec A, Meert J (2004) A "snowball Earth" climate triggered by continental break-up through changes in runoff. Science 428:303-306

Dott RH Jr (2003) The importance of Aeolian abrasion in supermature quartz sandstones and the paradox of weathering on vegetation-free landscapes. J Geol 111:387-405

Dupraz C, Reid RP, Braissant O, Decho AW, Norman RS, Visscher PY (2009) Processes of carbonate precipitation in modern microbial mats. Earth-Science Rev 96:141-162

Dutkiewicz A, Volk H, Ridley J, George S (2003) Biomarkers, brines, and oil in the Mesoproterozoic Roper Superbasin, Australia. Geology 31:981-984

Dymek RF, Klein C (1988) Chemistry, petrology and origin of banded iron-formation lithologies from the 3800 Ma Isua Supracrustal Belt, West Greenland. Precambrian Res 39:247-302

Ehrenfreund P, Charnley SB (2000) Organic molecules in the interstellar medium, comets, and meteorites. Ann Rev Astron Astrophys 38:427-483

El Goresy A, Gillet P, Chen M, Künstler F, Graup G, Stähle V (2001) In situ discovery of shock-induced graphite-diamond phase transition in gneisses from the Ries crater, Germany. Am Mineral 86:611-621

England GL, Rasmussen B, Krapez B, Groves DL (2002) Paleoenvironmental significance of rounded pyrite in siliciclastic sequences of the Late Archean Witwatersrand Basin: Oxygen-deficient atmosphere or hydrothermal alteration. Sedimentology 49:1133-1136

Evans DAD (2006) Proterozoic low orbital obliquity and axial-dipolar geomagnetic field from evaporite palaeolatitudes. Nature 44:51-55

Fairchild IJ (1993) Balmy shores and icy wastes: the paradox of carbonates associated with glacial deposits in Neoproterozoic times. *In:* Sedimentology Review 1. Wright VP (ed) Blackwell, Oxford, UK, p 1-16

Farquhar J, Bao H, Thiemens MH (2000) Atmospheric influence of Earth's earliest sulfur cycle. Science 289:756-758

Farquhar J, Peters M, Johnston DT, Strauss H, Masterson A, Wiechert U, Kaufman AJ (2007) Isotopic evidence for mesoarchean anoxia and changing atmospheric sulphur chemistry. Nature 449:706-709

Farquhar J, Savarino I, Airieau S, Thiemens MH (2001) Observations of wavelength-sensitive, mass-independent sulfur isotope effects during SO_2 photolysis: Implications for the early atmosphere. J Geophys Res 106:1-11

Fike DA, Grotzinger JP, Pratt LM, Summons RE (2006) Oxidation of the Ediacaran ocean. Nature 444:744-747

Frankel RB, Bazylinski DA (2003) Biologically induced mineralization by bacteria. Rev Mineral Geochem 54:95-114

Fredriksson K, Kerridge JF (1988) Carbonates and sulphates in CI chondrites: Formation by aqueous activity on the parent body. Meteoritics 23:35-44

Froelich PN, Klinkhammer GP, Bender ML, Luedtke NA, Heath GR, Cullen D, Dauphin P, Hammond D, Hartman B, Maynard V (1979) Early oxidation of organic matter in pelagic sediments of the eastern equatorial Atlantic: suboxic diagenesis. Geochim Cosmochim Acta 43:1075-1090

Furniss G, Rittel JF, Winston D (1998) Gas bubble and expansion crack origin of "molar-tooth" calcite structures in the Middle Proterozoic Belt Supergroup, western Montana. J Sediment Res 68:104-114

Gaffey MJ, Cloutis EA, Kelley MS, Reed KL (2002) Mineralogy of asteroids. *In:* Asteroids III. Bottke WF Jr, Cellino A, Paolicchi P (eds) University of Arizona Press, Tucson, p 183-204

Garrels RM, Christ CL (1965) Solutions, Minerals, and Equilibria. Harper & Row, New York

Genge MJ, Balme B, Jones AP (2001) Salt-bearing fumarole deposits in the summit crater of Oldoinyo Lengai, Northern Tanzania: Interactions between natrocarbonate lava and meteoric water. J Volcanology 106:111-122

Gillet P (1993) Stability of magnesite ($MgCO_3$) at mantle pressure and temperature conditions—a Raman-spectroscopic study. Am Mineral 78:1328-1331

Glein CR (2012) Theoretical and Experimental Studies of Cryogenic and Hydrothermal Organic Geochemistry. PhD thesis, Arizona State University, Tucson

Gold T (1999) The Deep Hot Biosphere. Copernicus, New York

Goldstein JI, Hewins RH, Romig AD Jr (1976) Carbides in lunar soils and rocks. 7th Lunar Science Conf Proc 1:807-818

Goodman EE, Kah LC (2007) Reassessing formation of Precambrian molar-tooth microspar: constraints from carbonate precipitation experiments. Geol Soc Am Abstracts with Programs 39:420

Gorjan P, Veevers JJ, Walter MR (2000) Neoproterozoic sulfur-isotope variation in Australia and global implications. Precambrian Res 100:151-179

Grady MM, Verchovsky AB, Franchi IA, Wright IP, Pillinger CT (2002) Light element geochemistry of the Tagish Lake CI2 chondrite: Comparison with CI1 and CM2 meteorites. Meteorit Planet Sci 37:713-735

Grew E, Hazen RM (2009) Evolution of the minerals of beryllium, a quintessential crustal element [Abstract]. Geol Soc Am Abstracts with Programs 41(7):99

Grew E, Hazen RM (2010a) Evolution of the minerals of beryllium, and comparison with boron mineral evolution [Abstract]. Geol Soc Am Abstracts with Programs 42(5):199

Grew E, Hazen RM (2010b) Evolution of boron minerals: Has early species diversity been lost from the geological record? [Abstract]. Geol Soc Am Abstracts with Programs 42(5):92
Grew ES, Bada JL, Hazen RM (2011) Borate minerals and origin of the RNA world. Origins Life Evol Biosph 41:307-316, doi 10.1007/s11084-101-9233-y
Grotzinger JP (1989) Facies and evolution of Precambrian depositional systems: emergence of the modern platform archetype. *In:* Controls on Carbonate Platform and Basin Development, Crevello PD, Wilson JJ, Sarg JF, Read JF (ed) SEPM Special Publication 44:79-106
Grotzinger JP, Kasting JF (1993) New constraints on Precambrian ocean composition. J Geol 101:235-243
Grotzinger JP, Knoll AH (1999) Stromatolites in Precambrian carbonates: Evolutionary mileposts or environmental dipsticks. Ann Rev Earth Planet Sci 27:313-358
Guo H, Du Y, Kah LC, Huang J, Hu C, Huang H (2012) Isotopic composition of organic and inorganic carbon from the Mesoproterozoic Jixian Group, North China: Implications for biological and oceanic evolution. Precambrian Res dx.doi.org/10.1016/j.precamres.2012.09.023
Halverson GP (2005) A Neoproterozoic chronology. *In:* Neoproterozoic Geobiology and Paleobiology. Topics in Geobiology, 27. Xiao S, Kaufman AJ (eds) Kluwer, New York, p 231-271
Halverson GP, Hoffman PF, Schrag DP, Maloof AC, Rice AHN (2005) Toward a Neoproterozoic composite carbon-isotope record. Geol Soc Am Bull 117:1-27
Hanneman RE, Strong HM, Bundy FP (1967) Hexagonal diamonds in meteorites: implications. Science 255:995-997
Hardie LA (1996) Secular variation in seawater chemistry: An explanation for the coupled secular variation in the mineralogies of marine limestones and potash evaporites over the past 600 m.y. Geology 24:279-283
Hardie LA (2003) Secular variations in Precambrian seawater chemistry and the timing of Precambrian aragonite seas and calcite seas. Geology 31:785-788
Harper DAT (2006) The Ordovician biodiversification: Setting an agenda for marine life. Palaeogeo, Palaeoclim, Palaeoecol 232:148-166
Harper EM, Palmer TJ, Alphey JR (1997) Evolutionary response by bivalves to changing Phanerozoic sea-water chemistry. Geol Mag 134:403–407, doi: 10.1017/S0016756897007061
Harrison TM, Blichert-Toft J, Müller W, Albarede F, Holden P, Mojzsis SJ (2005) Heterogeneous Hadean hafnium: evidence of continental crust at 4.4 to 4.5 Ga. Science 310:1947-1950
Hayes JM, Waldbauer JR (2006) The carbon cycle and associated redox processes through time. Phil Trans Roy Soc Lon 361:931-950
Hazen RM, Bekker A, Bish DL, Bleeker W, Downs RT, Farquhar J, Ferry JM, Grew ES, Knoll AH, Papineau DF, Ralph JP, Sverjensky DA, Valley JW (2011) Needs and opportunities in mineral evolution research. Am Mineral 96:953-963
Hazen RM, Downs RT, Jones AP, Kah L (2013) Carbon mineralogy and crystal chemistry. Rev Mineral Geochem 75:7-46
Hazen RM, Eldredge N (2010) Themes and variations in complex systems. Elements 6:43-46
Hazen RM, Ewing RC, Sverjensky DA (2009) Evolution of uranium and thorium minerals. Am Mineral 94:1293-1311
Hazen RM, Golden J, Downs RT, Hystad G, Grew ES, Azzolini D, Sverjensky DA (2012) Mercury (Hg) mineral evolution: A mineralogical record of supercontinent assembly, changing ocean geochemistry, and the emerging terrestrial biosphere. Am Mineral 97:1013-1042
Hazen RM, Papineau D, Bleeker W, Downs RT, Ferry JM, McCoy TJ, Sverjensky DA, Yang H (2008) Mineral evolution. Am Mineral 93:1693-1720
Hedges JI, Keil RG (1995) Sedimentary organic matter preservation: An assessment and speculative synthesis. Marine Chem 49:81-139
Helgeson HC, Delaney JM, Nesbitt HW, Bird DK (1978) Summary and critique of the thermodynamic properties of rock-forming minerals. Am J Sci 278A:1-229
Herd CDK, Blinova A, Simkus DN, et al. (2011) Origin and evolution of prebiotic organic matter as inferred from the Tagish Lake meteorite. Science 332:1304-1307
Higgins JA, Fischer WW, Schrag DP (2009) Oxygenation of the oceans and sediments: Consequences for the seafloor carbonate factory. Earth Planet Sci Let 284:25-33
Hoffman PF, Kaufman AJ, Halverson GP, Schrag DP (1998) A Neoproterozoic snowball Earth. Science 281:1342-1346
Hoffman PF, Prave AR (1996) A preliminary note on a revised subdivision and regional correlation of the Otavi Group based on glaciogenic diamictites and associated cap dolostones. Commun Geol Survey Namibia 11:77-82
Hoffman PF, Schrag DP (2000) Snowball Earth. Sci Am January 2000:68-75
Holland HD (1984) The Chemical Evolution of the Oceans and Atmosphere. Princeton University Press, Princeton, NJ
Holland HD (2002) Volcanic gases, black smokers, and the great oxidation event. Geochim Cosmochim Acta 66:3811-3826

Holland HD, Rye R (1997) Evidence in pre-2.2 Ga Paleosols for the early evolution of atmospheric oxygen and terrestrial biota; discussion and reply. Geology 25:857-859

Hopkins M, Harrison TM, Manning CE (2008) Low heat flow inferred from >4 Gyr zircons suggests Hadean plate boundary interactions. Nature 456:493:496, doi: 10.1038/nature07465

Hopkins M, Harrison TM, Manning CE (2010) Constraints on Hadean geodynamics from mineral inclusions in >4 Ga zircons. Earth Planet Sci Lett 298:367-376

Hurtgen MT, Arthur MA, Halverson GP (2005) Neoproterozoic sulfur isotopes, the evolution of microbial sulfur species, and the burial efficiency of sulfide as sedimentary pyrite. Geology 33:41-44

Hutchinson RW (1973) Volcanogenic sulfide deposits and their metallogenic significance. Econ Geology 68:1223-1246

Isley AE, Abbott DH (1999) Plume-related mafic volcanism and the deposition of banded iron formation. J Geophys Res 104:15461–15477

Ivanov BA, Deutsch A (2002) The phase diagram of $CaCO_3$ in relation to shock compression and decomposition. Phys Earth Planet Int 129:131-143

Jacobsen S (2001) Gas hydrates and deglaciations. Nature 412:691-693

Jacobsen SB, Pimentel-Klose MR (1988) A Nd isotopic study of the Hamersley and Michipicoten banded iron formations: the source of REE and Fe in Archean oceans. Earth Planet Sci Lett 87:29-44

Jakosky B, Anbar A, Taylor J, Lucey P (2004) Astrobiology Science Goals and Lunar Exploration. NASA Astrobiology Institute White Paper, 14 April 2004:1-17

James NP, Narbonne GM, Sherman AB (1998) Molar-tooth carbonates: shallow subtidal facies of the Mid- to Late Proterozoic. J Sediment Res 68:716-722

Jarosewich E (1990) Chemical analysis of meteorites: A compilation of stony and iron meteorite analyses. Meteoritics 25:323-327

Javaux EJ, Knoll AH, Walter MR (2004) TEM evidence for eukaryotic diversity in mid-Proterozoic oceans. Geobiology 2:121-132

Johnston D, Wing B, Farquhar J, Kaufman A, Strauss H, Lyons T, Kah L, Canfield D (2005) Active microbial sulfur disproportionation in the Mesoproterozoic. Science 310:1477-1479

Johnston DT, Wolfe-Simon F, Pearson A, Knoll AH (2009) Anoxygenic photosynthesis modulated Proterozoic oxygen and sustained Earth's middle age. Proc Natl Acad Sci USA 106:16925-16929

Jolliffe AW (1955) Geology and iron ores of Steep Rock Lake. Econ Geol 50:373-398

Jones AP (2007) The mineralogy of cosmic dust: astromineralogy. Eur J Mineral 19:771-782

Jones AP, Claeys P, Heuschkel S (2000) Impact melting of carbonates from the Chicxulub crater. Lecture Notes in Earth Sciences 91:343-361

Jones AP, Genge M, Carmody L (2013) Carbonate melts and carbonatites. Rev Mineral Geochem 75:289-322

Kah LC, Bartley JK (2011) Protracted oxygenation of the Proterozoic biosphere. Int Geol Rev 53:1424-1442

Kah LC, Bartley JK, Frank TD, Lyons TW (2006) Reconstructing sea-level change from the internal architecture of stromatolite reefs: an example from the Mesoproterozoic Sulky Formation, Dismal Lakes Group, arctic Canada. Canadian J Earth Sci 43:653-669

Kah LC, Bartley JK, Stagner AF (2009) Reinterpreting a Proterozoic enigma: *Conophyton–Jacutophyton* stromatolites of the Mesoproterozoic Atar Group, Mauritania. Int Assoc Sediment Spec Pub 41:277-295

Kah LC, Bartley JK, Teal DA (2012) Chemostratigraphy of the Late Mesoproterozoic Atar Group, Taoudeni Basin, Mauritania: Muted isotopic variability, facies correlation, and global isotopic trends. Precambrian Res 200-203:82-103

Kah LC, Knoll AH (1996) Microbenthic distribution in Proterozoic tidal flats: environmental and taphonomic considerations. Geology 24:79-82

Kah LC, Lyons TW, Frank TD (2004) Evidence for low marine sulphate and the protracted oxygenation of the Proterozoic biosphere. Nature 431:834-838

Kah LC, Riding R (2007) Mesoproterozoic carbon dioxide levels inferred from calcified cyanobacteria. Geology 35:799-802

Kappler A, Pasquero C, Konhauser KO, Newman DK (2005) Deposition of Banded Iron Formations by photoautotrophic Fe(II)-oxidizing bacteria. Geology 33:865-868

Kasting JF (1987) Theoretical constraints on oxygen and carbon dioxide concentrations in the Precambrian atmosphere. Precambrian Res 34:205-229

Kasting JF (2001) The rise of atmospheric oxygen. Science 293:819-820

Kasting JF, Eggler DH, Raeburn SP (1993) Mantle redox evolution and the oxidation state of the Archean atmosphere. J Geol 101:245-257

Kasting JF, Siefert JL (2002) Life and the evolution of Earth's atmosphere. Science 296:1066-1068

Kaufman AJ, Johnston DT, Farquhar J, Masterson AL, Lyons TW, Bates S, Anbar AD, Arnold GL, Garvin J, Buick R (2007) Late Archean biospheric oxygenation and atmospheric evolution. Science 317:1900-1903

Kaufman AJ, Xiao S (2003) High CO_2 levels in the Proterozoic atmosphere estimated from analyses of individual microfossils. Nature 425:279-282

Kendall B, Reinhard CT, Lyons TW, Kaufman AJ, Poulton SW, Anbar AD (2010) Pervasive oxygenation along late Archaean ocean margins. Nature Geosci 3:647-652

Kennedy MJ (1996) Stratigraphy, sedimentology, and isotope geochemistry of Australian Neoproterozoic postglacial cap dolostones; deglaciation, $\delta^{13}C$ excursions, and carbonate precipitation. J Sediment Petrol 66:1050-1064

Kennedy MJ, Droser M, Mayer LM, Pevear D, Mrofka D (2006) Late Precambrian oxygenation; inception of the clay mineral factory. Science 311:1446-1449

Kennedy MJ, Runnegar B, Prave AR, Hoffmann KH, Arthur MA (1998) Two or four Neoproterozoic glaciations? Geology 26:1059-1063

Kenney JF, Shnyukov YF, Krayishkin VA, Tchebanenko II, Klochko VP (2001) Dismissal of claims of a biological connection for natural petroleum. Energia 22:26-34

Kenward PA, Goldstein RH, Gonzalez LA, Roberts JA (2009) Precipitation of low temperature dolomite from an anaerobic microbial consortium: the role of methanogenic Archaea. Geobiology 7:556-565

Kerridge JF, Bunch TE (1979) Aqueous activity on asteroids: Evidence from carbonaceous chondrites. *In:* Asteroids. Garrels T (ed) University of Arizona Press, Tucson, p 745-764

Kimberley MM (1989) Exhalative origins of iron formations. Ore Geol Rev 5:13-145

Kirschvink JL (1992) A paleogeographic model for Vendian and Cambrian time. *In:* The Precambrian Biosphere. Schopf JW, Klein C (ed) Cambridge University Press, New York, p 51-52

Klein C (1974) Greenalite, stilpnomelane, minnesotaite, crocidolite, and carbonates in very low-grade metamorphic Precambrian iron-formation. Canadian Mineral 12:475-498

Klein C (2005) Some Precambrian banded iron-formations (BIFs) from around the world: Their age, geologic setting, mineralogy, metamorphism, geochemistry, and origin. Am Mineral 90:1473-1499

Knoll AH (2003) Biomineralization and evolutionary history. Rev Mineral Geochem 54:329-356

Knoll AH, Carroll SB (1999) Early animal evolution: emerging views from comparative biology and geology. Science 284:2130-2137

Knoll AH, Javaux EJ, Hewitt D, Cohen P (2006) Eukaryotic organisms in Proterozoic oceans. Philos Trans R Soc London, Biol Sci 361:1023-1038

Konhauser KO, Amskold L, Lalonde SV, Posth NR, Kappler A, Anbar A (2007) Decoupling photochemical Fe(II) oxidation from shallow-water BIF deposition. Earth Planet Sci Lett 258:87-100

Konhauser KO, Hamade T, Raiswell R, Morris RC, Ferris FG, Southam G, Canfield DE (2002) Could bacteria have formed the Precambrian banded iron-formations? Geology 30:1079-1082

Kraft S, Knittle E, Williams Q (1991) Carbonate stability in the Earth's mantle—a vibrational spectroscopic study of aragonite and dolomite at high-pressures and temperatures. J Geophysl Res Solid Earth 96:17997-18009

Kranz SA, Wolf-Gladrow D, Nehrke G, Langer G, Rost B (2010) Calcium carbonate precipitation induced by the growth of the marine cyanobacterium *Trichodesmium*. Limnology and Oceanography 55:2563-2569

Krivovichev SV (2010) The concept of mineral evolution in Russian mineralogical literature (1978-2008). 20th General Meeting of the IMA (IMA2010), Budapest, Hungary, August 21-27, Acta Mineralogica-Petrographica Abstract Series, Szeged 6:763

Kump LR, Barley ME (2007) Increased subaerial volcanism and the rise of atmospheric oxygen 2.5 billion years ago. Nature 448:1033-1036

Kump LR, Kasting JF, Barley ME (2001) Rise of atmospheric oxygen and the "upside down" Archean mantle. Geochem Geophys Geosystems 2:#2000GC000114

Kutcherov VG, Bendiliani NA, Alekseev VA, Kenney JF (2002) Synthesis of hydrocarbons from minerals at pressure up to 5 GPa. Proc Russian Acad Sci 387:789-792

Kvenvolden KA (1995) A review of the geochemistry of methane in natural gas hydrate. Org Geochem 23:997-1008

Kwok S (2009) Organic matter in space: from star dust to the Solar System. Astrophys Space Sci 319:5-21

LaBerge GL (1973) Possible biological origin of Precambrian iron-formations. Econ Geol 68:1098-1109

Langenhorst F, Deutsch A (2012) Shock metamorphism of minerals. Elements 8:31-36

Langenhorst F, Shafranovsky GI, Masaitis VL, Koivisto M (1999) Discovery of impact diamonds in a Fennoscandian crater and evidence for their genesis by solid-state transformation. Geology 27:747-750

Lathe R (2006) Early tides: Response to Varga et al. Icarus 180:277-280

Leclerc J, Weber F (1980) Geology and genesis of the Moanda manganese deposits. *In:* Geology and Geochemistry of Manganese, volume 2. Varentsov IM, Grasselly G (ed) E Schweizerbart'sche Verlagsbuchhandlung, Stuttgart, p 89-109

Lee MR, Torney C, Owen AW (2007) Magnesium-rich intralensar structures in schizochroal trilobites eyes. Palaeontology 50:1031-1037

Lepot K, Benzerara K, Brown GE, Philippot P (2008) Microbially influenced formation of 2,724-million-year-old stromatolites. Nature Geosci 1:1-4

London D (2008) Pegmatites. Can Mineral Special Publication 10, 347 p

Lowe DR (1980) Stromatolites 3,400-Myr old from the Archean of Western Australia. Nature 284:441-443
Lowenstam HA, Weiner S (1989) On Biomineralization. Oxford University Press, Oxford
Lowenstein TK, Timofeeff MN, Brennan ST, Hardie LA, Demicco RV (2001) Oscillations in Phanerozoic seawater chemistry: Evidence from fluid inclusions. Science 294:1086-1088
Lyons TW, Anbar AD, Severmann S, Scott C, Gill BC (2009) Tracking euxinia in the ancient ocean: a multiproxy perspective and Proterozoic case study. Ann Rev Earth Planet Sci 37:507-534
Lyons TW, Gellatly AM, McGoldrick PJ, Kah LC (2006) Proterozoic sedimentary exhalative (SEDEX) deposits and links to evolving global ocean chemistry. Geol Soc Am Memoir 198:169-184
MacPherson GJ (2007) Calcium-aluminum-rich inclusions in chondritic meteorites. *In:* Treatise on Geochemistry, Volume 1. Holland HD, Turekian KK (ed) p 201-246
Madhusudhan N, Lee K, Mousis O (2012) A possible carbon-rich interior in Super-Earth 55 Cancri e. Astrophys J Lett 759:L40, doi: 10.1088/2041-8205/759/2/L40
Manning CE, Shock EL, Sverjensky D (2013) The chemistry of carbon in aqueous fluids at crustal and upper-mantle conditions: experimental and theoretical constraints. Rev Mineral Geochem 75:109-148
Martinez I, Perez EMC, Matas J, Gillet P, Vidal G (1998) Experimental investigation of silicate-carbonate system at high pressure and high temperature. J Geophys Res-Solid Earth 103:5143-5163
Martins Z, Alexander CMO'D, Orzechowska GE, Fogel ML, Ehrenfreund P (2007) Indigenous amino acids in primitive CR meteorites. Meteorit Planet Sci 42:2125-2136
Masaitis VI, Futergendler SI, Gnevushev MA (1972) Diamonds in impactites of the Popigai meteor crater. All-Union Mineral Soc Proc 1:108-112
Mason B (1967) Extraterrestrial mineralogy. Am Mineral 52:307-325
Mayer LM, Schtik LL, Hardy KR, Wagai R, McCarthy J (2004) Organic matter in small mesopores in sediments and soils. Geochim Cosmochim Acta 68:3863-3872
McCollom TM (2013) Laboratory simulations of abiotic hydrocarbon formation in Earth's deep subsurface. Rev Mineral Geochem 75:467-494
McCoy TL (2010) Mineralogical evolution of meteorites. Elements 6:19-24, doi: 10.2113/gselements.6.1.19
McCoy TL, Mittlefehldt DW, Wilson L (2006) Asteroid differentiation. *In:* Meteorites and the Early Solar System II. Lauretta DS, McSween HY Jr (ed) University of Arizona Press, Tucson, p 733-746
McMillan M, Downs RT, Stein H, Zimmerman A, Beitscher B, Sverjensky DA, Papineau D, Armstrong J, Hazen RM (2010) Molybdenite mineral evolution: A study of trace elements through time. Geol Soc Am Abstracts with Programs 42 (5):93
Melezhik VA, Fallick AE, Makarikhin VV, Lyubtsov VV (1997) Links between Palaeoproterozoic palaeogeography and rise and decline of stromatolites: Fennoscandian Shield. Precambrian Res 82:311-348
Merz MUE (1992) The biology of carbonate precipitation by cyanobacteria. Facies 26:81-102
Messenger S, Keller LP, Stadermann FJ, Walker RM, Zinner E (2003) Samples of stars beyond the solar system: Silicate grains in interplanetary dust. Science 300:105-108
Messenger S, Sandford S, Brownlee D (2006) The population of starting materials available for solar system construction. *In:* Meteorites and the Early Solar System II. Lauretta DS, McSween HY Jr (ed) University of Arizona Press, Tucson, p 187-207
Meyer C (1985) Ore metals through geologic history. Science 227:1421-1428
Moore SE, Ferrell RE, Abaron P (1992) Diagenetic siderite and other ferroan carbonates in a modern subsiding sequence. J Sed Petrol 62:357-366
Mostefouai S, Hoppe P (2004) Discovery of abundant in situ silicate and spinel grains from red giant stars in a primitive meteorite. Astrophys J 613:L149-L152
Mozley PS (1989) Relation between depositional environment and the elemental composition of early diagenetic siderite. Geology 17:704-706
Murakami T, Utsinomiya S, Imazu Y, Prasadi N (2001) Direct evidence of late Archean to early Proterozoic anoxic atmosphere from a product of 2.5 Ga old weathering. Earth Planet Sci Lett 184:523-528
Nash JT, Granger HC, Adams SS (1981) Geology and concepts of genesis of important types of uranium deposits. Econ Geol 75th Anniversary Volume, p 63-116
Ni H, Keppler H (2013) Carbon in silicate melts. Rev Mineral Geochem 75:251-287
Nittler LR (2003) Presolar stardust in meteorites: Recent advances and scientific frontiers. Earth Planet Sci Lett 209:259-273
Ohmoto H, Watanabe Y, Kumazawa K (2004) Evidence from massive siderite beds for a CO_2-rich atmosphere before ~1.8 billion years ago. Nature 429:395-399
Olson JM, Blankenship RE (2004) Thinking about the evolution of photosynthesis. Photosynthesis Res 80:373-386
Ono S, Eigenbrode JL, Pavlov AA, Kharecha P, Rumble D III, Kasting JF, Freeman KH (2003) New insights into Archean sulfur cycle from mass-independent sulfur isotope records from the Hamersley Basin, Australia. Earth Planet Sci Lett 213:15-30

Palmer T, Wilson M (2004) Calcite precipitation and dissolution of biogenic aragonite in shallow Ordovician calcite seas. Lethaia 37:417, doi: 10.1080/00241160410002135
Papineau D (2010) Mineral environments of the earliest Earth. Elements 6:25-30
Papineau D, Mojzsis SJ, Coath CD, Karhu JA, McKeegan KD (2005) Multiple sulfur isotopes of sulfides from sediments in the aftermath of Paleoproterozoic glaciations. Geochim Cosmochim Acta 69:5033-5060
Papineau D, Mojzsis SJ, Schmitt AK (2007) Multiple sulfur isotopes from Paleoproterozoic Huronian interglacial sediments and the rise of atmospheric oxygen. Earth Planet Sci Lett 255:188-212
Paris F, Bottom B, Lapeyrie F (1996) In vitro weathering of phlogopite by ectomycorrhizal fungi. Plant and Soil 179:141-150
Pierrehumbert RT (2004) High levels of atmospheric carbon dioxide necessary for the termination of global glaciation. Nature 429:646-648
Planavsky NJ, McGoldrick P, Scott CT, Chao L, Reinhard CT, Kelly AE, Xuelei C, Bekker A, Love GD, Lyons TW (2011) Widespread iron-rich conditions in the mid-Proterozoic ocean. Nature 477:448-451
Pollock MD, Kah LC, Bartley JK (2006) Morphology of molar-tooth structures in Precambrian carbonates: influence of substrate rheology and implications for genesis. J Sediment Res 76:310-323
Porter SM (2007) Seawater chemistry and early carbonate biomineralization. Science 316:1302, doi: 10.1126/science.1137284
Poulton SW, Canfield DE, Fralick P (2004) The transition to a sulfidic ocean ~1.84 billion years ago. Nature 431:173-177
Preis W, Gamsjäger H (2002) Solid-solute phase equilibria in aqueous solution. XVI. Thermodynamic properties of malachite and azurite - predominance diagrams for the system Cu^{2+}-H_2O-CO_2. J Chem Thermodyn 34:631-650
Pruss SB, Finnegan S, Fischer WW, Knoll AH (2010) Carbonates in skeleton-poor seas: new insights from Cambrian and Ordovician strata of Laurentia. Palaios 25:73-84
Rasmussen B, Buick R (1999) Redox state of the Archean atmosphere: Evidence from detrital heavy minerals in ca.3250-2750 Ma sandstones from the Pilbara Craton, Australia. Geology 27:115-118
Rasmussen B, Fletcher IR, Brocks JJ, Kilburn MR (2008) Reassessing the first appearance of eukaryotes and cyanobacteria. Nature 455:1101-1104
Ridgwell AJ (2005) A mid-Mesozoic revolution in the regulation of ocean chemistry. Marine Geol 217:339-357
Ridgwell AJ, Kennedy MJ, Caldeira K (2003) Carbonate deposition, climate stability, and Neoproterozoic ice ages. Science 302:859-862
Ridgwell AJ, Zeebe RE (2005) The role of the global carbonate cycle in the regulation and evolution of the Earth system. Earth Planet Sci Lett 234:299-315
Riding R (2002) Microbial carbonates: the geological record of calcified bacterial-algal mats and biofilms. Sedimentology 47:179-214
Riding R (2006a) Cyanobacterial calcification, carbon dioxide concentrating mechanisms, and Proterozoic-Cambrian changes in atmospheric composition. Geobiology 4:299-316
Riding R (2006b) Microbial carbonate abundance compared with fluctuations in metazoan diversity over geological time. Sediment Geol 185:229-238
Riding R, Liang L (2005) Geobiology of microbial carbonates: metazoan and seawater saturation state influences on secular trends during the Phanerozoic. Palaeogeo Palaeoclim Palaeoecol 219:101-115
Ries JB (2010) Review: geological and experimental evidence for secular variation in seawater Mg/Ca (calcite-aragonite seas) and its effect on biological calcification. Biogeosciences 7:2795-2849
Ronov AB, Migdisov AA, Barskaya NV (1969) Tectonic cycles and regularities in the development of sedimentary rocks and paleogeographic environments of sedimentation of the Russian platform (an approach to a quantitative study). Sedimentology 13:179-212
Rose NM, Rosing MT, Bridgwater D (1996) The origin of metacarbonate rocks in the Archaean Isua Superacrustal Belt, West Greenland. Am J Sci 296:1004-1044
Ross AJ, Steele A, Fries MD, Kater L, Downes H, Jones AP, Smith CL, Jenniskens PM, Zolensky ME, Shadded MH (2011) MicroRaman spectroscopy of diamond and graphite in Almahata Sitta and comparison with other ureilites. Meteorit Planet Sci 46:364-378
Roy S (2006) Sedimentary manganese metallogenesis in response to the evolution of the Earth system. Earth Sci Rev 77:273-305
Rubin AE (1997a) Mineralogy of meteorite groups. Meteorit Planet Sci 32:231-247
Rubin AE (1997b) Mineralogy of meteorite groups: An update. Meteorit Planet Sci 32:733-734
Rumble D III, Duke EF, Hoering TC (1986) Hydrothermal graphite in New Hampshire: Evidence of carbon mobility during regional metamorphism. Geology 14:452-455
Rumble D III, Hoering TC (1986) Carbon isotope geochemistry of graphite vein deposits from New Hampshire, U. S. A. Geochem Cosmochim Acta 50:1239-1247
Runnegar B (1987) The evolution of mineral skeletons. *In:* Origin, Evolution, and Modern Aspects of Biomineralization in Plants and Animals. Crick RE (ed) Plenum, New York, p 75-94

Russell SS, Pillinger CT, Arden JW, Lee MR, Ott U (1992) A new type of meteoritic diamond in the enstatite chondrite Abee. Science 256:206-209

Ruzicka A, Snyder GA, Taylor LA (1999) Giant impact hypothesis for the origin of the Moon: A critical review of some geochemical evidence. *In:* Planetary Petrology and Geochemistry. Snyder GA, Neal CR, Ernst WG (ed) Geological Society of America, Boulder, Colorado, p 121-134

Rye R, Kuo PH, Holland HD (1995) Atmospheric carbon dioxide concentration before 2.2 billion years ago. Nature 378:603-605

Sagan C, Chyba C (1997) The early faint Sun paradox: Organic shielding of ultraviolet-labile greenhouse gases. Science 276:1217-1221

Sandberg PA (1983) An oscillating trend in Phanerozoic nonskeletal carbonate mineralogy. Nature 305:19-22

Sandberg PA (1985) Nonskeletal aragonite and pCO_2 in the Phanerozoic and Proterozoic. AGU Monograph 32:585-594

Sangster DF (1972) Precambrian volcanogenic massive sulfide deposits in Canada: A review. Geol Surv Canada Paper 72-22:1-43

Schmitt-Kopplin P, Gabelica Z, Gougeon RD, Fekete A, Kanawati B, Harir M, Gebefuegi I, Eckel G, Hertkorn N (2010) High molecular diversity of extraterrestrial organic matter in Murchison meteorite revealed 40 years after its fall. Proc Natl Acad Sci USA 107:2763-2768, doi: 10.1073/pnas.0912157107

Schwartzman DW, Volk T (1989) Biotic enhancement of weathering and the habitability of Earth. Nature 340:457-460

Sephton MA (2002) Organic compounds in carbonaceous meteorites. Natl Prod Rep 19:292-311

Sephton MA, Hazen RM (2013) On the origins of deep hydrocarbons. Rev Mineral Geochem 75:449-465

Servais T, Harper DAT, Li J, Munnecke A, Owen AW, Sheehan PM (2009) Understanding the Great Ordovician biodiversification Event (GOBE): Influences of paleogeography, paleoclimate, and paleoecology. GSA Today 19:4-10

Seto Y, Hamane D, Nagai T, Fujino F (2008) Fate of carbonates within oceanic plates subducted to the lower mantle, and a possible mechanism of diamond formation. Phys Chem Miner 35:223–229, doi: 10.1007/s00269-008-0215-9

Shen Y, Canfield DE, Knoll AH (2002) Middle Proterozoic ocean chemistry: evidence from the McArthur Basin, Northern Australia. Am J Sci 302:81-109

Shirey SB, Cartigny P, Frost DJ, Keshav S, Nestola F, Nimis P, Pearson DG, Sobolev NV, Walter MJ (2013) Diamonds and the geology of mantle carbon. Rev Mineral Geochem 75:355-421

Shirey SB, Kamber BS, Whitehouse MJ, Mueller PA, Basu AR (2008) A review of the isotopic and trace element evidence for mantle and crustal processes in the Hadean and Archean: Implications for the onset of plate tectonic subduction. *In:* When Did Plate Tectonics Start on Earth? Condie KC, Pease V (ed) Geological Society of America Special Paper 440, Boulder, Colorado, p 1-30

Shirey SB, Richardson SH (2011) Start of the Wilson cycle at 3 Ga shown by diamonds from the subcontinental mantle. Science 333:434-436, doi: 10.1126/science.1206275

Shock EL, Sassani DC, Betz H (1997a) Uranium in geologic fluids: Estimates of standard partial molal properties, oxidation potentials and hydrolysis constants at high temperatures and pressures. Geochim Cosmochim Acta 61:4245-4266

Shock EL, Sassani DC, Willis M, Sverjensky DA (1997b) Inorganic species in geologic fluids: Correlations among standard molal thermodynamic properties of aqueous cations, oxyanions, acid oxyanions, oxyacids and hydroxide complexes. Geochim Cosmochim Acta 61:907-950

Shukolyukov A, Lugmair GW (2002) ^{60}Fe—Light my fire. Meteoritics 27:289

Silver PG, Behn MD (2008) Intermittent plate tectonics. Science 319:85-88

Simkiss K, Wilbur KM (1989) Biomineralization. Academic Press, Elsevier, Amsterdam

Smithies RH, Champion DC (2000) The Archean high-Mg diorite suite: Links to tonalite-trondhjemite-granodiorite magmatism and implications for early Archean crustal growth. J Petrol 41:1653-1671

Smithies RH, Champion DC, Cassidy KF (2003) Formation of Earth's early Archaean continental crust. Precambrian Res 127:89-101

Stanley SM, Hardie LA (1998) Secular oscillations in the carbonate mineralogy of reef-building and sediment-producing organisms driven by tectonically forced shifts in seawater chemistry. Paleogeogr Paleoclimatol Paleoecol 144:3-19

Stroud RM, Nittler LR, Alexander CMO'D (2004) Polymorphism in presolar Al_2O_3 grains from asymptotic giant branch stars. Science 305:1455-1457

Strutt RJ (1910) Measurements of the rate at which helium is produced in thorianite and pitchblende, with a minimum estimate of their antiquity. Proc Royal Soc London A84:379-388

Sumner DW (1997) Carbonate precipitation and oxygen stratification in late Archean seawater as deduced from facies and stratigraphy of the Gamohaan and Frisco Formations, Transvaal Supergroup, South Africa. Am J Sci 297:455-487

Sumner DY, Grotzinger JP (1996a) Herringbone calcite: Petrography and environmental significance. J Sed Res 66:419-429
Sumner DY, Grotzinger JP (1996b) Were kinetics of Archean calcium carbonate precipitation related to oxygen concentration? Geology 24:119-122
Sumner DY, Grotzinger JP (2004) Implications for Neoarchaean ocean chemistry from primary carbonate mineralogy of the Campbellrand-Malmani Platform, South Africa. Sedimentology 51:1273-1299
Sverjensky DA. Lee N (2010) The Great Oxidation Event and mineral diversification. Elements 6:31-36
Tazaki K (2005) Microbial formation of a halloysite-like mineral. Clays Clay Miner 55:224-233
Thompson CK, Kah LC (2012) Sulfur isotope evidence for widespread euxinia and a fluctuating oxycline in Early to middle Ordovician greenhouse oceans. Palaeogeo, Palaeoclim, Palaeoecol 313:189-214
Thompson JB, Ferris FG (1990) Cyanobacterial precipitation of gypsum, calcite, and magnesite from natural alkaline lake water. Geology 18:995-998
Tkachev AV (2011) Evolution of metallogeny of granitic pegmatites associated with orogens throughout geological time. *In:* Granite-Related Ore Deposits. Sial AN, Bettencourt JS, De Campos CP, Ferreira VP (ed), Geological Society of London Special Publications 350:7-23
Tomlinson KY, Davis DW, Stone D, Hart T (2003) U-Pb age and Nd isotopic evidence for Archean terrane development and crustal recycling in the south-central Wabigoon Subprovince,Canada. Contrib Mineral Petrol 144:684-702
Tonks WB, Melosh HJ (1993) Magma ocean formation due to giant impacts. J Geophys Res 98:5319-5333
Touboul M, Kleine T, Bourdon B, Plame H, Wieler R (2007) Late formation and prolonged differentiation of the Moon inferred from W isotopes in lunar metals. Nature 450:1206-1209
Tourre SA, Sumner DY (2000) Geochemistry of herringbone calcite from an ancient Egyptian quarry. Geol Soc Am, Abstracts with Programs 32:215
Towe KM (2002) The problematic rise of Archean oxygen. Science 295:798-799
Ueno Y, Yamada K, Yoshida N, Maruyama S, Isozaki Y (2006) Evidence from fluid inclusions for microbial methanogenesis in the early Archaean era. Nature 440:516-519
Ueshima M, Tazaki K (1998) Bacterial bio-weathering of K-feldspar and biotite in granite. Clay Sci Japan 38:68-92
Van Kranendonk MJ (2006) Volcanic degassing, hydrothermal circulation and the flourishing of early life on Earth: new evidence from the Warrawoona Group, Pilbara Craton, Western Australia. Earth Sci Rev 74:197-240
Van Kranendonk MJ (2007) A review of the evidence for putative Paleoarchean life in the Pilbara Craton. *In:* Earth's Oldest Rocks. Van Kranendonk MJ, Smithies RH, Bennet V (ed), Developments in Precambrian Geology 15, Elsevier, Amsterdam, p 855-896
Van Kranendonk MJ (2011) Onset of plate tectonics. Science 333:413-414
Varga P, Rybicki KR, Denis C (2006) Comment on the paper "Fast tidal cycling and the origin of life" by Richard Lathe. Icarus 180:274-276
Veizer J, Bell K, Jansen SL (1992) Temporal distribution of carbonatites. Geology 20:1147-1149
Vollmer C, Hoppe P, Brenker FE, Holzapfel C (2007) Stellar $MgSiO_3$ perovskite: A shock-transformed silicate found in a meteorite. Astrophys J 666:L49-L52
Wadhwa M, Srinivasan G, Carlson RW (2006) Timescales of planetary differentiation in the early solar system. *In:* Meteorites in the Early Solar System II. Lauretta DS, McSween HY Jr (ed) University of Arizona Press, Tucson, p 715-731
Walter MR (1994) The earliest life on Earth: clues to finding life on Mars. *In:* Early Life on Earth. Bengtson S (ed), Nobel Symposium 84:270-286. Columbia University Press, New York
Walter MR, Buick R, Dunlop JSR (1980) Stromatolites 3,400-3,500 Myr old from the North-Pole area, Western-Australia. Nature 284:443-445
Weisberg MK, McCoy TJ, Krot AN (2006) Systematics and evolution of meteorite classification. *In:* Meteorites and the Early Solar System II. Lauretta DS, McSween HY Jr (ed), University of Arizona Press, Tucson, p 19-52
Wenk HR, Bulakh A (2004) Minerals: Their Constitution and Origin. Cambridge University Press, New York
Westall F, de Vries ST, Nijman W, Rouchon V, Orberger B, Pearson V, Watson J, Verchovsky A, Wright I, Rouzaud JN, Marchesini D, Severine A (2006) The 3.466 Ga "Kitty's Gap Cheil", an early Archean microbial ecosystem. Geol Soc Am Special Paper 405:105-131
Wilkinson BH, Givens KR (1986) Secular variation in abiotic marine carbonates: Constraints on Phanerozoic atmospheric carbon dioxide contents and oceanic Mg/Ca ratios. J Geol 94:321-333
Wilks ME, Nisbet EG (1988) Stratigraphy of the Steep Rock Group, northwest Ontario: a major Archaean unconformity and Archaean stromatolites. Canadian J Earth Sci 25:370-391
Williams JP, Blitz L, McKee CF (2000). The structure and evolution of molecular clouds: from clumps to cores to the IMF. Protostars and Planets IV. University of Arizona Press, Tucson, p 97

Workman RK, Hart SR (2005) Major and trace element composition of the depleted MORB mantle (DMM). Earth Planet Sci Lett 231:53-72

Young GM (1976) Iron-formation and glaciogenic rocks of the Rapitan Group, Northwest Territories, Canada. Precambrian Res 3:137-158

Young GM (1995) Are Neoproterozoic glacial deposits preserved on the margins of Laurentia related to the fragmentation of two supercontinents? Geology 23:153-156

Zegers TE (2004) Granite formation and emplacement as indicators of Archean tectonic processes. *In:* The Precambrian Earth: Tempos and Events. Eriksson PG, Altermann W, Nelson DR, Mueller WU, Catuneau O (ed), Elsevier, New York, p 103-118

Zerkle AL, Claire MW, Domagal-Goldman SD (2012) A bistable organic-rich atmosphere on the Neoarchaean Earth. Nature Geosci 5:359-363

Zhabin AG (1981) Is there evolution of mineral speciation on Earth? Doklady Earth Science Sections 247:142-144

Zhu YF, Ogasawara Y (2002) Carbon recycled into deep Earth: Evidence from dolomite dissociation in subduction-zone rocks. Geology 30:947-950

Reviews in Mineralogy & Geochemistry
Vol. 75 pp. 109-148, 2013

The Chemistry of Carbon in Aqueous Fluids at Crustal and Upper-Mantle Conditions: Experimental and Theoretical Constraints

Craig E. Manning

Department of Earth and Space Sciences
University of California, Los Angeles
Los Angeles, California 90095, U.S.A.

manning@ess.ucla.edu

Everett L. Shock

School of Earth and Space Exploration
Arizona State University
Tempe, Arizona 85287-1404, U.S.A.

Everett.Shock@asu.edu

Dimitri A. Sverjensky

Department of Earth and Planetary Sciences
The Johns Hopkins University
Baltimore, Maryland 21218, U.S.A.

sver@jhu.edu

INTRODUCTION

Carbon can be a major constituent of crustal and mantle fluids, occurring both as dissolved ionic species (e.g., carbonate ions or organic acids) and molecular species (e.g., CO_2, CO, CH_4, and more complex organic compounds). The chemistry of dissolved carbon changes dramatically with pressure (*P*) and temperature (*T*). In aqueous fluids at low *P* and *T*, molecular carbon gas species such as CO_2 and CH_4 saturate at low concentration to form a separate phase. With modest increases in *P* and *T*, these molecular species become fully miscible with H_2O, enabling deep crustal and mantle fluids to become highly concentrated in carbon. At such high concentrations, carbon species play an integral role as solvent components and, with H_2O, control the mobility of rock-forming elements in a wide range of geologic settings. The migration of carbon-bearing crustal and mantle fluids contributes to Earth's carbon cycle; however, the mechanisms, magnitudes, and time variations of carbon transfer from depth to the surface remain least understood parts of the global carbon budget (Berner 1991, 1994; Berner and Kothavala 2001).

Here we provide an overview of carbon in crustal and mantle fluids. We first review the evidence for the presence and abundance of carbon in these fluids. We then discuss oxidized and reduced carbon, both as solutes in H_2O-rich fluids and as major components of miscible CO_2-CH_4-H_2O fluids. Our goal is to provide some of the background needed to understand the role of fluids in the deep carbon cycle.

1529-6466/13/0075-0005$00.00 DOI: 10.2138/rmg.2013.75.5

Carbon in aqueous fluids of crust and mantle

Numerous lines of evidence indicate that carbon may be an important component of crustal and mantle fluids. Fluid inclusions provide direct samples of carbon-bearing fluids from a range of environments. Carbon species in fluid inclusions include molecular gas species (CO_2, CH_4), carbonate ions, and complex organic compounds, including petroleum (Roedder 1984). Carbon-bearing fluid inclusions occur in all crustal metamorphic settings, but they have also been reported in samples derived from mantle depths, including nearly pure CO_2 inclusions in olivine in mantle xenoliths (Roedder 1965; Deines 2002), inclusions in ultrahigh-pressure metamorphic minerals exhumed from mantle depths (Fu et al. 2003b; Frezzotti et al. 2011), and carbon-bearing fluid inclusions in diamonds from depths corresponding to more than 5 GPa (Navon et al. 1988; Schrauder and Navon 1993).

The formation of carbon-bearing minerals in fluid-flow features such as veins and segregations are *prima facie* indications of carbon transport by deep fluids. Environments in which carbonate veins have been observed range from shallow crustal settings to rocks exhumed from subduction zones (Gao et al. 2007) and, rarely, mantle xenoliths (Demeny et al. 2010). Graphite is also widely observed as a vein mineral, most famously perhaps in the Borrowdale graphite deposit of the Lake District in the United Kingdom (e.g., Barrenechea et al. 2009). The occurrence of C-bearing minerals in metamorphic veins is consistent with the observation that the C content of metamorphic rocks decreases with increasing metamorphic grade. For example, pelagic clay lithologies ("pelites") progressively decarbonate during metamorphism: whereas the global average oceanic sediment has 3.01 wt% CO_2, low-grade metapelites have an average of 2.31 wt% CO_2, and high-grade metapelites average 0.22 wt% CO_2 (Shaw 1956; Plank and Langmuir 1998). The decarbonation correlates with dehydration, clearly demonstrating that prograde metamorphic reactions liberate a fluid phase containing both H_2O and carbon as components. Similarly, the development of calc-silicate skarns in carbonate lithologies (Einaudi et al. 1981), in which fluid flow induces replacement of carbonate minerals (chiefly calcite) by silicates and oxides, requires liberation of carbon to water-rich fluids. Finally, spring waters discharging from active metamorphic terranes commonly contain carbon derived from depth (Irwin 1970; Chiodini et al. 1995; Chiodini et al. 1999; Becker et al. 2008; Wheat et al. 2008).

Sources of carbon in aqueous fluids of the crust and mantle

The carbon that is incorporated into deep fluids is derived from two sources. It may be liberated from the host rocks during fluid-rock reaction ("internal sources"), or it may be introduced by mixing with other fluids ("external sources").

Internal carbon sources. Oxidized carbon is incorporated into rocks by primary accumulation and crystallization processes, and by secondary weathering, alteration, or cementation processes. The dominant primary internal source of oxidized carbon is sedimentary. Pure limestone generated by accumulation of biomineralized calcite (shells, etc.) contains 44 wt% CO_2, whereas dolomite contains 48 wt% CO_2. Varying amounts of siliciclastic detritus found in "impure" carbonates lowers the CO_2 concentration. The carbonate compensation depth limits the accumulation of carbon in pelagic sediments. But even deep-ocean sedimentary packages contain at least some CO_2: the global average composition of oceanic sediment entering subduction zones is 3.01 ± 1.44 wt% CO_2 (Plank and Langmuir 1998). In addition, calcite is one of the most common cements found in sandstone. Carbonate minerals are rare products of magmatic crystallization of silicate magmas, though they occur as primary phases in C-rich magmas such as carbonatites and kimberlites (Jones et al. 2013). Carbonate minerals have been observed as inclusions in silicate minerals in mantle xenoliths (e.g., McGetchin and Besancon 1973).

Near Earth's surface, secondary processes can enrich igneous, metamorphic, and sedimentary lithologies in CO_2 by weathering and alteration. On continents, rock weathering extracts CO_2 from the atmosphere to produce secondary carbonate minerals (Urey 1952; Berner et al. 1983). In submarine settings, seafloor alteration processes lead to significant CO_2 uptake by secondary carbonation of basaltic crust. For example, fresh mid-ocean ridge basalt (MORB) is estimated to contain ~0.15 wt% CO_2, but during alteration and seafloor weathering the CO_2 content of the upper 600 m of the oceanic crust may increase to ~3 wt%, and the average CO_2 gain by the entire crustal section is elevated to ~0.4 wt% (Staudigel 2003). Abundant carbonate veins in oceanic serpentinites are observed in core, dredge hauls, and ophiolites, indicating that altered oceanic mantle rocks likewise contain significant carbon (e.g., Thompson et al. 1968; Thompson 1972; Bonatti et al. 1974, 1980; Trommsdorff et al. 1980; Morgan and Milliken 1996; Schrenk et al. 2013).

The most common source of reduced carbon is buried organic material found in sedimentary lithologies. Carbon in sedimentary basins is present in a variety of species and phases that span a substantial range of redox states. Familiar organic compounds found in sedimentary basins include the fossil fuels such as coal, petroleum, and natural gas, which typically have their origins in the transformation of detrital organic remains of life (Sephton and Hazen 2013). The biomolecules that accumulate with mineral grains in sediments and sedimentary rocks are those compounds that are most resistant to microbial modification. The most refractory compounds are membrane molecules of microbes and lignin molecules from plants, which can be transformed into petroleum or coal, respectively, if the subsurface geologic conditions are conducive. Through a complex series of reactions, these compounds may transform to graphite during crustal metamorphism. This graphite can be an important source for carbon in metamorphic fluids.

Condensed zero-valent and reduced carbon occurs in mantle rocks in a variety of forms (Mathez et al. 1984). It is found as a free phase as graphite or diamond, or as carbide minerals such as moissanite, cementite or other Fe-C compounds (Dasgupta and Hirschmann 2010; Shiryaev et al. 2011; Hazen et al. 2013). Small amounts of carbon may dissolve in mantle minerals (Tingle and Green 1987; Tingle et al. 1988; Keppler et al. 2003; Shcheka et al. 2006; Ni and Keppler 2013) or coat grain surfaces (Mathez 1987; Pineau and Mathez 1990; Mathez and Mogk 1998).

External carbon sources. The carbon in a system experiencing fluid-mineral interaction need not be solely internally derived from the local rock host. Fluids carrying carbon from external sources may mix with an otherwise carbon-free fluid. At least at depths above the brittle-ductile transition, meteoric waters drawn downward by hydrothermal or metamorphic circulation may carry atmosphere-derived carbon. Magmas also represent a potentially important carbon source. Carbon in volcanic gases typically occurs as CO_2; reduced species such as CO and CH_4 are very low in abundance (Symonds et al. 1994; Burton et al. 2013). Pre-eruptive CO_2 contents of the main types of mafic magmas are 2000-7000 ppm in ocean island basalts and ~1500 ppm in normal MORB (Marty and Tolstikhin 1998; Gerlach et al. 2002; Oppenheimer 2003; Dasgupta 2013; Ni and Keppler 2013). Andesite exhibits a wide range of pre-eruptive CO_2, from below detection to 2500 ppm (Wallace 2005), and CO_2 is typically below detection in silicic magmas such as dacites and rhyolites (Oppenheimer 2003). Carbonatite and carbonated silicate magmas, though rare, carry substantial carbon and may act as a carbon source where they trigger production of more common magmas (Dasgupta and Hirschmann 2006; Jones et al. 2013).

Although inferred pre-eruptive carbon contents of the more common magma types are generally low, molecular carbon species are strongly partitioned into the vapor phase when magmas reach saturation. This fractionation means that substantial carbon may be lost prior to entrapment of melt inclusions or liberation of volcanic gas, both of which form the basis for estimates of the above volatile abundances. In many cases, concentration estimates are therefore simply lower bounds. This factor may be particularly important in convergent margins and orogenic belts. For example, Blundy et al. (2010) proposed early saturation of CO_2 and more CO_2-rich arc magmas than previously assumed. The occurrence of magmatic calcite inclusions in granitoids is a test of this idea. Audétat et al. (2004) describe magmatic calcite inclusions in quartz and apatite in a quartz monzodioritic dike at Santa Rita, New Mexico. Calcite on the liquidus in granitic systems requires crystallization at depths of at least 10 km (Swanson 1979; Audétat et al. 2004), at conditions of high CO_2 partial pressure. The Santa Rita example suggests that, at least locally, very high carbon contents may in fact occur in felsic systems. Rare carbonate-bearing scapolite that is rich in meionite component [$Ca_4Al_6Si_6O_{24}(CO_3)$] has been reported from a range of volcanic and plutonic rock types (Goff et al. 1982; Mittwede 1994; Smith et al. 2008).

Carbon contents may also be particularly high in alkaline and peralkaline magmatic systems due to elevated carbonate solubility (Koster van Groos and Wyllie 1968). Alkali carbonate/bicarbonate-rich fluids have been reported from numerous granitic pegmatites (Anderson et al. 2001; Sirbescu and Nabelek 2003a,b; Thomas et al. 2006, 2011). Evidence is chiefly the presence of fluid and melt inclusions containing carbonate daughter minerals such as nahcolite (Na_2CO_3), zabuyelite (Li_2CO_3), and even potassium carbonate (K_2CO_3). In a detailed study, Thomas et al. (2011) report evidence for pegmatite emplacement from a three-phase fluid system of hydrous carbonate melt, a hydrous carbonate-saturated silicate melt, and CO_2-rich vapor. Total carbonate species concentrations in the vapor phase may exceed 30-40 wt%. These observations demonstrate that magmatic systems represent an important, though highly variable, source of carbon in the geologic environment through which they pass.

Finally, mantle degassing may provide an important source of carbon (Burton et al. 2013; Dasgupta 2013). Evidence for mantle fluids in deep environments is typically obscured by more voluminous fluids sourced from crustal rocks. However, fluid inclusions in mantle xenoliths record evidence for reduced carbon species, including CH_4, CO, and, potentially, COS (Melton et al. 1972; Melton and Giardini 1974; Murck et al. 1978; Bergman and Dubessy 1984; Tomilenko et al. 1998). In addition, carbonate-metasomatized shear zones of the deep crust display mantle-like C, Sr, and He isotope ratios, leading to the inference that components of the fluids that deposited the carbonates were initially of mantle origin (Baratov et al. 1984; Lapin et al. 1987; Stern and Gwinn 1990; Dahlgren et al. 1993; Dunai and Touret 1993; Oliver et al. 1993; Wickham et al. 1994).

OXIDIZED CARBON IN AQUEOUS FLUIDS AT HIGH *P* AND *T*

A vast body of experimental and theoretical work has shown that in pure H_2O at ambient conditions and along the liquid-vapor saturation curve, species of oxidized carbon dissolved in pure H_2O include carbonate ion (CO_3^{2-}), bicarbonate ion (HCO_3^-), and dissolved CO_2 ($CO_{2,aq}$; Fig. 1). A fourth possible species, "true" carbonic acid (H_2CO_3; Fig. 1), has been isolated as a pure gas and solid (e.g., Terlouw et al. 1987), but decomposes rapidly in H_2O, such that the reaction

$$H_2CO_3 = CO_{2,aq} + H_2O \tag{1}$$

proceeds far to the right; for example, at room T and P, H_2CO_3 concentration is about ~0.1% of $CO_{2,aq}$ (Loerting et al. 2000; Ludwig and Kornath 2000; Tossell 2006; England et al. 2011). Detection of these low concentrations of H_2CO_3 in aqueous solutions has now been convincingly achieved (Falcke and Eberle 1990; Soli and Byrne 2002; Adamczyk et al. 2009). Nevertheless, because of its very low concentration, geochemists conventionally treat the carbon present in both hydrated neutral species as $CO_{2,aq}$.

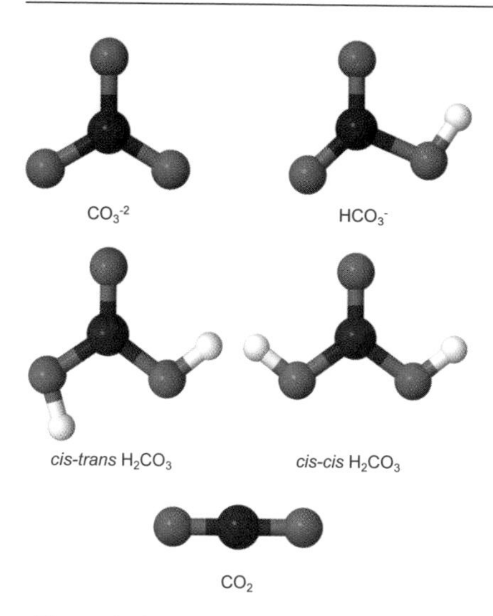

Figure 1. Gas-phase structures of the main oxidized-carbon species found in deep aqueous fluids. Carbon atoms are black, oxygen atoms gray, and hydrogen atoms white. The carbonate ion (CO_3^{2-}) has trigonal planar structure. The C-O bond distance is 0.131 nm and the bond angle is 120°. One of the three C-O bonds is a double bond. In bicarbonate (HCO_3^-), the hydration of an oxygen atom lengthens the corresponding C-O bond and all bond angles rotate slightly to accommodate. The *cis-cis* carbonic acid (H_2CO_3) structure is more stable than the *cis-trans* variant (e.g., Mori et al. 2011); the unhydrated oxygen shares a double bond with carbon. Carbon dioxide (CO_2) is a linear molecule with double C-O bonds that are 0.16 nm in length.

The predominant oxidized carbon species interact via two stepwise dissociation reactions. The first involves generation of bicarbonate from $CO_{2,aq}$ and a solvent H_2O molecule:

$$CO_{2,aq} + H_2O = HCO_3^- + H^+ \tag{2}$$

The second stepwise dissociation reaction is

$$HCO_3^- = CO_3^{2-} + H^+ \tag{3}$$

Figure 2a shows that at ambient conditions, neutral pH of H_2O lies between pK_2 and pK_3, where pK_i is the negative logarithm of the equilibrium constant K of reaction i. Thus, bicarbonate will often be the predominant species when pH is fixed independently of the carbon system.

Natural crustal and mantle solutions are complex, and contain substantial dissolved metal cations. These may interact with carbonate ions to form ion pairs such as $NaCO_3^-$, $CaCO_3^\circ$, or $CaHCO_3^+$. But dissolved oxidized carbon chemistry will vary strongly with geologic environment even in dilute aqueous solutions, because the pK values and the equilibrium constant for H_2O dissociation are strong functions of P and T. Thus, as illustrated in Figure 2b, the predominant species can be expected to change in deep crustal and mantle settings.

Aqueous fluids at high P and T

Experimental constraints on homogeneous systems. Whereas there is a voluminous literature on aqueous carbonate ion speciation at ambient conditions and along the liquid-vapor saturation curve of H_2O, there have been few direct studies of homogenous aqueous carbonate systems at high P and T. This lack is chiefly due to the experimental challenges posed by working at these conditions. Read (1975) appears to have been the first experimentalist to examine aqueous carbonate equilibria at pressures greater than a few hundred atmospheres. Extending earlier work by Ellis (1959a) and Ryzhenko (1963), he used electrical conductivity measurements to 250 °C and 0.2 GPa (= 2 kbar) to determine the equilibrium constant for reaction (2):

$$K = \frac{a_{HCO_3^-}\, a_{H^+}}{a_{CO_2,aq}\, a_{H_2O}} \tag{4}$$

The results revealed that K rises with increasing P at constant T, but drops with increasing T at constant P. Thus, reaction (2) is driven to the right on isothermal compression, but to the left on isobaric heating.

Kruse and Franck (1982) compressed $KHCO_3$ solutions at up to 300 °C and 50 MPa, and used Raman spectroscopy to show that CO_3^{2-} is favored relative to HCO_3^-. Frantz (1998)

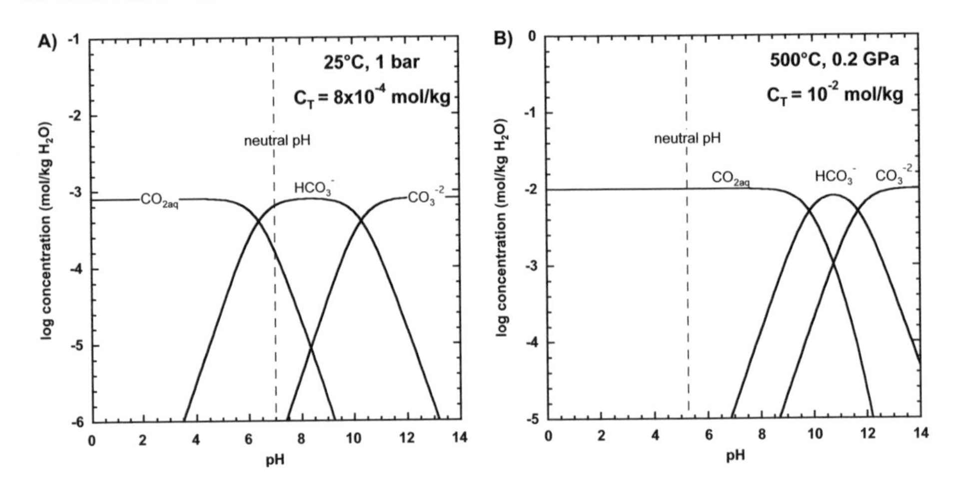

Figure 2. Variation in the abundances of the main oxidized carbon species with pH at 25 °C, 1 bar (A), and 500 °C, 0.2 GPa (B). Calculations assume unit activity coefficients of all species; data are from Shock et al (1989, 1997b). Total carbon concentration (C_T) in (A) is 8×10^{-4} molal, the global average riverine bicarbonate concentration (Garrels and Mackenzie 1971). In (B), C_T is set to 10^{-2} molal. Vertical dashed line shows neutral pH at each set of conditions. Comparison of (A) and (B) highlight that the pH range over which HCO_3^- is stable decreases as *P* and *T* rise.

used a similar approach on 1 molal K_2CO_3 and $KHCO_3$ solutions. He studied solutions along two isobars of 0.1 and 0.2 GPa, to 550 °C. When adjusted for Raman scattering cross-section ratios, the results support isobaric decreases in HCO_3^- relative to $CO_{2,aq}$ and CO_3^{2-} (Fig. 3). Martinez et al. (2004) studied 0.5 and 2 m K_2CO_3 solutions in a diamond cell. Raman spectra in runs to 400 °C and 0.03 GPa indicated the presence of CO_3^{2-}, but not HCO_3^-. They inferred that HCO_3^- stability decreases significantly at high pressure. Thus, it appears that bicarbonate ion becomes less stable relative to $CO_{2,aq}$ and carbonate ion as *P* and *T* rise to crustal and mantle conditions.

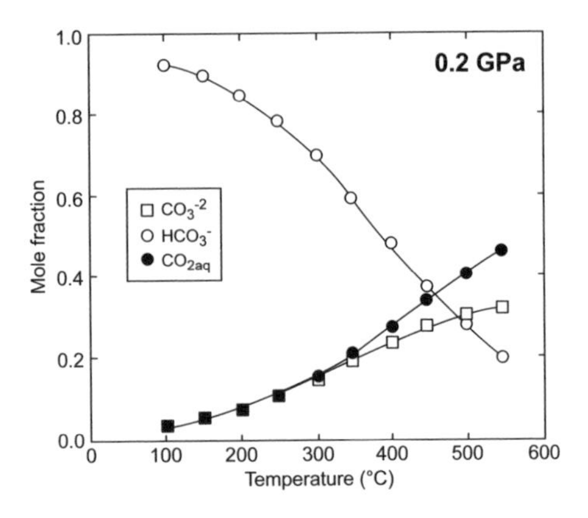

Figure 3. Mole fraction of total dissolved carbonate, bicarbonate, and $CO_{2,aq}$ in 1 molal $KHCO_3$ solutions at 0.2 GPa (Frantz 1998). The relative concentration of bicarbonate decreases isobarically with rising temperature.

None of the spectroscopic studies yielded evidence for metal-carbonate ion pairing, although the solutions were relatively dilute. In addition, no evidence has been found for carbonic acid, consistent with results at ambient conditions (e.g., Davis and Oliver 1972); however, Falcke and Eberle (1990) found evidence of H_2CO_3 at high ionic strengths. The significantly higher ionic strengths of high *P-T* solutions (Manning 2004) means that this species may yet be significant at these conditions. Finally, there is not yet any evidence of other species in high *P-T* aqueous solutions. For example, it has been suggested that dicarbonate ion ($C_2O_5^{2-}$) may be produced by reaction of carbonate ion with H_2O:

$$2CO_3^{2-} + H_2O = C_2O_5^{2-} + 2OH^- \quad (5)$$

(Zeller et al. 2005). Although dicarbonate ions may form when CO_2 dissolves into carbonate melts via

$$CO_2 + CO_3^{2-} = C_2O_5^{2-} \tag{6}$$

(Claes et al. 1996), there is as yet no evidence for this species in relevant geological fluids at crustal or mantle conditions.

Experimental studies of carbonate mineral solubility in H_2O. The most voluminous source of information on oxidized carbon species in aqueous solutions comes from studies of the solubility of carbonate minerals in H_2O and mixed solvents. In general, experimental work has focused on the solubility of divalent metal carbonates, chiefly calcite, because carbonate minerals involving monovalent or trivalent cations are highly soluble or require extremely acidic pH, respectively (Rimstidt 1997). The solubility of carbonate minerals in H_2O is strongly dependent on pH, regardless of *P* and *T* (Fig. 4). However, when carbonate minerals dissolve in initially pure H_2O, the solute products shift pH to a value that is more alkaline than neutral pH at the conditions of interest. Hence, all carbonate minerals contribute alkalinity to crustal and mantle fluids. We focus first on calcite in H_2O, then other minerals in H_2O, then carbonate minerals in more complex solutions.

Numerous studies investigated calcite solubility in H_2O (CO_2-free or equilibrated with the atmosphere) at low *P* and *T* along the H_2O liquid-vapor saturation curve and up to ~0.1 GPa (Wells 1915; Frear 1929; Schloemer 1952; Morey 1962; Segnit 1962; Plummer 1982). The results showed that calcite dissolution in pure H_2O is congruent, and that solubility decreases with increasing temperature isobarically and along the H_2O boiling curve—the well-known "reverse solubility" effect for calcite.

There are only 3 studies of calcite solubility in H_2O at higher *P* and *T* appropriate for metamorphism and mantle fluids (Fig. 5). Walther and Long (1986) reported initial results at 0.1-0.3 GPa, 300-600 °C. They observed that solubility decreased with increasing temperature at 0.1-0.2 GPa; however at 0.3 GPa solubility was constant or increased slightly with temperature. Fein and Walther (1989) later showed that the solubilities of Walther and Long (1986) were too low because they used less accurate post-experiment analytical procedures.

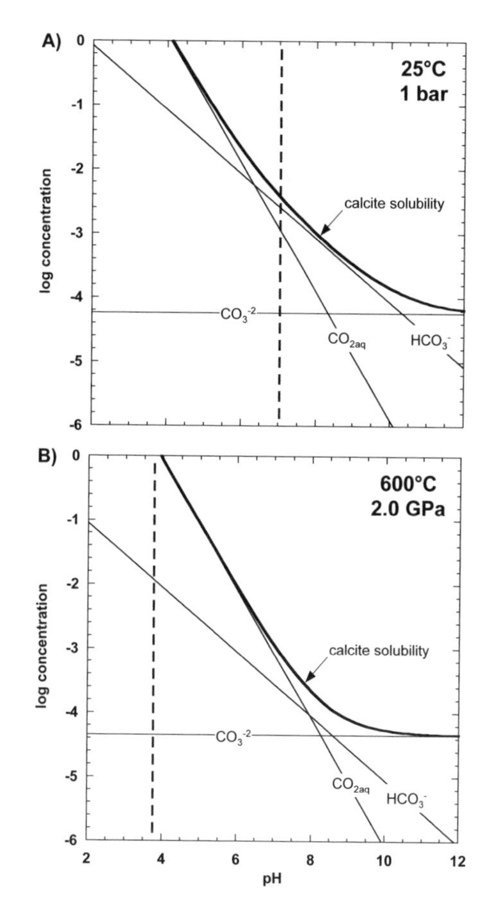

Figure 4. Calcite solubility in pure H_2O at 25 °C, 1 bar (A) and 600 °C, 2.0 GPa (B), as a function of pH. Calcite solubility (bold lines) is the sum of the concentrations of each constituent carbonate species (light lines). The general form of the solubility curves and the solubility at high pH are very similar. However the increase in the equilibrium constant for H_2O dissociation with *P* and *T* leads to an increase in calcite solubility in near-neutral fluids.

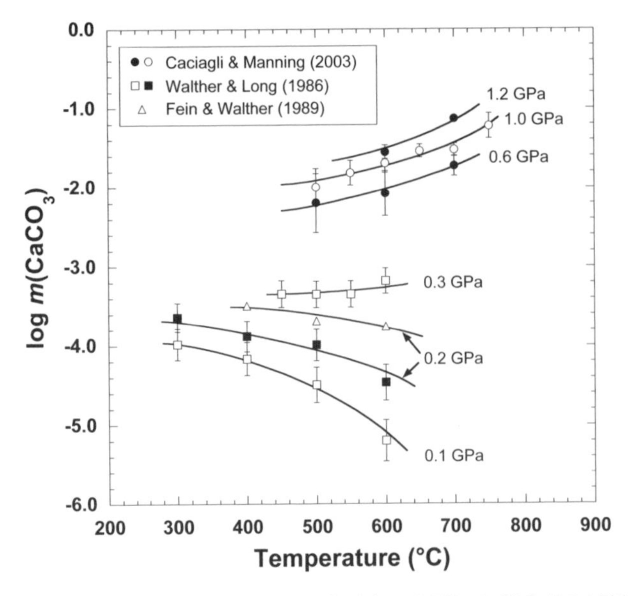

Figure 5. Summary of high-pressure determinations of calcite solubility in H_2O. Solubilities in molality.

However, the revised results yield similar solubility trends at 0.2 GPa (Fig. 5). Caciagli and Manning (2003) extended these studies to deep crustal and mantle pressures (Fig. 5). Their data confirm that at high pressure calcite exhibits an isobaric rise in solubility with temperature.

Calcite transforms to aragonite at high pressure (Boettcher 1968; Johannes 1971); no studies have yet investigated aragonite solubility directly, though Caciagli and Manning (2003) extrapolated their results into the aragonite stability field. Lennie (2005) found that, at ambient *T*, ikaite ($CaCO_3{\cdot}6H_2O$) was stable with respect to calcite + H_2O; however, solubility data were not obtained.

Although solubilities of a wide range of metal carbonate minerals in H_2O have been investigated at low pressures and temperatures (below 0.1 GPa), high-pressure studies are rare. Sanchez-Valle et al. (2003) combined hydrothermal diamond anvil cell (HDAC) methods with synchrotron X-ray fluorescence spectroscopy to determine $SrCO_3$ solubility in H_2O to 3.6 GPa and 525 °C. Strontianite was used because a high atomic number was required to obtain sufficiently favorable detection limits. Heating/compression runs in the HDAC revealed that strontianite solubility increases with pressure and temperature. Maximum solubility at 3.3-3.6 GPa and 475-525 °C was ~0.2 mol/kg H_2O. Siderite ($FeCO_3$) dissolution in H_2O has been studied to 400 °C and 1.13 GPa in the HDAC (Marocchi et al. 2011), but the acquired Raman spectra were not used to determine quantitative solubility values.

The paucity of data on the dissolution of simple carbonate minerals in pure H_2O at the high pressures presents serious challenges for testing thermodynamic models of carbonate ions and mineral solubility (see below). However, the hydrothermal piston-cylinder methods (e.g., Caciagli and Manning 2003) in parallel with hydrothermal diamond anvil cell methods (Sanchez-Valle et al. 2003) hold promise for generating such data in the near future.

Experimental studies of calcite solubility in NaCl-H_2O. Studies at $P < 0.1$ GPa show that calcite solubility in H_2O increases with addition of NaCl (Ellis 1963; Malinin 1972). Fein and Walther (1989) extended these studies to higher pressure of 0.2 GPa and temperature up to 600 °C. They showed that with addition of up to 0.1 m NaCl, calcite solubility increased consistent with formation of $CaCl^+$ in solution (Fig. 6). Newton and Manning (2002a) conducted experiments on calcite solubility in H_2O-NaCl at 600-900 °C, 1.0 GPa, and to highly concentrated NaCl solutions approaching halite saturation. They showed that calcite solubility increased with NaCl at all investigated conditions (Fig. 7). The pressure, temperature, and composition dependence of calcite molality (m_{CaCO_3}) are described by

$$m_{CaCO_3} = \left[-0.051 + 1.65\times10^{-4}T + X_{NaCl}^2 \exp\left(-3.071 + 4.749\times10^{-6}T^2\right)\right]\left(0.76 + 0.024P\right) \quad (7)$$

with P in kbar (1 kbar = 0.1 GPa) and T in kelvins. The solubility increase with temperature and salinity is so great that critical mixing of NaCl-rich hydrous carbonate liquid and $CaCO_3$-rich saline solution was proposed at 1.0 GPa at about 1000 °C and NaCl mole fraction (X_{NaCl}) of ~0.4.

The enhancement of solubility increases dramatically with temperature at constant X_{NaCl}. Experiments at 0.6, 1.0, and 1.4 GPa revealed a slight increase with pressure (~20%) at fixed X_{NaCl}. Moreover, Newton and Manning's solubility data display a simple dependence on the

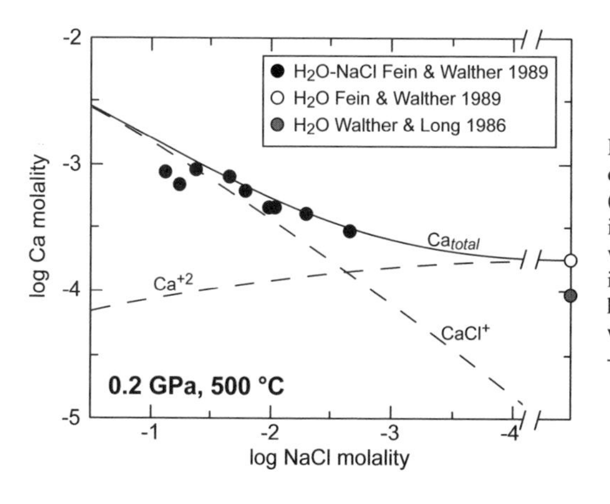

Figure 6. Dependence of calcite solubility on NaCl concentration at 500 °C, 0.2 GPa (after Fein and Walther 1989). Calcite solubility, expressed as total dissolved Ca, increases with increasing NaCl. The corresponding rise in total chloride results in Ca-Cl ion pairing; however, no evidence of Na-carbonate pairing was observed. Solubility in pure H_2O plots at $-\infty$, to right of break in scale.

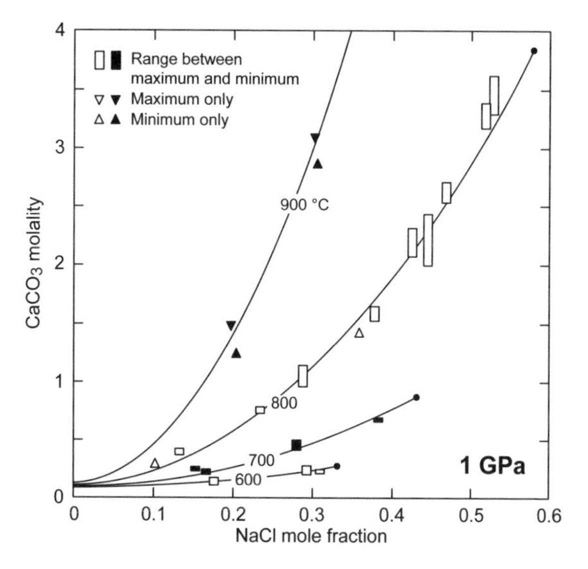

Figure 7. Experimentally determined $CaCO_3$ molality at 1.0 GPa, as a function of NaCl mole fraction (after Newton and Manning 2002a). Vertical size of rectangles reflects the range between maximum and minum solbility from a The 600, 700, and 800 °C curves are extrapolated to halite saturation (filled circles) in the system NaCl-H_2O (Aranovich and Newton 1996), ignoring dissolved $CaCO_3$.

square of the NaCl mole fraction. This result has important implications for the nature of carbon and other solutes. An increase in solubility over the entire investigated range shows that the solute species are not hydrous; if they were, their concentrations would be required to decline with rising NaCl in concert with decreasing H_2O activity. These concentrated solutions behave in a manner similar to molten salts (Newton and Manning 2002a, 2005; Tropper and Manning 2007), in which the activity of NaCl is proportional to the square of its mole fraction (Aranovich and Newton 1996), and solute mixing is nearly ideal. This result led to the inference that the dissolution reaction, written in terms of predominant species, was

$$\underset{calcite}{CaCO_3} + NaCl = CaCl^+ + NaCO_3^- \tag{8}$$

Further discussion of the chemistry of saline brines can be found in Newton and Manning (2010).

Experimental studies of calcite solubility in CO_2-H_2O. Miller (1952), Ellis (1959b), Sharp and Kennedy (1965), and Malinin and Kanukov (1972) showed that, at low fixed P and T, calcite solubility in H_2O increases to a maximum and then declines with increasing X_{CO_2}. Fein and Walther (1987) extended these studies to 0.2 GPa, 550 °C, and found similar behavior. At 0.2 GPa, calcite solubility reaches a maximum at X_{CO_2} = 0.025 to 0.05 and then decreases (Fig. 8). Fein and Walther suggested that this solubility behavior arises from the tradeoff between increasing formation of bicarbonate and decreasing H_2O activity as X_{CO_2} increased.

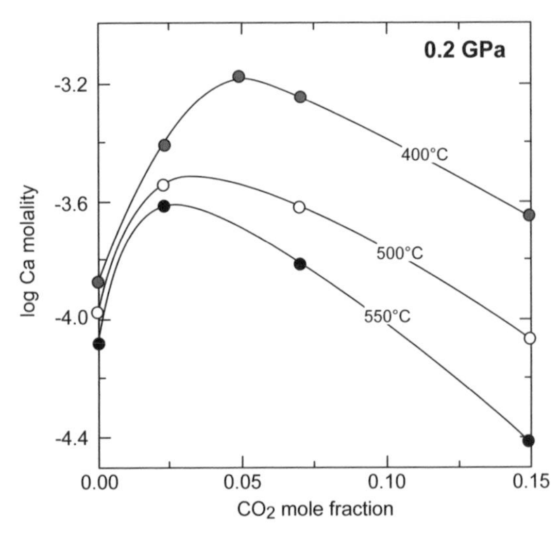

Figure 8. Calcite solubility in CO_2-H_2O fluid as a function of CO_2 mole fraction at 0.2 GPa (Fein and Walther 1987). Addition of CO_2 to pure H_2O initially causes solubility to increase, but solubility maximizes and then declines with more CO_2. Fein and Walther (1987) attribute this behavior to a decrease in H_2O activity.

Thermodynamics of oxidized carbon in dilute aqueous systems. Two fundamentally different approaches to treat oxidized dissolved carbon species have evolved in the literature, depending on the application. The first approach, which could be termed a component approach, is commonly used for the treatment of metamorphic fluids and fluids in the mantle (Anderson and Crerar 1993; Zhang and Duan 2009). It treats CO_2 as a component for which the partial molal Gibbs free energy ($\overline{G}_{CO_2;P,T}$) is expressed by

$$\overline{G}_{CO_2;P,T} = \overline{G}^0_{CO_2;P=1.0,T} + RT\ln\chi_{CO_2;P,T}X_{CO_2;P,T}P_{CO_2;P,T} \tag{9}$$

where the first term on the right-hand side represents the standard partial molal Gibbs energy of CO_2 ($\overline{G}^0_{CO_2;P=1.0,T}$). The standard state is the hypothetical ideal gas at the temperature of interest and 1 atmosphere. The second term on the right-hand side represents the connection from the standard state to the real fluid. In this term, $\chi_{CO_2;P,T}$ represents the fugacity coefficient, $X_{CO_2;P,T}$ the mole fraction and $P_{CO_2;P,T}$ the partial pressure of CO_2, all at the pressure and temperature of interest.

This approach ignores the detailed speciation of the dissolved carbon dioxide. There is no concept of pH in the model. Advantages of this approach are that the first term on the right-hand side of Equation (9) can be easily calculated from experimental data and statistical mechanics, and that it permits consideration of the full compositional range along the CO_2-H_2O binary (see below). Most of the labor involved in this approach is associated with evaluating the second term on the right-hand side. Specifically, very intensive experimental and theoretical efforts are needed to derive the fugacity coefficients for pure fluids and mixtures of fluids (Jacobs and Kerrick 1981b; Kerrick and Jacobs 1981; Holland and Powell 1991; Zhang et al. 2007; Zhang and Duan 2009). Without any experimental data it is difficult to implement this approach.

The second approach could be called the speciation approach, commonly used in the aqueous geochemistry of hydrothermal fluids. Here, all known dissolved carbon species are treated explicitly and the dependence on pH and concentrations of other components can be evaluated. For example, for the dissolved carbonate ion, we treat the partial molal Gibbs free energy of formation of the ion ($\Delta\overline{G}^0_{f,CO_3^{2-};P,T}$) by defining

$$\Delta\overline{G}_{f,CO_3^{2-};P,T} = \Delta\overline{G}^0_{f,CO_3^{2-};P_r,T_r} + RT\ln\gamma_{CO_3^{2-};P,T}m_{CO_3^{2-};P,T} \tag{10}$$

where $\Delta\overline{G}^0_{f,CO_3^{2-};P_r,T_r}$ represents the standard partial molal free energy of formation at reference pressure and temperature, and $\gamma_{CO_3^{2-};P,T}$ and $m_{CO_3^{2-};P,T}$ represent the activity coefficient and molality, respectively, of the aqueous carbonate ion. Here the standard state is the hypothetical 1.0 M solution referenced to infinite dilution at the temperature and pressure of interest. The advantage of this approach is that ions are treated explicitly, enabling pH variations to be analyzed as a result of interactions with silicate mineral assemblages that control activity ratios such as $a_{Ca^{2+}} / a^2_{H^+}$. Disadvantages of this approach are that only species whose existence is known can be included in the model, and its applicability is limited to H_2O-rich solutions and conditions where H_2O solvent properties are well known.

The most difficult and uncertain aspect of evaluating Equation (10) is the calculation of the first term on the right-hand side: the standard Gibbs free energy of formation of the ion. Considerable effort has been expended in developing methods to calculate this term for many types of aqueous species including metal complexes and biomolecules (Shock and Helgeson 1988, 1990; Shock et al. 1989, 1997a, 1997b; Shock 1995; Amend and Helgeson 1997; Sverjensky et al. 1997; Plyasunov and Shock 2001; Richard 2001; Dick et al. 2006). In contrast, the second term on the right-hand side can be approximated—at least for water-rich fluids—by a variety of models, which can be used to evaluate the aqueous ion activity coefficients. Experimental and theoretical studies relevant to these two approaches are summarized below.

As an example of the species approach we focus on the aqueous CO_3^{2-} ion. The standard Gibbs free energy of formation of the ion at elevated pressure and temperature can be calculated from

$$\Delta\overline{G}^0_{f,CO_3^{2-};P,T} = \Delta\overline{G}^0_{f,CO_3^{2-};P_r,T_r} + f(P,T) + \omega_{CO_3^{2-}}\left(\frac{1}{\varepsilon_{H_2O}} - 1\right) \tag{11}$$

where $\Delta\overline{G}^0_{f,CO_3^{2-};P_r,T_r}$ again represents the free energy at the reference pressure and temperature (Shock et al. 1997b). The function $f(P,T)$ is a complex function of equation-of-state coefficients of the carbonate ion associated with the non-solvation part of the Gibbs free energy change, and $\omega_{CO_3^{2-}}$ represents the equation-of-state coefficient for the carbonate ion associated with the solvation part of the Gibbs free energy change, which also depends on the dielectric constant of pure water (ε_{H_2O}). Equation-of-state coefficients have been developed for hundreds of aqueous species and are part of the data file for the programs SUPCRT92 (Johnson et al. 1992) and CHNOSZ (Dick et al. 2008).

Equation (11) emphasizes that the dielectric constant at pressure and temperature is a critical parameter for evaluating the standard Gibbs energies of aqueous species at high pressures and temperatures. However, extensive sets of experimental values of ε_{H_2O} have only been measured to about 0.5 GPa and 550 °C (Heger et al. 1980). A synthesis of these values together with estimates from the Kirkwood equation (Pitzer 1983) is incorporated in SUPCRT92 which enables prediction to 0.5 GPa and 1000 °C (Shock et al. 1992). A more recent synthesis, also based on the Kirkwood equation has been extrapolated to 1.0 GPa and 1,200 K (Fernandez et al. 1997). The limitation of SUPCRT92 calculations to 0.5 GPa has long been a severe roadblock to the application of quantitative aqueous geochemistry to deep crustal and mantle conditions. Methods for making calculations at higher pressures up to 3.0 GPa are summarized below.

Much progress has been made in recent years by extrapolating experimental solubilities and individual equilibrium constants to pressures substantially greater than the 0.5 GPa limitation of SUPCRT92 (Manning 1994, 1998, 2004; Caciagli 2003; Dolejs and Manning 2010). Empirical extrapolations can be carried out by taking advantage of the approximate linearity of many equilibrium constants when plotted in terms of the log$K$ versus the log ρ_{H_2O} at a given temperature, where ρ_{H_2O} is the density of water. An example for the solubility product of calcite is shown in Figure 9. Extrapolation of equilibrium constants as in Figure 9, together with other aqueous phase equilibria involving dissolved oxidized carbon species enables prediction of solubility and aqueous speciation involving aqueous ions at elevated pressure and temperature without using Equations (10) and (11).

Evaluation of the aqueous ion activity coefficients in Equation (10) at pressures and temperatures to 0.5 GPa and 550 °C is carried out using the extended Debye-Hückel equation for NaCl-bearing fluids (Helgeson et al. 1981). For example, the activity coefficient of the carbonate ion is calculated according to

$$\log \gamma_{CO_3^{2-}} = -\frac{Z_{CO_3^{2-}}^2 A_\gamma \bar{I}^{0.5}}{a_{NaCl}^0 B_\gamma \bar{I}^{0.5}} + b_{\gamma,NaCl}\bar{I} \tag{12}$$

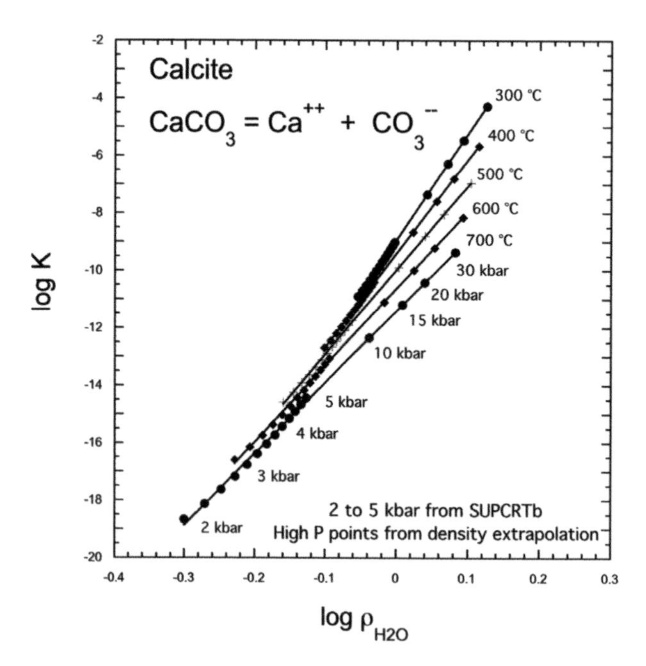

Figure 9. Predicted equilibrium constants for the solubility product of calcite. Values at 2-5 kbar (1 kbar = 0.1 GPa) calculated from SUPCRT92 (Johnson et al. 1992); values at higher pressure linearly extrapolated. The solubility product is lower than calcite solubility in H_2O due to speciation of carbon in the aqueous phase.

where Z refers to the charge on the carbonate ion (−2), A_γ and B_γ refer to Debye-Hückel coefficients that are properties of pure water (Helgeson and Kirkham 1974a,b); a^0_{NaCl} refers to the ion-size parameter for NaCl solutions, I refers to the ionic strength, and $b_{\gamma,NaCl}$ represents the extended term parameter for NaCl solutions. Using smoothed values of the dielectric constant from Helgeson and Kirkham (1974a), Franck et al. (1990), and Fernandez et al. (1997), it is possible to obtain estimates of A_γ and B_γ (Fig. 10). Values of the extended term parameter $b_{\gamma,NaCl}$ are subject to substantial uncertainty at elevated pressures and temperatures, but can be

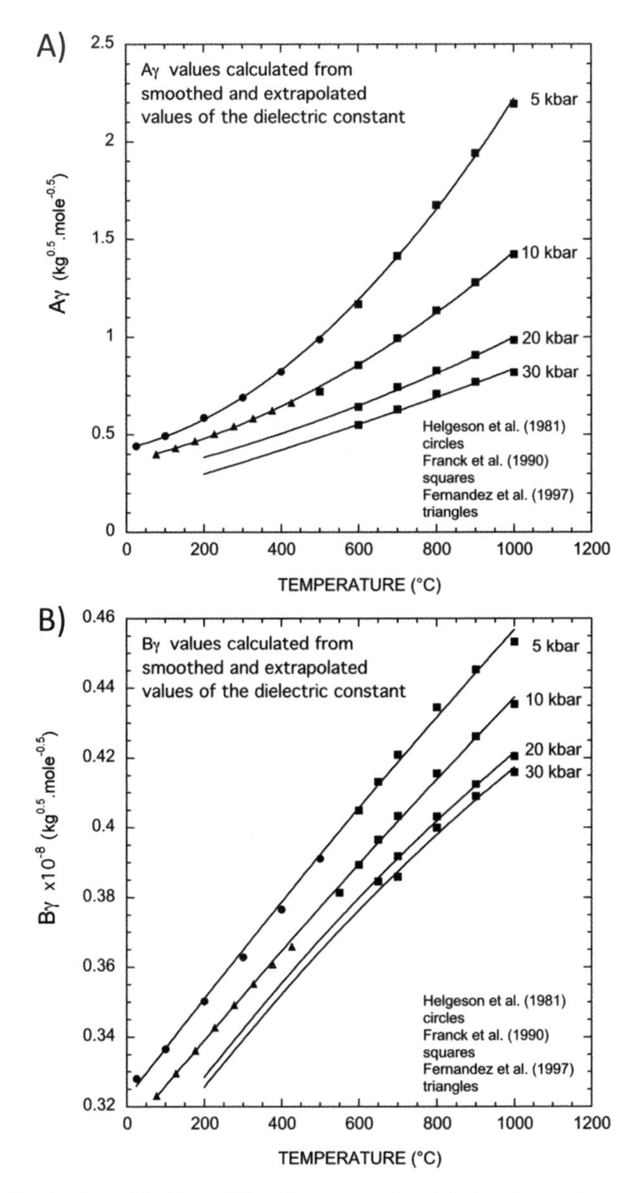

Figure 10. Predicted values of the Debye-Hückel parameters A_γ and B_γ, based on estimates of the dielectric constant of water from Helgeson et al. (1974a, 1981), Franck et al. (1990) and Fernandez et al. (1997) (1 kbar = 0.1 GPa).

estimated from correlation with the density of water to give the preliminary estimates shown in Figure 11.

Taken together, equilibrium constants such as those shown in Figure 10 and Debye-Hückel parameters such as those shown in Figures 10 and 11 can be used to estimate the solubilities of carbonate minerals in water. As an example, the solubility of calcite is shown in Figure 12 at 500 °C as a function of pressure. The predicted solubility of calcite can be compared with a variety of published experimental studies (Walther 1986; Fein 1989; Caciagli 2003). It can be seen that the predicted curve agrees well with the data from Fein and Walther (1989), but is higher than the data from Walther and Long (1986), and is in agreement with the lowest solubilities reported by Caciagli and Manning (2003). The steep increase in solubility between 0.1 and 0.5 GPa noted by Caciagli and Manning (2003) tapers off to a much slower rate of increase with pressure approaching 3.0 Gpa at 500 °C. In this range of pressures, the predominant reaction controlling the solubility of calcite is given by

$$\underset{calcite}{CaCO_3} + H_2O = Ca^{+2} + 2OH^- + CO_{2,aq} \quad (13)$$

The calculated pH values vary from 7.8 at 0.1 GPa to 5.4 at 3.0 GPa, yet less than 5% of the dissolved carbon is present as HCO_3^- and CO_3^{2-} in the fluids.

It should be emphasized that the theoretical curve shown in Figure 12 is completely predicted based on the established standard Gibbs free energy of formation of calcite (Berman 1988), the estimated equilibrium constants for

$$H_2O = H^+ + OH^- \quad (14)$$

and reactions (2), (3), and (13), extrapolated to elevated pressures as described above with the density model and the estimated aqueous ion activity coefficients discussed above. As noted above, the extrapolations for the log*K* values were based on SUPCRT92 predictions to 0.5 GPa using published equations of state for aqueous CO_2, HCO_3^-, and CO_3^{2-} (Shock et al. 1989, 1997). However, the density extrapolation for the water dissociation reaction (Eqn. 13) was based on an independent summary of experimental values (Marshall and Franck 1981).

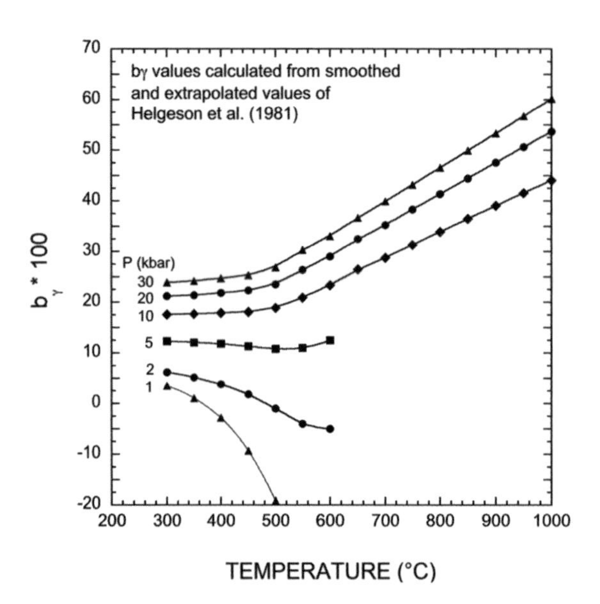

Figure 11. Predicted values of the extended term parameter ($b_{\gamma,NaCl}$) based on preliminary extrapolations of the values given in Helgeson et al. (1981) using empirical correlations with the density of water (1 kbar = 0.1 GPa).

CO_2-H_2O mixing and miscibility

Above a temperature of about 350 °C and pressures higher than the H_2O liquid-vapor saturation curve, H_2O and CO_2 are fully miscible (Fig. 13). Full miscibility therefore occurs at most conditions relevant for studies of the deep carbon cycle—much of the crust and all of the mantle—and fluids may be much richer in oxidized carbon than in the more "dilute" systems discussed above. However, complexities arise in fluids with additional components. While there is a substantial literature devoted to the thermodynamic behavior of the binary CO_2-H_2O system, much remains to be done to improve our understanding of more compositionally relevant fluids expected in deep environments.

Experimental and theoretical constraints on the CO_2-H_2O binary. The mixing of CO_2 and H_2O is highly nonideal (Eqn. 9), so considerable effort has been dedicated to generating experimental and theoretical constraints for use in petrology. Experimental studies take two approaches. A variety of methods has been used to obtain direct measurement of volumes of mixing in homogeneous CO_2-H_2O fluids (Todheide and Franck 1963; Takenouchi and Kennedy 1964; Sterner and Bodnar 1991; Seitz and Blencoe 1999). Experimental challenges generally limit these studies to low pressure. Studies utilizing synthetic fluid inclusions hold promise for

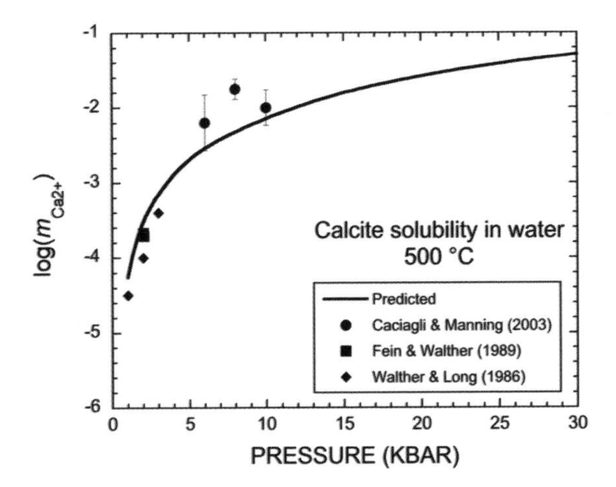

Figure 12. Predicted solubility of calcite in water as a function of pressure at 500 °C. The calculations were carried out using equilibrium constants for carbon-bearing species extrapolated using the density of water and aqueous activity coefficients calculated using Equation (12) and the Debye-Hückel parameters in Figures 10 and 11 (1 kbar = 0.1 GPa) (see text).

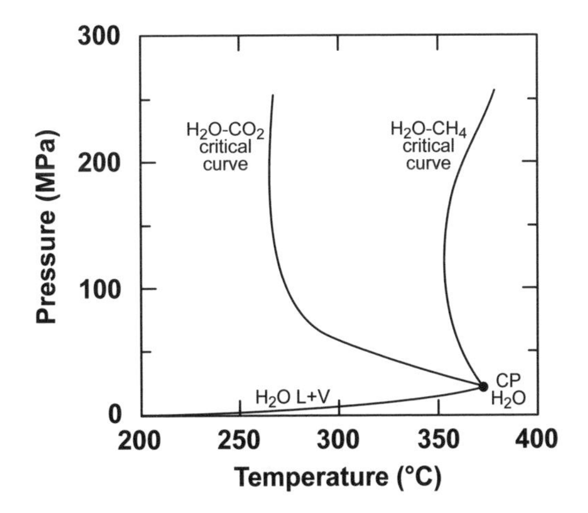

Figure 13. Pressure-temperature diagram showing critical curves for H_2O-CO_2 and H_2O-CH_4 mixing, after Liebscher (2010). The boiling curve for H_2O (H_2O L+V) terminates at the critical point (CP H_2O). At pressure higher than CP H_2O and assuming equilibrium, the binary systems H_2O-CO_2 and H_2O-CH_4 exist as two phases at temperature lower than the critical curve, or as a single fluid phase at temperature higher than the critical curve. In the two-phase field, an H_2O-rich liquid coexists with a C-rich vapor. The composition of the fluid varies along each binary critical curve.

extending *PVT* measurements to higher pressures (e.g., Sterner and Bodnar 1991; Frost and Wood 1997). An alternative is study of displacement of mineral equilibria as a function of fluid composition in mixed volatile systems. Building on pioneering work by Greenwood, Metz, Skippen, and co-workers (e.g., Greenwood 1967; Metz and Trommsdorff 1968; Skippen 1971), a number of investigators have provided high-pressure constraints on the mixing properties of the CO_2-H_2O binary (Eggert and Kerrick 1981; Jacobs and Kerrick 1981a; Aranovich and Newton 1999). Theoretical approaches also offer promise (Brodholt and Wood 1993; Destrigneville et al. 1996).

The experimental data and theoretical results serve as constraints for equations of state of CO_2-H_2O fluid mixtures. The literature on this topic is so large that even its review papers are too numerous to list completely. Useful compilations, contributions and critiques can be found in Ferry and Baumgartner (1987), Duan et al. (1996), and Gottschalk (2007).

CO_2-H_2O-NaCl and other complex fluids. Addition of NaCl to CO_2-H_2O yields an extensive region of immiscibility in the ternary system (e.g., Heinrich 2007). Experimental studies in this system were reviewed by Liebscher (2007). Extension of low-*P* experiments (Gehrig et al. 1979; Anovitz et al. 2004; Aranovich et al. 2010) to 0.5 GPa and beyond (Kotelnikov and Kotelnikova 1990; Johnson 1991; Frantz et al. 1992; Joyce and Holloway 1993; Gibert et al. 1998; Shmulovich and Graham 1999, 2004) indicates the presence of a miscibility gap between low density CO_2-H_2O rich vapor and NaCl-H_2O-rich brine at a wide range of crustal and mantle *P* and *T*. Experimental studies on $CaCl_2$-CO_2-H_2O yield similar results (Zhang and Frantz 1989; Shmulovich and Graham 2004). Equations of state for ternary system have been constructed based on experiment and theory (Bowers and Helgeson 1983a,b; Duan et al. 1995), allowing computation of simple phase equilibria involving one or two fluid phases.

Figure 14a shows an example of the isothermal, isobaric ternary at 0.9 GPa and 800 °C. A single, miscible fluid phase occurs in the H_2O-rich portion of the system. Phase separation

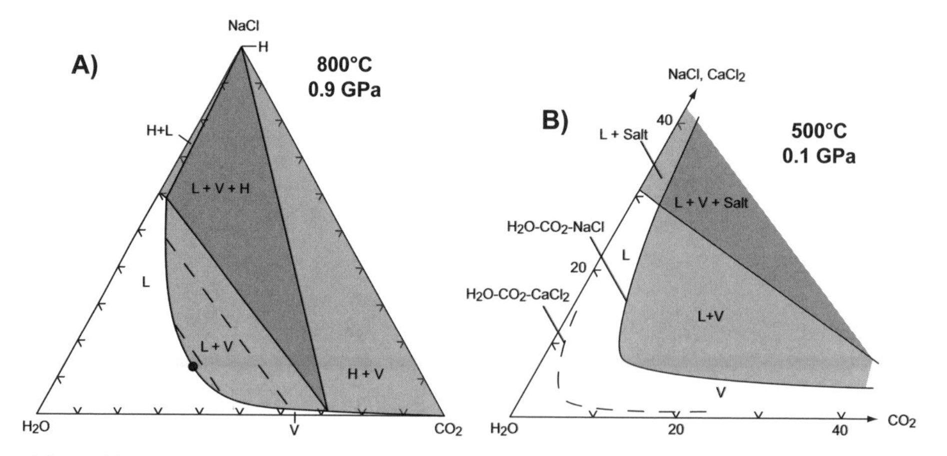

Figure 14. (A) Phase relations in the system H_2O-CO_2-NaCl at 0.9 GPa, 800 °C (Shmulovich and Graham 1999), in mol%. Abbreviations: H, halite, L, liquid, V, vapor. The unshaded region marks stability of a single fluid phase, which varies continuously from liquid like (L) to vapor-like (V). Light shading denotes fields where two phases coexist at equilibrium: liquid-vapor (L + V), halite-liquid (H + L), and halite-vapor (H + V); selected coexisting compositions are indicated by dashed tie lines in the L + V field. Dark shading denotes the liquid-vapor-halite (L + V+ H) field, in which the compositions of all three phases are fixed at the apices. (B) Comparison of relations in the systems H_2O-CO_2-NaCl (solid lines, shading as in A) and H_2O-CO_2-$CaCl_2$ (dashed line) at 0.1 GPa, 500 °C (Liebscher 2010). The two-phase field is larger for $CaCl_2$ than for NaCl.

occurs at < 65 mol% H_2O, yielding an NaCl-rich brine and a CO_2-rich vapor. At these conditions, the brine phase attains halite saturation at ~60 mol% NaCl, whereas halite saturates in the vapor phase at only ~1 mol% NaCl in fluids with > 72 mol% CO_2. As illustrated in Figure 14b, replacement of NaCl by $CaCl_2$ significantly expands the miscibility gap, all else equal.

Variations in the topology of the ternary system are illustrated in Figure 15, which shows that the miscibility gap is expected to persist for a wide range of crustal and mantle conditions, even for low-salinity CO_2-H_2O fluids. Newton and Manning (2002) hypothesized that, in the presence of calcite, H_2O-NaCl fluids may coexist with a hydrous, saline $CaCO_3$ liquid at 800-1000 °C, 1.0 GPa (Fig. 16). In general, extensive immiscibility in the NaCl-CO_2-H_2O ternary system has fundamental importance for carbon transport in the crust and mantle. It requires that, in relatively H_2O-poor systems, CO_2 is largely partitioned into a low-density vapor phase that may move separately from a dense, saline brine (Touret 1985; Skippen and Trommsdorff 1986; Trommsdorff and Skippen 1986; Newton et al. 1998; Heinrich et al. 2004).

Shmulovich et al. (2006) reported experimental data and a model for quartz solubility in CO_2-H_2O fluids containing different salts. However, there is currently insufficient data to constrain models that include the properties of ternary fluids in this system and ionic species. Though preliminary models have been constructed (e.g., Duan et al. 1995), more accurate and complete data and models are needed for treating complex reactive flow problems in crustal and mantle settings.

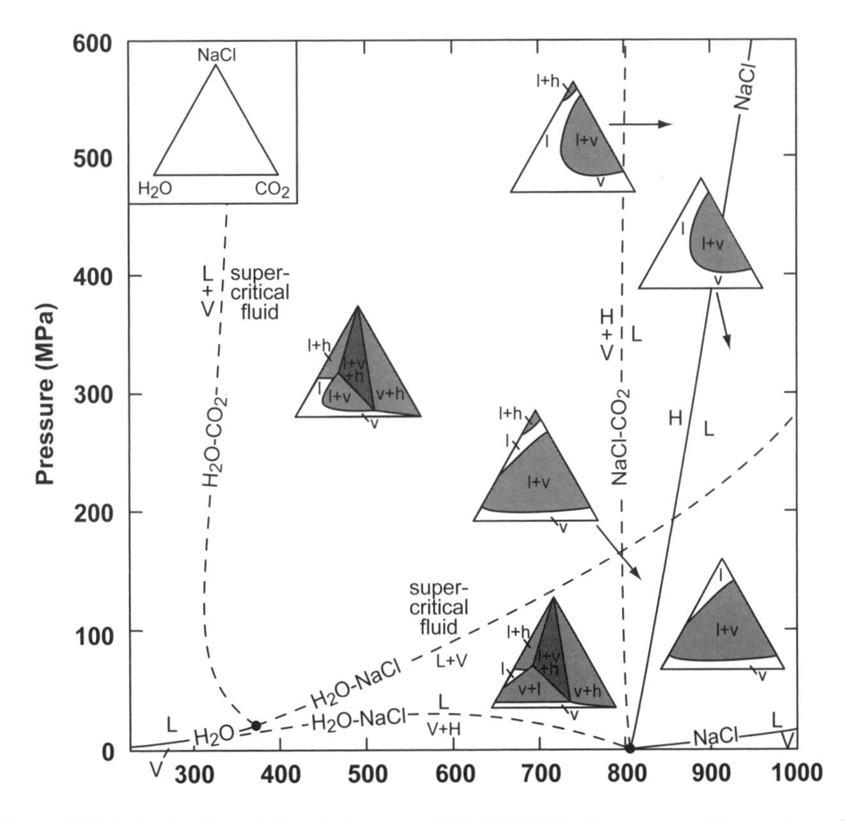

Figure 15. Variation in phase relations in the system H_2O-CO_2-NaCl with pressure and temperature, after Heinrich et al. (2004). Inset shows apices of the ternary compositional space. Abbreviations: L, liquid; V, vapor; H, halite. Shading as in Figure 14. Tie-lines omitted for clarity.

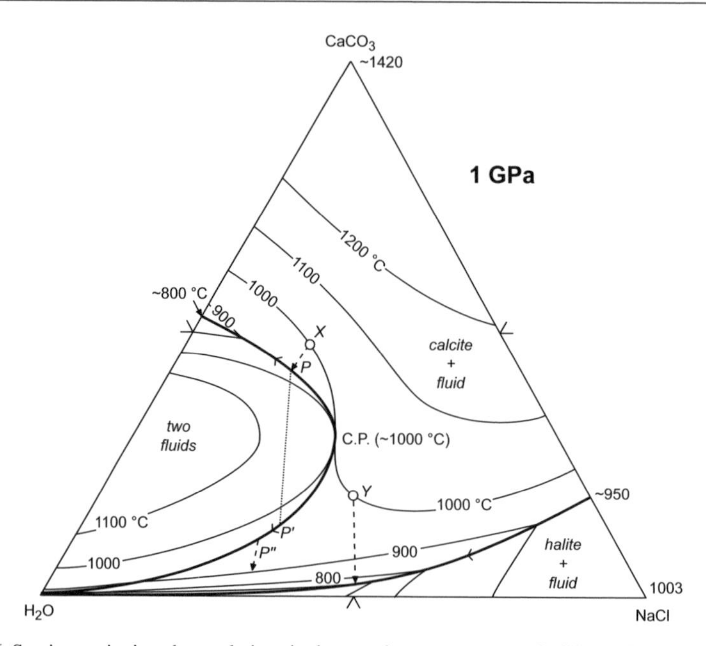

Figure 16. Semi-quantitative phase relations in the pseudo-ternary system $CaCO_3$-NaCl-H_2O at 1 GPa (in mol%, after Newton and Manning 2002a). Newton and Manning hypothesized critical mixing between a hydrous, NaCl-bearing carbonate liquid and a $CaCO_3$-rich brine at ~1000 °C and 30 mol% NaCl, 25 mol% $CaCO_3$ (CP). A hydrous carbonate liquid in equilibrium with calcite at 1000 °C (composition *X*) will, upon cooling to 950 °C (point *P*), exsolve a concentrated salt solution (point *P′*). Both fluids will crystallize calcite upon further cooling over a narrow temperature interval. At *P″*, only the $CaCO_3$-rich brine phase remains, and deposits most of solute $CaCO_3$ in cooling only 200 °C further. A calcite-saturated fluid that contains more NaCl (composition *Y*), but cooling from the same starting temperature of 1000 °C, will avoid the two-fluid region and deposit nearly all of its substantial dissolved carbonate in isobaric cooling past 800 °C.

It is generally assumed that non-polar molecules such as CO_2 are poor solvents. However, it has been hypothesized that CO_2-rich fluids are responsible for element metasomatism in several geologic contexts. For example, petrologic observations of some granulites require a fluid phase with low H_2O activity that is capable of dissolving and redistributing important major and trace elements. Newton et al. (1980) argued that a silicate melt could not explain observations, and proposed a CO_2-rich fluid instead. Although supported by evidence from CO_2-rich fluid inclusions, petrologic considerations were problematic (Lamb and Valley 1987; Yardley and Valley 1997). The requirement that such a fluid was responsible for metasomatic transfer of presumed low-solubility elements such as alkalis and Th poses further challenges. Similar proposals have been made for REE and other element metasomatism in mantle xenoliths (Berkesi et al. 2012).

The immiscibility between brine and vapor in the H_2O-CO_2-NaCl system even at very high *P* and *T* potentially solves this conundrum (Touret 1985; Newton et al. 1998; Newton and Manning 2010). In this hypothesis, metasomatism takes place in the presence of a two-phase fluid. The CO_2-rich vapor phase has wetting properties that lead to its selective entrapment as fluid inclusions, while the brine phase is a powerful solvent responsible for observed metasomatism (Gibert et al. 1998). Both phases contain CO_2, but partitioning between the phases differs depending on *P*, *T*, and composition. Such two-phase systems may be more important in crustal metamorphism than has previously been appreciated.

Mineral solubility and solute structure in CO_2-H_2O fluids. The presence of CO_2 strongly influences mineral solubility and material transport by crustal and mantle aqueous fluids

(Walther and Helgeson 1980). Because CO_2 is non-polar (Fig. 1), its addition to H_2O lowers the solvent dielectric constant at constant P and T (Walther and Schott 1988; Walther 1992), indicating that dilution of H_2O by CO_2 reduces the number of H_2O molecules that solvate solutes dissolved in the fluid.

The disruption of ion hydration by CO_2 in high P-T fluids was studied by Evans et al. (2009) using EXAFS on RbBr-H_2O-CO_2 fluids to 579 °C and 0.26 GPa. The fluids were trapped as synthetic fluid inclusions in corundum. In CO_2-free solutions, the number of nearest neighbor oxygen atoms (in hydrating H_2O molecules) decreased from 6 ± 0.6 to 1.4 ± 0.1 as T increased from 20 to 534 °C, and P to ~0.3 GPa. At $X_{CO_2} = 0.08$, Evans et al. (2009) infer decreases of 16 and 22% in the number of nearest-neighbor oxygen atoms at 312 and 445 °C, respectively. This decrease suggests that CO_2 addition leads to a reduced extent of ion hydration.

In their study of calcite solubility in H_2O-CO_2 fluids, Fein and Walther (1987) found that at constant pressure and temperature, calcite solubility initially increases with increasing X_{CO_2} (Fig. 8). It reaches a maximum at $X_{CO_2} = 0.025$ to 0.05 and then decreases. Taking the calcite dissolution reaction to be

$$\underset{calcite}{CaCO_3} + 2H^+ = Ca^{+2} + CO_{2,aq} + H_2O \tag{15}$$

they inferred that the initial isothermal, isobaric solubility increase is due to reaction (2) progressing to the right. Although H_2O activity declines with increasing CO_2, the rise in CO_2 evidently counters the drop in H_2O activity (Eqn. 9). Fein and Walther (1987) suggested that the decline in calcite solubility is due to diminished extent of hydration of Ca^{+2} in solution.

Similar effects lead to a decrease in quartz solubility with increasing X_{CO_2}. Solute silica does not form complexes with CO_2, so changes in H_2O activity are chiefly responsible for solubility variations with X_{CO_2} at a given P and T. Walther and Orville (1983) suggested that H_2O solvation of dissolved silica could be assessed by writing the quartz dissolution reaction as:

$$\underset{quartz}{SiO_2} + nH_2O = \underset{solute\ complex}{Si(OH)_4 \cdot (n-2)H_2O} \tag{16}$$

The equation includes two moles of H_2O as hydroxyl and n–2 moles of molecular, solvation H_2O per mole of solute silicon. At constant P and T, the number of H_2O of solvation can be determined from quartz solubility data by the relation $n = d\log a_{Si(OH)_4 \cdot (n-2)H_2O} \,/\, d\log a_{H_2O}$. Early work lacked sufficient precision to derive the silica hydration state. Walther and Orville (1983) concluded that, assuming all silica dissolved as monomeric $Si(OH)_4$, $n \approx 4$ (i.e., two molecular H_2O of solvation) for many conditions, though uncertainties in n were at least ±1. Later studies yielded n of 3.5 or 2 (Shmulovich et al. 2006; Akinfiev and Diamond 2009), or suggested that n decreases with increasing X_{CO_2} at constant P and T (Newton and Manning 2000; Shmulovich et al. 2001).

Improved accuracy of this approach has been attained by taking advantage of new and more precise equations of state for H_2O-CO_2 fluids (e.g., Aranovich and Newton 1999) and by accounting explicitly for aqueous silica polymerization (e.g., Zotov and Keppler 2000, 2002; Newton and Manning 2002b, 2003). Newton and Manning (2009) determined quartz solubility in H_2O-CO_2 fluids at 800 °C and 1.0 GPa. Using experimentally constrained models of H_2O activity and mixing of silica monomers and dimers, they determined that n = 4.0 for their experiments and all previous high-quality data at different P and T. They also obtained n = 7.0 for silica dimers. These results indicate that, regardless of silica species, there are two solvating H_2O molecules per Si in H_2O-CO_2 fluids for a large range of crustal metasomatic processes. This result is somewhat surprising because it implies that the hydration state of aqueous silica does not change over a wide range of X_{CO_2}. Evidently, within the limits of X_{CO_2} studied so far, the decline in silica solubility with increasing CO_2 concentration means that there is always sufficient molecular H_2O to supply the two H_2O molecules associated with each dissolved Si. The simple model for silica solubility in mixed H_2O-CO_2 fluids aids study of crustal mass transfer (e.g., Ferry et al. 2011).

REDUCED CARBON IN AQUEOUS FLUIDS AT HIGH *P* AND *T*

CH_4 and CO solubility in H_2O

Gas solubilities in H_2O generally decrease with rising temperature, but this is only true up to a certain temperature along the vapor-liquid saturation (boiling) curve for H_2O. Solubility minima vary among gases (Shock et al. 1989; Shock and McKinnon 1993; Plyasunov and Shock 2001). In the case of methane, the solubility minimum occurs at about 100 °C as shown in Figure 17. It can also be seen in Figure 17 that there is increasing experimental disagreement in the solubility of methane with increasing temperature, and the scatter seems to maximize at higher temperatures. The curve in Figure 17 is calculated with thermodynamic data and revised HKF equation-of-state parameters from Plyasunov and Shock (2001), who also used experimental high-temperature enthalpy of solution data and partial molal heat capacity data for $CH_{4,aq}$ in their regression procedure. Note that the calculated equilibrium constant for methane dissolution is an order of magnitude less negative at 300 °C than at the solubility minimum at 100 °C, suggesting that significant amounts of methane can be lost from solution in cooling portions of hydrothermal systems.

Carbon monoxide is similar in solubility to CH_4, as indicated by the curves in Figure 18, and both gases are considerably less soluble than CO_2. Note that the minimum log*K* for the CO dissolution reaction is a little lower in temperature than that for CH_4, but that both minima are at considerably lower temperatures than that for CO_2, which occurs at about 175 °C. Although the

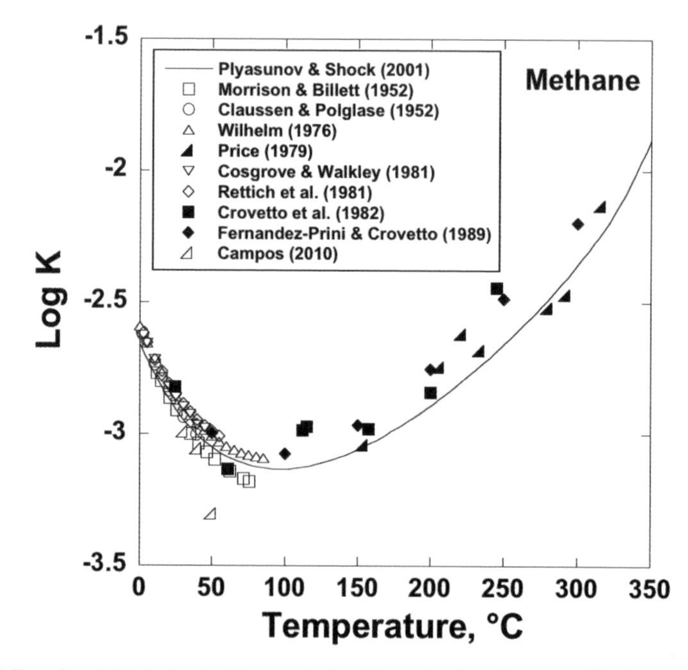

Figure 17. Experimental data (symbols) and calculated (curve) values for the equilibrium constant for the reaction $CH_{4,g} = CH_{4,aq}$ along the H_2O boiling curve.

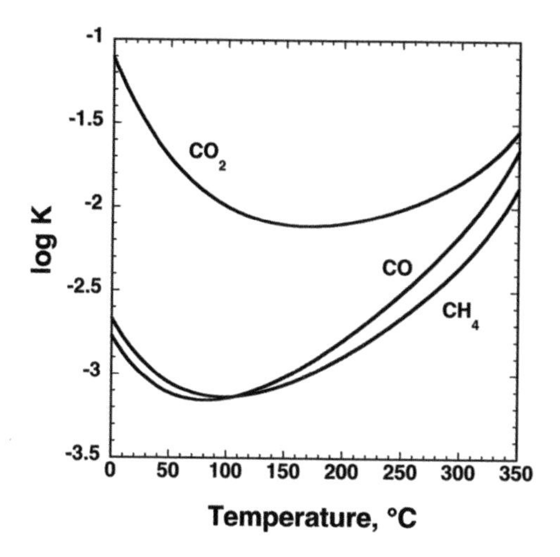

Figure 18. Equilibrium constants for dissolution of CO_2, CO, and CH_4 in water along the H_2O boiling curve. The curve for CH_4 is the same as that shown in Figure 17, and the curves for CO_2 and CO are similarly constrained by experimental data (Shock et al. 1989; Shock and McKinnon 1993). At similar fugacities, CO and CH_4 would exhibit similar solubilities, which would be lower than the corresponding solubility of CO_2.

curves differ by about 1.5 log units at 0 °C, they tend to converge with increasing temperature and are within 0.7 log units of one another at 350 °C. As a result, there are highly variable consequences for exsolution of gases during cooling of hydrothermal fluids.

As with CO_2, CH_4 and CO become fully miscible with H_2O at T greater than about 350 °C and modest pressure (Fig. 13). Thus, in most crustal and mantle contexts, binary fluids are stable as a single phase for all compositions. Numerous equations of state based on experiment and theory have been derived to describe the behavior of these fluids. Gottschalk (2007) provides a recent review of the topic.

Further similarities between CH_4-H_2O and CO_2-H_2O (and likely CO-H_2O as well) can be found in the role played by the addition of NaCl. As in the CO_2-H_2O system discussed above, addition of NaCl leads to unmixing into two fluid phases, a H_2O-rich brine, and a CH_4-rich vapor (Fig. 19). Generally, it can be expected that the similarity in CH_4-H_2O and CO_2-H_2O mixing properties should lead to very similar geochemical behavior of the molecular carbon species in aqueous solutions. For example, it can be anticipated that the effect of CH_4 on quartz solubility is similar to that of CO_2 as described by Newton and Manning (2009), all else equal.

Kinetic inhibition of CH_4 formation

In many crustal fluids, coexisting CO_2 and CH_4 are not in equilibrium with each other (e.g., Janecky and Seyfried 1986; Shock 1988, 1990; Charlou et al. 1998, 2000; McCollom and Seewald 2001). Disequilibrium is most pronounced at temperatures below about 500 °C (Shock 1990, 1992), which are relevant to low-grade metamorphism, hydrothermal systems, sedimentary basins, and subduction zones. The underlying reasons for this disequilibrium state lie in the difficulty of breaking bonds and transferring the eight electrons required for CO_2 and CH_4 to react reversibly. Consequently, stable equilibrium between CO_2 and CH_4 is in many cases attained only at high P and T. In many hydrothermal systems, and throughout sedimentary basins, CH_4 is so slow to form that it persists in concentrations that are far from equilibrium with redox conditions determined by coexisting mineral assemblages.

As shown in Figure 20, the kinetic inhibition of CO_2-CH_4 equilibration produces a "window of opportunity" for metastable persistence of a wide array of aqueous organic compounds at conditions where C-O-H fluids at equilibrium would otherwise consist almost entirely of CO_2, CH_4, H_2O, and H_2. Although organic compounds are less stable than CO_2 or CH_4 depending on the prevailing oxidation-reduction conditions, the sluggish kinetics allow them to persist

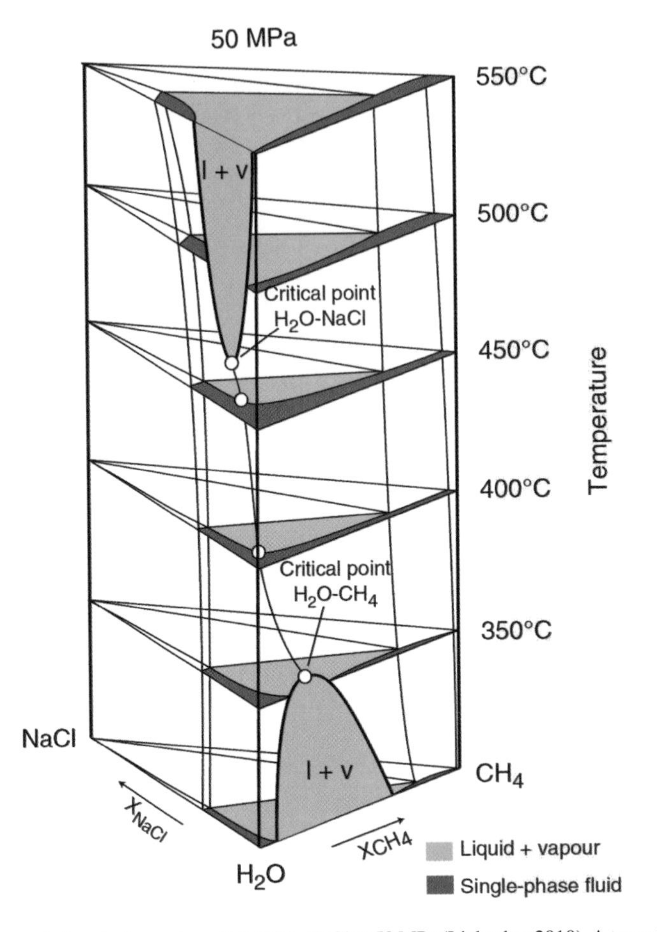

Figure 19. Phase relations in the systems H_2O-CH_4-NaCl at 50 MPa (Liebscher 2010). A two-phase field in which NaCl-rich aqueous brine coexists with CH_4-rich vapor is persists over a wide range of temperature. [Reproduced from Liebscher (2010) *Geofluids*, Vol. 10, p. 3-19, by permission of John Wiley & Sons.]

within this window for millions to billions of years. This metastability is why petroleum, coal, kerogen, bitumen, and other forms of organic matter persevere in sedimentary and low-grade metamorphic rocks, and why organic compounds are often encountered in hydrothermal systems. Slow reaction rates also help explain why there are organic compounds present in carbonaceous meteorites billions of years after their formation through complex processes in the condensing solar nebula and on the parent bodies of the meteorites. In fact, metastable organic compounds are relatively abundant throughout the Solar System, especially in the outer reaches populated by icy satellites and comets, which appear to have extensive inventories of organic compounds.

Even though organic compounds persist in metastable states, they continue to react in response to thermodynamic driving forces. In many cases, reactions among organic compounds reach metastable equilibrium states, in which the organic compounds attain ratios consistent with equilibrium constants for reactions written among them. These metastable states were first recognized in data from natural systems (Shock 1988, 1989, 1994; Helgeson et al. 1993), which led to experimental tests first with hydrocarbons (Seewald 1994) and then with organic

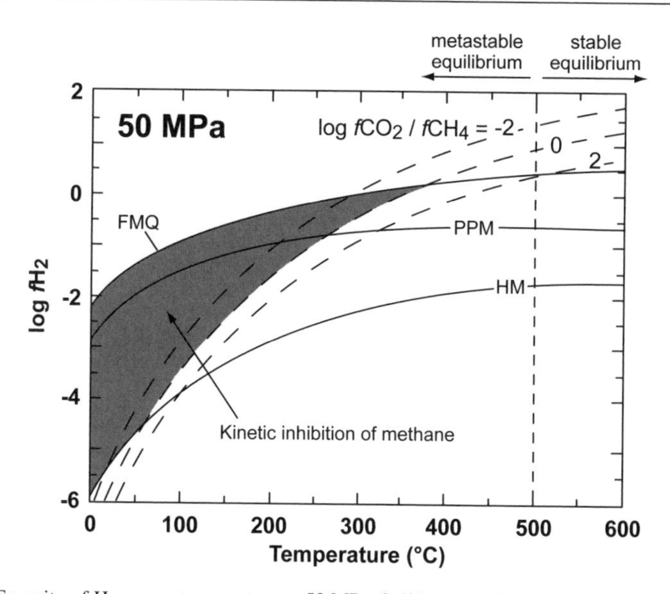

Figure 20. Fugacity of H_2 versus temperature at 50 MPa. Solid curves show values of log f_{H_2} buffered by the mineral equilibria fayalite-quartz-magnetite (FMQ), pyrite-pyrrhotite-magneite (PPM), and hematite-magnetite (HM). Dashed curves correspond to contours of log (f_{CO_2}/f_{CH_4}) equal to 2, 0, and −2. Dotted vertical line at 500 °C demarcates hypothetical boundary between stable equilibrium in the C-H-O system ($T > 500$°C), and kinetic inhibition of CO_2 reduction to CH_4 where metastable equilibrium states between CO_2 and aqueous organic compounds may prevail ($T < 500$ °C). Shaded area corresponds to the region where synthesis of aqueous organic compounds in metastable states may be most easily detected. After Shock (1992).

acids, alcohols, and ketones (Seewald 2001; McCollom and Seewald 2003a,b; see McCollom 2013). Recently it was shown that many conversions between alkanes and alkenes, alkenes and alcohols, and alcohols and ketones are reversible reactions at temperatures and pressures of upper-crustal hydrothermal systems (Yang et al. 2012; Shipp et al. in press). Abundances of organic compounds in sedimentary basins suggest that, in some cases, metastable equilibrium states also include CO_2 (Shock 1988, 1989, 1994; Helgeson et al. 1993), leading to the hypothesis that abiotic organic synthesis in hydrothermal systems proceeds from CO_2 to organic compounds (Shock 1990, 1992; Shock and Schulte 1998; McCollom and Seewald 2007; Shock and Canovas 2010; McCollom 2013).

Reduced carbon and aqueous fluids at high *P* and *T*

Burial metamorphism of organic compounds. Resilient compounds derived from plants (lignin) and microbes (long-chain carboxylic acids) progressively transform on burial. Because of the slow pace of methane formation, organic acids and other dissolved components that might not otherwise persist are added to coexisting fluid (and gas if present). There are many pathways for these transformations, and a host of products. Oxidation of organic carbon and decarboxylation of preexisting carboxyl groups in organic matter yield CO_2. Small organic acids are released during petroleum and coal formation (followed, perhaps, by their decarboxylation). Hydrolytic disproportionation processes transform hydrocarbons. Alkanes, alkenes, alcohols, ketones, aldehydes, and other compounds are generated through aqueous organic transformation reactions. Finally, methane and other small organic compounds are released during the conversion of lignin to coal. While much attention has been focused on these transformations in the context of the origin and evolution of coal and petroleum deposits, the same processes operate on even small concentrations of organic matter that are insufficient to generate large deposits.

Significantly, it is only in the last few decades that H_2O has been recognized as a reactant or product involved in many of these organic transformations. This view challenges the traditional assumption that water is a passive participant in the physical movement of organic compounds in sedimentary basins (Hoering 1984; Shock 1988, 1989, 1994; Helgeson et al. 1993, 2009; Lewan 1997; Lewan and Ruble 2002; Lewan and Roy 2011; Reeves et al. 2012).

Lignin, which makes up plants' structural parts (e.g., wood), has evolved to be resistant to attack. The biosynthesis of lignin by plants starts with three aromatic alcohol building blocks (monolignols) that are derived from phenylalanine (see structures in Fig. 21). Biosynthetic processes in plant cells generate polymers of these alcohols through processes that are slowly being revealed (Davin and Lewis 2005). In the polymerization process, the overall stoichiometry of lignin structures largely reflects that of the monolignols, with only slight modifications. As a consequence, the model lignin molecules shown in Figure 21 plot near the locations of

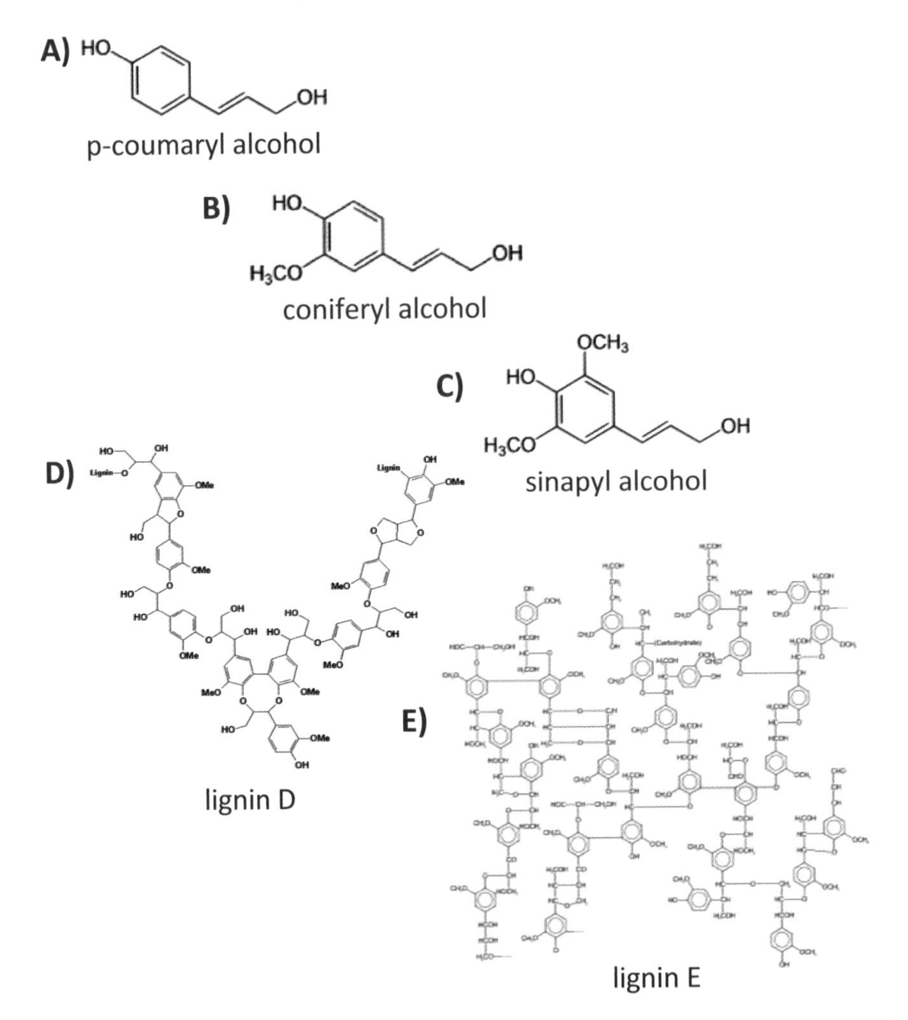

Figure 21. Structures of aromatic alcohols (A-C) that are the building blocks of lignin, and two model lignin structures (D-E). Lignin D has the overall stoichiometry of $C_{90}H_{93}O_{33}$; that of E is $C_{272}H_{290}O_{88}$. Both are plotted in the ternary diagram shown in Figure 22.

the monolignols in the C-H-O ternary diagram shown in Figure 22. Taken together, the open symbols in Figure 22 represent the plant material available for incorporation into sedimentary basins.

In the case of lignin, burial in sedimentary basins leads to the generation of the low-rank coal lignite. Plotting compositions of lignin and lignite coal in Figure 22 reveals that the transformation of lignin to lignite is accompanied by a decrease in the relative abundance of hydrogen. The most direct pathway for hydrogen loss is by the liberation of methane. Note that a vector drawn from the location where methane plots in the diagram connects the compositions of buried plant material with those of lignite 1, the lowest ranking coal composition. Where it occurs, removal of methane from lignin pushes the residual solid composition toward lignite. Reference to Figure 21 shows that numerous methoxy (-OCH_3) groups are present in models for lignin, suggesting that methanol might also be a product of coal formation. The vector extending from methanol shows that lignin would evolve toward the composition of bituminous coal. Any combination of methane and methanol production during lignin transformation drives the residual solid material into the compositional range bounded by the arrowheads of the two solid vectors (Fig. 22). The methane and methanol produced will be dissolved in aqueous fluids that coexist with this organic transformation process, and some of the methane can also exist as natural gas.

Higher coal ranks—bituminous coal and anthracite—plot near the carbon apex of the ternary diagram shown in Figure 22. The transformation of lignite into these higher ranks of coal requires the removal of oxygen, which can occur in several ways, some more likely

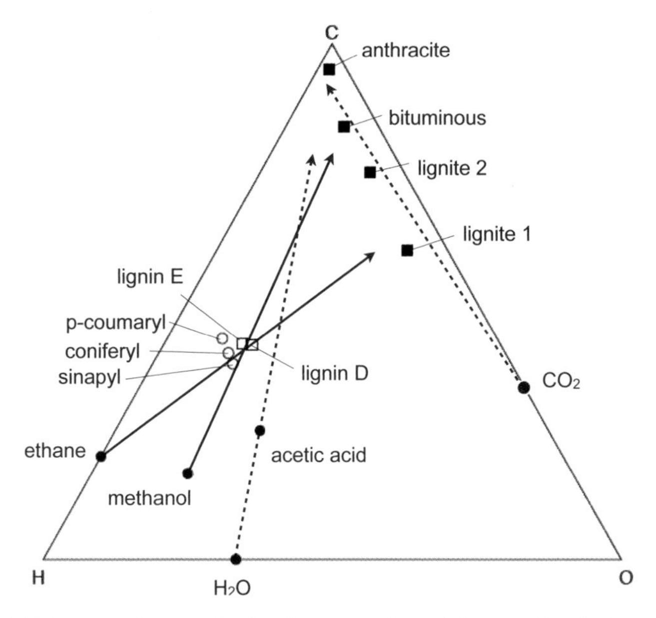

Figure 22. C-H-O ternary diagram showing the compositions of the monolignols from which lignin is synthesized by plants, the model lignin fragments shown in Fig. 21, and the composition of various coal compositions. Solid arrows show how loss of methane would drive the trajectories of the remaining solid organic matter as its composition transforms toward that of lignite coal. Increasing rank of coal from lignite to bituminous to anthracite requires loss of O and H, which can be accomplished by dehydration, or by release of acetic acid as indicated by the dashed arrow. Combinations of these vectors outline the transition of lignin to coal in sedimentary basins, and require loss of methane and water or acetic acid, or perhaps other small organic solutes.

than others. One is the release of O_2; however, this process seems unlikely in highly reduced sedimentary basins, because if O_2 were released it would likely react immediately with reduced constituents in the system (e.g., reduced forms of S, Fe, N, or C). Effective ways to remove oxygen during elevation of coal rank are the production of CO_2 and/or H_2O, shown by dashed vectors in Figure 22. In the case of dehydration, numerous hydroxyl groups are candidates for this process (Fig. 21); however, acetic acid lies on this same vector, making it difficult to resolve the effects of dehydration from those resulting from acetic acid production. Fluids associated with some coal deposits are enriched in acetic and other small organic acids (Fisher and Boles 1990). In summary, the transformation reactions leading to coal accompany the alteration of lignin in any sedimentary rock. They lead to the release of methanol, acetic acid, other small organic solutes, and methane to sedimentary basin fluids.

Major resilient constituents of microbial membranes are long-chain carboxylic acids. It is widely thought that during the organic transformations that accompany burial and heating of sediments, these acids undergo decarboxylation. When a carboxyl group is cleaved from the rest of the molecule, CO_2 and long-chain alkanes are generated (Cooper and Bray 1963; Jurg and Eisma 1964; Robinson 1966; Kvenvolden and Weiser 1967; Smith 1967; Shimoyama and Johns 1971, 1972; Philippi 1974; Snape et al. 1981; Kissin 1987; Hunt et al. 2002). Tie lines in Figure 23 show how several carboxylic acids (indicated by carbon numbers) can decarboxylate into CO_2 and an alkane. As examples, acetic acid (2 in Fig. 23) produces CO_2 and methane via

$$CH_3COOH \rightarrow CH_4 + CO_2 \tag{17}$$

and hexadecanoic acid (16 in Fig. 23) would yield CO_2 and pentadecane via

$$C_{15}H_{31}COOH \rightarrow C_{15}H_{32} + CO_2 \tag{18}$$

The products of decarboxylation reactions lie along the C-O and C-H binaries of the ternary diagram in Figure 23. This geometry emphasizes that decarboxylation is a disproportionation reaction, because in each case the products are one compound with bulk carbon oxidation state that is more oxidized (CO_2) and one with carbon that is more reduced (an alkane) than in the reacting acid.

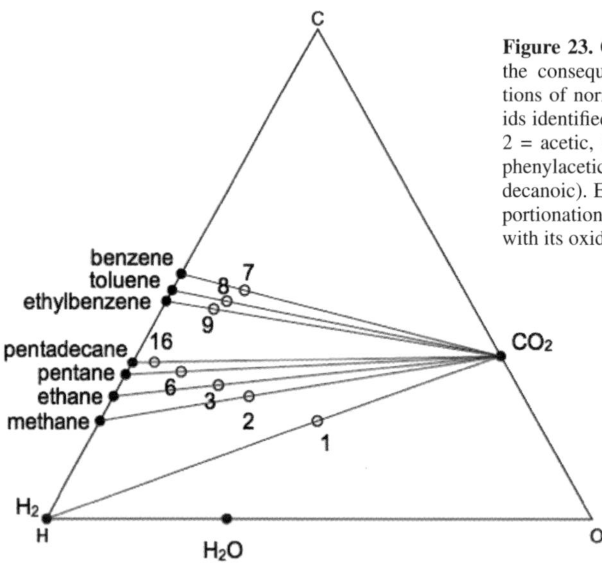

Figure 23. C-H-O ternary diagram showing the consequences of decarboxylation reactions of normal and aromatic carboxylic acids identified by carbon number (1 = formic, 2 = acetic, 6 = hexanoic, 7 = benzoic, 8 = phenylacetic, 9 = hydrocinnamic, 16 = hexadecanoic). Each decarboxylation is a disproportionation, and tie-lines connect each acid with its oxidized and reduced products.

The production of hydrocarbons and CO_2 by acid decarboxylation is advantageous to the geochemist, because organic acids are far more soluble in water than are hydrocarbons, and it is easier to model the chemistry and transport of aqueous acids as hydrocarbon precursors than it is to model the hydrocarbons themselves. Recognition of this situation led researchers to investigate the rates of carboxylic acid decarboxylation. Initial results showed highly variable rates depending on the composition of the experimental container used (Kharaka et al. 1983; Palmer and Drummond 1986). Palmer and Drummond (1987) determined that the least catalytic surface to use is gold, and that has been the standard for subsequent work. Early studies focused on acetic acid, which is among the most abundant organic acids found in oilfield brines and other deep fluids; rates were determined by quantifying its decreasing concentration (Kharaka et al. 1983). Subsequently, researchers attempted to show mass balance in their experiments, with the hope of demonstrating equal concentrations of methane and CO_2 generated by the decarboxylation reaction (Eqn. 17). In some cases it could be argued that roughly similar abundances of methane and CO_2 were produced, giving some confidence that decarboxylation rates could be established (Palmer and Drummond 1986; Bell and Palmer 1994; Bell et al. 1994). However, the composition of the experimental solution, the presence of minerals, and the presence or absence of a gas headspace in the experiments also allowed processes other than decarboxylation to occur. These variables were explored in detail by McCollom and Seewald (2003a,b), who used mineral assemblages to control the oxidation state of their experiments, and maintained careful control of mass balance.

When put to the test, the elegant simplicity of decarboxylation runs into difficulties in explaining the fate of organic acids in geologic fluids. As revealed by analysis of natural samples and laboratory experiments, the transformations undergone by aqueous organic compounds are more complicated. McCollom and Seewald (2003b) showed, for example, that the decarboxylation reaction could explain the fate of acetic acid in the presence of the assemblage pyrite + pyrrhotite + magnetite at 325 °C and 35 MPa, but that at the same conditions, the mineral assemblages hematite + magnetite or hematite + magnetite + pyrite drove the oxidation of carbon in acetic acid, probably via the overall reaction

$$CH_3COOH + 2H_2O \rightarrow 2CO_2 + 4H_2 \tag{19}$$

McCollom and Seewald (2003b) also showed that, at the same conditions and in the presence of the same mineral assemblages, valeric acid (C_4H_9COOH) transformed by a variety of reaction pathways. One such pathway was the degradation of valeric acid to formic acid and butene (C_4H_8) in the presence of hematite + magnetite, implying an overall reaction such as

$$C_4H_9COOH \rightarrow C_4H_8 + HCOOH \tag{20}$$

At the high P and T of these experiments, formic acid rapidly converts to CO_2 and H_2 (McCollom and Seewald 2001, 2003a; Seewald et al. 2006) via

$$HCOOH \rightarrow CO_2 + H_2 \tag{21}$$

At more reduced conditions consistent with the pyrite + pyrrhotite + magnetite assemblage, McCollom and Seewald (2003b) showed that butene would be rapidly reduced to butane (C_4H_{10}), via

$$C_4H_8 + H_2 \rightarrow C_4H_{10} \tag{22}$$

which, together with the conversion of formic acid to CO_2 (Eqn. 21), would produce the misleading appearance of direct decarboxylation.

As indicated by these reactions and reference to Figure 23, the generation of formic acid is not colinear with a carboxylic acid reactant and the alkane product containing one less carbon. As an example, a line connecting pentane (C_5H_{12}) and formic acid (HCOOH) would not pass through hexanoic acid ($C_5H_{11}COOH$; 6 in Fig. 23). Instead, hexanoic acid plots above this line, implying that the addition of H_2 is required in the overall reaction, which is indeed the case as given by

$$C_5H_{11}COOH + H_2 \rightarrow C_5H_{12} + HCOOH \quad (23)$$

If formic acid breakdown followed Equation (21), then H_2 would be consumed and produced in the overall process, again giving the misleading impression of direct decarboxylation.

In principle, during transformation of membrane biomolecules it would also be possible for other C-C bonds to be broken, allowing the formation of acetic, propanoic or other small carboxylic acids together with correspondingly shorter alkanes. The extent of decarboxylation, deformylation, and other reactions could then determine the fates of these acids, which tend to accumulate in oil-field brines and other aqueous fluids co-produced with petroleum (Shock 1988, 1989, 1994; Helgeson et al. 1993; Seewald 2001, 2003). The mechanisms of decarboxylation reactions at the molecular level are currently under investigation (Glein and Shock, unpublished data).

Reduced carbon in aqueous fluids at greater depths. Graphite is the dominant crystalline form of reduced carbon at conditions of the middle and lower crust and upper mantle. It may be produced by a variety of pathways, such as metamorphism of organic matter, reduction of carbonate minerals, partial melting, or infiltration of externally derived carbon-bearing fluids (Nokleberg 1973; Andreae 1974; Perry and Ahmad 1977; Rumble et al. 1977; Wada et al. 1994; Luque et al. 1998). The composition of fluids coexisting with graphite-bearing rocks varies strongly with *P*, *T*, fluid composition, and oxygen fugacity (French 1966; Ohmoto and Kerrick 1977; Huizenga 2001, 2011; Huizenga and Touret 2012). A simple way to explore variations in graphite solubility is to track variations in the graphite saturation surface in the C-O-H ternary system (Fig. 24). Each ternary diagram in Figure 24 is constructed at a fixed *P* and *T*, but f_{O_2} varies with composition within the plot. The carbon content in a graphite-saturated fluid

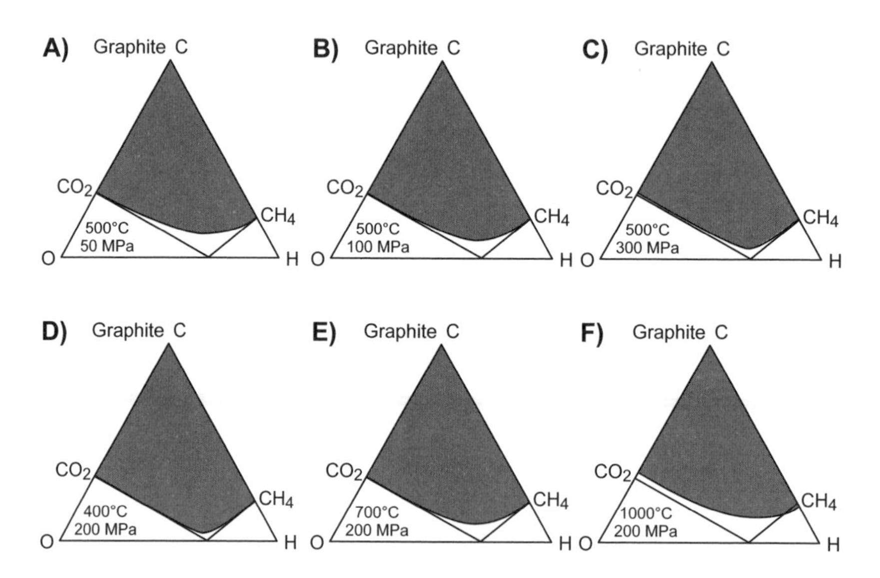

Figure 24. Predicted solubility of graphite in C-O-H fluids at a range of conditions of crustal metamorphism. Shaded region in each ternary (A-F) denotes the graphite + fluid stability field. Graphite solubility varies with fluid composition along the lower, curved boundary of the two-phase field and can be surprisingly high even in H_2O-rich fluids. Assumes ideal mixing of molecular species; after Spear (1993) and Ferry and Baumgartner (1987).

corresponds to the curve passing from the C-O binary to the C-H binary. Figure 24 illustrates that graphite-saturated fluid composition is variable at fixed *P* and *T*: graphite solubility is lower in CH_4-rich fluid than in CO_2-rich fluid, and minimum graphite solubilities occur in initially pure H_2O. Comparison of the ternaries also confirms that graphite solubility varies with *P* and *T*. An important conclusion to be drawn from the ternary diagrams is that, at many conditions, the solubility of graphite, nominally a refractory phase, can be surprisingly high.

In general, devolatilization of graphite-bearing rocks will produce dissolved, reduced carbon at a concentration limited by the solubility of graphite at the ambient f_{O_2}. The relative abundance of carbon and its distribution among dominant carbon-bearing molecular components such as CH_4, CO_2, and COS species, will vary strongly with *P* and *T*. For example, in the presence of graphite at relatively reducing conditions, a C-O-H-S fluid coexisting with the model mineral assemblage biotite + K-feldspar + pyrite + pyrrhotite will have total carbon concentration and species abundances that are strong functions of *T* at constant *P* (Fig. 25; Ferry and Baumgartner, 1987). Carbon concentrations are higher at oxidizing conditions, where virtually all carbon is present as CO_2.

Where reducing conditions prevail in the lower crust and upper mantle, the predominant dissolved form of carbon is generally CH_4. This conclusion is supported by fluid inclusions containing CH_4 from regional metamorphic terranes (e.g., Sisson and Hollister 1990; Huff and Nabelek 2007). Interestingly, CH_4 appears to be quite common as a solute in subduction-zone settings (Zheng et al. 2000, 2003; Fu et al. 2001, 2002, 2003a,b; Shi et al. 2005).

Numerous studies have shown that CH_4 (and hydrocarbons) can form at high pressures and temperatures (Kenney et al. 2002; Kutcherov et al. 2002; Scott et al. 2004; Kolesnikov et al. 2009; Sharma et al. 2009; Marocchi et al. 2011). But in many settings where CH_4 is recorded as a component of metamorphic fluids, the temperature of entrapment does not exceed 500 °C, which raises the question of why it has formed despite the kinetic inhibition of methane formation at these temperatures. Rates of abiogenic methanogenesis may be increased by catalysis involving mineral surfaces in the host rock. Much geochemical research on heterogeneous catalysis has been motivated by observations that abundant reduced gases

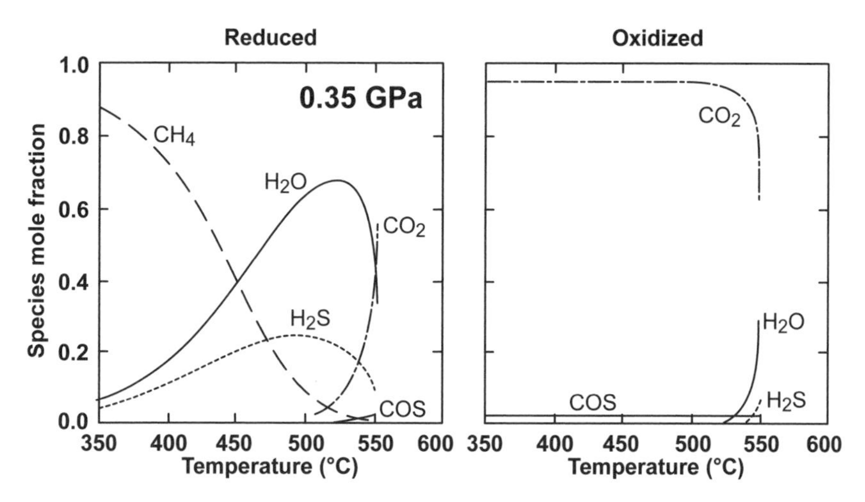

Figure 25. Fluid composition and species abundances in equilibrium with biotite + K-feldspar + pyrite + pyrrhottite + graphite, as a function of temperature at 0.35 GPa. Fluid composition and speciation varies strongly from relatively reducing (left) to oxidizing (right) conditions. Assumes ideal mixing of molecular species; after Spear (1993) and Ferry and Baumgartner (1987).

are produced during serpentinization of olivine-rich ultramafic rocks, with metal oxides, Fe-Ni alloys, and Fe-Ni sulfides as byproducts (Schrenk et al. 2013). A range of experimental studies has demonstrated the effectiveness of these catalytic effects on accelerating the rate of alkanogenesis (Horita and Berndt 1999; Foustoukos and Seyfried 2004; Fu et al. 2008).

CONCLUDING REMARKS

Our understanding of the chemistry of aqueous carbon is advanced at conditions relevant for the shallow geologic environments; however, much remains to be done to extend this framework to the higher-pressure systems relevant to Earth's deep carbon cycle. Most progress has been made in development and application of equations of state for molecular fluids. Unfortunately, this simple framework is inadequate for treating many problems of mass transfer, where other species (ions, metal carbonates, organic acids, etc.) must be taken into account. The problem is compounded by limits in the application of HKF theory to $\leq$ 0.5 GPa, and relatively few experimental investigations at the requisite high pressures. This situation is especially true for reduced carbon species: even if such species are metastable with respect to CH_4, the kinetic inhibition of CH_4 formation means that they may be important in many deep systems of interest.

Nevertheless, recent advances hold promise for progress in this field. Developments of hydrothermal piston-cylinder and hydrothermal diamond-anvil cell approaches to mineral solubility and fluid characterization now allow for robust, routine experimental work at high P and T. In addition, recognition of the comparative simplicity of H_2O behavior at high P and T has led to exploitation of simple correlations of mineral solubility and homogeneous equilibria with H_2O density. These advances are opening for the first time the realm of deep fluid flow to robust aqueous geochemical methods. The combined experimental and theoretical avenues thus promise new insights into the terrestrial deep carbon cycle in the coming years.

ACKNOWLEDGMENTS

We thank John Brodholt, Robert Newton and Codi Lazar for thorough reviews. Pam Hill and Edwin Schauble are thanked for jmol assistance. This work was supported by grants from the National Science Foundation (EAR-1049901 to CEM), the University of California Lab Research Program (CEM), the Department of Energy (DOE Grant DE-FG02-96ER-14616 to DAS), and the Deep Carbon Observatory.

REFERENCES

Adamczyk K, Prémont-Schwarz M, Pines D, Pines E, Nibbering ETJ (2009) Real-time observation of carbonic acid formation in aqueous solution. Science 326:1690-1694

Amend JP, Helgeson HC (1997) Group additivity equations of state for calculating the standard molal thermodynamic properties of aqueous organic species at elevated temperatures and pressures. Geochim Cosmochim Acta 61:11-46

Anderson AJ, Clark AH, Gray S (2001) The occurrence and origin of zabuyelite (Li_2CO_3) in spodumene-hosted fluid inclusions: implications for the internal evolution of rare-element granitic pegmatites. Can Mineral 39:1513-1527

Anderson GM, Crerar DA (1993) Thermodynamics in Geochemistry: The Equilibrium Model. Oxford University Press, Oxford

Andreae MO (1974) Chemical and stable isotope composition of high-grade metamorphic rocks from Arendal-area, southern Norway. Contrib Mineral Petrol 47:299-316

Anovitz LM, Labotka TC, Blencoe JG, Horita J (2004) Experimental determination of the activity-composition relations and phase equilibria of H_2O-CO_2-NaCl fluids at 500 °C, 500 bars. Geochim Cosmochim Acta 68:3557-3567

Aranovich LY, Newton RC (1996) H_2O activity in concentrated NaCl solutions at high pressures and temperatures measured by the brucite-periclase equilibrium. Contrib Mineral Petrol 125:200-212

Aranovich LY, Newton RC (1999) Experimental determination of CO_2-H_2O activity-composition relations at 600-1000 °C and 6-14 kbar by reversed decarbonation and dehydration reactions. Am Mineral 84:1319-1332

Aranovich LY, Zakirov IV, Sretenskaya NG, Gerya TV (2010) Ternary system H_2O-CO_2-NaCl at high T-P parameters: An empirical mixing model. Geochem Int 48:446-455

Audétat A, Pettke T, Dolejs D (2004) Magmatic anhydrite and calcite in the ore-forming quartz-monzodiorite magma at Santa Rita, New Mexico (USA): genetic constraints on porphyry-Cu mineralization. Lithos 72:147-161

Baratov RB, Gnutenko NA, Kuzemko VN (1984) Regional carbonization connected with the epi-Hercynian tectogenesis in the southern Tien Shan. Dokl Akad Nauk 274:124-126

Barrenechea JF, Luque FJ, Milward D, Ortega L, Beyssac O, Rodas M (2009) Graphite morphologies from the Borrowdale deposit (NW England, UK): Raman and SIMS data. Contrib Mineral Petrol 158:37-51

Becker JA, Bickle MJ, Galy A, Holland TJB (2008) Himalayan metamorphic CO_2 fluxes: Quantitative constraints from hydrothermal springs. Earth Planet Sci Lett 265:616-629

Bell JLS, Palmer DA (1994) Experimental studies of organic acid decomposition. *In*: Organic Acids in Geological Processes Pittman ED, Lewan MD (eds) Springer-Verlag, Berlin, p 227-269

Bell JLS, Palmer DA, Barnes HL, Drummond SE (1994) Thermal decomposition of acetate: III. Catalysis by mineral surfaces. Geochim Cosmochim Acta 58:4155-4177

Bergman SC, Dubessy J (1984) CO_2-CO fluid inclusions in a composite peridotite xenolith - implications for upper mantle oxygen fugacity. Contrib Mineral Petrol 85:1-13

Berkesi M, Guzmics T, Szabo C, Dubessy J, Bodnar RJ, Hidas K, Ratter K (2012) The role of CO_2-rich fluids in trace element transport and metasomatism in the lithospheric mantle beneath the Central Pannonian Basin, Hungary, based on fluid inclusions in mantle xenoliths. Earth Planet Sci Lett 331-332:8-20

Berman RG (1988) Internally-consistent thermodynamic data for minerals in the system Na_2O-K_2O-CaO-MgO-FeO-Fe_2O_3-Al_2O_3-SiO_2-TiO_2-H_2O-CO_2. J Petrol 29:445-522

Berner RA (1991) A model for atmospheric CO_2 over Phanerozoic time. Am J Sci 291:339-376

Berner RA (1994) GEOCARB II: a revised model of atmospheric CO_2 over Phanerozoic time. Am J Sci 294:56-91

Berner RA, Kothavala Z (2001) GEOCARB III: a revised model of atmospheric CO_2 over Phanerozoic time. Am J Sci 301:182-204

Berner RA, Lasaga AC, Garrels RM (1983) The carbonate-silicate geochemical cycle and its effect on atmospheric carbon dioxide over the past 100 million years. Am J Sci 283:641-683

Blundy J, Cashman KV, Rust A, Witham F (2010) A case for CO_2-rich arc magmas. Earth Planet Sci Lett 290:289-301

Boettcher AL, Wyllie PJ (1968) The calcite-aragonite transition measured in the system CaO-CO_2-H_2O. J Geol 76:314-330

Bonatti E, Emiliani C, Ferrara G, Honnorez J, Rydell H (1974) Ultramafic-carbonate breccias from equatorial mid-Atlantic ridge. Marine Geol 16:83-102

Bonatti E, Lawrence JR, Hamlyn PR, Breger D (1980) Aragonite from deep-sea ultramafic rocks. Geochim Cosmochim Acta 44:1207-1214

Bowers TS, Helgeson HC (1983a) Calculation of the thermodynamic and geochemical consequences of nonideal mixing in the system H_2O-CO_2-NaCl on phase relations in geologic systems: equation of state for H_2O-CO_2-NaCl fluids at high pressures and temperatures. Geochim Cosmochim Acta 47:1247-1275

Bowers TS, Helgeson HC (1983b) Calculation of the thermodynamic and geochemical consequences of nonideal mixing in the system H_2O-CO_2-NaCl on phase-relations in geologic systems - metamorphic equilibria at high-pressures and temperatures. Am Mineral 68:1059-1075

Brodholt J, Wood B (1993) Molecular-dynamics simulations of the properties of CO_2-H_2O mixtures at high-pressures and temperatures. Am Mineral 78:558-564

Burton MR, Sawyer GM, Granieri D (2013) Deep carbon emissions from volcanoes. Rev Mineral Geochem 75:323-354

Caciagli NC, Manning CE (2003) The solubility of calcite in water at 6-16 kbar and 500-800 °C. Contrib Mineral Petrol 146:275-285

Campos CEPS, Penello JR, Pessoa FLP, Uller AMC (2010) Experimental measurement and thermodynamic modeling for the solubility of methane in water and hexadecane. J Chem Eng Data 55:2576-2580

Charlou JL, Donval JP, Douville E, Jean-Baptiste P, Radford-Knoery J, Fouquet Y, Dapoigny A, Stievenard M (2000) Compared geochemical signatures and the evolution of Menez Gwen (37°50′N) and Lucky Strike (37°17′N) hydrothermal fluids, south of the Azores Triple Junction on the Mid-Atlantic Ridge. Chem Geol 171:49-75

Charlou JL, Fouquet Y, Bougault H, Donval JP, Etoubleau J, Jean-Baptiste P, Dapoigny A, Appriou P, Rona PA (1998) Intense CH_4 plumes generated by serpentinization of ultramafic rocks at the intersection of the 15°20′N fracture zone and the Mid-Atlantic Ridge. Geochim Cosmochim Acta 62:2323-2333

Chiodini G, Frondini F, Kerrick DM, Rogie J, Parello F, Peruzzi L, Zanzari AR (1999) Quantification of deep CO_2 fluxes from Central Italy. Examples of carbon balance for regional aquifers and of soil diffuse degassing. Chem Geol 159:205-222

Chiodini G, Frondini F, Ponziani F (1995) Deep structures and carbon-dioxide degassing in central Italy. Geothermics 24:81-94

Claes P, Thirion B, Glibert J (1996) Solubility of CO_2 in the molten Na_2CO_3-K_2CO_3 (42 mol%) eutectic mixture at 800 °C. Electrochim Acta 41:141-146

Claussen WF, Polglase MF (1952) Solubilities and structures in aqueous aliphatic hydrocarbon solutions. J Am Chem Soc 74:4817-4819

Cooper JE, Bray EE (1963) A postulated role of fatty acids in petroleum formation. Geochim Cosmochim Acta 27:1113-1127

Cosgrove BA, Walkley J (1981) Solubilities of gases in H_2O and ^2H_2O. J Chromatogr 216:161-167

Crovetto R, Fernandez-Prini R, Japas ML (1982) Solubilities of inert gases and methane in H_2O and D_2O in the temperature range of 300 to 600 K. J Chem Phys 76:1077-1086

Dahlgren S, Bogoch R, Magaritz M, Michard A (1993) Hydrothermal dolomite marbles associated with charnockitic magmatism in the Proterozoic Bamble Shear Belt, south Norway. Contrib Mineral Petrol 113:394-408

Dasgupta R (2013) Ingassing, storage, and outgassing of terrestrial carbon through geologic time. Rev Mineral Geochem 75:183-229

Dasgupta R, Hirschmann MM (2006) Melting in the Earth's deep upper mantle caused by carbon dioxide. Nature 440:659-662

Dasgupta R, Hirschmann MM (2010) The deep carbon cycle and melting in Earth's interior. Earth Planet Sci Lett 298:1-13

Davin LB, Lewis NG (2005) Lignin primary structures and dirigent sites. Curr Opin Biotechnol 16:407-415

Davis AR, Oliver BG (1972) A vibrational-spectroscopic study of the species present in CO_2-H_2O system. J Solution Chem 1:329-338

Deines P (2002) The carbon isotope geochemistry of mantle xenoliths. Earth Sci Rev 58:247-278

Demeny A, Dallai L, Frezzotti ML, Vennemann TW, Embey-Isztin A, Dobosi G, Nagy G (2010) Origin of CO_2 and carbonate veins in mantle-derived xenoliths in the Pannonian Basin. Lithos 117:172-182

Destrigneville CM, Brodholt JP, Wood BJ (1996) Monte Carlo simulation of H_2O-CO_2 mixtures to 1073.15 K and 30 kbar. Chem Geol 133:53-65

Dick JM (2008) Calculation of the relative metastabilities of proteins using the CHNOSZ software package. Geochem Trans 9:10, doi: 10.1186/1467-4866-9-10

Dick JM, LaRowe DE, Helgeson HC (2006) Temperature, pressure, and electrochemical constraints on protein speciation: Group additivity calculation of the standard molal thermodynamic properties of ionized unfolded proteins. Biogeosciences 3:311-336

Dolejs D, Manning CE (2010) Thermodynamic model for mineral solubility in aqueous fluids: theory, calibration and application to model fluid-flow systems. Geofluids 10:20-40

Duan ZH, Moller N, Weare JH (1995) Equation of state for the NaCl-H_2O-CO_2 system - prediction of phase-equilibria and volumetric properties. Geochim Cosmochim Acta 59:2869-2882

Duan ZH, Moller N, Weare JH (1996) A general equation of state for supercritical fluid mixtures and molecular dynamics simulation of mixture PVTX properties. Geochim Cosmochim Acta 60:1209-1216

Dunai TJ, Touret JLR (1993) A noble gas study of a granulite sample from the Nilgiri Hills, southern India: implications for granulite formation. Earth Planet Sci Lett 119:271-281

Eggert RG, Kerrick DM (1981) Metamorphic equilibria in the siliceous dolomite system - 6 kbar experimental data and geologic implications. Geochim Cosmochim Acta 45:1039-1049

Einaudi MT, Meinert LD, Newberry RJ (1981) Skarn deposits. *In*: Economic Geology Seventy-Fifth Anniversary Volume. Skinner BJ (ed) The Economic Geology Publishing Company, New Haven, p 317-391

Ellis AJ (1959a) The effect of pressure on the first dissociation constant of "carbonic acid". J Chem Soc 1959:3689-3699

Ellis AJ (1959b) The solubility of calcite in carbon dioxide solutions. Am J Sci 257:354-365

Ellis AJ (1963) The solubility of calcite in sodium chloride solutions at high temperatures. Am J Sci 261:259-267

England AH, Duffin AM, Schwartz CP, Uejio JS, Prendergast D, Saykally RJ (2011) On the hydration and hydrolysis of carbon dioxide. Chem Phys Lett 514:187-195

Evans K, Gordon RA, Mavrogenes JA, Tailby N (2009) The effect of CO_2 on the speciation of RbBr in solution at temperatures of 579 °C and pressures of 0.26 GPa. Geochim Cosmochim Acta 73:2631-2644

Falcke H, Eberle SH (1990) Raman spectroscopic identification of carbonic acid. Water Res 24:685-688

Fein JB, Walther JV (1989) Calcite solubility and speciation in supercritical NaCl-HCl aqueous fluids. Contrib Mineral Petrol 103:317-324

Fein JB, Wlather JV (1987) Calcite solubility in supercritical CO_2-H_2O fluids. Geochim Cosmochim Acta 51:1665-1673

Fernandez DP, Goodwin ARH, Lemmon EW, Levelt Sengers JMH, Williams RC (1997) A formulation for the static permittivity of water and steam at temperatures from 238 K to 873 K at pressures up to 1200 MPa, including derivatives and Debye-Hückel coefficients. J Phys Chem Ref Data 26:1125-1166

Fernandez-Prini R, Crovetto R (1989) Evaluation of data on solubility of simple apolar gases in light and heavy water at high temperature. J Phys Chem Ref Data 18:1231-1243

Ferry JM, Baumgartner L (1987) Thermodynamic models of molecular fluids at the elevated pressures and temperatures of crustal metamorphism. Rev Mineral 17:323-365

Ferry JM, Ushikubo T, Valley JW (2011) Formation of forsterite by silicification of dolomite during contact metamorphism. J Petrol 52:1619-1640

Fisher JB, Boles JR (1990) Water-rock interaction in Tertiary sandstones, San Joaquin basin, California, U.S.A.: Diagenetic controls on water composition. Chem Geol 82:83-101

Foustoukos DI, Seyfried WE (2004) Hydrocarbons in hydrothermal vent fluids: The role of chromium-bearing catalysts. Science 304:1002-1005

Franck EU, Rosenzweig S, Christoforakos M (1990) Calculation of the dielectric constant of water to 1000 °C and very high pressures. Ber Bunsen Ges Phys Chem 94:199-203

Frantz JD (1998) Raman spectra of potassium carbonate and bicarbonate aqueous fluids at elevated temperatures and pressures: comparison with theoretical simulations. Chem Geol 152:211-225

Frantz JD, Popp RK, Hoering TC (1992) The compositional limits of fluid immiscibility in the system H_2O-CO_2-NaCl as determined with the use of synthetic fluid inclusions in conjunction with mass-spectrometry. Chem Geol 98:237-255

Frear GL, Johnston DH (1929) The solubility of calcium carbonate (calcite) in certain aqueous solutions at 25 °C. J Am Chem Soc 51:2082-2092

French BM (1966) Some geological implications of equilibrium between graphite and a C-H-O gas at high temperatures and pressures. Rev Geophys 4:223-253

Frezzotti ML, Selverstone J, Sharp ZD, Compagnoni R (2011) Carbonate dissolution during subduction revealed by diamond-bearing rocks from the Alps. Nature Geosci 4:703-706

Frost DJ, Wood BJ (1997) Experimental measurements of the properties of H_2O-CO_2 mixtures at high pressures and temperatures. Geochim Cosmochim Acta 61:3301-3309

Fu B, Touret JLR, Zheng YF (2001) Fluid inclusions in coesite-bearing eclogites and jadeite quartzite at Shuanghe, Dabie Shan (China). J Metamorph Geol 19:529-545

Fu B, Touret JLR, Zheng YF (2003a) Remnants of premetamorphic fluid and oxygen isotopic signatures in eclogites and garnet clinopyroxenite from the Dabie-Sulu terranes, eastern China. J Metamorph Geol 21:561-578

Fu B, Touret JLR, Zheng YF, Jahn BM (2003b) Fluid inclusions in granulites, granulitized eclogites and garnet clinopyroxenites from the Dabie-Sulu terranes, eastern China. Lithos 70:293-319

Fu B, Zheng YF, Touret JLR (2002) Petrological, isotopic and fluid inclusion studies of eclogites from Sujiahe, NW Dabie Shan (China). Chem Geol 187:107-128

Fu Q, Foustoukos DI, Seyfried WE (2008) Mineral catalyzed organic synthesis in hydrothermal systems: An experimental study using time-of-flight secondary ion mass spectrometry. Geophys Res Lett 35, doi: 10.1029/2008GL033389

Gao J, John T, Klemd R, Xiong XM (2007) Mobilization of Ti-Nb-Ta during subduction: evidence from rutile-bearing dehydration segregations and veins hosted in eclogite, Tianshan, NW China. Geochim Cosmochim Acta 71:4974-4996

Garrels RM, Mackenzie FT (1971) Evolution of Sedimentary Rocks. Norton, New York

Gehrig M, Lentz H, Franck EU (1979) Thermodynamic properties of water-carbon dioxide-sodium chloride mixtures at high temperatures and pressures. *In*: High Pressure Science and Technology. Timmerhaus KD, Barber MS (eds) Plenum, New York, p 534-542

Gerlach TM, McGee KA, Elias T, Sutton AJ, Doukas MP (2002) Carbon dioxide emission rate of Kilauea Volcano: Implications for primary magma and the summit reservoir. J Geophys Res 107, doi: 10.1029/2001jb000407

Gibert F, Guillaume D, Laporte D (1998) Importance of fluid immiscibility in the H_2O-NaCl-CO_2 system and selective CO_2 entrapment in granulites: experimental phase diagram at 5-7 kbar, 900 °C and wetting textures. Eur J Mineral 10:1109-1123

Goff F, Arney BH, Eddy AC (1982) Scapolite phenocrysts in a latite dome, northwest Arizona, USA. Earth Planet Sci Lett 60:86-92

Gottschalk M (2007) Equations of state for complex fluids. Rev Mineral Geochem 65:49-97

Greenwood HJ (1967) Wollastonite - stability in H_2O-CO_2 mixtures and occurence in a contact-metamorphic aureole near Salmo British Columbia, Canada. Am Mineral 52:1669-1680

Hazen RM, Downs RT, Jones AP, Kah L (2013) Carbon mineralogy and crystal chemistry. Rev Mineral Geochem 75:7-46

Heger K, Uematsu M, Franck EU (1980) The Static Dielectric Constant of Water at High Pressures and Temperatures to 500 MPa and 550°C. Ber Bunsen Ges Phys Chem 84:758-762

Heinrich W (2007) Fluid immiscibility in metamorphic rocks. Rev Mineral Geochem 65:389-430

Heinrich W, Churakov SS, Gottschalk M (2004) Mineral-fluid equilibria in the system CaO-MgO-SiO_2-H_2O-CO_2-$NaCl$ and the record of reactive fluid flow in contact metamorphic aureoles. Contrib Mineral Petrol 148:131-149

Helgeson HC, Kirkham DH (1974a) Theoretical prediction of the thermodynamic behavior of aqueous electrolytes at high pressures and temperatures: I. summary of the thermodynamic/electrostatic properties of the solvent. Am J Sci 274:1089-1198

Helgeson HC, Kirkham DH (1974b) Theoretical prediction of the thermodynamic behavior of aqueous electrolytes at high pressures and temperatures: II. Debye-Hückel parameters for activity coefficients and relative partial molal properties. Am J Sci 274:1199-1261

Helgeson HC, Kirkham DH, Flowers GC (1981) Theoretical prediction of the thermodynamic behavior of aqueous electrolytes at high pressures and temperatures: IV. Calculation of activity coefficients, osmotic coefficients, and apparent molal and standard and relative partial molal properties to 600°C and 5 kb. Am J Sci 281:1249-1516

Helgeson HC, Knox AM, Owens CE, Shock EL (1993) Petroleum, oil-field waters, and authigenic mineral assemblages - are they in metastable equilibrium in hydrocarbon reservoirs. Geochim Cosmochim Acta 57:3295-3339

Helgeson HC, Richard L, McKenzie WF, Norton DL, Schmitt A (2009) A chemical and thermodynamic model of oil generation in hydrocarbon source rocks. Geochim Cosmochim Acta 73:594-695

Hoering TC (1984) Thermal reactions of kerogen with added water, heavy water and pure organic substances. Organic Geochem 5:267-278

Holland TJB, Powell R (1991) A compensated-Redlich-Kwong (CORK) equation for volumes and fugacities of CO_2 and H_2O in the range 1 bar to 50 kbar and 100-1600 °C. Contrib Mineral Petrol 109:265-273

Horita J, Berndt ME (1999) Abiogenic methane formation and isotopic fractionation under hydrothermal conditions. Science 285:1055-1057

Huff TA, Nabelek PI (2007) Production of carbonic fluids during metamorphism of graphitic pelites in a collisional orogen - An assessment from fluid inclusions. Geochim Cosmochim Acta 71:4997-5015

Huizenga JM (2001) Thermodynamic modelling of C-O-H fluids. Lithos 55:101-114

Huizenga JM (2011) Thermodynamic modelling of a cooling C-O-H fluid-graphite system: implications for hydrothermal graphite precipitation. Mineral Dep 46:23-33

Huizenga JM, Touret JLR (2012) Granulites, CO_2 and graphite. Gondwana Res 22:799-809

Hunt JE, Philp RP, Kvenvolden KA (2002) Early developments in petroleum geochemistry. Organic Geochem 9:1025-1052

Irwin WP (1970) Metamorphic waters from the Pacific tectonic belt of the west coast of the United States. Science 168:973-975

Jacobs GK, Kerrick DM (1981a) Devolatilization equilibria in H_2O-CO_2 and H_2O-CO_2-$NaCl$ fluids: an experimental and thermodynamic evaluation at elevated pressures and temperatures. Am Mineral 66:1135-1153

Jacobs GK, Kerrick DM (1981b) Methane: An equation of state with application to the ternary system H_2O-CO_2-CH_4. Geochim Cosmochim Acta 45:607-614

Janecky DR, Seyfried WE (1986) Hydrothermal serpentinization of peridotite within the oceanic-crust - experimental investigations of mineralogy and major element chemistry. Geochim Cosmochim Acta 50:1357-1378

Johannes WP, Puhan D (1971) The calcite-aragonite transition, reinvestigated. Contrib Mineral Petrol 31:28-38

Johnson EL (1991) Experimentally determined limits for H_2O-CO_2-$NaCl$ immiscibility in granulites. Geology 19:925-928

Johnson JW, Oelkers EH, Helgeson HC (1992) SUPCRT92: A software package for calculating the standard molal thermodynamic properties of minerals, gases, aqueous species, and reactions from 1 to 5000 bar and 0 to 1000 °C. Comput Geosci 18:899-947

Jones AP, Genge M, Carmody L (2013) Carbonate melts and carbonatites. Rev Mineral Geochem 75:289-322

Joyce DB, Holloway JR (1993) An experimental-determination of the thermodynamic properties of H_2O-CO_2-$NaCl$ fluids at high-pressures and temperatures. Geochim Cosmochim Acta 57:733-746

Jurg JW, Eisma E (1964) Petroleum hydrocarbons: generation from fatty acid. Science 144:1451-1452

Kenney JF, Kutcherov VA, Bendeliani NA, Alekseev VA (2002) The evolution of multicomponeint systems at high pressures: VI. The thermodynamic stability of the hydrogen-carbon system: The genesis of hydrocarbons and the origin of petroleum. Proc Natl Acad Sci USA 99:10976-10981

Keppler H, Wiedenbeck M, Shcheka SS (2003) Carbon solubility in olivine and the mode of carbon storage in the Earth's mantle. Nature 424:414-416

Kerrick DM, Jacobs GK (1981) A modified Redlich-Kwong equation for H_2O, CO_2, and H_2O-CO_2 mixtures at elevated pressures and temperatures. Am J Sci 281:735-767

Kharaka YK, Carothers WW, Rosenbauer RJ (1983) Thermal decarboxylation of acetic acid: Implications for origin of natural gas. Geochim Cosmochim Acta 47:397-402

Kissin YV (1987) Catagenesis and composition of petroleum: Origins of *n*-alkanes and isoalkanes in petroleum crudes. Geochim Cosmochim Acta 51:2445-2457

Kolesnikov A, Kutcherov VG, Goncharov AF (2009) Methane-derived hydrocarbons produced under upper-mantle conditions. Nature Geosci 2:566-570

Koster van Groos AF, Wyllie PJ (1968) Liquid immiscibility in the join $NaAlSi_3O_8$-Na_2CO_3-H_2O and its bearing on the genesis of carbonatites. Am J Sci 266:932-967

Kotelnikov AR, Kotelnikova ZA (1990) Experimental-study of phase state of the system H_2O-CO_2-NaCl by method of synthetic fluid inclusions in quartz. Geokhimiya 1990:526-537

Kruse R, Franck EU (1982) Raman spectra of hydrothermal solutions of CO_2 and HCO_3 at high temperatures and pressures. Ber Bunsen Ges Phys Chem 86:1036-1038

Kutcherov VG, Bendeliani NA, Alekseev VA, Kenney JF (2002) Synthesis of hydrocarbons from minerals at pressures up to 5 GPa. Dokl Phys Chem 387:328-330

Kvenvolden KA, Weiser D (1967) A mathematical model of a geochemical process: Normal paraffin formation from normal fatty acids. Geochim Cosmochim Acta 31:1281-1309

Lamb WM, Valley JW (1987) Post-metamorphic CO_2-rich inclusions in granulites. Contrib Mineral Petrol 96:485-495

Lapin AV, Ploshko VV, Malyshev AA (1987) Carbonatites of the Tatar deep-seated fault zone, Siberia. Int Geol Rev 29:551-567

Lennie AR (2005) Ikaite ($CaCO_3 \cdot 6H_2O$) compressibility at high water pressure: a synchrotron X-ray diffraction study. Mineral Mag 69:325-335

Lewan MD (1997) Experiments on the role of water in petroleum formation. Geochim Cosmochim Acta 61:3691-3723

Lewan MD, Roy S (2011) Role of water in hydrocarbon generation from Type-I kerogen in Mahogany oil shale of the Green River Formation. Organic Geochem 42:31-41

Lewan MD, Ruble TE (2002) Comparison of petroleum generation kinetics by isothermal hydrous and nonisothermal open-system pyrolysis. Organic Geochem 33:1457-1475

Liebscher A (2007) Experimental studies in model fluid systems. Fluid-Fluid Interactions 65:15-47

Liebscher A (2010) Aqueous fluids at elevated pressure and temperature. Geofluids 10:3-19

Loerting T, Tautermann C, Kroemer RT, Kohl I, Hallbrucker A, Mayer E, Liedl KR (2000) On the Surprising Kinetic Stability of Carbonic Acid (H_2CO_3). Angew Chem Int Ed 39:891-894

Ludwig R, Kornath A (2000) In spite of the chemist's belief: Carbonic acid is surprisingly stable. Angew Chem Int Ed 39:1421-1423

Luque FJ, Pasteris JD, Wopenka B, Rodas M, Barrenechea JF (1998) Natural fluid-deposited graphite: Mineralogical characteristics and mechanisms of formation. Am J Sci 298:471-498

Malinin SD, Kanukov AB (1972) The solubility of calcite in homogeneous H_2O-NaCl-CO_2 systems in the 200-600 degrees C temperature interval. Geochem Int 8:668-679

Manning CE (1994) The solubility of quartz in H_2O in the lower crust and upper-mantle. Geochim Cosmochim Acta 58:4831-4839

Manning CE (1998) Fluid composition at the blueschist-eclogite transition in the model system Na_2O-MgO-Al_2O_3-SiO_2-H_2O-HCl. Schweiz Mineral Petrol Mitt 78:225-242

Manning CE (2004) The chemistry of subduction-zone fluids. Earth Planet Sci Lett 223:1-16

Marocchi M, Bureau H, Fiquet G, Guyot F (2011) In-situ monitoring of the formation of carbon compounds during the dissolution of iron(II) carbonate (siderite). Chem Geol 290:145-155

Marshall WL, Franck EU (1981) Ion product of water substance, 0-1000 °C, 1-10,000 bars: new international formulation and its background. J Phys Chem Ref Data 10:295-304

Martinez I, Sanchez-Valle C, Daniel I, Reynard B (2004) High-pressure and high-temperature Raman spectroscopy of carbonate ions in aqueous solution. Chem Geol 207:47-58

Marty B, Tolstikhin IN (1998) CO_2 fluxes from mid-ocean ridges, arcs and plumes. Chem Geol 145:233-248

Mathez EA (1987) Carbonaceous matter in mantle xenoliths - composition and relevance to the isotopes. Geochim Cosmochim Acta 51:2339-2347

Mathez EA, Dietrich VJ, Irving AJ (1984) The geochemistry of carbon in mantle peridotites. Geochim Cosmochim Acta 48:1849-1859

Mathez EA, Mogk DM (1998) Characterization of carbon compounds on a pyroxene surface from a gabbro xenolith in basalt by time-of-flight secondary ion mass spectrometry. Am Mineral 83:918-924

McCollom TM (2013) Laboratory simulations of abiotic hydrocarbon formation in Earth's deep subsurface. Rev Mineral Geochem 75:467-494

McCollom TM, Seewald JS (2001) A reassessment of the potential for reduction of dissolved CO_2 to hydrocarbons during serpentinization of olivine. Geochim Cosmochim Acta 65:3769-3778

McCollom TM, Seewald JS (2003a) Experimental constraints on the hydrothermal reactivity of organic acids and acid anions: I. Formic acid and formate. Geochim Cosmochim Acta 67:3625-3644

McCollom TM, Seewald JS (2003b) Experimental study of the hydrothermal reactivity of organic acids and acid anions: II. Acetic acid, acetate, and valeric acid. Geochim Cosmochim Acta 67:3645-3664

McCollom TM, Seewald JS (2007) Abiotic synthesis of organic compounds in deep-sea hydrothermal environments. Chem Rev 107:382-401

McGetchin TR, Besancon JR (1973) Carbonate inclusions in mantle-derived pyropes. Earth Planet Sci Lett 18:408-410

Melton CE, Giardini AA (1974) Composition and significance of gas released from natural diamonds from Africa and Brazil. Am Mineral 59:775-782

Melton CE, Giardini AA, Salotti CA (1972) Observation of nitrogen, water, carbon-dioxide, methane and argon as impurities in natural diamonds. Am Mineral 57:1518-1523

Metz P, Trommsdorff V (1968) On phase equilibria in metamorphosed siliceous dolomites. Contrib Mineral Petrol 18:305-309

Miller JP (1952) A portion of the system calcium carbonate-carbon dioxide-water, with geological implications. Am J Sci 250:161-203

Mittwede SK (1994) Primary scapolite in a granitic pegmatite, western Cherokee County, South Carolina. Can Mineral 32:617-622

Morey GW (1962) The action of water on calcite, magnesite and dolomite. Am Mineral 47:1456-1460

Morgan JK, Milliken KL (1996) Petrography of calcite veins in serpentinized peridotite basement rocks from the Iberia Abyssal Plain, sites 897 and 899: Kinematic and environmental implications. Proc Ocean Drilling Prog, Sci Results 149:553-558

Mori T, Suma K, Sumiyoshi Y, Endo Y (2011) Spectroscopic detection of the most stable carbonic acid, cis-cis H_2CO_3. J Chem Phys 134, doi:10.1063/1.3532084

Morrison TJ, Billett F (1952) The salting-out of non-electrolytes. Part II. The effect of variations in non-electrolyte. J Chem Soc 1952:3819-3822, doi:10.1039/jr9520003819-3822

Murck BW, Burruss RC, Hollister LS (1978) Phase-equilibria in fluid inclusions in ultramafic xenoliths. Am Mineral 63:40-46

Navon O, Hutcheon ID, Rossman GR, Wasserburg GJ (1988) Mantle-derived fluids in diamond micro-inclusions. Nature 335:784-789

Newton RC, Aranovich LY, Hansen EC, Vandenheuvel BA (1998) Hypersaline fluids in Precambrian deep-crustal metamorphism. Precambrian Res 91:41-63

Newton RC, Manning CE (2002a) Experimental determination of calcite solubility in H_2O-NaCl solutions at deep crust/ upper mantle pressures and temperatures: Implications for metasomatic processes in shear zones. Am Mineral 87:1401-1409

Newton RC, Manning CE (2002b) Solubility of silica in equilibrium with enstatite, forsterite, and H_2O at deep crust/upper mantle pressures and temperatures and an activity-concentration model for polymerization of aqueous silica. Geochim Cosmochim Acta 66:4165-4176

Newton RC, Manning CE (2003) Activity coefficient and polymerization of aqueous silica at 800 °C, 12 kbar, from solubility measurements on SiO_2-buffering mineral assemblages. Contrib Mineral Petrol 146:135-143

Newton RC, Manning CE (2005) Solubility of anhydrite, CaSO4, in NaCl-H2O solutions at high pressures and temperatures: applications to fluid-rock interaction. J Petrol 46:701-716

Newton RC, Manning CE (2009) Hydration state and activity of aqueous silica in H_2O-CO_2 fluids at high pressure and temperature. Am Mineral 94:1287-1290

Newton RC, Manning CE (2010) Role of saline fluids in deep-crustal and upper-mantle metasomatism: insights from experimental studies. Geofluids 10:58-72

Newton RC, Smith JV, Windley BF (1980) Carbonic metamorphism, granulites and crustal growth. Nature 288:45-50

Ni H, Keppler H (2013) Carbon in silicate melts. Rev Mineral Geochem 75:251-287

Nokleberg WJ (1973) CO_2 as a source of oxygen in metasomatism of carbonates. Am J Sci 273:498-514

Ohmoto H, Kerrick D (1977) Devolatilization equilibria in graphitic systems. Am J Sci 277:1013-1044

Oliver NHS, Cartwright I, Wall VJ, Golding SD (1993) The stable isotope signature of kilometer-scale fracture-dominated metamorphic fluid pathways, Mary Kathleen, Australia. J Metamorph Geol 11:705-720

Oppenheimer C (2003) Volcanic degassing. *In*: The Crust, Vol. 3. Treatise on Geochemistry. Rudnick RL Holland HD, Turekian KK (eds) Elsevier-Pergamon, Oxford, p 123-166

Palmer DA, Drummond SE (1986) Thermal decarboxylation of acetate. Part I. The kinetics and mechanism of reaction in aqueous solution. Geochim Cosmochim Acta 50:813-823

Palmer DA, Drummond SE (1987) Thermodynamics of the formation of ferrous acetate complexes. J Electrochem Soc 134:C506-C506

Perry EC Jr, Ahmad SN (1977) Carbon isotope composition of graphite and carbonate minerals from 3.8-AE metamorphosed sediments, Isukasia, Greenland. Earth Planet Sci Lett 36:280-284

Philippi GT (1974) The influence of marine and terrestrial source material on the composition of petroleum. Geochim Cosmochim Acta 38:947-966

Pineau F, Mathez EA (1990) Carbon isotopes in xenoliths from the Hhualalai volcano, Hawaii, and the generation of isotopic variability. Geochim Cosmochim Acta 54:217-227

Pitzer KS (1983) Dielectric constant of water at very high temperature and pressure. Proc Natl Acad Sci USA 80:4575-4576

Plank T, Langmuir CH (1998) The chemical composition of subducting sediment and its consequences for the crust and mantle. Chem Geol 145:325-394

Plummer LN, Busenberg E (1982) The solubilities of calcite, aragonite and vaterite in CO_2-H_2O solutions between 0 and 90 °C, and an evaluation of the aqueous model for the system $CaCO_3$-CO_2-H_2O. Geochim Cosmochim Acta 46:1011-1040

Plyasunov AV, Shock EL (2001) Correlation strategy for determining the parameters of the revised Helgeson-Kirkham-Flowers model for aqueous nonelectrolytes. Geochim Cosmochim Acta 65:3879-3900

Price LC (1979) Aqueous solubility of methane at elevated pressures and temperatures. AAPG Bull 63:1527-1533

Read AJ (1975) The first ionization constant of carbonic acid from 25 to 250 °C and to 2000 bar. J Solution Chem 4:53-70

Reeves EP, Seewald JS, Sylva SP (2012) Hydrogen isotope exchange between n-alkanes and water under hydrothermal conditions. Geochim Cosmochim Acta 77:582-599

Rettich TR, Handa YP, Battino R, Wilhelm E (1981) Solubility of gases in liquids. 13. High-precision determination of Henry's constants for methane and ethane in liquid water at 275 to 328 K. J Phys Chem 85:3230-3237

Richard L (2001) Calculation of the standard molal thermodynamic properties as a function of temperature and pressure of some geochemically important organic sulfur compounds. Geochim Cosmochim Acta 65:3827-3877

Rimstidt JD (1997) Gangue mineral transport and deposition. *In*: Geochemistry of Hydrothermal Ore Deposits. 3rd ed. Barnes HL (ed) John Wiley and Sons, New York, p 487-515

Robinson R (1966) The origins of petroleum. Nature 212:1291-1295

Roedder E (1965) Liquid CO_2 inclusions in olivine-bearing nodules and phenocrysts from basalts. Am Mineral 50:1746-1782

Roedder E (1984) Fluid inclusions. Rev Mineral 12:1-644

Rumble D, III, Hoering TC, Grew ES (1977) The relation of carbon isotopic composition to graphitization of carbonaceous materials from the Narragansett Basin, Rhode Island. Carnegie Inst Wash Yearbook 76:623-625

Ryzhenko BN (1963) Dissociation constant values of carbonic acid at elevated temperatures. Dokl Akad Nauk 149:639-641

Sanchez-Valle C, Martinez I, Daniel I, Philippot P, Bohic S, Simonovici A (2003) Dissolution of strontianite at high P-T conditions: an in-situ synchrotron X-ray fluorescence study. Am Mineral 88:978-985

Schloemer VH (1952) Hydrothermale Untersuchungen über das System CaO-MgO-CO_2-H_2O. Neues Jarhb Mineral Monatsch 1952:129-135

Schrauder M, Navon O (1993) Solid carbon dioxide in a natural diamond. Nature 365:42-44

Schrenk MO, Brazelton WJ, Lang SQ (2013) Serpentinization, carbon, and deep life. Rev Mineral Geochem 75:575-606

Scott HP, Hemley RJ, Mao HK, Herschbach DR, Fried LE, Howard WM, Bastea S (2004) Generation of methane in the Earth's mantle: In situ high pressure-temperature measurements of carbonate reduction. Proc Natl Acad Sci USA 101:14023-14026

Seewald JS (1994) Evidence for metastable equilibrium between hydrocarbons under hydrothermal conditions. Nature 370:285-287

Seewald JS (2001) Aqueous geochemistry of low molecular weight hydrocarbons at elevated temperatures and pressures: Constraints from mineral buffered laboratory experiments. Geochim Cosmochim Acta 65:1641-1664

Seewald JS (2003) Organic-inorganic interactions in petroleum-producing sedimentary basins. Nature 426:327-333

Seewald JS, Zolotov MY, McCollom T (2006) Experimental investigation of single carbon compounds under hydrothermal conditions. Geochim Cosmochim Acta 70:446-460

Segnit ER, Holland HD, Biscardi CJ (1962) The solubility of calcite in aqueous solutions -I. The solubility of calcite in water between 75° and 200° at CO_2 pressures up to 60 atm. Geochim Cosmochim Acta 26:1301-1331

Seitz JC, Blencoe JG (1999) The CO_2-H_2O system. I. Experimental determination of volumetric properties at 400°C, 10-100 MPa. Geochim Cosmochim Acta 63:1559-1569

Sephton MA, Hazen RM (2013) On the origins of deep hydrocarbons. Rev Mineral Geochem 75:449-465

Sharma A, Cody GD, Hemley RJ (2009) In situ diamond-anvil cell observations of methanogenesis at high pressures and temperatures. Energy Fuels 23:5571-5579

Sharp WE, Kennedy GC (1965) The system CaO-CO_2-H_2O in the two-phase region calcite+aqueous solution. J Geol 73:391-403

Shaw DM (1956) Geochemistry of pelitic rocks. 3. Major elements and general geochemistry. Bull Geol Soc Am 67:919-934

Shcheka SS, Wiedenbeck M, Frost DJ, Keppler H (2006) Carbon solubility in mantle minerals. Earth Planet Sci Lett 245:730-742

Shi GU, Tropper P, Cui WY, Tan J, Wang CQ (2005) Methane (CH_4)-bearing fluid inclusions in the Myanmar jadeitite. Geochem J 39:503-516

Shimoyama A, Johns WD (1971) Catalytic conversion of fatty acids to petroleum-like paraffins and their maturation. Nature-Phys Sci 232:140-144

Shimoyama A, Johns WD (1972) Formation of alkanes from fatty acids in the presence of $CaCO_3$. Geochim Cosmochim Acta 36:87-91

Shipp J, Gould I, Herckes P, Shock EL, Williams L, Hartnett H (in press) Organic functional group transformations in water at elevated temperature and pressure: Reversibility, reactivity, and mechanisms. Geochim Cosmochim Acta

Shiryaev AA, Griffin WL, Stoyanov E (2011) Moissanite (SiC) from kimberlites: Polytypes, trace elements, inclusions and speculations on origin. Lithos 122:152-164

Shmulovich KI, Graham CM (1999) An experimental study of phase equilibria in the system H_2O-CO_2-NaCl at 800 °C and 9 kbar. Contrib Mineral Petrol 136:247-257

Shmulovich KI, Graham CM (2004) An experimental study of phase equilibria in the systems H_2O-CO_2-$CaCl_2$ and H_2O-CO_2-NaCl at high pressures and temperatures (500-800 °C, 0.5-0.9 GPa): geological and geophysical applications. Contrib Mineral Petrol 146:450-462

Shmulovich KI, Yardley BWD, Graham C M (2006) Solubility of quartz in crustal fluids: experiments and general equations for salt solutions and H_2O-CO_2 mixtures at 400-800°C and 0.1-0.9 GPa. Geofluids 6:154-167

Shock E, Canovas P (2010) The potential for abiotic organic synthesis and biosynthesis at seafloor hydrothermal systems. Geofluids 10:161-192

Shock EL (1988) Organic-acid metastability in sedimentary basins. Geology 16:886-890

Shock EL (1989) Corrections to "Organic-acid metastability in sedimentary basins". Geology 17:572-573

Shock EL (1990) Geochemical constraints on the origin of organic compounds in hydrothermal systems. Origins Life Evol Biosphere 20:331-367

Shock EL (1992) Chemical environments of submarine hydrothermal systems. Orig Life Evol Biosph 22:67-107

Shock EL (1994) Application of thermodynamic calculations to geochemical processes involving organic acids. *In*: The Role of Organic Acids in Geological Processes. Lewan M, Pittman E (eds) Springer-Verlag, p 270-318

Shock EL (1995) Organic acids in hydrothermal solutions: Standard molal thermodynamic properties of carboxylic acids and estimates of dissociation constants at high temperatures and pressures. Am J Sci 295:496-580

Shock EL, Helgeson HC (1988) Calculation of the thermodynamic and transport properties of aqueous species at high pressures and temperatures: correlation algorithms for ionic species and equation of state predictions to 5 kb and 1000 °C. Geochim Cosmochim Acta 52:2009-2036

Shock EL, Helgeson HC (1990) Calculation of the thermodynamic and transport properties of aqueous species at high pressures and temperatures: Standard partial molal properties of organic species. Geochim Cosmochim Acta 54:915-945

Shock EL, Helgeson HC, Sverjensky DA (1989) Calculation of the thermodynamic and transport properties of aqueous species at high pressures and temperatures: standard partial molal properties of inorganic neutral species. Geochim Cosmochim Acta 53:2157-2183

Shock EL, McKinnon WB (1993) Hydrothermal processing of cometary volatiles - applications to Triton. Icarus 106:464-477

Shock EL, Oelkers EH, Johnson JW, Sverjensky DA, Helgeson HC (1992) Calculation of the thermodynamic and transport properties of aqueous species at high pressures and temperatures: Effective electrostatic radii to 1000°C and 5 kb. Faraday Soc Trans 88:803-826

Shock EL, Sassani DC, Betz H (1997a) Uranium in geologic fluids: Estimates of standard partial molal properties, oxidation potentials and hydrolysis constants at high temperatures and pressures. Geochim Cosmochim Acta 61:4245-4266

Shock EL, Sassani DC, Willis M, Sverjensky DA (1997b) Inorganic species in geologic fluids: correlations among standard molal thermodynamic properties of aqueous ions and hydroxide complexes. Geochim Cosmochim Acta 61:907-950

Shock EL, Schulte MD (1998) Organic synthesis during fluid mixing in hydrothermal systems. J Geophys Res Planets 103:28513-28527

Sirbescu MLC, Nabelek PI (2003a) Crustal melts below 400 °C. Geology 31:685-688

Sirbescu MLC, Nabelek PI (2003b) Crystallization conditions and evolution of magmatic fluids in the Harney Peak Granite and associated pegmatites, Black Hills, South Dakota—evidence from fluid inclusions. Geochim Cosmochim Acta 67:2443-2465

Sisson VB, Hollister LS (1990) A fluid-inclusion study of metamorphosed pelitic and carbonate rocks, south-central Maine. Am Mineral 75:59-70
Skippen G, Trommsdorff V (1986) The influence of NaCl and KCl on phase-relations in metamorphosed carbonate rocks. Am J Sci 286:81-104
Skippen GB (1971) Experimental data for reactions in siliceous marbles. J Geol 79:457-481
Smith GC, Holness MB, Bunbury JM (2008) Interstitial magmatic scapolite in glass-bearing crystalline nodules from the Kula Volcanic Province, Western Turkey. Mineral Mag 72:1243-1259
Smith HM (1967) The hydrocarbon constituents of petroleum and some possible lipid precursors. J Am Oil Chem Soc 44:680-690
Snape CE, Stokes BJ, Bartle KD (1981) Identification o straight-chain fatty acids in coal extracts and their geochemical relation with straight-chain alkanes. Fuel 60:903-908
Soli AL, Byrne RH (2002) CO_2 system hydration and dehydration kinetics and the equilibrium CO_2/H_2CO_3 ratio in aqueous NaCl solution. Marine Chem 78:65-73
Spear FS (1993) Metamorphic Phase Equilibria and Pressure-Temperature-Time Paths. Mineralogical Society of America, Washington, D. C.
Staudigel H (2003) Hydrothermal alteration processes in the oeanic crust. *In*: The Crust, Vol. 3. Treatise on Geochemistry. Rudnick RL Holland HD, Turekian KK (eds) Elsevier-Pergamon, Oxford, p 511-535
Stern RJ, Gwinn CJ (1990) Origin of Late Precambrian intrusive carbonates, eastern desert of Egypt and Sudan: C, O and Sr isotope evidence. Precambrian Res 46:259-272
Sterner SM, Bodnar RJ (1991) Synthetic fluid inclusions. X: Experimental-determination of P-V-T-X properties in the CO_2-H_2O system to 6 kb and 700 °C. Am J Sci 291:1-54
Sverjensky DA, Shock EL, Helgeson HC (1997) Prediction of the thermodynamic properties of aqueous metal complexes to 1000 °C and 5 kb. Geochim Cosmochim Acta 61:1359-1412
Swanson SE (1979) The effect of CO_2 on phase equilibria and crystal growth in the system $KAlSi_3O_8$-$NaAlSi_3O_8$-$CaAl_2Si_2O_8$-SiO_2-H_2O-CO_2 to 8000 bars. Am J Sci 279:703-720
Symonds RB, Rose WI, Bluth GJS, Gerlach TM (1994) Volcanic-gas studies: methods, results, and applications. Rev Mineral 30:1-66
Takenouchi S, Kennedy GC (1964) Binary system H_2O-CO_2 at high temperatures + pressures. Am J Sci 262:1055-1074
Terlouw JK, Lebrilla CB, Schwarz H (1987) Thermolysis of NH_4HCO_3—A simple route to the formation of free carbonic acid (H_2CO_3) in the gas phase. Angew Chem Int Ed 26:354-355
Thomas R, Davidson P, Schmidt C (2011) Extreme alkali bicarbonate- and carbonate-rich fluid inclusions in granite pegmatite from the Precambrian Rønne granite, Bornholm Island, Denmark. Contrib Mineral Petrol 161:315-329
Thomas R, Webster JD, Davidson P (2006) Understanding pegmatite formation: the melt and fluid inclusion approach. *In*: Melt Inclusions in Plutonic Rocks. Webster JD (ed) Mineralogical Association of Canada, p 189-210
Thompson G (1972) Geochemical study of some lithified carbonate sediments from deep-sea. Geochim Cosmochim Acta 36:1237-1253
Thompson G, Bowen VT, Melson WG, Cifelli R (1968) Lithified carbonates from deep-sea of equatorial Atlantic. J Sed Pet 38:1305-1312
Tingle TN, Green HW (1987) Carbon solubility in olivine - implications for upper mantle evolution. Geology 15:324-326
Tingle TN, Green HW, Finnerty AA (1988) Experiments and observations bearing on the solubility and diffusivity of carbon in olivine. J Geophys Res 93:15289-15304
Todheide K, Franck EU (1963) Das Zweiphasengebiet und die kritische Kurve im System Kohlendioxyd-Wasser bei hohen Drucken. Ber Bunsen Ges Phys Chem 67:836-836
Tomilenko AA, Chepurov AI, Pal'Yanov YN, Shebanin AP, Sobolev NV (1998) Hydrocarbon inclusions in synthetic diamonds. Eur J Mineral 10:1135-1141
Tossell JA (2006) H_2CO_3 and its oligomers: Structures, stabilities, vibrational and NMR spectra, and acidities. Inorg Chem 45:5961-5970
Touret JLR (1985) Fluid regime in southern Norway: the record of fluid inclusions. *In*: The Deep Proterozoic Crust in the North Atlantic Provinces. Tobi AC, Touret JLR (eds) Reidel, Dordecht, p 517-549
Trommsdorff V, Evans BW, Pfeifer HR (1980) Ophicarbonate rocks - metamorphic reactions and possible origin. Arch Sci 33:361-364
Trommsdorff V, Skippen G (1986) Vapor loss (boiling) as a mechanism for fluid evolution in metamorphic rocks. Contrib Mineral Petrol 94:317-322
Tropper P, Manning CE (2007) The solubility of fluorite in H_2O and H_2O-NaCl at high pressure and temperature. Chem Geol 242:299-306
Urey HC (1952) The Planets, Their Origin and Development. Yale University Press, New Haven

Wada H, Tomita T, Matsuura K, Iuchi K, Ito M, Morikiyo T (1994) Graphitization of carbonaceous matter during metamorphism with references to carbonate and pelitic rocks of contact and regional metamorphisms, Japan. Contrib Mineral Petrol 118:217-228

Wallace PJ (2005) Volatiles in subduction zone magmas: concentrations and fluxes based on melt inclusion and volcanic gas data. J Volcan Geotherm Res 140:217-240

Walther JV (1992) Ionic association in H_2O-CO_2 fluids at mid-crustal conditions. J Metamorph Geol 10:789-797

Walther JV, Helgeson HC (1980) Description and interpretation of metasomatic phase relations at high pressures and temperatures: 1. equilibrium activities of ionic species in nonideal mixtures of CO_2 and H_2O. Am J Sci 280:575-606

Walther JV, Long MI (1986) Experimental determination of calcite solubilities in supercritical H_2O. Int Symp Water-Rock Interaction 5:609-611

Walther JV, Schott J (1988) The dielectric constant approach to speciation and ion pairing at high temperature and pressure. Nature 332:635-638

Wells RC (1915) The solubility of calcite in water in contact with the atmosphere and its variation with temperature. J Wash Acad Sci 5:617-622

Wheat CG, Fryer P, Fisher AT, Hulme S, Jannasch H, Mottl MJ, Becker K (2008) Borehole observations of fluid flow from South Chamorro Seamount, an active serpentinite mud volcano in the Mariana forearc. Earth Planet Sci Lett 267:401-409

Wickham SM, Janardhan AS, Stern RJ (1994) Regional carbonate alteration of the crust by mantle-derived magmatic fluids, Tamil Nadu, South India. J Geol 102:379-398

Wilhelm E, Battino R, Wilcock RJ (1977) Low-pressure solubility of gases in liquid water. Chem Rev 77:219-262

Yang Z, Gould IR, Williams L, Hartnett H, Shock EL (2012) The central role of ketones in reversible and irreversible hydrothermal organic functional group transformations. Geochim Cosmochim Acta 98:48-65

Yardley BWD, Valley JW (1997) The petrologic case for a dry lower crust. J Geophys Res 102:12,173-12,185

Zeller KP, Schuler P, Haiss P (2005) The hidden equilibrium in aqueous sodium carbonate solutions - evidence for the formation of dicarbonate anion. Eur J Inorg Chem 1:168-172

Zhang C, Duan ZH (2009) A model for C-O-H fluid in the Earth's mantle. Geochim Cosmochim Acta 73:2089-2102

Zhang C, Duan ZH, Zhang Z (2007) Molecular dynamics simulation of the CH_4 and CH_4-H_2O systems up to 10 GPa and 2573 K. Geochim Cosmochim Acta 71:2036-2055

Zhang YG, Frantz JD (1989) Experimental-determination of the compositional limits of immiscibility in the system $CaCl_2$-H_2O-CO_2 at high-temperatures and pressures using synthetic fluid inclusions. Chem Geol 74:289-308

Zheng YF, Gong B, Li YL, Wang ZR, Fu B (2000) Carbon concentrations and isotopic ratios of eclogites from the Dabie and Sulu terranes in China. Chem Geol 168:291-305

Zheng YF, Gong B, Zhao ZF, Fu B, Li YL (2003) Two types of gneisses associated with eclogite at Shuanghe in the Dabie terrane: carbon isotope, zircon U-Pb dating and oxygen isotope. Lithos 70:321-343

Zotov N, Keppler H (2000) In-situ Raman spectra of dissolved silica species in aqueous fluids to 900 °C and 14 kbar. Am Mineral 85:600-604

Zotov N, Keppler H (2002) Silica speciation in aqueous fluids at high pressures and high temperatures. Chem Geol 184:71-82

Reviews in Mineralogy & Geochemistry
Vol. 75 pp. 149-181, 2013

Primordial Origins of Earth's Carbon

Bernard Marty

Centre de Recherches Pétrographiques et Géochimiques, CNRS,
Université de Lorraine, BP 20,
54220 Vandoeuvre les Nancy Cedex, France

bmarty@crpg.cnrs-nancy.fr

Conel M. O'D. Alexander

Department of Terrestrial Magnetism
Carnegie Institution of Washington
5241 Broad Branch Road, NW
Washington, DC 20015-1305, U.S.A.

alexander@dtm.ciw.edu

Sean N. Raymond

Univ. Bordeaux and CNRS
Laboratoire d'Astrophysique de Bordeaux, UMR 5804,
F-33270, Floirac, France.

rayray.sean@gmail.com

INTRODUCTION

It is commonly assumed that the building blocks of the terrestrial planets were derived from a cosmochemical reservoir that is best represented by chondrites, the so-called chondritic Earth model. This view is possibly a good approximation for refractory elements (although it has been recently questioned; e.g., Caro et al. 2008), but for volatile elements, other cosmochemical reservoirs might have contributed to Earth, such as the solar nebula gas and/or cometary matter (Owen et al. 1992; Dauphas 2003; Pepin 2006). Hence, in order to get insights into the origin of the carbon in Earth, it is necessary to compare: (i) the elemental abundances and isotopic compositions of not only carbon, but also other volatile elements in potential cosmochemical "ancestors," and (ii) the ancestral compositions with those of terrestrial volatiles. This approach is the only one that has the potential for understanding the origin of the carbon in Earth but it has several intrinsic limitations. First, the terrestrial carbon budget is not well known, and, for the deep reservoir(s) such as the core and the lower mantle, is highly model-dependent (Dasgupta 2013; Wood et al. 2013). Second, the cosmochemical reservoir(s) that contributed volatile elements to proto-Earth may not exist anymore because planet formation might have completely exhausted them (most of the mass present in the inner solar system is now in Venus and Earth). Third, planetary formation processes (accretion, differentiation, early evolution of the atmospheres) might have drastically modified the original elemental and isotopic compositions of the volatile elements in Earth. Despite these limitations, robust constraints on the origin(s) of the carbon in Earth can be deduced from comparative planetology of volatile elements, which is the focus of this chapter.

Carbon in the cloud of gas and dust from which the solar system formed was probably mainly in the forms of gaseous CO and of organic-rich carbonaceous dust (e.g., Zubko et al.

1529-6466/13/0075-0006$00.00 DOI: 10.2138/rmg.2013.75.6

2004). Other forms of carbon, such as CH_4, volatile organics, and diamonds, were also probably present but in much smaller proportions. In primitive meteorites, the depletion of carbon relative to refractory elements and normalized to the solar composition indicates that indeed a large fraction of this element was not in refractory phase(s) and had a volatile element behavior like hydrogen, nitrogen, and noble gases (e.g., Anders and Grevesse 1989; Lodders 2003). Furthermore, in primitive meteorites, the total amounts of carbon, nitrogen, and trapped noble gases tend to correlate (Otting and Zärhinger 1967; Kerridge 1985). Thus, to first order the noble gases can be taken as proxies for the behavior of carbon during planetary building events, keeping in mind that under specific conditions, such as low oxygen fugacities and in the presence of metal, carbon and nitrogen might have behaved more like refractory elements than like highly volatile elements, such as noble gases.

In this chapter, we first discuss the origin of the carbon isotopes in the universe and in the solar system. We present estimates of the elemental and isotopic compositions of carbon and, when necessary, of other volatile elements in various solar system reservoirs. We then discuss the latest estimates of the carbon content and isotopic compositions in the different terrestrial reservoirs. From the comparison between these contrasting inventories, we discuss the various possible processes of delivery of volatiles to Earth. Finally, the early terrestrial carbon cycle is introduced.

CARBON IN THE UNIVERSE

Nucleosynthesis of carbon and stellar evolution

Only hydrogen, helium, and some lithium were created in significant amounts in the Big Bang (Table 1). All other elements were formed by nucleosynthesis in stars (e.g., Burbidge et al. 1957; Truran and Heger 2003; Meyer and Zinner 2006), except for most lithium, beryllium, and boron, which are largely the products of fragmentation (spallation) of heavier nuclei by energetic cosmic rays in the interstellar medium. The build-up of elements (and isotopes) heavier than helium once star formation began in galaxies is often referred to as galactic chemical evolution (GCE). Stars range in mass from the smallest brown dwarfs with mass of ~0.012 solar masses (M_o) to super massive stars of >100 M_o.

Table 1. The principal nucleosynthetic formation reactions and sources of the isotopes of hydrogen, carbon and nitrogen.

	Nucleosynthesis	Sources
^{1}H	Big Bang	Big Bang
D	Big Bang	Big Bang
^{12}C	He-burning	SN, AGB
^{13}C	CNO cycle	AGB
^{14}N	CNO cycle	AGB
^{15}N	CNO cycle	Novae, SN, AGB

SN=supernovae, AGB=asymptotic giant branch stars.

Nucleosynthesis of ever-heavier isotopes is the prime energy source for stars and it proceeds in stages that require increasing temperatures and pressures with each new stage. The first stages of nucleosynthesis, deuterium ($D + {}^1H = {}^3He$) and lithium (${}^7Li + {}^1H = 2\,{}^4He$) burning, occur even as a star is still forming/contracting (pre-main sequence). Brown dwarfs, the lowest mass stars (0.012-0.08 M_o), do not get beyond the deuterium-burning stage. Higher mass stars join the main sequence, where they spend most of their lives when they begin burning ^1H in their cores. Hydrogen burning converts ^1H to ^4He via proton-proton chain and CNO cycle reactions. In the CNO cycle, carbon, nitrogen, and oxygen isotopes inherited from earlier generations of stars "catalyze" the conversion of ^1H to ^4He. The isotopes ^{13}C, ^{14}N, and ^{17}O are important intermediate products of the CNO cycle.

The main sequence life of a star ends when it has exhausted all the ^{1}H in its core. In <10 M_0 stars, hydrogen burning continues in a shell above the core and the star becomes a red giant branch (RGB) star. Massive convection at this transition brings hydrogen-burning products to the stellar surface, most notably ^{13}C, ^{14}N, and ^{17}O. Eventually, contraction of the core and mass added to it from the hydrogen-burning shell produce conditions that enable the next stage, core helium burning. In helium burning, three ^{4}He nuclei are fused together in the triple-alpha reaction, to form ^{12}C. Further alpha addition leads to the formation of ^{16}O, but reaction rates beyond this are very slow. During core helium burning, the entire star contracts and it leaves the RGB. However, once helium is exhausted in the core, shell burning resumes and the star joins the asymptotic giant branch (AGB). For stars of ~3-10 M_0, there is a second dredge-up of hydrogen-burnt material into the envelope of the star. The $^{12}C/^{13}C$ ratios in the envelopes of most stars at the beginning of the AGB are <10, but once alternating phases of hydrogen and helium burning in shells above the exhausted core become established, ^{12}C-rich helium-burnt material is periodically mixed into the envelope. For ≤4 M_0 stars, the result of adding this material to the envelope is a rapid increase in both the $^{12}C/^{13}C$ and the C/O ratios. Eventually, so much ^{12}C is added that the C/O ratio of the envelope, which was initially less than one, exceeds one. For 4-10 M_0 stars, conditions at the base of the envelope allow for some hydrogen burning, which destroys much of the added ^{12}C, so that the $^{12}C/^{13}C$ ratio remains low and the C/O ratio stays below one.

The chemistry of the envelope during the AGB is important because the AGB stars begin to lose their envelopes in massive stellar winds, the development of which is driven at least in part by dust formation. The chemistry of envelopes dictates the types of dust that form—envelopes with C/O < 1 are dominated by crystalline and amorphous silicates and oxides, with most of the carbon tied up in CO, while those with C/O > 1 are dominated by SiC and graphite and/or amorphous carbon, and most of the oxygen is in CO. These AGB winds are important sources of freshly synthesized material, particularly ^{13}C and ^{17}O, and dust for the interstellar medium (ISM).

Eventually, the entire envelope is lost, leaving behind the inert core, now a white dwarf, which is dominated in most cases by ^{12}C and ^{16}O. The cores of the most massive AGB stars may experience carbon burning, which converts ^{12}C to ^{20}Ne and ^{24}Mg, producing an O-Ne-Mg white dwarf. A white dwarf is not always the final state of an intermediate mass star. If it is part of a close binary, it can gain mass from its less evolved companion. The increase in temperature and pressure at the surface as material is accreted eventually results in a nova explosion driven by runaway CNO cycle reactions ($^{12}C/^{13}C < 5$). However, not all the accreted material is blown off in a nova explosion and, if after repeated nova events the mass of a white dwarf eventually exceeds ~1.4 M_0, it will explode as a type Ia supernova.

Once 9-25 M_0 stars leave the main sequence, their initial evolution is not unlike that of intermediate mass ones, with the development of a supergiant phase and dredge-up that enriches the envelope in CNO-cycle products. However, there is never any dredge-up of helium-burnt material, so the C/O < 1. Significant mass loss and dust formation is observed to occur during the supergiant phase, but the envelope is never entirely lost before the star becomes a type II supernova. The internal structure of the pre-supernova star is a series of shells that have experienced ever more advanced stages of nucleosynthesis with increasing depth. These include a relatively massive, carbon-rich helium-burning shell, but the majority of the shells are oxygen-rich. Two of the major nucleosynthetic products in type II supernovae are ^{12}C and ^{16}O. Further nucleosynthesis occurs during the explosion, producing the elements heavier than iron.

For the most massive stars, (> 25 M_0), mass loss during the supergiant phase is so vigorous that the envelope is lost, exposing hydrogen-burnt material directly at the surface. With continued mass loss, the products of hydrogen burning, such as ^{14}N, become highly overabundant. The star is now a Wolf-Rayet star. Further mass loss reveals helium-burnt material, rich in ^{12}C and

^{16}O (C/O > 1), and it is only at this stage that dust formation is observed. The ultimate fate of a WR star is a type Ib supernova that produces a similar array of elements/isotopes to type IIs.

Galactic chemical evolution

The lifetimes of stars vary dramatically with their masses. Stars that are only slightly less massive than the Sun have lifetimes that are of the order of the age of the Universe, > 10^{10} years. These low-mass stars will not have contributed to GCE and will not be discussed further here. Intermediate and massive stars (> 1 M_o) have lifetimes that range from ~10^{10} years at the low mass end to only a few million years at the high mass end. Early in the history of the Galaxy, massive stars (their winds and type II and Ib supernovae) would have been the main sources of elements/isotopes heavier than helium. The main products would have been isotopes like ^{12}C and ^{16}O, so-called primary isotopes because they can be built up from the primary products of the Big Bang (hydrogen and helium). With time, and after several generations of massive star formation, the longer-lived intermediate mass stars would have started dying and isotopes like ^{13}C and ^{17}O would have begun to build up. Evidence for this GCE can be found in long-lived low-mass stars, some of which can have very low abundances of elements heavier than helium (low metallicity). Because rates of star formation decrease with increasing radius from the Galactic center, the effects of GCE can also be seen in the carbon and oxygen isotopic compositions of the interstellar medium, with $^{12}C/^{13}C$ and $^{16}O/^{17}O$ ratios increasing more-or-less monotonically with radial distance.

Carbon in the interstellar medium and the presolar molecular cloud

Once injected into the ISM, stellar material is subject to a number of energetic processes that modifies it. These events include supernova-driven shock waves, cosmic rays, and UV irradiation. In the diffuse ISM, these energetic processes mean that complex molecules cannot form and dust is either damaged or destroyed. Most rock-forming elements are condensed in amorphous silicate dust, while carbon is distributed between CO and a poorly characterized carbonaceous dust (Zubko et al. 2004; Draine and Li 2007). In the higher densities of molecular clouds, most of the material is shielded from shock waves and UV irradiation. The interiors of molecular clouds are also much denser and colder than the diffuse ISM—temperatures may reach 10 K or less. As a result, all but the most volatile material is condensed in icy dust grains. Despite these temperatures, a complex chemistry can occur in molecular clouds in the gas phase, on grain surfaces, and even within grains (Herbst and van Dishoeck 2009). This chemistry is driven to a large extent by cosmic rays that can penetrate the clouds—activation barriers for reactions between cosmic ray generated ions and other gaseous species are essentially zero, allowing for reactions at even these very low temperatures. Because the reaction temperatures are so low, extreme isotopic fractionations can be generated. Deuterium, for instance, can be enriched by several orders of magnitude in water and simple organic molecules.

Carbon content and isotopic composition of the solar nebula

Roughly 99.8% of the total mass of the present solar system is in the Sun. Thus, the general assumption is that the Sun's bulk composition must closely resemble the average composition of the material from which the solar system formed. Except for the brief, early episode that destroyed most of the deuterium and lithium, the outer regions of the Sun are unaffected by nucleosynthesis in the stellar interior (e.g., Geiss and Bochsler 1982). Therefore, spectroscopic measurements of the Sun's photosphere combined with hydrodynamic models of its atmosphere can be used to determine most elemental and even some isotopic abundances (Asplund et al. 2009). To estimate the primordial solar or "cosmic" composition, small corrections to the present-day photospheric abundances are needed to account for gravitational settling and diffusion of elements that deplete the photosphere over the lifetime of the solar system (Turcotte et al. 1998). Solar photospheric abundances have been measured for most elements, although they continue to be refined (Asplund et al. 2009). In particular, since 2002 there has been a significant revision downwards in the abundances of carbon, nitrogen, and oxygen. The composition of the solar wind and Jupiter's atmosphere can also provide useful constraints for the abundances of some elements and isotopes. Tables 2 and 3 list the abundances and isotopic compositions of a restricted number of elements that are particularly relevant here. Figure 1 compares the isotope variations of hydrogen and nitrogen in the solar system, which show a close affinity between primitive meteorites and Earth.

To first order, the solar composition resembles the compositions of young stars in the local neighborhood, as well as the composition of the local interstellar medium (e.g., Nieva and Przybilla 2011). There is scatter in the compositions of other stars in the vicinity of the Sun. In part, this diversity reflects uncertainties in the measurements, but it also reflects spatial and temporal variations in the compositions of the interstellar material from which the stars formed (Kobayashi and Nakasato 2011). The composition of the interstellar medium is expected to vary in time and space as a result of the infall of material onto the Galactic disk, local "pollution" by highly evolved mass-losing stars (e.g., supernovae, novae, and AGB stars), and because of an overall gradient in star formation rates. Nevertheless, the solar composition is a useful and commonly used reference with which to compare the compositions of other astronomical objects.

Remarkably, the composition of one group of primitive meteorites, the CI chondrites, closely resembles the non-volatile (i.e., not including hydrogen, carbon, nitrogen, oxygen, and the noble gases) photospheric abundances. In fact, the resemblance is so close that the

Table 2. The abundances (wt%) of hydrogen, carbon, and nitrogen in various reservoirs and objects.

	Protosolar[1]	**Earth**[2]	**Halley**[3]	**CI-CM**[4]
H	71.54	3.0×10^{-2}	6.0	1.2
He	27.03			1.5×10^{-6}
C	0.25	5.2×10^{-2}	18	3.65
N	7.3×10^{-2}	1.7×10^{-4}	0.8	0.15
Ne	1.3×10^{-3}	7.6×10^{-10}		3.55×10^{-8}
Ar	7.8×10^{-5}	5.5×10^{-9}		1.58×10^{-7}
Kr	1.2×10^{-7}	4.6×10^{-10}		6.25×10^{-9}
Xe	1.8×10^{-8}	6.0×10^{-11}		1.74×10^{-8}

[1] Asplund et al. (2009). [2] Marty (2012). [3] Delsemme (1991) and assuming that the organics has a composition like Halley CHON particles (Kissel and Krueger 1987). [4] Alexander et al. (2012) and Marty (2012) for noble gases; averages of two carbonaceous chondrites, Murchison (CM) and Orgueil (CI).

Table 3. Key isotopic ratios for hydrogen, carbon and nitrogen in various objects.

	Protosolar	**Earth**	**Meteorites : chondrites**	**Comets : Oort cloud**	**Comet : 103P/Hartley2 Kuiper belt**
D/H (10^{-6})	20 (±3)[1]	149[11]	120-408[2,6,10]	319 (±59)[4]	161(±24)[4]
$^{13}C/^{12}C$ (10^{-2})	<1.0[5]	1.12	1.10-1.14 (±0.02)[6,10]	1.08 (±0.10)[9]	1.05 (±0.14)[7]
$^{15}N/^{14}N$ (10^{-3})	2.27(±0.03)[8]	3.67	3.51-5.00 (±0.2)[6,10]	6.80 (±1.25)[9]	6.45 (±0.90)[7]

[1]Geiss and Gloeckler (2003). [2] Robert (2003). [3]Bockelée-Morvan et al. (2008). [4]Hartogh et al. (2011), only H_2O. [5]Hashizume et al. (2004). [6]Kerridge (1985). [7]Meech et al. (2011), only CN. [8]Marty et al. (2011). [9]Manfroid et al. (2009), only CN and HCN. [10]Alexander et al. (2012). [11]Lécuyer et al. (1998).

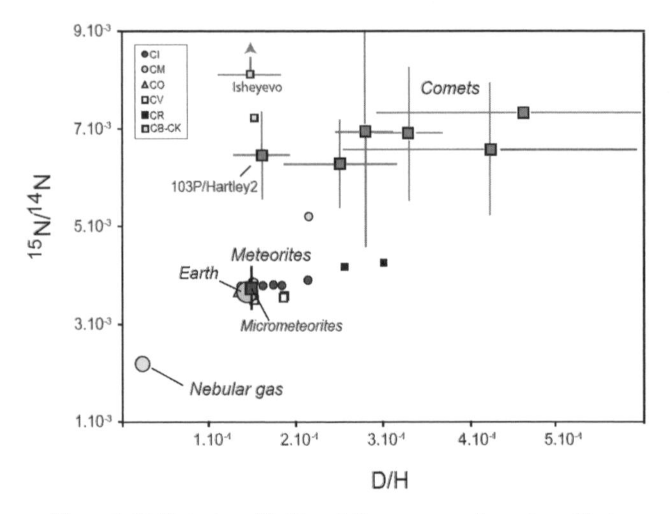

Figure 1. Stable isotope (H, N) variations among solar system objects and reservoirs (adapted from Marty 2012; data sources therein).

appropriately scaled CI abundances of many elements and most isotopes are preferred over the photospheric abundances because they can be measured more accurately (e.g., Lodders 2003; Asplund et al. 2009). The general assumption that the bulk isotopic compositions of the elements in CI chondrites are solar is not valid for the most volatile elements, such as hydrogen, carbon, nitrogen, oxygen, and the noble gases. This is because physical and chemical processes in the solar nebula produced several carriers of these elements (e.g., H_2, H_2O, CO, CH_4, NH_3, and N_2, as well as dust/ice) with different volatilities that were physically and isotopically fractionated relative to one another, so that the CI chondrites did not accrete all of these carriers when they formed. Because solar (or CI) is the starting composition from which all solar system materials are assumed to have ultimately evolved, the compositional variations of solar system objects, particularly meteorites, are often expressed as deviations relative to CI. Here we also use a CI-CM average composition for volatile abundances, since CM-type meteorites such as Murchison are also remarkably primitive and volatile rich, present analogies with micrometeorites (another potentially important source of terrestrial volatiles, see below), and most CMs have experienced less aqueous alteration than CIs.

Volatile abundances and isotope compositions in comets with special reference to carbon

The volatile species emitted by comets are dominated by water, followed by CO, CO_2, NH_3, and simple organic molecules (Bockelée-Morvan et al. 2004; Mumma and Charnley 2012). To date, there have been no unambiguous remote detections of noble gases in comets. Helium and neon have been measured in cometary matter from comet Wild 2 (or 81P/Wild 2) that was returned to Earth by the Stardust mission (Marty et al. 2008). The isotopic composition of neon more closely resembles that of neon trapped in chondrites (phase "Q" neon) rather than that of solar wind neon, and the helium and neon abundances were found to be very high and only matched by lunar regolith that had been irradiated by the solar wind (SW) for long periods of time. These observations suggest that the light noble gases in Stardust matter were implanted by solar irradiation early in the formation of the solar system, provided that phase Q neon is indeed derived from solar neon due to isotope fractionation during implantation, as suggested by some models (see Ott 2002, for a review). Hence implantation of solar gases could be a significant source for trace gases, but probably not for carbon that is predominantly in organic compounds.

Based on the data obtained by the VEGA-1 spacecraft as it flew past comet Halley, much of the carbon in that comet is in C-, H-, O-, and N-rich (CHON) dust particles (Kissel and Krueger 1987; Jessberger et al. 1988). The C/Mg ratio in Halley dust is 6-12 times that of CI chondrites (Jessberger et al. 1988), significantly higher than for the particles captured by the Stardust mission to comet 81P/Wild 2 (Brownlee et al. 2006). This difference could be largely due to the high relative velocity (6 km/s) between the Stardust spacecraft and the Wild 2 particles at the time they were captured. The particles were captured in a very low-density medium, but the large kinetic energies of the cometary particles would have favored the preservation of only the more competent and refractory materials. Nevertheless, variations in composition between comets cannot be ruled out.

For comet Halley, Delsemme (1991) estimated a bulk composition of 43 wt% water, 26 wt% organics, and 31 wt% silicates. If most of the organics is in CHON-like material (Kissel and Krueger 1987) that is ~70 wt% carbon, this would mean that Halley-like comets have ~18 wt% carbon (Table 1). Greenberg (1998) arrived at a similar composition for comets in general using a range of constraints, including the Halley data. This carbon content would correspond to a bulk H/C molar ratio for Halley of about four, not so different from ratios observed in volatile-rich CI and CM carbonaceous chondrites (molar H/C of 3-13).

Interplanetary dust particles

Interplanetary dust particles (IDPs) are small particles that are collected in the upper atmosphere (Bradley 2003). Most IDPs are < 50-100 μm across. Larger particles, known as cluster IDPs, tend to be very fragile and fragment on the collectors. Their small size and/or fluffy nature means that IDPs slow down relatively gently during atmospheric entry. Consequently, unlike their larger and/or denser counterparts that at least partially melt (cosmic spherules and micrometeorites) and often vaporize, IDPs are less severely heated on entering the atmosphere. Dynamical arguments suggest that IDPs have both cometary and asteroidal sources (e.g., Messenger et al. 2003a).

Most IDPs that have been studied have grossly (factors of 2-3) chondritic elemental compositions, except for volatile elements that can be lost during atmospheric entry heating or are subject to terrestrial contamination (Schramm et al. 1989; Thomas et al. 1993; Kehm et al. 2002). Extraterrestrial particles with non-chondritic compositions occur in collections, but these objects are more difficult to distinguish from terrestrial contaminants and so are rarely studied. The chondritic particles fall into two morphological groups—porous (CP-) and smooth (CS-) particles. The CP-IDPs are largely anhydrous, composed mostly of submicron crystalline and amorphous silicates, metal, sulfide, and organic matter. The silicate compositions tend to be very heterogeneous. Because of their very fine-grained nature, high degree of chemical disequilibrium, and high concentration of organic matter and presolar grains, CP-IDPs are thought to be the most primitive solar system materials available for study. In the CS-IDPs, the silicates are dominated by hydrous minerals (predominantly serpentines and clay minerals). Although not identical, the CS-IDPs resemble the matrices of the aqueously altered chondrites. Because of their resemblance to chondrites and the general belief that comets did not experience aqueous alteration, it is widely assumed that CS-IDPs are derived from asteroids. On the other hand, there are no meteorites that resemble CP-IDPs, which, along with their apparently very primitive nature, have led most researchers to conclude that CP-IDPs are from comets. A cometary origin of at least some CP-IDPs is supported by the high abundances of isotopically anomalous organic matter and presolar grains in them (e.g., Floss et al. 2006; Busemann et al. 2009; Duprat et al. 2010).

On average, the abundance of carbon in IDPs is ~12 wt% (Thomas et al. 1993), roughly three times that in CI chondrites, although there is considerable variation. Most of this carbon is in organic matter. Messenger et al. (2003b) reviewed the hydrogen, nitrogen and carbon isotopic compositions of IDPs. Their bulk carbon isotopic compositions fall within the terrestrial range, but their average isotopic composition of $\delta^{13}C = -45‰$ is lighter than any bulk meteorites or the bulk Earth (although for Earth the bulk carbon isotope composition is not precisely known, see below). Hydrogen and nitrogen isotopic compositions vary enormously within and between IDPs. The most striking features of the hydrogen and nitrogen in IDPs are their often very large deuterium and ^{15}N enrichments. These enrichments are poorly correlated despite generally being associated with organic matter (see also: Aléon et al. 2003; Keller et al. 2004; Floss et al. 2006; Busemann et al. 2009). The hydrogen and nitrogen isotopic variations are similar to those seen in the more primitive chondrites (e.g., Floss and Stadermann 2009) and their organic matter (e.g., Busemann et al. 2006) when they are analyzed at similar spatial scales to IDPs.

Meteorites

With the rare exceptions of Lunar and Martian falls, meteorites are fragments of asteroids from the asteroid belt located between the orbits of Mars and Jupiter. They are broadly divided into primitive chondrites, and achondrites that have undergone melting and differentiation on their parent bodies (e.g., Scott and Krot 2003). Achondrites generally have low abundances of carbon and other volatiles (hydrogen, nitrogen, and noble gases), presumably because their carriers were largely destroyed during melting. The notable exceptions are the ureilites that contain up to ~5-7 wt% in poorly graphitized carbon and the shock-produced high-pressure carbon allotropes diamond and lonsdaleite. The ureilites are a potent potential source of Earth's carbon. However, their CI-CM-normalized hydrogen and nitrogen contents are much lower than for carbon. Later on, we will argue that the relative abundances of Earth's volatiles are CI-CM-like. Hence, if ureilites were a major source of Earth's carbon, sources that were strongly depleted in carbon relative to the other volatiles must be found. To date, no such sources have been identified. Consequently, the ureilites notwithstanding, here we will concentrate on the chondrites as the most likely sources of Earth's volatiles.

Historically, the chondrites have been divided into three classes based on their compositions and mineralogies (ordinary, carbonaceous, and enstatite). These three classes in turn have been subdivided into a number of groups: ordinary chondrites into H, L, and LL; carbonaceous chondrites into CI, CM, CR, CV, CO, CB, CH, and CK; and enstatite chondrites into EH and EL. The name carbonaceous chondrite is a historical one and is a bit misleading because some ordinary and enstatite chondrites contain more carbon than some carbonaceous chondrites (Table 4). The chondrite classification scheme is still evolving as more meteorites are found—two new classes (R and K chondrites) have been identified, and a number of individual meteorites do not belong to any recognized group.

After formation, the chondrites experienced secondary modification (thermal metamorphism and aqueous alteration) on their parent bodies. A petrographic classification scheme for secondary processes divides the chondrites into 6 types—types 3 to 6 reflect increasing extent of thermal metamorphism, and types 3 to 1 reflect increasing degrees of aqueous alteration. By convention, the chemical classification is followed by the petrologic one (e.g., CI1, CM2, CV3).

The chondrites are principally made up of three components—chondrules, refractory inclusions, and fine-grained matrix—whose relative abundances vary widely (Brearley and Jones 1998). Refractory inclusions and chondrules are high-temperature objects, and normally contain little or no carbon. Almost all the carbon in the most primitive chondrites is found in their matrices, as are the other most volatile elements and presolar circumstellar grains. The matrix abundances of these materials in all chondrites are similar to those in CI chondrites (Huss 1990; Huss and Lewis 1994; Alexander et al. 2007).

The main carbonaceous component in chondrites is organic, but aqueously altered meteorites often also contain carbonate (Table 4). The organics in chondrites also account for a significant fraction of the hydrogen and most of the nitrogen in the bulk chondrites, as well as

Table 4. The range of abundances and isotopic compositions of carbon in bulk and in the directly determined major components—insoluble organic material (IOM) and carbonates (carb)—in the main chondrite groups. Shock-heated meteorites were not included in the calculation of the ranges. Tagish Lake is a carbonaceous chondrite that has affinities with the CM and CI chondrites, and fell on January 18th, 2000.

		Min. C (wt%)	Max. C (wt%)	Min. $\delta^{13}C$ (‰)	Max. $\delta^{13}C$ (‰)	Ref.
CI						
	Bulk	3.5	3.9	−14.7	−4.3	1
	IOM	2	2.3	−17.0	−17.0	2
	Carb.	0.014	0.420	47.8	61.0	3
CM						
	Bulk	1.2	3.1	−19.9	3.7	1
	IOM	0.74	1.3	−34.2	−8.4	2,4
	Carb.	0.023	0.55	3.0	67.8	3
CR						
	Bulk	1	2.7	−11.5	0.0	1
	IOM	0.34	1.5	−26.6	−20.3	2,4
	Carb.	0.0023	0.048	49.9	65.4	3
Tagish Lake						
	Bulk	4	4.1	9.4	14.0	1
	IOM	1.7	1.9	−14.3	−13.1	5
	Carb.		1.3		67.6	6
CV						
	Bulk	0.27	1.1	−17.1		1,7
	IOM	0.12	0.68	−15.4	−6.8	2,4
CO						
	Bulk	0.12	0.84		−7.1	1,7
	IOM	0.04	0.48	−13.9	−4.5	2,4
	Bulk					
OC						
	Bulk	0.01	0.59	−22.6		1,8
	IOM	0	0.36	−23.7	−10.4	2,4
	Carb.	0	1493	−6.4		3
EC						
	Bulk	0.29	0.7	−14.1	−4.1	9

[1]Alexander et al. (2012). [2]Alexander et al. (2007). [3]Grady et al. (1988). [4]Alexander et al. (2010). [5]Alexander et al. (2012). [6]Grady et al. (2002). [7]Kerridge (1985). [8]Grady et al. (1982). [9]Grady et al. (1988).

being associated with the carrier of most of the noble gases. Carbon is also present in "primitive" meteorites as nanodiamonds that can make up to ~1,000 ppm of the matrix in chondrites (Huss 1990). In bulk, the nanodiamonds contain a trace noble gas component (so-called Xenon-HL) that is clearly of nucleosynthetic origin, so at least some of them are circumstellar. However, nanodiamonds have bulk carbon and nitrogen bulk isotopic compositions that are solar (e.g., Ott et al. 2012), and a dual origin (solar system and inheritance from the presolar molecular cloud) cannot be ruled out. Other trace carbonaceous components include circumstellar SiC and graphite that formed around supernovae, novae, and AGB stars prior to the formation of the solar system (e.g., Nittler 2003).

In all chondrite groups, the carbon content decreases with increasing thermal metamorphism, hence the low minimum carbon contents of CV, CO, and ordinary chondrites. Generally, little or no carbon remains by the type 3-4 transition, and if any remains much of it may be terrestrial contamination (Alexander et al. 1998). The CK and R chondrites have all been heavily metamorphosed, which is why they have not been included in Table 4. Only in the highly reduced enstatite chondrites is carbonaceous material preserved to higher metamorphic grades (Grady et al. 1986; Alexander et al. 1998). Aqueous alteration can also modify this organic matter, but the effects tend to be subtler (Alexander et al. 2007, 2010; Herd et al. 2011). Hence, to understand the origins of this organic material, here we concentrate on the most primitive chondrites (CI, CR, CM, and Tagish Lake) available to us.

The organic matter in chondrites—relationship to IDPs, comets, and ISM

The organic matter in chondrites can be divided into soluble (SOM) and insoluble (IOM) fractions (Gilmour 2003; Pizzarello et al. 2006). Even in the most primitive meteorites, the concentrations of traditionally defined SOM, that are soluble in typical solvents (e.g., water, methanol, toluene, etc.), is no more than a few hundred parts per million. The SOM is composed of a very complex suite of compounds that include amino acids, carboxylic acids, and polycyclic aromatic hydrocarbons (PAHs). The IOM and carbonates are usually assumed to make up the bulk (>75%) of the carbon in chondrites. However, the true IOM, that which remains insoluble even after demineralization with acids, only comprises about half of the bulk carbon in primitive chondrites (Table 4). The remainder is in a poorly understood material that is insolvent until hydrolyzed by acids, but it has a composition that is probably not very different from the IOM (e.g., Alexander et al. 2012). For this discussion, it will be assumed that the IOM and the hydrolysable material are closely related.

The most primitive meteoritic IOM has an elemental composition ($C_{100}H_{75}N_4O_{15}S_4$; Alexander et al. 2007) that is similar to the average composition ($C_{100}H_{80}N_4O_{20}S_2$) of comet Halley CHON particles (Kissel and Krueger 1987). The IOM appears to be composed of small PAHs decorated and cross-linked by short, highly-branched aliphatic material (e.g. Cody et al. 2002). At least in terms of hydrogen, nitrogen, and carbon isotopes, the bulk compositions and range of compositions seen in the most primitive meteoritic IOM (Busemann et al. 2006) also resembles the IOM in the most primitive IDPs (Aléon et al. 2000; Messenger 2000; Brownlee et al. 2006; Busemann et al. 2009). The large deuterium and ^{15}N isotopic enrichments in IOM in chondrites and IDPs are thought to be the result of ion-molecule reactions and other processes in the ISM (Robert and Epstein 1982; Yang and Epstein 1983; Messenger 2000; Aléon et al. 2003) and/or the outer solar system (Aikawa et al. 2002; Gourier et al. 2008). There are variations in the isotopic composition of IOM within and between chondrites and IDPs, but these mostly reflect parent body processing in chondrites (Alexander et al. 2007, 2010; Herd et al. 2011) and atmospheric entry heating of IDPs (Bockelée-Morvan et al. 2004). The organic particles in the Stardust samples also have hydrogen and nitrogen isotopic anomalies (Brownlee et al. 2006; McKeegan et al. 2006; Matrajt et al. 2008; De Gregorio et al. 2010). The hydrogen isotope anomalies in the Stardust samples are more subdued than in the most primitive chondrites and IDPs, but this difference may reflect modification during capture. A further link between the organics in chondrites, IDPs, and comets are the presence of organic nanoglobules in all three (Flynn et al. 2006; Garvie and Buseck 2006; De Gregorio et al. 2010; Matrajt et al. 2012). Thus, the organics in meteorites, CP-IDPs, and comets appear to be related despite their parent bodies' very different formation conditions and locations.

The bulk carbon content of CI chondrites, which is largely in organic material, represents ~10% of the carbon available (assuming a solar composition for the primordial solar system). The carbon contents of comet Halley and the parent bodies of IDPs may be much higher (30-100% of available carbon), although the uncertainties are large (e.g., Alexander 2005). Nevertheless, whether the organic matter was inherited from the protosolar molecular cloud or formed in the solar system, the formation mechanism was relatively efficient.

THE SOLAR SYSTEM: DYNAMICS

Despite its apparently well-ordered orbital architecture, the solar system's history is thought to have been dynamic. Indeed, there are at least two phases of evolution during which Earth could very well have been prevented from forming or have been destroyed. Our understanding of these phases comes from models of orbital dynamics and should thus not be interpreted as absolute truth. Indeed, Alexander et al. (2012) have questioned some of the predictions of the "Nice" and "Grand Tack" models that will be described below. Nonetheless, it is important to realize that simpler models systematically fail to reproduce the observed orbital architecture of the solar system (e.g., Raymond et al. 2009).

During the first dangerous phase, interactions with the gaseous protoplanetary disk may have caused Jupiter to migrate inward to just 1.5 AU, in the immediate vicinity of Earth's formation zone, before "tacking" and migrating back outward beyond 5 AU (the "Grand Tack" model of Walsh et al. 2011). This excursion of Jupiter into the inner solar system may explain the relatively small mass of Mars. Had the timing of events, particularly of Saturn's orbital migration relative to Jupiter's, been different, Jupiter might well have continued migrating inward and decimated Earth's building blocks. In the second dangerous phase the giant planets are thought to have undergone a dynamical instability that re-arranged their orbital architecture and caused the late heavy bombardment (LHB; the instability is referred to as the "Nice" model of Gomes et al. 2005; Tsiganis et al. 2005; and Morbidelli et al. 2007). Although this did cause a large increase in the impact flux throughout the solar system, it was an extremely weak instability: the eccentricity distribution of extra-solar planets suggests that between 50 and 90% of all planetary systems with giant planets undergo much stronger instabilities that would likely have destroyed Earth, or at least provoked collisions between the terrestrial planets (Raymond et al. 2010, 2011). Even in the current epoch, chaotic dynamics driven by gravitational interactions between planets allow for the possibility of a dynamical instability and future collisions between the terrestrial planets (Laskar and Gastineau 2009).

We now review the current paradigm of the formation history of the solar system, focusing on the growth of the terrestrial planets and how they acquired their volatiles, notably their carbon. For a more detailed account of the physics of planet formation, the reader is directed to recent reviews on the subject, notably Papaloizou and Terquem (2006) and Morbidelli et al. (2012).

The solar system formed from a disk of gas and dust orbiting the young Sun, sometimes called the "solar nebula" or "protosolar nebula." This disk is assumed to have had the same bulk composition as the Sun, but the composition of the gas and dust phases was not uniform throughout the disk. Driven by both stellar and viscous heating, the disk was hotter closer to the Sun and cooler farther away. Given the temperature-dependent condensation sequence, any given species remains in the gas phase interior to an orbital radius that corresponds to its condensation temperature for the relevant, very low pressure, and should condense beyond that radius. For example, interior to the "snow line," where the temperature is ~170 K, water will remain in the vapor phase, but beyond the snow line it condenses as ice. Likewise, each species in a disk has its own condensation line, exterior to which it will condense and interior to which it remains in the gas phase. The composition of solids that formed within the disk thus reflects the local temperature and the disk's composition; this confluence is called the *condensation sequence* (Pollack et al. 1994; Lodders 2003).

The dominant carbon-bearing condensable species in disks are CH_4 and CO (e.g., Dodson-Robinson et al. 2009), although, as noted in the preceding section, carbon-rich dust that is not necessarily seen astronomically may constitute another important, possibly dominant, source of carbon. Polycyclic aromatic hydrocarbons (PAHs) are observed to be abundant in the interstellar medium (Tielens 2008) and in carbonaceous meteorites (see previous section), and may thus constitute an important source of carbon (Kress et al. 2010). Two important orbital radii in terms of carbon abundance have been proposed: a radius analogous to the snow line beyond which carbon can condense—referred to as the "tar line" by Lodders (2004); and the radius interior to which the PAHs that were initially presented in the disk are destroyed by thermally-driven reactions—dubbed the "soot line" by Kress et al. (2010). However, complex, refractory organics like those seen in meteorites, IDPs and comets do not form spontaneously via condensation of CH_4, CO, etc. at low temperatures. Fischer-Tropsch-type (FTT) synthesis can occur if the appropriate catalysts are present, but there is no evidence that the organics in meteorites, for instance, formed by this mechanism (e.g., Alexander et al. 2007). The IOM does start to break down at temperatures of 300-400 °C, thus a more appropriate definition for a "tar line" would be the isotherm beyond which the IOM-like material is stable in the solar nebula.

Protoplanetary disks are not static, but evolve in time. The gaseous component of the disk is dissipated on a timescale of a few million years, as inferred from observations of infrared excesses in the spectra of young stars (Haisch et al. 2001; Meyer et al. 2008). As the disk dissipates it cools, such that the location of the various condensation fronts move inward in time (Sasselov and Lecar 2000; Ciesla and Cuzzi 2006; Dodson-Robinson et al. 2009). The exact positions of the condensation fronts depend on the disk's detailed temperature and pressure structure, which in turn is determined by poorly constrained physical characteristics, such as the viscosity and the opacity (e.g., Garaud and Lin 2007). In addition, dust particles do not remain on static orbits within the disk but migrate radially due to drag forces and pressure, although models suggest that dust migration probably does not produce large pileups (Hughes and Armitage 2010).

Terrestrial planets and giant planet cores form from the dust component of the disk. Giant planets subsequently accrete massive gaseous atmospheres directly from the gaseous component of the disk. Given the strong constraint that gaseous protoplanetary disks only survive for a few million years (Haisch et al. 2001), and the equally strong isotopic constraints that Earth's accretion lasted for at least 11 to 30 m.y. (Yin et al. 2002; Kleine et al. 2002) and possibly closer to 50 to 100 m.y. (Touboul et al. 2007), it appears that gas giant planets form faster than the much smaller terrestrial planets. This difference is counter-intuitive, given that the giant planets' cores are thought to be more massive than Earth (Guillot 2005), and may be explained by the simple increase in the amount of condensable material in the giant planet-forming part of the disk because it lies beyond the snow line (Stevenson and Lunine 1988; Kokubo and Ida 2002). Although the model we present is the current paradigm, we note that there exists an alternate "top-down" model for giant planet formation that invokes gravitational instability in the gas disk (Boss 1997).

Planet formation occurs in a series of dynamical steps. First, dust grains agglomerate to form mm- to cm-sized pebbles or possibly even m-sized boulders via low-velocity collisions (e.g., Blum and Wurm 2008). Next, planetesimals form from the pebbles/boulders, probably by hydrodynamical processes that efficiently concentrate these particles to a large enough degree to create gravitationally-bound clumps (Chiang and Youdin 2010; and references therein). The sizes of planetesimals is debated: some formation and collisional models suggest that planetesimals are "born big," with radii of hundreds to a thousand kilometers (Johansen and Youdin 2007; Cuzzi et al. 2008, 2010; Morbidelli et al. 2009; Chambers 2010), while other models argue in favor of much smaller, sub-km sized planetesimals (Weidenschilling 2011).

The next phase in the growth of solid bodies is the accretion of planetary embryos from planetesimals. A planetesimal that grows larger than its neighbors can rapidly increase its collisional cross section due to gravitational focusing and undergo runaway accretion (Safronov and Zvjagina 1969; Greenberg et al. 1978; Wetherill and Stewart 1993). Runaway accretion slows to so-called "oligarchic growth" as the orbits of nearby planetesimals are excited by the growing embryo (Ida and Makino 1992; Kokubo and Ida 1998). This process is thought to produce a population of embryos with comparable masses: roughly lunar to Mars mass in the terrestrial planet region, and ~Earth mass in the giant planet region (Kokubo and Ida 2002).

A giant planet core needs to reach at least 5 to 10 Earth masses before it can efficiently accrete gas from the disk (Ikoma et al. 2000), but the growth of giant planet cores from Mars- to Earth-mass embryos remains poorly understood. In this size range, embryos are far more efficient at scattering planetesimals than accreting them, and standard growth models fail to produce giant planet cores in the lifetime of the gaseous disk (Thommes et al. 2003). However, two proposed mechanisms may help to solve this problem. First, embryos are massive enough to have thin gaseous envelopes, which act to enhance their collisional cross section by a factor of up to ten or more in radius (Inaba and Ikoma 2003). Second, embryos probably experience large-scale orbital migration via both the back-reaction from planetesimal scattering (Fernandez and Ip 1984; Kirsh et al. 2009) and via tidal interactions with the gaseous protoplanetary disk, often referred to as "type 1" migration (Goldreich and Tremaine 1980). Hydrodynamical models show that there exist locations in the disk where type 1 migration is convergent (Lyra et al. 2010; Paardekooper et al. 2011). A combination of these mechanisms may allow relatively small embryos to rapidly grow into full-fledged giant planet cores (e.g., Levison et al. 2010; Horn et al. 2012), although this remains an area of active research.

Giant planet cores accrete gas from the disk at a rate that is limited by their ability to cool and contract (i.e., their atmospheric opacity), thus freeing up space within their Hill sphere for additional gas (e.g., Hubickyj et al. 2005). Thus, gas accretion initially proceeds at a relatively slow pace; the combination of this time lag and the disk's limited lifetime may explain why Neptune-sized planets are so much more common than Jupiter-sized planets among exoplanets (e.g., Howard et al. 2010). Once the mass in a core's gaseous envelope is comparable to the solid core mass, gas accretion enters a runaway phase and the planet becomes a gas giant in just ~10^5 years (Mizuno 1980). As the planet accretes, it clears an annular gap in the disk, which then constrains its accretion rate and final mass (e.g., Lissauer et al. 2009).

Once a giant planet clears a gap in the disk its orbital evolution is inextricably linked to the disk dynamics. This behavior occurs because the gap prevents gas exterior to the planet's orbit from interacting viscously with gas interior to the planet's orbit. As the disk viscously evolves and spreads radially, the vast majority of the gas spirals inward onto the star, and a small fraction of the mass spreads to large orbital radii to balance the angular momentum budget (Lynden-Bell and Pringle 1974). In the giant planet forming part of the disk, the gas flow is generally inward such that a gas giant is essentially dragged inward on the disk's viscous timescale of 10^{5-6} years in a process called "type 2" migration (Lin and Papaloizou 1986).

Of course, orbital migration of both type 1 and type 2 only occurs in the presence of the massive but short-lived gaseous component of the disk. When the disk dissipates, the gas giants have necessarily reached close to their final masses and orbits, but the most dynamic phase of terrestrial planet formation is in full swing.

The final phase in the growth of the terrestrial planets involves the collisional accumulation of planetesimals and planetary embryos. Embryos initially remain isolated from one another as they grow by accreting planetesimals and stay on low-eccentricity orbits via dynamical friction (Kokubo and Ida 1998). Embryos eventually start to interact with each other once the local surface density in planetesimals and embryos is comparable (Kenyon and Bromley 2006). Late-stage terrestrial accretion is thus characterized by slow growth by planetesimal-embryo impacts and punctuated growth by giant embryo-embryo impacts (Wetherill 1985). Each planet's feeding zone spreads outward and widens in time during this chaotic phase of strong gravitational scattering (Raymond et al. 2006; Fig. 2).

A planet's feeding zone determines the mixture of material that condensed in different regions and at different temperatures, and thus the planet's final composition. The temperature

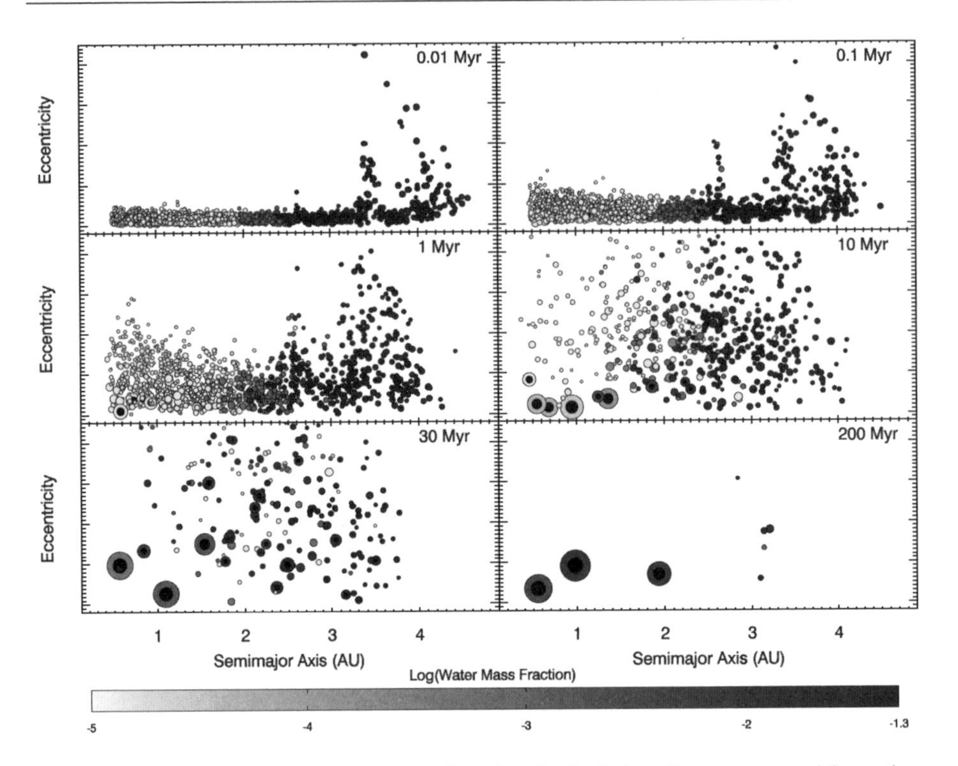

Figure 2. Six snapshots in time of the orbital configuration of a simulation of late-stage terrestrial accretion from Raymond et al. (2006). The size of each body is proportional to its relative physical size and scales as its mass$^{1/3}$. Jupiter was also included in the simulation on a circular orbit at 5.5 AU (not shown in plot). The color of each particle corresponds to its water content: particles inside 2 AU started the simulation dry, particles that started from 2 to 2.5 AU have 0.1% water by mass, and particles that started beyond 2.5 AU have 5% water by mass; during collisions the water mass fraction is a calculated using a simple mass balance. In this particular simulation three terrestrial planets formed, as did a few particles trapped in the asteroid belt. The two inner planets are reasonably good Venus and Earth analogues in terms of their masses and orbits, but the outer planet is roughly nine times more massive than the real Mars. The Earth analogue in this simulation accreted a large volume of water of roughly ten times Earth's current water budget, although water loss during impacts was not accounted for. Recall that we expect carbon delivery to Earth to have followed the same dynamical pathway as water delivery.

at 1 AU is usually thought to have been too hot throughout the disk phase for water to condense (Boss 1998), although some disk models yield low enough temperatures at 1 AU to allow water to condense (Woolum and Cassen 1999; Sasselov and Lecar 2000). Although it is possible that water vapor could be adsorbed onto grains at 1 AU (Muralidharan et al. 2009), it is generally thought that water must have been delivered to Earth by impacts with icy bodies that formed at cooler temperatures. Primitive outer asteroidal material (C-complex) is a likely source of Earth's water (Morbidelli et al. 2000; Raymond et al. 2004, 2007). C-complex asteroids are thought to be the sources of the carbonaceous chondrites, and of them CI and CM chondrites have compositions that most closely resemble the volatile elemental and isotopic composition of Earth (Alexander et al. 2012). And it is during an advanced accretion phase that Earth accreted the bulk of its water (Fig. 2). As we will discuss below, the same arguments hold for Earth's carbon.

Measurements of isotopic chronometers, especially the hafnium-tungsten system, in meteorites and Earth rocks tell us that Earth's last giant impact occurred at roughly 11 to 150 Ma

(Kleine et al. 2002; Yin et al. 2002; Touboul et al. 2007). This event is thought to have been an impact with a ~Mars-sized body, which spun out a disk of material to form the Moon (Benz et al. 1986; Canup and Asphaug 2001). Mars, on the other hand, appears not to have undergone any giant impacts after just a few Ma (Dauphas and Pourmand 2011). Mars may therefore represent a remnant embryo, while Earth is a fully-grown planet that underwent many giant impacts.

After the end of the giant impact phase there remains a population of remnant planetesimals. The final sweep up of these planetesimals occurs by gravitational scattering by the planets, and these planetesimals end up either colliding with the Sun, a terrestrial or giant planet, or getting scattered outward into interstellar space. The fraction of planetesimals that collide with the terrestrial planets provided the so-called "Late Veneer," which is characterized by the incorporation of highly-siderophile elements (HSEs) into the planets' mantles (Kimura et al. 1974). Given that HSEs are "iron-loving," they should have been sequestered into the core during planetary differentiation, and so their presence in the mantle indicates that they were accreted after the last core formation event or at least after the time when there was significant core-mantle re-equilibration. The "Late Veneer" phase represents the tail end of accretion, and the source region of "Late Veneer" impactors is constrained by comparing the bulk mantle composition with meteorites (e.g., Dauphas et al. 2004; Burkhardt et al. 2011). At even later times, a very small amount of material was added to Earth during and after the late heavy bombardment (Gomes et al. 2005, Bottke et al. 2012).

Figure 2 shows the evolution of a dynamical simulation of late stage terrestrial accretion from Raymond et al. (2006). In the simulation, accretion transitions from the primordial accretion to the "Late Veneer" after the last giant impact that was energetic enough to stimulate core-mantle equilibration (and thus reset the hafnium-tungsten chronometer). In this simulation the last giant impact occurred at ~60 Ma for an Earth analog planet (and 20 Ma for a Venus analog).

Accretion simulations like the one from Figure 2 have succeeded in reproducing a number of characteristics of the actual terrestrial planets, such as their number and approximate total mass (Wetherill 1990) and their low orbital eccentricities and inclinations (Raymond et al. 2006, 2009; O'Brien et al. 2006; Morishima et al. 2008, 2010). In addition, these simulations were able to explain the origin and isotopic signature of Earth's water (Morbidelli et al. 2000; Raymond et al. 2007).

However, these simulations failed to produce realistic Mars analogs. Planets at Mars' orbital distance were systematically a few to ten times more massive than the real Mars (0.11 Earth masses; Wetherill, 1991; Raymond et al. 2006, 2009; O'Brien et al. 2006). Improvements in numerical resolution and the inclusion of additional physical effects have been unable to solve the Mars problem (Morishima et al. 2010).

The only successful solution to the Mars problem came from a change in the "initial conditions." The aforementioned simulations started from disks of planetesimals and embryos that stretched from a few tenths of an AU out to 4 to 5 AU, just inside Jupiter's current orbit. Mars formed big simply because there was too much mass in the Mars' feeding zone and no mechanism to clear out that mass. However, Hansen (2009) showed that if the terrestrial planets only formed from a narrow annulus of embryos from 0.7 to 1 AU, a small Mars is produced naturally as an edge effect. Earth and Venus are big because they formed within the annulus of embryos, and Mars and Mercury are small because they were scattered beyond the annulus, thus limiting their growth. In fact, Hansen (2009) was able to reproduce the masses and orbital configurations of all of the terrestrial planets. The only problem was that his initial conditions were completely ad-hoc.

Walsh et al. (2011) proposed a new model called the "Grand Tack," which provides a solution to the Mars problem while remaining consistent with the large-scale evolution of the solar system (Fig. 3). The model relies on the fact that, because the giant planets form much faster than the terrestrial planets, they can sculpt the disk of terrestrial embryos.

The Grand Tack model starts late in the gas disk phase, when Jupiter was fully formed but Saturn had not yet grown to its final size. Jupiter migrated via the type 2 mechanism inward from its formation zone, assumed to be at a few AU, perhaps just beyond the snow line. Meanwhile, Saturn started to accrete gas and migrated inward rapidly, catching up with Jupiter. Hydrodynamical simulations show that Saturn is naturally trapped in the 2:3 mean motion resonance with Jupiter (Pierens and Nelson 2008). When the two planets are locked in this configuration, their direction of migration is switched to outward (Masset and Snellgrove 2001;

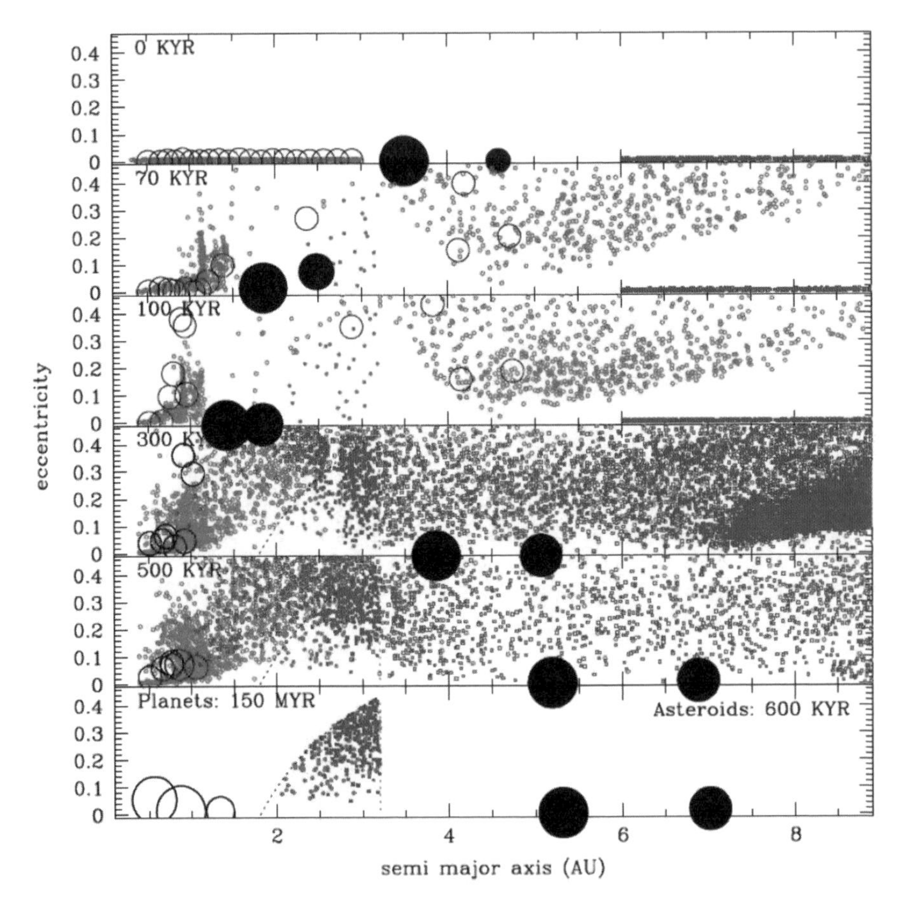

Figure 3. Snapshots in time of a simulation that represents the Grand Tack model (Walsh et al. 2011). The simulation starts with three parts: (i) an inner disk of planetary embryos (open circles) and dry (red) planetesimals; (ii) a fully-formed Jupiter at 3.5 AU and Saturn's core trapped in 2:3 resonance (large black dots); and (iii) a pristine outer disk of water-rich planetesimals. Jupiter and Saturn migrate inward together while Saturn continues to grow by accreting gas from the disk. Once Saturn reaches its final mass the two planets' migration reverses direction and they migrate outward until the disk dissipates and leaves them stranded at 5.4 and 7.1 AU. (See Walsh et al. 2011 for a discussion of plausible orbital histories of Jupiter and Saturn). The inner disk is truncated at 1 AU and the terrestrial planets that formed are similar to the actual ones, including Mars. Water and carbon are delivered to Earth in the form of outer disk planetesimals that "overshoot" the asteroid belt as they are scattered inward by Jupiter during the giant planets' outward migration. They can be seen in the 300 and 500 kyr panels as the blue symbols at low semi-major axis and generally high eccentricity.

Morbidelli and Crida, 2007; Pierens and Nelson 2008; Pierens and Raymond 2011). Jupiter and Saturn migrate outward until the disk dissipates; stranding them close to their current orbits.

If Jupiter's "tack"—i.e., change in the direction of its migration—occurred at ~1.5 AU, then the inner disk of embryos is naturally truncated at about 1 AU. The terrestrial planets that form from this disk are a quantitative match to the real ones (Walsh et al. 2011). Although Jupiter migrates through the asteroid belt twice (once going inward, once outward), it is repopulated by two distinct sources. The inner belt contains planetesimals that originated interior to Jupiter's orbit and were subsequently scattered outward during Jupiter's inward migration, and then back inward during Jupiter's outward migration. The efficiency with which inner disk planetesimals are implanted into the final asteroid belt is only ~0.1%, but this is actually in very good agreement with the current belt. The outer asteroid belt contains planetesimals that originated beyond Jupiter's orbit, either between the giant planets or beyond the ice giants' orbits, and were scattered inward during outward migration and implanted with an efficiency of ~1%. Thus, in the context of the Grand Tack model, S-complex asteroids represent leftovers from the terrestrial planet region and C-complex asteroids are leftovers from the giant planet region.

Water delivery to Earth occurs in the Grand Tack model in the form of C-complex material that is scattered inward past the asteroid belt during Jupiter's outward migration (Walsh et al. 2011). For every C-complex planetesimal that was implanted into the main belt, roughly 10 to 20 planetesimals were scattered onto orbits with roughly the same semi-major axis but with eccentricities high enough to intercept the inner annulus of embryos. This C-complex material therefore represents a "pollution" of the inner annulus of embryos by C-complex planetesimals at the few percent level. Accretion simulations including this tail of C-complex material show that it does indeed deliver about the right amount of water to Earth. In the context of the Grand Tack model, carbon was delivered to Earth by C-complex material, along with the water and other volatiles (see next section).

To conclude this section, we emphasize that although studies have mainly focused on the delivery of water to Earth, the delivery of carbon represents a problem of comparable magnitude. Carbon chemistry in protoplanetary disks is complex, and the dominant species may have been CH_4, CO, refractory carbonaceous dust, and/or perhaps free PAHs. However, since carbonaceous chondrite meteorites are thought to represent C-complex asteroids, and since these represent the probable source of water on Earth, carbon probably followed the same path as water in its delivery to Earth.

CLUES TO THE ORIGIN OF CARBON ON EARTH

Terrestrial carbon inventory

Making an inventory of terrestrial carbon is not an easy task. The sizes of the main surface reservoirs of carbon (carbonates and reduced carbon in the biosphere and in hydrocarbons) are well documented (Dasgupta 2013; Hazen et al. 2013a; Jones et al. 2013; Manning et al. 2013; Ni and Keppler 2013). In contrast, the inventory of the deep carbon reservoirs is not well constrained (Dasgupta 2013; Shirey et al. 2013; Wood et al. 2013), so that estimates require calibration relative to other geochemical proxies and are somewhat model-dependent. Estimates of bulk Earth carbon content vary by over an order of magnitude (e.g., 50 ppm carbon for Zhang and Zindler 1993; ≥500 ppm carbon for Marty 2012). These values are nevertheless low compared to those of potential cosmochemical contributors (comets: 180,000 ppm; carbonaceous chondrites: 36,000 ppm, Table 1). Below we review the carbon content of the major terrestrial reservoirs.

Earth's surface. The major reservoirs of carbon at the surface of our planet are carbonates in marine and continental sedimentary rocks, and reduced carbon in the biosphere and in fossil hydrocarbons. Carbonates occur as continental massive units from ancient oceanic platforms, as oceanic sediments, and as veins and alteration phases in the oceanic crust (Hazen et al. 2013a,b). The amount of reduced carbon is lower than the carbonate reservoir by about a factor of ~4 to 5. Current estimates of the total amount of carbon at Earth's surface are around 7-11 × 10^{21} moles (Ronov and Yaroshevskhiy 1976; Javoy et al. 1982; Holser et al. 1988; Hayes and Waldbauer 2006). Carbonates have $\delta^{13}C$ values of around 0‰ relative to PDB (the standard PDB is a cretaceous carbonate) and terrestrial organic carbon is more depleted in ^{13}C, with values around −30‰, which is a signature of its biogenic origin. Consequently, the bulk $\delta^{13}C$ of the surface inventory is around −5 to −10‰. The composition of this surface carbon may not have varied greatly in the geological past, since ancient sedimentary rocks have carbonate and organic carbon values in the same range (Hayes and Waldbauer 2006). This bulk value is also surprisingly comparable to estimates of the bulk mantle value of −5‰; the surface inventory is a mixture of two pools of carbon with different isotopic compositions, one of which being of presumably biogenic origin. This similarity and the constancy of the bulk surface value with time seems to indicate that the relative proportions of carbonates and reduced carbon have been about the same throughout the geological history, which has important implications both for the exchange of carbon between Earth's surface and deep interior, and for the size of the biosphere through time.

Mantle. The carbon content of mantle-derived material is generally very low because carbon is extensively outgassed from lavas. Thus, current estimates of the mantle carbon content are indirect and based on the calibration of carbon to specific geochemical tracers whose geochemical cycles are well constrained. Here we review current estimates of the carbon content of the mantle. All of them are based on the assumption that carbon is incompatible during partial melting of the mantle source; that is, it goes quantitatively to the surface during magma generation and volcanism.

The highest rate of magma production takes place along divergent plate margins at mid-ocean ridges, which are also the locus of most of mantle degassing. Thus, most estimates are for the mantle source of such magmas, referred in the literature as the depleted mantle (DM). The rare isotope of helium, ^{3}He, is of primordial origin in Earth, that is it was trapped within Earth as it formed from a cosmochemical reservoir and it is still degassing from the mantle at present. Its abundance in the atmosphere is low as helium escapes to space. The flux of ^{3}He from the mantle has been quantified from excesses of ^{3}He in deep-sea waters (knowing the residence time of the latter). The original estimate of 1000 ± 300 moles ^{3}He/yr (Craig et al. 1975) has been recently revised downward to 527 ± 106 moles/yr (Bianchi et al. 2010). The C/^{3}He molar ratio measured in mid-ocean ridge basalts (Marty and Jambon 1987; Marty and Tolstikhin 1998) and in mid-ocean ridge hydrothermal vents is on average 2.2 ± 0.6 × 10^{9}. Together with an average partial melting rate of 12 ± 4% for mid-ocean ridge basalts (MORBs) and a total magma production rate of 21 km^3/yr at all mid-ocean ridges, one obtains a carbon content of the MORB mantle source of 27 ± 11 ppm carbon. Another approach is to analyze samples from the mantle that are not degassed, or for which degassing fractionation can be corrected for, and to calibrate carbon relative to a non-volatile trace element, such as niobium. Saal et al. (2002) and Cartigny et al. (2008) estimated the CO_2/Nb ratio of the DM at 240 and 530, respectively, resulting in estimated carbon contents for the DM of 19 ppm and 44 ppm, respectively. Hirschmann and Dasgupta (2009) and Salters and Stracke (2004) estimated the carbon content of the DM at 14 ± 3 ppm and 16 ± 9 ppm, respectively, using similar approaches. All these estimates suggest a carbon content for the DM of ~20 to 30 ppm, close to the bulk Earth estimate of Zhang and Zindler (1993) of 50 ppm.

Noble gases released at centers of hot spot volcanism (e.g., Hawaii, Yellowstone, Réunion) that are fed by mantle plumes have isotopic compositions suggesting that they come from a less degassed region of the mantle. Thus, the DM is unlikely to represent the bulk Earth inventory. ^{40}Ar is produced by the radioactive decay of ^{40}K, with a half-life of 1.25 Ga, and the total amount of radiogenic ^{40}Ar produced over 4.5 Ga from terrestrial potassium can be readily computed. A known fraction of ^{40}Ar is now in the atmosphere and the complementary amount of radiogenic ^{40}Ar trapped in silicate Earth is obtained by mass balance. For a bulk silicate Earth (BSE) potassium content of 280 ± 120 ppm (2σ) (Arevalo et al. 2009), about half of the ^{40}Ar produced over Earth's history is in the atmosphere, and the other half is therefore still trapped in silicate Earth (Allègre et al. 1996; Ozima and Podosek 2002). The carbon content of the BSE can be scaled to ^{40}Ar using the composition of gases and rocks from mantle plume provinces following the method presented in Marty (2012), which gives 765 ppm carbon for the BSE (with a large uncertainty of 420 ppm at the 2σ level when all errors are propagated).

An independent estimate of the BSE carbon content can be obtained using primordial noble gases. Extinct and extant radioactivities producing noble gas isotopes indicate that most of the non-radiogenic noble gases, such as ^{36}Ar, are in the atmosphere and their mantle abundances are order(s) of magnitude lower. The mean $C/^{36}Ar$ ratio of carbonaceous chondrites is $5.6 \pm 2.0 \times 10^7$ (C and ^{36}Ar data from Otting and Zähringer 1967) and is not significantly different from the ratio for the much less volatile-rich enstatite chondrites ($5.9 \pm 4.2 \times 10^7$, computed with data from Otting and Zähringer 1967). Assuming a chondritic source and from the ^{36}Ar atmospheric inventory, one can compute a bulk Earth carbon content that is 920 ± 330 ppm, which is within the uncertainties of the carbon content computed above using a different approach.

These independent estimates suggest that the terrestrial carbon content of the Earth is within 500 to 1000 ppm—a C content that corresponds to the contribution of about 2 ± 1 wt% carbonaceous chondrite (CI-CM) material to a dry proto-Earth (Fig. 4). If this estimate is correct, there are several important implications for the origin of volatiles on Earth:

(i) There exists large reservoir(s) of carbon in Earth that are not yet documented. The occurrence of gas-rich regions of the mantle is supported by the isotopic compositions of mantle plume-derived noble gases (helium, neon) isotope, which point to a mantle reservoir that is less degassed that the DM one;

(ii) Most (>90%) of the carbon is not at Earth's surface, but is trapped in bulk silicate Earth, contrary to the case of noble gases that are mostly in the atmosphere;

(iii) The contribution of ~2 wt% CI material for carbon and other volatile elements (Fig. 4) is higher than the so-called "Late Veneer" of about 0.3 wt% (range: 0.1-0.8 wt%) of chondritic material to account for the mantle inventory of platinum group elements that never equilibrated with metal, presumably because they were added after the main episodes of core formation (Kimura et al. 1974).

Carbon in the core? From experimental data, it has been proposed that carbon may be present in the core at the percent level (Wood 1993; Dasgupta and Walker 2008; Wood et al. 2013). A high carbon content of the core would change dramatically the carbon inventory of Earth, and also the bulk carbon isotope composition of our planet, since carbon stored in the core would be isotopically different/fractionated with respect to mantle carbon. For instance, a carbon content of 2% (Wood 1993) in the core corresponds to a BSE content of 6600 ppm carbon. If the source of this carbon was chondritic, then it should have been accompanied by ~ 6×10^{16} moles of ^{36}Ar (using the chondritic $C/^{36}Ar$ from the preceding subsection), which is an order of magnitude higher than the current estimate of the terrestrial inventory. One could argue that the noble gases were largely lost to space from the primitive atmosphere and that only the reactive volatiles, such as carbon and, presumably, nitrogen and hydrogen, were quantitatively trapped in Earth. However, the same argument against extensive sequestration of carbon into the core also applies to the halogens (Sharp et al. 2009). The chlorine content of bulk Earth is 1.1×10^{23} g, or 3.6×10^{-7} mol $^{35}Cl/g$. Because Cl is not siderophile (Sharp et al. 2009), chlorine is in the mantle, in the crust, and in the oceans. The average carbonaceous chondrite $^{35}Cl/^{12}C$

ratio is ~200. The equivalent carbon concentration of a chondritic Earth is 800 ppm, in fact remarkably comparable to the carbon content computed in the preceding subsection from the ^{40}Ar budget and from the $C/^{36}Ar$ ratio. The isotopic composition of terrestrial chlorine is similar to that of chondrites and there is, therefore, no reason to suspect that chlorine has been lost to space in early times. Whether or not the amount of carbon in the core is low is an open question that requires further metal-silicate partitioning experimental studies involving not only carbon, but also nitrogen, hydrogen, and noble gases.

Volatile (C-H-N-noble gas) elemental and isotopic constraints

Carbon isotopes are the primary source of information on the origin of this element in Earth, but the message is ambiguous since carbon has only two isotopes and, because it is a light element, the extent of its fractionation can be high and can superimpose isotope fractionation effects to source heterogeneities. The stable isotope compositions of other volatile elements, such as hydrogen (water) and nitrogen and the associated noble gases, are important supplementary sources of information on the origin of terrestrial carbon, and are discussed in this section.

The range of $\delta^{13}C$ values of mantle carbon, based on the analysis of diamonds, MORBs, mantle plume rocks, and volcanic gases is thought to be around $-5 \pm 3‰$ (Deines 1980; Des Marais and Moore 1984; Javoy et al. 1986; Marty and Zimmermann 1999). Lighter (more negative) values have been reported for a few MORB glasses and volcanic gases and have been interpreted as being due to isotope fractionation during degassing (Javoy et al. 1986). Lighter isotopic compositions in some of diamonds have generally been interpreted as due to isotopic fractionation or incorporation of recycled organic carbon (Dienes 1980; Cartigny et al. 1998). Because a large amount of carbon may be sequestrated in the core, a firm carbon isotope budget of Earth is not yet possible. However, the isotope fractionation of carbon between metal and silicate at the very high temperatures and pressures that prevailed during core formation are unlikely to have been larger than a few per mil.

Terrestrial carbon might have been supplied by the following cosmochemical reservoirs (see also section above):

(i) *The protosolar nebula.* Solar-like neon is present in the mantle and indicates that a solar-like source supplied a fraction of the terrestrial volatile elements. Whether such a source also supplied carbon to Earth is an open question. From the analysis of solar wind implanted in lunar regolith grains, Hashizume et al. (2004) proposed that the solar wind isotopic composition is light, with an upper limit for $\delta^{13}C$ of $-105 \pm 20‰$. The solar wind may be enhanced in the light isotopes due to Coulomb drag, so that the solar value could be heavier than this limit by about 20-30‰ per mass unit for light elements like carbon, nitrogen, or oxygen (Bodmer and Bochsler 1998). Thus the solar carbon isotope composition could still be light, possibly lighter than $-80‰$. If this estimate is correct, then the presence of solar carbon at depth should be reflected by a light carbon component for samples of deep mantle origin, like those associated with mantle plumes. Such an isotopically light carbon component has yet to be observed: mantle carbon shows values of around $-5‰$ whatever the source is, either MORB-like or plume-like. D/H and $^{15}N/^{14}N$ isotope systematics could only allow for a small (< 10%) fraction of terrestrial hydrogen to be of solar origin (Marty 2012; Alexander et al. 2012).

(ii) *Comets.* The D/H ratios of these bodies are generally 2-3 times the terrestrial value (Bockelée-Morvan et al. 1998, 2008, Mumma and Charnley 2012), which led to the consensus view that ocean water, and, by extension other major volatiles, like carbon and nitrogen, cannot be derived solely from comets. Recently, a Jupiter family comet, presumably originating from the Kuiper belt, has been shown to have a terrestrial-like water D/H (Hartogh et al. 2011), although Alexander et al. (2012) argue that the bulk D/H of this comet may be significantly higher if it contains Halley-like abundances of organic matter and this organic matter has a

D/H ratio like that in the most primitive meteorite organics. $^{15}N/^{14}N$ ratios of comets that have been measured to date are all much higher ($\delta^{15}N > 800‰$, where the reference is atmospheric nitrogen; Arpigny et al. 2003; Bockelée-Morvan et al. 2008; Mumma and Charnley 2012) than the range of terrestrial values ($\delta^{15}N$ from −30 to +40‰ for the most extreme end-members), which would preclude a major cometary contribution for terrestrial nitrogen. However, there may be a caveat: the nitrogen isotopic compositions have only been measured in cometary CN and HCN, and the composition of cometary NH_3, N_2 (if quantitatively trapped), and refractory organic matter are not known.

(iii) *Chondrites*. Their $\delta^{13}C$ compositions are generally negative, with values of around 0 to −15‰ (Table 4), which encompass the terrestrial carbon isotopic composition and, therefore, support a chondritic origin for this element. The hydrogen and nitrogen isotope ratios of Earth are best matched by the bulk CI chondrites, followed by the CMs, with a small input of solar material (Marty 2012; Alexander et al. 2012).

There is one notable exception for a chondritic source for terrestrial volatiles, and that is the isotopic composition of atmospheric xenon. Atmospheric xenon isotopes and, to a lesser extent, krypton isotopes are enriched by 3-4% per amu (about 1% per amu for krypton), relative to chondritic or solar compositions. Xenon is also depleted in its elemental abundance by a factor of about 20 relative to other noble gases in the atmosphere (compared to a chondritic abundance pattern). This fact, known as the xenon paradox, has been regarded as the result of early escape processes specific to the physical properties of xenon (Pepin 1991; Tolstikhin and O'Nions 1994; Dauphas 2003). Recently, Pujol et al. (2011) reported xenon isotopic compositions in Archaean rocks that are intermediate between atmospheric and chondritic ones, and suggested a long-term, non-thermal escape of atmospheric xenon through time due to ionization by solar UV light. It is not known whether such a process could have affected carbon at Earth's surface. Its effect might have been limited if carbon is indeed mostly sequestrated in the mantle. The possibility of xenon loss from the atmosphere after Earth's forming events has implications for the timing of early atmospheric evolution and of Earth's formation (see below).

Inferences on the nature of Earth's building blocks

According to our estimate, the delivery of volatile elements, including carbon, to Earth required the contribution of 1 to 3 wt% of "wet" material, having probable affinities with CI, or possibly CM, chondrites (Fig. 4). By comparison, platinum group elements, which are in chondritic proportions in the mantle, require a lower contribution of about 0.3 wt% (range: 0.1-0.8 wt%) of chondritic material after core formation. This difference suggests that the delivery of volatile elements was not a late veneer event and was already on-going during terrestrial differentiation. The depletion of nitrogen in the BSE (Fig. 4), if due to nitrogen trapping in the core, also requires early accretion of volatiles. This point is important for understanding the nature of the building blocks of our planet. The ruthenium and molybdenum isotope signatures of most meteorite groups correlate (Dauphas et al. 2004; Burkhardt et al. 2011) and show enrichments in s-process isotopes relative to Earth's. Molybdenum is moderately siderophile, while ruthenium is highly siderophile. Thus, mantle molybdenum was predominantly accreted during the main stage of planetary growth, as were volatile elements, while the ruthenium now present in the mantle was delivered as part of a "Late Veneer." Thus, the respective cosmochemical reservoirs of these deliveries must have had comparable isotopic compositions during and after Earth's accretion. While most meteorite groups show enrichments in s-process molybdenum and ruthenium isotopes relative to Earth, the exceptions are the enstatite and CI (Orgueil) chondrites for molybdenum and, presumably, for ruthenium, which both have terrestrial-like molybdenum isotopic patterns. Hence making Earth from a cosmochemical reservoir having an enstatite-like isotope composition (which has the appropriate oxygen isotope signature) with addition of "wet" material akin to CI-like material is not only compatible with volatile element compositions, but also could have provided a way to oxidize the composition of Earth.

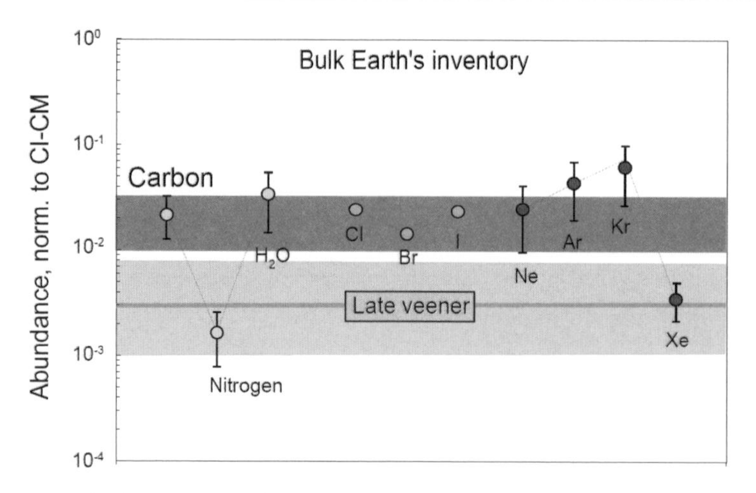

Figure 4. Normalized abundance of volatile elements in bulk Earth (data from Marty 2012). Each volatile molar abundance is divided by the mass of Earth (5.97×10^{27} g) and normalized by the corresponding content in CI-CM chondrites. Note that the terrestrial abundance pattern is close to chondritic and corresponds to the contribution of ~2 ± 1% carbonaceous chondrite material. Such contribution is in excess of the "Late Veneer" one, corresponding to the amount of highly siderophile elements in the mantle.

Importantly, if the volatile elements were accreted before the end of terrestrial differentiation as we argue here, a major fraction of them must have survived the Moon-forming impact and other giant impacts that built early Earth. An open question of this scenario is the behavior of carbon and other volatile elements during core formation if they were delivered, not as a late veneer, but during terrestrial differentiation. Much progress is expected from experimental high-pressure, high-temperature work on volatile elements in the presence of metal and silicate.

Is cosmic dust a major source of terrestrial volatiles?

Micrometeorites (MMs) and interplanetary dust particles (IDPs) are potentially important suppliers of terrestrial volatiles because: (i) the D/H distribution of MMs' values presents a frequency peak that coincides with that of terrestrial water (Engrand et al. 1999); (ii) at present, their mass flux of 10,000 to 40,000 tons/yr (Love and Brownlee 1993; Engrand and Maurette 1998) exceeds by far that of meteorites (~8 tons/yr; Bland et al. 1996); and (iii) "cosmic" dust may be rich in carbon (e.g., Matrajt et el. 2003). Marty et al. (2005) estimated that < 10% atmospheric nitrogen could have been supplied by cosmic dust after Earth's building events since 4.4 Ga ago, with essentially the correct nitrogen isotopic composition, based on the analysis of nitrogen in MMs and using the lunar surface record of ET contribution to planetary surfaces. The impact on the carbon inventory of our planet might have been even more limited since nitrogen is apparently depleted by one order of magnitude in Earth relative to carbon (Fig. 4).

From a dynamical point of view, IDPs do not represent a plausible source for the bulk of Earth's carbon. During planetary growth, dust is efficiently accreted by planetesimals and planetary embryos (Leinhardt et al. 2009; Lambrechts and Johansen 2012) such that its lifetime during the main phases of planetary accretion is short. Dust is re-generated by collisions, but is swept up by larger bodies on a timescale that is short compared with the radial drift timescale (Leinhardt et al 2009). It is only after the "dust has settled" in terms of planet formation, and the density of planetesimals and surviving embryos (a.k.a., terrestrial planets) has decreased sufficiently in the inner solar system, that dust particles are able to drift over large radial distances and thus deliver volatiles such as carbon and water to Earth. These arguments would break down if the snow-/soot-line swept in to ~1 AU during the last stretch of the gaseous protoplanetary disk phase (e.g., Sasselov and Lecar 2000), producing volatile-rich dust grains close to Earth's orbit. However, it has not been demonstrated that such a scenario could deliver the requisite amount of volatiles to Earth.

Timing of volatile delivery and retention. The time constraints for this epoch are scarce and somewhat model dependent. Inner planet bodies of the size of Mars accreted in a few Ma (Dauphas and Pourmand 2011). It takes longer to make an Earth-size planet, but how long is a matter of debate. The ^{182}Hf-^{182}W extinct radiochronometer permits one to explore metal-silicate differentiation of planetary bodies, given the half-life of ^{182}Hf (9 Ma) and the lithophile/siderophile nature of Hf/W. Earth has a different W isotope composition compared to chondrites, which allows one to set the last episode of metal-silicate differentiation, presumably the formation of the core, at 11 to 30 Ma after start of solar system condensation (e.g., Yin et al. 2002; Kleine et al. 2002). The final stage of core formation has been attributed by many to the Moon-forming impact, after which no similarly catastrophic collisions would have taken place. Touboul et al. (2007) have proposed a longer timeframe (60-120 Ma), since they could not find any W isotopic difference between Earth and Moon. They argued that, because the terrestrial and lunar mantles have different Hf/W ratios, the absence of any W isotopic difference between the two bodies implies that their last major episode of differentiation (the Moon-forming impact) took place after ^{182}Hf was decayed, in practice, $\geq$ 50 Ma. However, this assumption would not hold if Earth and the Moon's mantles have a similar Hf/W ratio, which could have been homogenized between the terrestrial and lunar mantles during or in the aftermaths of the Moon-forming impact.

Independent time constraints for Earth's formation come from the notion of atmospheric closure using the extinct radioactivity of ^{129}I (this isotope decays to ^{129}Xe with a half-life of 16 Ma). The initial abundance of ^{129}I can be estimated from that of terrestrial iodine (the stable ^{127}I isotope) and from the amount of ^{129}Xe in the atmosphere in excess of the stable Xe isotope composition. This mass balance yields a closure age of about 120 Ma after the start of solar system formation for the atmosphere (Wetherill 1975), which is also that of bulk Earth since the mantle inventory of radiogenic ^{129}Xe is small compared to the atmospheric one (e.g., Ozima and Podosek 2002). Combining this radiochronometric system with the other relevant extinct radioactivity of ^{244}Pu ($T_{1/2}$ = 82 Ma) that fissions to $^{131\text{-}136}$Xe yields a similar age of ~80 Ma (Kunz et al. 1998), since this age range is mostly determined by the decay constant of ^{129}I. However, if xenon has been escaping for a long period of time (Pujol et al. 2011), then the closure ages have to be corrected for such loss and yield < 50 Ma (range 35-50 Ma) for the "age" of the atmosphere and therefore for the main episodes of volatile delivery. Such timing is more consistent with the early age of lunar formation initially proposed on the basis of W differences between chondrites and Earth. This closure age may represent the time when the Earth retained quantitatively its most volatile elements.

In the context of the Grand Tack model (Walsh et al. 2011) the accretion of volatile-rich (C-complex) material occurs mostly in the first ~50 Ma of Earth's history (Fig. 5). In dynamical simulations, Earth analogues accrete a total of 2 to 3% of carbon-rich material, mainly from small planetesimals, with a tail that extends beyond 100 Ma (O'Brien et al. 2010) In contrast, Earth analogues in simulations with fixed giant planet orbits (e.g., the one in Fig. 2) tend to accrete far more C-complex material, typically 5 to 20% (Raymond et al. 2007, 2009).

Volatile delivery is not perfectly efficient. Some water is lost during accretion due to impact heating for very high-velocity impacts such as those from comets (e.g., Svetsov 2007). Giant impacts may also strip a fraction of water from the growing Earth (Canup and Pierazzo 2006), although the entire atmosphere must be completely removed by an impact before a large fraction of Earth's water can be stripped (Genda and Abe 2005). However, the question of volatile retention merits additional study as only a fraction of the relevant circumstances have been carefully explored.

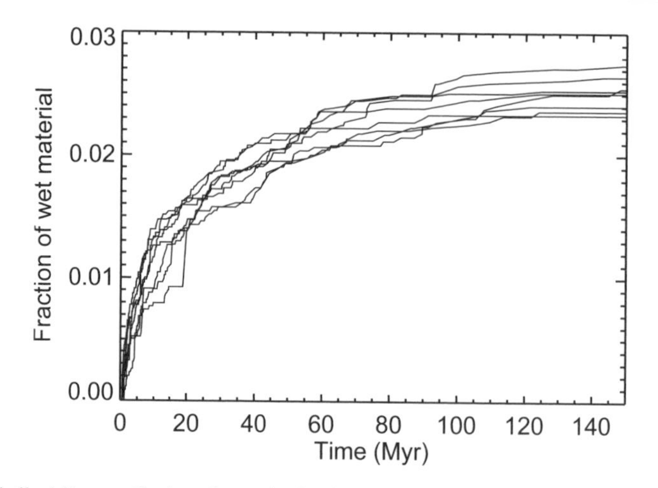

Figure 5. Volatile delivery to Earth analogues in simulations of terrestrial planet formation in the context of the Grand Tack model. Earth analogues are defined here simply as simulated planets with masses of 0.7 to 1.3 Earth masses and orbital semi-major axes between 0.8 and 1.25. The plot shows the fraction of "wet" (carbon-rich C-complex) material accreted onto each simulated planet as a function of time. Each Earth analogue accreted 2-3% of C-complex material in the form of planetesimals that were scattered inward during Jupiter and Saturn's early outward migration but that were accreted on a much longer timescale of tens to 100 Ma.

CARBON TRAPPING IN EARTH

Earth experienced very high temperatures during accretion, which might have peaked with the Earth-Moon forming event (Cameron 1997; Pahlevan and Stevenson 2007). Under these conditions, the retention of volatile elements may have been problematic. Noble gases are inert by nature and are presently concentrated in the atmosphere. Chemically reactive volatiles, such as carbon, nitrogen, and water, are able to change speciation and solubility depending on redox conditions: at low f_{O_2} (below IW) carbon, nitrogen, and hydrogen solubility in silicates increases drastically (e.g., Kadik et al. 2011), suggesting that significant amounts of these elements could have been retained in a magma ocean under reducing conditions.

From dynamical simulations, contributions from wet planetesimals akin to carbonaceous chondrites are a natural consequence of the evolution of the inner solar system. Such contributions are not likely to have occurred before the late phases of the gaseous protoplanetary disk phase, and more likely towards the last stages of terrestrial accretion, starting after a few Ma and lasting for several tens of Ma. At that time, the size of Earth was likely to have exceeded half of its present size, and in-falling wet planetesimals were quantitatively degassing upon impact (Lange and Ahrens 1982, 1986), generating a CO_2-rich steam atmosphere and probably local episodes of intense magmatism. Because the solubility of CO_2 is low in molten silicates, only a small fraction of impact-generated CO_2 could have been trapped in silicates by equilibrium dissolution, and other means of sequestering carbon in the mantle need to be found. At this stage of Earth formation, it is unlikely that reducing conditions below IW could have persisted for long intervals of time, particularly because in-falling bodies were oxidized, and the placing of significant amounts of carbon into reduced phases, such as graphite or diamond, was unlikely.

The problem of trapping CO_2 from a steam atmosphere into Earth has been addressed by Sleep et al. (2001). After the Moon-forming impact, Earth's surface evolved towards a CO_2-rich steam atmosphere overlying liquid water in a geologically short period of time (few Ma) due to the bombardment by remnant planetesimals both wet and dry (see Fig. 5). An upper limit for the CO_2 partial pressure is taken to be the one corresponding to all terrestrial carbon being in the atmosphere in the form of CO_2. The atmospheric pressure of CO_2 corresponding to a carbon inventory of 2.5×10^{22} moles (Zhang and Zindler 1993) is 21.5 MPa (215 bar), or ~0.2 GPa (2 kbar) if one uses total bulk Earth carbon content estimated by Marty (2012; 2.6×10^{23} moles). The actual CO_2 partial pressure could have been much less. Impacts of volatile-rich bodies could have continued for 10^7 to 10^8 years, while transfer of CO_2 from the atmosphere and into the mantle may have occurred on a time scale of ~1 Ma or less, corresponding to the duration of a magma ocean episode induced by a large impact (Sleep et al. 2001). For illustration, if impactors all had sizes of 100 km, a typical size for primordial asteroids, ~20,000 of them with 3-4% carbon would have supplied the required 2 wt% of Earth's mass to supply terrestrial volatiles. Over 30 Ma, this effect would have led to an average CO_2 partial pressure of only about 100 bars. In these conditions, most water would have been liquid, with a H_2O water pressure of only a few bars.

Based on analogies with present-day mid-ocean ridge systems, Sleep et al. (2001) have argued that atmospheric CO_2 precipitated as carbonates in the basaltic crust through hydrothermal circulation. Because of the hotter thermal regime of Earth, convection was much faster and the magmas were less sluggish, resulting in efficient hydrothermal trapping of carbon in the newly created crust. Foundering of crustal blocks would then transfer carbon into the mantle. Part of "subducted" carbon would then be stored at great depth, escaping recirculation to the surface and degassing. It is possible that, due to the pressure dependence of oxygen fugacity, such recycled carbon could then have been sequestrated in high-pressure phases. A critical unknown for this model is the efficiency of transfer into the mantle, since a hotter regime may also favor degassing of foundering blocks by analogy with modern hot subduction zones. Another crucial question is whether the subducted carbon could escape shallow convection and be transferred deeper into the mantle. According to Sleep et al. (2001), CO_2 trapping in the oceanic crust and transfer to the mantle would have to have been efficient in order for the temperature at Earth's surface to allow liquid water to be stable in on the order of 10^6 to 10^8 years, allowing perhaps the development of biological activity in the 50 to 100 °C temperature range.

Several areas of research that to be explored, such as the speciation of carbonates during their transfer into the mantle and the possible existence of carbon-rich fluids in the mantle lasting from this epoch, the direct injection at mantle depth of carbon-rich material during impacts, and the fact that all carbon present in Earth was probably not at the surface at the same time. We are only opening the magic book of Earth's formation and early evolution.

ACKNOWLEDGEMENTS

This study is funded by the European Research Council under the European Community's Seventh Framework Programme (FP7/2007-2013 Grant Agreement no. [267255] to BM). SNR thanks the CNRS's PNP program, the Conseil Regional d'Aquitaine, and the Virtual Planetary Laboratory lead team of the NASA Astrobiology Institute. CA was partially funded by the NASA Astrobiology Institute and NASA Cosmochemistry Grant NNX11AG67G. We thank Sara Russell, Adrian Jones, and Matthew Genge for comments and suggestions, and Robert Hazen for careful editing. CRPG contribution n° 2198.

REFERENCES

Aikawa Y, van Zadelhoff GJ, van Dishoeck EF, Herbst E (2002) Warm molecular layers in protoplanetary disks. Astron Astrophys 386:622-632

Aléon J, Engrand C, Robert F, Chaussidon M (2000) Clues on the origin of interplanetary dust particles from the isotopic study of their hydrogen-bearing phases. Geochim Cosmochim Acta 65:4399-4412

Aléon J, Robert F, Chaussidon M, Marty B (2003) Nitrogen isotopic composition of macromolecular organic matter in interplanetary dust particles. Geochim Cosmochim Acta 67:3773-3783
Alexander CMO'D (2005) Re-examining the role of chondrules in producing the volatile element fractionations in chondrites. Meteorit Planet Sci 40:943-965
Alexander CMO'D, Bowden R, Fogel ML, Howard KT, Herd CDK, Nittler LR (2012) The provenances of asteroids, and their contributions to the volatile inventories of the terrestrial planets. Science 337:721-723
Alexander CMO'D, Fogel M, Yabuta H, Cody GD (2007) The origin and evolution of chondrites recorded in the elemental and isotopic compositions of their macromolecular organic matter. Geochim Cosmochim Acta 71:4380-4403
Alexander CMO'D, Newsome SD, Fogel ML, Nittler LR, Busemann H, Cody GD (2010) Deuterium enrichments in chondritic macromolecular material – Implications for the origin, evolution of organics, water and asteroids. Geochim Cosmochim Acta 74:4417-4437
Alexander CMO'D, Russell, SS, Arden JW, Ash RD, Grady MM, Pillinger CT (1998) The origin of chondritic macromolecular organic matter: a carbon and nitrogen isotope study. Meteorit Planet Sci 33:603-622
Allègre CJ, Hofmann AW, O'Nions RK (1996) The argon constraints on mantle structure. Geophys Res Lett 23:3555-3557
Anders E, Grevesse N (1989) Abundances of the elements: meteoritic and solar. Geochim Cosmochim Acta 53:197-214
Arevalo R, McDonough WF, Luong M (2009) The K/U ratio of the silicate Earth: insights into mantle composition, structure and thermal evolution. Earth Planet Sci Lett 278:361-369
Arpigny C, Jehin E., Manfroid J, Hutsemékers D, Schulz R, Stüwe JA, Zucconi JM, Ilyin I (2003) Anomalous nitrogen isotope ratio in comets. Science 301:1522-1524
Asplund M, Grevesse N, Sauval AJ, Scott P (2009) The chemical composition of the sun. Annu Rev Astron Astrophys 47:481-522
Benz W Slattery WL Cameron AGW (1986) The origin of the Moon and the single impact hypothesis. I. Icarus 66:515-535
Bianchi D, Sarmiento JL, Gnanadesikan A, Key RM, Schlosser P, Newton R (2010) Low helium flux from the mantle inferred from simulations of oceanic helium isotope data. Earth Planet Sci Lett 297:379-386
Bland PA, Smith TB, Jull AJT, Berry FJ, Bevan AWR, Cloudt S, Pillinger CT (1996) The flux of meteorites to the Earth over the last 50 000 years. Month Not Royal Astron Soc 283:551-565
Blum J, Wurm G (2008) The growth mechanisms of macroscopic bodies in protoplanetary disks. Annu Rev Astron Astrophys 46:21-56
Bockelée-Morvan D, Biver N, Jehin E, Cochran AL, Wiesemeyer H, Manfroid J, Hutsemekers D, Arpigny C, Boissier J, Cochran W, Colom P, Crovisier J, Milutinovic N, Moreno R, Prochaska JX, Ramirez I, Schulz R, Zucconi JM (2008) Large excess of heavy nitrogen in both hydrogen cyanide and cyanogen from comet 17P/Holmes. Astrophys J 679:L49-L52
Bockelée-Morvan D, Crovisier J, Mumma MJ, Weaver HA (2004) The composition of cometary volatiles. *In:* Comets 2. Festou MC, Keller U, Weaver HA (eds) University of Arizona Press, Tucson p 391-423
Bockelée-Morvan D, Gautier D, Lis DC, Young K, Keene J, Phillips T, Owen T, Crovisier J, Goldsmith PF, Bergin EA, Despois D, Wootten A (1998) Deuterated water in comet C 1996 B2 (Hyakutake) and its implications for the origin of comets. Icarus 133:147-162
Bodmer P, Bochsler P (1998) The helium isotopic ratio in the solar wind and ion fractionation in the corona by inefficient Coulomb drag. Astron Astrophys 337:921-927
Boss AP (1997) Giant planet formation by gravitational instability. Science 276:1836-1839
Boss AP (1998) Temperatures in protoplanetary disks. Annu Rev Earth Planet Sci 26:53-80
Bottke WF, Vokrouhlicky D, Minton D, Nesvorny D, Morbidelli A, Brasser R, Simonson B, Levison HF (2012) An Archaean heavy bombardment from a destabilized extension of the asteroid belt. Nature 485:78-81
Bradley JP (2003) Interplanetary dust particles. *In:* Treatise on Geochemistry. Meteorites, Comets and Planets. Davis AM (ed) Elsevier-Pergamon, Oxford p 689-712
Brearley AJ, Jones RH (1998) Chondritic meteorites. Rev Mineral Geochem 36:3-001 - 3-398
Brownlee D, Tsou P, Aléon J, Alexander CMO'D, Araki, T, Bajt S, Baratta GA, Bastien R, Bland P, Bleuet P, Borg, J, Bradley A, Brenker F, Brennan S, Bridges JC, Browing ND, Brucato JR, Bullock E, Burchell MJ et al. (2006) Comet 81P/Wild 2 under a microscope. Science 314:1711-1716
Burbidge EM, Burbidge GM, Fowler GM (1957) Synthesis of the elements in stars. Rev Modern Phys 29:547-650
Burkhardt C, Kleine T, Oberli F, Pack A, Bourdon B, Wieler R (2011) Molybdenum isotope anomalies in meteorites: constraints on solar nebula evolution and origin of the Earth. Earth Planet Sci Lett 312:390-400
Busemann H, Nguyen AN, Cody GD, Hoppe P, Kilcoyne ALD, Stroud RM, Zega TJ, Nittler LR (2009) Ultra-primitive interplanetary dust particles from the comet 26P/Grigg-Skjellerup dust stream collection. Earth Planet Sci Lett 288:44-57

Busemann H, Young AF, Alexander CMO'D Hoppe P, Mukhopadhyay S, Nittler LR (2006) Interstellar chemistry recorded in organic matter from primitive meteorites. Science 312:727-730

Cameron AGW (1997) The origin of the Moon and the single impact hypothesis. V. Icarus 126:126-137

Canup RM, Asphaug E (2001) Origin of the Moon in a giant impact near the end of the Earth's formation. Nature 412:708-712

Canup RM, Pierazzo E (2006) Retention of water during planet-scale collisions. 37th Ann Lunar Planet Sci #2146

Caro G, Bourdon B, Halliday AN, Quitté G. (2008) Super-chondritic Sm/Nd ratios in Mars, the Earth and the Moon. Nature 452:336-339

Cartigny P, Harris JW, Javoy M (1998) Subduction-related diamonds? - The evidence for a mantle-derived origin from coupled $\delta^{13}C$-$\delta^{15}N$ determinations. Chem Geol 147:147-159

Cartigny P, Harris JW, Javoy M (2001) Diamond genesis, mantle fractionations and mantle nitrogen content: a study of $\delta^{13}C$-N concentrations in diamonds. Earth Planet Sci Lett 185:85-98

Cartigny P, Pineau F, Aubaud C, Javoy M (2008) Towards a consistent mantle carbon flux estimate: Insights from volatile systematics (H_2O/Ce, δD, CO_2/Nb) in the North Atlantic mantle (14° N and 34° N). Earth Planet Sci Lett 265:672-685

Chambers JE (2010) Planetesimal formation by turbulent concentration. Icarus 208:505-517

Chiang E, Youdin AN (2010) Forming planetesimals in solar and extrasolar nebulae. Annu Rev Earth Planet Sci 38:493-522

Ciesla FJ, Cuzzi JN (2006) The evolution of the water distribution in a viscous protoplanetary disk. Icarus 181:178-204

Cody GD, Alexander CMO'D, Tera F (2002) Solid state (^1H and ^{13}C) NMR spectroscopy of the insoluble organic residue in the Murchison meteorite: a self-consistent quantitative analysis. Geochim Cosmochim Acta 66:1851-1865

Craig H, Clarke WB, Beg MA (1975) Excess ^3He in deep waters on the East Pacific Rise. Earth Planet Sci Lett 26:125-132

Cuzzi JN, Hogan RC, Shariff K (2008) Toward planetesimals: dense chondrule clumps in the protoplanetary nebula. Astrophys J 687:1432-1447

Cuzzi JN, Hogan RC, Bottke WF (2010) Towards initial mass functions for asteroids and Kuiper Belt Objects. Icarus. 208:518-538

Dasgupta R (2013) Ingassing, storage, and outgassing of terrestrial carbon through geologic time. Rev Mineral Geochem 75:183-229

Dasgupta R, Walker D (2008) Carbon solubility in core melts in a shallow magma ocean environment and distribution of carbon between the Earth's core and the mantle. Geochim Cosmochim Acta 72:4627-4641

Dauphas N (2003) The dual origin of the terrestrial atmosphere. Icarus 165:326-339

Dauphas N, Davis AM, Marty B, Reisberg L (2004) The cosmic molybdenum-ruthenium isotope correlation. Earth Planet Sci Lett 226:465-475

Dauphas N, Pourmand A (2011) Hf-W-Th evidence for rapid growth of Mars and its status as a planetary embryo. Nature 473:489-492

De Gregorio BT, Stroud RM, Nittler LR, Alexander CMO'D, Kilcoyne ALD, Zega TJ (2010) Isotopic anomalies in organic nanoglobules from Comet 81P/Wild 2:Comparison to Murchison nanoglobules and isotopic anomalies induced in terrestrial organics by electron irradiation. Geochim Cosmochim Acta 74:4454-4470

Deines P (1980) The carbon isotopic composition of diamonds - relationship to diamond shape, color, occurrence and vapor composition. Geochim Cosmochim Acta 44:943-961

Delsemme AH (1991) Nature and history of the organic compounds in comets - an astrophysical view. *In:* Comets in the Post-Halley Era. Newburn Jr RL, Neugebauer M, Rahe J (eds) Kluwer Academic Press, Dordrecht p 377-428

Des Marais DJ, Moore JG (1984) Carbon and its isotopes in mid-oceanic basaltic glasses. Earth Planet Sci Lett 62:43-57

Dodson-Robinson SE, Willacy K, Bodenheimer P, Turner NJ, Beichman CA (2009) Ice lines, planetesimal composition and solid surface density in the solar nebula. Icarus 200:672-693

Draine BT, Li A (2007) Infrared emission from interstellar dust. IV. The silicate-graphite-PAH model in the post-Spitzer era. Astrophys J 657:810-837

Duprat J, Dobrica E, Engrand C, Aleon J, Marrocchi Y, Mostefaoui S, Meibom A, Leroux H, Rouzaud JN, Gounelle M, Robert F (2010) Extreme deuterium excesses in ultracarbonaceous micrometeorites from central Antarctic snow. Science 328:742-745

Engrand C, Deloule E, Robert F, Maurette M, Kurat G (1999) Extraterrestrial water in micrometeorites and cosmic spherules from Antarctica: An ion microprobe study. Meteorit Planet Sci 34:773-786

Engrand C, Maurette M (1998) Carbonaceous micrometeorites from Antarctica. Meteorit Planet Sci 33:565-580

Fernandez JA, Ip WH (1984) Some dynamical aspects of the accretion of Uranus and Neptune: the exchange of orbital angular-momentum with planetesimals. Icarus 58:109-120

Floss C, Stadermann FJ (2009) High abundances of circumstellar and interstellar C-anomalous phases in the primitive CR3 chondrites QUE 99177 and MET 00426. Astrophys J 697:1242-1255

Floss C, Stadermann FJ, Bradley JP, Dai ZR, Bajt S, Graham G, Lea AS (2006) Identification of isotopically primitive interplanetary dust particles: A NanoSIMS isotopic imaging study. Geochim Cosmochim Acta 70:2371-2399

Flynn GJ, Bleuet P, Borg J, Bradley JP, Brenker FE, Brennan S, Bridges J, Brownlee DE, Bullock ES, Burghammer M, Clark BC, Dai ZR, Daghlian CP, Djouadi Z, Fakra S, Ferroir T, Floss C, Franchi IA, Gainsforth Z, Gallien J-P, Gillet P, et al. (2006) Elemental compositions of comet 81P/Wild 2 samples collected by Stardust. Science 314:1731-1735

Garaud P, Lin DNC (2007) The effect of internal dissipation and surface irradiation on the structure of disks, and the location of the snow line around Sun-like stars. Astrophys J 654:606-624.

Garvie LAJ, Buseck PR (2006) Carbonaceous materials in the acid residue from the Orgueil carbonaceous chondrite meteorite. Meteorit Planet Sci 41:633-642

Geiss J, Bochsler P (1982) Nitrogen isotopes in the solar system. Geochim Cosmochim Acta 46:529-548

Geiss J, Gloeckler G (2003) Isotopic composition of H, He and Ne in the protosolar cloud. Space Sci Rev 106:3-18

Genda H, Abe Y (2005) Enhanced atmospheric loss on protoplanets at the giant impact phase in the presence of oceans. Nature 433:842-844

Gilmour I (2003) Structural and isotopic analysis of organic matter in carbonaceous chondrites. *In:* Treatise on Geochemistry. Meteorites, Comets and Planets. Davis AM (ed) Elsevier-Pergamon, Oxford p 269-290

Goldreich P, Tremaine S (1980) Disk-satellite interactions. Astrophys J 241:425-441

Gomes R, Levison HF, Tsiganis K, Morbidelli A (2005) Origin of the cataclysmic Late Heavy Bombardment period of the terrestrial planets. Nature 435:466-469

Gourier D, Robert F, Delpoux O, Binet L, Vezin H, Moissette A, Derenne S (2008) Extreme deuterium enrichment of organic radicals in the Orgueil meteorite: revisiting the interstellar interpretation? Geochim Cosmochim Acta 72:1914-1923

Grady MM, Swart PK, Pillinger CT (1982) The variable carbon isotopic composition of type 3 ordinary chondrites. J Geophys Res 87:A289-A296

Grady MM, Verchovsky AB, Franchi IA, Wright IP, Pillinger CT (2002) Light element geochemistry of the Tagish Lake CI2 chondrite: comparison with CI1 and CM2 meteorites. Meteorit Planet Sci 37:713-735

Grady MM, Wright IP, Carr LP, Pillinger CT (1986) Compositional differences in enstatite chondrites based on carbon and nitrogen stable isotope measurements. Geochim Cosmochim Acta 50:2799-2813

Grady MM, Wright IP, Swart PK, Pillinger CT (1988) The carbon and oxygen isotopic composition of meteoritic carbonates. Geochim Cosmochim Acta 52:2855-2866

Greenberg JM (1998) Making a comet nucleus. Astron Astrophys 330:375-380

Greenberg R, Wacker JF, Hartmann WK, Chapman CR (1978) Planetesimals to planets - numerical simulation of collisional evolution. Icarus 35:1-26

Guillot T (2005) The interiors of giant planets: models, and outstanding questions. Annu Rev Earth Planet Sci 33:493-530

Haisch KE, Lada EA, Pina RK, Telesco CM, Lada CJ (2001) A mid-infrared study of the young stellar population in the NGC 2024 cluster. Astron J 121:1512-1521

Hansen BMS (2009) Formation of the terrestrial planets from a narrow annulus. Astrophys J 703:1131-1140

Hartogh P, Lis DC, Bockelee-Morvan D, de Val-Borro M, Biver N, Kuppers M, Emprechtinger M, Bergin EA, Crovisier J, Rengel M, Moreno R, Szutowicz S, Blake GA (2011) Ocean-like water in the Jupiter-family comet 103P/Hartley 2. Nature 478:218-220

Hashizume K, Chaussidon M, Marty B, Terada K (2004) Protosolar carbon isotopic composition: implications for the origin of meteoritic organics. Astrophys J 600:480-484

Hayes JF, Waldbauer JR (2006) The carbon cycle and associated redox processes through time. Phil Trans Royal Soc London, B361:931-950

Hazen RM, Downs RT, Jones AP, Kah L (2013a) Carbon mineralogy and crystal chemistry. Rev Mineral Geochem 75:7-46

Hazen RM, Downs RT, Kah, L, Sverjensky D (2013b) Carbon mineral evolution. Rev Mineral Geochem 75:79-107

Herbst E, van Dishoeck EF (2009) Complex organic interstellar molecules. Annu Rev Astron Astrophys 47:427-480

Herd CDK, Blinova A, Simkus DN, Juang Y, Tarozo R, Alexander CMO'D, Gyngard F, Nittler LR, Cody GD, Fogel ML, Kebukawa Y, Kilcoyne ALD, Hlits RW, Slater GF, Glavin DP, Dworkin JP, Callahan MP, Elsila JE, De Gregorio BT, Stroud RM (2011) Origin and evolution of prebiotic organic matter as inferred from the Tagish Lake meteorite. Science 332:1304-1307

Hirschmann MM, Dasgupta R (2009) The H/C ratios of Earth's near-surface and deep reservoirs, and consequences for deep Earth volatile cycles. Chem Geol 262:4-16

Holser WT, Schidlowski M, MacKenzie FT, Maynard JB (1988) Geochemical cycles of carbon, and sulfur. *In:* Chemical cycles in the evolution of the Earth. Gregor MC, Garrels RM, Mackenzie FT, Maynard JB (eds) Wiley, New York, p 5-59.

Horn B, Lyra W, Mac Low MM, Sandor Z (2012) Orbital migration of interacting low-mass planets in evolutionary radiative turbulent models. Astrophys J 750:34-42

Howard AW, Johnson JA, Marcy GW, Fischer DA, Wright JT, Bernat D, Henry GW, Peek KMG, Isaacson H, Apps K, Endl M, Cochran WD, Valenti JA, Anderson J, Piskunov NE (2010) The California planet survey. I. Four new giant exoplanets. Astrophys J 721:1467-1481

Hubickyj O, Bodenheimer P, Lissauer JJ (2005) Accretion of the gaseous envelope of Jupiter around a 5-10 Earth-mass core. Icarus 179:415-431

Hughes ALH, Armitage PJ (2010) particle transport in evolving protoplanetary disks: implications for results from Stardust. Astrophys J 719:1633-1653

Huss GR (1990) Ubiquitous interstellar diamond and SiC in primitive chondrites: abundances reflect metamorphism. Nature 347:159-162

Huss GR, Lewis RS (1994) Noble gases in presolar diamonds. II. Component abundances reflect thermal processing. Meteorit 29:811-829

Ida S, Makino JI (1992) N-body simulation of gravitational interaction between planetesimals and a protoplanet .1. Velocity distribution of planetseimals. Icarus 96:107-120

Ikoma M, Nakazawa K, Emori H (2000) Formation of giant planets: dependences on core accretion rate and grain opacity. Astrophys J 537:1013-1025

Inaba S, Ikoma M (2003) Enhanced collisional growth of a protoplanet that has an atmosphere. Astron Astrophys 410:711-723

Javoy M, Pineau F, Allègre CJ (1982) Carbon geodynamic cycle. Nature 300:171-173

Javoy M, Pineau F, Delorme H (1986) Carbon and nitrogen isotopes in the mantle. Chem Geol 57:41-62

Jessberger EK, Christoforidis A, Kissel J (1988) Aspects of the major element composition of Halley's dust. Nature 332:691-695

Johansen A, Youdin A (2007) Protoplanetary disk turbulence driven by the streaming instability: nonlinear saturation and particle concentration. Astrophys J 662:627-641

Jones AP, Genge M, Carmody L (2013) Carbonate melts and carbonatites. Rev Mineral Geochem 75:289-322

Kadik AA, Kurovskaya NA, Ignat'ev YA, Kononkova NN, Koltashev VV, Plotnichenko VG (2011) Influence of oxygen fugacity on the solubility of nitrogen, carbon and hydrogen in $FeO-Na_2O-SiO_2-Al_2O_3$ melts in equilibrium with metallic iron at 1.5 GPa and 1400 °C. Geochem Int 49:429-438

Kehm K, Flynn GJ, Sutton SR, Hohenberg CM (2002) Combined noble gas and trace element measurements on individual stratospheric interplanetary dust particles. Meteorit Planet Sci 37:1323-1335

Keller LP, Messenger S, Flynn GJ, Clemett S, Wirick S, Jacobsen C (2004) The nature of molecular cloud material in interplanetary dust. Geochim Cosmochim Acta 68:2577-2589

Kenyon SJ, Bromley BC (2006) Terrestrial planet formation. I. The transition from oligarchic growth to chaotic growth. Astron J 131:1837-1850

Kerridge F (1985) Carbon, hydrogen and nitrogen in carbonaceous chondrites: abundances, and isotopic compositions in bulk samples. Geochim Cosmochim Acta 49:1707-1714

Kimura K, Lewis RS, Anders E (1974) Distribution of gold and rhenium between nickel-iron and silicate melts - Implications for abundance of siderophile elements on Earth and Moon. Geochim Cosmochim Acta 38:683-701

Kirsh DR, Duncan M, Brasser R, Levison HF (2009) Simulations of planet migration driven by planetesimal scattering. Icarus 199:197-209

Kissel J, Krueger FR (1987) The organic component in dust from comet Halley as measured by the PUMA mass spectrometer on board Vega 1. Nature 326:755-760

Kleine T, Munker C, Mezger K, Palme H (2002) Rapid accretion and early core formation on asteroids and the terrestrial planets from Hf-W chronometry. Nature 418:952-955

Kobayashi C, Nakasato N (2011) Chemodynamical simulations of the Milky Way galaxy. Astrophys J 729:16, doi: 10.1088/0004-637X/729/1/16

Kokubo E, Ida S (1998) Oligarchic growth of protoplanets. Icarus 131:171-178

Kokubo E, Ida S (2002) Formation of protoplanet systems and diversity of planetary systems. Astrophys J 581:666-680

Kress ME, Tielens A, Frenklach M (2010) The "soot line": destruction of presolar polycyclic aromatic hydrocarbons in the terrestrial planet-forming region of disks. Adv Space Res 46:44-49

Kunz J, Staudacher T, Allègre CJ (1998) Plutonium-fission xenon found in the Earth's mantle. Science 280:877-880

Lambrechts M, Johansen A (2012) Rapid growth of gas-giant cores by pebble accretion. Astron Astrophys 544:A32

Lange MA, Ahrens TJ (1982) The evolution of an impact-generated atmosphere. Icarus 51:96-120

Lange MA, Ahrens TJ (1986) Shock-induced CO_2 loss from $CaCO_3$ - Implications for early planetary atmospheres. Earth Planet Sci Lett 77:409-418
Laskar J, Gastineau M (2009) Existence of collisional trajectories of Mercury, Mars and Venus with the Earth. Nature 459:817-819
Lécuyer, C, Gillet P, Robert F (1998) The hydrogen isotope composition of seawater and the global water cycle. Chem Geol 145:249-261
Leinhardt ZM, Richardson DC, Lufkin G, Haseltine J (2009) Planetesimals to protoplanets - II. Effect of debris on terrestrial planet formation. Mon Not R Astron Soc 396:718-728
Levison HF, Thommes E, Duncan MJ (2010) Modelling the formation of giant planet cores: I. Evaluating key processes. Astron J 139:1297-1314
Lin DNC, Papaloizou J (1986) On the tidal interaction between protoplanets and the protoplanetary disk. 3. Orbital migrations of protoplanets. Astrophys J 309:846-857
Lissauer JJ, Hubickyj O, D'Angelo G, Bodenheimer P (2009) Models of Jupiter's growth incorporating thermal and hydrodynamic constraints. Icarus 199:338-350
Lodders K (2003) Solar system abundances and condensation temperatures of the elements. Astrophys J 591:1220-1247
Lodders K (2004) Jupiter formed with more tar than ice. Astrophys J 611:587-597
Love S, Brownlee DE (1993) A direct measurement of the terrestrial mass accretion rate of cosmic dust. Science 262:550-553
Lynden-Bell D, Pringle JE (1974) Evolution of viscous disks and origin of nebular variables. Mon Not R Astron Soc 168:603-637
Lyra W, Paardekooper SJ, Mac Low MM (2010) Orbital migration of low-mass planets in evolutionary radiative models: avoiding catastrophic infall. Astrophys J 715:L68-L73.
Manfroid J, Jehin E, Hutsemekers D Cochran A, Zucconi JM, Arpigny C, Schulz R, Stuwe JA, Ilyin I_ (2009) The CN isotopic ratios in comets. Astron Astrophys 503:613-U354.
Manning CE, Shock EL, Sverjensky D (2013) The chemistry of carbon in aqueous fluids at crustal and upper-mantle conditions: experimental and theoretical constraints. Rev Mineral Geochem 75:109-148
Marty B (2012) The origins and concentrations of water, carbon, nitrogen and noble gases on Earth. Earth Planet Sci Lett 313-314:56-66
Marty B, Chaussidon M, Wiens RC, Jurewicz AJG, Burnett DS (2011) A ^{15}N-poor isotopic composition for the solar system as shown by Genesis solar wind samples. Science 332:1533-1536
Marty B, Jambon A (1987) $C/^3He$ in volatile fluxes from the solid Earth: implications for carbon geodynamics. Earth Planet Sci Lett 83:16-26
Marty B, Palma RL, Pepin RO, Zimmermann L, Schlutter DJ, Burnard PG, Westphal AJ, Snead CJ, Bajt S, Becker RH, Simones JE (2008) Helium and neon abundances and compositions in cometary matter. Science 319:75-78
Marty B, Robert P, Zimmermann L (2005) Nitrogen and noble gases in micrometeorites. Meteorit Planet Sci 40:881-894
Marty B, Tolstikhin IN (1998) CO_2 fluxes from mid-ocean ridges, arcs and plumes. Chem Geol 145:233-248
Marty B, Zimmermann L (1999) Volatiles (He, C, N, Ar) in mid-ocean ridge basalts: assessment of shallow-level fractionation and characterization of source composition. Geochim Cosmochim Acta 63:3619-3633
Masset F, Snellgrove M (2001) Reversing type II migration: resonance trapping of a lighter giant protoplanet. Mon Not R Astron Soc 320:L55-L59
Matrajt G et al. (2008) Carbon investigation of two Stardust particles: a TEM, NanoSIMS, and XANES study. Meteorit Planet Sci 43:315-334
Matrajt G, Messenger S, Brownlee D, Joswiak D (2012) Diverse forms of primordial organic matter identified in interplanetary dust particles. Meteorit Planet Sci 47:525-549
Matrajt G, Taylor S, Flynn G, Brownlee D, Joswiak D (2003) A nuclear microprobe study of the distribution and concentration of carbon and nitrogen in Murchison and Tagish Lake meteorites, Antarctic micrometeorites, and IDPs: implications for astrobiology. Meteorit Planet Sci 38:1585-1600
McKeegan KD, Aléon J, Bradley J, Brownlee D, Busemann H, Butterworth A, Chaussidon M, Fallon S, Floss C, Glimor J, Gounelle M, Graham G, Guan Y, Heck PR, Hoppe P, Hutcheon ID, Huth J, Ishii H, Ito M, Jacobsen SB, et al. (2006) Isotopic compositions of cometary matter returned by Stardust. Science 314:1724-1728
Meech KJ, A'Hearn MF, Adams JA, Bacci P, Bai J, Barrera L, Battelino M, Bauer JM, Becklin E, Bhatt B, Biver N, Bockelée-Morvan D, Bodewits D, Böhnhardt H, Boissier J, Bonev BP, Borghini W, Brucato JR, et al. (2011) EPOXI; Comet 103P/Hartley 2 observations from a worldwide campaign. Astrophys J 734:9-16
Messenger S (2000) Identification of molecular-cloud material in interplanetary dust particles. Nature 404:968-971
Messenger S, Keller LP, Stadermann FJ, Walker RM, Zinner E (2003a) Samples of stars beyond the solar system: silicate grains in interplanetary dust. Science 300:105-108

Messenger S, Stadermann FJ, Floss C, Nittler LR, Mukhopadhyay S (2003b) Isotopic signatures of presolar materials in interplanetary dust. Space Sci Rev 106:155-172

Meyer BS, Zinner E (2006) Nucleosynthesis. *In:* Meteorites and the Early Solar System II. Lauretta DS, McSween Jr HY (eds) Univ. Arizona Press, Tucson p 69-108

Meyer MR, Carpenter JM, Mamajek EE, Hillenbrand LA, Hollenbach D, Moro-Martin A, Kim JS, Silverstone MD, Najita J, Hines DC, Pascucci I, Stauffer JR, Bouwman J, Backman DE (2008) Evolution of mid-infrared excess around sun-like stars: constraints on models of terrestrial planet formation. Astrophys J 673:L181-L184

Mizuno H (1980) Formation of the giant planets. Prog Theor Phys 64:544-557

Morbidelli A, Chambers J, Lunine JI, Petit JM, Robert F, Valsecchi GB, Cyr KE (2000) Source regions and timescales for the delivery of water to the Earth. Meteorit Planet Sci 35:1309-1320

Morbidelli A, Bottke WF, Nesvorny D, Levison HF (2009) Asteroids were born big. Icarus 204:558-573

Morbidelli A, Crida A (2007) The dynamics of Jupiter and Saturn in the gaseous protoplanetary disk. Icarus 191:158-171

Morbidelli A, Lunine JI, O'Brien DP, Raymond SN, Walsh KJ (2012) Building terrestrial planets. Annu Rev Earth Planet Sci 40:251-275

Morbidelli A, Tsiganis K, Crida A, Levison HF, Gomes R (2007) Dynamics of the giant planets of the solar system in the gaseous protoplanetary disk and their relationship to the current orbital architecture. Astron J 134:1790-1798

Morishima R, Schmidt MW, Stadel J, Moore B (2008) Formation and accretion history of terrestrial planets from runaway growth through to late time: implications for orbital eccentricity. Astrophys J 685:1247-1261

Morishima R, Stadel J, Moore B (2010) From planetesimals to terrestrial planets: n-body simulations including the effects of nebular gas and giant planets. Icarus 207:517-535

Mumma MJ, Charnley SB (2012) The chemical composition of comets - Emerging taxonomies and natal heritage. Annu Rev Astron Astrophys 49:471-524

Muralidharan K, Stimpfl M, de Leeuw NH, Deymier PA, Runge K, Drake MJ (2009) Water in the inner solar system: insights from atomistic and electronic-structure calculations. Meteorit Planet Sci 44:A136

Ni H, Keppler H (2013) Carbon in silicate melts. Rev Mineral Geochem 75:251-287

Nieva MF, Przybilla N (2011) Fundamental parameters of "normal" B stars in the solar neighborhood. *In:* Active OB Stars: Structure, Evolution, Mass-Loss, and Critical Limits. Neiner C (ed) IAU Symposium Proceedings Series. Cambridge University Press, Cambridge p 566-570

Nittler LR (2003) Presolar stardust in meteorites: recent advances and scientific frontiers. Earth Planet Sci Lett 209:259-273

O'Brien DP, Morbidelli A, Levison HF (2006) Terrestrial planet formation with strong dynamical friction. Icarus 184:39-58

O'Brien DP, Walsh KJ, Morbidelli A, Raymond SN, Mandel AM, Bond JC (2010) Early giant planet migration in the solar system: Geochemical and cosmochemical implications for terrestrial planet formation. DPS meeting #42, Bull Am Astron Soc 42: 948

Ott U (2002) Noble gases in meteorites - Trapped components. Rev Mineral Geochem 47:71-100

Ott U, Besmehn A, Farouqi K, Hallmann O, Hoppe P, Kratz KL, Melber K, Wallner A (2012) New attempts to understand nanodiamond stardust. Publ Astron Soc Australia 29:90-97

Otting W, Zähringer J (1967) Total carbon content and primordial rare gases in chondrites. Geochim Cosmochim Acta 31:1949-1956

Owen T, Bar-Nun A, Kleinfeld I (1992) Possible cometary origin of heavy noble gases in the atmospheres of Venus, Earth and Mars. Nature 358:43-46

Ozima M, Podosek FA (2002) Noble Gas Geochemistry. Cambridge University Press, Cambridge

Paardekooper S-J, Baruteau C, Kley W (2011) A torque formula for non-isothermal Type I planetary migration - II. Effects of diffusion. Mon Not R Astron Soc 410:293-303

Pahlevan K, Stevenson DJ (2007) Equilibration in the aftermath of the lunar-forming giant impact. Earth Planet Sci Lett 262:438-449

Papaloizou JCB, Terquem C (2006) Planet formation and migration. Rep Prog Phys 69:119-180

Pepin RO (1991) On the origin and early evolution of terrestrial planet atmospheres and meteoritic volatiles. Icarus 92:2-79

Pepin RO (2006) Atmospheres on the terrestrial planets: clues to origin and evolution. Earth Planet Sci Lett 252:1-14

Pierens A, Nelson RP (2008) Constraints on resonant-trapping for two planets embedded in a protoplanetary disc. Astron Astrophys 482:333-340

Pierens A, Raymond SN (2011) Two phase, inward-then-outward migration of Jupiter and Saturn in the gaseous solar nebula. Astron Astrophys 533:14, http://dx.doi.org/10.1051/0004-6361/201117451

Pizzarello S, Cooper GW, Flynn GJ (2006) The nature and distribution of the organic material in carbonaceous chondrites and interplanetary dust particles. *In:* Meteorites and the Early Solar System II. Lauretta DS, McSween Jr HY (eds) Univ Arizona Press, Tucson p 625-651
Pollack JB, Hollenbach D, Beckwith S, Simonelli DP, Roush T, Fong W (1994) Composition and radiative properties of grains of molecular clouds and accretion disks. Astrophys J 421:615-639
Pujol M, Marty B, Burgess R (2011) Chondritic-like xenon trapped in Archean rocks: a possible signature of the ancient atmosphere. Earth Planet Sci Lett 308:298-306
Raymond SN, Armitage PJ, Gorelick N (2010) Planet-planet scattering in planetesimal disks. II. Predictions for outer extrasolar planetary systems. Astrophys J 711:772-795
Raymond SN, Armitage PJ, Moro-Martin A, Booth M, Wyatt MC, Armstrong JC, Mandell AM Selsis F, West AA (2011) Debris disks as signposts of terrestrial planet formation. Astron Astrophys 530, doi: 10.1051/0004-6361/201116456
Raymond SN, O'Brien DP, Morbidelli A, Kaib NA (2009) Building the terrestrial planets: constrained accretion in the inner Solar System. Icarus 203:644-662
Raymond SN, Quinn T, Lunine JI (2004) Making other earths: dynamical simulations of terrestrial planet formation and water delivery. Icarus 168:1-17
Raymond SN, Quinn T, Lunine JI (2006) High-resolution simulations of the final assembly of Earth-like planets I. Terrestrial accretion and dynamics. Icarus 183:265-282
Raymond SN, Quinn T, Lunine JI (2007) High-resolution simulations of the final assembly of Earth-like planets. 2. Water delivery and planetary habitability. Astrobiol 7:66-84
Robert F (2003) The D/H ratio in chondrites. Space Sci Rev 106:87-101
Robert F, Epstein S (1982) The concentration and isotopic composition of hydrogen, carbon and nitrogen in carbonaceous chondrites. Geochim Cosmochim Acta 46:81-95
Ronov AB, Yaroshevskhiy AA (1976) A new model for the chemical structure of the Earth's crust. Geochem Int 13:89-121
Saal AE, Hauri EH, Langmuir CH, Perfit MR (2002) Vapour undersaturation in primitive mid-ocean-ridge basalt and the volatile content of Earth's upper mantle. Nature 419:451-455
Safronov VS, Zvjagina EV (1969) Relative sizes of largest bodies during accumulation of planets. Icarus 10:109-115
Salters VJM, Stracke A (2004) Composition of the depleted mantle. Geochem Geophys Geosyst 5:Q05B07, doi:10.1029/2003GC000597
Sasselov DD, Lecar M (2000) On the snow line in dusty protoplanetary disks. Astrophys J 528:995-998
Schramm LS, Brownlee DE, Wheelock MM (1989) Major element composition of stratospheric micrometeorites. Meteoritics 24:99-112
Scott ERD, Krot AN (2003) Chondrites and their components. *In:* Treatise on Geochemistry. Meteorites, Comets and Planets. Davis AM (ed) Elsevier-Pergamon, Oxford p 143-200
Sharp ZD, Draper DS, Agee CB (2009) Core-mantle partitioning of chlorine and a new estimate for the hydrogen abundance of the Earth. 40th Lunar Planet Sci Conf, Houston, CD-ROM, abstr #1209
Shirey SB, Cartigny P, Frost DJ, Keshav S, Nestola F, Nimis P, Pearson DG, Sobolev NV, Walter MJ (2013) Diamonds and the geology of mantle carbon. Rev Mineral Geochem 75:355-421
Sleep NH, Zahnle K, Neuhoff PS (2001) Initiation of clement surface conditions on the earliest Earth. Proc Natl Acad Sci USA 98:3666-3672
Stevenson DJ, Lunine JI (1988) Rapid formation of Jupiter by diffusive redistribution of water vapor in the solar nebula. Icarus 75:146-155
Svetsov VV (2007) Atmospheric erosion and replenishment induced by impacts of cosmic bodies upon the Earth and Mars. Solar System Res 41:28-41
Thomas KL, Blanford GE, Keller LP, Klöck W, McKay DS (1993) Carbon abundance and silicate mineralogy of anhydrous interplanetary dust particles. Geochim Cosmochim Acta 57:1551-1566
Thommes EW, Duncan MJ, Levison HF (2003) Oligarchic growth of giant planets. Icarus 161:431-455
Tielens AGGM (2008) Interstellar polycyclic aromatic hydrocarbon molecules. Annu Rev Astron Astrophys 46:289-337
Tolstikhin IN, O'Nions RK (1994) The Earth missing xenon - a combination of early degassing and rare gas loss from the atmosphere. Chem Geol 115:1-6
Touboul M, Kleine T, Bourdon B, Palme H, Wieler R (2007) Late formation and prolonged differentiation of the Moon inferred from W isotopes in lunar metals. Nature 450:1206-1209
Truran Jr JW, Heger A (2003) Origin of the Elements. *In:* Treatise on Geochemistry. Meteorites, Comets and Planets. Davis AM (ed) Elsevier-Pergamon, Oxford p 1-15
Tsiganis K, Gomes R, Morbidelli A, Levison HF (2005) Origin of the orbital architecture of the giant planets of the Solar System. Nature 435:459-461
Turcotte S, Richer J, Michaud G, Iglesias CA, Rogers FJ (1998) Consistent solar evolution model including diffusion and radiative acceleration effects. Astrophys J 504:539-558

Walsh KJ, Morbidelli A, Raymond SN, O'Brien DP, Mandell AM (2005) A low mass for Mars from Jupiter's early gas-driven migration. Nature 475:206-209
Weidenschilling SJ (2011) Initial sizes of planetesimals and accretion of the asteroids. Icarus 214:671-684
Wetherill GW (1975) Radiometric chronology of early Solar-System. Annu Rev Nucl Particle Sci 25:283-328
Wetherill GW (1985) Occurrence of giant impacts during the growth of the terrestrial planets. Science 228:877-879
Wetherill GW (1990) Formation of the Earth. Annu Rev Earth Planet Sci 18:205-256
Wetherill GW (1991) Occurrence of Earth-like bodies in planetary systems. Science 253:535-538
Wetherill GW, Stewart GR (1993) Formation of planetary embryos - effects of fragmentation, low relative velocity, and independent variation of eccentricity and inclination. Icarus 106:190-209
Wood BJ (1993) Carbon in the core. Earth Planet Sci Lett 117:593-607
Wood BJ, Li J, Shahar A (2013) Carbon in the core: its influence on the properties of core and mantle. Rev Mineral Geochem 75:231-250
Woolum DS, Cassen P (1999) Astronomical constraints on nebular temperatures: implications for planetesimal formation. Meteorit Planet Sci 34:897-907
Yang J, Epstein S (1983) Interstellar organic-matter in meteorites. Geochim Cosmochim Acta 47:2199-2216
Yin Q, Jacobsen SB, Yamashita K, Blichert-Toft J, Telouk P, Albarede F (2002) A short timescale for terrestrial planets from Hf-W chronometry. Nature 418:949-952
Zhang Y, Zindler A (1993) Distribution and evolution of carbon and nitrogen in Earth. Earth Planet Sci Lett 117:331-345
Zubko V, Dwek E, Arendt RG (2004.) Interstellar dust models consistent with extinction, emission, and abundance constraints. Astrophys J 152:211-249

Reviews in Mineralogy & Geochemistry
Vol. 75 pp. 183-229, 2013

Ingassing, Storage, and Outgassing of Terrestrial Carbon through Geologic Time

Rajdeep Dasgupta

Department of Earth Science
Rice University
6100 Main Street, MS 126
Houston, Texas 77005, U.S.A.

Rajdeep.Dasgupta@rice.edu

INTRODUCTION

Earth is unique among the terrestrial planets in our solar system in having a fluid envelope that fosters life. The secrets behind Earth's habitable climate are well-tuned cycles of carbon (C) and other volatiles. While on time-scales of ten to thousands of years the chemistry of fluids in the atmosphere, hydrosphere, and biosphere is dictated by fluxes of carbon between near surface reservoirs, over hundreds of millions to billions of years it is maintained by chemical interactions of carbon between Earth's interior, more specifically the mantle, and the exosphere (Berner 1999). This is because of the fact that the estimated total mass of C in the mantle is greater than that observed in the exosphere (Sleep and Zahnle 2001; Dasgupta and Hirschmann 2010) and the average residence time of carbon in the mantle is between 1 and 4 Ga (Sleep and Zahnle 2001; Dasgupta and Hirschmann 2006). But how did Earth's mantle attain and maintain the inventory of mantle carbon over geologic time? And is the residence time of carbon in the mantle, as constrained by the present-day fluxes, a true reflection of carbon ingassing and outgassing rates throughout Earth's history? Also, when in the planet's history did its mantle carbon inventory become established and how did it change through geologic time? The answers to these questions are important because of carbon's importance in a number of fields of Earth sciences, such as the thermal history of Earth [e.g., trace-volume carbonated melt may extract highly incompatible heat-producing elements from great depths (Dalou et al. 2009; Dasgupta et al. 2009b; Grassi et al. 2012)], the internal differentiation of the mantle and core [carbon influences element partitioning in both carbonate-silicate (e.g., Blundy and Dalton 2000; Dasgupta et al. 2009b; Dalou et al. 2009; Grassi et al. 2012) and metallic systems (Chabot et al. 2006; Hayden et al. 2011; Buono et al. submitted)], long-term evolution of climate (e.g., Kasting and Catling 2003; Hayes and Waldbauer 2006), and origin and evolution of life. Carbon also plays a role in the inner workings of the planetary mantle through mobilizing hydrogen from nominally anhydrous silicate minerals (Dasgupta et al. 2007a) and by stabilizing incipient carbonated melts in the mantle (e.g., Eggler 1976; Wyllie 1977; Dalton and Presnall 1998; Dasgupta and Hirschmann 2006; Dasgupta et al. 2013b). In the absence of carbon-induced melting and efficient extraction of such mobile melt, the presence of "water" in mineral structures controls the viscosity and the creep behavior of the mantle and determines whether a planet would "live" long (e.g., Earth), with systematic solid state convection that sustains plate tectonic cycles or "die" shortly after its birth (e.g., Mars and Venus), without a lasting tectonic cycle. Similarly, the presence of trace amounts of carbonated melt in Earth's upper mantle may be critical in regulating asthenospheric processes and stabilizing plate tectonics (Eggler 1976; Dasgupta et al. 2007a; Hirschmann 2010). Finally, any deep Earth carbon that is sequestered in the metallic core also influences the physical properties of inner and outer core (e.g., Poirier

1529-6466/13/0075-0007$00.00 DOI: 10.2138/rmg.2013.75.7

1994; Hillgren et al. 2000; Mookherjee et al. 2011; Wood et al. 2013) and may have played a role in the dynamics of core formation, timing of core crystallization, and dynamics of magma ocean processes, including core-mantle segregation (Dasgupta et al. 2009a).

Because the abundance and mode of storage of mantle carbon are central to carbon's role in global geodynamics and controlling variables influencing carbon's distribution in other terrestrial reservoirs (crust plus atmosphere system and core), it is critical to constrain the processes that modulated the mantle carbon inventory through time. In this chapter, ingassing, outgassing, and storage mechanisms of terrestrial carbon, from the time period of early planetary differentiation and magma ocean in the Hadean Eon to the plate tectonic cycles of modern Earth through Phanerozoic, are reviewed (Fig. 1). Thus this chapter will primarily be in two parts: (1) inheritance of mantle carbon (focusing on Hadean processes) and (2) retention of mantle carbon (focusing on plate-tectonic cycling from Archean to present). The terms "ingassing" and "outgassing" are going to be used in a general sense, i.e., any mechanisms that introduce carbon to the mantle will be grouped as a carbon "ingassing" or acquisition mechanism and any mechanisms that remove carbon from the mantle will be listed as a carbon "outgassing" mechanism. The terms ingassing and outgassing are not used to imply that carbon is present in form of a gaseous phase.

CARBON INHERITANCE — MAGMA OCEAN CARBON CYCLE

How did Earth's mantle acquire or inherit its carbon budget and when did the inheritance take place? The origin of carbon on Earth, similar to many other volatile elements, is debated (Marty et al. 2013). The building blocks of Earth, which is traditionally thought to be carbonaceous chondrites, contain several weight percent of carbon depending on the sub-type (2.7-4.4 wt% with group average of 3.52 ± 0.48 wt%) (Anders and Grevesse 1989; Lodders 2003; Lodders 2010). Although carbonaceous chondrite is commonly used as the model for the terrestrial building block because of their compositional similarities with the solar photosphere (Lodders 2003), geochemical arguments for other types of chondrites and non-chondritic compositions, likely with quite different and lower abundance of carbon, also exist in modern literature (e.g., Alard et al. 2000; Javoy et al. 2010; Caro 2011; Warren 2011; Campbell and O'Neill 2012). Further, despite having concentration, in carbonaceous chondrite, on the order of weight percent, owing to very low condensation temperatures for carbon-bearing gases, ices, and other solid phases (e.g., CO, CH_4, and graphite; 41-626 K) carbon is thought to be largely lost during Earth's accretion (Abe 1997; Genda and Abe 2003). A similar decrease in the carbon abundance is also observed by comparing composition of solar photosphere to CI-type chondrites, where the latter incorporate only 10% of the photospheric carbon (Lodders 2003). Indeed, most of the estimates of bulk Earth carbon fall below 0.1 wt% (McDonough 2003; Marty 2012), although higher values (0.09-0.37 wt%) are also proposed in earlier literature (Trull et al. 1993). Despite the fact that this bulk Earth carbon budget and the timing as to when this amount was acquired remain highly uncertain it is not implausible, however, that Earth inherited this amount from chondritic materials during accretion (Morbidelli et al. 2012). Assuming that some finite concentration of carbon was inherited from the time of planetary accretion, in the following the possible fate of such carbon during early Earth differentiation—core formation and magma ocean processes is reviewed.

Magma ocean carbon cycle during core formation

A number of studies discussed the importance and various scenarios of carbon fractionation in a magma ocean (Kuramoto and Matsui 1996; Kuramoto 1997; Dasgupta and Walker 2008; Hirschmann and Dasgupta 2009; Dasgupta and Hirschmann 2010; Hirschmann 2012; Dasgupta et al. 2013a) (Fig. 1). This initial fractionation is important as it must have had an effect on Earth's early thermal and dynamical evolution, its geochemical differentiation, its path to an

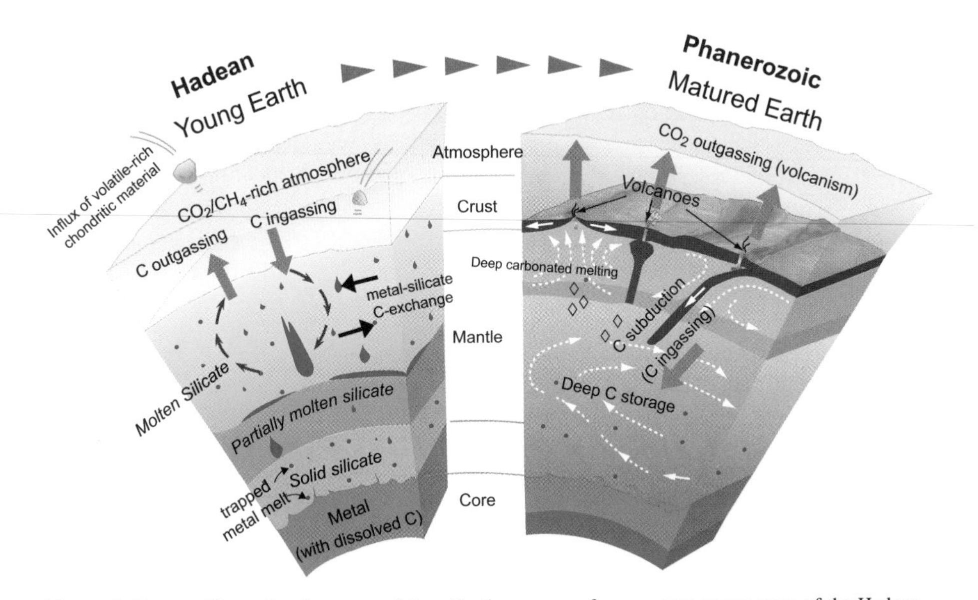

Figure 1. Cartoon illustrating the range of deep-Earth processes from magma ocean stage of the Hadean Eon to plate tectonic framework of the modern world that likely influenced the long-term carbon cycle. This chapter reviews Earth's deep carbon cycle in light of these processes. The early Earth processes cartoon (left), although shown together, may be temporally separated. For example, if carbon exchange between the proto-atmosphere and the magma ocean had an important role in the later evolution of the mantle, such a process was required to continue after the core formation had ceased (see text for details).

equable climate, and development of prebiotic chemistry. Magma ocean processes must have also set the initial distribution of carbon and conditions for further development of Earth's deep carbon cycle and have forced Earth to evolve differently compared to the planet Mars (Kuramoto 1997). The key information on the fate of carbon during early Earth differentiation is how the element was partitioned between various reservoirs, viz. core, mantle (magma ocean), and proto-atmosphere.

Equilibrium fractionation of carbon between the core and mantle and the contribution of core towards bulk Earth carbon inventory. Several studies over the last three decades have attempted to constrain the partitioning of siderophile elements between metallic and silicate melts in magma ocean environments, but similar experiments constraining volatile element fractionation remained limited (e.g., Li and Agee 1996; Okuchi 1997). Fractionation of elements, such as siderophile elements, between metallic core and silicate mantle during core formation was largely responsible for setting the elemental distribution in the terrestrial reservoirs and setting the stage for crust and mantle geochemistry to evolve. The same fractionation process likely also had influenced the bulk distribution of the terrestrial volatiles in general and carbon in particular (Kuramoto and Matsui 1996; Kuramoto 1997; Dasgupta and Walker 2008; Dasgupta et al. 2013a). If the entire bulk Earth carbon budget participated in the metal-silicate fractionation, and if $D_C^{metal/silicate}$ (partition coefficient of carbon between metal and silicate = mass fraction of carbon in metal melt/ mass fraction of carbon in silicate melt) is <1, then carbon would be concentrated mostly in the outer silicate layer. On the other hand, if $D_C^{metal/silicate}$ >>1 then bulk carbon of the planet would mostly be sequestered into the core and hence would have much less impact on the long-term carbon cycle. If carbon really behaved as a highly siderophile element, a late volatile-rich veneer or other post-core segregation processes would likely be responsible for bringing carbon to the planet. Alternatively, in the event carbon preferentially partitioned into the core-forming liquid, a less extreme iron affinity for carbon ($D_C^{metal/silicate}$ >1) might have allowed Earth's molten silicate to retain enough carbon in dissolved form to explain Earth's present-day carbon budget of the mantle. In order to answer these questions, the knowledge of $D_C^{metal/silicate}$, over the plausible range of depth-temperature-oxygen fugacity-melt compositions is necessary. However, studies investigating partitioning of carbon between metallic and silicate melt are scarce (Dasgupta and Walker 2008; Dasgupta et al. 2012; Dasgupta et al. 2013a).

In the absence of any experimental measurements of $D_C^{metal/silicate}$, Dasgupta and Walker (2008) bracketed the possible range of $D_C^{metal/silicate}$ assuming that (1) the present-day inventory of mantle carbon (30-1100 ppm C), as derived from the concentration of the same in mantle derived magmas, was produced by equilibrium partitioning of carbon between metallic and silicate melt in a magma ocean, (2) the entire present-day core mass of 32.3% equilibrated with the whole mantle melt of 67.7%, and (3) a maximum of ~6-7 wt% C can be dissolved into the metallic melt at magma ocean conditions. This analysis suggested that with bulk Earth carbon of 730 ppm (McDonough 2003), the bulk core will end up acquiring 0.25 ± 0.15 wt% C. If more recent bulk Earth carbon estimate of 530 ± 210 ppm (Marty 2012) is used, the core carbon estimate would be slightly lower. However, it is important to note that Marty's (2012) bulk Earth carbon estimate assumes that there is no carbon in the core and distributes the bulk silicate Earth (BSE) carbon content to the whole Earth mass (5.98×10^{27} g). But the mass balance calculation of Dasgupta and Walker (2008) showed that Earth's core can store as much as $(4.8 \pm 2.9) \times 10^{24}$ g C, i.e., the metallic core alone can contribute as much as 803 ppm C to bulk Earth. The analysis of Dasgupta and Walker (2008), although simple, highlighted the importance of possible carbon storage in the core towards the total inventory of bulk Earth carbon.

More recent experimental effort generated direct measurements of $D_C^{metal/silicate}$ at shallow magma ocean conditions (Dasgupta et al. 2012; Dasgupta et al. 2013a). These experiments at 1-5 GPa and 1500-2100 °C at oxygen fugacity (f_{O_2}) 1.5 to 2 log units below the iron-wüstite buffer (~IW−1.5 to IW−2.0) explored carbon partitioning between Fe-rich metallic melt and mafic-ultramafic silicate melts (non-bridging oxygen over tetrahedrally coordinated cations, NBO/T of 0.9-2.8). These studies demonstrated that carbon indeed behaves as a strongly siderophile element ($D_C^{metal/silicate}$ varying between ~5500 and >150) and metal affinity of carbon increases with increasing pressure and decreases with increasing temperature, silicate melt depolymerization, extent of hydration, and oxygen fugacity. Despite these experimental measurements, more work needs to be done to fully explore the effect of various key intensive variables on $D_C^{metal/silicate}$. In particular, experiments will need to quantify $D_C^{metal/silicate}$ relevant for magma oceans that are quite deep (Li and Agee 1996; Chabot and Agee 2003; Righter 2011) and more reduced (Wood et al. 2006). As pointed out by Dasgupta et al. (2013a), future experiments will also have to take into account more complex metallic alloy liquid chemistry including the presence of sulfur, silicon, and oxygen and a known fugacity of other trace gases such as H_2O and H_2 that can influence the speciation of carbon in silicate melts. But because the conditions of core-mantle equilibration continue to be a subject of active debate (e.g., Righter 2011; Rubie et al. 2011), it is worth exploring the predictions based on already constrained values of $D_C^{metal/silicate}$ on the relative carbon budget of core and the mantle.

Figure 2, modified from Dasgupta et al. (2013a), shows the predicted concentration of carbon in the core and the BSE as a function of bulk Earth carbon that participated in the core-mantle fractionation in a magma ocean. For this calculation, it was assumed that the present-day mass of Earth's core (32.3% by weight of Earth) equilibrated with the whole molten mantle, i.e., the concentration of carbon in the core, C_C^{core} and the concentration of carbon in the mantle, C_C^{mantle} are related to the bulk Earth carbon, C_C^0 and $D_C^{metal/silicate}$ by the following equations:

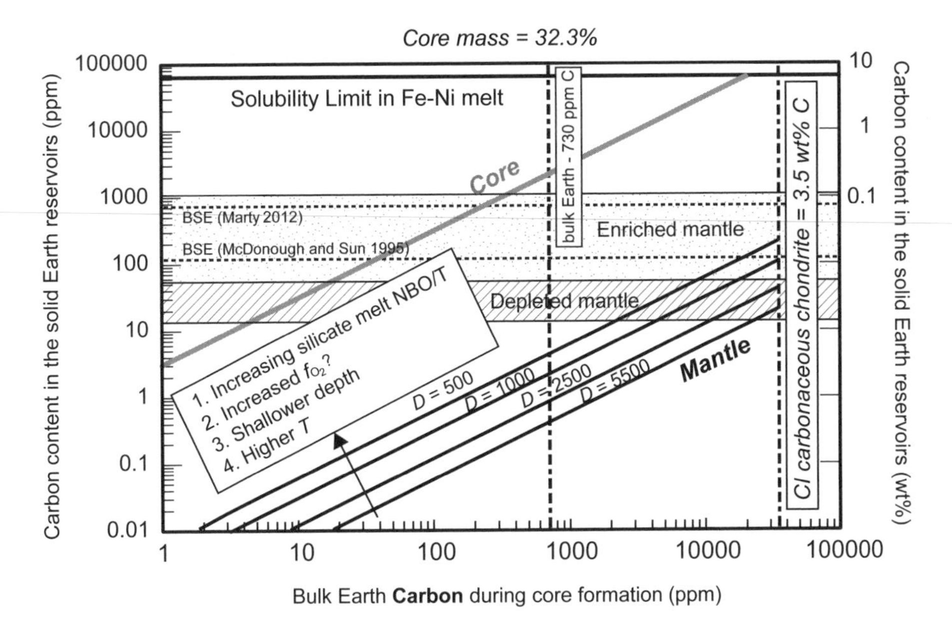

Figure 2. The effect of equilibrium partitioning of carbon between metallic and silicate melt in a magma ocean on the inventory of mantle and core carbon, modified after Dasgupta et al. (2013a). This calculation assumes that the inventory of carbon in the BSE and the core reflects perfect equilibration of the present-day mass of the core (32.3%) with the mass of silicate Earth. Carbon concentration of deep Earth reservoirs (core: grey curve and mantle: black curves) are plotted as a function of bulk Earth carbon that participated in the core-mantle fractionation event. The mantle and core concentrations are plotted for four different $D_C^{metal/silicate}$ values of 5500, 2500, 1000, and 500, which may cover all the possible conditions of core-mantle equilibration. The higher values are appropriate for magma ocean that is deep, dry, and more reduced. Plotted for comparison are the estimated range of modern Earth mantle carbon contents [depleted mantle similar to the source regions of mid-ocean ridge basalts: ~50-200 ppm CO_2 (Marty and Tolstikhin 1998; Saal et al. 2002; Cartigny et al. 2008); enriched mantle similar to the source regions of intraplate ocean island basalt or off-axis seamounts: up to 4000 ppm CO_2; (Pineau et al. 2004)], primarily based on volcanic CO_2 flux and the estimate of BSE carbon content of 120 ppm (McDonough and Sun 1995) and ~765 ppm (Marty 2012). Also plotted for reference are the carbon abundance in CI chondrite of ~3.5 wt% (Anders and Grevesse 1989; Lodders 2003), estimated bulk Earth carbon content of 730 ppm (McDonough 2003), and carbon solubility limit for Fe-rich alloy melt at high temperature and low pressure (Dasgupta and Walker 2008; Nakajima et al. 2009; Siebert et al. 2011; Dasgupta et al. 2013a).

$$C_C^{core} = \frac{C_C^0 - 0.677C_C^{mantle}}{0.323} \tag{1}$$

$$C_C^{mantle} = \frac{C_C^0}{0.323D_C^{metal/silicate} + 0.677} \tag{2}$$

Given that magma ocean metal-silicate equilibration likely took place over a range of P-T-f_{O_2} (Wood et al. 2006), in Figure 2 the estimated carbon content of the core and the mantle residual to core formation are plotted for $D_C^{metal/silicate}$ varying from 5500 to 500. The higher $D_C^{metal/silicate}$ values likely capture carbon fractionation in a very deep magma ocean under more reduced conditions, with very dry magma ocean (low fugacity of H_2 and H_2O), or for evolved silicate magma compositions and the lower $D_C^{metal/silicate}$ values capture carbon fractionation scenarios in shallower or wet magma oceans, with extreme temperatures, and more primitive (higher degree of melt depolymerization) compositions.

Figure 2 adopted from Dasgupta et al. (2013a) suggests that the average carbon content of Earth's present-day depleted mantle can be matched by equilibrium core-mantle fractionation of carbon if bulk Earth carbon during core formation was ≥0.4-3.5 wt% (≥4000-35000 ppm C). In order to generate enriched mantle domains similar to those of some of the ocean island basalt (OIB) source regions (50-1000 ppm C in the source), a minimum bulk Earth carbon of 2.1 wt% and in excess of 3.5 wt% is necessary. The abundance of bulk Earth carbon during core-mantle segregation is uncertain, however. Many estimates of bulk Earth carbon suggest values ≤1000 ppm (McDonough 2003; Marty 2012). Owing to volatility of carbon during the accretion process, carbon is thought to be largely lost from the chondritic building blocks. If the available bulk Earth carbon was only ~730 ppm (McDonough, 2003), for the plotted range of $D_C^{metal/silicate}$, the molten mantle residual to core formation could only possess 0.4-4.5 ppm C (1.5-16.5 ppm CO_2). This value is only 3-33% of the present-day mantle budget of carbon with 50 ppm bulk CO_2 (~14 ppm C). If the modern bulk mantle C content is closer to the carbon for enriched mantle domains (e.g., 500 ppm CO_2 or 136 ppm C), the contribution of carbon that was left behind after core formation will be even smaller. In this latter case or if the primitive mantle C content estimate of 120 ppm, after McDonough and Sun (1995), is used, the equilibrium core formation will leave behind a molten silicate that contains only 0.3-3.8% of the present-day mantle carbon. If the BSE carbon content of ~765 ppm (Marty 2012) is used then >99% of silicate Earth C needs to be derived from post core-segregation processes. Hence, whatever the exact carbon content of the present-day BSE is, such concentration is in excess with respect to what is predicted by magma ocean chemical equilibration.

The analysis presented above clearly suggests that, owing to siderophile nature of carbon in core-forming magma ocean conditions, only a small fraction of the present-day mantle or BSE carbon can be primordial, which dates back to the Hadean Eon prior to the last giant impact. In other words, most of the primordial carbon, which dates back prior to core formation time and not degassed to atmosphere or lost to space, was likely sequestered into the metallic core. The measured range of $D_C^{metal/silicate}$ and bulk Earth carbon of 730 ppm (McDonough, 2003) make the bulk core store ~0.23 wt% C. This concentration of carbon in the bulk core (1.932×10^{27} g) makes a total of 4.44×10^{24} g C, or ~744 ppm C for the whole Earth. This calculation underscores the importance of knowing the carbon abundance that participated in the core-formation processes, because with core contribution of >700 ppm C for bulk Earth, the estimate of bulk Earth carbon can at least double. For example, if the estimate of BSE carbon content of 765 ± 300 ppm C is normalized with respect to the whole Earth, a bulk Earth carbon content of 530 ± 210 ppm C results (Marty 2012). But with 0.23 wt% carbon in the core, the bulk Earth carbon content becomes as much as 1277 ± 210 ppm C. The estimate of 0.23 wt% C in the core derived from experimental partition coefficients of carbon between metallic and silicate melt (Dasgupta et al. 2013a) is similar to the earlier estimates obtained based on mass balance approach (McDonough 2003; Dasgupta and Walker 2008). The value of ~0.2 wt% carbon in the core is also consistent with the Pb-isotopic age of Earth, which seems to require strong partitioning of Pb into the core throughout Earth's accretion and Pb behaves as a strongly siderophile element only when the carbon content of the metallic liquid is small and far from saturation (Wood and Halliday 2010). However, although these geochemical estimates are all converging towards only a modest carbon content of the bulk core, some earlier work suggested carbon core content as much as 1.2 wt% (Yi et al. 2000). The work of Wood et al. (2013) estimated a similar but somewhat higher abundance of carbon in the core than those derived from currently available partitioning data. These authors (Wood et al. 2013) suggested, based on the C/S ratio of accreting materials and the effect of S on siderophile element partitioning between metal and silicate, that the core contains ~0.6 wt% carbon.

With the metallic core taking possession of Earth's carbon during the Hadean eon, an explanation of the modern mantle carbon budget requires some later replenishment event. This is the essence of the "excess" mantle carbon paradox. If perfect core-mantle equilibration is conceived, then the only way to avoid this is for bulk Earth to have as much carbon as carbonaceous chondrite (Fig. 2). However, even if there was 3.5 wt% carbon taking part in core-mantle fractionation, the residual silicate liquid would not retain enough carbon to match the BSE estimate of 765 ± 300 ppm C as proposed by Marty (2012). Also, with as much as 3.5 wt% C available, carbon could end up being the sole light element in the core, reaching the Fe-rich melt saturation value of ~8-10 wt% (Dasgupta and Walker 2008; Lord et al. 2009; Nakajima et al. 2009; Siebert et al. 2011; Dasgupta et al. 2013a) and perhaps precipitating graphite or diamond in the mantle. But given the low condensation temperatures of relevant carbon compounds, it is extremely unlikely that weight percent level carbon was acquired from chondritic building blocks. With the magma ocean being depleted in carbon instantly after core segregation, other processes to bring back carbon to the present-day BSE budget need to be invoked. What can some of those processes be?

Heterogeneous accretion and imperfect metal-silicate equilibration during core formation: the fate of carbon. If the dynamics of core-mantle separation are considered in detail, the possibility of perfect equilibration between metallic and silicate melt in a magma ocean occurs when the core melt is thoroughly emulsified in the magma ocean. In other words, sinking metal melt droplets had to have the dimensions smaller than the length scale of diffusive equilibration for elements of interest. Metal droplet diameters of ~1 cm is thought to be required for efficient equilibration with silicate melts. This likely happened when undifferentiated objects accreted or when small impactors (differentiated or undifferentiated) collided with proto-Earth, setting the equilibrium fractionation of carbon and other siderophile elements. But the geochemistry of Hf-W isotope systematics (Halliday 2004) and also those of Ni, Co, and W (Rubie et al. 2011) seem to require imperfect equilibration. Dahl and Stevenson (2010) showed that if the dynamics of accretion of large, differentiated impactors are considered, it is possible to have scenarios where the core of the giant impactor merges directly with the proto-Earth core, bypassing major chemical interaction with Earth's mantle. If this was the case, then the core:mantle mass ratio relevant for equilibrium partitioning of carbon (and any other elements) could be very different than their relative present-day masses. Dahl and Stevenson (2010) argued that only 1-20% of the cumulative core mass might have equilibrated with the magma ocean when late, differentiated, large impactors collided with proto-Earth. In this specific case, one could envision the most extreme cases of core to mantle masses (~1:210 to 1:10.5) relevant for equilibration. Dasgupta et al. (2013a) showed that with such mass ratios and for a bulk Earth carbon of 730 ppm, the magma ocean could retain as much as ~1.5 (20% of core mass equilibrating with the magma ocean and with $D_C^{metal/silicate}$ of 5500) to 217 ppm (1% of cumulative core mass equilibrating with the magma ocean with $D_C^{metal/silicate}$ of 500) carbon in dissolved form (Fig. 2). Interestingly, in the latter case, i.e., if only 1-2% of the metallic core mass equilibrated with the entire magma ocean, mass balance predicts core liquid to dissolve 10-15 wt% C. But such concentration could be higher than the solubility of carbon alloy in liquid at magma ocean conditions (Fig. 3; (Dasgupta and Walker 2008; Nakajima et al. 2009; Siebert et al. 2011; Dasgupta et al. 2013a). Thus the most extreme form of core-mantle disequilibrium—a very small fraction of alloy liquid equilibrating with a large mass of molten silicate—could have forced over-saturation of graphite or diamond during core segregation. In this scenario, the diamond or graphite would float from the segregating core liquid and contribute to the overlying silicate magma ocean carbon budget. Thus one might argue that this extremely imperfect core-mantle equilibration scenario could eliminate the "excess carbon" in the mantle problem by letting the magma ocean possess sizeable carbon content in solution and by forcing the segregating alloy liquid leave behind graphite or diamond. This model is similar to the one proposed by Hirschmann (2012). It should be noted that the carbon content of Earth's core becomes quite uncertain in this scenario, because the final make-up of the core depends on the composition and relative masses of the cores of different impactors. Dasgupta et al. (2013a) argued this scenario may be unlikely; because when the core of a large impactor merges directly

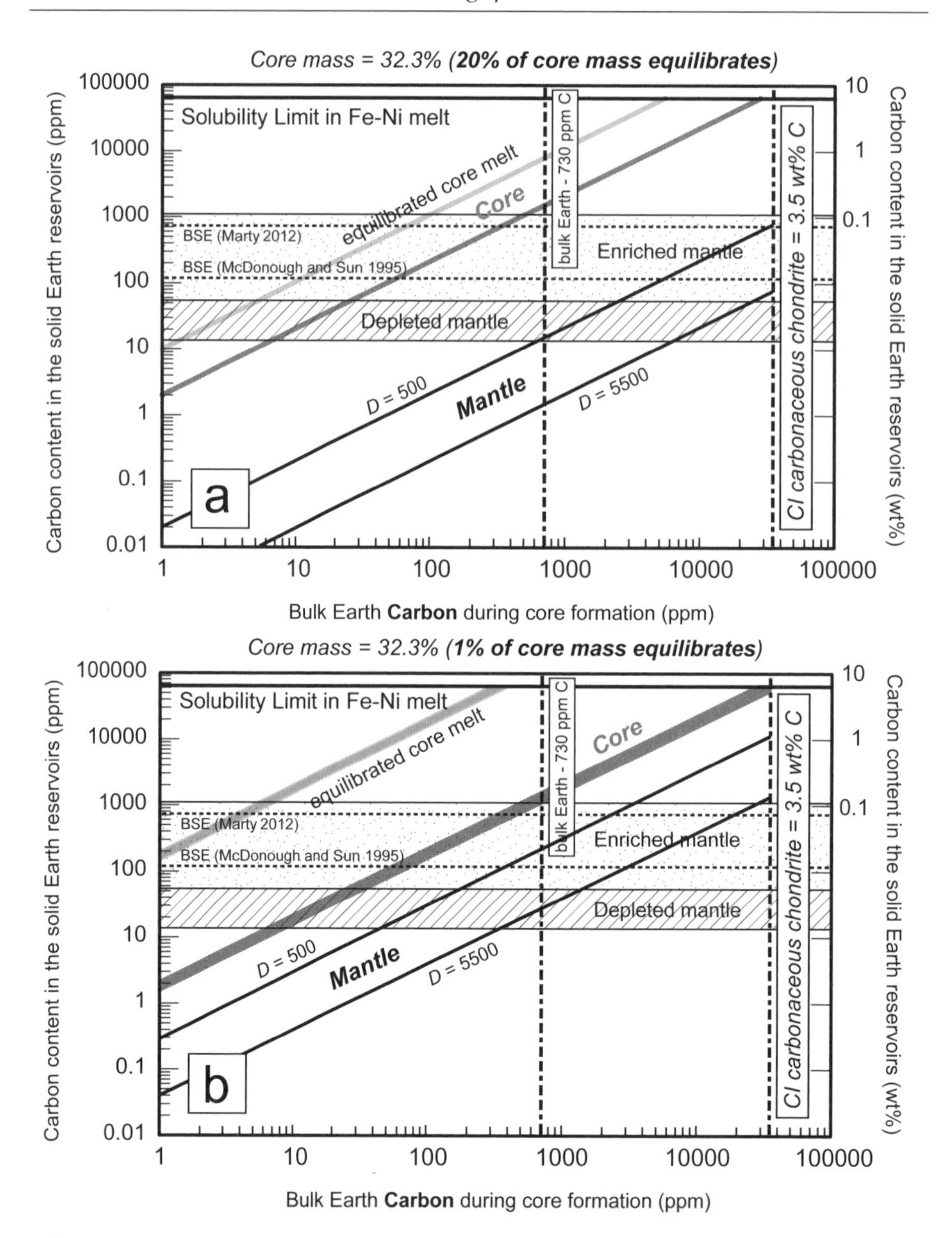

Figure 3. The effect of incomplete alloy melt-silicate melt equilibrium on the inventory of mantle and core carbon. The calculation in this figure assumes that between 20% (a) and 1% (b) of the present-day core mass equilibrated with the silicate magma ocean. The fraction of metallic alloy melt mass that equilibrated with the mantle was taken from Dahl and Stevenson (2010), who specifically modeled the impact of an already differentiated object with the proto-Earth, for the case where very limited metal-silicate equilibration takes place before the core of the impactor merges with the core of the proto-Earth. Shown for reference from Figure 2 are the present day mantle carbon content estimates for depleted and enriched sources, the BSE carbon content estimates, bulk Earth carbon content estimate of McDonough (2003), and carbon content of CI-type meteorite. For the sake of clarity, calculations are shown for only two $D_C^{metal/silicate}$ values of 500 and 5500. In general, if only a small fraction of the segregating alloy liquid equilibrates with the magma

(caption continued on facing page)

with the core of proto-Earth, the latter fractions of accreting metallic melt may not equilibrate with the entire silicate mantle mass. In particular, the other side of the planet and/ or the lower part of the solid mantle, would not provide storage for carbon (Keppler et al. 2003; Shcheka et al. 2006). Unmolten silicates will also not allow extreme metallic to silicate melt mass ratio for chemical equilibration (Golabek et al. 2008) necessary to force metallic liquid into carbon saturation. Hence, even if most of the terrestrial volatiles, including carbon, were delivered to Earth during the late stage of magma ocean and accretion (Morbidelli et al. 2012), it is unclear whether the very low ratio of alloy: silicate melt necessary for carbon saturation of segregating alloy liquid was ever achieved. Moreover, it has been argued that thorough emulsification of differentiated impactors' cores is achieved (conditions required for equilibrium separation of the core and the mantle) when dynamics of the impact process are considered (Kendall and Melosh 2012). Finally, even if thorough emulsification of the cores of large impactors did not take place the terrestrial magma ocean, metal-silicate equilibration with respect to carbon might have still taken place. Dasgupta et al. (2013a) pointed out that the length scale of diffusion for carbon in alloy melt is longer than that for common siderophile elements (Goldberg and Belton 1974; Dobson and Wiedenbeck 2002). Hence, if some modest fraction of carbon was available during the main stage of accretion and alloy-magma ocean equilibration, carbon partitioning in a system with alloy:silicate melt ratio not too different from the respective present-day masses mantle and core likely took place.

Magma ocean carbon cycle after core formation

Magma ocean storage capacity of carbon. Although the above analysis shows that the bulk of the carbon available in magma ocean environment during core formation was likely partitioned into the core, it also suggests the ability of silicate magma to keep in solution some finite fraction of carbon. This observation is unlike some previous suggestions where carbon was thought to be perfectly incompatible in silicate magma during core-mantle equilibration (Kuramoto and Matsui 1996; Kuramoto 1997; Yi et al. 2000). Given the uncertainty in bulk Earth carbon in general and the concentration prevalent during the Hadean time in particular, it is important to evaluate the storage capacity of carbon of the terrestrial magma ocean so that the limit to which magma ocean could supply carbon to the crystallizing mantle can be evaluated. The knowledge of magma ocean storage capacity of carbon is also important so that the conditions of appearance of various carbon-rich accessory phases (e.g., graphite/diamond, Fe-carbide, silicon carbide, C-rich alloy melt/ sulfide-rich melt with dissolved carbon, carbonated melt, and crystalline carbonate) can be constrained as a function of evolving temperature and oxygen fugacity of the planet.

The storage capacity of carbon in silicate liquid depends strongly on the dominant speciation or bonding environment of carbon, which in turn is controlled by a number of key variables including pressure, temperature, melt composition, and the last but not the least, the fugacity of oxygen, f_{O_2} and fugacity of other volatile species, such as f_{H_2O} and f_{H_2} that may impact carbon dissolution (Hirschmann 2012; Dasgupta et al. 2013a). Based on the spectroscopic studies of natural and experimental glasses at the conditions relevant for the basalt genesis of modern

---

(continuation of Figure 3 caption)
ocean, the latter is more likely to retain a good fraction of Earth's bulk carbon. It can be noted that with bulk Earth carbon of 730 ppm, the magma ocean can achieve as much as 27-217 ppm C (b), which overlaps with the estimates of present-day mantle budget. Also, in this case the equilibrating fraction of alloy melt may reach C-saturation (~8 wt% C) and hence release graphite/diamond. Flotation of these diamonds may elevate the magma ocean C-content thus perhaps even attaining the BSE value of ~765 ppm C (Marty 2012). However, it is argued in the text that this extreme case of incomplete core-mantle equilibration is unlikely. The bulk core carbon content estimation (dark grey lines) assumes that the unequilibrated fraction of the segregating metallic melt contained no carbon.

Earth, carbon is known to dissolve in mafic-ultramafic melts chiefly as carbonate anions bonded to network modifiers such as Ca^{2+}, Mg^{2+}, and Fe^{2+} (Blank and Brooker 1994; Dixon 1997; Brooker et al. 2001; Morizet et al. 2002; Lesne et al. 2011). However, CO_2 solubility in silicate liquids decreases with decreasing oxygen fugacity (Pawley et al. 1992; Thibault and Holloway 1994; Morizet et al. 2010; Mysen et al. 2011) and at conditions as reduced as Fe-metal saturation (~IW), carbonates may not be the dominant carbon species of interest for reduced magma oceans. Thus Equilibria of C-O-H fluids as a function of oxygen fugacity suggest that at f_{O_2} at or near core-forming conditions (IW–4 to IW+1), the dominant fluid species of interest are CH_4, CO, and H_2 rather than CO_2 and H_2O (e.g., Holloway and Jakobsson 1986; Zhang and Duan 2009). Although the number of studies investigating the solubility and dissolution of carbon in silicate melt at similarly reduced conditions are scarce, the available studies suggest that CO, CH_4 (or other higher order C-H molecules or groups, such as alkyne group (C≡C-H) bonded to silicate network), metal carbonyl groups, and possible structural units such as Si-C (Kadik et al. 2004; Kadik et al. 2006; Mysen and Yamashita 2010; Mysen et al. 2011; Wetzel et al. 2012; Dasgupta et al. 2013a) are all important species to be considered for carbon in reduced melts. Therefore, the topic of interest is the storage capacity of carbon species at reduced magma ocean conditions.

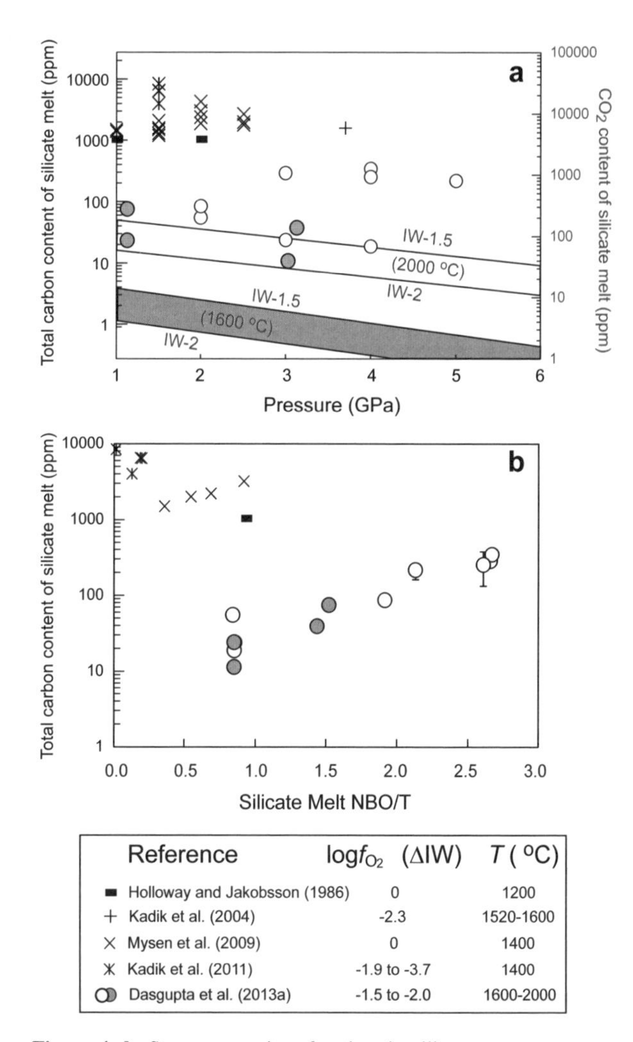

Figure 4a,b. Storage capacity of carbon in silicate magma at core-forming conditions (f_{O_2} ~IW–4 to IW), modified after Dasgupta et al. (2013a). (a) Total carbon content in silicate melts as a function of pressure. A range of silicate compositions saturated with CH_4-H_2 fluid and/or graphite and at temperatures of 1200-2000 °C is plotted. Also shown for reference are the CO_2 solubilities in a tholeiitic basalt at graphite saturation at 1600 and 2000 °C and between f_{O_2} similar to IW–2 to IW–1.5, following the model of Holloway (1992). (b) Estimated concentration of total dissolved carbon in silicate melt as a function of melt depolymerization index NBO/T (Mysen 1991) for the studies plotted in (a).

Figure 4 compiles the available experimental data of carbon solubility in silicate melts at conditions reduced enough to be relevant for core-mantle equilibration (IW–1 to IW–4;

Wood et al. 2006) or early stages of cooling magma ocean, soon after core formation (up to IW+1). Both C-rich vapor saturated (Holloway and Jakobsson 1986; Jakobsson and Holloway 1986; Mysen et al. 2009) and vapor-absent, graphite-saturated (Kadik et al. 2004; Kadik et al. 2011; Dasgupta et al. 2012, 2013a) experiments are included. The key features of dissolved carbon content in basaltic silicate melt (mostly at graphite saturation) are its increase with increasing temperature and melt depolymerization index as expressed by NBO/T, and decrease with increasing pressure. The dependence of total carbon solubility on pressure and temperature mimics what is expected for CO_2 solubility at graphite saturation (Holloway et al. 1992; Hirschmann and Withers 2008) (Fig. 4c). However, the total carbon content in tholeiites, at a f_{O_2} < IW, remains as much as an order of magnitude higher than the value expected for carbon dissolved only as carbonates (Dasgupta et al. 2013a). If total dissolved carbon content in tholeiitic melt saturated with CH_4-H_2 fluid reported by Holloway and Jakobsson (1986) is appropriate (Fig. 4a,b), then the contribution of the reduced or hydrogenated species to the total carbon content may be even higher. Furthermore, when CO_2 solubility predictions of reduced, graphite-saturated melts are compared with total dissolved carbon content at graphite saturation in more depolymerized (peridotitic, komatiitic) melts (Dasgupta et al. 2013a), it is observed that the difference between the two are even more dramatic (Fig. 4a); i.e., the depolymerized melts may have up to three orders of magnitude higher carbon content than that expected from CO_2 solubility alone. This latter discrepancy may owe in part to the fact that the model of Holloway et al. (1992) was developed for a tholeiitic basalt and hence does not apply to more MgO-rich and depolymerized melt compositions. Indeed the total carbon contents at graphite saturation, obtained for tholeiitic compositions, are consistently lower compared to the ones for peridotitic, komatiitic, and alkali basaltic compositions in the study of Dasgupta et al. (2013a). This observation suggests that the carbon solubility at graphite saturation is higher for more depolymerized melts (Fig. 4a) and thus the lat-

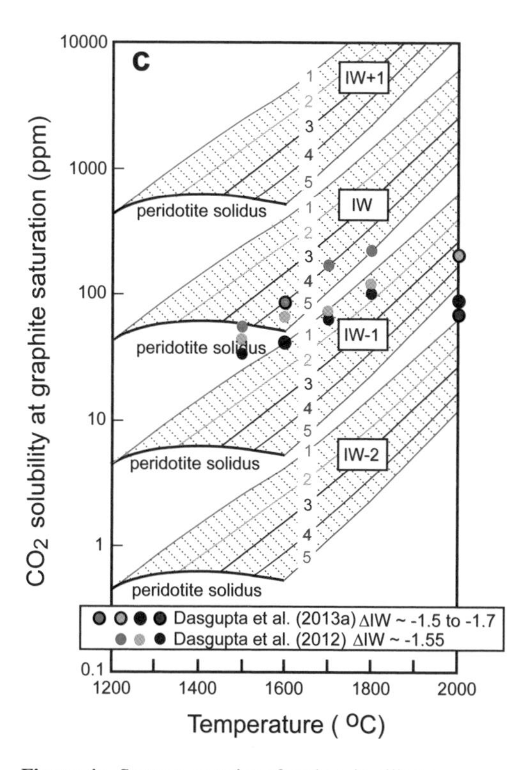

Figure 4c. Storage capacity of carbon in silicate magma at or near core-forming conditions (f_{O_2} ~IW–2 to IW+1). (c) CO_2 solubility in a tholeiitic basalt at graphite saturation as a function of temperature at 1-5 GPa and reduced conditions compared to the total dissolved carbon measured in a number of studies on tholeiitic compositions (Dasgupta et al. 2012; Dasgupta et al. 2013a) at similarly reducing conditions. The data and calculated lines are color-coded by pressure - 1 GPa - green, 2 GPa - orange, 3 GPa - black, 4 GPa - blue, and 5 GPa - red. This plot shows that the total carbon content at graphite saturation for tholeiitic basalts increases with increasing temperature and decreases with increasing pressure and these systematics are similar to what is expected for CO_2 solubility from thermodynamic calculations. However, the experimentally measured total dissolved carbon content at a given oxygen fugacity is significantly higher than that of the carbon in the form of dissolved carbonates alone, which highlights the role of other reduced species and f_{H_2} and f_{H_2O} in the magma ocean storage capacity of carbon.

ter are more appropriate for magma ocean carbon saturation limits. More work in calibrating the CO_2 solubility in Fe-Mg-rich depolymerized basalts at graphite saturation and at reduced (IW−2 to IW+1) conditions will be needed to evaluate what fraction of the total dissolved carbon in the terrestrial magma ocean was carbonated species and what fraction was metal carbonyl complexes and hydrogenated species, such as methane. Also, CO_2 storage capacity of mafic-ultramafic melts increases with f_{O_2} and f_{H_2O} (Eggler and Rosenhauer 1978; Holloway et al. 1992; Dasgupta et al. 2013a), so it is also going to be important to constrain the relative proportion of carbonated versus hydrogenated and other reduced species as a function of changing f_{O_2} of the magma ocean.

Magma ocean-atmosphere interaction and ingassing of carbon? The pressure and temperature dependence of storage capacity of carbon (either as CO_2 or as total dissolved carbon) at graphite saturation for the reduced magma ocean has important implications for the nature of interaction between magma ocean and proto atmosphere and for the fate of carbon during the crystallization of magma ocean.

A plausible mechanism for the BSE gaining carbon after core formation is through interaction of the magma ocean with the Hadean atmosphere (Hirschmann 2012; Dasgupta et al. 2013a). The negative effect of pressure on total carbon/CO_2 solubility at graphite saturation, as shown in the preceding section, makes this process viable. The depth dependence of CO_2 and total carbon solubility in mafic magma at graphite saturation and near iron-wüstite buffer (Fig. 4c) (Holloway et al. 1992; Hirschmann and Withers 2008; Dasgupta et al. 2013a) suggests that the carbon storage capacity of magma ocean at shallow depths is greater and diminishes at the expense of graphite/diamond or carbide-rich metallic melt at greater depths. In this scenario, a magma ocean could dissolve carbon through interaction with an early C-rich atmosphere and precipitate diamond/metal carbide melt in its deeper parts as convection brought a batch of magma down to greater depths. Precipitation and sequestration of carbon-rich phases at depth would lead to C-depletion of magma and thus upon upwelling would be able to dissolve more carbon/CO_2 from the atmosphere (Fig. 5b). This cycling might have served as an efficient mechanism of magma ocean ingassing, bringing the mantle inventory of carbon up to the present-day value or to match the suggested value of BSE well before magma ocean crystallization (Hirschmann 2012; Dasgupta et al. 2013a). The time scale over which this carbon ingassing process likely took place is unclear at present. The chemical evolution of Earth's primitive atmosphere is controversial (e.g., Kasting 1993; Zahnle et al. 2007; Zahnle et al. 2010) but several authors suggest persistence of a CO_2-rich atmosphere either through degassing of an earlier magma ocean (Elkins-Tanton 2008) or via condensation of a liquid water ocean, which could allow the residual atmosphere to be composed chiefly of CO_2 (Zahnle et al. 2007). CO_2-rich atmosphere could have existed when redox state of the shallow magma ocean was somewhat more oxidized than that imposed by saturation of metallic core liquid, i.e., when dissolved C-species in magma at shallow depths was dominantly CO_2 rather than methane or other reduced species. A study on oxidation state of Hadean zircon suggests that such an environment was created as early as only within a few million years after core formation (Trail et al. 2011). Thus if magma ocean carbon ingassing requires the presence of a CO_2-rich atmosphere, then such a scenario was possibly created within few tens of million years after core formation.

But what if the carbon in the atmosphere existed as reduced gases such as methane? Zahnle et al. (2010) showed that an atmosphere generated by impact degassing would tend to have a composition reflective of the impacting bodies (rather than the mantle of the proto Earth), and these impactors tend to be strongly reducing and volatile-rich. A consequence is that, although CO- or methane-rich atmospheres are not necessarily stable as steady states, they quite likely have existed as long-lived transients, after a number of impacts. The other authors have also argued that Earth retained a significant portion of the impactors' atmosphere. For example, Genda and Abe (2003) studied a planet with a solid surface and found that after an impact

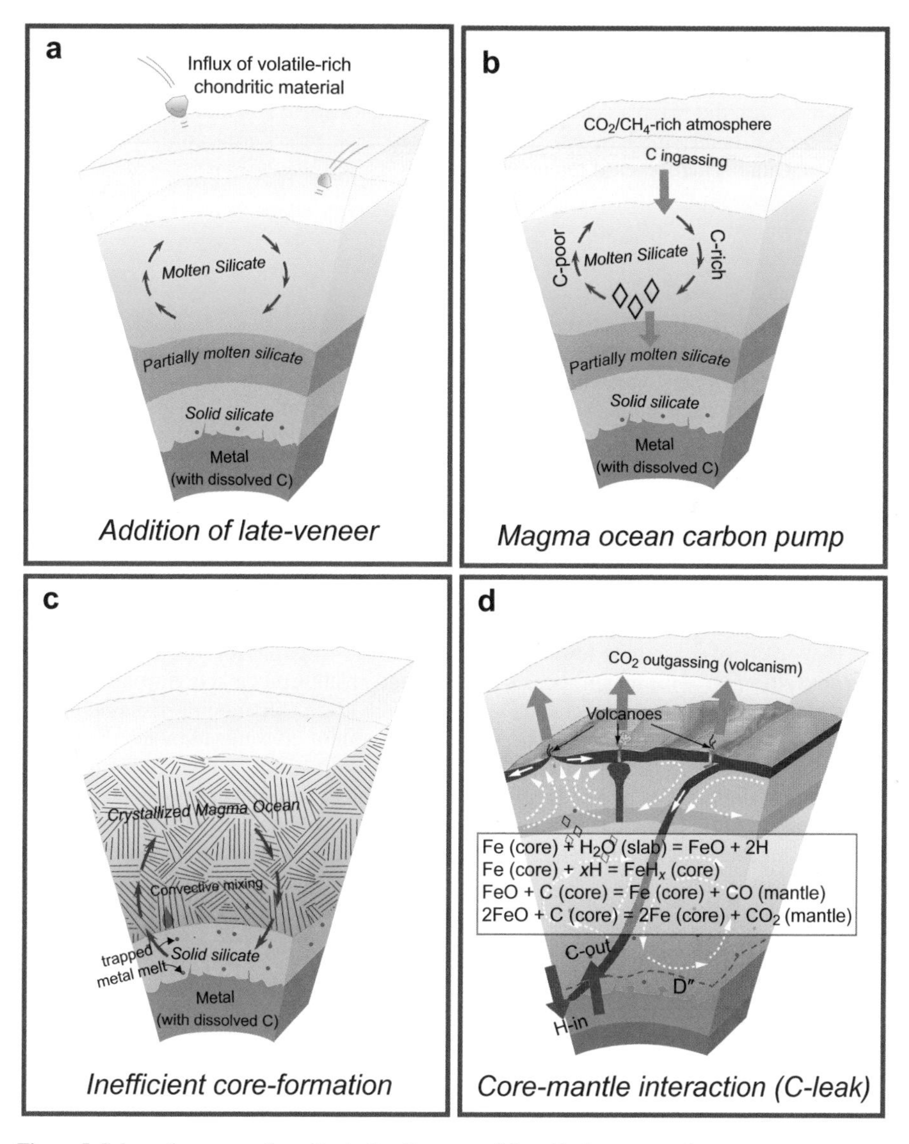

Figure 5. Schematic cross sections illustrating the range of deep Earth processes from magma ocean stage of the Hadean Eon to the plate tectonic framework of the modern world, which may explain the "excess carbon" in the mantle. (a) Addition of volatile-rich materials to Earth after core formation has ceased. (b) Interaction between a magma ocean and carbon-rich proto-atmosphere to increase the C-budget of the magma ocean. (c) Convective mixing of C-rich lower mantle materials with the entire mantle, after the crystallization of the magma ocean. (d) Core-mantle interactions and possible back-release of C-bearing fluids/melts aided by the presence of "water"; introduction of water to core-mantle boundary regions may be aided by subduction.

of a Mars-sized body, Earth retained 70% of its atmosphere; if the impactor had only 10% Earth mass, Earth retained 80 to 90% of its atmosphere. Proto-Earth, therefore, could have had a more massive atmosphere following the impact than it had before and this atmosphere could have been more reduced. Planets with liquid surfaces are thought to lose far more of their atmospheres during large impacts because of enhanced surface motion (Elkins-Tanton 2012). During the terrestrial magma ocean stage, large impacts thus would have removed large fractions of, but not the entire, atmosphere (Genda and Abe 2005). This atmosphere would then re-equilibrate with the magma ocean, causing magma to dissolve volatiles up to the solubility limits. Therefore, being struck by one or more reduced impactors, Earth could have inherited a methane-rich atmosphere, even if temporarily.

Could the magma ocean carbon ingassing operate with atmosphere being methane-rich rather than CO_2-rich? The depth dependence of methane solubility is ambiguous at present (Mysen et al. 2009; Ardia et al. 2011) but Dasgupta and co-workers (2012, 2013a) suggest that at graphite saturation, the total dissolved carbon that comprise more than one hydrogenated species of carbon (e.g., methane, methyl group, alkyne group) along with carbonates, also show a negative relationship with pressure. Hence, although somewhat speculative at this stage, magma ocean-atmosphere interaction may have acted as a carbon pump not only for a CO_2-rich atmosphere but also for a reduced (CH_4-bearing?) atmosphere.

Magma ocean crystallization and the fate of carbon. The above discussion suggests that although the core might have taken possession of most of Earth's carbon, a magma ocean-proto atmosphere interaction likely created an opportunity to restore the BSE carbon budget. Then the obvious next step to consider is the fate of the dissolved carbon in the magma ocean. The fate of dissolved carbon during crystallization of the terrestrial magma ocean depends on the Earth's temperature-f_{O_2} evolution. Because carbon solubility in reduced magma ocean drops precipitously with falling temperature (Fig. 4c), the storage capacity near the peridotite solidus can be 1-2 orders of magnitude lower than at temperatures relevant for magma oceans. Thus during crystallization of the magma ocean, diamond and graphite precipitation would take place as the peridotite solidus is approached. The temperature of precipitation of diamond/graphite in a cooling magma ocean would also depend on the depth of crystallization and the fugacity of hydrogen and water. Owing to decreased solubility of carbon as a function of increasing pressure, greater depths would see precipitation of diamond at a higher temperature. Thus if the crystallization of a well-mixed magma ocean proceeds within a narrow range of relatively reduced f_{O_2}, the lower part would tend to be more carbon-rich. Carbon enrichment in the lower part of the crystallizing magma ocean could be enhanced if magma ocean carbon ingassing also was efficient.

Carbon addition by late veneer? The late veneer, which has to be delivered after the latest stage of core formation, is widely considered to have added the refractory highly siderophile elements to silicate Earth (e.g., Chou 1978; Wänke and Dreibus 1988). A late veneer has also been invoked as a possible source for bulk Earth's or silicate Earth's water (Owen and Bar-Nun 1995; Morbidelli et al. 2000; Dauphas and Marty 2002; Albarede 2009; Marty 2012; Fig. 5a). It was also argued to be the main source of carbon and sulfur in silicate Earth (Yi et al. 2000). Addition of Earth's volatile elements by the late veneer has been suggested partly because of the premise that Earth may have lost most of its volatiles to space following a giant impact (Abe 1997; Genda and Abe 2003) and that the budget of terrestrial volatiles would therefore have to be reintroduced. But the proposition that BSE carbon is chiefly introduced by a "late veneer" is not entirely satisfactory.

In order to evaluate whether a "late veneer" could provide the bulk of Earth's carbon (and other volatiles), it is critical to evaluate how much of what types of materials were added and whether those additions satisfy the geochemistry of other siderophile elements. For example, the identical W isotopic composition of the Moon and the BSE has been argued to limit the amount of material that can be added as a late veneer to Earth after the giant impact to less than 0.3 ± 0.3 wt% of Earth's mass of ordinary chondrite or less than 0.5 ± 0.6 wt% CI-type carbonaceous chondrite based on their known W isotopic compositions (Halliday 2008). Similarly, the explanation of the mantle abundance of highly siderophile elements such as Pt and Pd seems to limit the addition of chondritic veneer to ~0.7 wt% (Holzheid et al. 2000). Abundance levels as low as 0.003 × Earth mass and as high as 0.007 × Earth mass have been argued by other studies as well (Morgan et al. 2001; Drake and Righter 2002). Addition of 0.3-0.7% of CI-type chondrite with 3.5 wt% C (Lodders 2003) would add 105-245 ppm C (385-898 ppm CO_2; Fig. 6) to Earth after core formation had occurred. Although this range of carbon concentration may be sufficient to explain the carbon geochemistry of mantle domains as enriched as those of most plume source regions or the BSE estimate of McDonough and Sun (1995), it is distinctly less than the BSE carbon abundance of 765 ± 300 ppm as estimated by Marty (2012). Thus it may be difficult to invoke the addition of CI-type chondritic material to bring carbon levels of the post-veneer mantle to the levels needed to match BSE without increasing the highly siderophile elements rather more than they are seen to be in "excess."

Invoking CI-type carbonaceous chondrite as the agent of volatile and siderophile element delivery to the BSE is also problematic for the Os isotopic composition of the mantle, which matches that expected from a veneer with a Re/Os ratio like that of ordinary chondrites (Walker et al. 2002). Moreover, the Ru and Mo isotopic compositions of meteorites are correlated (Dauphas et al. 2004), and the BSE appears to be different from the composition of CI chondrites and more similar to ordinary chondrites (Drake 2005). One can go through a similar exercise to evaluate whether the ordinary chondrites can deliver the right concoction of

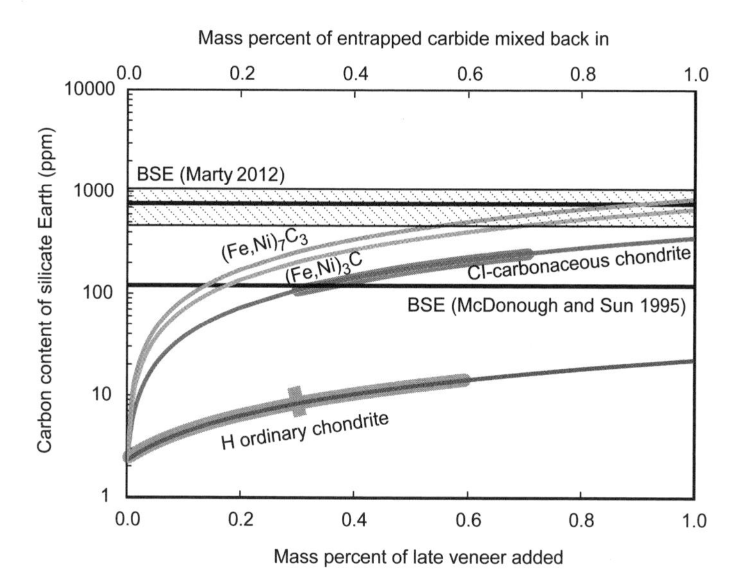

Figure 6. The effect of addition of chondritic late veneer and back-mixing of Fe-rich carbide from inefficient core formation on the bulk mantle carbon budget. The suggested range of CI-carbonaceous chondrite addition as required by the refractory siderophile element budget of the mantle and W-isotopic composition of the Earth-Moon system is highlighted in grey. The mean and standard deviation of the suggested value of ordinary chondrite addition are also highlighted in grey. Shown for reference are the BSE carbon estimate given by McDonough and Sun (1995) and Marty (2012). Plausible BSE carbon abundance of as much as 900-3700 ppm, i.e., even higher than the estimate of Marty (2012), exists in literature (Trull et al. 1993; Wood 1995; Wood et al. 1996) but are not plotted here for simplicity. The figure shows that if BSE carbon content of Marty (2012) or any other higher value is appropriate then the addition of chondritic "late veneer" cannot be the sole mechanism of bringing carbon to Earth. If on the other hand C-bearing core melt trapped in the lower mantle undergoes fractional crystallization, then some mantle domains could be Fe-carbide-bearing. Plotted, for example, are enrichments predicted for mixing such mantle domains containing stoichiometric $(Fe,Ni)_3C$ or $(Fe,Ni)_7C_3$ carbides. Admixing ~0.6 to 1.0 wt% of Fe-rich carbide can potentially elevate the magma ocean residual to core formation to the carbon level similar to that of BSE.

refractory siderophile elements and volatiles, including carbon. Halliday (2008) argued that in order to satisfy the W-isotopic composition of the BSE, no more than 0.3 ± 0.3 wt% of Earth's mass of H ordinary chondrite can be added. Thus with ordinary chondrite C content of 0.2 wt% (Moore and Lewis 1967), only ~6 ppm of carbon can be added through late veneer and hence cannot be a chief delivery agent to the BSE carbon budget.

Finally, if the BSE volatiles derive chiefly from a chondritic late veneer, it is also difficult to explain why the present-day H/C ratio of BSE (Hirschmann and Dasgupta 2009) is distinctly higher than all known chondritic materials (Kerridge 1985). Furthermore, carbon isotopic compositions of CI chondritic materials ($\delta^{13}C$~ −15 to −7‰) are distinctly lighter than the average carbon isotope composition of Earth's mantle ($\delta^{13}C$~ −5‰) (Kerridge 1985; Deines 2002). Therefore, any late delivery mechanisms may need to resort to more than one source that may be geochemically different than the known meteorites (Raymond et al. 2004; Raymond et al. 2006; Raymond et al. 2007; Albarede 2009); for example, carbonaceous chondrite and comets both may contribute differentially to the budget of carbon and water on Earth. In summary, although some parts of BSE carbon's delivery through a late veneer cannot be ruled out, it appears unlikely that late influx of chondritic materials is the primary process to elevate the carbon budget of the mantle after core formation.

Inefficient core formation and deep carbon storage. Inefficient core formation has been invoked as a plausible mechanism for explaining the "excess" siderophile element abundance in the mantle by a number of authors (e.g., Jones and Drake 1986; Newsom and Sims 1991). Dasgupta et al. (2013a) suggested that a similar mechanism can be surmised to explain the "excess" carbon in the mantle. In this model, metallic liquid would drain inefficiently to join the core, so some small fraction of melt would be trapped in the solid mantle matrix, providing a source of carbon in the mantle (Figs. 1 and 5c). Interfacial energies between carbon-bearing metallic alloy melts and mantle minerals are not constrained at present but observed dihedral angles between many other metallic melt compositions and solid silicate minerals of >60° and mostly >90°-100° support such hypothesis (Minarik et al. 1996; Shannon and Agee 1996; Terasaki et al. 2007; Mann et al. 2008; Terasaki et al. 2008). Trapped Fe-rich, carbon-bearing metallic melt will provide a source for both refractory siderophile elements and carbon in the mantle. According to this suggestion of Dasgupta et al. (2013a), Earth's lower mantle could start off being more carbon-rich and solid state convection after magma ocean solidification would bring up the lower mantle material with accessory Fe-rich alloy, Fe-rich carbide, or C-bearing metallic alloy liquid (Fig. 5c). Dasgupta et al. (2013a) suggested that the exact chemical form of carbon storage in the solid lower mantle will depend on the phase relations of Fe-(±Ni)-C bearing metallic systems and the mantle adiabat of interest. Using the study of Lord et al. (2009) on Fe-C binary, Dasgupta et al. (2013a) noted that the key observations are: (1) with increasing pressure from 10 to 50 GPa, the solubility of carbon in crystalline-Fe likely diminishes and thus the probability of iron carbide stability increases; and (2) depending on the base of the magma ocean, the trapped core melt in the solid matrix at relatively shallower depths may exist as a molten alloy whereas at greater depths it will exist as either iron metal + cohenite ($Fe + Fe_3C$) or iron + Eckstrom-Adcock carbide ($Fe + Fe_7C_3$) assemblages.

The contribution of carbon dissolved in alloy or as Fe-rich carbide to the total budget of upper mantle carbon depends critically on its composition. If the trapped metallic melt undergoes batch freezing and has only modest carbon content, as will be the case if it is derived from equilibrium partitioning in a low bulk carbon environment, then its input to the present-day, upper mantle carbon inventory will be small. For example, if trapped metal with ~0.2 wt% C is only 0.1-1% of the lower mantle, then such a mantle domain will have only 2-20 ppm C (~7-73 ppm CO_2). It is not implausible, however, as argued above that in some mantle domains, that the trapped metallic liquid precipitates as Fe-rich carbide (e.g., Fe_7C_3 or Fe_3C solid solutions). Carbide formation is facilitated by the Fe-C eutectic composition becoming increasingly carbon-poor with increasing pressure (Wood 1993; Lord et al. 2009) and carbide-saturated Fe-alloy becoming poorer in carbon with increasing pressure (Walker et al. accepted). If separation of carbide from the remaining metallic liquid takes place, perhaps aided by the presence of silicate partial melts, then some lower mantle domains can be quite carbon-rich. If the lower mantle domain contains 0.1-1 wt% of Fe_7C_3, then such parcels can deliver ~84-840 ppm C (~310-3100 ppm CO_2) and convective mixing of only 10% of such mantle mass can contribute >30-300 ppm CO_2 to the Earth's depleted upper mantle (Fig. 6). It can be seen from Figure 6 that trapped carbide fractions have greater leverage, compared to delivery of late chondritic veneer, in supplying carbon to a depleted mantle. Another way to have the trapped core liquid be rich in carbon is to have carbon partitioning with a relatively low alloy:magma ocean mass ratio. For example, if only 20% of the core mass equilibrated with the whole mantle mass magma ocean, then the equilibrated alloy liquid could have had carbon content as high as 1 wt% (Fig. 3) even with a bulk Earth carbon as low as 730 ppm. Therefore, if volatile delivery was delayed to the latter part of accretion and if low alloy:magma ocean ratio was the norm for carbon fractionation, trapped core liquid can be quite C-rich.

If core-forming alloy liquid indeed provided some initial carbon, then the carrier phase of interest can be much more complex than that captured by the Fe-Ni-C system alone. For example, what if the segregating liquid alloy is also sulfur bearing? In such a case, phase relations of Fe-Ni-C-S systems will have to be considered. In that case diamond saturation may be achieved, owing to lower solubility of carbon in sulfide-rich melt (see the later discussion on carbon storage in the present-day mantle). Similarly, if the alloy liquid of interest is Si-bearing, phase relations in the Fe-Ni-Si-C system may come into play. In the latter case, crystallization of moissanite (SiC) in a reduced lower mantle cannot be ruled out. Although rarely found, moissanites have been reported as inclusions in natural diamonds (Leung et al. 1990; Leung 1990). If segregating alloy liquid crystallizes diamond and/or moissanite in the lower mantle, mixing of such mantle materials will be far more effective in bringing the bulk mantle carbon budget up.

Transfer of carbon in trapped reduced phases by convective and oxidative processes. If core-forming liquid trapped in the lower mantle mineral matrix is the chief source of carbon in Earth's mantle in general, and in present-day Earth's upper mantle in particular, then some form of oxidative process is necessary for carbon to be mobilized from these reduced accessory phases. The following end-member reactions can be posited for such oxidative release of carbon:

$$\underset{carbide}{2Fe_3C} + \underset{px/pvskt}{4FeSiO_3} + 5O_2 = \underset{carbonate}{2FeCO_3} + \underset{ol/wad/ring}{4Fe_2SiO_4} \quad (3)$$

$$\underset{carbide}{Fe_7C_3} + \underset{px/pvskt}{30FeSiO_3} + \underset{gt/pvskt}{13Fe^{3+}{}_2O_3} = \underset{carbonate}{3FeCO_3} + \underset{ol/wad/ring}{30Fe_2SiO_4} \quad (4)$$

where carbide reflects either a true Fe-rich carbide phase or a C-bearing component in the Fe-rich metal alloy and $Fe^{3+}{}_2O_3$ reflects a component in garnet (*gt*) or perovskite (*pvskt*), recognizing the highest Fe^{3+} uptake by these silicate phases (Frost et al. 2004; Rohrbach et al. 2011). Fe_2SiO_4 also reflects an end-member component in olivine (*ol*), wadsleyite (*wad*), or ringwoodite (*ring*) whereas $FeCO_3$ reflects either a component in a crystalline carbonate or a component in a carbonitic or carbonated silicate melt. Furthermore, oxidation of metal alloy or metal carbide likely does not directly generate carbonate or CO_2; more likely it proceeds through an intermediate stage of metal oxidation and diamond/graphite (*dia/gr*) precipitation (Frost and McCammon 2008) such as:

$$\underset{carbide/alloy}{Fe_7C_3} + \underset{in\ gt/pvskt}{7Fe_2O_3} = \underset{dia/gr}{3C} + \underset{in\ silicates/ferropericlase}{21FeO} \quad (5)$$

$$\underset{carbide/alloy}{Fe_3C} + \underset{n\ gt/pvskt}{3Fe_2O_3} = \underset{gr/dia}{C} + \underset{in\ silicates/ferropericlase}{9FeO} \quad (6)$$

followed by oxidation of graphite/diamond through reactions of the sort:

$$\underset{gr/dia}{C} + \underset{ol/wad/ring}{Fe_2SiO_4} + O_2 = \underset{carbonate}{FeCO_3} + \underset{px/pvskt}{FeSiO_3} \quad (7)$$

$$\underset{gr/dia}{C} + \underset{px/pvskt}{3FeSiO_3} + \underset{in\ gt/pvskt}{2Fe_2O_3} = \underset{carbonate}{FeCO_3} + \underset{ol/wad/ring}{3Fe_2SiO_4} \quad (8)$$

Again, all of the above reactions are shown with both oxygen and Fe^{3+} as oxidants. Rohrbach and Schmidt (2011) showed that the transition from a metal and diamond-bearing mantle to a carbonated mantle takes place at ~250 km depth for a geotherm relevant to the present-day sub-ridge environment. If this depth was relevant for the mantle soon after magma ocean crystallization, then oxidative release of carbon from metal alloy and/ or carbide would have taken place somewhat above the mantle transition zone. However, the proportion of metal alloy/ carbide that might be physically entrapped in solid silicate matrix of the lower mantle can be more than the metal proportion of ~1 wt% expected by Fe^{2+} disproportionation in the present-day lower mantle (Frost et al. 2004) and much more than 0.1 wt% supposedly present in the modern deep upper mantle. Similarly, the composition of the trapped metal and/or metal carbide phase may have much higher Fe/Ni ratio compared to those formed by iron disproportionation. Hence the buffering capacity of such metal alloy/carbide-rich mantle may be different than what is predicted for present-day Earth's metal-poor mantle below ~250 km (Rohrbach et al. 2007, 2011). Future experiments will need to constrain the conditions of redox reactions involving variable metal alloy compositions.

CARBON RETENTION: MODULATING MANTLE CARBON BUDGET THROUGH THE WILSON CYCLE

The preceding discussion sheds some light on the path for attaining the inventory of mantle carbon in general and that of the upper mantle in particular. It was argued that it is possible to conceive a set of processes that yield carbon content similar to that of the present-day budget by late Hadean to early Archean time. If Earth's mantle attained a carbon content similar to the BSE inventory early in its life, how is such a budget maintained and processed in the light of the ongoing differentiation of the planet for more than 4 Ga? While magma ocean processes dominated the proceedings of deep carbon cycle in the Hadean, the carbon cycle post magma ocean crystallization was modulated by the thermal vigor of solid state convection and plate tectonic cycles. In this section, the current thinking of Earth's deep carbon cycle in the plate tectonic framework of mantle differentiation is reviewed. It will be shown that the mantle inventory of carbon may have gone through definite changes as a function of time, between the Archean Eon and the present-day and that key processes that affected such changes are the efficiency of crustal carbon recycling and the efficiency of CO_2 release through partial melting in oceanic provinces, both of which are controlled by the evolving thermal state of Earth's mantle at various tectonic settings. For magmatic outgassing of carbon, the key parameter of interest is the depth of onset of decompression melting and the extent of melting; the deeper and higher the extent of partial melting of the mantle, greater is the efficiency of carbon liberation (e.g., Dasgupta and Hirschmann, 2006; Dasgupta et al. 2013b). For recycling of crustal carbon, the key factor is the relative positions of decarbonation and decarbonation melting reactions of the lithospheric lithologies and depth-temperature paths of downgoing lithospheric materials. If at shallow depths (e.g., sub-arc depths), the former are cooler compared to the latter then one would predict inefficient deep subduction of carbon with most carbon being stripped off from the downgoing slab and released back to the exosphere by arc magmatism. If, on the other hand, up to the sub-arc depths the downgoing lithologies experience temperatures cooler than the key decarbonation reactions, then carbon hosted in lithosphere is expected to avoid the magmatic release in the volcanic arcs and participate in the deeper cycling and mantle processes.

Carbon cycle in an ancient Earth with greater thermal vigor: an era of more efficient outgassing?

A number of naturally relevant composition studies based on ancient rock records suggest that Earth's mantle was hotter in the Archean through Proterozoic time (Herzberg et al. 2010; Lee et al. 2010). While the average mantle potential temperatures (T_P) relevant for modern Earth is estimated to be ~1300-1400 °C (e.g., Ita and Stixrude 1992; McKenzie et al. 2005; Herzberg et al. 2007; Herzberg and Asimow 2008), basalt genesis from non-arc settings at 2.5-3.0 Ga requires that the potential temperatures were at least 1500-1600 °C. If Archean komatiites are considered, then the relevant mantle potential temperatures would be at least 1700 °C (Herzberg et al. 2010; Lee et al. 2010). The question then is how deep was the onset of decompression melting in the Archean-Proterozoic time. Decompression melting beneath mid-oceanic ridges, which particularly controls the release of carbon and other highly incompatible elements, is the depth of carbonated peridotite solidus. Figure 7 shows a compilation of experimental data that

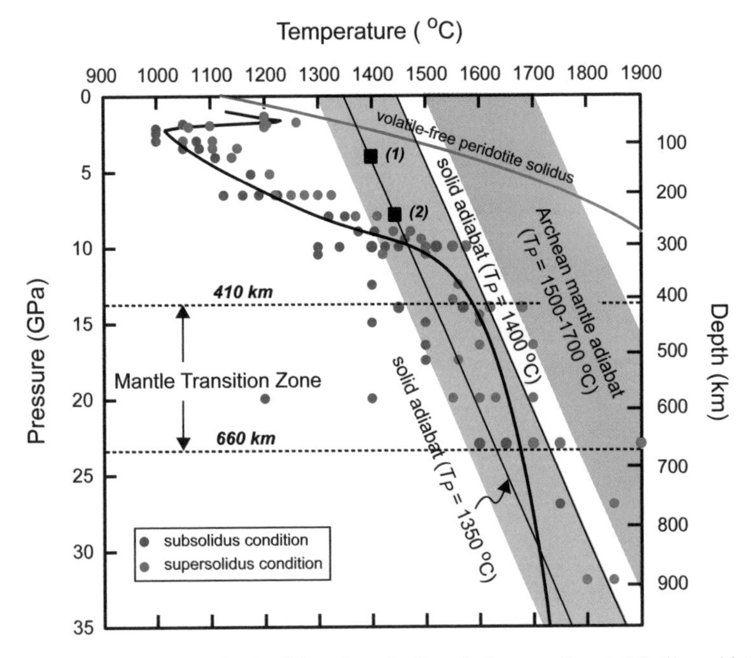

Figure 7. Experimentally constrained solidus of nominally anhydrous, carbonated fertile peridotite. The experiments include those from Falloon and Green (1989) at 1.4-3.5 GPa, Dasgupta and Hirschmann (2006) at 3.0-10.0 GPa, Dasgupta and Hirschmann (2007a) at 6.6 GPa, Ghosh et al. (2009) at 10.0-20.0 GPa, Litasov and Ohtani (2009) at 16.5-32.0 GPa, and Rohrbach and Schmidt (2011) at 10-23 GPa. The shape of the solidus over 3-10 GPa pressure range is constrained by the experiments of Dasgupta and Hirschmann (2006), but the exact location is adjusted to that expected for a peridotite with ppm level CO_2 after Dasgupta and Hirschmann (2007a). Subsolidus and supersolidus experiments are given by solid blue and open red circles, respectively. The preferred solidus from 2 to 35 GPa is fitted with a polynomial. Unlike in the study of Dasgupta and Hirschmann (2010), where the solidus at >10 GPa was parameterized by correcting for the bulk compositional differences between the studies performed at lower (≤10 GPa; natural fertile peridotite compositions) and higher (≥10 GPa; fertile peridotite with excess alkalis or peridotite in the simple system CMASN+CO_2) pressures, here the solidus at >10 GPa is parameterized by using the brackets from the study of Rohrbach and Schmidt (2011). Also shown for reference (black boxes on the 1350 °C mantle adiabat) are the conditions of redox melting as proposed by (1) Stagno and Frost (2010) and (2) Rohrbach and Schmidt (2011). The green and saffron shaded bands indicate approximate location of the solid mantle adiabats of the modern mantle (T_P = 1300-1450 °C) and the Archean mantle (T_P = 1500-1700 °C), respectively.

constrains the solidus of carbonated fertile peridotite solidus. At depths in excess of ~70 km, the carbonated peridotite solidus is 350-500 °C cooler than that of volatile-free solidus. The shape of the solidus is such that it intersects the solid mantle adiabat for T_P of 1350 °C at ~10 GPa. Thus, in modern Earth's upwelling carbonated mantle, melting initiates at a minimum depth of ~300 km. Another curious feature of the fertile carbonated peridotite solidus is that it changes slope sharply at ~10 GPa and remains similar to the solid mantle adiabat with T_P of ~1350-1400 °C at higher pressures. Hence adopting any hotter mantle adiabat for the Archean and part of the Proterozoic (Herzberg et al. 2010) would predict that the entire mantle may have possibly remained above the carbonated peridotite solidus. The implication is that the mantle can be very efficiently processed through the carbonated peridotite solidus.

But is the hotter thermal state of the mantle the only factor affecting the depth of onset of partial melting and carbon release? In order to use the depth of intersection of solid mantle adiabats and carbonated peridotite solidus as the depth of first melting in the adiabatically upwelling mantle, carbon in peridotite is required to be stored as mineral carbonates (e.g., Falloon and Green 1989; Dalton and Presnall 1998; Dasgupta and Hirschmann 2006; Dasgupta and Hirschmann 2007a). However, carbon in the mantle can be stored not only as mineral carbonates (magnesite and dolomite solid solution for mantle peridotite) but also as graphite or diamond (Eggler and Baker 1982; Eggler 1983; Luth 1993; Luth 1999; Dasgupta and Hirschmann 2010; Rohrbach and Schmidt 2011; Dasgupta et al. 2013b). Therefore, the depth of onset of melting must vary as a function of the mode of carbon storage. Two factors critically affect whether carbon in the mantle at depths is being stored in carbonates or in one of the reduced phases: (1) the locations of the equilibrium reactions that constrain the stability of carbonate versus graphite/diamond in P-T-f_{O_2} space, and (2) the oxygen fugacity profile of the mantle as a function of depth.

(1) Equilibrium reactions. Carbonate-carbon (graphite:G or diamond:D) equilibria involving upper mantle silicates at subsolidus conditions have long been constrained (Eggler and Baker 1982; Eggler 1983; Luth 1993).

$$\text{EMOG/D:}\quad \underset{enstatite}{Mg_2Si_2O_6} + \underset{magnesite}{2MgCO_3} = \underset{olivine}{2Mg_2SiO_4} + \underset{graphite}{2C} + 2O_2 \tag{9}$$

$$\text{EDDOG/D:}\quad \underset{enstatite}{2Mg_2Si_2O_6} + \underset{dolomite}{CaMg(CO_3)_2} = \underset{diopside}{CaMgSi_2O_6} + \underset{olivine}{2Mg_2SiO_4} + \underset{graphite}{2C} + O_2 \tag{10}$$

Reaction (9) also can be rewritten to constrain the oxygen fugacity at which magnesite coexists with diamond, clinoenstatite ($MgSiO_3$), and wadsleyite/ ringwoodite (Mg_2SiO_4) at transition zone depths (Stagno et al. 2011).

(2) Depth versus oxygen fugacity. The depth versus effective oxygen fugacity profile depends on the chemical species that are responsible for controlling the oxygenation potential of mantle assemblages. Silicate equilibria involving the exchange of Fe^{3+} and Fe^{2+} between silicate minerals could be responsible for controlling the apparent f_{O_2} (Gudmundsson and Wood 1995; Rohrbach et al. 2007; Frost and McCammon 2008) and in this case the chemical form of carbon will rely on the intersection of the iron redox state imposed f_{O_2} and the carbonate-carbon (graphite/diamond) transformation reactions. However, mantle f_{O_2} may also be buffered by carbon-carbonate equilibria (Eggler 1983; Luth 1999), in which case the stability of carbon versus carbonate and f_{O_2} versus depth are not controlled by the energetics of Fe^{3+} incorporation into silicate minerals such as garnet or perovskite (Gudmundsson and Wood 1995; McCammon 2005; Frost and McCammon 2008) but by the EMOD-type equilibrium (Equation 9). Although many studies have favored the Fe^{3+}-Fe^{2+} equilibria as the independent variable affecting effective f_{O_2} at depth, whether carbon-carbonate equilibrium buffers the f_{O_2} instead remains an open question. In particular, if the Hadean and Archean mantle was carbon-rich (owing to inefficient core formation or through magma ocean-atmosphere interaction or by addition of late-veneer as discussed previously), then the buffering capacity of Fe^{3+}-Fe^{2+} in silicates might have been overwhelmed by that involving carbon (Luth 1999). Not having continental crusts of sufficient volume, the exospheric storage of carbon may have been limited and the entire of budget of BSE carbon might have been in the mantle (Marty and Jambon 1987). In fact, if the BSE carbon content (765-3700 ppm) estimated by Marty (2012) or by earlier works (Trull et al. 1993) applied to the whole mantle of the Hadean or Archean Eons, then carbon-carbonate equilibria would indeed control the f_{O_2} of the deep mantle. If this was the case then the first melting of the upwelling mantle would take place very deep in the mantle at the carbonated peridotite solidus.

Inefficient subduction of carbon in the Archean and Proterozoic?

The timing of initiation and the nature of tectonics involving ancient subduction zones are actively debated. The mineral assemblages included in diamonds spanning the past 3.5 billion years suggest that the subduction may have initiated at ~3.0 Ga (Shirey and Richardson 2011). Much older subduction-type environments have even been suggested based on thermobarometric estimates of the inclusions in 4.02-4.19 Ga zircons (Hopkins et al. 2008; Hopkins et al. 2010). If some form of lithospheric recycling initiated in the late Hadean to early Archean, then what was the thermal vigor of such ancient recycling processes and how did it evolve through the Archean and Proterozoic Eons? This question is crucial because in order to predict the subduction potential of carbon, the thermal profile of the downgoing plate as a function of depth needs to be known. Geodynamic predictions suggest that subduction may have initiated with T_P being as much as 175-200 °C hotter than of the modern Earth (Sizova et al. 2010). However, the increase in mantle potential temperature should not be directly reflected in the slab surface temperature. Hotter mantle would generate thicker crust and faster plate velocity, both of which would offset the effect of hotter mantle wedge temperature and therefore the slab surface temperature-depth trajectories of ancient subduction may not be much hotter. These geodynamic considerations suggest that the temperature during subduction may have been ~87-100 °C (50% of the potential temperature) hotter (van Keken personal communication). In addition to the geodynamic predictions, the estimates of ancient geothermal gradient based on thermobarometry of Precambrian rock records also exist (e.g., Nakajima et al. 1990; Möller et al. 1995; Komiya et al. 2002; Brown 2006; Moyen et al. 2006; Mints et al. 2010; Saha et al. 2010). In Figure 8, following the approach of Dasgupta and Hirschmann (2010) and Tsuno and Dasgupta (2011), the plausible depth-temperature trajectories experienced by subducting rocks during the Archean-Proterozoic are compared with the experimental constraints on decarbonation reactions (CO_2-rich fluid or carbonated melt liberation) of subducting lithologies. Subduction zone thermal conditions based both on geodynamic and petrologic constraints are presented. It can be noticed that the estimates derived from the rock records suggest higher subduction zone temperatures than those predicted by geodynamic modeling. For simplicity only one plausible *P-T* trajectory of ancient subduction that is hotter than Cascadia subduction zone by 100 °C is plotted. This estimate is most likely at the upper end of all the ranges slab-surface temperatures, given that most other modern subduction zones are cooler than that of Cascadia. Therefore, the difference between the geodynamic predictions and natural rock records is likely even more than what appears in Figure 8.

The compositions and the relative proportions of carbon-bearing rock types during this ancient time-period are poorly constrained and in Figure 8 it is assumed that all the three dominant carbonated lithologies that subduct in modern Earth—that is carbonated ocean floor sediments, carbonated altered basalt, and carbonated peridotite (ophicarbonate)—participated in the subduction-type environment of deep time, although their relative importance in terms of the flux of carbon in the downgoing slab may have been different than what has been estimated for modern subduction zones globally (Sleep and Zahnle 2001; Jarrard 2003; Dasgupta and Hirschmann 2010).

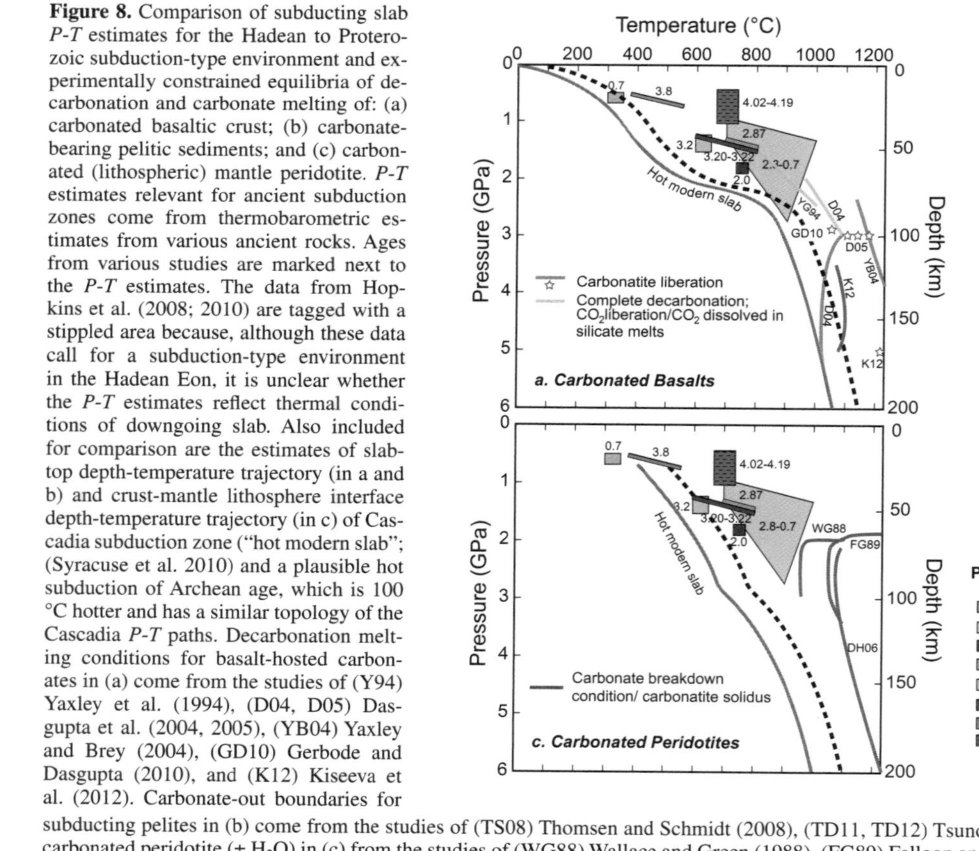

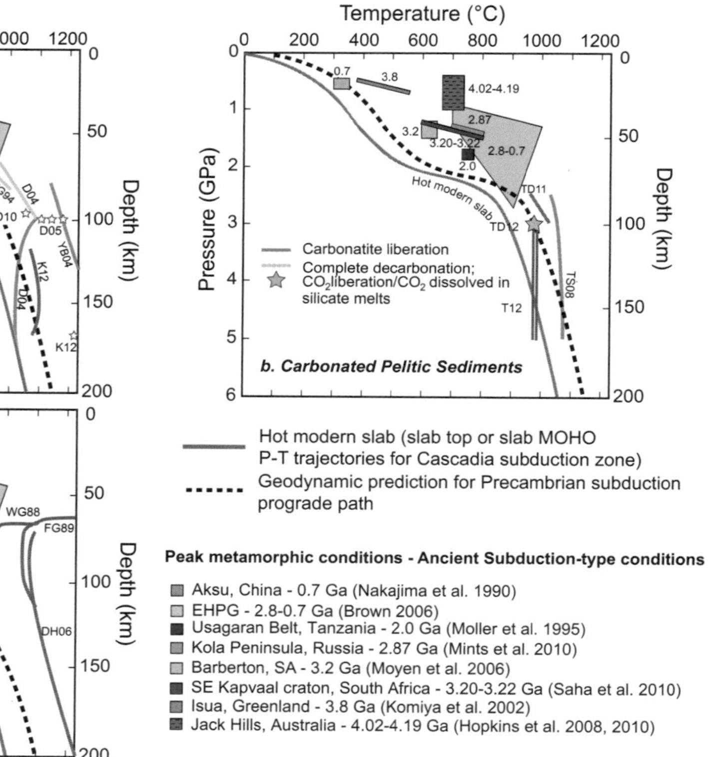

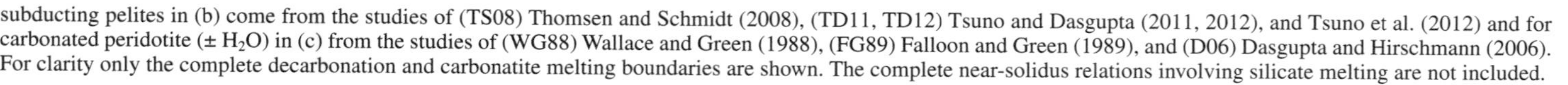

Figure 8. Comparison of subducting slab *P-T* estimates for the Hadean to Proterozoic subduction-type environment and experimentally constrained equilibria of decarbonation and carbonate melting of: (a) carbonated basaltic crust; (b) carbonate-bearing pelitic sediments; and (c) carbonated (lithospheric) mantle peridotite. *P-T* estimates relevant for ancient subduction zones come from thermobarometric estimates from various ancient rocks. Ages from various studies are marked next to the *P-T* estimates. The data from Hopkins et al. (2008; 2010) are tagged with a stippled area because, although these data call for a subduction-type environment in the Hadean Eon, it is unclear whether the *P-T* estimates reflect thermal conditions of downgoing slab. Also included for comparison are the estimates of slab-top depth-temperature trajectory (in a and b) and crust-mantle lithosphere interface depth-temperature trajectory (in c) of Cascadia subduction zone ("hot modern slab"; (Syracuse et al. 2010) and a plausible hot subduction of Archean age, which is 100 °C hotter and has a similar topology of the Cascadia *P-T* paths. Decarbonation melting conditions for basalt-hosted carbonates in (a) come from the studies of (Y94) Yaxley et al. (1994), (D04, D05) Dasgupta et al. (2004, 2005), (YB04) Yaxley and Brey (2004), (GD10) Gerbode and Dasgupta (2010), and (K12) Kiseeva et al. (2012). Carbonate-out boundaries for subducting pelites in (b) come from the studies of (TS08) Thomsen and Schmidt (2008), (TD11, TD12) Tsuno and Dasgupta (2011, 2012), and Tsuno et al. (2012) and for carbonated peridotite (± H_2O) in (c) from the studies of (WG88) Wallace and Green (1988), (FG89) Falloon and Green (1989), and (D06) Dasgupta and Hirschmann (2006). For clarity only the complete decarbonation and carbonatite melting boundaries are shown. The complete near-solidus relations involving silicate melting are not included.

Figure 8b shows that on average, at any given pressure, carbonated pelitic sediments have the lowest decarbonation and carbonate melting temperatures, thus they are most prone to losing their carbon inventory during subduction. Carbonated ocean floor basalts or their derivatives, carbonated eclogite, have intermediate decarbonation or carbonate melting temperatures and carbonated peridotites have the highest carbonate melting solidi. Thus if a slab surface *P-T* trajectory hotter by ~100 °C than the hottest subduction zone of modern Earth (Cascadia) is considered, then compositions similar to carbonated silicate sediments of the present-day ocean floor will be completely carbon-free by carbonate melting or decarbonation by depths of ~100-150 km. Whereas if metamorphic conditions recorded in ancient crustal assemblages rightly capture the prograde subduction path of the Precambrian, then sediment decarbonation may have been completed as shallow as 80-100 km. These depth estimates are well within modern Earth sub-arc depths of 72-179 km (Syracuse and Abers 2006).

With *P-T* conditions similar to those recorded in the Precambrian rock record, altered ocean floor basalts would undergo complete decarbonation by 80-120 km (Yaxley and Green 1994; Dasgupta et al. 2004; Yaxley and Brey 2004; Dasgupta et al. 2005; Gerbode and Dasgupta 2010; Kiseeva et al. 2012). Whereas if slab-top temperatures were only up to 100 °C hotter than the hottest subduction zone of the present era, then carbonated basaltic compositions, which stabilize calcitic crystalline carbonate in the eclogite (Yaxley and Green 1994; Yaxley and Brey 2004; Dasgupta et al. 2005), would survive significant decarbonation in the Precambrian sub-arc depths (Fig. 8a). Under similar conditions, basaltic eclogites with more magnesian crystalline carbonate (dolomite to magnesite solid solution), would undergo complete decarbonation and carbonate melting by 125-170 km depth (Dasgupta et al. 2004; Dasgupta et al. 2005; Gerbode and Dasgupta 2010).

The fate of carbon hosted in oceanic mantle lithosphere is somewhat unclear, partly because of the fact that compositions that derive from serpentinized and carbonated depleted peridotite have not been extensively studied experimentally. Also, compositions representative of serpentinized carbonated peridotite and their dehydrated products (fluid-absent, carbonated peridotite) are poorly constrained. In terms of the phase relations of altered lithosphereic mantle, what exist are thermodynamic calculations relevant for low-pressure devolatilization of ophicarbonates (Kerrick and Connolly 1998) and melting and decarbonation relations of more fertile peridotite compositions under hydrous (Wallace and Green 1988) or nominally anhydrous (Falloon and Green 1989; Dasgupta and Hirschmann 2006; Dasgupta and Hirschmann 2007a) conditions. If these phase relations are a reasonable approximation of high-pressure behavior of carbonated lithosphere, then subducting mantle likely had the best prospect of carrying carbon down deep. This conclusion follows because peridotitic lithologies provide the greatest thermal stability of crystalline carbonates and they go down along the coolest paths during subduction (Fig. 8c).

Although Precambrian subduction can be argued to be hotter on average than modern Earth, what remains uncertain is the time scale over which slab thermal structure evolved to something similar to that of the present day. Figure 9a presents a conjectural projection of all the various temperature estimates for subduction-type environments between the Archean Eon and Neoproterozoic Era to a constant depth of ~70 km. The different *P-T* estimates are projected to a single pressure of ~2 GPa, using the topology of the slab surface *P-T* path estimated for the present-day subduction zone in Cascadia. Although for comparison with experimental data of decarbonation/melting, slab temperatures at somewhat higher pressures are necessary, P = 2 GPa is preferred to avoid significant extrapolation of the available thermobarometric data, which record peak metamorphic temperatures at $P \leq 2.5$ GPa and mostly < 2 GPa. However, Figure 9a does shed some light on the extent of excess temperature that the downgoing crustal rocks of Precambrian likely have suffered compared to those that subduct in modern Earth. It shows that slab temperatures were likely hotter by as much as 100 °C as recently as 1 Ga even

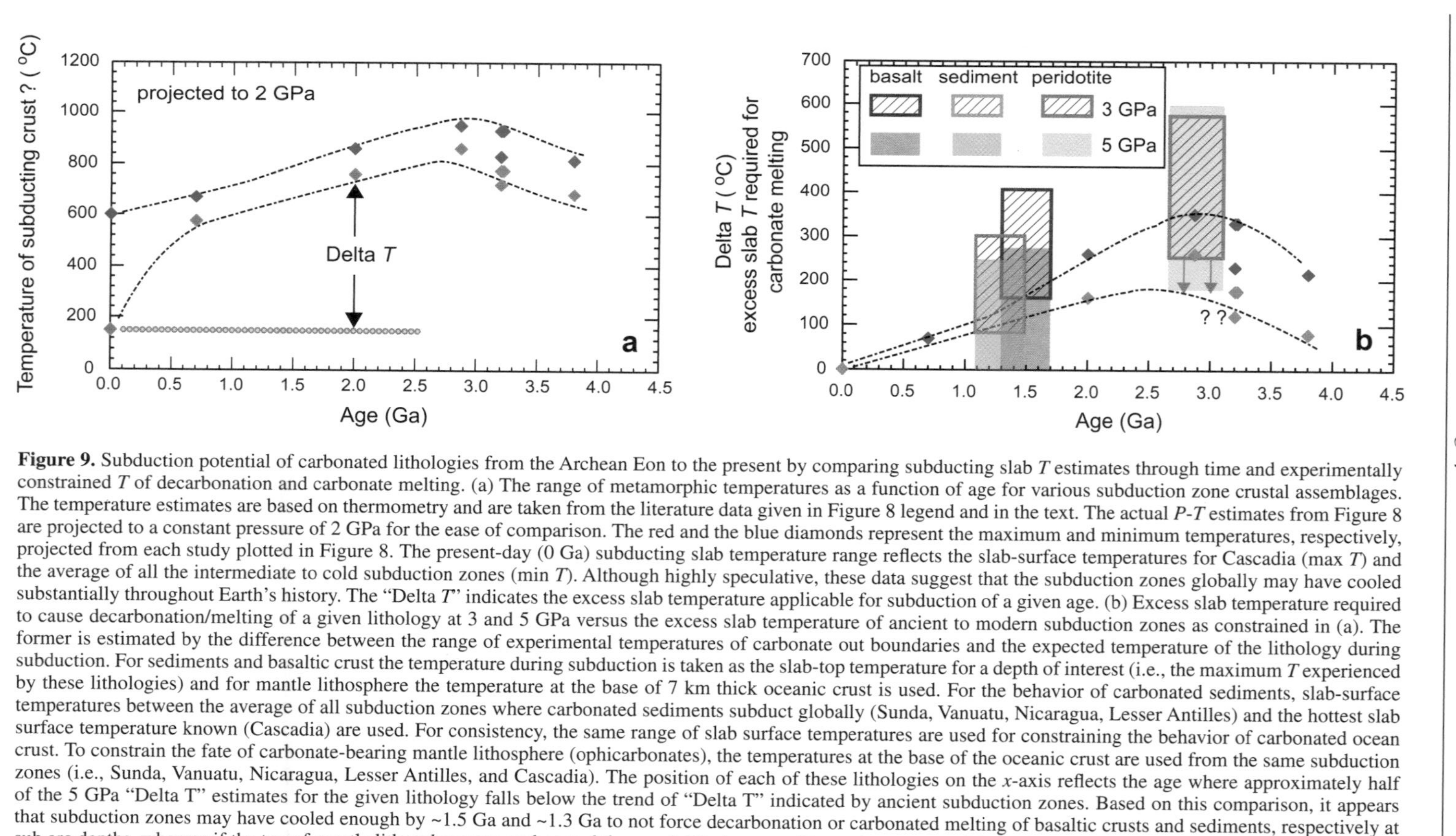

Figure 9. Subduction potential of carbonated lithologies from the Archean Eon to the present by comparing subducting slab *T* estimates through time and experimentally constrained *T* of decarbonation and carbonate melting. (a) The range of metamorphic temperatures as a function of age for various subduction zone crustal assemblages. The temperature estimates are based on thermometry and are taken from the literature data given in Figure 8 legend and in the text. The actual *P-T* estimates from Figure 8 are projected to a constant pressure of 2 GPa for the ease of comparison. The red and the blue diamonds represent the maximum and minimum temperatures, respectively, projected from each study plotted in Figure 8. The present-day (0 Ga) subducting slab temperature range reflects the slab-surface temperatures for Cascadia (max *T*) and the average of all the intermediate to cold subduction zones (min *T*). Although highly speculative, these data suggest that the subduction zones globally may have cooled substantially throughout Earth's history. The "Delta *T*" indicates the excess slab temperature applicable for subduction of a given age. (b) Excess slab temperature required to cause decarbonation/melting of a given lithology at 3 and 5 GPa versus the excess slab temperature of ancient to modern subduction zones as constrained in (a). The former is estimated by the difference between the range of experimental temperatures of carbonate out boundaries and the expected temperature of the lithology during subduction. For sediments and basaltic crust the temperature during subduction is taken as the slab-top temperature for a depth of interest (i.e., the maximum *T* experienced by these lithologies) and for mantle lithosphere the temperature at the base of 7 km thick oceanic crust is used. For the behavior of carbonated sediments, slab-surface temperatures between the average of all subduction zones where carbonated sediments subduct globally (Sunda, Vanuatu, Nicaragua, Lesser Antilles) and the hottest slab surface temperature known (Cascadia) are used. For consistency, the same range of slab surface temperatures are used for constraining the behavior of carbonated ocean crust. To constrain the fate of carbonate-bearing mantle lithosphere (ophicarbonates), the temperatures at the base of the oceanic crust are used from the same subduction zones (i.e., Sunda, Vanuatu, Nicaragua, Lesser Antilles, and Cascadia). The position of each of these lithologies on the *x*-axis reflects the age where approximately half of the 5 GPa "Delta T" estimates for the given lithology falls below the trend of "Delta T" indicated by ancient subduction zones. Based on this comparison, it appears that subduction zones may have cooled enough by ~1.5 Ga and ~1.3 Ga to not force decarbonation or carbonated melting of basaltic crusts and sediments, respectively at sub arc depths, whereas if the top of mantle lithosphere was carbonated then such lithology would remain unsusceptible to decarbonation even in the early Archean Eon.

if the hottest subduction zone of modern Earth is used as a reference (Fig. 9). If the trend of excess slab temperature versus age is followed, then it appears that many carbonated sediment and carbonated ocean floor basalt compositions likely suffered sub-arc depth decarbonation and carbonated partial melting in all of the late Hadean Eon, the Archean Eon, and perhaps at least the first half of the Proterozoic Eon. Only as recently as 1.2-1.5 Ga does subduction appear to be cold enough to allow significant subduction of crustal carbonate beyond sub-arc depths of 100-170 km. However, depending on the choice of carbonated crust compositions, some deep subduction likely took place even earlier. Ophicarbonates could, however, have subducted even during the Archean-Proterozoic time as the *P-T* trajectories of mantle lithosphere, which are cooler than the overlying crust, likely remained significantly below the solidi temperatures of carbonated ($\pm H_2O$) peridotite (Fig. 9b). If the most recent estimate of serpentinites-hosted carbon content (500-1000 ppm C; primarily in the form of sea-water carbonates and smaller fractions of organic carbon; (Alt et al. 2012) is considered as a relevant estimate for the ancient subduction, then with serpentinites making up 10% of the 10-km thick layer of suboceanic mantle, an estimated $4.8\text{-}9.6 \times 10^{12}$ g C/yr would be added to the subducting assemblage. Considering the present-day input of carbon through subduction of altered oceanic crust alone is ~6.1×10^{13} g C/yr, the contribution of carbon hosted in the mantle section of the Archean Eon would have been about an order of magnitude lower. Future studies will have to constrain the abundance of mantle-hosted carbon in the Archean and Proterozoic lithospheric sections in order evaluate the potential role of mantle subduction in deep carbon cycle of the Archean Eon.

Another possibility of somewhat enhanced subduction of carbonate in the Precambrian is that, owing to higher solidi of carbonated peridotite with respect to carbonated basalt and sediments, some fraction of released CO_2/carbonate melt may freeze in the mantle wedge peridotite immediately above the downgoing slab. If this was the case, this hanging wall peridotite can be dragged down into the deeper mantle along with the subducting slab. It is unclear how much carbon may get locked into peridotite in this fashion, because crystalline carbonate stability in compositions relevant for carbonated slab melt influxed peridotite is unconstrained. Likewise, despite the thermal hindrance to early Earth carbonate subduction, subduction of reduced carbon may have been another mechanism for carbon ingassing in the Hadean, as graphite/diamond has no notable effect on the solidus of dry mantle lithologies. This idea gains support from the fact that a significant fraction of near-surface carbon could have been graphite or other reduced phases in early Earth (Catling et al. 2001).

The above discussion suggests that, owing to somewhat more efficient decarbonation of the subducting crusts at relatively shallow depths, the flux of CO_2 through Archean volcanic arcs likely was higher and deep subduction of carbon may have been hindered. In addition, owing to Earth's much hotter conditions, the subduction cycles may have also been episodic in nature for most part of the Precambrian (O'Neill et al. 2007; Moyen and van Hunen 2012), thus making subduction introduction of crustal carbon to the mantle even a less reliable ingassing process.

Petrologic solution to Faint Young Sun Paradox — the role of arc volcanism. A corollary of the discussion in the preceding section is that during the Hadean to Mesoproterozoic time the majority of Earth's carbon that did not segregate to the core, or was lost to space during accretion, was likely restricted to the exosphere (crust + atmosphere) with only a limited cycling involving the shallow mantle wedge. The convecting mantle's carbon-poor state would be exacerbated by the fact that carbon outgassing may have been more efficient with deeper carbonated melting. This prediction of more efficient subduction decarbonation (along with deep carbonated melting beneath oceanic volcanic centers) or at least less systematic subduction must have had important implications for the Faint Young Sun Paradox (Sagan and Mullen 1972; Feulner 2012). This paradox describes the apparent contradiction between geologic evidence of liquid water on Earth's surface as early as Hadean time and the astrophysical expectation that the Hadean Sun was ~30% less luminous than today at ~4.6 Ga (Sagan and

Mullen 1972) and remained at least 10% less luminous as recently as 1.5 Ga (Gough 1981). If atmospheric chemistry was the same as that of modern Earth, with Sun's output being only 70-90% of that of the modern epoch, Earth's surface temperature (T_s) would remain <273 K (Budyko 1969; Kasting and Catling 2003; Zahnle 2006; Zahnle et al. 2007), thus preventing water to exist in the liquid form as recently as 1.5-2.0 Ga (Kasting and Catling 2003). One of the ways to get around this problem is to have higher concentrations of atmospheric CO_2 than modern Earth's (preindustrial) nominal value of ~300 ppmv (Owen et al. 1979; Walker et al. 1981; Kuhn and Kasting 1983). However, although higher partial pressure of CO_2 in the ancient terrestrial atmosphere is a popular model, little has been discussed in literature about the source of excess volcanic CO_2 (Feulner 2012).

The comparison between the estimated temperatures of Precambrian subduction zones and the high-pressure experimental phase relations of crustal decarbonation suggests that significant CO_2 liberation at volcanic arcs likely helped to sustain higher CO_2 content in the exosphere for at least 2.5 billion years, between ~4 Ga and 1.5 Ga (Fig. 9b). Although the role of volcanic CO_2 as the main greenhouse gas required to offset the lower solar luminosity has been speculated before, the analyses presented here suggests that it specifically was CO_2 outgassing at ancient arcs that likely provided the excess CO_2. But how much carbon was released to atmosphere through arc volcanism during the Archean Eon? Similar to the present-day, carbonated basaltic crust likely was the chief source of crustal carbon entering ancient subduction zones. In present-day Earth, carbonate precipitated as veins and present in vesicles in the upper volcanics of ocean floor basalt amounts to ~0.3 wt% CO_2 of the 7 km thick ocean crust (Alt and Teagle 1999; Alt 2004). For a subduction rate of 3 km^2/yr, relevant for the Phanerozoic Eon (Reymer and Schubert 1984), this equates to subduction of 6.1×10^{13} g of C/yr (2.2×10^{14} g of CO_2/yr) (Dasgupta and Hirschmann 2010): all of which could have potentially been released in the Archean Eon. With greater vigor of hydrothermal activity and faster sea-floor spreading, the carbonation rate of the Archean ocean crust and flux of carbon to subduction zones could have been more extreme, however. For example, the study of Nakamura and Kato (2004) on the top 500 m of early Archean (3.46 Ga) hydrothermally altered basaltic rocks exposed near the Marble Bar area of the eastern Pilbara Craton, Western Australia, suggests that the carbon flux to subduction zones at this time period was ~4.6×10^{14} g of C/yr (~1.7×10^{15} g of CO_2/yr). Shibuya and co-workers (2012) investigated the Cleaverville area of Pilbara Craton and constrained carbonation of the top 4000 m of the middle Archean MORB-like greenstone composition. This latest study yielded an even more extreme flux of subducting carbon of ~1.8×10^{15} g of C/yr (6.6×10^{15} g of CO_2/yr). Again, all of this carbon being outgassed at volcanic arcs would yield at least 1-2 orders of magnitude higher flux than the total flux of modern volcanic CO_2 emission, with all magmatic centers combined.

Punctuation of the deep carbon cycle of the Precambrian by supercontinent formation? The preceding section suggests that breakdown and outgassing of recycled crustal carbonates at volcanic arcs were perhaps much more frequent in most of the Precambrian and this process may have implications for the Faint Young Sun paradox. If this was the case, does it mean there was very limited deep ingassing of crustal carbonates through the Precambrian? Were the subduction zones hot throughout the Archean and early Proterozoic Eons? The thermal consequence of supercontinent aggregation and disaggregation suggests otherwise. Formation of a supercontinent elevates the T_P underneath the continents owing to heat transfer inefficiency of thick, stagnant continental lithosphere relative to thinner, subducting oceanic lithosphere (Coltice et al. 2009; Lenardic et al. 2011). Moreover, if a supercontinent is surrounded by subduction zones, to keep the average T_P constant, an increase in T_P beneath continent is accompanied by a drop in T_P beneath oceans. Lenardic et al. (2011) showed that the change in mantle thermal state would result from the insulating effect of continental lithosphere no longer being communicated to the suboceanic mantle via thermal mixing. Thus supercontinent formation and break-up in the Precambrian (Condie 2004) might have caused the mantle wedge

T_P, and hence the near-surface slab temperatures, to fluctuate. The implication for subduction of crustal carbonate is that recycling of carbon beyond sub-arc depths may be promoted during the stability of a supercontinent with lower mantle wedge potential temperatures. Supercontinent break-up, on the other hand, would lead to a transient burst of enhanced convective vigor driven by a strong lateral thermal gradient beneath continents and oceans and corresponding increases in the T_P of the mantle wedge. This change, in turn, would trigger breakdown of slab carbonates, causing massive CO_2 outgassing at arcs. However, slower plate velocity during this supercontinent stage may offset the effect of a cooler mantle wedge and may lead to warming of the slab-surface during subduction. This effect would then be more effective in triggering slab decarbonation. More work, using both rock records and geodynamic modeling, is needed to constrain the thermal structures of the Precambrian subduction zones through time, as there may be short-term oscillations from hot to moderately hot or relatively cold subductions, modulating the efficiency of carbonate recycling. This effect is particularly important because inefficient ingassing of CO_2 may have helped the planet to recover from hard snow-ball state as recent as Neoproterozoic (Kirschivink 1992).

Carbon ingassing in modern Earth

Recycling. A number of studies have discussed Earth's carbon cycle, taking into consideration the cycle involving the planet's interior relevant for the Phanerozoic Eon in general and that of modern Earth in particular (Kerrick 2001; Sleep and Zahnle 2001; Hayes and Waldbauer 2006; Dasgupta and Hirschmann 2010). Thus, here only the salient features of modern Earth's deep carbon ingassing via recycling of near-surface rocks are reviewed, highlighting the new observations and drawing attention to potential new directions of study.

Dasgupta and Hirschmann (2010) suggested that the total input of carbon into subduction zones is (6.1-11.4) × 10^{13} g of C/yr, which includes an oceanic mantle lithosphere contribution of 3.6 × 10^{13} g of C/yr (assuming ophicarbonates contain ~11 wt% CO_2 and the top 100 m of the mantle lithosphere is composed of a pervasive ophicarbonate layer). However, if this value is revised with bulk serpentinite content of only 500-1000 ppm C (Alt et al. 2012) and the estimate of serpentinite proportion as preferred by the same authors (Alt et al. 2012), mantle subduction does not contribute more than (~0.5-1.0) × 10^{13} g C/yr. Thus the revised estimate of the present-day subduction input of carbon, combining all lithologies, becomes (5.4-8.8) × 10^{13} g C/yr. Although the global outgassing flux of carbon through intraplate volcanism is largely unconstrained, if such flux is small (given much smaller mass flux of intraplate volcanism), the present-day subduction input of carbon may be more than the outgassing flux combining ridges, hotspots, and arcs. With mantle inventory of carbon being the topic of interest, the obvious question then becomes that how much of this carbon goes past sub-arc depths of 72-173 km (Syracuse and Abers 2006) and thus likely participates in a much longer time scale cycle involving the deep mantle.

Similar to the Archean-Proterozoic times, carbon ingassing flux of the present-day mantle is also being shaped primarily by thermal vigor of the plate boundary processes, i.e., thermal structure of modern subduction zones. The preceding discussion and Figures 8 and 9 indicate that although carbon ingassing by crustal recycling likely was hindered by a hotter Earth in the Hadean Eon, in the Archean Eon and a significant portion of the Proterozoic Eon slab-surface temperatures likely became cooler than the average decarbonation and carbonate melting solidi temperatures by the Mesoproterozoic to Neoproterozoic era. Indeed, the comparison of *P-T* paths of downgoing slabs estimated for modern Earth subduction zones (van Keken et al. 2002; Syracuse et al. 2010) (Fig. 10), combined with petrologic constraints on carbonate/CO_2-bearing lithospheric assemblages, has prompted many investigators to conclude that at present-day carbon subducts deep into Earth with only minor outgassing via arc environments (Yaxley and Green 1994; Kerrick and Connolly 1998; Molina and Poli 2000; Kerrick 2001; Kerrick and Connolly 2001; Dasgupta et al. 2004; Dasgupta et al. 2005; Thomsen and Schmidt

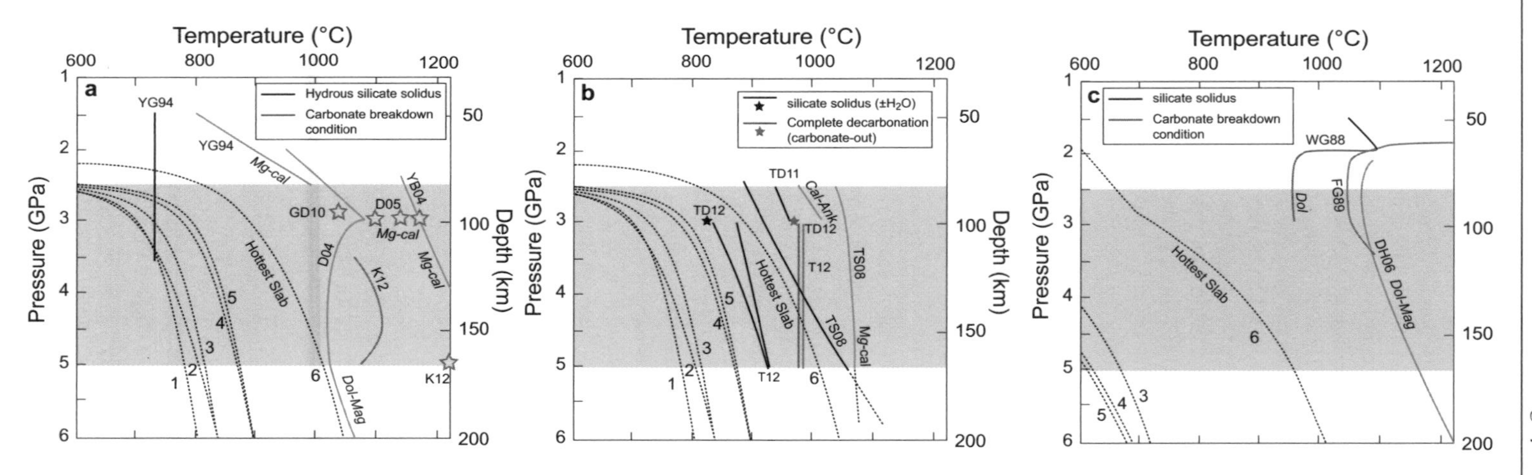

Subducting slab P-T conditions

1 - Sunda 2 - Antilles 3 - South Vanuatu
4 - Nicaragua 5 - North Vanuatu 6 - North Cascadia

Hottest slab - slab-top trajectory for Cascadia

Sub-arc depth range

Carbonate mineral abbreviations

Mg-cal - magnesian calcite solid solution
Dol-Mag - Dolomite-magnesite solid solution
Cal-Ank - Calcite-ankerite solid solution

Figure 10. Subduction efficiency of carbon in present day Earth, based on the comparison of subducting slab *P-T* paths worldwide (Syracuse et al. 2010) and the available experimental boundaries of decarbonation and melting of carbonated basalt (a), carbonated pelite (b), and carbonated peridotite (c). Experiments that constrain the decarbonation and melting of subducting lithologies are the same as those plotted in Figure 8. The subducting slab *P-T* paths (dashed lines with numbers next to them) plotted in (a) and (b) are those for the top surface of the slabs, given that both sedimentary and secondary carbonates in oceanic basalt reside mostly within top 500 m of the slab. The subducting slab *P-T* paths in (c) are those for the base of the 7 km thick oceanic crust or at the top surface of the mantle lithosphere. The slab-surface *P-T* paths in (b), marked with 1 through 5, are specifically for those subduction zones where ocean-floor sediments are known to have a distinct carbonate fraction (Plank and Langmuir 1998). The gray shaded region marks the range of sub-arc depths for all modern subduction zones combined (Syracuse and Abers 2006). Carbonate mineral major element compositions stable along the phase boundaries are marked (see legend to the left). Lithologies that stabilize dolomite solid solution or dolomite-magnesite solid solution on average have lower carbonate-out boundaries compared to those that stabilize calcitic solid solution.

2008; Dasgupta and Hirschmann 2010; Tsuno and Dasgupta 2011; Tsuno and Dasgupta 2012). If this is the case then a significant portion of the original input of (5.4-8.8) × 10^{13} g C/yr could be recycled deep into the mantle. However, the arc flux of CO_2 [(1.8-3.7) × 10^{13} g C/yr; (Sano and Williams 1996)] demands that up to ~20-70% of the original carbon input to subduction zones be returned to the atmosphere by arc magmatism. In addition, fluxes associated with metamorphic and hydrothermal fluids not generally included in volcanic flux estimate may call for even greater extent of CO_2 transport from slab to the surface. Furthermore, the proposition that primary arc magmas might be more CO_2-rich than previously thought (Blundy et al. 2010) and intrusives associated with arc volcanos may also contain carbon, suggest that arc flux of CO_2 may be larger than previously estimated. Thus, there is an apparent contradiction between the petrologic prediction of carbonate stability in the subducting slab and measured flux of carbon in arc volcanic centers. The following are some of the processes that can potentially explain this paradox.

1. A significant decarbonation of crustal carbonates (in altered basalt and in sediments) may take place by influx of water derived from extraneous sources, such as underlying serpentinite sections (Gorman et al. 2006). The chief uncertainty in this process is pervasiveness of the fluid flushing through basaltic crust and sediment at sub-arc depths. If serpentinized zones are created via bending faults in the outer rise, then serpentinites may be spatially restricted (Ranero et al. 2003) and thus fluids liberated from them may not flush through the overlying crust pervasively but remain in channels. In such a scenario, decarbonation may not be significant.

2. The hydrated top portions of the subducting slab, including sediments, basaltic crust, and mantle lithospheric section, may detach from the downgoing slab and rise to wedge mantle as Rayleigh-Taylor instabilities (e.g., Gerya and Yuen 2003; Castro and Gerya 2008). If this happens then slab carbonates may experience hotter wedge temperature, thereby undergoing melting or decarbonation and releasing CO_2.

3. Owing to density contrast alone and not aided by hydration, downgoing sedimentary packages may undergo diapiric rise into the mantle wedge (Currie et al. 2007; Behn et al. 2011). Signature of sediment partial melt and sedimentary carbon in arc volcanics thus may be derived by sediment decarbonation and melting in the mantle wedge, as proposed by the studies of Tsuno and Dasgupta (2011, 2012), Tsuno et al. (2012), and Behn and co-workers (2011).

4. Based on fluid inclusions trapped in diamonds of ultra-high pressure rocks of Western Alps that preserved dissolved bicarbonate and carbonate ions and crystals of carbonate in the inclusions, it has been suggested that the transfer of carbon from slab to wedge takes place by carbonate dissolution in the fluid at sub-arc depths, rather than by decarbonation or melting (Frezzotti et al. 2011).

5. A sizeable portion of the global arc flux may derive from metamorphic decarbonation of crustal carbonates in the overriding plate through interaction with arc magmas. For example, Mount Etna [1.3 × 10^{13} g CO_2; (Allard et al. 1991)] and Vesuvius, which erupt through thick sections of crustal carbonates, make up 20% of the global CO_2 arc out flux. The presence of skarn xenoliths and heavy carbon isotope chemistry of fumarolic gases in Mt. Etna and Vesuvius do suggest sedimentary carbonate assimilation at arc crust (e.g., Iacono-Marziano et al. 2009). Although this non-mantle CO_2 outflux tends to be more important at old continental arcs, owing to the possible presence of crustal carbonates, even in the juvenile continental crust this contribution could be more than previously recognized. If global arc CO_2 flux is increased significantly by this form of crustal carbonates in the overriding plate, then one important corollary would be that a greater percentage of trench input of carbon actually gets recycled deep into the mantle.

The relative contributions of these various mechanisms to the outgassing flux of carbon in modern subduction zones are unconstrained at present. Also unconstrained is how much CO_2 can be carried in siliceous hydrous partial melts of sediments, which appears to be a an unavoidable agent of mass transfer in subduction zones. With all recent estimates of slab-surface temperatures being hotter than the fluid-present solidi of subducting crust and sediments (van Keken et al. 2002; Syracuse et al. 2010; Cooper et al. 2012) and the flux of hydrous fluid perhaps being available from breakdown of hydrous phases in subjacent lithologies (e.g., serpentinites), it becomes critical to know the carrying capacity of CO_2 in high-pressure hydrous siliceous melts (Ni and Keppler 2013).

Acquisition of mantle carbon via interaction of subducted 'water' and metallic core? All possible processes for explaining the "excess" carbon in the mantle paradox that were discussed before are early Earth processes that could have restored the mantle carbon budget well before the initiation of plate-tectonics. However, as suggested by Dasgupta et al. (2013a), the initiation of plate tectonic cycles may have provided another pathway for elevating mantle carbon over geologic time, i.e., in addition to introducing surficial carbon by recycling. They proposed that if subduction of oceanic plate brings water to the core-mantle boundary (CMB) regions, it may be possible that water, expelled from breakdown of hydrous phases or contained in hydrous phases, reacts with metallic alloy melt from outer core to form Fe-hydrides, FeH_x (Okuchi 1997; Okuchi 1998; Terasaki et al. 2012); Fig. 5d). Oxygen, being less soluble in the metallic core and released in the process, then may react with carbon in the core to form CO_2, CO, or $FeCO_3$, following a set of reactions such as:

$$Fe + H_2O = FeO + 2H \tag{11}$$

$$Fe + xH = FeH_x \tag{12}$$

$$FeO + C = Fe + CO \tag{13}$$

$$2FeO + C = 2Fe + CO_2 \tag{14}$$

$$FeO + CO_2 = FeCO_3 \tag{15}$$

Excess hydrogen, not dissolved in the metallic core, may also react with carbon to release hydrogenated species such as methane, following the reaction:

$$C + 2H_2 = CH_4 \tag{16}$$

It is also possible that such reaction stabilizes other multi-light element alloys, such as $FeSiH_x$ (Terasaki et al. 2011), following the reaction such:

$$FeSi + xH = FeSiH_x \tag{17}$$

The produced CO, CO_2, hydrocarbons, or carbonate most likely forms a component in a melt phase in D″. Such C-bearing melt may be released from the core, eventually causing the mantle carbon concentration to increase. Similar to the model of inefficient core formation, this possibility could also make the lower mantle more carbon-rich. One caveat to this speculation is that while the reaction with subducted hydrous phases and the core may release carbon to the mantle, if a subducted slab carries carbon down to the CMB regions, then subducted carbon may also be lost to core. Therefore, if both recycling and core-slab interactions contribute to modern Earth's mantle carbon budget, then slabs in the CMB graveyard need to be hydrous but carbon-poor.

Carbon in modern Earth's mantle: recycled versus primordial? With deep carbonate subduction being quite efficient at least over the Phanerozoic Eon, and maybe for the past 1 billion year, one may wonder whether the present-day inventory of mantle carbon is chiefly recycled or primordial. For example, if mantle outgassing rate at oceanic ridges is controlled by deep carbonated melting at a depth exceeding 300 km (Dasgupta and Hirschmann 2006), then the residence time of mantle carbon can be as short as 1 Ga, thus requiring a significant fraction of the present-day mantle carbon to be recycled. Whereas if redox freezing (as graphite, diamond, C-bearing metal alloy, or metal carbide) suppresses deep melting-induced efficient liberation of carbon, then the residence time of carbon in the mantle may approach 4 Ga, requiring limited input from Phanerozoic subduction to make up its inventory.

Carbon isotopic composition of mantle-derived samples shed light on the possible origin of present-day mantle carbon. The average value of $\delta^{13}C$ of mantle carbon is ~ –5 (Deines 2002), with more negative excursions [i.e., carbon that is isotopically lighter, up to –25‰; (Stachel et al. 2005)] typically associated with graphite and diamonds (e.g., Schulze et al. 1997; Walter et al. 2011). Although the origin of "light" carbon in the mantle is debated (Cartigny 2005), a leading hypothesis is that the light mantle carbon ($\delta^{13}C < -10‰$) results from subduction of the isotopically light organic carbon fraction of altered oceanic crust, sediments, or mantle lithosphere (Stachel et al. 2005). However, it still remains an unsolved mystery why most of the mantle carbon presents $\delta^{13}C \sim -5‰$, given that the dominant mass flux of carbon via carbonate subduction injects $\delta^{13}C \sim -1‰$ (Coltice et al. 2004; Stachel et al. 2005). In particular, in the context of ancient versus modern processes shaping the carbon chemistry of the mantle as discussed in this chapter, the study of Satish-Kumar et al. (2011) re-opens the debate on the origin of the carbon isotope compositional variability of the mantle. Satish-Kumar and co-workers (2011) presented experimental evidence of carbon isotope fractionation between Fe-carbide melt (C-saturated metallic-Fe melt) and graphite/diamond and showed that Fe-carbide melt prefers isotopically light, ^{12}C-enriched carbon over graphite/diamond. Although it remains unclear whether Earth's early magma ocean was graphite-saturated and whether the C-isotope fractionation between silicate magma ocean and C-poor metal melt would yield a similar ^{12}C-enriched core melt, it is plausible that carbon derived from metallic melt generates in part the light isotopic signature observed in many deep diamonds. Therefore, if inefficient core formation leaves behind a small fraction of C-bearing metallic melt, such phases may contribute towards the carbon isotopic heterogeneity of mantle samples and in particular may provide the light carbon in the mantle. Furthermore, carbon isotopic fractionation between metal and silicate melts might have established the $\delta^{13}C$ near the average mantle value of –5‰ and potentially explain the difference between CI chondrites ($\delta^{13}C$ –7 to –15‰) (Kerridge 1985) and that of Earth's mantle.

Break-up of Pangea and perturbation of the Phanerozoic deep carbon cycle. Although present-day estimates of slab thermal profiles suggest only a limited decarbonation of downgoing lithosphere at sub-arc depths, any increase in arc mantle potential temperatures would disrupt such balance of deep carbon cycling and force carbon to be released at sub-arc depths and thus outgassed by arc volcanism. Were there any time periods in the Phanerozoic Eon when the subduction zones could have become hotter? Formation and break-up of supercontinent Pangea might have created such a scenario. Thermal mixing in the mantle soon after the supercontinent break-up could have caused the arc source mantle to heat-up by as much as 200 °C (Lenardic et al. 2011). As a consequence, the slab surface temperatures beneath arcs might have increased by ~100 °C. With such a change in slab temperatures within a short period of time, carbonate melting and decarbonation of subducting crusts are expected and cause shut-down of the deep carbon influx for perhaps a few tens of million years. Very short residence times of subducted carbon and release through arc volcanism thus might have played a critical role in the Cretaceous and early Paleogene (~140-50 Ma) greenhouse climate. Greater abundance of continental arcs over island arcs at this time period (Lee et al. 2013) might have also caused excess CO_2 to be released through metamorphic decarbonation and assimilation of sedimentary carbonates in the arc crust as observed in Mt Vesuvius and Etna of modern Earth (Iacono Marziano et al. 2008; Iacono-Marziano et al. 2009; Lee et al. 2013).

Stable forms of carbon in the modern mantle and carbon outgassing

With deep ingassing appearing to be efficient for most part of the last ~1.5 Ga, the question becomes what are the stable forms of carbon in Earth's modern mantle (Oganov et al. 2013)? Moreover, how do the stable hosts of carbon vary as a function of depth and various tectonic settings? Again, a number of studies reviewed the mineralogic and petrologic constraints on carbon storage in the mantle (Luth 1999; Dasgupta and Hirschmann 2010), thus this section is aimed at highlighting only some of the new insights based on recent experimental observations.

Carbonated melt in the convecting oceanic mantle. Owing to very limited solubility of carbon in the silicate minerals of Earth's mantle (Keppler et al. 2003; Shcheka et al. 2006), carbon's presence in the mantle is controlled chiefly by the thermodynamic stability of various C-bearing accessory phases and dynamic and thermodynamic stability of C-bearing fluids and melts. In shallow oceanic mantle beneath ridges, oxidized form of carbon exists and carbonated melt—either carbonatite or carbonated silicate melt (Dalton and Presnall 1998; Gudfinnsson and Presnall 2005; Dasgupta and Hirschmann 2006; Dasgupta et al. 2007a, 2013b)—could dominate the budget of mantle carbon to ~300 km (Fig. 7). However, it is also plausible that carbonated melts freeze by a redox process at the expense of diamond/graphite (e.g., Eqns. 7 and 8) at depths shallower than ~300 km (the depth of intersection of carbonated peridotite solidus and the mantle adiabat of T_P ~1350 °C) (Stagno and Frost 2010; Rohrbach and Schmidt 2011; Dasgupta et al. 2013b). The depth of reduction of carbonated melt to diamond/graphite is debated at present. Stagno and Frost (2010) suggest such destabilization of carbonated melt as shallow as 100-150 km, whereas the study of Rohrbach and Schmidt (2011) places this depth to ~250 km (Fig. 6), based on the experimental observation that majorite (significant host of Fe^{3+}) becomes stable at such a depth and saturation of Fe-Ni alloy is achieved (Rohrbach et al. 2007; Rohrbach et al. 2011). The carbon release depth of 300 km (by carbonated melting) versus as shallow as ~100-150 km (by redox melting) must have critical difference for the outgassing rate of carbon in the modern mantle. For example, Dasgupta et al. (2013b) showed that with redox transformation of carbonate to diamond at ~250 km depth, the first melt in the upwelling mantle globally will be a carbonated silicate melt of kimberlitic affinity rather than carbonatite *sensu stricto*. This conclusion rests on the experimental observation that carbonate-silicate melt mixing is favored at greater depth and hence the stability field of carbonated silicate melt expands over the field of true carbonatite. Dasgupta et al. (2013b) argued that if the depth-oxygen fugacity evolution is taken into account, carbonated silicate melt with 15-25 wt% CO_2 and modest amount of water is likely a more important agent releasing incompatible trace elements and fluids to the exosphere.

Even though carbonate melt/mineral stability may be compromised at the face of iron disproportionation-induced redox freezing, locally carbonate-rich regions (perhaps regions that are affected by subduction of carbonates) may persist to even lower mantle depths (Biellmann et al. 1993; Stagno et al. 2011). In such a scenario the topology of carbonated peridotite solidus could apply to the entire depth range plotted in Figure 7, i.e., through mantle transition zone to lower mantle. Experimental determination of the solidus slope of carbonated peridotite at transition zone and lower mantle depths, combined with the knowledge of the same in the upper mantle, suggests possibly an interesting case of carbonatite generation in the deep mantle. Figure 7 shows that if a mantle adiabat for T_P of 1350 °C is followed, then there could be a "double-crossing" of the carbonated peridotite solidus, one at ~10 GPa and another at ~30 GPa. The exact depth of the latter intersection remains somewhat uncertain owing to lack of experimental constraints on the solidus for relevant compositions at >25 GPa. However, the key point is that the "double-crossing" of the adiabat and the solidus suggests that carbonatite generation not only can take place at ~300 km depth by adiabatic decompression of the mantle, but also at lower mantle depth of ~800-900 km by compression or convective down-welling of the mantle transition zone materials. A corollary of the solidus-adiabat double-crossing is that a large fraction of the transition zone may be below the carbonatite solidus and thus may have carbon stored as magnesite, provided the oxygen fugacity is >IW+2 (Stagno et al. 2011). Thus, trace carbonatite melt may affect the seismic structure of the deep upper mantle and also shallow lower mantle but perhaps not the mantle transition zone.

Carbon and carbonates in the continental lithospheric mantle. While oxidized form of carbon in asthenospheric upper mantle is chiefly a melt phase, continental lithospheric mantle can contain stable crystalline carbonates. Figure 11 shows that in the shallow continental lithospheric mantle to a depth of ~100-220 km (geotherms corresponding to surface heat flux of 40-50 mW m^{-2}), carbon can exist as crystalline carbonate, i.e., magnesite and dolomite solid solution (Fig. 11) and only at greater depths do the continental shield geotherms cross the carbonated peridotite solidus and hence magnesio-carbonatite melt (Dasgupta and Hirschmann 2007b) becomes stable in the lithospheric mantle. If the carbonated peridotite domain is also hydrous and amphibole-bearing, then the solidus can be slightly lower (Wallace and Green 1988) and magnesite/dolomite may be restricted to slightly shallower depths. Although carbonatitic melt may be stable at depths deeper than 100-220 km, diminishing oxygen fugacity with depth, as recorded in the $Fe^{3+}/\Sigma Fe$ of garnets in cratonic mantle xenoliths and determined based on the oxy-thermobarometry calibration of Gudmundsson and Wood (1995), suggests that graphite/diamond may become stable over crystalline carbonate or carbonated melt at 100-150 km depth (Woodland and Koch 2003; McCammon and Kopylova 2004; Stagno and Frost 2010; Yaxley et al. 2012). If this is the case, then the stability of primary carbonatitic melt in Archean cratonic environments may be restricted within a narrow window of ~1050-1100 °C

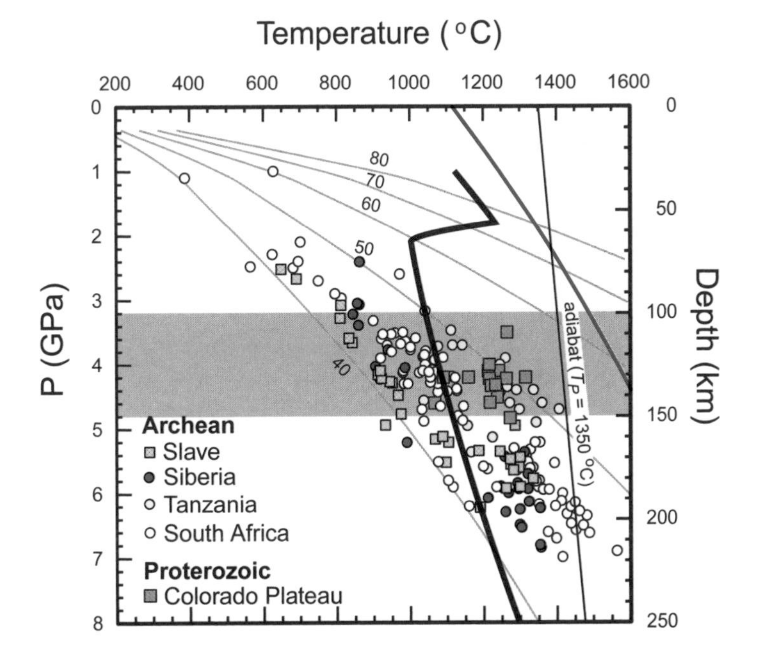

Figure 11. Comparison of carbonated fertile peridotite solidus (thick black line) from Figure 6 with geotherms based on xenoliths of the Archean and Proterozoic cratonic peridotite. Steady-state conductive geotherms are for a surface heat flux of 40-80 mW m^{-2}. Shown for reference are volatile-free peridotite solidus (red line) from Hirschmann (2000) and proposed depth range of carbonate to graphite/diamond transition (gray horizontal band; (Stagno and Frost 2010; Yaxley et al. 2012) based on skiagite activity in garnet and experimental oxy-barometry calibration of Gudmundsson and Wood (1995). The xenolith geotherm data are from Lee et al. (2011).

and 100-150 km. Somewhat deeper production of carbonated melt in the cratons is possible, however, with metasomatically oxidized and more carbon-rich domains, either through redox or carbonated melting (e.g., Foley 2008). Although carbonate melting may be restricted for greater thickness of thermal boundary layer in the Archean cratons, elevated geotherms similar to those of the Proterozoic (surface heat flux of ~50 mW m^{-2}) and Phanerozoic terrains (surface heat flux of >60-80 mW m^{-2}) would stabilize carbonatitic melt as shallow as 70-100 km (Fig. 11), depths at which conversion from carbonate to graphite likely does not occur.

Deep storage of carbon — deep upper mantle to lower mantle. Although oxidized phases such as CO_2-rich fluid, carbonatite, and carbonated silicate melt are the chief hosts of carbon in the shallow (< 200-250 km) upper mantle, the dominant storage mechanism of carbon changes at greater depths. In a mantle poor in carbon, and thus oxygen fugacity being controlled by the Fe^{2+}-Fe^{3+} exchange and energetics of Fe^{3+} incorporation in mantle silicates, mantle likely becomes metal-saturated as shallow as 250±30 km and remains such at greater depths (Frost and McCammon 2008; Rohrbach et al. 2011). A consequence of this environment is the equilibrium reaction of carbon with Fe-Ni metal and mantle storage of the former in Fe-rich alloy, Fe-rich carbides, and as diamond (Frost and McCammon 2008; Dasgupta and Hirschmann 2010; Buono et al. submitted). Albeit rare, natural inclusions of Fe-carbide, Fe_3C and metallic-Fe in mantle-derived diamonds (Sharp 1966; Jacob et al. 2004; Kaminsky and Wirth 2011) validate such experimental predictions. But what further experimental constraints can be provided in terms of the relative roles of alloy versus carbides versus diamond versus metallic liquid as the host of deep carbon? The exact phase of interest that hosts carbon depends on the bulk composition in the Fe-C (±Ni ±S) system, and the *P-T* condition in question. The concentration of Fe-Ni metal is predicted to vary between 0.1 and 1.0 wt% with ~0.1 wt% metal alloy being stable from the deep upper mantle through the transition zone and ~1 wt% metal alloy being stable almost throughout the lower mantle. The alloy composition is also predicted to vary from 39 wt% Fe - 61 wt% Ni at ~250 km depth and 88 wt% Fe - 12 wt% Ni at the lower mantle (Frost and McCammon 2008). The current knowledge of the effect of Ni on the high-pressure phase relations in the Fe-C system is limited thus the prediction from Fe-C binary phase diagrams are discussed first. In Figure 12, the estimated Fe-C phase diagrams at pressures of 10 and 50 GPa from the study of Lord et al. (2009) are presented and compared with the mantle temperatures expected at such depths. The key observation from Figures 12a and 12b are: (a) at the base of the upper mantle and for an average MORB source carbon content not exceeding 30 ppm C (equivalent of 110 ppm CO_2) all the carbon will be dissolved in an alloy ± a metallic melt phase; (b) for cohenite to be a phase of interest, the deep upper mantle needs to contain carbon in excess of ~50 ppm; and (c) for diamond to be an equilibrium phase, carbon content in excess of ~90 ppm is needed. Although the detailed phase diagram of the Fe-Ni-C system as a function of pressure is not well constrained, recent experimental work at 3 and 6 GPa suggests that Ni behaves as an incompatible element in the cohenite-Fe-Ni-C(±S) liquid system (Buono et al. submitted), suggesting that Ni does not stabilize cohenite and Ni lowers the melting point of cohenite. Hence, it is expected that if the metal is Ni-rich at the base of the upper mantle and throughout the transition zone then the Fe-Ni-C melt will be a more dominant carbon host than predicted from the Fe-C phase diagram alone. Moreover, as sulfur in the mantle is present almost entirely as sulfides—given the oxygen fugacity of the mantle (e.g., Jugo et al. 2010)—equilibrium phase relations and geochemistry of the Fe-(±Ni)-C-S system also become relevant to constrain the equilibrium carbon-bearing phase. Sulfide-metal-metal carbide connection in the mantle is also evident from common association of pyrrhotite or troilite with iron carbide and Fe-rich metal alloy phases in inclusions in diamonds (Sharp 1966; Jacob et al. 2004). In Figure 13, the plausible range of Fe-(±Ni)-C-S composition relevant for the mantle subsystem is shown. Also shown are the available near-liquidus phase relations for Fe-Ni-C-S, Fe-Ni-C, and Fe-C-S bulk compositions over the pressure range of 3 to 6 GPa (Dasgupta et al. 2009a; Buono et al. submitted). It can be observed, with reasonable extrapolation of the experimental

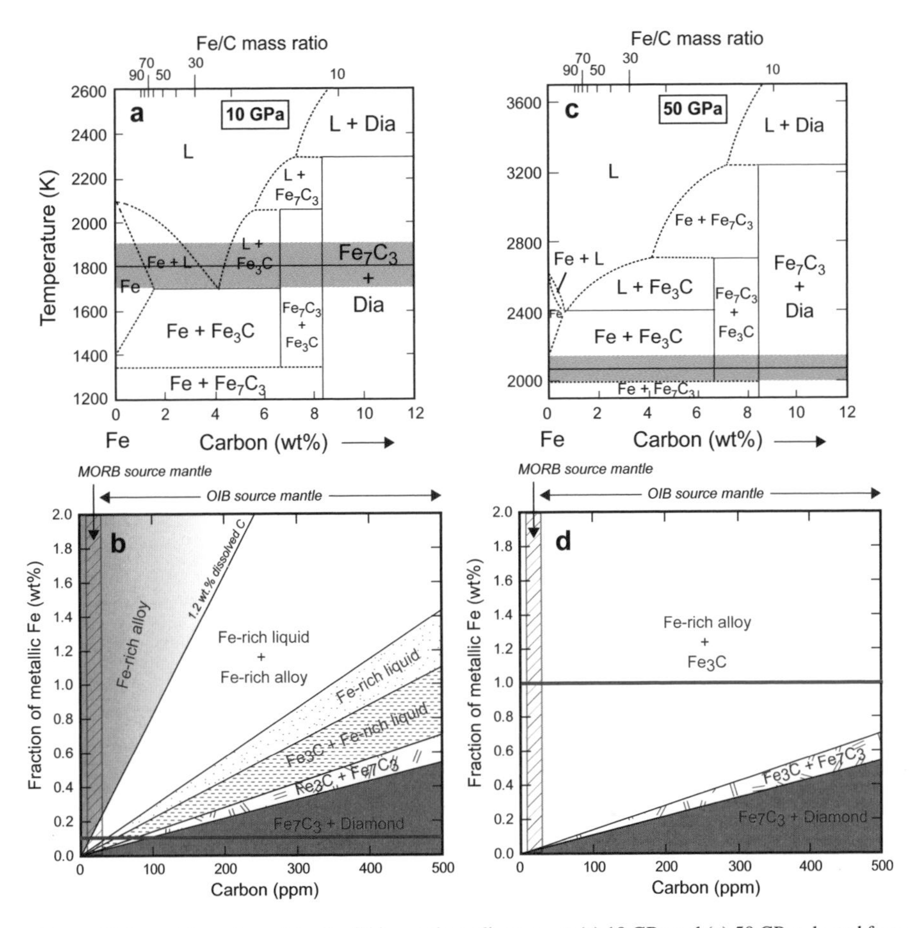

Figure 12. Iron-rich portion of the Fe-C binary phase diagrams at (a) 10 GPa and (c) 50 GPa adapted from the study of Lord et al. (2009). Also marked along the top *x*-axis are the Fe/C mass ratios for various bulk compositions. The shaded band with lines in the middle in (a) and (c) indicate the expected temperatures of the mantle along an average mantle adiabat derived from Brown and Shakland (1981) at these depths. (b) and (d), based on the phase diagram (a) and (c) respectively, show the expected phase assemblages in Fe-Ni metal alloy in the mantle (in weight percent) and carbon content (in ppm) of the mantle space. (b) and (d) thus present the expected reduced carbon-bearing phases in the deep upper mantle and lower mantle, respectively as a function of metallic alloy and carbon content. The red horizontal lines in (b) and (d) mark the expected mass fraction of alloy at these mantle depths as suggested by Frost and McCammon (2008). Reduced solubilities of carbon in carbide-saturated alloy at higher pressures suggest that carbide-saturation of the mantle takes place even for a carbon depleted MORB-source mantle at lower mantle depths, whereas for a similarly carbon-poor deep upper mantle, the entire carbon inventory could be dissolved in a metallic alloy phase (see text for details). The plot in (d) also suggests that unless a small amount of Ni has a large influence on the Fe-C phase diagram, diamond stability in the lower mantle requires the mantle to be extremely carbon-rich or much poorer in metal alloy fraction.

phase boundaries, that the sub-ridge adiabatic temperatures may yield an Fe-Ni-C-S melt at depth of first metal saturation, i.e., 250 ± 30 km. If the mantle adiabatic gradient relevant for intraplate ocean island sources is considered then Fe-Ni-C or Fe-Ni-C-S melt will likely be stable throughout the deep upper mantle, possibly even throughout the transition zone. Furthermore, although carbon-bearing Fe-Ni-S melt likely is stable in the deep upper mantle and transition zone, the involvement of sulfur may help and be necessary for stabilizing diamond at these

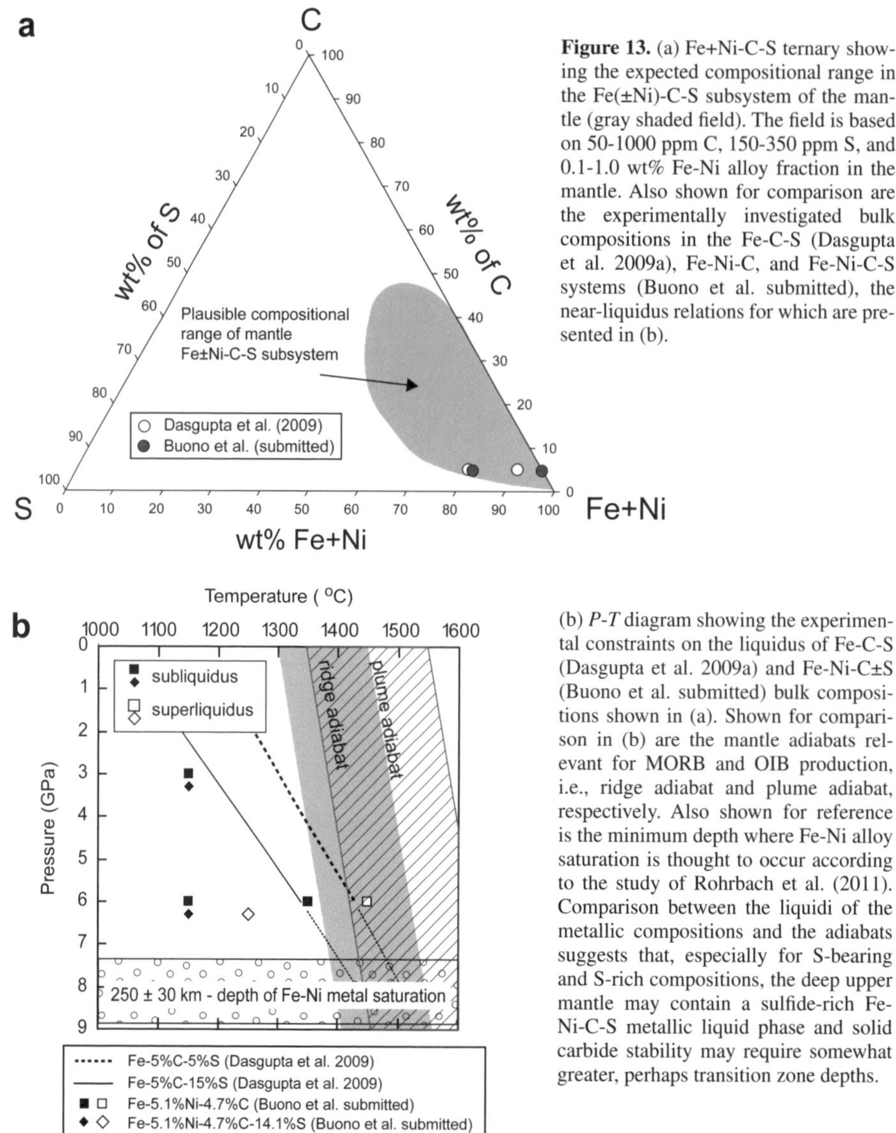

Figure 13. (a) Fe+Ni-C-S ternary showing the expected compositional range in the Fe(±Ni)-C-S subsystem of the mantle (gray shaded field). The field is based on 50-1000 ppm C, 150-350 ppm S, and 0.1-1.0 wt% Fe-Ni alloy fraction in the mantle. Also shown for comparison are the experimentally investigated bulk compositions in the Fe-C-S (Dasgupta et al. 2009a), Fe-Ni-C, and Fe-Ni-C-S systems (Buono et al. submitted), the near-liquidus relations for which are presented in (b).

(b) *P-T* diagram showing the experimental constraints on the liquidus of Fe-C-S (Dasgupta et al. 2009a) and Fe-Ni-C±S (Buono et al. submitted) bulk compositions shown in (a). Shown for comparison in (b) are the mantle adiabats relevant for MORB and OIB production, i.e., ridge adiabat and plume adiabat, respectively. Also shown for reference is the minimum depth where Fe-Ni alloy saturation is thought to occur according to the study of Rohrbach et al. (2011). Comparison between the liquidi of the metallic compositions and the adiabats suggests that, especially for S-bearing and S-rich compositions, the deep upper mantle may contain a sulfide-rich Fe-Ni-C-S metallic liquid phase and solid carbide stability may require somewhat greater, perhaps transition zone depths.

depths. Owing to strong non-ideality of mixing in the Fe-(±Ni)-C-S systems (Wang et al. 1991; Corgne et al. 2008; Dasgupta et al. 2009a), S-rich Fe-Ni-S melt may repel carbon and force diamond saturation. Data available in Fe-Ni-C-S systems suggest that diamond nucleation may be aided by the presence of a Fe-Ni sulfide melt (Shushkanova and Litvin 2008; Zhimulev et al. 2012) because C-solubility in Fe-alloy melt is known to diminish with increasing sulfur content (Ohtani and Nishizawa 1986; Tsymbulov and Tsmekhman 2001; Zhimulev et al. 2012). Thus, although Figure 12 suggests that deep upper mantle diamond only forms from a C-rich (>100 ppm C or ~370 ppm CO_2) mantle, S-rich systems may lead to diamond formation at lower C content. However, more work on sulfide-metal-carbide system is necessary to constrain the relative contributions to various C-bearing phases in detail.

The scenario of carbon storage in the lower mantle could be different (Fig. 12c,d). This expected change is aided by the following. (1) The metal content in the lower mantle can be as much as 1 wt% (Frost and McCammon 2008) and hence, unless the lower mantle is more carbon-rich compared to the deep upper mantle, the effective Fe/C ratios of the metal subsystem become higher. (2) The composition of the metal phase is Fe-rich and Ni-poor, hence the Fe-C binary exerts more control on the mode carbon storage. (3) The solubility of C in carbide-saturated metallic alloy is expected to diminish strongly as a function of pressure (Lord et al. 2009; Walker et al. accepted). The estimated rate of movement of the Fe-C eutectic as a function of P to C-poor compositions suggests that the C-solubility in metal alloy may become negligible by 50 GPa (Fig. 12c; Lord et al. 2009). A consequence of all of these observations is that a vast region of lower mantle may display coexistence of cohenite and Fe-rich metallic alloy for Fe/C ratios relevant for both MORB and ocean island basalt source regions. This assemblage is distinct from the mode of carbon storage in metallic systems at upper mantle and transition zone depths, where carbide may not be a stable phase for carbon-poor MORB source compositions (≤110 ppm CO_2). It can also be noted in Figure 12d that, if the lower mantle contains ~1 wt% Fe-rich metal alloy, then in an equilibrium scenario the stability of diamond and Fe_7C_3 is unexpected unless the mantle domains are extremely carbon-rich (>700-1000 ppm C). More experimental work in the Fe-Ni-C-S system at lower mantle depths will be needed to refine these predictions and gain further insight into deep mantle carbon storage in reduced phases.

CONCLUDING REMARKS

In response to the evolving thermal vigor and fugacity of oxygen and other volatiles, the behavior and fate of carbon and its partitioning between various layers of Earth (atmosphere, silicate Earth, and metallic core) varied through geologic time. During the first few tens of millions of years of the Hadean Eon, carbon that was not lost to space and participated in the core-forming magma ocean processes appears to have partitioned strongly into the metallic core. This fractionation would have made core the largest terrestrial reservoir of carbon. However, uncertainty in the core carbon budget remains owing to poor constraints on bulk Earth carbon in a magma ocean environment, the lack of consensus on the extent of metal-silicate equilibration, the composition of the impactors (including the composition of the core of the differentiated impactors), and the lack of availability of experimental data on partition coefficient of carbon in a deep, reduced magma ocean environment, involving a multi-component metallic alloy melt. With almost all carbon of the chondritic building blocks being lost to space, restricted to nascent atmosphere, or sunk to the core, the silicate liquid mantle soon after core segregation was probably carbon-poor. Processes such as late bombardment of volatile-rich material, entrapped C-bearing metallic liquid in the pore spaces of lower mantle solids, and ingassing from a C-rich atmosphere had opportunities to replenish mantle carbon by the end of the Hadean Eon. In fact, the mantle may have even attained carbon concentrations higher than that sampled by present-day oceanic volcanism at the end of the Hadean Eon. Greater thermal vigor of convection in the Archean Eon likely caused deeper and greater volume of carbonated melt generation and thus efficient outgassing of carbon ensued. The thermal state of the crustal recycling zone also likely was hotter until ~1.5 Ga; such slab thermal structure along with inconsistent subduction cycle likely hindered deep ingassing of surficial carbon. Massive release of CO_2 at Archean and Paleoproterozoic volcanic arcs thus may have supplied the necessary dose of greenhouse gas in the atmosphere to offset the dimmer early Sun and help sustain liquid water on Earth's surface. With secular cooling of Earth's mantle, the global systematics of ingassing and outgassing appear to have changed by the Meso- to Neoproterozoic Eras. The depth-temperature paths of subduction zones became amenable to transport of crustal carbonates past arc-magmatic processing depth, setting up systematic deep ingassing of carbon. A possible short-term disruption of carbon ingassing and outgassing pattern may be caused by the mantle thermal state, owing to formation and break-up of supercontinents.

It remains unclear how the subducted carbon over the past 1-2 billion years distributed itself in the mantle, but with subduction going past the mantle transition zone, it is not unlikely that subducted carbon of the Proterozoic and Phanerozoic Eons is somewhat concentrated in the transition zone and lower mantle. It has also been speculated that deeply subducted water, carried in hydrous phases, may react with outer core alloy melt to trigger release of some core carbon. While ingassing of crustal carbon appear efficient in modern Earth, the petrology and dynamics of subduction zone processes including formation of sediment diapirs and fluid fluxed hydrous melting need to be closely evaluated to reconcile the modern arc flux of CO_2. With cycling of carbonates deep into the modern mantle globally, the storage mechanisms at depth becomes a key issue. Carbonate minerals and carbonatitic melt are stable at the shallow part of continental lithospheric mantle, while carbonated silicate melts are stable at the asthenospheric mantle. In subduction-influenced, carbon-rich, oxidized domains, carbonatitic melt may even be stable at the deep upper mantle through the top of the lower mantle. However, oxidized forms of carbon at depths greater than the deep upper mantle may be an exception and reduced phases such as Fe-Ni-S-C melt, Fe-Ni alloy, and Fe-rich carbides are the phases of interest with increasing depth, while diamond becomes a stable phase only in C-rich and/or S-rich mantle domains. Further work in multi-component metallic system will be required to understand the full spectrum of reduced carbon storage in the mantle. It also needs to be explored whether there are remnants of metallic alloy/carbide melt left behind from the event of core formation and whether such phases preserve very different compositions than those expected from iron disproportionation in silicates and subsequent alloy-C reaction.

ACKNOWLEDGMENTS

This chapter benefited from formal reviews by Hilary Downes and the editors Adrian Jones and Robert Hazen. The chapter also benefited from various discussions the author had with Dave Walker, Cin-Ty Lee, Terry Plank, Adrian Lenardic, Mark Harrison, Peter van Keken, and Norm Sleep. Han Chi, Kyusei Tsuno, Megan Duncan, and Antonio S. Buono are thanked for influencing the views of the author through their recent and ongoing work and by providing some of their unpublished data. Cyril Aubaud and Pierre Cartigny are thanked for providing the seed cartoon of the Earth's internal structure that motivated Figure 1. The author received support from NSF grant EAR-0911442 and OCE-0841035 and from a Packard Fellowship for Science and Engineering.

REFERENCES

Abe Y (1997) Thermal and chemical evolution of the terrestrial magma ocean. Phys Earth Planet Inter 100(1-4):27-39, doi: 10.1016/S0031-9201(96)03229-3

Alard O, Griffin WL, Lorand JP, Jackson SE, O'Reilly SY (2000) Non-chondritic distribution of the highly siderophile elements in mantle sulphides. Nature 407(6806):891-894

Albarede F (2009) Volatile accretion history of the terrestrial planets and dynamic implications. Nature 461(7268):1227-1233

Allard P, Carbonnelle J, Dajlevic D, Bronec JL, Morel P, Robe MC, Maurenas JM, Faivre-Pierret R, Martin D, Sabroux JC, Zettwoog P (1991) Eruptive and diffuse emissions of CO_2 from Mount Etna. Nature 351(6325):387-391, doi: 10.1038/351387a0

Alt JC (2004) Alteration of the upper oceanic crust: mineralogy, chemistry, and processes. *In*: Hydrogeology of the Oceanic Lithosphere. Vol. Davis EE, Elderfield H (eds) Cambridge University Press, p 497-535

Alt JC, Garrido CJ, Shanks Iii WC, Turchyn A, Padrón-Navarta JA, López Sánchez-Vizcaíno V, Gómez Pugnaire MT, Marchesi C (2012) Recycling of water, carbon, and sulfur during subduction of serpentinites: A stable isotope study of Cerro del Almirez, Spain. Earth Planet Sci Lett 327-328(0):50-60, doi: 10.1016/j.epsl.2012.01.029

Alt JC, Teagle DAH (1999) The uptake of carbon during alteration of ocean crust. Geochim Cosmochim Acta 63:1527-1535

Anders E, Grevesse N (1989) Abundances of the elements: meteoritic and solar. Geochim Cosmochim Acta 53:197-214

Ardia P, Withers AC, Hirschmann MM (2011) Methane solubility under reduced conditions in a haplobasaltic liquid. *In:* 42nd Lunar and Planetary Science Conference, Woodlands, Texas, USA. LPI Contribution No. 1608, p. 1659

Behn MD, Kelemen PB, Hirth G, Hacker BR, Massonne H-J (2011) Diapirs as the source of the sediment signature in arc lavas. Nat Geosci 4(9):641-646, doi: 10.1038/ngeo1214

Berner RA (1999) A new look at the long-term carbon cycle. GSA Today 9(11):1-6

Biellmann C, Gillet P, Guyot F, Peyronneau J, Reynard B (1993) Experimental evidence for carbonate stability in the Earth's lower mantle. Earth Planet Sci Lett 118:31-41

Blank JG, Brooker RA (1994) Experimental studies of carbon dioxide in silicate melts: solubility, speciation, and stable carbon isotope behavior. Rev Mineral 30:157-186

Blundy J, Cashman KV, Rust A, Witham F (2010) A case for CO_2-rich arc magmas. Earth Planet Sci Lett 290(3-4):289-301, doi: 10.1016/j.epsl.2009.12.013

Blundy JD, Dalton J (2000) Experimental comparison of trace element partitioning between clinopyroxene and melt in carbonate and silicate systems, and implications for mantle metasomatism. Contrib Mineral Petrol 139:356-371

Brooker RA, Kohn SC, Holloway JR, McMillan PF (2001) Structural controls on the solubility of CO_2 in silicate melts Part II: IR characteristics of carbonate groups in silicate glasses. Chem Geol 174:241-254

Brown JM, Shankland TJ (1981) Thermodynamic parameters in the Earth as determined from seismic profiles. Geophys J Int 66(3):579-596

Brown M (2006) Duality of thermal regimes is the distinctive characteristic of plate tectonics since the Neoarchean. Geology 34(11):961-964

Budyko MI (1969) The effect of solar radiation variations on the climate of the Earth. Tellus 21(5):611-619

Buono AS, Dasgupta R, Lee CT-A, Walker D (submitted) Siderophile element partitioning between cohenite and liquid in Fe-Ni-S-C systems and implications for geochemistry of planetary cores and mantles. Geochim Cosmochim Acta

Campbell IH, O'Neill HSC (2012) Evidence against a chondritic Earth. Nature 483(7391):553-558

Caro G (2011) Early silicate earth differentiation. Annu Rev Earth Planet Sci Lett 39(1):31-58

Cartigny P (2005) Stable isotopes and the origin of diamond. Elements 1(2):79-84

Cartigny P, Pineau F, Aubaud C, Javoy M (2008) Towards a consistent mantle carbon flux estimate: Insights from volatile systematics (H_2O/Ce, δD, CO_2/Nb) in the North Atlantic mantle (14° N and 34° N). Earth Planet Sci Lett 265:672-685

Castro A, Gerya TV (2008) Magmatic implications of mantle wedge plumes: Experimental study. Lithos 103(1-2):138-148, doi: 10.1016/j.lithos.2007.09.012

Catling DC, Zahnle KJ, McKay C (2001) Biogenic methane, hydrogen escape, and the irreversible oxidation of early Earth. Science 293(5531):839-843, doi: 10.1126/science.1061976

Chabot NL, Agee CB (2003) Core formation in the Earth and Moon: New experimental constraints from V, Cr, and Mn. Geochim Cosmochim Acta 67:2077-2091

Chabot NL, Campbell AJ, Jones JH, Humayoun M, Vern Lauer HJ (2006) The influence of carbon on trace element partitioning behavior. Geochim Cosmochim Acta 70:1322-1335

Chou C-L (1978) Fractionation of siderophile elements in the Earth's upper mantle. Proc 9th Lunar Planet Sci Conf 1:219-230

Coltice N, Bertrand H, Rey P, Jourdan F, Phillips BR, Ricard Y (2009) Global warming of the mantle beneath continents back to the Archaean. Gond Res 15(3-4):254-266, doi: 10.1016/j.gr.2008.10.001

Coltice N, Simon L, Lecuyer C (2004) Carbon isotope cycle and mantle structure. Geophys Res Lett 31:L05603, doi: 10.1029/2003GL018873

Condie KC (2004) Supercontinents and superplume events: distinguishing signals in the geologic record. Phys Earth Planet Inter 146(1-2):319-332, doi: 10.1016/j.pepi.2003.04.002

Cooper L, Ruscitto DM, Plank T, Wallace P, Syracuse EM, Manning CE (2012) Global variations in H_2O/Ce I: slab surface temperatures beneath volcanic arcs. Geochem Geophys Geosyst Q03024, doi: 10.1029/2011GC003902

Corgne A, Wood BJ, Fei Y (2008) C- and S-rich molten alloy immiscibility and core formation of planetesimals. Geochim. Cosmochim. Acta 72:2409-2416

Currie CA, Beaumont C, Huismans RS (2007) The fate of subducted sediments: A case for backarc intrusion and underplating. Geology 35(12):1111-1114

Dahl TW, Stevenson DJ (2010) Turbulent mixing of metal and silicate during planet accretion — And interpretation of the Hf-W chronometer. Earth Planet Sci Lett 295(1-2):177-186

Dalou C, Koga KT, Hammouda T, Poitrasson F (2009) Trace element partitioning between carbonatitic melts and mantle transition zone minerals: Implications for the source of carbonatites. Geochim Cosmochim Acta 73(1):239-255
Dalton JA, Presnall DC (1998) Carbonatitic melts along the solidus of model lherzolite in the system CaO-MgO-Al_2O_3-SiO_2-CO_2 from 3 to 7 GPa. Contrib Mineral Petrol 131:123-135
Dasgupta R, Buono A, Whelan G, Walker D (2009a) High-pressure melting relations in Fe-C-S systems: implications for formation, evolution, and structure of metallic cores in planetary bodies. Geochim Cosmochim Acta 73:6678-6691, doi: 10.1016/j.gca.2009.08.001
Dasgupta R, Chi H, Shimizu N, Buono A, Walker D (2012) Carbon cycling in shallow magma oceans of terrestrial planets constrained by high P-T experiments. *In:* 43rd Lunar and Planetary Science Conference, Woodlands, Texas, USA. LPI Contribution No 1659, p 1767
Dasgupta R, Chi H, Shimizu N, Buono A, Walker D (2013a) Carbon solution and partitioning between metallic and silicate melts in a shallow magma ocean: implications for the origin and distribution of terrestrial carbon. Geochim Cosmochim Acta 102:191-212, doi: 10.1016/j.gca.2012.10.011
Dasgupta R, Hirschmann MM (2006) Melting in the Earth's deep upper mantle caused by carbon dioxide. Nature 440:659-662
Dasgupta R, Hirschmann MM (2007a) Effect of variable carbonate concentration on the solidus of mantle peridotite. Am Mineral 92:370-379
Dasgupta R, Hirschmann MM (2007b) A modified iterative sandwich method for determination of near-solidus partial melt compositions. II. Application to determination of near-solidus melt compositions of carbonated peridotite. Contrib Mineral Petrol 154:647-661
Dasgupta R, Hirschmann MM (2010) The deep carbon cycle and melting in Earth's interior. Earth Planet Sci Lett 298:1-13, doi: 10.1016/j.epsl.2010.06.039
Dasgupta R, Hirschmann MM, Dellas N (2005) The effect of bulk composition on the solidus of carbonated eclogite from partial melting experiments at 3 GPa. Contrib Mineral Petrol 149:288-305
Dasgupta R, Hirschmann MM, McDonough WF, Spiegelman M, Withers AC (2009b) Trace element partitioning between garnet lherzolite and carbonatite at 6.6 and 8.6 GPa with applications to the geochemistry of the mantle and of mantle-derived melts. Chem Geol 262:57-77, doi: 10.1016/j.chemgeo.2009.02.004
Dasgupta R, Hirschmann MM, Smith ND (2007a) Water follows carbon: CO_2 incites deep silicate melting and dehydration beneath mid-ocean ridges. Geology 35:135-138, doi: 10.1130/G22856A.1
Dasgupta, R, Hirschmann MM, Smith ND (2007b) Partial melting experiments of peridotite + CO_2 at 3 GPa and genesis of alkalic ocean island basalts. J Petrol 48:2093-2124
Dasgupta R, Hirschmann MM, Withers AC (2004) Deep global cycling of carbon constrained by the solidus of anhydrous, carbonated eclogite under upper mantle conditions. Earth Planet Sci Lett 227:73-85
Dasgupta R, Mallik A, Tsuno K, Withers AC, Hirth G, Hirschmann MM (2013b) Carbon-dioxide-rich silicate melt in the Earth's upper mantle. Nature 493:211-215
Dasgupta R, Walker D (2008) Carbon solubility in core melts in a shallow magma ocean environment and distribution of carbon between the Earth's core and the mantle. Geochim Cosmochim Acta 72:4627-4641, doi: 10.1016/j.gca.2008.06.023
Dauphas N, Davis AM, Marty B, Reisberg L (2004) The cosmic molybdenum-ruthenium isotope correlation. Earth Planet Sci Lett 226(3-4):465-475
Dauphas N, Marty B (2002) Inference on the nature and the mass of Earth's late veneer from noble metals and gases. J Geophys Res 107(E12):5129, doi: 10.1029/2001JE001617
Deines P (2002) The carbon isotope geochemistry of mantle xenoliths. Earth Sci Rev 58(3-4):247-278
Dixon JE (1997) Degassing of alkalic basalts. Am Mineral 82:368-378
Dobson DP, Wiedenbeck M (2002) Fe- and C-self-diffusion in liquid Fe_3C to 15 GPa. Geophys Res Lett 29, doi: 10.1029/2002GL015536
Drake MJ (2005) Origin of water in the terrestrial planets. Meteor Planet Sci 40(4):519-527
Drake MJ, Righter K (2002) Determining the composition of the Earth. Nature 416(6876):39-44
Eggler DH (1976) Does CO_2 cause partial melting in the low-velocity layer of the mantle? Geology 4:69-72
Eggler DH (1983) Upper mantle oxidation state: Evidence from olivine-orthopyroxene-ilmenite assemblages. Geophys Res Lett 10(5):365-368
Eggler DH, Baker DR (1982) Reduced volatiles in the system C O H: implications to mantle melting, fluid formation, and diamond genesis. *In*: Advances in Earth and Planetary Sciences. Vol 12. Akimoto S-I, Manghnani MH (eds), p 237-250
Eggler DH, Rosenhauer M (1978) Carbon dioxide in silicate melts: II. Solubilities of CO_2 and H_2O in $CaMgSi_2O_6$ (diopside) liquids and vapors at pressures to 40 kb. Am J Sci 278(1):64-94
Elkins-Tanton LT (2008) Linked magma ocean solidification and atmospheric growth for Earth and Mars. Earth Planet Sci Lett 271(1-4):181-191
Elkins-Tanton LT (2012) Magma oceans in the inner solar system. Annu Rev Earth Planet Sci 40(1):113-139, doi: 10.1146/annurev-earth-042711-105503

Falloon TJ, Green DH (1989) The solidus of carbonated, fertile peridotite. Earth Planet Sci Lett 94:364-370
Feulner G (2012) The faint young Sun problem. Rev Geophys 50(2):364-370, doi: 10.1016/0012-821X(89)90153-2
Foley SF (2008) Rejuvenation and erosion of the cratonic lithosphere. Nat Geosci 1:503-510
Frezzotti ML, Selverstone J, Sharp ZD, Compagnoni R (2011) Carbonate dissolution during subduction revealed by diamond-bearing rocks from the Alps. Nat Geosci 4(10):703-706
Frost DJ, Liebske C, Langenhorst F, McCammon CA, Tronnes RG, Rubie DC (2004) Experimental evidence for the existence of iron-rich metal in the Earth's lower mantle. Nature 428:409-412
Frost DJ, McCammon CA (2008) The Redox State of Earth's Mantle. Annual Rev Earth Planet Sci 36(1):389-420
Genda H, Abe Y (2003) Survival of a proto-atmosphere through the stage of giant impacts: the mechanical aspects. Icarus 164(1):149-162
Genda H, Abe Y (2005) Enhanced atmospheric loss on protoplanets at the giant impact phase in the presence of oceans. Nature 433(7028):842-844
Gerbode C, Dasgupta R (2010) Carbonate-fluxed melting of MORB-like pyroxenite at 2.9 GPa and genesis of HIMU ocean island basalts. J Petrol 51(10):2067-2088
Gerya TV, Yuen DA (2003) Rayleigh-Taylor instabilities from hydration and melting propel 'cold plumes' at subduction zones. Earth Planet Sci Lett 212(1-2):47-62
Ghosh S, Ohtani E, Litasov KD, Terasaki H (2009) Solidus of carbonated peridotite from 10 to 20 GPa and origin of magnesiocarbonatite melt in the Earth's deep mantle. Chem Geol 262:17-28, doi: 10.1016/j.chemgeo.2008.12.030
Golabek GJ, Schmeling H, Tackley PJ (2008) Earth's core formation aided by flow channelling instabilities induced by iron diapirs. Earth Planet Sci Lett 271(1-4):24-33
Goldberg D, Belton G (1974) The diffusion of carbon in iron-carbon alloys at 1560 °C. Metallurg Mat Trans B 5(7):1643-1648
Gorman PJ, Kerrick DM, Connolly JAD (2006) Modeling open system metamorphic decarbonation of subducting slabs. Geochem Geophys Geosyst 7:Q04007, doi: 10.1029/2005GC001125
Gough DO (1981) Solar interior structure and luminosity variations. Sol Phys 74(1):21-34
Grassi D, Schmidt MW, Günther D (2012) Element partitioning during carbonated pelite melting at 8, 13 and 22 GPa and the sediment signature in the EM mantle components. Earth Planet Sci Lett 327-328(0):84-96
Gudfinnsson G, Presnall DC (2005) Continuous gradations among primary carbonatitic, kimberlitic, melilititic, basaltic, picritic, and komatiitic melts in equilibrium with garnet lherzolite at 3-8 GPa. J Petrol 46:1645-1659
Gudmundsson G, Wood BJ (1995) Experimental tests of garnet peridotite oxygen barometry. Contrib Mineral Petrol 119:56-67
Halliday AN (2004) Mixing, volatile loss and compositional change during impact-driven accretion of the Earth. Nature 427(6974):505-509
Halliday AN (2008) A young Moon-forming giant impact at 70-110 million years accompanied by late-stage mixing, core formation and degassing of the Earth. Phil Trans Royal Soc 366(1883):4163-4181
Hayden LA, Van Orman JA, McDonough WF, Ash RD, Goodrich CA (2011) Trace element partitioning in the Fe-S-C system and its implications for planetary differentiation and the thermal history of ureilites. Geochim Cosmochim Acta 75(21):6570-6583
Hayes JF, Waldbauer JR (2006) The carbon cycle and associated redox processes through time. Phil Trans Royal Soc London B361:931-950
Herzberg C, Asimow PD (2008) Petrology of some oceanic island basalts: PRIMELT2.XLS software for primary magma calculation. Geochem Geophys Geosys 9:Q09001, doi: 10.1029/2008GC002057
Herzberg C, Asimow PD, Arndt N, Niu Y, Lesher CM, Fitton JG, Cheadle MJ, Saunders AD (2007) Temperatures in ambient mantle and plumes: Constraints from basalts, picrites, and komatiites. Geochem Geophys Geosyst 8:Q02006, doi: 10.1029/2006GC001390
Herzberg C, Condie K, Korenaga J (2010) Thermal history of the Earth and its petrological expression. Earth Planet Sci Lett 292(1-2):79-88
Hillgren VJ, Gessmann CK, Li J (2000) An experimental perspective on the light element in Earth's core. *In*: Origin of the Earth and Moon. Canup RM, Righter K (eds) The University of Arizona Press, Tucson, p 245-263
Hirschmann MM (2000) The mantle solidus: experimental constraints and the effect of peridotite composition. Geochem Geophys Geosys 1:2000GC000070
Hirschmann MM (2010) Partial melt in the oceanic low velocity zone. Phys Earth Planet Inter 179(1-2):60-71
Hirschmann MM (2012) Magma ocean influence on early atmosphere mass and composition. Earth Planet Sci Lett 341-344(0):48-57
Hirschmann MM, Dasgupta R (2009) The H/C ratios of Earth's near-surface and deep reservoirs, and consequences for deep Earth volatile cycles. Chem Geol 262:4-16, doi: 10.1016/j.chemgeo.2009.02.008

Hirschmann MM, Withers AC (2008) Ventilation of CO_2 from a reduced mantle and consequences for the early Martian greenhouse. Earth Planet Sci Lett 270(1-2):147-155

Holloway JR, Jakobsson S (1986) Volatile Solubilities in Magmas: Transport of Volatiles from Mantles to Planet Surfaces. J Geophys Res 91(B4):505-508, doi: 10.1029/JB091iB04p0D505

Holloway JR, Pan V, Gudmundsson G (1992) High-pressure fluid-absent melting experiments in the presence of graphite; oxygen fugacity, ferric/ferrous ratio and dissolved CO_2. Eur J Mineral 4(1):105-114

Holzheid A, Sylvester P, O'Neill HSC, Rubie DC, Palme H (2000) Evidence for a late chondritic veneer in the Earth's mantle from high-pressure partitioning of palladium and platinum. Nature 406(6794):396-399

Hopkins M, Harrison TM, Manning CE (2008) Low heat flow inferred from >4 Gyr zircons suggests Hadean plate boundary interactions. Nature 456(7221):493-496

Hopkins MD, Harrison TM, Manning CE (2010) Constraints on Hadean geodynamics from mineral inclusions in >4 Ga zircons. Earth Planet Sci Lett 298(3-4):367-376

Iacono Marziano G, Gaillard F, Pichavant M (2008) Limestone assimilation by basaltic magmas: an experimental re-assessment and application to Italian volcanoes. Contrib Mineral Petrol 155(6):719-738

Iacono-Marziano G, Gaillard F, Scaillet B, Pichavant M, Chiodini G (2009) Role of non-mantle CO_2 in the dynamics of volcano degassing: The Mount Vesuvius example. Geology 37(4):319-322

Ita J, Stixrude L (1992) Petrology, elasticity, and composition of the mantle transition zone. J Geophys Res 97:6849-6866

Jacob DE, Kronz A, Viljoen KS (2004) Cohenite, native iron and troilite inclusions in garnets from polycrystalline diamond aggregates. Contrib Mineral Petrol 146:566-576

Jakobsson S, Holloway JR (1986) Crystal-liquid experiments in the presence of a C-O-H fluid buffered by graphite + iron + wustite: Experimental method and near-liquidus relations in basanite. J Volcan Geotherm Res 29(1-4):265-291

Jarrard RD (2003) Subduction fluxes of water, carbon dioxide, chlorine, and potassium. Geochem Geophys Geosyst 4:8905, doi: 10.1029/2002GC000392

Javoy M, Kaminski E, Guyot F, Andrault D, Sanloup C, Moreira M, Labrosse S, Jambon A, Agrinier P, Davaille A, Jaupart C (2010) The chemical composition of the Earth: Enstatite chondrite models. Earth Planet Sci Lett 293(3-4):259-268

Jones JH, Drake MJ (1986) Geochemical constraints on core formation in the Earth. Nature 322(6076):221-228

Jugo PJ, Wilke M, Botcharnikov RE (2010) Sulfur K-edge XANES analysis of natural and synthetic basaltic glasses: Implications for S speciation and S content as function of oxygen fugacity. Geochim Cosmochim Acta 74:5926-5938

Kadik A, Pineau F, Litvin Y, Jendrzejewski N, Martinez I, Javoy M (2004) Formation of carbon and hydrogen species in magmas at low oxygen fugacity. J Petrol 45:1297-1310

Kadik AA, Kurovskaya NA, Ignat'ev YA, Kononkova NN, Koltashev VV, Plotnichenko VG (2011) Influence of Oxygen Fugacity on the Solubility of Nitrogen, Carbon, and Hydrogen in FeO-Na_2O-SiO_2-Al_2O_3 Melts in Equilibrium with Metallic Iron at 1.5 GPa and 1400°C. Geochem Int 49:429-438

Kadik AA, Litvin YA, Koltashev VV, Kryukova EB, Plotnichenko VG (2006) Solubility of hydrogen and carbon in reduced magmas of the early Earth's mantle. Geochem Int 44:33-47

Kaminsky FV, Wirth R (2011) Iron carbide inclusions in lower-mantle diamond from Juina, Brazil. Can Mineral 49(2):555-572

Kasting J (1993) Earth's early atmosphere. Science 259(5097):920-926

Kasting JF, Catling D (2003) Evolution of a habitable planet. Ann Rev Astron Astrophys 41(1):429-463

Kendall JD, Melosh HJ (2012) Fate of iron cores during planetesimal impacts. *In:* 43rd Lunar and Planetary Science Conference, Woodlands, Texas, USA. LPI Contribution No 1659, p 2699

Keppler H, Wiedenbeck M, Shcheka SS (2003) Carbon solubility in olivine and the mode of carbon storage in the Earth's mantle. Nature 424:414-416

Kerrick DM (2001) Present and past nonanthropogenic CO_2 degassing from the solid earth. Rev Geophys 39(4):565-585

Kerrick DM, Connolly JAD (1998) Subduction of ophicarbonates and recycling of CO_2 and H_2O. Geology 26(4):375-378

Kerrick DM, Connolly JAD (2001) Metamorphic devolatilization of subducted oceanic metabasalts: implications for seismicity, arc magmatism and volatile recycling. Earth Planet Sci Lett 189:19-29

Kerridge JF (1985) Carbon, hydrogen and nitrogen in carbonaceous chondrites: Abundances and isotopic compositions in bulk samples. Geochim Cosmochim Acta 49(8):1707-1714

Kirschivink JL (1992) Late Proterozoic low-latitude global glaciation: the snowball Earth. *In*: The Proterozoic Biosphere: A Multidisciplinary Study. Schopf JW, Klein C (eds) Cambridge University Press, p 51-52

Kiseeva ES, Yaxley GM, Hermann J, Litasov KD, Rosenthal A, Kamenetsky VS (2012) An experimental study of carbonated eclogite at 3.5-5.5 GPa—Implications for silicate and carbonate metasomatism in the cratonic mantle. J Petrol 53:727-759, doi: 10.1093/petrology/egr078

Komiya T, Hayashi M, Maruyama S, Yurimoto H (2002) Intermediate-P/T type Archean metamorphism of the Isua supracrustal belt: Implications for secular change of geothermal gradients at subduction zones and for Archean plate tectonics. Am J Sci 302(9):806-826

Kuhn WR, Kasting JF (1983) Effects of increased CO_2 concentrations on surface temperature of the early Earth. Nature 301(5895):53-55

Kuramoto K (1997) Accretion, core formation, H and C evolution of the Earth and Mars. Phys Earth Planet Inter 100:3-20

Kuramoto K, Matsui T (1996) Partitioning of H and C between the mantle and core during the core formation in the Earth: Its implications for the atmospheric evolution and redox state of early mantle. J Geophys Res 101:14909-14932

Lee C-TA, Luffi P, Chin EJ (2011) Building and destroying continental mantle. Ann Rev Earth Planet Sci 39(1):59-90

Lee C-TA, Luffi P, Höink T, Li J, Dasgupta R, Hernlund J (2010) Upside-down differentiation and generation of a 'primordial' lower mantle. Nature 463:930-933, doi: 910.1038/nature08824

Lee C-TA, Shen B, Slotnick B, Liao K, Dickens G, Yokoyama Y, Lenardic A, Dasgupta R, Jellinek M, S. LJ, Schneider T, Tice M (2013) Continent-island arc fluctuations, growth of crustal carbonates, and long-term climate change. Geosphere 9, doi: 10.1130/GES00822.1

Lenardic A, Moresi L, Jellinek AM, O'Neill CJ, Cooper CM, Lee CT (2011) Continents, supercontinents, mantle thermal mixing, and mantle thermal isolation: Theory, numerical simulations, and laboratory experiments. Geochem Geophys Geosyst 12(10):Q10016, doi: 10.1029/2011GC003663

Lesne P, Scaillet B, Pichavant M, Beny J-M (2011) The carbon dioxide solubility in alkali basalts: an experimental study. Contrib Mineral Petrol 162(1):153-168, doi: 10.1007/s00410-010-0585-0

Leung I, Guo W, Friedman I, Gleason J (1990) Natural occurrence of silicon carbide in a diamondiferous kimberlite from Fuxian. Nature 346(6282):352-354

Leung IS (1990) Silicon carbide cluster entrapped in a diamond from Fuxian, China. Am Mineral 75(9-10):1110-1119

Li J, Agee CB (1996) Geochemistry of mantle-core differentiation at high pressure. Nature 381:686-689

Litasov KD, Ohtani E (2009) Solidus and phase relations of carbonated peridotite in the system $CaO-Al_2O_3-MgO-SiO_2-Na_2O-CO_2$ to the lower mantle depths. Phys Earth Planet Int:doi: 10.1016/j.pepi.2009.07.008

Lodders K (2003) Solar system abundances and condensation temperatures of the elements. Astrophys J 591:1220-1247, doi: 10.1086/375492

Lodders K (2010) Solar system abundances of the elements. *In*: Principles and Perspectives in Cosmochemistry: Lecture Notes of the Kodai School on 'Synthesis of Elements in Stars'. Vol. Goswami A, Reddy BE (eds) Springer-Verlag, Berlin Heidelberg, p 379-417

Lord OT, Walter MJ, Dasgupta R, Walker D, Clark SM (2009) Melting in the Fe-C system to 70 GPa. Earth Planet Sci Lett 284:157-167

Luth RW (1993) Diamonds, eclogites, and oxidation state of the Earth's mantle. Science 261:66-68

Luth RW (1999) Carbon and carbonates in the mantle. *In*: Mantle Petrology: Field Observations and High Pressure Experimentation: A Tribute to Francis R. (Joe) Boyd. Vol 6. Fei Y, Bertka, C. M., Mysen, B. O. (ed) The Geochemical Society, p 297-316

Mann U, Frost DJ, Rubie DC (2008) The wetting ability of Si-bearing liquid Fe-alloys in a solid silicate matrix—percolation during core formation under reducing conditions? Phys Earth Planet Inter 167(1-2):1-7

Marty B (2012) The origins and concentrations of water, carbon, nitrogen and noble gases on Earth. Earth Planet Sci Lett 313-314(0):56-66

Marty B, Alexander CMO'D, Raymond SN (2013) Primordial origins of Earth's carbon. Rev Mineral Geochem 75:149-181

Marty B, Jambon A (1987) $C/^3He$ in volatile fluxes from the solid Earth: implications for carbon geodynamics. Earth Planet Sci Lett 83:16-26

Marty B, Tolstikhin IN (1998) CO_2 fluxes from mid-ocean ridges, arcs and plumes. Chem Geol 145:233-248

McCammon CA (2005) Mantle oxidation state and oxygen fugacity: constraints on mantle chemistry, structure, and dynamics. *In*: Earth's Deep Mantle: Structure, Composition, and Evolution. Vol 160. van der Hilst RD, Bass JD, Matas J, Trampert J (eds) American Geophysical Union, Washington D. C., p 221-242

McCammon CA, Kopylova MG (2004) A redox profile of the Slave mantle and oxygen fugacity control in the cratonic mantle. Contrib Mineral Petrol 148:55-68

McDonough WF (2003) Compositional model for the Earth's core. *In*: The Mantle and Core. Vol 2. Carlson RW (ed) Elsevier-Pergamon, Oxford, p 547-568

McDonough WF, Sun S-s (1995) The composition of the Earth. Chem Geol 120:223-253

McKenzie D, Jackson J, Priestley K (2005) Thermal structure of oceanic and continental lithosphere. Earth Planet Sci Lett 233:337-349

Minarik WG, Ryerson FJ, Watson EB (1996) Textural entrapment of core-forming melts. Science 272(5261):530-533

Mints MV, Belousova EA, Konilov AN, Natapov LM, Shchipansky AA, Griffin WL, O'Reilly SY, Dokukina KA, Kaulina TV (2010) Mesoarchean subduction processes: 2.87 Ga eclogites from the Kola Peninsula, Russia. Geology 38(8):739-742

Molina JF, Poli S (2000) Carbonate stability and fluid composition in subducted oceanic crust: an experimental study on H_2O-CO_2-bearing basalts. Earth Planet Sci Lett 176:295-310

Möller A, Appel P, Mezger K, Schenk V (1995) Evidence for a 2 Ga subduction zone: Eclogites in the Usagaran belt of Tanzania. Geology 23(12):1067-1070

Mookherjee M, Nakajima Y, Steinle-Neumann G, Glazyrin K, Wu X, Dubrovinsky L, McCammon C, Chumakov A (2011) High-pressure behavior of iron carbide (Fe_7C_3) at inner core conditions. J Geophys Res 116(B4):B04201

Moore CB, Lewis CF (1967) Total carbon content of ordinary chondrites. J Geophys Res 72(24):6289-6292

Morbidelli A, Chambers J, Lunine JI, Petit JM, Robert F, Valsecchi GB, Cyr KE (2000) Source regions and timescales for the delivery of water to the Earth. Meteor Planet Sci 35(6):1309-1320

Morbidelli A, Lunine JI, O'Brien DP, Raymond SN, Walsh KJ (2012) Building terrestrial planets. Ann Rev Earth Planet Sci 40(1):251-275

Morgan JW, Walker RJ, Brandon AD, Horan MF (2001) Siderophile elements in Earth's upper mantle and lunar breccias: Data synthesis suggests manifestations of the same late influx. Meteor Planet Sci 36(9):1257-1275

Morizet Y, Brooker RA, Kohn SC (2002) CO_2 in haplo-phonolite melt: solubility, speciation and carbonate complexation. Geochim Cosmochim Acta 66(10):1809-1820

Morizet Y, Paris M, Gaillard F, Scaillet B (2010) C-O-H fluid solubility in haplobasalt under reducing conditions: An experimental study. Chem Geol 279(1-2):1-16

Moyen J-F, van Hunen J (2012) Short-term episodicity of Archaean plate tectonics. Geology doi: 10.1130/G32894.32891

Moyen J-Fo, Stevens G, Kisters A (2006) Record of mid-Archaean subduction from metamorphism in the Barberton terrain, South Africa. Nature 442(7102):559-562

Mysen BO (1991) Volatiles in magmatic liquids. *In*: Physical Chemistry of Magma. Advances in Physical Geochemistry, Vol. 9. Perchuk LL, Kushiro I (eds) Cambridge University Press, New York, p 435-476

Mysen BO, Fogel ML, Morrill PL, Cody GD (2009) Solution behavior of reduced COH volatiles in silicate melts at high pressure and temperature. Geochim Cosmochim Acta 73(6):1696-1710

Mysen BO, Kumamoto K, Cody GD, Fogel ML (2011) Solubility and solution mechanisms of C-O-H volatiles in silicate melt with variable redox conditions and melt composition at upper mantle temperatures and pressures. Geochim Cosmochim Acta 75(20):6183-6199

Mysen BO, Yamashita S (2010) Speciation of reduced C-O-H volatiles in coexisting fluids and silicate melts determined in-situ to ~1.4 GPa and 800 °C. Geochim Cosmochim Acta 74(15):4577-4588

Nakajima T, Maruyama S, Uchiumi S, Liou JG, Wang X, Xiao X, Graham SA (1990) Evidence for late Proterozoic subduction from 700-Myr-old blueschists in China. Nature 346(6281):263-265

Nakajima Y, Takahashi E, Toshihiro S, Funakoshi K (2009) "Carbon in the core" revisited. Phys Earth Planet Inter 174:202-211, doi: 10.1016/j.pepi.2008.05.014

Nakamura K, Kato Y (2004) Carbonatization of oceanic crust by seafloor hydrothermal activity and its significance as a CO_2 sink in the Early Archean. Geochim Cosmochim Acta 68:4595-4618

Newsom HE, Sims KWW (1991) Core Formation During Early Accretion of the Earth. Science 252(5008):926-933

Ni H, Keppler H (2013) Carbon in silicate melts. Rev Mineral Geochem 75:251-287

Oganov AR, Hemley RJ, Hazen RM, Jones AP (2013) Structure, bonding, and mineralogy of carbon at extreme conditions. Rev Mineral Geochem 75:47-77

Ohtani H, Nishizawa T (1986) Calculation of Fe-C-S ternary phase diagram. Trans ISIJ 26:655-663

Okuchi T (1997) Hydrogen partitioning into molten iron at high pressure: implications for Earth's core. Science 278:1781-1784

Okuchi T (1998) The melting temperature of iron hydride at high pressures and its implications for the temperature of the Earth's core. J Phys Condens Matter 10:11595-11598

O'Neill C, Lenardic A, Moresi L, Torsvik TH, Lee CTA (2007) Episodic Precambrian subduction. Earth Planet Sci Lett 262(3-4):552-562

Owen T, Bar-Nun A (1995) Comets, impacts, and atmospheres. Icarus 116(2):215-226

Owen T, Cess RD, Ramanathan V (1979) Enhanced CO_2 greenhouse to compensate for reduced solar luminosity on early Earth. Nature 277(5698):640-642

Pawley AR, Holloway JR, McMillan PF (1992) The effect of oxygen fugacity on the solubility of carbon-oxygen fluids in basaltic melt. Earth Planet Sci Lett 110(1-4):213-225

Pineau F, Shilobreeva S, Hekinian R, Bidiau D, Javoy M (2004) Deep-sea explosive activity on the Mid-Atlantic Ridge near 34° 50′ N: a stable isotope (C, H, O) study. Chem Geol 211:159-175

Plank T, Langmuir CH (1998) The geochemical composition of subducting sediment and its consequences for the crust and mantle. Chem Geol 145:325-394

Poirier J-P (1994) Light elements in the Earth's outer core: a critical review. Phys Earth Planet Inter 85:319-337
Ranero CR, Phipps Morgan J, McIntosh K, Reichert C (2003) Bending-related faulting and mantle serpentinization at the Middle America trench. Nature 425(6956):367-373
Raymond SN, Quinn T, Lunine JI (2004) Making other earths: dynamical simulations of terrestrial planet formation and water delivery. Icarus 168(1):1-17
Raymond SN, Quinn T, Lunine JI (2006) High-resolution simulations of the final assembly of Earth-like planets. 1. Terrestrial accretion and dynamics. Icarus 183(2):265-282
Raymond SN, Quinn T, Lunine JI (2007) High-resolution simulations of the final assembly of Earth-like planets. 2. Water delivery and planetary habitability. Astrobiology 7:66-84
Reymer A, Schubert G (1984) Phanerozoic addition rates to the continental crust and crustal growth. Tectonics 3(1):63-77, doi: 10.1029/TC003i001p00063
Righter K (2011) Prediction of metal-silicate partition coefficients for siderophile elements: An update and assessment of PT conditions for metal-silicate equilibrium during accretion of the Earth. Earth Planet Sci Lett 304(1-2):158-167
Rohrbach A, Ballhaus C, Golla-Schindler U, Ulmer P, Kamenetsky VS, Kuzmin DV (2007) Metal saturation in the upper mantle. Nature 449:456-458
Rohrbach A, Ballhaus C, Ulmer P, Golla-Schindler U, Schonbohm D (2011) Experimental evidence for a reduced metal-saturated upper mantle. J Petrol 52(4):717-731
Rohrbach A, Schmidt MW (2011) Redox freezing and melting in the Earth's deep mantle resulting from carbon-iron redox coupling. Nature 472(7342):209-212
Rubie DC, Frost DJ, Mann U, Asahara Y, Nimmo F, Tsuno K, Kegler P, Holzheid A, Palme H (2011) Heterogeneous accretion, composition and core-mantle differentiation of the Earth. Earth Planet Sci Lett 301(1-2):31-42
Saal AE, Hauri E, Langmuir CH, Perfit MR (2002) Vapour undersaturation in primitive mid-ocean-ridge basalt and the volatile content of Earth's upper mantle. Nature 419:451-455
Sagan C, Mullen G (1972) Earth and Mars: Evolution of atmospheres and surface temperatures. Science 177(4043):52-56
Saha L, Hofmann A, Xie H, Hegner E, Wilson A, Wan Y, Liu D, Kröner A (2010) Zircon ages and metamorphic evolution of the Archean Assegaai-De Kraalen granitoid-greenstone terrain, southeastern Kaapvaal Craton. Am J Sci 310(10):1384-1420
Sano Y, Williams SN (1996) Fluxes of mantle and subducted carbon along convergent plate boundaries. Geophys Res Lett 23:2749-2752, doi: 10.1029/96gl02260
Satish-Kumar M, So H, Yoshino T, Kato M, Hiroi Y (2011) Experimental determination of carbon isotope fractionation between iron carbide melt and carbon: ^{12}C-enriched carbon in the Earth's core? Earth Planet Sci Lett 310(3-4):340-348
Schulze DJ, Valley JW, Viljoen KS, Stiefenhofer J, Spicuzza M (1997) Carbon isotope composition of graphite in mantle eclogites. J Geol 105(3):379-386
Shannon MC, Agee CB (1996) High pressure constraints on percolative core formation. Geophys Res Lett 23(20):2717-2720
Sharp WE (1966) Pyrrhotite: a common inclusion in South African diamonds. Nature 211(5047):402-403
Shcheka SS, Wiedenbeck M, Frost DJ, Keppler H (2006) Carbon solubility in mantle minerals. Earth Planet Sci Lett 245:730-742
Shibuya T, Tahata M, Kitajima K, Ueno Y, Komiya T, Yamamoto S, Igisu M, Terabayashi M, Sawaki Y, Takai K, Yoshida N, Maruyama S (2012) Depth variation of carbon and oxygen isotopes of calcites in Archean altered upperoceanic crust: Implications for the CO_2 flux from ocean to oceanic crust in the Archean. Earth Planet Sci Lett 321-322(0):64-73
Shirey SB, Richardson SH (2011) Start of the Wilson cycle at 3 ga shown by diamonds from subcontinental mantle. Science 333(6041):434-436
Shushkanova A, Litvin Y (2008) Diamond formation in sulfide pyrrhotite-carbon melts: Experiments at 6.0-7.1 GPa and application to natural conditions. Geochem Int 46(1):37-47
Siebert J, Corgne A, Ryerson FJ (2011) Systematics of metal-silicate partitioning for many siderophile elements applied to Earth's core formation. Geochim Cosmochim Acta 75(6):1451-1489
Sizova E, Gerya T, Brown M, Perchuk LL (2010) Subduction styles in the Precambrian: Insight from numerical experiments. Lithos 116(3-4):209-229
Sleep NH, Zahnle K (2001) Carbon dioxide cycling and implications for climate on ancient earth. J Geophys Res 106:1373-1399
Stachel T, Brey GP, Harris JW (2005) Inclusions in sublithospheric diamonds: glimpses of deep Earth. Elements 1(2):73-78
Stagno V, Frost DJ (2010) Carbon speciation in the asthenosphere: Experimental measurements of the redox conditions at which carbonate-bearing melts coexist with graphite or diamond in peridotite assemblages. Earth Planet Sci Lett 300(1-2):72-84

Stagno V, Tange Y, Miyajima N, McCammon CA, Irifune T, Frost DJ (2011) The stability of magnesite in the transition zone and the lower mantle as function of oxygen fugacity. Geophys Res Lett 38(19):L19309

Syracuse EM, Abers GA (2006) Global compilation of variations in slab depth beneath arc volcanoes and implications. Geochem Geophys Geosyst 7:Q05017, doi: 10.1029/2005GC001045

Syracuse EM, van Keken PE, Abers GA (2010) The global range of subduction zone thermal models. Phys Earth Planet Inter 183:73-90, doi: 10.1016/j.pepi.2010.02.004

Terasaki H, Frost DJ, Rubie DC, Langenhorst F (2007) Interconnectivity of Fe-O-S liquid in polycrystalline silicate perovskite at lower mantle conditions. Phys Earth Planet Inter 161(3-4):170-176

Terasaki H, Frost DJ, Rubie DC, Langenhorst F (2008) Percolative core formation in planetesimals. Earth Planet Sci Lett 273(1-2):132-137

Terasaki H, Ohtani E, Sakai T, Kamada S, Asanuma H, Shibazaki Y, Hirao N, Sata N, Ohishi Y, Sakamaki T, Suzuki A, Funakoshi K-i (2012) Stability of Fe-Ni hydride after the reaction between Fe-Ni alloy and hydrous phase (δ-AlOOH) up to 1.2 Mbar: Possibility of H contribution to the core density deficit. Phys Earth Planet Inter 194-195(0):18-24

Terasaki H, Shibazaki Y, Sakamaki T, Tateyama R, Ohtani E, Funakoshi K-i, Higo Y (2011) Hydrogenation of FeSi under high pressure. Am Mineral 96(1):93-99

Thibault Y, Holloway JR (1994) Solubility of CO_2 in a Ca-rich leucitite: effects of pressure, temperature, and oxygen fugacity. Contrib Mineral Petrol 116:216-224

Thomsen TB, Schmidt MW (2008) Melting of carbonaceous pelites at 2.5-5.0 GPa, silicate-carbonatitie liquid immiscibility, and potassium-carbon metasomatism of the mantle. Earth Planet Sci Lett 267:17-31

Trail D, Watson EB, Tailby ND (2011) The oxidation state of Hadean magmas and implications for early Earth's atmosphere. Nature 480(7375):79-82

Trull T, Nadeau S, Pineau F, Polve M, Javoy M (1993) C-He systematics in hotspot xenoliths: implications for mantle carbon contents and carbon recycling. Earth Planet Sci Lett 118:43-64

Tsuno K, Dasgupta R (2011) Melting phase relation of nominally anhydrous, carbonated pelitic-eclogite at 2.5-3.0 GPa and deep cycling of sedimentary carbon. Contrib Mineral Petrol 161:743-763

Tsuno K, Dasgupta R (2012) The effect of carbonates on near-solidus melting of pelite at 3 GPa: Relative efficiency of H_2O and CO_2 subduction. Earth Planet Sci Lett 319-320(0):185-196

Tsuno K, Dasgupta R, Danielson L, Righter K (2012) Flux of carbonate melt from deeply subducted pelitic sediments - geophysical and geochemical implications for the source of Central American volcanic arc. Geophys Res Lett 37:L16307, doi: 10.1029/2012GL052606

Tsymbulov LB, Tsmekhman LS (2001) Solubility of carbon in sulfide melts of the system Fe-Ni-S. Russ J Appl Chem 74:925-929

van Keken PE, Kiefer B, Peacock SM (2002) High-resolution models of subduction zones: Implications for mineral dehydration reactions and the transport of water into the deep mantle. Geochem Geophys Geosyst 3:1056, doi: 10.1029/2001GC000256

Walker D, Dasgupta R, Li J, Buono A (accepted) Nonstoichiometry and growth of some Fe-carbides. Contrib Mineral Petrol

Walker JCG, Hays PB, Kasting JF (1981) A negative feedback mechanism for the long-term stabilization of Earth's surface temperature. J Geophys Res 86(C10):9776-9782

Walker RJ, Horan MF, Morgan JW, Becker H, Grossman JN, Rubin AE (2002) Comparative ^{187}Re-^{187}Os systematics of chondrites: Implications regarding early solar system processes. Geochim Cosmochim Acta 66(23):4187-4201

Wallace ME, Green DH (1988) An experimental determination of primary carbonatite magma composition. Nature 335:343-346

Walter MJ, Kohn SC, Araujo D, Bulanova GP, Smith CB, Gaillou E, Wang J, Steele A, Shirey SB (2011) Deep mantle cycling of oceanic crust: evidence from diamonds and their mineral inclusions. Science 334(6052):54-57

Wang C, Hirama J, Nagasaka T, Ban-Ya S (1991) Phase equilibria of liquid Fe-S-C ternary system. ISIJ Int 31:1292-1299

Wänke H, Dreibus G (1988) Chemical composition and accretion history of terrestrial planets. Phil Trans R Soc Lond 325:545-557

Warren PH (2011) Stable-isotopic anomalies and the accretionary assemblage of the Earth and Mars: A subordinate role for carbonaceous chondrites. Earth Planet Sci Lett 311(1-2):93-100

Wetzel DT, Jacobsen SD, Rutherford MJ, Hauri EH, Saal AE (2012) The solubility and speciation of carbon in lunar picritic magmas. *In:* 43rd Lunar and Planetary Science Conference, Woodlands, Texas, USA. LPI Contribution No. 1659, p 1535

Wood BJ (1993) Carbon in the core. Earth Planet Sci Lett 117:593-607

Wood BJ (1995) Storage and recycling of H_2O and CO_2 in the earth. AIP Conf Proc 341(1):3-21

Wood BJ, Halliday AN (2010) The lead isotopic age of the Earth can be explained by core formation alone. Nature 465(7299):767-770

Wood BJ, Pawley A, Frost DR (1996) Water and carbon in the Earth's mantle. Philos Trans R Soc London 354:1495-1511

Wood BJ, Walter MJ, Wade J (2006) Accretion of the Earth and segregation of its core. Nature 441:825-833

Wood BJ, Li J, Shahar A (2013) Carbon in the core: its influence on the properties of core and mantle. Rev Mineral Geochem 75:231-250

Woodland AB, Koch M (2003) Variation in oxygen fugacity with depth in the upper mantle beneath the Kaapvaal craton, South Africa. Earth Planet Sci Lett 214:295-310

Wyllie PJ (1977) Peridotite-CO_2-H_2O, and carbonatitic liquid in the upper asthenosphere. Nature 266:45-47

Yaxley GM, Berry AJ, Kamenetsky VS, Woodland AB, Golovin AV (2012) An oxygen fugacity profile through the Siberian Craton — Fe K-edge XANES determinations of $Fe^{3+}/\Sigma Fe$ in garnets in peridotite xenoliths from the Udachnaya East kimberlite. Lithos 140-141(0):142-151

Yaxley GM, Brey GP (2004) Phase relations of carbonate-bearing eclogite assemblages from 2.5 to 5.5 GPa: implications for petrogenesis of carbonatites. Contrib Mineral Petrol 146:606-619

Yaxley GM, Green DH (1994) Experimental demonstration of refractory carbonate-bearing eclogite and siliceous melt in the subduction regime. Earth Planet Sci Lett 128:313-325

Yi W, Halliday AN, Alt JC, Lee D-C, Rehkämper M, Garcia MO, Langmuir CH, Su Y (2000) Cadmium, indium, tin, tellurium, and sulfur in oceanic basalts: Implications for chalcophile element fractionation in the Earth. J Geophys Res 105(B8):18927-18948

Zahnle K, Arndt N, Cockell C, Halliday A, Nisbet E, Selsis F, Sleep N (2007) Emergence of a habitable planet. Space Sci Rev 129(1):35-78

Zahnle K, Schaefer L, Fegley B (2010) Earth's Earliest Atmospheres. Cold Spring Harb Perspect Biol 2(10):a004895, doi: 10.1101/cshperspect.a004895

Zahnle KJ (2006) Earth's Earliest Atmosphere. Elements 2(4):217-222

Zhang C, Duan Z (2009) A model for C-O-H fluid in the Earth's mantle. Geochim Cosmochim Acta 73(7):2089-2102

Zhimulev E, Chepurov A, Sinyakova E, Sonin V, Pokhilenko N (2012) Diamond crystallization in the Fe-Co-S-C and Fe-Ni-S-C systems and the role of sulfide-metal melts in the genesis of diamond. Geochem Int 50(3):205-216

Reviews in Mineralogy & Geochemistry
Vol. 75 pp. 231-250, 2013

8

Carbon in the Core: Its Influence on the Properties of Core and Mantle

Bernard J. Wood

Department of Earth Sciences
University of Oxford
Oxford OX1 3AN, United Kingdom

bernie.wood@earth.ox.ac.uk

Jie Li

Department of Earth and Environmental Sciences
University of Michigan
Ann Arbor, Michigan 48109, U.S.A.

jackieli@umich.edu

Anat Shahar

Geophysical Laboratory
Carnegie Institution of Washington
Washington, DC 20015, U.S.A.

ashahar@ciw.edu

INTRODUCTION

Earth's core is known to be metallic, with a density of about 9.90 Mg·m^{-3} at the core-mantle boundary and as such is substantially denser than the surrounding mantle (5.56 Mg·m^{-3} at the core-mantle boundary; Dziewonski and Anderson 1981). Comparison with cosmic abundances suggests that the core is predominantly Fe with around 5% Ni (Allègre et al. 1995; McDonough 2003) and 8-12% of one or more light elements (Birch 1952). The latter conclusion comes from the observation that the core is appreciably less dense than pure Fe or Fe-Ni alloys under any plausible core temperature conditions (Stevenson 1981). The nature of the light element (or elements) has been the subject of considerable speculation, because of its bearing on Earth's overall bulk composition, the conditions under which the core formed, the temperature regime in the core, and possible ongoing interactions between core and mantle. Any element with substantially lower atomic number than iron ($z = 26$) would have the required effect on core density, but it must also be of high cosmic abundance and it must be soluble in liquid Fe under both the conditions of core formation and those of the outer core. A review of the likely contributors to the core density deficit (Wood 1993) concluded that S and C were the most likely candidate elements and acknowledged that Si, which is extensively soluble in Fe at low pressures, could also conceivably be present. More recently, arguments have been put forward in favor of H (Okuchi 1997) and O (Rubie et al. 2004) as major "light" elements in the core. Although the presence of any of these other elements would not exclude C from the core, dissolution of most of them in liquid Fe require specific compositions of accreting planetesimals and specific conditions of core formation. In order to place constraints on core composition, therefore, it is necessary to consider the cosmochemical abundances of the elements and the process of core formation in the early solar system.

1529-6466/13/0075-0008$00.00 DOI: 10.2138/rmg.2013.75.8

Earth's earliest history was marked by accretion from protoplanetary materials and segregation of the core within about 35 m.y. (Kleine et al. 2002; Yin et al. 2002) and formation of the Moon by giant impact approximately 100 m.y. after the origin of the solar system. During this primary differentiation, all elements were distributed between the Fe-rich metallic phase and the silicate mantle according to their partition coefficients D_i ($D_i = [i]_{metal}/[i]_{silicate}$). The net result is that the mantle is relatively depleted in those (siderophile) elements with high D_i, which partitioned strongly into the core and enriched in lithophile elements with low D values. These qualitative observations are placed in context by the observation that Earth's mantle has strong compositional affinities with chondritic meteorites (Allègre et al 1995; McDonough and Sun 1995). Figure 1 shows the abundances of a large number of elements in silicate Earth compared to those in CI chondrites plotted against the temperature at which 50% of the element would condense from a gas of solar composition. Refractory lithophile elements (those which condense at highest temperature) are present in the mantle in approximately chondritic proportions, which implies that all refractory elements are present in bulk Earth (core plus mantle) in approximately chondritic proportions. In contrast, silicate Earth is depleted in volatile elements relative to CI chondrites with a decreasing relative abundance with decreasing condensation temperature. Siderophile elements are partitioned into the core and, in the case of refractory elements, their concentrations in the metal phase may be estimated by mass balance by assuming overall chondritic abundance in bulk Earth (McDonough 2003). In contrast, the abundances of volatile elements such as S, C, and Si in the core are more difficult to estimate because of the non-chondritic bulk Earth ratios of these elements. Nevertheless, plausible bounds may be placed on their concentrations, as discussed below.

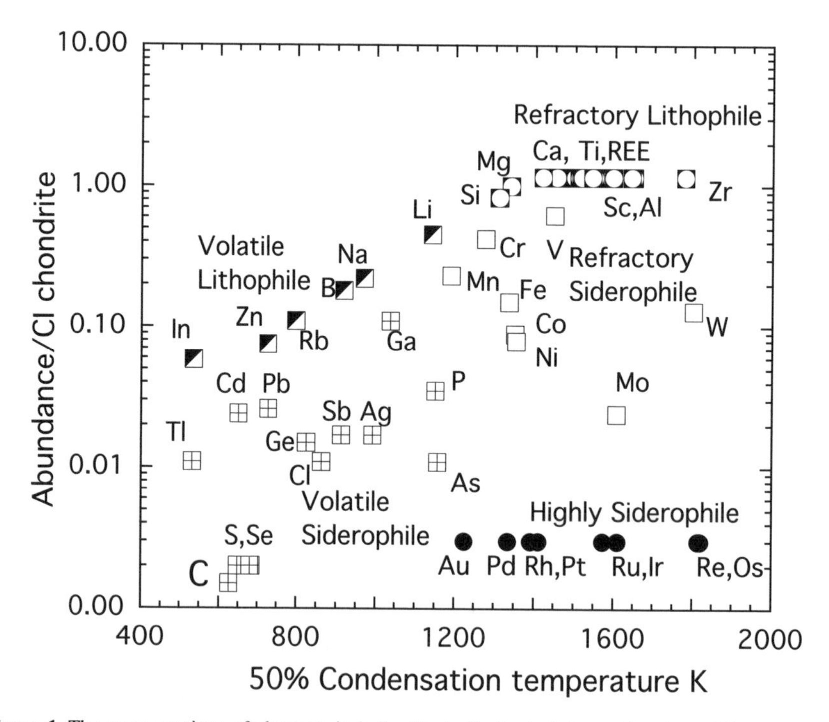

Figure 1. The concentrations of elements in bulk silicate Earth plotted as a function of the temperature at which 50% of the element would be condensed from a gas of solar composition (Lodders 2003). The elemental concentrations expressed as a ratio of abundance in Earth to CI chondrites have been normalized to Mg = 1.0.

The Si content of the core was estimated by Allègre et al. (1995) by assuming that Si, which is slightly depleted in silicate Earth, is in chondritic ratio to the bulk Earth contents of refractory lithophile elements. This plausible upper bound yields 7.3% Si in the core. In order to make an estimate of the S content of the core, it was assumed (Dreibus and Palme 1996) that the overall concentration of S in bulk Earth is in the same ratio to that in CI chondrites as elements of similar condensation temperature such as Zn. Then, by assuming that Zn is completely lithophile, Dreibus and Palme (1996) estimated that virtually all of Earth's S must be in the core and that, to give the whole Earth the same fraction of CI abundance as that of Zn, the core must contain about 1.7 wt% sulfur. The situation for carbon is more complex because it is much more difficult to estimate the bulk Earth concentration of this element than it is to estimate S abundance. The reason is that the condensation temperature of C is difficult to define and carbon behavior during accretion is also likely to have been quite complex. In Figure 1, we show a condensation temperature for carbon of 626 K, which is the estimate of Lodders (2003) based on an assumed equilibrium condensation of carbon to graphite but with kinetic inhibition of hydrocarbon formation. If complete equilibrium is assumed, with hydrocarbon species such as CH_4 forming, then the condensation temperature of carbon should correspond to that of methane ices at 41 K. Thus, there, is a wide range of potential condensation temperatures for carbon, which depend on the extent of kinetic inhibition of hydrocarbon formation, known to be significant, in the solar nebula. The result is that attempting to correlate carbon abundance with that of a lithophile element of "similar volatility" is impossible. Other cosmochemical arguments must be employed.

One of the short-lived radionuclides that was present at the beginning of the solar system, ^{107}Pd, decays to stable ^{107}Ag with a half-life of 6.5 m.y. Figure 1 shows that relatively volatile Ag is more abundant in silicate Earth, relative to CI chondrites, than is refractory Pd; that is, Pd/Ag of silicate Earth is subchondritic. This observation means that the $^{107}Ag/^{109}Ag$ ratio of silicate Earth would also be expected to be subchondritic since much of the current ^{107}Ag was produced from ^{107}Pd. In fact, despite the strong depletion of silicate Earth in Pd due to its highly siderophile nature, silicate Earth has a $^{107}Ag/^{109}Ag$ ratio the same as that of CI chondrites (Schönbächler et al. 2010). This similarity must mean that the Ag present in silicate Earth had its $^{107}Ag/^{109}Ag$ ratio established in a body (not silicate Earth) with chondritic Pd/Ag ratio (Schönbächler et al. 2010). In order to retain its chondritic $^{107}Ag/^{109}Ag$ ratio this silver must have mainly been added to Earth after ^{107}Pd was extinct so that no more ^{107}Ag was being produced. At this point, ~30 m.y. after the beginning of the solar system, Earth was already ~90% accreted. The implication is that Ag (and elements of similar volatility) was accreted to Earth late in accretionary history. Schönbächler et al. (2010) model the timing of this event as the first ~87% of accretion occurring from volatile-poor (Ag-poor) material and with the last 13% corresponding to volatile-rich (Ag-bearing) material perhaps related to the Moon-forming giant impact. This model explains the Ag isotopic composition of silicate Earth and would require that Ag and other volatile elements such as S and C were accreted from a volatile-rich (CI-like) impactor late in accretionary history. If this idea is correct, then we can estimate the C content of Earth by assuming that Earth has the C/S ratio of volatile-rich chondrites. This value would place an upper bound on the C content of the core of 1.1 wt% if we assume that the core contains 1.7 wt% sulfur and Earth has a C/S ratio the same as that of CI chondrites.

CARBON ISOTOPES AND CARBON CONTENT OF THE CORE

An additional constraint on the carbon content of the core can be gained by considering carbon isotope ratios in Earth and comparing them with other solar system bodies. Carbon has two stable isotopes ^{12}C (98.9%) and ^{13}C (1.1%). The $^{13}C/^{12}C$ ratio is a convenient way to express the separation of the two isotopes relative to an internationally accepted standard, Pee

Dee Belemnite (PDB). Deviations in this ratio are expressed in delta notation, where $\delta^{13}C = [(^{13}C/^{12}C)_{sample}/(^{13}C/^{12}C)_{PDB} - 1] \times 1000$.

Traditionally, carbon isotopes have been used to trace the movement and cycling of carbon between the atmosphere, oceans, and shallow subsurface environments. Experimental and theoretical work on isotope fractionation at high pressures and temperatures in the mantle has focused principally on the fractionation between graphite and diamond and a gas (such as CO_2; Bottinga 1969; Javoy et al. 1978). Since high temperatures cause decreases in equilibrium stable isotope fractionations, it was assumed for decades that carbon isotope fractionation under deep Earth conditions would be negligible. However, Peter Deines in an elegant paper (Deines 2002) summarized all the carbon isotope data to date and hypothesized that there might be "as yet unknown high-temperature fractionation" processes characterized by large fractionations.

The carbon isotopic signature of all whole rock mantle xenoliths and separated minerals is bimodal with sharp peaks at $\delta^{13}C = -5$ and -25‰ (Fig. 2). It has become accepted (see Deines 2002 for a review) that the peak at −5‰ represents the primitive mantle signature (that is, the signature that silicate Earth has had since its formation), and that the peak at −25‰ represents the incorporation of subducted organic material from the surface. Organic material on the surface is depleted in ^{13}C and through plate tectonic processes makes its way to the upper mantle, mixing with the carbon present, hence creating reservoirs with a −25‰ signature. Thus, it is generally accepted that bulk silicate Earth has a carbon isotope signature of about $\delta^{13}C = -5$‰ (Fig. 3). Figure 3 shows a comparison of silicate Earth carbon isotopic composition with those of other planetary and asteroidal bodies. As can be seen, all of the other solar system bodies for which we have data have more negative $\delta^{13}C$ than Earth. In Figure 3 Mars is represented by measurements of magmatic phases in the SNC meteorites (Grady et al. 2004), while Vesta is represented by measurements of HED meteorites (Grady et al. 1997). A large range in $\delta^{13}C$ is exhibited by chondrites, as shown in Figure 3 with values obtained from CO, CV, CI, and CM chondrites. We have excluded enstatite chondrites because of bulk compositional and Si isotopic evidence that they can only have provided a small fraction of the material present in bulk Earth (Fitoussi and Bourdon 2012). We also neglected the values for soluble organic matter and carbonates from the CI and CM chondrites, as they are a small percentage of the

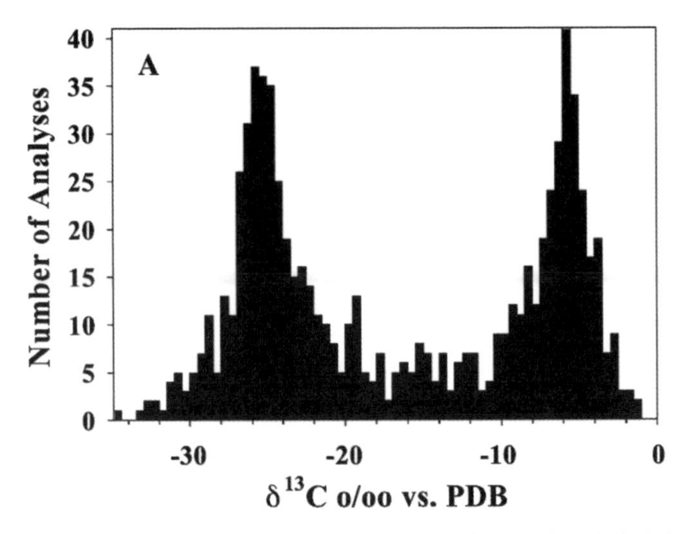

Figure 2. Figure from Deines (2002) illustrating the carbon isotopic fractionation of all whole rock mantle xenoliths and separated minerals from these xenoliths.

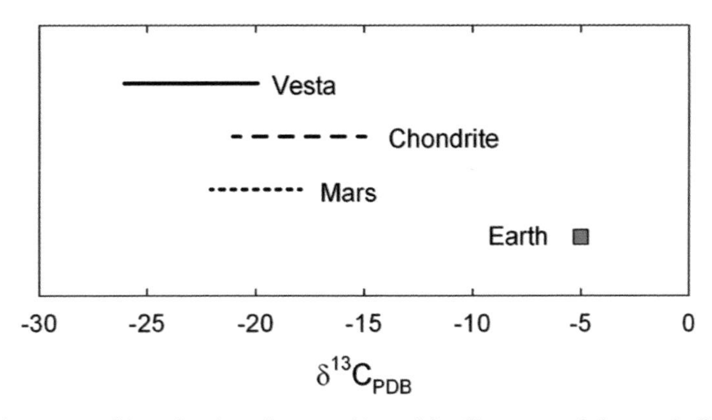

Figure 3. A summary of the carbon isotopic compositions of the silicate parts of planetary bodies as determined from HED meteorites (Vesta), SNC meteorites (Mars), and Earth compared to CI, CM, CO, and CV chondrites (Chondrite). See text for more detail.

carbon in meteorites and are not believed to be representative of the carbon-bearing material accreted to the Earth (Grady and Wright 2003).

It can readily be seen in Figure 3 that silicate Earth is an outlier, that is, that the supposed primitive Earth signature is in fact quite different from those of meteorites, and therefore of other planetary and asteroidal bodies. Current models of planet formation suggest that as they began to grow, Earth and the other terrestrial planets swept up volatile-poor material in fairly narrow feeding zones (Chambers 2005). Later in accretion, larger asteroidal and protoplanetary bodies from further out in the solar system were scattered into the inner solar system, causing violent impacts and delivering the more volatile elements to Earth and Mars (Chambers 2005). In support of these models, silver isotopic evidence, discussed above, suggests addition of CI-chondrite-like material to Earth late-on in its accretionary history (Schönbächler et al. 2010). A corollary of this accretionary scenario is that Earth, Mars, and the asteroids all received their volatiles, including carbon, from a similar source. This conclusion is in agreement with the suggestion that similarities between the isotopic compositions of H and N in Earth, Mars and chondritic meteorites indicate that these bodies all received their isotopic signatures from a common reservoir, which was isotopically well mixed (Marty 2012; Marty et al. 2013). We will therefore assume that the carbon isotopic compositions of Earth, Mars, Vesta, and chondrites were initially the same and that observed differences arose from differences in the accretionary processes on the different bodies.

The simplest way to explain the differences among Earth, Mars, Vesta, and chondrites is that Earth's primitive signature is not $\delta^{13}C = -5‰$ but similar to the other bodies at ~ $-20‰$ (Grady et al. 2004; Mikhail et al. 2010). The single largest event in Earth's history that changed its composition significantly is core formation (Fig. 1) and carbon is a strongly siderophile element. While Earth, Mars, and Vesta have all undergone core formation, and therefore might be expected to show similar carbon isotopic fractionations, there are important differences in the conditions attending their differentiation that could cause differences in the compositions of their cores. We will discuss these differences after we have addressed the potential for carbon isotopic fractionation due to core formation on Earth. Previous work on silicon (Georg et al. 2007; Shahar et al. 2011) has shown that core formation can impart a stable isotope signature to the planet, even at the very high temperatures necessary for segregation of liquid metal from liquid silicate. So would it be plausible that core formation could change the carbon isotopic signature of the entire silicate part of the planet by 15‰? And if so, what would that mean for the composition of the core?

In order to estimate the effect core formation could have on Earth's carbon isotope budget, it is necessary to know the isotopic fractionation factor for carbon partitioning between liquid metal and liquid silicate. In most cases of interest the fractionation factor has been measured experimentally, or can be estimated from natural samples or from theoretical calculations. However, for carbon none of these avenues is currently available: high-temperature experiments have only been done in the Fe-C system (Satish-Kumar et al. 2011); theoretical calculations have only been published for graphite, diamond, and carbon dioxide (Bottinga 1969; Polyakov and Kharlashina 1995); and iron meteorites show only a high-temperature carbon isotope fractionation in the Fe-C system. Therefore, due to the absence of data, we will estimate the potential for isotopic fractionation by using the Fe-C system as an analog for core formation.

Satish-Kumar et al. (2011) determined the carbon isotope fractionation between molten iron-carbon alloy and graphite as a function of temperature—the first study to address experimentally high-temperature (>1300 °C) carbon isotope fractionation. In this study, the authors find that ^{12}C preferentially incorporates into the metallic phase, leaving a ^{13}C-enriched signature in the graphite or diamond (depending on the pressure of the experiment). The fractionation factor $\Delta^{13}C$ follows the normal temperature dependence:

$$\Delta^{13}\mathrm{C}(_{gr/dia\text{-FeC}}) = 8.85\left\{\frac{10^6}{T^2}\right\} + 0.99 \quad (1)$$

where

$$\Delta^{13}\mathrm{C}(_{gr/dia-\mathrm{FeC}}) = \delta^{13}C_{gr/dia} - \delta^{13}C_{metal} \quad (2)$$

Iron meteorite data on a similar system reflect the same fractionation direction (Deines and Wickman 1975), where graphite is more enriched in ^{13}C than coexisting cohenite [$(Fe,Ni)_3C$]. These studies imply that if Earth has carbon in its core, it is enriched in ^{12}C, and has created a higher $^{13}C/^{12}C$ ratio in the mantle during core formation (Grady et al. 2004; Mikhail et al. 2010). This fractionation is in the right direction to explain an Earth that began with a $\delta^{13}C$ similar to other planetary bodies (−20‰) and evolved to −5‰ after core formation. There are two possible end-member models for core formation and carbon segregation to the metal: (1) A single stage of mantle-core equilibrium with initial $\delta^{13}C$ of the system of −20‰, and (2) Continuous core formation from the same system with Rayleigh isotopic fractionation.

The single stage model (1) leads to the following mass balance:

$$\delta^{13}\mathrm{C}_{BSE}X_\mathrm{C} + \delta^{13}\mathrm{C}_{core}(1 - X_\mathrm{C}) = \delta^{13}\mathrm{C}_{bulk\ earth} \quad (3)$$

where $\delta^{13}C_{BSE}$ is the carbon isotopic composition of bulk silicate Earth, X_C is the fraction of Earth's carbon that is in the silicate portion of the planet, $\delta^{13}C_{core}$ is the isotopic composition of the core, and $\delta^{13}C_{bulk\ earth}$ is the isotopic composition of chondrites (−20‰). It is easily shown that ($\delta^{13}C_{BSE} - \delta^{13}C_{core}$) must be greater than 15‰ in order for this mass balance to work with a BSE value of −5‰ and an initial bulk Earth value of −20‰. According to the measured fractionations of Equation (1), an isotopic fractionation factor greater than 15‰ could only apply at temperatures below 800 K, which are implausibly low for liquid metal coexisting with liquid silicate. Therefore, a single stage of segregation can be excluded unless it is found that metal-silicate carbon isotope fractionations are much larger than those in the Fe-C system.

For model (2), continuous core extraction with Rayleigh fractionation, the relevant equation is:

$$\delta^{13}\mathrm{C}_{BSE} - \delta^{13}\mathrm{C}_{bulk} = \Delta^{13}\mathrm{C}(_{core\text{-}BSE})\ln F \quad (4)$$

where F is the fraction of the original carbon remaining in the mantle. We obtain an F of 0.05 if the fractionation factor on the right hand side is −5‰ and an F of 0.014 if the fractionation factor is −3.5‰. Equation (1) (note the change of sign in $\delta^{13}C$) indicates that this would correspond to temperatures of 1500-1900 K, which are plausible if rather low given the high temperatures (~2500-3500 K) inferred by many authors to have attended terrestrial core formation (Wood et al. 2006; Rudge et al. 2010; Rubie et al. 2011). The result in terms of C content of the core varies widely because of the range of estimates of C content of bulk silicate Earth. The value of 120 ppm given by McDonough and Sun (1995), for example, leads to a core carbon content of 0.5-2.0 wt%, which is in reasonable accord with the cosmochemically derived value. However, a more recent estimate of carbon content of bulk silicate Earth of 765±300 ppm (Marty 2012; see also Dasgupta 2013) leads, assuming Rayleigh fractionation, to implausibly large C contents of the core (up to 15%). In the latter case a hybrid Rayleigh-equilibrium model would be required to yield lower, more plausible, carbon contents of the core. We conclude that the carbon isotopic composition of silicate Earth is consistent with core formation having fractionated carbon isotopes but that it is not currently possible to constrain the processes involved.

Given the hypothesis of a large shift in $\delta^{13}C$ of BSE from −20‰ to −5‰ due to core formation, we must now consider why Vesta and Mars, which both have cores, do not also exhibit shifts in their carbon isotopic signatures relative to the chondritic reference. The answer for Vesta may simply be a matter of the pressure of core formation. In order for significant amounts of C to dissolve in the liquid metal forming the core, oxygen fugacities must be close to (i.e., not too far above) those for equilibrium with graphite or diamond, otherwise all carbon will be present as oxidized species CO and CO_2. At atmospheric pressure equilibrium between graphite and C-O-H-S gas requires oxygen fugacities at least 6 $\log f_{O_2}$ units below the Fe-FeO (IW) buffer. Under these conditions, silicate melt coexisting with metal contains only a few hundred ppm of FeO (Wood BJ, unpublished data). The fact that Vesta was much more oxidized than this during differentiation is apparent from the FeO contents of eucrites, which are 2 orders of magnitude richer in FeO, corresponding to at least 4 log units higher in oxygen fugacity and hence far above graphite saturation. With increasing pressure graphite equilibrium moves to higher relative oxygen fugacities so that at 1.5 GPa it lies 3-4 log units above IW (Wood et al. 1990). However pressures within Vesta are only a few atmospheres so the low-pressure data are relevant and we can exclude conditions close to carbon saturation during core formation. Carbon should not, therefore, have been strongly partitioned into Vesta's core, an observation that also explains the low C contents of magmatic iron meteorites. Silicate Vesta then should not show any carbon isotopic difference from chondrites, as observed. The situation is much more complex for Mars because pressures within Mars during core formation must have been sufficiently high to stabilize graphite or diamond at oxygen fugacities above the IW buffer, which is the maximum conceivable during core segregation. If the explanation for silicate Earth's isotopic signature is correct then either Mars had core formation close to model (1), single-stage equilibrium, or most of the carbon in Mars' mantle was added to the planet after core formation had ceased—the so-called "late veneer." W isotopic data for Mars indicates that this planet grew very rapidly, being 50% accreted within 1.8 m.y. (Dauphas and Pourmand 2011), so single-stage core formation may approximate the differentiation process on Mars, but not on Earth, which grew 10 times more slowly (Kleine et al. 2002). In addition, recent data on the highly siderophile elements show that a "late veneer" of chondritic material after cessation of core formation occurred on Mars as well as Earth (Dale et al. 2012) so it is possible that this late veneer dominates carbon in silicate Mars, but not in the more slowly-accreting silicate Earth. In either case it is clear that the carbon present in silicate Mars has a very different history from that found in silicate Earth.

From the above discussion we can conclude that the carbon isotopic signature of silicate Earth is consistent with Earth's core containing ~1 wt% carbon, provided the silicate-metal fractionation factors for this element are close to those of graphite-liquid Fe carbide and that the core was extracted from the mantle under conditions approaching Rayleigh fractionation.

DENSITY AND PHASE DIAGRAM CONSTRAINTS ON THE CARBON CONTENT OF THE CORE

The preceding discussion, based on the estimated S content of the core and the isotopic composition of carbon in silicate Earth, leads us to the conclusion that the core may contain ~1% C. The question now is whether or not concentrations at this level are testable seismically. The presence of low atomic number elements in Earth's core was initially inferred from seismic observations and mineral physics measurements. These data showed that the core is less dense than pure iron under the corresponding pressure-temperature conditions by 5-8% in the outer liquid part and 2-5% in the inner solid part (Birch 1952; Stevenson 1981; Jeanloz 1990; Poirier 1994; Hillgren et al. 2000; Anderson and Isaak 2002; Li and Fei 2003; Komabayashi and Fei 2010).

Based on long extrapolations of the equation of state data available at the time, Wood (1993) proposed that Fe_3C was likely to be the first phase to crystallize out of an Fe-S-C liquid under core conditions, even for C contents <1% and hence could form the inner core. Recent experimental work on the Fe-C phase diagram, however, suggests that Fe_7C_3 would be a stronger candidate for the inner core than Fe_3C since it probably replaces Fe_3C to become the liquidus phase at core pressures (Lord et al. 2009; Nakajima et al. 2009; Oganov et al. 2013). An inner core dominated by iron carbide (either Fe_3C or Fe_7C_3) would be the largest reservoir of carbon in Earth, dwarfing the combined budget of known carbon in the atmosphere and crust by an order of magnitude. In order to test the hypothesis of a carbide-rich inner core we need, of course, to know the liquidus phase of the Fe-C, Fe-C-S-(Si) and related systems at core pressures.

The Fe-C phase diagram

Wood (1993) predicted that the eutectic composition in the system Fe-C becomes more iron-rich with increasing pressure to approach the iron end member at inner core pressures. As pressure increases, the stability field of Fe_3C was expected to expand relative to iron, with its melting point increasing more rapidly to substantially exceed that of iron at core pressures. Consequently, Fe_3C was predicted to be the first phase to crystallize from a liquid core containing sulfur and a small amount of carbon (< 1 wt%). This liquidus behavior is the basis of the prediction of carbide as the principal phase in the inner core.

The phase diagram of the Fe-C system at 1 atmosphere has been extensively studied because of its applications in the steel industry (Chipman 1972; Zhukov 2000). The binary system has a eutectic point between iron and Fe_3C at 4.1 wt% carbon (Fig. 4).

A number of experimental studies have aimed at measuring the effect of pressure on the eutectic composition in the Fe-C system using the large volume press apparatus. Some, but not all, of the data support Wood's (1993) prediction that the eutectic composition becomes poorer in carbon with increasing pressure (Hirayama et al. 1993; Nakajima et al. 2009). For example, the reported eutectic composition near 5 GPa is at 3.2 wt% (Strong and Chrenko 1971), 3.6 wt% (Nakajima et al. 2009), and 4.8 wt% carbon (Chabot et al. 2008), as opposed to 4.1% at atmospheric pressure. A much higher pressure study of Lord et al. (2009), however, supports the predicted shift in the eutectic composition with pressure. Lord et al. (2009) used a novel X-ray radiographic technique to measure the eutectic melting behavior of Fe-Fe_3C mixtures up to 70 GPa. They found a rapid drop in the carbon content of the eutectic composition at pressures above 20 GPa, based on which they predicted that the eutectic composition is nearly pure iron by ~50 GPa. Extrapolation of the eutectic data beyond 70 GPa can be inferred to indicate a shift back in the eutectic composition to higher carbon contents (Lord et al. 2009), but this estimate is speculative. What does appear to be established is that there is a trend to ~50 GPa of increasing stability of the carbide at a eutectic of decreasing carbon contents.

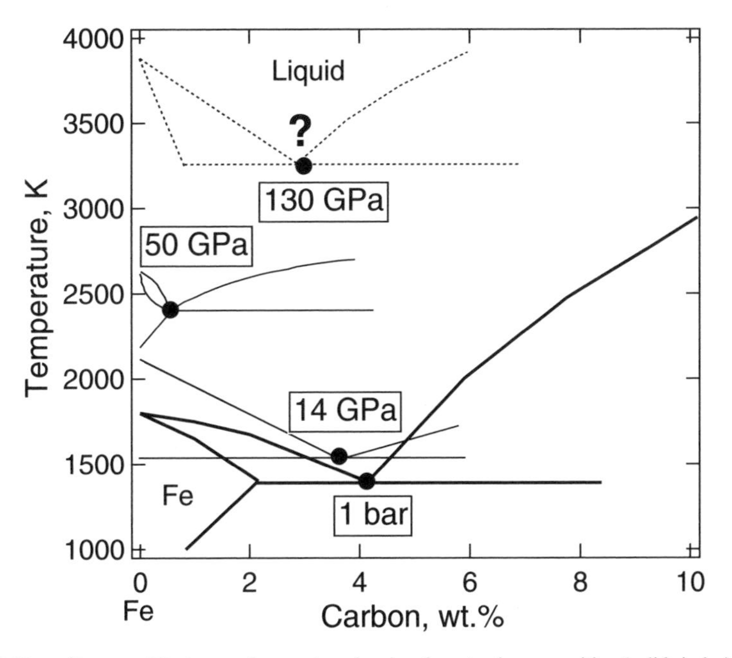

Figure 4. Phase diagram of the iron-carbon system showing the eutectic composition (solid circles) at different pressures. Note the decrease in observed carbon content at the eutectic to 50 GPa and that the 130 GPa diagram is an extrapolation.

There are strong interactions between carbon and sulfur dissolved in liquid Fe such that there is an extremely large miscibility gap in the Fe-S-C liquid system at 1 atm. Although this miscibility gap shrinks with increasing pressure up to 20 GPa (Wang et al. 1991; Corgne et al. 2008; Dasgupta et al. 2009), the net effect of adding sulfur is to increase the activity coefficient for C dissolved in Fe-rich liquids (Wood 1993). This change means that addition of S should decrease the carbon contents of liquids saturated in Fe_3C or Fe_7C_3. The implication is that, at core pressures, metallic liquids containing several % S may crystallize carbide first even if the carbon content of the liquid is very low (< 1%).

Densities of iron carbides

As discussed above, the presence of low atomic number ("light") elements in Earth's core was initially inferred from seismic observations and mineral physics measurements. Although a wide range of mixtures of Si, O, C, and S have been proposed as possible constituents of the "light" element in the outer core, the inner core is most likely to be an Fe-alloy or a compound of Fe with one of these low atomic number elements. The suggestion by Wood (1993) that Fe_3C would be the stable liquidus phase at very low C contents of the liquid outer core and that its density should be close to that observed, has stimulated a number of studies of the effects of compression on densities and elastic properties of iron carbides.

The natural form of Fe_3C, cohenite, occurs in iron meteorites (Ringwood 1960; Hazen et al. 2013). At 1 atm and 300 K, Fe_3C has an orthorhombic structure (space group *Pnma*, *Z* = 4). Its density (7.68 g/cc) is 2.5% smaller than that of iron in the bcc structure (7.88 g/cc), corresponding to ~ 0.4% density reduction for each 1 wt% carbon. Synchrotron-based X-ray diffraction (XRD) studies have shown that the crystal structure of Fe_3C remains stable to at least 187 GPa and 1500 K (Rouquette et al. 2008; Ono and Mibe 2010; Sata et al. 2010) and possibly to 356 GPa and 5520 K (Tateno et al. 2010).

On the basis of estimated equation of state (EOS) parameters, Wood (1993) suggested that Fe_3C approaches the observed density of the inner core under the appropriate high-pressure and high-temperature conditions. Compression measurements (Scogt et al. 2001; Li et al. 2002) support Wood's estimates, with best-fit isothermal EOS parameters of K_0 = 174 ± 6 GPa, K_0' = 4.8 ± 0.8 (using a neon pressure medium to 32 GPa), and K_0 = 175 ± 4 GPa, K_0' = 5.2 ± 0.3 (using a methanol-ethanol-water mixture combined with laser annealing). In contrast, based on first-principles calculations, Vočadlo et al. (2002) excluded Fe_3C as a major inner-core-forming phase. Their calculations suggest that Fe_3C transforms from ferromagnetic to non-magnetic at ~ 60 GPa and 0 K (Vočadlo et al. 2002; Mookherjee 2011; Oganov et al. 2013). A pressure-induced loss of magnetism at 300 K was also observed at ~ 25 GPa using X-ray emission spectroscopy (XES; Lin et al. 2004), at 9 GPa using X-ray magnetic circular dichroism (XMCD; Duman et al. 2005), and at 20-30 or 4.3-6.5 GPa using synchrotron Mössbauer spectroscopy (Prakapenka et al. 2004; Gao et al. 2008). The magnetic collapse was found to cause abrupt reduction in volume (Ono and Mibe 2010) and compressibility (Vočadlo et al. 2002; Lin et al. 2004; Sata et al. 2010), which may make it difficult to reconcile the density and elastic properties of Fe_3C with those of the inner core. It should be noted, however that the pressure of magnetic collapse in Fe_3C is highly uncertain, ranging from ~5 GPa to ~55 GPa (Lin et al. 2004; Duman et al. 2005; Gao et al. 2008; Ono and Mibe 2010). The discrepancies likely arise from the different methods used to detect the transition. Alternatively, it may reflect variable amounts of non-hydrostatic stress in samples compressed in neon, NaCl, or no pressure medium. Furthermore, as can be seen from Figure 5, the equation of state of Fe_3C has been determined to 200 GPa

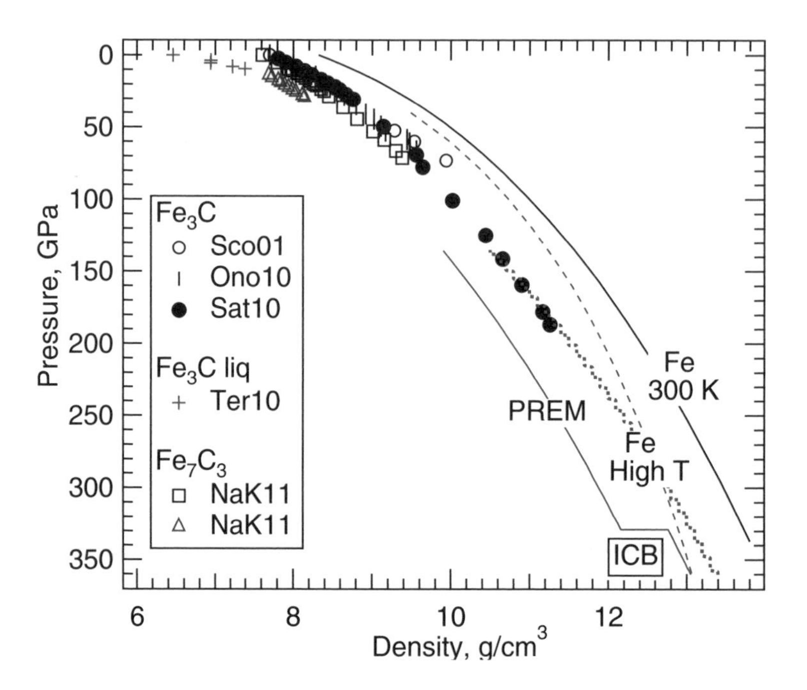

Figure 5. Densities of iron, iron carbides, and Earth's core as a function of pressure. Data sources: PREM (Dziewonski and Anderson 1981). Dashed line: Shocked Fe along a Hugoniot (Brown 2001). Solid line: statically compressed hcp Fe at 300 K (Mao et al. 1990). Note slightly higher densities in a more recent paper (Dewaele et al. 2006). Dotted line: Statically compressed hcp Fe at 5000 K (Dubrovinsky et al. 2000; Komabayashi and Fei 2010). Solid Fe_3C at 300 K (Ono and Mibe 2010; Sata et al. 2010; Scott et al. 2001) (vertical bars, solid circles and open circles respectively). Liquid Fe_3C (Terasaki et al. 2010). Solid Fe_7C_3 at 300 K (squares) and high temperature (triangles) (Nakajima et al. 2011).

at 300K (Sata et al. 2010) and the change in compressibility and volume associated with the suggested magnetic transition was found to be very small. Comparison of the density of Fe_3C with that of the inner core based on the EOS parameters of Sata et al. (2010) indicate that Fe_3C cannot be excluded as the principal inner core phase (Fig. 5).

Based on phase equilibrium measurements, it has recently been suggested that Fe_7C_3 is a more likely candidate for Earth's inner core than is Fe_3C (Lord et al. 2009; Nakajima et al. 2009). At 1 atm and 300 K, Fe_7C_3 adopts a hexagonal or orthorhombic structure (Fang et al. 2009; Oganov et al. 2013). In the more stable hexagonal structure (space group $P6_3mc$, $Z = 2$), the density of Fe_7C_3 (7.61 g/cc) is 3.4% smaller than bcc iron (7.88 g/cc), corresponding to ~0.4% density reduction for each 1 wt% carbon, which is nearly identical to that of Fe_3C. A synchrotron X-ray diffraction study found that the crystal structure of Fe_7C_3 remains stable up to 71.5 GPa and 1973 K, although anomalous compression behavior was observed at 18 GPa and 300 K, which is attributed to a ferromagnetic to paramagnetic transition (Nakajima et al. 2011). A theoretical study predicted magnetic collapse at ~67 GPa, causing a small increase in the bulk modulus (Mookherjee et al. 2011). A consideration of the measured densities of Fe_7C_3 (Fig. 5) show, however that, like Fe_3C, this phase could have the appropriate density for the inner core under inner core conditions.

In summary, EOS data available to date indicate that either Fe_3C or Fe_7C_3 could have the observed density of the inner core under inner core conditions. In order to discriminate between these and other phases it will be necessary to obtain density data under pressure-temperature conditions closer to those of the inner core. It should be noted however, that, even if carbides are shown not to be present in the inner core, there could still be substantial amounts of carbon in the liquid outer core.

Sound velocities of Fe, Fe_3C and those of the inner core

Apart from the density of the inner core, the solid phase or phases present must have appropriate sound velocities and anisotropy. Compared to the Preliminary Earth Reference Model (PREM) the compressional wave velocity (v_P) of Fe is somewhat higher and has a greater dependence on density while the shear wave velocity (v_S) is much higher but has a similar dependence on density to Earth (Fig. 6). In addition, seismic studies have shown that compressional waves travel through the inner core at velocities that are faster by 3-4% in the polar direction than in the equatorial plane (Creager 1992; Tromp 1993). Recent observations suggest shear-wave velocity in the inner core is also anisotropic (Wookey and Helffrich 2008). A candidate inner core component must therefore match the depth-dependent sound velocities from Earth's center to the inner core radius and also account, in large measure, for the compressional- and shear-wave anisotropy in the inner core. At ambient conditions the v_S of Fe_3C is 3.0 km/s according to ultrasonic interferometry and nuclear inelastic X-ray scattering (NRIXS) measurements (Dodd et al. 2003; Gao et al. 2008). The results on v_P vary from 5.2 km/s (ultrasonic; Dodd et al. 2003), 5.89 km/s (NRIXS; Gao et al. 2008), to 6.10 km/s by inelastic X-ray scattering (IXS; Fiquet et al. 2009). For comparison, the v_P and v_S of bcc Fe are 5.8 and 3.1 km/s, respectively (Lübbers et al. 2000; Fiquet et al. 2001; Mao et al. 2001).

At high pressures and 300 K, NRIXS data up to 50 GPa on Fe_3C show that v_P and v_S increase linearly with density (Gao et al. 2008). Upon magnetic transition near 5 GPa, the slope for v_S becomes significantly shallower, however, leading to an extrapolated v_S that is considerably smaller than that of iron under inner core pressures and plausibly in better agreement with PREM (Fig. 6). Furthermore, one set of measurements on Fe_3C to 47 GPa and 1450 K shows that, at high temperature v_S deviates from the linear relationship at 300 K towards lower values, potentially matching the anomalously low v_S in the inner core under the relevant pressure and temperature conditions (Gao et al. 2011). The results provide further support for Fe_3C as a major component of the inner core. In addition, experimental and computational data on Fe_3C

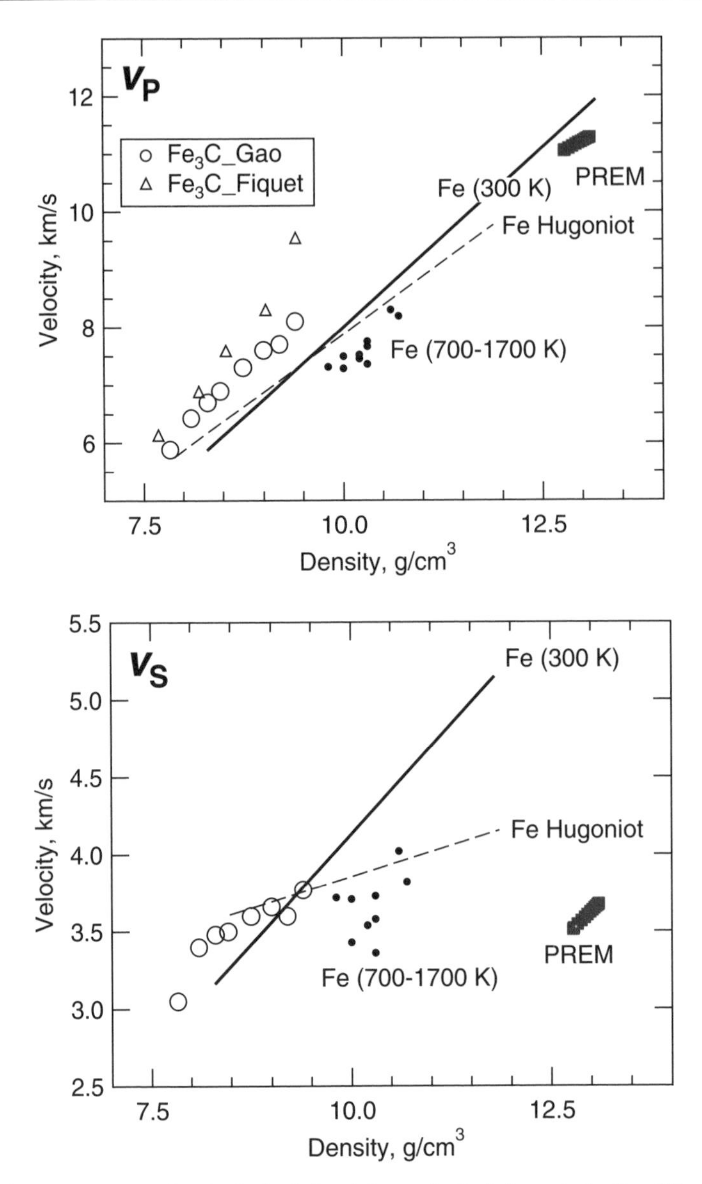

Figure 6. Aggregate compressional velocity (v_P) and shear velocity (v_S) versus density of Fe and Fe_3C in comparison with the PREM model. Data sources: Dashed line refers to shocked Fe along a Hugoniot (Duffy and Ahrens 1992). Statically compressed Fe at 300 K (solid line; Mao et al. 2001). Statically compressed Fe at high temperature (solid circles, Lin et al. 2005). NRIXS data on Fe_3C at 300 K (open circles, Gao et al. 2008, 2009). IXS data on Fe_3C at 300 K (triangles, Fiquet et al. 2009). PREM (Dziewonski and Anderson 1981).

indicate strong anisotropy in sound velocity under ambient conditions, providing yet another piece of supporting evidence for Fe_3C as a plausible candidate for the inner core (Nikolussi et al. 2008; Gao et al. 2009). In contrast to these results, however, the v_P derived from IXS data up to 68 GPa are considerably larger than the NRIXS results, suggesting that 1 wt% carbon in

Fe-Ni alloy (as opposed to 6.67% carbon in Fe_3C) could be sufficient to explain the difference in compressional wave velocity between PREM and experiments (Fiquet et al. 2009).

In summary, phase equilibrium experiments are consistent with crystallization of carbide (Fe_3C or Fe_7C_3) on the liquidus at very low C contents of a Fe-C liquid at pressures of 50 GPa and higher (Fig. 4). The stability of carbide or of C-bearing Fe alloy on the liquidus is enhanced by the addition of sulfur to the core because of the raised activity coefficient of C when S is added to the Fe-rich metallic liquid. Although density data to core pressures, available v_P and v_S data, and measured elastic anisotropy of Fe_3C are all broadly consistent with the inner core being dominantly Fe_3C or Fe_7C_3, some v_P data are more consistent with an Fe-Ni alloy containing a smaller amount of carbon. More high-temperature data at pressures approaching those of the inner core are required to refine these conclusions further.

CARBON IN THE CORE AND SIDEROPHILE ELEMENTS IN THE MANTLE

A final way of constraining the concentrations of "light" elements in the core is to return to the observation that, as the core was extracted from the mantle, siderophile elements were partitioned between metal and silicate according to their partition coefficients D_i ($D_i = [i]_{metal}/[i]_{silicate}$). Since alloys of iron metal with carbon, sulfur, silicon and other light elements are non-ideal solutions, the identities and concentrations of the light elements in the core will have, in some cases, marked effects on how strongly siderophile elements have partitioned into the core. It has been shown, for example, that addition of 4-5% carbon to liquid Fe changes the behavior of Pb from siderophile to lithophile because of strong Pb-C interactions in the metal (Wood and Halliday 2010). In contrast, W has a strong affinity for small amounts of C dissolved in the metal and becomes more siderophile as the C content of the metal increases (Wade et al. 2012). These effects have the potential to enable us to constrain the light element concentrations in the core from the concentrations of siderophile elements in the mantle. Wade et al. (2012) have shown, for example, that addition of sulfur to the metal has the potential to change the relative behavior of Mo and W from the prediction that W was more strongly partitioned into the core than Mo to the observed behavior, which implies that Mo was more siderophile during core formation than W. We consider now whether or not similar arguments can be used to constrain the C content of the core.

In order to consider the effects of S on the partitioning of Mo and W between the core and the mantle, we used a simple model of Earth accretion initially proposed several years ago (Wade and Wood 2005). Wade and Wood (2005) used a model of continuous accretion and core segregation in which every increment of metal added to the core equilibrated with a well-mixed mantle reservoir. They then showed that the experimental data on metal-silicate partitioning of Ni, Co, V, Mn, and Si would result in the observed concentrations of these elements in the mantle provided pressures of metal-silicate equilibration increased and Earth became progressively more oxidized as it grew. The model has been refined several times, most recently by addition of terms for the interactions between Si and the other elements dissolved in the metal (Tuff et al. 2011) and by the inclusion of a comprehensive database for Mo and W (Wade et al. 2012). As a basis we use the version of the accretionary model presented by Tuff et al. (2011), which makes explicit provision for Silicon-*i* interactions in the metal where *i* is any minor element of interest. This model generates the observed partitioning behavior of Co, Ni, W, Cr, V, Nb, and Si as the pressure of metal segregation increases from 0 to 36 GPa and Earth becomes more oxidized.

In order to expand the accretionary model to take account of the effects of S and C on the composition of the mantle and core, we have adopted Sulfur-*i* and Carbon-*i* interactions from the Steelmaking data sourcebook (Steelmaking 1988), which are used in conjunction with the epsilon model of activity coefficients in the metal:

$$\ln \gamma_{\mathrm{Fe}}^{met} = \sum_{i=2}^{N} \varepsilon_i^i (x_i + \ln(1-x_i)) - \sum_{j=2}^{N-1} \sum_{k=j+1}^{N} \varepsilon_j^k x_j x_k \left(1 + \frac{\ln(1-x_j)}{x_j} + \frac{\ln(1-x_k)}{x_k}\right)$$
$$+ \sum_{i=2}^{N} \sum_{\substack{k=2 \\ k \neq i}}^{N} \varepsilon_i^k x_i x_k \left(1 + \frac{\ln(1-x_k)}{x_k} - \frac{1}{1-x_i}\right) + \frac{1}{2} \sum_{j=2}^{N-1} \sum_{k=j+1}^{N} \varepsilon_j^k x_j^2 x_k^2 \left(\frac{1}{(1-x_j)} + \frac{1}{(1-x_k)} - 1\right)$$
$$- \sum_{i=2}^{N} \sum_{\substack{k=2 \\ k \neq i}}^{N} \varepsilon_i^k x_i^2 x_k^2 \left(\frac{1}{1-x_i} + \frac{1}{1-x_k} + \frac{x_i}{2(1-x_i)^2} - 1\right) \tag{5}$$

and

$$\ln \gamma_i^{met} = \ln \gamma_{\mathrm{Fe}}^{met} + \ln \gamma_i^{\circ} - \varepsilon_i^i \ln(1-x_i)$$
$$- \sum_{\substack{k=2 \\ k \neq i}}^{N} \varepsilon_i^k x_k \left(1 + \frac{\ln(1-x_k)}{x_k} - \frac{1}{1-x_k}\right)$$
$$+ \sum_{\substack{k=2 \\ k \neq i}}^{N} \varepsilon_i^k {x_k}^2 x_i \left(\frac{1}{1-x_i} + \frac{1}{1-x_k} + \frac{x_i}{2(1-x_i)^2} - 1\right) \tag{6}$$

In Equations (5) and (6), x_i refers to the mole fraction of i in the metal, γ_i^{met} is the activity coefficient of i, γ_i° is the activity coefficient of i infinitely dilute in liquid Fe, and ε_i^j is the interaction parameter between elements i and j. The interested reader may follow our activity calculations using the online calculator at *http://www.earth.ox.ac.uk/research/groups/experimental_petrology/tools*. All relevant ε_i^j terms are given by Wade et al. (2012) except for those involving carbon. The later, taken from the steelmaking data sourcebook are $\varepsilon_{\mathrm{C}}^{\mathrm{Mo}}$ of −6.15 and $\varepsilon_{\mathrm{C}}^{\mathrm{W}}$ of −6.64. We incorporated these values, together with those of the other elements of interest, in the model and found that the effects of adding carbon on core-mantle partitioning of Co, Ni, Cr, V, Nb, and Si are quite small but that carbon can have large effects on W and Mo. We therefore concentrated on these two elements and made versions of the model, all of which generated the correct mantle concentrations of Co, Ni, Cr, V, Nb, and Si.

For partitioning of W and Mo between a liquid peridotitic mantle and liquid Fe-rich metal we used the expressions given by Wade et al. (2012):

$$\log\left(K_D^{\mathrm{W}}\right)_{wt} = 1.85 - \frac{6728}{T} - \frac{77P}{T} \quad \pm 0.24 \tag{7}$$

The uncertainty corresponds to one standard error of the fit to a large number of experiments. Core-mantle partitioning (D_{W}) for tungsten is defined relative to that for Fe (well-known for Earth as a whole) in the expression for K_D:

$$\left(K_D^{\mathrm{W}}\right)_{wt} = \frac{\left(D_{\mathrm{W}}^{wt} \gamma_{\mathrm{W}}^{met}\right)}{\left(D_{\mathrm{Fe}}^{wt} \gamma_{\mathrm{Fe}}^{met}\right)^3}$$

Note that, for pure Fe $\gamma_{\mathrm{Fe}}^{met}$ is 1.0 while $\gamma_{\mathrm{W}}^{met}$ is 3.0 (at 1873 K; *http://www.earth.ox.ac.uk/research/groups/experimental_petrology/tools*). The D_{Fe} term is cubed to take account of a +6 oxidation state for W in the silicate.

For Mo the corresponding equation is:

$$\log K_D^{\mathrm{Mo}} = 1.44 - \frac{143}{T} - \frac{167P}{T} \quad \pm 0.19 \tag{8}$$

Where in this case the D_{Fe} term in the denominator of K_D is squared to take account of a +4 oxidation state of Mo in the silicate and γ_{Mo}^{met} in pure Fe is 1. Note that the oxidation states account for the partitioning behavior of Mo and W as a function of oxygen fugacity at oxygen fugacities down to at least 3.3 $\log f_{O_2}$ units below the IW buffer.

Figure 7 shows the results of the calculated effects of sulfur on the overall partitioning of W and Mo between mantle and core. As stated above we used the accretionary model described by Tuff et al. (2011) and Wade et al. (2012) with increasing oxidation state of silicate Earth as the planet grew. Sulfur was added to the core only during the last 13% of accretion in order to be consistent with the results of Schönbächler et al. (2010). As can be seen from Figure 7, addition of 2% S (rounded-up from the cosmochemical estimate of 1.7%) to the core (as 15% S in the last 13% of accreted metal) dramatically increases partitioning of Mo into the core and pushes calculated partitioning towards the range of probable partitioning behavior in Earth. Sulfur has little effect on W partitioning. Addition of carbon (Fig. 8) enhances partitioning of both Mo and W into the metal, such that ~0.6% carbon in the metal (as 5% C in the last 13% of segregated core) would enable the core-mantle partitioning of both Mo and W to reach the expected values, consistent with the observed concentrations of these elements in silicate Earth. Addition of 1% carbon would, however take both elements out of the expected range of core-mantle partitioning. Nevertheless, values > 0.5% carbon in the core are possible provided accretion of C was more protracted, over the last 25-30% of accretion for example.

CONCLUSIONS

We have used several distinct approaches to the problem of constraining the carbon content of Earth's core. The cosmochemical approach is firstly to consider the sulfur content of the core.

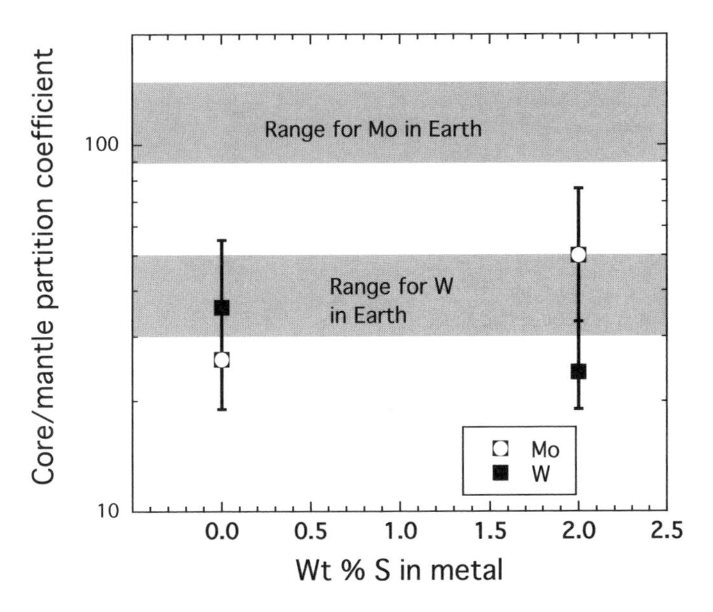

Figure 7. This figure illustrates the effect of adding 2% S in the metal on the core-mantle partitioning of W and Mo. The calculations are based on an accretionary model, which correctly reproduces the observed core-mantle partitioning of V, Cr, Nb, Co, Ni and Si (see text). S, which affects W and Mo much more than the other elements, was incorporated into the metal during the last 13% of accretion, consistent with Ag isotopic data on silicate Earth (Schönbächler et al. 2010).

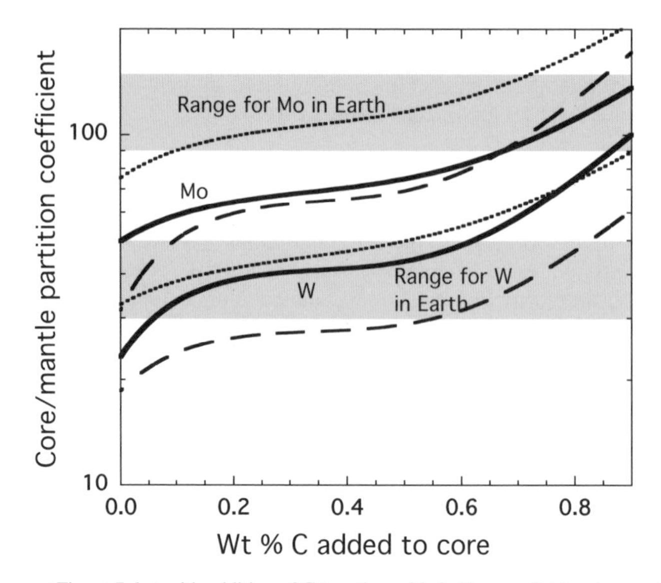

Figure 8. Same as Figure 7, but with addition of C together with S. Upper solid line is overall core/mantle partitioning of Mo and lower solid line equivalent partitioning of W. Short dashed and long dashed lines are uncertainty limits based on experimental uncertainties (Eqns. 7 and 8). In order to match observed core-mantle partitioning of Mo and W we need to add ~0.6% C to the core during the last 13% of accretion. The amount of C added can be increased if accretion of S and C is extended over the last 25-30% of accretion rather than the final 13% (see text).

Dreibus and Palme (1996) noted the decrease in abundance of lithophile elements in silicate Earth with decreasing condensation temperature (Fig. 1). They then correlated the observed silicate Earth abundance of S with that of Zn, a relatively lithophile element of virtually identical condensation temperature to S. Then, by assuming that S should, in the absence of a core, be present in the same ratio to Zn in the silicate Earth as in CI chondrites, they calculated the amount of sulfur "missing" from silicate Earth. This "missing" S was assumed to be present in the core, giving a concentration in the latter reservoir of 1.7 wt%. Our cosmochemical estimate of a maximum of 1.1% carbon in the core was obtained from this S concentration by noting that the Ag isotopic composition of silicate Earth is most readily reconciled with the observed value if volatile elements such as Ag, S, and C were added in approximately CI chondritic ratios to refractory elements late in Earth's accretion (Schönbächler et al. 2010). The CI chondritic ratio of C to S gives 1.1% C in the core.

Silicate Earth is profoundly different in C isotope ratios from other asteroidal and planetary bodies. Silicate Earth has a δ^{13}C of approximately −5‰, substantially higher than those of Mars, Vesta, and carbonaceous chondrite meteorites whose δ^{13}C values are all approximately −20‰. We show that, given plausible isotopic fractionations between liquid metal and liquid silicate, the difference between Earth and other bodies is explicable by strong partitioning of C into Earth's core. The major requirements would be that C was extracted to the core by a process approximating Rayleigh Fractionation and that the core contains ~1% C. The latter estimate can be greatly improved by better experimental data on metal-silicate fractionation of carbon isotopes and more precise estimates of the carbon content of bulk silicate Earth.

A third approach to constraining the core's carbon content is provided by available data on the densities, elastic properties, and phase equilibria of potential core materials under the pressure-temperature regime of the core. Phase equilibrium studies show that the eutectic composition in the Fe-C system shifts towards the Fe side with increasing pressure and that, at 50 GPa, compositions with <1% C can precipitate a carbide (Fe_3C) on their liquidus. This behavior likely continues to higher pressures, although Fe_3C may be replaced by Fe_7C_3 as the liquidus carbide phase. The densities and elastic properties of Fe_3C and Fe_7C_3 have not been measured at core conditions; however, it is conceivable, based on available data and the known properties of the core, that one of these phases is the predominant carbon-bearing component of the solid inner core. Demonstration that carbide is present in the inner core would enable us to constrain the carbon content of the core more precisely. Current data suggest, however, that a core carbon content of about 1% is feasible.

Our final method for constraining the carbon content of the core was to consider the effects of S and C on the partitioning of siderophile elements between core and mantle during Earth's accretion. We applied a simplified accretionary model in which Earth's growth was accompanied by oxidation of the mantle and extraction of the liquid core continuously at monotonically increasing pressure. This model generates the observed mantle concentrations of a number of weakly- and moderately-siderophile elements (V, Cr, Nb, Co, Ni, Si). The core-mantle partitioning of W and Mo were found to be very sensitive to the S and C contents of the metal. If we assume that ~2% S is present in the core and that this S was added during the last 13% of accretion (Schönbächler et al. 2010), then the current W and Mo contents of the mantle would be consistent with ~0.6% carbon in the core.

We conclude that all 4 approaches to carbon as a component of the "light" element in the core lead to an upper likely concentration of ~1% of this element in the core.

ACKNOWLEDGMENTS

BJW acknowledges the support of the NERC (U.K.) and the European Research Council for his research into core-mantle partitioning. JL acknowledges support of NSF grants EAR-102379, EAR-1291881, and DOE CI JL 2008-05246 ANTC. AS acknowledges the support of NSF grant EAR 0948131.

REFERENCES

Allègre CJ, Poirier J-P, Humler E, Hofmann AW (1995) The chemical composition of the Earth. Earth Planet Sci Lett 134(3-4):515-526

Anderson OL, Isaak DG (2002) Another look at the core density deficit of Earth's outer core. Phys Earth Planet Inter 131(1):19-27

Birch F (1952) Elasticity and constitution of the Earth's interior. J Geophys Res 57:227-286

Bottinga Y (1969) Calculated fractionation factors for carbon and hydrogen isotope exchange in system calcite-carbon dioxide-graphite-methane-hydrogen-water vapor. Geochim Cosmochim Acta 33:49-60

Brown JM (2001) The equation of state of iron to 450 GPa: Another high pressure solid phase? Geophys Res Lett 28(22):4339-4342

Chabot NL, Campbell AJ, McDonough WF, Draper DS, Agee CB, Humayun M, Watson HC, Cottrell E, Saslow SA (2008) The Fe-C system at 5 GPa and implications for Earth's core. Geochim Cosmochim Acta 72(16):4146-4158

Chambers JE (2005) Planet formation. *In:* Treatise on Geochemistry. Vol 2. Meteorites, Comets and Planets. Davies AM (*ed*) Elsevier, Amsterdam, The Netherlands, p 461-475

Chipman J (1972) Thermodynamics and phase-diagram of Fe-C system. Metall Trans 3(1):55-64

Corgne A, Keshav S, Wood BJ, McDonough WF, Fei Y (2008) Metal-silicate partitioning and constraints on core composition and oxygen fugacity during Earth accretion. Geochim Cosmochim Acta 72:574-589

Creager KC (1992) Anisotropy of the inner core from differential travel-times of the phases Pkp and Pkikp. Nature 356(6367):309-314

Dale CW, Burton KW, Greenwood RC, Gannoun A, Wade J, Wood BJ, Pearson DG (2012) Late accretion on the earliest planetesimals revealed by the highly siderophile elements. Science 336:72-75

Dasgupta R (2013) Ingassing, storage, and outgassing of terrestrial carbon through geologic time. Rev Mineral Geochem 75:183-229

Dasgupta R, Buono A, Whelan G, Walker D (2009) High-pressure melting relations in Fe-C-S systems: Implications for formation, evolution, and structure of metallic cores in planetary bodies. Geochim Cosmochim Acta 73(21):6678-6691
Dauphas N, Pourmand A (2011) Hf-W-Th evidence for rapid growth of Mars and its status as a planetary embryo. Nature 473:489-493
Deines P (2002) The carbon isotope geochemistry of mantle xenoliths. Earth Sci Rev 58(3-4):247-278
Deines P, Wickman FE (1975) Contribution to stable carbon isotope geochemistry of iron-meteorites. Geochim Cosmochim Acta 39(5):547-557
Dewaele A, Loubeyre P, Occelli F, Mezouar M, Dorogokupets PI, Torrent M (2006) Quasihydrostatic equation of state of iron above 2 Mbar. Phys Rev Lett 97(21), doi:10.1103/PhysRevLett.97.215504
Dodd SP, Saunders GA, Cankurtaran M, James B, Acet M (2003) Ultrasonic study of the temperature and hydrostatic-pressure dependences of the elastic properties of polycrystalline cementite (Fe_3C). Phys Status Solidi A 198(2):272-281
Dreibus G, Palme H (1996) Cosmochemical constraints on the sulfur content in the Earth's core. Geochim Cosmochim Acta 60:1125-1130
Dubrovinsky LS, Saxena SK, Tutti F, Rekhi S, LeBehan T (2000) In situ X-ray study of thermal expansion and phase transition of iron at multimegabar pressure. Phys Rev Lett 84(8):1720-1723
Duffy TS, Ahrens TJ (1992) Hugoniot sound velocities in metals with applications to the Earth's inner core. *In:* High Pressure Research: Applications to Earth and Planetary Sciences (Geophysical Monograph 67). Syono MMY (ed) Terra Scientific, Tokyo, Japan, p 353-361
Duman E, Acet M, Wassermann EF, Itie JP, Baudelet F, Mathon O, Pascarelli S (2005) Magnetic instabilities in Fe_3C cementite particles observed with FeK-edge X-ray circular dichroism under pressure. Phys Rev Lett 94(7), doi:10.1103/PhysRevLett.94.075502
Dziewonski AM, Anderson DL (1981) Preliminary reference Earth model. Phys Earth Planet Inter 25:297-356
Fang CM, van Huis MA, Zandbergen HW (2009) Structural, electronic, and magnetic properties of iron carbide Fe7C3 phases from first-principles theory. Phys Rev B 80(22), doi:10.1103/PhysRevB.80.224108
Fiquet G, Badro J, Gregoryanz E, Fei YW, Occelli F (2009) Sound velocity in iron carbide (Fe_3C) at high pressure: Implications for the carbon content of the Earth's inner core. Phys Earth Planet Inter 172(1-2):125-129
Fiquet G, Badro J, Guyot F, Requardt H, Krisch M (2001) Sound velocities in iron to 110 gigapascals. Science 291(5503):468-471
Fitoussi C, Bourdon B (2012) Silicon isotope evidence against an enstatite chondrite Earth. Science 335(6075):1477-1480
Gao LL, Chen B, Lerche M, Alp EE, Sturhahn W, Zhao JY, Yavas H, Li J (2009) Sound velocities of compressed Fe_3C from simultaneous synchrotron X-ray diffraction and nuclear resonant scattering measurements. J Synchrotron Radiat 16:714-722
Gao LL, Chen B, Wang J, Alp EE, Zhao J, Lerche M, Sturhahn W, Scott HP, Huang F, Ding Y, Sinogeikin SV, Lundstrom CC, Bass JD, Li J (2008) Pressure-induced magnetic transition and sound velocities of Fe(3)C: Implications for carbon in the Earth's inner core. Geophys Res Lett 35(17), doi:10.1029/2008GL034817
Gao LL, Chen B, Zhao JY, Alp EE, Sturhahn W, Li J (2011) Effect of temperature on sound velocities of compressed Fe_3C, a candidate component of the Earth's inner core. Earth Planet Sci Lett 309(3-4):213-220
Georg RB, Halliday AN, Schauble EA, Reynolds BC (2007) Silicon in the Earth's core. Nature 447(7148):1102-1106
Grady MM, Verchovsky AB, Wright IP (2004) Magmatic carbon in Martian meteorites: attempts to constrain the carbon cycle on Mars. Int J Astrobiol 3:117-124
Grady MM, Wright IP (2003) Elemental and isotopic abundances of carbon and nitrogen in meteorites. Space Sci Rev 106(1-4):231-248
Grady MM, Wright IP, Pillinger CT (1997) Carbon in howardite, eucrite and diogenite basaltic achondrites. Meteorit Planet Sci 32(6):863-868
Hazen RM, Downs RT, Jones AP, Kah L (2013) Carbon mineralogy and crystal chemistry. Rev Mineral Geochem 75:7-46
Hillgren VJ, Gessmann CK, Li J (2000) An experimental perspective on the light element in Earth's core. *In:* Origin of the Earth and Moon. Canup RM, Righter K (eds) University of Arizona Press, p 245-263
Hirayama Y, Fujii T, Kurita K (1993) The melting relation of the system, iron and carbon at high-pressure and its bearing on the early-stage of the Earth. Geophys Res Lett 20(19):2095-2098
Javoy M, Pineau F, Ilyama I (1978) Experimental-determination of isotopic fractionation between gaseous CO_2 and carbon dissolved in tholeiitic magma-preliminary study. Contrib Mineral Petrol 67:35-39
Jeanloz R (1990) The nature of the Earth's core. Annu Rev Earth Planet Sci 18:357-386
Kleine T, Munker C, Mezger K, Palme H (2002) Rapid accretion and early core formation on asteroids and the terrestrial planets from Hf-W chronometry. Nature 418(6901):952-955

Komabayashi T, Fei YW (2010) Internally consistent thermodynamic database for iron to the Earth's core conditions. J Geophys Res-Solid Earth 115, doi:10.1029/2009JB006442

Li J, Fei Y (2003) Experimental constraints on core composition. *In:* Treatise on Geochemistry. Vol 2. Holland HD, Turekian KK (eds) Elsevier Ltd, Amsterdam, p 521-546

Li J, Mao HK, Fei Y, Gregoryanz E, Eremets M, Zha CS (2002) Compression of Fe_3C to 30 GPa at room temperature. Phys Chem Miner 29(3):166-169

Lin JF, Struzhkin VV, Mao HK, Hemley RJ, Chow P, Hu MY, Li J (2004) Magnetic transition in compressed Fe_3C from x-ray emission spectroscopy. Phys Rev B 70(21), doi:10.1103/PhysRevB.70.212405

Lodders K (2003) Solar system abundances and condensation temperatures of the elements. Astrophys J 591:1220-1247

Lord OT, Walter MJ, Dasgupta R, Walker D, Clark SM (2009) Melting in the Fe-C system to 70 GPa. Earth Planet Sci Lett 284(1-2):157-167

Lübbers R, Grunsteudel HF, Chumakov AI, Wortmann G (2000) Density of phonon states in iron at high pressure. Science 287(5456):1250-1253

Mao HK, Xu J, Struzhkin VV, Shu J, Hemley RJ, Sturhahn W, Hu MY, Alp EE, Vočadalo L, Alfè D, Price GD, Gillan MJ, Schwoerer-Böhning M, Häusermann D, Eng P, Shen G, Giefers H, Lübbers R, Wortmann G (2001) Phonon density of states of iron up to 153 gigapascals. Science 292(5518):914-916

Mao HK, Wu Y, Chen LC, Shu JF, Jephcoat AP (1990) Static compression of iron to 300 Gpa and $Fe_{0.8}Ni_{0.2}$ Alloy to 260 Gpa - implications for composition of the core. J Geophys Res Solid Earth 95(B13):21737-21742

Marty B (2012) The origins and concentrations of water, carbon, nitrogen and noble gases on Earth. Earth Planet Sci Lett 313-314:56-66

McDonough WF (2003) Compositional model for the Earth's core. *In:* Treatise on Geochemistry. Vol 2. The Mantle and Core. Carlson RW (ed) Elsevier-Pergamon, Oxford p 547-568

McDonough WF, Sun S-s (1995) The composition of the Earth. Chem Geol 120(3-4):223-253

Mikhail S, Jones AP, Basu S, Milledge HJ, Dobson DP, Wood I, Beard A, Guillermier C, Verchovsky AB, Franchi IA (2010) Carbon isotope fractionation betwee Fe-carbide and diamond; a light C isotope reservoir in the deep Earth and Core? EOS Transactions AGU, Fall Meeting Supplementary Abstract U21A-0001

Mookherjee M (2011) Elasticity and anisotropy of Fe_3C at high pressures. Am Mineral 96(10):1530-1536

Mookherjee M, Nakajima Y, Steinle-Neumann G, Glazyrin K, Wu XA, Dubrovinsky L, McCammon C, Chumakov A (2011) High-pressure behavior of iron carbide (Fe_7C_3) at inner core conditions. J Geophys Res Solid Earth 116, doi:10.1029/2010JB007819

Nakajima Y, Takahashi E, Sata N, Nishihara Y, Hirose K, Funakoshi K, Ohishi Y (2011) Thermoelastic property and high-pressure stability of Fe_7C_3: Implication for iron-carbide in the Earth's core. Am Mineral 96(7):1158-1165

Nakajima Y, Takahashi E, Suzuki T, Funakoshi K (2009) "Carbon in the core" revisited. Phys Earth Planet Inter 174(1-4):202-211

Nikolussi M, Shang SL, Gressmann T, Leineweber A, Mittemeijer E, Wang Y, Liu ZK (2008) Extreme elastic anisotropy of cementite, Fe_3C: First-principles calculations and experimental evidence. Scripta Mater 59(8):814-817

Oganov AR, Hemley RJ, Hazen RM, Jones AP (2013) Structure, bonding, and mineralogy of carbon at extreme conditions. Rev Mineral Geochem 75:47-77

Okuchi T (1997) Hydrogen partitioning into molten iron at high pressure: Implications for Earth's core. Science 278(5344):1781-1784

Ono S, Mibe K (2010) Magnetic transition of iron carbide at high pressures. Phys Earth Planet Inter 180(1-2):1-6

Poirier JP (1994) Light-elements in the Earth's outer core - a critical-review. Phys Earth Planet Inter 85(3-4):319-337

Polyakov VB, Kharlashina NN (1995) The use of heat-capacity data to calculate carbon-isotope fractionation between graphite, diamond, and carbon-dioxide - a new approach. Geochim Cosmochim Acta 59(12):2561-2572

Prakapenka VB, Shen G, Sturhahn W, Rivers ML, Sutton SR, Uchida T (2004) Crystal structure and magnetic properties of Fe_3C at high pressures and high temperatures. EOS Transactions AGU, Fall Meeting Supplementary Abstract MR43A-0874

Ringwood AE (1960) Cohenite as a pressure indicator in iron meteorites. Geochim Cosmochim Acta 20(2):155-158

Rouquette J, Dolejs D, Kantor IY, McCammon CA, Frost DJ, Prakapenka VB, Dubrovinsky LS (2008) Iron-carbon interactions at high temperatures and pressures. Appl Phys Lett 92(12), doi:10.1063/1.2892400

Rubie DC, Frost DJ, Mann U, Asahara Y, Nimmo F, Tsuno K, Kegler P, Holzheid A, Palme H (2011) Heterogeneous accretion, composition and core-mantle differentiation of the Earth. Earth Planet Sci Lett 301:31-42

Rubie DC, Gessmann CK, Frost DJ (2004) Partitioning of oxygen during core formation on the Earth and Mars. Nature 429:58-61

Rudge JF, Kleine T, Bourdon B (2010) Broad bounds on Earth's accretion and core formation constrained by geochemical models. Nature Geoscience 3:439-443

Sata N, Hirose K, Shen GY, Nakajima Y, Ohishi Y, Hirao N (2010) Compression of FeSi, Fe_3C, $Fe_{0.95}O$, and FeS under the core pressures and implication for light element in the Earth's core. J Geophys Res-Solid Earth 115, doi:10.1029/2009JB006975

Satish-Kumar M, So H, Yoshino T, Kato M, Hiroi Y (2011) Experimental determination of carbon isotope fractionation between iron carbide melt and carbon: C-12-enriched carbon in the Earth's core? Earth Planet Sci Lett 310(3-4):340-348

Schönbächler M, Carlson RW, Horan MF, Mock TD, Hauri EH (2010) Heterogeneous accretion and the moderately volatile element budget of Earth. Science 328:884-887

Scott HP, Williams Q, Knittle E (2001) Stability and equation of state of Fe_3C to 73 GPa: Implications for carbon in the Earth's core. Geophys Res Lett 28(9):1875-1878

Shahar A, Hillgren VJ, Young ED, Fei YW, Macris CA, Deng LW (2011) High-temperature Si isotope fractionation between iron metal and silicate. Geochim Cosmochim Acta 75(23):7688-7697

Steelmaking (1988) Steelmaking Data Sourcebook. Gordon and Breach, New York

Stevenson DJ (1981) Models of the Earth's Core. Science 214:611-619

Strong HM, Chrenko RM (1971) Further studies on diamond growth rates and physical properties of laboratory-made diamond. J Phys Chem 75(12):1838-1843

Tateno S, Hirose K, Ohishi Y, Tatsumi Y (2010) The structure of iron in Earth's inner core. Science 330(6002):359-361

Terasaki H, Nishida K, Shibazaki Y, Sakamaki T, Suzuki A, Ohtani E, Kikegawa T (2010) Density measurement of Fe_3C liquid using X-ray absorption image up to 10 GPa and effect of light elements on compressibility of liquid iron. J Geophys Res-Solid Earth 115, doi:10.1029/2009JB006905

Tromp J (1993) Support for anisotropy of the Earths inner-core from free oscillations. Nature 366(6456):678-681

Tuff J, Wood BJ, Wade J (2011) The effect of Si on metal-silicate partitioning of siderophile elements and implications for the conditions of core formation. Geochim Cosmochim Acta 75:673-690

Vočadlo L, Brodholt J, Dobson DP, Knight KS, Marshall WG, Price GD, Wood IG (2002) The effect of ferromagnetism on the equation of state of Fe_3C studied by first-principles calculations. Earth Planet Sci Lett 203(1):567-575

Wade J, Wood BJ (2005) Core formation and the oxidation state of the Earth. Earth Planet Sci Lett 236:78-95

Wade J, Wood BJ, Tuff J (2012) Metal-silicate partitioning of Mo and W at high pressures and temperatures: Evidence for late accretion of sulfur to the Earth. Geochim Cosmochim Acta 85:58-74

Wang C, Hirama J, Nagasaka T, Shiro BY (1991) Phase-equilibria of liquid Fe-S-C ternary-system. ISIJ Int 31(11):1292-1299

Wood BJ (1993) Carbon in the core. Earth Planet Sci Lett 117:593-607

Wood BJ, Bryndzia LT, Johnson KE (1990) Mantle oxidation state and its relationship to tectonic environment and fluid speciation. Science 248(4953):337-345

Wood BJ, Halliday AN (2010) The lead isotopic composition of the Earth can be explained by core formation alone. Nature 465:767-770

Wood BJ, Walter MJ, Wade J (2006) Accretion of the Earth and segregation of its core. Nature 441:825-833

Wookey J, Helffrich G (2008) Inner-core shear-wave anisotropy and texture from an observation of PKJKP waves. Nature 454(7206):873-U824

Yin QZ, Jacobsen SB, Yamashita K, Blichert-Toft J, Telouk P, Albarede F (2002) A short timescale for terrestrial planet formation from Hf-W chronometry of meteorites. Nature 418(6901):949-952

Zhukov A (2000) Once more about the Fe-C phase diagram. Metal Sci Heat Treat 42(1):42-43

Reviews in Mineralogy & Geochemistry
Vol. 75 pp. 251-287, 2013

9

Carbon in Silicate Melts

Huaiwei Ni and Hans Keppler

Bayerisches Geoinstitut
95440 Bayreuth, Germany

Hans.Keppler@uni-bayreuth.de

INTRODUCTION

Silicate melts are the main agent for transporting carbon from Earth's interior to the surface. The carbon concentration in the atmosphere and the size of the carbon reservoir in oceans, sediments, and biomass are ultimately controlled by the balance between carbon removal through weathering, burial in sediments, and subduction on one hand and volcanic degassing on the other hand (e.g., Berner 1994). Carbon emissions from volcanoes may have ended the Neoproterozoic "snowball-Earth" glaciation (Hoffman et al. 1998) and they have been invoked as a potential mechanism that could link flood basalt eruptions to mass extinction events (Beerling 2002).

In Earth's deep interior, the strong partitioning of carbon into silicate melts relative to solid minerals may contribute to melting in the seismic low-velocity zone of the upper mantle and in the transition zone (e.g., Dasgupta and Hirschmann 2010; Keshav et al. 2011). The formation of some highly silica-undersaturated melts is likely related to the effects of carbon dioxide on melting in the mantle (Brey and Green 1975). While carbon is usually less abundant than water in magmas erupting at Earth's surface, the lower solubility of carbon dioxide (either molecular or as CO_3^{2-}) in silicate melts implies that it is primarily carbon dioxide that controls the nucleation of bubbles, which is an important aspect of eruption dynamics (e.g., Holloway 1976; Papale and Polacci 1999). In the lower mantle, more reduced carbon species may be dominant in silicate melts, which may behave in a different way than carbon dioxide (e.g., Kadik et al. 2004). Data on the solubility and speciation of carbon in silicate melts (in a broad sense, covering superliquidus liquids, supercooled liquids and glasses) and its effect on melt properties are therefore essential for understanding a wide range of phenomena in the Earth system.

CARBON SOLUBILITY IN SILICATE MELTS

Under typical redox conditions in the present crust and upper mantle (ΔQFM ranging from −2 to +5; Wood et al. 1990; McCammon 2005) and the intrinsic conditions in experimental pressure vessels (generally above QFM; e.g., Jakobsson 1997; Tamic et al. 2001), thermodynamic calculations reveal that the majority of carbon is present as carbon dioxide in geological fluids (Pawley et al. 1992; Holloway and Blank 1994; Manning et al. 2013). Correspondingly, the carbon dissolved in a silicate melt coexisting with such a fluid is predominately in the form of either molecular CO_2 or the carbonate group (CO_3^{2-}), depending on temperature, pressure, and melt composition. However, under more reduced conditions such as in the Archean or at greater depths of the modern Earth, CH_4 together with some CO may prevail in C-O-H fluids (Ballhaus 1995; Kump and Barley 2007; Manning et al. 2013), and they dissolve into silicate melts differently. The solubility of carbon species (CO_2, CO, or CH_4) in a variety of silicate melts under a broad range of pressure and temperature conditions

1529-6466/13/0075-0009$00.00 DOI: 10.2138/rmg.2013.75.9

is crucial for understanding the degassing of Earth and other terrestrial planets, the formation of the atmosphere, as well as the petrogenesis of various igneous rocks.

In an earlier volume of the RiMG series published nearly two decades ago, Blank and Brooker (1994) and Holloway and Blank (1994) presented two excellent reviews on the solubility of CO_2 in silicate melts. Lately, Moore (2008) discussed some experimental and modeling aspects related to this topic. Since 1994, extensive new experimental data have been reported (e.g., Dixon et al. 1995; Brooker et al. 1999, 2001a,b; Botcharnikov et al. 2005, 2006; Lesne et al. 2011). Molecular dynamics simulations have recently been applied to the investigation of CO_2 solubility (Guillot and Sator 2011), which are particularly needed for pressures beyond 3 GPa. There have also been some other studies focusing on reduced conditions (e.g., Mysen et al. 2009; Morizet et al. 2010). Furthermore, considerable efforts have been made on the development of general CO_2 solubility models (e.g., Papale 1997, 1999; Papale et al. 2006). All post-1994 studies devoted to carbon solubility in silicate melts are listed in Table 1. Data from these studies are summarized in Online Supplementary Table 1.

Below, we will first discuss the work on CO_2 solubility in nominally anhydrous melts (the systems with CO_2 being the only volatile component), which is followed by a review of the studies on CO_2 solubility in hydrous melts (the systems with binary volatiles CO_2-H_2O), and followed by a summary on the solubility of C-O-H fluids under reduced conditions (the systems with ternary volatiles).

CO_2 solubility in nominally anhydrous melts

Experimental method. Solubility experiments are performed by saturating some melt with CO_2 at high pressure (P) and temperature (T) and then quenching the melt to a glass. The CO_2 content of the quenched glass is then measured by FTIR (Fourier transform infrared spectroscopy), SIMS (secondary ion mass spectrometry), or some other method and is assumed to represent the CO_2 solubility in the melt at given P and T. This method obviously can only be used for systems where melts can be quenched into homogeneous glasses, without quench crystallization and without bubble nucleation during quench. For this reason, CO_2 solubility data for non-quenchable utramafic melts do not exist and direct solubility measurements at deep mantle pressures > 10 GPa, where quenching melts to glasses becomes very difficult, are not available either. At pressures below 1 GPa, the liquidus temperatures of many water-free silicate melts are so high that measuring CO_2 solubility requires experiments in internally-heated gas pressure vessels equipped with a rapid-quench device.

The starting materials in CO_2 solubility experiments are either chips/powders of nominally anhydrous glass + silver oxalate or a mixture of oxides and carbonates, enclosed in noble metal capsules such as platinum or $Au_{80}Pd_{20}$. Silver oxalate $Ag_2C_2O_4$ decomposes upon heating to metallic silver and pure CO_2. It is the most commonly used source of CO_2 in high-pressure experiments and has largely replaced various organic compounds, such as oxalic acid, that were used in early studies. In addition, direct loading of gaseous CO_2 into the capsule has also been practiced (Botcharnikov et al. 2006, 2007). At superliquidus temperature (or a lower temperature at which the melt is metastable) and high pressure (see Moore 2008 for a discussion of various pressure vessels), it is expected that a silicate melt reaches equilibrium with a pure CO_2 fluid within a time frame of minutes to days. However, hydrogen can diffuse into the system through capsules, as evidenced by dissolved H_2O in quenched melts. If special care is taken in the experimental procedures, the dissolved H_2O can be controlled to below 4000 ppm or even 1000 ppm (Brooker et al. 1999). H_2O contents at such level are believed to have only a minor effect on the dissolution behavior of CO_2 (Stolper et al. 1987). In some circumstances, carbon (e.g., arising from a graphite heater in a piston-cylinder apparatus) may also gain access to the system, but this phenomenon can be avoided through the improvement of experimental design (Brooker et al. 1999). Hydrogen diffusing into the sample capsule may not only introduce an additional, unwanted component, but it may also reduce CO_2 to other carbon species. Reduction may also be enhanced, perhaps through catalytic effects, by the capsule material, such as pure Pt. In order to retain carbon in the +4 oxidation state in such experiments, the oxygen fugacity therefore has to be kept high and diffusion of hydrogen into the capsule has to be suppressed as far as possible. In studies of Fe-bearing melts, the problem of iron loss to Pt can be alleviated by pre-saturating Pt capsules with Fe. Under high temperature and pressure, small amounts of melt components, such as alkali, may dissolve into the fluid phase. Despite all these experimental complexities, it is possible to produce a silicate melt of designated composition coexisting with a fluid with CO_2 mole fraction X_{CO_2} > 90%, which can be treated approximately as a system with a single volatile component CO_2.

Analytical techniques. The CO_2 concentration in the glass quenched from a CO_2-saturated run gives the CO_2 solubility of a melt at a specific pressure and temperature, if the concentration remained unchanged during quenching. Under equivalent conditions, CO_2 solubility is one to two orders of magnitude lower than the solubility of H_2O. The low solubility of CO_2 and the limitation of then available analytical techniques (e.g., weight loss) caused early experimental studies to center on GPa level pressures. β-track autoradiography was once used frequently in carbon analysis (e.g., Mysen et al. 1975, 1976), but later it was found to yield inaccurate carbon concentration (Tingle and Aines 1988; Blank and Brooker 1994). FTIR has become by far the most frequently adopted analytical technique (Table 1) because it non-destructively probes both CO_2 and H_2O, measures concentrations down to ppm level with high accuracy, and delivers information about carbon speciation. The molar absorptivities of FTIR bands (2350 cm^{-1} for molecular CO_2 and the doublets within 1350-1650 cm^{-1} for CO_3^{2-}) for each specific melt composition need to be pre-calibrated by absolute methods (such as manometry or bulk carbon analyzers). Different authors may report CO_2 solubility based on different molar absorptivities, which is a major source of inconsistency between different data sets. Combined with stepped heating, it is plausible to separate adsorbed carbon, CO_2 from vesicles in the glass, and the actual dissolved carbon (e.g., Jendrzejewski et al. 1997). In addition to FTIR and the bulk analytical methods, SIMS and NMR (nuclear magnetic resonance) have also been occasionally used for carbon analysis (Pan et al. 1991; Thibault and Holloway 1994; Brooker et al. 1999; Behrens et al. 2004a).

Pressure effect. CO_2 solubility increases with increasing CO_2 fugacity and therefore with increasing pressure. Figure 1a shows that to the first order approximation, CO_2 solubility in rhyolite melt (at 1123-1323 K) and basalt melt (at 1473-1573 K) is proportional to pressure at $P < 0.7$ GPa with a common slope of about 0.57 ppm CO_2/bar (despite the fact that CO_2 speciation in quenched rhyolite and basalt glass is quite different). Therefore, approximately

$$C \propto P \tag{1}$$

where C is CO_2 solubility in ppm or wt% and P is pressure. The approximate proportionality between C and P may apply to pressures as high as 4.0 GPa, although the slope varies for different melts (Fig. 1b).

Phenomenologically, the proportionality between CO_2 solubility and pressure, which has been used for empirical fitting of CO_2 solubility (e.g., Liu et al. 2005), resembles Henry's law. But from a thermodynamic perspective, CO_2 must have the same chemical potential in the fluid phase and in the melt phase. The CO_2 chemical potential in the fluid is directly related to the fugacity f_{CO_2} (instead of pressure), and that in the melt is directly related to the activity (instead of mole fraction) of CO_2. There are many equations of state for pure CO_2 fluid available as summarized in Gottschalk (2007). For consistency we have adopted the EoS from Duan et al. (1992) to calculate f_{CO_2}, which is not much different (within 10%) from the reported fugacities in the original papers based on the EoS from Kerrick and Jacobs (1981) or earlier work. When CO_2 solubility is plotted against f_{CO_2}, their correlation deviates significantly from

Table 1. Studies on carbon solubility in silicate melts (since 1994).

Year	Authors	Melt	Fluid	*T* (°C)	*P* (bar)	Analytical Method
Data						
1994	Holloway & Blank	Basanite	CO_2	1200-1400	1000-20000	Bulk analyzer + SIMS
1995	Dixon et al.	Basalt	CO_2-H_2O	1200	310-980	FTIR
1997	Jakobsson	Andesite	CO_2-H_2O	1400	10000	FTIR + Bulk analyzer
1997	Jendrzejewski et al.	Basalt	CO_2	1200-1300	250-1950	Manometry + FTIR
1999	Brooker et al.	SNA	CO_2	1450-1700	10000-35000	Bulk analyzer + FTIR +NMR
2000	Paonita et al.	Rhyolite, Basalt	CO_2-H_2O-He	1130-1160	1120-2150	Estimated from Papale (1999) model
2001a,b	Brooker et al.	SNAC, MSNAC, Andesite, Phonolite, Nephelinite, Melilitite	CO_2	1175-1600	2000-27000	Bulk analyzer + FTIR
2001	Tamic et al.	Rhyolite	CO_2-H_2O	800-1100	2000-5000	FTIR
2002	King & Holloway	Andesite	CO_2-H_2O	1300	10000	FTIR
2002	Morizet et al.	Haplophonolite	CO_2	1300-1550	10000-25000	Bulk analyzer + FTIR
2004a	Behrens et al.	Dacite	CO_2-H_2O	1250	1000-5000	SIMS + FTIR
2004	Kadik et al.	Basalt	C-O-H	1520-1600	37000	MS
2005	Botcharnikov et al.	Basalt, Alkali Basalt	CO_2-H_2O	1150-1200	2000-5000	FTIR
2006	Botcharnikov et al.	Andesite	CO_2-H_2O	1100-1300	2000-5000	FTIR
2007	Botcharnikov et al.	Andesite	CO_2-H_2O-Cl	1200	2000	FTIR
2009	Behrens et al.	Phonotephrite	CO_2-H_2O	1200-1250	2000-5000	FTIR
2009	Mysen et al.	SN	C-O-H	1400	10000-25000	Bulk analyzer
2010	Morizet et al.	Haplobasalt	C-O-H	1250	2000-3000	FTIR
2010	Shishkina et al.	Basalt	CO_2-H_2O	1250	500-5000	FTIR
2011	Lesne et al.	Alkali Basalt	CO_2-H_2O	1200	269-2059	FTIR
2011	Vetere et al.	Trachyandesite	CO_2-H_2O	1250	500-4000	FTIR + Bulk analyzer
2011	Webster et al.	Rhyolite	CO_2-H_2O-S	897-1100	1990-2040	FTIR
2011	Stanley et al.	Basalt	CO_2	1400-1625	10000-25000	FTIR
2011	Guillot & Sator	Rhyolite, Basalt, Kimberlite	CO_2	1200-2000	1000-150000	Molecular dynamics
2012	Iacono-Marziano et al.	Alkali basalt, Lamproite,-kamafugite	CO_2-H_2O	1200	485–4185	FTIR
Model						
1997	Dixon	Basalt, Basanite, Leucitite, Melilitite	CO_2	1200-1400	0-20000	
1997	Papale	Various	CO_2	900-1600	0-30000	
1999	Papale	Various	CO_2-H_2O	700-1600	0-5000	
2002	Newman & Lowenstern	Rhyolite, Basalt	CO_2-H_2O	700-1500	0-5000	
2005	Liu et al.	Rhyolite	CO_2-H_2O	700-1200	0-5000	
2006	Papale et al.	Various	CO_2-H_2O	700-1600	0-10000	

For studies before 1994, see Table A1 and Table A2 in Blank and Brooker (1994). SNA = SiO_2-Na_2O-Al_2O_3 melts; SNAC = SiO_2-Na_2O-Al_2O_3-CaO melts; MSNAC = MgO-SiO_2-Na_2O-Al_2O_3-CaO melts.

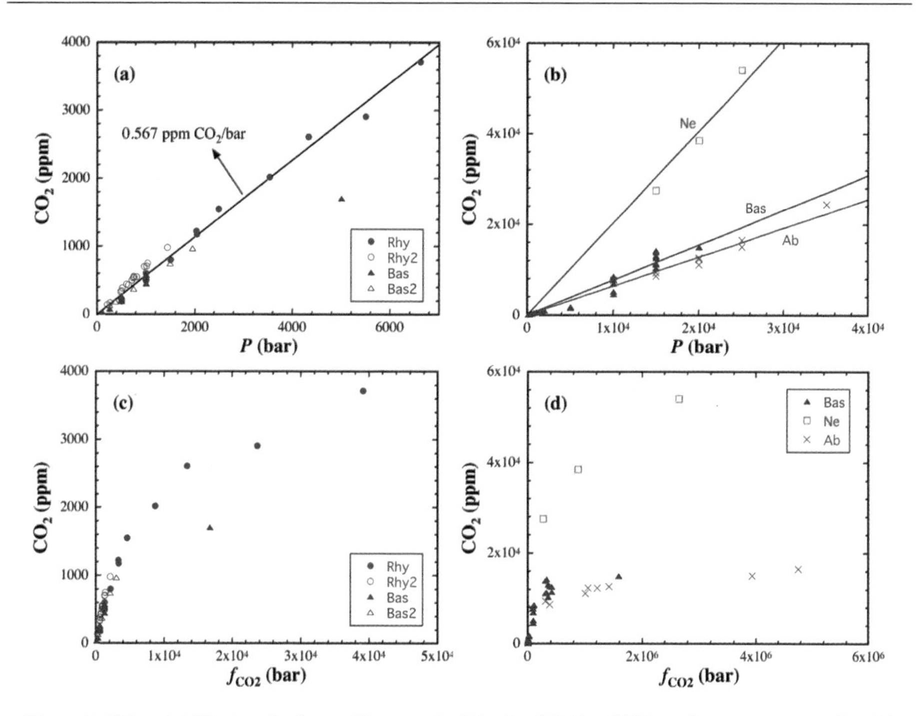

Figure 1. CO_2 solubility in anhydrous silicate melts (dissolved H_2O < 5000 ppm) versus pressure (in a,b) or CO_2 fugacity (in c,d). Data sources: Rhy = rhyolite melt at 1323 K (Fogel and Rutherford 1990); Rhy2 = rhyolite melt at 1123 K (Blank 1993); Bas = basalt melt at 1473 K (Stolper and Holloway 1988; Mattey 1991; Pan et al. 1991; Pawley et al. 1992); Bas2 = basalt melt at 1473-1573 K (Jendrzejewski et al. 1997); Ne = nepheline melt at 1973 K (Brooker et al. 1999); Ab = albite melt at 1723-1898 K (Stolper et al. 1987; Brooker et al. 1999). CO_2 fugacities are calculated from the equation of state for pure CO_2 fluid (Duan et al. 1992).

a proportional relationship at f_{CO_2} > 0.4 GPa (Fig. 1c,d). This nonlinearity can be attributed to the pressure dependence of CO_2 activity in the melt, which appears to compensate the non-ideal behavior of a real CO_2 fluid and cause the apparent proportionality between solubility and pressure.

Temperature effect. The influence of temperature on CO_2 solubility is less well constrained than the pressure influence. Blank and Brooker (1994) showed that after excluding the controversial data obtained using β-track autoradiography, CO_2 solubility decreases with increasing temperature, e.g., for rhyolite melt and albite melt (Fig. 2). Nevertheless, CO_2 solubility appears to be insensitive to temperature in basalt melt, and it even increases with increasing temperature in nepheline melt (Fig. 2). Furthermore, the direction of temperature effect may be reversed upon pressure change (Botcharnikov et al. 2005).

Composition effect. Compared to pressure and temperature effects, the dependence of CO_2 solubility on melt composition is more complex and more difficult to constrain. At ~1500 K and 0.2 GPa, CO_2 solubility increases weakly from rhyolite melt to dacite melt to andesite melt, but basalt melt and rhyolite melt have the lowest solubility (~1000 ppm) in the calc-alkaline series (Fig. 3a). On the other hand, CO_2 solubility increases nearly three-fold from basalt melt to alkali basalt melt to phonotephrite melt, all three of which have similar silica content. Therefore, CO_2 solubility appears to increase with increasing melt alkalinity, which is evidently demonstrated by a good correlation between solubility and total alkali content

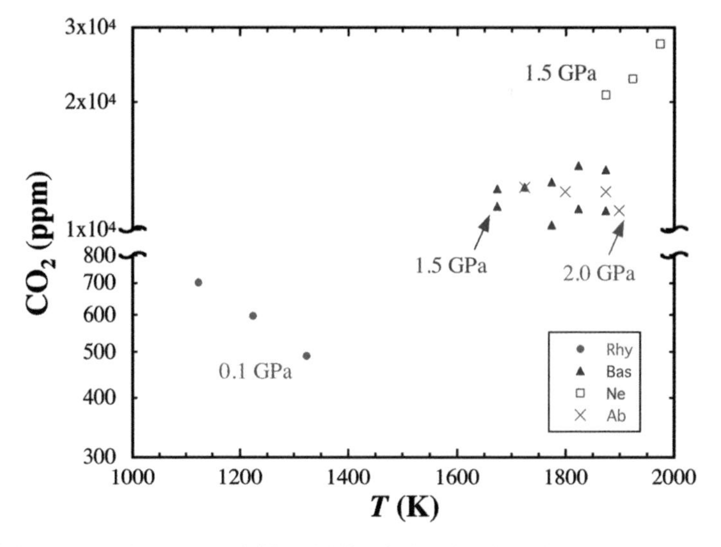

Figure 2. Temperature dependence of CO_2 solubility (in logarithmic scale) in anhydrous silicate melts (dissolved H_2O < 5000 ppm). Data sources: Rhy = rhyolite melt at 0.1 GPa (Fogel and Rutherford 1990; Blank 1993); Bas = basalt melt at 1.5 GPa (Pan et al. 1991); Ne = nepheline melt at 1.5 GPa (Brooker et al. 1999); Ab = albite melt at 2.0 GPa (Stolper et al. 1987; Brooker et al. 1999).

(Fig. 3b). The variation of CO_2 solubility with melt composition can be much more dramatic than illustrated in Figure 3. For example, Brooker et al. (2001a) showed that CO_2 solubility in a synthetic SNAC melt (35 wt% SiO_2, 10.5 wt% Al_2O_3, 49 wt% CaO, 5.5 wt% Na_2O) at 1500 K and 0.2 GPa can be as high as 8 wt%, which is about 80 times the solubility in basalt melt under equivalent conditions.

Molecular dynamics calculations. Most experimental solubility data fall within the pressure range of 0.01-3.0 GPa, with the only two exceptions being that Mysen et al. (1976) reported one datum (7.7 wt% CO_2) at 1898 K and 4.0 GPa in a melilitite melt and Brooker et al. (1999) reported one datum (2.44 wt% CO_2) at 1723 K and 3.5 GPa in albite melt. In principle solubility measurements could be extended to higher pressures by switching from a piston-cylinder apparatus to a multivanvil press, but there are numerous practical challenges. Complementarily, the solubility behavior at pressures > 3.0 GPa can now be investigated by molecular dynamics simulations.

Guillot and Sator (2011) performed the first molecular dynamics study on CO_2 solubility in silicate melts. A supercritical CO_2 fluid was numerically equilibrated with three natural melts at 1473-2273 K and 0.1-15 GPa. They found that CO_2 solubility increases more rapidly with the rise of pressure than predicted by the proportionality relationship, and that solubility decreases with increasing temperature (Fig. 4). The compositional dependence is weak among rhyolite melt, basalt melt and kimberlite melt at pressures below 8 GPa. They showed that their simulation results at low pressure are broadly consistent with experimental data.

Solubility model. Efforts were initially made (e.g., Stolper et al. 1987; Fogel and Rutherford 1990) to construct CO_2 solubility models for specific melt compositions, i.e., only the pressure (or fugacity) and temperature dependences were incorporated in the models. If CO_2 is treated as a simple component (i.e., CO_2 speciation in the melt phase is not accounted for), we consider the heterogeneous reaction

$$CO_2\ (\mathit{fluid}) = CO_2\ (\mathit{melt}) \tag{2}$$

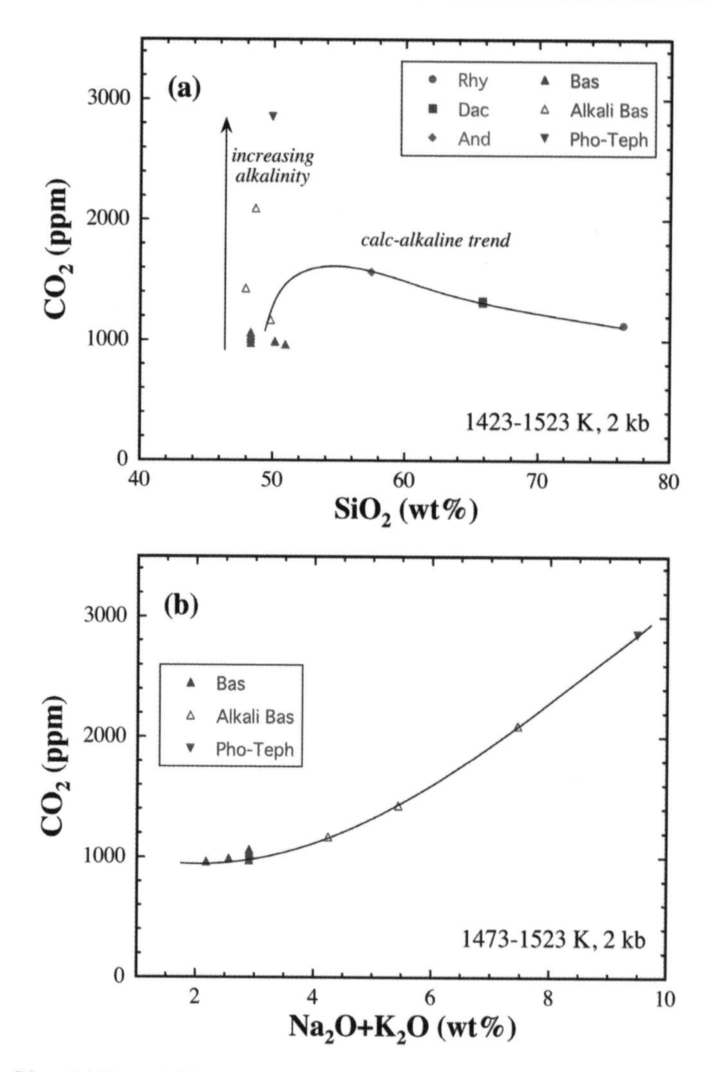

Figure 3. CO_2 solubility at 1423-1523 K and 0.2 GPa versus melt composition in anhydrous melts (the mole fraction of H_2O in the coexisting fluid phase is less than 0.1). Data sources: Rhy = rhyolite melt (Fogel and Rutherford 1990); Dac = dacite melt (Behrens et al. 2004a); And = andesite melt (Botcharnikov et al. 2006); Bas = basalt melt (Jendrzejewski et al. 1997; Botcharnikov et al. 2005; Shishkina et al. 2010); Alkali Bas = alkali basalt melt (Lesne et al. 2011); Pho-Teph = phonotephrite melt (Behrens et al. 2009). The curves are drawn to guide the eye.

The equilibrium constant K of the above reaction can be defined as

$$K = \frac{X_{CO_2}}{f_{CO_2}} \tag{3}$$

where f_{CO_2} is the CO_2 fugacity in the fluid phase and X_{CO_2} is the mole fraction of CO_2 in the melt phase. Note that unlike the unambiguous definition of CO_2 content of the melt in ppm or wt%, there are multiple ways defining the mole fraction of CO_2. If we limit our scope for treating a specific melt, the following definition is recommended,

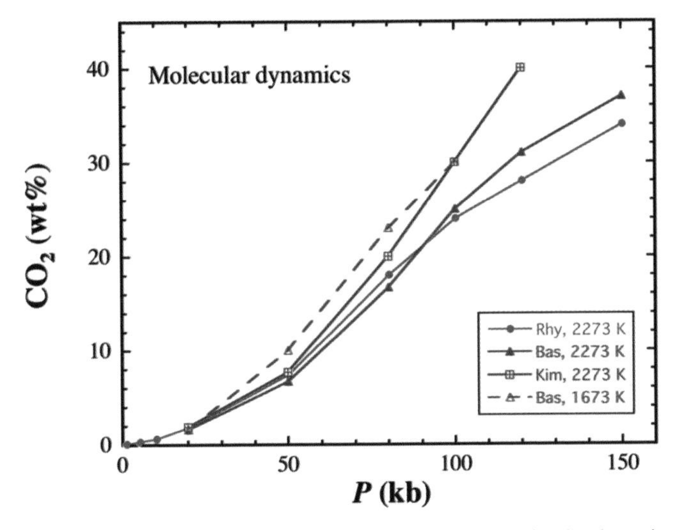

Figure 4. CO_2 solubility in anhydrous melts at 1673-2273 K based on molecular dynamics simulations. Rhy = rhyolite melt; Bas = basalt melt; Kim = kimberlite. Modified after Guillot and Sator (2011).

$$X_{CO_2} = \frac{\left(\dfrac{C}{44.01}\right)}{\left(\dfrac{C}{44.01} + \dfrac{(100-C)}{M}\right)} \tag{4}$$

where C is the CO_2 content of the melt in wt%, and 44.01 and M are the molecular weight of CO_2 and the formula weight of the volatile-excluded melt on a single oxygen basis (e.g., M = 32.78 g/mol for albite melt). Basically X_{CO_2} is proportional to C under realistic situations.

The equilibrium constant K varies with both pressure and temperature,

$$\left(\frac{\partial \ln K}{\partial P}\right)_{T,X} = -\frac{V}{RT} \tag{5a}$$

$$\left(\frac{\partial \ln K}{\partial T}\right)_{P,X} = \frac{\Delta H}{RT^2} \tag{5b}$$

where V is the partial molar volume of CO_2 in the melt, R is the gas constant, and ΔH is the molar dissolution enthalpy of CO_2 (ΔH is negative as the dissolution process is exothermic). If one sets the value of the equilibrium constant to be K_0 for a reference state (P_0, T_0), by combining Equation (3) and Equations (5a,b) we obtain

$$X_{CO_2} = f_{CO_2} K_0 \exp\left(\int_{P_0}^{P} -\frac{V}{RT}\mathrm{d}P + \int_{T_0}^{T} \frac{\Delta H}{RT^2}\mathrm{d}T\right) \tag{6}$$

With a further assumption that the pressure and temperature dependences of V and ΔH can be neglected, Equation (6) is simplified to the following form:

$$X_{CO_2} = f_{CO_2} K_0 \exp\left[\frac{V(P_0 - P)}{RT} + \frac{\Delta H}{R}\left(\frac{1}{T_0} - \frac{1}{T}\right)\right] \tag{7}$$

The above model correctly predicts a progressively gentler slope towards high P in solubility vs. fugacity plots (compare Fig. 1c,d) as well as a negative temperature dependence of CO_2 solubility. After designating a reference state (such as 1500 K and 1 atm) and selecting a valid equation of state of CO_2 fluid to calculate f_{CO_2}, one can fit experimental data $X(P, T)$ to Equation (7) and extract the three free parameters of K_0, V and ΔH (e.g., Fogel and Rutherford 1990).

Based on literature solubility data for melt compositions of basalt, basanite, leucitite and melilitite, Dixon (1997) presented a simple solubility model in which CO_2 solubility at 1473 K and 0.1 GPa is linearly correlated with the mole fractions of various cations. A more ambitious attempt was carried out by Papale (1997)—he compiled a large data set (263 data points) for various melt compositions and proposed a general CO_2 solubility model. The melt phase was regarded as a mixture of 11 oxide components SiO_2-TiO_2-Al_2O_3-Fe_2O_3-FeO-MnO-MgO-CaO-Na_2O-K_2O-CO_2 (hence the definition of CO_2 mole fraction was different from that in Eqn. 4). Papale (1997) presented a strict thermodynamic model based on the equivalence of CO_2 fugacity in the coexisting melt and fluid phases. However, he assumed that the partial molar volume of CO_2 in the melt coincides with that of pure CO_2 fluid, which was problematic and was discarded in their later models (Papale 1999; Papale et al. 2006). Compared to Equation (6), instead of extracting V and ΔH from fitting experimental data, the treatment in Papale (1997) is equivalent to assigning the volume of pure CO_2 fluid for V and assigning the enthalpy difference $H(P, T) - H(P_0, T)$ of pure CO_2 fluid for ΔH (at least mathematically). Also, K_0 in Equation (6) was replaced by a term related to the activity coefficient of CO_2 (γ) in the melt, which was again related to the excess Gibbs free energy of the melt (G^E). G^E was attributed to the interaction between oxide components according to the regular solution theory (here the compositional dependence weighs in). The interaction energy terms involving CO_2 were assumed to be pressure-dependent and were constrained by the experimental data.

The model of Papale (1997) reproduced 198 out of 263 solubility data to within 30%. But this model relied heavily on those early data involving β-track measurements. The large number of fitting parameters (22 in total not counting the interaction parameters between non-volatile components and not counting the designated reference pressure) compared to the limited number of melt compositions is also a cause for concern. Furthermore, the invariably positive temperature dependence (due to the replacement of dissolution enthalpy with $H(P, T) - H(P_0, T)$ of pure CO_2 fluid) is inconsistent with experimental observation. Special care should be taken when applying the Papale (1997) solubility model to pressures > 0.5 GPa and melts that are not covered by the data set used for modeling.

CO_2 solubility in hydrous melts

H_2O and CO_2 are considered to be the two most important volatile components in natural silicate melts. Numerous experiments have been performed to equilibrate a melt with a binary CO_2-H_2O fluid (Table 1), in which H_2O is loaded in the form of water or oxalic acid ($H_2C_2O_4$ or $H_2C_2O_4{\cdot}2H_2O$) (e.g., Dixon et al. 1995; Tamic et al. 2001). Even in some experiments where H_2O was not deliberately added to the system, careful examination of the quenched products indicated the presence of H_2O in both the melt phase and the fluid phase (e.g., Lesne et al. 2011).

Fluid composition analysis. The obtained CO_2 solubility in melt is meaningful only if the composition of the coexisting fluid can be characterized. One method is to separate the CO_2 from the H_2O of the fluid phase according to their different boiling point and then measure the amount of each component with gravimetry or manometry (e.g., Dixon et al. 1995; Jakobsson 1997; Shishkina et al. 2010). Another way of determining fluid composition is based on mass balance considerations together with the measurements of volatiles in the starting material and in the quenched melt (e.g., King and Holloway 2002).

CO_2-H_2O solubility data. Studies on CO_2-H_2O solubility often involve a series of experiments performed at the same pressure, temperature, and melt composition (on volatile-free basis) but with varying CO_2/H_2O ratios. The early experiments at pressures at GPa level (e.g., Mysen 1976) observed an initial increase in CO_2 solubility with the addition of H_2O and explained this with the depolymerization effect on the melt by H_2O. However, the measurements were made by β-track autoradiography, and the composition of the quenched fluid was unknown. By contrast, Blank (1993) and Dixon et al. (1995) showed for rhyolite melt and basalt melt at < 0.1 GPa that H_2O only causes a dilution effect lowering CO_2 fugacity and hence reducing CO_2 solubility in the melt. At a given temperature and pressure, CO_2 solubility should be proportional to f_{CO_2} (Henrian behavior; Eqn. 3 with K being a constant) and should also be proportional to CO_2 fraction in the fluid phase (here denoted as Y_{CO_2}) because the CO_2 fugacity coefficient does not vary significantly with CO_2/H_2O at temperatures far above 1000 °C and pressures not higher than a few GPa.

Experimental results for a variety of silicate melts obtained at 0.2-1.0 GPa are summarized in Figure 5. The behavior of CO_2 solubility at these intermediate pressures appears to be in the middle of the behavior at higher pressures (Mysen 1976) and that at lower pressures (Blank 1993; Dixon et al. 1995). Only one study on andesite melt at 1.0 GPa (King and Holloway 2002) observed a positive correlation between the dissolved CO_2 and the dissolved H_2O (Fig. 5b), in agreement with Mysen (1976). All the other studies generally showed a negative correlation between CO_2 and H_2O within each individual data set, but CO_2 solubility flattens out when the concentration of dissolved H_2O becomes sufficiently low (Fig. 5a,b). At pressures of 0.5 GPa or higher, CO_2 solubility is nearly constant over a broad H_2O concentration range. Plots of CO_2 solubility versus CO_2 fraction in the fluid phase or CO_2 fugacity (Fig. 5c-f) demonstrate a nonlinear correlation (the deviation from linearity enlarges as pressure increases), which suggests an evident non-Henrian behavior. In addition, the data of rhyolite, dacite, andesite, and basalt suggest that the compositional effect becomes more pronounced at higher pressure (Fig. 5a,c,e). At 1.0 GPa, CO_2 solubility in icelandite (ferroandesite) melt is roughly three times larger than that in andesite melt (Fig. 5b-d).

There have also been a few studies with an extra volatile component such as He, Cl, or S in addition to CO_2 and H_2O (Paonita et al. 2000; Botcharnikov et al. 2007; Webster et al. 2011), but these studies typically emphasize the effect of CO_2 on the solubility of the extra volatile component rather than CO_2 solubility itself.

CO_2-H_2O solubility models. Based on several experimental studies and the assumption of Henrian behavior for CO_2-H_2O dissolution, Newman and Lowenstern (2002) developed a program VolatileCalc that can be used in Excel to calculate the solubility of CO_2-H_2O in rhyolite melt and basalt melt at 700-1500 °C and less than 0.5 GPa. However, the CO_2-H_2O solubility data in Tamic et al. (2001) showed marked deviation from the predictions by VolatileCalc. Liu et al. (2005) provided an easy-to-use empirical expression of CO_2 solubility in rhyolite melt as a function of the partial pressure of H_2O and that of CO_2, applicable to 700-1200 °C and 0.5 GPa. Figure 6 presents the calculation results for rhyolite melt at 1373 K according to the VolatileCalc program and the model of Liu et al. (2005). The feature of CO_2 solubility first increasing with H_2O concentration in Figure 6b may be an artifact due to some inconsistency between the CO_2 solubility data of Fogel and Rutherford (1990) and the CO_2-H_2O solubility data of Tamic et al. (2001), based on which the model of Liu et al. (2005) was constructed.

Papale (1999) extended his previous model on pure CO_2 or pure H_2O solubility (Papale 1997) to the solubility of two-component CO_2-H_2O fluids in 12-component (10 oxides + 2 volatiles) silicate melts. Papale et al. (2006) updated the model of Papale (1999) by adding a large amount of new CO_2-H_2O solubility data and discarding the pre-1980 CO_2 solubility data (mostly from β-track autoradiography). The highlight of Papale (1999) and Papale et al. (2006) was again the treatment of compositional dependences. Unlike Papale (1997), the

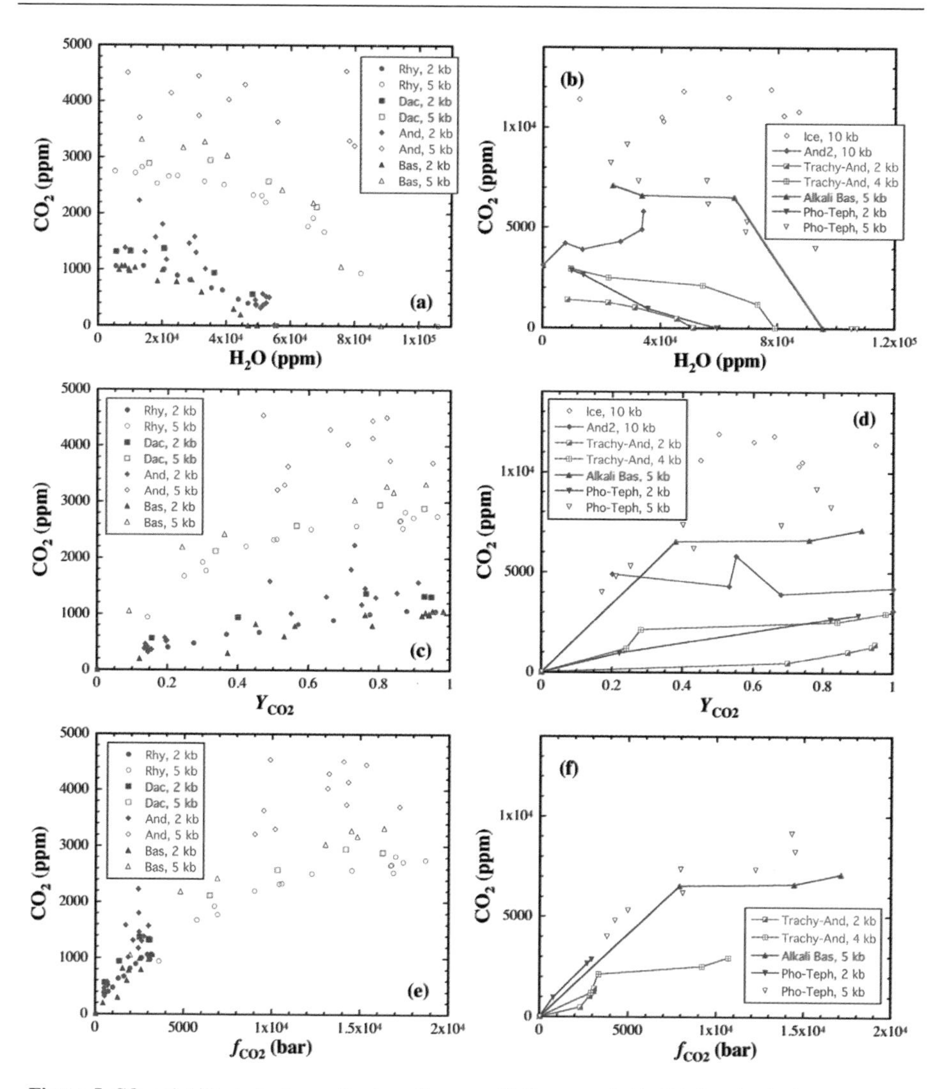

Figure 5. CO_2 solubility in hydrous silicate melts versus H_2O content (a,b), CO_2 fraction in the coexisting fluid (c,d), and CO_2 fugacity (e,f). Data sources: Rhy = rhyolite melt at 1373 K (Tamic et al. 2001); Dac = dacite melt at 1523 K (Behrens et al. 2004a); And = andesite melt at 1373-1573 K (Botcharnikov et al. 2006, 2007); Bas = basalt melt at 1473-1523 K (Botcharnikov et al. 2005; Shishkina et al. 2010); Ice = icelandite melt at 1673 K (Jakobsson 1997); And2 = andesite melt at 1573 K (King and Holloway 2002); Trachy-And = trachyandesite (shoshonite) melt at 1523 K (Vetere et al. 2011); Alkali Bas = alkali basalt melt at 1423 K (Botcharnikov et al. 2005); Pho-Teph = phonotephrite melt at 1473-1523 K (Behrens et al. 2009). CO_2 fugacities are derived from the equation of state for CO_2-H_2O fluids (Duan and Zhang 2006).

partial molar volume of CO_2 in the melt was assumed to be a 10-parameter (Papale 1999) and 3-parameter (Papale et al. 2006) function of pressure and temperature. Fitting of the experimental data indicated that the interaction between CO_2 and H_2O in the melt contributes negligibly to the excess Gibbs free energy of the melt. Papale et al. (2006) showed that their model reproduces most of the 173 CO_2 solubility data and 84 CO_2-H_2O solubility data (in terms of saturation pressure) within 25% relative.

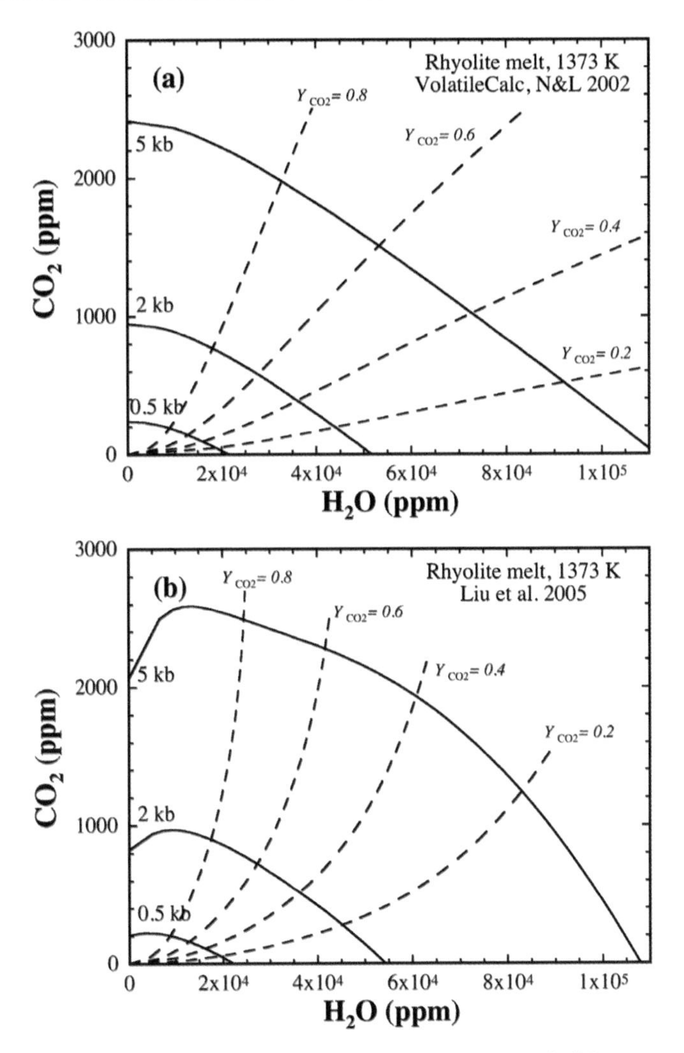

Figure 6. CO_2-H_2O solubility in rhyolite melt at 1373 K based on (a) the VolatileCalc program of Newman and Lowenstern (2002); and (b) the empirical model of Liu et al. (2005). The solid curves represent isobaric saturation, and the dashed curves represent fixed CO_2 mole fractions in the coexisting fluid phase.

The thermodynamic model of Papale et al. (2006) has indicated non-Henrian behavior of CO_2-H_2O solubility in rhyolite melt and basalt melt at high pressure (Fig. 7), in contrast with the Henrian behavior predicted by VolatileCalc (Fig. 6a). Based on Papale et al. (2006), for rhyolite melt at 0.5 GPa, CO_2 solubility first increases with dissolved H_2O in the melt; but for basalt melt at 0.5 GPa, CO_2 solubility decreases rapidly with the initial addition of H_2O. Later experimental data have demonstrated limited success of the Papale et al. (2006) model. Shishkina et al. (2010) showed that this model always underestimates volatile saturation pressure (in other words, the model overestimates CO_2-H_2O solubility) for basalt melt, opposite to the estimation by the VolatileCalc program (Fig. 8a). Furthermore, the non-ideality of CO_2-H_2O solubility in basalt melt is smaller than predicted by the Papale et al. (2006) model (Shishkina et al. 2010). There is also large disparity between the data for alkali basalt melt

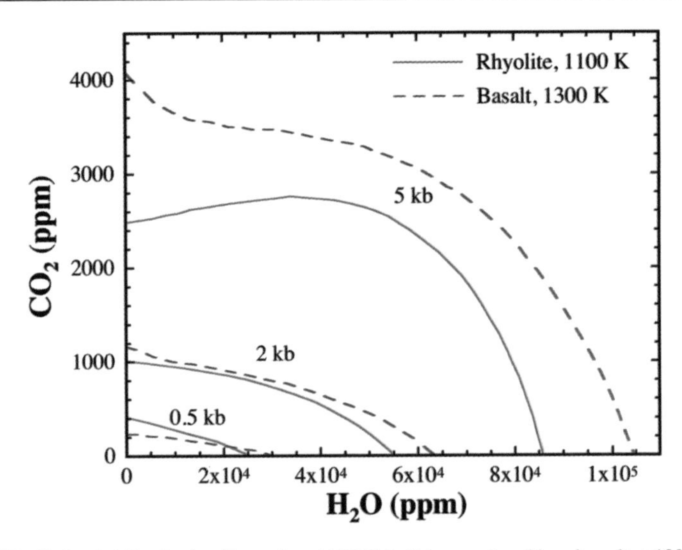

Figure 7. CO_2-H_2O solubility in rhyolite melt at 1100 K (solid curves) and basalt melt at 1300 K (dashed curves) based on the thermodynamic model of Papale et al. (2006).

(Lesne et al. 2011) and the model. Vetere et al. (2011) showed that the deviation between their data for trachyandesite melt at 0.05 GPa and 0.2 GPa and the model of Papale et al. (2006) is significant, but the data at 0.4 GPa agree with the model (Fig. 8b). The Papale et al. (2006) model may therefore need recalibration by new experimental data.

Solubility of C-O-H fluids under reduced conditions

The Archean Earth or certain deep regions in Earth's mantle today may be reduced enough for CO or CH_4 to be significant components in C-O-H fluids (Ballhaus 1995; Kump and Barley 2007; Frost and McCammon 2008). These species are expected to have different dissolution behavior in silicate melts.

CO-CO₂ solubility. At constant temperature and pressure, the CO/CO_2 ratio of a C-O fluid increases with decreasing oxygen fugacity until graphite saturation is reached (Fig. 9a). Under graphite saturation,

$$C(s) + CO_2(f) = 2CO(f) \tag{8}$$

where "*s*" and "*f*" denote the solid graphite phase and the fluid phase, respectively. The proportions of the two species calculated from the equilibrium constant of the above reaction vary significantly with temperature and pressure (Fig. 9b), which can be checked with post-experiment analysis on the fluid phase or the fluid inclusions in quenched glass (Pawley et al. 1992).

Eggler et al. (1979) showed for several melts that CO-CO_2 solubility is comparable (lower at subliquidus temperatures but higher at superliquidus temperatures) with the solubility of pure CO_2, and argued that CO is also soluble in silicate melts. However, their data were acquired by the potentially unreliable β-track autoradiography. On the contrary to Eggler et al. (1979), Pawley et al. (1992) demonstrated that at < 0.15 GPa the dissolved carbon (in the form of CO_3^{2-}) in a basalt melt is proportional to CO_2 fugacity, and hence concluded that the role of CO is limited to diluting CO_2 (i.e., the solubility of CO in the melt is much smaller than that of CO_2).

Solubility of reduced C-O-H fluids. Holloway and Jakobsson (1986) and Jakobsson and Holloway (1986) investigated the solubility of C-O-H fluids in several silicate melts at 1.0-2.5

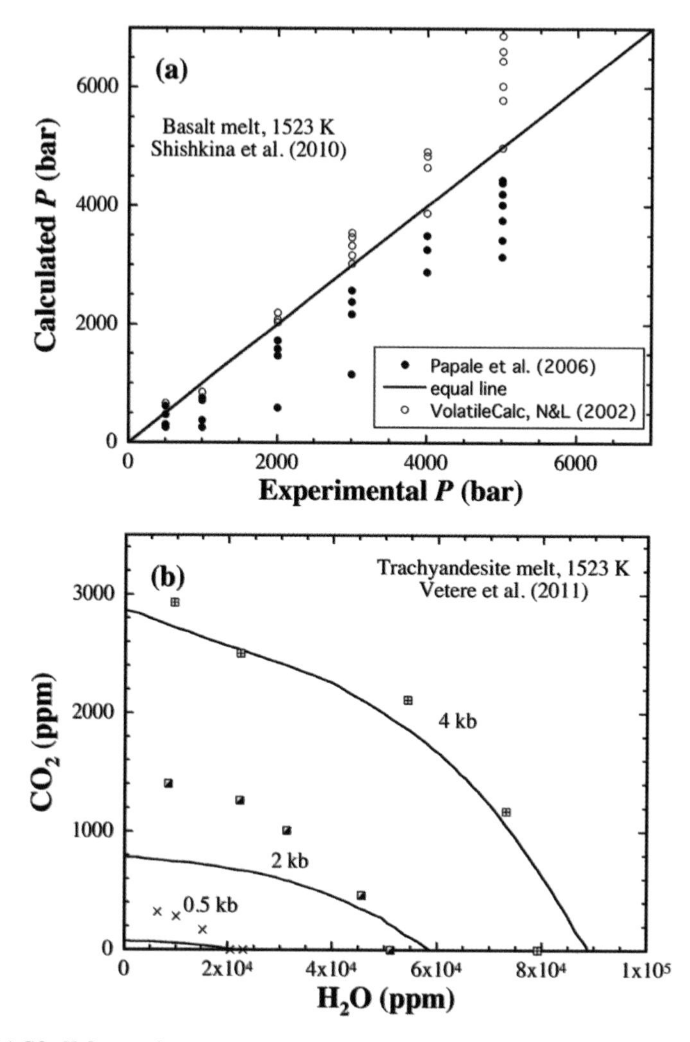

Figure 8. (a) CO_2-H_2O saturation pressure calculated from the Papale et al. (2006) model (solid circles) and the VolatileCalc program (open circles) of Newman and Lowenstern (2002) versus experimental pressure for basalt melt at 1523 K. After Shishkina et al. (2010). (b) Experimental CO_2-H_2O solubility in trachyandesite melt at 1523 K compared to the Papale et al. (2006) model (solid curves). After Vetere et al. (2011).

GPa under graphite saturation at the iron-wüstite (IW) buffer. For a given temperature and pressure, these two constraints (carbon activity is one and oxygen fugacity is at IW buffer) together with the mass balance constraint and the equilibrium constants of the following reactions,

$$CO + \frac{1}{2}O_2 = CO_2 \tag{9a}$$

$$H_2 + \frac{1}{2}O_2 = H_2O \tag{9b}$$

$$CO + 3H_2 = CH_4 + H_2O \tag{9c}$$

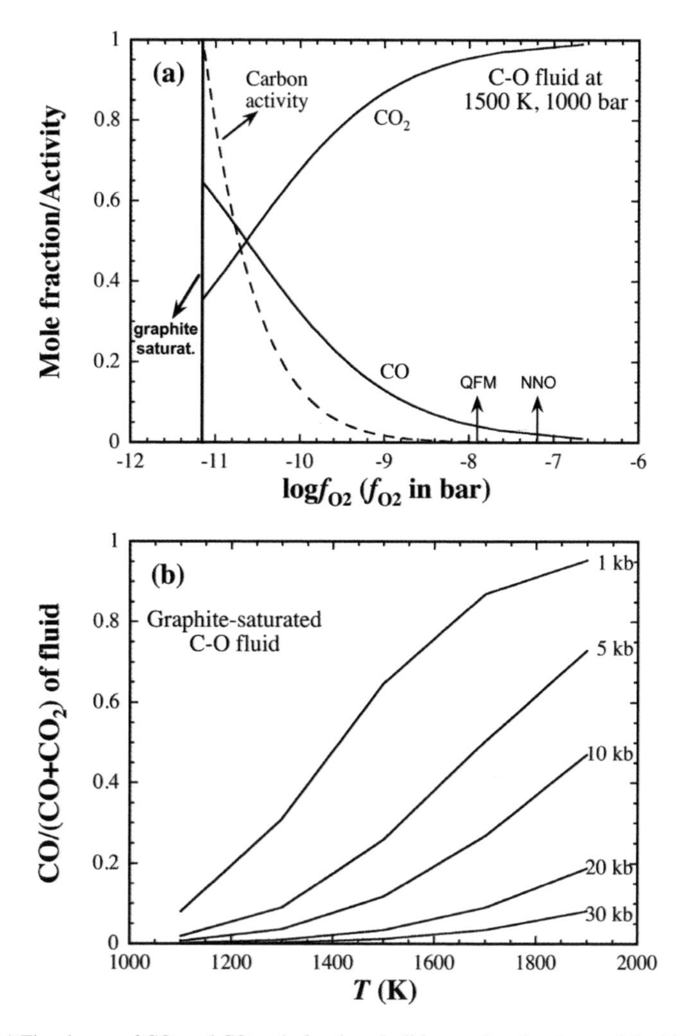

Figure 9. (a) The change of CO_2 and CO mole fractions (solid curves) and carbon activity (dashed curve) with oxygen fugacity for C-O fluid at 1500 K and 0.1 GPa. Oxygen buffers of NNO and QFM are indicated in short vertical arrows. CO_2 fraction approaches minimum at graphite saturation (carbon activity equals 1). (b) CO fraction of graphite-saturated fluid at various temperatures and pressures. Calculations are based on the standard thermodynamic data in the JANAF table (Chase 1998) and the fugacity coefficients using the GFluid program (Zhang and Duan 2010).

fix the composition of the fluid (the fractions of CH_4, CO, CO_2, H_2, O_2, and H_2O). Under the experimental conditions of Holloway and Jakobsson (1986) and Jakobsson and Holloway (1986), the dominating volatile species are CH_4 and H_2O, with minor amounts of H_2 and CO (Fig. 10). Jakobsson and Holloway (1986) suggested that CO is more soluble than CO_2 and CH_4 in silicate melts. However, their conclusion was purely based on quadrupole mass spectrometry. There was no FTIR or Raman evidence for the presence of molecular CO in the melt. Furthermore, high CO solubility appears to be inconsistent with low-pressure studies (Pawley et al. 1992; Morizet et al. 2010).

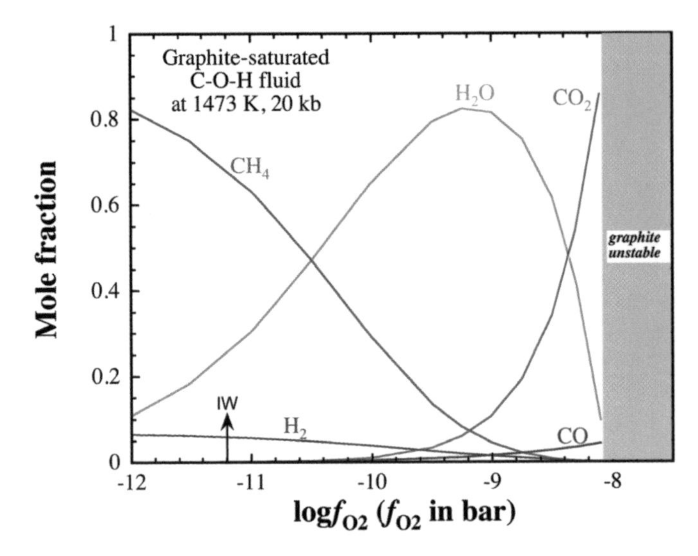

Figure 10. Mole fractions of various species in C-O-H fluid at 1473 K, 2.0 GPa and graphite saturation, calculated with the GFluid program (Zhang and Duan 2010). Oxygen buffer of iron-wüstite (IW) is indicated in short vertical arrow.

Kadik et al. (2004) also investigated the dissolution of C-O-H species in ferrobasalt melt at 3.7 GPa and 1800-1873 K under graphite saturation and IW buffer conditions. Their IW buffer fixed hydrogen fugacity instead of oxygen fugacity because it was placed outside the capsule, hence ΔIW with respect to oxygen was about −2.4. The system was so reduced that an iron metallic phase formed at equilibrium. Strictly speaking, their experiments were not really a solubility study under our definition as a fluid phase was absent. They concluded that the dissolved carbon (1600 ppm as C) should mostly be either atomic or amorphous based on the lack of observable bands in FTIR and Raman spectroscopy. Mysen et al. (2009) showed that CH_4 is quite soluble (2000-5000 ppm as CH_4) in Na_2O-SiO_2 melts at 10-2.5 GPa and 1673 K as molecular CH_4 or possibly species containing C≡C–H bonds. However, the composition of the coexisting fluid in their experiments was not examined, and may not be a binary oxygen-free CH_4+H_2 mixture as assumed. Under extremely reduced conditions (ΔIW ranging from −3.7 to −5.6 with respect to oxygen fugacity), Kadik et al. (2006, 2010, 2011) reported that the carbon dissolved in graphite-saturated iron-bearing melts at 1.5-4.0 GPa and 1673-1873 K was present as CH_4 species, based on Raman spectroscopy. Note that similar to Kadik et al. (2004), there was no free C-O-H fluid phase in coexistence with the melt.

Morizet et al. (2010) used Ar-H_2 gas as the pressure medium of internally-heated pressure vessels to produce intermediate reduced conditions (ΔQFM within ±2.6), under which CO_2, H_2O, and CO are the major species in the fluid phase. Based on a series of experiments at 1523 K and 0.2-0.3 GPa, they concluded that CH_4 and CO are essentially insoluble in haplobasalt melt.

CARBON SPECIATION IN SILICATE MELTS

Spectroscopic information on speciation

Infrared and Raman spectroscopy. Infrared and Raman spectroscopy are two types of vibrational spectroscopy, i.e., they probe the interaction of electromagnetic radiation with vibrations that occur in a molecule or in a condensed phase. The information obtained from both types of spectroscopy can be complementary, as due to different selection rules, some vibrations may only be detected in the infrared spectrum, while others are only Raman active. The normal modes (independent vibrations) of molecular CO_2 are shown in Figure 11 and Table 2, and those of the carbonate ion (CO_3^{2-}) are shown in Figure 12 and and Table 3. For the linear CO_2 molecule containing 3 atoms, there are $3 \times 3 - 5 = 4$ independent vibrations. Both the symmetric bending vibration at 667 cm^{-1} and the antisymmetric stretching vibration at 2349 cm^{-1} are infrared active; however, in silicate glasses, there is strong absorption from the glass matrix in the frequency range of the bending vibration, so that normally it cannot be observed. The antisymmetric stretching vibration is prominent in the spectra of glasses containing molecular CO_2 (Fig. 13), as the corresponding extinction coefficients are high and the absorption from the glass matrix in this frequency region is negligible. The symmetric stretching vibration near 1337 cm^{-1} may be observed in Raman spectra of CO_2-bearing glasses. Fortuitously, this frequency is nearly twice the frequency of the symmetric bending vibration at 667 cm^{-1}. In this situation, Fermi resonance may occur, where the first overtone of the bending vibration gains intensity by interaction with the symmetric stretching vibration. Therefore, a pair of bands ("Fermi diade") may be observed in the Raman spectrum of CO_2 gas, while usually only one band near 1382 cm^{-1} is seen in the spectra of glasses containing molecular CO_2 (Fig. 14). This can be useful to distinguish CO_2 in gas bubbles from CO_2 dissolved in the glass (e.g., Brooker et al. 1999).

For the carbonate group, there are $3 \times 4 - 6 = 6$ independent vibrations. As with molecular CO_2, the bending vibrations are usually hidden by the absorbance of the glass matrix in the infrared spectrum. However, the ν_3 antisymmetric stretching vibration is very intense in the infrared spectrum and only slightly overlaps with background absorption from the glass matrix. As noted in Table 3, this vibration is twofold degenerate, i.e., there are two physically different vibrations (with atoms vibrating in different directions), which in the undistorted CO_3^{2-} group have the same frequency. This is obvious from Figure 12: Rotating the image showing the movement of atoms during the antisymmetric stretching vibration by 120° or 240° produces vibrations of the same type, but with individual atoms moving in different directions. A more thorough analysis shows that of these three vibrations, only two are independent, the third one can be produced as a combination of the other two vibrations. The degeneracy of these

Table 2. Internal vibrations of the CO_2 molecule

Mode	Selection rule	Frequency (cm^{-1})
ν_1 symmetric stretch	Raman	1337
ν_2 symmetric bend	IR	667
ν_3 asymmetric stretch	IR	2349

The ν_2 symmetric bending vibration is two-fold degenerate, i.e., there are two physically different vibrations, which in a free molecule have the same frequency for symmetry reasons.

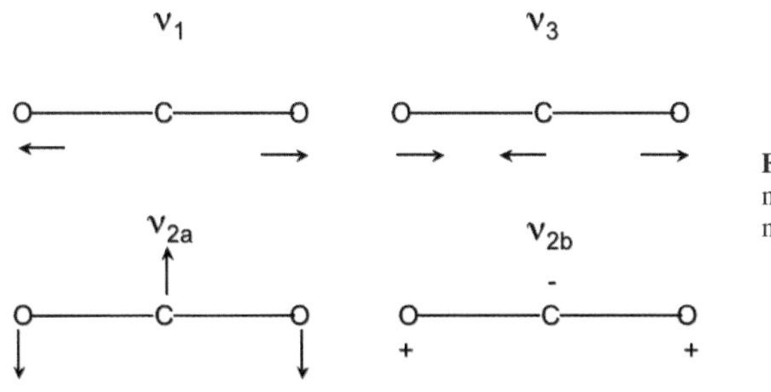

Figure 11. Normal modes of the CO_2 molecule

Table 3. Internal vibrations of the carbonate group

Mode	Symmetry	Selection rule	Frequency (cm^{-1})
ν_1 symmetric stretch	A_1'	Raman	1063
ν_2 out-of-plane bend	A_2'	IR	879
ν_3 asymmetric stretch	E′	IR + Raman	1415
ν_4 in-plane bend	E′	IR + Raman	680

The ν_3 and ν_4 vibrations are both two-fold degenerate, i.e., there are two physically different vibrations each, which in an undistorted carbonate group have the same frequency for symmetry reasons. Symmetry types are for trigonal D_{3h} symmetry.

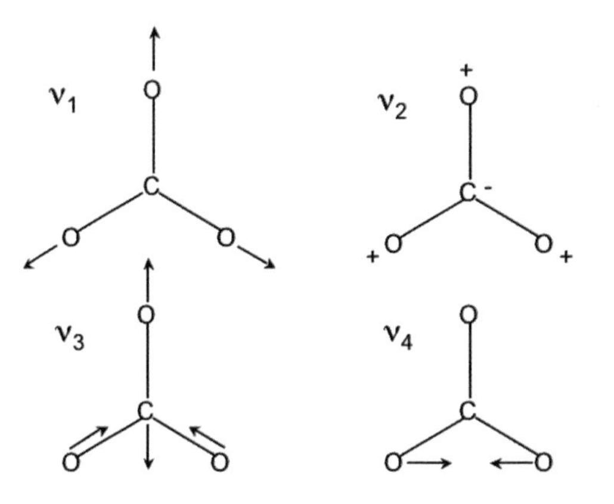

Figure 12. Normal modes of the carbonate group (CO_3^{2-}), for ideal trigonal D_{3h} geometry.

vibrations, however, will disappear if the environment of the carbonate group is asymmetric, e.g., if one of the oxygen atoms is more strongly bonded than the others. In such a situation, the two vibrations will have different frequencies and the asymmetric stretching vibration in the infrared spectrum will split into two components. This effect is observed in carbonate-bearing silicate glasses (Fig. 13) and the type of splitting observed contains valuable information about the environment of the carbonate group. Note that whatever the distortion in the environment of the carbonate group is, only one band (in a symmetric environment) or two bands (in a distorted environment) may be produced by one type of carbonate group. If there are more bands, there must be more structurally different carbonate groups. In theory, two infrared bands could also be produced by two structurally different carbonate groups, each residing in a symmetric environment. In such a situation, however, the Raman spectrum should also show two symmetric stretching bands at different frequencies and separate carbonate peaks should also be observed in the NMR spectrum. The most prominent band in the Raman spectrum of carbonate is the symmetric stretching vibration at 1063 cm^{-1}, which, however, often overlaps with the Si-O stretching vibrations of the glass matrix (Fig. 14).

Infrared spectroscopy measures absorption of infrared radiation, while Raman spectroscopy measured light scattering, often in the visible range. In general, infrared spectra can be more easily quantified than Raman spectra, since absorbance can routinely be measured with an accuracy of 1% relative or better. For this reason and because the infrared bands of the antisymmetric stretching vibrations of CO_2 and carbonate are well separated from the absorbance of the glass matrix, infrared spectroscopy has been used extensively to study the speciation of carbon in silicate glasses. Moreover, it has also been widely used as an analytical tool to derive CO_2 contents from the measured absorbance for CO_2 and carbonate using the Lambert Beer law $A = \varepsilon\, c\, d$, where A is linear or integral absorbance, ε is the molar extinction

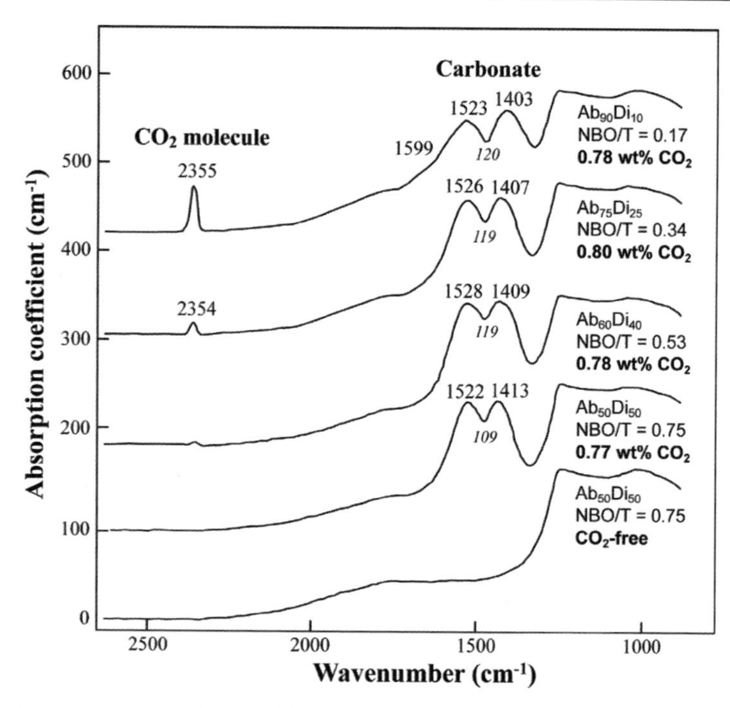

Figure 13. Infrared spectra of some CO_2-bearing glasses along the albite ($NaAlSi_3O_8$) - diopside ($CaMgSi_2O_6$) join. Bulk CO_2 content is about 0.8 wt%. While molecular CO_2 and carbonate coexist in the albite-rich compositions, the more diopside-rich compositions only contain carbonate. Note the splitting of the carbonate band into two components due to a low-symmetry environment. After Konschak (2008).

coefficient, c is concentration (in mol/l), and d is the sample thickness. Extinction coefficients of both molecular CO_2 and carbonate are matrix-dependent and need to be calibrated against some other analytical method that measures absolute carbon. Extinction coefficients for the antisymmetric stretching vibrations of molecular CO_2 and of carbonate are compiled in Table 4.

NMR spectroscopy. ^{13}C NMR (nuclear magnetic resonance) spectroscopy probes the chemical environment of a ^{13}C nucleus by measuring the energy required to change the orientation of this nucleus in a very strong external magnetic field (several Tesla). The local field seen by the nucleus will be shielded by the surrounding electron shell. Therefore, there is a chemical shift of the absorption frequency depending on the chemical environment of the nucleus; this shift is usually given as shift in ppm (i.e., 10^{-6}) relative to a standard (TMS, tetramethylsilane). In a solid material, such as a glass, dipolar interactions between neighboring nuclei will tend to broaden NMR peaks to such an extent that no structural information can be obtained. This effect can be suppressed by rapid (kHz) rotation of the sample at the magic angle (54°44′) relative to the magnetic field. This MAS (magic angle spinning) technique is therefore routinely used to acquire ^{13}C NMR spectra of carbon-bearing glasses (Fig. 15). Compared to infrared spectroscopy, ^{13}C NMR has been less frequently used to study carbon in glasses, however, it can yield useful complementary information. One particular advantage of ^{13}C NMR is that it is intrinsically quantitative; the areas of the NMR peaks are directly proportional to species abundance, or in other words, the intensity ratio of two peaks directly gives the abundance ratio of the corresponding species. Moreover, all carbon species in a sample will be detected, even if they do not possess any infrared active bands or bands that are only weak or overlap with other bands in the infrared spectrum.

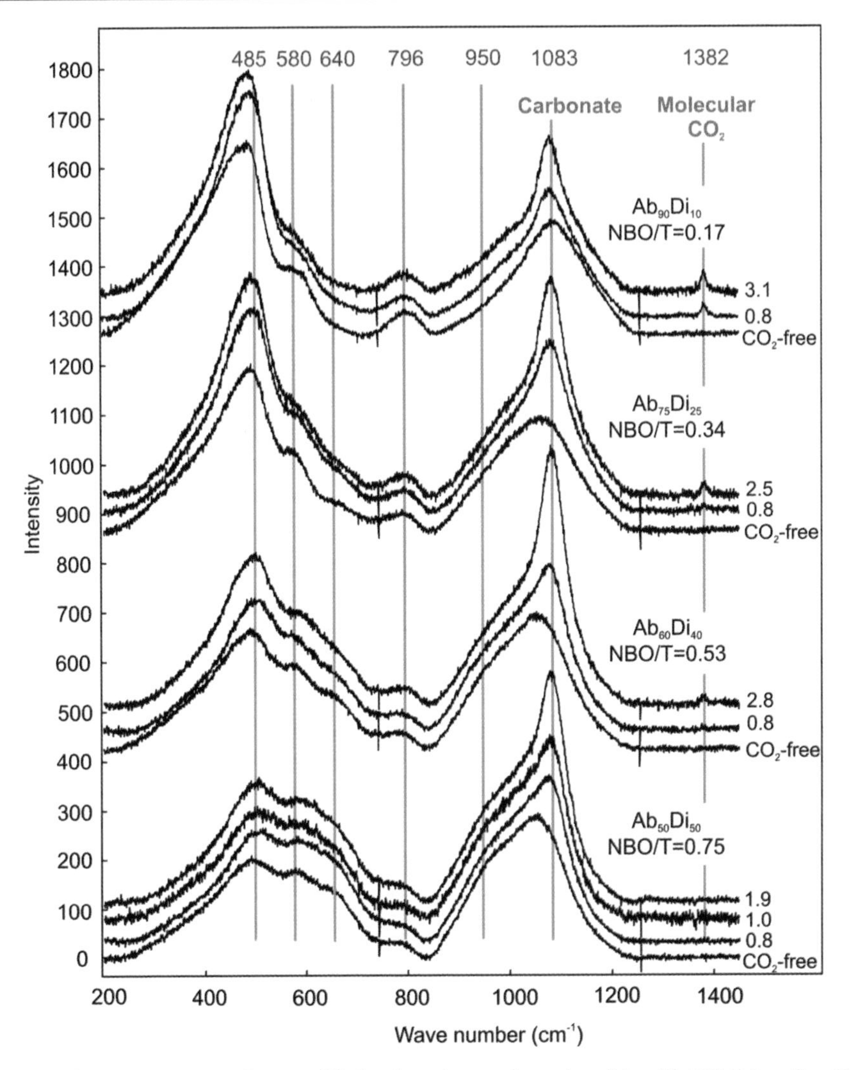

Figure 14. Raman spectra of some CO_2-bearing glasses along the albite ($NaAlSi_3O_8$) - diopside ($CaMgSi_2O_6$) join. For each glass composition, spectra with different bulk CO_2 content (in wt%) are shown. After Konschak (2008).

Carbon speciation in silicate glasses

Variation of carbonate and molecular CO_2 as function of glass composition. Wyllie and Tuttle (1959) noted that the differences in CO_2 solubility between felsic and mafic or ultramafic melts are likely due to the formation of carbonate in the latter compositions. Indeed, infrared and Raman spectroscopic studies of glasses later showed (e.g., Brey 1976; Fine and Stolper 1986; Stolper et al. 1987) that all CO_2 is dissolved as carbonate in basaltic glasses, while rhyolite, albite and other silica-rich glasses contain molecular CO_2 coexisting with at most minor amounts of carbonate. In andesite and phonolite glasses, molecular CO_2 and carbonate coexist (e.g., Brooker et al. 2001b). If melt compositions are expressed as NBO/T (i.e., non-bridging oxygen atoms per tetrahedron), increasing NBO/T or depolymerization of the melt

Table 4. Infrared extinction coefficients ε of CO_2 and carbonate group in silicate glasses

Composition	Reference	integral ε of CO_2 (L mol^{-1} cm^{-2})	linear ε of CO_2 (L mol^{-1} cm^{-1})	integral ε of CO_3^{2-} (L mol^{-1} cm^{-2})	linear ε of CO_3^{2-} (L mol^{-1} cm^{-1})	carbonate band (cm^{-1}) for linear ε
Albite – jadeite	Fine & Stolper (1985)	25 200 ± 1200	945 ± 45	24 100 ± 1900* 16 800 ± 1500 *	200 ± 15 235 ± 20	1610 1375
Albite	Nowak et al. (2003)	18 000 ± 1000		42 250 ± 2000		
Rhyolite	Behrens et al. (2004b)		1214 ± 78			
Dacite	Nowak et al. (2003)	16 000 ± 1000		40 100 ± 2000		
Dacite	Behrens et al. (2004a)		830		170	1530
Icelandite	Jakobsson (1997)			54450 27730* 27700*	 180 190	 1500 1420
Phonotephrite	Behrens et al. (2009)				308 ± 110	1430
Basanite	Dixon & Pan (1995)			60 000 ± 1700	283 ± 8	1525 and 1425
Ca-rich leucitite	Thibault & Holloway (1994)				340 ± 20	1515
Basalt**	Fine and Stolper (1986)			69500 ±3000	375 ± 20	1515 and 1435
Basalt	Shishkina et al. (2011)				317 ± 23	1430
Ferrobasalt	Stanley et al. (2011)			81500 ± 1500		
Shoshonite	Vetere et al. (2011)				356 ± 18	1430

Integral extinction coefficients are calibrated using the peak area, while linear extinction coefficients refer to peak height. For carbonate, there are usually two bands in the range of 1350 - 1650 cm^{-1}. The integral extinction coefficients refer to the integral under both bands, unless otherwise noted, while the linear extinction coefficients refer to the height of one band, which is specified in the last column of the table. All data for molecular CO_2 refer to the band near 2351 cm^{-1}. * these integral extinction coefficients refer to the individual carbonate bands specified in the last column. **this calibration includes basalt, diopside and a calcium aluminosilicate glass

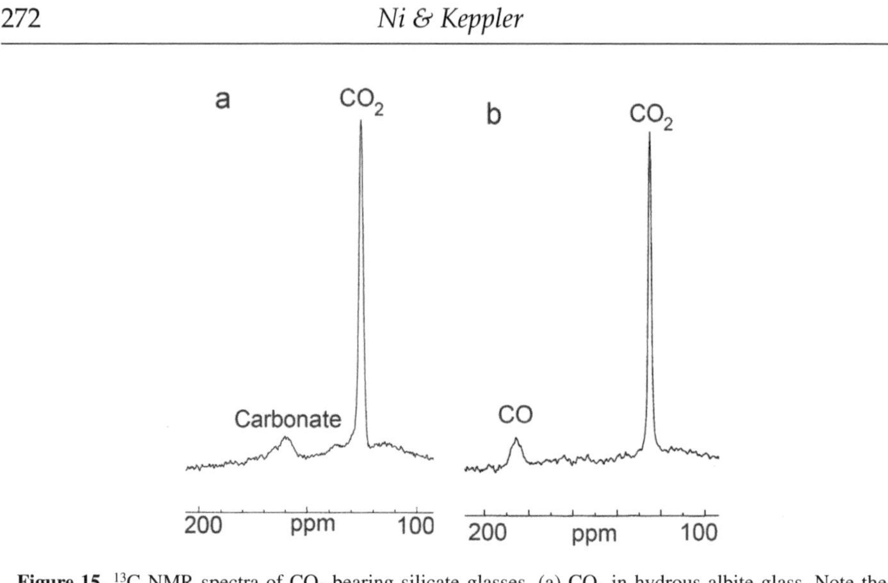

Figure 15. ^{13}C NMR spectra of CO_2-bearing silicate glasses. (a) CO_2 in hydrous albite glass. Note the asymmetry of the carbonate peak, suggesting the presence of more than one carbonate species. (b) CO_2 in rhyolite glass containing some CO and a trace of carbonate. Spectra courtesy of Simon Kohn.

favors the formation of carbonate in the glasses at the expense of molecular CO_2. This effect is illustrated in the infrared spectra of some glasses along the albite ($NaAlSi_3O_8$) - diopside ($CaMgSi_2O_6$) join in Figure 13. The corresponding Raman spectra are shown in Figure 14. Note that in the infrared spectrum, the degeneracy of the antisymmetric stretching vibration of carbonate is lifted, producing two bands separated by 109-120 cm^{-1}. On the other hand, there is one single symmetric stretching vibration of carbonate at 1083 cm^{-1} in the Raman spectrum (overlapping with the Si-O stretching vibrations), which confirms that these glasses contain one single type of carbonate group in some asymmetric environment. For the albite-rich compositions, both the Raman and infrared spectra show only one band for molecular CO_2, the antisymmetric stretching vibration in the infrared spectra at 2355 cm^{-1} and the symmetric stretching vibration at 1382 cm^{-1} in the Raman spectra. Data from ^{13}C NMR spectra (Fig. 15) are generally consistent with the speciation models derived from infrared spectra.

Given that the relative abundance of molecular CO_2 and carbonate appear to depend on the availability on non-bridging oxygen atoms in the glass, one may write equilibrium of the type:

$$CO_2 + O^{2-} = CO_3^{2-} \tag{10}$$

where "O^{2-}" stands for a non-bridging oxygen atom (e.g., Eggler and Rosenhauer 1978). The equilibrium constant for this reaction would then predict that the carbonate/CO_2 ratio should increase linearly with the activity of NBO in the melt or glass. However, for a given glass composition, the carbonate/CO_2 ratio should be independent of the bulk CO_2 concentration, which agrees with observation if glasses are produced under otherwise identical conditions. Note that the above equation also implies that a non-bridging oxygen atom belonging to some tetrahedrally coordinated ion such as Si^{4+} or Al^{3+} is being incorporated into the carbonate group, that is, the carbonate group may become attached to the silicate network of the glass or melt. This idea is adopted in many speciation models (e.g., Brooker et al. 2001b) and appears to be consistent with the result from molecular dynamics (Guillot and Sator 2011).

The degree of polymerization or the NBO/T ratio is, however, certainly not the only parameter that controls the carbonate/CO_2 ratio in glasses. Brooker et al. (1999) studied glasses along the join $NaAlO_2$-SiO_2, which should all be fully polymerized (NBO/T = 0).

However, while there is mostly molecular CO_2 and very little carbonate in albite ($NaAlSi_3O_8$) glass, carbonate is prominent and molecular CO_2 nearly absent in nepheline ($NaAlSiO_4$) glass. Moreover, replacing 2 Na^+ by 1 Ca^{2+} appears to strongly enhance carbonate at the expense of molecular CO_2 (Brooker et al. 2001b).

The nature of the carbonate groups. The splitting of the ν_3 asymmetric stretching vibration of carbonate in glasses as seen in infrared spectra (Fig. 13) suggests that the carbonate group is in some asymmetric environment. For most natural glass compositions (basalt, andesite, phonolite), the two band components are separated by 80-100 cm^{-1} (Brooker et al. 2001b), similar to the splitting observed in glasses of the albite-diopside join in Figure 13. However, a wide range of splittings $\Delta\nu_3$ have been observed in different synthetic glass systems and sometimes several distinct carbonate species coexist (Brooker et al. 1999, 2001b). (1) Very large splittings (215-295 cm^{-1}) can be observed in fully polymerized sodium aluminosilicate melts, e.g., along the albite-nepheline join. (2) In alkali silicate glasses, two carbonate groups may coexist, one with $\Delta\nu_3 \approx 300$ cm^{-1} and one with $\Delta\nu_3 \approx 35$ cm^{-1}. (3) Adding small amounts of Mg and particularly of Ca to fully polymerized glasses causes an abrupt change in the environment of the carbonate group. For Mg, new bands with $\Delta\nu_3 \approx 168$ cm^{-1} appear, while for Ca, bands with $\Delta\nu_3 \approx 80$ cm^{-1}, similar to those observed in natural glasses, become predominant in the spectra.

The observation that the splitting of the ν_3 bands of the carbonate group in natural glass composition is very similar to that observed upon addition of Ca to various base compositions strongly suggest that in these glasses, the carbonate ion is somehow associated with the Ca^{2+} ion. Moreover, the fact that the corresponding bands appear already when a small amount of Ca^{2+} is added implies that the association between Ca^{2+} and carbonate in the glass is very stable. Brooker et al. (2001b) suggested that the carbonate in Ca-bearing systems with a typical $\Delta\nu_3 \approx 80$ cm^{-1} is related to a carbonate group close to a Ca^{2+} ion and attached via a non-bridging oxygen atom to a silicate or aluminate tetrahedron. The carbonate groups with very large $\Delta\nu_3$ in fully polymerized sodium aluminosilicate systems may form bridges between two tetrahedra, while some peralkaline glasses, where $\Delta\nu_3$ is negligible, may contain carbonate groups not attached to any NBO and surrounded only by alkali ions.

The band assignments made above are plausible and in general agreement with predictions from molecular orbital calculations (Kubicki and Stolper 1995). However, the observed $\Delta\nu_3$ strictly is only a measure of the distortion in the environment of the carbonate group and by itself does not imply chemical bonding to a specific ion. Similar splittings as observed in glasses can sometimes be seen in crystalline carbonates. In simple, calcite-structure carbonates, there is only one ν_3 band (at 1435 cm^{-1} for calcite and at 1450 cm^{-1} for magnesite; White 1974). Small splittings are seen in alkali carbonates, such as Na_2CO_3 (1413 and 1425 cm^{-1}; White 1974), while much larger splittings occur in double carbonates such as shortite $Na_2Ca_2(CO_3)_3$. Shortite contains two crystallographically distinct carbonate groups (Dickens et al. 1971), yielding a total of four infrared bands at 1522, 1481, 1453, and 1410 cm^{-1} (White 1974). Both carbonate groups in the shortite structure are bonded to two Ca ions and one Na ion in the plane of the carbonate group and the resulting asymmetry in the environment is believed to cause the splitting of ν_3 (Taylor 1990). A relatively large splitting of the carbonate band ($\Delta\nu_3$ ≈ 100 cm^{-1}) has also been observed for scapolite, although the carbonate group in this mineral is not attached to a silicate tetrahedron (Papike and Stephenson 1966).

In some early studies, changes in the Si-O stretching region of the Raman spectra upon dissolution of CO_2 in the glass were interpreted in terms of CO_2 solubility mechanisms. Mysen and Virgo (1980a,b) suggested that CO_2 depolymerizes albite and anorthite glasses, while it polymerizes diopside and $NaCaAlSi_2O_7$ glass. However, the changes in the Raman spectra are generally very subtle (Fig. 14) and the models for the deconvolution and assignment of individual band components are not unique. Furthermore, while it is plausible that carbonate is associated with some cations, in particular with Ca^{2+} in Ca-bearing glasses, there is little spectroscopic evidence that would suggest the formation of cation-carbonate complexes, in the sense of stable, molecule-like units. Indeed, the position and splitting of carbonate bands observed in most glasses is well within the range of parameters observed for crystalline carbonates or minerals such as scapolite (see above), where carbonate is coordinated by some alkali or alkaline earth cations, but where discrete molecule-like carbonate-cation complexes do not exist. Molecular dynamics studies of CO_2 in silicate melts also failed to find evidence for such complexes (Guillot and Sator 2011).

The nature of molecular CO_2 dissolved in silicate glass. The antiymmetric stretching frequency of molecular CO_2 in glasses (near 2350 cm^{-1}, see also Fig. 13) is very close to the values observed to gaseous CO_2 (2348 cm^{-1}), implying a generally similar geometry of the molecule and only weak interactions with the host glass matrix. This is consistent with the very slight difference in ^{13}C chemical shift between pure CO_2 gas (124.2 ppm) and CO_2 in silicate glasses (125 ppm; Kohn et al. 1991; Brooker et al. 1999; Morizet et al. 2002). The CO_2 band observed in the glass, however, does not show any rotational fine structure, implying that the molecule cannot rotate freely. Also, the intensity ratio of the Fermi diade in the Raman spectra of the glasses is very different from gaseous CO_2 (Brooker et al. 1999) and the infrared extinction coefficients vary considerably with glass composition (Table 4). These observations suggest that there must be some, although weak, interaction between the CO_2 molecule and the matrix. Molecular orbital calculations suggest that the CO_2 molecule in silicate glasses has a slightly bent geometry, with a O-C-O angle of 168-179° (Tossel 1995; Kubicki and Stolper 1995). The molecular dynamics model of Guillot and Sator (2011) suggest that in silicate melts, the CO_2 molecules are not randomly distributed through the melt, but preferentially located near oxygen atoms, with a preference for non-bridging oxygen atoms.

Other carbon species in glasses. Carbon monoxide (CO) has sometimes been detected as a minor species in glasses prepared under reducing conditions. In the ^{13}C NMR spectra (Fig. 15b), it may occur as a minor peak at 183 ppm (Brooker et al. 1999). Under extremely reduced conditions, carbon may be present in atomic or amorphous form (Kadik et al. 2004) or even in the form of CH_4 species (Kadik et al. 2006, 2010; Mysen et al. 2009).

Equilibrium carbon speciation in silicate melts

Annealing experiments. Brey (1976) noted that infrared spectra of quenched glasses show only carbonate bands for depolymerized compositions, while in albite glass molecular CO_2 occurs. Brey (1976) suggested that in the albite melts at high temperature, CO_2 was also dissolved as carbonate and reverted to molecular CO_2 upon quenching. Stolper et al. (1987) noted that the ratio of carbonate to molecular CO_2 in quenched albite glasses appeared to depend slightly on run temperature, with higher temperatures shifting the equilibrium towards carbonate. However, the structure of glasses represents only the structure of the melt at the glass transformation temperature T_g. Above T_g, structural relaxation is so fast that it cannot be preserved during quenching. Accordingly, it is unlikely that variations in melt structure as function of run temperatures could be directly observed in quenched glasses. Nowak et al. (2003) later suggested that the variations observed by Stolper et al. (1987) may be the result of subtle variations in water content that affect the glass transformation temperature.

Direct evidence for the true temperature dependence of carbon species was provided by Morizet et al. (2001) and Nowak et al. (2003), who carried out annealing experiments of CO_2-bearing glasses below the glass transformation temperature. In both studies, it was observed that increasing annealing temperature shifts the equilibrium towards molecular CO_2 and not towards carbonate, as previously assumed. These studies also indicate that CO_2 speciation is decoupled from the relaxation of the bulk glass structure, i.e., the equilibrium between molecular CO_2 and carbonate can be reset at temperatures where relaxation of the bulk glass structure is not expected to occur.

Morizet et al. (2001) annealed CO_2-bearing, fully polymerized jadeite glasses at 400 to 575 °C in a 1-atm furnace for variable run durations and quenched the samples rapidly to room temperature. They found that in the experiment at 575 °C, the ratio of molecular CO_2 to carbonate first increases sharply for annealing times of less than one hour and then apparently reached some equilibrium value, while at 400 and 450 °C, the CO_2/carbonate ratio decreased. From their data, they concluded that the equilibrium between molecular CO_2 and carbonate shifts towards molecular CO_2 at high temperature. For jadeite glass and melt, they estimated standard state thermodynamic data for the speciation reaction (10) of $\Delta H = -17$ (+4/−8) kJ mol^{-1} and $\Delta S = -24$ (+6/−9) J mol^{-1} K^{-1}. Morizet et al. (2001) also give kinetic data for the rate constants of the interconversion between CO_2 and carbonate. For the forward reaction (10) they find an activation energy of 68 (+3/−31) kJ mol^{-1} and for the reverse reaction of 86 (+1 /−69) kJ mol^{-1}.

The annealing experiments of Morizet et al. (2001) were carried out at 1 atm, where CO_2 should ultimately exsolve from the glass. Therefore, it is conceivable that they do not fully represent (metastable) thermodynamic equilibrium. However, Nowak et al. (2003) carried out similar annealing experiments under pressure in an internally-heated gas pressure vessel using albite and dacite glasses and obtained results that broadly agree with those of Morizet et al. (2001). Some of the results of Nowak et al. (2003) are shown in Figure 16. Annealing CO_2-bearing albite and dacite glass below the glass transformation temperature appears to reset the equilibrium between molecular CO_2 and carbonate such that with increasing temperature, molecular CO_2 becomes more abundant and carbonate decreases. Notably, Nowak et al. (2003) could also show reversibility of the CO_2 speciation equilibrium, that is, the speciation observed in a dacite glass at 879 K was the same for glasses first annealed at 973 K or at 789 K. For the fully polymerized albite glass, they obtained $\Delta H = -12$ (±3) kJ mol^{-1} and ΔS =

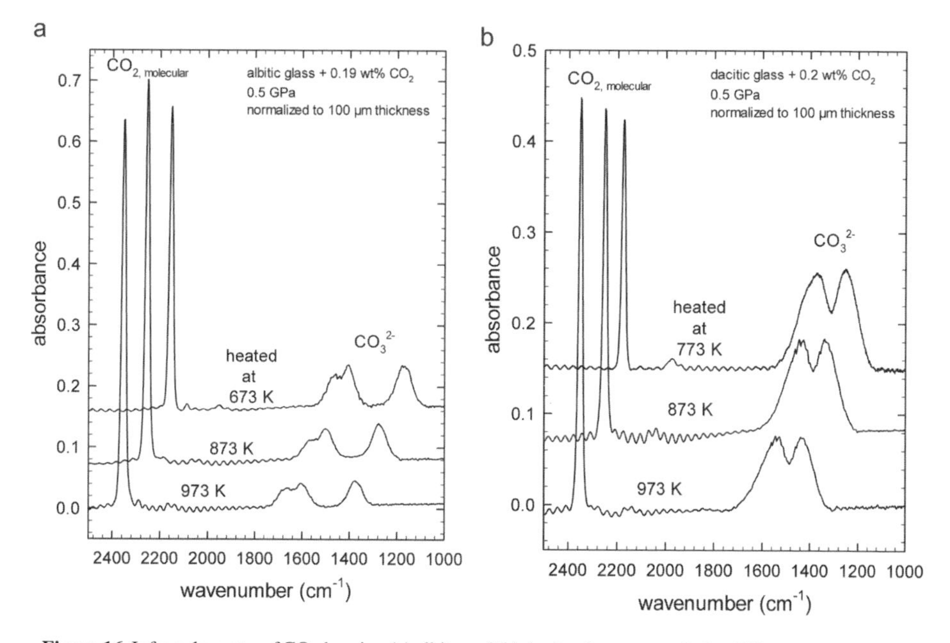

Figure 16. Infrared spectra of CO_2-bearing (a) albite and (b) dacite glasses annealed at different temperatures. Note the increase of molecular CO_2 and the decrease of carbonate with increasing temperature. Spectra courtesy of Marcus Nowak.

−23 (±2) J mol^{-1} K^{-1}, similar to the data for the fully polymerized jadeite glass reported by Morizet et al. (2001). However, for the slightly depolymerized dacite glass, they observed a significantly higher reaction enthalpy ($\Delta H = -29$ (±3) kJ mol^{-1}; $\Delta S = -32$ (±2) J mol^{-1} K^{-1}).

In situ high-temperature FTIR spectroscopy. Direct, *in situ* infrared spectroscopic measurements of CO_2 speciation in a range of silicate melts were reported by Konschak (2008) and by Konschak and Keppler (2009). These experiments are very difficult, as they require temperatures in excess of 1000 °C, which is at the limit of the externally-heated diamond anvil cells used for the measurements. Moreover, blackbody emission from the cell and from the sample becomes so strong in the mid-infrared region at high temperatures that spectra cannot be measured with a conventional infrared source anymore; a synchrotron infrared source is required. Despite these experimental difficulties, however, the results obtained by these *in situ* studies are in very good agreement with the annealing experiments by Morizet et al. (2001) and by Nowak et al. (2003). With increasing temperature, equilibrium (10) in the melt shifts towards molecular CO_2 and the enthalpy of the reaction increases with the depolymerization of the melt.

Figure 17 shows typical high-temperature FTIR spectra of CO_2-bearing phonolite glass as measured in an externally heated diamond anvil cell. Up to the glass transformation temperature of 700 °C, the absorbances of both the molecular CO_2 and of the carbonate decrease. This effect is due to a reduced population of the vibrational ground state with increasing temperature and can be quantitatively modeled by a Boltzmann distribution (Konschak 2008). Beyond the glass transformation temperature, however, the intensity of the band of molecular CO_2 increases again while the carbonate band nearly vanishes, indicating a conversion of carbonate to molecular CO_2 with increasing temperature. If these data are converted to species concentrations, the equilibrium constant for reaction (10) may be calculated (Fig. 18). For dacite melt, these measurements yield $\Delta H = -42$ (±12) kJ mol^{-1} and $\Delta S = -38$ (±14) J mol^{-1} K^{-1}, within error quite comparable to the value reported by Nowak et al. (2003) for annealing experiments on dacite glasses of the same composition (NBO/T = 0.09). The somewhat higher value for the enthalpy may either be a pressure effect or it may indicate that annealing below the glass transformation temperature does not completely relax CO_2 speciation. For a phonolite melt (NBO/T = 0.14), Konschak (2008) obtained $\Delta H = -65$ (±20) kJ mol^{-1}and $\Delta S =$ −51 (±20) J mol^{-1} K^{-1}. These data, together with those of Morizet et al. (2001) and Nowak et al. (2003) indicate a systematic increase of ΔH of reaction (10) with NBO/T (Fig. 19). By linear regression of ΔS and ΔH, as a function of NBO/T, Konschak (2008) constructed a model that predicts the equilibrium between molecular CO_2 and carbonate over a wide range of temperatures and compositions. As shown in Figure 20, the equilibrium constant, which for a model of ideal mixing of oxygen atoms is virtually identical with the molar carbonate/molecular CO_2 ratio, is strongly dependent on temperature. At 1000-1200 °C, there is indeed a major difference in CO_2 speciation between rhyolite (NBO/T $\approx$ 0) and basalt (NBO/T = 0.5-1), with carbonate prevailing in low-temperature basaltic melts. However, at higher temperatures near 1500 °C, this difference in speciation nearly disappears and molecular CO_2 is the predominant carbon species. This implies that the dependence of CO_2 solubility on melt composition should become less pronounced at higher temperatures, in general agreement with measurements and results from molecular dynamics simulations (Fig. 4).

The results from annealing experiments and from *in situ* measurements outlined here are in good agreement with the recent molecular dynamics simulation by Guillot and Sator (2011). They found that molecular CO_2 is present even in mafic and ultramafic melts at superliquidus conditions and the fraction of total carbon dissolved as molecular CO_2 increases with temperature, while it decreases with pressure. Molecular CO_2 is only loosely associated with melt structure (Fig. 21a). On the other hand, the carbonate groups are preferentially associated with non-bridging oxygen atoms while the association of CO_3^{2-} with bridging oxygen is also present (Fig. 21b).

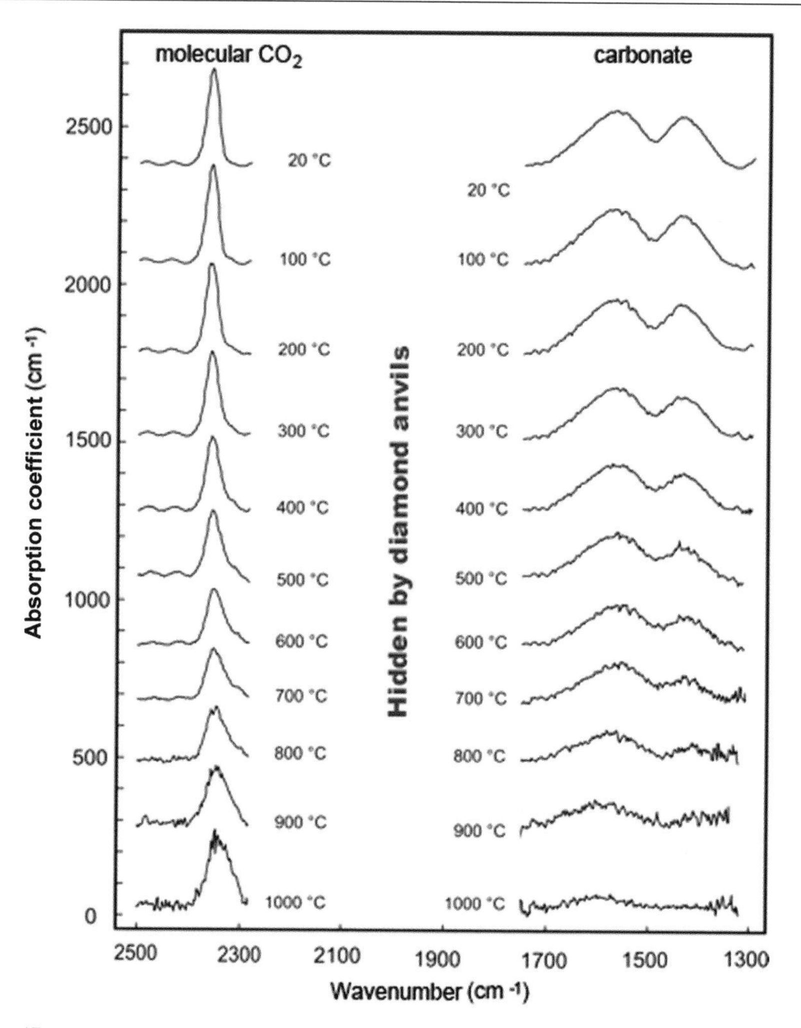

Figure 17. High-temperature *in situ* FTIR spectra of dacite melt with 1.6 wt% CO_2. Pressure increases from 5 GPa at room temperature to 14 GPa at 1000 C. Spectra were measured at the synchrotron source ANKA in Karlsruhe, Germany. After Konschak (2008).

PHYSICAL PROPERTIES OF CARBON-BEARING SILICATE MELTS

Viscosity and electrical conductivity

Brearley and Montana (1989) observed in high-pressure falling-sphere experiments that CO_2 slightly reduced the viscosity of albite melt, while the effect on sodium melilitite melt was negligible. White and Montana (1990) observed that 0.5 wt% CO_2 slightly decreases the viscosity of sanidine melt at 1.5-2 GPa and 1500 °C. Bourgue and Richet (2001) reported that the viscosity of a potassium silicate liquid with 56.9 mol% SiO_2 decreases by two orders of magnitude upon addition of 3.5 wt% CO_2 at 750 K and 1 atm. However, the effect of 1 wt% CO_2 at 1500 K on the viscosity of the same melt is almost negligible. More recently, Morizet et al. (2007) found that dissolved CO_2 has little or no effect on the glass transformation temperature of phonolite and jadeite glasses, implying a negligible effect of CO_2 on viscosity. Ni et al. (2011) found that 0.5 wt% CO_2 has virtually no effect on the electrical conductivity of basaltic

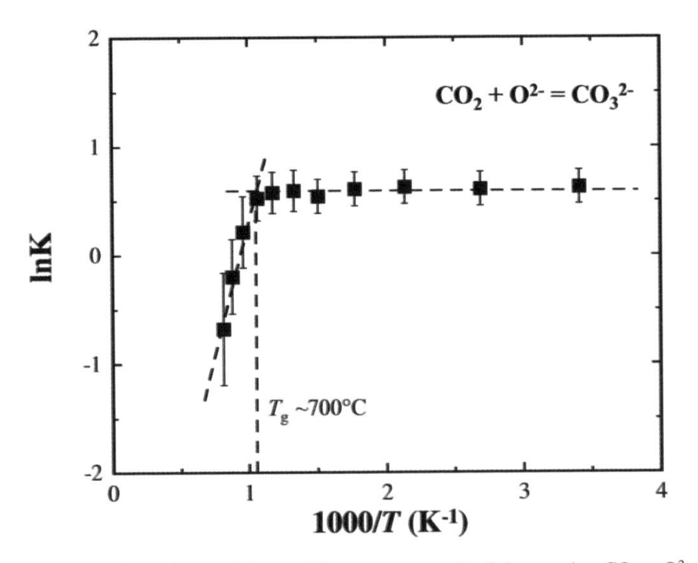

Figure 18. Temperature dependence of the equilibrium constant K of the reaction $CO_2 + O^{2-} = CO_3^{2-}$ in dacite melt from *in situ* FTIR measurements. Below the glass tranformation temperature T_g of about 700 °C, species concentrations are unchanged (out of equilibrium due to slow reaction). From Konschak (2008).

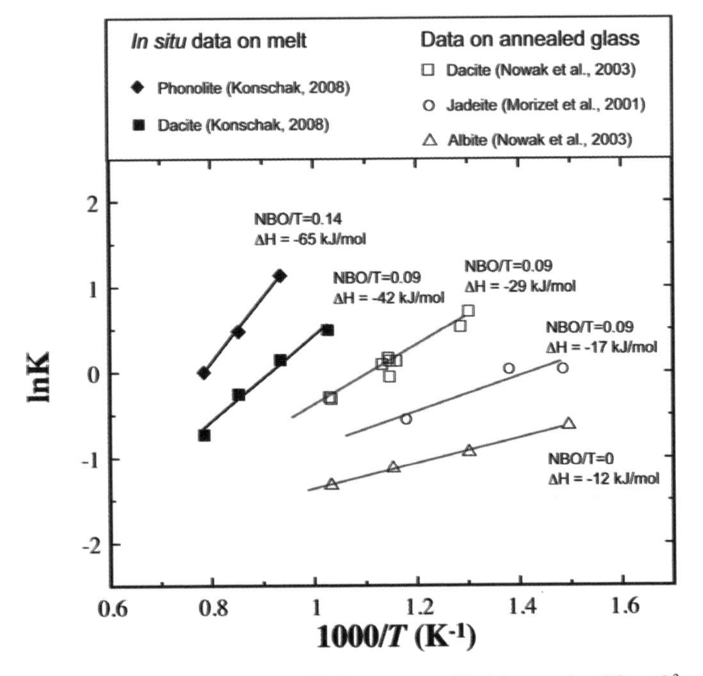

Figure 19. Temperature dependence of the equilibrium constant K of the reaction $CO_2 + O^{2-} = CO_3^{2-}$ for various melt compositions, from annealing experiments (open symbols) and *in situ* measurements (solid symbols). Parameter NBO/T and reaction enthalpy (ΔH) are indicated for each composition. From Konschak (2008).

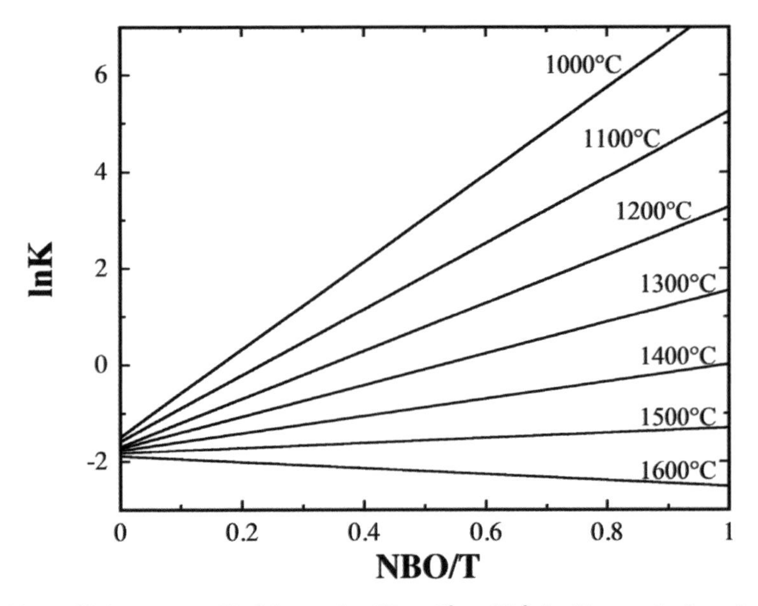

Figure 20. Equilibrium constant K of the reaction $CO_2 + O^{2-} = CO_3^{2-}$ in silicate melts for various temperatures and melt compositions as represented by the NBO/T parameter. For a model of ideal mixing of oxygen atoms, this constant nearly equals the molar carbonate/CO_2 ratio. Curves are calculated from a linear relationship between ΔS and ΔH of the reaction and NBO/T. From Konschak (2008).

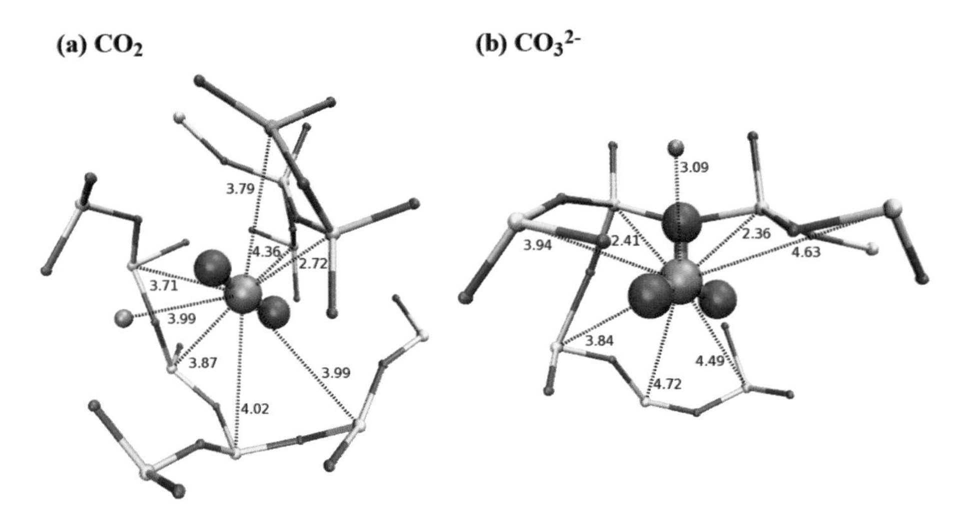

Figure 21. Snapshots of representative (a) molecular CO_2 and (b) carbonate group dissolved in basalt melt at 2273 K and 2.0 GPa from molecular dynamics simulation. Carbon atoms are the big blue balls, oxygen atoms are in red, silicon atoms are in yellow, aluminum atom is in green, and Ca atoms are the small blue balls. The numbers indicate carbon-cation distances in angstroms (only those cations < 5 angstroms away from the carbon atom are shown). Courtesy of Bertrand Guillot.

melts. The latter observation is in line with evidence from *in situ* spectroscopy (Konschak 2008) and molecular dynamics calculations (Guillot and Sator 2011) suggesting that molecular CO_2 becomes increasingly abundant with the rise of temperature. Generally, the effect of CO_2 on the transport properties of silicate melts is likely negligible, except perhaps for melts containing several wt% CO_2.

Density and molar volume

The dissolution of a light component such as CO_2 in a silicate melt will reduce density. At low pressures, this effect will be small due to the low bulk solubility of CO_2 in most melts. However, it may become very significant at deep mantle pressures. Using the sink-float method, Ghosh et al. (2007) determined the density of a basaltic melt with 5 wt% CO_2 at 2575 K and 19.5 GPa. From this measurement they derived a partial molar volume of CO_2 of 21.0 ± 1.0 cm^3/mol. Liu and Lange (2003) measured the partial molar volume of $CaCO_3$ in carbonate melts at 1 atm and obtained a partial molar volume of CO_2 in carbonate melts of 25.8 cm^3/mol. They estimated that the partial molar volume of CO_2 in alkaline, strongly depolymerized silicate melts should be 19 cm^3/mol or larger, assuming that CO_2 is dissolved in these melts as carbonate species similar to those occurring in carbonatite melts. Bourgue and Richet (2001) measured a partial molar volume of CO_2 of 25.6 ± 0.8 cm^3/mol in potassium silicate glasses at room temperature and observed that the presence of CO_2 has no effect on the thermal expansion coefficient. The thermodynamic analysis of the pressure dependence of CO_2 solubility in silicate melts suggests molar volumes in the order of 21-29 cm^3/mol (Lange 1994). These data are in broad agreement with the molecular dynamics simulations of Guillot and Sator (2011).

Diffusivity of carbon

The diffusion of carbon component in silicate melts has recently been reviewed by Zhang et al. (2007) and Zhang and Ni (2010), to which the readers are directed for a thorough discussion of the relevant studies before 2007. Here we first give a short summary largely based on Zhang and Ni (2010), which is then followed by an introduction of more recent developments. The early ^{14}C tracer diffusivity data by Watson (1991) and Watson et al. (1982) obtained by β-track autoradiography probably contained large errors because (a) β-track autoradiography cannot measure accurate carbon concentration (see the section on carbon solubility in this review); and more importantly (b) the β-particle range (in the order of 100 μm) turns out to be much higher than originally expected (i.e., the measured profiles carry significant broadening effects), as pointed out by Mungall (2002). All the later studies investigate CO_2 chemical transport and measure diffusion profiles with the more reliable FTIR microspectroscopy (Table 5). One important finding from these studies is that CO_2 diffusivity does not depend much on melt composition despite the fact that the speciation of CO_2 component can be very different (e.g., a higher fraction of molecular CO_2 in rhyolite melt than in basalt melt). Assuming the diffusion is dominated by molecular CO_2 (neutral and smaller in size), Nowak et al. (2004) attributed the weak compositional dependence to increasing molecular CO_2 diffusivity from rhyolite melt to basalt melt combined with decreasing proportion of molecular CO_2 (i.e., these two effects approximately cancel each other). Because of the scarcity of CO_2 diffusivity data, the similarity between Ar diffusivity and CO_2 diffusivity is exploited by Zhang et al. (2007) to derive the following model for apparent total CO_2 diffusivity (as well as Ar diffusivity) in rhyolite to basalt melts:

$$\ln D_{\text{total CO}_2} = -13.99 - \frac{17367 + 1944.8P}{T} + \frac{(855.2 + 271.2P)}{T} C_{\text{H}_2\text{O}} \tag{11}$$

where D is total CO_2 diffusivity in m^2/s (note that this total CO_2 diffusivity is different from molecular CO_2 diffusivity), T is absolute temperature, P is pressure in GPa, and C_{H_2O} is the total dissolved H_2O in wt%. This model, applicable within 673-1773 K, 0-1.5 GPa, and 0-5

Table 5. Studies on carbon diffusion in silicate melts

Year	Authors	Melt	Species	T (°C)	P (bar)	Analytical Method
Data						
1982	Watson et al.	SNA*, Haplobasalt	^{14}C	800-1500	500-18000	β-track mapping
1991	Watson	Rhyolite, Dacite	^{14}C	800-1100	10000	β-track mapping
1990	Fogel & Rutherford	Rhyolite	CO_2	1050	1000-2490	FTIR
1991	Zhang & Stolper	Basalt	CO_2	1300	10000	FTIR
1993	Blank	Rhyolite	CO_2	450-1050	500-1050	FTIR
2002	Sierralta et al.	Albite, Albite + Na_2O	CO_2	1250	5000	FTIR
2003	Liu	Dacite	CO_2	638	970	FTIR
2004	Nowak et al.	Rhyolite, Dacite, Andesite, Basalt Hawaiite	CO_2	1350	5000	FTIR
2005	Baker et al.	Trachyte	CO_2	1100-1300	10000-12000	FTIR
2010	Spickenbom et al.	SNA*	CO_2	1100-1350	5000	FTIR
2011	Guillot & Sator	Rhyolite, Basalt, Kimberlite	CO_2	1200-2000	20000-100000	Molecular dynamics
Model						
2007	Zhang et al.	Rhyolite to basalt	CO_2	500-1500	0-10000	

* SNA means SiO_2-Na_2O-Al_2O_3 melts.

wt% H_2O, predicts a positive H_2O effect and a negative pressure effect. It also implies that CO_2 concentration has no effect on CO_2 diffusivity, in contrast with the rapid increase of H_2O diffusivity with increasing H_2O concentration (Shaw 1974; Ni and Zhang 2008; Ni et al. 2009a,b).

In the last couple of years two new studies on CO_2 diffusion in silicate melts have been published, one experimental and the other computational. Spickenbom et al. (2010) showed that CO_2 diffusivity varies insignificantly for $Ab_{70}Qz_{30}$ melt to jadeite melt, but it increases by about a factor of 3 from albite melt to a soda-rich melt with 63.95 wt% SiO_2, 18.65 wt% Al_2O_3, and 17.4 wt% Na_2O (Fig. 22a), in accordance with Sierralta et al. (2002). Guillot and Sator (2011) performed the first molecular dynamics simulations to obtain CO_2 diffusivity at 2-10 GPa and 1473-2273 K from the mean square displacements of carbon atoms. They also found that CO_2 diffusivity increases notably with the degree of melt depolymerization (to a lesser extent from rhyolite melt to basalt melt than from basalt melt to kimberlite melt), as shown in Figure 22b using NBO/T as the index for the degree of melt depolymerization. Furthermore, their computed molecular CO_2 diffusivities support the explanation by Nowak et al. (2004) for the limited variation of total CO_2 diffusivity from rhyolite melt to basalt melt. One major difference between Guillot and Sator (2011) and the experimental studies is about the diffusivity of CO_3^{2-}. At 2273 K and 2.0 GPa, they found that CO_3^{2-} diffusivity, which is comparable to oxygen diffusivity in basalt and kimberlite melts, is lower than molecular CO_2 diffusivity by only a factor of 2-3, whereas previously CO_3^{2-} diffusivity was always assumed to be negligible (Nowak et al. 2004). However, Guillot and Sator (2011) appear to have overestimated total CO_2 diffusivities by roughly one order of magnitude.

FUTURE DIRECTIONS

After decades of experimental studies, we now have a reasonably good understanding of carbon solubility and speciation in silicate melts in Earth's crust and uppermost mantle. However, experimental data on the behavior of carbon in melts at the higher pressure regimes of the deeper upper mantle, the transition zone, and the lower mantle are lacking. An interesting, yet unexplored possibility is the conceivable occurrence of complete miscibility between CO_2 and alkaline silicate melts at very high pressures and temperatures. The behavior of carbon under reducing conditions is generally poorly explored. Moreover, there are no data on carbon solubility in peridotitic melts, and in particular in peridotitic melts under reducing conditions, which would be essential for understanding the behavior of carbon in a magma ocean. Such data would be essential for constraining the initial distribution of carbon in Earth. Many of these problems will likely require molecular dynamic simulations or a further advancement in *in situ* experimental methods for studying carbon speciation and solubility (Oganov et al. 2013). In Earth's upper mantle, carbon dioxide in general appears to have less influence on the physical properties of silicate melts than water, but its effects need to be better quantified. How reduced carbon species may modify melt properties also needs to be examined, and will provide additional challenges to experimental and theoretical geoscientists.

ACKNOWLEDGMENTS

We thank Zhigang Zhang for the program calculating CO_2 fugacity of C-O-H fluids, Bertrand Guillot for the microscopic pictures of carbon species, and Paolo Papale for discussion. A formal review by David Dobson has improved the manuscript.

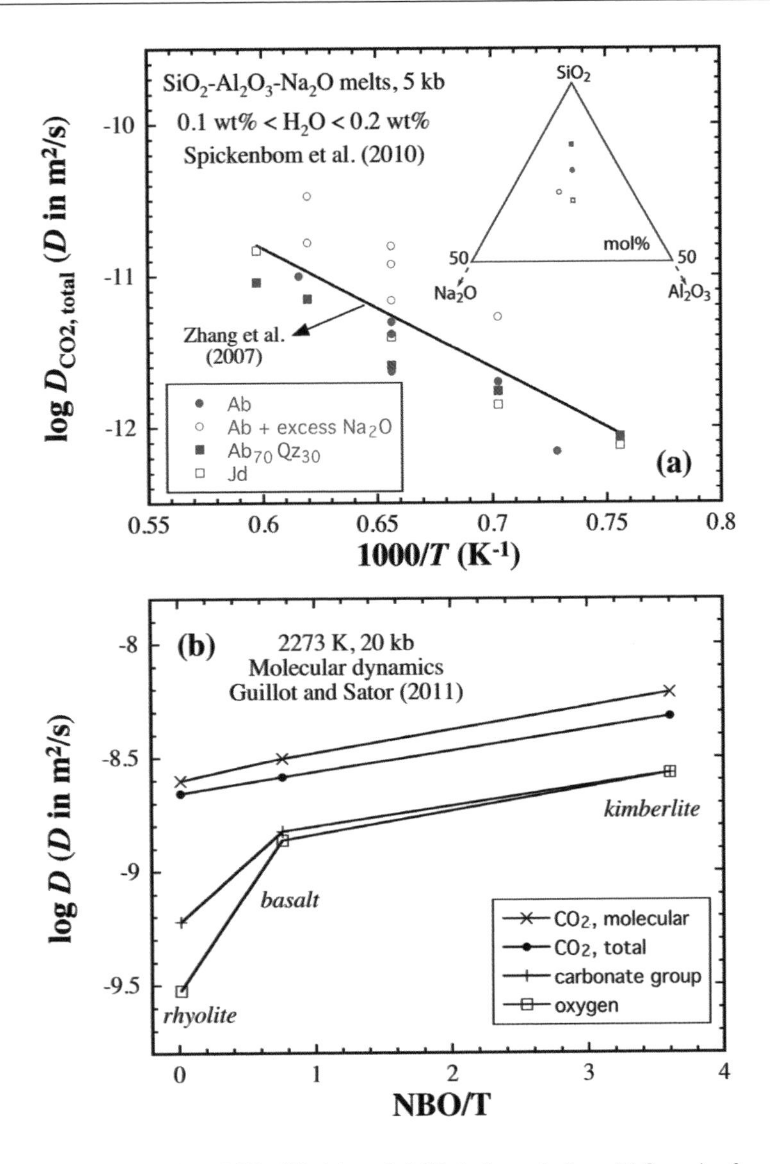

Figure 22. (a) Experimental total CO_2 diffusivity at 0.5 GPa in four anhydrous (H_2O ranging from 0.1 to 0.2 wt%) SiO_2-Al_2O_3-Na_2O melts with the inlet showing the loci of melts in compositional space. Modified after Spickenbom et al. (2010). Calculations using the composition-independent model of Zhang et al. (2007) at 0.15 wt% H_2O are shown for comparison (solid line). (b) Molecular dynamics-derived diffusivity of total CO_2, molecular CO_2, CO_3^{2-}, and oxygen versus NBO/T of three melts at 2273 K and 2.0 GPa. After Guillot and Sator (2011).

REFERENCES

Baker DR, Freda C, Brooker RA, Scarlato P (2005) Volatile diffusion in silicate melts and its effects on melt inclusions. Ann Geophys 48:699-717

Ballhaus C (1995) Is the upper mantle metal-saturated? Earth Planet Sci Lett 132:75-86

Beerling D (2002) CO_2 and the end-Triassic mass extinction. Nature 415:386-387

Behrens H, Misiti V, Freda C, Vetere F, Botcharnikov RE, Scarlato P (2009) Solubility of H_2O and CO_2 in ultrapotassic melts at 1200 and 1250 °C and pressure from 50 to 500 MPa. Am Mineral 94:105-120

Behrens H, Ohlhorst S, Holtz F, Champenois M (2004a) CO_2 solubility in dacitic melts equilibrated with H_2O-CO_2 fluids: implications for modeling the solubility of CO_2 in silicic melts. Geochim Cosmochim Acta 68:4687-4703

Behrens H, Tamic N, Holtz F (2004b) Determination of the molar absorption coefficient for the infrared absorption band of CO_2 in rhyolitic glasses. Am Mineral 89:301-306

Berner RA (1994) 3GEOCARB II: A revised model for atmospheric CO_2 over Phanerozoic time. Am J Sci 294:56-91

Blank JG (1993) An experimental investigation of the behavior of carbon dioxide in rhyolitic melt. PhD dissertation, California Institute of Technology, Pasadena, CA

Blank JG, Brooker RA (1994) Experimental studies of carbon dioxide in silicate melts: solubility, speciation, and stable carbon isotope behavior. Rev Mineral 30:157-186

Botcharnikov RE, Behrens H, Holtz F (2006) Solubility and speciation of C-O-H fluids in andesitic melt at T=1100-1300 °C and P=200 and 500 MPa. Chem Geol 229:125-143

Botcharnikov RE, Freise M, Holtz F, Behrens H (2005) Solubility of C-O-H mixtures in natural melts: new experimental data and application range of recent models. Ann Geophys 48:633-646

Botcharnikov RE, Holtz F, Behrens H (2007) The effect of CO_2 on the solubility of H_2O-Cl fluids in andesitic melt. Eur J Mineral 19:671-680

Bourgue E, Richet P (2001) The effects of dissolved CO_2 on the density and viscosity of silicate melts: a preliminary study. Earth Planet Sci Lett 193:57-68

Brearley M, Montana A (1989) The effect of CO_2 on the viscosity of silicate liquids at high pressure. Geochim Cosmochim Acta 53:2609-2616

Brey G (1976) CO_2 solubility and solubility mechanisms in silicate melts at high pressures. Contrib Mineral Petrol 57: 215-221

Brey G, Green DH (1975) The role of CO_2 in the genesis of olivine melilitite. Contrib Mineral Petrol 49:93-103

Brooker RA, Kohn SC, Holloway JR, McMillan PF (2001a) Structural controls on the solubility of CO_2 in silicate melts. Part I: bulk solubility data. Chem Geol 174:225-239

Brooker RA, Kohn SC, Holloway JR, McMillan PF (2001b) Structural controls on the solubility of CO_2 in silicate melts. Part II: IR characteristics of carbonate groups in silicate glasses. Chem Geol 174:241-254

Brooker RA, Kohn SC, Holloway JR, McMillan PF, Carroll MR (1999) Solubility, speciation and dissolution mechanisms for CO_2 in melts on the $NaAlO_2$-SiO_2 join. Geochim Cosmochim Acta 63:3549-3565

Chase MW Jr (1998) NIST-JANAF Thermochemical Tables. Fourth Edition. J Phys Chem Ref Data, Monograph 9

Dasgupta R, Hirschmann, MM (2010) The deep carbon cycle and melting in Earth's interior. Earth Planet Sci Lett 298:1-13

Dickens B, Hyman A, Brown WE (1971) Crystal structure of $Ca_2Na_2(CO_3)_3$ (shortite). J Res Nat Bur Stand 75A:129-135

Dixon JE (1997) Degassing of alkalic basalts. Am Mineral 82:368-378

Dixon JE, Pan V (1995) Determination of the molar absorptivity of dissolved carbonate in basanitic glass. Am Mineral 80:1339-1342

Dixon JE, Stolper EM, Holloway JR (1995) An experimental study of water and carbon dioxide solubilites in Mid-Ocean Ridge basaltic liquids. Part I: calibration and solubility models. J Petrol 36:1607-1631

Duan Z, Moller N, Weare JH (1992) An equation of state for the CH_4-CO_2-H_2O system: I. Pure systems from 0 to 1000 °C and 0 to 8000 bar. Geochim Cosmochim Acta 56:2605-2617

Duan Z, Zhang Z (2006) Equation of state of the H_2O, CO_2, and H_2O-CO_2 systems up to 10 GPa and 2573.15 K: molecular dynamics simulations with *ab inito* potential surface. Geochim Cosmochim Acta 70:2311-2324

Eggler DH, Mysen BO, Hoering TC, Holloway JR (1979) The solubility of carbon monoxide in silicate melts at high pressures and its effect on silicate phase relations. Earth Planet Sci Lett 43:321-330

Eggler DH, Rosenhauer M (1978) Carbon dioxide in silicate melts: II. Solubilities of CO_2 and H_2O in $CaMgSi_2O_6$ (diopside) liquids and vapors at pressures to 40 kbar. Am J Sci 278:64-94

Fine G, Stolper E (1985) The speciation of carbon dioxide in sodium aluminosilicate glasses. Contrib Mineral Petrol 91:105-121

Fine G, Stolper E (1986) Dissolved carbon dioxide in basaltic glasses: concentrations and speciation. Earth Planet Sci Lett 76:263-278

Fogel RA, Rutherford MJ (1990) The solubility of carbon dioxide in rhyolitic melts: a quantitative FTIR study. Am Mineral 75:1311-1326

Frost DJ, McCammon CA (2008). The redox state of the Earth's mantle. Ann Rev Earth Planet Sci 36:389-420

Ghosh S, Ohtani E, Litasov K, Suzuki A, Sakamaki T (2007) Stability of carbonated magma at the base of Earth's upper mantle. Geophys Res Lett 34:L22312, doi:10.1029/2007GL031349

Gottschalk M (2007) Equations of state for complex fluids. Rev Mineral Geochem 65:49-97
Guillot B, Sator N (2011) Carbon dioxide in silicate melts: a molecular dynamics simulation study. Geochim Cosmochim Acta 75:1829-1857
Hoffman PF, Kaufmann AJ, Halverson GP, Schrag DP (1998) A Neoproterozoic snowball Earth. Science 281:1342-1346
Holloway JR (1976) Fluids in the evolution of granitic magmas: consequences of finite CO_2 solubility. Geol Soc Am Bull 87:1513-1518
Holloway JR, Blank JG (1994) Application of experimental results to C-O-H species in natural melts. Rev Mineral Geochem 30:187-230
Holloway JR, Jakobsson S (1986) Volatile solubilities in magmas: transport of volatiles from mantles to planet surfaces. J Geophys Res 91:D505-D508
Iacono-Marziano G, Morizet Y, Le Trong E, Gaillard F (2012) New experimental data and semi-empirical parameterization of H_2O-CO_2 solubility in mafic melts. Geochim Cosmochim Acta 97:1-23
Jakobsson S (1997) Solubility of water and carbon dioxide in an icelandite at 1400 °C and 10 kilobars. Contrib Mineral Petrol 127:129-135
Jakobsson S, Holloway JR (1986) Crystal-liquid experiments in the presence of a C-O-H fluid buffered by graphite + iron + wustite: experimental method and near-liquidus relations in basanite. J Volcanol Geotherm Res 29:265-291
Jendrzejewski N, Trull TW, Pineau F, Javoy M (1997) Carbon solubility in Mid-Ocean Ridge basaltic melt at low pressures (250-1950 bar). Chem Geol 138:81-92
Kadik AA, Kurovskaya NA, Ignat'ev YA, Kononkova NN, Koltashev VV (2010) Influence of oxygen fugacity on the solubility of carbon and hydrogen in FeO-Na_2O-SiO_2-Al_2O_3 melts in equilibrium with liquid iron at 1.5 GPa and 1400 °C. Geochem Int 48:953-960
Kadik AA, Kurovskaya NA, Ignat'ev YA, Kononkova NN, Koltashev VV, Plotnichenko VG (2011) Influence of oxygen fugacity on the solubility of nitrogen, carbon and hydrogen in FeO-Na_2O-SiO_2-Al_2O_3 melts in equilibrium with metallic iron at 1.5 GPa and 1400 °C. Geochem Int 49:429-438
Kadik AA, Litvin YA, Koltashev VV, Kryukova EB, Plotnichenko VG (2006) Solubility of hydrogen and carbon in reduced magmas of the early Earth's mantle. Geochem Int 44:33-47
Kadik AA, Pineau F, Litvin YA, Jendrzejewski N, Martinez I, Javoy M (2004) Formation of carbon and hydrogen species in magmas at low oxygen fugacity. J Petrol 45:1297-1310
Kerrick DM, Jacobs GK (1981) A modified Redlich-Kwong equation for H_2O, CO_2, and H_2O-CO_2 mixtures at elevated pressures and temperatures. Am J Sci 281:735-767
Keshav S, Gudfinnsson GH, Presnall DC (2011) Melting phase relations of simplified carbonated peridotite at 12-26 GPa in the systems CaO-MgO-SiO_2-CO_2 and CaO-MgO-Al_2O_3-SiO_2-CO_2: highly calcic magmas in the transition zone of the Earth. J Petrol 52:2265-2291
King PL, Holloway JR (2002) CO_2 solubility and speciation in intermediate (andesitic) melts: the role of H_2O and composition. Geochim Cosmochim Acta 66:1627-1640
Kohn SC, Brooker RA, Dupree R (1991) ^{13}C MAS NMR: A method for studying CO_2 speciation in glasses. Geochim Cosmochim Acta 55:3879-3884
Konschak A (2008) CO_2 in Silikatschmelzen. Ph. D. dissertation, University of Bayreuth, Germany
Konschak A, Keppler H (2009) A model for CO_2 solubility in silicate melts. Geochim Cosmochim Acta 73:A 680
Kubicki JD, Stolper EM (1995) Structural roles of CO_2 and $[CO_3]^{2-}$ in fully polymerized, sodium aluminosilcate melts and glasses. Geochim Cosmochim Acta 59:683-698
Kump LR, Barley ME (2007) Increased subaerial volcanism and the rise of atmospheric oxygen 2.5 billion years ago. Nature 448:1033-1036
Lange RA (1994) The effect of H_2O, CO_2 and F on the density and viscosity of silicate melts. Rev Mineral 30:331-369
Lesne P, Scaillet B, Pichavant M, Beny J-M (2011) The carbon dioxide solubility in alkali basalts: an experimental study. Contrib Mineral Petrol 162:153-168
Liu Q, Lange RA (2003) New density measurements on carbonate liquids and the partial molar volume of the $CaCO_3$ component. Contrib Mineral Petrol 146:370-381
Liu Y (2003) Water in rhyolitic and dacitic melts. Ph.D. dissertation, University of Michigan
Liu Y, Zhang Y, Behrens H (2005) Solubility of H_2O in rhyolitic melts at low pressures and a new empirical model for mixed H_2O-CO_2 solubility in rhyolitic melts. J Volcanol Geotherm Res 143:219-235
Manning CE, Shock EL, Sverjensky D (2013) The chemistry of carbon in aqueous fluids at crustal and upper-mantle conditions: experimental and theoretical constraints. Rev Mineral Geochem 75:109-148
Mattey DP (1991) Carbon dioxide solubility and carbon isotope fractionation in basaltic melt. Geochim Cosmochim Acta 55:3467-3473
McCammon C (2005) The paradox of mantle redox. Science 308:807-808

Moore G (2008) Interpreting H_2O and CO_2 contents in melt inclusions: constraints from solubility experiments and modeling. Rev Mineral Geochem 69:333-361

Morizet Y, Brooker RA, Kohn SC (2002) CO_2 in haplo-phonolite melt: solubility, speciation and carbonate complexation. Geochim Cosmochim Acta 66:1809-1820

Morizet Y, Kohn SC, Brooker RA (2001) Annealing experiments on CO_2-bearing jadeite glass: an insight into the true temperature dependence of CO_2 speciation in silicate melts. Mineral Mag 65:701-707

Morizet Y, Paris M, Gaillard F, Scaillet B (2010) C-O-H fluid solubility in haplobasalt under reducing conditions: an experimental study. Chem Geol 279:1-16

Morizet, Y, Nichols ARL, Kohn SC, Brooker RA, Dingwell DB (2007) The influence of H_2O and CO_2 on the glass transition temperature: insights into the effects of volatiles on magma viscosity. Eur J Mineral 19:657-669

Mungall JE (2002) Empirical models relating viscosity and tracer diffusion in magmatic silicate melts. Geochim Cosmochim Acta 66:125-143

Mysen BO (1976) The role of volatiles in silicate melts: solubility of carbon dioxide and water in feldspar, pyroxene, and feldspathoid melts to 30 kb and 1625 °C. Am J Sci 276:969-996

Mysen BO, Arculus RJ, Eggler DH (1975) Solubility of carbon dioxide in melts of andesite, tholeiite, and olivine nephelininte composition to 30 kbar pressure. Contrib Mineral Petrol 53:227-239

Mysen BO, Eggler DH, Seitz MG, Holloway JR (1976) Carbon dioxide in silicate melts and crystals. Part I. Solubility measurements. Am J Sci 276:455-479

Mysen BO, Fogel ML, Morrill PL, Cody GD (2009) Solution behavior of reduced C-O-H volatiles in silicate melts at high pressure and temperature. Geochim Cosmochim Acta 73:1696-1710

Mysen BO, Virgo D (1980a) Solubility mechanism of carbon dioxide in silicate melts: a Raman spectroscopic study. Am Mineral 65:885-899

Mysen BO, Virgo D (1980b) The solubility behavior of CO_2 in melts on the join $NaAlSi_3O_8$-$CaAl_2Si_2O_8$-CO_2 at high pressures and temperatures: A Raman spectroscopic study. Am Mineral 65:1166-1175

Newman S, Lowenstern JB (2002) VolatileCalc: a silicate melt-H_2O-CO_2 solution model written in Visual Basic for excel. Computat Geosci 28:597-604

Ni H, Behrens H, Zhang Y (2009b) Water diffusion in dacitic melt. Geochim Cosmochim Acta 73:3642-3655

Ni H, Keppler H, Behrens H (2011) Electrical conductivity of hydrous basaltic melts: implications for partial melting in the upper mantle. Contrib Mineral Petrol 162:637-650

Ni H, Liu Y, Wang L, Zhang Y (2009a) Water speciation and diffusion in haploandesitic melts at 743-873 K and 100 MPa. Geochim Cosmochim Acta 73:3630-3641

Ni H, Zhang Y (2008) H_2O diffusion models in rhyolitic melt with new high pressure data. Chem Geol 250:68-78

Nowak M, Porbatzki D, Spickenbom K, Diedrich O (2003) Carbon dioxide speciation in silicate melts: a restart. Earth Planet Sci Lett 207:131-139

Nowak M, Schreen D, Spickenbom K (2004) Argon and CO_2 on the race track in silicate melts: a tool for the development of a CO_2 speciation and diffusion model. Geochim Cosmochim Acta 68:5127-5138

Oganov AR, Hemley RJ, Hazen RM, Jones AP (2013) Structure, bonding, and mineralogy of carbon at extreme conditions. Rev Mineral Geochem 75:47-77

Pan V, Holloway JR, Hervig RL (1991) The pressure and temperature dependence of carbon dioxide solubility in tholeiitic basalt melts. Geochim Cosmochim Acta 55:1587-1595

Paonita A, Gigli G, Gozzi D, Nuccio PM, Trigila R (2000) Investigation of the He solubility in H_2O-CO_2 bearing silicate liquids at moderate pressure: a new experimental method. Earth Planet Sci Lett 181:595-604

Papale P (1997) Modeling of the solubility of a one-component H_2O or CO_2 fluid in silicate liquids. Contrib Mineral Petrol 126:237-251

Papale P (1999) Modeling of the solubility of a two-component $H_2O + CO_2$ fluid in silicate liquids. Am Mineral 84:477-492

Papale P, Moretti R, Barbato D (2006) The compositional dependence of the saturation surface of $H_2O + CO_2$ fluids in silicate melts. Chem Geol 229:78-95

Papale P, Polacci M (1999) Role of carbon dioxide in the dynamics of magma ascent in explosive eruptions. Bull Volcanol 60:585-594

Papike JJ, Stephenson NC (1966) The crystal structure of mizzonite, a calcium- and carbonate-rich scapolite. Am Mineral 51:1014-1027

Pawley AR, Holloway JR, McMillan PF (1992) The effect of oxygen fugacity on the solubility of carbon-oxygen fluids in basaltic melt. Earth Planet Sci Lett 110:213-225

Shaw HR (1974) Diffusion of H_2O in granitic liquids: I. Experimental data; II. Mass transfer in magma chambers. *In:* Geochemical Transport and Kinetics. Hofmann AW, Giletti BJ, Yoder HS, Yund RA (eds) Carnegie Inst. Washington Publ., Washington, DC, pp 139-170

Shishkina TA, Botcharnikov RE, Holtz F, Almeev RR, Portnyagin MV (2010) Solubility of H_2O- and CO_2-bearing fluids in tholeiitic basalts at pressures up to 500 MPa. Chem Geol 277:115-125
Sierralta M, Nowak M, Keppler H (2002) The influence of bulk composition on the diffusivity of carbon dioxide in Na aluminosilicate melts. Am Mineral 87:1710-1716
Spickenbom K, Sierralta M, Nowak M (2010) Carbon dioxide and argon diffusion in silicate melts: insights into the CO_2 speciation in magmas. Geochim Cosmochim Acta 74:6541-6564
Stanley BD, Hirschmann MM, Withers AC (2011) CO_2 solubility in Martian basalts and Martian atmospheric evolution. Geochim Cosmochim Acta 75:5987-6003
Stolper EM, Fine G, Johnson T, Newman S (1987) Solubility of carbon dioxide in albitic melt. Am Mineral 72:1071-1085
Stolper EM, Holloway JR (1988) Experimental determination of the solubility of carbon dioxide in molten basalt at low pressure. Earth Planet Sci Lett 87:397-408
Tamic N, Behrens H, Holtz F (2001) The solubility of H_2O and CO_2 in rhyolitic melts in equilibrium with a mixed CO_2-H_2O fluid phase. Chem Geol 174:333-347
Taylor WR (1990) The dissolution mechanism of CO_2 in aluminosilicate melts - infrared spectroscopic constraints on the cationic environment of dissolved $[CO_3]^{2-}$. Eur J Mineral 2:547-563
Thibault Y, Holloway JR (1994) Solubility of CO_2 in a Ca-rich leucitite: effects of pressure, temperature, and oxygen fugacity. Contrib Mineral Petrol 116:216-224
Tingle TN, Aines RD (1988) Beta track autoradiography and infrared spectroscopy bearing on the solubility of CO_2 in albite melt at 2 GPa and 1450 °C. Contrib Mineral Petrol 100:222-225
Tossell JA (1995) Calculation of the ^{13}C NMR shieldings of the CO_2 complexes of aluminosilicates. Geochim Cosmochim Acta 59: 1299-1305
Vetere F, Botcharnikov RE, Holtz F, Behrens H, de Rosa R (2011) Solubility of H_2O and CO_2 in shoshonitic melts at 1250 °C and pressures from 50 to 400 MPa: implications for Campi Flegrei magmatic systems. J Volcanol Geotherm Res 202:251-261
Watson EB (1991) Diffusion of dissolved CO_2 and Cl in hydrous silicic to intermediate magmas. Geochim Cosmochim Acta 55:1897-1902
Watson EB, Sneeringer MA, Ross A (1982) Diffusion of dissolved carbonate in magmas: experimental results and applications. Earth Planet Sci Lett 61:346-358
Webster JD, Goldoff B, Shimizu N (2011) C-O-H-S fluids and granitic magma: how S partitions and modifies CO_2 concentrations of fluid-saturated felsic melt at 200 MPa. Contrib Mineral Petrol 162:849-865
White BS, Montana A (1990) The effect of H_2O and CO_2 on the viscosity of sanidine liquid at high pressures. J Geophys Res 95:15683-15693
White WB (1974) The carbonate minerals. *In:* The Infrared Spectra of Minerals. Farmer VC (ed) Mineralogical Society, London, p 227-284
Wood BJ, Bryndzia LT, Johnson KE (1990) Mantle oxidation state and its relationship to tectonic environment and fluid speciation. Science 248:337-345
Wyllie PJ, Tuttle OF (1959) Effect of carbon dioxide on the melting of granite and feldspars. Am J Sci 257:648-655
Zhang C, Duan Z (2010) GFluid: an Excel spreadsheet for investigating C-O-H fluid composition under high temperatures and pressures. Computat Geosci 36:569-572
Zhang Y, Ni H (2010) Diffusion of H, C, and O components in silicate melts. Rev Mineral Geochem 72:171-225
Zhang Y, Stolper EM (1991) Water diffusion in a basaltic melt. Nature 351:306-309
Zhang Y, Xu Z, Zhu M, Wang H (2007) Silicate melt properties and volcanic eruptions. Rev Geophys 45:RG4004, doi:10.1029/2006RG000216

Reviews in Mineralogy & Geochemistry
Vol. 75 pp. 289-322, 2013

Carbonate Melts and Carbonatites

Adrian P. Jones

Earth Sciences, University College London
Gower Street, London WC1E 6BT, United Kingdom

adrian.jones@ucl.ac.uk

Matthew Genge

Earth Sciences and Engineering, Imperial College London
S Kensington, London, SW7 2AZ, United Kingdom

m.genge@imperial.ac.uk

Laura Carmody

Earth Sciences, University College London
Gower Street, London WC1E 6BT, United Kingdom

(presently at Earth and Planetary Sciences, University of Tennessee
Knoxville, Tennessee 37996, U.S.A.)

lcarmod1@utk.edu

INTRODUCTION

Carbonatites are familiar to students of petrology as rare igneous rocks formed predominantly of carbonate, whose only modern expression is a single active volcano that erupts strongly alkaline carbonate lavas with no direct match in Earth's geological record (see Lengai movie in the electronic version of this chapter or on the MSA RiMG website). Based on their Sr-Nd-Pb isotopic data, stable isotopic compositions, noble gases, and experimental phase equilibria, they are derived from the mantle, showing almost no sign of contamination by the crust.

As liquids, carbonate melts have remarkable physical properties, which set them apart from the alkaline silicate melts with which they are often temporally associated. They show very high solubilities of many elements considered rare in silicate magmas, and they have the highest known melt capacities for dissolving water and other volatile species like halogens at crustal pressures. They are highly efficient transport agents of carbon from the mantle to the crust, remaining mobile over extraordinary ranges of temperature, and their very low viscosity should enhance connectivity along grain boundaries in the mantle where they are implicated in geochemical enrichment processes related to metasomatism.

Most carbonatites have unambiguous origins in the mantle and the limit to their depth is not known, but the likelihood that they may exist in the lower mantle (Kaminsky et al. 2009, 2012; Stoppa et al. 2009) needs to be appraised since they may exert a fundamental control on the mobility and long-term storage of deep carbon in Earth. Ultimately the stability of carbonate melt is an extension of the stability of carbonate minerals (Hazen et al. 2013a,b) subject critically to the mantle oxidation state (Luth 1993; Frost and McCammon 2008); carbonate-melts have also been predicted in the oceanic low-velocity zone and deep mantle (Hauri et al. 1993; Presnall and Gudfinnsson 2005) by laboratory petrology experiments (Wyllie 1995). Much remains to be discovered about carbonate melts at very high-pressures.

1529-6466/13/0075-0010$00.00 DOI: 10.2138/rmg.2013.75.10

VIDEO: Erupting Lengai volcano. For readers of the electronic version of this chapter, the video can be activated by either clicking on the play button or the image above. The video shows black natrocarbonatite lava spewing from one of several white "hornito" structures at the summit of Oldoinyo Lengai volcano in the East African Rift, Tanzania. This carbonatite lava is the lowest temperature (< 600 °C) and lowest viscosity lava known, and is seen dissolving a channel for itself before it rapidly overflows the rim of the crater. The last part of the movie shows a standing wave caused by a small obstruction near the crater rim. Besides the scientists' voices, the soundtrack records the strange clinking and water-like murmurs of the lava itself. The movie was filmed by Tobias Fischer, University of New Mexico during a 2005 Oldoinyo Lengai expedition led by Pete Burnard and Bernard Marty, CNRS, France. The volcano has a recent history typified by passive eruption of natrocarbonatite as shown in the movie, interspersed with larger dangerous explosive eruptions every few decades. The last major eruption in 2007-8 destroyed all of the crater hornitos in the movie and excavated a deep new summit crater; it was preceded by tremors measuring up to 6 on the Richter scale.

Beyond the current solid media experiments with piston cylinder, multi-anvil press, and diamond-anvil cell (e.g., calcite-dolomite-aragonite; Kraft et al. 1991), we may look to carbonate inclusions in diamonds (Kaminsky et al. 2001, 2009; Brenker et al. 2007; Kaminsky 2012) and high-pressure shock wave environments for clues, including shocked carbonate from impact craters, which can show isotopic shifts ($\delta^{13}C$ vary 5 per mil; Martinez et al. 1994, 1995; Jones et al. 2000a). Another unique property of carbonate melts is their high electrical conductivity—up to three orders of magnitude greater than silicate melts and five orders of magnitude higher than hydrated mantle material (see section *"Occurrence of carbonatities"*). Consequently, carbonate melts have been invoked to explain deep regions of the mantle asthenosphere characterized by anomalous conductivity. The presence of low volume (0.1%) carbonate melts are contenders to explain electrically conductive mantle regions, previously thought to be caused by silicate melts or water-bearing olivine (Gaillard et al. 2008).

Several books and reviews have been written about the systematic geochemistry and mineralogy and origin of carbonatites (Tuttle and Gittins 1966; Bell 1989; Bell et al. 1998; Jones 2000; Mitchell 2005; Woolley and Church 2005; Woolley and Kjarsgaard 2008a; Downes et al. 2012). Fundamental understanding of carbonatites was largely achieved in the last century when engineering developments in technology enabled experimental petrology to unlock the secrets of how carbonatites actually form, including their important connections with water, enabling early formative predictions about the stability of carbonate minerals in the upper mantle; the significance of free CO_2 and H_2O in the mantle transition zone; the derivation of kimberlitic and carbonatitic melts; and mantle metasomatism (Wyllie and Tuttle 1960, 1962; Wyllie and Huang 1976). The purpose of this review is to provide a framework for understanding carbonate melts, to highlight their potential role in providing vertical connectivity and pathways for deep carbon to be transferred from the mantle to Earth's surface, and to illustrate how further technological advances are still required to answer even the most basic questions about the abundance and mobility of carbon in Earth's deep interior.

CARBONATE MELTS

Physical properties

Very low magmatic temperatures and very low viscosity are striking features of alkali-carbonatite lavas at low pressure, for example at the sole active volcano Oldoinyo Lengai in Tanzania (Treiman and Schedl 1983; Krafft and Keller 1989; Dawson et al. 1990; Oppenheimer 1998). The low carbonate melt viscosity was first measured accurately in experiments using in-situ synchrotron radiation to track rapidly falling spheres (Dobson et al. 1996) and is compared with other data (Wolff 1994; Jones et al. 1995a) in Table 1. We note that only natural high *PT* iron-rich melts in Earth's core may approach such extremely low viscosities (Dobson et al. 2000).

Calculated physical properties by Genge et al. (1995b) predict that $CaCO_3$ melt densities increase from 2000 kg m^{-3} at $P = 0.1$ GPa to 2900 kg m^{-3} at $P = 10.0$ GPa, suggesting carbonate melts are significantly more compressible than silicate melts. Estimates of the constant pressure heat capacity of 1.65-1.90 J g^{-1} K^{-1}, isothermal compressibilities of 0.012-0.01 $\times$ 10^{-10} Pa^{-1} and thermal expansivities of 1.886-0.589 $\times$ 10^{-4} K^{-1} were also calculated (Genge et al. 1995b). Self-diffusion coefficients qualitatively suggest that $CaCO_3$ melts have very low viscosities at high-pressures to 11 GPa.

Although no other atomic simulations of carbonate melts have been performed, quantum mechanical *ab initio* evolutionary models, which allow the most energetically stable atomic

Table 1. Carbonate melt physical data measured *in situ* using synchrotron X-ray falling sphere method and calculations for pressures up to 5.5 GPa. Source references [1] Dobson et al. (1996), [2] Genge et al. (1995b) [3] Wolff (1994).

Composition	Pressure (GPa)	*T* (°C)	Density (g/cm^3)	Comment	Viscosity (PaS)	Reference
$K_2Mg(CO_3)_2$	atm	500	2.262			[1]
$K_2Ca(CO_3)_2$	atm	859	2.058			
$MgCO_3$ = Mc	atm		2.30			[2]
$K_2Mg(CO_3)_2$	3.00	800			0.036	[1]
	3.00	900			0.022	[1]
	5.50	1200			0.006	[1]
$K_2Ca(CO_3)_2$ = Kc	2.50	950	2.75		0.032	[1]
	2.50	1150	2.58		0.018	[1]
	4.00	1050	2.80		0.023	[1]
$Mc_{.25}Kc_{.75}$	2.00	1250			0.065	[1]
K_2CO_3	4.00	1500	3.10		0.023	[1]
RE-Carbonatite	3.00	530	4.10		0.155	[1]
$CaCO_3$				Thermal expansivity		[2]
Ca-carbonatite	atm	800		calculated	0.08	[3]
Natrocarbonatite	atm	800		calculated	0.008	[3]

structures to be predicted, suggest that radical transformation of carbonate mineral structures occur at lower mantle pressures (Oganov et al. 2008, 2013). Simulations predict that the stable $MgCO_3$ and $CaCO_3$ phases at pressures >82 GPa and >19 GPa respectively are dominated by corner-sharing CO_4 tetrahedra networks, with those of $CaCO_3$ adopting a β-cristobalite structure, and $MgCO_3$ adopting a pyroxene-like structure at pressures >110 GPa. Such predictions are supported experimentally by recovered Mg-Fe carbonates with polymerized structures (Boulard et al. 2011). The formation of CO_4^{4-} at high-pressure was predicted (Genge 1994) because it is compatible with the similar electronic configurations of C and Si, which satisfy the requirements of both sp^2 (trigonal) and sp^3 (tetrahedral) hybridization. Transformation from carbonate structures to tetrahedral CO_4 groups, however, requires the breaking of *pp* bonds and the formation of antibonding orbitals and is likely to be associated with significant activation energy. Metastable carbonate phases, therefore, are likely to be present across the transition region of the mantle at which such transformations occur. Indeed *ab initio* models predict that numerous energetically similar metastable phases exist for $MgCO_3$ (Oganov et al. 2013).

Atomic structure of carbonate melts

The atomic structures of carbonate melts have been little studied in comparison to the structure of silicate melts, but are fundamental in controlling their physical and chemical behavior in natural systems. Carbonate melts are ionic liquids consisting of carbonate CO_3^{2-} molecular anions and metal cations that interact principally due to coulombic interactions and are thus very different from silicate melts, which have network structures characterized by polymerization (Mysen 1983). Ionic carbonate melts have been considered to be structureless with no definite association between metal cations and carbonate molecules (Treiman and Schedl 1983). However the combined evidence from phase relations of carbonates, the solubility of metals in carbonate liquids, and the spectroscopy of carbonate glasses and atomic simulations, suggests that carbonate liquids have structure at scales larger than their component molecular groups.

Carbonate melts as ionic liquids

The ionic nature of carbonate melts and their inability to polymerize to form network structures is their most fundamental property and is a consequence of the electronic structure and intra-molecular bonding of the carbonate ion. Consideration of the electronic structure of Si^{4+} and C^{4+} demonstrates that the outer shells of both atoms have identical electron occupation, that is, Si^{4+} ($3s^2 3p^2$) and C^{4+} ($2s^2 2p^2$), and hence similar bonding characteristics might be expected. However, the differing electronegativities of Si and C, 1.9 and 2.6 respectively (Pauling 1960), result in Si-O bonds that are less polarized than C-O bonds, with a 50% ionic character and localized charge distribution on oxygen atoms. The small ionic radius Si^{4+} of 0.34 dictates tetrahedral coordination with oxygen, Si^{4+} readily adopts sp^3 hybrid covalent bonds, whereas C^{4+} is less restricted by the requirements of close packing and adopts an sp^2 hybridization in order to reduce columbic interaction of the oxygen atoms. In both C^{4+} and Si^{4+} the adoption of hybrid bonding orbitals is facilitated by excitation of an electron from an *s* orbital to occupy an empty *p* orbital. A consequence of the sp^2 hybridization of the carbonate ion is the formation of sp^2s bonds between C and O and the formation of two *pp* bonds (Fig. 1) above and below the plane of the molecule by interaction of C and O *p*-orbitals. Not only does the presence of a *pp* bond result in a double bond, shared over the three C-O bonds, but it also leaves only lone pair *p* orbital per oxygen orientated in the plane of the molecule. Hence, unlike SiO_4^{4-} tetrahedra, the CO_3^{2-} trigonal group has no unpaired orbitals available for covalent bonding, and is hence, unable to polymerize.

Cation electronegativity (χ)

The dissociation temperatures of crystalline unary carbonates indicate that metal cations exert an important control on carbonate melt structure since they control the stability of the carbonate ion. Carbonate dissociation temperatures decrease with increasing electronegativity

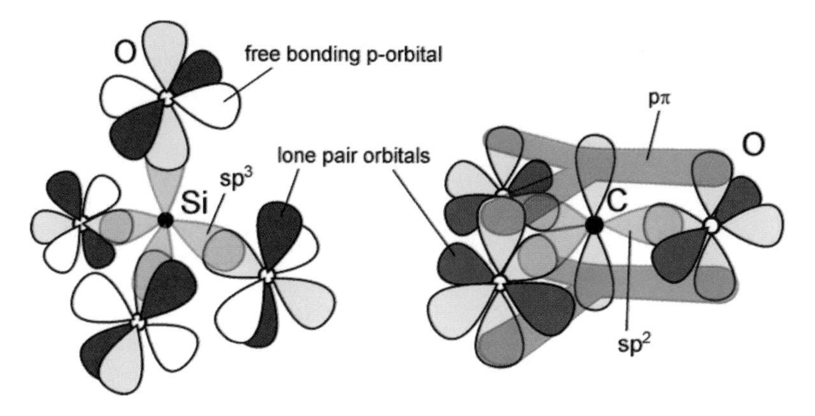

Figure 1. Showing the electronic configurations and molecular bonding of SiO_4 and CO_3.

(χ) of the metal cation for any particular polytype (Fig. 2; Weast 1972). Since χ relates to the ability of an atom to attract electrons, the variation in dissociation temperatures of crystalline carbonates with χ suggests that intra-molecular bond strengths are moderated by charge delocalization on CO_3^{2-} molecules due to the influence of nearby metal cations. Similar control by metal composition on CO_3^{2-} intra-molecular bond strengths within carbonate melts will be an important control on dissociation reactions that determine the abundance of carbonate ions available for complexation with metal cations (Genge 1994).

Different dissociation temperature-χ trends are observed for carbonates with different structures. Aragonite structure carbonates with 8-fold coordinated metal cations have higher dissociation temperatures than 6-fold coordinated calcite structure carbonates. The dependence of C-O bond strength on metal cation site coordination suggests that carbonate melt structure is likely to be as important in the dissociation of carbonate ions as composition. Transition metal carbonates have higher dissociation temperatures than predicted by χ of their metal cations, indicating that delocalization of intra-molecular bonding electrons is not the only control on carbonate dissociation (Fig. 2). Transition metals, however, are capable of coordinate bonding with carbonate ions, utilizing the lone pair orbitals of the carbonate ion to donate electrons to empty metal ligands. Coordinate bonding results in charge redistribution that will reduce delocalization of electrons from carbonate ion molecular orbitals, resulting in increased bond

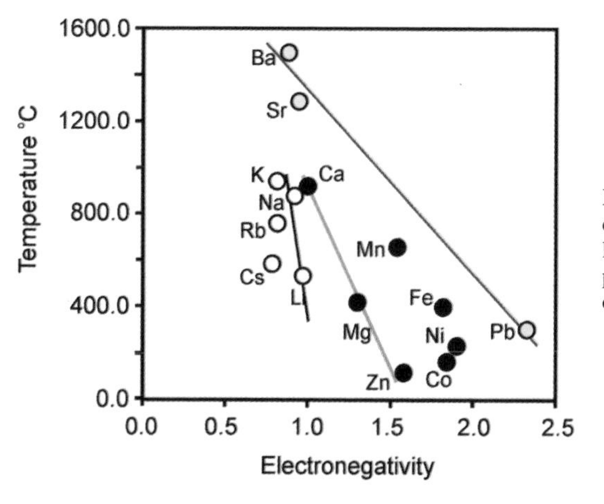

Figure 2. Showing the dissociation temperatures of crystalline carbonates at 1 atm compared with the electronegativity of metal cations.

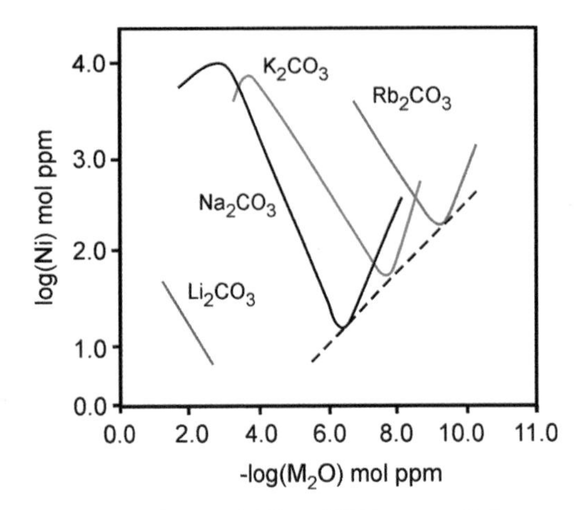

Figure 3. The solubility of NiO in unary alkali carbonate melts. After Orfield and Shores (1989).

strengths and greater carbonate stability. Carbonate ion dissociation is thus very important in the nature of speciation of components within carbonate melts.

Speciation

The solubility of metal oxides in carbonate melts can provide constraints on carbonate melt structure, and several studies have been performed on alkali carbonates at atmospheric pressure for use in molten carbonate fuel cells (MCFCs) focused on transition elements/oxides like NiO (Orfield and Shores 1988, 1989) and alkaline earth metal ceramics (Lessing et al. 1988). Solubility experiments for transition and alkaline earths in alkaline carbonate melts show solution as basic oxides or acid carbonate complexes (Fig. 3). Basic dissolution occurs at low abundances of the added metal oxide while carbonate dissolution is dominant at higher abundances. Illustrative reaction mechanisms for dissolution of NiO are shown below:

Basic	Acid
$NiO + O^{2-} = NiO_2^{2-}$	$NiO + CO_3^{2-} = NiCO_3 + O^{2-}$
$NiO + CO_3^{2-} = NiO_2^{2-} + CO_2$	$NiO + 2CO_3^{2-} = Ni(CO_3)^{2-} + O^{2-}$

Further details of speciation related to dissolution mechanisms in alkaline carbonate melts like Na_2CO_3 are provided by Orfield and Shores (1988), while Doyon et al. (1987) show that dependence on carbonate activity within the melt, oxygen fugacity, and partial pressure of CO_2 will also be important factors in controlling the speciation of metals. Orfield and Shore (1989) report experimentally derived NiO solubility in binary Na_2CO_3-K_2CO_3 melts, which exhibits significant divergences from ideal mixing behavior, and they provide additional data for Rb_2CO_3 and K_2CO_3 melts (Fig. 3).

Studies of MCFCs show that water can influence carbonate melt structure through coupled reactions with carbonate ions (Lu and Selman 1989) with a wide range of possible reactions prevalent under different oxygen fugacity and partial pressure of CO_2:

$$CO_3^{2-} + H_2O = CO_2 + 2OH^-$$

$$OH^- + CO_2 = CO_3^{2-} + H^+$$

$$3H_2 + CO = CH_4 + H_2O$$

The solubility of species in competition with CO_3^{2-} for metal cations, for example P, which in silicate melts is stabilized by divalent cations (Mysen et al. 1981) or OH^- are likely to decrease with increasing metal-carbonate complexation. Experimental studies on P solubility in $CaCO_3$ melts (Baker and Wyllie 1992) demonstrate that solubility is reduced with increasing partial pressure of CO_2 and decreasing temperature, consistent with increased formation of metal-carbonate associations in the melt (Genge et al. 1995b). The solubility of cations and molecular species within carbonate melts imply that these are present either as complexes with carbonate ions or as oxide and hydrate complexes, but the solubility data do not provide constraints on the sites or degree of order of these structural components.

Carbonate glasses

Two carbonate melt systems are known to quench to glasses at 0.1 GPa under laboratory conditions: (1) $MgCO_3$-K_2CO_3 (Faile et al. 1963; Ragone et al. 1966) and (2) $La(OH)_3$-$Ca(OH)_2$-$CaCO_3$-CaF_2-$BaSO_4$ (Jones and Wyllie 1983) and provide a means of investigating melt structure directly by spectroscopy. The phase relations of both systems are shown in (Fig. 4). Glass is a supercooled liquid and forms by cooling through the glass transition, a second order phase transition during which the translational and vibrational motions of molecular groups become restricted. Crucial in the formation of glasses is their failure to crystallize during cooling below the liquidus/solidus; thus high viscosity melts and those with low melting temperatures are most likely to form glasses due to the dependence of both crystallization and nucleation on diffusion (Turnbull 1956). The majority of glasses, therefore, form from melts with network structures, such as silicate melts, in which covalently bonded polymers resist rearrangement. Although both carbonate systems that form glass have low eutectic temperatures, they are likely to have low melt viscosities (Dobson et al. 1996). Glass formation in these systems is also not restricted only to low temperature melts. The formation of these carbonate glasses from ionic liquids is, therefore, anomalous and implies that an extended structural association occurs between molecular component groups. Infrared and Raman spectra of carbonate glasses (Genge et al. 1995a) indicate at least two structural populations of CO_3^{2-} (Fig. 5, Table 2); one vibrational frequency like common 6-fold coordinated carbonate, and a highly asymmetric site with large vibrational splitting of its ν_3 mode. The existence of two general structural populations of carbonate ion in the glasses may correspond to a flexible pseudo-network structure where alkaline earth elements (Ca and Mg) act as bridging cations linking carbonate groups by ionic bonds and having similar coordination to equivalent crystalline carbonates, while other components, such as K, act as network modifiers that support ring-structures within the flexible network (Genge et al. 1995b). The structure of carbonate melts is, therefore, envisaged as a network of metal-carbonate complexes held open by modifying species. In this context asymmetric carbonate sites were suggested to represent carbonate ions with non-bridging oxygen atoms without first neighbor bridging cations.

The spectral activity of the O-H stretching region within hydrous La-bearing glasses furthermore suggests that water exists both as molecular H_2O and OH, interacting variably with carbonate ions and as metal complexes occupying relatively high symmetry sites in these glasses. The presence of bicarbonate groups, however, is prohibited by the absence of their character-

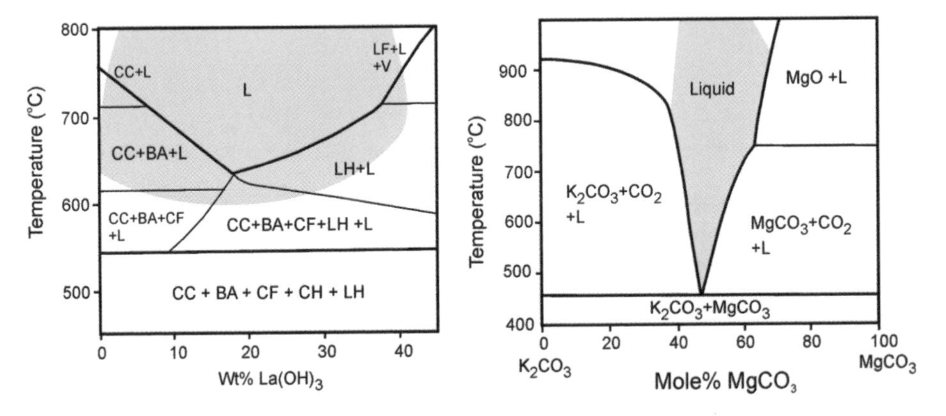

Figure 4. Phase relations in $La(OH)_3$-$Ca(OH)_2$-$CaCO_3$-CaF_2-$BaSO_4$ and $MgCO_3$-K_2CO_3 glass forming systems showing regions of melt which quench to glass (shaded areas). Abbreviations: CC-calcite, BA-barites, CF-fluorite, LH - lanthanum hydroxide, L - liquid. After Jones and Wyllie (1986) and Dobson (1995).

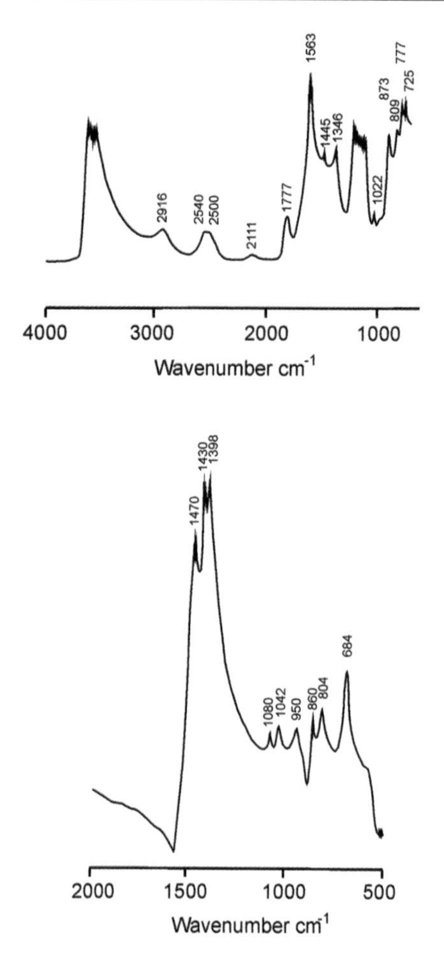

Figure 5. Background subtracted infrared reflectance spectra of La-bearing (upper) and Mg-K carbonate glass (lower); (Genge et al. 1995a).

istic O-H stretching frequencies. The absence of a molecular CO_2 ν_3 stretching mode from the infrared spectra of both carbonate glasses suggests, by comparison with IR spectra for silicate melts (Fine and Stolper 1986) very low concentrations of CO_2 in carbonate melts. In summary, IR and Raman spectroscopy of carbonate glasses, implies that for these restricted compositions, the carbonate melt structure comprises metal-carbonate complexes where a transient pseudo-network of ring structures with interstitial spaces occupied by modifying species and molecular groups (Genge et al. 1995b).

Atomic simulation of carbonates

Molecular dynamics simulations (MDS) of $CaCO_3$ from atmospheric pressure to 11.5 GPa (Genge et al. 1995b) suggest that $CaCO_3$ melts have the closest similarity to calcite structures, rather than aragonite, with similar bond lengths (Table 3) and broadly 6-fold Ca coordination (Table 4). The presence of second-nearest neighbor peaks, however, in particular for Ca-O and C-C suggest a degree of medium-range order consistent with associative metal-carbonate complexes (Fig. 6). Snapshots of melt structure (Fig. 7) indicate the occurrence of local density fluctuations, relating to spatial associations of carbonate groups and metal cations. The results of atomic simulations of $CaCO_3$ liquids are, therefore, broadly compatible with the implications of infra-red studies of carbonate glasses since they exhibit carbonate ions linked into a flexible network by ionic bonding to metal cations.

CARBONATITES

Understanding the geological context of carbonatites observed as volcanic products and magmatic rocks now at Earth's surface should not only consider their connections with other families of carbon-bearing igneous rocks like nephelinites, melilitites, and kimberlites, but also the direct evidence provided by experiments on their synthetic counterparts as carbonate melts. This section summarizes the current status of carbonatites.

Carbonatites are commonly defined as magmatic rocks with high modal abundance of carbonate minerals (>50 wt%) and geochemistry typified by high abundances of Sr, Ba, P and the light rare-earth elements (LREE) (Nelson et al. 1988). They have been subdivided (Fig. 8) on the basis of their dominant modal carbonate mineral, such as calcite-, or dolomite-carbonatites and on their corresponding major element geochemistry with Mg-, Ca, Fe- and REE-carbonatites (Woolley 1982; Le Bas 1987; Woolley and Kempe 1989). In parallel, a process-related classification would divide them into two groups: *primary carbonatites* and

Table 2. Infra-red absorption, reflectance and Raman frequencies with band assignments from carbonate glasses in cm^{-1} (Sharma and Simons 1980; Genge et al. 1995a). Symbols relate to the form of the band s-shoulder, b-broad, i-intense, w-weak.

$La(OH)_3$-$Ca(OH)_2$-$CaCO_3$-CaF_2-$BaSO_4$ Glass			
Raman	**Reflectance**	**Absorption**	**Assignment**
623(w)			CO_3 u_4 out of plane bend
690 (w)			
722 (w)	725 (w)		
	777 (w)		
	809 (w)		
870 (w)	873 (i)		CO_3 u_2 in plane bend
999 (i)	1022 (w)	999 (i)	CO_3 u_3 stretch
1123 (w)			SO_4 stretch
1300 (w)	1346 (i)		CO_3 u_3 stretch
1437 (i)			
1452 (s)			
1508 (w)			
1570 (b)	1563 (i)		
	1777 (i)	1770 (i)	$2u_2$ or u_1+u_4
1945 (i)			
2211 (i)	2111 (w)	2130 (w)	u_1+u_2
2521 (i)	2500 (w)	2504 (i)	u_1+u_3
	2540 (w)	2549 (i)	
2929 (i)	2916 (w)	2920 (i)	$2u_3$
3100 (b)			O-H stretch
	3550 (b)	3550 (b)	

$MgCO_3$-K_2CO_3 Glass			
Raman	**Reflectance**	**Absorption**	**Assignment**
690 (w)	684 (i)	621 (w)	
720 (w)	804 (w)	690 (w)	CO_3 u_4 out of plane bend
		724 (w)	
		804 (i)	
	860 (w)	872 (w)	CO_3 u_2 in plane bend
	950 (w)		
1053 (i)	1042 (i)	1060 (i)	CO_3 u_1 stretch
1072 (i)	1080 (s)	1075 (s)	
1387 (w)	1398 (s)		CO_3 u_3 stretch
1447 (i)	1430 (i)		
	1470 (s)		
1525 (w)			
		1745 (s)	$2u_2$ or u_1+u_4
		2455 (w)	u_1+u_3
		2560 (w)	
		3100 (b)	O-H stretch

carbothermal residua (Mitchell 2005). In this scheme, primary carbonatites can be further divided into groups of magmatic carbonatites associated with nephelinite, melilitite, kimberlite, and specific mantle-derived silicate magmas, formed by partial melting, whereas carbothermal residua carbonatites form as low-temperature fluids rich in CO_2, H_2O, and fluorine.

Table 3. Atomic separations (in nm) for simulated $CaCO_3$ melt at 1600 K and 0.06 GPa, and simulated crystalline polymorphs.

i-j	$CaCO_3$ melt	Calcite	Aragonite
O-O	0.2434	0.2251	0.2102
O-O(1)	0.3200	0.3210	0.3320
O-C	0.1364	0.1310	0.1276
O-Ca	0.2327	0.2320	0.2401
C-Ca	0.3424	0.3202	0.2953
C-C	0.2541	0.4002	0.2844
Ca-Ca	0.4173	0.4043	0.3902

Table 4. First-neighbor coordination numbers for $CaCO_3$ melts simulated at 1700 K at different simulated pressures.

Pressure (GPa)	0.067	3.27	11.56
Density (Kg m^{-3})	2090	2500	3000
O-O	2.912	3.157	4.848
O-C	2.826	2.853	2.544
O-Ca	5.128	5.316	6.320
C-Ca	1.433	1.410	1.911
C-C	5.434	5.420	5.006
Ca-Ca	6.310	6.210	6.102

Occurrence of carbonatites

Tectonic setting of carbonatites. Primarily, carbonatites are located within stable, intraplate settings, over half of which are in Africa, often occurring in peripheral regions to orogenic belts showing an apparent link to orogenic events or plate separation (Garson et al. 1984; Le Bas 1987; Bell 1989; Veizer et al. 1992). Carbonatite concentrations are also associated with topographic swells up to 1000 km across (Le Bas 1971; Srivastava et al. 1995). The occurrence of carbonatites in continental crust has perpetuated interpretation of their geochemistry in terms of a genetic connection (Bell and Blenkinsop 1987) and they have also been variously related to mantle plumes and large igneous provinces (LIPs); for thematic reviews see Gwalani et al (2010). A recent survey of the most complete world database shows striking lithological control with repeated activation of old carbonatites in Archaean-aged crust. This distribution precludes a direct link with mantle plumes and favors a fundamental link to the same underlying mantle source of carbon, which manifests in kimberlites (Woolley and Bailey 2012). The surface manifestation of carbonatites may belie their distribution in the underlying mantle, where some of their extreme physical properties, such as very high conductivity (Fig. 9) make them sensitive to remote geophysical testing (Gaillard et al. 2008). Understanding their deep origins must explain the growing number of carbonatites reported from unconventional tectonic associations, i.e., not continental rifts, but including oceanic islands, ophiolites, shear zones, deep subduction zones (Woolley 1991; Coltorti et al. 1999; Moine et al. 2004a; Rajesh and Arai 2006; Walter et al. 2008; Nasir 2011) and even connections to ultra-high pressure (UHP) metamorphic terranes (Attoh and Nude 2008).

Carbonatites overlying oceanic lithosphere are rare (Silva et al. 1981; Kogarko 1993; Hoernle et al. 2002; Jørgensen and Holm 2002), but discovery of carbonatite melt as interstitial

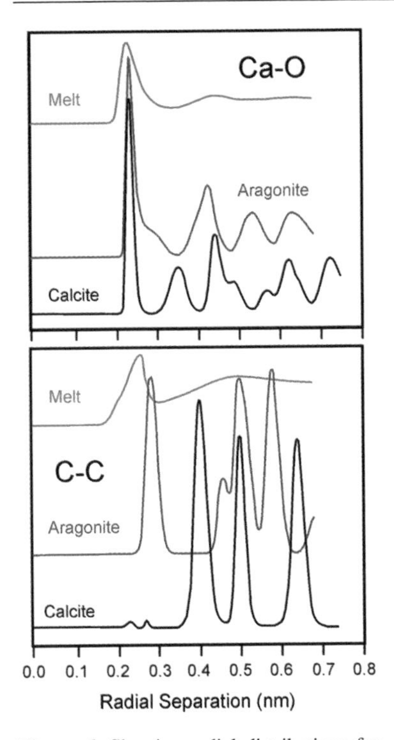

Figure 6. Showing radial distributions for $CaCO_3$ melt, calcite and aragonite from MDS simulations (Genge et al. 1995b).

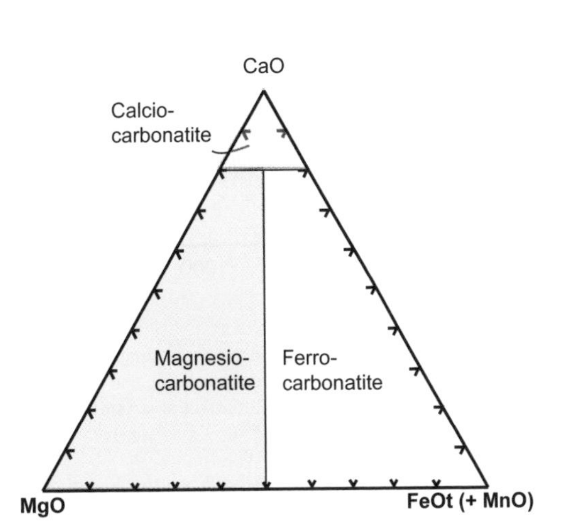

Figure 8. Carbonatite classification diagram; redrawn after (Woolley and Kempe 1989). Note: ferrocarbonatite can also be rich in REE.

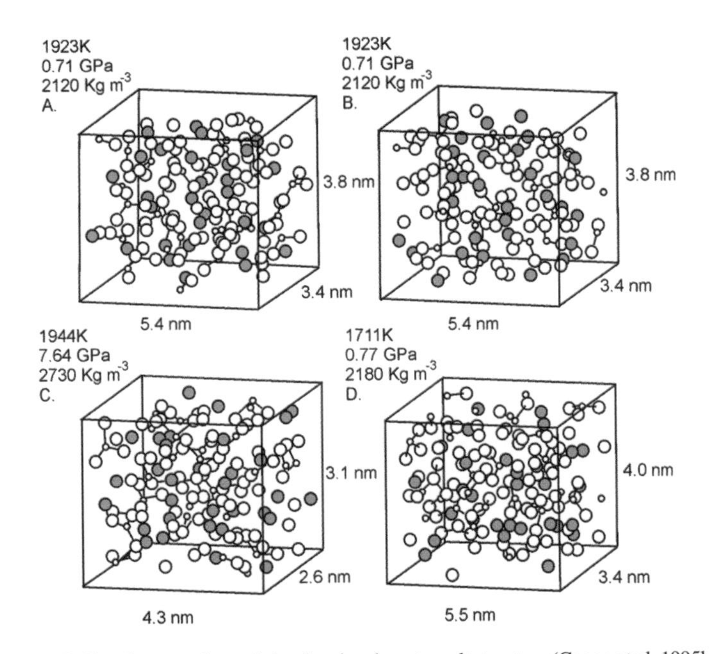

Figure 7. Showing snapshots of simulated carbonate melt structure (Genge et al. 1995b).

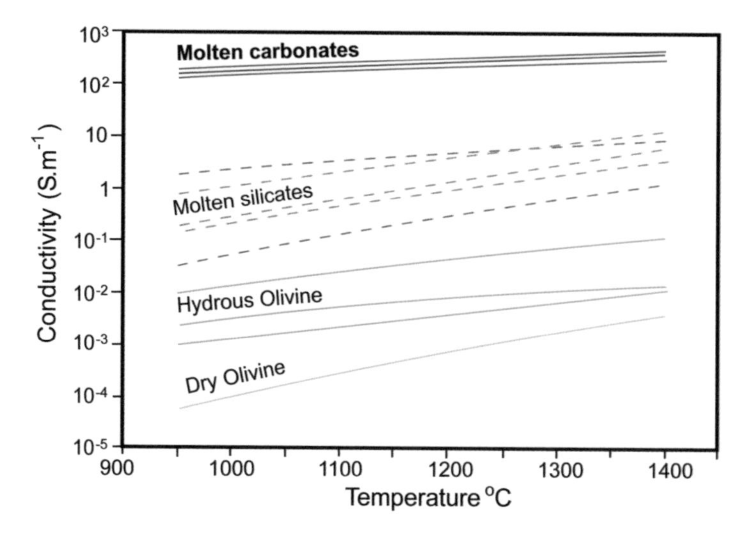

Figure 9. Electrical conductivity versus temperature shows very high values for molten carbonates compared with molten silicates, hydrous olivine and dry olivine mineral phases stable in the mantle as a function of temperature. Drawn after Gaillard et al. (2008). See also Sifre and Gaillard (2012).

pockets within dunite xenoliths from the Kerguelen Archipelago (Moine et al. 2004b) indicates that oceanic carbonatites may be more widespread. Geographical areas of carbonatitic activity are often very long-lived and, where unrelated to the migration of mantle plumes, a direct relationship with underlying lithosphere is likely (Genge 1994; Woolley and Bailey 2012), yet others are thought to be related to deep mantle plumes, for example Kola Peninsula (Marty et al. 1998), Canary Islands (Widom et al. 1999), Cape Verde Islands (Holm et al. 2006), Brazil (Toyoda et al. 1994), Deccan Traps (Simonetti et al. 1998), and Greenland (Larsen and Rex 1992). New regions with carbonatites are still being discovered, for example in the Middle East including Saudi Arabia, United Arab Emirates, and ophiolite-related carbonatite in Oman (Woolley 1991). A remarkable series of books dedicated to reviewing the systematic geographical distribution with individual maps of all carbonatites sourced from thousands of references, is provided by Woolley and has been used to publish a comprehensive global map and database of all known carbonatites (Woolley and Kjarsgaard 2008a).

Temporal distribution of carbonatites. Globally from just 56 known in 1987 (Fig. 10), there are now ~527 carbonatite occurrences, of which 49 are extrusive, ranging in age from Archaean to present (Woolley and Church 2005; Woolley and Kjarsgaard 2008a). The most commonly reported oldest dated carbonatite and associated silicate rocks are from Phalaborwa carbonatite in South Africa at 2063 to 2013 Ma (Masaki et al. 2005) and Siilinjarvi, Finland circa 2047 million years old (Puustinen 1972; Woolley and Kempe 1989). The Siilinjarvi age is typical, and based upon K-Ar dates of 1790±30 to 2030±30 Ma on phlogopite, 2530±45 Ma on richterite, and 2260±42 Ma on actinolite (Puustinen 1972), as well as 1850±40 Ma and 2280±40 Ma on phlogopite and richterite respectively from the main carbonatite. However, older ages are also discussed for Siilinjarvi, with reports of U-Pb dating of zircon from sövite indicating an age of 2580±200 Ma (Patchett et al. 1982) along with unpublished Sm-Nd data supporting an age of 2600 Ma (Basu et al. 1984).

Of the known extrusive carbonatites, 41 are calcio-carbonatites, 7 are dolomitic carbonatites, and only one extrusive carbonatite is alkaline natrocarbonatite (Woolley and Church 2005). Carbonatites tend not to occur as single rock units but rather as a suite in association with

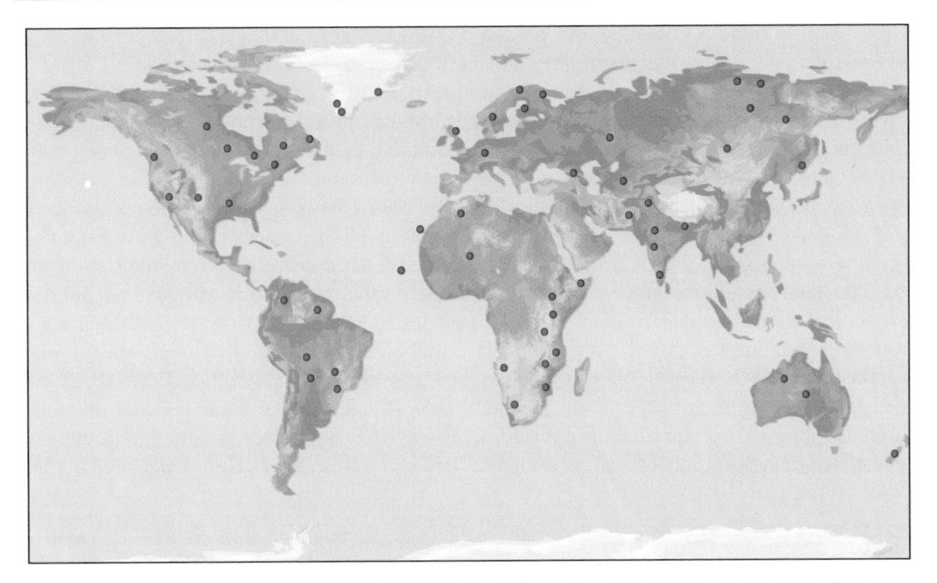

Figure 10. Global distribution of carbonatites from Le Bas (1987); this early compilation shows 56 carbonatites distributed across all continents, but mostly associated with continental modern and ancient rifting. In just over 25 years the number has transformed to over 500 today, culminating in a new metabase called "Carbonatite occurrences of the world: map and database" compiled by Woolley and Kjarsgaard (2008a), which is held at the Site of the Natural Resources in Canada, and is free to access at *http://geoscan.ess.nrcan.gc.ca/starweb/geoscan/servlet.starweb.*

alkaline silicate rocks, including a wide variety of ultramafic to felsic silicate igneous rocks from dunites to syenites, as in the Siberian Massif (Egorov 1970; Kogarko and Zartman 2007; Vladykin 2009), with only ~20% of carbonatites occurring without associated silicate rocks (Woolley and Kjarsgaard 2008a).

Despite the occurrence of carbonatites throughout the majority of geological periods since the Archean, their increase in frequency with decreasing age, and episodic clustering of activity, has led to the suggestion that conditions required for carbonatite formation are becoming more widespread (Woolley 1989). Alternatively Veizer et al (1992) argue that the apparent increase is a preservation artifact, with crustal erosion and preferential recycling of orogenic-related carbonatites, with a geodynamic half-life of carbonatite bodies close to ~445 m.y. Carbonatite rocks (orogenic and anorogenic) are relatively easily weathered in comparison to silicate rocks, so bias towards young ages in the geological record might be expected. However the argument for carbonatite age bias due to preservation presented by Viezer et al (1992) has been further debated through the consideration that cratonic material in which carbonatites are concentrated and not readily subducted; therefore, the backward projection through Precambrian time of the modern pace of recycling by subduction is not valid. Instead, Woolley and Bailey (2012) argue that the concentration of carbonatite material in late Archean cratonic regions is the result of re-opening and use of pre-existing lesions during plate movements. This activity is thought to occur in episodes, with some areas showing up to five events, with gaps of millions of years between episodes.

Geochemistry of carbonatites

Carbonatites occur as intrusive, extrusive and hydrothermal or replacement bodies that contain more than 50 vol% primary igneous carbonate minerals derived from carbonate magma (Streckeisen 1980) and with less than 20% SiO_2 (Le Maitre 2002). A summary classification is provided in Table 5 (Woolley and Kempe 1989; Le Maitre 2002). The category of REE-carbonatites have no formal definition, and they can be associated with variable Ca:Mg:Fe carbonatites and widely varying grain sizes and textures from fine-grained (Bayan Obo, Inner Mongolia, China), to pegmatitic (Kangankunde, Malawi), and porphyritic (Mountain Pass, California). Their modal REE-minerals may contribute notable colors in hand specimen either yellow, caused by bastnasite, synchisite, and REE-fluocarbonates, or vivid green, caused by monazite and REE phosphates (Wall and Mariano 1996). We suggest a whole-rock value of >1% RE_2O_3 as a working definition of REE carbonatite. Higher values (>5%) have been used by some mining geologists (Castor 2008); however, these deposits are often also rich in iron and thus have been termed ferrocarbonatite. The field relations of REE carbonatites and their association with elevated Th and U suggest primary mechanisms of crystal fractionation of carbonatite magma associated with secondary enrichment by volatile-rich metasomatic fluids (Le Bas et al. 2007; Yang et al. 2011), although relationships with alkaline silicate rocks may need further study as in the case of Mountain Pass (Castor 2008). Their residual magmatic nature is supported by experiments in mixed Ca-Ba-Sr-REE carbonate systems, which show the persistence of melts at low crustal pressures to very low-temperatures (Jones and Wyllie 1983, 1986; Wyllie et al. 1996).

Major element signatures. Carbonatites have high abundances of Sr, Ba, P, and light rare earth elements (LREEs), often > 3 orders of magnitude higher than those of chondritic meteorites or bulk Earth, and show a negative Zr and Hf anomaly (Nelson et al. 1988). Their chemical composition makes them powerful chemical probes for understanding the mantle, because the effects of crustal contamination are minimized. The average chemical compositions for carbonatite (Bell 1989) show that concentrations of Si, Ti, Mn, Ba, Fe, and F increase through the series calciocarbonatites – magnesiocarbonatites – ferrocarbonatites, but this sequence is unlikely to represent a simple crystal fractionation series (Gaspar and Wyllie 1984; Le Bas 1987). Al, Na, K, Sr, and P are variable throughout the carbonatite divisions (Table 6), with the exception of natrocarbonatite, which is dominated by Na_2O and K_2O.

Natrocarbonatite contains up to ~40 wt% (Na_2O+K_2O) with very low SiO_2, TiO_2, and Al_2O_3, high amounts of CaO and CO_2, and considerable BaO, SrO, P_2O_5, SO_3, Cl, F, and MnO in comparison to silicate igneous rocks (Ridley and Dawson 1975). Natrocarbonatite lavas erupted during the last ~50 years (especially in 1960-1966, 1988-1993, and 2006-2008) have similar compositions and their detailed geology and regional context are comprehensively reviewed in a monograph by Dawson (2008).

Table 5. Carbonatite nomenclature extended here from (Woolley and Kempe 1989)* to include Rare Earth (RE)-carbonatite and natrocarbonatite; FeO^T is total iron, RE_2O_3 = total REE oxides.

Class	Sub-division	Chemical Characteristic
Calciocarbonatite*	Sövite (coarse-grained); Alvikite (medium-to fine- grained)	CaO/(CaO+FeO+MgO > 0.80
Dolomite carbonatite	Beforsite	(Ca,Mg)-rich
Ferrocarbonatite*	—	(FeO^T + MnO) > MgO
Magnesiocarbonatite*	—	MgO > (FeO + MnO)
Rare earth carbonatite	Variable grain sizes modal REE minerals	RE_2O_3 > 1% wt
Natrocarbonatite	Lava at Oldoinyo Lengai volcano	(Na_2O + K_2O) > (CaO+MgO+FeO)

Table 6. Major element compositions of carbonatites (wt%).

	1	2	3	4	5	6	7
SiO_2	0.05	0.88	0.16	6.12	3.24	0.83	—
TiO_2	0.01	0.18	0.07	0.68	0.00	0.07	—
Al_2O_3	0.11	0.37	0.17	1.31	0.20	0.65	—
Fe_2O_3	0.41	2.62	4.04	7.55	11.50	11.00	—
MnO	0.48	0.39	0.41	0.75	5.18	5.53	1.56
MgO	0.48	0.31	0.67	12.75	10.74	0.36	19.0
CaO	14.43	53.60	51.20	29.03	25.85	43.60	28.8
Na_2O	33.89	0.09	0.25	0.14	—	0.05	—
K_2O	8.39	0.03	0.01	0.79	—	0.06	—
P_2O_5	0.93	3.18	1.52	2.66	1.27	0.42	—
CO_2	30.53	38.38	39.50	37.03	32.62	30.42	—
F	2.71	0.06	—	0.09	—	—	—
Cl	3.81	Trace	—	—	—	—	—
SO_3	2.88	—	—	0.89	0.49	—	—
SrO	1.35	0.23	0.10	0.01	0.73	0.07	1.10
BaO	1.26	0.08	0.17	0.11	2.48	>4.0	—
REE	0.1	0.05	0.3	—	2.82	1.5	—

Notes: Analyses 1-6 from Humphreys et al. (2010) and Le Bas (1987). Analysis 7 from Bailey (1989). 1. Natrocarbonatite from Oldoinyo Lengai; 2. Sövite dyke from Tundulu, Malawi; 3. Alvikite cone sheet, Homa Mountain, Western Kenya; 4. Berforsite (dolomite carbonatite) dyke, Alnö, Sweden; 5. Ferrocarbonatite with high Mg, Kangankunde, Malawi; 6. Ferrocarbonatite (low Mg), Homa Mountain, Western Kenya; 7. Magnesio-carbonatite, Rufunsa, Zambia.

Trace elements. Several "trace" elements achieve major levels in carbonatites. Steep light rare-earth element (LREE)-enriched patterns are typical (Fig. 11) and can occur as modal REE-minerals attractive for economic mining. Through the series of carbonatites from magnesiocarbonatites and calciocarbonatites to ferrocarbonatites, trace elements such as Co, Cr, Ni, and V decrease whereas the REE are most abundant in ferrocarbonatites and REE-carbonatites, often accompanied by U and Th. Few experiments have determined quantitative element partitioning in carbonate melts, but data exist for Zr, REE, and P (Jones and Wyllie 1983, 1986; Woolley and Kempe 1989; Jones et al. 1995b; Klemme and Meyer 2003; Wall and Zaitsev 2004; Ruberti et al. 2008; Xu et al. 2010). Natrocarbonatite lava contains high concentrations of LREEs with La/Yb and U/Th ratios being amongst the highest of all terrestrial lavas (Dawson 2008, and references therein).

Local chemical variations occur within the natrocarbonatites, with aphyric, phenocryst-poor lavas having a more fractionated pattern with enrichments in Mn, Mg, Fe, V, Ba, Rb, Nb, Y, and several volatile species in comparison to phenocryst-rich lavas and water contents are low (Keller and Krafft 1990). The 1993 natrocarbonatite lavas of Oldoinyo Lengai in Tanzania are particularly enriched in Ba, Cs, K, Mo, U, and LREEs relative to primitive mantle with slight variation between the lava types, i.e., spheroid-free or spheroid-bearing (Simonetti et al. 1997). Natrocarbonatite lava can have Ba/Sr > 0.7, which is the inverse to most primary calcite carbonatites, and Au contents are anomalously higher than continental crust. Natrocarbonatite also occurs as quenched melt in pyroclastic eruptions typified by proximal lapilli beds and tuffs (Church and Jones 1994) and natrocarbonatite tephra may have occurred at other volcanoes in the East African Rift, although there is controversy about whether such ash extends to the area of early hominid footprints at Laetoli (Hay 1983; Barker and Milliken 2008).

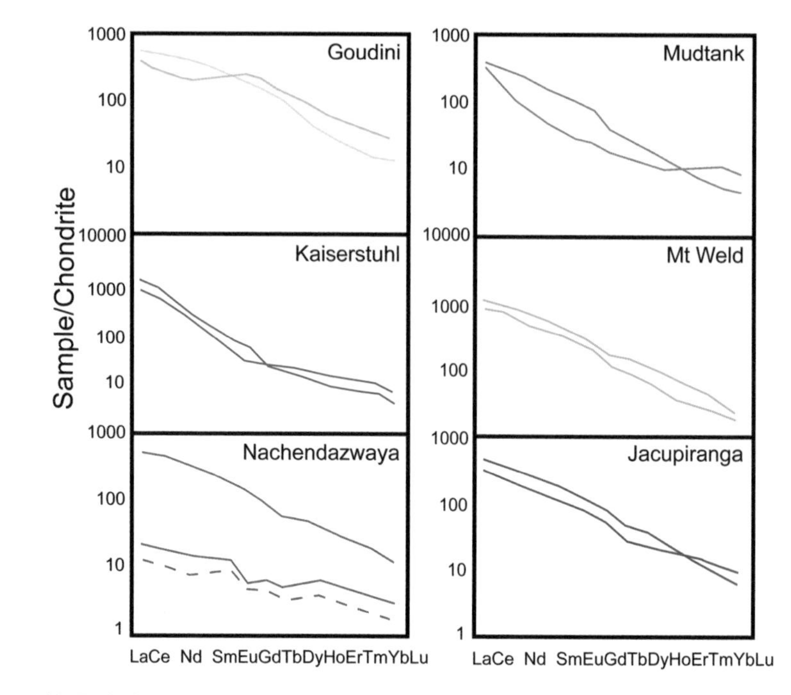

Figure 11. Typical chondrite-normalized REE concentrations of carbonatites from Goudini, Limpopo; Kaiserstuhl, Germany; Nachendazwaya, Tanzania; Mudtank and Mt Weld, Australia; Jacupiranga, Brazil, drafted after (Nelson et al. 1988) and see also (Lottermoser 1990; Currie et al. 1992; Huang et al. 1995; Verwoerd 2008). Notice the prominent light-REE enriched patterns, and absence of significant anomalies for Europium, consistent with mantle origins.

Carbonatite mineral deposits

Carbonatites contain minerals enriched in several key elements of immense economic interest, which include the REE ("strategic metals") niobium, uranium, and tantalum, and to a lesser extent, iron, copper, phosphorus, fluorite, barite, platinum group elements, silver, and gold (Richardson and Birkett 1996; Hornig-Kjarsgaard 1998). Political and economic aspects of their sustainability as producers of REE have recently been re-evaluated (Franks 2011) and currently hinge on large-scale REE carbonatite mineralization of the large Bayan Obo deposit in China (Le Bas et al. 1992, 2007; Smith and Henderson 2000; Yang et al. 2011), with high prices driving resource exploration in some countries such as Malawai (e.g., Kangankunde, AR Woolley, personal communication 2012). Carbonatites have been considered to host the majority of the world's niobium, with up to 10% of the western world's niobium once sourced from Niobec mine in the Oka carbonatite, Quebec (Scales 1989), where extreme compositional variation of pyrochlore group minerals may have been caused by magma mixing (Zurevinski and Mitchell 2004).

Table 7 shows data for significant carbonatite mineral deposits showing their potential, including estimated reserves and grades, chemical affinity and mineralogy, for both primary magmatic and secondary metasomatic deposits (Richardson and Birkett 1996). Secondary enrichment is often a result of remobilization of Nb and REEs by hydrothermal or carbothermal fluids enriched in F and CO_2 (Le Bas 1987; Smith and Henderson 2000; Fan et al. 2005). The REE-enriched nature of carbonatites has been linked to the preferential transport of REE by molecular CO_3 complexes in the melt during immiscible separation between coexisting silicate

Table 7. Mined carbonatite deposits with reserves (Megatons) and grade of interest for REE, Nb, CaF_2 (fluorite) and P_2O_5 (as phosphate minerals); for a comprehensive review of ore grades and tonnages for market conditions in the mid 1990's see Richardson and Birkett (1996).

Deposit	Reserve and Grade	Comments
Oka Carbonatite, Quebec	112.7 Mt at 0.44% Nb_2O_5 23.8 Mt at 0.2-0.5% REO	Hydrothermal REE mineralization especially pyrochlore
Phalaborwa, South Africa*	600 Mt at 7% P_2O_5 286 at 0.69% Cu 2.16 Mt REO	Banded carbonatite contains Cu sulfides, magnetite and baddeleyite.
Bayan Obo, Inner Mongolia	37 Mt at 6% REO; 1 Mt at 0.1% Nb	Largest mined REE deposit
Amba Dongar, India	11.6 Mt at 30% CaF_2	Ore associated with fenite units between carbonatite and country rock
Panda Hill, Tanzania	113 Mt at 0.3% Nb_2O_5	Disseminated pyrochlore, apatite, magnetite in sövite plug.

* Alternate spelling for Phalaborwa is Palabora.

and carbonate melts, resulting in an increased La/Lu ratio in the carbonatite relative to silicate melt (Cullers and Medaris 1977).

Isotopic signatures of carbonatites

Radiogenic isotope ratios. Although carbonatites are volumetrically insignificant compared with silicate igneous rocks, their widespread distribution on most continents coupled with their variation in age provides constraints on the evolution of the sub-continental mantle through time. Young carbonatites share significant isotopic similarities with young oceanic island basalts (OIB; Bell and Tilton 2001). Thus, alkaline silicate magmas and carbonatites in the important East African Rift, including the active carbonatite volcano Oldoinyo Lengai, lie close to the mixing line HIMU-EM1 (Fig. 12) identified in OIB. This mixing may represent either a lithospheric or a deeper mantle sub-lithospheric signature related to a mantle plume (Bell et al. 1998; Bell and Tilton 2001). In general carbonatites contain very low concentrations of Pb, far below crustal levels, offering a clear distinction from crustal carbonates. Their Pb isotopic compositions are not contaminated by crust, and can be used to probe the isotope geochemical signature of unseen deep carbon mantle reservoirs.

A linear array for covariation of Nd-Sr isotope data (Table 8) lies on the East African Carbonatite Line (EACL) and has often been thought to show binary mixing between the mantle reservoirs of HIMU (high $^{238}U/^{204}Pb$ thought to be the results of recycled ancient, altered oceanic crust) and EM1 ("enriched mantle 1," caused by the recycling of continental crust or lithosphere; Kalt et al. 1997). However, young carbonatites from the central Italian rift (Stoppa and Principe 1998; Stoppa et al. 2005, 2009) and Mesozoic carbonatites from Shandong China (Fig. 13) have much higher $^{87}Sr/^{86}Sr$ interpreted to reflect the controlling influence of metasomatized lithospheric mantle (Ying et al. 2004; Woolley and Bailey 2012).

Stable isotope ratios. Early studies of carbon and oxygen isotope ratios of carbonatites focused on coarse-grained intrusive carbonatites associated with alkaline silicate rocks in western Germany; Alnö, Sweden; and Colorado, USA, to define *primary igneous carbonatite* (Taylor et al. 1967) reproduced in Figure 14. By observing the origins of divergent trends in oxygen isotopes affected by secondary fluid/hydrothermal alteration, a common origin could be defined. Thus, the co-variation of carbon and oxygen isotopes were used to define the "box" in the range $\delta^{13}C$ −3.1 to −7.7 and $\delta^{18}O$ +5.3 to +8.4 for *primary igneous carbonatite*.

Table 8. Range of Sr and Nd isotope ratios for all rock units found at Oldoinyo Lengai natrocarbonatite volcano, and a range of carbonatite complexes with indicative key references.

Complex	^{87}Sr / ^{86}Sr	^{143}Nd / ^{144}Nd	Reference
Oldoinyo Lengai, Tanzania	0.70437-0.70445	0.51259-0.51263	(Bell and Dawson 1995)
Homa Bay, Kenya	0.70502	0.51244	(Bell and Blenkinsop 1987)
Panda Hill, Tanzania	0.70423	0.51249	(Bell and Blenkinsop 1987; Bell and Dawson 1995)
Igaliko, Greenland	0.70267-0.70380	0.51191-0.51206	(Pearce and Leng 1996)
Laiwa-Zibo, China	0.7095-0.7106	0.51155-0.51174	(Ying et al. 2004)
Il'mensky-Vishnevogorsky, Russia	0.70356-0.70470	0.51190-0.51231	(Nedosekova et al. 2009)

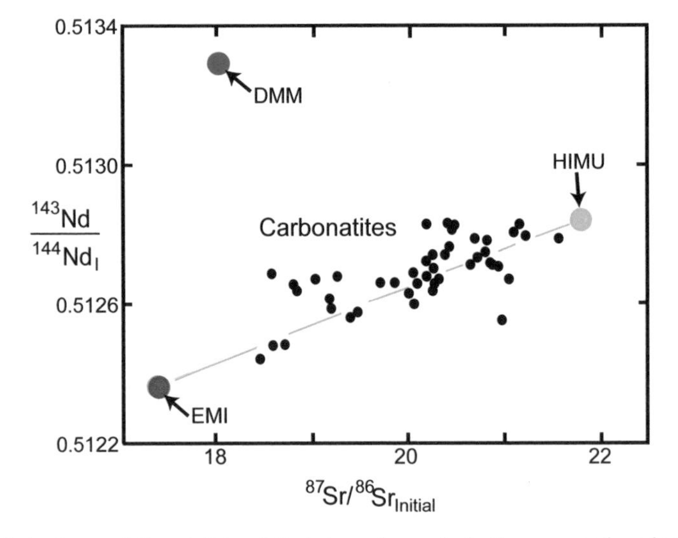

Figure 12. Isotopic covariation of Nd and Sr isotopes for geologically young carbonatites, plotted as $^{143}Nd/^{144}Nd$ Initial versus $^{87}Sr/^{86}Sr$ Initial, including positions of end member isotope reservoirs known as HIMU, DMM and EMI. After Bell and Tilton (2001). See also Bell and Blenkinsop (1987).

Subsequent studies of carbonatites show these features to be global in nature (Pineau et al. 1973; Suwa et al. 1975; Horstmann and Verwoerd 1997) including a more restricted range for fresh natrocarbonatite: $\delta^{13}C$ −6.3 to −7.1 and $\delta^{18}O$ +5.8 to +6.7 (Keller and Hoefs 1995; Zaitsev and Keller 2006). Natrocarbonatite undergoes atmospheric alteration towards heavier isotopes (Deines 1989) with $\delta^{18}O$ rapidly increasing to ~ +24‰ and $\delta^{13}C$ values changing less to −1.5‰ (Keller and Zaitsev 2006; Zaitsev and Keller 2006). Scarce representatives of apparent "oceanic" carbonatites from La Palma and Fuerteventura (Canary Islands, Spain), have $\delta^{18}O$ of +13‰ and appear to have been overprinted by fluid alteration (Hoernle et al. 2002; Demény et al. 2008). The role of isotope fractionation has often been neglected, but this can have a major influence on deep-carbon systems in general, including carbonate melts and carbonatites. Isotope fractionation has been considered to be a significant process in (mantle) carbonate systems (Deines 1968, 1970, 2004) and early experiments to 3 GPa showed C isotope fractionation between CO_2 vapor and carbonate melt (Mattey et al. 1990). Mass-independent fractionation of oxygen isotopes is also known from thermal decomposition of carbonates

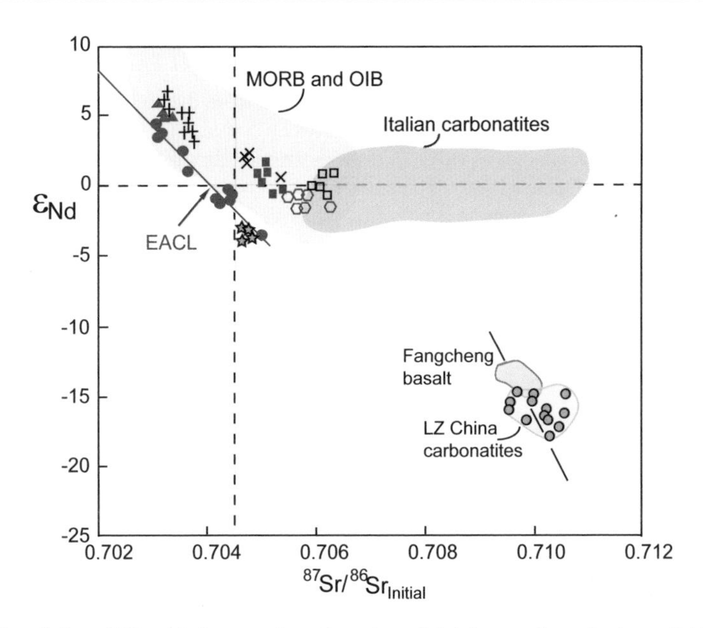

Figure 13. Covariation of Nd and Sr isotopes for carbonatites: global data set shows 3 primary fields (after Ying et al. 2004): (1) proximity to the EACL = East Africa Carbonatite Line (Bell and Blenkinsop 1987) (2) locus for extension to high Sr for Italian carbonatites (pale green field; Stoppa and Woolley 1997) and (3) steeper trend for Chinese Fangcheng carbonatite (Ying et al. 2004). Individual symbols show carbonatites from Africa (blue circles; Bell and Blenkinsop 1987), Magnet Cove (black crosses; Bell and Blenkinsop 1987), Pakistan (black/orange stars; Tilton et al. 1998), oceanic (triangles; Hoernle et al, 2002), Walloway Australia (black X's; Nelson et al. 1988), Jacupiranga Brazil (solid red squares; Huang et al. 1995), Amba Dongar India (blue hexagons; Simonetti et al. 1995), Vulture calcio-carbonatites Italy (open squares; Rosatelli et al. 2007), Laiwa-Zibu carbonatites from Fangcheng China (orange circles; Ying et al. 2004). Also shown are global basalt fields for MORB+OIB (pale yellow; Hoffman 1997), and Fangcheng basalts (pale blue; Ying et al. 2004).

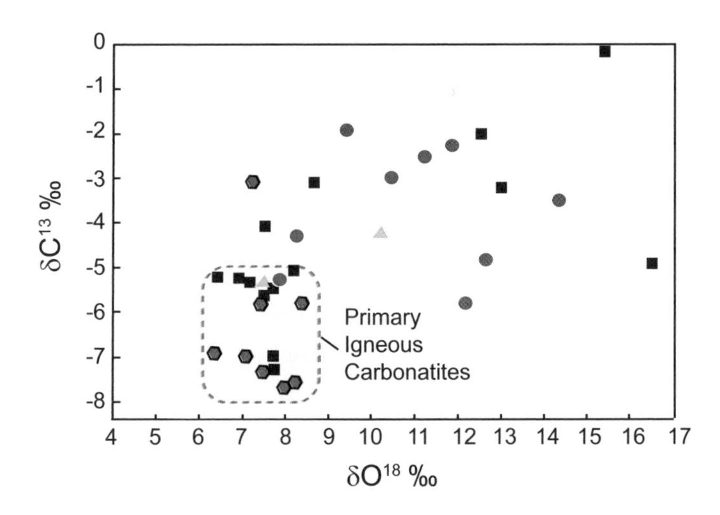

Figure 14. Stable isotope ($\delta^{18}O$ vs. $\delta^{13}C$) for global carbonatite complexes. Highlighted region of carbonatites thought to be unaffected by deuteric or hydrothermal alteration labeled "primary igneous carbonatite; after (Taylor et al. 1967). Symbols represent carbonatites from Africa (blue circles), Iron Hill (green triangles), Laacher See Germany (red hexagons), and Alnö Sweden (black squares).

(Miller et al. 2002) and there is a measurable effect of thermal decarbonation on stable isotope composition of carbonates (Sharp et al. 2003).

The mantle range for oxygen isotopes, as determined from the study of chondrites, mantle xenoliths and basalts, is relatively restricted to between 5 and 6‰, slightly offset from bulk peridotite silicate minerals such as olivine, orthopyroxene, clinopyroxene, and garnet, which average 7 to 8‰ $\delta^{18}O$ (Deines 1989). The carbon isotope signature of the mantle is more difficult to determine due to the apparent previous under-estimation of high-temperature isotope fractionation for carbon in the mantle (Mikhail et al. 2011), and also because there is large uncertainty over the average concentration of carbon in the mantle (Marty et al. 2013; Shirey et al. 2013; Wood et al. 2013). A large variability is observed in the isotopic compositions of meteorites, xenoliths, and basalts with respect to carbon, but carbonatites seem to have a relatively restricted range with a mode around -5‰. This value is more positive than average mantle diamond and chondritic meteorite isotope signatures, and may result from an unknown enrichment mechanism during melt formation from an isotopically averaged source region (Deines 2002).

GENESIS OF CARBONATITE MAGMAS

The three main theories for the origin of carbonatites are essentially:

1. Residual melts of fractionated carbonated nephelinite or melilitite (Gittins 1989; Gittins and Jago 1998).
2. Immiscible melt fractions of CO_2-saturated silicate melts (Freestone and Hamilton 1980; Amundsen 1987; Kjarsgaard and Hamilton 1988, 1989; Brooker and Hamilton 1990; Kjarsgaard and Peterson 1991; Church and Jones 1995; Lee and Wyllie 1997; Dawson 1998; Halama et al. 2005; Brooker and Kjarsgaard 2011).
3. Primary mantle melts generated through partial melting of CO_2-bearing peridotite (Wallace and Green 1988; Sweeney 1994; Harmer and Gittins 1998; Harmer et al. 1998; Ying et al. 2004).

Combinations of these three theories are also popular; for example carbonatite liquids generated by deep melting of carbonated eclogite in the upper mantle infiltrate overlying peridotite to produce silica under-saturated carbonate-bearing melts, which then penetrate the crust and evolve or un-mix (Yaxley and Brey 2004). Carbonatites have also been considered to be generated in the lithospheric mantle as partial melts rising rapidly above a hot ascending mantle plume. If these mantle carbonate melts stall, for example owing to thermal death, they generate carbonate-melt metasomatism in the mantle (Wyllie 1995). As the much hotter center of the plume approaches, melting is induced in the metasomatic horizon and results in generation of the carbonatite melts that are observed on the surface (Bizimis et al. 2003). Although the plume model is quite attractive, recent recognition of strong and repeated lithospheric controls in the compilation of global carbonatite ages are thought to argue against a direct connection to mantle plumes (Woolley and Bailey 2012).

Alternative models for natrocarbonatite petrogenesis have included anatexis of metasomatized basement (Morogan and Martin 1985), incorporation of trona sediments [$Na_3(CO_3)(HCO_3)\cdot 2H_2O$], remobilization of carbonate material already present on the volcano (Church and Jones 1995, and references therein), and condensation from a co-magmatic fluid (Nielsen and Veksler 2001, 2002).

Recently discovered provinces of Neogene- to Quaternary-aged carbonatite-silicate volcanism in Europe (Italy, France, and Spain) are providing important new lines of research for the mantle origins of extrusive carbonatite volcanism (Stoppa and Principe 1998; Stoppa et al. 2005). These new studies are exciting because, unlike carbonatites in general, the young volcanoes have transported mantle xenoliths, which can help constrain their source regions or wallrocks. Derived geotherms can also constrain transport pathways for deep carbon (Jones et al. 2000b; Downes et al. 2002; Bailey and Kearns 2012). These silico-carbonatite volcanic rocks also preserve high-pressure carbonate minerals like aragonite ($CaCO_3$; Hazen et al. 2013a) as early high-pressure inclusions in olivine (Fo_{87}) considered to have crystallized in the mantle at depths of >100 km (Humphreys et al. 2010, 2012).

Carbonate melt metasomatism

The low viscosity and chemical composition of carbonatites and carbonate melts (Table 1) makes them excellent metasomatic agents, and there are both experimental and natural examples. High P_2O_5/TiO_2 or variable solubility in synthetic carbonate melts was used to suggest metasomatized harzburgite by carbonatite melts (Ryabchikov et al. 1989; Baker and Wyllie 1992) and carbonate metasomatism has been inferred to occur in the lower mantle from experiments at 20-24.5 GPa and 1600-2000 °C (Gasparik and Litvin 2002). Thus, while transfer of mobile silicate (basaltic) melts may adequately account for the chemistry of many spinel peridotites and pyroxenites, highly mobile carbonate melts are believed to have played a pivotal role in the formation of apatite pyroxenites/wehrlites (converted peridotites), carbonate-bearing peridotites (reacted wallrock) and metasomatized mantle xenoliths in continental terrains (O'Reilly and Griffin 2000). Carbonate-metasomatism is observed in some eclogitic xenoliths (Pyle and Haggerty 1994, 1997) and massive eclogite terrains, implying an active role for metasomatic carbonate fluids in subduction zones (Selverstone et al. 1992). "Subduction" experiments with carbonated eclogite at 5-10 GPa, corresponding to approximately 150-300 km depth in the mantle show, that the carbonatitic solidus in eclogite is located at 4 GPa higher in pressure than in the peridotitic system; first carbonate partial melts can be calcium-rich [Ca/(Ca+Fe+Mg) ~ 0.80] in contrast to those produced by melting of carbonated peridotite [Ca/(Ca+Fe+Mg) ~ 0.50] (Hammouda 2003) possibly also influenced by the role of garnet in carbonated eclogite (Knoche et al. 1999). The Th/U ratio may be sensitive to mantle metasomatism related to carbonate or silicate melts where clinopyroxene is stable (Foley et al. 2001) and ultrapotassic mantle metasomatism in East Africa can be notably rich in fluorine (Edgar et al. 1994; Rosatelli et al. 2003) and may result during emplacement of immiscible silicate-carbonatite magmas (Rosatelli et al. 2003).

Melt inclusions in deep volcanic minerals and xenoliths from continental rift systems commonly demonstrate the effects of metasomatism caused by carbonate melts (Seifert and Thomas 1995; Jones et al. 2000b; Downes et al. 2002; Woolley and Bailey 2012). Carbonate melt metasomatism in lherzolite has also been related to derivation from kimberlite transport in the mantle (Bodinier et al. 2004). Sometimes direct evidence for high-pressure carbonate-melts occurs trapped as inclusions in minerals, such as in ultramafic xenoliths from Kerguelen related to oceanic upper mantle domains (Schiano et al. 1994), and globally as carbonatitic alkaline hydrous fluid inclusions in diamond (Navon et al. 1988; Guthrie et al. 1991; Schrauder and Navon 1993, 1994; Izraeli et al. 2001; Tomlinson et al. 2005), which in addition to common inorganic carbon-species C-O-H volatiles also may contain hydrocarbons and nitrogen (Tomilenko et al. 1997).

The percolation of carbonatitic material through olivine-rock matrix (i.e., mantle peridotite) has been proposed to occur at a rate of several millimeters per hour by a process of dissolution-precipitation (Hammouda and Laporte 2000). These infiltration rates are orders of magnitude higher than those previously found for basalt infiltration in mantle lithologies. However, it could be argued that such quick percolation may result in short residence times, which may inhibit chemical interaction and metasomatism to take place (Dalou et al. 2009). Transient mineral reactions to armor veins may greatly extend metasomatic pathways while depleting fluids in silicate-compatible and hydrous components (Jones et al. 1983; Menzies et al. 1987; Rudnick et al. 1993; Menzies and Chazot 1995; Dawson 2002). Evidence for carbonatite-related metasomatism has been repeatedly recognized in the East African rift region, as manifest in fenitized rocks and glimmerites with abundant phlogopite, pyroxene, and amphibole (Rhodes and Dawson 1975; Dawson and Smith 1988, 1992; Rudnick et al. 1993; Dawson et al. 1995), as well as across the globe in Italy (Rosatelli et al. 2007; Stoppa et al. 2008, 2009; Stoppa and Woolley, 1997), Australia (Andersen et al. 1984), Canary Islands (Frezzotti et al. 2002), Grande Comore in the Indian Ocean (Coltorti et al. 1999), and Greenland (Ionov et al. 1993).

Carbonate melt crystallization of diamond

Natural diamondiferous carbonatites are very rare but are known to occur at Chagatai, Uzbekistan, and a few other diamond-bearing carbonatites are likely related to the crystallization of carbonate-kimberlite (Litvin et al. 2001, 2003; Bobrov et al. 2004), where it is thought that the diamond is syngenetic (Palyanov et al. 1994). In principle, carbonate melts, which are over-saturated with respect to dissolved carbon and sometimes visibly bear graphite, are highly efficient diamond-forming systems under the thermodynamic conditions of diamond stability (Borzdov et al. 1999; Bobrov et al. 2004; Kogarko et al. 2010). Several observations of carbonate as inclusions in mantle diamond (Guthrie et al. 1991) lead to models of diamond formation involving carbonated melts in the mantle (Hammouda 2003). Carbonate melt inclusions have also been identified associated with diamond in continental crust from high pressure metamorphic rocks (De Corte et al. 2000; Korsakov et al. 2005), interpreted as very deep subduction (Korsakov and Hermann 2006) with unknown significance of recently recognized lonsdaleite (hexagonal diamond) in crustal rocks (Godard et al. 2011). Diamond is readily synthesized experimentally from simple carbonates (Pal'yanov et al. 1999a, 1999b, 2002a, 2002b), from alkali potassic carbonate (Shatskii et al. 2002), from hydrous-carbonate (Sokol et al. 2000), from carbonate with sulfide (Spivak et al. 2008), from mixed carbonate-alkali-halide systems (Tomlinson et al. 2004), and from mixed carbonate-silicate (kimberlitic) melts (Litvin 2003) to produce artificial diamond rock or "diamondite" with polycrystalline diamond (Litvin and Spivack 2003). Experiments show that "superdeep" diamond can crystallize rapidly from simple Mg-carbonate at pressures up to 20 GPa (Tomlinson et al. 2011). Thus, carbonate melts have been suggested as a medium for the efficient crystallization of diamonds in Earth's mantle, broadly consistent with global distribution of carbonatitic fluids trapped during the growth of natural "coated stone" type mantle diamond (Navon et al. 1988; Tomlinson et al. 2005; Klein-BenDavid et al. 2007) although quantitative details of a genetic connection between carbonate melts and fluids in the mantle are virtually unknown.

Magmas related to carbonate melts

For at least half a century field observations of alkaline silicate-carbonatite systems have suggested a variety of petrogenetic pathways between carbonatites and alkaline silicate magmas are preserved in surface volcanic products at least from the Proterozoic (Harmer 1999; Andersen 2008) through to Recent volcanism (Le Bas 1987). The associations have been tested with increasing sophistication as experimental technology has advanced. Pioneering phase equilibria studies delineated the important role of water (Wyllie and Tuttle 1962). More recent experiments quantified partial melting of mantle lherzolite using major components CaO-MgO-Al_2O_3-SiO_2-CO_2 with a CO_2 content of 0.15 wt%, which showed a continuous change in melt composition approximating to carbonatitic through to kimberlitic melts over the minor melting range 0-1%, initially to a few GPa pressure (Dalton and Presnall 1998), to 3-8 GPa (Gudfinsson and Presnall 2005), and to 6-10 GPa pressure (Brey et al. 2008), where additional preconditioning of the mantle by metasomatism is implicated. In general these experimental studies have avoided alkaline-bearing or indeed alkaline-rich and fluid-bearing systems (but see Litvin et al. 1998; Gasparik and Litvin 2002; Safonov et al. 2007; Brooker and Kjarsgaard

2011) and are still lacking in complexity, and for example, understanding of oxygen fugacity compared with natural systems.

Direct experiments with natural carbonatitic rocks and minerals suffer from a host of different problems, but they can provide indicative results (e.g., Bobrov et al. 2004). Carbonatites may share geochemical trace element signatures with kimberlites (Hornig-Kjarsgaard 1998; Le Roex et al. 2003) and associations with other families of ultramafic rocks such as lamprophyres where CO_2-rich ocelli (Huang et al. 2002) might imply arrested unmixing of immiscible liquids. Round spherical textures, such as ocelli in rocks are sometimes ambiguous (Brooker and Hamilton 1990) and coexisting silicate and carbonatite lapilli have been also used to argue against a genetic association (Andersen 2008). The composition of primary kimberlite from the Slave Craton, based on samples of aphanitic kimberlite from the Jericho kimberlite pipe, Northwest Territories, Canada, have minimum CO_2 contents (10-17 wt%) and geochemistry suggesting carbonatite affinities (Price et al. 2000). Potassic silicate glass occurs with calcite carbonatite in lapilli from extrusive carbonatites at Rangwa Caldera Complex, Kenya (Rosatelli et al. 2003). Group II mantle kimberlites have been shown to be experimentally related to carbonate melts in their mantle source region (Ulmer and Sweeney 2002) and metasomatized mantle xenoliths can show, for example, geochemical signatures transitional between kimberlites and carbonatites (Jones 1989). Detailed potential relationships between distinct families of silicate rocks and carbonatites are given in dedicated books and reviews (Le Bas 1987; Woolley and Kempe 1989; Woolley 2003; Mitchell 2005; Woolley and Kjarsgaard 2008b), and others have blurred the boundaries between carbonatite and kimberlite volcanism (Sparks et al. 2009) as evident in early experiments (Wyllie and Huang 1976).

Other indicative experiments of upper mantle carbonate melts have investigated liquid immiscibility (Brooker and Hamilton 1990; Lee and Wyllie 1996), the transition from carbonate to silicate melts in the simplified CaO-MgO-SiO_2-CO_2 system (Moore and Wood 1998), mixed carbonate-chloride-silicate systems (Safonov et al. 2007), dissociation of carbonate in low pressure halide melts (Combes et al. 1977; Cherginets and Rebrova 2002, 2003) and the melting behavior of mantle carbonate-phlogopite lherzolite (Thibault et al. 1992).

FUTURE RESEARCH

Fundamental gaps remain in our knowledge of the petrogenesis of carbonate melts in the deep mantle, which cloud our understanding of the possible origins of carbonatites, including their relationships to mantle silicate rocks and to the deepest known mantle magmas, kimberlites.

Carbonatites at high-pressure

New geological discoveries that provide insights on carbonatites are still being made, and can be very instructive. For example, carbonatite-silicate volcanism carrying mantle xenoliths, carbonatites in major shear zones, and carbonatites associated with ophiolite belts are providing new tests and ideas for conventional research into mantle-derived carbonatitic melts (Tilton et al. 1998; Nasir et al. 2003; Stoppa et al. 2005; Nasir 2006, 2011; Rajesh and Arai 2006; Attoh and Nude 2008; Humphreys et al. 2010; Bailey and Kearns 2012). However, we still lack quantitative understanding of high-pressure behavior in several critical areas; for example high-pressure behavior of mixed C-O-H fluids to calculate carrying capacities over the range of compositions transitional between carbonate-silicate mantle systems, and how these operate in fluid-rich pathways escaping from subduction zones (Stalder et al. 1998). What have been the roles of carbonate melts in the distribution of carbon throughout Earth's history, and how do these roles relate to changing mantle dynamics and different plate tectonic settings? Key will be to map out the phase stability of carbonates from the crust into the mantle to connect observations from the deepest natural samples (inclusions in diamond) with predictions from theory and experiment (Berg 1986; Biellmann et al. 1993a, 1993b; Brenker 2005; Brenker et al. 2006, 2007). Particularly relevant will be to understand the relative stabilities of low-pressure carbonate and newly discovered families of high-pressure tetracarbonates.

Melt structure of tetracarbonates?

The transformation of carbonate minerals to "tetracarbonates" represents potentially the most significant step-forward in the understanding of the solid reservoirs of carbon at deeper mantle pressures (Oganov et al. 2013), in particular since experimental results suggest the transformation may involve a range of carbon-bearing phases, including nanodiamond (Boulard et al. 2011). The implications for carbonate melt structures, if these are present under lower mantle geotherms, is intriguing to consider. Changes in the density and viscosity in silicate melts occur corresponding to phase changes in the equivalent solid phase due to increases in coordination (Genge 1994; Karki and Stixrude 2010). Thus, jadeite and albite melts display increases in density in both solid and quenched glass phases between 1.0-1.5 GPa and decreases in viscosity relating to the plagioclase to garnet phase transition (Karki and Stixrude 2010). We might therefore expect changes in behavior in carbonate melts corresponding to phase transitions in their crystalline solids with increases in coordination and close packing under compression. Crystalline $CaCO_3$ transforms from calcite to aragonite structures at ~2 GPa (Suito et al. 2001); however, MDS simulations of $CaCO_3$ melts suggest no equivalent increase in Ca coordination from 6- to 8-fold (Genge et al. 1995b). The large compressibility of $CaCO_3$ melts predicted from these simulations, however, would imply that coordination increases are inevitable at higher pressures (>11 GPa). If transformation of $CaCO_3$ melt structure to a tetrahedral carbonate occurs at lower mantle pressures then a dramatic change in melt properties would be expected owing to the ability of CO_4 to form polymerizable networks. If the theoretical increases in carbonate melt viscosity at high pressures are verified (Jones and Oganov 2010) this behavior would fundamentally inhibit mobility of carbonate melts in the lower mantle and transform previous conceptions (Jones and Oganov 2009; Stoppa et al. 2009). Significant changes in the chemical properties of carbonate melts dramatically change their solubility for metal cations. The site of transformation of tetracarbonate to carbonate magmas in the lower mantle could, therefore, be extremely significant for storage of deep mantle carbon, and we might envisage, for example, precipitation of solid phases from rising carbonate melts.

ACKNOWLEDGMENTS

We are grateful to Alan Woolley, for providing a review of an early draft and to the Natural History Museum London for access to additional specimens and research facilities. The groundwork for this review was supported through graduate studentship training awards at University College London provided by NERC.

REFERENCES

Amundsen HEF (1987) Evidence of liquid immiscibility in the upper mantle. Nature 327:692-695

Andersen T (2008) Qassiarssuk Complex, Gardar Rift, southwest Greenland. Can Mineral 46(4):933-950

Andersen T, O'Reilly SY, Griffin WL (1984) The trapped fluid phase in upper mantle xenoliths from Victoria, Australia: implications for mantle metasomatism. Contrib Mineral Petrol 88(1):72-85

Attoh K, Nude PM (2008) Tectonic significance of carbonatite and ultrahigh-pressure rocks in the Pan-African Dahomeyide suture zone, southeastern Ghana. J Geol Soc 297:217-231

Bailey DK (1989) Carbonate-rich melts from the mantle in the volcanoes of south-east Zambia. Nature 338:415-418

Bailey DK, Kearns S (2012) New forms of abundant carbonatite–silicate volcanism: recognition criteria and further target locations. Mineral Mag 76:271-284

Baker MB, Wyllie PJ (1992) High-pressure apatite solubility in carbonate-rich liquids: implications for mantle metasomatism. Geochim Cosmochim Acta 56(9):3409-3422

Barker DS, Milliken KL (2008) Cementation of the footprint tuff, Laetoli, Tanzania. Can Mineral 46(4):831-841

Basu AR, Goodwin AM, Tatsumoto M (1984) Sm-Nd study of Archean alkalic rocks from the Superior Province of the Canadian Shield. Earth Planet Sci Lett 70(1):40-46

Bell DR, Schmitz MD, Janney PE (2003) Mesozoic thermal evolution of the southern African mantle lithosphere. Lithos 71(2-4):273-287

Bell K (1989) Carbonatites: Genesis and Evolution. Unwin Hyman, London

Bell K, Blenkinsop J (1987) Nd and Sr isotopic compositions of East African carbonatites: Implications for mantle heterogeneity. Geology 15(2):99-102

Bell K, Dawson JB (1995) Nd and Sr Isotope Systematics of the Active Carbonatite Volcano, Oldoinyo Lengai. *In:* Carbonatite Volcanism: Oldoinyo Lengai and the Petrogenesis of Natrocarbonatites. Bell K, Keller J (eds) IAVCEI Proceedings in Volcanology 4:101-112

Bell K, Kjarsgaard BA, Simonetti A (1998) Carbonatites: into the twenty first century. J Petrol 39(11-12):1839-1845

Bell K, Tilton GR (2001) Nd, Pb and Sr isotopic compositions of East African carbonatites: evidence for mantle mixing and plume inhomogeneity. J Petrol 42(10):1927-1945

Berg GW (1986) Evidence for carbonate in the Mantle. Nature 324(6092):50-51

Biellmann C, Gillet P, Guyot F, Peyronneau J, Reynard B (1993a) Experimental-evidence for carbonate stability in the Earths lower mantle. Earth Planet Sci Lett 118(1-4):31-41

Biellmann C, Guyot F, Gillet P, Reynard B (1993b) High-pressure stability of carbonates: quenching of calcite-II, high-pressure polymorph of $CaCO_3$. Eur J Mineral 5(3):503-510

Bizimis M, Salters VJM, Dawson JB (2003) The brevity of carbonatite sources in the mantle: evidence from Hf isotopes. Contrib Mineral Petrol 145:281-300

Bobrov AV, Litvin YA, Divaev FK (2004) Phase relations and diamond synthesis in the carbonate-silicate rocks of the Chagatai Complex, Western Uzbekistan: Results of experiments at P = 4-7 GPa and T = 1200-1700 degrees C. Geochem Int 42(1):39-48

Bodinier JL, Menzies MA, Shimizu N, Frey FA, McPherson E (2004) Silicate, hydrous and carbonate metasomatism at Lherz, France: Contemporaneous derivatives of silicate melt-harzburgite reaction. J Petrol 45(2):299-320

Borzdov YM, Sokol AG, Pal'yanov YN, Kalinin AA, Sobolev NV (1999) Studies of diamond crystallization in alkaline silicate, carbonate and carbonate-silicate melts. Doklady Akademii Nauk 366(4):530-533

Boulard E, Gloter A, Corgne A, Antonangeli D, Auzende A-L, Perrillat J-P, Guyot Fo, Fiquet G (2011) New host for carbon in the deep Earth. Proc Natl Acad Sci USA 108(13):5184-5187

Brenker FE (2005) Detection of a Ca-rich lithology in the Earth's deep (>300 km) convecting mantle. Earth Planet Sci Lett 236:579-587

Brenker FE, Vollmer C, Vincze L, Vekemans B, Szymanski A, Janssens K, Szaloki I, Nasdala L, Joswig W, Kaminsky F (2006) CO_2-recycling to the deep convecting mantle. Geochim Cosmochim Acta 70(18, Supplement 1):A66-A66

Brenker FE, Vollmer C, Vincze L, Vekemans B, Szymanski A, Janssens K, Szaloki I, Nasdala L, Joswig W, Kaminsky F (2007) Carbonates from the lower part of transition zone or even the lower mantle. Earth Planet Sci Lett 260(1-2):1-9

Brey GP, Bulatov VK, Girnis AV, Lahaye Y (2008) Experimental melting of carbonated peridotite at 6-10 GPa. J Petrol 49(4):797-821

Brooker RA, Hamilton DL (1990) Three-liquid immiscibility and the origin of carbonatites. Nature 346:459-462

Brooker RA, Kjarsgaard BA (2011) Silicate-carbonate liquid immiscibility and phase relations in the system SiO_2-Na_2O-Al_2O_3-CaO-CO_2 at 0.1-2.5 GPa with applications to carbonatite genesis. J Petrol 52(7-8):1281-1305

Castor SB (2008) The Mountain Pass Rare Earth carbonatite and associated ultrapotassic rocks, California. Can Mineral 46(4): 779-806

Cherginets VL, Rebrova TP (2003) On carbonate ion dissociation in molten alkali metal halides at approximate to 800 degrees C. J Chem Eng Data 48(3):463-467

Cherginets VL, Rebroya TP (2002) Dissociation of carbonate ions in molten chlorides of alkali metals. Russian J Phys Chem 76(1):118-120

Church AA, Jones AP (1994) Hollow natrocarbonatite lapilli from the 1992 eruption of Oldoinyo-Lengai, Tanzania. J Geol Soc 151:59-63

Church AA, Jones AP (1995) Silicate-Carbonate Immiscibility at Oldoinyo-Lengai. J Petrol 36(4):869-889

Coltorti M, Bonadiman C, Hinton RW, Siena F, Upton BGJ (1999) Carbonatite metasomatism of the oceanic upper mantle: Evidence from clinopyroxenes and glasses in ultramafic xenoliths of Grande Comore, Indian Ocean. J Petrol 40:133-165

Combes R, Feys R, Tremillon B (1977) Dissociation of carbonate in molten NaCl-KCl. J Electroanal Chem 83:383-385

Cullers RL, Medaris G (1977) Rare earth elements in carbonatite and cogenetic alkaline rocks: Examples from Seabrook Lake and Callander Bay, Ontario. Contrib Mineral Petrol 65(2):143-153

Currie KL, Knutson J, Temby PA (1992) The Mud Tank carbonatite complex, central Australia—an example of metasomatism at mid-crustal levels. Contrib Mineral Petrol 109(3):326-339
Dalou C, Koga KT, Hammouda T, Poitrasson F (2009) Trace element partitioning between carbonatitic melts and mantle transition zone minerals: Implications for the source of carbonatites. Geochim Cosmochim Acta 73(1):239-255
Dalton JA, Presnall DC (1998) The continuum of primary carbonatitic-kimberlitic melt compositions in equilibrium with lherzolite: Data from the system CaO-MgO-Al_2O_3-SiO_2-CO_2 at 6 GPa. J Petrol 39(11-12):1953-1964
Dawson JB (1998) Peralkaline nephelinite-natrocarbonatite relationships at Oldoinyo Lengai, Tanzania. J Petrol 39(11-12):2077-2094
Dawson JB (2002) Metasomatism and partial melting in upper mantle peridotite xenoliths from the Lashaine volcano, northern Tanzania. J Petrol 43(9):1749-1777
Dawson JB (2008) The Gregory Rift Valley and Neogene-Recent Volcanoes of Northern Tanzania. Geological Society London, London
Dawson JB, Pinkerton H, Norton GE, Pyle DM (1990) Physico-chemical properties of alkali-carbonatite lavas, Tanzania. Geology 18:260-263
Dawson JB, Smith JV (1988) Metasomatized and veined upper-mantle xenoliths from Pello Hill, Tanzania: evidence for anomalously-light mantle beneath the Tanzanian sector of the East African Rift Valley. Contrib Mineral Petrol 100(4):510-527
Dawson JB, Smith JV (1992) Potassium loss during metasomatic alteration of mica pyroxenite from Oldoinyo Lengai, northern Tanzania: contrasts with fenitization. Contrib Mineral Petrol 112(2):254-260
Dawson JB, Smith JV, Steele IM (1995) Petrology and mineral chemistry of plutonic igneous xenoliths from the carbonatite volcano, Oldoinyo Lengai, Tanzania. J Petrol 36(3):797-826
De Corte K, Korsakov A, Taylor WR, Cartigny P, Ader M, De Paepe P (2000) Diamond growth during ultrahigh-pressure metamorphism of the Kokchetav Massif, northern Kazakhstan. Island Arc 9(3):428-438
Deines P (1968) The carbon and oxygen isotopic composition of carbonates from a mica peridotite dike near Dixonville, Pennsylvania. Geochim Cosmochim Acta 32(6):613-625
Deines P (1970) The carbon and oxygen isotopic composition of carbonates from the Oka carbonatite complex, Quebec, Canada. Geochim Cosmochim Acta 34(11):1199-1225
Deines P (1989) Stable isotope variations in Carbonatites. *In:* Carbonatites: Genesis and Evolution. Bell K (ed) Unwin Hyman, London, p 301-359
Deines P (2002) The carbon isotope geochemistry of mantle xenoliths. Earth Sci Rev 58(3-4):247-278
Deines P (2004) Carbon isotope effects in carbonate systems. Geochim Cosmochim Acta 68(12):2659-2679
Deméńy A, Casillas R, Ahijado A, de La Nuez J, Andrew Milton J, Nagy G (2008) Carbonate xenoliths in La Palma: Carbonatite or alteration product? Chemie Erde 68(4):369-381
Dobson DP (1995) Synthesis of carbonate melts at high pressure and the origin of natural kimberlites. PhD Dissertation, University College London, London.
Dobson DP, Crichton WA, Vočadlo L, Jones AP, Wang YB, Uchida T, Rivers M, Sutton SR, Brodholt JP (2000) In situ measurement of viscosity of liquids in the Fe-FeS system at high pressures and temperatures. Am Mineral 85:1838-1842
Dobson DP, Jones AP, Rabe R, Sekine T, Kurita K, Taniguchi T, Kondo T, Kato T, Shimomura O, Urakawa S (1996) In-situ measurement of viscosity and density of carbonate melts at high pressure. Earth Planet Sci Lett 143(1-4):207-215
Downes H, Kostoula T, Jones AP, Thirlwall MF (2002) Geochemistry and Sr-Nd isotope composition of peridotite xenoliths from Monte Vulture, south central Italy. Contrib Mineral Petrol 144:78-92
Downes H, Wall F, Demeny A, Szabo C (2012) Continuing the carbonatite controversy: preface. Mineral Mag 76:255-257
Doyon JD, Gilbert T, Davies G, Paetsch L (1987) NiO solubility in mixed alkali/alkaline earth carbonates. J Electrochem Soc 134(12):3035-3038
Edgar AD, Lloyd FE, Vukadinovic D (1994) The role of fluorine in the evolution of ultrapotassic magmas. Mineral Petrol 51(2-4):173-193
Egorov LS (1970) Carbonatites and ultrabasic-alkaline rocks of the Maimecha-Kotui region, N. Siberia. Lithos 3(4):341-359
Faile S, Roy DM, Tuttle OF (1963) The Preparation, properties and structure of carbonate glasses. Interim Report, October 1, 1962-August 1, 1963. Other Information: Orig. Receipt Date: 31-DEC-63, p 22. Penn State University, Pennsylvania.
Fan H-R, Hu F-F, Wang K-Y, Xie Y-H, Mao J, Bierlein FP (2005) Aqueous-carbonic-REE fluids in the giant Bayan Obo deposit, China: implications for REE mineralization. (Mineral Deposit Research: Meeting the Global Challenge) Springer-Verlag, Berlin Heidelberg, p 945-948
Fine G, Stolper E (1986) Dissolved carbon dioxide in basaltic glasses: concentrations and speciation. Earth Planet Sci Lett 76(3-4):263-278

Foley SF, Petibon CM, Jenner GA, Kjarsgaard BA (2001) High U/Th partitioning by clinopyroxene from alkali silicate and carbonatite metasomatism: an origin for Th/U disequilibrium in mantle melts? Terra Nova 13(2):104-109
Franks SM (2011) Rare Earth Minerals: Policies and Issues. Nova Science, Hauppage, NY.
Freestone IC, Hamilton DL (1980) The role of liquid immiscibility in the genesis of carbonatites — An experimental study. Contrib Mineral Petrol 73(2):105-117
Frezzotti ML, Andersen T, Neumann ER, Simonsen SL (2002) Carbonatite melt-CO_2 fluid inclusions in mantle xenoliths from Tenerife, Canary Islands: A story of trapping, immiscibility and fluid-rock interaction in the upper mantle. Lithos 64(3-4):77-96
Frost DJ, McCammon CA (2008) The redox state of Earth's mantle. Ann Rev Earth Planet Sci 36(1):389-420
Gaillard F, Malki M, Iacono-Marziano G, Pichavant M, Scaillet B (2008) Carbonatite melts and electrical conductivity in the asthenosphere. Science 322(5906):1363-1365
Garson MS, Coats JS, Rock NMS, Deans T (1984) Fenites, breccia dykes, albitites, and carbonatitic veins near the Great Glen Fault, Inverness, Scotland. J Geol Soc 141(4):711-732
Gaspar JC, Wyllie PJ (1984) The alleged kimberlite-carbonatite relationship - evidence from ilmenite and spinel from Premier and Wesselton Mines and the Benfontein Sill, South-Africa. Contrib Mineral Petrol 85(2):133-140
Gasparik T, Litvin YA (2002) Experimental investigation of the effect of metasomatism by carbonatic melt on the composition and structure of the deep mantle. Lithos 60(3-4):129-143
Genge MJ (1994) The structure of carbonate melts and implications for the petrogenesis of carbonatite magmas. PhD Dissertation, University College London, London.
Genge MJ, Jones AP, Price GD (1995a) An infrared and Raman study of carbonate glasses: implications for the structure of carbonatite magmas. Geochim Cosmochim Acta 59(5):927-937
Genge MJ, Price GD, Jones AP (1995b) Molecular dynamics simulations of $CaCO_3$ melts to mantle pressures and temperatures: implications for carbonatite magmas. Earth Planet Sci Lett 131(3-4):225-238
Gittins J (1989) The origin and evolution of carbonatite magmas. *In:* Carbonatites: Genesis and Evolution, Bell K (ed) Unwin Hyman, London, p 580-600
Gittins J, Jago BC (1998) Differentiation of natrocarbonatite magma at Oldoinyo Lengai volcano, Tanzania. Mineral Mag 62:759-768
Godard G, Frezzotti ML, Palmeri R, Smith DC (2011) Origin of high-pressure disordered metastable phases (lonsdaleite and incipiently amorphized quartz) in metamorphic rocks: Geodynamic shock or crystal-scale overpressure? *In* Ultrahigh Pressure Metamorphism, 1st Edition: 25 Years After the Discovery of Coesite and Diamond. Dobrzhinetskaya L, Cuthbert S, Faryad W (eds) Elsevier, Amsterdam, p 125-148
Gudfinsson GH, Presnall DC (2005) Continuous gradations among primary carbonatitic, kimberlitic, melilititic, basaltic, picritic, and komatiitic melts in equilibrium with garnet lherzolite at 3-8 GPa. J Petrol 46(8):1645-1659
Guthrie GD, Veblen DR, Navon O, Rossman GR (1991) Sub-micrometer fluid inclusions in turbid-diamond coats. Earth Planet Sci Lett 105(1-3):1-12
Halama R, Vennemann T, Siebel W, Markl G (2005) The Gronnedal-Ika carbonatite-syenite complex, South Greenland: carbonatite formation by liquid immiscibility. J Petrol 46(1):191-217
Hammouda T (2003) High-pressure melting of carbonated eclogite and experimental constraints on carbon recycling and storage in the mantle. Earth Planet Sci Lett 214(1-2):357-368
Hammouda T, Laporte D (2000) Ultrafast mantle impregnation by carbonatite melts. Geology 28(3):283-285
Harmer RE (1999) The petrogenetic association of carbonatite and alkaline magmatism: constraints from the Spitskop Complex, South Africa. J Petrol 40(4):525-548
Harmer RE, Gittins J (1998) The Case for Primary, Mantle-derived Carbonatite Magma. J Petrol 39(11-12):1895-1903
Harmer RE, Lee CA, Eglington BM (1998) A deep mantle source for carbonatite magmatism; evidence from the nephelinites and carbonatites of the Buhera district, SE Zimbabwe. Earth Planet Sci Lett 158:131-142
Hauri E, Shimizu N, Dieu JJ, Hart SR (1993) Evidence for hotspot-related carbonatite metasomatism in the oceanic upper mantle. Nature 365:221-227
Hay RL (1983) Natrocarbonatite tephra of Kerimasi volcano, Tanzania. Geology 11(10):599-602
Hazen RM, Downs RT, Jones AP, Kah L (2013a) Carbon mineralogy and crystal chemistry. Rev Mineral Geochem 75:7-46
Hazen RM, Downs RT, Kah, L, Sverjensky D (2013b) Carbon mineral evolution. Rev Mineral Geochem 75:79-107
Hoernle K, Tilton G, Le Bas M, Duggen S, Garbe-Schönberg D (2002) Geochemistry of oceanic carbonatites compared with continental carbonatites: mantle recycling of oceanic crustal carbonate. Contrib Mineral Petrol 142(5):520-542
Hoffman AW (1997) Mantle geochemistry: the message from oceanic volcanism. Nature 385:219-229

Holm PM, Wilson JR, Christensen BP, Hansen L, Hansen SL, Hein KM, Mortensen AK, Pedersen R, Plesner S, Runge MK (2006) Sampling the Cape Verde mantle plume: Evolution of melt compositions on Santo Antåo, Cape Verde Islands. J Petrol 47(1):145-189

Hornig-Kjarsgaard I (1998) Rare earth elements in sovitic carbonatites and their mineral phases. J Petrol 39(11-12):2105-2121

Horstmann UE, Verwoerd WJ (1997) Carbon and oxygen isotope variations in southern African carbonatites. J African Earth Sci 25(1):115-136

Huang Y-M, Hawkesworth CJ, van Calsteren P, McDermott F (1995) Geochemical characteristics and origin of the Jacupiranga carbonatites, Brazil. Chem Geol 119(1-4):79-99

Huang Z, Liu C, Xiao H, Han R, Xu C, Li W, Zhong K (2002) Study on the carbonate ocelli-bearing lamprophyre dykes in the Ailaoshan gold deposit zone, Yunnan Province. Sci China Ser D Earth Sci 45(6):494-502

Humphreys ER, Bailey DK, Hawkesworth CJ, Wall F, Najorka J, Rankin A (2010) Aragonite in olivine from Calatrava, Spain—Evidence for mantle carbonatite melts from >100 km depth. Geology 38:911-914

Ionov DA, Dupuy C, O'Reilly SY, Kopylova MG, Genshaft YS (1993) Carbonated peridotite xenoliths from Spitsbergen: implications for trace element signature of mantle carbonate metasomatism. Earth Planet Sci Lett 119(3):283-297

Izraeli ES, Harris JW, Navon O (2001) Brine inclusions in diamonds: a new upper mantle fluid. Earth Planet Sci Lett 187(3-4):323-332

Jones AP (1989) Upper mantle enrichment by kimberlitic or carbonatitic magmatism. *In:* Carbonatites: Genesis and Evolution. Bell K (ed) Unwin Hyman, London, p 448-463

Jones AP (2000) Carbonatite thematic set. Mineral Mag 64(4):581-582

Jones AP, Claeys P, Heuschkel S (2000a) Impact melting of carbonates from the Chicxulub crater. *In:* Impacts and the Early Earth. 91. Gilmour I, Koeberl C (eds) Spinger-Verlag, Berlin, p 343-361

Jones AP, Dobson DP, Genge M (1995a) Comment on physical-properties of carbonatite magmas inferred from molten-salt data, and application to extraction patterns from carbonatite-silicate magma chambers - discussion. Geol Mag 132(1):121-121

Jones AP, Kostoula T, Stoppa F, Woolley AR (2000b) Petrography and mineral chemistry of mantle xenoliths in a carbonate-rich melilititic tuff from Mt. Vulture volcano, southern Italy. Mineral Mag 64(4):593-613

Jones AP, Oganov A (2009) Superdeep carbonate melts in the Earth. Goldschmidt Conference, p A603, Vienna, Austria.

Jones AP, Oganov A (2010) Carbon rich melts in the Earth's deep mantle. *In:* Diamond Workshop Bressanone. F Nestol (ed) *http://www.univie.ac.at/Mineralogie/EMU/media/NESTOLA/Jones_1.pdf*

Jones AP, Smith JV, Dawson JB (1983) Mantle metasomatism in 14 veined peridotites from Bultfontein Mine, South Africa. J Geol 90:435-453

Jones AP, Wall F, Williams CT (1995b) Rare Earth Minerals: Chemistry, Origin and Ore Deposits. Springer-Verlag, Berlin

Jones AP, Wyllie PJ (1983) Low temperature glass quenched from synthetic rare earth carbonatite: implications for the origin of the Mountain Pass deposit, California. Econ Geol 78:1721-1723

Jones AP, Wyllie PJ (1986) Solubility of rare earth elements in carbonatite magmas, indicated by the liquidus surface in $CaCO_3$-$Ca(OH)_2$-$La(OH)_3$ at 1 kbar pressure. Appl Geochem 1(1):95-102

Jørgensen JØ, Holm PM (2002) Temporal variation and carbonatite contamination in primitive ocean island volcanics from Sao Vicente, Cape Verde Islands. Chem Geol 192:249-267

Kalt A, Hegner E, Satir M (1997) Nd, Sr, and Pb isotopic evidence for diverse lithospheric mantle sources of East African Rift carbonatites. Tectonophysics 278(1-4):31-45

Kaminsky F (2012) Mineralogy of the lower mantle: A review of 'super-deep' mineral inclusions in diamond. Earth Sci Rev 110:127-147

Kaminsky F, Wirth R, Schreiber A, Thomas R (2009) Nyerereite and nahcolite inclusions in diamond: evidence for lower-mantle carbonatitic magmas. Mineral Mag 73(5):797-816

Kaminsky, Kaminsky F, Zakharchenko, Zakharchenko O, Davies, Davies R, Griffin, Griffin W, Khachatryan B, Khachatryan-Blinova G, Shiryaev, Shiryaev A (2001) Superdeep diamonds from the Juina area, Mato Grosso State, Brazil. Contrib Mineral Petrol 140(6):734-753

Karki BB, Stixrude L (2010) Viscosity of $MgSiO_3$ liquid at mantle conditions and implications for early magma ocean. Science 328:740-742

Keller J, Hoefs J (1995) Stable isotope characteristics of recent natrocarbonatites from Oldoinyo Lengai. *In:* Carbonatite Volcanism: Oldoinyo Lengai and the Petrogenesis of Natrocarbonatites. Bell K, Keller J (eds) IAVCEI Proceedings in Volcanology 4:113-123

Keller J, Krafft M (1990) Effusive natrocarbonatite activity of Oldoinyo Lengai, June 1988. Bull Volcanol 52(8):629-645

Keller J, Zaitsev AN (2006) Calciocarbonatite dykes at Oldoinyo Lengai, Tanzania: the fate of natrocarbonatite. Can Mineral 44(4):857-876

Kjarsgaard B, Hamilton DL (1988) Liquid immiscibility and the origin of alkali-poor carbonatites. Mineral Mag 52:43-55

Kjarsgaard B, Hamilton DL (1989) The genesis of carbonatites by immiscibility. *In:* Carbonatites: Genesis and Evolution. Bell K (ed) Unwin Hyman, London, p 388-404
Kjarsgaard B, Peterson T (1991) Nephelinite-carbonatite liquid immiscibility at Shombole volcano, East Africa: Petrographic and experimental evidence. Mineral Petrol 43(4):293-314
Klein-BenDavid O, Izraeli ES, Hauri E, Navon O (2007) Fluid inclusions in diamonds from the Diavik mine, Canada and the evolution of diamond-forming fluids. Geochim Cosmochim Acta 71(3):723-744
Klemme S, Meyer HP (2003) Trace element partitioning between baddeleyite and carbonatite melt at high pressures and high temperatures. Chem Geol 199(3-4):233-242
Knoche R, Sweeney RJ, Luth RW (1999) Carbonation and decarbonation of eclogites: the role of garnet. Contrib Mineral Petrol 135(4):332-339
Kogarko LN (1993) Geochemical characteristics of oceanic carbonatites from Cape Verde Islands. South African J Geol 96:119-125
Kogarko LN, Ryabchikov ID, Divaev FK, Wall F (2010) Regime of carbon compounds in carbonatites in Uzbekistan: evidence from carbon isotopic composition and thermodynamic simulations. Geochem Int 11:1053-1063
Kogarko LN, Zartman RE (2007) A Pb isotope investigation of the Guli massif, Maymecha-Kotuy alkaline-ultramafic complex, Siberian flood basalt province, Polar Siberia. Mineral Petrol 89(1):113-132
Korsakov AV, Hermann J (2006) Silicate and carbonate melt inclusions associated with diamonds in deeply subducted carbonate rocks. Earth Planet Sci Lett 241(1-2):104-118
Korsakov AV, Vandenabeele P, Theunissen K (2005) Discrimination of metamorphic diamond populations by Raman spectroscopy (Kokchetav, Kazakhstan). Spectrochim Acta A 61:2378-2385
Krafft M, Keller J (1989) Temperature measurements in carbonatite lava lakes and flows from Oldoinyo Lengai, Tanzania. Science 245(4914):168-170
Kraft S, Knittle E, Williams Q (1991) Carbonate stability in the Earths mantle - a vibrational spectroscopic study of aragonite and dolomite at high-pressures and temperatures. J Geophys Res 96(B11):17997-18009
Larsen LM, Rex DC (1992) A review of the 2500 Ma span of alkaline-ultramafic, potassic and carbonatitic magmatism in West Greenland. Lithos 28(3-6):367-402
Le Bas MJ (1971) Per-alkaline volcanism, crustal swelling, and rifting. Nature Phys Sci 230(12):85-87
Le Bas MJ (1987) Nephelinites and carbonatites. Geol Soc London Special Pub 30(1):53-83
Le Bas MJ, Kellere J, Wall F, Williams CT, Peishan Z (1992) Carbonatite dykes at Bayan Obo, Inner Mongolia, China. Mineral Petrol 46:195-228
Le Bas MJ, SYang XM, Taylor RN, Spiro B, Milton JA, Peishan Z (2007) New evidence from a calcite-dolomite carbonatite dyke for the magmatic origin of the massive Bayan Obo ore-bearing dolomite marble, inner Mongolia, China. Mineral Petrol 91:281-307
Le Maitre RW (2002) Igneous Rocks: A Classification and Glossary of Terms. Cambridge University Press, Cambridge, UK
Le Roex AP, Bell DR, Davis P (2003) Petrogenesis of group I kimberlites from Kimberley, South Africa: Evidence from bulk-rock geochemistry. J Petrol 44(12):2261-2286
Lee WJ, Wyllie PJ (1996) Liquid immiscibility in the join $NaAlSi_3O_8$-$CaCO_3$ to 2 center dot 5 GPa and the origin of calciocarbonatite magmas. J Petrol 37(5):1125-1152
Lee WJ, Wyllie PJ (1997) Liquid immiscibility between nephelinite and carbonatite from 1.0 to 2.5 GPa compared with mantle melt compositions. Contrib Mineral Petrol 127(1):1-16
Lessing PA, Yang ZZ, Miller GR, Yamada H (1988) Corrosion of metal oxide ceramics in molten lithium-potassium carbonates. J Electrochem Soc 135(5):1049-1057
Litvin YA (2003) Mantle genesis of diamond in carbonate-silicate-carbon melts of variable chemistry: Evidence from high-pressure experiments. Geochim Cosmochim Acta 67(18):A256-A256
Litvin YA, Chudinovskikh LT, Zharikov VA (1998) Crystallization of diamond in the system $Na_2Mg(CO_3)_2$-$K_2Mg(CO_3)_2$-C at 8-10 GPa. Doklady Akad Nauk 359(5):668-670
Litvin YA, Jones AP, Beard AD, Divaev FK, Zharikov VA (2001) Crystallization of diamond and syngenetic minerals in melts of diamondiferous carbonatites at Chagatai Massif, Uzbekistan; Experiment at 7.0 GPa. Dokl Earth Sci 318A:1066-1069
Litvin YA, Spivack A (2003) Rapid growth of diamondite at the contact between graphite and carbonate melt: Experiments at 7.5-8.5 GPa. Dokl Earth Sci 391A:888-891
Litvin YA, Spivak AV, Matveev YA (2003) Experimental study of diamond formation in the molten carbonate-silicate rocks of the Kokchetav metamorphic complex at 5.5-7.5 GPa. Geochem Int 41(11):1090-1098
Lottermoser BG (1990) Rare-earth element mineralisation within the Mt. Weld carbonatite laterite, Western Australia. Lithos 24(2):151-167
Lu SH, Selman JR (1989) Hydrogen oxidation in molten carbonate: mechanistic analysis of potential sweep data J Electrochem Soc 136:1063-1072
Luth RW (1993) Diamonds, eclogites, and the oxidation-state of the Earth's mantle. Science 261:66-68
Martinez I, Agrinier P, Scharer U, Javoy M (1994) A SEM ATEM and stable-isotope study of carbonates from the Haughton Impact Crater, Canada. Earth Planet Sci Lett 121(3-4):559-574

Martinez I, Deutsch A, Scharer U, Ildefonse P, Guyot F, Agrinier P (1995) Shock recovery experiments on dolomite and thermodynamical calculations of impact-induced decarbonation. J Geophys Res Solid Earth 100(B8):15465-15476

Marty B, Tolstikhin I, Kamensky IL, Nivin V, Balaganskaya E, Zimmermann J-L (1998) Plume-derived rare gases in 380 Ma carbonatites from the Kola region (Russia) and the argon isotopic composition in the deep mantle. Earth Planet Sci Lett 164(1-2):179-192

Marty B, Alexander CMO'D, Raymond SN (2013) Primordial origins of Earth's carbon. Rev Mineral Geochem 75:149-181

Masaki Y, Yuka H, Naoko N, Hirogo K (2005) Rb-Sr, Sm-Nd ages of the Phalaborwa Carbonatite Complex, South Africa. Polar Geosci 18:101-113

Mattey DP, Taylor WR, Green DH, Pillinger CT (1990) Carbon isotopic fractionation between CO_2 vapor, silicate and carbonate melts: an experimental-study to 30 Kbar. Contrib Mineral Petrol 104(4):492-505

Menzies M, Chazot G (1995) Fluid processes in diamond to spinel facies shallow mantle. J Geodyn 20(4):387-415

Menzies MA, Rogers NW, Tindle A, Hawkesworth CJ (1987) Metasomatic and enrichment processes in lithospheric peridotites, an effect of asthenosphere-lithosphere Interaction. *In:* Mantle Metasomatism. Menzies MA, Hawkesworth CJ (eds) Academic Press, London, p 313-385

Mikhail S, Shahar A, Hunt S, A, Verchovsky AB, Jones AP (2011) An experimental investigation of the pressure effect on stable isotope fractionation at high temperature: Implications for mantle processes and core formation in celestial bodies. Lunar Planet Sci Conf Proc 42:1376.

Miller MF, Franchi IA, Thiemens MH, Jackson TL, Brack A, Kurat G, Pillinger CT (2002) Mass-independent fractionation of oxygen isotopes during thermal decomposition of carbonates. Proc Natl Acad Sci USA 99(17):10988-10993

Mitchell RH (2005) Carbonatites and carbonatites and carbonatites. Can Mineral 43(6):2049-2068

Moine BN, Gregoire M, O'Reilly SY, Delpech G, Sheppard SMF, Lorand JP, Renac C, Giret A, Cottin JY (2004a) Carbonatite melt in oceanic upper mantle beneath the Kerguelen Archipelago. Lithos 75(1-2):239-252

Moine BN, Gregoire M, O'Reilly SY, Delpech G, Sheppard SMF, Lorand JP, Renac C, Giret A, Cottin JY (2004b) Carbonatite melt in oceanic upper mantle beneath the Kerguelen Archipelago. Lithos 75:239-252

Moore KR, Wood BJ (1998) The transition from carbonate to silicate melts in the CaO-MgO- SiO_2-CO_2 system. J Petrol 39(11-12):1943-1951

Morogan V, Martin RF (1985) Mineralogy and partial melting of fenitized crustal xenoliths in the Oldoinyo Lengai carbonatitic volcano, Tanzania. Am Mineral 70:1114-1126

Mysen BO (1983) The structure of silicate melts. Annu Rev Earth Planet Sci 11(1):75-97

Mysen BO, Ryerson FJ, Virgo D (1981) The structural role of phosphorus in silicate melts. Am Mineral 66:106-117

Nasir S (2006) Geochemistry and petrology of Tertiary volcanic rocks and related ultramafic xenoliths from the central and eastern Oman Mountains. Lithos 90:249-270

Nasir S (2011) Petrogenesis of ultramafic lamprophyres and carbonatites from the Batain Nappes, eastern Oman continental margin. Contrib Mineral Petrol 161:47-74

Nasir S, Hanna S, Al-Hajari S (2003) The petrogenetic association of carbonatite and alkaline magmatism: Constrains from the Masfut-Rawda Ridge, Northern Oman Mountains. Mineral Petrol 77:235-258

Navon O, Hutcheon ID, Rossman GR, Wasserburg GJ (1988) Mantle-derived fluids in diamond micro-inclusions. Nature 335:784-789

Nedosekova IL, Vladykin NV, Pribavkin SV, Bayanova TB (2009) The Il'mensky-Vishnevogorsky Miaskite-Carbonatite Complex, the Urals, Russia: Origin, ore resource potential, and sources. Geol Ore Deposits 51:139-161

Nelson DR, Chivas AR, Chappell BW, McCulloch MT (1988) Geochemical and isotopic systematics in carbonatites and implications for the evolution of ocean-island sources. Geochim Cosmochim Acta 52:1-17

Nielsen TF, Veksler IV (2002) Is natrocarbonatite a cognate fluid condensate? Contrib Mineral Petrol 142(4):425-435

Nielsen TFD, Veksler IV (2001) Oldoinyo Lengai natrocarbonatite revisited: a cognate fluid condensate? J Afr Earth Sci 32(1):A27-A28

Oganov AR, Hemley RJ, Hazen RM, Jones AP (2013) Structure, bonding, and mineralogy of carbon at extreme conditions. Rev Mineral Geochem 75:47-77

Oganov A, Ono S, Ma Y, Glass CW, Garcia A (2008) Novel high-pressure structures of $MgCO_3$, $CaCO_3$ and CO_2 and their role in Earth's lower mantle. Earth Planet Sci Lett 273(1-2):38-47

Oppenheimer C (1998) Satellite observation of active carbonatite volcanism at Ol Doinyo Lengai, Tanzania. Int J Remote Sensing 19(1):55-64

O'Reilly SY, Griffin WL (2000) Apatite in the mantle: implications for metasomatic processes and high heat production in Phanerozoic mantle. Lithos 53(3-4):217-232

Orfield ML, Shores DA (1988) Solubility of NiO in molten Li_2CO_3-Na_2CO_3-K_2CO_3, and Rb_2CO_3 at 910 °C. J Electrochem Soc 135:1662-1672
Orfield ML, Shores DA (1989) The solubility of NiO in binary mixtures of molten carbonates. J Electrochem Soc 136(10):2862-2866
Pal'yanov YN, Khokhryakov AF, Borzdov YM, Doroshev AM, Tomilenko AA, Sobolev NV (1994) Inclusions in synthetic diamonds. Dokl Akad Nauk 338(1):78-80
Pal'yanov YN, Sokol AG, Borzdov YM, Khokhryakov AF (2002a) Fluid-bearing alkaline carbonate melts as the medium for the formation of diamonds in the Earth's mantle: An experimental study. Lithos 60(3-4):145-159
Pal'yanov YN, Sokol AG, Borzdov YM, Khokhryakov AF, Shatsky AF, Sobolev NV (1999a) The diamond growth from Li_2CO_3, Na_2CO_3, K_2CO_3 and Cs_2CO_3 solvent-catalysts at P = 7 GPa and T = 1700-1750 degrees C. Diamond Relat Mater 8(6):1118-1124
Pal'yanov YN, Sokol AG, Borzdov YM, Khokhryakov AF, Sobolev NV (1999b) Diamond formation from mantle carbonate fluids. Nature 400(6743):417-418
Pal'yanov YN, Sokol AG, Borzdov YM, Khokhryakov AF, Sobolev NV (2002b) Diamond formation through carbonate-silicate interaction. Am Mineral 87:1009-1013
Patchett JP, Kuovo O, Hedge CE, Tatsumoto M (1982) Evolution of continental crust and mantle heterogeneity: Evidence from Hf isotopes. Contrib Mineral Petrol 78(3):279-297
Pauling L (1960) The Nature of the Chemical Bond and the Structure of Molecules and Crystals. Cornell University Press, New York
Pearce NJG, Leng MJ (1996) The origin of carbonatites and related rocks from the Igaliko Dyke Swarm, Gardar Province, South Greenland: field, geochemical and C-O-Sr-Nd isotope evidence. Lithos 39:21-40
Pineau F, Javoy M, Allegre CJ (1973) Etude systématique des isotopes de l'oxygène, du carbone et du strontium dans les carbonatites. Geochim Cosmochim Acta 37:2363-2377
Presnall DC, Gudfinnsson GH (2005) Carbonate-rich melts in the oceanic low-velocity zone and deep mantle. Geol Soc Am Spec Pap 388:207-216
Price SE, Russell JK, Kopylova MG (2000) Primitive Magma From the Jericho Pipe, N.W.T., Canada: Constraints on primary kimberlite melt chemistry. J Petrol 41(6):789-808
Puustinen K (1972) Richterite and actinolite from the Siilinjarvi carbonatite complex, Finland. Bull Geol Soc Finland 44:83-86
Pyle JM, Haggerty SE (1994) Silicate-carbonate liquid immiscibility in upper-mantle eclogites: Implications for natrosilicic and carbonatitic conjugate melts. Geochim Cosmochim Acta 58:2997-3011
Pyle JM, Haggerty SE (1997) Eclogites and the metasomatism of eclogites from the Jagersfontein Kimberlite: Punctuated transport and implications for alkali magmatism. Geochim Cosmochim Acta 62(7):1207-1231
Ragone SE, Datta RK, Roy DM, Tuttle OF (1966) The system potassium carbonate-magnesium carbonate. J Phys Chem 70(10):3360-3361
Rajesh VJ, Arai S (2006) Baddeleyite-apatite-spinel-phlogopite (BASP) rock in Achankovil Shear Zone, South India, as a probable cumulate from melts of carbonatite affinity. Lithos 90(1-2):1-18
Rhodes JM, Dawson JB (1975) Major and trace element chemistry of peridotite inclusions from the Lashaine volcano, Tanzania. Phys Chem Earth 9:545-557
Richardson DG, Birkett TC (1996) Carbonatite-associated deposits. *In:* Geology of Canadian Mineral Deposit Types. Eckstrand OR, Sinclair WD, Thorpe RI (eds) Geological Survey of Canada, Ottawa, p 541-558
Ridley WI, Dawson JB (1975) Lithophile trace element data bearing on the origin of peridotite xenoliths, ankaramite and carbonatite from Lashaine volcano, N. Tanzania. Phys Chem Earth 9:559-569
Rosatelli G, Wall F, Le Bas MJ (2003) Potassic glass and calcite carbonatite in lapilli from extrusive carbonatites at Rangwa Caldera Complex, Kenya. Mineral Mag 67(5):931-955
Rosatelli G, Wall F, Stoppa F (2007) Calcio-carbonatite melts and metasomatism in the mantle beneath Mt. Vulture (Southern Italy). Lithos 99(3-4):229-248
Ruberti E, Enrich GER, Gomes CB, Comin-Chiarmonti P (2008) Hydrothermal REE fluorocarbonate mineralization at Barra do Itapirapua, a multiple stockwork carbonatite, Southern Brazil. Can Mineral 46(4):901-914
Rudnick RL, McDonough WF, Chappell BW (1993) Carbonatite Metasomatism in the Northern Tanzanian Mantle: Petrographic and Geochemical Characteristics. Earth Planet Sci Lett 114(4):463-475
Ryabchikov ID, Baker M, Wyllie PJ (1989) Phosphate-bearing carbonatite melts equilibrated with mantle lherzolites at 30-kbar. Geokhimiya (5):725-729
Safonov OG, Perchuk LL, Litvin YA (2007) Melting relations in the chloride-carbonate-silicate systems at high-pressure and the model for formation of alkalic diamond-forming liquids in the upper mantle. Earth Planet Sci Lett 253:112-128
Scales M (1989) Niobec: one-of-a-kind mine. Can Min J 110:43-46
Schiano P, Clocchiatti R, Shimizu N, Weis D, Mattielli N (1994) Cogenetic silica-rich and carbonate-rich melts trapped in mantle minerals in Kerguelen ultramafic xenoliths: implications for metasomatism in the oceanic upper-mantle. Earth Planet Sci Lett 123(1-4):167-178

Schrauder M, Navon O (1993) Solid carbon-dioxide in a natural diamond. Nature 365:42-44
Schrauder M, Navon O (1994) Hydrous and carbonatitic mantle fluids in fibrous diamonds from Jwaneng, Botswana. Geochim Cosmochim Acta 58:761-771
Seifert W, Thomas R (1995) Silicate-carbonate immiscibility: A melt inclusion study of olivine melilitite and wehrlite xenoliths in tephrite from the Elbe zone, Germany. Chemie Erde 55(4):263-279
Selverstone J, Franz G, Thomas S, Getty S (1992) Fluid variability in 2-GPa eclogites as an indicator of fluid behavior during subduction. Contrib Mineral Petrol 112(2-3):341-357
Sharma SK, Simons B (1980) Raman study of K_2CO_3-$MgCO_3$ glasses. Carnegie Institute. Washington Yearbook 79:322-326, *http://www.archive.org/details/yearbookcarne79197980carn*
Sharp ZD, Papike JJ, Durakiewicz T (2003) The effect of thermal decarbonation on stable isotope compositions of carbonates. Am Mineral 88:87-92
Shatskii AF, Borzdov YM, Sokol AG, Pal'yanov YN (2002) Phase formation and diamond crystallization in carbon-bearing ultrapotassic carbonate-silicate systems. Geol Geofiz 43(10):940-950
Shirey SB, Cartigny P, Frost DJ, Keshav S, Nestola F, Nimis P, Pearson DG, Sobolev NV, Walter MJ (2013) Diamonds and the geology of mantle carbon. Rev Mineral Geochem 75:355-421
Sifre D, Gaillard F (2012) Electrical conductivity measurements on hydrous carbonate melts at mantle pressure. European Mineralogical Conference 1:EMC2012-502, *http://meetingorganizer.copernicus.org/EMC2012/EMC2012-502.pdf*
Silva LC, Le Bas MJ, Robertson AHF (1981) An oceanic carbonatite volcano on Santiago, Cape Verde Islands. Nature 294:644-645
Simonetti A, Bell K, Shrady C (1997) Trace- and rare-earth-element geochemistry of the June 1993 natrocarbonatite lavas, Oldoinyo Lengai (Tanzania): Implications for the origin of carbonatite magmas. J Volcanol Geotherm Res 75:89-106
Simonetti A, Bell K, Viladkar SG (1995) Isotopic data from the Amba Dongar carbonatite complex, west-central India; evidence for an enriched mantle source. Chem Geol 122:185-198
Simonetti A, Goldstein SL, Schmidberger SS, Viladkar SG (1998) Geochemical and Nd, Pb, and Sr isotope data from Deccan alkaline complexes: Inferences for mantle sources and plume-lithosphere interaction. J Petrol 39(11-12):1847-1864
Smith MP, Henderson P (2000) Preliminary fluid inclusion constraints on fluid evolution in the Bayan Obo Fe-REE-Nb deposit, Inner Mongolia, China. Econ Geol 95(7):1371-1388
Sokol AG, Tomilenko AA, Pal'yanov YN, Borzdov YM, Pal'yanova GA, Khokhryakov AF (2000) Fluid regime of diamond crystallisation in carbonate-carbon systems. Euro J Mineral 12(2):367-375
Sparks RSJ, Brooker RA, Field M, Kavanagh J, Schumacher JC, Walter MJ, White J (2009) The nature of erupting kimberlite melts. Nature 112:429-438
Spivak AV, Litvin YA, Shushkanova AV, Litvin VYu, Shiryaev AA (2008) Diamond formation in carbonate-silicate-sulfide-carbon melts: Raman- and IR-microspectroscopy. Euro J Mineral 20:341-347
Srivastava RK, Rajesh K, Hall RP (1995) Tectonic Setting of Indian Carbonatites. *In:* Magmatism in Relation to Diverse Tectonic Settings. Srivastava RK, Rajesh K, Chandra R (eds) Balkema, Rotterdam, p 135-154
Stalder R, Foley SF, Brey GP, Horn I (1998) Mineral aqueous fluid partitioning of trace elements at 900-1200 degrees C and 3.0-5.7 GPa: New experimental data for garnet, clinopyroxene, and rutile, and implications for mantle metasomatism. Geochim Cosmochim Acta 62(10):1781-1801
Stoppa F, Jones A, Sharygin V (2009) Nyerereite from carbonatite rocks at Vulture volcano: implications for mantle metasomatism and petrogenesis of alkali carbonate melts. Cent Euro J Geosci 1(2):131-151
Stoppa F, Principe C (1998) High energy eruption of carbonatitic magma at Mt. Vulture (Southern Italy): The Monticchio Lakes Formation. J Volc Geotherm Res 80(1-2):137-153
Stoppa F, Rosatelli G, Wall F, Jeffries T (2005) Geochemistry of carbonatite: silicate pairs in nature: a case history from Central Italy. Lithos 85:26-47
Stoppa F, Woolley AR (1997) The Italian carbonatites: field occurrence, petrology and regional significance. Mineral Petrol 59:43-67
Streckeisen A (1980) Classification and nomenclature of volcanic rocks, lamprophyres, carbonatites and melilitic rocks. IUGS Subcomission on the Systematics of Igneous Rocks. Geol Rundsch 69:194-207
Suito K, Namba J, Horikawa T, Taniguchi Y, Sakurai N, Kobayashi M, Onodera A, Shimomura O, Kikegawa T (2001) Phase relations of $CaCO_3$ at high pressure and high temperature. Am Mineral 86(9):997-1002
Suwa BK, Oana S, Wada H, Osaki S (1975) Isotope geochemistry and petrology of African carbonatites. Phys Chem Earth 9:735-745
Sweeney RJ (1994) Carbonatite melt compositions in the Earth's mantle. Earth Planet Sci Lett 128(3-4):259-270
Taylor HP Jr, Frechen J, Degens ET (1967) Oxygen and carbon isotope studies of carbonatites from the Laacher See District, West Germany and the Alnö District, Sweden. Geochim Cosmochim Acta 31(3):407-430
Thibault Y, Edgar AD, Lloyd FE (1992) Experimental investigation of melts from a carbonated phlogopite lherzolite: Implications for metasomatism in the continental lithospheric mantle. Am Mineral 77(7-8):784-794

Tilton GR, Bryce JG, Mateen A (1998) Pb-Sr-Nd isotope data from 30 and 300 Ma collision zone carbonatites in north-west Pakistan, J Petrol 39:1865-1871

Tomilenko AA, Chepurov AI, Palyanov YN, Pokhilenko LN, Shebanin AP (1997) Volatile components in the upper mantle (from data on fluid inclusions). Geol Geofiz 38(1):276-285

Tomlinson E, De Schrijver I, De Corte K, Jones AP, Moens L, Vanhaecke F (2005) Trace element compositions of submicroscopic inclusions in coated diamond: A tool for understanding diamond petrogenesis. Geochim Cosmochim Acta 69(19):4719-4732

Tomlinson E, Jones A, Milledge J (2004) High-pressure experimental growth of diamond using C-K_2CO_3-KCl as an analogue for Cl-bearing carbonate fluid. Lithos 77(1-4):287-294

Tomlinson EL, Howell D, Jones AP, Frost DJ (2011) Characteristics of HPHT diamond grown at sub-lithopshere conditions (10-20 GPa). Diamond Relat Mater 20(1):11-17

Toyoda K, Horiuchi H, Tokonami M (1994) Dupal anomaly of Brazilian carbonatites: Geochemical correlations with hotspots in the South Atlantic and implications for the mantle source. Earth Planet Sci Lett 126(4):315-331

Treiman AH, Schedl A (1983) Properties of carbonatite magma and processes in carbonatite magma chambers. J Geol 91(4):437-447

Turnbull D (1956) Phase changes. *In:* Solid State Physics 3. Sietz F, Turnbull D (eds) Academic Press, New York, p 225-306

Tuttle OF, Gittins J (eds) (1966) Carbonatites. Interscience Publishers, New York

Ulmer P, Sweeney RJ (2002) Generation and differentiation of group II kimberlites: Constraints from a high-pressure experimental study to 10 GPa. Geochim Cosmochim Acta 66(12):2139-2153

Veizer J, Bell K, Jansen SL (1992) Temporal distribution of carbonatites. Geology 20:1147-1149

Verwoerd WJ (2008) The Goudini carbonatite complex, South Africa: a re-appraisal. Can Mineral 46(4):825-830

Vladykin NV (2009) Potassium alkaline lamproite-carbonatite complexes: petrology, genesis, and ore reserves. Russian Geol Geophys 50(12):1119-1128

Wall F, Mariano AN (1996) Rare earth minerals in carbonatites: A discussion centred on the Kangankunde carbonatite, Malawi. *In:* Rare Earth Minerals: Chemistry Origin and Ore Deposits. Jones AP, Wall F, Williams CT (eds) Chapman and Hall, London, p 193-225

Wall F, Zaitsev AN (2004) Phoscorites and Carbonatites: From Mantle to Mine. Mineralogical Society, London

Wallace ME, Green DH (1988) An experimental determination of primary carbonatite magma composition. Nature 335:343-346

Walter MJ, Bulanova GP, Armstrong LS, Keshav S, Blundy JD, Gudfinnsson G, Lord OT, Lennie AR, Clark SM, Smith CB, Gobbo L (2008) Primary carbonatite melt from deeply subducted oceanic crust. Nature 454(7204):622-625

Weast R (1972) CRC Handbook of Chemistry and Physics. CRC Press, Boca Raton, Florida

Widom E, Hoernle KA, Shirey SB, Schminke H-U (1999) Os isotope systematics in the Canary Islands and Madeira: Lithospheric contamination and mantle plume signatures. J Petrol 40(2):279-296

Wolff JA (1994) Physical properties of carbonatite magmas inferred from molten salt data, and application to extraction patterns from carbonatite–silicate magma chambers. Geol Mag 131:145-153

Wood BJ, Li J, Shahar A (2013) Carbon in the core: its influence on the properties of core and mantle. Rev Mineral Geochem 75:231-250

Woolley AR (1982) A discussion of carbonatite evolution and nomenclature, and the generation of sodic and potassic fenites. Mineral Mag 46:7-13

Woolley AR (1991) Extrusive carbonatites from the Uyaynah Area, United Arab Emirates. J Petrol 32:1143-1167

Woolley AR (2003) Igneous silicate rocks associated with carbonatites: their diversity, relative abundances and implications for carbonatite genesis. Periodico di Mineralogia 72:9-17

Woolley AR, Bailey DK (2012) The crucial role of lithospheric structure in the generation and release of carbonatites: geological evidence. Mineral Mag 76:259-270

Woolley AR, Church AA (2005) Extrusive carbonatites: A brief review. Lithos 85(1-4):1-14

Woolley AR, Kempe DRC (1989) Carbonatites: Nomenclature, average chemical compositions and element distribution. *In* Carbonatites: Genesis and Evolution. Bell K (ed) Unwin Hyman, London, p 1-14

Woolley AR, Kjarsgaard BA (2008a) Carbonatite occurrences of the world: map and database. Geological Survey of Canada Open File # 5796, 28 p

Woolley AR, Kjarsgaard BA (2008b) Paragenetic types of carbonatite as indicated by the diversity and relative abundances of associated silicate rocks: evidence from a global database. Can Mineral 46(4):741-752

Wyllie PJ (1995) Experimental petrology of upper-mantle materials, process and products. J Geodyn 20(4):429-468

Wyllie PJ, Huang WL (1976) Carbonation and melting reactions in the system CaO-MgO-SiO_2-CO_2 at mantle pressures with geophysical and petrological applications. Contrib Mineral Petrol 54:79-107

Wyllie PJ, Jones AP, Deng J (1996) Rare earth elements in carbonate-rich melts from mantle to crust. *In:* Rare Earth Minerals: Chemistry, Origin and Ore deposits. Jones AP, Wall F, Williams CT (eds) Chapman and Hall, London, p 77-104

Wyllie PJ, Tuttle OF (1960) The system $CaO-CO_2-H_2O$ and the origin of carbonatites. J Petrol 1:1-46

Wyllie PJ, Tuttle OF (1962) Carbonatitic lavas. Nature 194:1269

Xu C, Wang L, Song W, Wu M (2010) Carbonatites in China: a review for genesis and mineralization. Geosci Frontiers 1:105-114

Yang K-F, Fan H-R, Santosh F, Hu FF, Wang K-Y (2011) Mesoproterozoic carbonatitic magmatism in the Bayan Obo deposit, Inner Mongolia, North China: constraints for the mechanism of super accumulation of rare earth elements. Ore Geol Rev 40(1):122-131

Yaxley GM, Brey GP (2004) Phase relations of carbonate-bearing eclogite assemblages from 2.5 to 5.5 GPa: implications for petrogenesis of carbonatites. Contrib Mineral Petrol 146(5):606-619

Ying J, Zhou X, Zhang H (2004) Geochemical and isotopic investigation of the Laiwu–Zibo carbonatites from western Shandong Province, China, and implications for their petrogenesis and enriched mantle source. Lithos 75(3-4):413-426

Zaitsev AN, Keller J (2006) Mineralogical and chemical transformation of Oldoinyo Lengai natrocarbonatites, Tanzania. Lithos 91(1-4):191-207

Zurevinski SE, Mitchell RH (2004) Extreme compositional variation of pyrochlore-group minerals at the Oka carbonatite complex, Quebec: evidence of magma mixing? Can Mineral 42:1159-1168

Reviews in Mineralogy & Geochemistry
Vol. 75 pp. 323-354, 2013

Deep Carbon Emissions from Volcanoes

Michael R. Burton

*Istituto Nazionale di Geofisica e Vulcanologia
Via della Faggiola, 32
56123 Pisa, Italy*

burton@pi.ingv.it

Georgina M. Sawyer

*Laboratoire Magmas et Volcans, Université Blaise Pascal
5 rue Kessler, 63038 Clermont Ferrand, France
and
Istituto Nazionale di Geofisica e Vulcanologia
Via della Faggiola, 32
56123 Pisa, Italy*

Domenico Granieri

*Istituto Nazionale di Geofisica e Vulcanologia
Via della Faggiola, 32
56123 Pisa, Italy*

INTRODUCTION: VOLCANIC CO_2 EMISSIONS IN THE GEOLOGICAL CARBON CYCLE

Over long periods of time (~Ma), we may consider the oceans, atmosphere and biosphere as a single exospheric reservoir for CO_2. The geological carbon cycle describes the inputs to this exosphere from mantle degassing, metamorphism of subducted carbonates and outputs from weathering of aluminosilicate rocks (Walker et al. 1981). A feedback mechanism relates the weathering rate with the amount of CO_2 in the atmosphere via the greenhouse effect (e.g., Wang et al. 1976). An increase in atmospheric CO_2 concentrations induces higher temperatures, leading to higher rates of weathering, which draw down atmospheric CO_2 concentrations (Berner 1991). Atmospheric CO_2 concentrations are therefore stabilized over long timescales by this feedback mechanism (Zeebe and Caldeira 2008). This process may have played a role (Feulner et al. 2012) in stabilizing temperatures on Earth while solar radiation steadily increased due to stellar evolution (Bahcall et al. 2001). In this context the role of CO_2 degassing from the Earth is clearly fundamental to the stability of the climate, and therefore to life on Earth. Notwithstanding this importance, the flux of CO_2 from the Earth is poorly constrained. The uncertainty in our knowledge of this critical input into the geological carbon cycle led Berner and Lagasa (1989) to state that it is the most vexing problem facing us in understanding that cycle.

Notwithstanding the uncertainties in our understanding of CO_2 degassing from Earth, it is clear that these natural emissions were recently dwarfed by anthropogenic emissions, which have rapidly increased since industrialization began on a large scale in the 18th century, leading to a rapid increase in atmospheric CO_2 concentrations. While atmospheric CO_2 concentrations have varied between 190-280 ppm for the last 400,000 years (Zeebe and Caldeira 2008), human activity has produced a remarkable increase in CO_2 abundance, particularly in the last 100 years, with concentrations reaching ~390 ppmv at the time of writing. This situation highlights

1529-6466/13/0075-0011$00.00 DOI: 10.2138/rmg.2013.75.11

the importance of understanding the natural carbon cycle, so that we may better determine the evolution of the anthropogenic perturbation.

The principle elements of the multifaceted and complex geological carbon cycle are summarized in Figure 1. The main sources of carbon are active and inactive volcanism from arcs and rift zones and metamorphism of crustal carbonates. The main sinks for geological carbon are silicate weathering and carbonation of oceanic crust. Knowledge of both the total magnitude of carbon ingassing during subduction and carbon released from volcanism and metamorphism would allow quantification of the evolution and relative distribution of volatiles in the crust and mantle (Dasgupta and Hirschmann 2010).

The main focus of this work is the role of volcanism in producing CO_2 in the atmosphere and oceans. Volcanic CO_2 sources can be divided into several categories, direct and diffuse degassing from active arc and rift volcanoes, diffuse degassing from inactive volcanoes and regional diffuse degassing from intrusive plutonic structures with associated crustal metamorphism. We focus here on non-eruptive degassing because, as shown below, continuous emission of CO_2 from multiple sources appears to dominate short-lived eruptive emissions from point sources.

CO_2 released directly from active volcanoes has three main sources, CO_2 dissolved in the mantle, recycled CO_2 from subducted crustal material (e.g., Marty and Tolstikhin 1998) and decarbonation of shallow crustal material (e.g., Troll et al. 2012). Separating the relative proportions of mantle and crustal carbon is possible through investigation of the isotopic composition of emitted carbon (e.g., Chiodini et al. 2011) and is increasingly important given that during eruptions magmatic intrusions may interact with crustal material, strongly enhancing the CO_2 output of the volcanic system (Troll et al. 2012), at least temporarily. The magnitude of diffuse mantle CO_2 can also be identified isotopically in mixed metamorphic and magmatic gases using Carbon (Chiodini et al. 2011) or Helium isotopes as a proxy for deep mantle sources in both major fault systems (Pili et al. 2011) and crustal tectonic structures (Crossey et al. 2009).

Our current estimates of volcanic carbon emissions are poorly constrained due to a lack of direct measurements. Measuring CO_2 in subaerial volcanic plumes is a challenge, because while CO_2 typically makes up ~10 mol% of volcanic gas emissions (the majority of which is normally water vapor), mixing with the atmosphere rapidly dilutes the volcanic CO_2 signature. Nevertheless, technological advances and an increase in the number of volcanoes studied have

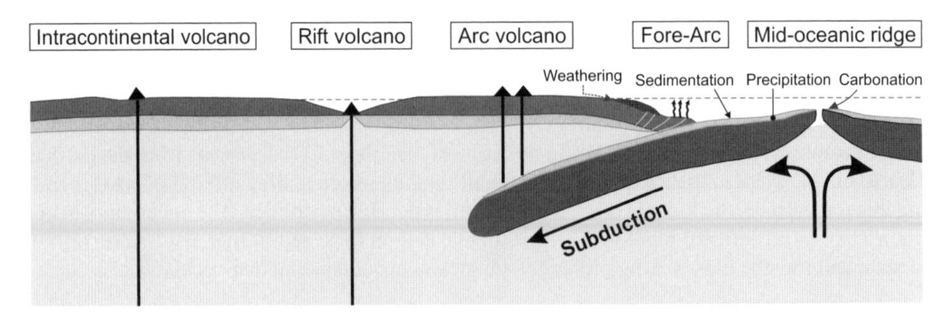

Figure 1. Schematic diagram showing the main sources and sinks for volcanic CO_2 on Earth within the geological carbon cycle. CO_2 is released at MORs during crustal genesis, but it is also absorbed into the newly formed crust in high temperature hydrothermal reactions. Carbonates precipitate directly into oceanic crust and collect in sediments before the subducting slab is carried under the mantle wedge. Volcanism then produces CO_2 emissions from the fore-arc (through cold seeps), arc volcanism, rift volcanism, intracontinental volcanoes and hotspots. The CO_2 emitted into the atmosphere reacts in weathering reactions with silicate rocks, carrying C back into the ocean where the geological carbon cycle is eventually closed through carbon sequestration into the subducting slab material.

greatly increased our knowledge of volcanic CO_2 fluxes over the last 10 years. One of the main goals of this work is to update global geological CO_2 flux estimates (e.g., Kerrick 2001; Mörner and Etiope 2002; Fischer 2008) using the recently acquired volcanic CO_2 flux data.

Diffuse CO_2 degassing from both volcanic and tectonic structures is a large contributor to the global geological CO_2 emission, but is difficult to measure due to the large areal extent that may be in play, and the large number of degassing sites throughout the globe. Measuring the CO_2 degassing rates into volcanic lakes and from submarine volcanism have significant technical challenges. In the following we review the state of the art of volcanic CO_2 measurements and present a catalogue of reported, quantified emissions from geological sources. These measurements are then extrapolated to produce estimates of the global volcanic CO_2 flux. These estimates are compared with previously published estimates of total CO_2 emissions, silicate weathering rates and the rate of carbon consumption during subduction. We then examine the dynamic role of CO_2 within magmatic systems and the magnitude of CO_2 released during eruptions.

Carbon species in Earth degassing

CO_2 is not the only carbon-containing molecule emitted from the Earth. In order of decreasing emissions, CO_2, CH_4, CO and OCS all contribute to the total carbon budget. Mörner and Etiope (2002) estimated that the global emission of CO_2 from Earth degassing was ~600 million tonnes of CO_2 per year (Mt/yr, 1 Mt = 10^{12} g), with ~300 Mt/yr produced from subaerial volcanism, and another 300 Mt/yr produced from non-volcanic inorganic degassing, mostly from tectonically active areas (Chiodini et al. 2005). For comparison, Cadle (1980) estimated that volcanic activity produces 0.34 Mt/yr of CH_4. Mud volcanoes in Azerbijan were estimated to produce ~1 Mt/yr of CH_4, however the global flux from mud volcanism is not known. Hydrocarbon seepage of CH_4 globally is estimated to produce between 8 and 68 Mt/yr (Hornafius et al. 1999). Etiope et al. (2008) estimated that global CH_4 emissions from geological sources to be 53 Mt/yr, a significant proportion of the geological C output.

CO is emitted directly from volcanoes, with a CO_2/CO ratio that varies between ~10 and ~1000 depending on the oxygen fugacity and temperature of the fluid co-existing with melt prior to outgassing. OCS is also directly emitted within volcanic plumes, but in even smaller relative amounts than CO, typically 1000-10,000:1 for CO_2:OCS (Mori and Notsu 1997; Burton et al. 2007a; Oppenheimer and Kyle 2008; Sawyer et al. 2008a). OCS is the most abundant S bearing gas species in the atmosphere, contributes to stratospheric sulfuric acid aerosol generation (Crutzen 1976) and is an efficient greenhouse gas (Brühl et al. 2012). Its budget is dominated by emissions from oceans and anthropogenic processes. From a total global output of ~1.3 Mt/yr of OCS only 0.03 Mt is estimated to arise from volcanism (Watts 2000). CS_2 is the final trace carbon gas emission from volcanoes, with a similar flux and chemistry to OCS.

While recent estimates of geological emissions of CH_4 (Etiope et al. 2008) clearly indicate that these emissions are significant compared with geological CO_2 on the global scale, in this work we focus on CO_2 emissions, and use the most recent volcanic CO_2 flux data to update the measured global volcanic CO_2 flux.

METHODS FOR MEASURING GEOLOGICAL CO_2 EFFLUX

Ground-based measurements of volcanic plumes

Directly quantifying volcanic CO_2 fluxes in the atmosphere is challenging due to the relatively abundant concentration of background CO_2, currently ~390 ppm. To put this in context, a strong volcanic CO_2 source such as Mt. Etna, Italy, produces a gas plume where 1 km downwind the average concentration of volcanic CO_2 is only ~4 ppm (based on calculations with VOLCALPUFF plume dispersal model, Barsotti et al. 2008). Thus, an *in situ* measurement 1 km downwind needs to resolve a mere 1% excess CO_2 compared with the background concentration. Most volcanoes emit less CO_2 than Mt. Etna, so this is an optimistic scenario. This difficulty has led most researchers to focus on measurements of the volcanic emissions close to the source, using *in situ* and proximal remote sensing techniques. In such measurements, volcanic CO_2 fluxes are normally determined by measuring the ratio of volcanic CO_2 to another volcanic gas, typically SO_2 due to the ease with which its flux can be measured (Oppenheimer et al. 2011), and then calculating the CO_2 flux as the product of the CO_2/SO_2 ratio and the SO_2 flux. The objective of the majority of the following techniques in the context of quantifying CO_2 fluxes is therefore the determination of the CO_2/SO_2 ratio in the volcanic gas.

An exception to this combined CO_2/SO_2 and SO_2 flux approach was adopted by Marty and Le Cloarec (1992) who utilized global volcanic flux estimates of ^{210}Po and $^{210}Po/CO_2$ ratios measured in direct sampling (see below) to derive estimates of global CO_2 fluxes.

Direct sampling of a volcanic gas can be achieved with the use of Giggenbach bottles (Giggenbach and Goguel 1989), where high temperature fumarolic gases are collected in an alkaline solution for later laboratory analysis. This approach allows both bulk and trace gas species to be quantified, but the fact that the most abundant gas component, H_2O, can condense in the tube leading to the alkali solution bottle means that H_2O is challenging to quantify, and therefore absolute concentrations of the other species can be difficult to define. Air contamination is difficult to avoid, and can further increase the difficulty in determining the original volcanic gas concentrations. Such measurements require working in extremely close proximity to the degassing vent, and are ideally performed on the hottest and most highly pressurized emissions (to avoid air contamination), making their collection challenging and potentially hazardous. In addition, some of the most voluminous volcanic gas sources release very little volatiles from fumaroles, instead the bulk of the emission is open-vent degassing from craters. These plume emissions are impossible to sample without air contamination with such an approach.

The MultiGas approach (Shinohara 2005) has greatly simplified the measurement of volcanic CO_2/SO_2 ratios, allowing automatic, unattended analysis of volcanic plumes for extended periods of time (Aiuppa et al. 2007). This instrument combines a near-infrared spectroscopic measurement of CO_2 and H_2O concentrations with a solid-state chemical sensor for quantification of SO_2 and H_2S. The relatively low cost, low power requirement and ease of use of the instrument make it probably the most convenient and cost-effective way of determining *in situ* CO_2/SO_2 ratios available today. Some potential errors can arise, however, due to the different response time of the CO_2 and SO_2 sensors. Typical response times for the near-infrared optical technique used to measure CO_2 is ~1 s, while response times for SO_2 chemical sensors is typically longer, e.g., ~13-31 s for an Alphasense chemical sensor (Roberts et al. 2012). This means that fast changes in chemical composition or concentration are challenging to capture, however through the use of longer integration times problems arising from diverse sensor response times can be avoided. Quickly changing gas concentrations could instead be captured in theory with an optically based SO_2 measurement.

Remote sensing measurements of CO_2 amounts can be performed with infrared spectroscopy if the volcanic gas concentration is sufficiently high above the background atmospheric CO_2 amount. The first such pioneering infrared measurements of volcanic CO_2 amounts were conducted remarkably early, in 1969 by Naughton et al. (1969), during a lava fountain on Kīlauea, Hawai'i. Since then, open-path Fourier transform infrared (OP-FTIR) spectroscopy (Mori and Notsu 1997) has become a well-utilized tool to measure *in situ* volcanic gas compositions. Modern FTIR spectrometers are light (~8 kg), relatively low power (~30 W) and require no cryogenic cooling (e.g., La Spina et al. 2010), yet allow the simultaneous measurement of many volcanic gases, including H_2O, CO_2, SO_2, HCl, HF, CO OCS and SiF_4. Measurements of volcanic CO_2 require an infrared light source, either an infrared lamp (Burton et al. 2000) or hot volcanic rocks (Allard et al. 2005; Burton et al. 2007a; Sawyer et al. 2008a). Infrared radiation is absorbed by volcanic gases before being measured with the infrared spectrometer and the resulting spectra can be analyzed (Burton et al. 2000, 2007a) to produce relative abundances of volcanic H_2O, CO_2, SO_2 and several other gases. Smith et al (2011) concluded that such analytical approaches could deliver accuracies of ~5% in CO_2 amounts. The greatest challenge in performing OP-FTIR measurements is obtaining a suitable source of IR radiation, with sufficient volcanic gas between the source and the spectrometer. Recent innovations in applying the remote-controlled mode at the summit of Stromboli (La Spina et al. 2013) show that it can be used for high temporal resolution monitoring of multiple gas sources.

CO_2 emissions from fumarole fields can be determined through the introduction of a known flux of a tracer gas, such as SF_6, and then measuring the volcanic CO_2/SF_6 ratio in the downwind gas emission. Mori et al. (2001) successfully used this approach to measure CO_2 flux emissions from fumarolic vents on Izu-Oshima (Japan), Kirishima (Japan) and Teide (Canary Islands, Spain).

A recent innovation has been the use of portable mass spectrometers to determine *in situ* volcanic gas compositions (Diaz et al. 2010). This approach was used successfully before and after the 5th January 2010 eruption of Turrialba (Costa Rica), revealing significant changes in CO_2, SO_2 and He concentrations.

Volcanic SO_2 flux measurements

As described above the determination of CO_2 flux requires a further step after measurement of CO_2/SO_2 ratios, multiplication with an SO_2 flux. Several assessments of arc volcanic CO_2 emissions have been produced (e.g., Hilton et al. 2002; Fischer and Marty 2005) by interpreting volcanic SO_2 inventories (e.g., Andres and Kasgnoc 1998; Halmer et al. 2002) Since errors on the SO_2 flux propagate into the CO_2 flux we briefly examine here the methods used to measure SO_2 flux, together with their associated errors. The SO_2 flux is much easier to measure directly than the CO_2 flux for two reasons: firstly, SO_2 is not present in the unpolluted troposphere and secondly, SO_2 has a convenient, relatively strong absorption band in the ultraviolet, easily accessible using scattered sunlight as a source. This has allowed the creation of automatic networks of UV scanners, that permit volcanic SO_2 fluxes to be monitored (Edmonds et al. 2003; Burton et al. 2009; Galle et al. 2010; Oppenheimer et al. 2011). Typically in the literature the greatest quoted source of error in SO_2 flux measurements derives from wind speed estimates, and this error is normally indicated to be ~20-30%. A recent innovation in ground-based measurements of SO_2 fluxes is the SO_2 imaging camera, which uses an imaging sensor sensitive to the UV and optical filters to produce specific sensitivity to SO_2 (e.g., Mori and Burton 2006). This approach has the potential to correct implicitly for wind velocity but it suffers potentially from cross-talk between SO_2 and volcanic ash or aerosol.

Recent work on subtle radiative transfer issues relating to ground-based UV SO_2 flux measurements highlight that there could be large, previously ignored, errors associated with light dilution (Kern et al. 2012). This is a process where light scattering from below the volcanic plume enters the instrument, diluting the light which passed through the volcanic plume from above, resulting in a net underestimation in the SO_2 flux, of up to 90%. The true significance and importance of this effect, and the number of published SO_2 fluxes that require re-analysis, has yet to be evaluated. It should be noted therefore that estimates of global CO_2 flux based on published SO_2 flux data may be subject to revision, depending on the impact of light dilution on SO_2 flux measurements.

In addition to ground-based measurements of SO_2 flux, satellite-based measurements are often used, working in both the ultraviolet (e.g., OMI, SCIAMACHY) and infrared (e.g., MODIS, ASTER, IASI) wavelengths. These instruments can produce maps of SO_2 abundance with a repeat time of ~days. With such a repeat rate the same plume is normally not observed in two different images, and it is therefore not possible to perform a simple cross-correlation to determine plume velocity and therefore flux from images of SO_2 abundance. The conversion from SO_2 abundance image to a quantitative degassing rate is therefore not trivial, because both the age and velocity of the plume at each point in the image must be derived using an independent method (e.g., Merucci et al. 2011). An additional challenge with satellite-based measurements of SO_2 flux is that the sensitivity to SO_2 decreases in the lower atmosphere, such that low-lying volcanoes are difficult to measure unless they are in eruption, and even higher altitude volcanoes require a relatively large SO_2 degassing rate to be reliably quantified (Carn et al. in press).

Measurements of CO_2 flux using the combination of CO_2/SO_2 ratios and SO_2 fluxes therefore reflect uncertainties in both measurements, typically estimated to be 10-20% and 25-30% respectively. As mentioned above, recent work in radiative transfer analysis highlights the potentially important, but largely overlooked, role that light dilution may play in producing potentially significant underestimates of SO_2 fluxes from UV measurements (Kern et al. 2012). Notwithstanding these drawbacks, at the current time these data are the main constraints available for subaerial deep CO_2 output from volcanoes.

Recent technological advances have allowed such integrated CO_2/SO_2, SO_2 flux measurements to be fully automated for the first time, allowing real-time monitoring of CO_2 fluxes. This is particularly important because the low solubility of CO_2 in magmas means that deep, pre-eruptive, magmatic intrusions may be heralded at the surface by increases in CO_2 flux. Aiuppa et al. (2011) reported the first time series of CO_2 flux collected using combined MultiGas and SO_2 flux networks (Burton et al. 2009) on Stromboli volcano. These revealed distinct oscillations in CO_2 emissions, with periods of relatively high CO_2 degassing followed by periods of low CO_2 degassing, producing a steady average CO_2 degassing rate of ~550 t/d. Such a pattern suggests a steady state supply of CO_2 which is modulated by gas accumulation / permeability / magma supply processes within the magma feeding system. Interestingly, more intense explosive activity was observed after a period of intense CO_2 degassing, opening the possibility of using such observations to forecast explosive volcanic activity at this volcano.

Airborne measurements of volcanic plumes

Two main approaches have been used to measure volcanic CO_2 fluxes from the air. A direct method consists of flying an *in situ* CO_2 analyzer (typically a closed-path near-infrared spectrometer) in a raster or ladder traverse across the cross-section of a volcanic plume (Gerlach et al. 1997; Werner et al. 2008). The resulting data can be interpolated to produce a CO_2 concentration map, which can be integrated over the cross-sectional area of the plume and multiplied with wind speed to produce a CO_2 flux. This approach has the advantage of being a direct measurement of the CO_2 emissions, however each measurement requires ~1 hour and therefore very stable wind conditions are required in order to avoid errors in the flux calculation.

The second airborne approach combines plume traverse measurements with an ultraviolet spectrometer to derive SO_2 flux with *in situ* measurement of the CO_2/SO_2 ratio provided with a closed-path FTIR spectrometer, sampling ambient air as the plane flies through the plume (Gerlach et al. 1998). This is probably the most robust methodology currently available for measuring CO_2 fluxes, as the SO_2 flux analysis can be performed close enough to the plume that light dilution is insignificant. Flying has its own challenges, however, due to the technical constraints involved in performing measurements on a vibrating platform, as well as the costs associated with flight time and difficulties presented from flying within the volcanic plume.

Airborne measurements of volcanic CO_2 may also be achieved by viewing infrared radiance from the ground through a volcanic plume with a hyperspectral radiometer, and this has been successfully demonstrated on Kīlauea, Hawai'i using the AVIRIS hyperspectral imager (Spinetti et al. 2008). Results obtained with AVIRIS agreed well with ground-based measurements of CO_2 emissions, validating the method.

More recently, similar measurements to those performed by Gerlach et al. (1998) have been conducted from an unmanned aerial vehicle platform (UAV) (McGonigle et al. 2008). Such an approach is appealing due to the significantly reduced cost and risk, as well as increased accessibility. However, the legal framework for conducting such measurements is complex and varies greatly between countries, making general take-up of such methodologies so far quite limited. Future longer distance stand-off measurements may be possible using larger UAVs offering an intermediate option to space-based satellite retrievals of air column soundings.

Space-based measurements of volcanic plumes

Global measurements of volcanic CO_2 emissions would be the ideal approach to quantifying the global volcanic deep CO_2 budget. The most promising observing platform for volcanic CO_2 is the aptly named Orbiting Carbon Observatory (OCO) (Crisp et al. 2004). The rocket carrying the OCO failed to reach orbit when launched in 2009, and a new launch with a replacement satellite is currently planned for 2014. The OCO utilizes a single telescope to feed light to spectrometers which measure the columnar abundance of both O_2 and CO_2, using absorption bands at wavelengths of 0.67 micron for O_2, and 1.61 and 2.06 micron for CO_2. The purpose of the O_2 column amount measurement is to normalize the CO_2 column measurement to an average CO_2 mole fraction in ppmv. The final error on the average CO_2 concentration in the column is 0.3 wt%. The footprint of the OCO will be 1 km by 1.5 km at nadir, and will have a repeat observation period of 16 days.

A simple calculation of the CO_2 emission from a strong emitter such as Mt. Etna allows a direct estimation of the feasibility of such measurements with OCO. Etna degasses CO_2 at an average rate of 16,000 t/d or ~190 kg s^{-1} (Allard et al. 1991; Aiuppa et al. 2006, 2008; La Spina et al. 2010). Assuming an optimal geometry in which the gas source was at one edge of a single OCO footprint and the entire plume contained within the 1.5 km length of the pixel, with a windspeed of 5 m s^{-1} the maximum age of CO_2 in the footprint would be ~300 seconds, and the total volcanic CO_2 mass would be ~57,000 tonnes. Converting this CO_2 mass to molecules and averaging over the OCO nadir footprint area produces a vertical column amount of ~5 × 10^{19} molecules cm^{-2}. The atmospheric vertical column amount of CO_2 at the average altitude of Etna is ~6 × 10^{21} molecules cm^{-2} and therefore the volcanic signal would be ~0.8 % of the atmospheric column, which is above the 0.3% error limit of OCO. A slower wind would produce a higher volcanic CO_2 amount, while a less optimal geometry would decrease the relative contribution from the volcano. OCO therefore has the potential for measuring passive CO_2 emissions from Mt. Etna, in optimal conditions. During eruptions the CO_2 emission rate would increase, allowing for easier detection. The majority of degassing volcanoes are less productive than Mt. Etna, however, and would present a challenge for detection from OCO, unless they were undergoing an eruption.

Ground-based measurements of diffuse deep CO_2

Significant amounts of diffuse CO_2 are released from active volcanic areas, not only during eruptions but also during quiescent periods. This volcanic CO_2 discharge occurs over the flanks of the volcanic edifice as diffuse soil emanations (Allard et al. 1991; Baubron et al. 1990), and adds to the voluminous and more obvious degassing from fumaroles and summit craters. Many CO_2 soil flux measurement techniques have been applied to quantify these gases and include both direct and indirect methods (e.g., Reiners 1968; Kucera and Kirkham 1971; Kanemasu et al. 1974; Parkinson 1981).

The indirect methods are based on the determination of the CO_2 concentration gradient in the soil (Camarda et al. 2006). These methods can be applied only if the transport of the gas is dominated by the diffusion and some properties of the medium are known. Direct methods require dynamic or static procedures, whether or not a flux of air is used to extract gas from the soil. The dynamic procedures require some corrections depending on the physical properties of the soil in the measurement point and on the design of the instrumental apparatus. Furthermore, all dynamic procedures are affected by overpressurization or depressurization depending upon the magnitude of the air flux chosen by the operator and according to Kanemasu et al. (1974) results are strongly affected by the physical modifications induced by pumping under different flux regimes. Camarda et al. (2006) present a demonstration of the indirect method applied to measuring CO_2 fluxes from Vulcano (Italy), and show that with low pumping rates the sensitivity of the method to soil permeability is reduced.

The accumulation chamber method (or closed-chamber method) is a direct, static method originally used in agricultural sciences to determine soil respiration (Parkinson 1981) and then successfully adapted to measure CO_2 soil flux of volcanological interest by Tonani and Miele (1991). This method is based on the measurement of the CO_2 concentration increase inside an open-bottomed chamber of known volume, inverted on the soil surface. The initial rate of change of the concentration is proportional to the CO_2 flux (Tonani and Miele 1991; Chiodini et al. 1996). The method does not require either assumptions about soil characteristics or the regime of the flux (advective/diffusive). The method has been tested by several authors under controlled laboratory conditions and provides reproducibility of 10% (Chiodini et al. 1998). In a field reproducibility test of the method, carried out at two points with high and low CO_2 flux, Carapezza and Granieri (2004) found an uncertainty of 12% for high fluxes and 24% for low fluxes. Multiple measurements performed by ground-based methodologies allow a mapping of CO_2 flux and an estimation of the total CO_2 release by use of interpolation algorithms (Cardellini et al. 2003; Chiodini et al. 2005).

Eddy covariance or alternately Eddy correlation (EC) is a micrometeorological technique (e.g., Baldocchi 2003) recently proposed as a method to monitor volcanic CO_2 emissions (Werner et al. 2000, 2003; Anderson and Farrar 2001; Lewicki et al. 2008). The basis of the EC is the calculation of the flux at the surface through the covariance between the fluctuations of the vertical component of the wind and the fluctuations of the gas concentration in atmosphere. The EC provides advantage of being an automated, time-averaged and area-integrated technique with a spatial scale significantly larger (square meter to square kilometer) than that of the ground-based methods (e.g., accumulation chamber). However the volcanic environment is often too heterogeneous for EC application, as suggested by the theory underlying EC, in terms of spatial and temporal variability of surface fluxes and morphology of the measuring field.

Diffusive degassing of deep CO_2 in tectonically active areas

Since the early work of Irwin and Barnes (1980), it has become clear that a close relationship exists between active tectonic areas and anomalous crustal emissions of CO_2. Due to their high crustal permeability, faults act as preferential pathways for the upward migration and eventual release of deep gases to the aquifers or directly to the atmosphere. Regional aquifers located in areas of high CO_2 flux can dissolve most or part of the deeply generated gas because the relatively high solubility of CO_2 in water. A carbon mass balance in the involved aquifer can be used to obtain an estimation of the amount of CO_2 dissolved by groundwater. However, the large range of $^{13}\delta C_{CO_2}$ observed in such aquifers suggests that carbon can derive from multiple sources: atmospheric C, biogenic C, carbonate minerals derived C and deeply derived C. Therefore, an approach by coupling groundwater chemistry with hydrologic and isotopic data has to be applied in order to differentiate shallow versus deep sources. Chiodini et al. (2004) showed that in the tectonically active area of the Italian Apennines, approximately 40% of the inorganic carbon in the groundwater derives from magmatic sources. This observation suggests that there may be significant amounts of magmatic CO_2 released in tectonic areas, perhaps a similar order of magnitude as subaerial volcanic degassing (Chiodini et al. 2004). Indeed, in volcanic areas, the dissolved CO_2 in groundwater can be a significant component of the total CO_2 flux at the volcano (e.g., Rose and Davisson 1996; Sorey et al. 1998; Inguaggiato et al. 2012).

Submarine measurements

Volatile emissions from the axis of mid-ocean ridges (MORs) in the form of black smokers are dramatic examples of submarine deep carbon emissions, however they tend to be short-lived and unpredictable, making collection of gas samples challenging. Nevertheless measurements of the composition and flux of such emissions have been conducted at dozens of sites (Kelley et al. 2004), using a wide range of gas collection techniques from piloted and remotely controlled submersible craft, allowing later analysis of gas samples in the laboratory. Direct measurements have been performed on only a fraction of the world's MOR, and therefore previous work on the fluxes of CO_2 from MOR has focused on quantifying emissions relative to a better-constrained global production parameter. These have included crustal production rates and the C content of the mantle (Gerlach 1989, 1991; Javoy and Pineau 1991; Holloway 1998; Cartigny et al. 2001; Saal et al. 2002), global mantle ^3He flux (Corliss et al. 1979; Des Marais and Moore 1984; Marty and Jambon 1987; Sarda and Graham 1990; Graham and Sarda 1991; Marty and Zimmerman 1999), hydrothermal fluid flux (Elderfield and Schultz 1996) and the $CO_2/^3He$ ratio in hydrothermal plumes. The latter two require estimates of global ^3He fluxes, which are produced primarily at MORs (Allard et al. 1992).

An important aspect of MOR volcanism is that during the process of formation CO_2 reacts with hot rock, sequestering CO_2. In addition, dissolved carbonate in seawater reacts progressively within the shallowest ~60 m oceanic crust, producing steadily higher carbonate concentrations with increasing crustal age. Measurements of drill cores of the upper oceanic crust allowed Alt and Teagle (1999) to quantify the magnitude of the CO_2 sink produced by crust reactions as of the magnitude of 150 Mt/yr CO_2. This is of similar magnitude to the MOR CO_2 flux of 97 ± 40 Mt/yr CO_2, indeed it is probable that reactions in the oceanic crust absorb more CO_2 than is emitted from MORs.

As well as emissions from the main axis of the MOR, degassing takes place on the flanks of the ridge, driving circulation of seawater through the crust. Sansone et al. (1998) sampled gas emissions from the eastern flank of the Juan de Fuca ridge with the Alvin deep sea vessel, both directly and through inverted funnels to concentrate the gas flow into titanium gas-tight samplers (Massoth et al. 1989). Further samples were collected using titanium syringe samples (Von Damm et al. 1985). Gas samples were acidified and then vacuum-extracted at sea with a glass/stainless-steel vacuum line. The total gas volume was determined with high precision capacitance manometers, and the extracted gas from each sample was sealed in break-seal glass ampoules for analysis ashore with gas chromatography and mass spectrometry.

Gas emissions also occur from active submarine arc volcanoes. Lupton et al. (2008) measured gas output from eleven volcanoes along the Mariana and Tonga-Kermadec arcs with remote controlled vehicles during three expeditions. Four of these volcanoes were found to produce distinct gas and liquid CO_2 gas emissions, together with hydrothermal emissions from the main vents. Vent emissions were sampled using seawater-filled titanium alloy gas-tight bottles connected via tubing to a Ti sampling snout inserted directly into the vent. A valve was then opened and hydraulic pressure filled the bottle with a sample of vent gas. On the ship, samples were acidified and transferred under vacuum ampoules made of Pyrex and low-He permeability alumino-silicate glass for later laboratory analysis. Collection of liquid CO_2 droplets was challenging due to the ~1000 fold expansion of liquid CO_2 when converted to CO_2 gas at 1 atm pressure, necessitating the use of a small volume Ti gas-tight bottle. Gas bubbles were collected with a plastic cylinder with relief valve that was placed over the emission until filled.

Understanding of the carbon balance in a subduction zone requires knowledge of the amount of carbon entering the zone within the subducting slab, CO_2 loss from main arc volcanism and back arc, and the submarine fore-arc (see Fig. 1). This latter was measured from seeps in the Central America subduction zone by Füri et al. (2010). Seep fluids were collected over a 12 month period at the submarine segment of the Costa Rica fore-arc margin using 1/8 inch diameter copper tubing attached to a submarine flux meter operating in continuous pumping mode to measure CO_2 and CH_4 fluxes. Temporal variations during the sampling period were revealed by cutting the copper tubing in 0.4 m sections under vacuum and extracting the stored volatile samples in the laboratory for isotope ratios and compositions.

Volcanic lakes are significant, but previously unrecognized (Pérez et al. 2011) contributors to global deep CO_2 budgets. CO_2 gas emissions from volcanic lakes are in the form of both diffuse degassing from the lake surface and bubbling (Mazot and Taran 2009). Lake CO_2 emissions were therefore measured with a floating gas accumulation chamber with an in-built NIR sensor to measure CO_2 concentrations. Conversion of CO_2 concentrations to fluxes was made using simultaneous measurements of pressure and temperature.

REPORTED MEASUREMENTS OF DEEP CARBON FLUXES

Subaerial volcanism

During the Holocene, ~1500 volcanoes on land erupted and in recorded history there have been 550 known eruptions. Typically 50-70 volcanoes erupt explosively each year and ~500 produce a gas emission either through hydrothermal systems or open-vent degassing (Siebert and Simkin 2002). Of these, only a small fraction have had their CO_2 flux measured directly, however the number of measured volcanoes has greatly increased in recent years. In this section we first present published data on measured volcanic CO_2 emission rates from active volcanoes, diffuse degassing of volcanic areas and tectonically active areas, followed by volcanic lakes and submarine emissions. We conclude by producing a global sum of CO_2 emission rates.

We report in Table 1 all known volcanic plume CO_2 flux measurements from persistently degassing volcanoes. We have chosen data in which CO_2 fluxes were measured either by near simultaneous measurement of CO_2/SO_2 ratios and SO_2 / gas flux, or direct measurement of the CO_2 flux. This was done because the number of volcanoes for which CO_2 flux has been measured accurately has greatly increased in the last years, with 40 new measurements reported since 2000. We performed a simple average of the reported CO_2 flux measurements for each volcano to produce Table 2, a summary which allows the total CO_2 flux from 33 measured volcanic gas plumes to be calculated as 59.7 Mt/yr. We note that this measured flux by itself is higher than the maximum estimated for global passive degassing from Williams et al. (1992), highlighting the fundamental importance of direct measurements of volcanic CO_2 fluxes in quantifying the volcanic CO_2 inventory.

In Table 3, we report diffuse CO_2 fluxes from historically active volcanoes, which have been verified through isotopic analysis to be of magmatic origin. This list is not an exhaustive collation of all diffuse CO_2 degassing measurements, but relfects the most updated or complete diffuse CO_2 flux measured at each volcano. The total CO_2 flux from the 30 measured volcanoes is 6.4 Mt/yr, including emissions from diffuse soil degassing and those measured in groundwater.

CO_2 fluxes from tectonic structures, hydrothermal systems or inactive volcanic areas are reported in Table 4, distinguishing between measured (or estimated) fluxes from soils from those dissolved in groundwaters. The two major measured contributors to the total tectonic, hydrothermal and inactive volcano CO_2 flux of 66 Mt/yr are tectonic degassing in Italy (10 Mt/yr, Chiodini et al. 2004) and hydrothermal emissions from Yellowstone (8.6 Mt/yr, Werner and Brantley 2003). We also include in this list an estimate of the total CO_2 flux produced by hydrothermal activity in Indonesia-Philippines (1.8 Mt/yr, Seward and Kerrick 1996) and

Table 1. Volcanic plume CO_2 fluxes from persistently degassing volcanoes (alphabetically ordered by country)

Volcano	Location	CO_2 Flux (t/d)	CO_2 Flux (Mt/yr)	Method	Date	Reference
Erebus	Antarctica	1,930	0.70	Airborne Li-COR	Dec 1997, Dec 1999, Jan 2001	Wardell et al. (2004)
Erebus	Antarctica	1,330	0.49	Ground-based OP-FTIR and scanning UV spectrometer	Dec 2004	Oppenheimer & Kyle (2008)
Villarrica	Chile	477	0.17	Ground-based OP-FTIR and airborne UV spectrometer traverses	Mar 2009	Sawyer et al. (2011)
Galeras	Colombia	1,020	0.37	High temp. summit fumarole sampling and COSPEC	1989-1995	Zapata et al. (1997)
Nyiragongo	DR Congo	95,500	34.86	Direct sampling with video footage of plume dimensions and rise rate	1959, 1972	Le Guern (1987)
Nyiragongo	DR Congo	9,320	3.40	Ground-based OP-FTIR and UV spectrometer vehicle traverses	May/Jun 2005, Jan 2006	Sawyer et al. (2008a)
Sierra Negra	Ecuador (Galápagos)	394	0.14	Ground-based multi-gas sensor and UV spectrometer walking traverses	Jun-Jul 2006	Padron et al. (2012)
Erta Ale	Ethiopia	54	0.02	Based on heat budget calculations	1971, 1973, 1974	Le Guern et al. (1979)
Erta Ale	Ethiopia	60	0.02	Ground-based OP-FTIR and UV spectrometer walking traverses	15 Oct 2005	Sawyer et al. (2008b)
Grímsvötn	Iceland	532	0.19	Sampling of subglacial crater lake	1954-1991	Agustdottir & Brantley (1994)
Merapi	Indonesia	240	0.09	High temp. dome fumarole sampling and routine SO_2 flux measurements		Toutain et al. (2009)
Etna	Italy	35,000	12.8	Plume sampling or airborne IR analysers and ground-based and airborne COSPEC;	1977-1984	Allard et al. (1991)

Table 1 (continued). Volcanic plume CO_2 fluxes

Volcano	Location	CO_2 Flux (t/d)	CO_2 Flux (Mt/yr)	Method	Date	Reference
Etna	Italy	9,000	3.29	Ground-based multi-gas (Voragine and NE) and UV spectrometer vehicle traverses	Sept 2004-Sept 2005	Aiuppa et al. (2006)
Etna	Italy	5,090	1.86	Ground-based multi-gas (Voragine) and UV spectrometer vehicle traverses	May 2005-Nov 2006	Aiuppa et al. (2008)
Stromboli	Italy	4,350	1.59	Fumarole and quiescent plume sampling and airborne COSPEC	1980-1993	Allard et al. (1994)
Stromboli	Italy	1,073	0.39	Ground-based OP-FTIR automated UV spectrometer network	9 Apr 2002 (IR) 2006 (UV)	Burton et al. (2007a,b)
Stromboli	Italy	550	0.20	Ground-based multi-gas sensor and automated UV spectrometer network	Sept 2008-Jul 2010	Aiuppa et al. (2011)
Vulcano	Italy	420	0.15	Ground-based multi-gas measurement and UV spectrometer vehicle traverses	Dec 2004	Aiuppa et al. (2005)
Vulcano	Italy	170	0.06	UAV mounted electrochemical sensors and UV spectrometer	Apr 2007	McGonigle et al. (2008)
Vulcano	Italy	362	0.13	High temp. fumarole sampling and UV spectrometer vehicle traverses	Sept 2007	Inguaggiato et al. (2012)
Miyakejima	Japan	14,500	5.29	Airborne Li-COR, SO_2 electrochemical sensor or pulsed fluorescence SO_2 analyzer and COSPEC	2000-2001	Shinohara et al. (2003)
Satsuma-Iwojima	Japan	100	0.04	High temp. fumarole sampling and COSPEC;	Oct 1999	Shimoike et al. (2002)
Popocatépetl	Mexico	9,000	3.29	Airborne Li-COR ladder surveys	Jun 1995	Gerlach et al. (1997)
Popocatépetl	Mexico	40,000	14.6	Airborne Li-COR ladder surveys	1996-1998	Delgado et al. (1998)
Popocatépetl	Mexico	38,000	13.87	Passive OP-FTIR and COSPEC	Feb 1998	Goff et al. (2001)

Soufrière Hills	Montserrat	1,468	0.54	Ground-based multi-gas sensor and automated UV spectrometer network	Jul 2008	Edmonds et al. (2010)
White Island	New Zealand	950	0.35	High temp. fumarole sampling and airborne COSPEC	1982-1984	Rose et al. (1986)
White Island	New Zealand	2,610	0.95	Airborne Li-COR ladder survey	Jan 1998	Wardell et al. (2001)
Masaya	Nicaragua	2,940	1.07	Ground-based OP-FTIR and COSPEC vehicle traverses	1998-1999	Burton et al. (2000)
Masaya	Nicaragua	930	0.34	Ground-based OP-FTIR and UV spectrometer vehicle traverses	Mar 2009	Martin et al. (2010)
Kudryavy	Russia	50	0.02	Fumarole sampling and ground-based COSPEC	Aug 1995	Fischer et al. (1998)
Bezymianny	Russia	990	0.37	Dome fumarole sampling and FLYSPEC ground-based scanning and airborne traverses	Aug 2007, Jul 2009	Lopez et al. (in press)
Gorely	Russia	660	0.24	Ground-based multi-gas measurement and UV camera	6 Sept 2011	Aiuppa et al. (2012)
Oldoinyo Lengai	Tanzania	6,630	2.42	Airborne Li-COR measurements	Jun 1994	Brantley & Koepenick (1995)
Augustine	USA	1,760	0.64	Restored gas samples and COSPEC	1986-1987	Symonds et al. (1992)
Douglas	USA	trace	trace	Airborne Li-COR measurements	2000-2006	Doukas & McGee (2007)
Griggs	USA	not detected	not detected	Airborne Li-COR measurement	1 Jul 2002	Doukas & McGee (2007)
Iliamna	USA	131	0.05	Airborne Li-COR measurements	1996-2005	Doukas & McGee (2007)
Kīlauea, Pu'u 'Ō'ō	USA	3,950	1.45	Airborne closed-path FTIR	15 Feb 1984, 4 Mar 1984	Greenland et al. (1985)
Kīlauea, Pu'u 'Ō'ō	USA	300	0.11	Airborne Li-COR, closed-path FTIR and COSPEC	19 Sept 1995	Gerlach et al. (1998)
Kīlauea, Pu'u 'Ō'ō	USA	396	0.14	Airborne Visible/Infrared Imaging Spectrometer (AVIRIS)	26 Apr 2000	Spinetti et al. (2008)

Table 1 (continued). Volcanic plume CO_2 fluxes

Volcano	Location	CO_2 Flux (t/d)	CO_2 Flux (Mt/yr)	Method	Date	Reference
Kīlauea summit	USA	1,600	0.58	Airborne closed-path FTIR	13 Feb 1984	Greenland et al. (1985)
Kīlauea summit	USA	8,500	3.10	Ground-based Li-COR, closed-path FTIR and COSPEC	20 Sept 1995, 20 Oct 1998, 6 May 1999	Gerlach et al. (2002)
Kīlauea summit	USA	4,900	1.79	Ground-based Li-COR, Interscan electrochemical SO_2 analyzer and COSPEC;	Jun-Jul 2003	Hager et al. (2008)
Mageik	USA	341	0.12	Airborne Li-COR measurements	2000-2006	Doukas & McGee (2007)
Martin	USA	56	0.02	Airborne Li-COR measurements	1998-2006	Doukas & McGee (2007)
Mt. Baker	USA	187	0.07	Airborne Li-COR measurement	13 Sept 2000	McGee et al. (2001)
Mt. Baker	USA	150	0.05	Airborne Li-COR measurement	2007	Werner et al. (2009)
Peulik	USA	not detected	not detected	Airborne Li-COR measurement	24 May 1998	Doukas & McGee (2007)
Reboubt	USA	18	0.01	Airborne Li-COR measurements	1997-2005	Doukas & McGee (2007)
Spurr	USA	633	0.23	Airborne Li-COR measurements	2004-2006	Doukas & McGee (2007)
Spurr Crater Peak	USA	334	0.12	Airborne Li-COR measurements	1996-2006	Doukas & McGee (2007)
Ukinrek Maars	USA	187	0.07	Airborne Li-COR measurement	24 May 1998	Doukas & McGee (2007)
Veniaminof	USA	not detected	not detected	Airborne Li-COR measurement	2 Aug 2003	Doukas & McGee (2007)
Ambrym	Vanuatu	20,000	7.30	Multi-gas sensor & airborne UV spectrometer traverses	2007	Allard et al. (2009)
Yasur	Vanuatu	840	0.31	Ground-based multi-gas sensor and UV spectrometer vehicle traverses	21 Oct 2007	Métrich et al. (2011)

Table 2. Mean volcanic plume CO_2 fluxes from persistently degassing volcanoes (ordered by CO_2 flux)

Volcano	Country	CO_2 Flux (t/d)	CO_2 Flux (Mt/yr)
Nyiragongo	DR Congo	52,410	19.13
Popocatépetl	Mexico	29,000	10.59
Ambrym	Vanuatu	20,000	7.30
Etna	Italy	16,363	5.97
Miyakejima	Japan	14,500	5.29
Oldoinyo Lengai	Tanzania	6,630	2.42
Kīlauea	USA	6,549	2.39
Stromboli	Italy	1,991	0.73
Masaya	Nicaragua	1,935	0.71
White Island	New Zealand	1,780	0.65
Augustine	USA	1,760	0.64
Erebus	Antarctica	1,630	0.59
Soufrière Hills	Montserrat	1,468	0.54
Galeras	Colombia	1,020	0.37
Bezymianny	Russia	990	0.36
Spurr	USA	967	0.35
Yasur	Vanuatu	840	0.31
Gorely	Russia	660	0.24
Grímsvötn	Iceland	532	0.19
Villarrica	Chile	477	0.17
Sierra Negra	Ecuador (Galápagos)	394	0.14
Mageik	USA	341	0.12
Vulcano	Italy	317	0.12
Merapi	Indonesia	240	0.09
Ukinrek Maars	USA	187	0.07
Mt. Baker	USA	169	0.06
Iliamna	USA	131	0.05
Satsuma-Iwojima	Japan	100	0.04
Erta Ale	Ethiopia	57	0.02
Martin	USA	56	0.02
Kudryavy	Russia	50	0.02
Redoubt	USA	18	0.01
Douglas	USA	trace	trace
	Total	**163,562**	**59.70**

Table 3. Diffuse CO_2 emissions from historically active volcanoes (alphabetically ordered by country)

Volcano	Country	CO_2 soil gas flux (t/yr)	Dissolved CO_2 flux (t/yr)	Reference
Erebus	Antarctica	14,600		Wardell et al. (2003)
Sierra Negra	Ecuador (Galápagos)	220,825		Padron et al. (2012)
Santa Ana	El Salvador	59,130		Salazar et al. (2004)
Nea Kameni	Greece	5,621		Chiodini et al. (1998)
Nisyros	Greece	24,784		Caliro et al. (2005)
Hengill volcanic system	Iceland	165,345		Hernández et al. (2012)
Krafla geothermal system	Iceland	84,000		Ármannsson et al. (2007)
Reykjanes volcanic sys.	Iceland	12,660		Óskarsson & Fridriksson (2011)
Merapi	Indonesia	78,475		Toutain et al. (2009)
Etna	Italy	1,000,000	250,000	D'Alessandro et al. (1997)
Ischia	Italy	468,940	9,461	Pecoraiano et al. (2005)
Pantelleria	Italy	361,000	34,000	Favara et al. (2001)
Solfatara, Campi Flegrei	Italy	556,260		Chiodini et al. (2001)
Stromboli	Italy	82,125		Carapezza & Federico (2000)
Vesuvio	Italy	55,115		Frondini et al. (2004)
Vulcano	Italy	41,975	2,190	Inguaggiato et al. (2012)
Miyakejima (Oyama)	Japan	43,618		Hernández et al.(2001a)
Satsuma-Iwojima	Japan	7,300		Shimoike et al. (2002)
Showa-Shinzan	Japan	3,760		Hernández et al. (2006)
Usu	Japan	60,712		Hernández et al.(2001b)
Popocatépetl	Mexico	not detected		Varley & Armienta (2001)
Cerro Negro	Nicaragua	1,022,000		Salazar et al. (2001)
Masaya caldera	Nicaragua	630,720		Pérez et al (2000)
Masaya, Comalito	Nicaragua	6,935		Chiodini et al. (2005)
Rabaul	Papua New Guineau	876,000		Pérez et al. (1998)
Furnas	Portugal		9,358	Cruz et al. (1999)
Oldoinyo Lengai	Tanzania	36,432		Koepenick et al. (1996)
Teide	Tenerife	38,836	64,605	Hernández et al. (2000); Marrero et al. (2008)
Lassen	USA	35,000	7,600	Rose & Davisson (1996)
Mt. Shasta	USA		8,500	Rose & Davisson (1996)
Ukinrek Maars	USA	11,863	1,095	Evans et al. (2009)
	Total (t/yr)	**6,004,031**	**386,809**	
	Total (Mt/yr)	**6.00**	**0.39**	

Table 4. CO_2 emissions from tectonic, hydrothermal or inactive volcanic areas (alphabetically ordered by country)

Area or vent	Country	CO_2 soil gas flux (t/yr)	Dissolved CO_2 flux (t/yr)	Reference
Tengchong Cenozoic volcanic field	China		3,580	Cheng et al. (2012)
Albani Hills	Italy	26,840	157,960	Chiodini & Frondini (2001)
Bossoleto, Siena	Italy	3,500		Mörner & Etiope (2002)
Caldara di Manziana	Italy	73,000		Chiodini et al. (1999)
Campanian degassing structure	Italy		3,080,000	Chiodini et al. (2004)
Castiglioni, Siena	Italy	4,400		Mörner & Etiope (2002)
Latera	Italy	127,750		Chiodini et al. (2007)
Mefite d'Ansanto	Italy	730,000		Chiodini et al. (2010)
Naftìa Lake area	Italy	72,217		Giammanco et al. (2007)
Pienza	Italy	4,015		Rogie et al. (2000)
Poggio dell'Ulivo	Italy	73,000		Chiodini et al. (1999)
Rapalano Cecilia	Italy	17,520		Rogie et al. (2000)
Rapalano Mofete Diambra	Italy	35,040		Rogie et al. (2000)
San Sisto	Italy	21,600		Italiano et al. (2000)
Selvena	Italy	6,205		Rogie et al. (2000)
Telese	Italy	20,000		Italiano et al. (2000)
Tuscan Roman degassing structure	Italy		6,160,000	Chiodini et al. (2004)
Umbertide	Italy	5,840		Rogie et al. (2000)
Ustica	Italy	260,000		Etiope et al. (1999)
Hakkoda	Japan	27,010		Hernández et al. (2003)
Taupo	New Zealand		440,000	Seward & Kerrick (1996)
Yangbajain	Tibet	50,370		Chiodini et al. (1998)
Mammoth Mountain	USA	189,800	14,600	Sorey et al. (1998)
Mt Washington and Belknap Crater	USA		2,400	James at al. (1999)
Mt Jefferson	USA		8,000	James at al. (1999)
Mt Bachelor	USA		1,800	James at al. (1999)
Salton Trough	USA		44,000	Kerrick et al. (1995)
Three Sisters	USA		4,400	James at al. (1999)
Yellowstone	USA	8,580,000		Werner & Brantley (2003)
Indonesia-Philippines		1,800,000		Seward & Kerrick (1996)
Subaerial Pacific rim		44,000,000		Seward & Kerrick (1996)
	Total (t/yr)	**56,128,107**	**9,916,740**	
	Total (Mt/yr)	**56.13**	**9.92**	

the subaerial Pacific rim (44 Mt/yr, Seward and Kerrick 1996). These estimates are based on extrapolations from the CO_2 emissions observed from the 150 km long Taupo Volcanic Zone (New Zealand) to the 18,000 km long Pacific Rim.

The global emissions of CO_2 from volcanic lakes were recently assessed by Pérez et al. (2011), who pointed out that volcanic lakes had not been included in previous estimates of global geological carbon efflux (e.g., Kerrick et al. 2001; Mörner and Etiope 2002). They found that CO_2 emissions increased with increasing acidity in volcanic lakes, reflecting the acidity of the volcanic gas discharge. Measurements were conducted on 32 volcanic lakes which were divided into three types of water based on pH, alkali, neutral and acid. Average flux per unit area for each type was then used to calculate a global volcanic lake estimate, extrapolating to an estimated number of volcanic lakes in the world (769). This number of lakes is greater than the number of lakes reported in the literature (138) by a factor which reflects the regional under-sampling between actual lakes and lakes reported in the scientific literature. This methodology assumes that the average acidity-emission rate relationship in the 32 measured lakes is a faithful average representation of the global lake population. A further estimate was produced by defining 4 populations in the measured data set based on frequency, emission rate and lake size and extrapolating to all 769 volcanic lakes. The combination of these two approaches yielded a global volcanic lake CO_2 emission of 117 ± 19 Mt/yr, of which 94 Mt/yr is attributed to magmatic degassing.

Submarine volcanism

There are three main submarine sources of CO_2, MOR, arc volcanoes and fore-arc degassing (which appears to be dominated by CH_4 emissions (Füri et al. 2010)). Global emissions from MOR have been determined by various authors, as reported in Table 5. The large spread of MOR fluxes, from 4.4 to 792, reflects uncertainties in the dissolved contents of C and global ^3He fluxes. Marty and Tolstikhin (1998) performed a careful examination of the $CO_2/^3He$ ratios used and determined a median value of 2.2×10^9 with standard deviation of 0.7×10^9. Using a ^3He flux of 1000 ± 250 mol/yr (Farley et al. 1995) they derived a MOR CO_2 flux of 97 ± 40 Mt/yr CO_2. The more recent determination of MOR CO_2 flux from Resing et al. (2004) who measured $CO_2/^3He$ in MOR hydrothermal plumes of 55 ± 33 Mt/yr is in reasonable agreement with that estimate.

While measurements of CO_2 release from the cooler flanks of MORs and submarine arc volcanoes increase in number each year, global estimates of submarine CO_2 emissions are extremely difficult to make. The large areal extent and our relatively poor knowledge of the submarine surface suggests that there is ample opportunity for unknown or unrecognized active volcanism (e.g., cold liquid CO_2 emissions, Lupton et al. 2008), but at the current time it is not possible to make quantitative estimates of the global CO_2 emissions from such sources.

INVENTORIES OF GLOBAL VOLCANIC DEEP CARBON FLUX: IMPLICATIONS FOR THE GEOLOGICAL CARBON CYCLE

Estimates of global deep carbon emission rates

In Table 6, we summarize the measured fluxes from the subaerial sources and MOR, and attempt to extrapolate from these measurements to global estimates of the CO_2 flux for each source. In the case of volcanic plume passive degassing the GVN catalogue (Siebert and Simkin 2002) indicates that there are ~150 such actively degassing volcanoes on Earth. While our catalogue of 33 CO_2 flux plume measurements (Table 2: total flux 59.7 Mt/yr) is significantly larger than previously collated, it reflects only 22% of the total number of active volcanoes. While our current compilation includes some large emitters, suggesting that the major sources have been already identified, we highlight how this total flux has increased due

Table 5. MOR global CO_2 flux (Mt/yr)

Method	Min	Max	Reference
Crustal production rates and C content of mantle	10	35	Gerlach (1989)
	22	39.6	Gerlach (1991)
	572	748	Javoy & Pineau (1991)
	128	255	Holloway (1998)
	176	792	Cartigny et al. (2001)
	28.6	41	Saal et al. (2002)
$CO_2/^3He$ in MORB glass, global mantle ^3He flux	75	119	Marty & Jambon (1987)
	18	44	Sarda & Graham (1990)
	106	264	Graham & Sarda (1991)
	66	119	Marty & Zimmerman (1999)
$CO_2/^3He$ in MOR fluids, global mantle ^3He flux	44	70	Corliss et al. (1979)
	20	57	Des Marais & Moore (1984)
Hydrothermal fluid flux and composition	4.4	53	Elderfield & Schultz (1996)
$CO_2/^3He$ in plumes	22	88	Resing et al. (2004)
Summary	**4.4**	**792**	

Table 6. Summary of measured volcanic CO_2 fluxes and estimated global emissions (Mt/yr)

Source	Measured CO_2 flux	N° measured	N° global	% global	Estimated Global CO_2 flux	Ref.
Volcanic plume passive degassing	59.7	33	~150	22	271	[1]
Diffuse emissions from historically active volcanoes	6.4	30	~550	5.5	117	[1]
Emissions from tectonic, hydrothermal or inactive volcanic areas	66	—	—	—	>66	[1]
Volcanic lakes	6.7	32	769	4.2	94	[2]
MOR	97	—	—		97	[3]
			Total		637	
			Total (no MOR)		540	

References: [1] This work; [2] Perez et al. (2011); [3] Marty and Tolstikhin (1998)

to previously unrecognized large emissions from e.g., Ambrym (Vanuatu). Further large, but as yet unquantified, sources of CO_2 emissions may be present in Papua New Guinea, the Banda Sunda arc and the Vanuatu island chain. We therefore conclude that the clearest and probably most accurate way to extrapolate from the current catalogue of plume emissions to a global estimate is through a linear extrapolation. Extrapolating from the measured 33 to an estimated 150 plume-creating, passively degassing volcanoes we estimate that the global plume CO_2 flux is ~271 Mt/yr (see Table 6).

The total number of historically active volcanoes reported by GVN is ~550, and 30 (5.5%) of these have had diffuse CO_2 soil degassing fluxes quantified, as reported in Table 3, for a total of 6.4 Mt/yr. Extrapolating to a global flux, assuming a similar distribution of fluxes in the unmeasured fluxes as seen in those measured, produces a total of 117 Mt/yr from diffuse degassing from the flanks of historically active volcanoes (Table 6).

The total diffuse CO_2 flux from inactive volcanoes, hydrothermal and tectonic structures reported in Table 4 is more challenging to extrapolate to a global scale. Our current constraints on the tectonic CO_2 flux comes almost entirely from the work of Chiodini et al. (2004) who examined actively degassing tectonic structures in Italy. The abundance of such structures on Earth is unknown, and this therefore represents a source of great uncertainty in estimates of total deep carbon flux. This uncertainty makes it challenging to sensibly extrapolate to a global estimate of tectonic CO_2 fluxes, and therefore we use only reported fluxes, and highlight the possibility that the true total may be significantly larger. Emissions from hydrothermal systems estimated by Seward and Kerrick (1996) are already extrapolated to cover a significant proportion of the volcanically active surface of the Earth. We therefore use the total presented in Table 4 for the CO_2 emissions from tectonic, hydrothermal and inactive volcanoes as a lower limit for the global emission of CO_2 from these sources.

Summing the extrapolated passive plume, diffuse degassing, lake degassing global estimates and emissions from inactive, hydrothermal and tectonic structures produces a total subaerial volcanic flux of 540 Mt/yr, and a global emission (including MOR emissions) of 637 Mt/yr (Table 6). Thus global volcanic CO_2 fluxes are only ~1.8% of the anthropogenic CO_2 emission of 35,000 Mt per year (Friedlingstein et al. 2010).

Comparison with previous estimates of subaerial volcanic CO_2 flux

There have been several papers which estimate the global CO_2 flux, as shown in Table 7. Our update of the global volcanic CO_2 flux, 637 Mt/yr, is larger than the maximum suggested by Marty and Tolstikhin (1998) of 440 Mt/yr. This is in part because CO_2 emissions from volcanic lakes were not addressed in that work. The total subaerial flux we calculate of 540 Mt/year is also higher than that proposed by Mörner and Etiope (2002), due primarily to the improvement in measurements of persistently degassing volcanoes. We note that Mörner and Etiope (2002) included the fluxes from single eruptive events from Pinatubo (1991) and Mt. St. Helens (1980) in their inventory of volcanoes contributing to the annual global CO_2 flux. Other papers cited in Table 7 appear to have significantly underestimated the global subaerial CO_2 flux, primarily due to a lack of field measurements.

Table 7. Global volcanic subaerial CO_2 flux (Mt/yr)

CO_2 flux	Reference
79	Gerlach (1991)
145	Varekamp et al. (1992)
66	Allard (1992)
88	Marty and Le Cloarec (1992)
65	Williams et al. (1992)
136	Sano and Williams (1996)
242	Marty and Tolstikhin (1998)
99	Kerrick (2001)
300	Mörner and Etiope (2002)
540	This work

Balancing CO_2 emission rates with weathering and subduction rates

In the absence of a continual supply of CO_2 from volcanic and tectonic degassing the CO_2 content of the atmospheres and oceans would be gradually depleted through CO_2 removal by weathering (Gerlach 1991). The fact that, instead, pre-industrial CO_2 concentrations are relatively stable suggest a balance between CO_2 removal by weathering and CO_2 supply by Earth degassing on timescales of ~0.5 Ma (Walker et al. 1987; Berner 1991). Over such timescales weathering of carbonates has no impact on removal of atmospheric CO_2, because

HCO_3^- supplied to the ocean by carbonate weathering releases the CO_2 it captured from the atmosphere during calcite precipitation (Berner 1991). Therefore in the long timescale of the geological carbon cycle CO_2 emissions from geological sources should balance consumption from silicate weathering and oceanic crust alteration. Gaillardet et al. (1999) found that CO_2 consumption from continental silicate weathering was 515 Mt/yr, which matches well with our estimates of subaerial volcanic CO_2 degassing (540 Mt/yr). However, inclusion of 300 Mt/yr CO_2 released by metamorphism (Morner and Etiope 2002) produces a total lithospheric subaerial CO_2 emission of 840 Mt/yr. This is larger than the current estimates of silicate weathering, suggesting that, assuming steady-state, weathering rates might be slightly higher to absorb all the emitted CO_2.

Dasgupta and Hirschmann (2010) calculated the total ingassing of CO_2 into subduction zones from the combination of three lithologies in the subducting slab, altered oceanic crust, sediments and mantle, producing an estimated CO_2 consumption rate of 403 Mt/yr. We note that this is lower than the CO_2 consumption rate due to silicate weathering (515 Mt/yr, Gaillerdet et al. 1999), which is slightly inconsistent (but within uncertainties of such estimates), as the eventual destiny of CO_2 consumed by silicate weathering will be deposition on the seafloor or precipitation within the oceanic crust. In order to maintain steady-state quantities of CO_2 in the exosphere this consumption should be balanced by the total emission from MORs, subaerial degassing and metamorphism (calculated here to be 937 Mt/yr). Given the likely underestimate in our total lithospheric CO_2 emissions arising from lack of knowledge of tectonic degassing, it appears reasonable to conclude that the ingassing rate may be an underestimate. However, it is clear that there are large uncertainties in both sums.

THE ROLE OF DEEP CARBON IN VOLCANIC ACTIVITY

Original CO_2 contents of magma

The volatile content of magmas, together with their evolving viscosity and degassing behavior during ascent, helps to determine whether a volcano will be quietly degassing or violently erupting for a given magma input rate. Models of magma dynamics require knowledge of original volatile content in order to reproduce physically accurate processes occurring during an eruption and in quiescent phases. Furthermore, knowledge of the original volatile contents of magmas allows calculation of the magma mass required to produce an observed gas flux, permitting quantitative comparison of fluxes with geophysical and volcanological observations. Measurements of original volatile contents are therefore of great interest.

Melt inclusions (MIs) provide records of original volatile contents, through analysis of pockets of melt trapped inside growing crystals during magma ascent or storage, and such studies have been carried out on many eruption products. However, the presence of a separate fluid phase at the moment of inclusion entrapment will produce an underestimate in the concentrations of dissolved volatiles. Wallace (2005) concluded that no melt inclusions sample arc magmas undegassed with respect to CO_2. Blundy et al. (2010) used MI measurements from Mt. St. Helens to show that the dissolved volatile contents of shallow magmas were strongly affected by CO_2-rich fluids rising from magmas at greater depth, concluding that the original CO_2 contents of arc magmas was likely to be significantly higher than that recorded in MIs. Using inferred CO_2 contents in arc andesites and dacites of 1.5 wt% they calculate that the CO_2 contents of parental, mantle-derived basalts would contain 0.3 wt%. This relatively high CO_2 content is in agreement with previous estimates of volatile contents of arc magmas (Wallace 2005).

Original CO_2 contents of magmas can also be estimated by assuming a steady-state condition for a persistently degassing volcano, and comparing the observed CO_2 flux together with the flux of a more soluble gas species whose original volatile content has been well-characterized, such as SO_2. Gerlach et al. (2002) performed such a calculation for Kīlauea volcano, Hawai'i, concluding that the bulk CO_2 concentration required to match magma input rates and CO_2 output rates was 0.70 wt%. Dixon and Clague (2001) measured a dissolved CO_2 content of at a depth of 1500-2000 m at the Loihi seamount in the Hawai'i chain. While the CO_2 concentrations were very low, the samples contained fluid-filled vesicles with high CO_2 contents, which allowed a bulk CO_2 concentration of up to 0.63 wt% to be determined, in fair agreement with the estimate from Gerlach et al. (2002). A recent study (Barsanti et al. 2009) introduced a more complex note to the examination of original CO_2 contents at Hawai'i, with a statistical analysis of MI CO_2 and H_2O concentrations which revealed distinct magma batches some of which could contain 2-6 wt% CO_2. High CO_2 contents of magmas feeding Etna and Stromboli have also been proposed based on degassing mass balance calculations, with amounts ranging between 1.6 and 2.2 wt% (Spilliaert et al. 2006; Burton et al. 2007a). These few available estimates of original CO_2 contents from mass balance determinations open the possibility that CO_2 contents of magmas feeding active volcanoes are in general higher than is expected based on CO_2 contents of melt inclusions.

Importance of a deep exsolved volatile phase on magma dynamics and eruptive style

The presence of a CO_2-rich volatile phase at great pressure can strongly affect the dynamics of magma ascent and eruption, because the style and intensity of eruptive activity is controlled in part by the distribution of gas phase in a magma during eruption (Eichelberger et al. 1986; Jaupart and Vergniolle 1988). Persistently degassing volcanoes can release vast amounts of gas at the surface non-explosively, implying storage within the crust of large volumes of degassed magma (Crisp 1984, Francis et al. 1993) via magma convection (Kazahaya et al. 1994).

Exsolved volatiles can ascend from depth, accumulating in foams that can produce Strombolian activity (e.g., Menand and Phillips 2007; Jaupart and Vergniolle 1988). Gas can stream through magma from depth to the surface (Wallace et al. 2005), as surmised to occur at Soufrière Hills volcano, Montserrat (Edmonds et al. 2010) and Stromboli volcano (Aiuppa et al. 2010). Perhaps most importantly of all however, exsolved gas accumulation can produce powerful explosive eruptions. The eruption of Pinatubo in 1991 (Pallister et al. 1992) was one of the most violent in recorded history. It produced a much greater mass of S than was to be expected from dissolved S contents and the volume of erupted material, suggesting the presence of a voluminous pre-eruptive gas phase (Wallace and Gerlach 1994), likely produced from basaltic underplating crystallizing as anhydrite (Matthews et al. 1992) which triggered the eruption (Pallister et al. 1992).

MAGNITUDE OF ERUPTIVE DEEP CARBON EMISSIONS

It is useful to compare the CO_2 emission rates for subaerial volcanism of 540 Mt/yr reported in Table 7 with a single large eruption such as the ~5 km^3 eruption of Mt. Pinatubo in 1991, producing ~50 Mt of CO_2 (Gerlach et al. 2011), equivalent to merely ~5 weeks of global subaerial volcanic emissions. The Pinatubo 1991 syn-eruptive emission is therefore dwarfed by the time-averaged continuous CO_2 emissions from global volcanism. Indeed, the present day CO_2 emission rate from the lake filling the crater formed during the eruption of Pinatubo is 884 t/d (Perez et al. 2011), suggesting that in the 31 years since that eruption ~10 Mt of CO_2 has been produced, ~20% of that emitted during the eruption.

Using the volumes of erupted material produced by the three largest eruptions of the last 200 years (Self et al. 2006) we may estimate their CO_2 emissions, assuming a similar erupted volume to CO_2 emission amount to that estimated for Pinatubo (10 Mt CO_2 per km^3 erupted, equivalent to ~1 wt% CO_2 content). The eruption of Tambora (Indonesia) in 1815 is estimated to have produced 30 km^3 of products (Self et al. 2006), with an inferred output of 300 Mt of CO_2. Krakatua in 1883 (Indonesia) and Katmai-Novarupta in 1912 (Alaska) each produced 12 km^3, and ~120 Mt of CO_2. The total CO_2 output of the four largest eruptions in the last 200 years is therefore ~600 Mt of CO_2, slightly less than we estimate for subaerial volcanic degassing in a single year, and therefore only 0.6% of the amount of gas released through continuous volcanic activity in the same time period. It therefore appears that the continuous degassing of active and inactive volcanoes dominates the short-lived paroxysmal emissions produced in large eruptions.

Crisp (1984) calculated that the average eruption rate from volcanoes over the last 300 years was 0.1 km^3 magma per year, which with ~1 wt% CO_2 content suggests an annual output of ~1 Mt CO_2, only 0.2% of the estimated annual subaerial CO_2 emissions. This demonstrates that degassing of unerupted magma dominates degassing of erupted lava on the planet, and emphasizes the fundamental role that unerupted magmatic intrusions must have in contributing to the global volcanic CO_2 flux. Such intrusions may produce unexpectedly high CO_2 emissions if they interact with crustal carbonates (Troll et al. 2012). This important process could be assessed quantitatively if a method could be developed for measuring volcanic $^{12}C/^{13}C$ ratios in the field.

SUMMARY

In recent years, measurements of CO_2 flux from volcanoes and volcanic areas have greatly increased, particularly on persistently degassing volcanoes, of which ~22% have had their CO_2 flux quantified. Notwithstanding this progress, it is clear that the CO_2 emissions from the majority of volcanic sources are still unknown. Using the available data from plume measurements from 33 degassing volcanoes we determine a total CO_2 flux of 59.7 Mt/yr. Extrapolating this to ~150 active volcanoes produces a total of 271 Mt/yr CO_2. Extrapolation of the measured 6.4 Mt/yr of CO_2 emitted from the flanks of 30 historically active volcanoes to all 550 historically active volcanoes produces a global emission rate of 117 Mt/yr. Perez et al. (2011) calculated the global emission from volcanic lakes to be 94 Mt/yr CO_2. The sum of these fluxes produces an updated estimate of the global subaerial volcanic CO_2 flux of 474 Mt/yr. Emissions from tectonic, hydrothermal and inactive volcanic areas contribute a further 66 Mt/yr to this total (Table 6), producing a total subaerial volcanic emission of 540 Mt/yr. An extrapolation to a global estimate is not straightforward for tectonic-related degassing, as the number of areas which produce such emissions is not known. Given the fact that ~10 Mt/yr is produced by Italy alone it is possible that the global total is significant, and this merits further investigation. We highlight also that the magnitude of CO_2 emissions from both cold and hot non-MOR submarine volcanic sources are currently effectively unknown.

Our subaerial volcanic CO_2 flux matches well with estimates of CO_2 removal rates of 515 Mt/yr due to silicate weathering, which, over timescales of 0.5 Ma, should balance lithospheric CO_2 emissions. However, inclusion of the metamorphic CO_2 flux of 300 Mt/yr calculated by Morner and Etiope (2002) produces a total subaerial lithospheric flux of 840 Mt/yr, suggesting that, assuming steady-state, weathering rates might be slightly higher in order to absorb all the CO_2 emitted from the lithosphere.

The global subaerial CO_2 flux we report is higher than previous estimates, but remains insignificant relative to anthropogenic emissions, which are two orders of magnitude greater at 35,000 Mt/yr (Friedlingstein et al. 2010). Nevertheless, it is clear that uncertainties in volcanic CO_2 emission rates remain high and significant upward revisions of the lithospheric CO_2 flux cannot be ruled out. This uncertainty also limits our understanding of global volcanic carbon budgets and the evolution of the distribution of CO_2 between the crust and the mantle. Furthermore, with the notable exception of continuous CO_2 flux monitoring at a handful of volcanoes we have very little data with which to assess CO_2 flux variations across different timescales. It is clear that there is much further work to be done surveying CO_2 emissions from both active and inactive volcanoes.

Continuous global CO_2 emissions from passively degassing volcanoes over timescales longer than a few months dominate CO_2 emissions produced by relatively short-lived eruptions. Nevertheless, we highlight that dramatic CO_2 emissions may occur during magmatic intrusion events, and that the sporadic and short-term nature of field measurements to date may lead to such events being missed. To this end, robust field-portable instruments capable of measuring $^{12}C/^{13}C$ ratios in volcanic CO_2 emissions would be of great utility in order to distinguish CO_2 produced during metamorphism of crustal carbonates from magmatic CO_2.

ACKNOWLEDGMENTS

We thank Adrian Jones and Bob Hazen for the opportunity to contribute to this volume. Adrian Jones' and Tamsin Mather's constructive reviews of an earlier version of this work are greatly appreciated. Sara Barsotti is thanked for her simulations of CO_2 concentrations in the Etna plume with VOLCALPUFF. MB acknowledges support from ERC project CO2Volc 279802.

REFERENCES

Ágústsdóttir AM, Brantley SL (1994) Volatile fluxes integrated over four decades at Grimsvötn volcano, Iceland. J Geophys Res 99:9505-9522, doi: 10.1029/93JB03597

Aiuppa A, Bertagnini A, Metrich N, Moretti R, Di Muro A, Liuzzo M, Tamburello G (2010) A model of degassing for Stromboli volcano. Earth Planet Sci Lett 295(1-2):195-204

Aiuppa A, Burton M, Allard P, Caltabiano T, Giudice G, Gurrieri S, Liuzzo M, Salerno G (2011) First observational evidence for the CO_2-driven origin of Stromboli's major explosions. Solid Earth 2(2):135-142, doi: 10.5194/se-2-135-2011

Aiuppa A, Federico C, Giudice G, Gurrieri S (2005) Chemical mapping of a fumarole field: La Fossa Crater, Vulcano Island (Aeolian Islands, Italy). Geophys Res Lett 32:L13309, doi: 10.1029/2005GL023207

Aiuppa A, Federico C, Giudice G, Gurrieri S, Liuzzo M, Shinohara H, Favara R, Valenza M (2006) Rates of carbon dioxide plume degassing from Mount Etna volcano. J Geophys Res Solid Earth 111:B09207, doi: 10.1029/2006JB004307

Aiuppa A, Giudice G, Gurrieri S, Liuzzo M, Burton M, Caltabiano T, McGonigle AJS, Salerno G, Shinohara H, Valenza M (2008) Total volatile flux from Mount Etna. Geophys Res Lett 35(24):L24302, doi: 10.1029/2008GL035871

Aiuppa A, Giudice G, Liuzzo M, Tamburello G, Allard P, Calabrese S, Chaplygin I, McGonigle AJS, Taran Y (2012) First volatile inventory for Gorely volcano, Kamchatka. Geophys Res Lett 39:L06307, doi: 10.1029/2012GL051177

Aiuppa A, Moretti R, Federico C, Giudice G, Gurrieri S, Liuzzo M, Papale P, Shinohara H, Valenza M (2007) Forecasting Etna eruptions by real-time observation of volcanic gas composition. Geology 35:1115-1118, doi: 10.1130/G24149A.1

Allard P (1992) Global emissions of helium-3 by subaerial volcanism. Geophys Res Lett 19:1479–1481

Allard P, Aiuppa A, Bani P, Metrich N, Bertagnini A, Gauthier PG, Parello F, Sawyer GM, Shinohara H, Bagnato E, Mariet C, Garaebiti E, Pelletier B (2009) Ambrym basaltic volcano (Vanuatu Arc): volatile fluxes, magma degassing rate and chamber depth. AGU Fall Meeting 2009, abstract #V24C-04

Allard P, Burton MR, Mure F (2005) Spectroscopic evidence for a lava fountain driven by previously accumulated magmatic gas. Nature 433(7024):407-410, doi: 10.1038/nature03246

Allard P, Carbonnelle J, Dajlevic D, Le Bronec J, Morel P, Robe MC, Maurenas JM, Faivre-Pierret R, Martin D, Sabroux JC, Zettwoog P (1991) Eruptive and diffuse emissions of CO_2 from Mount Etna. Nature 351:387-391, doi: 10.1038/351387a0

Allard P, Carbonnelle J, Metrich N, Loyer H, Zettwoog P (1994) Sulphur output and magma degassing budget of Stromboli volcano. Nature 368(6469):326-330, doi: 10.1038/368326a0

Alt JC, Teagle DAH (1999) The uptake of carbon during alteration of ocean crust. Geochim Cosmochim Acta 63:1527–1535

Anderson DE, Farrar CD (2001) Eddy covariance measurement of CO_2 flux to the atmosphere from an area of high volcanogenic emissions, Mammoth Mountain, California. Chem Geol 177(1-2):31-42, doi: 10.1016/S0009-2541(00)00380-6

Andres RJ, Kasgnoc AD (1998) A time-averaged inventory of subaerial volcanic sulfur emissions. J Geophys Res 103:25251-25261, doi: 10.1029/98JD02091

Ármannsson H, Fridriksson T, Wiese F, Hernández P, Pérez N (2007) CO_2 budget of the Krafla geothermal system, NE-Iceland. *In:* Bullen TD, Wang Y (eds) Water–Rock Interaction. Taylor & Francis Group, London, p 189-192

Bahcall JN, Pinsonneault MH, Basu S (2001) Solar models: Current epoch and time dependences, neutrinos, and helioseismological properties. Astrophys J 555:990, doi: 10.1086/321493

Baldocchi DD (2003) Assessing the eddy covariance technique for evaluating carbon dioxide exchange rates of ecosystems: past, present and future. Global Change Bio 9(4):479-492, doi: 10.1046/j.1365-2486.2003.00629.x

Barsanti M, Papale P, Barbato D, Moretti R, Boschi E, Hauri E, Longo A (2009) Heterogeneous large total CO_2 abundance in the shallow magmatic system of Kilauea volcano, Hawaii. J Geophys Res 114:B12201, doi: 10.1029/2008JB006187

Barsotti S, Neri A, Scire JS (2008) The VOL-CALPUFF model for atmospheric ash dispersal: 1. Approach and physical formulation. J Geophys Res 113:B03208, doi: 10.1029/2006JB004623

Baubron JC, Allard P, Toutain JP (1990) Diffuse volcanic emissions of carbon dioxide from Vulcano Island, Italy. Nature 344(6261):51-53, doi: 10.1038/344051a0

Berner RA (1991) A model for atmospheric CO_2 over Phanerozoic time. Am J Sci 291:339-376, doi: 10.2475/ajs.291.4.339

Berner RA, Lasaga AC (1989) Modeling the geochemical carbon cycle. Sci Am 260:74-81

Blundy J, Cashman KV, Rust A, Witham F (2010) A case for CO_2-rich arc magmas. Earth Planet Sci Lett 290:289-301, doi: 10.1016/j.epsl.2009.12.013

Brantley SL, Koepenick KW (1995) Measured carbon dioxide emissions from Oldoinyo Lengai and the skewed distribution of passive volcanic fluxes. Geology 23(10):933-936, doi: 10.1130/0091-7613(1995)023<0933:MCDEFO>2.3.CO;2

Brühl C, Lelieveld J, Crutzen PJ, Tost H (2012) The role of carbonyl sulphide as a source of stratospheric sulphate aerosol and its impact on climate. Atmos Chem Phys 12:1239-1253, doi: 10.5194/acp-12-1239-2012

Burton M, Allard P, Murè F, La Spina A (2007a) Depth of slug-driven strombolian explosive activity. Science 317:227-230

Burton MR, Caltabiano T, Mure F, Salerno G, Randazzo D (2009) SO_2 flux from Stromboli during the 2007 eruption: Results from the FLAME network and traverse measurements. J Volcanol Geotherm Res 182:214-220, doi: 10.1016/j.jvolgeores.2008.11.025

Burton MR, Mader HM, Polacci M (2007b) The role of gas percolation in quiescent degassing of persistently active basaltic volcanoes. Earth Planet Sci Lett 264:46-60, doi: 10.1016/j.epsl.2007.08.028

Burton MR, Oppenheimer C, Horrocks LA, Francis PW (2000) Remote sensing of CO_2 and H_2O emission rates from Masaya volcano, Nicaragua. Geology 28(10):915-918, doi: 10.1130/0091-7613(2000)28<915:RSOCAH>2.0.CO;2

Cadle RD (1980) A comparison of volcanic with other fluxes of atmospheric trace gas constituents. Rev Geophys Space Phys 18:746-752

Caliro S, Chiodini G, Galluzzo D, Granieri D, La Rocca M, Saccorotti G, Ventura G (2005) Recent activity of Nisyros volcano (Greece) inferred from structural, geochemical and seismological data. Bull Volcanol 67(4):358-369, doi: 10.1007/s00445-004-0381-7

Camarda M, Gurrieri S, Valenza M (2006) CO_2 flux measurements in volcanic areas using the dynamic concentration method: Influence of soil permeability. J Geophys Res 111:B05202, doi: 10.1029/2005JB003898

Carapezza ML, Federico C (2000) The contribution of fluid geochemistry to the volcano monitoring of Stromboli. J Volcanol Geotherm Res 95(1-4):227-245, doi: 10.1016/S0377-0273(99)00128-6

Carapezza ML, Granieri D (2004) CO_2 soil flux at Vulcano (Italy): comparison between active and passive methods. Appl Geochem 19(1):73-88, doi: 10.1016/S0883-2927(03)00111-2

Cardellini C Chiodini G, Frondini F (2003) Application of stochastic simulation to CO_2 flux from soil: Mapping and quantification of gas release. J Geophys Res Solid Earth 108:2425, doi: 10.1029/2002JB002165

Carn SA, Krotkov NA, Yang K, Krueger AJ (in press) Measuring global volcanic degassing with the Ozone Monitoring Instrument (OMI). Spec Publ Geol Soc London

Cartigny P, Jendrzejewksi N, Pineau F, Petit E, Javoy M (2001) Volatile (C, N, Ar) variability in MORB and the respective roles of mantle source heterogeneity and 25 degassing: the case study of the Southwest Indian ridge. Earth Planet Sci Lett 194:241-257

Cheng ZH, Guo ZF, Zhang ML, Zhang LH (2012) CO_2 flux estimations of hot springs in the Tengchong Cenozoic volcanic field, Yunnan Province, SW China. Acta Petrol Sinica 28(4):1217-1224

Chiodini G, Baldini A, Barberi F, Carapezza ML, Cardellini C, Frondini F, Granieri D, Ranaldi M (2007) Carbon dioxide degassing at Latera caldera (Italy): Evidence of geothermal reservoir and evaluation of its potential energy. J Geophys Res Solid Earth 112:B12204, doi: 10.1029/2006JB004896

Chiodini G, Caliro S, Aiuppa A, Avino R, Granieri D, Moretti M, Parello F (2011) First $^{13}C/^{12}C$ isotopic characterisation of volcanic plume CO_2. Bull Volcanol 73:531-542, doi: 10.1007/s00445-010-0423-2

Chiodini G, Cardellini C, Amato A, Boschi E, Caliro S, Frondini F, Ventura G (2004) Carbon dioxide Earth degassing and seismogenesis in central and southern Italy. Geophys Res Lett 31(7):L07615, doi: 10.1029/2004GL019480

Chiodini G, Cioni R, Guidi M, Raco B, Marini L (1998) Soil CO_2 flux measurements in volcanic and geothermal areas. Appl Geochem 13:543-552

Chiodini G, Frondini F (2001) Carbon dioxide degassing from the Albani Hills volcanic region, Central Italy. Chem Geol 177(1-2):67-83, doi: 10.1016/S0009-2541(00)00382-X

Chiodini G, Frondini F, Cardellini C, Granieri D, Marini L, Ventura G (2001) CO_2 degassing and energy release at Solfatara volcano, Campi Flegrei, Italy. J Geophys Res Solid Earth 106:16,213-16,221, doi: 10.1029/2001JB000246

Chiodini G, Frondini F, Kerrick DM, Rogie J, Parello F, Peruzzi L, Zanzari AR (1999) Quantification of deep CO_2 fluxes from Central Italy. Examples of carbon balance for regional aquifers and of soil diffuse degassing. Chem Geol 159(1-4):205-222, doi: 10.1016/S0009-2541(99)00030-3

Chiodini G, Frondini F, Raco B (1996) Diffuse emission of CO_2 from the Fossa crater, Vulcano Island (Italy). Bull Volcanol 58:41-50

Chiodini G, Granieri D, Avino R, Caliro S, Costa A, Minopoli C, Vilardo G (2010) Non-volcanic CO_2 Earth degassing: Case of Mefite d'Ansanto (southern Apennines), Italy. Geophys Res Lett 37:L11303, doi: 10.1029/2010GL042858

Chiodini G, Granieri D, Avino R, Caliro S, Costa A, Werner C (2005) Carbon dioxide diffuse degassing and estimation of heat release from volcanic and hydrothermal systems, J Geophys Res Solid Earth 110:B08204, doi: 10.1029/2004JB003542

Corliss JB, Dymond J, Forgon LI, Edmond JM, von Herzen RP, Ballard RD, Green K, Williams D, Bainbridge A, Crane K, van Andel TH (1979) Submarine thermal springs on the Galàpagos Rift. Science 203:1073-1083

Crispa D, Atlasb RM, Breonc F-M, Browna LR, Burrowsd JP, Ciaisc P, Connore BJ, Doneyf SC, Fungg IY, Jacobh DJ, Milleri CE, O'Brienj D, Pawsonb S, Randersonk JT, Raynerj P, Salawitcha RJ, Sandera SP, Sena B, Stephensm GL, Tansn PP, Toona GC, Wennbergk PO, Wofsyh SC, Yungk YL, Kuangk Z, Chudasamaa B, Spraguea G, Weissa B, Pollocko R, Kenyonp D, Schrollp S (2004) The orbiting carbon observatory (OCO) mission. Trace Constituents in the Troposphere and Lower Stratosphere: Adv Space Res 34(4), doi: 10.1016/j.asr.2003.08.062

Crisp JA (1984) Rates of magma emplacement and volcanic output. J Volcanol Geotherm Res 20(3-4):177-211, doi: 10.1016/0377-0273(84)90039-8

Crossey LJ, Karlstrom KE, Springer AE Newell D, Hilton DR, Fischer T (2009) Degassing of mantle-derived CO_2 and He from springs in the southern Colorado Plateau region-Neotectonic connections and implications for groundwater systems. Geol Soc Am Bull 121(7-8):1034-1053, doi: 10.1130/B26394.1

Crutzen PJ (1976) The possible importance of COS for the sulfate layer of the stratosphere. Geophys Res Lett 3:73-76

Cruz JV, Coutinho RM, Carvalho MR, Oskarsson N, Gislason SR (1999) Chemistry of waters from Furnas volcano, São Miguel, Azores: fluxes of volcanic carbon dioxide and leached material. J Volcanol Geotherm Res 92:151-167

D'Alessandro W, Giammanco S, Parello F, Valenza M (1997) CO_2 output and $\delta^{13}C(CO_2)$ from Mount Etna as indicators of degassing of shallow asthenosphere. Bull Volcanol 58:455-458

Dasgupta R, Hirschmann MM (2010) The deep carbon cycle and melting in Earth's interior. Earth Planet Sci Lett (Frontiers) 298:1-13

Delgado H, Piedad-Sanchez N, Galvan L, Julio T, Alvarez M, Cardenas L (1998) CO_2 flux measurements at Popocatepetl volcano: II Magnitude of emissions and significance. EOS Trans/Supplement 79, no 10, p F926

Des Marais DJ, Moore JG (1984) Carbon and its isotopes in mid-oceanic basaltic glasses. Earth Planet Sci Lett 69:43-57, doi: 10.1016/0012-821X(84)90073-6

Diaz JA, Pieri D, Arkin CR, Gore E, Griffin TP, Fladeland M, Bland G, Soto C, Madrigal Y, Castillo D, Rojas E, Achi S (2010) Utilization of in situ airborne MS-based instrumentation for the study of gaseous emissions at active volcanoes. Int J Mass Spectrom 295(3):105-112, doi: 10.1016/j.ijms.2010.04.013

Dixon JE, Clague DA (2001) Volatiles in basaltic glasses from Loihi seamount, Hawaii: evidence for a relatively dry plume component. J Petrol 42:627-654

Doukas MP, McGee KA (2007) A compilation of gas emission-rate data from volcanoes of Cook Inlet (Spurr, Crater Peak, Redoubt, Iliamna, and Augustine) and Alaska Peninsula (Douglas, Fourpeaked, Griggs, Mageik, Martin, Peulik, Ukinrek Maars, and Veniaminof), Alaska, from 1995-2006. US Geol Soc Open-File Report 2007-1400

Edmonds M, Aiuppa A, Humphreys M, Moretti R, Giudice G, Martin RS, Herd RA, Christopher T (2010) Excess volatiles supplied by mingling of mafic magma at an andesite arc volcano. Geochem Geophys Geosyst 11:Q04005, doi: 10.1029/2009GC002781

Edmonds M, Herd RA, Galle B, Oppenheimer CM (2003) Automated, high time-resolution measurements of SO_2 flux at Soufriere Hills Volcano, Montserrat. Bull Volcanol 65:578-586, doi: 10.1007/s00445-003-0286-x

Eichelberger JC, Carrigan CR, Westrich HR, Shannon JR (1986) Non-explosive volcanism. Nature 323:598-602, doi: 10.1038/323598a0

Elderfield H, Schultz A (1996) Mid-ocean ridge hydrothermal fluxes and the chemical composition of the ocean. Annu Rev Earth Planet Sci 24:191-224

Etiope G Beneduce P, Calcara M, Favali P, Frugoni F, Schiattarella M, Smriglio G (1999) Structural pattern and CO_2-CH_4 degassing of Ustica Island, Southern Tyrrhenian basin. J Volcanol Geotherm Res 88:291-304

Etiope G, Milkov AV, Derbyshire E (2008) Did geologic emissions of methane play any role in Quaternary climate change? Global Planet Change 61(1-2):79-88, doi:10.1016/j.gloplacha.2007.08.008

Evans WC, Bergfeld D, McGimsey RG, Hunt AG (2009) Diffuse gas emissions at the Ukinrek Maars, Alaska: Implications for magmatic degassing and volcanic monitoring. Appl Geochem 24(4):527-535, doi: 10.1016/j.apgeochem.2008.12.007

Farley KA, Maier-Reimer E, Schlosser P, Broecker WS (1995). Constraints on mantle ^3He fluxes and deep-sea circulation from an oceanic general circulation model. J Geophys Res 100:3829-3839, doi: 10.1029/94JB02913

Favara R, Giammanco S, Inguaggiato S, Pecoraino G (2001) Preliminary estimate of CO_2 output from Pantelleria Island volcano (Sicily, Italy): evidence of active mantle degassing. Appl Geochem 16:883-894

Feulner G (2012) The faint young Sun problem. Rev Geophys 50:RG2006, doi: 10.1029/2011RG000375

Fischer TP (2008) Fluxes of volatiles (H_2O, CO_2, N-2, Cl, F) from arc volcanoes. Geochem J 42(1):21-38

Fischer TP, Giggenbach WF, Sano Y, Williams SN (1998) Fluxes and sources of volatiles discharged from Kudryavy, a subduction zone volcano, Kurile Islands. Earth Planet Sci Lett 160(1-2):81-96, doi: 10.1016/S0012-821X(98)00086-7

Fischer TP, Marty B (2005) Volatile abundances in the sub-arc mantle: insights from volcanic and hydrothermal gas discharges. J Volcanol Geotherm Res 140:205-216

Francis PW, Oppenheimer C, Stevenson D (1993) Endogenous growth of persistently active volcanoes. Nature 366:554-557, doi: 10.1038/366554a0

Friedlingstein P, Houghton RA, Marland G, Hackler J, Boden TA, Conway TJ, Canadell JG, Raupach MR, Ciais P, Le Quéré C (2010) Update on CO_2 emissions. Nat Geosci 3(12):811-812, doi: 10.1038/ngeo1022

Frondini F, Chiodini G, Caliro S, Cardellini C, Granieri D, Ventura G (2004) Diffuse CO_2 degassing at Vesuvio, Italy. Bull Volcanol 66(7):642-651, doi: 10.1007/s00445-004-0346-x

Füri ED, Hilton R, Tryon MD, Brown KM, McMurtry GM, Brückmann W, Wheat CG (2010) Carbon release from submarine seeps at the Costa Rica fore-arc: Implications for the volatile cycle at the Central America convergent margin. Geochem Geophys Geosyst 11:Q04S21, doi: 10.1029/2009GC002810

Gaillardet J, Dupre B, Louvat P, Allegre CJ (1999) Global silicate weathering and CO_2 consumption rates deduced from the chemistry of the large rivers. Chem Geol 159(1-4):3-30

Galle B, Johansson M, Rivera C, Zhang Y, Kihlman M, Kern C, Lehmann T, Platt U, Arellano S, Hidalgo S (2010) Network for Observation of Volcanic and Atmospheric Change (NOVAC)—A global network for volcanic gas monitoring: Network layout and instrument description. J Geophys Res 115:D05304, doi: 10.1029/2009JD011823

Gerlach TM (1989) Degassing of carbon dioxide from basaltic magma at spreading centers: II. Mid-oceanic ridge basalts. J Volcanol Geotherm Res 39:231-232

Gerlach TM (1991) Present-day CO_2 emissions from volcanoes. EOS Trans, Am Geophys Union 72 (23):249-255

Gerlach TM (2011) Volcanic versus anthropogenic carbon dioxide. EOS Trans, 92(24):201-208

Gerlach TM, Delgado H, McGee KA, Doukas MP, Venegas JJ, Cardenas L (1997) Application of the LI-COR CO_2 analyzer to volcanic plumes: A case study, volcan Popocatepetl, Mexico, June 7 and 10, 1995. J Geophys Res Solid Earth 102:8005-8019, doi: 10.1029/96JB03887

Gerlach TM, McGee KA, Elias T, Sutton AJ, Doukas MP (2002) Carbon dioxide emission rate of Kilauea Volcano: Implications for primary magma and the summit reservoir. J Geophys Res Solid Earth 107:2189, doi: 10.1029/2001JB000407

Gerlach TM, McGee KA, Sutton AJ, Elias T (1998) Rates of volcanic CO_2 degassing from airborne determinations of SO_2, emission rates and plume CO_2/SO_2: Test study at Pu'u 'O'o cone, Kilauea volcano, Hawaii. Geophys Res Lett 25(14):2675-2678, doi: 10.1029/98GL02030

Giammanco S, Parello F, Gambardella B, Schifano R, Pizzu S, Galante G (2007) Focused and diffuse effluxes of CO_2 from mud volcanoes and mofettes south of Mt. Etna (Italy). J Volcanol Geotherm Res 165:46-63, doi: 10.1016/j.jvolgeores.2007.04.010

Giggenbach WF, Goguel RL (1989) Methods for the collection and analysis of geothermal and volcanic water and gas samples. Department of Scientific and Industrial Research, Chemistry Division, Report 2401

Goff F, Love SP, Warren RG, Counce D, Obenholzner J, Siebe C, Schmidt SC (2001) Passive infrared remote sensing evidence for large, intermittent CO_2 emissions at Popocatepetl volcano, Mexico. Chem Geol 177(1-2):133-156, doi: 10.1016/S0009-2541(00)00387-9

Graham D, Sarda P (1991) Reply to comment by TM Gerlach on "Mid-ocean ridge popping rocks: implications for degassing at ridge crests". Earth Planet Sci Lett 105: 568-573

Greenland LP, Rose WI, Stokes JB (1985) An estimate of gas emissions and magmatic gas content from Kilauea volcano. Geochim Cosmochim Acta 49(1):125-129, doi: 10.1016/0016-7037(85)90196-6

Hager SA, Gerlach TM, Wallace PJ (2008) Summit CO_2 emission rates by the CO_2/SO_2 ratio method at Kilauea Volcano, Hawai'i, during a period of sustained inflation. J Volcanol Geotherm Res 177(4):875-882, doi: 10.1016/j.jvolgeores.2008.06.033

Halmer MM, Schmincke HU, Graf HF (2002) The annual volcanic gas input into the atmosphere, in particular into the stratosphere: A global data set for the past 100 years. J Volcanol Geotherm Res 115:511-528

Hernández P, Pérez N, Salazar J, Sato M, Notsu K, Wakita H (2000) Soil gas CO_2, CH_4, and H_2 distribution in and around Las Cañadas caldera, Tenerife, Canary Islands, Spain. J Volcanol Geotherm Res 103:425-438

Hernández PA , Pérez NM, Fridriksson T, Egbert J, Ilyinskaya E, Thárhallsson A, Ívarsson G, Gíslason G, Gunnarsson I, Jónsson B, Padrón E, Melián G, Mori T, Notsu K (2012) Diffuse volcanic degassing and thermal energy release from Hengill volcanic system, Iceland. Bull Volcanol 74:2435-2448, doi: 10.1007/s00445-012-0673-2

Hernández PA, Notsu K, Okada H, Mori T, Sato M, Barahona F, Pérez NM (2006) Diffuse emission of CO_2 from Showa-Shinzan, Hokkaido, Japan: A sign of volcanic dome degassing. Pure Appl Geophys 163(4):869-881, doi: 10.1007/s00024-006-0038-x

Hernández PA, Notsu K, Salazar JM, Mori T, Natale G, Okada H, Virgili G, Shimoike Y, Sato M, Pérez NM (2001b) Carbon dioxide degassing by advective flow from Usu volcano, Japan. Science 292:83-86

Hernández PA, Notsu K, Tsurumi M, Mori T, Ohno M, Shimoike Y, Salazar J, Pérez NM (2003) Carbon dioxide emissions from soils at Hakkoda, north Japan. J Geophys Res 108:2210, doi: 10.1029/2002JB001847

Hernández PA, Salazar JM, Shimoike Y, Mori T, Notsu K, Pérez NM (2001a) Diffuse emission of CO_2 from Miyakejima volcano. Japan Chem Geol 177:175-185

Hilton DR, Fischer TP, Marty B (2002) Noble gases and volatile recycling at subduction zones. Rev Mineral Geochem 47:319-370

Holloway JR (1998) Graphite-melt equilibria during mantle melting: constraints on CO_2 in MORB magmas and the carbon content of the mantle. Chem Geol 147:89-97

Hornafius JS, Quigley D, Luyendyk BP (1999) The world's most spectacular marine hydrocarbon seeps (Coal Oil Point, Santa Barbara Channel, California): quantification of emissions. J Geophys Res 104:20703-20711

Inguaggiato S, Mazot A, Diliberto IS, Inguaggiato C, Madonia P, Rouwet D, Vita F (2012) Total CO_2 output from Vulcano island (Aeolian Islands, Italy). Geochem Geophys Geosyst 13:Q02012, doi: 10.1029/2011GC003920

Irwin WP, Barnes I (1980) Tectonic relations of carbon dioxide discharges and earthquakes. J Geophys Res 85:3115-3121, doi: 10.1029/JB085iB06p03115

Italiano F, Martelli M, Martinelli G, Nuccio PM (2000) Geochemical evidence of melt intrusions along lithospheric faults of the Southern Apennines, Italy: geodynamic and seismogenic implications. J Geophys Res 105:13569-13578

James ER, Manga M, Rose TP (1999) CO_2 degassing in the Oregon cascades. Geology 27(9):823-826, doi: 10.1130/0091-7613(1999)027<0823:CDITOC>2.3.CO;2

Jaupart C, Vergniolle S (1988) Laboratory models of Hawaiian and Strombolian eruptions. Nature 331:58-60, doi: 10.1038/331058a0

Javoy M, Pineau F (1991) The volatiles record of a bpoppingQ rock from the Mid-Atlantic ridge at 148N: chemical and isotopic composition of a gas trapped in the vesicles. Earth Planet Sci Lett 107:598-611

Kanemasu ET, Powers WL, Sij JW (1974) Field chamber measurements of CO_2 flux from soil surface. Soil Sci 118(4):233-237, doi: 10.1097/00010694-197410000-00001

Kazahaya K, Shinohara H, Saito G (1994) Excessive degassing of Izu-Oshima volcano: Magma convection in a conduit. Bull Volcanol 56:207-216, doi: 10.1007/BF00279605

Kelley DS, Lilley MD, Fruh-Green GL (2004) Volatiles in submarine environments: Food for life. *In:* The Subseafloor Biosphere at Mid-Ocean Ridges. Geophys Monogr Ser 144, Wilcock W, DeLong EF, Kelley DS, Baross JA, Craig Cary S (eds) AGU, Washington, D.C., p 167-190

Kern C, Deutschmann T, Werner C, Sutton AJ, Elias T, Kelly PJ (2012) Improving the accuracy of SO_2 column densities and emission rates obtained from upward-looking UV-spectroscopic measurements of volcanic plumes by taking realistic radiative transfer into account. J Geophys Res 117:D20302, doi: 10.1029/2012JD017936

Kerrick DM (2001) Present and past non-anthropogenic CO_2 degassing from the solid earth. Rev Geophys 39(4):565-585

Kerrick DM, McKibben MA, Seward TM, Caldeira K (1995) Convective hydrothermal CO_2 emission from high heat flow regions. Chem Geol 121:285-293

Koepenick KW, Brantley SL, Thompson JM, Rowe GL, Nyblade AA, Moshy C (1996) Volatile emissions from the crater and flank of Oldoinyo Lengai, Tanzania. J Geophys Res 101:13819-13830

Kucera CL, Kirkham DR (1971) Soil respiration studies in tall grass prairie in Missouri. Ecology 52(5):912-915, doi: 10.2307/1936043

La Spina A, Burton M, Harig R, Murè F, Rausch P, Jordan M, Caltabiano T (2013) New insights into volcanic processes at Stromboli from Cerberus, a remote-controlled open-path FTIR scanner. J Volcanol Geotherm Res 249:66-76

La Spina A, Burton M, Salerno GG (2010) Unravelling the processes controlling gas emissions from the central and northeast craters of Mt. Etna. J Volcanol Geotherm Res 198(3-4):368-376

Le Guern F (1987) Mechanism of energy transfer in the lava lake of Niragongo (Zaire), 1959-1977. J Volcanol Geotherm Res 31(1-2):17-31, doi: 10.1016/0377-0273(87)90003-5

Le Guern F, Carbonnelle J, Tazieff H (1979) Erta 'Ale lava lake: heat and gas transfer to the atmosphere. J Volcanol Geotherm Res 6:27–48

Lewicki JL, Fischer ML, Hilley GE (2008) Six-week time series of eddy covariance CO_2 flux at Mammoth Mountain, California: Performance evaluation and role of meteorological forcing. J Volcanol Geotherm Res 171(3-4):178-190, doi: 10.1016/j.jvolgeores.2007.11.029

Lopez T, Ushakov S, Izbekov P, Tassi F, Cahill C, Neill O, Werner C (in press) Constraints on magma processes, subsurface conditions, and total volatile flux at Bezymianny Volcano in 2007–2010 from direct and remote volcanic gas measurements. J Volcanol Geotherm Res, doi: 10.1016/j.jvolgeores.2012.10.015

Lupton J, Lilley M, Butterfield D, Evans L, Embley R, Massoth G, Christenson B, Nakamura K, Schmidt M (2008) Venting of a separate CO_2-rich gas phase from submarine arc volcanoes: Examples from the Mariana and Tonga-Kermadec arcs. J Geophys Res 113:B08S12, doi: 10.1029/2007JB005467

Marrero R, López DL, Hernández PA , Pérez NM (2008) Carbon dioxide discharged through the Las Cañadas Aquifer, Tenerife, Canary Islands. Pure Appl Geophys 165:147-172, doi: 10.1007/s00024-007-0287-3

Martin RS, Sawyer GM, Spampinato L, Salerno GG, Ramirez C, Ilyinskaya E, Witt MLI, Mather TA, Watson IM, Phillips JC, Oppenheimer C (2010) A total volatile inventory for Masaya Volcano, Nicaragua. J Geophys Res Solid Earth 115:B09215, doi: 10.1029/2010JB007480

Marty B, Jambon A (1987) $C/^3He$ in volatile fluxes from the solid earth: implications for carbon geodynamics. Earth Planet Sci Lett 83:16-26

Marty B, Le Cloarec MF (1992) Helium-3 and CO_2 fluxes from subaerial volcanoes estimated from polonium-210 emissions. J Volcanol Geotherm Res 53:67-72

Marty B, Tolstikhin IN (1998) CO_2 fluxes from mid-ocean ridges, arcs and plumes. Chem Geol 145:233-248

Marty B, Zimmerman L (1999) Volatiles (He, C, N, Ar) in mid-ocean ridge basalts: assessment of shallow-level fractionation and characterization of source composition. Geochim Cosmochim Acta 63(21):3619-3633, doi: 10.1016/S0016-7037(99)00169-6

Massoth GJ, Milburn HB, Hammond SR, Butterfield DA, McDuff RE, Lupton JE (1989) The geochemistry of submarine venting fluids at Axial Volcano, Juan de Fuca Ridge: New sampling methods and a VENTS program rationale. *In:* Global Venting, Midwater, and Benthic Ecological Processes. National Undersea Research Program, Research Report 88-4. De Luca MP, Babb I (eds) US Department of Commerce: National Oceanic and Atmospheric Administration, p 29-59

Matthews SJ, Jones AP, Bristow CS (1992) A simple magma-mixing model for sulfur behavior in calc-alkaline volcanic rocks: mineralogical evidence from Mount Pinatubo 1991 eruption. J Geol Soc London 149: 863-866

Mazot A, Taran Y (2009) CO_2 flux from the volcanic lake of El Chichón (Mexico): Geofís Int 48(1):73-83

McGee KA, Doukas MP, Gerlach TM (2001) Quiescent hydrogen sulfide and carbon dioxide degassing from Mount Baker, Washington. Geophys Res Lett 28(23):4479-4482, doi: 10.1029/2001GL013250

McGonigle AJS, Aiuppa A, Giudice G, Tamburello G, Hodson AJ, Gurrieri S (2008) Unmanned aerial vehicle measurements of volcanic carbon dioxide fluxes. Geophys Res Lett 35(6):L06303, doi: 10.1029/2007GL032508

Menand T, Phillips HC (2007) Gas segregation in dykes and sills. J Volcanol Geotherm Res 159(4):393-408, doi: 10.1016/j.jvolgeores.2006.08.003

Merucci L, Burton M, Corradini S, Salerno GG (2011) Reconstruction of SO_2 flux emission chronology from space-based measurements. J Volcanol Geotherm Res 206(3-4):80-87, doi: 10.1016/j.jvolgeores.2011.07.002

Métrich N, Allard P, Aiuppa A, Bani P, Bertagnini A, Shinohara H, Parello F, Di Muro A, Garaebiti E, Belhadj O, Massare D (2011) Magma and volatile supply to post-collapse volcanism and block resurgence in Siwi Caldera (Tanna Island, Vanuatu Arc). J Petrol 52(6):1077-1105, doi: 10.1093/petrology/egr019

Mori T, Burton MR (2006) The SO_2 camera: A simple, fast and cheap method for ground-based imaging of SO_2 in volcanic plumes. Geophys Res Lett 33(24):L24804, doi: 10.1029/2006GL027916

Mori T, Hernández PA, Salazar JML, Pérez NM, Notsu K (2001) An in situ method for measuring CO_2 flux from volcanic-hydrothermal fumaroles. Chem Geol 177(1-2):85-99, doi: 10.1016/S0009-2541(00)00384-3

Mori T, Notsu K (1997) Remote CO, COS, CO_2, SO_2, HCl detection and temperature estimation of volcanic gas. Geophys Res Lett 24(16):2047-2050, doi: 10.1029/97GL52058

Mörner NA, Etiope G (2002) Carbon degassing from the lithosphere. Global Planet Change 33:185-203

Naughton JJ, Derby JV, Glover RB (1969) Infrared measurements on volcanic gas and fume: Kilauea eruption, 1968. J Geophys Res 74:3273-3277

Oppenheimer C, Kyle PR (2008) Probing the magma plumbing of Erebus volcano, Antarctica, by open-path FTIR spectroscopy of gas emissions. J Volcanol Geotherm Res 177(3):743-754, doi: 10.1016/j.jvolgeores.2007.08.022

Oppenheimer C, Scaillet B, Martin RS (2011) Sulfur degassing from volcanoes: source conditions, surveillance, plume chemistry and earth system impacts. *In:* Sulfur in magmas and melts: its importance for natural and technical processes. Rev Mineral Geochem 73:363-422, doi: 10.2138/rmg.2011.73.13

Óskarsson F, Fridriksson T (2011) Reykjanes production field. Geochemical monitoring in 2010. ÍSOR Report 2011/050, Iceland GeoSurvey, Reykjavík, p 51

Padron E, Hernández PA, Pérez NM, Toulkeridis T, Melian G, Barrancos J, Virgili G, Sumino H, Notsu K (2012) Fumarole/plume and diffuse CO_2 emission from Sierra Negra caldera, Galapagos archipelago. Bull Volcanol 74(6):1509-1519, doi: 10.1007/s00445-012-0610-4

Pallister JS, Hoblitt RP, Reyes AG (1992) A basalt trigger for the 1991 eruptions of Pinatubo volcano. Nature 356(6368):426-428, doi: 10.1038/356426a0

Parkinson KJ (1981) An improved method for measuring soil respiration in the field. J Appl Ecol 18:221-228

Pecoraino G, Brusca L, D'Alessandro W, Giammanco S, Inguaggiato S, Longo M (2005) Total CO_2 output from Ischia Island volcano (Italy). Geochem J 39:451-458

Pérez N, Melian G, Salazar J, Saballos A, Alvarez J, Segura F, Hernández P, Notsu K (2000) Diffuse degassing of CO2 from Masaya Caldera, Nicaragua, Central America. EOS Trans, Am Geophys Union 81(48), Fall Meet Suppl, 2000

Pérez NM et al. (2011) Global CO_2 emission from volcanic lakes. Geology 39:235-238, doi: 10.1130/G31586.1

Pérez NM, Nakai S, Wakita H, Notsu K, Talai B (1998) Anomalous diffuse degassing of Helium-3 and CO_2 related to the active ring-fault structure at Rabaul Caldera, Papua New Guinea. Abstract AGU Fall Meet, V12B-11

Pili E, Kennedy BM, Conrad ME, Gratier J-P (2011) Isotopic evidence for the infiltration of mantle and metamorphic CO_2-H_2O fluids from below in faulted rocks from the San Andreas Fault system. Chem Geol 281(3-4):242-252, doi: 10.1016/j.chemgeo.2010.12.011

Reiners WA (1968) Carbon dioxide evolution from the floor of three Minnesota forests. Ecology 49:471-483

Resing JA, Lupton JE, Feely RA, Lilley MD (2004) CO_2 and ^3He in hydrothermal plumes: implications for mid-ocean ridge CO_2 flux. Earth Planet Sci Lett 226:449-464

Roberts TJ, Braban CF, Oppenheimer C, Martin RS, Freshwater RA, Dawson DH, Griffiths PT, Cox RA, Saffell JR, Jones RL (2012) Electrochemical sensing of volcanic gases, Chem Geol 332-333:74-91, doi: 10.1016/j.chemgeo.2012.08.027

Rogie JD, Kerrick DM, Chiodini G, Frondini F (2000) Flux measurements of nonvolcanic CO_2 emission from some vents in central Italy. J Geophys Res Solid Earth 105:8425-8445, doi: 10.1029/1999JB900430

Rose TP, Davisson ML (1996) Radiocarbon in hydrologic systems containing dissolved magmatic carbon dioxide. Science 273(5280):1367-1370, doi: 10.1126/science.273.5280.1367

Rose WI, Chuan RL, Giggenbach WF, Kyle PR, Symonds RB (1986) Rates of sulfur dioxide and particle emissions from White Island volcano, New Zealand, and an estimate of the total flux of major gaseous species. Bull Volcanol 48(4), doi: 181-188

Saal AE, Hauri EH, Langmuir CH, Perfit MR (2002) Vapour undersaturation in primitive mid-ocean-ridge basalt and the volatile content of earth's upper mantle. Nature 419:451-455

Salazar JML, Hernández PA, Pérez NM, Melian G, Alvarez J, Segura F, Notsu K (2001) Diffuse emission of carbon dioxide from Cerro Negro volcano, Nicaragua, Central America. Geophys Res Lett 28(22):4275-4278, doi: 10.1029/2001GL013709

Salazar JML, Hernández PA, Pérez NM, Olmos R, Barahona F, Cartagena R, Soriano T, Lopez DL, Sumino H, Notsu K (2004) Spatial and temporal variations of diffuse CO_2 degassing at the Santa Ana–Izalco–Coatepeque volcanic complex, El Salvador, Central America. Geol Soc Am Special Papers 375:135-146, doi: 10.1130/0-8137-2375-2.135

Sano Y, Williams SN (1996) Fluxes of mantle and subducted carbon along convergent plate boundaries. Geophys Res Lett 23(20):2749-2752

Sansone FJ, Mottl MJ, Olson EJ, Wheat CG, Lilley MD (1998) CO_2-depleted fluids from mid-ocean ridge-flank hydrothermal springs. Geochim Cosmochim Acta 62(13):2247-2252, doi: 10.1016/S0016-7037(98)00135-5

Sarda SP, Graham D (1990) Mid-ocean ridge popping rocks: implications for degassing at ridge crests. Earth Planet Sci Lett 97:268-289

Sawyer GM, Carn SA, Tsanev VI, Oppenheimer C, Burton MR (2008a) Investigation into magma degassing at Nyiragongo volcano, Democratic Republic of the Congo. Geochem Geophys Geosyst 9:Q02017, doi: 10.1029/2007GC001829

Sawyer GM, Oppenheimer C, Tsanev VI, Yirgu G (2008b) Magmatic degassing at Erta 'Ale volcano, Ethiopia. J Volcanol Geotherm Res 178:837-846, doi: 10.1016/j.jvolgeores.2008.09.017

Sawyer GM, Salerno GG, Le Blond J, Martin RS, Spampinato L, Roberts T, Tsanev VI, Mather TA, Witt MLI, Oppenheimer C (2011) Gas and particle emissions from Villarrica Volcano, Chile. J Volcanol Geotherm Res 203(1-2):62-75, doi: 10.1016/j.jvolgeores.2011.04.003

Self S (2006) The effects and consequences of very large explosive eruptions. Philos Trans R Soc A 364:2073-2097, doi: 10.1098/rsta.2006.1814

Seward TM, Kerrick DM (1996) Hydrothermal CO_2 emission from the Taupo Volcanic Zone, New Zealand. Earth Planet Sci Lett 139(1-2):105-113, doi: 10.1016/0012-821X(96)00011-8

Shimoike Y, Kazahaya K, Shinohara H (2002) Soil gas emission of volcanic CO_2 at Satsuma-Iwojima volcano, Japan. Earth Planet Sci Lett 54(3):239-247

Shinohara H (2005) A new technique to estimate volcanic gas composition: plume measurements with a portable multi-sensor system. J Volcanol Geotherm Res 143(4):319-333

Shinohara H, Kazahaya K, Saito G, Fukui K, Odai M (2003) Variation of CO_2/SO_2 ratio in volcanic plumes of Miyakejima: Stable degassing deduced from heliborne measurements. Geophys Res Lett 30(5):1208, doi: 10.1029/2002GL016105

Siebert L, Simkin T (2002) Volcanoes of the world: an illustrated catalogue of Holocene volcanoes and their eruptions. Smithsonian Institution, Global Volcanism Program Digital Information Series, GVP-3, (*http://www.volcano.si.edu/world/*)

Smith TEL, Wooster MJ, Tattaris M, Griffith DWT (2011) Absolute accuracy and sensitivity analysis of OP-FTIR retrievals of CO_2, CH_4 and CO over concentrations representative of "clean air" and "polluted plumes". Atmos Meas Tech 4:97-116, doi: 10.5194/amt-4-97-2011

Sorey ML, Evans WC, Kennedy BM, Farrar CD, Hainsworth LJ, Hausback B (1998) Carbon dioxide and helium emissions from a reservoir of magmatic gas beneath Mammoth Mountain, California. J Geopys Res Solid Earth 103:15,503-15,323, doi: 10.1029/98JB01389

Spilliaert N, Allard P, Métrich N, Sobolev AV (2006) Melt inclusion record of the conditions of ascent, degassing, and extrusion of volatile-rich alkali basalt during the powerful 2002 flank eruption of Mount Etna (Italy). J Geophys Res 111:B04203, doi: 10.1029/2005JB003934

Spinetti C, Carrere V, Buongiorno MF, Sutton AJ, Elias T (2008) Carbon dioxide of Pu'u'O'o volcanic plume at Kilauea retrieved by AVIRIS hyperspectral data. Remote Sens Environ 112(6):3192-3199, doi: 10.1016/j.rse.2008.03.010

Symonds RB, Reed MH, Rose WI (1992) Origin, speciation, and fluxes of trace-element gases at Augustine volcano, Alaska – Insights into magma degassing and fumarolic processes. Geochim Cosmochim Acta 56(2):633-657, doi: 10.1016/0016-7037(92)90087-Y

Tonani F, Miele G (1991) Methods for measuring flow of carbon dioxide through soil in volcanic setting. *In:* International Conference on Active Volcanoes and Risk Mitigation, IAVCEI, Naples, Italy, 27 Aug. to 1 Sept.

Toutain J-P, Sortino F, Baubron J-C, Richon P, Surono, Sumarti S, Nonell A (2009) Structure and CO_2 budget of Merapi volcano during inter-eruptive periods. Bull Volcanol 71(7):815-826, doi: 10.1007/s00445-009-0266-x

Troll VR, Hilton DR, Jolis EM, Chadwick JP, Blythe LS, Deegan FM, Schwarzkopf LM, Zimmer M (2012) Crustal CO_2 liberation during the 2006 eruption and earthquake events at Merapi volcano, Indonesia. Geophys Res Lett 39:L11302, doi: 10.1029/2012GL051307

Varekamp JC, Kreulen R, Poorter RPE, Van Bergen MJ (1992) Carbon sources in arc volcanism, with implications for the carbon cycle. Terra Nova 4(3)363-373, doi: 10.1111/j.1365-3121.1992.tb00825.x

Varley NR, Armienta MA (2001) The absence of diffuse degassing at Popocatépetl volcano, Mexico. Chem Geol 177:157-173

Von Damm KL, Edmond JM, Grant G, Measures CI, Walden B, Weiss RF (1985) Chemistry of submarine hydrothermal solutions at 21°N, East Pacific Rise. Geochim Cosmochim Acta 49:2197-2220

Walker JCG, Hays PB, Kasting JF (1981) A negative feedback mechanism for the long-term stabilization of Earth's surface temperature. J Geophys Res 86:9776-9782, doi: 10.1029/JC086iC10p09776

Wallace PJ (2005) Volatiles in subduction zone magmas: concentrations and fluxes based on melt inclusion and volcanic gas data. J Volcanol Geotherm Res 140(1-3):217-240, doi: 10.1016/j.jvolgeores.2004.07.023

Wallace PJ, Gerlach TM (1994) Magmatic vapor source for sulfur-dioxide released during volcanic-eruptions - evidence from Mount Pinatubo. Science 265(5171):497-499

Wang WC, Yung YL, Lacis AA, Mo T, Hansen JE (1976) Greenhouse effects due to man-made perturbations of trace gases. Science 194(4266): 685-690

Wardell LJ, Kyle PR, Campbell AR (2003) Carbon dioxide emissions from fumarolic ice towers, Mount Erebus volcano, Antarctica. *In:* Volcanic Degassing. Oppenheimer C, Pyle DM, Barclay J (eds) Geol Soc London, Special Publications 213:213-246

Wardell LJ, Kyle PR, Chaffin C (2004) Carbon dioxide and carbon monoxide emission rates from an alkaline intra-plate volcano: Mt. Erebus, Antarctica. J Volcanol Geotherm Res 131(1-2):109-121, doi: 10.1016/S0377-0273(03)00320-2

Wardell LJ, Kyle PR, Dunbar N, Christenson B (2001) White Island volcano, New Zealand: carbon dioxide and sulfur dioxide emission rates and melt inclusion studies. Chem Geol 177(1-2):187-200, doi: 10.1016/S0009-2541(00)00391-0

Watts SF (2000) The mass budgets of carbonyl sulfide, dimethyl sulfide, carbon disulfide and hydrogen sulfide. Atmos Environ 34:761-779

Werner C, Brantley S (2003) CO_2 emissions from the Yellowstone volcanic system. Geochem Geophys Geosyst 4:1061, doi: 10.1029/2002GC000473

Werner C, Evans WC, Poland M, Tucker DS, Doukas MP (2009) Long-term changes in quiescent degassing at Mount Baker Volcano, Washington, USA; Evidence for a stalled intrusion in 1975 and connection to a deep magma source. J Volcanol Geotherm Res 186:379-386, doi: 10.1016/j.jvolgeores.2009.07.006

Werner C, Hurst T, Scott B, Sherburn S, Christenson BW, Britten K, Cole-Baker J, Mullan B (2008) Variability of passive gas emissions, seismicity, and deformation during crater lake growth at White Island Volcano, New Zealand, 2002–2006. J Geophys Res 113:B01204, doi: 10.1029/2007JB005094

Werner C, Wyngaard JC, Brantley SL (2000) Eddy correlation measurement of hydrothermal gases. Geophys Res Lett 27(18):2925-2928, doi: 10.1029/2000GL011765

Williams SN, Schaefer SJ, Calvache ML, Lopez D (1992) Global carbon dioxide emission to the atmosphere by volcanoes. Geochim Cosmochim Acta 56(4):1765-1770, doi: 10.1016/0016-7037(92)90243-C

Zapata GJA, Calvache VML, Cortés JGP, Fischer TP, Garzon VG, Gómez MD, Narvaez ML, Ordónez VM, Ortega EA, Stix J, Torres CR, Williams SN (1997) SO_2 fluxes from Galeras volcano, Colombia, 1989-1995: progressive degassing and conduit obstruction of a Decade Volcano. J Volcanol Geotherm Res 77:195-208

Zeebe RE, Caldeira K (2008) Close mass balance of long-term carbon fluxes from ice-core CO_2 and ocean chemistry records. Nature Geosci 1:312-315

Reviews in Mineralogy & Geochemistry
Vol. 75 pp. 355-421, 2013

12

Diamonds and the Geology of Mantle Carbon

Steven B. Shirey

*Department of Terrestrial Magnetism, Carnegie Institution of Washington
5241 Broad Branch Road, NW, Washington, DC 20015, U.S.A.*

shirey@dtm.ciw.edu

Pierre Cartigny

*Laboratoire de Géochimie des Isotopes Stables de l'Institut de Physique du Globe de Paris, UMR 7154, Université Paris Denis-Diderot,
PRES Sorbonne Paris-Cité, Office n°511, 1 rue Jussieu,75005 Paris, France*

Daniel J. Frost

Bayerisches Geoinstitut, Universität Bayreuth, D-95440 Bayreuth, Germany

Shantanu Keshav

*Geosciences Montpellier, University of Montpellier 2
CNRS & UMR 5243, Montpellier, France*

Fabrizio Nestola, Paolo Nimis

Department of Geosciences, University of Padua, Via Gradenigo 6, 1-35131 Padova, Italy

D. Graham Pearson

*Department of Earth and Atmospheric Sciences, University of Alberta,
1-26 Earth Sciences Building, Edmonton, Alberta, Canada T6G 2E3*

Nikolai V. Sobolev

*V.S. Sobolev Institute of Geology and Mineralogy, Siberian Branch of
Russian Academy of Sciences, 630090 Novosibirsk 90, Russia*

Michael J. Walter

School of Earth Sciences, University of Bristol, Bristol BS8 1RJ, United Kingdom

INTRODUCTION TO DIAMOND CHARACTERISTICS

Introduction

Earth's carbon, derived from planetesimals in the 1 AU region during accretion of the Solar System, still retains similarities to carbon found in meteorites (Marty et al. 2013) even after 4.57 billion years of geological processing. The range in isotopic composition of carbon on Earth versus meteorites is nearly identical and, for both, diamond is a common, if volumetrically minor, carbon mineral (Haggerty 1999). Diamond is one of the three native carbon minerals on Earth (the other two being graphite and lonsdaleite). It can crystallize throughout the mantle below about 150 km and can occur metastably in the crust. Diamond is a rare mineral, occurring at the part-per-billion level even within the most diamondiferous volcanic host rock although some

1529-6466/13/0075-0012$00.00 DOI: 10.2138/rmg.2013.75.12

rare eclogites have been known to contain 10-15% diamond. As a trace mineral it is unevenly distributed and, except for occurrences in metamorphosed crustal rocks, it is a xenocrystic phase within the series of volcanic rocks (kimberlites, lamproites, ultramafic lamprohyres), which bring it to the surface and host it. The occurrence of diamond on Earth's surface results from its unique resistance to alteration/dissolution and the sometimes accidental circumstances of its sampling by the volcanic host rock. Diamonds are usually the chief minerals left from their depth of formation, because intact diamondiferous mantle xenoliths are rare.

Diamond has been intensively studied over the last 40 years to provide extraordinary information on our planet's interior. For example, from the study of its inclusions, diamond is recognized as the only material sampling the "very deep" mantle to depths exceeding 800 km (Harte et al. 1999; McCammon 2001; Stachel and Harris 2009; Harte 2010) although most crystals (~95%) derive from shallower depths (150 to 250 km). Diamonds are less useful in determining carbon fluxes on Earth because they provide only a small, highly variably distributed sample that is usually not directly related to the host magma. One major achievement in our understanding of diamond formation both in Earth's mantle and metamorphic rocks is the increasing evidence for its formation from a mobile C-bearing phase, commonly referred to as "C-O-H-bearing fluid or melt". *These free fluids give diamond the remarkable ability to track carbon mobility in the deep mantle, as well as mantle mineralogy and mantle redox state and hence a unique ability to follow the path and history of the carbon from which diamond is composed. Thus, diamond truly occupies a unique position in any discussion of the igneous and metamorphic aspects of Earth's carbon cycle.* Beyond simply providing deep samples, diamond studies have revealed active geodynamics. These studies have pinpointed the initiation of subduction (Shirey and Richardson 2011), tracked the transfer of material through the mantle transition zone (Stachel et al. 2005; Walter et al. 2011), recorded the timing of ingress of fluids to the continental lithosphere (Richardson et al. 1984; Pearson et al. 1998; Shirey et al. 2004b), preserved carbonatitic fluids that trigger deep mantle melting (Schrauder and Navon 1994; Walter et al. 2008; Klein-BenDavid et al. 2009; Klein-BenDavid et al. 2010; Kopylova et al. 2010), captured the redox state of the mantle (McCammon et al. 2004; Stagno and Frost 2010; Rohrbach and Schmidt 2011), and provided samples of primordial noble gases (Wada and Matsuda 1998; Ozima and Igarashi 2000).

The present chapter does not attempt to review all aspects of diamond studies—for this an entire volume of *Reviews in Mineralogy and Geochemistry* would be required. For summaries on the various aspects of diamond mineralogy, geochemistry, and formation, the reader is referred to works by Harris and coworkers (1968, 1979, 1992), Sobolev and others (1977, 1990), Deines (1980), Gurney and others (2010), Pearson and coworkers (1999, 2003), Stachel et al. (2005), Cartigny (2005), Spetsius and Taylor (2008), Stachel and Harris (2008), Harte (2010), and Tappert and Tappert (2011). *Rather, this chapter will review the key observations and the current state of research on naturally occurring diamonds using modern methods of analysis as they apply to diamond formation, the source of carbon-bearing species in the mantle, the role of carbon during mantle melting, and the geologic history of the mantle with emphasis on the main difficulties to be unlocked in future studies.*

Types of diamond. For a cubic mineral of simple composition, diamond displays a remarkable range of properties. Diamond also displays a variety of shapes reflecting growth under variable conditions of supersaturation and resorption. These characteristics have been covered in recent popular books and articles on diamond (e.g., Harlow 1998; Harlow and Davies 2005; Spetsius and Taylor 2008; Tappert and Tappert 2011). For the geological purposes of tracing the history of C-bearing fluids it is important to consider the three main forms in which diamond occurs: polycrystalline, monocrystalline, and coated (Fig. 1). Polycrystalline diamond includes many subtypes of mantle-derived diamonds (e.g., framesite, bort, ballas), some of unknown origin (e.g., carbonado), and some impact diamonds (e.g., yakutite). Most

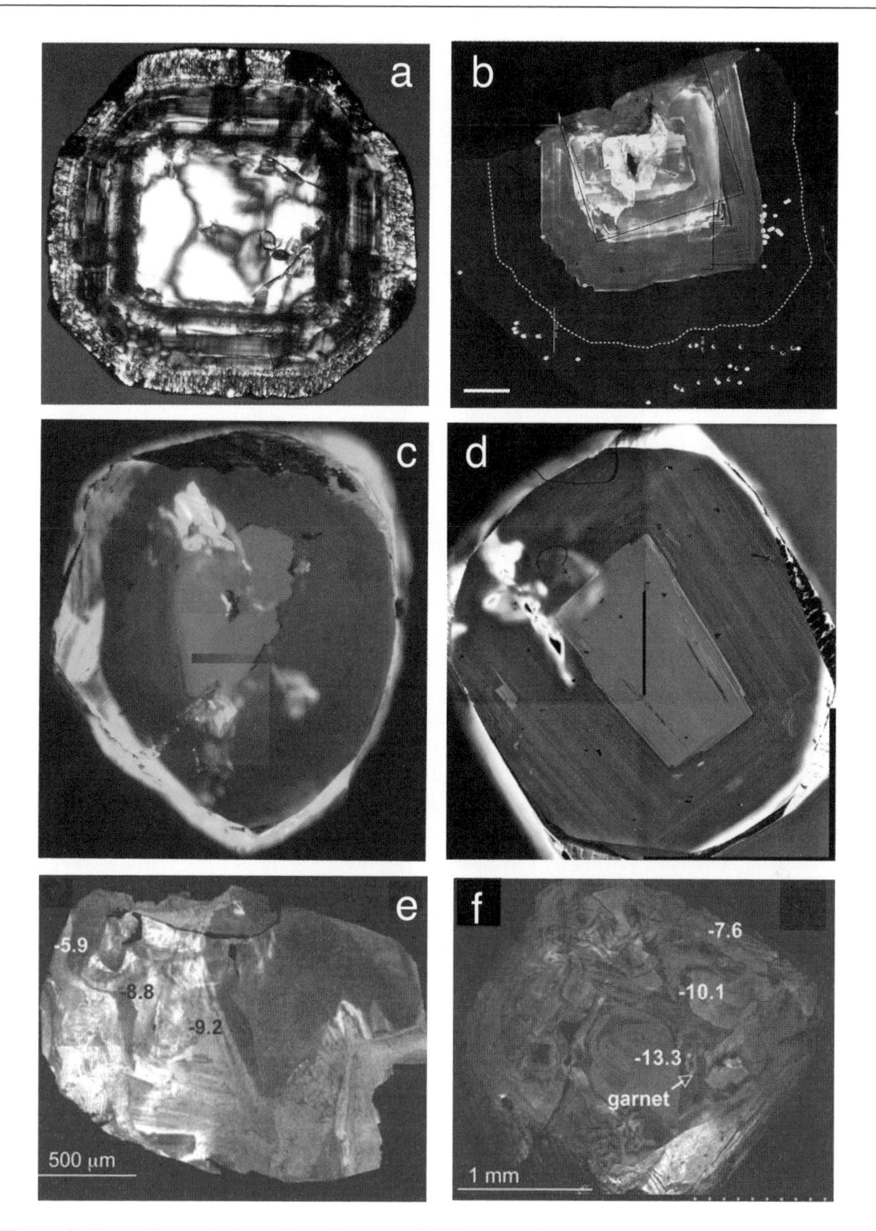

Figure 1. These diamond plates show the textural differences that can occur between coated diamonds (a,b) monocrystalline lithospheric diamonds (c,d), and monocrystalline sub-lithospheric diamonds (e,f). Coated diamond in (a) is an optical photomicrograph, plane light courtesy Ofra Klein-Bendavid. Diameter of diamond is 1 cm. Coated diamond (b) from the Congo (alluvial) is a catholuminescence (CL) image (Used by permission of Elsevier Limited, from Kopylova et al. (2010) *Earth and Planetary Science Letters*, Vol. 291, Fig. 1, p. 128). Dots indicate positions of analyzed inclusions, scale bar is 0.6 mm. Diamonds in (c) and (d) are from Orapa, Botswana. They are both about 6 mm across. The color CL images show multiple growth histories with significant resorption in (c) after a first stage of growth. Note the very thin growth rings in (d). Diamonds (e) and (f) are gray-scale CL images of sub-lithospheric diamonds from the Collier 4 kimberlite pipe, Juina field, Brazil (Used by permission of Springer, from Bulanova et al. (2010) *Contributions to Mineralogy and Petrology*, Vol. 160, Figs. 3d,e, p. 493). Diamond in (e) is about 3 mm on the long axis. Note the irregular zoning in both diamonds.

studied mantle-derived polycrystalline diamonds fall into two main categories: 1) framboids of diamond crystals sintered together without silicates (bort); and 2) diamondite, fine to medium-grained rocks composed of subequal amounts of silicate minerals (typically garnet and pyroxene but lacking olivine) and diamond. Monocrystalline diamond is micro (<0.5 mm) to macro (>0.5 mm) single-crystal diamond from which gem diamonds are cut and polished. Often these monocrystals display a complicated internal growth history with episodes of resorption and regrowth (Fig. 1) as well as simple composite forms (e.g., twin forms such as macles and intergrowths). Monocrystalline diamonds have been an important source of inclusions for study and the ages of these diamonds have been determined via geochronology on their inclusions as Proterozoic to Archean. Coated diamonds are a special case of of mixed polycrystalline-monocrystalline diamond, where monocrystals have been overgrown by a thick, cloudy, polycrystalline coat laden with microinclusions of fluid. If the coat is composed of rods or blades of diamond (Fig. 1) it will exhibit a fibrous structure and be termed a fibrous diamond. These coats are believed to grow during transport in the kimberlite and thus represent young, new diamond growth surrounding often ancient diamond (e.g., Boyd et al. 1994). More research to date has been on monocrystals because they are more available for study, are less complicated morphologically, and are the most robust hosts for mineral inclusions.

Diamond parental and host rocks. Diamonds are chiefly carried to Earth's surface in only three rare types of magmas: kimberlite, lamproite, and lamprophyre (e.g., Gurney et al. 2010). Of the three types, kimberlites are by far the most important, with several thousand known, of which some 30% are diamondiferous. A similar percentage of the several hundred known lamproites is diamondiferous and diamonds are occasionally recorded from ultramafic lamprophyres. Lamproites are next in importance to kimberlites because they host the world's largest diamond mine, Argyle (Australia), and notable diamond occurrences in the United States and India. Lamprophyres currently are only of petrological interest as hosting the oldest known diamonds, which occur in Wawa, Ontario. In general, these magma types are derived by small amounts of melting deep within the mantle, are relatively volatile (H_2O, CO_2, F, or Cl) and MgO-rich, erupt rapidly, and are not oxidizing. In nearly all cases of magmatically-hosted diamonds, Archean to Proterozoic diamonds are carried by Phanerozoic to younger (Cretaceous/Tertiary) kimberlitic volcanic rocks (Pearson and Shirey 1999; Gurney et al. 2010). The composition of kimberlitic magmas can vary widely depending on the relative proportions disaggregated mantle xenoliths, phenocryst phases such as olivine, assimilated country rock, the ratio of H_2O to CO_2 in the volatile phase, and the extent of interaction with metasomatic minerals in the subcontinental lithospheric mantle. Of great importance for diamond petrogenesis the distinction between Group I (GI) and Group II (GII) kimberlites as diamond carriers. Generally, GI kimberlites contain normal (i.e., non-metasomatic) mantle minerals and initial isotopic compositions for Sr, Nd, Hf, and Pb that are indicative of equilibration chiefly with the convecting mantle, whereas GII kimberlites contain micaceous and metasomatic minerals and extreme isotopic compositions for Sr, Nd, Hf, and Pb that are indicative of equilibration with the metasomatized subcontinental lithospheric mantle (e.g., Smith 1983).

Within the mantle, eclogite and peridotite are the main parental rocks of diamonds, as the loose monocrystalline diamonds seen in kimberlite are considered to have been released from eclogitic or peridotitic hosts by alteration and mechanical disaggregation during sampling by the kimberlitic magma in the lithosphere or early transport by the kimberlite (Kirkley et al. 1991; Harlow 1998). Both diamondiferous and diamond-free eclogites often survive transport by kimberlite, whereas diamondiferous peridotite is exceptionally rare and nearly all peridotite xenoliths are diamond-free. The physical distribution of diamond in eclogite has been studied recently by CAT-scan techniques, which have revealed that diamonds are often found in between the major silicate phases in their host eclogite and often along pathways where metasomatic fluids traveled (Keller et al. 1999; Anand et al. 2004). Presumably diamond in peridotite has a similar textural relationship to its major silicates; however, the ready reaction of CO_2-rich diamond-forming fluids with abundant magnesian silicates to form friable magnesite ($MgCO_3$) along silicate grain boundaries promotes disaggregation of the xenoliths and release of the diamonds, thus destroying the textural relationship with their host.

In the crust, diamonds are found directly within their host lithologies, which have been exhumed by the orogenic process of continental collision (Ogasawara 2005; Dobrzhinetskaya 2012; Schertl and Sobolev 2012). These occurrences are typically in carbonate-bearing rocks, and/or those that have been subjected to the flow of water/carbonate-bearing metamorphic fluids as can be seen by infrared spectroscopy showing fluid inclusions with carbonates, silicates, hydroxyl groups, and water (e.g., de Corte et al. 1998). In the Kokchetav massif, northern Kazakhstan, for example (Sobolev and Shatsky 1990; Claoue-Long et al. 1991), these hosts include garnet-biotite gneisses and schists, which make up 85 vol% of the rocks with the remainder being dolomite, Mg-calcite, garnet, and clinopyroxene in different proportions. Other high-pressure terranes contain diamond as multiphase inclusions coexisting with minerals such as garnet, K-clinopyroxene, magnesite, high-Si phengite, and coesite. In many cases, these inclusions appear to be shielded from retrogression by garnet, zircon, or kyanite, showing the significance of the host minerals as containers of UHP mineral inclusions and their significance in the search for further diamondiferous metamorphic rocks. In the Erzebirge terrane, Germany, diamond occurs within all three host minerals in a muscovite-quartz-feldspar rock (Massonne 2003). In the Western Gneiss terrane, Norway, diamond is enclosed exclusively with spinel in garnet that occurs within garnet-websterite pods (Dobrzhinetskaya et al. 1995; van Roermund et al. 2002), whereas in the Dabie Shan and North Quaidem areas, China, diamond occurs with coesite and jadeite within garnet or zircon in eclogite, garnet pyroxenite, and jadeite (Shutong et al. 1992; Song et al. 2005).

Diamond formation also occurs during the high pressures and temperatures produced when an extraterrestrial body impacts Earth's surface (impact diamonds). Given the short time scales, diamond formation occurs within either C-rich targets (graphite to diamond solid-state transition) or impact melts (Hough et al. 1995; Koeberl et al. 1997). Impact diamonds can reach up to 1 cm size in the well-studied Popigai impact crater (Koeberl et al. 1997) but diamonds are typically much smaller (submicron to millimeter in size). Lonsdaleite, the hexagonal-form of *sp*3-bonded carbon (Hazen et al. 2013), is characteristic of impact diamonds. Although somewhat rare in nature, impact diamond and lonsdaleite could actually be spread over Earth's entire surface, as illustrated by their occurrence at the K/T boundary layer. Impact diamonds are comparatively little studied and have been mostly used to identify the occurrence of large impacts (Hough et al. 1997).

Diamond distribution in Earth. At depths below about 150-200 km along both continental and oceanic geothermal gradients, the entire Earth, including the base of the continental lithosphere and the convecting mantle beneath the lithosphere-asthenosphere boundary, is in the diamond stability field (e.g., Stachel et al. 2005; Fig. 2). Locally, given the typically low solubility of C in mantle silicate and oxides, there is the potential to crystallize diamond within a large volume of Earth's mantle in the presence of a free C-bearing phase such as methane or carbonate. But under the conditions of typical plume-related magmatism and relatively slow transport to the surface, diamonds will re-equilibrate, either graphitize or more likely oxidize. With few exceptions diamonds are erupted only by kimberlite and lamproite magma and such volcanism is rare. Thus, the true amount of diamond crystallizing at depth in Earth below the lithosphere is not known but could be much larger than what has erupted with kimberlite. When these diamonds are exhumed from below the continental lithosphere, they are referred to as *superdeep* or *sub-lithospheric* diamonds, whereas if they are derived from within the continental lithosphere they are referred to as lithospheric (Fig. 2). Crustal diamonds that occur in high-pressure metamorphic terranes are known as "ultra-high-pressure metamorphic" or UHPM diamonds for the amazingly high pressures that they signify for crustal conditions. Note

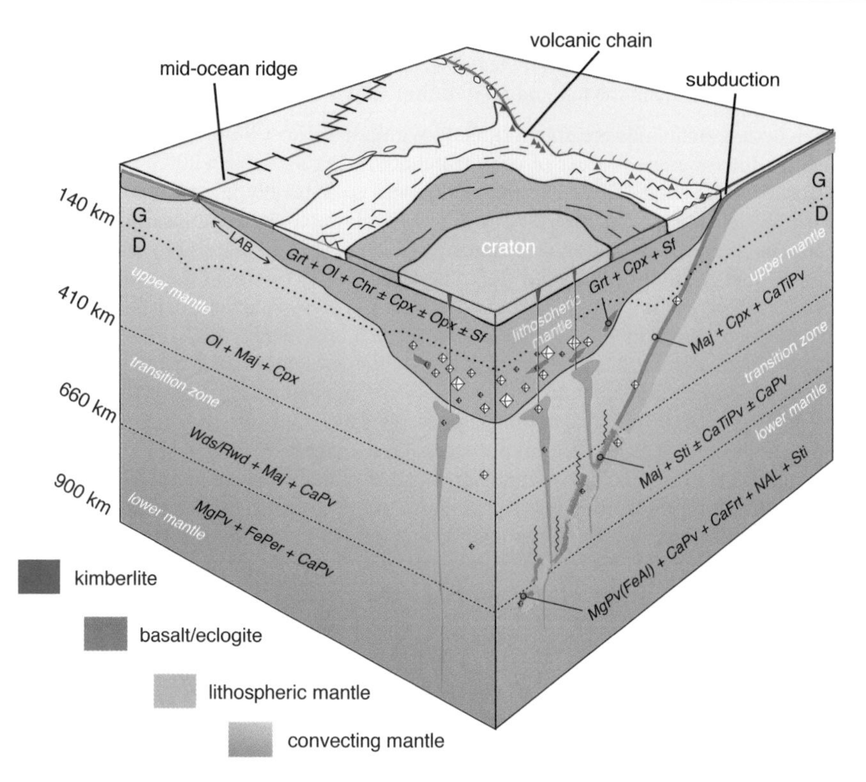

Figure 2. Block diagram showing the basic relationship between a continental craton, its lithospheric mantle keel and diamond stable regions in the keel, and the convecting mantle. Under the right f_{O_2}, diamonds can form in the convecting mantle, the subducting slab, and the mantle keel. Figure redrawn from an original by Tappert and Tappert (2011) with additions. G=graphite, D=diamond, LAB=lithosphere/asthenosphere boundary. Mineral assemblage information and abbreviations defined in Tables 1 and 2. These assemblages give the expected inclusions to be found in peridotitic or ultramafic (left) and eclogitic or basaltic (right) rock compositions.

that all mantle diamonds form at higher pressures than UHPM diamonds. The point here is that diamond is a good tracer mineral for carbon throughout the mantle and crust.

Geologic setting for diamond formation. At Earth's surface, macro-diamonds are highly unevenly distributed, being found primarily within erupted kimberlite with a direct association to stable Archean continental nuclei (e.g., Harlow and Davies 2005; Fig. 3). Beneath the seismically stable, old, interior portions of cratons, the lithospheric mantle extends from about 40 km depth down to perhaps 250-300 km, whereas under the oceans it is thinner and extends from about 40 km to only about 110 km (e.g., Jordan 1975, 1978; Ritsema et al. 2004). Because of the downward protruding shape and the long-term attachment of this mantle to the continental crust of the craton, this portion of mantle has taken the term "mantle keel" (Fig. 2). Archean lithospheric keels are more melt-depleted and deeper than Proterozoic and younger continental mantle keels and their lowermost reaches are squarely in the diamond stability field (e.g., Haggerty 1999; Fig. 2). They are the most diamond-friendly regions of Earth and their depth is thought to facilitate the production of kimberlitic magma by deepening the onset of melting of carbonated mantle. This connection produces an amazingly strong association between diamondiferous kimberlite and the oldest interior portions of continental nuclei having the most melt-depleted mantle keels—a relationship known as "Clifford's Rule" (e.g., Kennedy

1964; Clifford 1966). This association is so strong that it has proved to be the most essential consideration in diamond exploration; its corollary is the complete absence of macrocrystalline diamonds in the oceanic mantle.

Diamonds found in kimberlite provide a recent snapshot of their occurrence at depth in the mantle because most kimberlites are relatively young. Nonetheless, since the diamonds are ancient (e.g., billions of years older than the kimberlite) their compositions can record the active geological processes that initially placed the diamondiferous fluids into the lithosphere. *Thus diamonds alone can provide a record of the carbon cycle older than the oldest oceanic lithosphere (e.g., >200 Ma), especially in the Archean and Proterozic, when major changes in Earth's geodynamics, crustal growth, and atmospheric chemistry occurred.* The oldest diamonds yet dated are thought to be intimately associated with initial production of the depleted mantle keel itself either in the plume upwelling and attendant melting that is responsible for the extensive depletion (e.g., Aulbach et al. 2009b), or in some closely-associated recycling (e.g., Westerlund et al. 2006) and advective thickening (e.g., Jordan 1975, 1978). They potentially track some of the oldest carbon-bearing fluids released by Earth. Once formed, the keel is not just a passive player in diamond genesis. During continental collision the keel can capture eclogite (e.g., Shirey and Richardson 2011) and trap fluids emanating from any underthrusted oceanic lithosphere (e.g., Aulbach et al. 2009b). During the supercontinent cycle, orogenesis around ancient continental nuclei may permit marginal subduction to repeatedly add diamond-forming fluids generated from tectonic processes near the edges of cratonic blocks (e.g., Richardson et al. 2004; Aulbach et al. 2009a) or orogenesis may even rework the lithosphere beneath mobile belts, in some cases recycling the extant carbon that is part of much older mantle lithosphere (e.g., Smit et al. 2010). Diamonds formed in association with continental tectonism record geologic processes from the deepest portions of the continents and can be a key to understanding the stabilization of the continents. Persistent, sub-lithospheric magmatism can also be a source of heat and fluids to add diamonds to the interior of cratons from below. The continental keels provide the only evidence for the source, timing, and geological causes of such ancient deep carbon.

Crustal diamonds from ultra-high-pressure (UHP) metamorphic terranes record the fate of carbon trapped at much shallower levels in the lithosphere. Diamonds found in these crustal settings are often cubic and microcrystalline and occur with metamorphic mineral assemblages that can be used to trace the diamond-forming reactions. Here we have the converse of Clifford's Rule in that these diamonds are chiefly forming in Paleozoic to Cenozoic orogenic belts, where extreme conditions of continental tectonic instability have allowed the crustal section to be buried to the diamond stability field and later exhumed (e.g., Ogasawara 2005; Dobrzhinetskaya 2012; Fig. 3). Although the study of Sumino et al. (2011) highlighted the occurrence of some mantle-derived rare gases, there is no doubt that, for the vast majority of UHP diamonds, the carbon reservoir is crustal, apparently isolated from mantle carbon. The crustal source of carbon is evident from the association of the diamonds with metasedimentary protoliths, the unusual chemical composition of their fluids (e.g., Hwang et al. 2005, 2006), and their high-N content (up to 1 wt%) or heavy $\delta^{15}N$ (Cartigny 2005, 2010) that are not found among mantle-derived diamonds.

Microscale components in diamonds

The advent of microanalytical techniques, such as secondary ion mass spectrometry (SIMS), laser ablation inductively-coupled plasma mass spectrometry (LA-ICPMS), focused ion beam (FIB) lift out, high-resolution transmission electron microscopy (HRTEM), scanning electron microscopy (SEM), and high-intensity light sources (synchrotron), have revolutionized the ability to look at the diamond itself even beyond the way that the electron microprobe (EPMA) revolutionized mineral analysis in the 1960's. Not only can ever smaller inclusions be found and imaged, but their chemical and isotopic compositions can be determined in some instances.

Elemental substitution. The substitution of elements into the diamond structure has long been an area of study because of its effect on the gem qualities, and hence value, of natural diamond and its effect on the physio-chemical properties of diamond. The first quantitative measurements of trace elements in diamonds were published by Fesq et al (1975). More than 60% of elements in the periodic table can be found in diamond but chiefly it is only nitrogen, boron, hydrogen, silicon, and nickel that substitute into the diamond structure (e.g., Field 1992; Gaillou et al. 2012) in routinely measurable quantities. Nitrogen is the main substitutional diamond impurity (Kaiser and Bond 1959) and, for historical reasons (Robertson et al. 1934), it forms the basis of diamond classification into so-called Type I (nitrogen-bearing) and Type II (nitrogen so low as to be thought of as essentially nitrogen-free) diamond. With modern instrumentation it proves possible to detect traces of N in diamonds that previously would have been termed Type II. Pearson et al. (2003) suggest defining type II as <20 ppm N, but this value may decrease in the future. In diamond, nitrogen occurs as different N-bearing centers, the most abundant (A, B, C defects) being the basis of the diamond classification into IaA, IaB, and Ib diamond respectively (e.g., Harlow 1998; Breeding and Shigley 2009). A second-order diffusion process (Chrenko et al. 1977; Evans and Qi 1982) leads C-centers (single substitution N-defect) originally present in the diamond matrix to migrate to form A-defects (N-pairs) and subsequently B-defect (cluster of 4 N-atoms around a vacancy). This difference can best be ascertained by infrared (IR) spectroscopy (e.g., Breeding and Shigley 2009). The abundance of N-bearing diamonds vary from one locality to the other. From published data, about 70% of diamonds contain > 20 ppm nitrogen and are classified as Type Ia, most (99.9%) being mixtures of IaA and IaB. Early studies investigated the potential of N-aggregation to date mantle residence time of diamond, but appeared to be so sensitive to temperature (Evans and Harris 1989) to make N-aggregation is actually a better thermometer (Taylor et al. 1990). Consistency in diamond N-aggregation state occurs, however. Metamorphic diamonds have short residence times in the crust (typically a few Ma; Finnie et al. 1994) at rather low temperatures (<1000 °C) and therefore display low-aggregation states (Ib-IaA diamonds). Fibrous diamonds and the coats of coated diamonds, being related to the kimberlite magmatism, also have short residence times (close to 1 Ma; Boyd et al. 1987; Navon et al. 1988), but at higher (i.e., mantle) temperatures and consistently display higher N-aggregation states (99% are IaA diamonds, plus some rare Ib-IaA diamonds; for review see Cartigny 2010 and references therein). Finally, having spent

---

Figure 3 (*on facing page*). Diamond localities of the world in relation to Archean cratons and classified as to whether they are kimberlite-hosted and from mantle keels (lithospheric), kimberlite-hosted and from the convecting mantle (superdeep), of surface origin (i.e., weathered out of original host; alluvial), from ultra-high pressure crustal terranes (UHP), or formed by the shock of meteorite impact (impact). The crustal age/craton basemap is modified from Pearson and Wittig (2008). Diamond locality information from Tappert et al. (2009), Harte (2010), Harte and Richardson (2011), Tappert and Tappert (2011), Dobrzhinetskaya (2012), and information from the authors. Localities as follows: (1) Diavik, Ekati, Snap Lake, Jericho, Gahcho Kue, DO-27; (2) Fort a la Corne; (3) Buffalo Hills; (4) State Line; (5) Prairie Creek; (6) Wawa; (7) Victor; (8) Renard; (9) Guaniamo; (10) Juina/Sao Luis; (11) Arenapolis; (12) Coromandel, Abaete, Canasta; (13) Chapad Daimantina; (14) Boa Vista; (15) Koidu; (16) Kankan; (17) Akwatia; (18) Tortiya; (19) Aredor; (20) Bangui; (21) Mbuji-Mayi; (22) Camafuca, Cuango, Catoca; (23) Mavinga; (24) Mwadui; (25) Luderitz, Oranjemund, Namaqualand; (26) Orapa/Damtshaa, Lhetlakane, Jwaneng, Finsch; (27) Murowa, Venetia, The Oaks, Marsfontein, Premier, Dokolwayo, Roberts Victor, Letseng-la-Terae, Jagersfontein, Koffiefontein, Monastery, Kimberley (Bultfontein, Kimberley, DeBeers, Dutoitspan, Kamfersdam, Wesselton); (28) Kollur; (29) Majhgawan/Panna; (30) Momeik; (31) Theindaw; (32) Phuket; (33) West Kalimantan; (34) South Kalimantan; (35) Springfield Basin, Eurelia/Orroro, Echunga; (36) Argyle, Ellendale, Bow River; (37) Merlin; (38) Copetown/Bingara; (39) Mengyin; (40) Fuxian; (41) Mir, 23rd Party Congress, Dachnaya, Internationalskaya, Nyurbinskaya; (42) Aykhal, Yubileynaya, Udachnaya, Zarnitsa, Sytykanskaya, Komsomolskaya; (43) Ural Mts.; (44) Arkhangelsk; (45) Kaavi-Kuopio; (46) W Alps; (47) Moldanubian; (48) Norway; (49) Rhodope; (50) Urals; (51) Kokchetav; (52) Qinling; (53) Dabie; (54) Sulu; (55) Kontum; (56) Java; (57) New England Fold Belt; (58) Canadian Cordillera; (59) Lappajarvi; (60); Reis; (61) Zapadnaya; (62) Popigai; (63) Sudbury; and (64) Chixculub.

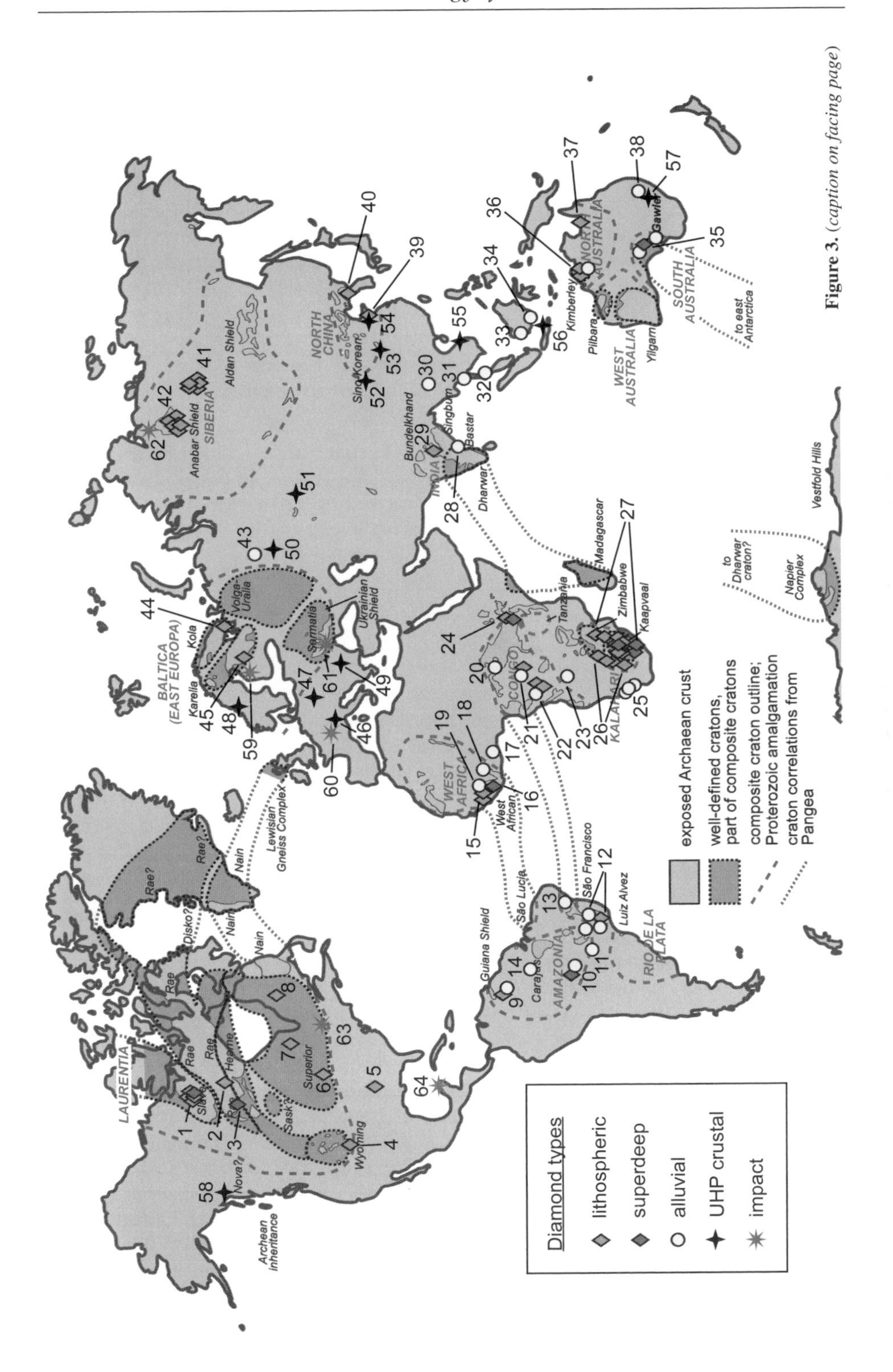

Figure 3. *(caption on facing page)*

billions of years at mantle temperatures, xenocrystal eclogitic and peridotitic diamonds show higher aggregation states (>99.9% are IaA-IaB diamonds). Transition zone and lower mantle are usually deprived in nitrogen (Type II) but the few type I deep diamonds are characterized by highly aggregated nitrogen (IaB diamond), reflecting their high-temperature environment (Stachel et al. 2002 and references therein).

For determining the source of the carbon in the fluids/melts that have crystallized diamond, —an essential aspect of carbon cycle research—the $^{13}C/^{12}C$ and $^{15}N/^{14}N$ isotopic compositions, and N abundance have emerged as the most important measurements (Cartigny 2005; and see section on "*Stable isotopic compositions and formation of diamonds*"). These measurements can be done in bulk by gas source, isotope ratio mass spectrometry (IRMS) at high precision and accuracy or on individual spots by SIMS in spatial relationship to diamond growth zones at lower levels of precision and accuracy. Fractionation of both the C and N isotopic compositions occurs with diamond growth and active debate centers on the extent to which fractionation versus source reservoir composition controls the isotopic composition of a diamond (Cartigny et al. 2003; Stachel and Harris 2009).

Trace element analysis by LA-ICPMS recently has emerged as a powerful new tool for relating diamonds to the fluids/melts from which they have grown (see below). But at current sensitivity and blank levels the trace elements analyzed by LA-ICPMS are in the micro-inclusions of cloudy and fibrous diamonds and are not direct constituents of the clear, gem-quality diamond lattice.

Fluid and micro-mineral inclusions. For purposes of discussion, a distinction is being made here between discrete micro-inclusions located usually in the interior of gem-quality monocrystalline diamonds and suited for individualized X-ray, thermobarometric, chemical, and isotopic study (discussions in the "*Inclusions Hosted in Diamonds*" section) and dispersed clusters and clouds of nano- to micro-inclusions that have been included during coated diamond growth or exsolved from melt after diamond crystallization. Fibrous and coated diamonds can occur as single-crystal cubes (fibrous cuboids) or as the thick outer rim or coat on clear octahedral cores. Some diamonds oscillate between fluid-poor (gem-quality, layer-by-layer growth) and fluid-rich, fibrous growth (e.g., Fig 1, top), whereas other have a center of fluid-rich fibrous growth that transforms into fluid-poor, gem diamond outwards (so-called "cloudy diamonds"). The first published report of fluids in fibrous diamonds was made by Chrenko et al. (1967) and early studies of their growth structures were made by Custers (1950) and Kamiya and Lang (1964).

Fibrous diamonds generally make up less than 1% of mine production but can comprise as much as 8% at some mines such as Jwaneng, Botswana (Harris 1992), 90% in Mbuji Mayi Zaire, and 50% in Sierra Leone (see Boyd et al. 1994 for review). Typically, the compositions within any one diamond are uniform but significant variations exist between individual diamonds. Fibrous diamonds have been the active research focus of Navon and coworkers for many years (e.g., Navon et al. 1988; Weiss et al. 2008, 2011), as summarized by Pearson et al. (2003). The fluid compositions are typically measured for major elements by electron microprobe (Navon et al. 1988; totals are typically <5% and are normalized to 100%) or FTIR (Weiss et al. 2008), whereas INAA (Schrauder et al. 1996) and ICPMS (Resano et al. 2003; Tomlinson et al. 2005, 2006; Zedgenizov et al. 2007; Rege et al. 2008, 2010; McNeill et al. 2009; Tomlinson and Mueller 2009; Klein-BenDavid et al. 2010) are the preferred methods for trace element determinations. In addition, radiogenic isotope characteristics can now be measured by novel applications (e.g., combustion, off-line laser ablation) of standard analytical techniques (Akagi and Masuda 1988; Klein-BenDavid et al. 2010). The fluids range in composition from carbonatitic to hydrous and silicic end-members, with intermediate compositions (Schrauder and Navon 1994; Fig. 4). The carbonatitic fluid is rich in carbonate, CaO, FeO, MgO, and P_2O_5 with a magnesio-carbonatitic end-member (Klein-BenDavid et al. 2009; Kopylova et al. 2010)

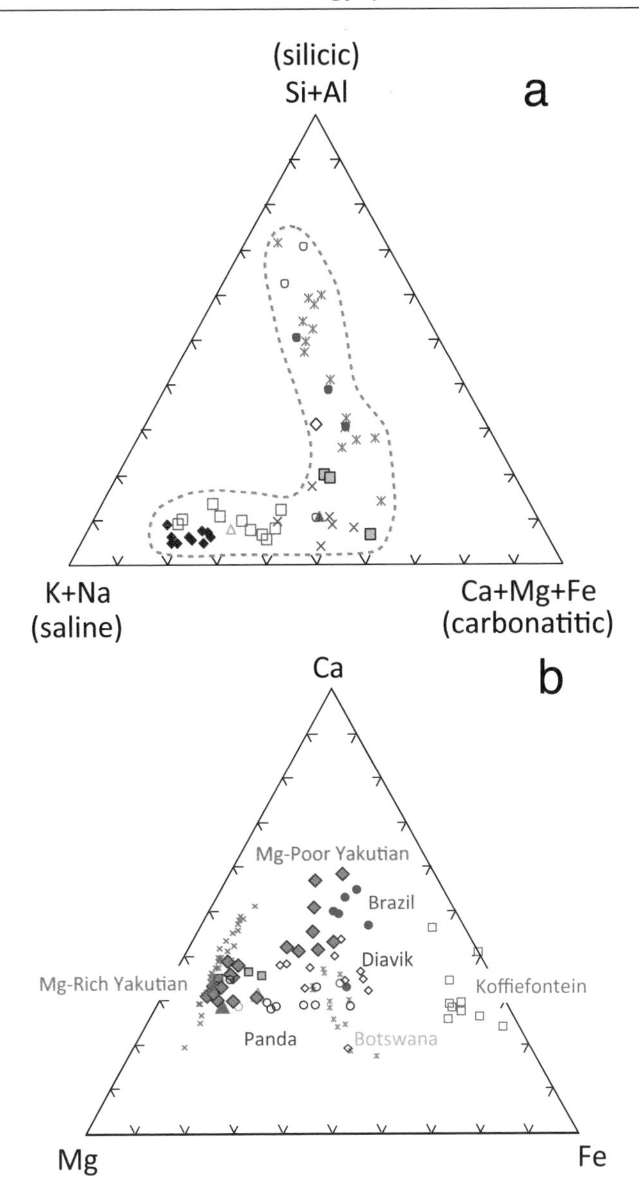

Figure 4. Composition of fluids in diamonds from worldwide locations. Data from the work of Klein-BenDavid (2004, 2007a, 2009), Izraeli (2001), and Tomlinson (2006, 2009). Note the clear delineation of three end-members (a), the large compositional variability (b), and that some localities have specific differences in their Fe/Mg (b).

whereas the hydrous fluid is rich in SiO_2, Al_2O_3 (Schrauder and Navon 1994). K_2O contents are high in both fluid types. In contrast, fluid inclusions from cloudy diamonds contain much higher Cl contents and are classified as brines, being distinct from the other fluid types found in fibrous diamonds (Izraeli et al. 2001). The brines carry very little SiO_2 (3-4 wt%), possibly because of the low water content restricting the solvating capacity of the fluid. Recent work over the last decade has discovered fluids in diamonds (Fig. 4) from some localities (Diavik, Udachnaya,

Kankan) that are carbonatitic and with high-Mg content (10 to 15 wt%; Klein-BenDavid et al. 2009; Kopylova et al. 2010; Weiss et al. 2011). These fluids are capable of being in equilibrium with carbonated peridotite and thus represent a more primitive end-member perhaps existing deeper in the lithosphere and associated with the fluids in proto-kimberlitic magmas (Klein-BenDavid et al. 2009). Their high alkali (K_2O > 10 wt%) and Cl contents are thought to be too elevated to be completely primary and may suggest assimilation of these components (Klein-BenDavid et al. 2009). Recent experimental work (e.g., Bureau et al. 2012) has confirmed these studies of natural diamond by showing that hydrous silicate melt and aqueous fluid can coexist in regions of active diamond growth, though at higher pressures and temperatures these fluids become one supercritical fluid. These fluids crystallize cloudy and fibrous diamond with complex mixed-phase inclusions (Bureau et al. 2012).

The concentration of incompatible elements of varying geochemical affinity (K, Na, Br, Rb, Sr, Zr, Cs, Ba, Hf, Ta, Th, U, and the light rare earth elements) in the fluid inclusions from fibrous diamonds is much higher than in typical mantle-derived magmas and phenocryst-hosted melt inclusions (Schrauder et al. 1996; Weiss et al. 2008, 2011; Tomlinson and Mueller 2009; Klein-BenDavid et al. 2010). Recent advancements in LA-ICPMS techniques (e.g., Klein-BenDavid et al. 2010) have permitted enough improvements in trace element measurements so that even Sr, Nd, and Pb isotopic compositions can now be determined (Fig. 5). The concentrations of

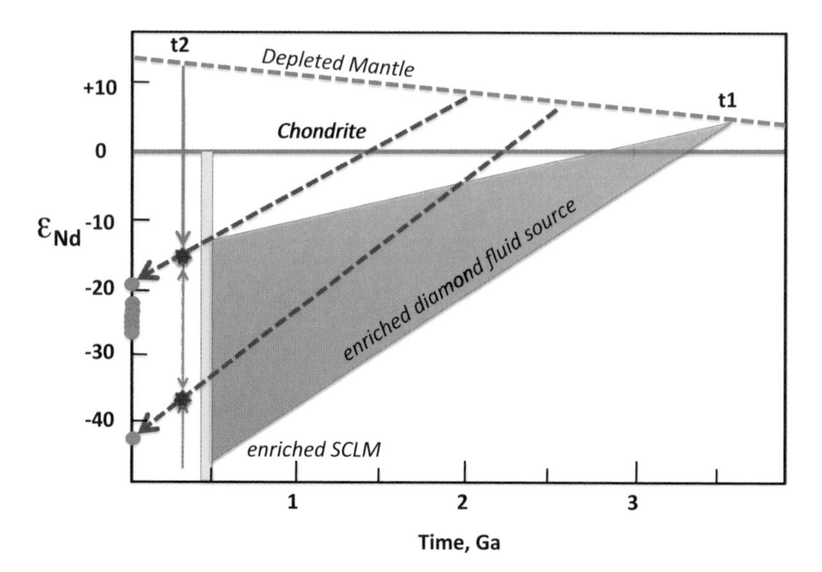

Figure 5. Nd isotope evolution diagram based on data in Klein-BenDavid et al. (2010; and unpublished) on fibrous diamonds from Botswana obtained by closed-system laser ablation ICP-MS (McNeill et al. 2009). Nd evolution illustrated is meant to be typical of the multi-stage history required explain not only the Nd data but the concurrently obtained Pb and Sr isotopic compositions and trace element data (not shown, but see Klein-BenDavid et al. 2010) on the same mass of diamond. Initial development of the enriched reservoir, at t1, requires early separation of the enriched source from either a chondritic or depleted mantle reservoir sometime in the Archean (as required by elevated $^{207}Pb/^{204}Pb$ at modest $^{206}Pb/^{204}Pb$). A spectrum of incompatible element enrichment was generated at this time, leading to a wide range in Nd isotope compositions over 2 to 3 Ga. A second event (t2), in the Phanerozoic, may coincide with diamond formation and involved mixing (as depicted by arrows) between a fluid derived from a more depleted mantle composition (the convecting mantle) and the highly-enriched mantle source generated between t1 and t2. This process created diamond-fluids with mixed compositions (circles) whose model ages (dashed lines) give ages younger than t1 and do not have direct age significance. The timing of stage t2 is constrained to be in the Phanerozoic by Pb isotope systematics and the un-aggregated nitrogen present in all fluid-rich fibrous diamonds.

most trace elements decrease by a factor of two from the carbonate-rich fluids to the hydrous fluids, with the high-Mg carbonatitic fluids among the highest trace element content (e.g., Weiss et al. 2011). Rare earth element (REE) contents of the fluid inclusions are higher than those of kimberlites and lamproites but the fluids show very similar levels of extreme light REE enrichment over the heavy REE typical of these rock types. Negative anomalies in Nb, Sr, Zr, Hf, and Ti are typical (e.g., McNeill et al. 2009; Tomlinson and Mueller 2009; Weiss et al. 2009, 2011; Klein-BenDavid et al. 2010). These features suggest that the fluids may be related to carbonatite- or kimberlite-like magmas in displaying a dominant metasomatic component in their source and fractionation of carbonate plus other phases such as rutile, and zircon. The amount of fractionation of these latter minerals must be small; otherwise middle REE and heavy REE systematics of the fluids would show distinctive fractionation effects. Early Sr isotope studies (Akagi and Masuda 1988) and carbon isotope measurements (Boyd et al. 1987, 1992) supported a link between the fluids in fibrous diamonds and kimberlite-like magmas. But recent, more complete isotopic work (Klein-BenDavid et al. 2010) has shown that fluids in fibrous diamonds have Sr isotopic compositions too radiogenic and Nd isotopic compositions too unradiogenic to be related just to the kimberlite hosting the diamonds. Generation of these fluids must also involve lithospheric components. These fluids are thought to have been derived from mixing between an ancient component derived from the breakdown of micaceous phases in the lithosphere and carbonatitic-kimberlitic fluids from beneath the lithosphere (Klein-BenDavid et al. 2010). Nonetheless, the incompatible-element-rich nature of the fluids in fibrous diamonds illustrate that carbonate-rich and hydrous deep mantle fluids are efficient carriers of incompatible elements. Despite the wide range of Sr and Nd isotopic compositions (Klein-BenDavid et al. 2010; Fig. 5), the C and N isotopic compositions of all fibrous diamonds measured so far (summarized in Cartigny 2005; additional data in Klein-BenDavid et al. 2010) is very restricted and falls close to the canonical mantle values of −5‰, indicating a dominant source within the convecting mantle.

Most fibrous/fluid-rich diamonds and diamond coats are thought to have formed near the age of kimberlite eruption and hence have not experienced protracted mantle residence times on the basis of their unaggregated nitrogen (IaA diamonds) and the similarity of fluids to kimberlitic fluids. Smith et al. (2012) have recently analyzed fluids within fibrous diamonds from Wawa, Ontario, where the diamonds are thought to have been emplaced at Earth's surface a minimum of 2.7 Ga ago. These authors find trace element systematics that are very similar to the fluids in "modern" fibrous diamonds. This study is strong evidence for a similar parentage for these late Archean diamonds and those fibrous diamonds of today, and provides supporting evidence for a dominant style/source of diamond formation from the late Archean onwards as proposed by Shirey and Richardson (2011).

Transmission electron microscopy coupled with sample removal by focused ion beam lift-out (TEM, FIB; Wirth 2004, 2009) is increasingly being applied to the study of fluid- and inclusion-rich diamonds (e.g., Dobrzhinetskaya et al. 2005, 2006, 2007; Klein-BenDavid et al. 2006, 2007b; Logvinova et al. 2008; Jacob et al. 2011). Klein-BenDavid et al. (2006) found abundant solid inclusions of carbonates, halides, apatite, and high-Si mica. The TEM results, together with the narrow range of compositions measured by EPMA, along with the volatiles evident from infrared (IR) spectroscopy, suggest that the micro-inclusions trapped a uniform, dense, supercritical fluid and that the included nano-phases grew as secondary phases during cooling. Kvasnytsya and Wirth (2009) found nano-inclusions of typical lherzolitic silicates in micro-diamonds from the Ukraine, plus a number of likely epigenetic phases, such as graphite and K-richterite. Fe-Sn oxides were also reported, whose primary origin is unclear. The same study also revealed abundant nano-inclusions of carbonate, ilmenite, rutile, apatite, sylvite, and low-Si mica in fibrous microdiamonds. This assemblage is consistent with formation of these diamonds from a carbonatitic to a slightly silicic melt, rich in alkali and volatile components, as proposed for other fibrous diamonds. A single polycrystalline diamond aggregate from Orapa was studied by Jacob et al. (2011), who found a syngenetic micro- and nano-inclusion suite of magnetite, pyrrhotite, omphacite, garnet, rutile, and C-O-H fluid. This assemblage led the authors to propose a novel redox reaction between carbonatitic melt and the sulfide-bearing eclogite during diamond crystallization.

While fibrous and cloudy diamonds with their characteristically high abundances of a variety of inclusions were studied first, these techniques are being increasingly applied to clearer, monocrystalline stones as the only way to look at rare, exceedingly tiny inclusions in superdeep diamonds (e.g., Brenker et al. 2002, 2007; Wirth et al. 2007, 2009; Kaminsky et al. 2009; Kaminsky and Wirth 2011). Certain suites of superdeep diamonds have yielded a dazzling array of nano-inclusions although whether they are syngenetic remains to be proven. Some inclusions, such as halides, anhydrite, phlogopite, or hydrous aluminosilicate (Wirth et al. 2007, 2009) are unexpected in the mantle unless they have been introduced by deep recycling of volatile-enriched slabs. Other superdeep inclusions, such as iron carbide and nitrocarbide (Kaminsky and Wirth 2011), are providing primary evidence for the redox shifts due to the disproportionation of Fe^{3+} into perovskite, whereas nyerereite and nahcolite (Kaminsky et al. 2009) suggest the existence of primary carbonatite associated with diamond formation (e.g., Walter et al. 2008). One expects that these techniques will be applied with more frequency to typical lithospheric monocrystalline diamonds of low inclusion abundance. In the end, it seems clear that such TEM and microanalytical studies will provide the answers to the questions of the speciation of diamond-forming fluids and the oxygen fugacity of the mantle regions in which diamond forms.

Internal textures in diamonds

Diamonds show no growth zonation in visible light but display it in polarized light, photoluminescence, and cathodoluminesce (CL). The best technique for observations is CL, which has been widely applied to diamonds. In CL a beam of electrons generated in either an electron probe or a Luminoscope™ (a microscope-mounted CL instrument) can excite photons through electron transfer. In diamonds this technique is best accomplished on polished plates, which must be oriented perpendicular to one of the {110} axes (e.g., not parallel to {100} or {111}; Bulanova et al. 2005) to cut across growth faces. Nitrogen, the major diamond impurity, is the chief activator of CL in diamond. The technique has been widely applied to diamonds and dramatic images displaying a variety of growth textures (e.g., see Fig. 1) have been presented by Bulanova (1996), Kaminsky and Khachatryan (2004), Spetsius and Taylor (2008), and Tappert and Tappert (2011).

Lithospheric diamond textures. Despite the irregular forms that exist (macles, bort, etc.) most monocrystalline lithospheric diamonds have an internal structure that is roughly concentric. The zoning patterns are characterized by two chief features: 1) extremely thin oscillations between stronger and weaker luminescence (Fig. 1c); and, 2) alternating episodes of resorption and overgrowth on top of the resorption (Fig. 1d). Both features strongly support the idea that diamond grows from an aqueous fluid and/or low-viscosity melt with an aqueous component (Bureau et al. 2012) rather than a solid medium such as graphite. Growth from graphite is not only energetically unfavorable for monocrystalline diamonds (Stachel and Harris 2009), but growth from graphite would likely not produce the internal diamond textures observed such as the fine oscillations (e.g., rapid change in N content) or periods of resorption between periods of growth. Growth of polycrystalline diamonds directly from graphite is feasible, however (Irifune et al. 2004). The growth history revealed by CL is extremely important to the interpretation of individual mineral inclusions (e.g., Pearson et al. 1999a,b; Westerlund et al. 2006) and C and N isotopic composition changes during diamond growth (e.g., Boyd et al. 1987; Cartigny et al. 2001; Stachel and Harris 2009; Smart et al. 2011). In combination with electron backscatter diffraction (EBSD) and the FIB-SEM, CL can even be used to study the 3-dimensional growth zonation around inclusions in diamond. In general, the ability to analyze individual inclusions and analyze C and N isotopes *in situ* by SIMS has made it essential to use such techniques to guide the placement of *in situ* analyses.

Superdeep diamond textures. The textures revealed in sub-lithospheric diamonds are strikingly different than those seen in lithospheric diamonds (Hayman et al. 2005; Bulanova et al. 2010; Araujo et al. 2013). External morphologies of such diamonds are not polycrystalline in the same way that lithospheric diamonds can be, yet they rarely form euhedral monocrystals either. In CL, regular concentric zonation is rare. Instead, these diamonds (Figs. 1e,f) are characterized by multiple growth centers, non-concentric zonation of a blocky nature, and even what appears to be deformation texture; in short they display almost polycrystalline internal structures. The major difference between sub-lithospheric diamonds compared to lithospheric diamonds is that they grow at much higher pressure and temperature and in a mantle that is actively convecting, whereas lithospheric diamonds grow in a mantle host that is not convecting. At present, it remains speculation as to whether these textural differences are caused by the dramatic differences in the nature of the host mantle or by the possibility that some growth from solid graphite (e.g., Irifune et al. 2004) is favored by the much higher *P-T* conditions and deformation.

DIAMOND FORMATION

Diamond formation in the mantle is generally considered to be a metasomatic process (e.g., Haggerty 1999; Stachel et al. 2005). The agents of metasomatism, most likely supercritical fluids or melts, react with the mantle rocks in which they infiltrate, and diamond crystallizes as a consequence of the reduction of carbon via redox reactions, simple examples of which are:

$$CO_2 = C + O_2 \tag{1}$$

$$CH_4 + O_2 = C + 2H_2O \tag{2}$$

As such, the speciation of carbon and the formation of diamond will be intimately associated with the oxidation state of mantle rocks, which is likely controlled by Fe^0-Fe^{2+}-Fe^{3+} components in silicate minerals, metals, and melts (e.g., Rohrbach et al. 2007, 2011; Frost and McCammon 2008; Rohrbach and Schmidt 2011). A synthesis of diamond formation includes a modern view of mantle oxidation state, carbon speciation in peridotitic and eclogitic mantle rocks, diamond growth within a framework provided by experimental and thermodynamic observations, as well as by the stable isotopic compositions of diamonds. Discussion is arranged separately for lithospheric versus sub-lithospheric diamonds because of their different and unique nature.

Experimental and thermodynamic constraints of growth in the lithospheric mantle

Oxygen fugacity in the cratonic mantle. The majority of diamonds sampled at the surface originated in cratonic lithosphere at depths less than 250 km and so the prevailing oxygen fugacity in cratonic mantle peridotite is important to consider. At the pressures of diamond stability, garnet is stable in peridotite. The oxygen fugacity at which garnet peridotite xenoliths from cratonic mantle equilibrated can be determined using the oxy-thermobarometer calibrated by Gudmundsson and Wood (1995) that employs the garnet-olivine-orthopyroxene (GOO) equilibrium,

$$\text{GOO} \qquad \underset{garnet}{2Fe_3Fe_2^{3+}Si_3O_{12}} = \underset{olivine}{4Fe_2SiO_4} + \underset{opx}{2FeSiO_3} + O_2 \tag{3}$$

(see also Woodland and Peltonen 1999). Figures 6a and 6c show f_{O_2} determinations made using this oxy-thermobarometer for garnet-bearing xenoliths from the Kaapvaal craton (see also Frost and McCammon 2008) from equilibration pressures and temperatures corresponding to the diamond stability field, determined from further thermometry and barometry equilibria (Luth et al. 1990; Woodland and Koch 2003; Creighton et al. 2009; Lazarov et al. 2009). We note here

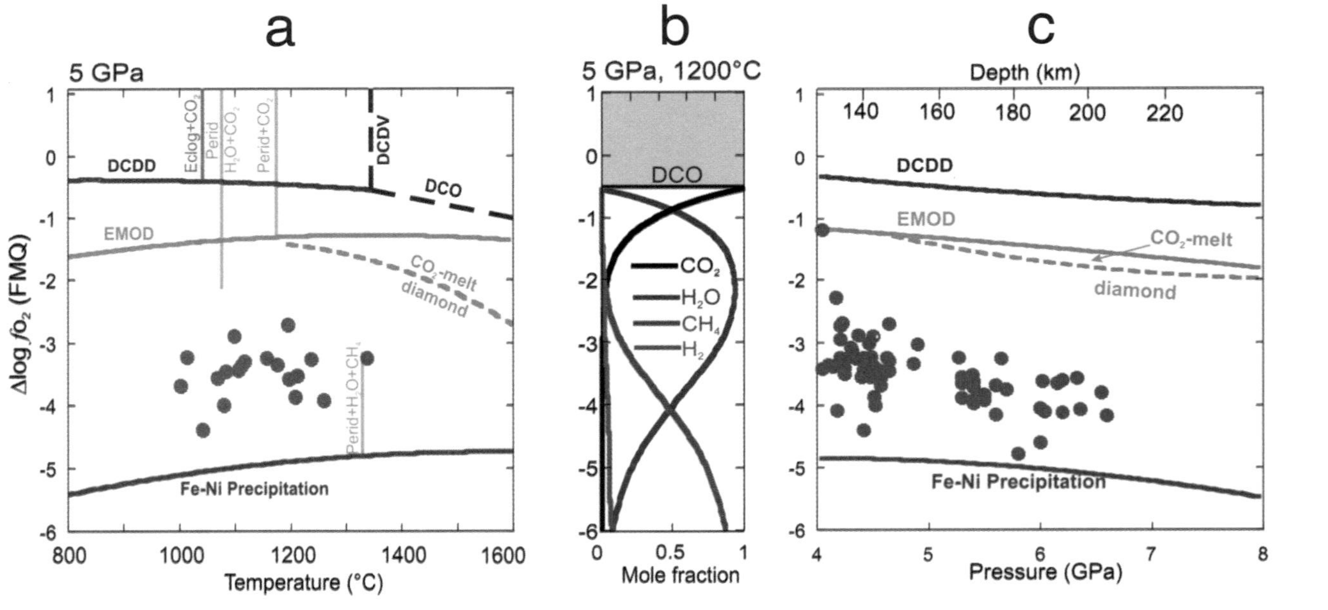

Figure 6. The oxygen fugacity relative to FMQ estimated for garnet peridotite xenoliths (red circles) compared with reactions that determine the speciation of carbon over the pressure and temperature range of cratonic diamond formation. (a) The oxygen fugacity of mantle xenoliths with equilibration pressures of 5 ± 0.5 GPa as a function of temperature. The curve for the EMOD buffer (solid orange) describes the f_{O_2} of the equilibrium between diamond and magnesite in peridotitic rocks, while the diamond CO_2-melt curve (dashed orange) describes the same equilibrium for molten carbonate. Curves describing the stability of carbonate with respect to diamond and CO_2 in eclogite rocks are shown in blue as defined in the text. The Fe-Ni precipitation curve marks the f_{O_2} where Ni-rich Fe metal will start to exsolve from peridotite minerals, as calculated using the procedure described by O'Neill and Wall (1987). Vertical lines in grey show melting temperatures for peridotite saturated with H_2O and CO_2 (perid+H_2O+CO_2; Foley et al. 2009), carbonated peridotite (perid+CO_2; Dasgupta and Hirschmann 2006), and reduced peridotite saturated with H_2O and CH_4 (perid+H_2O+CH_4; Taylor and Green 1988). The blue vertical line shows the carbonated eclogite solidus (Eclog+CO_2) of Dasgupta et al. (2004). The vertical extent of these curves is an approximation of the f_{O_2} range over which such melts are stable. (b) The speciation of a C-O-H fluid in equilibrium with diamond calculated at 5 GPa and 1200 °C using the equation of states from Belonoshko and Saxena (1992) following the procedure of Holloway (1987). Diamond is unstable above the horizontal DCO buffer line. (c) The f_{O_2} of mantle xenoliths as a function of pressure or depth. Carbon speciation curves are the same as in A) but are calculated along the average geothermal gradient recorded by the xenoliths.

that the oxy-thermobarometer of Gudmundsson and Wood (1995) has not been tested experimentally at pressures above 3 GPa, so some unquantified uncertainty exists in garnet peridotite oxygen fugacity estimates due to extrapolation of thermodynamic data beyond the tested range. In Figure 6a and 6c oxygen fugacity is reported relative to the fayalite-magnetite-quartz (FMQ) buffer reaction,

$$\text{FMQ} \qquad \underset{fayalite}{3Fe_2SiO_4} + O_2 = \underset{magnetite}{2Fe_3O_4} + \underset{quartz}{3SiO_2} \tag{4}$$

i.e., $\Delta \log f_{O_2}$ (FMQ), which removes some of the temperature and pressure dependence inherent in all f_{O_2} dependent equilibria (e.g., Frost et al. 1988). The majority of xenolith samples plot in the range between −2.5 and −4.5 $\Delta \log f_{O_2}$ (FMQ). There is a slight f_{O_2} pressure dependence to the xenolith samples, which mainly arises from the volume change of Equation (3) that has the tendency to drive the determined oxygen fugacities to lower levels at high pressures (e.g., Frost and McCammon 2008). Consequently, some of the highest-pressure xenolith samples shown in Figure 1c, which record the lowest oxygen fugacities, are relatively fertile peridotites with garnet $Fe^{3+}/\Sigma Fe$ ratios that are among the highest for these samples. Many xenolith samples in fact record oxygen fugacities close to the Fe-Ni precipitation curve. This curve marks the f_{O_2} where Ni-rich Fe alloy will start to precipitate out of mantle silicates as a consequence of reduction of iron oxide, which can be calculated following procedures described in O'Neill and Wall (1987). The curve has a f_{O_2} close to the iron-wüstite buffer (IW),

$$\text{IW} \qquad \underset{iron}{2Fe} + O_2 = \underset{wüstite}{2FeO} \tag{5}$$

and marks an effective lower bound in mantle f_{O_2} because significant amounts of FeO would be required to reduce from the silicates before the f_{O_2} could pass substantially below this curve. The important point here is that at pressures where diamond is stable, cratonic lithosphere is likely to have a prevailing f_{O_2} that is reducing enough that carbon can exist as diamond.

Carbon speciation in peridotite. The highest oxygen fugacity at which diamond could form within carbonated peridotite assemblages can be described in simplified terms by the reaction,

$$\text{EMOD} \qquad \underset{enstatite}{MgSiO_3} + \underset{magnesite}{MgCO_3} = \underset{olivine}{Mg_2SiO_4} + \underset{diamond}{C} + O_2 \tag{6}$$

which is referred to by the mineral acronym EMOD (enstatite-magnesite-olivine-diamond) (Eggler and Baker 1982; Luth 1993). The f_{O_2} buffered by this equilibrium is shown in Figures 6a and 6c. Curves plotted in Figure 6c are calculated along the average geothermal gradient for the Archean lithosphere recorded by the xenolith samples, and EMOD falls between −2 and −0.5 $\Delta \log f_{O_2}$ (FMQ). At oxygen fugacities above EMOD diamonds are unstable in peridotite rocks with respect to magnesite, and below EMOD carbonate minerals are unstable. As shown, the vast majority of mantle xenoliths plot firmly in the diamond stability field with respect to EMOD. The EMOD buffer can be calculated using thermodynamic data as in Figure 6 (Holland and Powell 2011) and recent experiments that have measured the f_{O_2} of this buffer using independent redox sensitive equilibria are reasonably consistent with such calculations (Stagno and Frost 2010).

Stagno and Frost (2010) measured the f_{O_2} of the equilibrium between diamond and carbonate melt for a peridotite assemblage as a function of pressure and temperature. The f_{O_2} of this diamond and CO_2-bearing melt equilibria was found to evolve to lower values compared to the extrapolated EMOD buffer as temperatures increase, as shown in Figure 6a. This lowering of the f_{O_2} arises not only from the difference in thermodynamic properties between the mineral and melt phases, but also results from the dilution of the carbonate melt by silicate components at high temperature. Experiments confirm that the dilution of the carbonate melt component by solutes other than silicates can also drive the f_{O_2} of the diamond melt equilibria to lower levels (Stagno and Frost 2010). H_2O is likely important in this role, as can be seen from the following calculations of C-O-H fluid speciation, but other solutes such as brines and phosphates would also act in a similar way.

Figure 6b shows the speciation of a C-O-H fluid calculated at 5 GPa and 1200 °C as a function of f_{O_2}, assuming an ideal mixing model (Holloway 1987; Belonoshko and Saxena 1992). This speciation calculation considers only the fluid phase and ignores the potential reaction of fluid species with silicate minerals to produce carbonates or volatile-rich melts for example. However, it provides a framework to examine likely volatile speciation as a function of f_{O_2}, even if the predicted volatile species would in reality be components in other phases. At oxygen fugacities compatible with the grey region in Figure 6b, diamond is unstable with respect to CO_2 fluid. Diamond can only form below the f_{O_2} defined by the DCO (diamond-carbon-oxygen) buffer equilibrium,

$$\text{DCO} \qquad C + O_2 = CO_2 \qquad (7)$$

The DCO buffer, as shown in Figure 6b, is generally above but within 1 log unit of EMOD at lithosphere conditions. At oxygen fugacities below the DCO buffer diamond remains stable but the equilibrium C-O-H fluid phase evolves to become more H_2O-rich. At approximately 2 log units below DCO the so-called "water maximum" occurs, where almost pure H_2O fluid is in equilibrium with diamond. At oxygen fugacities below the water maximum, concentrations of CH_4 in the fluid phase, and to a lesser extent H_2, start to increase. As shown in Figure 6a by the grey vertical melting curves, carbonate melts are likely to form at temperatures in the region of 1200 °C, but pure carbonates are incompatible with the f_{O_2} of mantle xenoliths as described above. As also shown in Figure 6a, the H_2O-CO_2 peridotite solidus is depressed to low temperatures (Foley et al. 2009) but H_2O-rich carbonate melts are also unlikely to be stable at the f_{O_2} recorded by most xenoliths. The majority of samples in fact record oxygen fugacities compatible with the existence of H_2O-CH_4 fluids, which are likely to form fluids rather than melts in the mantle due to the smaller melting point depression associated with these species (Jakobsson and Holloway 1986; Taylor and Green 1988). In a recent experimental study where the fluid in quenched experimental samples was analyzed using gas chromatography to quantify fluid speciation, however, much higher concentrations of H_2 compared to CH_4 were identified in reduced gas mixtures produced at 6.3 GPa and 1400-1600 °C (Sokol et al. 2009).

The majority of peridotite samples record an f_{O_2} consistent with the stability of H_2O-CH_4 fluids if the calculations are informative (Fig. 6), or possibly H_2O-H_2 fluids if the experiments are a better guide. In either case, these results do not mean that diamonds necessarily formed from such fluids. Instead, diamond crystallization can occur as a consequence of redox gradients that exist when metasomatic melts or fluids infiltrate mantle peridotite. Fluids that are either more oxidizing or more reducing than the fluid that would be stable at the oxygen fugacity of mantle peridotite would be expected to crystallize diamond. For example, oxidized carbonate-rich melts or high-density fluids could crystallize diamond by reduction of CO_2 component (as in Eqn. 1), whereas diamond growth from reducing fluids could occur by oxidation of CH_4 (e.g., Eqn. 2). In these cases oxygen is absorbed or supplied by local re-adjustment of Fe^{2+}-Fe^{3+} equilibria in mantle minerals. Evidence for such reactions were reported by McCammon et al. (2001), who noted significant zonation in the $Fe^{3+}/\Sigma Fe$ ratios determined for garnets in mantle xenoliths from the Wesselton kimberlite, consistent with the passage of liquids that have metasomatized and oxidized only the outer rim of garnet grains.

Carbon speciation in eclogite. A large proportion of cratonic diamonds and diamond inclusions are associated with eclogitic rocks and minerals. Currently, however, there is no calibrated oxythermobarometer that can be used to determine the f_{O_2} of eclogitic rocks. Luth (1993) performed experiments on model carbonated eclogite and concluded that the equilibrium,

DCDD $$CaMg(CO_3)_2 + 2SiO_2 = CaMgSi_2O_6 + 2C + 2O_2 \quad (9)$$
dolomite coesite diopside diamond

which has the mineral acronym DCDD (dolomite-coesite-diopside-diamond), would control the stability of carbonate minerals and diamond in eclogitic rocks (see also Luth 1999). As shown in Figure 6, DCDD is approximately 1 log unit above EMOD, implying that the diamond stability field is larger with respect to f_{O_2} in eclogitic rocks. The larger stability field would imply that carbonate-bearing melts or fluids stable within peridotite rocks could be reduced to diamond on entering eclogites, even if the f_{O_2} remained essentially constant. This change in controlling equilibria may be a factor in the close association between diamonds and eclogitic xenoliths, and the prevalence of eclogitic inclusions in certain suites of lithospheric diamonds.

Diamond formation in the lithospheric mantle (experimental results). The oxygen fugacity structure of the lithospheric mantle discussed above indicates that diamond is the likely form of carbon within deep lithospheric mantle. Fluids and melts have long been favored as potential growth media for diamonds, but it is a challenge to deduce the exact conditions of growth from diamonds themselves due to their elemental purity. However, mineral and fluid inclusions, trace impurity chemistry (e.g., N, H, and other trace elements) and growth morphology, can provide important information for interpreting growth history. Diamond nucleation and growth experiments have a long and glorious history, having been motivated by the importance of diamonds in both industry and academia, and provide an important context for observations from natural diamonds (Hazen 1999). A brief synopsis of relevant experiments will contribute to understanding diamond growth in the mantle.

Diamond synthesis directly from C-O or C-O-H fluids has been studied experimentally at pressures generally appropriate for the lithosphere (e.g., ~5-8 GPa), although experimental temperatures tend to be higher than lithospheric because of the chemically simple systems. In general, results show that diamonds can grow from a wide range of fluid compositions at oxygen fugacities at or below the DCO buffer (Hong et al. 1999; Akaishi et al. 2000, 2001; Kumar et al. 2000; Pal'yanov et al. 2000; Sokol et al. 2001b, 2009; Sun et al. 2001; Yamaoka et al. 2002). These studies attest to the fecundity of CO_2-rich fluids, CO_2-H_2O fluids, graphite-H_2O fluid, and CH_4-rich fluids as media for nucleation and growth of diamond. Nucleation and growth is apparently enhanced in H_2O-rich fluids but inhibited in H_2-rich fluids (Sokol et al. 2009).

Diamond growth can occur directly by reduction of carbonate components in minerals and melts, and numerous experiments show that carbonated fluids and melts provide productive diamond-forming media (e.g., Pal'yanov et al. 1998; Sato et al. 1999; Sokol et al. 2000, 2001a, 2004; Arima et al. 2002; Sokol and Pal'yanov 2004; Spivak and Litvin 2004). For example, at 7.7 GPa diamonds can form readily either from molten Ca- and Mg-carbonate (Sato et al. 1999; Arima et al. 2002) or from solid carbonate in equilibrium with a reduced, CH_4-H_2O fluid (Yamaoka et al. 2002). Experiments also indicate enhanced diamond growth from dolomitic melts in the presence of fluids enriched in H_2O and CO_2 (Sokol et al. 2000), although Bataleva et al. (2012) found that CO_2-rich ferrous carbonate-silicate melt can be an effective waterless medium for diamond crystallization at 6.3 GPa. The addition of alkalies to carbonate-rich fluids and melts also yields fertile diamond growth media (Litvin et al. 1997, 1998a,b; Litvin and Zharikov 1999).

Although experiments in simplified C-O-H fluid and carbonated systems are essential and insightful, fluids and melts in the lithospheric mantle will react with silicate minerals in peridotite or eclogite, which can lead to a wide range of chemically diverse compositions as seen, for example, in fibrous diamonds. Experimental data show that an array of complex fluid and melt compositions involving C-O-H fluids, carbonates, chlorides, and silicates, reminiscent of those trapped in natural diamonds, can provide suitable diamond growth media. Pal'yanov et al (2002, 2005) show that reaction of carbonated fluids or melts with silicates can lead to diamond nucleation and growth. Alkaline-carbonate-silicate melts can be highly efficient for diamond formation, but nucleation and growth is apparently limited to specific compositional ranges (Shatsky et al. 2002). Litvin (2009) discusses how in the system Na_2O-K_2O-MgO-CaO-Al_2O_3-SiO_2-C at 8.5 GPa that, as silicate components dissolve into melts, a concentration exists beyond which diamond nucleation and growth is inhibited. According to experiments from model peridotitic-carbonate systems, the barrier may occur at ~30% dissolved silicate (Litvin et al. 2008), whereas in model eclogitic-carbonate a value closer to 50% dissolved silicate is indicated (Litvin and Bobrov 2008). As in simple systems, H_2O apparently enhances diamond crystallization in more complex alkali-chloride-carbonate-silicate-water systems. Diamonds can grow readily in volatile-rich kimberlitic magma (Arima et al. 1993), and a number of studies have verified diamond nucleation and growth in a range of alkali- and chloride-rich C-bearing systems (Pal'yanov et al. 2007b; Safonov et al. 2007, 2011; Pal'yanov and Sokol 2009). A common theme amongst these studies is the importance of fluid/melt composition in facilitating or inhibiting diamond nucleation and in determining growth mechanism and crystal form.

Sulfide inclusions, such as pyrrhotite, are common amongst inclusions in lithospheric diamonds. Bulanova et al (1998) identified the potential importance of sulfide melts, possibly immiscible with a volatile-rich silicate melt, in the nucleation and growth of diamond. Simple system experiments verified that carbon-saturated sulfide melts nucleate and crystallize cubo-octahedral diamond (Pal'yanov et al. 2001, 2006, 2009; Litvin et al. 2002), even if low carbon solubility indicates a limited role for sulfide as an agent of C dissolution and transport. Gunn and Luth (2006) suggest that FeS melt may dissolve sufficient oxygen such that carbonate in a coexisting melt could be reduced by a reaction such as:

$$MgCO_3 + MgSiO_3 = Mg_2SiO_4 + C + O_2 \quad (10)$$

where the oxygen is dissolved in the Fe-S-O melt. Whereas Palyanov et al (2007a) showed that when a carbonate component is involved, the role of sulfur may increase due to its important role as a reducing agent, as in the simplified reaction:

$$2FeS + CO_2 = 2FeO + S_2 + C \quad (11)$$

where CO_2 would be a component in a fluid or melt. Shushkanova and Litvin (2006) showed that at 6 GPa sulfide-carbonic melts are highly efficient diamond-forming media, and that formation of diamond polycrystals, reminiscent of natural diamondite and carbonado, can occur from highly C-oversaturated sulfide melts. In experiments in silicate-carbonate-sulfide systems, immiscible carbonate-silicate and sulfide melts form, and diamonds can nucleate and grow from either media (Shushkanova and Litvin 2008), although again the overall low solubility of C in sulfide would make it a less efficient diamond producer than coexisting carbonate.

Although rare, metallic iron, sometimes accompanied by wüstite, has been reported as inclusions in natural diamonds (e.g., Bulanova et al. 1998, 2010; Stachel et al. 1998a). Molten transition metals (e.g., Fe, Ni, Co) have long been noted for their utility as solvents for diamond growth (Bundy et al. 1955; Strong and Hanneman 1967; Sumiya et al. 2000). Given that the oxygen fugacity of the mantle may reach the metal saturation curve as described above, the potential for Fe-rich metallic melts as diamond-forming agents in the mantle is clear. Fedorov et al (2002) studied diamond crystallization from Fe-Ni-C melts and found diamond nucleation and growth at P-T conditions appropriate for the lithosphere. These authors found that either iron (iron-nickel) or wüstite can crystallize together with diamond, depending on the redox conditions, and that iron-carbon melts are stable over a range of f_{O_2} ranging from the stability field of iron to that of wüstite. Siebert et al. (2005) produced diamonds in experiments by reaction between carbonates and highly reducing Si-bearing iron metal phases. Thus, it is clear that metallic melts are excellent catalytic solvents for diamond growth, and if they occur in the mantle could be agents of diamond formation. What is not clear is why metallic iron inclusions would precipitate from liquids in equilibrium with diamond. According to calculated and experimental phase relations in the Fe-C system, there is no stability field for diamond +

Fe metal, as the intermediate carbide phases Fe_3C and Fe_7C_3 are stable throughout the pressure-temperature range of diamond stability in the mantle (Wood 1993; Lord et al. 2009; Nakajima et al. 2009; Oganov et al. 2013; Wood et al. 2013).

Experimental and thermodynamic constraints of growth in the sub-lithospheric mantle

Oxygen fugacity in the sub-lithospheric mantle. A number of studies have proposed that in a vertically isochemical mantle the oxygen fugacity will decrease with increasing pressure as a result of the stabilization of Fe_2O_3 over FeO components in modally abundant mantle minerals (O'Neill et al. 1993b; Ballhaus 1995; Frost et al. 2004; Rohrbach et al. 2007; Frost and McCammon 2008). Experiments that support this stabilization demonstrate high $Fe^{3+}/\Sigma Fe$ ratios in mineral phases from the deep upper mantle, transition zone and lower mantle, even when these minerals are equilibrated with iron metal. Rohrbach et al. (2007), for example, showed that at pressures above 10 GPa majoritic garnet contains over 20% of total Fe in the Fe^{3+} state in equilibrium with Fe metal. Wadsleyite, the main mineral in the transition zone, has been shown to contain about 2% Fe^{3+} at the same f_{O_2} (O'Neill et al. 1993a).

As can be seen in Figure 6a, many of the deepest lithospheric samples are not far displaced from the Fe-Ni precipitation curve where equilibrium with metal would start to occur. Therefore the f_{O_2} of the base of the upper mantle and transition zone are likely to be on average close to the IW oxygen buffer. Frost and McCammon (2008), for example, estimate that for an upper mantle Fe/O content, mantle oxygen fugacity would decrease with depth such that an FeNi-metal alloy would precipitate beginning at depths of about 250 km. Estimates for the bulk $Fe^{3+}/\Sigma Fe$ ratio of the upper mantle are less than 2% (Canil and O'Neill 1996; O'Neill et al. 1993a). In the transition zone where the dominant minerals are wadsleyite and majoritic garnet, the upper mantle Fe^{3+} content would be below that required for these minerals to be in equilibrium with Fe metal. Thus, for a mantle with constant Fe/O, the implication is that another source must produce the additional Fe^{3+} required for these iron-bearing minerals to be in equilibrium with iron metal (O'Neill et al. 1993b; Ballhaus 1995; Rohrbach et al. 2007). The additional source of Fe^{3+} is predicted to be disproportionation of FeO in mineral phases through the reaction:

$$3FeO = Fe_2O_3 + Fe \qquad (12)$$

In the lower mantle the modally predominant mineral is aluminous silicate perovskite, which has an $Fe^{3+}/\Sigma Fe$ content of over 50% when coexisting with metallic Fe (Frost et al. 2004). As in the case of the transition zone, if whole mantle convection occurs and the total oxygen content of the upper mantle is similar to the lower mantle (e.g., constant Fe/O), then metallic Fe-Ni alloy must precipitate in the lower mantle to provide sufficient Fe^{3+} for perovskite (Frost et al. 2004; Frost and McCammon 2008). The f_{O_2} of the lower mantle is therefore also likely to be at or below IW and anomalously large concentrations of Fe_2O_3 would have to arise from somewhere to raise the f_{O_2} above IW. Thus, the deep upper mantle and the entirety of the transition zone and lower mantle are expected to be reducing and metal-saturated.

Carbon speciation in the sub-lithospheric mantle. Several recent experimental studies have indicated that the f_{O_2} of buffering reactions between diamond and magnesite, which are analogs to EMOD in the transition zone and lower mantle rocks, are at approximately the same oxygen fugacity relative to IW as EMOD in the upper mantle (Rohrbach and Schmidt 2011; Stagno et al. 2011). For example, the reaction,

$$MgO + C + O_2 = MgCO_3 \qquad (13)$$

is the analog buffering reaction to EMOD throughout the lower mantle and lies between two and three log units above IW. Therefore, given no large perturbation in the bulk oxygen content, the deeper mantle should be in the diamond stability field.

On the basis of calculations of carbon speciation in the C-O-H system like those shown in Figure 6, the deep upper mantle, transition zone, and possibly the uppermost lower mantle could be in equilibrium with a CH_4-H_2O fluid, whereas in the deeper lower mantle equilibrium fluids may become H_2O-dominated (Frost and McCammon 2008). We stress again that such a fluid is calculated without consideration of equilibrium with silicate minerals, and with gross extrapolation of thermodynamic data. Experiments and first-principles calculations are needed to identify and quantify carbon species in more realistic deep mantle fluids and melts (e.g., Dasgupta 2013; Manning et al. 2013).

Diamond formation in the sub-lithospheric mantle. As described in the preceding sections, over much of its depth range, sub-lithospheric mantle is metal-saturated. If a free fluid phase were present in this reducing primitive mantle peridotite, calculations suggest it will be CH_4 and H_2O-rich. However, given the high-storage capacity of the deep upper mantle and transition zone for hydrogen (e.g., Hirschmann et al. 2009), a free fluid phase likely is not present, and carbon may be locked up in solid Fe(Ni) carbides and diamond. Given the low estimates for the C content of the primitive mantle (~50 to 200 ppm), carbides such as Fe_3C and Fe_7C_3 that are stable along a mantle adiabat (Wood 1993; Lord et al. 2009; Nakajima et al. 2009; Wood et al. 2013) can accommodate the entire primitive mantle carbon budget. Thus, in ambient primitive sub-lithospheric mantle, diamond may not be present.

We expect that, in general, sub-lithospheric diamonds likely formed by metasomatic processes involving reducing ambient mantle with C-bearing fluids or melts. In common with the lithospheric mantle, diamond crystallization may occur predominantly as a result of redox equilibria between a metasomatic melt and solid mantle phases. Mineral inclusions in sub-lithospheric diamonds, such as majorite, Ca-rich perovskite, and Mg-rich perovskite, provide evidence for diamond formation at depths throughout the deep upper mantle, transition, and the upper part of the lower mantle (see Stachel 2005; Harte 2010). An important goal of future research will be to link sub-lithospheric inclusion mineralogy and trace element composition to diamond fluid composition, deep mantle melt migration, and mantle redox in a way that can be related to mantle convection patterns (e.g., Walter et al. 2008).

Stable isotopic compositions and the formation of diamonds

C-isotopes of diamonds. More than four thousand carbon isotopic compositions on diamonds are available, including representative data from worldwide diamond mines including Siberia; Canada; Australia; Brazil; and West, East, and southern Africa (Botswana and South Africa), as well as diamonds formed in other usually less constrained contexts. Historically, these data stem from the combustion analyses of diamonds (e.g., Deines 1980; Galimov 1985; Cartigny et al. 2001; Cartigny 2005) but these bulk analyses are being rapidly augmented by somewhat less precise, *in situ* analyses made by SIMS (e.g., Hauri et al. 2002).

The distribution of $\delta^{13}C$-values of diamonds formed in Earth's mantle (Fig. 7) are usually divided into distinct diamond populations on the basis of several factors: (1) their eclogitic versus peridotitic paragenesis as inferred from their inclusions, (2) their relative crystallization ages as inferred by younger, often kimberlite-age fibrous coats of coated diamonds grown upon older, typically Archean and Proterozoic monocrystalline diamond cores, (3) their depth of origin in the lithosphere, transition zone or lower mantle as inferred from their inclusions, and (4) their morphology as monocrystalline versus polycrystalline aggregates (including boart, framesites), the latter whose formation age might be closely related to the kimberlite eruption age (e.g., Heaney et al. 2005). In addition, $\delta^{13}C$ values have been measured on (5) polycrystalline, sintered diamonds known as carbonado (Jeynes 1978), whose extra-terrestrial, mantle or crustal origin is unclear (see Cartigny 2010, and references therein), and (6) microdiamonds (typically < 500 microns) formed in crustal rocks subducted to pressures greater than 3.5 GPa along a cold, crustal geotherms. These latter two groups are considered petrogenetically distinct and will be discussed separately.

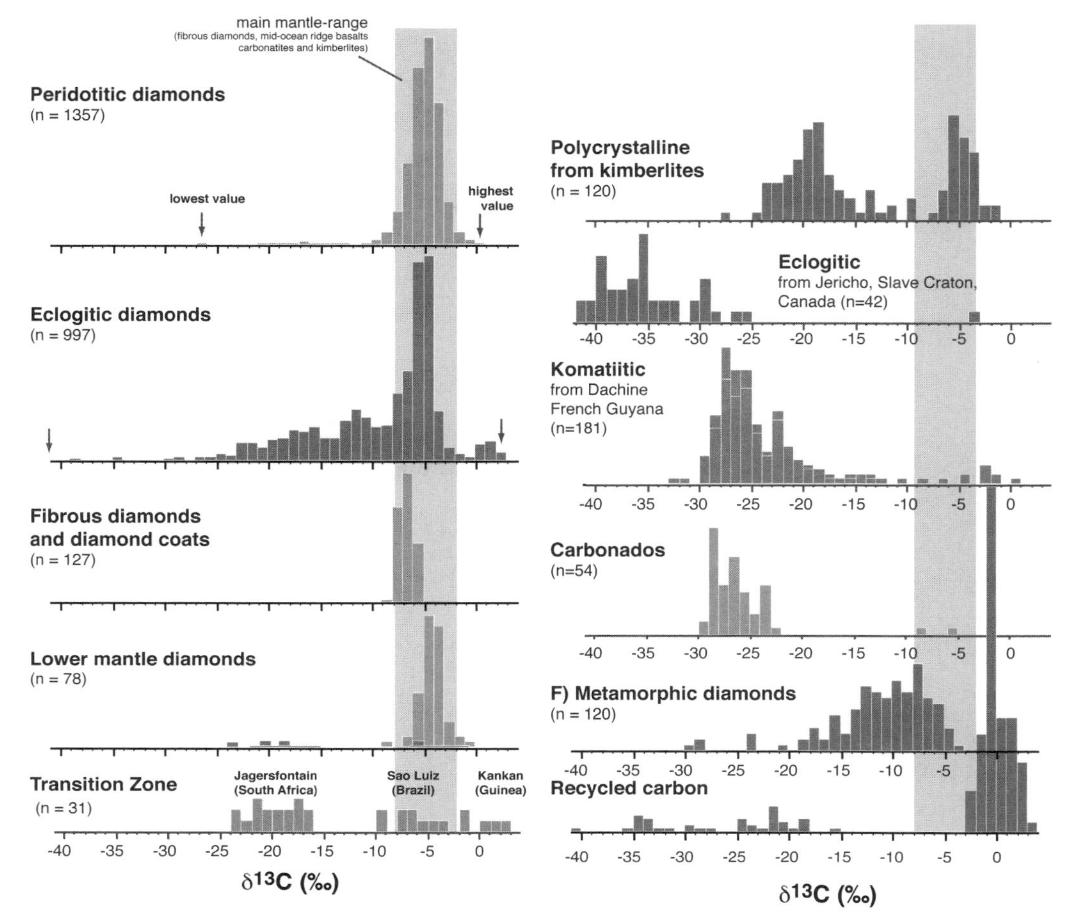

Figure 7. Carbon isotope distributions of worldwide diamonds from a different types of occurrences (e.g., see Fig. 3). Widely-studied peridotitic and eclogitic lithospheric diamonds (upper left) are compared to superdeep and coated diamonds as well as other diamond types. Monocrystalline diamonds unusual in their low $\delta^{13}C$ are represented by eclogitic diamonds from Jericho and diamonds (unknown paragenesis) from the Dachine metakomatiite or metalamprophyre. Polycrystalline diamonds, metamorphic UHP diamonds and carbonado diamonds whose origin remains unknown are also compared. The "main mantle range" defined by fibrous diamonds (i.e., kimberlite-related, see text for details), carbonatites, and carbonates from kimberlites. For references see Cartigny (2005, 2010).

Figure 7 is a summary diagram of the worldwide data set for the C isotopic composition of diamond. Several key points are clear from the distributions seen:

- Worldwide samples cover a large carbon isotopic composition ($\delta^{13}C$) ranging from –41 to +5‰, a range close to that displayed by sedimentary rocks.
- Approximately 72% are contained within a narrower interval of –8 to –2‰, centered on a value approximately –5 ± 1‰. This range is similar to the range displayed by other mantle-derived rocks such as mid-ocean ridge basalts, ocean island basalts, carbonatites, kimberlites.
- The distribution is continuous with a clear decrease in frequency on either side of a $\delta^{13}C$-value of about –5‰.
- The $\delta^{13}C$-distributions are significantly different between their respective growth environments. Peridotitic diamonds cover a narrower range of $\delta^{13}C$-values (from –26.4 to +0.2 ‰), than eclogitic diamonds (from –41.3 to +2.7‰), whilst both coated and lower mantle peridotitic diamonds show narrow ranges of values (–8.1 to –4.9‰) and (–8.5 to –0.5‰), respectively.
- Deep (from transition zone and lower mantle) eclogitic diamonds have variable abundances of negative or positive $\delta^{13}C$-values, (here defined as being below –10‰)
- Diamond formed in metamorphic rocks subducted at ultra-high pressures have $\delta^{13}C$-values ranging from –30 to –3‰, whereas carbonados are mostly from –32 to –25 ‰ and up to –5‰.

Individual mines commonly display $\delta^{13}C$-distributions similar to those illustrated by the worldwide distributions in that their peridotitic diamonds have the typical uni-modal, mantle-like C isotopic distribution and their eclogitic diamonds despite showing nearly the same mode show a strong negative skewness (Fig. 7). However there are several striking exceptions where unique uni-modal eclogitic compositions centered outside the typical mantle-like C values occur: $\delta^{13}C$ of –35‰ for the Jericho kimberlitic diamonds, Slave craton Canada (Fig. 7; De Stefano et al. 2009; Smart et al. 2011); $\delta^{13}C$ of –27‰ for the Dachine lamprophyric or komatiitic diamonds, French Guyana (Fig. 7; Cartigny 2010; Smith et al. 2012); $\delta^{13}C$ of –15‰ for the Guaniamo kimberlitic diamonds, Venezuela (Galimov et al. 1999; Kaminsky et al. 2000), $\delta^{13}C$ of –11‰ for the Argyle lamproitic diamonds, western Australia (Jaques et al. 1989) and $\delta^{13}C$ of +2‰ for placer diamonds from New South Wales, eastern Australia (Sobolev 1984). Rare peridotitic diamonds analyzed so far from these sources show typical mantle $\delta^{13}C$-values of around –5‰. Other localities such as Orapa, Botswana and Jagersfontain, South Africa (Deines et al. 1991, 1993) show a strong bimodal $\delta^{13}C$-distribution with a first peak at about –5‰ and a second at about –20‰ in both peridotitic and eclogitic diamonds. The important question of whether these carbon isotopic compositions are primordial, reflect mantle carbon, or are evidence for subducted carbon depends on the how C isotopic compositions fractionate and evolve in mantle fluids as they migrate through the mantle and diamonds crystallize.

Diamond C-isotopic variability, speciation of carbon, and carbon sources. *In-situ* analyses of single diamonds usually display a limited range ~3‰, which is more isotopically homogeneous in C-isotope composition than the range of all diamonds from the same mine. The trends in C-isotope composition with other tracers, such as N-content and N-isotopic composition, recorded either within a single diamond or within a diamond population, can help in elucidating the speciation (oxidized versus reduced) of carbon in the fluid/melt associated with diamond precipitation (Deines 1980; Thomassot et al. 2007). Carbon isotopic composition is sensitive to f_{O_2} because the reduced carbon in a diamond is depleted in ^{13}C, at isotope equilibrium, by 2-3‰ at T ~1000 °C compared to oxidized carbon (CO_2, carbonate). In comparison, diamond in equilibrium with a more reduced form of carbon such as methane is enriched in ^{13}C by about 1‰. The exact magnitude of isotope fractionation between diamond and coexisting fluid relies on theoretical calculations and experimental data (e.g., Deines 1980; Deines and Eggler 2009; Mysen et al. 2009), whose constraints on isotopic exchange are limited, at present, in temperature, pressure, and composition. Even with these uncertainties, there is no doubt that isotopic fractionation occurs as diamonds grow from fluids (see discussion in Thomassot et al. 2007; Smart et al. 2009) and must be considered in addition to C isotopic differences inherited from C sources.

Diamonds can form from reduced (methane; Eqn. 2) or oxidized (CO_2, carbonate; Eqns. 1, 9-12) carbon (e.g., Stachel and Harris 2009). A diamond or a diamond population displaying a histogram of $\delta^{13}C$-distribution with negative skewness from the mean mantle value and a linear relationship between $\delta^{13}C$ and logarithmic values of diamond N-contents has been suggested to be formed from methane (Fig. 8; Cartigny et al. 2001; Stachel and Harris 2009). The occurrence of CO_2 inclusions (Schrauder and Navon 1993) particularly in some eclogitic diamonds (Chinn et al. 1995; Cartigny et al. 1998), the occurrence of metasomatic inclusions (Leost et al. 2003), positive skewness of some peridotitic $\delta^{13}C$ distributions (Stachel and Harris 2009), and $\delta^{13}C$-N relationships in core-rim traverses of individual diamonds (Bulanova et al. 2002; Smart et al. 2011) provide evidence for diamond precipitation from oxidized carbon. Further evidence for formation from oxidized carbon is found in the fibrous coats of coated diamonds, which trap mantle fluids bearing carbonate nano-inclusions (Navon et al. 1988; see also Klein-BenDavid et al. 2010 and references therein). Even though fibrous diamonds often are homogeneous (typically 3‰; Cartigny et al. 2003) they can also document, sometimes within single growth zones, a consistent increase of 3‰ in $\delta^{13}C$-range starting from a value near −8 up to −5‰ (e.g., Boyd et al. 1992; Klein-BenDavid et al. 2010; and references therein). These observations are again consistent with precipitation from oxidized carbon. It is worth noting, however, that in most cases, and especially for sub-lithospheric diamonds, this direct inference cannot be made. In this case, additional tracers are needed to better characterize the proportion of diamond precipitating from reduced/oxidized carbon.

Diamond crystallization alone from its source melt/fluid seems unlikely to account for the largest range in $\delta^{13}C$ values of diamonds, because otherwise eclogitic and peridotitic diamonds

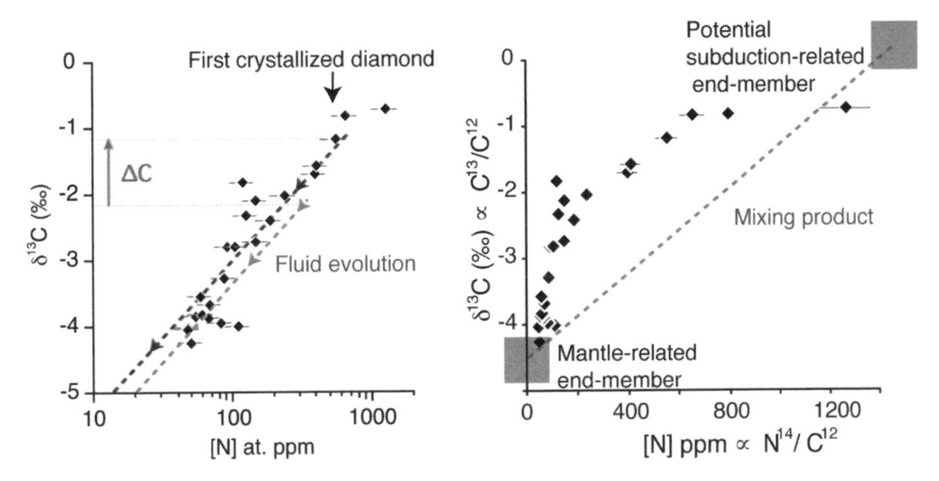

Figure 8. An illustration of the $\delta^{13}C$-N variability that can be produced during the crystallization of diamonds in a natural lithospheric peridotite. The trends are modeled to record diamond growth from methane of multiple diamonds that all found in the same xenolith. Each data point represents a single diamond. [Adapted with permission of Elsevier from Thomassot et al. (2007), *Earth Planet Sci Lett*, Vol. 257, Figs. 2,3, p. 366.]

would display similar distributions. This fundamental observation is the basis for suggesting that diamond growth can be a recorder of mantle $\delta^{13}C$-variability (Javoy et al. 1986; Galimov 1991) by being inherited from various mantle sources. There are two schools of thought. Following the early suggestion of Deines et al. (1993) that diamond would record primordial isotopic heterogeneity, the occurrence of primordial carbon has been suggested for a few diamonds from Kankan, that have a $\delta^{13}C$ value close to −3.5‰ (i.e., being neither isotopically heterogeneous nor depleted in the ^{13}C-isotope; Palot et al. 2012). While the existence of primordial mantle carbon reservoirs can explain the general range of diamond $\delta^{13}C$ (e.g., Haggerty 1999), it still fails to account for the distinct $\delta^{13}C$-distributions among the general population of eclogitic and peridotitic diamonds. The alternative, non-exclusive interpretation is that mantle $\delta^{13}C$-variability recorded in diamonds (i.e., primarily eclogitic) reflects the persistence of isotope variability resulting from the subduction of sedimentary carbon (e.g., Sobolev and Sobolev 1980). The basaltic crust of the oceanic lithosphere contains considerable organic carbon, which is isotopically light (low) in its $\delta^{13}C$ composition. The oceanic lithosphere also is hydrothermally altered and takes on anomalous compositions in $\delta^{18}O$. Subduction of this material and its conversion to eclogite, a common and ongoing geological process, can account for both the anomalous $\delta^{13}C$ of diamonds and $\delta^{18}O$ of eclogite xenoliths in kimberlite. The possibility that diamond might form from subducted carbon is usually also addressed from the study of diamond structural impurities and mineral inclusions.

A contribution to the differences in eclogitic and peridotitic $\delta^{13}C$ isotope distributions also could result from distinctly different evolution of oxidized-C-bearing fluids in eclogite versus peridotite (Cartigny et al. 1998, 2001). Differences in evolution are due to differences in fluid H_2O/CO_2-ratio and the extent of decarbonation reactions in olivine-free eclogitic compositions. Because carbonate decarbonation is associated with $^{13}C/^{12}C$ fractionation (CO_2 being ^{13}C-enriched) this additional process can increase C-isotope variability found in eclogitic diamonds. Although this model is based on the previous experimental work (Luth 1993; Knoche et al. 1999) and finds increasing support (e.g., Stachel and Harris 2009), it is worth noting that it relies on a limited series of experiments. In particular, experiments involving methane are still lacking. Future work is likely to highlight new reactions that might be relevant to better understand eclogitic diamond formation.

Nitrogen in diamonds: contents, speciation, and isotopic composition. Interest in studying N-isotopes in diamonds originates from the original identification that mantle nitrogen is deprived in the ^{15}N isotope compared to surface reservoirs such as the atmosphere (0‰) and the crust and its sediments, which are enriched in ^{15}N (Javoy et al. 1984). This view has since been supported by a wealth of data (Fig. 9). Analysis of fibrous diamonds, mid-ocean ridge basalts, and older mantle derived-samples such as peridotitic diamonds that show predominantly negative $\delta^{15}N$ values. In contrast, metasediments show enrichment the ^{15}N-isotope (positive $\delta^{15}N$) both in the present and the Archean; the amount of negative $\delta^{15}N$ values in sediments is very rare (as reviewed by Thomazo et al. 2009). Furthermore, during sediment subduction, if any devolatilization occurs it would preferentially release ^{14}N, leaving a further ^{15}N-enriched subducted material (Bebout and Fogel 1992; Busigny et al. 2003) as shown by data on metamorphic diamonds (Cartigny et al. 2004 and references therein). Although a limited dataset is available, altered ocean crust and oceanic lithosphere also display positive $\delta^{15}N$ (Philippot et al. 2007; Busigny et al. 2011). Nitrogen in sedimentary/crustal rocks occurs as ammonium ions substituting for potassium and therefore the behavior of nitrogen in subduction zones is expected to follow the fate of potassic minerals, some of which have been experimentally demonstrated to be stable to mantle depths (Watenphul et al. 2009, 2010). If eclogitic diamonds form from subducted carbon and nitrogen, they would be predicted to display positive $\delta^{15}N$-values, as seen in subducted metamorphic rocks (i.e., values distinctly higher than a homogeneous mantle with $\delta^{15}N$ isotopic composition near −5‰). This is certainly

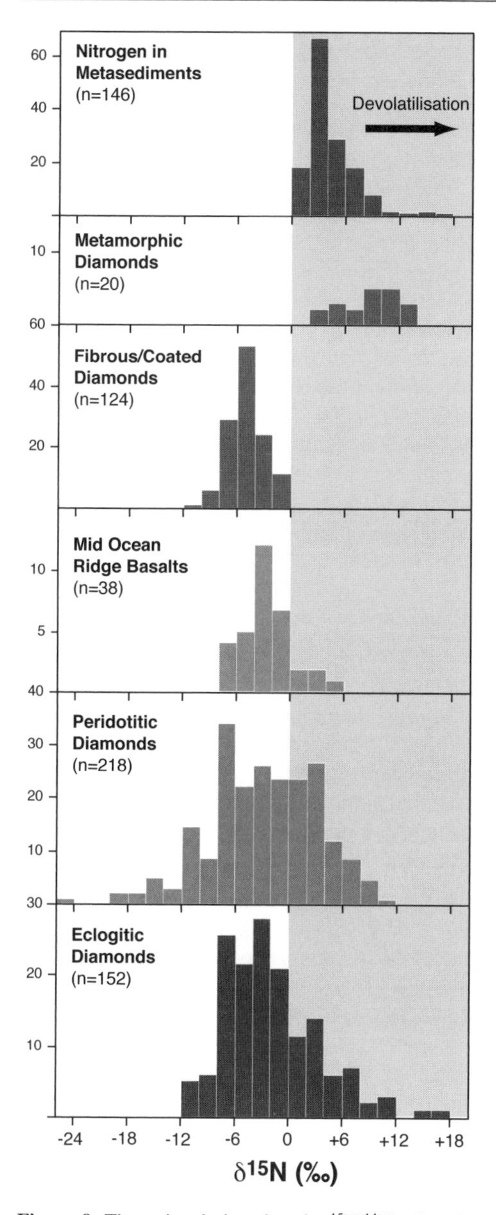

Figure 9. The rationale in using the $^{15}N/^{14}N$-values in diamonds lies in the distinct distributions displayed by surface and mantle reservoirs. Nitrogen in metasedimentary and metamorphic diamonds, metagabbros and metaophioloites (not shown) being enriched in ^{15}N compared to the mantle shown here by fibrous/coated diamond, mid-ocean ridge basalts (not shown) and peridotitic diamonds. The similarities of the $^{15}N/^{14}N$-distributions of eclogitic and peridotitic diamonds must be explained if the source of carbon and nitrogen in eclogitic diamond is to be largely related to subduction (see text for discussion). From Cartigny (2005).

the case for most of the deep diamonds from the mantle transition zone such as at Jagersfontein, South Africa and the Sao Luis-Juina fields, Brazil (Tappert et al. 2005a; Palot et al. 2012) and possibly at Dachine, French Guyana (Cartigny 2010; Smith et al. 2012). But many eclogitic diamonds worldwide show a $\delta^{15}N$-distribution similar to most peridotitic diamonds, pointing to a mantle origin of their nitrogen and therefore of their carbon (Fig. 9). About half of eclogitic diamonds with low-$\delta^{13}C$ values also show negative $\delta^{15}N$-values. These $\delta^{13}C$-$\delta^{15}N$-N co-variations have been argued to be inconsistent with simple mixing of subduction components (see Cartigny 2005 for review) and the data thus have been interpreted to reflect decarbonation reactions occurring in eclogites. However, the remarkably low-$\delta^{15}N$ compositions of the peridotitic diamonds from Pipe 50, China (Cartigny et al. 1997) suggests that variability in the N isotopic composition of mantle exists, and might mask the straightforward assignment of a positive $\delta^{15}N$ to recycled components involved in diamond formation. With a heterogeneous mantle, the $\delta^{13}C$-$\delta^{15}N$ mixing relationships become more complex. Evidence for subduction is clear from many studies: sulfur isotope of sulfide inclusions, oxygen isotopes of eclogitic silicate inclusions and eclogite nodules, and geological considerations. But subduction seems to require some decoupling of carbon from other elements, like N. This view can be reconciled by metasomatic processes during diamond formation. But it remains unclear, if carbon is a massively cycled element, why the presence of low $\delta^{13}C$ for recycled carbon in eclogitic diamonds remains rarer than the normal mantle-like $\delta^{13}C$. Perhaps the amount of C recycled form the mantle portion of the slab which should have $\delta^{13}C$ near −5‰ has been underestimated.

INCLUSIONS HOSTED IN DIAMONDS

Thermobarometry

Chemical thermobarometry. Chemical thermobarometers are expressions that allow one to retrieve the temperature (or pressure) of formation of a mineral species or mineral assemblage knowing the chemical compositions of the minerals and how these compositions are expected to vary with P (or T) of formation. In principle, thermobarometers are based on chemical equilibria between at least two mineral species, but approximate formulations can sometimes be devised that consider the composition of a single mineral, which is assumed to be in equilibrium with another phase capable of buffering its composition under certain P and T. Single-mineral thermobarometers are particularly useful for diamond studies, inasmuch as (1) most inclusions in diamonds are made of isolated mineral grains, (2) non-touching mineral grains included in the same diamond may have been incorporated at different times and P-T conditions and thus may not have been in equilibrium, and (3) polymineralic inclusions made of touching mineral grains presumably had enough time to re-equilibrate at depth after diamond formation during long-standing storage in the mantle. Single-mineral thermobarometry of monomineralic inclusions will provide an indication of the P-T of formation of the diamond, provided the inclusions are syngenetic, or had time to re-equilibrate completely during diamond-forming processes, and did not undergo any transformation afterwards. In contrast, two-mineral thermobarometry of touching inclusions may not necessarily provide the temperature of diamond formation, although the pressure estimate will still provide an indication of the depth of provenance of the diamond.

The reader is referred to Stachel and Harris (2008) for a review of P-T estimates for lithospheric diamonds based on single- and two-mineral thermobarometry of their inclusions. Estimates for over one thousand stones indicate that lithospheric diamonds can form at any depth within the appropriate P-T conditions for diamond stability, with a temperature mode around 1150-1200 °C. Rather than reflecting a favored condition for diamond formation, this mode may simply represent the most "probable" temperature range within the limits imposed by diamond stability, depth of the lithosphere and mantle adiabat under typical cratonic thermal regimes.

The reliability of some of the thermometers used by Stachel and Harris (2008) and by many previous workers, particularly those based on Fe-Mg exchange reactions between garnet and orthopyroxene, clinopyroxene or olivine, and the popular two-pyroxene thermometer of Brey and Köhler (1990), is questionable (Nimis and Grütter 2010). The accuracy of the widely used Opx-Grt barometers at $P > 5$ GPa and for orthopyroxenes with excess Na over Cr + Ti (i.e., about 13% of reported Opx inclusions) also remains to be explored in detail (Carswell et al. 1991; Nimis and Grütter 2012). In particularly unfavorable cases, errors can exceed 150-200 °C and 1.5 GPa (i.e., 45 km in estimated depth). Especially at a local scale and for specific inclusion populations, these errors may obscure possible heterogeneities in the vertical distribution of diamonds. Thermobarometry of eclogitic inclusions remains particularly problematic, because Fe-Mg exchange thermometry of Grt-Cpx pairs is affected by large uncertainties (at least ±100 °C), mostly owing to Fe^{3+} problems (Krogh Ravna and Paquin 2003), and fully satisfactory, well-tested barometers are not yet available (Fig. 10).

It is also worth noting that in some diamond suites non-touching inclusions apparently yield hotter conditions than touching inclusions and xenoliths from the same kimberlite (e.g., Meyer and Tsai 1976; Stachel et al. 1998b; Viljoen et al. 1999; Phillips et al. 2004) and that no relationship occurs between the non-touching inclusion temperatures and those deduced from the N-aggregation state of the diamond. This result suggests diamond formation during transient thermal perturbations, secular cooling of the lithosphere after diamond formation or, simply, disequilibrium. In many other cases, however, no discrepancy exists (Hervig et al.

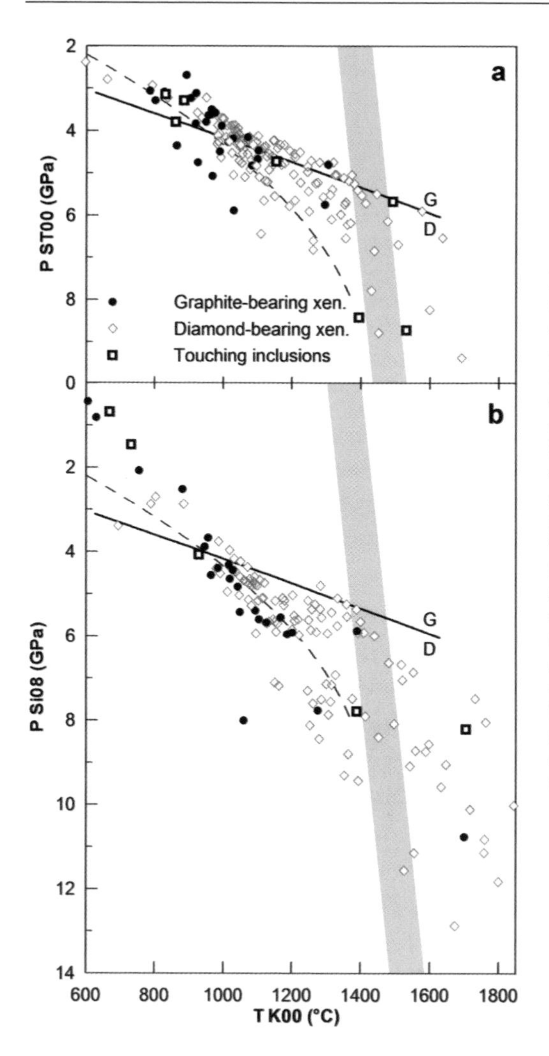

Figure 10. *P-T* estimates for eclogitic xenoliths and touching Grt-Cpx pairs included in eclogitic diamonds using a combination of the Fe-Mg exchange thermometer (K00; Krogh Ravna 2000) with two different versions of the Grt-Cpx barometer (ST00: Simakov and Taylor 2000; Si08: Simakov 2008). Both combinations produce a substantial overlap of diamond-bearing and graphite-bearing eclogites, a significant proportion of diamond-bearing samples in the graphite stability field, and variable proportions of excessive *T* and *P* values (much greater than expected for majorite-poor garnets), suggesting poor reliability of eclogite thermobarometry. Dashed line: conductive geotherms for a surface heat flow of 40 mW/m^2 after Pollack and Chapman (1977); black solid line: graphite-diamond boundary after Day (2012); grey band: *T* range for mantle adiabat based on mantle potential temperatures of 1300 to 1400 °C.

1980; Sobolev et al. 1997; Nimis 2002), suggesting that diamond-forming fluids were thermally equilibrated with the ambient mantle. This result implies that the lithospheric mantle had already cooled to a conductive thermal regime billions of years ago when the diamonds formed and that this thermal regime was comparable to that recorded in mantle xenoliths erupted during emplacement of the much younger host kimberlite (e.g., Cretaceous in the case of the Kaapvaal and Slave cratons). The possibility that these inclusions are all protogenetic and did not re-equilibrate completely during diamond crystallization should also be considered.

In view of the above complications, in the present review particular emphasis is placed on the results of single-mineral thermobarometry on monomineralic inclusions, because they have the best potential to reflect the true *P-T* conditions of diamond formation. Considering the wide range of pressures and temperatures under which diamonds may form, the potentially large influence of input *P* on *T* estimates (and vice versa), and the possible formation of diamond under perturbed or ancient thermal conditions, the most useful mineral species will be those that demonstrably allow sufficiently accurate retrieval of both *P* and *T* of formation. At present, the choice is restricted to diopsides belonging to the ultramafic paragenesis that can be assumed to be in equilibrium with garnet and orthopyroxene and allow application of the Cr-in-Cpx barometer and enstatite-in-Cpx thermometer of Nimis and Taylor (2000; Fig. 11a). Thermal re-equilibration of Cpx with Grt alone (± olivine) causes negligible effects on *P-T* estimates; therefore Cpx + Grt ± olivine polymineralic inclusions should provide the same information as monomineralic Cpx inclusions (Nimis 2002). The enstatite-in-Cpx thermometer has proved to be a very robust method (Nimis and Grütter 2010). The Cr-in-Cpx barometer has two limitations: first, tests against experiments indicated progressive underestimation at $P > 4.5$ GPa (up to ca. −0.8 GPa at $P = 7$ GPa; Nimis 2002); second, typical analytical uncertainties may propagate large errors for Cpx with low values of $a\mathrm{Cr} = \mathrm{Cr} - 0.81 \cdot \mathrm{Na} \cdot \mathrm{Cr}/(\mathrm{Cr} + \mathrm{Al})$ (atoms per formula unit), which is the main building block in the barometer formulation. Standard analytical conditions may result in errors exceeding 1-2 GPa for compositions with $a\mathrm{Cr} < 0.005$ (i.e., for about 10% of reported diopside inclusions). This error accounts well for the larger overall scatter in *P-T* points for inclusions using single-Cpx thermobarometry compared with Opx-Grt thermobarometry (Fig. 11a). Filtered *P-T* estimates confirm the mode around 1150-1200 °C, the distribution of most *P-T* values along typical cratonic geotherms, and the existence of a few "hot" inclusions approaching the mantle adiabat (Fig. 11a). Systematic shift of most

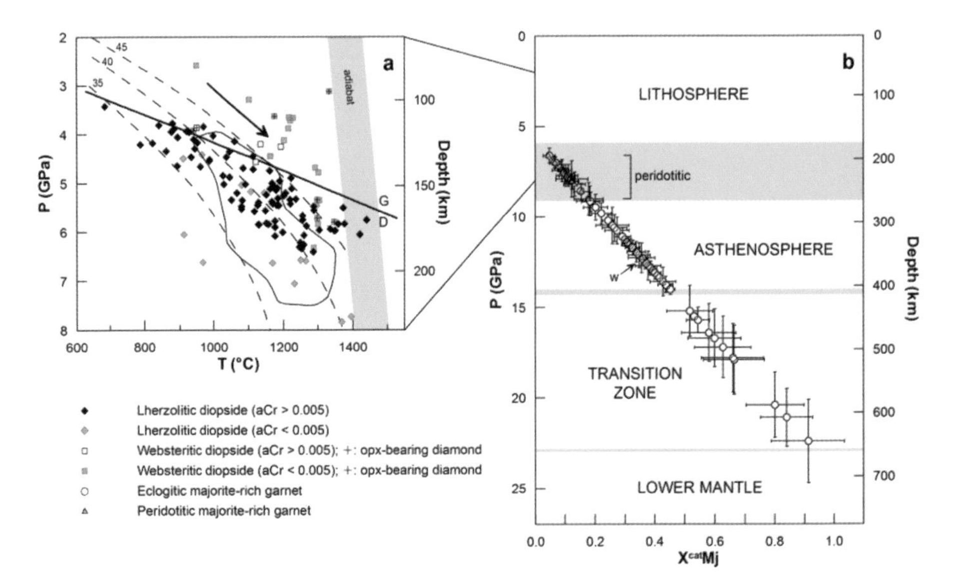

Figure 11. (a) *P-T* estimates for lherzolitic and websteritic Cpx inclusions (monomineralic and Opx-free polymineralic inclusions) in worldwide diamonds based on single-Cpx thermobarometry (Nimis and Taylor 2000). The estimates should correspond to the conditions of formation of the diamonds, with some *P* underestimation at $P > 4.5$ GPa. The scatter is considerably reduced if one excludes diopsides with aCr < 0.005, for which *P* estimates can be severely in error (see text for further explanation). Inclusions in websteritic diamonds containing also non-touching Opx inclusions are indicated with crossed-symbols. For all other websteritic inclusions equilibrium with Opx is not proved and *T* (and hence *P*) estimates may be strongly underestimated; the arrow shows the effect of an increase of *T* on calculated *P*. Outlined field encompasses *P-T* estimates for touching and non-touching inclusions from worldwide sources based on Opx-Grt thermobarometry (Harley 1984; Brey and Koehler 1990) after Stachel and Harris (2008). Conductive geotherms for different surface heat flows ($\mathrm{mW/m^2}$) after Pollack and Chapman (1977); graphite-diamond boundary after Day (2012). *T* range for mantle adiabat is based on mantle potential temperatures of 1300 to 1400 °C. (b) Relationships between molar fraction of majoritic components and *P* for isolated inclusions of majorite-rich Grt in worldwide diamonds based on majorite barometry (Collerson et al. 2010). Bracket indicates *P* range for non-wehrlitic peridotitic inclusions.

websteritic diopsides to lower *P* may be due to *T* underestimation owing to absence of Opx in the original assemblage or to poor reliability of the Cr-in-Cpx barometer for very low-Cr# compositions—the barometer calibration only included diopsides with Cr# = 0.09-0.44 (Nimis and Taylor 2000).

Majorite-rich garnet inclusions allow retrieval of *P* from the fraction of majoritic component (Collerson et al. 2010). Although *T* remains undetermined, the majorite barometer appears to be thermally and compositionally robust, thus allowing the minimum pressure of formation to be estimated. Even allowing for generous uncertainties, the results for isolated monomineralic inclusions indicate beyond any doubt that diamonds containing majoritic garnet could form at very great depth (Fig. 11b). Ranges of possible temperatures of formation may vary over several hundred °C, depending on the interpreted formational setting: <1200 °C for eclogitic diamonds formed in a subducting slab (Stachel et al. 2005), about 1250-1400 °C for peridotitic diamonds formed in a deep lithosphere (Pokhilenko et al. 2004), >1400 °C for diamonds formed in ascending mantle plumes (Davies et al. 2004; Bulanova et al. 2010). Although the abundance of diamonds with majorite-rich garnet inclusions decreases with increasing *P*, existing estimates indicate a more or less continuous spreading from the deep lithosphere to the deep transition zone (Fig. 11b). Such distribution supports a potential genetic link between many majoritic garnet-bearing diamonds and the rare super-deep diamonds with inclusions of interpreted lower-mantle origin (Stachel et al. 2005; Tappert et al. 2009; Walter et al. 2011). Harte (2010) cautioned that many of the relatively low-*P* majoritic inclusions may originally have formed at much greater depth. If they formed in a clinopyroxene-poor medium, the pyroxene can be consumed by dilution into the garnet with no further change in composition recorded with increasing depth. Furthermore, slow rise of the diamond during mantle upwelling may lead to subsolidus re-equilibration of the inclusion (e.g., clinopyroxene exsolution) to lower *P* (Harte and Cayzer 2007). Although special care is given when investigating the chemical composition of diamond inclusions, the possibility that cryptic exsolution has been overlooked in some of these inclusions should be considered.

Elastic methods for geobarometry of diamonds. Elastic methods provide a potential, generally non-destructive alternative to chemical thermobarometry for the evaluation of the pressure of formation of a diamond containing a monomineralic inclusion. These methods are based on the measurement of the "internal pressure" (hereafter P_i and also called "residual" or "remnant" pressure), that is the pressure exerted by the diamond on the inclusion when the diamond-inclusion pair is at room *P-T*. Such pressure can be retreived by using three different techniques: (1) microRaman spectroscopy (e.g., Izraeli et al. 1999; Sobolev et al. 2000; Nasdala et al. 2003; Barron et al. 2008); (2) strain birefringence analysis (Howell et al. 2010), and (3) single-crystal X-ray diffraction (Harris et al. 1970; Nestola et al. 2012). Combining the P_i with data on the thermoelastic parameters (i.e., volume bulk modulus and its pressure and temperature derivatives, volume thermal expansion, shear modulus) of the diamond and of the inclusion allows one to calculate an "isomeke", i.e., a curve in *P-T* space along which the volume of the inclusion is equal to the volume of the cavity within the diamond for a fixed value of P_i. Such line constrains the possible conditions under which the diamond and the inclusion formed. If *T* is known independently, e.g., from FTIR data, or the isomeke is not strongly dependent on *T*, then the *P* at the time of encapsulation of the inclusion can be determined.

Available estimates of the pressure of formation for coesite inclusions based on P_i data are generally much too low for diamond stability (Fig. 12). On the whole, *P* estimates for olivine are more acceptable, but they still straddle the graphite-diamond boundary, indicating again some *P* underestimation at least for some samples (Fig. 12). The limited success of elastic methods thus far indicates that either the diamonds did not behave in a solely elastic fashion or that thermoelastic data for the minerals are inaccurate (see the recent review by Howell et al. 2012). The potential applicability of elastic methods to inclusions of important minerals for which

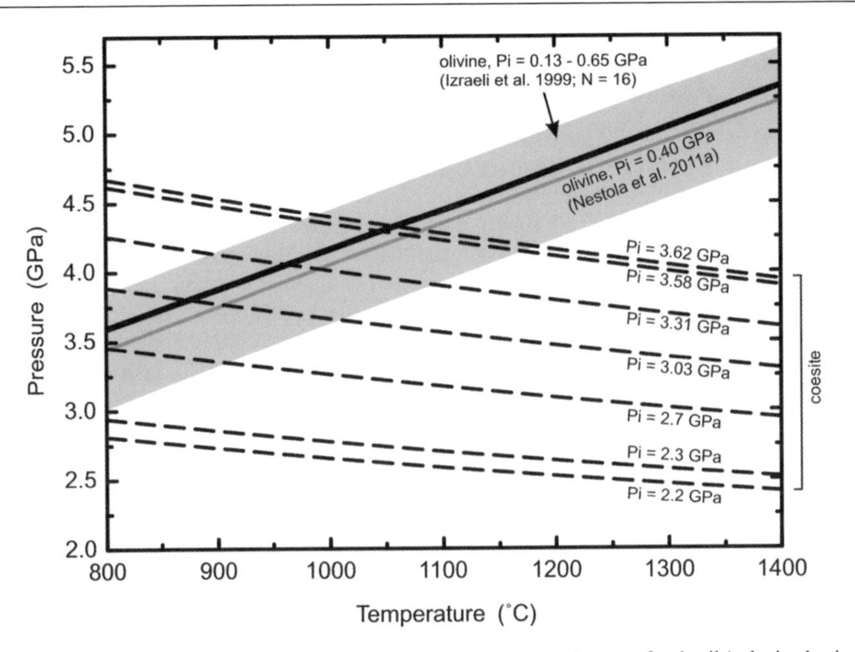

Figure 12. Isomekes for inclusions in diamonds based on P_i estimates (see text for details) obtained using different techniques and thermoelastic parameters as in Howell et al. (2010). Sources of P_i data: olivine - Izraeli et al. (1999; Raman, grey band), Nestola et al. (2011; X-ray diffraction, grey line); coesite - Sobolev et al. (2000; Raman), Nasdala et al. (2003; Raman), Barron et al. (2008; Raman), Howell et al. (2010; birefringence analysis). Black solid line: graphite-diamond boundary after Day (2012).

single-mineral chemical barometers do not exist (e.g., olivine, chromite, coesite) makes these methods worthy of further testing. However, re-assessment of thermoelastic parameters for the minerals included in diamond using state-of-the-art techniques and equipment is necessary before these methods can be considered trustworthy.

Geochemistry and age

Syngenesis or protogenesis? Mineral inclusions can be classified as protogenetic, syngenetic, or epigenetic according to the timing of their crystallization (earlier, contemporaneous, or later) with respect to that of their diamond host (Meyer 1987). Inclusions forming along fractures or made of alteration minerals after former syn- or protogenetic inclusions can be identified as epigenetic. Discrimination of syngenetic and protogenetic inclusions is less straightforward. Such distinction is important, because in the case of syngenesis any geological information extracted from the inclusion (e.g., *P-T* of formation, geochemical environment, age) would also unequivocally apply to its host diamond. A protogenetic inclusion would record conditions that existed before its encapsulation but this might range from geologically short to very long timescales. In the latter case a protogenetic inclusion could be unrelated to diamond formation. Demonstrably protogenetic inclusions would support models of diamond formation involving fluxes of C-bearing fluids through pre-existing mantle rocks and could help explain occurrences of isotopically different inclusions in the same generation of diamond (e.g., Thomassot et al. 2009). In cases of protogenicity, although absolutely accurate ages of diamond formation would not be obtained from the inclusions, a maximum age would be obtained and a general age pattern of diamond growth in a region of lithospheric mantle might still be evident (see section on "*Age systematics and isotopic compositions*").

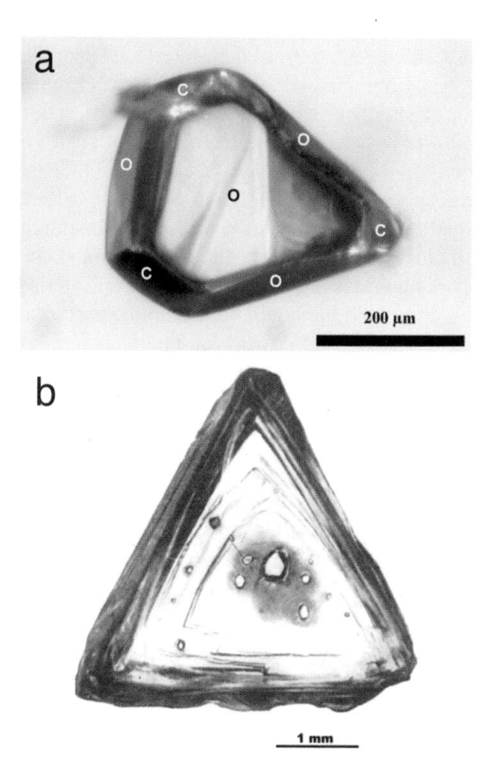

Figure 13. (a) Inclusion of olivine in diamond showing diamond-imposed, flattened cubo-octahedral habit (modified from Nestola et al. 2011). Faces of the dominant octahedral (o) and cubic (c) forms are indicated. (b) Diamond macle (twinned and flattened) with a number of olivine inclusions having major faces parallel to the octahedral diamond face (modified from Sobolev et al. 1972).

The most commonly used proof of syngenesis is the imposition of the morphology of the diamond on the inclusion (Fig. 13; e.g., Harris 1968; Sobolev et al. 1969, 1972; Sobolev 1977; Harris and Gurney 1979; Meyer 1985, 1987; Pearson and Shirey 1999; Sobolev et al. 2009). Compositional consistency with associated mineral inclusions is another important criterion as, for example, in the syngenetic low-Si mica inclusions that are documented in peridotite (U/P-type) and eclogitic (E-type) diamonds as phlogopite and biotite, respectively (Sobolev et al. 2009). The recognition of several inclusions of harzburgitic garnet with diamond-imposed morphology having trace element compositions indicative of multistage geochemical evolution has challenged the morphology criterion (Taylor et al. 2003). The observation that in many cases diamond growth zones, as revealed by cathodoluminescence studies, do not wrap around the inclusions is consistent with, although it does not prove, syngenesis (Bulanova 1995). An epitaxial relationship between an inclusion and its host would represent a more robust proof of syngenicity (e.g., Futergendler and Frank-Kamenetsky 1961; Harris 1968; Harris and Gurney 1979; Wiggers de Vries et al. 2011). Although some apparently recurrent crystallographic orientations with potential epitaxial significance have been found for some inclusions, such orientations are rarely determined and a systematic survey for the different mineral species is lacking or has been restricted to limited sets of samples (see review in Harris and Gurney 1979).

A more recent approach to investigating diamond-inclusion relationships relies on the combination of high-resolution techniques to better understand diamond growth, especially in relation to inclusions. For example Wiggers de Vries et al. (2011) applied the CL technique along with electron backscatter diffraction (EBSD), using FIB-SEM, to study the three-dimensional growth zonation around inclusions in diamond. EBSD orientation mapping revealed that three chromite inclusions in a single diamond studied by these authors have a potential epitaxial relation with the host, within ±0.4°. One of the chromite inclusions is surrounded by a non-luminescent CL halo that has apparent crystallographic morphology with symmetrically oriented pointed features. The CL halo has ~200 ppm Cr and ~75 ppm Fe and is interpreted to have a secondary origin as it overprints a major primary diamond growth structure. The diamond zonation adjacent to the chromite and the morphology of the inclusion records changes in the relative growth rates and habits at the diamond-chromite interface, thus supporting a syngenetic relationship.

A resolutive approach would be one that combines accurate measurement of the crystallographic orientations of the inclusion and its host with calculations of their interfacial energies in a number of possible reciprocal orientations. Crystallographic orientations can be determined with high accuracy and precision by *in situ*, non-destructive, single-crystal X-ray diffraction on the inclusions still trapped in their diamond hosts by adapting the methods developed for high-pressure studies of single crystals in diamond-anvil cells (e.g., Nestola et al. 2011). These methods overcome technical issues related to the accurate visual centering of the inclusions, a common difficulty in routine X-ray diffractometry, thus allowing investigation of diamonds with unfavorable morphology or with multiple inclusions. *Ab initio* quantum-mechanical calculations may then show whether any particular orientation is energetically favored and should hence be expected in the case of syngenesis. This combined methodology cannot be used routinely because it requires dedicated laboratories and equipment. In particular, interfacial energy calculations have never been performed on inclusions in diamond. Until a statistically significant number of crystallographic and interfacial energy data are produced, the classification of any inclusion as syngenetic based purely on morphological or crystallographic criteria should be considered with caution.

Inclusion type and paragenesis. Silicate inclusions in lithospheric diamonds are commonly classified into 2 dominant parageneses—peridotitic (P-type, with harzburgitic and lherzolitic members) and eclogitic (E-type). A minor websteritic paragenesis is present at some localities and a wherlitic paragenesis also can be tentatively identified (Stachel and Harris 2008 and references therein; Fig. 14, Table 1). For garnets, this classification is clearly resolved on the basis of Cr contents, with P-type garnets having > 1 wt% Cr_2O_3 (e.g., Schulze 1983; Fig. 14) and on the basis of Cr# (100Cr/[Cr + Al]) for clinopyroxenes where P-type clinopyroxenes (Cr-diopsides) have a Cr# of 7 to 10 (Stachel and Harris 2008). The websteritic inclusion suite is not as clearly defined and has been used to classify silicates with transitional mineral chemistry between P- and E-type paragenesis. For instance, Gurney et al. (1984) use this classification for garnets with Cr_2O_3 contents > 1 wt% that have abnormally low Mg#. Aulbach et al. (2002) have applied the websteritic classification to garnets and clinopyroxenes that have E-type chemical affinities (Cr_2O_3 in garnet generally <2.5%; low Cr# in clinopyroxene) and chemical traits implying coexistence with orthopyroxene. Grütter et al. (2004) distinguish websteritic garnets as having

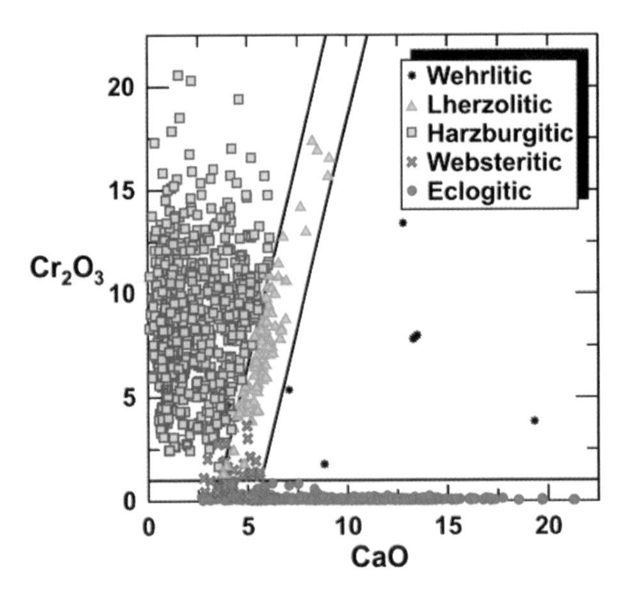

Figure 14. Garnet from a worldwide database on the classification in a plot Cr_2O_3 versus CaO (wt%) with compositional fields of Grütter et al. (2004). See text for discussion. (Figure and caption used by permission of Elsevier Limited, from Stachel and Harris (2008) *Ore Geology Reviews*, Vol. 34, Fig. 4, p. 8).

Table 1. Mineral inclusions in lithospheric diamonds and their associated parageneses (modified from Sobolev 1983).

Type	Mineral parageneses	Principal mineral inclusions in diamond	Specific compositional features	Polycrystalline	Xenoliths
ULTRAMAFIC (PERIDOTITIC)	harzburgite-dunite	**Cr-Prp** (Mg# >85; Ca# <15; Cr# >15), **Ol** (Fo_{92-94}), **Chr** (Cr# >85), **Sf** (Ni > 17 wt%), [Esk, Phl]	Ca-poor Cr-Prp	+	+
	lherzolite	**Prp** (Mg# >80; Ca# >15), **Ol** (Fo_{90-92}), **En**, **Cr-Di**, **Sf** (Ni >17 wt%), [Chr, Phl, Mgs, Ilm, Rt]	Prp with moderate CaO (4-7 wt%)	+	+
	wehrlite	**Cr-Prp** (Mg# >80; Ca# >20 +Ol (Fo_{92}), **Cr-Di**, **Sf** (Ni > 17 wt%), [Chr]	Ca-rich Cr-Prp (Ca# > 20)	+	–
WEBSTERITIC	websterite-pyroxenite	**Prp** (Mg# >70), **Cpx**, **Sf** (Ni 10-17 wt%), [En]	Absence of Ol	+	+
ECLOGITIC	eclogite	**Grt** (Ca# < 50), **Omp**, **Sf** (Ni < 10 wt%), [Rt, Phl (Bt), Dol]	Na in Grt K in Omp	+	+
	coesite eclogite	**Grt** (Ca# < 50), **Omp**, **Coe**, **Sf** (Ni < 10 wt%), [Rt]	Coe	–	+
	ilmenite eclogite	**Grt**(Ca# < 50), **Omp**, **Ilm**, **Sf** (Ni< 10 wt%), [Rt]	Ilm	+	–
	kyanite eclogite	**Grt** (Ca# < 50), **Omp**, **Ky**, **Sf** (Ni < 10 wt%), [Sa, Rt, Coe]	Ky	–	+
	corundum eclogite	**Grt** (Ca# < 50), **Omp**, **Crn**, **Sf** (Ni < 10 t%), [Ky, Rt]	Crn	–	+
	grospydite	**Grt** (Ca# >50), **Omp**, **Ky**, **Sf** (Ni < 10 wt%), [Crn, Rt]	Grt (Ca# > 50)	–	+

The more common mineral inclusions in diamond are shown in boldface; the rarer inclusion minerals are in normal typeface within brackets. Sulfides are common minerals in all diamond parageneses. Sometimes rare minerals like ferropericlase, moissanite and others may be present as inclusions in diamonds. (+) indicates the presence diamond type or rock associated with the paragenesis; whereas (–) indicates their absence. Abbreviations of minerals are after Whitney and Evans (2010) as follows: chrome-pyrope (Cr-Prp), olivine (Ol), chromite (Chr), sulfide (Sf), eskolaite (Esk), phlogopite (Phl), pyrope (Prp), enstatite (En), chrome-diopside (Cr-Di), magnesite (Mgs), ilmenite (Ilm), rutile (Rt), clinopyroxene (Cpx), garnet (Grt), omphacite (Omp), kyanite (Ky), sanidine (Sa), coesite (Coe), and corundum (Crn). Polycrystalline diamond includes boart, framesite and diamondite (see text for details). Mg# = 100Mg/(Mg+Fe); Ca# = 100Ca/(Ca+Mg+Fe+Mn); Cr# = 100Cr/(Cr+Al).

relatively low Cr_2O_3 (<2.5%) and CaO contents of <6 wt%. Sodium content can be useful as Sobolev and Lavrent'ev (1977) noted that E-type garnets contained elevated Na_2O (>0.1%)

The Cr_2O_3 versus CaO plot for garnets (Fig. 14) has developed into a central means of classifying garnets and is a key diamond exploration tool (Gurney and Switzer 1973; Sobolev et al. 1973; Schulze 1983; Gurney et al. 1984; Grütter et al. 2004). In this compositional space, Sobolev et al. (1973) first identified garnets that had not equilibrated with clinopyroxene as having low-Ca, high-Cr characteristics and designated these garnets as having come from highly-depleted harzburgitic to dunitic lithologies. In contrast, lherzolitic garnets occupy a distinct linear trend originating from circa 2.5 wt% Cr_2O_3 and 3 wt% CaO with a slope of ~ 0.3 CaO to 1 Cr_2O_3 (Fig. 14). Gurney (1984) defined a similar lherzolitic trend based on a line that separated 85% of diamond inclusion garnets with Cr_2O_3 >4 wt%, extending upwards in Cr-Ca space with a similar slope to the Sobolev trend. Grütter et al. (2004) proposed more precise boundaries to the lherzolitic field, which are largely the same as the early classifications as well as more clearly defining the eclogitic and websteritic compositional fields on this plot. Sobolev (1977) likewise distinguished high-Mg and Cr chromites (Fe/(Fe + Mg) <50%; Cr_2O_3>62 wt%) as characterizing diamond inclusion spinels and reflecting a harzburgite association.

Mirroring the approach to silicates, a basic subdivision of sulfides into P- and E-type chemistries, based predominately on Ni-content (Table 1) was proposed by Yefimova et al. (1983), although Deines and Harris (1995) showed a clear compositional overlap. Subsequent studies of sulfides for Re-Os dating, while adding to this continuum, have demonstrated that Os content (i.e. 3-200 ppb for E-types and ~2,000-30,000 ppb for P-types) is a more sensitive discriminant (e.g, Pearson and Shirey 1999).

The study of mineral inclusions in sublithospheric diamonds is in a youthful stage compared to the study of inclusions in lithospheric diamonds in part due to the rarity of specimens, small grain size, and difficulties in recognizing original high-pressure minerals from their low-pressure, retrograde assemblages (e.g., Table 2). Regardless, sublithospheric inclusions can be divided into ultramafic (peridotitic) and basaltic (eclogitic) types in a parallel manner to inclusions in lithospheric diamonds. Ultramafic types are characterized by high-pressure magnesium-rich phases such as Mg-perovskite, ringwoodite, wadsleyite, and olivine with ferro-periclase, majorite, and Ca-perovskite and their low pressure breakdown products (Table 2). Basaltic types are characterized by assemblages richer in basaltic components such as Ca, Al, Si, and Ti including majorite, clinopyroxene, CaTi-perovskite, Ca-perovskite, Ca-ferrite, stishovite, and the "new aluminum phase" (NAL; Table 2).

Trace elements. Pioneering work (Sobolev and Lavrent'ev 1971; Sobolev et al. 1972; Hervig et al. 1980), using elevated beam currents and extended counting times, allowed the measurement of a limited number of trace elements in silicates included in diamonds. The advent of the ion microprobe allowed the first analyses of a broad spectrum of trace elements, including the petrogenetically useful rare earth elements (REE), which indicate the extreme enrichment of incompatible elements in fluids that formed them (Shimizu and Richardson 1987). While laser-ablation ICPMS has the potential to make the same analyses (Davies et al. 2004), the significantly more destructive nature of this technique has meant that trace element studies focused almost exclusively on the ion-microprobe technique (Ireland et al. 1994; Shimizu and Sobolev 1995; Shimizu et al. 1997; Stachel and Harris 1997; Stachel et al. 1998b, 1999, 2000, 2004; Harte et al. 1999; Aulbach et al. 2002; Promprated et al. 2004; Tappert et al. 2005b).

Stachel et al. (2004) pointed out that, from a global perspective, peridotitic garnets included in diamonds from cratonic lithosphere show a full spectrum of REE patterns from those known as sinusoidal (i.e., they have positive slopes from La to Nd or even Eu, negative slopes in the middle REE and positive slopes again from Ho or Er to Lu; Fig. 15a), to those that are light REE depleted with flat middle REE- to heavy REE sections. The light REE enrichment

Table 2. Mineral inclusions in sublithospheric diamonds and their associated parageneses (modified from Harte 2010 and references therein, with additions).

Type	Mineral facies	Approximate Depth (km)	Indicative mineral assemblage	Retrograde transformations	Principal mineral inclusions in diamond
ULTRAMAFIC (PERIDOTITIC)	upper mantle	< 410	Ol+Maj+Cpx	Maj→Cpx+Grt	Ol, Grt, Cpx
	transition zone	410 - 660	Wds/Rwd+Maj±CaPv	Wds/Rwd→Ol, Maj→Cpx+Grt, CaPv→Wal	Ol, Grt, Cpx, Wal
	UM/LM boundary association	~ 660	Rwd+MgPv(low-Al)+FePer+Maj+CaPv	Rwd→Ol, MgPv→En, CaPv→Wal	Ol, En, FePer, Wal
	lower mantle	> 660	MgPv(Al)+FePer+CaPv	MgPv(Al)→En, CaPv→Wal	En(Al), FePer, Wal
BASALTIC (ECLOGITIC)	upper mantle	~ 300 - 450	Maj+Cpx±CaTiPv	CaTiPv→Pv+Wal	Grt, Cpx, Pv, Wal
	transition zone	450 - 600	Maj+Sti±CaTiPv±CaPv	Maj→Cpx+Grt, CaTiPv→Pv+Wal Sti→Qz/Coe	Grt, Cpx, Pv, Wal, SiO_2
	UM/LM boundary association	~600 - 750	Maj±CaTiPv±CaPv ±NAL+Sti	Maj→Cpx+Gar, NAL→Spl+Kls, Sti→Qz/Coe	NaAlPx, Gar, Cpx, Pv, Wal,Spl, Kls, SiO_2
	lower mantle	> ~750	MgPv(Fe, Al)+CaPv+CaFrt+NAL+Sti	MgPv→TAPP±En±Spl, NAL→Spl+Kls CaFrt→Spl+Nph, Sti→Qz/Coe	En, TAPP, Spl, Kls, Nph, SiO_2

The indicative mineral assemblages of the depth zones of sublithospheric inclusions in diamond are never seen due to exsolution and transformations to low-pressure phases as indicated. Furthermore, the rarity of these inclusions and their small size means that, in most cases, crystal structure is inferred from elemental composition. Therefore, the depth ranges of these diamonds with inclusions are approximate because they are derived from the principal mineral inclusions observed in the diamond at low pressure. UM= upper mantle, LM=lower mantle. Abbreviations of minerals are after Whitney and Evans (2010) as follows: olivine (Ol), majorite (Maj), clinopyroxene, (Cpx), wadsleyite (Wds), ringwoodite (Rwd), calcium perovskite (CaPv), magnesium perovskite (MgPv), ferropericlase (FePer), calciu-titianium perovskite (CaTiPv), stishovite (Sti), new aluminum phase (NAL), calcium ferrite (CaFrt), garnet (grt), walstromite (Wal), quartz (Qz), coesite (Coe), spinel (Spl), kalsilite (Kls), nepheline (Nph), enstatite (En) , perovskite (Pv), and tetragonal almandine pyrope phase (TAPP).

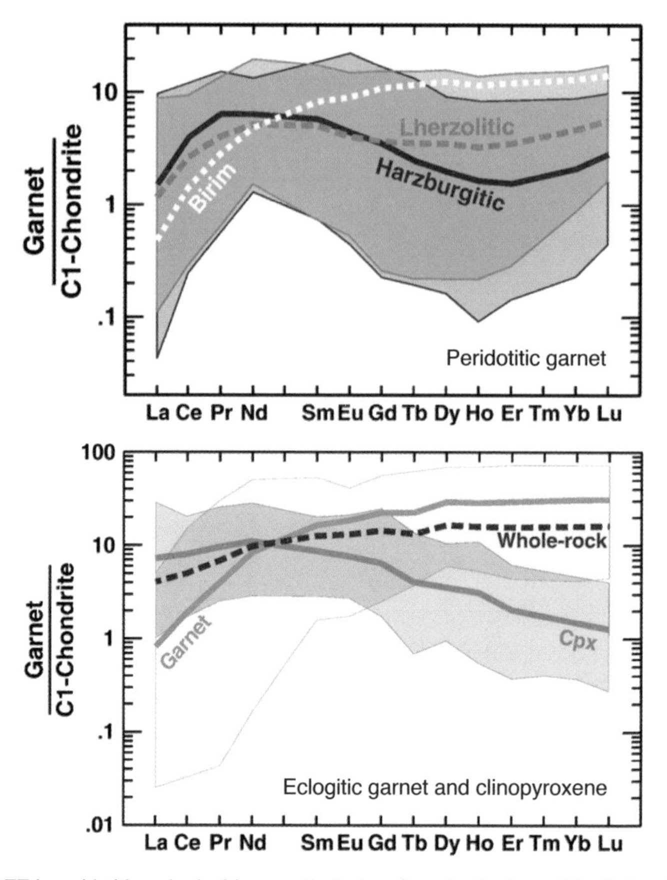

Figure 15. REE in peridotitic and eclogitic garnet inclusions from the database of Stachel et al. (2004) and normalized to C1-chondrite (McDonough and Sun 1995). The top diagram shows the compositional fields and average compositions of harzburgitic and lherzolitic garnets. The average composition of lherzolitic garnets from Birim diamonds (Akwatia, Ghana; Fig. 3) is shown to represent "normal" REE patterns generally (but not exclusively) restricted to the lherzolitic paragenesis. The strongly sinusoidal REE patterns typically observed for the harzburgitic paragenesis represent depleted garnets that have been re-enriched in light REE. This pattern requires a metasomatic agent with extremely high light REE to heavy REE, believed to indicate fluid metasomatism. Ranges and average REE contents of eclogitic garnet and clinopyroxene inclusions are shown in the bottom panel. Calculated REE pattern for eclogitic whole-rock (blue dashed line) assumes a modal garnet-clinopyroxene ratio of 1:1. (Figures and captions used by permission of Elsevier Limited, from Stachel and Harris (2008) *Ore Geology Reviews*, Vol. 34, Figs. 21 and 22, pp. 17 and 19).

and strongly sinusoidal patterns are generally restricted to very depleted (high-Cr, low-Ca), harzburgitic major element chemistries in diamond inclusions (Stachel et al. 2009) and garnets in diamond-bearing peridotites (Nixon 1987; Pearson et al. 1995; Shimizu et al. 1997; Stachel et al. 1998b; Klein-BenDavid and Pearson 2009) and this pattern has led to the suggestion that this signature is characteristic of the imprint left on wall-rocks by the passage of diamond-forming fluids. Garnets with strongly sinusoidal REE patterns often have Sr enrichments of 10 to >40 ppm (Pearson et al. 1995; Shimizu and Sobolev 1995) and these features, together with other characteristics, have led to the prevalent interpretation that these garnets equilibrated with C-O-H fluids (e.g., Stachel et al. 2004) or carbonatitic fluids/melts (Navon 1999) associated with diamond formation. The classic light REE depleted, typically more fertile lherzolitic garnets seem to have last equilibrated with a silicate melt. Stachel et al. (2004) see the continuum between these two end-member garnet types as resulting from a spread of melt-fluid compositions generated by fractional crystallization and reaction with lithospheric wall rocks. Burgess and Harte (2004) called such a process "percolative fractionation." Isotopic studies of diamond-forming fluids indicate that there are multiple fluid sources involved in diamond genesis (McNeill et al. 2009; Klein-BenDavid et al. 2010) and hence the "C-O-H fluid" designation is likely to encompass fluids of differing types. The co-variation of Zr with Y in garnets can also be used to characterize the different metasomatic interactions that mantle garnets may have experienced (Griffin et al. 1999). When garnets included in diamonds are examined in this way (Stachel and Harris 2008; their Fig. 22), the majority of harzburgitic garnets plot in the field for garnets that have experienced large amounts of melt depletion, with a clear trend of Zr enrichment that is usually associated with phlogopite metasomatism. This trend may be reflective of one of the sources of fluids being derived from mica-rich metasomes within the lithosphere (Klein-BenDavid et al. 2010). In contrast, lherzolitic garnets range from the melt-depleted field into the regions of coupled Zr and Y enrichment that signifies silicate melt metasomatism.

A subset of very depleted, high-Cr garnets with either very high equilibration temperatures (Buffalo Head Hills, Alberta; Banas et al. 2007) or majoritic (high-Na) compositions (Promprated et al. 2004) could indicate, in some locations, the possible presence of very deep (>300 km) lithospheric mantle in places, or derivation from detached slabs of basal lithosphere by kimberlites on route to the surface. Such occurrences, while rare, offer a valuable opportunity to better understand mantle geodynamics and further study is warranted.

Despite the plethora of analyses from many different cratons, the exact nature of the parental diamond melt-fluid to peridotitic diamonds remains unconstrained, because of the multiple stages of depletion and metasomatism that silicates included within diamonds have experienced. Difficulties in the geochemical interpretations include the assumption of equilibrium to calculate parental fluids/melts using trace element partitioning data and the likely lack of applicability of the partitioning data due to strong differences in distribution coefficients related to poorly constrained parental fluid compositions. Trace element partitioning data between lherzolitic and high-Cr harzburgitic garnets over a range of C-O-H and carbonatitic fluid/melt compositions, would greatly assist our understanding of P-type diamond formation, notwithstanding the technical difficulties of such experiments.

E-type silicates included in diamonds show much less trace element variability that P-type silicates. Eclogitic garnet inclusions have light REE depleted patterns that show broad similarities to garnets from crustal eclogites, with light REE ~ 1× chondritic abundances and heavy REE ~ 30× chondritic (Fig. 15b; e.g., Ireland et al. 1994; Taylor et al. 1996; Stachel et al. 1999, 2000, 2004). REE patterns for clinopyroxenes (omphacites) appear in broad equilibrium with the garnets, with light REE enrichment and heavy REE at ~ 1× chondritic. Ireland et al (1994) noted that the major and incompatible trace element compositions of eclogitic silicates included in diamonds were more depleted than host eclogite xenoliths and interpreted these compositions as reflecting the extraction of a TTG melt during subduction of an oceanic crustal precursor that experienced eclogite facies metamorphism. The observation of positive and negative Eu anomalies in both garnet and clinopyroxene inclusions supports an origin via oceanic crustal protoliths (e.g., Promprated et al. 2004; Stachel et al. 2004), in agreement with the widely accepted origin of most eclogite xenoliths erupted by kimberlites through cratons (e.g., MacGregor and Manton 1986; Jacob 2004).

While the parentage of the E-type silicates seems straightforward with the available data, the origin of the diamonds that surround E-type silicates is not. The simplest hypothesis for the origin of E-type diamonds would be via solid-state growth from a carbon-bearing crustal precursor. However, there is little evidence for the solid-state growth of most diamonds (see review in Stachel and Harris 2009) and increasing evidence for their metasomatic growth. The strongest evidence for a metasomatic origin for diamonds come from studies that document extreme chemical variations across multiple inclusions in the same diamond (Sobolev and Efimova 1998; Taylor et al. 1998; Keller et al. 1999; Bulanova et al. 2004), and relations between diamonds and host silicates in diamondiferous xenoliths revealed by microscopy (Spetsius et al. 2002; Spetsius and Taylor 2008) and X-ray micro-tomography (Keller et al. 1999; Taylor et al. 2003; Anand et al. 2004). The range in silicate inclusion equilibration temperatures indicated by diamond inclusion thermobarometry, from supra- to sub-solidus, has also been used to support a metasomatic origin for most diamonds under melt-dominated (eclogite and lherzolite) and C-O-H fluid-dominated (harzburgitic) conditions (Stachel and Harris 2008). A temporal variability has been suggested in this process, on the basis of lithospheric redox and C-isotope compositions, with Meso- to Paleoarchean diamonds possibly forming via reduction of methane-rich fluids permeating the lithosphere, whereas in post-Archean times, reduction of carbonate-rich melts better explains the C isotopic systematics (Stachel and Harris 2009). This view contrasts however with vanadium-scandium systematics (e.g., Canil 2002), which do not highlight any significant secular change in mantle oxygen fugacity.

Key aspects of these models require better understanding: 1) the temporal evolution of the lithospheric mantle redox state, 2) the mechanism of diamond formation via interaction of melt and C-O-H fluid with mantle wall rocks, and 3) the origin of the proposed metasomatic fluids—are there local sources of carbon that are remobilized over centimeters, or fluids streaming through the lithosphere on kilometer scale-lengths?

Age systematics and isotopic compositions. Currently, it is not possible to date monocrystalline diamonds by direct analysis of the diamond crystal. Fibrous diamonds, being related to the kimberlite magmatism, have an age very close to that of kimberlite. But with only poorly aggregated nitrogen, the exact residence time (million year time scales) and temperature(s) in the mantle remains unclear. As such, all viable ages produced so far have been obtained by the analysis of solid inclusions within diamonds that are assumed to be syngenetic (see above) with the diamonds; the reader is directed to the isotopic age dating review of Pearson and Shirey (1999) for more details. The first dating studies were performed on sulfide (Kramers 1979) and silicate (Richardson et al. 1984) inclusions and indicated the likely antiquity of the host diamonds. Both these groundbreaking studies were made by pooling together numerous (sometimes >100) inclusions because of analytical constraints. Subsequently, the Sm-Nd isotope system became the method of choice and yielded a number of isochron ages from suites of diamonds from southern Africa (Smith et al. 1991; Richardson et al. 1990, 1999, 2004), Siberia (Richardson and Harris 1997), and Western Australia (Richardson 1986).

Some studies have questioned the validity of ages obtained on composites of inclusions (see discussion in Navon 1999) and this uncertainty drove the need to make analyses on single inclusions. The Ar-Ar method was the first to be applied to single clinopyroxene inclusions in diamonds (Phillips et al. 1989; Burgess et al. 1992) because eclogitic omphacite contains sufficient potassium to allow age determinations. It soon became clear that unexposed, pristine inclusions are essential for this approach due to potential diffusion of radiogenic Ar to the silicate-diamond interface, yielding ages that range upwards from the kimberlite eruption age in cleaved diamonds (Burgess et al. 1992). Because of the potential for incorporation of the locally abundant ambient ^{40}Ar in the mantle (e.g., Pearson et al. 1998), the ages can be viewed as absolute maxima for encapsulation of the inclusion by the diamond. Orapa eclogitic pyroxenes analyzed by Burgess et al. (2004) gave ages of 906 to 1032 Ma, consistent with previously determined Sm-Nd ages, with a few samples yielding ages >2500 Ma, hinting at the presence of multiple diamond age populations in this kimberlite. These authors also found ages of 520 Ma for eclogitic omphacites from the Venetia kimberlite, indicating a population of diamonds that formed shortly before kimberlite eruption. In summary, Ar-Ar geochronology studies of eclogitic pyroxenes from southern Africa largely confirm the results of the Sm-Nd approach, indicating E-type diamond formation from the Neoarchean onwards.

Analysis of single inclusions in diamond has been most effectively realized using the Re-Os isotope system in sulfides. The relatively high Re and Os contents of sulfides from both E- and P-type parageneses, allow analyses of Os in the sub-picogram to nanogram range (Pearson et al. 1998, 1999b; Pearson and Shirey 1999). The focus on obtaining relatively large sulfides for analysis has lead to an apparent bias towards dating studies involving E-type diamonds (e.g., Pearson et al. 1998; Richardson et al. 2001, 2004; Shirey et al. 2004a, 2004b; Aulbach et al. 2009a) although P-type sulfide inclusions have been analyzed in some instances (Pearson et al. 1999a, 1999b; Westerlund et al. 2006; Smith et al. 2009; Smit et al. 2010).

The general picture of lithospheric diamond formation revealed by Re-Os dating is that there are multiple diamond ages within one kimberlite (e.g., Pearson et al. 1998; Richardson et al. 2004; Aulbach et al. 2009a) and that all E-type sulfide-bearing diamonds analyzed so far appear to have formed in the Neoarchean and later (Pearson et al. 1998; Pearson and Shirey 1999; Richardson et al. 2001, 2004; Shirey et al. 2004b, 2008; Aulbach et al. 2009a,b; Laiginhas et al. 2009; Shirey and Richardson 2011). In contrast, diamonds containing P-type sulfides are older, having started to form from the Mesoarchean onwards (Pearson et al. 1999a, 1999b; Westerlund et al. 2006; Aulbach et al. 2009b; Smit et al. 2010; Shirey and Richardson 2011), with the exception of a single Mesozoic diamond from Koffiefontein (Pearson et al. 1998) and one from Jagersfontein (Aulbach et al. 2009b).

In addition to these "mainstream" approaches, there has been a small number of studies using the U-Pb system on zircon (Kinny and Meyer 1994) and yimengite (Hamilton et al. 2003). Bulanova et al. (2004) also made Ar-Ar age determinations on yimengite inclusions in a diamonds from the Sese kimberlite, Zimbabwe, producing apparent ages from 538 to 892 Ma. These studies yielded relatively young formation ages. Such phases that may be related to the proto-kimberlite melts, and are part of a growing body of evidence, augmented by Re-Os isotopes (Pearson et al. 1998; Aulbach et al. 2009a), Sm-Nd and Ar-Ar studies in E-type clinopyroxenes (Richardson 1986; Burgess et al. 2004) and N-aggregation systematics, that indicate a proportion of gem diamond growth shortly before kimberlite eruption.

So far only 2 age determinations have been made on ultra-deep, sub-lithospheric diamonds. Bulanova et al. (2010) made a U-Pb ion probe determination of a Ca-silicate perovskite (re-equilibrated to walstromite) diamond inclusion from the Collier-4 kimberlite, Brazil that yielded an age of 107 ± 7 Ma, only 14 Ma older than the pipe emplacement age of 93 Ma. This study did not use matrix-matched standards and, as such, the date must be viewed as preliminary. However, the results are supported by nitrogen aggregation data for super deep diamonds from this pipe, which imply, for an assumed temperature of ~1500 °C, a maximum mantle residence time of <10 Ma (Bulanova et al. 2010). In contrast, a single sulfide inclusion in an ultra-deep diamond from Juina examined by Hutchinson et al. (2012) indicated a formation age likely to be significantly in excess of 500 Ma, considerably older than the circa 90-Ma pipe emplacement age. This sort of age (400 to 800 Ma) is seen for some radiogenic isotopic systems (e.g., Sm-Nd, Lu-Hf) in oceanic basalts and abyssal peridotites and thus is consistent with this diamond having grown deep in the convecting oceanic mantle. A similar conclusion can be drawn from the Sr and Nd isotopic compositions reported in majoritic garnet inclusions in diamonds from the mantle transition zone, also from the Brazilian craton (Sao Luis), which plot in the middle of the oceanic Sr and Nd isotopic array and distinctly different from the garnet inclusions from lithospheric diamonds (Harte and Richardson 2011). The scarcity of data from sulfides in ultra-deep diamonds has so far restricted age information on ultra-deep diamonds and this is certainly an area that will see more effort in the future.

Stable isotopic signatures in diamond inclusions. The origin of diamond can also be addressed from the study of O- and S-isotopes in silicate and sulfide inclusions ($n < 20$ and $n < 50$, respectively). Although fairly limited data are available, both provide evidence for the involvement of subduction-related material. The $^{18}O/^{16}O$-isotope variability ($\delta^{18}O$ from +4 to +16‰) of eclogitic diamond inclusions, contrasts with the mantle homogeneity displayed by peridotite xenoliths and diamond inclusions and compares with the known range measured in altered ocean crust (see Lowry et al. 1999; Schulze et al. 2003; Anand et al. 2004). Although often equilibrated at lower pressure and temperature, eclogite xenoliths have O-isotope compositions consistent with those in eclogitic diamond inclusions (see Jacob 2004 for review). Eclogitic sulfide inclusions also display variable $^{34}S/^{32}S$ ratios ($\delta^{34}S$ from −11 to +14‰) that compare well with sediments and altered oceanic crust, but it must be emphasized that the most recent studies did not reproduce such a large range of values (Farquhar et al. 2002; Thomassot et al. 2009 and references therein). Unambiguous evidence for the involvement of subduction-related sulfur in eclogitic diamond inclusions is brought from the recognition of mass-independent fractionations of sulfur isotopes (i.e., $\delta^{33}S \neq 0.5 \times \delta^{34}S$) within sulfide diamond inclusions (Fig. 16; Farquhar et al. 2002; Thomassot et al. 2009). The only known geologically relevant process to be associated with sulfur isotopic compositions with mass-independent fractionations is the UV-photolysis of sulfur-bearing molecules in an O_2-deprived composition: on Earth, such conditions were met during the Archean Eon (Farquhar et al. 2000). Therefore the identification of mass-independent fractionations of sulfur isotopes within sulfides eclogitic diamond inclusions (Farquhar et al. 2002), and their absence within peridotitic diamonds (Cartigny et al. 2009), demonstrate the occurrence of recycled Archean sedimentary sulfur in the former.

GEOLOGY OF MANTLE CARBON FROM DIAMONDS

Geodynamics, carbon mobility and reservoirs

Diamond—the mechanisms by which it crystallizes, the relationship it bears to the explosive kimberlite host magmas that deliver it, and the distribution of different diamond types in a geologic context—provide the key to understanding the carbon cycle in the deep mantle.

Continent assembly, plate tectonics, and ancient carbon recycling. As a consequence of old cratons having preserved mantle keels, the diamond record remains one of the prime ways to examine continental tectonics from mantle depths and avoid the later overprinting effects of metasomatism. It is also a main way to get some idea of Earth's ancient igneous carbon cycle.

Shirey and Richardson (2011) extended the observations of isotopic ages to all diamond formation ages so far determined and combined them with the diamond type to show that only P-type diamonds were forming in the lithosphere during the Paleoarchean to Mesoarchean, whereas E-type diamonds began to form post-3 Ga and became the prevalent type of diamond formed (Fig. 17). These authors went on to propose, based on the diamond inclusion geochronology database that includes both the Sm-Nd data on silicate inclusions and the Re-Os data on sulfides, that continental dispersion and subduction tectonics as we know it in the modern Earth—the Wilson cycle—initiated in the Neoarchean. The transition to an Earth dominated by lateral tectonics and subduction is not well explained at the present time, but the 3-Ga shift to eclogitic inclusions does correlate with a major change in crustal geologic style (Van Kranendonk 2010, 2011) and crustal growth mechanisms (Dhuime et al. 2012), which do support the Wilson Cycle onset conclusion. It is not known, for example, whether the appearance of eclogite simply signals a more effective capture of slabs by a buoyant keel in which they can be retained or the actual appearance of large slabs in the geologic record. If it is the latter, this change may have important implications for the nature of carbon-bearing fluids and their delivery to diamond-forming depths in the mantle. Using skewness in the carbon isotopic composition histogram of P-Type diamonds,

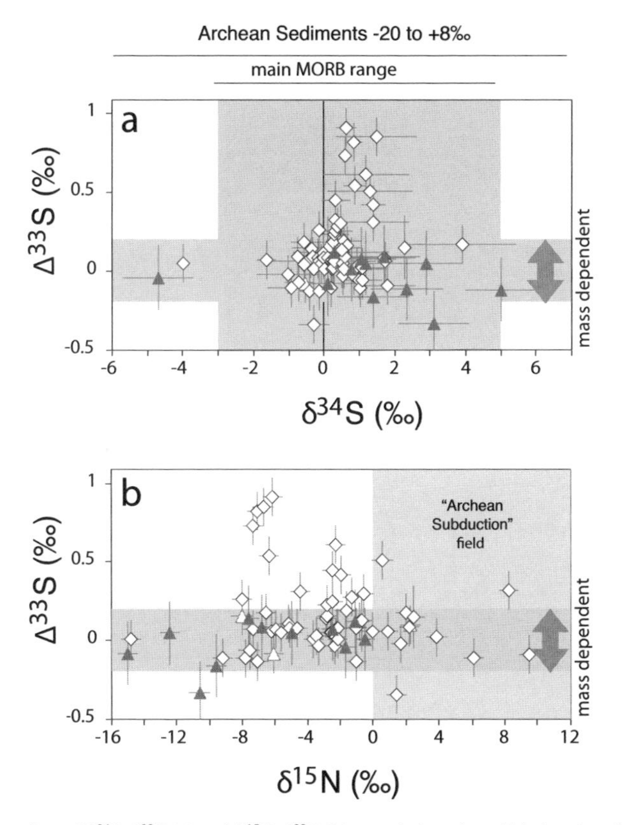

Figure 16. Illustration of $\delta^{34}S$-$\Delta^{33}S$ (a) and $\delta^{15}N$-$\Delta^{33}S$ (b) covariations in sulfide-bearing diamonds. Filled triangles and filled diamonds correspond to sulfides from peridotitic and eclogitic diamonds respectively from the Ekati mine, Canada (Panda kimberlite; Cartigny et al. 2009); open diamonds are eclogitic sulfides from the Jwaneng and Orapa kimberlites, Botswana (Farquhar et al. 2002; Thomassot et al. 2009). Note that almost no samples fall within the Archean subduction field. The inconsistency between the evidence for the occurrence for subucted sulfur in sulfide inclusion (from non zero $\Delta^{33}S$) and absence (from negative $\delta^{15}N$) could be reconciled considering metasomatic diamond formation enclosing a pre-existing sulfide although the N isotopic composition of the Archean mantle is poorly known (see text for details). Dashed line: conductive geotherms for a surface heat flow of 40 mW/m^2 after Pollack and Chapman (1977); black solid line: graphite-diamond boundary after Day (2012); grey band: T range for mantle adiabat based on mantle potential temperatures of 1300 to 1400 °C. (Used by permission of Elsevier Limited, from Cartigny et al. (2009) *Lithos*, Vol. 112S, Fig. 6, p. 861)

Stachel and Harris (2009) proposed that there may a change in the mechanism by which older diamonds form from methane oxidation in the Mesoarchean Era to the way that younger diamonds form by carbonate reduction in the Proterozoic Eon. If this observation is combined with the proposed onset of the Wilson Cycle, it could signify a change from the geodynamic processes that would favor primary mantle devolatilization and/or the outgassing of recycled reduced fluids, to the geodynamic processes that favor carbonate recycling via slab subduction.

Diamond inclusion populations and ages have also been linked to broad-scale regional lithosphere evolution. A number of authors have made the link between diamond formation events or pulses and large-scale thermo-tectonic events recorded in the craton crust (Shirey et al. 2002, 2004b; Richardson et al. 2004; Pearson and Wittig 2008; Richardson and Shirey 2008; Aulbach et al. 2009a,b). Shirey et al. (2002, 2004a,b) have noted that in southern Africa, mantle

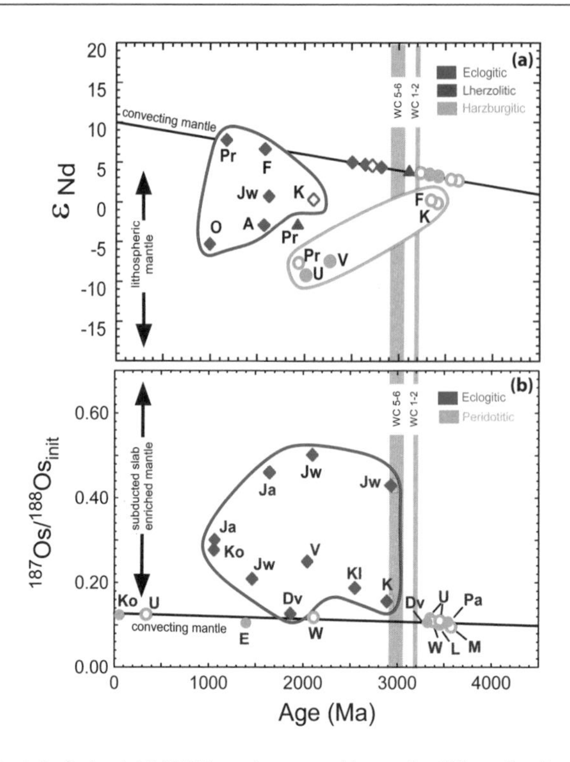

Figure 17. (a) Silicate inclusion initial Nd isotopic composition vs Sm-Nd age for diamonds of peridotitic (circles), lherzolitic (triangle), and eclogitic (diamonds) parageneses. Closed symbols are isochron studies for composites of garnet and clinopyroxene grains. Open symbols are model age studies for composites of garnet only. Unlabeled points on convecting mantle curve are mantle extraction ages extrapolated from labeled points. (b) Sulfide inclusion initial Os isotopic composition vs Re-Os age. Closed symbols are isochron studies; open symbols are model age studies for single grains. For isochron studies, mantle extraction ages extrapolated from labeled points are typically <100 million years (one scale division) older than the isochron age. WC 1-2 represents the Wilson Cycle rifting (stage 1-2) for the Pilbara craton (Van Kranendonk et al. 2010) whereas WC 5-6 represents the Wilson Cycle continental closure (stage 5-6) for the Kaapvaal craton. Locality abbreviations are as follows: (Pr) Premier, (O) Orapa, (A) Aryle, (Jw) Jwaneng, (F) Finsch, (Ko) Koffiefontein, (U) Udachnaya, (V) Venetia, (K) Kimberley pool, (E) Ellendale, (W) Wellington, (Dv) Diavik, (Kl) Klipspringer, (Pa) Panda, (M) Murowa, (L) Letseng. (Figure and caption used by permission of the American Association for Advancement of Science, from Shirey and Richardson (2011) *Science*, Vol. 333, Fig. 1, p. 434).

lithosphere with slower seismic P-wave velocity relative to the craton average correlates with a greater proportion of E-type versus P-type silicate inclusions in diamonds, a greater incidence of younger (Mesoproterozoic Era) Sm-Nd inclusion ages, a greater proportion of diamonds with light C-isotope compositions and a lower proportion of low-N diamonds (Fig. 18). This correlation was proposed to result from Proterozoic modification of Mesoarchean lithosphere by large-scale tectono-magmatic events in the Proterozoic Eon, which added new lherzolitic and eclogitic diamonds to an original harzburgitic inventory of diamonds. In this case, diamond-forming fluids equilibrated with pre-existing silicates and incorporated them as inclusions, retaining the mineralogical differences imparted by the sub-lithospheric magmatism of the Bushveld Complex, now retained as fossil seismic velocity differences (Fig. 18).

The craton-wide pattern for southern Africa displayed by sulfide-bearing diamonds is different than the pattern seen with silicate-bearing diamonds. This difference stems from the

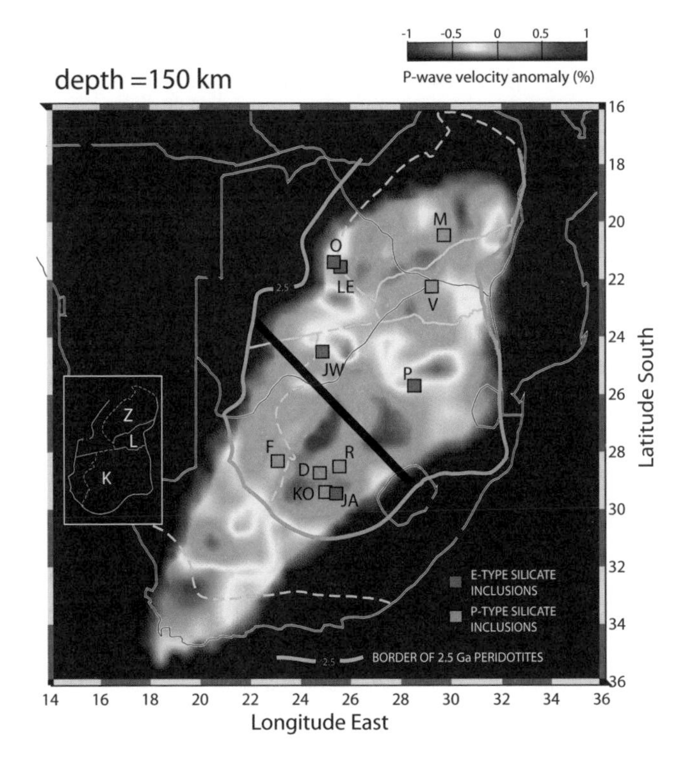

Figure 18. Tomographic image of the lithospheric mantle derived from seismic P-wave data at a depth of 150 km (e.g., James et al. 2001; Fouch et al. 2004). The Kaapvaal and Zimbabwe cratonic blocks and the Limpopo mobile belt (K,L, and Z of inset) collectively comprise the Kaapvaal-Zimbabwe craton. Bold green line indicates the outermost boundary of the Kaapvaal-Zimbabwe craton as defined by the break between Archean and Proterozoic Re-Os ages for peridotite xenoliths (Carlson et al. 1996; Carlson et al. 1999; Janney et al. 2010). Colored squares are diamond mines: red = predominantly eclogitic silicate inclusions (Letlhakane, LE; other abbreviations in caption to Fig. 17) and green squares = predominantly peridotitic silicate inclusions (Roberts Victor, R; other abbreviations in caption to Fig. 17). (Figure and caption used by permission of the American Association for Advancement of Science, from Shirey et al. (2002) *Science*, Vol. 297, Fig. 1, p. 1634).

close association of the sulfide-bearing diamonds with sulfide-containing eclogite that is the source of the diamond fluids and the occurrence of sulfides in different diamonds than those that contain silicate inclusions. The Kaapvaal craton was assembled from two independent continental blocks after 2.97 Ga (Schmitz et al. 2004). Westward-facing subduction underneath the western or "Kimberley" block (Schmitz et al. 2004) effectively made the western block the hanging wall for the fluids and sulfur carried by the oceanic slab (e.g., Aulbach et al. 2009b) and the western cratonic keel the recipient of any eclogite that could be incorporated. The surface distribution of diamond ages and types is a result of this process. West of the suture, all diamond mines contain 2.9 Ga diamonds whereas east of the suture 2.9 Ga diamonds are absent (Fig. 19). If younger ages occur it is either where the lithosphere was subject to subduction at its margin or where persistent, pervasive sub-lithospheric magmatism took place (e.g., Bushveld complex; Fig. 19). The systematic relationship of diamond age and type to the geologic processes that have affected the deepest parts of the continental lithosphere, besides being an excellent exploration model, makes diamond formation a premier tracer of the passage of ancient carbon-rich fluids.

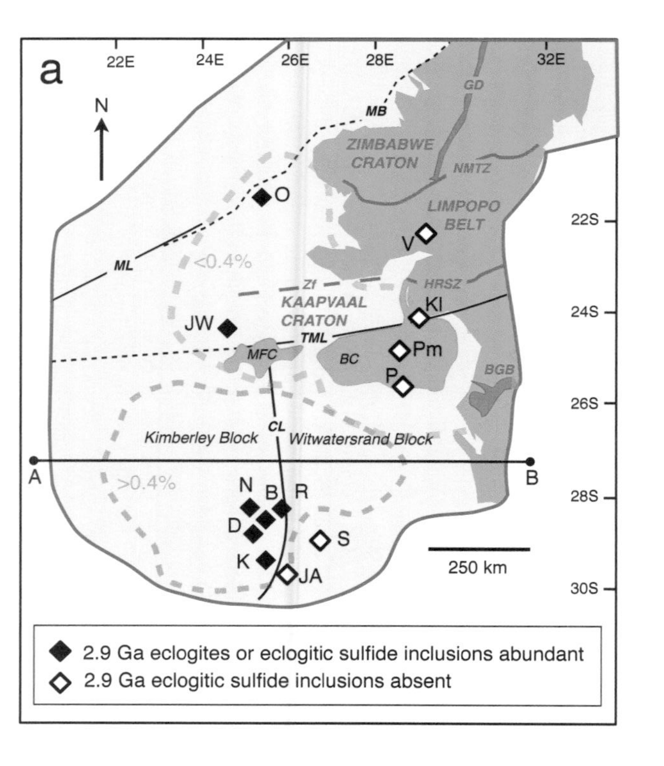

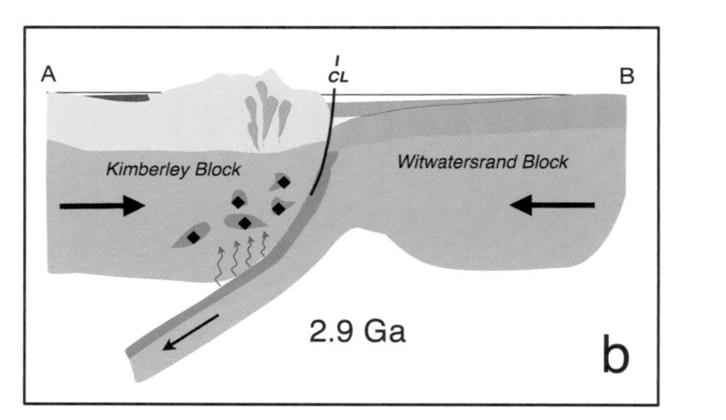

Figure 19. (a) Schematic map view of the Kaapvaal-Zimbabwe craton of southern Africa showing the localities where suites of eclogitic sulfide inclusions or whole rock eclogites have been studied and give 2.9 Ga ages (black diamonds) versus eclogitic sulfide suites where 2.9 Ga ages are absent (white diamonds) superposed on a geological basemap modified from McCourt et al. (2004) and Richardson et al. (2009). Note the assymmetry in age distribution. Data from the literature with locality abbreviations as in Figures 17, 18 with the addition of Star (S; Schmitz unpub data), Palmietgat (Pm; Simelane unpub data) and Klipspringer (Kl; Westerlund et al. 2004). Area of Archean outcrop in pink, Archean covered by supracrustal rocks in blue, outcrop of the Proterozoic Bushveld Complex (BC) and Molopo Farms Complex (MFC) in orange. The region of anomalously high P-wave velocity (>0.4%, blue dashed line) and low P-wave velocity (<0.4%, orange dashed line) is shown for the mantle lithosphere at 150 km depth (James et al. 2001; Fouch et al. 2004). The craton boundary (see sources in Fig. 18 caption) is shown by the red line, the suture between its two craton halves (Schmitz et al. 2004) by the black line labeled CL (Coleburg lineament). Other structural and igneous elements of the Kaapvaal-Zimbabwe craton (McCourt et al. 2004) as follows: Great Dyke (GD), Barberton greenstone belt (BGB), Thabazimbi-Murchison Line (TML), Hought River Shear Zone (HRSZ), Zoetfontein Fault (ZF), Northern Marginal Thrust Zone (NMTZ), southern margin of Magondi Belt (MB), Makgadikgadi Line (ML). (b) Cross section of craton along line A-B illustrating that the Kimberley Block was on the hangingwall for westward dipping subduction during continent collision at 2.9 Ga and received the bulk of diamond-forming fluids, sulfur, and eclogitic components at this time. Figure modified from Schmitz et al. (2004) and Aulbach et al. (2009b).

Lithospheric diamonds are old, typically 1.0 to 3.5 Ga, and crystallized during and after lithosphere construction. Both primordial mantle carbon and carbon sourced from subducted lithosphere could have participated in diamond-forming events in the lithosphere as a consequence of fluid and melt metasomatism. The thermodynamic and experimental observations described above are permissive of a large number of pathways for diamond crystallization, and from a wide compositional range of fluids, melts, and solids (Fig. 20a,b; see section on "*Diamond Formation*"). The considerable compositional range of mineral and fluid inclusions found in diamonds likely attests to many of these pathways being important in nature. What is clear from a mantle redox point of view is that diamond crystallization from fluids and melts in mantle lithosphere is an explicable and expected outcome of mantle metasomatism.

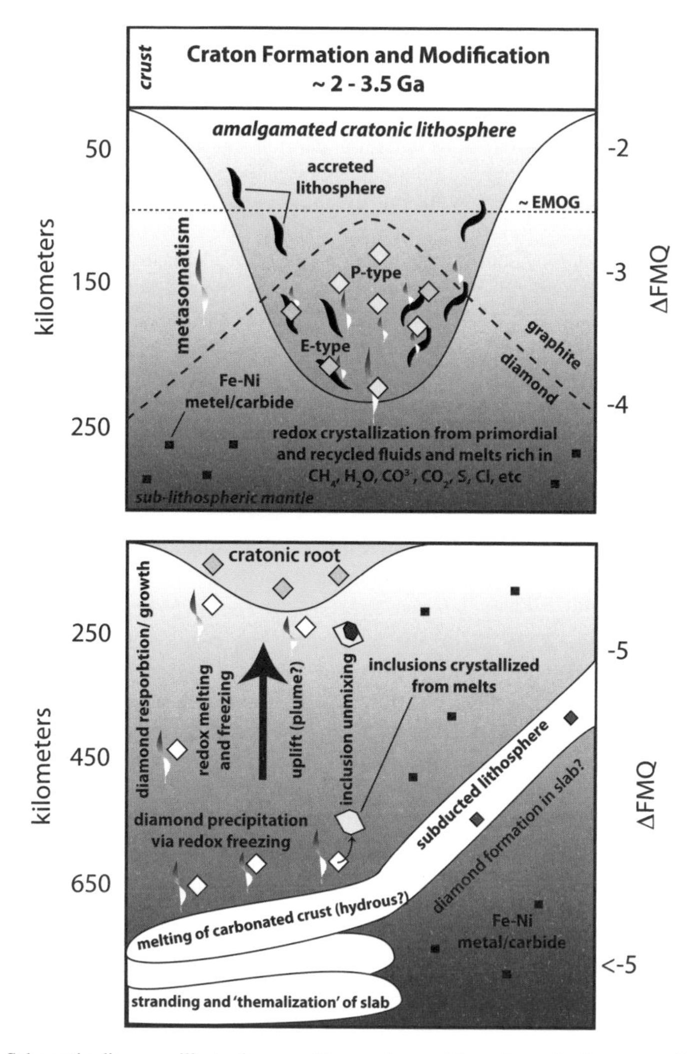

Figure 20. Schematic diagrams illustrating possible mantle conditions (e.g., depth, oxygen fugacity) and geodynamic settings in which lithopsheric and sub-lithospheric diamonds and their mineral inclusions are formed. The relative roles of mantle and subducted lithospheric protoliths, crystallization from partial melts, mantle redox, and mantle flow, can all be deduced and quantified through detailed, integrated studies of diamonds and their inclusions. See text for detailed discussion.

Deep carbon cycling with mantle convection: sub-lithospheric diamonds. Harte (2010) has suggested a link between the depth intervals over which sub-lithospheric inclusions most commonly occur (e.g., Table 2) based on geobarametry, and dehydration of hydrous or nominally anhydrous minerals in subducting lithospheric slabs. He postulated depth control by the position at which fluid or melt occurs. Harte suggested that dehydration of lawsonite in subducted mafic rocks provides a location for melt formation and the inclusion of the shallower (~300 km) majoritic inclusions. Deeper majoritic inclusions from the transition zone may occur as a consequence of dehydration melting at the wadsleyite-to-olivine transformation whereas the deepest diamonds possibly are related to dehydration of hydrous ringwoodite and dense hydrous Mg-silicates formed in subducted peridotites. A link between subducted protoliths and sub-lithospheric diamonds is strong in many cases. For example, Tappert et al. (2005a,b) showed that majorite garnets from Jagersfontein have Eu anomalies linking them to subducted oceanic crust, while the diamond hosts have extremely negative carbon isotope compositions possibly derived from a subducted carbon source. These authors postulate diamond formation by direct conversion from graphite in a subducted slab.

Many sub-lithospheric inclusions in natural diamonds provide ample evidence for the role of deep melts in their origin. Phase relations and abundances of incompatible trace elements show that effectively all reported inclusions in diamond interpreted as relic Ca-rich perovskite likely crystallized directly from melts derived from low extents of melting (Wang et al. 2000; Walter et al. 2008; Bulanova et al. 2010). The majority of majoritic garnet inclusions in sub-lithospheric diamonds could also have crystallized in equilibrium with low-degree, carbonate-rich melts (Keshav et al. 2005; Walter et al. 2008; Bulanova et al. 2010). Walter et al. (2008) and Bulanova et al. (2010) argue for a model involving diamond and inclusion co-precipitation from low-degree, carbonate-rich melts. These melts are envisioned to have originated within subducted oceanic lithosphere that became stranded in the deep transition zone or shallow lower mantle and thermalized with ambient mantle (Fig. 20b). Under such conditions, carbonated slab materials, including sediment, basalt, and peridotite, would potentially release low-degree, carbonate-rich melts, possibly hydrous, into the surrounding mantle. As illustrated in Figure 20b, the oxidized, carbonate-rich melts would be unstable in the ambient, reducing mantle, and when the highly mobile melts infiltrate the surrounding mantle, reaction with the mantle and reduction of carbonate results in diamond precipitation by "redox freezing" (Rohrbach and Schmidt 2011).

A feature of some inclusions in super-deep diamonds is unmixing of originally homogeneous phases into a composite of two or more phases (Table 2). Unmixing is common in majorite inclusions (Harte and Cayzer 2007) and occurs in Ti-rich Ca-perovskite (Walter et al. 2008; Bulanova et al. 2010). Walter et al. (2011) have argued that a suite of unmixed inclusions from Juina, Brazil, represents original Al-rich Mg-perovskite, CF-phase, and NAL-phase that formed in the lower mantle (Table 2). These observations indicate that the diamonds were transported upward by as much as hundreds of kilometers from their place of origin prior to incorporation into kimberlite magmas. The mechanism for upwelling beneath a craton is unclear, but could be related to a deep-seated mantle plume in the case of the sub-lithospheric mantle beneath Brazil (e.g., Harte and Cayzer 2007; Bulanova et al. 2010). Mantle that has undergone metasomatism and diamond formation via redox freezing would be locally more carbon-rich and more oxidizing than ambient mantle. Upon upwelling of such metasomatized mantle, carbonate will become stabilized at a depth that will depend on the carbon content and the ambient f_{O_2}. If this stabilization occurs, local oxidation of diamond to carbonate will drastically lower the solidus, resulting in "redox melting" (Fig. 20b; Taylor and Green 1988; Stagno and Frost 2010; Rohrbach and Schmidt 2011). Newly formed carbonated melts would then intrude more reducing mantle, and redox freezing could again occur. This process of repeated redox freezing, including melting, may in part explain the complex textures observed in many super-deep diamonds that include multiple growth centers, resorption and re-growth, and intense and complex zoning (Fig. 1, bottom).

Diamonds from the transition zone cover a large range in $^{13}C/^{12}C$ ratio (Fig. 7), yet display restricted and distinct ranges from one locality to another. For example, Kankan transition zone diamonds have higher $^{13}C/^{12}C$ ($\delta^{13}C$ ~1‰), whereas those from Jagerfontain are much lower $^{13}C/^{12}C$ ($\delta^{13}C$ ~ −20‰; Fig. 7). As these diamonds are eclogitic in nature, it remains unknown whether these values are characteristic of larger transition zone domains or only apply to local eclogitic regions. The link between diamond formation in the transition zone and the subduction factory has been emphasized by the unique C-isotope characteristics of the diamonds and the trace element patterns their included eclogitic garnets (e.g., Tappert et al. 2005b), and is essential for the "redox-freezing" model for super-deep diamond growth (see above). N-isotope geochemistry of most studied transition zone diamonds also support the subduction-factory link (Palot et al. 2012), although this link has been challenging to confirm with $^{15}N/^{14}N$ studies because of the preponderance sub-lithospheric diamonds that are Type-II (nitrogen-free). Low $^{13}C/^{12}C$ compositions are rarer in the studied lower mantle diamond population (Fig. 7; Pearson et al. 2003), which might suggest eclogitic material rarely reaches the lower mantle. Recently, though, a suite of low $^{13}C/^{12}C$ ($\delta^{13}C$ ~ −24‰) lower mantle diamonds containing a high-pressure basaltic mineral assemblage was studied, confirming that recycling can reach the lower mantle (Walter et al. 2011). In general, the complex growth pattern of super-deep diamonds identified by cathodoluminescence (Fig. 1; Araujo et al. 2013) is substantiated by C-isotope heterogeneity, illustrating multiple diamond growth events in a changing *P-T* environment (e.g., Bulanova et al. 2010; Palot et al. 2012).

Carbon mobility with melt: the diamond-kimberlite-carbonatite connection. With the exception of rare lamproite and lamprophyre, lithospheric and sub-lithospheric diamonds have been transported to the surface exclusively in rocks of kimberlitic composition (e.g., Gurney et al. 2010). Experimental studies and igneous petrology establish a petrogenetic link between kimberlite, carbonatite, and carbonated peridotite (Gudfinnsson and Presnall 2005; Keshav et al. 2005; Walter et al. 2008; Keshav et al. 2011; Russell et al. 2012; Jones et al. 2013). Where kimberlites originate and how they form are matters beyond simple diamond transport, for, although nearly all monocrystalline diamonds are much older (e.g., many tens of millions to billions of years older) than their host kimberlite, there is much that the study of kimberlites can contribute to understanding deep mantle fluids and melts in the region of diamond growth.

Kimberlites are rare but have been found on every continent and are associated with the cratonic portion that has a mantle lithospheric keel (e.g., Figs. 2, 3). Kimberlites are well known to have erupted more commonly in the Phanerozoic Eon than in the Pre-Cambrian (Gurney et al. 2010). The number of known kimberlite occurrences older than 1 Ga is fairly small, and although at the moment Archean kimberlites are not known, the presence of alluvial macro-diamonds in late Archean sediments indicates that such ancient kimberlites may indeed have existed (e.g., Gurney et al. 2010; Kopylova et al. 2011). Apparently the number of kimberlites that erupted globally increased dramatically around the Phanerozoic Eon (e.g., Smith et al. 1994; Heaman et al. 2004), although Tertiary kimberlites are much less abundant and only one Quaternary example has been reported (Dawson 1994).

Most kimberlites are not diamond bearing, but those that are must have originated at least as deep as the onset of diamond stability, which is at ~140 km, a depth that is consistent with results from thermobarometry on mantle xenoliths in kimberlites. The presence of sub-lithospheric diamonds in some kimberlites places kimberlite magma generation within the asthenosphere or deeper in the mantle, at least for these kimberlite pipes. The isotopic composition and mineralogy of Group I kimberlites also supports a sub-lithospheric origin, whereas the lithospheric affinity of Group II raises question of a shallower source that may be in the lithosphere for these kimberlites. Kimberlite magmas collect and transport a tremendous amount and variety of foreign material, including xenoliths and xenocrysts derived from both mantle and crustal sources. For this reason, establishing the bulk composition of primary kimberlite magma has long been problematic (e.g., Mitchell 2008). However, from rare hypabyssal aphanitic examples and from chemical re-constructions, a general consensus has emerged that kimberlites are silica-undersaturated (~15 to 35 wt% SiO_2) and MgO-rich (~20 to 35 wt%), and contain a high volatile component, possibly rich in both carbon (~5 to 20 wt%) and water (~5 to 10 wt%; see Mitchell 2008; Sparks et al. 2009). The lack of knowledge of the composition of primary kimberlite magma renders petrogenetic models for their origin non-unique. For example, their elemental enrichments and volatile-rich nature might indicate either very low-degree partial melting of cryptically metasomatized mantle (Dalton and Presnall 1998; Becker and Le Roex 2006) or higher-degree melting of pervasively veined mantle (Mitchell 1995, 2004).

Kimberlites also bear resemblance to carbonatitic rocks in terms of their high degree of silica undersaturation and enrichments in incompatible elements and volatiles, and for this reason there is a possible petrogenetic link between these magma types. On the basis of experimental melting phase relations of model carbonated peridotite, there is a continuum of compositions ranging from carbonatitic magmas (e.g., $SiO_2 < 5$ wt%) at the solidus to compositions akin to kimberlites at higher degrees of melting (Dalton and Presnall 1998; Gudfinnsson and Presnall 2005) at pressures at least up to up to 8 GPa. With this link to carbonatites, kimberlites themselves then would provide information about the deep carbon cycle by being generated from carbonatitic sources.

Another possible link is that carbonatitic melts may have been responsible for the metasomatic conditioning of the mantle source regions from which kimberlites form. Carbonatitic melts are probably highly mobile in the mantle (Minarik and Watson 1995; Hammouda and Laporte 2000; Jones et al. 2013) and are considered to be very effective metasomatic agents (Green and Wallace 1988; Hauri et al. 1993), influencing the mantle either chemically or modally. Melting of carbonate-metasomatized mantle, especially in the presence of water, may produce primary kimberlite magmas. Recently a model was developed whereby kimberlites form as the product of reaction of proto-carbonatitic melts with orthopyroxene-bearing mantle peridotite (Fig. 21). Dissolution of orthopyroxene into carbonatite increases the silica content, which at the same time decreases the solubility of volatiles (i.e., CO_2 and H_2O). Exsolution of the volatile phase provides buoyancy for the magma and a mechanism for rapid upward migration and incorporation of xenolithic material. Continued dissolution of orthopyroxene as the magma rises eventually changes the carbonatitic primary magma into a kimberlitic magma, and the progressively decreasing volatile solubility eventually leads to the rapid and explosive transport and emplacement of diamonds from the deep mantle to the shallow crust or surface (Fig. 21).

Carbon reservoirs: primordial versus recycled carbon. Carbon is a massively cycled element and it is likely that it is close to steady state in Earth's mantle (e.g., Javoy et al. 1982; Jambon 1994). In other words, the central question is not whether carbon is subducted but rather whether we can record the isotope heterogeneity introduced in Earth's mantle by subduction and distinguish it from the heterogeneity induced by intra-mantle processes (e.g., metasomatism), or even by mixing with a homogeneous (or heterogeneous) C-isotopic reservoir.

Furthermore, with mantle carbon having a long residence time, ca. 4.5 b.y. and other isotopic evidence for long-lived ancient mantle reservoirs (e.g., Jackson et al. 2010; Touboul et al. 2012), it is possible that a primordial C-reservoir might still exist and be sampled. The occurrence of low $\delta^{13}C$ in diamonds has been often ascribed to reflect a primordial heterogeneity of carbon in the mantle (Deines et al. 1991, 1993; Haggerty 1999) although this model, among others, fails to account for their occurrence among eclogitic rather than in peridotitic diamonds. The argument according to which mantle carbon might be isotopically heterogeneous was originally based on the similarities between the $\delta^{13}C$ distribution of carbon in iron meteorites and diamonds

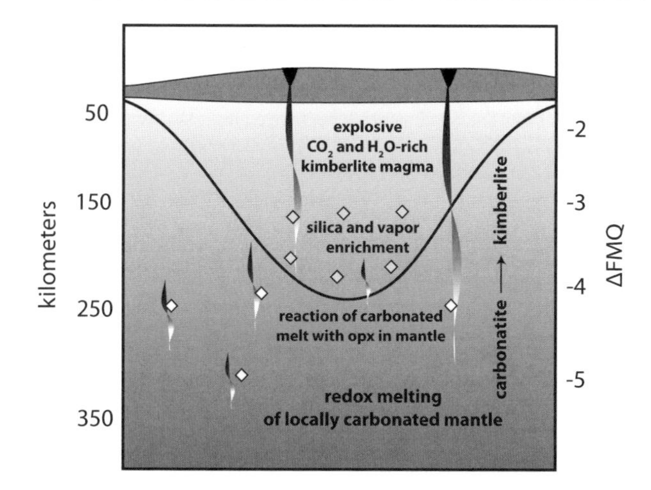

Figure 21. A schematic diagram illustrating the growth environment of lithospheric diamonds and their mineral inclusions. Studies show the importance of C-O-H-S metasomatic fluids in diamond growth, and the chemistry of mineral inclusions clearly reveals both peridotitic (P-type) and eclogitic (E-type) protoliths. The chemistry, age, and tectonic setting of the diamonds and their inclusions provide powerful constraints for models of the growth and evolution of cratonic mantle lithosphere. Carbonatite to kimberlite evolution after Russell et al. (2012).

but how such heterogeneities could survive mantle convection and homogenization during a period of a magma ocean remained unaddressed. In a model in which carbon is at steady state, it is anticipated that Earth's convective mantle (i.e., the reservoir from which the oceanic and continental crust are extracted and subducted ocean crust is recycled) and the primordial reservoir would have similar $\delta^{13}C$ close to −5‰. This similarity would also be the case for carbon exchanged between the surface and the mantle; degassed carbon and recycled carbon would display time-integrated $\delta^{13}C$ close to −5‰. In this context, the fact that ocean island basalts, mid-ocean ridge basalts, carbonatites, kimberlite, and diamonds have a $\delta^{13}C$ mode close to −5‰ is not inconsistent whatever the source of their carbon. The evidence for primordial carbon is therefore typically inferred from other systematics such as rare gases in ocean island basalts, but the respective contributions of carbon from the primordial and convective reservoirs remain to be established as their concentration in the primordial reservoir remains unknown. Additional evidence for primordial heterogeneity in diamonds has been suggested from three samples displaying low $\delta^{15}N$-values (< −25‰). In this case the $\delta^{13}C$-value close to −3.5 ‰ was suggested (Palot et al. 2012) but the size of the primordial reservoir and its carbon concentration cannot be addressed.

Several diamond populations (from the Dachine metakomatiite or metalamprophyre in French Guyana and from Jericho kimberlite in Northern Slave with $\delta^{13}C$-modes ~−28 ‰ and −38‰ respectively) are unique and difficult to interpret in the light of heterogeneity being either primordial, subducted, or mantle-related. This difficulty arises because there is almost no sediment with $\delta^{13}C$ as low as −40‰. Their preservation of such low $\delta^{13}C$ further requires the virtual absence of any (i.e., ^{13}C-enriched) carbonates in their sources, which is not consistent with observations in metamorphic rocks and sediments in general. Those odd C-isotope distributions might alternatively reflect an as yet unknown process(es), but we lack appropriate experimental work to investigate such a possibility.

Overall, the data suggest that the existence of a primordial carbon reservoir can be recognized and sampled from the study of diamonds, but its existence can only be established in the light of additional tracers (trace elements, radiogenic and other stable isotope systematics). Although usually at sub-ppm or ppb levels, diamonds contain many impurities that can be used in future studies to perhaps resolve the many open issues that have been highlighted in this review chapter.

OUTSTANDING QUESTIONS AND FUTURE WORK

The comprehensive, cross-disciplinary nature of this review identifies some of the areas where important unknowns in diamond research can be addressed with future work: (1) the quantitative partitioning of elements and fractionation of isotopes during diamond growth, (2) the co-genetic (or not) relationship of diamond to its host inclusions and the age of diamonds, (3) the recognition and significance of primordial carbon, primary mantle carbon, or subducted carbon in the composition of diamond, (4) the speciation of C in diamond-forming fluids and the processes that control the oxygen fugacity of the mantle, (5) the deepest diamonds, their ultra-high pressure inclusions and the geodynamic processes occurring in convecting the mantle, (6) the experimental simulation of diamond formation from a variety of mantle fluids and melts, and (7) the nanostructural characteristics of diamond as they relate to all aspects of diamond formation.

The expected answers to questions in these areas will lead to a new understanding of the conditions of diamond formation in the deep mantle, how diamond-forming and diamond-carrying melts interact with mantle peridotite, whether a significant reservoir of mantle carbon is primordial or recycled, and how carbon is transported and stored in the mantle now and in the past (as long as 3.5 billion years ago). Such integrated research on natural diamond has the potential to transform our knowledge about the sources of the surface volcanic flux of carbon, the connections between carbon in the biosphere to carbon in the deep mantle, the behavior of carbon in Earth's interior under extreme conditions, and the geodynamics of Earth's mantle.

ACKNOWLEDGMENTS

We appreciate the very thorough reviews of Galina Bulanova, Robert Hazen, and Chris Smith and the careful and patient, thorough editorial handling of Robert Hazen and Jodi Rosso. Discussions with Richard Carlson, Genet Duke, Jeffrey Harris, Erik Hauri, Sami Mikhail, and Bjorn Mysen contributed to the content, for which the authors alone are responsible.

The manuscript was completed with support from the following agencies and institutions: NSF (EAR1049992) to SBS, NASA (Astrobiology CAN 5) to the Carnegie Node, NERC (NE/J024821/1) to MJW, DFG (FR1555/5-1) to DJF, CNRS and IPGP support to PC, EU Marie Curie Grant (7th Program) to SK, CERC support to DGP, ERC Starting Grant 2012 (Agreement #307322) to FN and PN, and the resident institutions of the authors.

REFERENCES

Akagi T, Masuda A (1988) Isotopic and elemental evidence for a relationship between kimberlite and Zaire cubic diamonds. Nature 336:665-667

Akaishi M, Kumar M, Kanda H, Yamaoka S (2001) Reactions between carbon and a reduced C-O-H fluid under diamond-stable HP-HT condition. Diamond Relat Mater 10:2125-2130, doi: 10.1016/S0925-9635(01)00490-3

Akaishi M, Shaji Kumar MD, Kanda H, Yamaoka S (2000) Formation process of diamond from supercritical H_2O-CO_2 fluid under high pressure and high temperature conditions. Diamond Relat Mater 9:1945-1950, doi: 10.1016/S0925-9635(00)00366-6

Anand M, Taylor LA, Misra KC, Carlson WD, Sobolev NV (2004) Nature of diamonds in Yakutian eclogites: views from eclogite tomography and mineral inclusions in diamonds. Lithos 77:333-348, doi: 10.1016/j.lithos.2004.03.026

Araujo D, Gaspar JC, Bulanova GP, Smith CB, Walter MJ, Kohn SC, Hauri EH (2013) Diamonds from kimberlites and alluvial deposits from Juina, Brazil. J Geol Soc India 81: in press

Arima M, Kozai Y, Akaishi M (2002) Diamond nucleation and growth by reduction of carbonate melts under high-pressure and high-temperature conditions. Geology 30:691-694, doi: 10.1130/0091-7613(2002)030<0691:DNAGBR>2.0.CO;2

Arima M, Nakayama K, Akaishi M, Yamaoka S, Kanda H (1993) Crystallization of diamond from a silicate melt of kimberlite composition in high-pressure and high-temperature experiments. Geology 21:968-970, doi: 10.1130/0091-7613(1993)021<0968:CODFAS>2.3.CO;2

Aulbach S, Shirey SB, Stachel T, Creighton S, Muehlenbachs K, Harris JW (2009a) Diamond formation episodes at the southern margin of the Kaapvaal Craton; Re-Os systematics of sulfide inclusions from the Jagersfontein Mine. Contrib Mineral Petrol 157:525-540, doi: 10.1007/s00410-008-0350-9

Aulbach S, Stachel T, Creaser RA, Heaman LM, Shirey SB, Muehlenbachs K, Eichenberg D, Harris JW (2009b) Sulphide survival and diamond genesis during formation and evolution of Archaean subcontinental lithosphere; a comparison between the Slave and Kaapvaal Cratons. Lithos 112:747-757, doi: 10.1016/j.lithos.2009.03.048

Aulbach S, Stachel T, Viljoen KS, Brey GP, Harris JW (2002) Eclogitic and websteritic diamond sources beneath the Limpopo Belt; is slab-melting the link? Contrib Mineral Petrol 143:56-70, doi: 10.1007/s00410-001-0331-8

Ballhaus C (1995) Is the upper mantle metal-saturated? Earth Planet Sci Lett 132:75-86, doi: 10.1016/0012-821X(95)00047-G

Banas A, Stachel T, Muehlenbachs K, McCandless TE (2007) Diamonds from the Buffalo Head Hills, Alberta; formation in a non-conventional setting. Lithos 93:199-213, doi: 10.1016/j.lithos.2006.07.001

Barron LM, Mernagh TP, Barron BJ (2008) Using strain birefringence in diamond to estimate the remnant pressure on an inclusion. Aust J Earth Sci 55:159-165, doi: 10.1080/08120090701689332

Bataleva YV, Palyanov YN, Sokol AG, Borzdov YM, Palyanova GA (2012) Conditions for the origin of oxidized carbonate-silicate melts: Implications for mantle metasomatism and diamond formation. Lithos 128-131:113-125, doi: 10.1016/j.lithos.2011.10.010

Bebout GE, Fogel ML (1992) Nitrogen-isotope compositions of metasedimentary rocks in the Catalina Schist, California: Implications for metamorphic devolatilization history. Geochim Cosmochim Acta 56:2839-2849

Becker M, Le Roex AP (2006) Geochemistry of South African on- and off- craton, Group I and Group II kimberlites; petrogenesis and source region evolution. J Petrol 47:673-703, doi: 10.1093/petrology/egi089

Belonoshko A, Saxena S (1992) A unified equation of state for fluids of C-H-O-N-S-Ar composition and their mixtures up to very high-temperatures and pressures. Geochim Cosmochim Acta 56:3611-3626, doi: 10.1016/0016-7037(92)90157-E

Boyd SR, Mattey DP, Pillinger CT, Milledge HJ, Mendelssohn M, Seal M (1987) Multiple growth events during diamond genesis: an integrated study of carbon and nitrogen isotopes and nitrogen aggregation state in coated stones. Earth Planet Sci Lett 86:341-353

Boyd SR, Pillinger CT, Milledge HJ, Mendelssohn MJ, Seal M (1992) C and N isotopic composition and the infrared absorption spectra of coated diamonds: evidence for the regional uniformity of CO_2-H_2O rich fluids in lithospheric mantle. Earth Planet Sci Lett 109:633-644, doi: 10.1016/0012-821X(92)90066-5

Boyd SR, Pineau F, Javoy M (1994) Modelling the growth of natural diamonds. Chem Geol 116:29-42

Breeding CM, Shigley JE (2009) The "type" classification system of diamonds and its importance in gemology. Gems Gemology 45:96-111

Brenker FE, Stachel T, Harris JW (2002) Exhumation of lower mantle inclusions in diamond; a TEM investigation of retrograde phase transitions, reactions and exsolution. Earth Planet Sci Lett 198:1-9, doi: 10.1016/s0012-821x(02)00514-9

Brenker FE, Vollmer C, Vincze L, Vekemans B, Szymanski A, Janssens K, Szaloki I, Nasdala L, Joswig W, Kaminsky F (2007) Carbonates from the lower part of transition zone or even the lower mantle. Earth Planet Sci Lett 260:1-9, doi: 10.1016/j.epsl.2007.02.038

Brey GP, Koehler T (1990) Geothermobarometry in four-phase lherzolites; II, new thermobarometers, and practical assessment of existing thermobarometers. J Petrol 31:1353-1378

Bulanova GP (1995) The formation of diamond. J Geochem Explor 53:1-23, doi: 10.1016/0375-6742(94)00016-5

Bulanova GP, Griffin WL, Ryan CG (1998) Nucleation environment of diamonds from Yakutian kimberlites. Mineral Mag 62:409-419

Bulanova GP, Griffin WL, Ryan CG, Shestakova OY, Barnes SJ (1996) Trace elements in sulfide inclusions from Yakutia diamonds. Contrib Mineral Petrol 124:111-125

Bulanova GP, Muchemwa E, Pearson DG, Griffin BJ, Kelley SP, Klemme S, Smith CB (2004) Syngenetic inclusions of yimengite in diamond from Sese Kimberlite (Zimbabwe); evidence for metasomatic conditions of growth. Lithos 77:181-192, doi: 10.1016/j.lithos.2004.04.002

Bulanova GP, Pearson DG, Hauri EH, Griffin BJ (2002) Carbon and nitrogen isotope systematics within a sector-growth diamond from the Mir Kimberlite, Yakutia. Chem Geol 188:105-123, doi: 10.1016/s0009-2541(02)00075-x

Bulanova GP, Varshavsky AV, Kotegov VA (2005) A venture into the interior of natural diamond; genetic information and implications for the gem industry. J Gemmol (1986) 29:377-386

Bulanova GP, Walter MJ, Smith CB, Kohn SC, Armstrong LS, Blundy J, Gobbo L (2010) Mineral inclusions in sub-lithospheric diamonds from Collier 4 kimberlite pipe, Juina, Brazil; subducted protoliths, carbonated melts and primary kimberlite magmatism. Contrib Mineral Petrol 160:489-510, doi: 10.1007/s00410-010-0490-6

Bundy FP, Hall HT, Strong HM, Wentorf RH (1955) Man-made diamonds. Nature 176:51-55, doi: 10.1038/176051a0

Bureau H, Langenhorst F, Auzende A-L, Frost DJ, Esteve I, Siebert J (2012) The growth of fibrous, cloudy and polycrystalline diamonds. Geochim Cosmochim Acta 77:202-214, doi: 10.1016/j.gca.2011.11.016

Burgess R, Kiviets GB, Harris JW (2004) Ar/Ar age determinations of eclogitic clinopyroxene and garnet inclusions in diamonds from the Venetia and Orapa kimberlites. Lithos 77:113-124, doi: 10.1016/j.lithos.2004.03.048

Burgess R, Turner G, Harris JW (1992) $^{40}Ar/^{39}Ar$ laser probe studies of clinopyroxene inclusions in eclogitic diamonds. Geochim Cosmochim Acta 56:389-402, doi: 10.1016/0016-7037(92)90140-e

Burgess SR, Harte B (2004) Tracing lithosphere evolution through the analysis of heterogeneous G9-G10 garnets in peridotite xenoliths; II, REE chemistry. J Petrol 45:609-634, doi: 10.1093/petrology/egg095

Busigny V, Cartigny P, Philippot P (2011) Nitrogen isotopes in ophiolitic metagabbros: A re-evaluation of modern nitrogen fluxes in subduction zones and implication for the early Earth atmosphere. Geochim Cosmochim Acta 75:7502-7521, doi: 10.1016/j.gca.2011.09.049

Busigny V, Cartigny P, Philippot P, Ader M, Javoy M (2003) Massive recycling of nitrogen and other fluid-mobile elements (K, Rb, Cs, H) in a cold slab environment: evidence from HP to UHP oceanic metasediments of the Schistes Lustres nappe (western Alps, Europe). Earth Planet Sci Lett 215:27-42

Canil D (2002) Vanadium in peridotites, mantle redox and tectonic environments: Archean to present. Earth Planet Sci Lett, 195:75-90, doi: 10.1016/S0012-821X(01)00582-9

Canil D, O'Neill H (1996) Distribution of ferric iron in some upper-mantle assemblages. J Petrol 37:609-635

Carlson RW, Grove TL, de Wit MJ, Gurney JJ (1996) Program to study crust and mantle of the Archean craton in southern Africa. Eos, Trans Am Geophys Union 77:273-277

Carlson RW, Pearson DG, Boyd FR, Shirey SB, Irvine G, Menzies AH, Gurney JJ (1999) Re-Os systematics of lithospheric peridotites: implications for lithosphere formation and preservation. *In:* The J. B. Dawson Volume. Gurney JJ, Gurney JL, Pascoe MD, Richardson SH (eds) Red Roof Design, Cape Town, p 99-108

Carswell DA, Yardley BWD, Schumacher JC (1991) The garnet-orthopyroxene Al barometer; problematic application to natural garnet lherzolite assemblages. Mineral Mag 55:19-31

Cartigny P (2005) Stable isotopes and the origin of diamonds. Elements 1:79-84, doi: 10.2113/gselements.1.2.79

Cartigny P (2010) Mantle-related carbonados? Geochemical insights from diamonds from the Dachine komatiite (French Guiana). Earth Planet Sci Lett 296:329-339, doi: 10.1016/j.epsl.2010.05.015

Cartigny P, Boyd SR, Harris JW, Javoy M (1997) Nitrogen isotopes in peridotitic diamonds from Fuxian, China; the mantle signature. Terra Nova 9:175-179

Cartigny P, Chinn I, Viljoen KS, Robinson D (2004) Early proterozoic ultrahigh pressure metamorphism: Evidence from microdiamonds. Science 304:853-855

Cartigny P, Farquhar J, Thomassot E, Harris JW, Wing B, Masterson A, McKeegan K, Stachel T (2009) A mantle origin for Paleoarchean peridotitic diamonds from the Panda kimberlite, Slave Craton; evidence from ^{13}C, ^{15}N and $^{33,34}S$ stable isotope systematics. Lithos 112:852-864, doi: 10.1016/j.lithos.2009.06.007

Cartigny P, Harris JW, Javoy M (1998) Eclogitic diamond formation at Jwaneng: no room for a recycled component. Science 280:1421-1424

Cartigny P, Harris JW, Javoy M (2001) Diamond genesis, mantle fractionations and mantle nitrogen content: a study of $\delta^{13}C$-N concentrations in diamonds. Earth Planet Sci Lett 185:85-98

Cartigny P, Harris JW, Taylor A, Davies R, Javoy M (2003) On the possibility of a kinetic fractionation of nitrogen stable isotopes during natural diamond growth. Geochim Cosmochim Acta 67:1571-1576, doi: 10.1016/s0016-7037(03)00028-0

Chinn IL, Gurney JJ, Milledge JH, Taylor WR, Woods PA (1995) Cathodoluminescence properties of CO_2-bearing and CO_2-free diamonds from the George Creek K1 kimberlite dike. Int Geol Rev 37:254-258

Chrenko RM, McDonald RS, Darrow KA (1967) Infra-red spectra of diamond coat. Nature 213:474-476

Chrenko RM, Tuft RE, Strong HM (1977) Transformation of the state of nitrogen in diamond. Nature 270:141-144

Claoue-Long JC, Sobolev NV, Shatsky VS, Sobolev AV (1991) Zircon response to diamond-pressure metamorphism in the Kokchetav Massif, USSR. Geology 19:710-713

Clifford TN (1966) Tectono-metallogenic units and metallogenic provinces of Africa. Earth Planet Sci Lett 1:421-434

Collerson KD, Williams Q, Kamber BS, Omori S, Arai H, Ohtani E (2010) Majoritic garnet; a new approach to pressure estimation of shock events in meteorites and the encapsulation of sub-lithospheric inclusions in diamond. Geochim Cosmochim Acta 74:5939-5957, doi: 10.1016/j.gca.2010.07.005

Creighton S, Stachel T, Matveev S, Höfer H, McCammon C, Luth RW (2009) Oxidation of the Kaapvaal lithospheric mantle driven by metasomatism. Contrib Mineral Petrol 157:491-504, doi: 10.1007/s00410-008-0348-3

Custers JFH (1950) On the nature of the opal-like outer layer of coated diamonds. Am Mineral 35:51-58

Dalton JA, Presnall DC (1998) The continuum of primary carbonatitic-kimberlitic melt compositions in equilibrium with lherzolite; data from the system CaO-MgO-Al_2O_3-SiO_2-CO_2 at 6 GPa. J Petrol 39:1953-1964, doi: 10.1093/petrology/39.11.1953

Dasgupta R (2013) Ingassing, storage, and outgassing of terrestrial carbon through geologic time. Rev Mineral Geochem 75:183-229

Dasgupta R, Hirschmann M (2006) Melting in the Earth's deep upper mantle caused by carbon dioxide. Nature 440:659-662, doi: 10.1038/nature04612

Dasgupta R, Hirschmann M, Withers A (2004) Deep global cycling of carbon constrained by the solidus of anhydrous, carbonated eclogite under upper mantle conditions. Earth Planet Sci Lett 227:73-85, doi: 10.1016/j.epsl.2004.08.004

Davies RM, Griffin WL, O'Reilly SY, McCandless TE (2004) Inclusions in diamonds from the K14 and K10 kimberlites, Buffalo Hills, Alberta, Canada; diamond growth in a plume? Lithos 77:99-111, doi: 10.1016/j.lithos.2004.04.008

Dawson JB (1994) Quaternary kimberlitic volcanism on the Tanzania Craton. Contrib Mineral Petrol 116:473-485

Day HW (2012) A revised diamond-graphite transition curve. Am Mineral 97:52-62, doi: 10.2138/am.2011.3763

de Corte K, Cartigny P, Shatsky VS, Sobolev NV, Javoy M (1998) Evidence of fluid inclusions in metamorphic microdiamonds from the Kokchetav massif, northern Kazakhstan. Geochim Cosmochim Acta 62:3765-3773, doi: 10.1016/S0016-7037(98)00266-X

De Stefano A, Kopylova MG, Cartigny P, Afanasiev V (2009) Diamonds and eclogites of the Jericho Kimberlite (northern Canada). Contrib Mineral Petrol 158:295-315, doi: 10.1007/s00410-009-0384-7

Deines P (1980) The carbon isotopic composition of diamonds - relationship to diamond shape, color, occurrence and vapor composition. Geochim Cosmochim Acta 44:943-961

Deines P, Eggler DH (2009) Experimental determination of carbon isotope fractionation between $CaCO_3$ and graphite. Geochim Cosmochim Acta 73:7256-7274, doi: 10.1016/j.gca.2009.09.005

Deines P, Harris JW (1995) Sulfide inclusion chemistry and carbon isotopes of African diamonds. Geochim Cosmochim Acta 59:3173-3188, doi: 10.1016/0016-7037(95)00205-e

Deines P, Harris JW, Gurney JJ (1991) The carbon isotopic composition and nitrogen content of lithospheric and asthenospheric diamonds from the Jagersfontein and Koffiefontein kimberlites, South Africa. Geochim Cosmochim Acta 55:2615-2626

Deines P, Harris JW, Gurney JJ (1993) Depth-related carbon isotope and nitrogen concentration variability in the mantle below the Orapa kimberlite, Botswana, Africa. Geochim Cosmochim Acta 57:2781-2796

Dhuime B, Hawkesworth CJ, Cawood PA, Storey CD (2012) A change in the geodynamics of continental growth 3 billion years ago. Science 335:1334-1336, doi: 10.1126/science.1216066

Dobrzhinetskaya LF (2012) Microdiamonds — Frontier of ultrahigh-pressure metamorphism: A review. Gondwana Res 21:207-223, doi: 10.1016/j.gr.2011.07.014

Dobrzhinetskaya LF, Eide E, Larsen R, Sturt B, Tronnes R, Smith D, Taylor W, Posukhovat T (1995) Microdiamond in high-grade metamorphic rocks of the Western Gneiss Region, Norway. Geology 23:597-600, doi: 10.1130/0091-7613(1995)023<0597:MIHGMR>2.3.CO;2

Dobrzhinetskaya LF, Wirth R, Green HW (2005) Direct observation and analysis of a trapped COH fluid growth medium in metamorphic diamond. Terra Nova 17:472-477, doi: 10.1111/j.1365-3121.2005.00635.x

Dobrzhinetskaya LF, Wirth R, Green HW II (2007) A look inside of diamond-formaing media in deep subduction zones. Proc Natl Acad Sci USA 104:9128-9132, doi: 10.1073/pnas.0609161104

Dobrzhinetskaya LF, Wirth R, Green HW, II (2006) Nanometric inclusions of carbonates in Kokchetav diamonds from Kazakhstan; a new constraint for the depth of metamorphic diamond crystallization. Earth Planet Sci Lett 243:85-93, doi: 10.1016/j.epsl.2005.11.030

Eggler DH, Baker DR (1982) Reduced volatiles in the system C-O-H; implications to mantle melting, fluid formation, and diamond genesis. *In:* High Pressure Research in Geophysics, Vol. 12. Akimoto S, Manghnani MH (eds) Center for Academic Publications Japan, Tokyo, Japan, p 237-250

Evans T, Harris JW (1989) Nitrogen aggregation, inclusion equilibration temperatures and the age of diamonds. *In*: Kimberlites and Related Rocks. Proc 4th Int Kimberlite Conf, Perth, Australia Vol 2. Ross J et al. (eds) Blackwell, Cambridge, MA, p 991-996

Evans T, Qi Z (1982) The kinetics of the aggregation of nitrogen atoms in diamond. Proc R Soc London A 381:159-178

Farquhar J, Bao HM, Thiemens M (2000) Atmospheric influence of Earth's earliest sulfur cycle. Science 289:756-758

Farquhar J, Wing BA, McKeegan KD, Harris JW, Cartigny P, Thiemens MH (2002) Mass-independent sulfur of inclusions in diamond and sulfur recycling on early Earth. Science 298:2369-2372, doi: 10.1126/science.1078617

Fedorov II, Chepurov AA, Dereppe JM (2002) Redox conditions of metal-carbon melts and natural diamond genesis. Geochem J 36:247-253

Fesq HW, Bibby DM, Erasmus CS, Kable EJD, Sellschop JPF (1975) A comparative trace element study of diamonds from Premier, Finsch and Jagersfontein mines, South Africa. Phys Chem Earth 9:817-836

Field JE (1992) The Properties of Natural and Synthetic Diamond. Academic Press, New York

Finnie KS, Fisher D, Griffin WL, Harris JW, Sobolev NV (1994) Nitrogen aggregation in metamorphic diamonds from Kazakhstan. Geochim Cosmochim Acta 58:5173-5177

Foley S, Yaxley G, Rosenthal A, Buhre S, Kiseeva E, Rapp R, Jacob D (2009) The composition of near-solidus melts of peridotite in the presence of CO_2 and H_2O between 40 and 60 kbar. Lithos 112:274-283, doi: 10.1016/j.lithos.2009.03.020

Fouch MJ, James DE, VanDecar JC, Van Der Lee S (2004) Mantle seismic structure beneath the Kaapvaal and Zimbabwe Cratons. S Afr J Geol 107: 33-44

Frost BR, Lindsley DH, Andersen DJ (1988) Fe-Ti oxide-silicate equilibria; assemblages with fayalitic olivine. Am Mineral 73:727-740

Frost DJ, Liebske C, Langenhorst F, McCammon CA, Tronnes RG, Rubie DC (2004) Experimental evidence for the existence of iron-rich metal in the Earth's lower mantle. Nature 428:409-412, doi: 10.1038/nature02413

Frost DJ, McCammon CA (2008) The redox state of Earth's mantle. Annu Rev Earth Planet Sci 36:389-420, doi: 10.1146/annurev.earth.36.031207.124322

Futergendler SI, Frank-Kamenetsky VA (1961) Oriented inclusions of olivine, garnet and chromite in diamonds. Notes Mineral Soc Russia 90:230-236

Gaillou E, Post JE, Rost D, Butler JE (2012) Boron in natural type IIb blue diamonds; chemical and spectroscopic measurements. Am Mineral 97:1-18, doi: 10.2138/am.2012.3925

Galimov EM (1985) The relation between formation conditions and variations in isotope composition of diamonds. Geochem Int 22:118-142

Galimov EM (1991) Isotope fractionation related to kimberlite magmatism and diamond formation. Geochim Cosmochim Acta 55:1697-1708

Galimov EM, Sobolev NV, Efimova ES (1999) Carbon isotopic composition of Venezuela diamond. Dokl Akad Nauk 364:101-106

Green DH, Wallace ME (1988) Mantle metasomatism by ephemeral carbonatite melts. Nature 336:459-462, doi: 10.1038/336459a0

Griffin WL, Fisher NI, Friedman J, Ryan CG, O'Reilly SY (1999) Cr-pyrope garnets in the lithospheric mantle; I, Composition systematics and relations to tectonic setting. J Petrol 40:679-704, doi: 10.1093/petrology/40.5.679

Grütter HS, Gurney JJ, Menzies AH, Winter F (2004) An updated classification scheme for mantle-derived garnet, for use by diamond explorers. Lithos 77:841-857, doi: 10.1016/j.lithos.2004.04.012

Gudfinnsson GH, Presnall DC (2005) Continuous gradations among primary carbonatitic, kimberlitic, melilititic, basaltic, picritic, and komatiitic melts in equilibrium with garnet lherzolite at 3-8 GPa. J Petrol 46:1645-1659, doi: 10.1093/petrology/egi029

Gudmundsson G, Wood B (1995) Experimental tests of garnet peridotite oxygen barometry. Contrib Mineral Petrol 119:56-67, doi: 10.1007/BF00310717

Gunn SC, Luth RW (2006) Carbonate reduction by Fe-S-O melts at high pressure and high temperature. Am Mineral 91:1110-1116, doi: 10.2138/am.2006.2009

Gurney JJ, Harris JW, Rickard RS (1984) Minerals associated with diamonds from the Roberts Victor Mine In: KimberlitesII:The mantle and crust-mantle relationships. Kornprobst J (ed) Elsevier., Amsterdam, Netherlands (NLD), p 25-32

Gurney JJ, Helmstaedt HH, Richardson SH, Shirey SB (2010) Diamonds through time. Econ Geol 105:689-712, doi: 10.2113/gsecongeo.105.3.689

Gurney JJ, Switzer GS (1973) The discovery of garnets closely related to diamonds in the Finsch Pipe, South Africa. Contrib Mineral Petrol 39:103-116

Haggerty SE (1999) A diamond trilogy; superplumes, supercontinents, and supernovae. Science 285:851-860

Hamilton MA, Sobolev NV, Stern RA, Pearson DG (2003) SHRIMP U-Pb dating of a perovskite inclusion: evidence for a syneruption age for diamond, Sytykanskaya kimberlite pipe, Yakutia region, Siberia. Proc 8th Int Kimberlite Conf, Victoria, Canada FLA 0388

Hammouda T, Laporte D (2000) Ultrafast mantle impregnation by carbonatite melts. Geology 28:283-285, doi: 10.1130/0091-7613(2000)028<0283:UMIBCM>2.3.CO;2

Harley SL (1984) An experimental study of the partitioning of Fe and Mg between garnet and orthopyroxene. Contrib Mineral Petrol 86:359-373

Harlow GE (1998) The Nature of Diamonds. Cambridge University Press, Cambridge, UK
Harlow GE, Davies RM (2005) Diamonds. Elements 1:67-70, doi: 10.2113/gselements.1.2.67
Harris JW (1968) The recognition of diamond inclusions. Part I: syngenetic inclusions. Ind Diamond Rev 28:402-410
Harris JW (1992) Diamond geology. *In:* The Properties of Natural and Synthetic Diamond. Field JE (ed) Academic Press, New York, p 345-393
Harris JW, Gurney JJ (1979) Inclusions in diamond. *In:* The Properties of Diamond. Field JE (ed) Academic Press, London, UK, p 555-591
Harris JW, Milledge HJ, Barron THK, Munn RW (1970) Thermal expansion of garnets included in diamond. J Geophys Res 75:5775-5792
Harte B (2010) Diamond formation in the deep mantle; the record of mineral inclusions and their distribution in relation to mantle dehydration zones. Mineral Mag 74:189-215, doi: 10.1180/minmag.2010.074.2.189
Harte B, Cayzer N (2007) Decompression and unmixing of crystals included in diamonds from the mantle transition zone. Phys Chem Min 34:647-656, doi: 10.10007/s00269-007-0178-2
Harte B, Harris JW, Hutchison MT, Watt GR, Wilding MC (1999) Lower mantle mineral associations in diamonds from Sao Luiz, Brazil. *In:* Mantle Petrology; Field Observations and High-Pressure Experimentation: A Tribute to Francis R. (Joe) Boyd. Fei Y, Bertka CM, Mysen BO (eds) Geochemical Society—University of Houston, Department of Chemistry, Houston, Texas, p 125-153
Harte B, Richardson SH (2011) Mineral inclusions in diamonds track the evolution of a Mesozoic subducted slab beneath West Gondwanaland. Gondwana Res 21:236-245, doi: 10.1016/j.gr.2011.07.001
Hauri EH, Shimizu N, Dieu JJ, Hart SR (1993) Evidence for hotspot-related carbonatite metasomatism in the oceanic upper mantle. Nature 365:221-227, doi: 10.1038/365221a0
Hauri EH, Wang J, Pearson DG, Bulanova GP (2002) Microanalysis of $d^{13}C$, $d^{15}N$, and N abundances in diamonds by secondary ion mass spectrometry. Chem Geol 185:149-163
Hayman PC, Kopylova MG, Kaminsky FV (2005) Lower mantle diamonds from Rio Soriso (Juina area, Mato Grosso, Brazil). Contrib Mineral Petrol 149:430-445, doi: 10.1007/s00410-005-0657-8
Hazen RM (1999) The Diamond Makers. Cambridge University Press, Cambridge, UK
Hazen RM, Downs RT, Jones AP, Kah L (2013) Carbon mineralogy and crystal chemistry. Rev Mineral Geochem 75:7-46
Heaman LM, Kjarsgaard BA, Creaser RA (2004) The temporal evolution of North American kimberlites. Lithos 76:377-397, doi: 10.1016/j.lithos.2004.03.047
Heaney PJ, Vicenzi EP, De S (2005) Strange diamonds: the mysterious origins of carbonado and framesite. Elements 1:85-89, doi: 10.2113/gselements.1.2.85
Hervig RL, Smith JV, Steele IM, Gurney JJ, Meyer HOA, Harris JW (1980) Diamonds; minor elements in silicate inclusions; pressure-temperature implications. J Geophys Res 85:6919-6929
Hirschmann MM, Tenner T, Aubaud C, Withers AC (2009) Dehydration melting of nominally anhydrous mantle: the primacy of partitioning. Phys Earth Planet Inter 176:54-68, doi: 10.1016/j.pepi.2009.04.001
Holland TJB, Powell R (2011) An improved and extended internally consistent thermodynamic dataset for phases of petrological interest, involving a new equation of state for solids. J Metamorph Geol 29:333-383
Holloway JR (1987) Igneous fluids. Rev Mineral 17:211-233
Hong S, Akaishi M, Yamaoka S (1999) Nucleation of diamond in the system of carbon and water under very high pressure and temperature. J Cryst Growth 200:326-328
Hough RM, Gilmour I, Pillinger CT, Arden JW, Gilkess KWR, Yuan J, Milledge HJ (1995) Diamond and silicon carbide in impact melt rock from the Ries impact crater. Nature 378:41-44, doi: 10.1038/378041a0
Hough RM, Gilmour I, Pillinger CT, Langenhorst F, Montanari A (1997) Diamonds from the iridium-rich K-T boundary layer at Arroyo el Mimbral, Tamaulipas, Mexico. Geology 25:1019, doi: 10.1130/0091-7613(1997)025<1019:DFTIRK>2.3.CO;2
Howell D, Wood IG, Dobson DP, Jones AP, Nasdala L, Harris JW (2010) Quantifying strain birefringence halos around inclusions in diamond. Contrib Mineral Petrol 160:705-717, doi: 10.1007/s00410-010-0503-5
Howell D, Wood IG, Nestola F, Nimis P, Nasdala L (2012) Inclusions under remnant pressure in diamond: A multi-technique approach. Eur J Mineral 24:563-573, doi: 10.1127/0935-1221/2012/0024-2183
Hutchinson MT, Dale CW, Nowell FA, Laiginhas FA, Pearson DG (2012) Age constraints on ultra-deep mantle petrology shown by Juina diamonds. Proc 10th Int Kimberlite Conf, Bangalore, India: 10IKC-108
Hwang S-L, Chu H-T, Yui T-F, Shen P, Schertl H-P, Liou JG, Sobolev NV (2006) Nanometer-size P/K-rich silica glass (former melt) inclusions in microdiamond from the gneisses of Kokchetav and Erzgebirge massifs: Diversified characteristics of the formation media of metamorphic microdiamond in UHP rocks due to host-rock buffering. Earth Planet Sci Lett 243:94-106, doi: 10.1016/j.epsl.2005.12.015
Hwang S-L, Shen P, Chu H-T, Yui T-F, Liou JG, Sobolev NV, Shatsky VS (2005) Crust-derived potassic fluid in metamorphic microdiamond. Earth Planet Sci Lett 231:295-306, doi: 10.1016/j.epsl.2005.01.002
Ireland TR, Rudnick RL, Spetsius Z (1994) Trace elements in diamond inclusions from eclogites reveal link to Archean granites. Earth Planet Sci Lett 128:199-213, doi: 10.1016/0012-821x(94)90145-7

Irifune T, Kurio A, Sakamoto S, Inoue T, Sumiya H, Funakoshi K-i (2004) Formation of pure polycrystalline diamond by direct conversion of graphite at high pressure and high temperature. Phys Earth Planet Inter 143-144:593-600, doi: 10.1016/j.pepi.2003.06.004

Izraeli ES, Harris JW, Navon O (1999) Raman barometry of diamond formation. Earth Planet Sci Lett 173:351-360

Izraeli ES, Harris JW, Navon O (2001) Brine inclusions in diamonds; a new upper mantle fluid. Earth Planet Sci Lett 187:323-332

Jackson MG, Carlson RW, Kurz MD, Kempton PD, Francis D, Blusztajn J (2010) Evidence for the survival of the oldest terrestrial mantle reservoir. Nature 466:853, doi: 10.1038/nature09287

Jacob DE (2004) Nature and origin of eclogite xenoliths from kimberlites. Lithos 77:295-316, doi: 10.1016/j.lithos.2004.03.038

Jacob DE, Wirth R, Enzmann F, Kronz A, Schreiber A (2011) Nano-inclusion suite and high resolution micro-computed-tomography of polycrystalline diamond (framesite) from Orapa, Botswana. Earth Planet Sci Lett 308:307-316, doi: 10.1016/j.epsl.2011.05.056

Jakobsson S, Holloway J (1986) Crystal-liquid experiments in the presence of a C-O-H fluid buffered by graphite+iron+wustite: experimental-method and near-liquidus relations in basanite. J Volcan Geotherm Res 29:265-291, doi: 10.1016/0377-0273(86)90048-X

Jambon A (1994) Earth degassing and large-scale geochemical cycling of volatile elements. Rev Mineral 30:479-517

James DE, Fouch MJ, VanDecar JC, van der Lee S, Group KS (2001) Tectospheric structure beneath southern Africa. Geophys Res Lett 28:2485-2488

Janney P, Shirey S, Carlson R, Pearson D, Bell D, le Roex A, Ishikawa A, Nixon P, Boyd F (2010) Age, composition and thermal characteristics of South African off-craton mantle lithosphere: Evidence for a multi-stage history. J Petrol 51:1849-1890, doi: 10.1093/petrology/egq041

Jaques AL, Hall AE, Sheraton JW, Smith CB, Sun SS, Drew RM, Foudoulis C, Ellingsen K (1989) Composition of crystalline inclusions and C-isotopic composition of Argyle and Ellendale diamonds. *In:* Kimberlites and Related Rocks: Their Mantle/Crust Setting, Diamonds and Diamond Exploration, Vol 2. Blackwell Scientific, Australia, p 966-989

Javoy M, Pineau F, Allegre CJ (1982) Carbon geodynamic cycle. Nature 300:171-173, doi: 10.1038/300171a0

Javoy M, Pineau F, Delorme H (1986) Carbon and nitrogen isotopes in the mantle. Chem Geol 57:41-62

Javoy M, Pineau F, Demaiffe D (1984) Nitrogen and carbon isotopic composition in the diamonds of Mbuji Mayi (Zaïre). Earth Planet Sci Lett 68:399-412

Jeynes C (1978) Natural polycrystalline diamond. Ind Diamond Rev 1:14-23

Jones AP, Genge M, Carmody L (2013) Carbonate melts and carbonatites. Rev Mineral Geochem 75:289-322

Jordan TH (1975) Continental tectosphere. Rev Geophys 13:1-12

Jordan TH (1978) Composition and development of continental tectosphere. Nature 274:544-548, doi: 10.1038/274544a0

Kaiser W, Bond WL (1959) Nitrogen, a major impurity in common type I diamond. Phys Rev 115:857-863

Kaminsky F, Wirth R, Thomas R (2009) Nyerereite and nahcolite inclusions in diamond: evidence for lower-mantle carbonatitic magmas. Mineral Mag 73:797-816, doi: 10.1180/minmag.2009.073.5.797

Kaminsky FV, Khachatryan GK (2004) The relationship between the distribution of nitrogen impurity centres in diamond crystals and their internal structure and mechanism of growth. Lithos 77:255-271, doi: 10.1016/j.lithos.2004.04.035

Kaminsky FV, Wirth R (2011) Iron carbide inclusions in lower-mantle diamond from Juina, Brazil. Canad Min 49:555-572, doi: 10.3749/canmin.49.2.555

Kaminsky FV, Zakharchenko OD, Griffin WL, Channer DMDR, Khachatryan-Blinova GK (2000) Diamonds from the Guaniamo area, Venezuela. Can Mineral 38:1347-1370

Kamiya Y, Lang AR (1964) On the structure of coated diamonds. Philos Mag 11:347-356

Keller RA, Taylor LA, Snyder GA, Sobolev VN, Carlson WD, Bezborodov SM, Sobolev NV (1999) Detailed pull-apart of a diamondiferous eclogite xenolith; implications for mantle processes during diamond genesis. Proc 7th Int Kimberlite Conf 1:397-402

Kennedy WQ (1964) The structural differentiation of Africa in the Pan-African (±500 m.y.) tectonic episode. *In:* 8th Annual Report of the Research Institute of African Geology, p 48-49

Keshav S, Corgne A, Gudfinnsson GH, Bizimis M, McDonough WF, Fei Y (2005) Kimberlite petrogenesis; insights from clinopyroxene-melt partitioning experiments at 6 GPa in the CaO-MgO-Al_2O_3-SiO_2-CO_2 system. Geochim Cosmochim Acta 69:2829-2845, doi: 10.1016/j.gca.2005.01.012

Keshav S, Gudfinnsson GH, Presnall DC (2011) Melting phase relations of simplified carbonated peridotite at 12-26 GPa in the systems CaO-MgO-SiO_2-CO_2 and CaO-MgO-Al_2O_3-SiO_2-CO_2: Highly calcic magmas in the transition zone of the Earth. J Petrol 52:2265-2291, doi: 10.1093/petrology/egr048

Kinny PD, Meyer HOA (1994) Zircon from the mantle; a new way to date old diamonds. J Geol 102:475-481

Kirkley MB, Gurney JJ, Levinson AA (1991) Age, origin, and emplacement of diamonds; scientific advances in the last decade. Gems Gemology 27:2-25

Klein-BenDavid O, Izraeli ES, Hauri E, Navon O (2004) Mantle fluid evolution; a tale of one diamond. Lithos 77:243-253, doi: 10.1016/j.lithos.2004.04.003

Klein-BenDavid O, Izraeli ES, Hauri EH, Navon O (2007a) Fluid inclusions in diamonds from the Diavik Mine, Canada and the evolution of diamond-forming fluids. Geochim Cosmochim Acta 71:723-744, doi: 10.1016/j.gca.2006.10.008

Klein-BenDavid O, Logvinova AM, Schrauder M, Spetius ZV, Weiss Y, Hauri EH, Kaminsky FV, Sobolev NV, Navon O (2009) High-Mg carbonatitic microinclusions in some Yakutian diamonds; a new type of diamond-forming fluid. Lithos 112:648-659, doi: 10.1016/j.lithos.2009.03.015

Klein-BenDavid O, Pearson DG (2009) Origins of subcalcic garnets and their relation to diamond forming fluids; case studies from Ekati (NWT, Canada) and Murowa (Zimbabwe). Geochim Cosmochim Acta 73:837-855, doi: 10.1016/j.gca.2008.04.044

Klein-BenDavid O, Pearson DG, Nowell GM, Ottley C, McNeill JCR, Cartigny P (2010) Mixed fluid sources involved in diamond growth constrained by Sr-Nd-Pb-C-N isotopes and trace elements. Earth Planet Sci Lett 289:123-133, doi: 10.1016/j.epsl.2009.10.035

Klein-BenDavid O, Wirth R, Navon O (2006) TEM imaging and analysis of microinclusions in diamonds; a close look at diamond-growing fluids. Am Mineral 91:353-365, doi: 10.2138/am.2006.1864

Klein-BenDavid O, Wirth R, Navon O (2007b) Micrometer-scale cavities in fibrous and cloudy diamonds; a glance into diamond dissolution events. Earth Planet Sci Lett 264:89-103, doi: 10.1016/j.epsl.2007.09.004

Knoche R, Sweeney RJ, Luth RW (1999) Carbonation and decarbonation of eclogites: the role of garnet. Contrib Mineral Petrol 135:332-339, doi: 10.1007/s004100050515

Koeberl C, Masaitis VL, Shafranovsky GI, Gilmour I, Langenhorst F, Schrauder M (1997) Diamonds from the Popigai impact structure, Russia. Geology 25:967, doi: 10.1130/0091-7613(1997)025<0967:DFTPIS>2.3.CO;2

Kopylova M, Navon O, Dubrovinsky L, Khachatryan G (2010) Carbonatitic mineralogy of natural diamond-forming fluids. Earth Planet Sci Lett 291:126-137, doi: 10.1016/j.epsl.2009.12.056

Kopylova MG, Afanasiev VP, Bruce LF, Thurston PC, Ryder J (2011) Metaconglomerate preserves evidence for kimberlite, diamondiferous root and medium grade terrane of a pre-2.7 Ga Southern Superior protocraton. Earth Planet Sci Lett 312:213-225, doi: 10.1016/j.epsl.2011.09.057

Kramers JD (1979) Lead, uranium, strontium, potassium and rubidium in inclusion-bearing diamonds and mantle-derived xenoliths from southern Africa. Earth Planet Sci Lett 42:58-70

Krogh Ravna EJ (2000) The garnet-clinopyroxene Fe^{2+}-Mg geothermometer; an updated calibration. J Metamorphic Geol 18:211-219

Krogh Ravna EJ, Paquin J (2003) Thermobarometric methodologies applicable to eclogites and garnet ultrabasites. EMU Notes Mineral 5:229-259

Kumar M, Akaishi M, Yamaoka S (2000) Formation of diamond from supercritical H_2O-CO_2 fluid at high pressure and high temperature. J Crystal Growth 213:203-206

Kvasnytsya VM, Wirth R (2009) Nanoinclusions in microdiamonds from Neogenic sands of the Ukraine (Samotkan' Placer); a TEM study. Lithos 113:454-464, doi: 10.1016/j.lithos.2009.05.019

Laiginhas FA, Pearson DG, Phillips D, Burgess R, Harris JW (2009) Re-Os and $^{40}Ar/^{39}Ar$ isotope measurements of inclusions in alluvial diamonds from the Ural Mountains; constraints on diamond genesis and eruption ages. Lithos 112:714-723, doi: 10.1016/j.lithos.2009.03.003

Lazarov M, Woodland AB, Brey GP (2009) Thermal state and redox conditions of the Kaapvaal mantle: A study of xenoliths from the Finsch mine, South Africa. Lithos 112:913-923, doi: 10.1016/j.lithos.2009.03.035

Leost I, Stachel T, Brey GP, Harris JW, Ryabchikov ID (2003) Diamond formation and source carbonation: mineral associations in diamonds from Namibia. Contrib Mineral Petrol 145:15-24, doi: 10.1007/s00410-003-0442-5

Litvin Y (2009) The physicochemical conditions of diamond formation in the mantle matter: experimental studies. Russian Geol Geophys 50:1188-1200

Litvin Y, Bobrov A (2008) Experimental study of diamond crystallization in carbonate-peridotite melts at 8.5 GPa. Dokl Earth Sci 422:1167-1171, doi: 10.1134/S1028334X08070386

Litvin Y, Butvina V, Bobrov A, Zharikov V (2002) The first synthesis of diamond in sulfide-carbon systems: The role of sulfides in diamond genesis. Dokl Earth Sci 382:40-43

Litvin Y, Chudinovskikh L, Zharikov V (1997) Crystallization of diamond and graphite in the mantle alkaline-carbonate melts in the experiments at pressure 7-11 GPa. Dokl Akad Nauk 355:669-672

Litvin Y, Chudinovskikh LT, Zharikov VA (1998a) The growth of diamond on seed crystals in the $Na_2Mg(CO_3)_2$-$K_2Mg(CO_3)2$-C system at 8-10 Gpa. Dokl Earth Sci 359A:464-466

Litvin Y, Chudinovskikh LT, Zharikov VA (1998b) Crystallization of diamond in the $Na_2Mg(CO_3)_2$-$K_2Mg(CO_3)_2$-C system at 8-10 GPa. Dokl Earth Sci 359A:433-435

Litvin Y, Litvin V, Kadik A (2008) Study of diamond and graphite crystallization from eclogite-carbonatite melts at 8.5 GPa: the role of silicates in diamond genesis. Dokl Earth Sci 419:486-491

Litvin Y, Zharikov VA (1999) Primary fluid-carbonatitic inclusions in diamond simulating by the system K_2O-Na_2O-CaO-MgO-FeO-CO_2 as a diamond-producting medium in experiment at 7-9 GPa. Dokl Akad Nauk 367:397-401

Logvinova AM, Wirth R, Fedorova EN, Sobolev NV (2008) Nanometre-sized mineral and fluid inclusions in cloudy Siberian diamonds; new insights on diamond formation. Eur J Mineral 20:317-331, doi: 10.1127/0935-1221/2008/0020-1815

Lord OT, Walter MJ, Dasgupta R, Walker D, Clark SM (2009) Melting in the Fe-C system to 70 GPa. Earth Planet Sci Lett 284:157-167, doi: 10.1016/j.epsl.2009.04.017

Lowry D, Mattey DP, Harris JW (1999) Oxygen isotope composition of syngenetic inclusions in diamond from the Finsch Mine, RSA. Geochim Cosmochim Acta 63:1825-1836

Luth RW (1993) Diamonds, eclogites, and the oxidation state of the Earth's mantle. Science 261:66-68

Luth RW (1999) Carbon and carbonates in the mantle. Geochem Soc Spec Pub 6:297-316

Luth RW, Virgo D, Boyd FR, Wood BJ (1990) Ferric iron in mantle-derived garnets; implications for thermobarometry and for the oxidation state of the mantle. Contrib Mineral Petrol 104:56-72

MacGregor ID, Manton WI (1986) Roberts Victor eclogites; ancient oceanic crust. J Geophys Res 91:14,063-014,079, doi: 10.1029/JB091iB14p14063

Manning CE, Shock EL, Sverjensky D (2013) The chemistry of carbon in aqueous fluids at crustal and upper-mantle conditions: experimental and theoretical constraints. Rev Mineral Geochem 75:109-148

Marty B, Alexander CMO'D, Raymond SN (2013) Primordial origins of Earth's carbon. Rev Mineral Geochem 75:149-181

Massonne H (2003) A comparison of the evolution of diamondiferous quartz-rich rocks from the Saxonian Erzgebirge and the Kokchetav Massif: are so-called diamondiferous gneisses magmatic rocks? Earth Planet Sci Lett 216:347-364, doi: 10.1016/S0012-821X(03)00512-0

McCammon C (2001) Deep diamond mysteries. Science 293:813-814

McCammon CA, Griffin WL, Shee SR, O'Neill HSC (2001) Oxidation during metasomatism in ultramafic xenoliths from the Wesselton Kimberlite, South Africa; implications for the survival of diamond. Contrib Mineral Petrol 141:287-296

McCammon CA, Stachel T, Harris JW (2004) Iron oxidation state in lower mantle mineral assemblages; II, Inclusions in diamonds from Kankan, Guinea. Earth Planet Sci Lett 222:423-434, doi: 10.1016/j.epsl.2004.03.019

McCourt S, Kampunzu AB, Bagai Z, Armstrong RA (2004) The crustal architecture of Archaean terranes in Northeastern Botswana. S Afr J Geol 107:147-158

McDonough WF, Sun SS (1995) The composition of the Earth. Chem Geol 120:223-253, doi: 10.1016/0009-2541(94)00140-4

McNeill J, Pearson DG, Klein-Bendavid O, Nowell GM, Ottley CJ, Chinn I (2009) Quantitative analysis of trace element concentrations in some gem-quality diamonds. Journal Of Physics-Condensed Matter 21:364207, doi: 10.1088/0953-8984/21/36/364207

Meyer HOA (1985) Genesis of diamond; a mantle saga. Am Mineral 70:344-355

Meyer HOA (1987) Inclusions in diamond. *In:* Mantle Xenoliths. Nixon Peter H (ed) John Wiley & Sons, Chichester, United Kingdom, p 501-523

Meyer HOA, Tsai HM (1976) Mineral inclusions in diamond; temperature and pressure of equilibration. Science 191:849-851

Minarik WG, Watson EB (1995) Interconnectivity of carbonate melt at low melt fraction. Earth Planet Sci Lett 133:423-437, doi: 10.1016/0012-821x(95)00085-q

Mitchell RH (1995) Kimberlites, Orangeites, and Related Rocks. Plenum Press, New York

Mitchell RH (2004) Experimental studies at 5-12 GPa of the Ondermatjie hypabyssal kimberlite. Lithos 76:551-564, doi: 10.1016/j.lithos.2004.03.032

Mitchell RH (2008) Petrology of hypabyssal kimberlites; relevance to primary magma compositions. J Volcan Geotherm Res 174:1-8, doi: 10.1016/j.jvolgeores.2007.12.024

Mysen BO, Fogel ML, Morrill PL, Cody GD (2009) Solution behavior of reduced COH volatiles in silicate melts at high pressure and temperature. Geochim Cosmochim Acta 73:1696-1710, doi: 10.1016/j.gca.2008.12.016

Nakajima Y, Takahashi E, Suzuki T, Funakoshi K-I (2009) "Carbon in the core" revisited. Phys Earth Planet Int 174:202-211, doi: 10.1016/j.pepi.2008.05.014

Nasdala L, Brenker FE, Glinnemann J, Hofmeister W, Gasparik T, Harris JW, Stachel T, Reese I (2003) Spectroscopic 2D-tomography; residual pressure and strain around mineral inclusions in diamonds. Eur J Mineral 15:931-935, doi: 10.1127/0935-1221/2003/0015-0931

Navon O (1999) Diamond formation in the Earth's mantle. Proc 7th Int Kimberlite Conf 2:584-604

Navon O, Hutcheon ID, Rossman GR, Wasserburg GJ (1988) Mantle-derived fluids in diamond micro-inclusions. Nature 355:784-789

Nestola F, Merli M, Nimis P, Parisatto M, Kopylova M, Safonov OG, De Stefano A, Longo M, Ziberna L, Manghnani MH (2012) In-situ analysis of garnet inclusion in diamond using single-crystal X-ray diffraction and X-ray micro-tomography. Eur J Mineral 24:599-606, doi: 10.1127/0935-1221/2012/0024-2212

Nestola F, Nimis P, Ziberna L, Longo M, Marzoli A, Harris JW, Manghnani MH, Fedortchouk Y (2011) First crystal-structure determination of olivine in diamond; composition and implications for provenance in the Earth's mantle. Earth Planet Sci Lett 305:249-255, doi: 10.1016/j.epsl.2011.03.007

Nimis P (2002) The pressures and temperatures of formation of diamond based on thermobarometry of chromian diopside inclusions. Canad Mineral 40, Part 3:871-884

Nimis P, Grütter H (2010) Internally consistent geothermometers for garnet peridotites and pyroxenites. Contrib Mineral Petrol 159:411-427, doi: 10.1007/s00410-009-0455-9

Nimis P, Grütter H (2012) Discussion of "The applicability of garnet-orthopyroxene geobarometry in mantle xenoliths", by Wu C.-M. and Zhao G. (Lithos, v. 125, p. 1-9). Lithos 142:285-287, doi: 10.1016/j.lithos.2011.09.006

Nimis P, Taylor WR (2000) Single clinopyroxene thermobarometry for garnet peridotites; Part I, Calibration and testing of a Cr-in-Cpx barometer an an enstatite-in-cpx thermometer. Contrib Mineral Petrol 139:541-554, doi: 10.1007/s004100000156

Nixon PH (1987) Kimberlitic xenoliths and their cratonic setting. *In:* Mantle Xenoliths. Nixon PH (ed) John Wiley & Sons, Chichester, United Kingdom (GBR), p 215-246

O'Neill HSC, McCammon CA, Canil D, Rubie DC, Ross CR, II, Seifert F (1993a) Mossbauer spectroscopy of mantle transition zone phases and determination of minimum Fe^{3+} content. Am Mineral 78:456-460

O'Neill HSC, Rubie DC, Canil D, Geiger C, Ross CR (1993b) Ferric iron in the upper mantle and in transition zone assemblages: implications for relative oxygen fugacities in the mantle. *In:* Evolution of the Earth and Planets. Takahashi E, Jeanloz R, Rubie DC (eds) Monograph 74. American Geophysical Union, Washington, p 73-88

O'Neill HSC, Wall VJ (1987) The olivine-orthopyroxene-spinel oxygen geobarometer, the nickel precipitation curve, and the oxygen fugacity of the Earth's upper mantle. J Petrol 28:1169-1191

Oganov AR, Hemley RJ, Hazen RM, Jones AP (2013) Structure, bonding, and mineralogy of carbon at extreme conditions. Rev Mineral Geochem 75:47-77

Ogasawara Y (2005) Microdiamonds in ultrahigh-pressure metamorphic rocks. Elements 1:91-96

Ozima M, Igarashi G (2000) The primordial noble gases in the Earth; a key constraint on Earth evolution models. Earth Planet Sci Lett 176:219-232

Pal'yanov YN, Borzdov YM, Bataleva YV, Sokol AG, Palyanova GA, Kupriyanov IN (2007a) Reducing role of sulfides and diamond formation in the Earth's mantle. Earth Planet Sci Lett 260:242-256, doi: 10.1016/j.epsl.2007.05.033

Pal'yanov YN, Borzdov YM, Khokhryakov AF, Kupriyanov IN, Sobolev NV (2006) Sulfide melts-graphite interaction at HPHT conditions; implications for diamond genesis. Earth Planet Sci Lett 250:269-280, doi: 10.1016/j.epsl.2006.06.049

Pal'yanov YN, Borzdov YM, Kupriyanov I, Gusev V, Khokhryakov AF, Sokol AG (2001) High-pressure synthesis and characterization of diamond from a sulfur-carbon system. Diamond Relat Mater 10:2145-2152

Pal'yanov YN, Kupriyanov IN, Borzdov YM, Sokol AG, Khokhryakov AF (2009) Diamond crystallization from a sulfur-carbon system at HPHT conditions. Crystal Growth 9:2922-2926

Pal'yanov YN, Shatsky VS, Sobolev NV, Sokol AG (2007b) The role of mantle ultrapotassic fluids in diamond formation. Proc Nat Acad Sci USA 104:9122-9127, doi: 10.1073/pnas.0608134104

Pal'yanov YN, Sokol AG (2009) The effect of composition of mantle fluids/melts on diamond formation processes. Lithos 112:690-700, doi: 10.1016/j.lithos.2009.03.018

Pal'yanov YN, Sokol AG, Borzdov YM, Khokhryakov A, Sobolev NV (1998) Crystallization of diamond in the $CaCO_3$-C, $MgCO_3$-C and $CaMg(CO_3)_2$-C systems. Dokl Akad Nauk 363:230-233

Pal'yanov YN, Sokol AG, Borzdov YM, Khokhryakov AF, Sobolev NV (2002) Diamond formation through carbonate-silicate interaction. Am Mineral 87:1009-1013

Pal'yanov YN, Sokol AG, Khokhryakov AF, Pal'yanova GA, Borzdov YM, Sobolev NV (2000) Diamond and graphite crystallization in COH fluid at PT parameters of the natural diamond formation. Dokl Earth Sci 375A:1395-1398

Pal'yanov YN, Sokol AG, Tomilenko AA, Sobolev NV (2005) Conditions of diamond formation through carbonate-silicate interaction. Eur J Mineral 17:207-214, doi: 10.1127/0935-1221/2005/0017-0207

Palot M, Cartigny P, Harris JW, Kaminsky F, Stachel T (2012) Evidence for deep mantle convection and primordial heterogeneity from nitrogen and carbon stable isotopes in diamond. Earth Planet Sci Lett 357-358:179-193

Pearson DG, Canil D, Shirey SB (2003) Mantle samples included in volcanic rocks: xenoliths and diamonds. *In:* Treatise on Geochemistry: Vol. 2, The Mantle. Carlson RW (ed) Elsevier, New York, p 171-277

Pearson DG, Carlson RW, Shirey SB, Boyd FR, Nixon PH (1995) Stabilisation of Archaean lithospheric mantle; a Re-Os isotope study of peridotite xenoliths from the Kaapvaal Craton. Earth Planet Sci Lett 134:341-357, doi: 10.1016/0012-821x(95)00125-v
Pearson DG, Shirey SB (1999) Isotopic dating of diamonds. *In:* Application of Radiogenic Isotopes to Ore Deposit Research and Exploration. Lambert DD, Ruiz J (eds) Society of Economic Geologists, Boulder, CO, United States, p 143-171
Pearson DG, Shirey SB, Bulanova GP, Carlson RW, Milledge HJ (1999a) Dating and paragenetic distinction of diamonds using the Re-Os isotope system; application to some Siberian diamonds. Proc 7th Int Kimberlite Conf 2:637-643
Pearson DG, Shirey SB, Bulanova GP, Carlson RW, Milledge HJ (1999b) Re-Os isotope measurements of single sulfide inclusions in a Siberian diamond and its nitrogen aggregation systematics. Geochim Cosmochim Acta 63:703-711, doi: 10.1016/s0016-7037(99)00042-3
Pearson DG, Shirey SB, Harris JW, Carlson RW (1998) Sulphide inclusions in diamonds from the Koffiefontein kimberlite, S Africa; constraints on diamond ages and mantle Re-Os systematics. Earth Planet Sci Lett 160:311-326
Pearson DG, Wittig N (2008) Formation of Archaean continental lithosphere and its diamonds; the root of the problem. J Geol Soc London 165:895-914, doi: 10.1144/0016-76492008-003
Philippot P, Busigny V, Scambelluri M, Cartigny P (2007) Oxygen and nitrogen isotopes as tracers of fluid activities in serpentinites and metasediments during subduction. Mineral Petrol 91:11-24, doi: 10.1007/s00710-007-0183-7
Phillips D, Harris JW, Viljoen KS (2004) Mineral chemistry and thermobarometry of inclusions from De Beers Pool diamonds, Kimberley, South Africa. Lithos 77:155-179, doi: 10.1016/j.lithos.2004.04.005
Phillips D, Onstott TC, Harris JW (1989) $^{40}Ar/^{39}Ar$ laser-probe dating of diamond inclusions from Premier kimberlite. Nature 340:460-462, doi: 10.1038/340460a0
Pokhilenko NP, Sobolev NV, Reutsky VN, Hall AE, Taylor LA (2004) Crystalline inclusions and C isotope ratios in diamonds from the Snap Lake-King Lake kimberlite dyke system; evidence of ultradeep and enriched lithospheric mantle. Lithos 77:57-67, doi: 10.1016/j.lithos.2004.04.019
Pollack HN, Chapman DS (1977) On the regional variation of heat flow, geotherms, and lithospheric thickness. Tectonophysics 38:279-296
Promprated P, Taylor LA, Anand M, Floss C, Sobolev NV, Pokhilenko NP (2004) Multiple-mineral inclusions in diamonds from the Snap Lake/King Lake kimberlite dike, Slave Craton, Canada; a trace element perspective. Lithos 77:69-81, doi: 10.1016/j.lithos.2004.04.009
Rege S, Griffin WL, Kurat G, Jackson SE, Pearson NJ, O'Reilly SY (2008) Trace-element geochemistry of diamondite; crystallisation of diamond from kimberlite-carbonatite melts. Lithos 106:39-54, doi: 10.1016/j.lithos.2008.06.002
Rege S, Griffin WL, Pearson NJ, Araujo D, Zedgenizov D, O'Reilly SY (2010) Trace element patterns of fibrous and monocrystalline diamonds insights into mantle fluids. Lithos (Oslo) 118:313-337, doi: 10.1016/j.lithos.2010.05.007
Resano M, Vanhaecke F, Hutsebaut D, de Corte D, Moens L (2003) Possibilities of laser ablation-inductively-coupled plasma-mass spectrometry for diamond fingerprinting. J Anal At Spectrom 18:1238-1242
Richardson SH (1986) Latter-day origin of diamonds of eclogitic paragenesis. Nature 322:623-626
Richardson SH, Chinn IL, Harris JW (1999) Age and origin of eclogitic diamonds from the Jwaneng Kimberlite, Botswana. Proc 7th Int Kimberlite Conf 2:709-713
Richardson SH, Erlank AJ, Harris JW, Hart SR (1990) Eclogitic diamonds of Proterozoic age from Cretaceous kimberlites. Nature 346:54-56, doi: 10.1038/346054a0
Richardson SH, Gurney JJ, Erlank AJ, Harris JW (1984) Origin of diamonds in old enriched mantle. Nature 310:198-202
Richardson SH, Harris JW (1997) Antiquity of peridotitic diamonds from the Siberian Craton. Earth Planet Sci Lett 151:271-277, doi: 10.1016/s0012-821x(97)81853-5
Richardson SH, Pöml P, Shirey SB, Harris, JW (2009) Age and origin of peridotitic diamonds from Venetia, Limpopo Belt, Kaapvaal-Zimbabwe craton. Lithos 112: 785-792, doi: 10.1016/j.lithos.2009.05.017
Richardson SH, Shirey SB (2008) Continental mantle signature of Bushveld magmas and coeval diamonds. Nature 453:910-913, doi: 10.1038/nature07073
Richardson SH, Shirey SB, Harris JW (2004) Episodic diamond genesis at Jwaneng, Botswana, and implications for Kaapvaal Craton evolution. Lithos 77:143-154, doi: 10.1016/j.lithos.2004.04.027
Richardson SH, Shirey SB, Harris JW, Carlson RW (2001) Archean subduction recorded by Re-Os isotopes in eclogitic sulfide inclusions in Kimberley diamonds. Earth Planet Sci Lett 191:257-266, doi: 10.1016/s0012-821x(01)00419-8
Ritsema J, van Heijst H, Woodhouse J (2004) Global transition zone tomography. J Geophys Res Solid Earth 109:14, doi: 10.1029/2003JB002610
Robertson R, Fox JJ, Martin AE (1934) Two types of diamond. Philos Trans R Soc London A232:463-535

Rohrbach A, Ballhaus C, Golla-Schindler U, Ulmer P, Kamenetsky VS, Kuzmin DV (2007) Metal saturation in the upper mantle. Nature 449:456-458, doi: 10.1038/nature06183

Rohrbach A, Ballhaus C, Ulmer P, Golla-Schindler U, Schoenbohm D (2011) Experimental evidence for a reduced metal-saturated upper mantle. J Petrol 52:717-731, doi: 10.1093/petrology.egq101

Rohrbach A, Schmidt MW (2011) Redox freezing and melting in the Earth's deep mantle resulting from carbon-iron redox coupling. Nature 472:209-214, doi: 10.1038/nature09899

Russell JK, Porritt LA, Lavallee Y, Dingwell DB (2012) Kimberlite ascent by assimilation; fueled buoyancy. Nature 481:352-356, doi: 10.1038/nature10740

Safonov OG, Kamenetsky VS, Perchuk LL (2011) Links between carbonatite and kimberlite melts in chlorid-carbonate-silicate systems; experiments and application to natural assemblages. J Petrol 52:1307-1331, doi: 10.1093/petrology/egq034

Safonov OG, Perchuk LL, Litvin YA (2007) Melting relations in the chloride-carbonate-silicate systems at high-pressure and the model for formation of alkalic diamond-forming liquids in the upper mantle. Earth Planet Sci Lett 253:112-128, doi: 10.1016/j.epsl.2006.10.020

Sato K, Akaishi M, Yamaoka S (1999) Spontaneous nucleation of diamond in the system $MgCO_3$-$CaCO_3$-C at 7.7 GPa. Diamond Relat Mater 8:1900-1905

Schertl H-P, Sobolev NV (2012) The Kokchetav massif, Kazakhstan: "Type locality" of diamond-bearing UHP metamorphic rocks. J Asian Earth Sci, doi: 10.1016/j.jseaes.2012.10.032

Schmitz MD, Bowring SA, de Wit MJ, Gartz V (2004) Subduction and terrane collision stabilize the western Kaapvaal craton tectosphere 2.9 billion years ago. Earth Planet Sci Lett 222:363-376, doi: 10.1016/j.epsl.2004.03.036

Schrauder M, Koeberl C, Navon O (1996) Trace element analyses of fluid-bearing diamonds from Jwaneng, Botswana. Geochim Cosmochim Acta 60:4711-4724

Schrauder M, Navon O (1993) Solid carbon dioxide in a natural diamond. Nature 365:42-44

Schrauder M, Navon O (1994) Hydrous and carbonatitic mantle fluids in fibrous diamonds from Jwaneng, Botswana. Geochim Cosmochim Acta 58:761-771

Schulze DJ (1983) Graphic rutile-olivine intergrowths from South African kimberlites. Carnegie Inst Wash YearBook 82:343-346

Schulze DJ, Harte B, Valley JW, Brenan JM, Channer DMD (2003) Extreme crustal oxygen isotope signatures preserved in coesite in diamond. Nature 423:68-70, doi: 10.1038/nature01615

Shatsky AF, Borzdov YM, Sokol AG, Pal'yanov YN (2002) Phase formation and diamond crystallization in carbon-bearing ultrapotassic carbonate-silicate systems. Geol Geofiz. 43:940-950

Shimizu N, Richardson SH (1987) Trace element abundance patterns of garnet inclusions in peridotite-suite diamonds. Geochim Cosmochim Acta 51:755-758, doi: 10.1016/0016-7037(87)90085-8

Shimizu N, Sobolev NV (1995) Young peridotitic diamonds from the Mir kimberlite pipe. Nature 375:394-397, doi: 10.1038/375394a0

Shimizu N, Sobolev NV, Yefimova ES (1997) Chemical heterogeneities of inclusion garnets and juvenile character of peridotitic diamonds from Siberia. Russ Geol Geophys 38:356-372

Shirey SB, Harris JW, Richardson SH, Fouch MJ, James DE, Cartigny P, Deines P, Viljoen F (2002) Diamond genesis, seismic structure, and evolution of the Kaapvaal-Zimbabwe Craton. Science 297:1683-1686, doi: 10.1126/science.1072384

Shirey SB, Kamber BS, Whitehouse MJ, Mueller PA, Basu AR (2008) A review of the isotopic and trace element evidence for mantle and crustal processes in the Hadean and Archean; implications for the onset of plate tectonic subduction. Geol Soc Am Spec Paper 440:1-29, doi: 10.1130/2008.2440(01)

Shirey SB, Richardson SH (2011) Start of the Wilson cycle at 3 Ga shown by diamonds from subcontinental mantle. Science 333:434-436, doi: 10.1126/science.1206275

Shirey SB, Richardson SH, Harris JW (2004a) Age, paragenesis and composition of diamonds and evolution of the Precambrian mantle lithosphere of Southern Africa. S Afr J Geol 107:91-106

Shirey SB, Richardson SH, Harris JW (2004b) Integrated models of diamond formation and craton evolution. Lithos 77:923-944, doi: 10.1016/j.lithos.2004.04.018

Shushkanova AV, Litvin V (2008) Diamond nucleation and growth in sulfide-carbon melts: an experimental study at 6.0-7.1 GPa. Eur J Mineral 20:349-355

Shushkanova AV, Litvin YA (2006) Formation of diamond polycrystals in pyrrhotite-carbonic melt; experiments at 6.7 GPa. Dokl Earth Sci 409:916-920, doi: 10.1134/s1028334x06060183

Shutong X, Okay AI, Shouyuan J, Sengor AMC, Wen S, Yican L, Laili J (1992) Diamond from the Dabie Shan metamorphic rocks and its implication for tectonic setting. Science 256:80-82, doi: 10.1126/science.256.5053.80

Siebert J, Guyot F, Malavergne V (2005) Diamond formation in metal-carbonate interactions. Earth Planet Sci Lett 229:205-216, doi: 10.1016/j.epsl.2004.10.036

Simakov SK (2008) Garnet-clinopyroxene and clinopyroxene geothermobarometry of deep mantle and crust eclogites and peridotites. Lithos 106:125-136, doi: 10.1016/j.lithos.2008.06.013

Simakov SK, Taylor LA (2000) Geobarometry for mantle eclogites; solubility of Ca-Tschermaks in clinopyroxene. International Geol Rev 42:534-544

Smart KA, Chacko T, Stachel T, Muehlenbachs K, Stern RA, Heaman LM (2011) Diamond growth from oxidized carbon sources beneath the Northern Slave Craton, Canada: A δ^{13}-N study of eclogite-hosted diamonds from the Jericho kimberlite. Geochim Cosmochim Acta 75:6027-6047, doi: 10.1016/j.gca.2011.07.028

Smart KA, Heaman LM, Chacko T, Simonetti A, Kopylova M, Mah D, Daniels D (2009) The origin of high-MgO diamond eclogites from the Jericho Kimberlite, Canada. Earth Planet Sci Lett 284:527-537, doi: 10.1016/j.epsl.2009.05.020

Smit KV, Shirey SB, Richardson SH, le Roex AP, Gurney JJ (2010) Re/Os isotopic composition of peridotitic sulphide inclusions in diamonds from Ellendale, Australia; age constraints on Kimberley cratonic lithosphere. Geochim Cosmochim Acta 74:3292-3306, doi: 10.1016/j.gca.2010.03.001

Smith CB (1983) Pb, Sr and Nd isotopic evidence for sources of southern African Cretaceous kimberlites. Nature 304:51-54, doi: 10.1038/304051a0

Smith CB, Bulanova GP, Walter MJ, Kohn SC, Mikhail S, Gobbo L (2012) Origin of diamonds from the Dachine ultramafic, French Guyana. Proc 10th Int Kimberlite Conf Bangalore, India, 10IKC-97

Smith CB, Clark TC, Barton ES, Bristow JW (1994) Emplacement ages of kimberlite occurrences in the Prieska region, southwest border of the Kaapvaal Craton, South Africa. Chem Geol 113:149-169, doi: 10.1016/0009-2541(94)90010-8

Smith CB, Gurney JJ, Harris JW, Otter MB, Robinson DN, Kirkley MB, Jagoutz E (1991) Neodymium and strontium isotope systematics of eclogite and websterite paragenesis inclusions from single diamonds. Geochim Cosmochim Acta 55:2579-2590

Smith CB, Pearson DG, Bulanova GP, Beard AD, Carlson RW, Wittig N, Sims K, Chimuka L, Muchemwa E (2009) Extremely depleted lithospheric mantle and diamonds beneath the southern Zimbabwe Craton. Lithos 112:1120-1132, doi: 10.1016/j.lithos.2009.05.013

Smith EM, Kopylova MG, Nowell GM, Pearson DG, Ryder J (2012) Archean mantle fluids preserved in fibrous diamonds from Wawa, Superior craton. Geology 40:1071-1074, doi: 10.1130/G33231.1

Sobolev NV (1977) Deep-Seated Inclusions in Kimberlites and the Problem of the Composition of the Upper Mantle. American Geophysical Union, Washington, DC

Sobolev NV (1983) Parageneses of the diamonds and the problems of mineral formation in deep seated conditions. Zap Vses Miner Obshch 112:389-397

Sobolev NV (1984) Crystalline inclusions in diamonds from New South Wales, Australia. *In:* Kimberlite occurrence and Origin: A Basis for Conceptual Models in Exploration. Glover JE, Harris PG (eds) University of Western Australia, Perth, p 213-226

Sobolev NV, Botkunov AI, Bakumenko IT, Sobolev VS (1972) Crystalline inclusions with octahedral faces in diamonds. Dokl Akad Nauk 204:117-120

Sobolev NV, Efimova ES (1998) Compositional variations of chromite inclusions as an indicator of the zonation of diamond crystals. Dokl Earth Sci 359:163-166

Sobolev NV, Fursenko BA, Goryainov SV, Shu J, Hemley RJ, Mao H-K, Boyd FR (2000) Fossilized high pressure from the Earth's deep interior; the coesite-in-diamond barometer. Proc Natl Acad Sci USA 97:11875-11879, doi: 10.1073/pnas.220408697

Sobolev NV, Kaminsky FV, Griffin WL, Yefimova ES, Win TT, Ryan CG, Botkunov AI (1997) Mineral inclusions in diamonds from the Sputnik kimberlite pipe, Yakutia. Lithos 39:135-157, doi: 10.1016/s0024-4937(96)00022-9

Sobolev NV, Lavrent'ev Y, Pospelova LN, Sobolev EV (1969) Chrome pyropes from the diamonds of Yakutia. Dokl Akad Nauk 189:162-165

Sobolev NV, Lavrent'ev YG (1971) Isomorphic sodium admixture in garnets formed at high pressures. Contrib Mineral Petrol 21:1-12

Sobolev NV, Lavrent'ev YG, Pokhilenko NP, Usova LV (1973) Chrome-Rich Garnets from the Kimberlites of Yakutia and Their Parageneses. Contrib Mineral Petrol 40:39-52

Sobolev NV, Logvinova AM, Yefimova ES (2009) Syngenetic phlogopite inclusions in kimberlite-hosted diamonds: implications for role of volatiles in diamond formation. Russ Geol Geophys 50(12):1234-1248

Sobolev NV, Shatsky VS (1990) Diamond inclusions in garnets from metamorphic rocks: a new environment for diamond formation. Nature 343:742-746, doi: 10.1038/343742a0

Sobolev VS, Sobolev NV (1980) New evidence on subduction to great depths of the eclogitized crustal rocks (in Russian). Dokl Akad Nauk 250:683-685

Sobolev VS, Sobolev NV, Lavrent'yev YG (1972) Inclusions in diamond from diamond-bearing eclogite. Dokl Akad Nauk 207:164-167

Sokol AG, Borzdov YM, Pal yYN, Khokhryakov AF, Sobolev NV (2001a) An experimental demonstration of diamond formation in the dolomite-carbon and dolomite-fluid-carbon systems. Eur J Mineral 13:893-900

Sokol AG, Pal'yanov YN (2004) Diamond crystallization in fluid and carbonate-fluid systems under mantle P-T conditions; 2, An analytical review of experimental data. Geochem Int 42:1018-1032

Sokol AG, Pal'yanov YN, Khokhryakov AF, Borzdov YM (2001b) Diamond and graphite crystallization from C-O-H fluids under high pressure and high temperature conditions. Diamond Relat Mater 10:2131-2136

Sokol AG, Pal'yanov YN, Pal'yanova GA, Tomilenko AA (2004) Diamond crystallization in fluid and carbonate-fluid systems under mantle P-T conditions; 1, Fluid composition. Geochem Int 42:830-838

Sokol AG, Pal'yanova GA, Pal'yanov YN, Tomilenko AA, Melenevskiy VN (2009) Fluid regime and diamond formation in the reduced mantle; experimental constraints. Geochim Cosmochim Acta 73:5820-5834, doi: 10.1016/j.gca.2009.06.010

Sokol AG, Tomilenko AA, Pal'yanov YN, Borzdov YM, Pal'yanova GA, Khokhryakov AF (2000) Fluid regime of diamond crystallisation in carbonate-carbon systems. Eur J Mineral 12:367-375

Song S, Zhang L, Niu Y, Su L, Jian P, Liu D (2005) Geochronology of diamond-bearing zircons from garnet peridotite in the North Qaidam UHPM belt, Northern Tibetan Plateau: A record of complex histories from oceanic lithosphere subduction to continental collision. Earth Planet Sci Lett 234:99-118, doi: 10.1016/j.epsl.2005.02.036

Sparks RSJ, Brooker RA, Field M, Kavanagh J, Schumacher JC, Walter MJ, White J (2009) The nature of erupting kimberlite melts. Lithos 112:429-438, doi: 10.1016/j.lithos.2009.05.032

Spetsius ZV, Belousova EA, Griffin WL, O'Reilly SY, Pearson NJ (2002) Archean sulfide inclusions in Paleozoic zircon megacrysts from the Mir Kimberlite, Yakutia; implications for the dating of diamonds. Earth Planet Sci Lett 199:111-126, doi: 10.1016/s0012-821x(02)00539-3

Spetsius ZV, Taylor LA (2008) Diamonds of Yakutia: Photographic Evidence for their Origin. Tranquility Base Press, Lenoir City, Tennessee

Spivak AV, Litvin Y (2004) Diamond syntheses in multicomponent carbonate-carbon melts of natural chemistry: elementary processes and properties. Diamond Relat Mater 13: 482-487

Stachel T, Aulbach S, Brey GP, Harris JW, Leost I, Tappert R, Viljoen KS (2004) The trace element composition of silicate inclusions in diamonds; a review. Lithos 77:1-19, doi: 10.1016/j.lithos.2004.03.027

Stachel T, Brey GP, Harris JW (1998a) Rare and unusual mineral inclusions in diamonds from Mwadui, Tanzania. Contrib Mineral Petrol 132:34-47

Stachel T, Brey GP, Harris JW (2005) Inclusions in sub-lithospheric diamonds; glimpses of deep earth. Elements 1:73-87

Stachel T, Harris JW (1997) Diamond precipitation and mantle metasomatism; evidence from the trace element chemistry of silicate inclusions in diamonds from Akwatia, Ghana. Contrib Mineral Petrol 129:143-154, doi: 10.1007/s004100050328

Stachel T, Harris JW (2008) The origin of cratonic diamonds - constraints from mineral inclusions. Ore Geol Rev 34:5-32, doi: 10.1016/j.oregeorev.2007.05.002

Stachel T, Harris JW (2009) Formation of diamond in the Earth's mantle. J Phys Condens Mat 21:364206, doi: 10.1088/0953-8984/21/36/364206

Stachel T, Harris JW, Aulbach S, Deines P (2002) Kankan diamonds (Guinea) III; $\delta^{13}C$ and nitrogen characteristics of deep diamonds. Contrib Mineral Petrol 142:465-475

Stachel T, Harris JW, Brey GP (1999) REE patterns of peridotitic and eclogitic inclusions in diamonds from Mwadui (Tanzania). Proc 7th Int Kimberlite Conf 2:829-835

Stachel T, Harris JW, Brey GP, Joswig W (2000) Kankan diamonds (Guinea); II, Lower mantle inclusion parageneses. Contrib Mineral Petrol 140:16-27, doi: 10.1007/s004100000174

Stachel T, Harris JW, Muehlenbachs K (2009) Sources of carbon in inclusion bearing diamonds. Lithos 112:625-637, doi: 10.1016/j.lithos.2009.04.017

Stachel T, Viljoen KS, Brey G, Harris JW (1998b) Metasomatic processes in lherzolitic and harzburgitic domains of diamondiferous lithospheric mantle; REE in garnets from xenoliths and inclusions in diamonds. Earth Planet Sci Lett 159:1-12, doi: 10.1016/s0012-821x(98)00064-8

Stagno V, Frost DJ (2010) Carbon speciation in the asthenosphere; experimental measurements of the redox conditions at which carbonate-bearing melts coexist with graphite or diamond in peridotite assemblages. Earth Planet Sci Lett 300:72-84, doi: 10.1016/j.epsl.2010.09.038

Stagno V, Tange Y, Miyajima N, McCammon CA, Irifune T, Frost DJ (2011) The stability of magnesite in the transition zone and the lower mantle as function of oxygen fugacity. Geophys Res Lett 38:L19309, doi: 10.1029/2011GL049560

Strong HM, Hanneman RE (1967) Crystallization of diamond and graphite. J Chem Phys 46:3668-3676

Sumino H, Dobrzhinetskaya LF, Burgess R, Kagi H (2011) Deep-mantle-derived noble gases in metamorphic diamonds from the Kokchetav Massif, Kazakhstan. Earth Planet Sci Lett 307:439-449, doi: 10.1016/j.epsl.2011.05.018

Sumiya H, Toda N, Satoh S (2000) High-quality large diamond crystals. New Diamond Front Carbon Technol 10:233-251

Sun L, Wu Q, Wang WK (2001) Bulk diamond formation from graphite in the presence of C-O-H fluid under high pressure. High Pressure Res 21:159-173

Tappert R, Foden J, Stachel T, Muehlenbachs K, Tappert M, Wills K (2009) Deep mantle diamonds from South Australia; a record of Pacific subduction at the Gondwanan margin. Geology 37:43-46, doi: 10.1130/g25055a.1

Tappert R, Stachel T, Harris JW, Muehlenbachs K, Ludwig T, Brey GP (2005a) Diamonds from Jagersfontein (South Africa); messengers from the sub-lithospheric mantle. Contrib Mineral Petrol 150:505-522, doi: 10.1007/s00410-005-0035-6

Tappert R, Stachel T, Harris JW, Muehlenbachs K, Ludwig T, Brey GP (2005b) Subducting oceanic crust; the source of deep diamonds. Geology 33:565-568, doi: 10.1130/g21637.1

Tappert R, Tappert MC (2011) Diamonds in Nature: A Guide to Rough Diamonds. Springer Verlag, Berlin

Taylor LA, Anand M, Promprated P, Floss C, Sobolev NV (2003) The significance of mineral inclusions in large diamonds from Yakutia, Russia. Am Mineral 88:912-920

Taylor LA, Milledge HJ, Bulanova GP, Snyder GA, Keller RA (1998) Metasomatic eclogitic diamond growth; evidence from multiple diamond inclusions. Int Geol Rev 40:663-676

Taylor LA, Snyder GA, Crozaz G, Sobolev VN, Yefimova ES, Sobolev NV (1996) Eclogitic inclusions in diamonds; evidence of complex mantle processes over time. Earth Planet Sci Lett 142:535-551, doi: 10.1016/0012-821x(96)00106-9

Taylor WR, Green DH (1988) Measurement of reduced peridotite-C-O-H solidus and implications for redox melting of the mantle. Nature 332:349-352, doi: 10.1038/332349a0

Taylor WR, Jaques AL, Ridd M (1990) Nitrogen-defect aggregation characteristics of some Australian diamonds: Time-temperature constraints on the source regions of pipe and alluvial diamonds. Am Mineral 75:1290-1310

Thomassot E, Cartigny P, Harris JW, Lorand JP, Rollion-Bard C, Chaussidon M (2009) Metasomatic diamond growth: A multi-isotope study ^{13}C, ^{15}N, ^{33}S, ^{34}S of sulphide inclusions and their host diamonds from Jwaneng (Botswana). Earth Planet Sci Lett 282:79-90, doi: 10.1016/j.epsl.2009.03.001

Thomassot E, Cartigny P, Harris JW, Viljoen KS (2007) Methane-related diamond crystallization in the Earth's mantle: Stable isotope evidences from a single diamond-bearing xenolith. Earth Planet Sci Lett 257:362-371, doi: 10.1016/j.epsl.2007.02.020

Thomazo C, Pinti DL, Busigny V, Ader M, Hashizume K, Philippot P (2009) Biological activity and the Earth's surface evolution: Insights from carbon, sulfur, nitrogen and iron stable isotopes in the rock record. C R Palevol 8:665-678, doi: 10.1016/j.crpv.2009.02.003

Tomlinson EL, de Schrijver I, de Corte K, Jones AP, Moens L, Vanhaecke F (2005) Trace element compositions of submicroscopic inclusions in coated diamond; a tool for understanding diamond petrogenesis. Geochim Cosmochim Acta 69:4719-4732, doi: 10.1016/j.gca.2005.06.014

Tomlinson EL, Jones AP, Harris JW (2006) Co-existing fluid and silicate inclusions in mantle diamond. Earth Planet Sci Lett 250:581-595, doi: 10.1016/j.epsl.2006.08.005

Tomlinson EL, Mueller W (2009) A snapshot of mantle metasomatism; trace element analysis of coexisting fluid (LA-ICP-MS) and silicate (SIMS) inclusions in fibrous diamonds. Earth Planet Sci Lett 279:362-372, doi: 10.1016/j.epsl.2009.01.010

Touboul M, Puchtel IS, Walker RJ (2012) 182W Evidence for Long-Term Preservation of Early Mantle Differentiation Products. Science (New York, NY) 335:1065-1069, doi: 10.1126/science.1216351

Van Kranendonk M (2010) Two types of Archean continental crust: Plume and plate tectonics on early Earth. Am J Sci 310:1187-1209, doi: 10.2475/10.2010.01

Van Kranendonk MJ (2011) Onset of plate tectonics. Science 333:413-414, doi: 10.1126/science.1208766

Van Kranendonk MJ, Smithies RH, Hickman AH, Wingate MTD, Bodorkos S (2010) Evidence for Mesoarchean (similar to 3.2 Ga) rifting of the Pilbara Craton: The missing link in an early Precambrian Wilson cycle. Precamb Res 177:145-161, doi: 10.1016/j.precamres.2009.11.007

van Roermund H, Carswell D, Drury M, Heijboer T (2002) Microdiamonds in a megacrystic garnet websterite pod from Bardane on the island of Fjortoft, western Norway: Evidence for diamond formation in mantle rocks during deep continental subduction. Geology 30:959-962, doi: 10.1130/0091-7613(2002)030<0959:MIAMGW>2.0.CO;2

Viljoen KS, Phillips D, Harris JW, Robinson DH (1999) Mineral inclusions in diamonds from the Venetia kimberlites, Northern Province, South Africa. Proc 7th Int Kimberlite Conf 2:888-895

Wada N, Matsuda Ji (1998) A noble gas study of cubic diamonds from Zaire; constraints on their mantle source. Geochim Cosmochim Acta 62:2335-2345

Walter MJ, Bulanova GP, Armstrong LS, Keshav S, Blundy JD, Gudfinnsson G, Lord OT, Lennie AR, Clark SM, Smith CB, Gobbo L (2008) Primary carbonatite melt from deeply subducted oceanic crust. Nature 454:622, doi: 10.1038/nature07132

Walter MJ, Kohn SC, Araujo D, Bulanova GP, Smith CB, Gaillou E, Wang J, Steele A, Shirey SB (2011) Deep mantle cycling of oceanic crust; evidence from diamonds and their mineral inclusions. Science 334:54-57, doi: 10.1126/science.1209300

Wang W, Gasparik T, Rapp RP (2000) Partitioning of rare earth elements between $CaSiO_3$ perovskite and coexisting phases; constraints on the formation of $CaSiO_3$ inclusions in diamonds. Earth Planet Sci Lett 181:291-300

Watenphul A, Wunder B, Heinrich W (2009) High-pressure ammonium-bearing silicates: Implications for nitrogen and hydrogen storage in the Earth's mantle. Am Mineral 94:283-292, doi: 10.2138/am.2009.2995

Watenphul A, Wunder B, Wirth R, Heinrich W (2010) Ammonium-bearing clinopyroxene: A potential nitrogen reservoir in the Earth's mantle. Chem Geol 270:240-248, doi: 10.1016/j.chemgeo.2009.12.003

Weiss Y, Griffin WL, Bell DR, Navon O (2011) High-Mg carbonatitic melts in diamonds, kimberlites and the sub-continental lithosphere. Earth Planet Sci Lett 309:337-347, doi: 10.1016/j.epsl.2011.07.012

Weiss Y, Griffin WL, Elhlou S, Navon O (2008) Comparison between LA-ICP-MS and EPMA analysis of trace elements in diamonds. Chem Geol 252:158-168, doi: 10.1016/j.chemgeo.2008.02.008

Weiss Y, Kessel R, Griffin WL, Kiflawi I, Klein-BenDavid O, Bell DR, Harris JW, Navon O (2009) A new model for the evolution of diamond-forming fluids; evidence from microinclusion-bearing diamonds from Kankan, Guinea. Lithos 112:660-674, doi: 10.1016/j.lithos.2009.05.038

Westerlund KJ, Gurney JJ, Carlson RW, Shirey SB, Hauri EH, Richardson SH (2004) A metasomatic origin for late Archean eclogitic diamonds: Implications from internal morphology of diamonds and Re-Os and S isotope characteristics of their sulfide inclusions from the late Jurassic Klipspringer kimberlites. S Afr Jour Geol 107:119-130, doi: 10.2113/107.1-2.119

Westerlund KJ, Shirey SB, Richardson SH, Carlson RW, Gurney JJ, Harris JW (2006) A subduction wedge origin for Paleoarchean peridotitic diamonds and harzburgites from the Panda Kimberlite, Slave Craton; evidence from Re-Os isotope systematics. Contrib Mineral Petrol 152:275-294, doi: 10.1007/s00410-006-0101-8

Whitney DL, Evans BW (2010) Abbreviations for names of rock-forming minerals. Am Mineral 95:185-187

Wiggers de Vries DF, Drury MR, de Winter DAM, Bulanova GP, Pearson DG, Davies GR (2011) Three-dimensional cathodoluminescence imaging and electron backscatter diffraction; tools for studying the genetic nature of diamond inclusions. Contrib Mineral Petrol 161:565-579, doi: 10.1007/s00410-010-0550-y

Wirth R (2004) Focused ion beam (FIB); a novel technology for advanced application of micro- and nanoanalysis in geosciences and applied mineralogy. Eur J Mineral 16:863-876, doi: 10.1127/0935-1221/2004/0016-0863

Wirth R (2009) Focused Ion Beam (FIB) combined with SEM and TEM: Advanced analytical tools for studies of chemical composition, microstructure and crystal structure in geomaterials on a nanometre scale. Chem Geol 261:217-229, doi: 10.1016/j.chemgeo.2008.05.019

Wirth R, Kaminsky F, Matsyuk S, Schreiber A (2009) Unusual micro- and nano-inclusions in diamonds from the Juina Area, Brazil. Earth Planet Sci Lett 286:292, doi: 10.1016/j.epsl.2009.06.043

Wirth R, Vollmer C, Brenker F, Matsyuk S, Kaminsky F (2007) Inclusions of nanocrystalline hydrous aluminium silicate "Phase Egg" in superdeep diamonds from Juina (Mato Grosso State, Brazil). Earth Planet Sci Lett 259:384, doi: 10.1016/j.epsl.2007.04.041

Wood BJ (1993) Carbon in the core. Earth Planet Sci Lett 117:593-607, doi: 10.1016/0012-821x(93)90105-i

Wood BJ, Li J, Shahar A (2013) Carbon in the core: its influence on the properties of core and mantle. Rev Mineral Geochem 75:231-250

Woodland AB, Koch M (2003) Variation in oxygen fugacity with depth in the upper mantle beneath the Kaapvaal Craton, Southern Africa. Earth Planet Sci Lett 214:295-310, doi: 10.1016/s0012-821x(03)00379-0

Woodland AB, Peltonen P (1999) Ferric iron contents of garnet and clinopyroxene and estimated oxygen fugacities of peridotite xenoliths from the eastern Finland kimberlite province. Proc 7th Int Kimberlite Conf 2:904-911

Yamaoka S, Kumar M, Kanda H, Akaishi M (2002) Formation of diamond from $CaCO_3$ in a reduced C-O-H fluid at HP-HT. Diamond Relat Mater 11:1496-1504

Yefimova ES, Sobolev NV, Pospelova LN (1983) Vklyucheniya sul'fidov v almazakh i osobennosti ikh paragenezisa. Sulfide inclusions in diamonds and their paragenesis. Zap Vses Miner Obshch 112:300-310

Zedgenizov DA, Rege S, Griffin WL, Kagi H, Shatsky VS (2007) Composition of trapped fluids in cuboid fibrous diamonds from the Udachnaya Kimberlite; LAM-ICPMS analysis. Chem Geol 240:151-162, doi: 10.1016/j.chemgeo.2007.02.003

Reviews in Mineralogy & Geochemistry
Vol. 75 pp. 423-448, 2013

Nanoprobes for Deep Carbon

Wendy L. Mao and Eglantine Boulard

Department of Geological and Environmental Sciences
Stanford University
Stanford, California 94305-2115, U.S.A.

wmao@stanford.edu *boulard@stanford.edu*

"Who looks outside, dreams; who looks inside, awakes."
— Carl Jung

INTRODUCTION

Surficial observations reveal carbon in a great variety of organic, inorganic, and biological forms that subduct with descending slabs and rise and erupt in volcanoes. Due to the lack of experimental means for studying carbon under extreme deep Earth conditions, we have limited information on the density and bonding nature of carbon-bearing fluids, and virtually no information on the texture and porosity of fluid-rock assemblages. Our knowledge on some of the most fundamental questions surrounding the deep carbon cycle becomes increasingly tenuous as we move into the planet. For example, in what form do carbon-bearing materials exist deep within Earth (Oganov et al. 2013)? How does carbon move within the planet's deep interior (Dasgupta 2013)? To address these types of questions, we need to improve our understanding of carbon-bearing phases at the extreme pressure-temperature conditions existing in Earth. As the fourth most abundant element in the universe, the backbone of organic matter and major energy carriers, pure carbon forms a variety of allotropes including both crystalline and disordered structures such as diamond, graphite, graphene, buckyballs, nanotubes and glassy carbon with numerous and exciting potential in technological applications. Adding the more than 370 other carbon-bearing mineral species (Hazen et al. 2013), this represents a huge range of structures and bonding and fascinating (as well as complex) physics and chemistry. Currently much is unknown about the behavior of carbon-bearing phases at high pressures and temperatures. Experimental study of materials behavior at extreme conditions requires the ability to reach simultaneous high pressure-temperature conditions, and the development and implementation of a battery of micro/nanoscale probes to characterize samples. In addition, studying carbon brings its own set of complications and considerations.

In this chapter we first review some of the techniques for reaching ultrahigh pressures and temperatures, focusing on the laser-heated diamond anvil cell (DAC). We then discuss state-of-the-art *ex situ* techniques for studying quenched carbon-bearing samples with nanoscale resolution. Significant progress has been made to bring the study of the deep Earth on par with the capabilities available for surface studies, and we will discuss some of the *in situ* techniques most relevant to studying carbon. Finally we will look ahead to future developments and prospects for the experimental study of deep carbon.

SYNTHESIZING SAMPLES AT HIGH PRESSURES AND TEMPERATURES

The diamond anvil cell (DAC) is the main static pressure device for studying carbon-bearing phases at Earth's lower mantle and core conditions. Following its invention in 1959

1529-6466/13/0075-0013$00.00 DOI: 10.2138/rmg.2013.75.13

(Bassett 2009), decades of development have extended the pressure-temperature range of the laser-heated DAC so that it now covers conditions of the entire geotherm of the planet (Tateno et al. 2010).

High pressure

A DAC anvil consists of a small pressure-bearing tip (culet) on one end of a diamond that expands to a large base (table) at the other end. The force applied to the low-pressure table of opposing anvils is transmitted to the culet where the pressure intensification is inversely proportional to the culet:table area ratio. The maximum achievable pressures in an anvil device depend upon the anvil material and the anvil geometry. In practice all anvils, including diamond, deform elastically and develop a cup, and the cupping increases with pressure (Fig. 1). Eventually the rims of the two anvils touch, and the pressure ceases to rise with further increase of force. In order to extend the pressure limit, bevels or convex shapes are added to the culet to avoid the rims touching. Beveled anvils can reach 200-300 GPa before the bevel flattens. The highest pressure a diamond anvil can sustain is not a fixed number but rather a range with increasing probability of failure at higher pressures. Flat, unbeveled diamond anvils can safely and repeatedly reach 50-70 GPa, but beyond this the probability of diamond failure increases sharply. Optimally cut beveled anvils can reliably reach 100-200 GPa, but have the undesirable property that the anvils always develop ring cracks upon pressure release due to the irreversibility of the gasket flow. The failure rate of beveled anvils increases above 200 GPa, and experiments beyond 250 GPa are extremely challenging, especially when coupled with very high temperatures.

In a DAC, samples are compressed within a hole, which acts as a sample chamber within a gasket confinement ring. Many different types of gaskets as well as pressure transmitting media and thermal insulating layers can be used depending on many factors (e.g., the sample, pressure-temperature range, degree of hydrostaticity required, etc.). Many varieties of DACs have been designed for specific measurements. For example, a symmetric DAC is typically used for X-ray diffraction (XRD) studies measured along the compression axis and sample synthesis for *ex situ* studies (Fig. 2a). This cell is also compatible with laser-heating. For X-ray Raman spectroscopy, a panoramic DAC with large openings through the side will be used, since the incident X-ray beam is directed through the gasket (Fig. 2b). In this case, an X-ray transparent material, high-strength beryllium, is used as the gasket. In the case of X-ray tomography, one needs maximum access around the radial direction, which has led to the modification of a plate DAC to a new cross DAC (Fig. 2c). More in-depth discussion of DAC technology can be found in a number of review articles (e.g., Boehler 2005; Mao and Mao 2007).

High temperature

Temperatures in excess of 5000 K (Tateno et al. 2010) can be achieved for samples simultaneously under extreme pressure in DACs by heating with high-powered infrared lasers (Bassett 2001), and variations of double-sided laser heating systems are available in-house and at synchrotron X-ray beamlines for reaching simultaneous high pressures and temperatures (Shen et al. 2001). Temperature gradients can be minimized by creating a flat-top power profile and by sandwiching samples between two thermal insulating layers to minimize heat loss through the diamond anvils. Developments in portable laser-heating systems have made it possible to add the high-temperature dimension to high-pressure research at beamlines without dedicated laser-heating systems (Shen et al. 2001; Boehler et al. 2009). The laser-heating technique normally requires an opaque, laser-absorbing sample and does not couple with transparent samples. This problem has been overcome by using CO_2 lasers (Tschauner et al. 2001) or by adding laser absorbers (e.g., mixing inert laser absorbers for solid samples, or using a metallic foil with a high melting point as the laser absorber, which in effect turns into an internal furnace for fluid samples). A small 10-20 μm diameter hole is drilled into the foil, which is placed inside

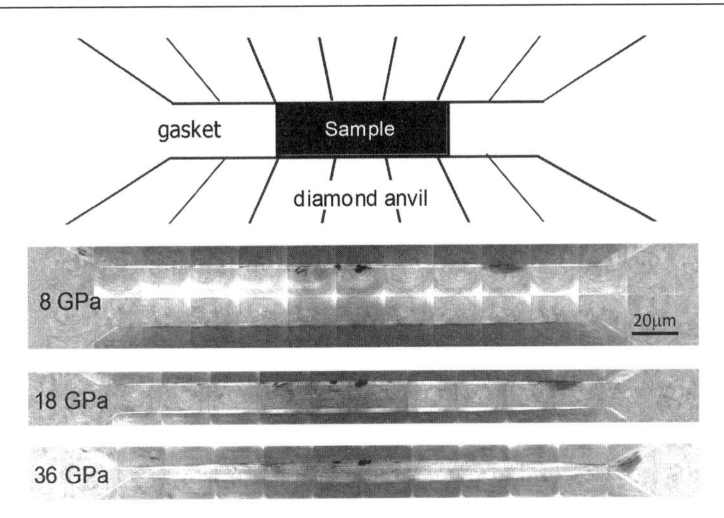

Figure 1. *Top*: Schematic diagram of DAC sample configuration focusing on sample and diamond culets. *Bottom panels*: Three X-ray radiographs collected upon increasing pressure, which illustrate the anvils deforming and cupping with pressure.

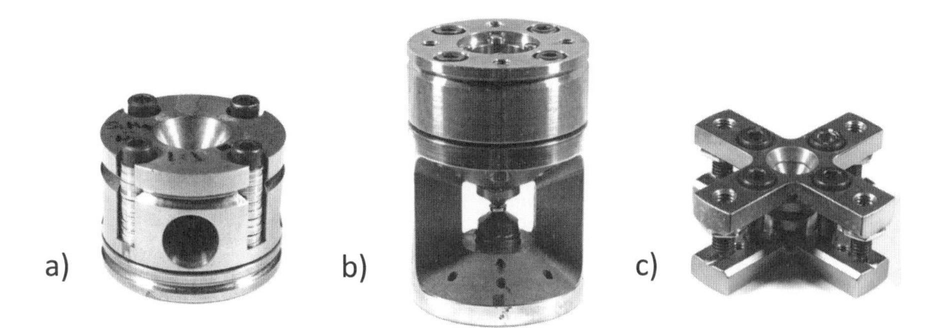

Figure 2. (a) Symmetric DAC; (b) Panoramic DAC with radial access; (c) cross DAC with radial access and axial access.

a gasket chamber filled with the transparent samples and then compressed to high pressures. An infrared YLF laser beam 30 μm in diameter is centered at the hole and heats the peripheral metal foil around the small hole; the heated part forms a donut-shaped furnace, effectively heating the transparent sample inside. Transparent samples such as CO_2 have been heated to high temperatures at high pressure, indicating the high heating efficiency of the internal donut furnace method (Santoro et al. 2004).

Spatial resolution

If the sample and textures are quenchable for *ex situ* experiments, measurements are generally not limited by the spatial resolution of available techniques (except for optical probes, where one can run into the diffraction limit below ~1 μm). The main challenge for *ex situ* analysis is being able to quench and recover the very small DAC samples. The potential of focused ion beam (FIB) techniques for preparing samples for many electron and X-ray probes is discussed in the next section. For *in situ* experiments, diamond is transparent below 5 eV, opaque between 5 eV and soft X-rays up to 5 keV, and is then transparent again to X-rays above

5 keV. Diamond is also transparent to neutrons but opaque to particles that require vacuum, such as electrons, ions, and protons. Non-penetrating radiation between 5 eV and 5 keV and techniques that require vacuum conditions are therefore incompatible with the high-pressure environment and are restricted to *ex situ* studies of quenched samples, or have to be replaced with equivalent *in situ* probes if available.

EX SITU TECHNIQUES

As discussed in the previous section, the laser heated DAC is the only static high-pressure tool available for reproducing the entire range of pressure and temperature conditions existing in Earth's mantle and core. To achieve these pressure and temperature conditions, the size of the samples has to be very small. Also, after recovery to ambient conditions, the samples are often very fragile; therefore their *ex situ* characterization necessitates special preparation such as focused ion beam (FIB). *Ex situ* techniques, as their name indicates, are conducted after the extreme pressure and temperatures that can induce structural changes, chemical reactions, and decomposition, have been released. The recovered sample may not always preserve all the characteristics of the high-pressure and temperature phases, so a comparison between the *ex situ* and *in situ* analyses is preferred when possible. However, *ex situ* analyses often offer more flexibility than *in situ* measurements and higher quality results. Numerous *ex situ* analytical techniques are available, which can provide a large range of information from the micron to the Ångstrom (= 0.1 nm) scale. Previous DAC studies have demonstrated the great potential in combining *in situ* analyses with *ex situ* analyses in order to obtain a detailed description of the run product (e.g., Irifune et al. 2005; Auzende et al. 2008; Fiquet et al. 2010).

In the following section we first discuss the use of scanning electron microscopy (SEM) to image the recovered samples and identify areas of interest for further *ex situ* study. Portions of interest are then prepared using a focused ion beam (FIB) instrument. Among the numerous *ex situ* probes available, we will describe transmission electron microscopy (TEM) and the associated analytical techniques and scanning transmission X-ray microscopy (STXM) and how they can be used to study deep carbon.

Sample preparation: FIB-SEM

Scanning Electron Microscopy (SEM). Scanning electron microscopy (SEM) is a convenient and ubiquitous imaging technique that does not require special sample preparation other than coating. Samples must be electrically conductive to prevent the accumulation of electrostatic charge at the surface. Nonconductive specimens tend to charge when scanned by the electron beam, and are therefore usually coated with an ultrathin coating, typically gold, gold/palladium alloy, platinum, or graphite. In an SEM, an electron beam, typically with energy between 0.2 to 40 keV, is focused on the surface of the sample by different electromagnetic lenses and condensers. The beam size is usually between 0.4 to 5 nm. The surface of the sample is scanned by this focused electron beam, and different detectors are used to record the emitted electrons or electromagnetic radiation that result from the interaction the electron beam with the sample. Low-energy secondary electrons form images sensitive to the topography of the sample. Since heavy elements (high atomic number, Z) backscatter electrons more strongly than light elements (low Z), backscattered electron are used to detect contrast between areas with different chemical compositions. Finally, X-rays emitted by the interaction of the electron beam with the sample are detected by X-ray energy dispersive spectroscopy (EDS) analyses, which give the composition of the observed area.

With its nanoscale resolution, the SEM represents a fast and non-destructive analytical method to observe directly recovered samples from high-pressure experiments. Electron beams do induce contamination in the form of carbonaceous deposits over the sample surface bombarded by the electron beam; however, this effect is limited to the surface of the sample and will not affect FIB cross-sections. The low voltage electron beam commonly used for SEM imaging (5 keV) generally will not damage the sample. Use of an SEM to study the texture of samples synthesized by piston cylinder and multi-anvil apparatuses is pervasive. For example, SEM images are commonly used to look for textural evidence of melting in the recovered sample (e.g., Rohrbach et al. 2007; Stagno and Frost 2010). This method was applied to determine the solidus of a carbonated peridote (Dasgupta and Hirschmann 2006). SEM is also very useful for imaging recovered samples from DAC experiments, which require high spatial resolution. SEM images and the chemical analyses allow identification of the laser-heated spot, which is usually a distinct circular area in the sample (Fig. 3) or other area of interest in the recovered samples. This identification is critical for the preparation of the sample for other *ex situ* analyses.

Focused Ion Beam (FIB). TEM requires very thin samples that are transparent to electrons, typically about 100 nm in thickness. Tripod polishing, ion milling, or ultramicrotomy can be used to prepare such thin sections. However, due to the small size and fragility of DAC samples these traditional sample preparation methods are very difficult to utilize and often lead to sample destruction. Moreover, only the FIB provides the opportunity to also image at the nanoscale, and thus the ability to carefully choose the region to be analyzed using the TEM. This selection is particularly critical when preparing a recovered DAC sample that has undergone large pressure and temperature gradients, and/or is compositionally heterogeneous. For example, with the FIB, one can extract a thin section in the center of the laser-heated area, minimizing thermal gradients in the observed area (Greaves et al. 2008; Auzende et al. 2008; Fiquet et al. 2010).

FIB milling of TEM sections has been exploited by the semiconductor industry over the past two decades, primarily to ensure quality control by TEM examination of silicon wafers (Anderson et al. 1992; Stevie et al. 1995). A FIB column is very similar to a SEM column, except that it uses a beam of gallium ions. When the high-energy gallium ions strike the sample, they sputter atoms from the surface. Thin sections from recovered DAC experiments can thus be prepared by milling into the samples. Secondary electrons or ions can be collected to form the

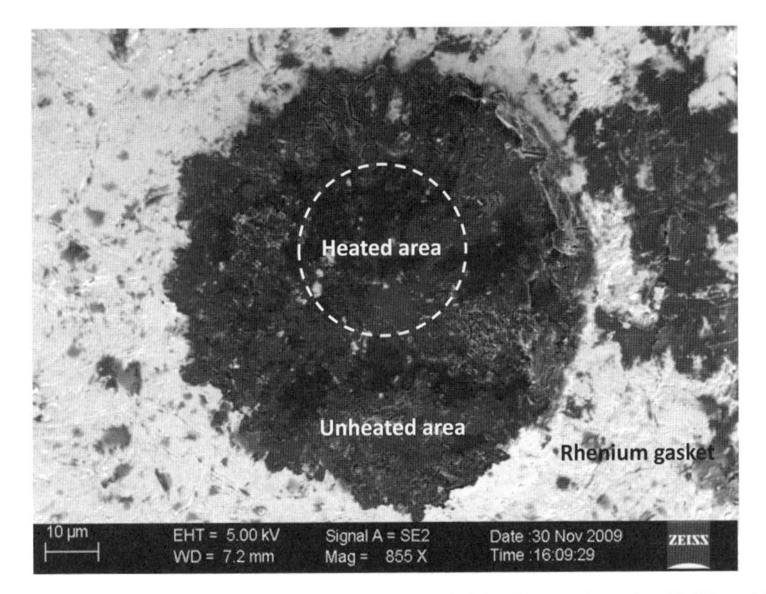

Figure 3. Secondary electron image of a recovered sample ($MgCO_3$+Pt) transformed at 85 GPa and 2400 K in DAC. The edge of the heated area can easily be discerned by textural changes.

FIB image. Most of the more recent instruments are dual beam, combining a FIB and a SEM column, the first one being used for milling and the second for imaging with a better resolution and without damaging the samples. Figure 4 shows the different steps for FIB preparation of a recovered high-pressure DAC sample: the SEM is used to select the heated area that will be cut (Fig. 3), first a thin section of about 1 micron thickness is milled (Fig. 4a) and extracted with a nanomanipulator (Fig. 4b) and then welded onto a TEM copper grid (Fig. 4c). Finally, the sample is thinned to about 100-nm thickness (Fig. 4d). One of the major issues in the use of FIB for preparation of samples for crystallography studies is the possible amorphization of the thin section during the thinning process. The use of a lower intensity ion beam during the thinning can help to minimize amorphization, but depending on the sample this can be problematic. Further details on the extraction of TEM thin sections by FIB are available in a number of references (Heaney et al. 2001; Marquardt and Marquardt 2012).

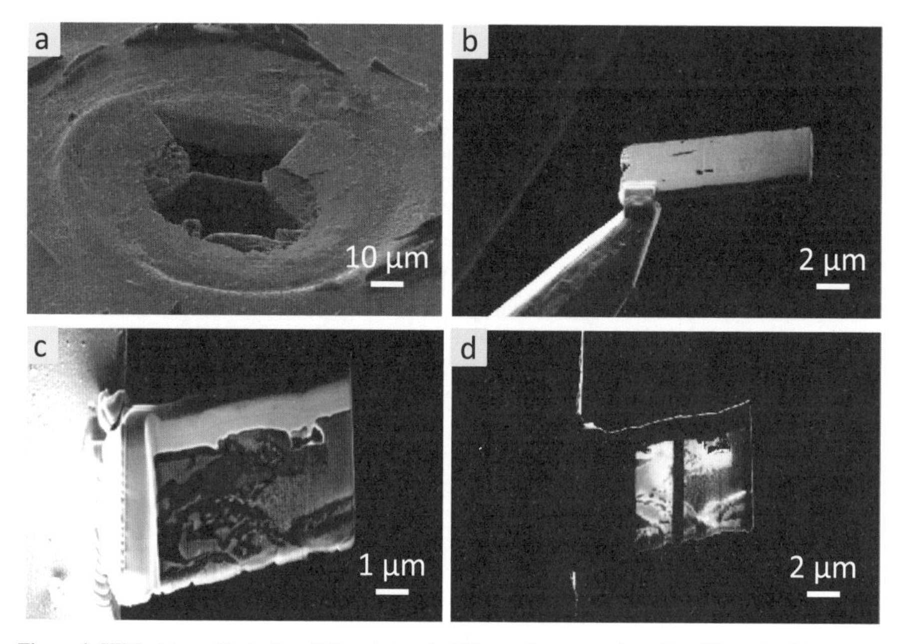

Figure 4. SEM pictures illustrating different steps in FIB sample preparation: (a) milling of a thick section from the recovered DAC sample; (b) extraction of this section using a nanomanipulator; (c) welding of the thick section on a TEM grid; d) after final thinning of the section welded to a copper TEM grid. For this last step, a low current Ga^+ beam was used in order to minimize damage, such as amorphization of the sample.

Characterization tools

Transmission electron microscopy (TEM). TEM is a powerful instrument for understanding reaction mechanisms and studying the distribution of phases resulting from chemical reactions. TEM gives the opportunity to image the thin sample section from the micron scale down to Ångstrom resolution. In addition, based on the electron-sample interactions, several chemical analyses can be conducted on the different phases. TEM can thus provide structural and chemical information on recovered DAC samples. In a TEM the electron beam is focused on the sample by several electromagnetic lenses. Unlike SEM, the image is a result of the interaction of the electron transmitted through the sample. To a first approximation, the image contrast is controlled by the absorption of electrons in the material, so it varies depending on the thickness of the sample and composition of the phases.

Given its ability to analyze multiple phases with high spatial resolution, TEM analysis is very complementary to *in situ* studies and is an especially powerful tool in the study of multi-phase reactions for which *in situ* XRD peaks may overlap. High-resolution TEM (HRTEM), also allows one to probe the presence of phases that exist in small quantities that cannot be detected by XRD. Identification of the different phases is also important when an unknown phase is present among other reaction products. For example the formation of the newly discovered high-pressure phase of $(Mg,Fe)CO_3$ is a result of a redox reaction that also leads to the formation of magnetite and nanodiamonds, which are easily identified by TEM (Figs. 5b,c; Boulard et al. 2011). The microstructure of the thin section observed by TEM provides information on the interaction between different phases, as well as textural and chemical zonation that results from reactions in the sample or may indicate a temperature or pressure gradient in the sample during compression and laser heating (e.g., Irifune et al. 2005). The ability to obtain atomic-scale resolution provides unique information on the nanostructure, which enables the study of deformation processes of minerals under pressure, observations of dislocations and epitaxial crystallization (Veblen et al. 1993; Mussi et al. 2010; Couvy et al. 2011).

Imaging can also be performed in scanning TEM (STEM), in which the electron beam is focused and scanned across the sample with a probe size of about 1 nm. STEM has the advantage over conventional TEM of delivering a lower electron dose (as the probe is rapidly scanned over the sample), thus facilitating the study of beam-sensitive materials. Figure 5a shows a high-angle annular dark-field (HAADF) STEM picture of a recovered sample. Contrast variations in a HAADF STEM picture can generally be explained by a change in sample thickness and/or electronic density of atoms (Z contrast). If the thickness of the FIB foil is constant in the observed area, contrast in the STEM picture can only be due to changes in the electron density, and hence to chemical variations; that is, light elements such as carbon appear dark while heavier elements are brighter. TEM is a comprehensive analytical instrument.

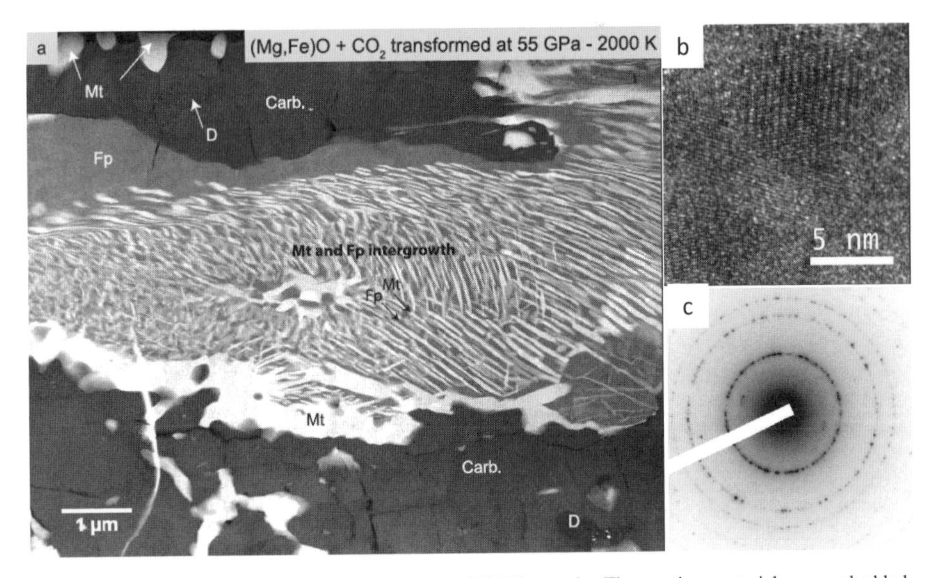

Figure 5. (a) HAADF-STEM image of a recovered DAC sample: The starting material was embedded within CO_2 and heated on both sides. The carbonation reaction front thus progressed from both sides, but did not affect the central part of the sample. In each of the reaction zones, the starting material is surrounded by Fe-bearing magnesite [$(Mg,Fe)CO_3$]. Magnetite is also present in the Fe-Mg carbonate areas in association with nanodiamonds; (b) TEM image of nano-diamonds; (c) electron diffraction from nano-diamonds. Modified after Boulard et al. (2012).

The interaction between the electron beam and the sample enables a variety of measurements such as electron diffraction, X-ray energy dispersive spectroscopy, and electron energy loss spectroscopy. Moreover, thanks to the high resolution of a TEM, it is possible to conduct these measurements on each phase in a multiphase assembly separately.

Electron diffraction. In this technique, a portion of the incident electrons that is transmitted through the specimen are elastically scattered, and generate an electron diffraction pattern, which provide information on the crystal structure. An aperture is used to select a precise area to be analyzed, which, thanks to the high-resolution of a TEM, can be as small as a nm-sized monocrystal. Electron diffraction patterns of a single crystal exhibit a periodic pattern over the entire observed area. Each spot on the diffraction pattern corresponds to diffraction from a different set of crystallographic planes in the same crystal. The spots are discrete and correspond to the points in the reciprocal lattice, their location and intensity provides information about the crystallographic structure of the phase being investigated. The distance between the central spot (direct transmitted beam) and a diffraction spot is related to the lattice spacing, and the angles between two diffraction spots is equal to the angle between the corresponding lattice planes. Electron diffraction can thus be used to determine lattice parameters and to help evaluate potential space groups for identification of known phases, as well as determination of the structure of an unknown phase. In this case, different diffraction patterns are taken at different zone axes by rotating the monocrystal. Although refining a low-symmetry structure necessitates the collection of a high number of electron diffraction images, high-symmetry crystal structures can be determined with relative ease. Diffraction patterns for polycrystalline materials are observed when the diffraction pattern is taken from the area containing multiple non-oriented crystals. Here the diffraction spots are then arranged into concentric rings (Fig. 5c).

Electron diffraction is a powerful tool for solving and refining structures and currently represents a promising technique for nanomaterials studies. This technique is, however, not commonly applied for the refinement of a high-pressure structure as it has two requirements. The high-pressure phase is quenchable back to room temperature and pressure, and the sample must not amorphize during FIB milling. For example, perovskite-structured silicates, which are synthesized at high pressure, are often observed as amorphous silicate phases in thin section (Auzende et al. 2008; Ricolleau et al. 2010). In the case of carbonates, the different high-pressure phases of magnesite ($MgCO_3$) were shown by XRD to be unquenchable at low temperature and pressure, and again were present as amorphous carbonate phases in the recovered sample (Isshiki et al. 2004; Irifune et al. 2005; Boulard et al. 2011). A crystalline phase was observed after high pressure-temperature transformation of siderite ($FeCO_3$); however, the structures observed by electron diffraction are different from those measured via *in situ* high-pressure and high-temperature XRD (Boulard et al. 2012).

High-pressure phases are by definition metastable and the energy deposited by the high-voltage electron beam (usually 200 keV) can cause amorphization while rotating the crystal to obtain diffraction patterns at different orientations.

X-ray energy dispersive spectroscopy (XEDS). Until recently, the determination of the chemical composition of lower-mantle phases was mainly indirectly inferred from the relationship of unit-cell volumes and composition by XRD (e.g., Mao et al. 1997). X-ray energy dispersive spectroscopy (variously called EDX and XEDS) can provide direct chemical analysis of each phase in a quenched sample. By imparting additional energy to samples, electrons from the incident beam excite an atom in a sample enough to eject an electron from its inner shells. The resulting electron vacancy is filled by an electron from an outer shell, emitting an X-ray with energy equal to the energy difference between the two electronic states. This emitted X-ray is characteristic of a particular element and the intensity of the peak is proportional to its abundance in the sample. Given the approximation that the section observed is of a constant thickness, each mineralogical phase of a specific composition will produce a different XEDS spectra. Qualitative analysis can easily be conducted by the identification of the different peaks recorded. In order to perform quantitative analyses, it is necessary to compare the analyses with the XEDS spectra of standards with known compositions that have been analyzed under the same analytical conditions. This technique was applied, for example, to determine the chemical composition of each phase in a MORB assemblage in order to determine the density of a MORB at different pressure and temperature conditions in Earth's mantle (Ricolleau et al. 2010). However, the X-ray emissions from light elements (e.g., boron, carbon, nitrogen, and oxygen) are weak, and not all instruments are equipped for analyzing these low-Z elements. Thus, in the case of the study of carbon-bearing phases, these chemical analyses are usually only qualitative for light element abundance. We will see in the next section that electron energy loss spectroscopy (EELS) is usually more accurate for the low-Z elements. XEDS can be conducted either on one specific phase by focusing the electron beam on a particular spot or it can be applied in a scanning mode. One XEDS spectrum is recorded at each step, which gives the opportunity to obtain element maps.

Electron energy loss spectroscopy (EELS). In addition to XEDS, EELS is a powerful and complementary method for chemical analyses. EELS is particularly well suited for the analysis of low concentration and low-Z elements such as carbon, while XEDS is more appropriate for heavy elements. The study of the fine structure of an EELS spectrum also provides element specific information on the environment and the chemical bonding of the atom. EELS involves measuring the energy loss of a fast electron beam that is inelastically scattered when passing through a sample. This technique necessitates a thin sample (<100 nm thickness). The loss of energy can occur in various ways that are measured in different portions of the spectrum. The low-energy loss region of the spectrum corresponds to the energy range of 0 to 50 eV. This region includes the zero loss peak at 0 eV that corresponds to electrons that have not interacted with the sample and peaks corresponding to the electrons that have interacted with the external shells of atoms. For energies higher than 50 eV, one can observe the core-loss peaks, formed by the interaction of electrons with the inner shells of an atom. Excitation of the atom by a transmitted electron gives rise to ionization edges in the energy-loss spectrum, which is equivalent to the absorption edges observed in X-ray absorption spectroscopy (XAS). The energy of excitation is specific to an element, so the identification of the atoms present can be easily determined by the edge energy. EELS is very sensitive and can be used to detect the presence of an element even at fairly low concentrations.

Core-level excitations provide unique spectroscopic information about the excited atom and its bonding states. Inner-shell excitation gives rise to ionization edges. The near-edge region, up to 30-40 eV above threshold, often shows very strong modulations, called fine structure. This so-called near-edge structure (ELNES) is highly sensitive to the nature of chemical bonds and to the local coordination around the excited atom. Fingerprint identification can be applied to determine in which form the element is present. Carbon exhibits numerous phases with distinct sp^3, sp^2, and sp hybridized bonds, which gives carbon a range of capability to bind with other elements. Figure 6 shows different carbon K edges for a number of carbon-bearing compounds. The different allotropes of carbon are easily distinguishable. Due to similarities in their bonding, amorphous carbon and graphite have similar edge shapes, with a lower energy peak at 285 eV and a second more intense feature at 290 eV. Diamond, however, has a very different edge shape with a peak maximum at 292.6 eV. The differences between the carbon K edges for graphite and diamond may be explained by the presence of sp^2 bonding in graphite, which results in a peak at 285 eV, identified as transitions to the π^* anti-bonding molecular orbital, and a second peak at 290 eV, due to transitions to σ^* orbitals (Weng et al. 1989; Brandes et al. 2008). In diamond, the bonding between the carbon atoms can be described in terms of tetrahedrally directed sp^3 hybrid orbitals, and the first peak is identified as arising from transitions to molecular orbitals of σ^* character (Weng et al. 1989). Between the extremes of sp^3-bonded diamond and sp^2-bonded graphite exists a large range of carbonaceous materials that have intermediate bond types, such

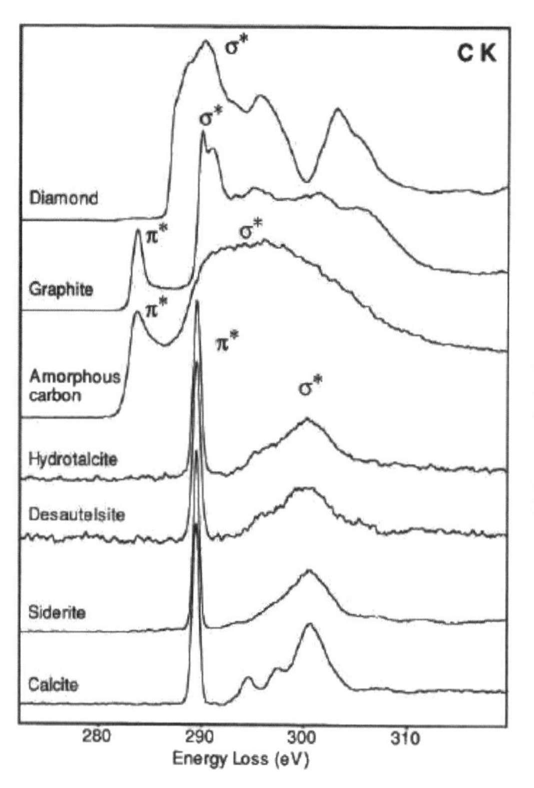

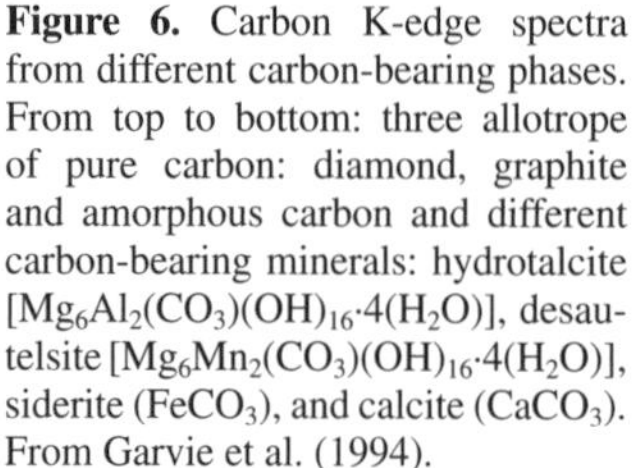
Figure 6. Carbon K-edge spectra from different carbon-bearing phases. From top to bottom: three allotrope of pure carbon: diamond, graphite and amorphous carbon and different carbon-bearing minerals: hydrotalcite [$Mg_6Al_2(CO_3)(OH)_{16}{\cdot}4(H_2O)$], desautelsite [$Mg_6Mn_2(CO_3)(OH)_{16}{\cdot}4(H_2O)$], siderite ($FeCO_3$), and calcite ($CaCO_3$). From Garvie et al. (1994).

as glassy carbon and evaporated amorphous carbon (Robertson 1986; Robertson and O'Reilly 1987). In the carbonate anion, CO_3^{2-}, a central carbon atom is bonded to three planar trigonal oxygen anions. The carbon K ELNES from the minerals containing the carbonate anion exhibit an edge shape consisting of a sharp initial peak at 290.3 eV and a broader, less intense, feature with a maximum at 301.3 eV. These features may be attributed to transitions to the unoccupied π^* and σ^* anti-bonding molecular orbitals of a CO_3^{2-} cluster (Hofer and Golob 1987; Garvie et al. 1994; Brandes et al. 2004; Schumacher et al. 2005).

If no reference spectra exist, *ab initio* calculations can be used for interpreting ELNES spectra. Figure 7a shows carbon K-edge EELS spectra collected in the untransformed and transformed regions for Fe-bearing magnesite compressed to 80 GPa and heated to 2300 K (Boulard et al. 2011). In the case of untransformed carbonate phase, the peak at 290.3 eV corresponds to planar CO_3^{2-} carbonate groups. The spectrum measured in the transformed area shows a main peak shifted at 290.7 eV and a smaller peak at 287.5 eV. As no reference EELS spectra exist for that new high-pressure phase, density functional theory (DFT) calculations of the unoccupied electronic density of state were performed in order to interpret the carbon K-edge measured (Fig. 7b). The magnesite density of state shows a narrow peak at ~5 eV above the fermi level that corresponds to the observed carbon K-edge and its "molecular" style CO_3^{2-} signature. The density of state obtained for the new high-pressure carbon-bearing phase does not show any molecular peak but rather a broad band at higher energy (from 7 to 11 eV). The geometry of CO_3^{2-} and $C_3O_9^{6-}$ rings can be seen on the right-hand side of Figure 7b. In order to see how the tetrahedral $C_3O_9^{6-}$ rings change during the decompression, structures have been relaxed at ambient conditions. The result is shown at the bottom. These structural changes have a strong influence on the unoccupied density of state. Indeed, as expected for less dense

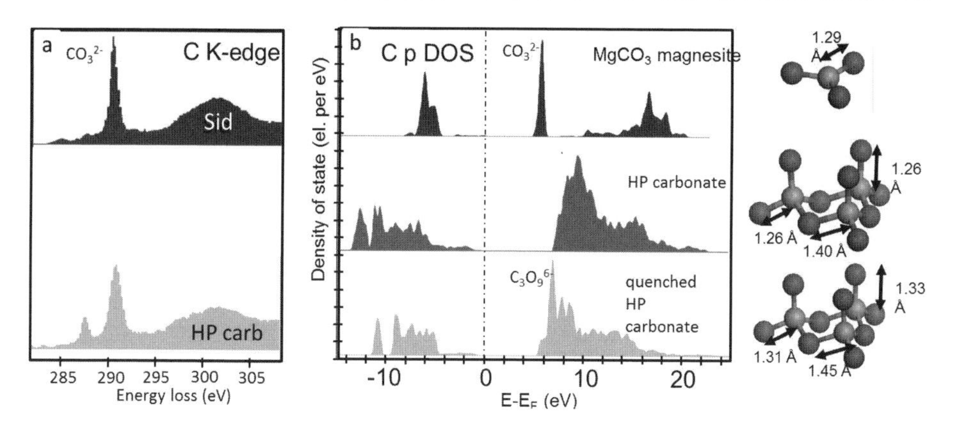

Figure 7. (a) Carbon K-edge spectra collected from a recovered sample of Fe-bearing magnesite compressed and heated to 80 GPa and 2300 K in the untransformed (top spectra) and transformed (bottom spectra) areas; (b) DFT calculations for electron density of state of the carbon atoms (*p* orbital symmetry) in different structures (schematics shown at right). This unoccupied density of state roughly corresponds to the excitation probed by EELS at the carbon K-edge (excitonic effects are neglected in the calculation). Modified after Boulard et al. (2011).

structures, the overall unoccupied density of state shifts to lower energy. Narrow peaks are present and correspond to tetrahedral $C_3O_9^{6-}$ rings, whose molecular signatures are similar to those observed in EELS in samples recovered from high-pressure experiments (i.e., a peak observed at around 290.7 eV).

When applied to 3*d* transition metals, EELS allows the determination of redox state of the element (e.g., Paterson and Krivanek 1990). The ferrous/ferric iron ratio, for example, can be estimated by various methods such as Mössbauer spectroscopy and electron probe microanalysis (EPMA); however, only EELS performed in TEM can provide a spatial resolution as small as a few nanometers. Calibrations have been performed in order to quantify the $Fe^{3+}/\Sigma Fe$ ratio in minerals by analyzing Fe $L_{2,3}$ ELNES (Van Aken et al. 1998; van Aken and Liebscher 2002). Determination of $Fe^{3+}/\Sigma Fe$ ratio can be applied to determine the oxygen fugacity of a mineral assemblage, and thus study at which f_{O_2} carbonates are reduced into diamond at mantle conditions (e.g., Stagno et al. 2011).

Core-loss peak intensity is proportional to the number of each element in the measured region. If the background for a particular absorption edge can be extrapolated and subtracted, the remaining core-loss intensity provides a quantitative estimate of the concentration of the corresponding element (e.g., Egerton 2009). This analysis allows the application of energy-selected imaging, which shows the spatial distribution of one element—the equivalent of an X-ray elemental map. When EELS is performed in the STEM mode, it is possible to obtain not only elemental maps but also maps of the distribution of different elemental forms (i.e., maps of carbon in carbonate groups versus carbon in the diamond structure). Finally elemental ratios can be calculated by measuring the different elemental ionization edges in the same window. EELS can thus also be applied to determine the chemical composition of a material (Egerton 1979; Aitouchen et al. 1997). While EELS provides absorption spectra with very high spatial resolution, high-energy resolution can be obtained by scanning transmission X-ray microscopy (STXM). These two analytical methods are thus highly complementary.

Scanning transmission X-ray microscopy (STXM). STXM is based on a synchrotron light source that provides X-ray absorption spectra. These measurements can be done directly on FIB thin sections, such as the one prepared for TEM observation. As an example of comparison of

STXM and EELS analysis, Figure 8 presents the images and spectra measured on a recovered sample of FeO+CO_2 transformed at 75 GPa and 2200 K. Typical spatial resolution is 30 nm (Hitchcock et al. 2008) in STXM, while STEM-EELS presents the advantage of a probe size of ~1 nm, which allow analysis of very small particles. However, STXM is far superior with regard to spectral resolution: up to 0.1 eV at 290 eV.

STXM measurements consist of focusing a monochromatic beam at one particular energy (between 200 and 2000 eV) and scanning the sample. Detectors positioned behind the sample measure the intensity of the beam transmitted on each step of the sample with a spatial resolution up to 30 nm. A series of transmitted pictures is collected at every 0.1 eV step on an energy range that covers the absorption edge of the element being analyzed (Fig. 9). This stack of transmission images gives 3D data: x, y, and the spectral dimension. In order to reconstitute an X-ray absorption spectrum, these pictures need to be aligned and converted into optical density. From there, an X-ray absorption spectrum can be extracted on each pixel of the picture.

Radiation damage is an issue for both TEM-EELS and STXM since both techniques use ionizing radiation. However, recent studies have demonstrated that X-ray microscopy produced

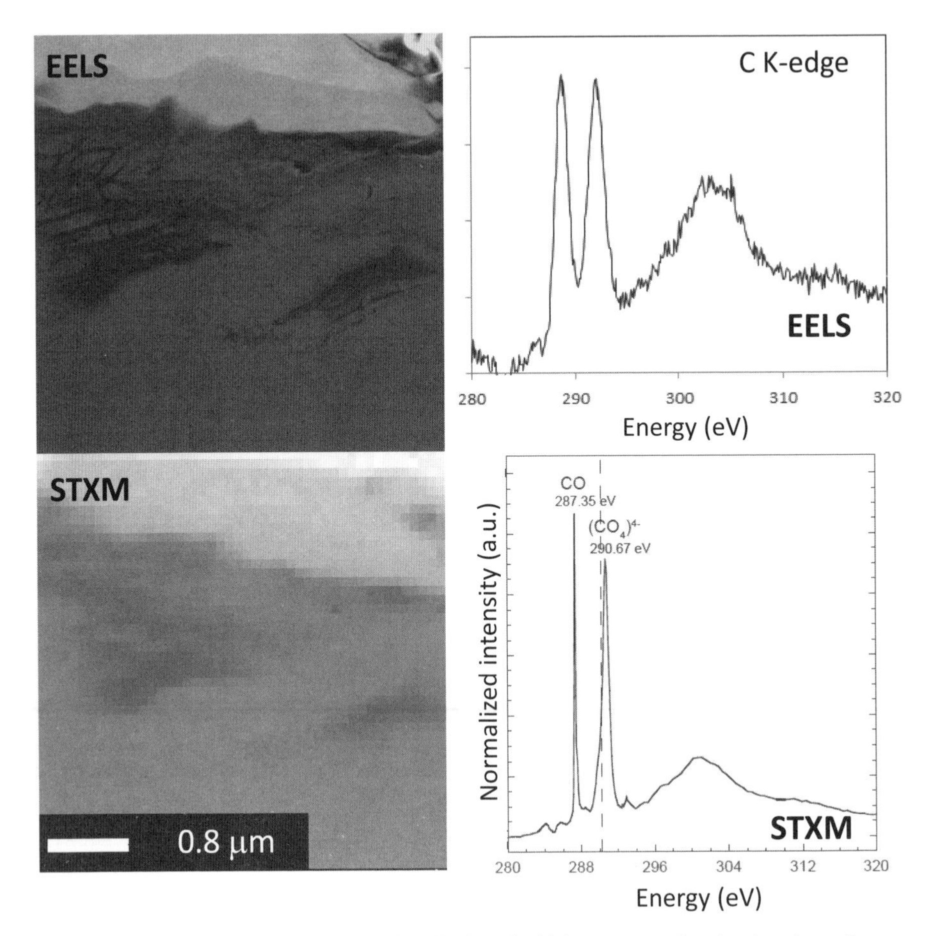

Figure 8. *Top*: STEM-EELS image and carbon K edge of a high-pressure carbon-bearing phase. *Bottom*: STXM image of the same area and carbon K edge of the same phase. While EELS provides absorption spectra at very high spatial resolution, higher energy resolution is obtained by STXM.

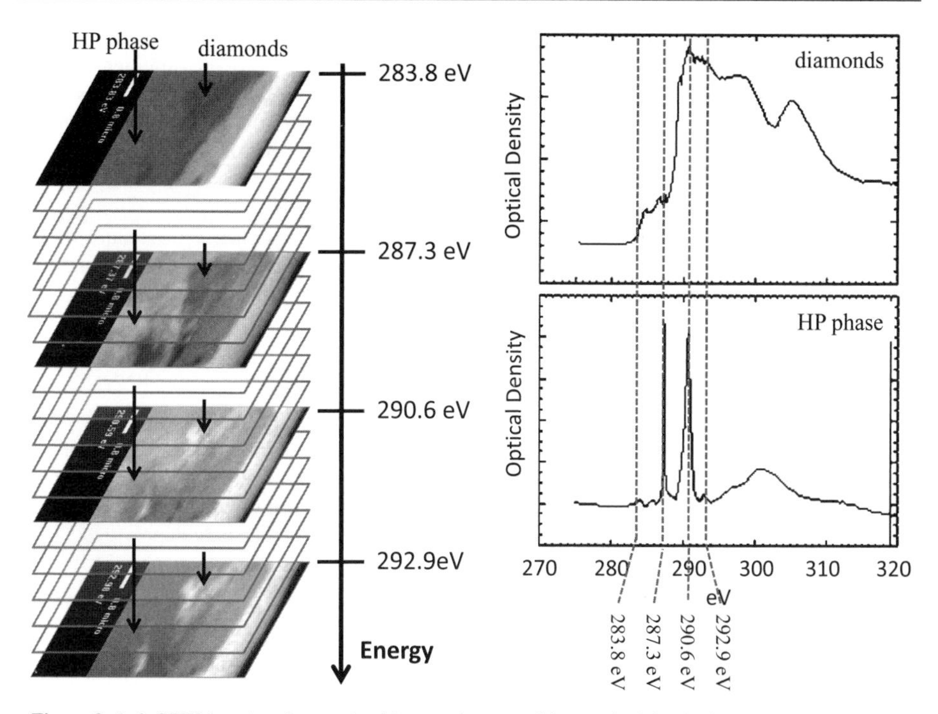

Figure 9. *Left*: STXM stacks of transmitted images (processed into optical density images) taken at 0.1 eV energy intervals, from which spectra can be extracted for each area. *Right*: Spectra for one area containing the high-pressure (HP) phase, corresponding to a new phase of $FeCO_3$. Spectra from a different area containing to nanodiamonds is shown at top.

far less damage to the sample than TEM-EELS (e.g., Hitchcock et al. 2008). STXM presents great advantages for the study of recovered samples from DAC experiments for which the small quantity of materials makes them more precious and metastablity of the high-pressure recovered phases can make them very fragile under electron beam. STXM has been applied primarily to study carbon in natural samples, such as carbon bearing materials in extraterrestrial sample or the relationships between mineral and organic chemical species (e.g., Lepot et al. 2008). However, few STXM studies have yet been conducted on recovered DAC samples. As shown in Figure 8, STXM have been used to highlight slight energy shifts in the carbon K-edge between a trigonally coordinated carbonate (which shows a molecular peak at 290.3 eV) and a high-pressure tetrahedrally coordinated carbon phase (Boulard et al. 2012). In contrast to EELS, STXM permits the precise determination of the energy of this molecular peak. These results demonstrate the exciting potential for using STXM in the *ex situ* study of high-pressure carbon-bearing phases.

IN SITU TECHNIQUES

The *ex situ* probes described in the previous section are extremely powerful, and highlight the need to be able to bring the study at deep Earth conditions on par with the capabilities available for quenched studies. The standard sub-micron probes using focused electrons (electron microscopies), ions (nanoSIMS), or surface contact (atomic force microscopy) require a low-pressure, near-vacuum environment that is incompatible with high-pressure experimental environments. Optical probes can access the high-pressure sample through the diamond windows but are restricted by the μm-scale diffraction limit of optical wavelengths. We would like to be able to conduct *in situ* high-pressure and temperature measurements, which currently can only be done at low pressure (e.g., TEM-like nano-crystallography, imaging and 3D tomography beyond the optical diffraction limit, and EELS-like probes for oxidation and bonding properties). *In situ* techniques are especially critical in the case of non-quenchable phases or dynamic studies where squeeze, cook, and look experiments are not sufficient. Pressure is an intensive parameter; the quality of *in situ* measurements at high pressure is controlled by the size of the analytical probe relative to the size of the sample, rather than the absolute sample size (Wang et al. 2010a). In the case of *in situ* techniques, the size of current X-ray probes has been a key limitation for resolving stress gradients, compositional heterogeneity, texture, and other characteristics at megabar (= 100 GPa) pressures, where significant differences occur on a sub-micron level. This problem can be partially remedied by using a nano/sub-μm sized beam to improve the spatial resolution of pressure gradients that are perpendicular to the beam.

The excellent transparency of single-crystal diamond windows to a wide range of electromagnetic radiation offers the potential for the development of numerous analytical probes for more comprehensive, *in situ* characterization of high pressure-temperature behavior. Synchrotron X-ray probes offer a number of advantages for high-pressure diamond anvil cell work. The high brightness and penetrating power enables many measurements which previously have been flux-limited (Mao and Mao 2007). The very short wavelength of high-energy X-rays also relaxes the constraint of the theoretical diffraction limit. In practice the focus size is restricted by the quality of the X-ray optics. A number of strategies exist for nanoscale focusing of X-rays with major progress in focusing with mirrors, lenses, and also lensless options; 100-600 nm-sized X-ray beams focused using K-B mirrors are well established at specialized nano-focusing beamlines (Wang et al. 2010a). Compound refractive lenses can focus to ~100nm, but are limited to higher energies due to absorption of X-ray by the lens. Fresnel zone plates can achieve tens of nm full-width half-maximum (FWHM) focusing resolution for hard X-rays (Als-Nielsen and McMorrow 2011), and have the advantage of operating over a wide range of energy.

In this section, we discuss *in situ* X-ray techniques for high-pressure work and their promise for deep carbon research. They include developments in diffraction (nanoscale XRD or nanoXRD) for studying structure, spectroscopy (e.g., X-ray Raman spectroscopy or XRS) for studying bonding, and imaging (e.g., nanoscale X-ray computed tomography or nanoXCT).

Nanoscale X-ray diffraction

Structural information on an atomic level is essential for understanding the properties of materials at high pressure. XRD has long been the bread-and-butter probe for *in situ* structural studies at high pressure. Compared with the 5-10 μm-sized X-ray beams that are now routine at high-pressure DAC X-ray diffraction synchrotron beamlines, nanoscale beams (i.e., those smaller than 1 μm) can resolve signals between sample and gasket, and pressure and temperature gradients. The use of nanoscale X-ray beams also enables selection of individual sub-micron crystals/phases in a heterogeneous sample and single-crystal XRD studies in nominally polycrystalline samples.

Resolving pressure and temperature gradients. Multi-megabar experiments are made possible by beveling the diamond anvils in order to sharpen the pressure gradient in the solid gasket that supports the peak pressure at the center of the culet. Achieving a very steep pressure gradient in the gasket and then placing the samples at the position where the pressure is maximum and the pressure gradient is zero is critical for reaching ultrahigh pressures (Fig. 10). Advances in ultrahigh-pressure technology thus critically depend upon the ability to resolve the pressure distribution and to determine the peak pressure (Wang et al. 2010a). The use of a nano-focused X-ray beam enables resolution of steep pressure gradients, and to detect peak pressures at sub-μm length scales. In laser-heating experiments, thermal gradients can also pose major issues. The flat-top portion of the laser spot should be much larger than the X-ray beam in

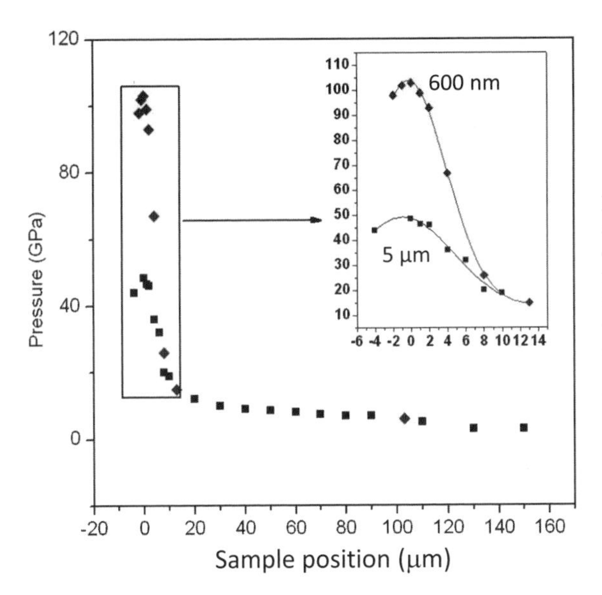

Figure 10. The pressure distribution as a function of sample position (radial distance from the center of anvil) determined with a ~5-μm (black squares) and a 600-nm (black diamonds) focused X-ray beam. A 20 GPa/μm gradient in 1-μm area at the peak pressure of 105 GPa can be observed. From Wang et al. (2010a).

order to minimize temperature gradients. With a smaller beam this essential condition is much easier to control.

Resolving multiple phases. Multiple samples have been studied in the same pressure chamber for direct comparison of their equations of state under the same pressure conditions, and natural rock specimens have been studied to simulate more realistic high-pressure geological environments (Lee et al. 2004). Such studies, while very promising, have been limited due to the overlap of diffraction peaks, which complicates analysis of the different phases involved. These problems can be overcome by using an X-ray beam that is an order of magnitude smaller, which could then distinguish between multiple samples and phases. This resolution is critical when studying possible reactions between carbon-bearing phases at mantle conditions. At lower pressures, carbon is found in accessory phases (e.g., CO_2 rich fluids, carbonates, hydrocarbons, graphite/diamond) and unlike H_2O does not dissolve in major silicate minerals. At higher pressures, carbon-bearing phases may undergo dramatic structure changes (e.g., coordination change in Mg-Fe carbonates) and react with silicate and oxide minerals, changing their structure (Boulard et al. 2011, 2012).

Enabling single-crystal studies on polycrystalline samples. XRD from single crystals can provide well-constrained, detailed structural information that is crucial for understanding the microscopic mechanisms of high-pressure phenomena and characterizing novel pressure-induced phase transitions. However, single crystals often break up into powders after reconstructive transitions or due to non-hydrostatic conditions. Without single crystals, we are forced to use various polycrystalline XRD methods, which do not give as definitive a result. These polycrystalline methods may assume a "good" powder sample; that is, statistically a nearly infinite number of randomly distributed crystallites, which thus yields a smooth and uniform XRD pattern. Alternatively, knowledge of the non-uniform sample texture is required. In reality, most new phases formed under multi-GPa pressures are neither a good powder nor a single crystal, and their XRD patterns are often spotty with a 5-μm X-ray probe leading to intensity data that are often insufficient for reliable Rietveld structure refinement. Whether a sample should be considered a single crystal or a powder depends upon the number of crystallites impinged by the X-ray beam (Fig. 11). It will be single-crystal XRD if the beam

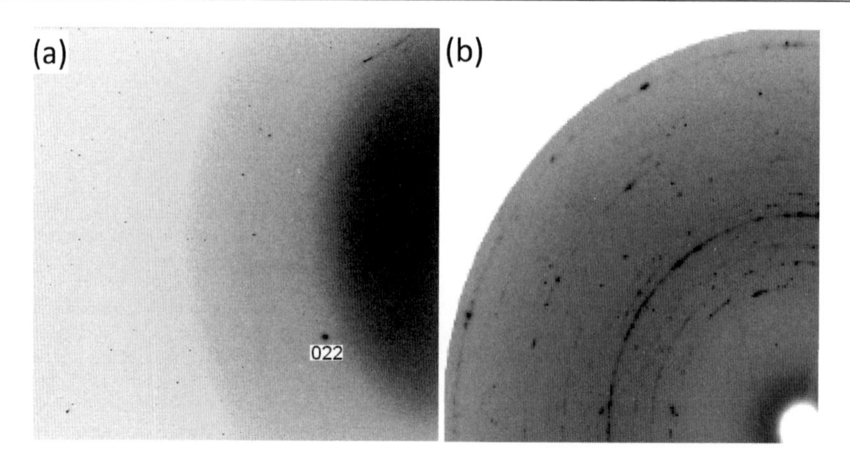

Figure 11. XRD patterns for the same $Mg_{0.6}Fe_{0.4}SiO_3$ post-perovskite sample measured at 142 GPa with 250-nm (a) and 5-μm (b) monochromatic X-ray beams. From Wang et al. (2010).

is smaller than the crystal, polycrystalline XRD if the beam covers a very large number of crystals, or spotty polycrystalline XRD if the beam covers an insufficient number of crystals. Reducing the X-ray probe size to a sub-micron focused X-ray beam enables us to carry out single-crystal studies even in the polycrystalline sample with grain sizes in the μm or sub-μm range (Wang et al. 2010a). This capability may be critical for structural determination of new carbon-bearing phases (e.g., high-pressure tetrahedrally-coordinated carbonates; Boulard et al. 2011), where the low scattering of carbon and multiple phases present make powder patterns difficult to interpret uniquely.

X-ray Raman spectroscopy

Carbon *K*-edge spectroscopy is a powerful tool for studying carbon bonding and chemistry, but is largely limited to being a surficial probe, owing to the low carbon K-edge energy (Rueff and Shukla 2010). This problem can be overcome by inelastic X-ray scattering. In high-pressure XRS, the high-energy incident X-ray penetrates the pressure vessel and reaches the sample. The scattered photon loses a portion of energy corresponding to the electron edge of the element of interest in the sample, and still is high enough in energy to exit the vessel to be registered on the analyzer-detector system. In XRS experiments, single-crystal analyzers collect scattered X-rays, and focus them to the detector in a nearly backscattering geometry, thus fixing the energy at the elastic line. The incident X-ray energy is scanned relative to the elastic line to determine the inelastic (Raman) shift. XRS features are relatively insensitive to momentum transfer (q), and thus the angle is set at an angle to optimize the intensity, and multiple analyzers can be used to increase the counting rate without concern for their differences in q.

XRS spectra of second-row elements from Li (56 eV) to O (543 eV) have been successfully observed at high pressures. A number of XRS studies have been conducted on a variety of carbon systems, which reveal interesting bonding changes *in situ* at high pressure. These studies, including graphite (Mao et al. 2003), C_{60} (Kumar et al. 2007a,b), benzene (Pravica et al. 2007), and glassy carbon (Lin et al. 2011), demonstrate the great promise of this technique for looking at deep carbon phases (Fig. 12). XRS has widened the reach of X-ray absorption spectroscopy (XAS) on low-Z samples, which was once limited to the soft X-ray range, to systems and sample conditions where the penetration capability of a hard X-ray probe is essential. In addition it has opened up the possibility to measure the spectra that are symmetry forbidden in XAS (Bergmann et al. 2003). This capability has resulted in rapid growth in XRS over the

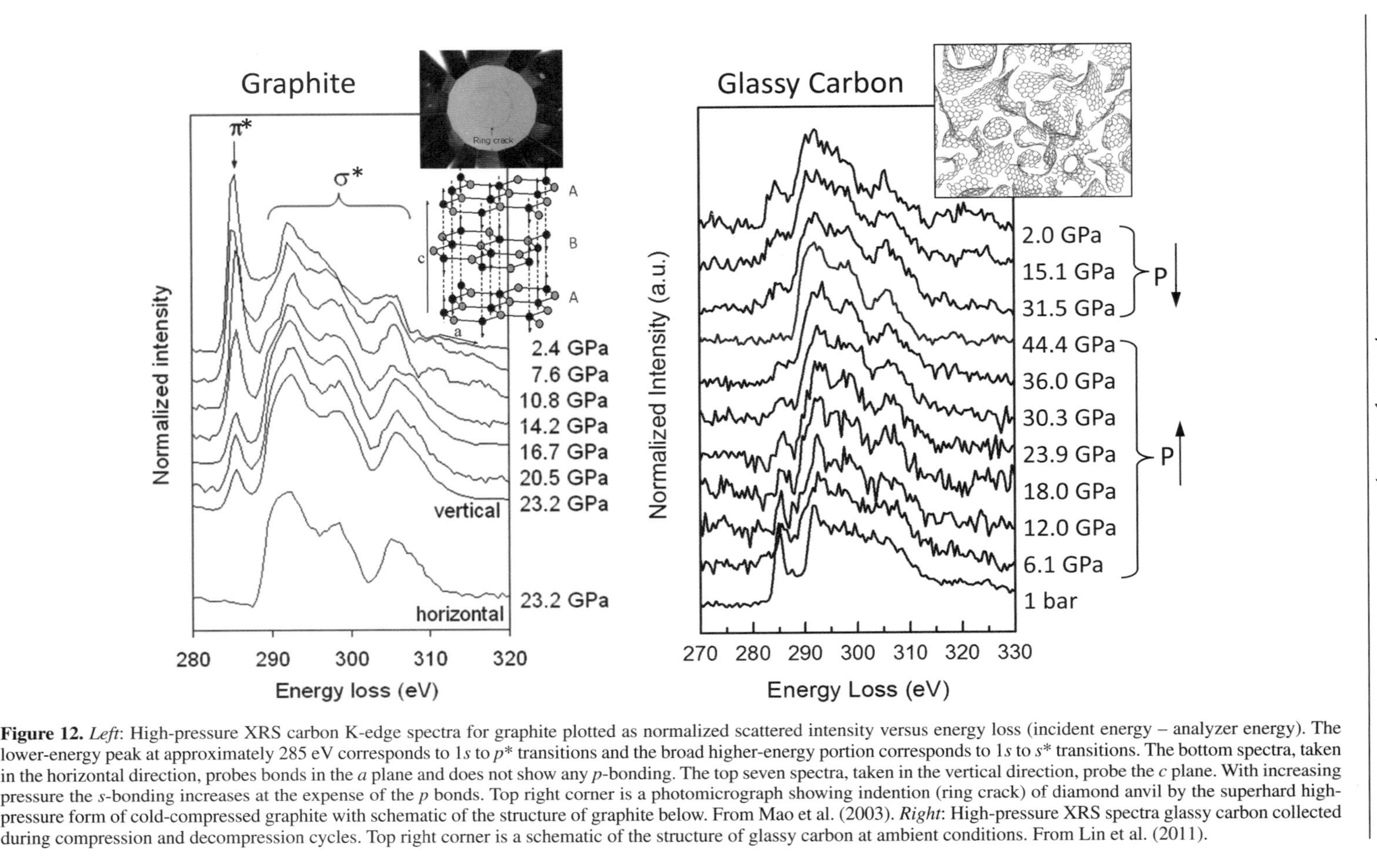

Figure 12. *Left*: High-pressure XRS carbon K-edge spectra for graphite plotted as normalized scattered intensity versus energy loss (incident energy – analyzer energy). The lower-energy peak at approximately 285 eV corresponds to 1*s* to *p** transitions and the broad higher-energy portion corresponds to 1*s* to *s** transitions. The bottom spectra, taken in the horizontal direction, probes bonds in the *a* plane and does not show any *p*-bonding. The top seven spectra, taken in the vertical direction, probe the *c* plane. With increasing pressure the *s*-bonding increases at the expense of the *p* bonds. Top right corner is a photomicrograph showing indention (ring crack) of diamond anvil by the superhard high-pressure form of cold-compressed graphite with schematic of the structure of graphite below. From Mao et al. (2003). *Right*: High-pressure XRS spectra glassy carbon collected during compression and decompression cycles. Top right corner is a schematic of the structure of glassy carbon at ambient conditions. From Lin et al. (2011).

last few years, yet given the large number of systems that could uniquely profit from XRS, the number of such studies has been very small when compared to conventional XAS. Progress has been hampered because only a few heavily oversubscribed instruments exist worldwide, and due to the diminutive XRS cross-section, in many systems very long scanning times are required to obtain a high-quality spectrum.

Currently, conducting XRS on carbon phases above 50 GPa is extremely challenging. The sample becomes very thin, dramatically reducing counts, and the signal to noise ratio also declines (more background scattering). These limitations are being overcome with improvements in a number of areas including those in DAC sample configuration (e.g., maintaining much thicker samples to higher pressures using novel composite gaskets; Wang et al. 2011), construction of new XRS instruments with more and improved analyzers to collect more solid angle and flux, improvements in X-ray optics and focusing, and brighter light sources.

X-ray imaging

X-ray computed tomography (XCT) is a powerful, non-destructive method for imaging the internal structure of a sample. The potential applications in wide-ranging fields for XCT have been recognized since its initial development and use in medical imaging. XCT has been used very successfully for cm-sized and larger geological samples with a resolution of up to 10 μm for in-house systems with micro-focused X-ray tubes, and has become a standard method for studying the texture of multiple phases (e.g., studying pore spaces in petroleum engineering; Mees et al. 2003). More recently at synchrotron sources, microXCT instruments with improved spatial resolution of ~2.5 μm have been used to look at natural samples (e.g., pumice clasts; Gualda et al. 2010), the connectivity of quenched basalt melt networks in peridotites using phase contrast (Zhu et al. 2011), and quenched core-forming melts within a silicate matrix (Watson and Roberts 2011). Measurements at high pressure have been conducted in a Drickamer press with rotating anvils to investigate the deformation of vitreous carbon and forsterite spheres in a FeS matrix (Wang et al. 2010b) and for direct determination of the volume of glasses (Lesher et al. 2009). DAC techniques have been used to study the equation of state of amorphous Se up to 10.7 GPa (Liu et al. 2008). For a typical sample size with dimensions of ~100 μm in a DAC at moderate pressures, 1-μm spatial resolution can only provide 10^{-2} resolution in volume, which is an order of magnitude lower than diffraction for crystalline materials. At higher pressure (>10 GPa), the sample thickness is dramatically reduced to tens or even less than 10 μm. In this case, μm-resolution tomography will not provide accurate volume measurements. To reach Earth's lower mantle and core conditions, the much higher pressures required and potential for heterogeneous samples require the development of much higher (nanoscale) spatial resolution to image the small, tens of microns-sized DAC samples and their much smaller features. One also has to consider the sample absorption at the particular X-ray energy used in order to ensure adequate absorption contrast.

Nanoscale XCT. The three basic components of an XCT system are an X-ray source, a detector, and a rotation system. The X-rays illuminate the sample and 2D radiographs are collected by the detector, which is positioned behind the sample. For synchrotron 3D XCT, the source and detector are fixed, and the sample sits on a rotation axis. A series of radiographs are taken over 180° around the single axis to yield a 3D XCT image, which can then be rotated around any axis and sliced at any angles and depth by software to reveal detailed internal information, while the sample is preserved without damage. For nanoXCT, a number of additional components are required to attain nanoscale resolution, the most important being the special X-ray focusing optics, which are typically micro Fresnel zone plates that can enable tens of nm resolution as discussed previously. For example, the full field X-ray microscope at Beamline 6-2 of the Stanford Synchrotron Radiation Lightsource is capable of 40-nm resolution from 4-14 keV over a 30-μm field of view (Fig. 13). The relatively large depth of focus provides full transmission imaging as opposed to surface only for electron microscopy. Spectroscopic imaging above and

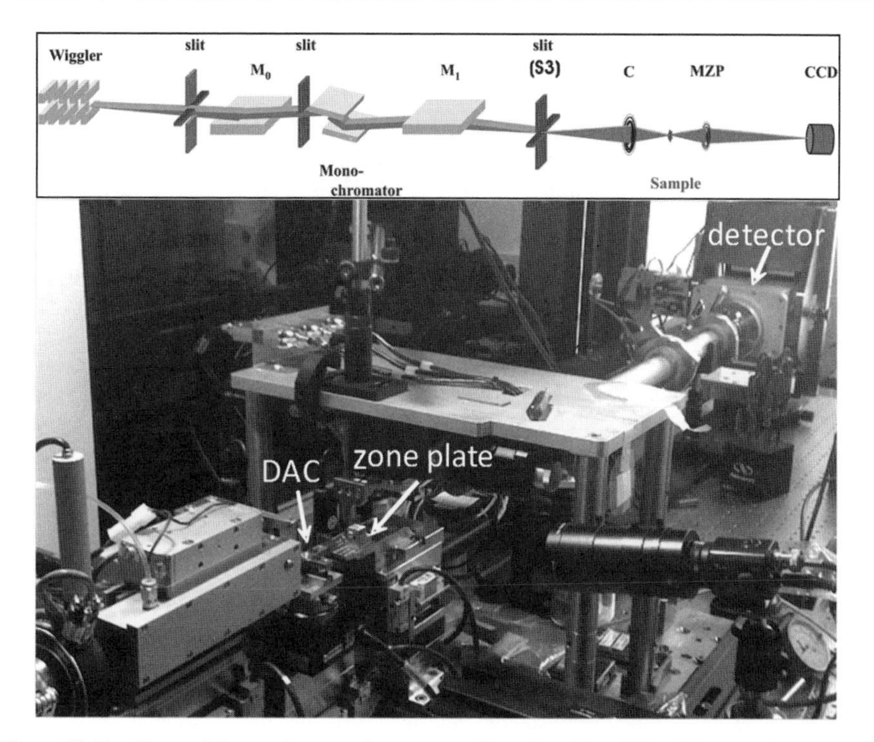

Figure 13. *Top*: Setup of X-ray microscope instrument at Beamline 6-2 at SSRL. The X-ray beam passes through vertical and horizontal slits, a vertically collimating mirror (M0) followed by vertical slits, monochromator, and toroidal mirror (M1) to focus the beam at the virtual source (S3). Capillary condenser (C) provides hollow cone illumination. Sample image is focused by micro zone plates (MZP) onto CCD detector, with optional phase ring. [Reprinted from Andrews et al. (2008) by permission of the publisher (Taylor & Francis Ltd, *http://www.tandf.co.uk/journals*).] *Bottom*: Photo of Xradia instrument with the cross DAC mounted at sample position and incident X-rays travel from bottom left to top right. The incident X-ray beam is directed through the sample via four side openings in DAC, which is then rotated to get the tomographic image.

below the absorption edge facilitates studies of nanoscale element distribution. NanoXCT has been used to image samples with biological, environmental, and materials applications, but has been applied only very recently to high-pressure mineral and rock studies (Andrews et al. 2009). In terms of lensless imaging approaches, coherent diffraction imaging and holography are also areas of active research where current resolution is limited to tens of nm due to limits in the coherent flux for coherent imaging (Miao et al. 2008) and the size of reference beam for holography (McNulty et al. 1992). Both of these techniques are very promising and can reach very high resolution, but have not been widely applied for high-pressure research.

The tens of nm spatial resolution of nanoscale X-ray imaging enables a number of exciting capabilities. It allows us to have accuracy in volume determinations for amorphous or poorly crystalline phases, which rivals that from X-ray diffraction of crystals. Reconstruction of multicomponent samples allows monitoring of reaction progress. In addition, we can make *in situ* observations of shape and textural changes. For deep carbon research, nanoXCT will allow researchers to garner three-dimensional X-ray tomographic images of carbon-bearing phases in pore spaces, grain boundaries, and channels within minerals and rocks in laser-heated diamond-anvil cells, and view the interaction of fluid CO_2 and hydrocarbons with solid rocks with clarity rivaling the XCT used by petroleum researchers to visualize the storage and flow of oil and natural gas in sediments at near-surface conditions. The low X-ray absorption of carbon-rich samples can lead to difficulties when working in absorption contrast. For single-phase equation of state determinations, a highly absorbing coating (e.g., Pt) can be deposited to improve X-ray contrast. If trying to image the shape and texture of heterogeneous samples, phase contrast can be used to highlight interfaces in the sample.

NanoXCT on quenched samples. Samples synthesized at high pressures and temperatures within a DAC can be recovered and then imaged using nanoXCT. In order to prepare the sample, a small portion (typically less than 20 μm × 20 μm in cross section) can be cut out using the FIB in much the same way as for the *ex situ* characterization methods. Figure 14 shows images from nanoXCT reconstructions of the detailed morphology of a sample composed of a molten Fe-alloy within a solid silicate matrix synthesized at moderate and very high pressures. For the small number of samples investigated so far, changes in the melt structure with pressure can already be seen. At the highest pressure studied (64 GPa), the shape of the melt changes and forms thin platelets. This morphology may be a reflection of very low dihedral angles, as the Fe-melt efficiently wets the surface of the silicate grains (note that the silicate at this pressure is in the perovskite structure). Alternatively, the change may result from the development of shape-preferred orientation. The change in connectivity from isolated Fe-alloy spheres to a connected Fe-alloy melt network with pressure can be clearly imaged without even having to calculate the dihedral angle. One can imagine investigating the interactions of carbon-rich fluids with surrounding rock by imaging the fluid portion relative to the solid grains.

These promising results demonstrate that nanoXCT on quenched samples can be conducted to 100 GPa and > 3000 K. An alternative method for analysis of quenched samples is slicing the sample with the FIB (or polishing off thin layers) and then analyzing the 2D images with SEM, repeating for sections through the sample, and reconstructing the 3D picture layer-by-layer (Bruhn et al. 2000) However, this procedure is destructive, time-consuming, and may suffer from sample-alteration by the high-energy electrons and ions. Nevertheless, FIB/SEM analysis after 3D nanoXCT measurement on quenched samples can provide critical benchmarks and calibrations for the spatial and chemical resolution of this new nanoXCT technique.

In situ nanoXCT at high pressures and temperatures. For *in situ* studies, two types of DACs have been developed that have been optimized for XCT through a beryllium gasket: a panoramic DAC that has been widely used for spectroscopic studies to >100 GPa, and a newly designed cross DAC that adopts the principle of the "plate DAC" (Boehler 2006; Figs. 2b,c)

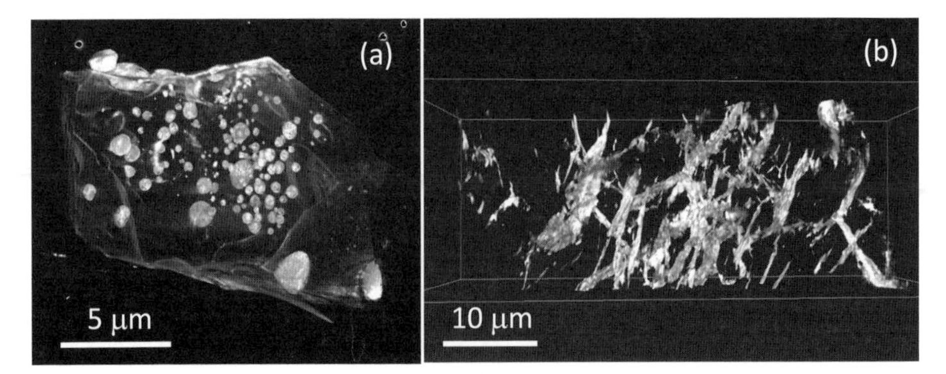

Figure 14. Images from 3D reconstruction of quenched samples composed of mixtures of molten Fe-rich liquids (bright white) in a silicate matrix (dark gray) at (a) pressure-temperature conditions of 8 GPa and 1800 °C, where the Fe-rich melt forms isolated spheres and at (b) higher pressure-temperature conditions of 64 GPa and 3000 °C, where the Fe-rich melt forms connected channels.

and whose functionality has been tested. Both DACs have wide radial access in and out of the equatorial plane when used with an X-ray-transparent high-strength beryllium gasket. For conducting TXM, the typical panoramic DAC has a 110-degree radial access and the cross DAC has 150-degree access. DACs need solid posts sufficiently strong in tension to support the opposing diamond anvils as they are pushed against each other. These posts block the X-rays and prevent access over a certain angular range so, while further improvements to the DAC design can increase the angular access, full 180-degree access is currently not possible. For large volume presses, an option is to use high-load thrust bearings to permit the rotation of the sample relative to the press, however, the need for high-precision alignment has prevented its implemented in DAC designs so far. For DACs, the "missing angle" problem can be partially mitigated with specialized reconstruction algorithms (Miao et al. 2005; Wang et al. 2012).

An application for *in situ* nanoXCT is the determination of the equations of state of amorphous materials. For crystalline materials, powder and single-crystal diffraction can be used to measure the equation of state with ~1.0×10^{-3} resolution. Owing to their lack of long range order, diffraction patterns from amorphous and liquid phases are often difficult to interpret and the position of broad diffraction maxima is not directly related to macroscopic density, as it is in the case of crystalline materials. Nano XCT can provide geometric information on samples regardless of their crystallinity. As a benchmark for this type of technique, the volume change with pressure for a crystalline Sn sample was determined using tomography with nanoscale spatial resolution TXM in a DAC (Wang et al. 2012). The results were consistent with diffraction results and had comparable error bars (Fig. 15), demonstrating the potential for the application of this method to a large variety of amorphous materials and liquids. Preliminary results on glassy carbon show that with proper sample preparation (in this case a coating of Pt to help with absorption contrast), one can image the volume of a low-Z amorphous material like glassy carbon (Fig. 16). One can also look at the shapes of multiple phases within heterogeneous samples (fluid-solid interactions) and study how these topologies change *in situ* at high pressure and variable temperature. Phase contrast can be employed to highlight interfaces between phases with similar X-ray scattering power.

Other contrast mechanisms. A successful XCT experiment requires careful consideration of sample absorption. If the absorption is too high, the sample is opaque, and if it is too low, the absorption contrast will be low. In addition to reaching higher pressures in a DAC, the sample size reduction from mm-sized in conventional and microXCT to μm-sized in nanoXCT has the advantage that the absorption of small samples can easily be optimized to electronic edges (e.g., for the Fe K-edge at 7.1 keV). Therefore, comparison of 3D XCT measurements collected above and below the Fe K-edge can be used for determination of 3D mapping of the distribution and partitioning of Fe, which has implications for the oxidation of the mantle and influences the oxidation state of deep carbon (Frost and McCammon 2008). By collecting the 3D XCT at small energy steps for from the pre-edge region through the electronic edge, it is possible to probe and map the oxidation state, spin state, and crystallographic site for Fe from its XANES signal. In terms of penetration of the DAC and collection of sufficient signal, we can access transition element *K*-edges and rare earth element *L*-edges, providing a map of coordination and oxidation states, as well as quantitative composition information.

Exciting developments in XRD and XRS have also been used for 3D tomographic reconstruction of carbon-bearing phases although the spatial resolution is still at the few μm to tens of μm range. Recent results at ambient pressure using XRS and XRD demonstrate that the differences in the chemical-bonding reflected in the carbon *K*-edge and the XRD patterns can be used to image diamond in graphite (Huotari et al. 2011). *In situ* high-pressure XRD computed tomography is also actively being developed at synchrotron sources and has been applied on carbon-bearing phases (Bleuet et al. 2008; Alvarez-Murga et al. 2011).

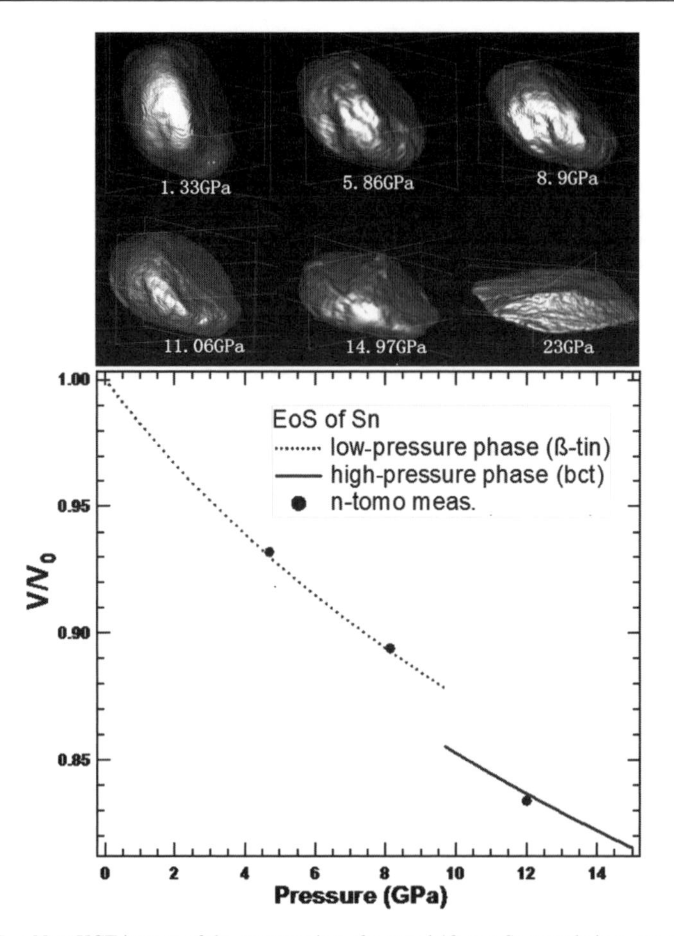

Figure 15. *Top*: NanoXCT images of the compression of a round 10-mm Sn sample in a panoramic DAC. With increasing pressure, one can follow the change in volume and shape to determine the equation of state and the shape preferred orientation. *Bottom*: Volumes measured as a function of pressure at 4.7, 8.1, and 12.0 GPa (the size of the black dots represents the estimated error bar). The equation of state of Sn for both the low-pressure and high-pressure phases are plotted together and demonstrate the excellent agreement between the XRD and nanoXCT results. From Wang et al. (2012).

CONCLUSIONS AND OUTLOOK

The new millennium has seen the rapid development of a battery of *ex situ* and *in situ* techniques that can be used for investigating deep carbon. The ultimate goal for experimental deep-carbon studies is to be able to determine the physics and chemistry of all possible carbon-bearing phases at conditions down to the base of the lower mantle. Although we still have a long way to go in terms of realizing this goal, researchers have made considerable progress. We have nanoscale diffraction and imaging tools in hand (and they are also being actively improved and developed), and nanoscale spectroscopy is on the horizon. With improvements in beamline instruments, insertion devices, and brighter sources, we may realize nanoscale resolution with X-ray Raman spectroscopy and other inelastic scattering techniques. The future is also bright (pardon the pun) for *in situ* synchrotron X-ray studies and also neutron studies with new and upgraded sources coming on-line. New ultrafast laser techniques and hard X-ray free electron

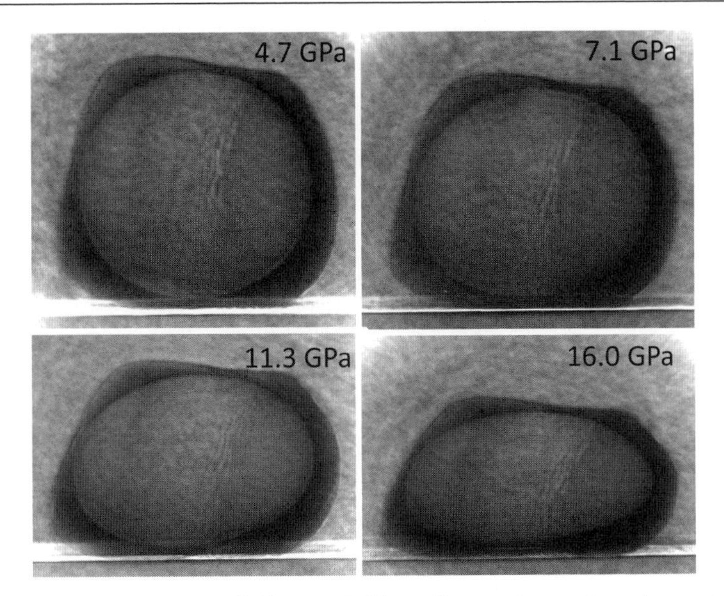

Figure 16. *In situ* high-pressure TXM images of a 15-mm diameter glassy carbon sphere coated with ~1-mm thick Pt compressed to four different pressures in a DAC.

lasers offer the possibility for future deep carbon studies with time-resolved measurements and the experimental study of the mechanisms and transformation pathways involved in phase transitions.

REFERENCES

Aitouchen A, Kihn Y, Zanchi G (1997) Quantitative EELS by spectrum parametrization. Microsc Microanal Microstruct 8:369-378

Als-Nielsen J, McMorrow D (2011) Elements of Modern X-ray Physics. John Wiley & Sons Inc, West Sussex

Alvarez-Murga M, Bleuet P, Marques L, Lepoittevin C, Boudet N, Gabarino G, Mezouar M, Hodeau J-L (2011) Microstructural mapping of C_{60} phase transformation into disordered graphite at high pressure, using X-ray diffraction microtomography. J Appl Crystallogr 44:163-171

Anderson R, Tracy B, Bravman JC (1992) Specimen Preparation for Transmission Electron Microscopy of Materials. III. Materials Research Society, Pittsburg, PA

Andrews JC, Brennan S, Liu Y, Pianetta P, Almeida EAC, van der Meulen MCH, Ishii H, Mester Z, Ouerdane L, Gelb J, Feser M, Rudati J, Tkachuk A, Yun W (2009) Full-field transmission X-ray microscopy for bio-imaging. J Phys: Conf Ser 186, doi: 10.1088/1742-6596/186/1/012081

Andrews JC, Brennan S, Patty C, Luening K, Pianetta P, Almeida E, van der Meulen MCH, Feser M, Gelb J, Rudati J, A. Tkachuk A, Yun WB (2008) A High Resolution, Hard X-ray Bio-imaging Facility at SSRL. Synchrotron Radiation News 21:17-26

Auzende A-L, Badro J, Ryerson FJ, Weber PK, Fallon SJ, Addad A, Siebert J, Fiquet G (2008) Element partitioning between magnesium silicate perovskite and ferropericlase: new insights into bulk lower-mantle geochemistry. Earth Planet Sci Lett 269:164-174

Bassett WA (2001) The birth and development of laser heating in diamond anvil cells. Rev Sci Instrum 72:1270-1272

Bassett WA (2009) Diamond anvil cell, 50th birthday. High Pressure Res 29(2):163-186.

Bergmann U, Groenzin H, Mullins OC, Glatzer P, Fetzer J, Cramer SP (2003) Carbon K-edge x-ray Raman spectroscopy support simple yet powerful description of aromatic hydrocarbon and asphaltenes. Chem Phys Lett 369:184-191

Bleuet P, Welcomme E, Dooryhee E, Susini J, Hodeau J-L, Walter P (2008) Probing the structure of heterogeneous diluted materials by diffraction tomography. Nat Mater 7:468-472

Boehler R (2005) Diamond cells and new materials. Mater Today 8:34-42

Boehler R (2006) New diamond cell for single-crystal x-ray diffraction. Rev Sci Instrum 77:115103

Boehler R, Musshoff HG, Ditz R, Aquilanti G, Trapananti A (2009) Portable laser-heating stand for synchrotron applications. Rev Sci Instrum 80:045103

Boulard E, Gloter A, Corgne A, Antonangeli D, Auzende A-L, Perrillat J-P, Guyot F, Fiquet G (2011) New host for carbon in the deep Earth. Proc Natl Acad Sci USA 108: 5184–5187

Boulard E, Menguy N, Auzende A-L, Benzerara K, Bureau H, Antonangeli D, Corgne A, Morard G, Siebert J, Perrillat J-P, Guyot F, Fiquet G (2012) Experimental investigation of the stability of Fe-rich carbonates in the lower mantle. J Geophys Res 117:1-15

Brandes JA, Cody G, Rumble D, Haberstroh P, Wirick S, Gelinas Y (2008) Carbon K-edge XANES spectromicroscopy of natural graphite. Carbon 46:1424-1434

Brandes JA, Lee C, Wakeham S, Peterson M, Jacobsen C, Wirick S, Cody G (2004) Examining marine particulate organic matter at sub-micron scales using scanning transmission X-ray microscopy and carbon X-ray absorption near edge structure spectroscopy. Mar Chem 92:107-121

Bruhn D, Groebner N, Kohlstedt DL (2000) An interconnected network of core-forming melts produced by shear deformation. Nature 403:883-886

Cody GD, Alexander CMO, Yabuta H, Kilcoyne ALD, Araki T, Ade H, Dera P, Fogel M, Militzer B, Mysen BO (2008) Organic thermometry for chondritic parent bodies. Earth Planet Sci Lett 272:446-455

Couvy H, Cordier P, Chen J (2011) Dislocation microstructures in majorite garnet experimentally deformed in the multi-anvil apparatus. Am Mineral 96:549-552

Dasgupta R (2013) Ingassing, storage, and outgassing of terrestrial carbon through geologic time. Rev Mineral Geochem 75:183-229

Dasgupta R, Hirschmann MM (2006) Melting in the Earth's deep upper mantle caused by carbon dioxide. Nature 440:659-62

Egerton RF (1979) K-shell ionization cross-sections for use in microanalysis. Ultramicroscopy 4:169-179

Egerton RF (2009) Electron energy-loss spectroscopy in the TEM. Rep Prog Phys 72:016502

Fiquet G, Auzende A-L, Siebert J, Corgne A, Bureau H, Ozawa H, Garbarino G (2010) Melting of peridotite to 140 gigapascals. Science 329:1516-1518

Frost DJ, McCammon CA (2008) The redox state of Earth's mantle. Annu Rev Earth Planet Sci 36:389-420

Garvie LAJ, Craven AJ, Brydson R (1994) Use of electron-energy loss near-edge fine structure in the study of minerals. Am Mineral 79:411-425

Greaves G, Jephcoat AP, Bouhifd MA, Donnelly SE (2008) A cross-sectional transmission electron microscopy study of iron recovered from a laser-heated diamond anvil cell. J Phys: Conf Ser 126:012047

Gualda GAR, Pamukcu AS, Claiborne LL, Rivers ML (2010) Quantitative 3D petrography using x-ray tomography. 3. Documenting accessory phases with differential absorption tomography. Geosphere 2010 6:782-792; doi: 10.1130/GES00568.1

Hazen RM, Downs RT, Jones AP, Kah L (2013) Carbon mineralogy and crystal chemistry. Rev Mineral Geochem 75:7-46

Heaney PJH, Vicenzi EP, Giannuzzi LA, Livi KJT (2001) Focused ion beam milling: a method of site-specific sample extraction for microanalysis of Earth and planetary materials. Am Mineral 86:1094-1099

Hitchcock AP, Dynes JJ, Johansson G, Wang J, Botton G (2008) Comparison of NEXAFS microscopy and TEM-EELS for studies of soft matter. Micron 39:741-748

Hofer F, Golob P (1987) New examples for near-edge fine structures in electron energy loss spectroscopy. Ultramicroscopy 21:379-383

Huotari S, Pylkkänen T, Verbeni R, Monaco G, Hämäläinen K (2011) Direct tomography with chemical-bond contrast. Nat Mater, doi: 10.1038/nmat3031.

Irifune T, Isshiki M, Sakamoto S (2005) Transmission electron microscope observation of the high-pressure form of magnesite retrieved from laser heated diamond anvil cell. Earth Planet Sci Lett 239:98-105

Isshiki M, Irifune T, Hirose K, Ono S, Ohishi Y, Watanuki T, Nishibori E, Takata M, Sakata M (2004) Stability of magnesite and its high-pressure form in the lowermost mantle. Nature 427:60-63

Kumar RS, Cornelius AL, Pravica MG, Nicol MF, Hu MY, Chow P (2007a) Bonding changes in single wall carbon nanotubes (SWCNT) on Ti and TiH_2 addition probed by X-ray Raman scattering. Diamond Rel Mater 16:1136-1139

Kumar RS, Pravica MG, Cornelius AL, Nicol MF, Hu MY, Chow P (2007b) X-ray Raman scattering studies on C_{60} fullerenes and multi-walled carbon nanotubes under pressure. Diamond Rel Mater 16:1250-1253

Lee KKM, O'Neill B, Jeanloz R (2004) Limits to resolution in composition and density in ultra high-pressure experiments on natural mantle-rock samples. Phys Earth Planet Inter 143-44:241-253

Lepot K, Benzerara K, Brown GE, Philippot P (2008) Microbially influenced formation of 2,724-million-year-old stromatolites. Nat Geosci 1:118-121

Lesher CE, Wang Y, Gaudio S, Clark A, Nishiyama N, Rivers ML (2009) Volumetric properties of magnesium silicate glasses and supercooled liquid at high pressure by X-ray microtomography. Phys Earth Planet Inter 174:292-301

Lin Y, Zhang L, Mao H-k, Chow P, Xiao Y, Baldini M, Shu J, Mao WL (2011) Amorphous diamond: a high-pressure superhard carbon allotrope. Phys Rev Lett 107:175504
Liu H, Wang L, Xiao X, De Carlo F, Feng J, Mao HK, Hemley RJ (2008) Anomalous high-pressure behavior of amorphous selenium from synchrotron x-ray diffraction and microtomography. Proc Natl Acad Sci USA 105:13229–13234
Mao HK, Mao WL (2007) 2.09 Theory and Practice - Diamond-Anvil Cells and Probes for High P-T Mineral Physics Studies. *In:* Price GD (ed) Treatise on Geophysics: Mineral Physics 2, p 231-268. Elsevier, Amsterdam
Mao HK, Shen G, Hemley RJ (1997) Multivariable dependence of Fe-Mg partitioning in the lower mantle. Science 278:2098-2100
Mao WL, Mao HK, Eng P, Trainor T, Newville M, Kao CC, Heinz DL, Shu J, Meng Y, Hemley RJ (2003) Bonding changes in compressed superhard graphite. Science 302:425-427
Marquardt H, Marquardt K (2012) Focused ion beam preparation and characterization of single-crystal samples for high-pressure experiments in the diamond-anvil cell. Am Mineral 97:299-304
McNulty I, Kirz J, Jacobsen C, Anderson EH, Howells MR, Kern DP (1992) High-resolution imaging by Fourier Transform X-ray holography. Science 256:1009-1012
Mees F, Swennen R, Geet MV, Jacobs P (eds) (2003) Applications of X-ray Computed Tomography in the Geosciences. Geological Society, London, Special Publications, Vol. 215
Miao J, Förster F, Levi O (2005) Equally sloped tomography with oversampling reconstruction. Phys Rev B 72:052103
Miao J, Ishikawa T, Shen Q, Earnest T (2008) Extending the methodology of x-ray crystallography to allow structure determination of non-crystalline materials, whole cells and single macromolecular complexes. Annu Rev Phys Chem 59:387-410
Mussi A, Cordier P, Mainprice D, Frost DJ (2010) Transmission electron microscopy characterization of dislocations and slip systems in K-lingunite: implications for the seismic anisotropy of subducted crust. Phys Earth Planet Inter 182:50-58
Oganov AR, Hemley RJ, Hazen RM, Jones AP (2013) Structure, bonding, and mineralogy of carbon at extreme conditions. Rev Mineral Geochem 75:47-77
Paterson JH, Krivanek O (1990) ELNES of 3d transition-metal oxides II Variations with oxidation state and crystal structure. Ultramicroscopy 32:319-325
Pravica M, Grubor-Urosevic O, Hu M, Chow P, Yulga B, Liermann P (2007) X-ray Raman spectroscopic study of benzene at high pressure. J Phys Chem B 111:11635
Ricolleau A, Perrillat J-P, Fiquet G, Daniel I, Matas J, Addad A, Menguy N, Cardon H, Mezouar M, Guignot N (2010) Phase relations and equation of state of a natural MORB: implications for the density profile of subducted oceanic crust in the Earth's lower mantle. J Geophys Res 115:B08202
Robertson J (1986) Amorphous carbon. Adv Phys 35:317-374
Robertson J, O'Reilly EP (1987) Electronic and atomic structure of amorphous carbon. Phys Rev B 35:2946-2957
Rohrbach A, Ballhaus C, Golla-Schindler U, Ulmer P, Kamenetsky VS, Kuzmin DV (2007) Metal saturation in the upper mantle. Nature 449:456-458
Rueff J-P, Shukla A. (2010) Inelastic x-ray scattering by electronic excitations under high pressure. Rev Modern Phys 82:847-896
Santoro M, Lin J-F, Mao HK, Hemley RJ (2004) In situ high P-T Raman spectroscopy and laser heating of carbon dioxide. J Chem Phys 121:2780-2787
Schumacher M, Christl I, Scheinost AC, Jacobsen C, Kretzschmar R (2005) Chemical heterogeneity of organic soil colloids investigated by scanning transmission X-ray microscopy and C-1s NEXAFS microspectroscopy. Environ Sci Technol 39:9094-9100
Shen G, Rivers ML, Wang Y, Sutton SR (2001) Laser heated diamond cell system at the Advanced Photon Source for in situ x-ray measurements at high pressure and temperature. Rev Sci Instrum 72:1273-1282
Stagno V, Frost DJ (2010) Carbon speciation in the asthenosphere: Experimental measurements of the redox conditions at which carbonate-bearing melts coexist with graphite or diamond in peridotite assemblages. Earth Planet Sci Lett 300:72-84
Stagno V, Tange Y, Miyajima N, McCammon CA, Irifune T, Frost DJ (2011) The stability of magnesite in the transition zone and the lower mantle as function of oxygen fugacity. Geophys Res Lett 38:1-5
Stevie FA, Shane TC, Kahora PM, Hull R, Bahnck D, Kannan VC, David E (1995) Applications of focused ion beams in microelectronics production, design and development. Surf Interface Anal 23:61-65
Tateno S, Hirose K, Ohishi Y, Tatsumi Y (2010) The structure of iron in Earth's inner core. Science 330:359-361
Tschauner O, Mao HK, Hemley RJ (2001) New transformations of CO_2 at high pressures and temperatures. Phys Rev Lett 87:075701
van Aken PA, Liebscher B (2002) Quantification of ferrous/ferric ratios in minerals: new evaluation schemes of Fe L_{23} electron energy-loss near-edge spectra. Phys Chem Mineral 29:188-200

van Aken PA, Liebscher B, Styrsa VJ (1998) Quantitative determination of iron oxidation states in minerals using Fe $L_{2,3}$-edge electron energy-loss near-edge structure spectroscopy. Phys Rev B 4:323-327

Veblen DR, Banfield JF, Guthrie GD, Heaney PJ, Ilton ES, Livi KJ, Smelik EA (1993) High-resolution and analytical transmission electron microscopy of mineral disorder and reactions. Science 260:1465-1472

Wang J, Yang W, Wang S, Xiao X, Carlo FD, Liu Y, Mao WL (2012) High pressure nano-tomography study using iteration method. J Appl Phys 111:112626

Wang L, Ding Y, Yang W, Liu W, Cai Z, Kung J, Shu J, Hemley RJ, Mao WL, Mao HK (2010a) Nanoprobe measurements of materials at megabar pressures. Proc Natl Acad Sci USA 107:6140-6145

Wang L, Yang W, Xiao Y, Liu B, Chow P, Shen G, Mao WL, Mao HK (2011) Application of a new composite cubic-boron nitride gasket assembly for high pressure inelastic x-ray scattering studies of carbon related materials. Rev Sci Instrum 82:073902

Wang Y, Lesher C, Fiquet G, Rivers ML, Nishiyama N, Siebert J, Roberts J, Morard G, Gaudio S, Clark A, Watson H, Menguy N, Guyot F (2010b) In situ high-pressure and temperature x-ray microtomographic imaging during large deformation: a new technique for studying mechanical behavior of multiphase composites. Geosphere 2011 7:40-53; doi: 10.1130/GES00560.1

Watson HC, Roberts JJ (2011) Connectivity of core forming melts: Experimental constraints from electrical conductivity and X-ray tomography. Phys Earth Planet Inter 186:172-182

Weng X, Rez P, Ma H (1989) Carbon E-shell near-edge structure: Multiple scattering and band-theory calculations. Phys Rev B 40:4175-4178

Zhu W, Gaetani GA, Fusseis F, Montési LGJ, De Carlo F (2011) Microtomography of partially molten rocks: three-dimensional melt distribution in mantle peridotite. Science 332:88-91

Reviews in Mineralogy & Geochemistry
Vol. 75 pp. 449-465, 2013

14

On the Origins of Deep Hydrocarbons

Mark A. Sephton

Earth Science & Engineering
Imperial College London, South Kensington Campus
London SW7 2AZ, United Kingdom

m.a.sephton@imperial.ac.uk

Robert M. Hazen

Geophysical Laboratory, Carnegie Institution of Washington
5251 Broad Branch Road NW
Washington, DC 20015, U.S.A.

rhazen@ciw.edu

INTRODUCTION

Deep deposits of hydrocarbons, including varied reservoirs of petroleum and natural gas, represent the most economically important component of the deep carbon cycle. Yet despite their intensive study and exploitation for more than a century, details of the origins of some deep hydrocarbons remain a matter of vocal debate in some scientific circles. This long and continuing history of controversy may surprise some readers, for the biogenic origins of "fossil fuels"—a principle buttressed by a vast primary scientific literature and established as textbook orthodoxy in North America and many other parts of the world—might appear to be settled fact. Nevertheless, conventional wisdom continues to be challenged by some scientists.

The principal objectives of this chapter are: (1) to review the overwhelming evidence for the biogenic origins of most known deep hydrocarbon reservoirs; (2) to present equally persuasive experimental, theoretical, and field evidence, which indicates that components of some deep hydrocarbon deposits appear to have an abiotic origin; and (3) to suggest future studies that might help to achieve a more nuanced resolution of this sometimes polarized topic.

BIOGENIC ORIGINS OF DEEP HYDROCARBONS

Types of hydrocarbons

Deep hydrocarbons include a rich diversity of organic chemical compounds in the form of petroleum deposits, including oil and gas in various reservoirs, bitumen in oil sands, coal and clathrate hydrates. The major gaseous hydrocarbons are the alkanes methane (natural gas, CH_4), ethane (C_2H_6), propane (C_3H_8), and butane (C_4H_{10}). Liquid components of petroleum include a complex mixture primarily of linear and cyclic hydrocarbons from C_5 to C_{17}, as well as numerous other molecular species, while solid hydrocarbons include such broad categories as paraffin waxes (typically from C_{18} to C_{40}). In addition, mature coal deposits sometimes hold a suite of unusual pure crystalline hydrocarbon phases and other organic minerals (see Hazen et al. 2013).

Water-based clathrates, also known as gas hydrates or clathrate hydrates, are an important emerging source of deep methane that deserve special notice in the context of deep hydrocarbons. These remarkable crystalline water-cage compounds, which form at low temperatures

1529-6466/13/0075-0014$00.00 DOI: 10.2138/rmg.2013.75.14

(< 0 °C) and elevated pressures (> 6 MPa), have potential applications both as a major methane source and as model materials for efficient energy storage (Buffett 2000; Boswell 2009; Koh et al. 2009, 2011). Several clathrate hydrate structure types feature a variety of cage sizes and shapes, depending on the size and shape of the incorporated gas molecule (see Hazen et al. 2013).

"Methane ice," by far the dominant natural clathrate hydrate mineral, forms in permafrost zones below a depth of ~130 meters and in marine sediments on the outer continental shelves (Max 2003; Guggenheim and Koster van Groos 2003; Koh et al. 2011). The extent of methane hydrate is remarkable, with total estimated methane storage of 2×10^{16} m^3 (Kvenvolden 1995; Milkov 2004). Methane ice thus represents a potential energy source that is orders of magnitude greater than the proven traditional natural gas reserves (Allison and Boswell 2007), and which may exceed the energy content of all known fossil fuel reserves (Kvenvolden 1995; Grace et al. 2008).

It is well established that methane-bearing hydrate clathrates arise from H_2O-CH_4 fluids subjected to low-temperature and high-pressure (Buffett 2000; Hyndman and Davis 1992; Max 2003), at which conditions an H_2O framework crystallizes around the CH_4 template molecules. However, the origins of the methane may be varied and are a continuing subject of debate (Abrajano et al. 1988; Horita and Berndt 1999; Fu et al. 2007; Chen et al. 2008). Subsurface methanogenic microorganisms represent one significant source of methane generated at relatively low temperatures (Chapelle et al. 2002; D'Hondt et al. 2004; Hinrichs et al. 2006; Jorgensen and Boetius 2007; Roussel et al. 2008; Schrenk et al. 2010, 2013; Mason et al. 2010; Menez et al. 2012; Colwell and D'Hondt 2013). Thermal cracking of hydrocarbons in petroleum at high temperature followed by migration of the methane to surficial environments is another important contributor to the formation of methane hydrates. Whether there are substantial additional abiotic sources from the lower crust and mantle is as yet unresolved (see below).

Diagenesis and kerogen formation

The more conventional view of petroleum formation is that it formed when selected aliquots of biomass from dead organisms were buried in a sedimentary basin and subjected to diagenesis through prolonged exposure to microbial decay followed by increasing temperatures and pressures. Oxygen-poor conditions, produced by exhaustion of local oxygen levels by biomass decay and often sustained by physical barriers to oxygen recharge, are obvious enhancers for fossil organic matter preservation and passage into the geosphere. The major organic components in life are large, high molecular weight entities and the most resistant of these units are preserved in sediments, augmented by cross-linking reactions that polymerize and incorporate smaller units into the complex network. The high molecular weight sedimentary organic matter is termed kerogen from the Greek for "wax former." It is worth noting that not all of life's organic matter is reflected in kerogen. Even under relatively favorable conditions less that 1% of the starting organism, representing the most resistant chemical constituents, may be preserved (Demaison and Moore 1980).

The chemistry of kerogen depends strongly on its contributing organisms and several different types are formed. Type I forms from mainly algae, Type II from a mixture of algae and land plants, and Type III primarily from land plants. Irrespective of kerogen type, increased temperature and pressure leads to thermal dissociation or "cracking" and produces petroleum. Owing to their intrinsic chemical constitution, Type I and Type II kerogens are predisposed to generate oil while Type III kerogens produce gas. At more extreme temperature and pressures, petroleum can undergo secondary cracking reactions that result in significant quantities of the smallest hydrocarbon molecule, methane.

Petroleum reservoirs often comprise porous and permeable sedimentary rock (Selley 1985). Under such conditions at Earth's surface organic matter is normally destroyed, so it follows that petroleum must have been introduced into the reservoir from elsewhere. The passage of petroleum from source rock to reservoir, driven largely by buoyancy, is termed migration. Oil, gas, and water are stratified in reservoirs based on density, indicating that the fluids have travelled vertically. For vertical migration to be arrested the reservoir must be capped by an impermeable seal.

ABIOTIC ORIGINS OF DEEP HYDROCARBONS

Deep gas theories

The hypothesis that at least some components of petroleum have a deep abiotic origin in the lower crust or mantle has a long history, with influential support and elaboration by Russian chemist and mineralogist Dimitri Mendeleev (Mendeleev 1877) and astronomer and mathematician Fred Hoyle (Hoyle 1955). The deep-Earth gas hypothesis proposes that abiogenic methane reflects a cosmic organic inheritance that is subsequently released by the mantle and migrates towards the surface utilizing weaknesses in the crust such as plate boundaries, faults, and sites of meteorite impacts. The deep sourced methane polymerizes en route to higher molecular weight hydrocarbons that ultimately form petroleum deposits. Members of the so-called "Russian-Ukrainian School" pursued this model with theory, experiments, and field observations (Kudyratsev 1951; Kenny 1996; Glasby 2006; Safronov 2009). Superficial support for this theory is provided by the increase in abundance of methane with depth in petroleum-containing basins. However, it is known that the higher temperatures associated with greater depths in Earth's subsurface promotes the cracking of high molecular weight hydrocarbons to produce lower molecular weight units, the ultimate product of which is methane. Hence, enhanced methane concentrations with depth are most effectively explained as an organic response to the geothermal gradient rather than closer proximity to a mantle source of methane. Yet the position that petroleum is primarily abiotic in origin is still held by some advocates (Gold 1992; Kenney et al. 2001). The theory, however, is not completely without merit and mounting evidence points to facile and potentially widespread abiotic synthesis of methane, and possibly some higher hydrocarbons, under specific geochemical conditions. Major issues exist however, over the scientific rigor of popularized deep gas concepts and the quantitative importance of the underlying processes to the world's hydrocarbon resources.

Thomas Gold and the "Deep Hot Biosphere"

Early 20th-century research of the Russian-Ukrainian School, though extensive, was published in Russian and until the 1990s was largely ignored in the West. The abiotic petroleum hypothesis first gained significant exposure outside the Soviet Union during the 1977 gasoline crisis, when Cornell astrophysicist Thomas (Tommy) Gold published an editorial in the *Wall Street Journal* (Gold 1977) in which he claimed that most deep hydrocarbons are generated abiotically in the mantle and migrate to the crust, where they act as an energy source for microbes producing a deep microbial ecosystem—the "Deep Hot Biosphere"— that may rival the surface biosphere in mass and volume. The organic remains of microbes in the deep hot biosphere represent the source of biological molecules in petroleum as a biological overprint onto an abiogenic organic mixture (Gold and Soter 1980, 1982; Gold 1992, 1999; see Hazen 2005). The proposal contains an implicit concession that the biological molecules in petroleum cannot be generated by polymerization of methane.

Echoing some of the previous arguments of the Russian-Ukrainian School, Gold presented several lines of evidence for abiotic petroleum, including (1) the presence of clearly abiotic hydrocarbons and other organic molecules in meteorites and on other solar system bodies; (2) plausible synthetic pathways for mantle hydrocarbon production; (3) the association of hydrocarbon deposits with helium and other trace gases, presumably from mantle sources; (4) the existence of extensive deep microbial communities that impart an overprint of biomarkers onto the abiotic petroleum; (5) the tendency of hydrocarbon reservoirs to occur at many depths in a single locality, implying an underlying deep source; (6) the distribution of hydrocarbon deposits related to underlying mantle structures; (7) the distribution of metals and other trace elements in petroleum, which correlate more strongly with chondrite compositions than presumed crustal sources; and (8) the occurrence of hydrocarbons in non-sedimentary formations such as crystalline rocks.

Evidence for abiotic hydrocarbon synthesis

While compelling evidence for the abiotic synthesis of petroleum is lacking, there is unambiguous experimental, theoretical, and field evidence for deep abiotic origins of some hydrocarbons (McCollom 2013).

Abiotic hydrocarbons in space. Carbon is the fourth most abundant element in the Cosmos and remote observations indicate that the Universe is replete with organic compounds produced without the influence of biology. Studies of the absorption spectra of dense molecular clouds reveal more than 150 small molecules (Ehrenfreund and Charnley 2000; Kwok 2009). The greatest amounts of hydrocarbons in our solar system are associated with the gas giant planets and their satellites, where significant amounts are present in atmospheres and on icy surfaces. The prevalence of non-biological hydrocarbons in the early stages of our solar system is evidenced by the chemistry of carbonaceous meteorites, which represent fragments of ancient asteroids, unchanged since shortly after the birth of the solar system. Carbonaceous meteorites contain percentage levels of organic matter that were generated in the absence of biology (Sephton 2002). The widespread presence of organic matter in the Cosmos suggests that no exotic mechanism therefore need be invoked to propose abiogenic hydrocarbons on Earth. In fact the biogenic origin of some hydrocarbons on Earth is, thus far, a unique observation in an abiotic hydrocarbon-rich Universe.

In contrast to our Cosmic environment, organic matter on Earth is associated primarily with biological processes. The distinction between biological and non-biological organic matter has held a long-term fascination for scientists and philosophers. For instance, all ancient mythologies invoke divine intervention and the use of some abstract vital force for the production of the organic compounds that constitute life (Fry 2000). Following the 7th and 6th centuries BC, such vitalist theories were often challenged by more materialistic concepts that assumed only differences in levels of organization between living and non-living entities. A key moment for materialism came in 1828, when Friedrich Wöhler produced the organic chemical urea $CO(NH_2)_2$ from inorganic ammonium cyanate (NH_4CNO), using a procedure now called the Wöhler synthesis (Wöhler 1828). The generation of an organic compound from inorganic material reflects the ability of non-biological processes to produce organic materials and it is an unavoidable conclusion that the origin of life on Earth was by definition a process of abiotic organic synthesis, e.g., (Lahav 1999; Wills and Bada 2000; Hazen 2005).

In general, organic compounds generated by abiotic reactions such as those that produced meteoritic organic matter are characterized by complete structural diversity; biological organic compounds produced by synthetic reactions directed by enzymes are notably specific in structure (Sephton and Botta 2005). Organic compounds in petroleum, by contrast, contain numerous structures in abundances that could not be produced by the random polymerization from methane.

Experimental hydrocarbon synthesis. Hydrocarbon synthesis is achievable under laboratory conditions and in shallow geological environments and several key reactions are notable.

1) The Fischer-Tropsch synthesis (Fischer and Tropsch 1926) involves the conversion of carbon monoxide or carbon dioxide to hydrocarbons:

$$nCO + (2n + 1)H_2 \rightarrow C_nH_{2n+2} + nH_2O$$

$$nCO_2 + (3n + 1)H_2 \rightarrow C_nH_{2n+2} + 2_nH_2O$$

The products of the Fischer-Tropsch synthesis are low molecular weight hydrocarbons and often form a Schulz-Flory distribution, where the log of the hydrocarbon concentration decreases linearly with increasing carbon numbers (Salvi and Williams-Jones 1997). The Fischer-Tropsch process has been proposed as a generator of hydrocarbons in hydrothermal vent settings (Horita and Berndt 1999; Foustoukos and Seyfried 2004; McCollom and Seewald 2006; Proskurowski et al. 2008; Proskurowski et al. 2008).

2) Simple thermal metamorphism of carbonates such as calcite ($CaCO_3$), dolomite [$CaMg(CO_3)_2$], and siderite ($FeCO_3$) at 400 °C in the presence of H_2 can produce low molecular weight hydrocarbons (Giardini and Salotti 1969), as in the following unbalanced reaction:

$$FeO + CaCO_3 + H_2O \rightarrow Fe_3O_4 + CH_4 + CaO$$

3) Thermal metamorphism of carbonates and graphite below 300-400 °C may generate methane (Holloway 1984), as displayed in the following unbalanced reaction:

$$Mg_3Si_4O_{10}(OH)_2 + CaCO_3 + C_{graphite} + H_2O \rightarrow CaMg(CO_3)_2 + SiO_2 + CH_4$$

4) Methane is common at mid-ocean ridges and serpentinization reactions are implicated (McCollom and Seewald 2001; Charlou et al. 2002; Kelley et al. 2001, 2005; Schrenk et al. 2013). Hydrogen forms during the hydration of olivine and can combine with carbon dioxide to form methane (Abrajano et al. 1990), as follows (unbalanced reaction):

$$(Fe,Mg)_2SiO_4 + H_2O + CO_2 \rightarrow Mg_3Si_2O_5(OH)_4 + Fe_3O_4 + CH_4$$

In addition, for those reactions that proceed under conditions relevant to Earth's near-surface, the influence of pressure may be an important factor for synthesis. This environmental control is especially pertinent to the theories of deep gas and a number of experimental studies at elevated temperatures and pressures relevant to deep crust and mantle conditions point to the possibility of deep abiotic hydrocarbon synthesis.

1) The generation of petroleum hydrocarbons up to $C_{10}H_{22}$ was demonstrated using solid iron oxide, marble, and water at temperatures of 1,500 °C and pressures above 3 GPa, corresponding to a depth of more than 100 km (Kenney et al. 2002; Kutcherov et al. 2002).

2) A heated diamond-anvil cell has been employed to explore the calcite-iron oxide-water system at mantle pressures and temperatures (Scott et al. 2004). Based on Raman spectroscopic and X-ray diffraction evidence, methane and possibly other light hydrocarbon species were successfully synthesized through the unbalanced reaction:

$$CaCO_3 + FeO + H_2O \rightarrow CH_4 + CaO + Fe_3O_4$$

The Scott et al. (2004) experiments represent the first attempts to test the Russian-Ukranian School hypothesis under conditions comparable to those found in the mantle. Subsequent experimentation has successfully repeated the observations (Sharma et al. 2009).

3) Pressures above 2 GPa and temperatures from 1000-1500 K appear to induce polymerization of methane to hydrocarbons (ethane, propane, and butane), molecular hydrogen, and graphite; the reaction is reversible under these conditions (Kolesnikov et al. 2009).

Although elegant and efficient reactions exist that can produce substantial amounts of hydrocarbons from simple precursors in the absence of biology, care must be taken when extrapolating data to natural settings. Convincing evidence must be found that the production mechanism is contributing. Lessons can be learned from meteoritic organic matter studies, where the attractive Fischer-Tropsch mechanism was supported even though ample evidence was available that this reaction was not contributing quantitatively significant amounts of material to the organic inventory (Sephton et al. 2001).

Hydrocarbons in minerals. Evidence has been presented for organic "protomolecule assembly" in igneous rocks (Freund et al. 2001). According to Freund's hypothesis, igneous melts inevitably incorporate carbon and other impurities at high temperatures. As minerals crystallize, they also incorporate a small amount of carbon, which concentrates along linear crystal defects and form alkanes. However, these results have been challenged by other workers (Keppler et al. 2003; Shcheka et al. 2006; see also Hazen 2005; Ni and Keppler 2013).

Abiotic hydrocarbons from hydrothermal vents. A number of authors have presented experimental, theoretical, and field evidence for the possible release of abiotic hydrocarbon species from hydrothermal vents (Simoneit et al. 2004; see McCollum 2013). Hydrocarbons have been collected from numerous field sites (Tingle et al. 1991; Holm and Charlou 2001; Charlou et al. 2002; Sherwood-Lollar et al. 2002, 2006; Simoneit et al. 2004; Loncke et al. 2004; Proskurowski et al. 2008; Konn et al. 2009; Lang et al. 2010). Observations of gas chemistry, abundance, and fluxes in various vent settings have provided an opportunity to assess quantitatively the relevance of proposed abiotic species to commercial oil and gas resources. Mean values of molar $CH_4/^3He$ in mantle-derived fluids imply that less than 200 ppm of abiotic gas are present in economically relevant reservoirs (Jenden et al. 1993).

These field observations of hydrocarbons from hydrothermal systems have been complemented by experimental studies that attempt to replicate natural conditions. A number of these experiments have demonstrated abiotic formation of methane and other organic compounds, particularly in studies that mimic serpentinization. The experiments underscore the importance of transition elements (i.e., Fe, Ni, Co, Cr, V, and Mn) in the hydrothermal synthesis of organic molecules by Fischer-Tropsch-type reactions (Heinen and Lauwers 1996; McCollom et al. 1999; Seewald 2001; McCollom and Seewald 2001, 2006, 2007; Cody 2004; Foustoukos and Seyfried 2004). These results suggest an intriguing connection between mineral-mediated organic synthesis and biological catalysts, many of which incorporate transition metal atoms at their active sites (Adams 1992; Beinert et al. 1997). Biochemical reactions promoted by enzymes that incorporate transition metals may thus represent reaction pathways that have survived and evolved from the prebiotic geochemical world.

DETERMINING SOURCE—CHEMICAL EVIDENCE

Pyrolysis experiments

The concept that petroleum could be generated by the thermal degradation of sedimentary organic matter is supported by centuries of shale utilization and experimentation. Shale with high organic carbon contents has been heated to produce oil for lighting since the 19th century, leading to inevitable suggestions that petroleum was formed from an analogous process in the subsurface. The hypothesis was confirmed by experiments revealing that substances similar to petroleum can be produced by pyrolysis (thermal decomposition without combustion) of a variety of animal substances (Engler 1912). Modern experiments create conditions that simulate subsurface maturation more closely by heating shale in the presence of water. These hydrous pyrolysis experiments generate products from potential source rocks that are physically and chemically similar to natural crude oils (Lewan et al. 1979).

Molecular biomarkers

Petroleum contains biomarkers that are the organic chemical remnants of once living organisms. The presence of biomarkers is a key indicator for an ultimately biological source of petroleum. Recognition of the first biomarker arose from the structural correlation between the biological pigment chlorophyll and its porphyrin degradation products in sediments and petroleum (Treibs 1934). Counter arguments have suggested that biomarkers are either contaminants introduced by microbes in the reservoir or by the solvation of biological structures during the upward migration of hydrocarbons originating from mantle sourced methane (Gold 1985). Yet such interpretations fail to account for the maturity information contained in the relative abundance of biomarker isomers. The maturity parameter concept relies on the relative instability of biological structures once buried and heated in the subsurface; the biological isomers are progressively transformed to more geologically stable configuration (Mackenzie et al. 1982). Such biomarker maturity parameters in petroleum are consistent with temperatures at which thermal degradation of kerogen would occur; migrating fluids that extract biomarkers from multiple depths would result in mixed maturity signals that do not correspond to the predicted maturity of the putative source rock. Moreover, the biomarker inventory in petroleum is not exclusively derived from Bacteria or Archaea that may survive in the subsurface and molecular fossils of surface-living eukaryotes are common, including those from algae (e.g., dinosterol from dinoflagellates; Boon et al. 1979) and land plants (e.g., oleanane from angiosperms; Moldowan et al. 1994).

Optical activity. A substance is said to be optically active if it rotates the plane of linearly polarized light. For rotation to occur the substance must contain chiral centers in which carbon atoms have four different groups attached. Differing arrangements of groups around the chiral carbon can produce mirror images that are non-superimposable. Optically pure compounds contain only one isomeric form and rotate plane-polarized light. The enzymatically directed synthesis of biological compounds leads to optically pure forms. Racemic mixtures contain more than one isomeric form in equal proportions and do not rotate plane-polarized light. The random synthesis associated with non-biological reactions typically leads to racemic mixtures, whereas petroleum tends to be optically active (Oakwood et al. 1952), reflecting contributions from preferred isomerical forms derived from biologically mediated reactions, specifically steranes and triterpenoids (Silverman 1971). As may be expected, the optical activity of petroleum increases during biodegradation as the relatively resistant chiral steranes and triterpenoids are concentrated (Winters and Williams 1969), and negatively correlates with thermal maturity as chiral centers are lost at high temperatures (Williams 1974) as the compound mixtures become racemic. The detection of optically active compounds in meteorites (Nagy et al. 1964; Nagy 1966), which contain non-biological organic matter, has been used as a criticism of the assignment of biology as a source of optically active compounds in petroleum. Yet some meteorite observations were subsequently attributed to analytical artefacts (Hayatsu 1965, 1966; Meinschein et al. 1966) and more recent analyses on better curated samples implied racemic or near racemic mixtures (Kvenvolden et al. 1970, 1971). Modern analyses have recognized some isomeric preference but not to the extent seen in biological materials and the observation is limited to certain compound classes, namely the structurally related amino and hydroxy acids and amino acid precursors (Pizzarello and Cronin 2000; Pizzarello et al. 2008).

Compound-specific studies of carbon isotopes. The carbon atoms that make up the skeleton of hydrocarbons contain two stable isotopes, ^{12}C and ^{13}C. The lighter carbon isotope is more reactive than its heavier counterpart and takes part in reactions more readily. The preferential incorporation of the lighter carbon isotope in reaction product leads to isotope fractionation and can reveal synthetic mechanisms. Abiotic hydrocarbons generally reveal an increase in ^{12}C with carbon number in accord with the kinetically controlled synthesis of higher molecular weight homologues from simpler precursors (Sephton and Gilmour 2001). Such trends are observed for gases released by the freeze thaw disaggregation of meteorites (Yuen et al. 1984) and by those produced in methane spark discharge experiments (DesMarais et al. 1981). By contrast the thermal cracking of high molecular weight biologically-derived hydrocarbons produces the opposite trend where a decrease in ^{12}C with carbon number is observed (DesMarais et al. 1981). Such patterns are common to the thermal cracking products of hydrocarbons in the laboratory and the gases emitted by hydrothermal systems. The gases recovered from petroleum reservoirs follow the cracking trend and do not support an origin by methane polymerization (Chung et al. 1988). The two trends are often proposed as a means of discriminating between gases generated by ultimately abiotic or biotic (followed by thermal degradation) processes and their utility is demonstrated in Figure 1. Carbon (and hydrogen) isotope ratios for hydrocarbons in the crystalline rocks of the Canadian Shield do imply an abiogenic source from the polymerization of methane, but the distinction between these and commercially relevant petroleum deposits have ruled out the presence of a globally significant abiogenic source of hydrocarbons (Sherwood-Lollar et al. 2002). The diagnostic potential of the carbon isotope with carbon number trends has been qualified, however, with recognized difficulties arising from gas mixing, diffusion, or oxidation (Sherwood-Lollar et al. 2008; Burruss and Laughrey 2010) alongside variable fractionation owing to different reaction pressures (Wei et al. 2012); biodegradation can also induce modifications to the original carbon isotope versus carbon number isotope profile (e.g., Prinzhofer et al. 2010; Fig. 1).

Thermochemical calculations of hydrocarbon stability. Thermodynamic calculations have been used to suggest that only methane and elemental carbon are stable at the relatively shallow depths and lower temperatures and pressures associated with petroleum reservoirs (Kenney et al. 2002). In contrast, depths of 100 km, pressures above 3 GPa, and temperatures of 900 °C have been proposed as conducive to hydrocarbon formation. To produce petroleum deposits the products need to be rapidly quenched requiring vertical migration at unlikely speed. Thermodynamic equilibrium is not inevitable and biology uses energy from processes such as photosynthesis to constantly maintain organic matter that is in thermodynamic disequilibrium

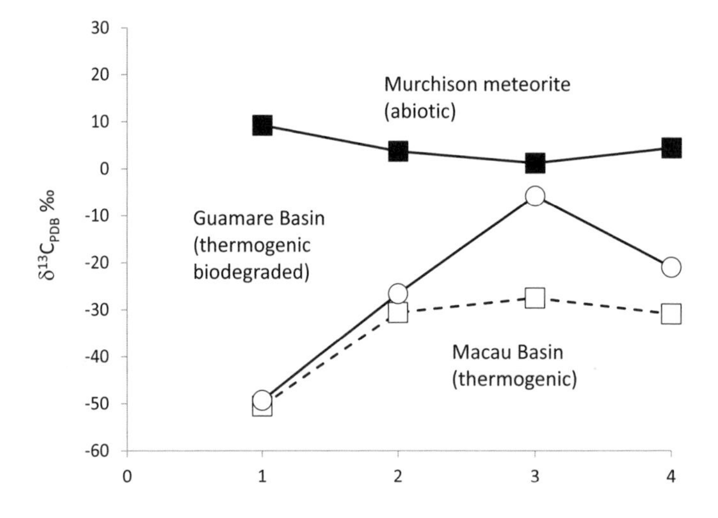

Figure 1. Carbon isotope ratios plotted against carbon number for straight chain hydrocarbons from meteorites (Yuen et al. 1984) and commercial gas fields (Prinzhofer et al. 2010). The trends for abiogenic and thermogenic hydrocarbons are almost mirror images; processes such as biodegradation can, however, disrupt these diagnostic patterns.

with its environment. It is an aliquot of this organic matter, dominated by the lipid components, that survives diagenesis and enters the geosphere to form kerogen. Many industrial processes (Cook and Sherwood 1991) and experimental studies (Lewan et al. 1979) have demonstrated how heating rocks rich in kerogens can produce petroleum.

Helium. The association of helium with hydrocarbons has been used to assess contributions from the mantle. Helium is often found in high concentrations in petroleum relative to the host rock. There are two significant sources of helium in crustal fluids: ^{3}He is a primordial isotope that was created in the Big Bang and incorporated into Earth as it formed, whereas ^{4}He is produced by radioactive decay of uranium and thorium and is constantly added to Earth's helium inventory (Mackintosh and Ballentine 2012). Helium isotopes can thus help to discriminate between sources. It has been controversially proposed that upwardly migrating hydrocarbons will entrain helium from uranium- and thorium-rich source rocks, resulting in its accumulation in petroleum reservoirs (Gold 1999). Yet constant ^{3}He/C ratios have been observed at mid-ocean ridges, reflecting mantle values (Marty et al. 1989). Helium is particularly evident near active fault systems (Hooker et al. 1985; O'Nions and Oxburgh 1988; Marty et al. 1992). Different ^{3}He/C values in petroleum reservoirs suggest that the main source of economic hydrocarbon accumulations is not the mantle. Hence, helium isotope ratios in petroleum reservoirs indicate a predominantly mantle provenance for this noble gas and flux calculations suggest that any associated mantle hydrocarbons would represent a minimal proportion of the total hydrocarbon reservoir (Ballentine et al. 1991; Jenden et al. 1993; Ballentine and Sherwood Lollar 2002).

DETERMINING SOURCE—GEOLOGIC EVIDENCE

Association with temperature and source rocks

Petroleum deposits are associated with areas in which source rocks have been subjected to elevated temperatures in the subsurface. This correlation was first recognized by a coal geologist who related the thermal transformation of coal deposits in a region with the appearance of petroleum reservoirs (Hunt 1853). The implication was that organic matter in fine-grained rocks would have been heated to similar extents as the coals and petroleum produced. Decades of subsequent research have led to the recognition of geochemical fingerprints; petroleum seeping from reservoirs reveals stable isotope ratios and molecular biomarkers that correlates it with with thermally matured source rocks (Seifert and Moldowan 1978). As petroleum leaves the source rock and begins migration it retains a molecular inheritance from the source rock, enabling the two to be correlated (Peters et al. 2005). Moreover, under thermal stress, the biological configurations of entrained molecules are transferred to more thermally stable forms (Mackenzie et al. 1982). The progress of transformation from biological to geological isomers can be monitored and correlated to temperature. Such maturity parameters indicate that oil can be related to source rocks that have achieved temperatures above 60 °C. Exploration in places where geochemical data indicates a mature source rock down-dip of the target area provides the greatest chance of success (Demaison 1984).

Faults and plate margins. A consequence of the abiogenic theory of hydrocarbon formation would be an association of petroleum deposits with the putative conduits for upward migration, namely convergent plate margins and major fault zones (Gold and Soter 1980). Where petroleum is present at plate boundaries however its presence can be attributed to other factors. In particular, plate boundaries are associated with high geothermal gradients that enhance the formation, migration, and entrapment of oil and gas (Klemme 1975).

Crystalline rock reservoirs. The association of hydrocarbons with igneous rocks has been taken as prima facie evidence for abiogenic synthesis. A number reservoirs are situated in igneous rocks and, although most are relatively small, some giant fields (recoverable reserves > 500 million barrels of oil equivalent) do exist (Schutter 2003). Yet the location of petroleum within igneous rocks is only circumstantial evidence because fluids can migrate over great distances (Bredehoeft and Norton 1990). Fundamentally, the accumulation of migrated petroleum in igneous rocks only requires the necessary porosity and permeability. Although pore spaces are not normally associated with igneous rocks, there are many ways to develop porosity and permeability and, in some cases, they may be more porous and permeable than the adjacent sediments (Schutter 2003). For instance, igneous rocks may have primary porosity, such as that associated with extrusive rocks. Igneous rocks may also develop secondary porosity from retrograde metamorphism or alteration by hydrothermal activity and fracturing as the rocks are cooled or weathered. Porosity in igneous rocks is remarkably robust and once produced resists compaction relatively well compared to sedimentary rock. When crystalline rock reservoirs are considered in a global context, it becomes clear that they represent less than 1% of the world's petroleum deposits. The remaining 99% points to an overwhelming correlation of petroleum with sedimentary rocks, indicating a likely origin from sedimentary organic matter (Ulmishek and Klemme 1991).

Perhaps the most celebrated example of a crystalline rock reservoir is the Siljin Ring Complex of Sweden, a 368 Ma meteorite impact site that features fractured granite county rock with hydrocarbon veins and methane seeps. From 1986 to 1992, Gold convinced the Swedish State Power Board to drill two exploratory wells in the granite—a 7-year, $40 million project that yielded a small, though tantalizing quantity of oil-like hydrocarbons. Proponents of deep abiotic origins for petroleum were convinced that these results vindicated the hypothesis (Gold 1999). Others disagreed and pointed to the extensive quantities of drilling oils and nearby sedimentary formations as the source of the recovered hydrocarbons, while many disappointed Swedish investors saw the project as a failure (Brown 1999; Hazen 2005).

Mantle oxidation state. For abiogenic hydrocarbons to be produced from the mantle, conditions must be reducing to prevent the oxidation of hydrogen-rich organic compounds such as methane. It has been suggested that the compositions of ultrabasic rocks, meteorites, comets, and planets indicate that Earth's mantle should be sufficiently reducing in nature. The presence of reducing agents such as hydrogen and metallic iron-nickel would enable the production and survival of hydrocarbons such as methane. Yet data suggest that Earth's upper mantle is too oxidized to allow the persistence of more than small amounts of methane or hydrogen (see Manning et al. 2013). Within surface geological features that communicate with the mantle, mid-ocean ridges tend to be relatively reduced compared to subduction zones (Wood et al. 1990), but the dominant gases observed are carbon dioxide and water. Major forms of carbon in the upper mantle are diamonds and carbonate (Wood et al. 1996) and methane is a minor constituent in fluid inclusions from the upper mantle (Roedder 1984; Apps and van der Kamp 1993). Gases from the mantle are transported to Earth's surface and mainly dissolved in magmas and gases released are composed of carbon dioxide, water, and sulfur dioxide or hydrogen sulfide. The concentrations of methane in mantle-derived gases thus cannot be the source of commercial gas fields.

SELECTED CASE STUDIES

The controversy surrounding the origins of hydrocarbon gases can be placed in context by reference to a number of case studies. The Lost City deep-sea hydrothermal vent is discussed by Schrenk et al. (2013) and the Siljin Ring Complex is reviewed above, but the two very different case studies of Mountsorrel in the United Kingdom and the Songliao Basin of China are summarized below.

Mountsorrel, United Kingdom

A classic case of abiotic-biotic hydrocarbon controversy is provided by the presence of hydrocarbons discovered in igneous rocks at Mountsorrel, Leicestershire, UK. Ordovician granodiorite rocks are cut by Carboniferous basaltic dikes. Liquid and solid hydrocarbons are observed along joints and fractures occurring parallel to the granodiorite-basalt contact but also cross cut the dykes in places, indicating that they postdate both intrusions (King 1959). Researchers used organic geochemical techniques to assign an abiotic source to the hydrocarbons based on the presence of an unresolved complex mixture during gas chromatographic analysis. The unresolved complex mixture implied structural diversity for the organic compounds in the igneous rocks (Ponnamperuma and Pering 1966)—data that were used to support the abiogenic hydrocarbon theory (Gold and Soter 1982). However, subsequent organic geochemical investigations point to the ability of biodegradation to remove alkanes and isoprenoidal alkanes, thus leaving an unresolved complex mixture behind (Wenger et al. 2002). More recent analyses of the Mountsorrel hydrocarbons revealed that the unresolved complex mixture contained abundant hopanes and steranes, pointing to an ultimately biological source probably from organic-rich Carboniferous rocks and likely introduced by hydrothermal fluids associated with the intrusion (Gou et al. 1987). Parnell (1988) believed this biodegradation mechanism to be a common occurrence and that effectively all hydrocarbons in igneous rocks in the UK were from sedimentary sources.

The Songliao Basin, China

The Songliao Basin in eastern China is amongst the largest Cretaceous rift basins in the world. Basement rocks comprise Precambrian to Palaeozoic metamorphic and igneous rocks and Paleozoic to Mesozoic granites (Pei et al. 2007). Mesozoic and Cenozoic sedimentary rocks overlie the basement (Gao et al. 1994) and include petroleum source rocks generated by two substantial lacustrine anoxic events (Hou et al. 2000). Significant attention has been paid to the presence of a reversed order for carbon isotope compositions of alkanes in natural gases in the Chinese gas reservoirs. It has been proposed that gases pass through faults in the crust, assisted by earthquakes, to accumulate in the Songliao basin. Data have been used to suggest that abiotic gases can contribute significantly to commercial gas reservoirs (Guo et al. 1997). However, although these reversed trends can result from the presence of abiotic gas, alternative mechanisms are available, including the mixing of humic and sapropelic gases, the mixing of gases from source rocks of different maturity, and the influence of biodegradation (Dai et al. 2004).

CONCLUSIONS: UNRESOLVED QUESTIONS IN THE ORIGINS OF DEEP HYDROCARBONS

This survey of the origins of deep hydrocarbons demonstrates that much is known regarding both the biotic origins of petroleum and the abiotic origins of some light hydrocarbons, notably methane. Nevertheless, uncertainties persist and much remains to be learned.

1. *What is the source of methane in clathrate hydrates?* The methane concentrated in clathrate hydrates might come from several different sources, including microbial methanogenesis, thermal cracking of biomass, and abiotic synthesis under varied conditions from relatively shallow serpentinization zones to the mantle. Is there a way to discriminate among these sources? Several workers (Wang and Frenklach 1994; Eiler and Schauble 2004; Ghosh et al. 2006) have suggested that the temperature of formation of methane and other light hydrocarbons might be deduced from the ratio of isotopologs, which are chemically identical molecules that differ in their combination of isotopes. Methane has two doubly substituted isotopologs: $^{12}CH_2D_2$ and $^{13}CH_3D$. Theoretical calculations suggest that measurements of the ratio of these two isotopologs could provide a sensitive indicator of the temperature at which a sample of methane formed—data that might distinguish between low-temperature methanogenesis, moderate temperature thermal cracking of biomass, or

high-temperature abiotic synthesis in the mantle. Such data is imminent and will be provided by newly developed mass spectrometers (Jones 2012).

2. *How can we better interrogate gases?* Stable isotope methods have provided valuable insights into the formation mechanisms of methane and its higher homologues, but can new methods uncover more information locked within deep gases? In contrast to the higher molecular weight hydrocarbon biomarkers, gases contain little structural information that can signpost origin. Dual isotopic methods (Butterworth et al. 2004) and position specific methods (Corso and Brenna 1997) can cast light onto the synthetic mechanisms that produced deep gas.
3. *What associations can reveal the source and history of deep gases?* The presence of gaseous consorts can often reveal the origin of hydrocarbons. The isotopic measurement of noble gases and hydrocarbons (Ballentine et al. 2002; Sherwood-Lollar and Ballentine 2009) illustrate how combined techniques can constrain the possible histories of gas mixtures. What new associations and chemical signals can be combined to build a body of circumstantial evidence to provide the most plausible interpretations?
4. *How will unconventional gas reservoirs change our understanding?* Much of the considerations associated with hydrocarbons and their sources involve conventional gas reservoirs and hydrothermal vents. Yet the natural gas industry is undergoing a revolution following technological advancements that allow the production of shale gas. This particular type of natural gas is hosted in a tight reservoir and has not appreciably migrated from its biogenic organic-rich source. The obscuring nature of processes such as biodegradation and source mixing will, presumably, be absent in such deposits. The chemical investigation of shale gas reservoirs will provide new and valuable baseline data to which previously acquired conventional reservoir information can be compared.

ACKNOWLEDGMENTS

We thank John Baross, Russell Hemley, Andrea Mangum, and Craig Schiffries for invaluable discussions and suggestions during the preparation of this manuscript. The authors gratefully acknowledge the Deep Carbon Observatory, the Alfred P. Sloan Foundation, the Carnegie Institution of Washington, the National Science Foundation, and NASA's Astrobiology Institute for support of this study.

REFERENCES

Abrajano TA, Sturchio NC, Bohlke JK, Lyon GL, Poreda RJ, Stevens CM (1988) Methane-hydrogen gas seeps, Zambales Ophiolite, Philippines: Deep or shallow origin? Chem Geol 71:211-222

Abrajano TA, Sturchio NC, Kennedy BM, Lyon GL, Muehlenbachs K, Bohlke JK (1990) Geochemistry of reduced gas related to serpentinization of the Zambales ophiolites, Philippines. Appl Geochem 5:625-630

Adams MWW (1992) Novel iron—sulfur centers in metalloenzymes and redox proteins from extremely thermophilic bacteria. *In:* Advances in Inorganic Chemistry. Richard C (ed) Academic Press, New York, p 341-396

Allison E, Boswell R (2007) Methane hydrate, future energy within our grasp, an overview. DOE Overview Document *http://www.fossil.energy.gov/programs/oilgas/hydrates/*

Apps JA, van der Kamp PC (1993) Energy gases of abiogenic origin in the Earth's crust. US Geol Surv Prof Paper 1570:81-132

Ballentine CJ, Burgess R, Marty B (2002) Tracing fluid origin, transport and interaction in the crust. Rev Min Geochem 47:539-614

Ballentine CJ, O'Nions RK, Oxburgh ER, Horvath F, Deak J (1991) Rare gas constraints on hydrocarbon accumulation, crustal degassing and groundwater flow in the Pannonian Basin. Earth Planet Sci Lett 105:229-246

Ballentine CJ, Sherwood-Lollar B (2002) Regional groundwater focusing of nitrogen and noble gases into the Hugoton-Panhandle giant gas field, USA. Geochim Cosmochim Acta 66:2483-2497
Beinert H, Holm RH, Münck E (1997) Iron-sulfur clusters: Nature's modular, multipurpose structures. Science 277:653-659
Boon JJ, Rijpstra WIC, De Lange F, De Leeuw JW, Yoshioka M, Shimizu Y (1979) Black Sea sterol - a molecular fossil for dinoflagellate blooms. Nature 277:125-127
Boswell R (2009) Is gas hydrate energy within reach? Science 325:957-958
Bredehoeft JD, Norton DL (1990) Mass and Energy Transport in the Deforming Earth's Crust: The role of Fluids in Crustal Processes. National Academy Press, Washington, DC
Brown A (1999) Upwelling of hot gas. Am Scientist 87:372-372
Buffett BA (2000) Clathrate hydrates. Ann Rev Earth Planet Sci 28:477-507
Burruss RC, Laughrey CD (2010) Carbon and hydrogen isotopic reversals in deep basin gas: Evidence for limits to the stability of hydrocarbons. Org Geochem 41:1285-1296
Butterworth AL, Aballain O, Chappellaz J, Sephton MA (2004) Combined element (H and C) stable isotope ratios of methane in carbonaceous chondrites. Mon Not Roy Astron Soc 347:807-812
Chapelle FH, O'Neill K, Bradley PM, Methe BA, Ciufo SA, Knobel LL, Lovley DR (2002) A hydrogen-based subsurface microbial community dominated by methanogens. Nature 415:312-315
Charlou JL, Donval JP, Fouquet Y, Jean-Baptiste P, Holm N (2002) Geochemistry of high H_2 and CH_4 vent fluids issuing from ultramafic rocks at the Rainbow hydrothermal field (36°14′ N, MAR). Chem Geol 191:345-359
Chen J, Lin L, Dong J, Zheng H, Liu G (2008) Methane formation from $CaCO_3$ reduction catalyzed by high pressure. Chinese Chem Lett 19:475-478
Chung HM, Gormly JR, Squires RM (1988) Origin of gaseous hydrocarbons in subsurface environments: Theoretical considerations of carbon isotope distribution. Chem Geol 71:97-104
Cody GD (2004) Transition metal sulfides and the origins of metabolism. Ann Rev Earth Planet Sci 32:569-599
Colwell FS, D'Hondt S (2013) Nature and extent of the deep biosphere. Rev Mineral Geochem 75:547-574
Cook AC, Sherwood NR (1991) Classification of oil shales, coals and other organic-rich rocks. Org Geochem 17:211-222
Corso TN, Brenna JT (1997) High-precision position-specific isotope analysis. Proc Natl Acad Sci USA 94:1049-1053
Dai J, Xia X, Qin S, Zhao J (2004) Origins of partially reversed alkane $\delta^{13}C$ values for biogenic gases in China. Org Geochem 35:405-411
Demaison GJ (1984) The generative basin concept. *In:* Petroleum Geochemistry and Basin Evaluation. Demaison GJ, Murris RJ (ed) American Association of Petroleum Geologists, Tulsa, Oklahoma, p 1-14
Demaison GJ, Moore GT (1980) Anoxic environments and oil source bed genesis. Bull Am Assoc Petrol Geol 64:1179-1209
DesMarais DJ, Donchin JH, Nehring NL, Truesdell AH (1981) Molecular carbon isotopic evidence for the origin of geothermal hydrocarbons. Nature 292:826-828
D'Hondt S, Jørgensen BB, Miller DJ, Batzke A, Blake R, Cragg BA, Cypionka H, Dickens GR, Ferdelman T, Hinrichs K-U, Holm NG, Mitterer R, Spivack A, Wang G, Bekins B, Engelen B, Ford K, Gettemy G, Rutherford SD, Sass H, Skilbeck CG, Aiello IW, Guèrin G, House CH, Inagaki F, Meister P, Naehr T, Niitsuma S, Parkes RJ, Schippers A, Smith DC, Teske A, Wiegel J, Padilla CN, Acosta JLS (2004) Distributions of microbial activities in deep subseafloor sediments. Science 306:2216-2221
Ehrenfreund P, Charnley SB (2000) Organic molecules in the interstellar medium, comets, and meteorites: A voyage from dark clouds to the early earth. Ann Rev Astron Astrophys 38:427-483
Eiler JM, Schauble E (2004) ^{18}O ^{13}C ^{16}O in Earth's atmosphere. Geochim Cosmochim Acta 68:4767-4777
Engler C (1912) Die Bildung der Haupt-Bestandteile des Erdöls. Petroleum (Berlin) 7:399-403
Fischer F, Tropsch H (1926) Über die direkte synthese von erdöl-kohlenwasserstoffen bei gewöhnlichtem druck. Berichte der Deutschen Chemischen Gesellschaft 59:830-831
Foustoukos DI, Seyfried WE (2004) Hydrocarbons in hydrothermal vent fluids: The role of chromium-bearing catalysts. Science 304:1002-1005
Freund F, Staple A, Scoville J (2001) Organic protomolecule assembly in igneous minerals. Proc Natl Acad Sci 98:2142-2147
Fry I (2000) The Emergence of Life on Earth. A Historical and Scientific Overview. Free Association Books, London
Fu Q, Sherwood Lollar B, Horita J, Lacrampe-Couloume G, Seyfried WE Jr (2007) Abiotic formation of hydrocarbons under hydrothermal conditions: Constraints from chemical and isotope data. Geochim Cosmochim Acta 71:1982-1998
Gao RQ, Zhang Y, Cui TC (1994) Cretaceous Petroleum Bearing Strata in the Songliao Basin: Petroleum Industry Press, Beijing, 1-333
Ghosh P, Adkins J, Affek H, Balta B, Guo W, Schauble EA, Schrag D, Eiler JM (2006) ^{13}C–^{18}O bonds in carbonate minerals: A new kind of paleothermometer. Geochim Cosmochim Acta 70:1439-1456

Giardini AA, Salotti CA (1969) Kinetics and relations in the calcite-hydrogen reaction and reactions in the dolomite-hydrogen and siderite-hydrogen systems. Am Mineral 54:1151-1172
Glasby GP (2006) Abiogenic origin of hydrocarbons: An historical overview. Resour Geol 56:83-96
Gold T (1977) Rethinking the origins of oil and gas. The Wall Street Journal 8 June 1977
Gold T (1985) The origin of natural gas and petroleum and the prognosis for future supplies. Ann Rev Energy 10:53-77
Gold T (1992) The deep, hot biosphere. Proc Natl Acad Sci USA 89:6045-6049
Gold T (1999) The Deep Hot Biosphere. Copernicus, New York
Gold T, Soter S (1980) The deep earth gas hypothesis. Sci Am 242:155-161
Gold T, Soter S (1982) Abiogenic methane and the origin of petroleum. Energy Exploration Exploitation 1:89-104
Gou X, Fowler MG, Comet PA, Manning DAC, Douglas AG, McEvoy J, Giger W (1987) Investigation of three natural bitumens from central England by hydrous pyrolysis and gas chromatography-mass spectrometry. Chem Geol 64:181-95
Grace J, Collett T, Colwell F, Englezos P, Jones E, Mansell R, Meekison JP, Ommer R, Pooladi-Darvish M, Riedel M, Ripmeester JA, Shipp C, Willoughby E (2008) Energy from gas hydrates—Assessing the opportunities and challenges for Canada. Report of the Expert Panel on Gas Hydrates, Council of Canadian Academies, September 2008
Guggenheim S, Koster van Groos AFK (2003) New gas-hydrate phase: Synthesis and stability of clay–methane hydrate intercalate. Geology 31:653-656
Guo Z, Wang X, Liu W (1997) Reservoir-forming features of abiotic origin gas in Songliao Basin. Science in China Series D: Earth Sciences 40:621-626
Hayatsu R (1965) Optical activity in the Orgueil meteorite. Science 149:443-447
Hayatsu R (1966) Artifacts in polarimetry and optical activity in meteorites. Science 153:859-861
Hazen RM (2005) Genesis: The Scientific Quest for Life's Origin. Joseph Henry Press, Washington, DC
Hazen RM, Downs RT, Jones AP, Kah L (2013) Carbon mineralogy and crystal chemistry. Rev Mineral Geochem 75:7-46
Heinen W, Lauwers AM (1996) Organic sulfur compounds resulting from the interaction of iron sulfide, hydrogen sulfide and carbon dioxide in an anaerobic aqueous environment. Orig Life Evol Biosph 26:131-150
Hinrichs K-U, Hayes JM, Bach W, Spivack AJ, Hmelo LR, Holm NG, Johnson CG, Sylva SP (2006) Biological formation of ethane and propane in the deep marine subsurface. Proc Natl Acad Sci 103:14684-14689
Holloway JR (1984) Graphite-CH_4-H_2O-CO_2 equilibria at low-grade metamorphic conditions. Geology 12:455-458
Holm NG, Charlou JL (2001) Initial indications of abiotic formation of hydrocarbons in the Rainbow ultramafic hydrothermal system, Mid-Atlantic Ridge. Earth Planet Sci Lett 191:1-8
Hooker PJ, O'Nions RK, Oxburgh ER (1985) Helium isotopes in North Sea gas fields and the Rhine rift. Nature 318:273-275
Horita J, Berndt ME (1999) Abiogenic methane formation and isotopic fractionation under hydrothermal conditions. Science 285:1055-1057
Hou D, Li M, Huang O (2000) Marine transgressional events in the gigantic freshwater lake Songliao: paleontological and geochemical evidence: Org Geochem 31:763-768
Hoyle F (1955) Frontiers of Astronomy. Heineman, London
Hunt TS (1853) Report on the Geology of Canada. Canadian Geological Survey Report: Progress to 1863
Hyndman RD, Davis EE (1992) A mechanism for the formation of methane hydrate and sea-floor bottom-simulating reflectors by vertical fluid expulsion. J Geophys Res-Solid Earth 97:7025-7041
Jenden PD, Hilton DR, Kaplan IR, Craig H (1993) Abiogenic hydrocarbons and mantle helium in oil and gas fields, in: Howell, D.G. (Ed.), The Future of Energy Gases - U.S. Geological Survey Professional Paper 1570. United States Government Printing Office, Washington, pp. 31-56
Jones N (2012) Source code: The methane race. Earth 57:40-45
Jorgensen BB, Boetius A (2007) Feast and famine--microbial life in the deep-sea bed. Nat Rev Micro 5:770-781
Kelley DS, Karson JA, Blackman DK, Fruh-Green GL, Butterfield DA, Lilley MD, Olson EJ, Schrenk MO, Roe KK, Lebon GT, Rivizzigno P (2001) An off-axis hydrothermal vent field near the Mid-Atlantic Ridge at 30 degrees N. Nature 412:145-149
Kelley DS, Karson JA, Früh-Green GL, Yoerger DR, Shank TM, Butterfield DA, Hayes JM, Schrenk MO, Olson EJ, Proskurowski G, Jakuba M, Bradley A, Larson B, Ludwig K, Glickson D, Buckman K, Bradley AS, Brazelton WJ, Roe K, Elend MJ, Delacour A, Bernasconi SM, Lilley MD, Baross JA, Summons RE, Sylva SP (2005) A serpentinite-hosted ecosystem: The Lost City hydrothermal field. Science 307:1428-1434

Kenney JF, Kutcherov VA, Bendeliani NA, Alekseev VA (2002) The evolution of multicomponent systems at high pressures: VI. The thermodynamic stability of the hydrogen–carbon system: The genesis of hydrocarbons and the origin of petroleum. Proc Natl Acad Sci 99:10976-10981

Kenney JF, Shnyukov YF, Krayishkin VA, Tchebanenko II, Klochko VP (2001) Dismissal of claims of a biological connection for natural petroleum. Energia 22:26-34

Keppler H, Wiedenbeck M, Shcheka SS (2003) Carbon solubility in olivine and the mode of carbon storage in the Earth's mantle. Nature 424:414-416

King RJ (1959) The mineralization of the Mountsorrel Granodiorite. Trans Leicester Lit Phil Soc 53:18-29

Klemme HD (1975) Geothermal gradients, heatflow and hydrocarbon recovery. *In:* Petroleum and Global Tectonics. Fischer AG, Judson S (ed) Princeton University Press, Princeton, New Jersey, p 251-304

Koh CA, Sloan ED, Sum AK, Wu DT (2011) Fundamentals and applications of gas hydrates. Ann Rev Chem Biomolec Eng 2:237-257

Koh CA, Sum AK, Sloan ED (2009) Gas hydrates: Unlocking the energy from icy cages. J Appl Phys 106:061101-061114

Kolesnikov A, Kutcherov VG, Goncharov AF (2009) Methane-derived hydrocarbons produced under upper-mantle conditions. Nature Geosci 2:566-570

Konn C, Charlou JL, Donval JP, Holm NG, Dehairs F, Bouillon S (2009) Hydrocarbons and oxidized organic compounds in hydrothermal fluids from Rainbow and Lost City ultramafic-hosted vents. Chem Geol 258:299-314

Kudyratsev NA (1951) Against the organic hypothesis of the origin of petroleum. Neftianoye Khozyaistvo 9:17-29

Kutcherov VG, Bendeliani NA, Alekseev VA, Kenney JF (2002) Synthesis of hydrocarbons from minerals at pressures up to 5 GPa. Doklady Phys Chem 387:328-331

Kvenvolden K, Lawless J, Pering K, Peterson E, Flores J, Ponnamperuma C, Kaplan IR, Moore C (1970) Evidence for extra-terrestrial amino acids and hydrocarbons in the Murchison meteorite. Nature 228:928-926

Kvenvolden KA (1995) A review of the geochemistry of methane in natural gas hydrate. Org Geochem 23:997-1008

Kvenvolden KA, Lawless JG, Ponnamperuma C (1971) Nonprotein amino acids in the Murchison meteorite. Proc Natl Acad Sci USA 68:86-490

Kwok S (2009) Organic matter in space: from star dust to the Solar System. Astrophys Space Sci 319:5-21

Lahav N (1999) Biogenesis. Oxford University Press, Oxford, UK

Lang SQ, Butterfield DA, Schulte M, Kelley DS, Lilley MD (2010) Elevated concentrations of formate, acetate and dissolved organic carbon found at the Lost City hydrothermal field. Geochim Cosmochim Acta 74:941-952

Lewan MD, Winters JC, McDonald JH (1979) Generation of oil-like pyrolysates from organic-rich shales. Science 203:897-899

Loncke L, Mascle J, Fanil Scientific P (2004) Mud volcanoes, gas chimneys, pockmarks and mounds in the Nile deep-sea fan (Eastern Mediterranean): Geophysical evidences. Mar Pet Geol 21:669-689

Mackenzie AS, Lamb NA, Maxwell JR (1982) Steroid hydrocarbons and the thermal history of sediments. Nature 295:223-226

Mackintosh SJ, Ballentine CJ (2012) Using $^{3}He/^{4}He$ isotope ratios to identify the source of deep reservoir contributions to shallow fluids and soil gas. Chem Geol 304-305:142-150

Manning CE, Shock EL, Sverjensky D (2013) The chemistry of carbon in aqueous fluids at crustal and upper-mantle conditions: experimental and theoretical constraints. Rev Mineral Geochem 75:109-148

Marty B, Jambon A, Sano Y (1989) Helium isotopes and CO_2 in volcanic gases of Japan. Chem Geol 76:25-40

Marty B, O'Nions RK, Oxburgh ER, Martel D, Lombardi S (1992) Helium isotopes in Alpine regions. Tectonophysics 206:71-78

Mason OU, Nakagawa T, Rosner M, Van Nostrand JD, Zhou J, Maruyama A, Fisk MR, Giovannoni SJ (2010) First investigation of the microbiology of the deepest layer of ocean crust. PLoS ONE 5:e15399

Max MD (2003) Natural Gas Hydrate in Oceanic and Permafrost Environments. Kluwer Academic Publishers, Dordrecht

McCollom TM (2013) Laboratory simulations of abiotic hydrocarbon formation in Earth's deep subsurface. Rev Mineral Geochem 75:467-494

McCollom TM, Ritter G, Simoneit BR (1999) Lipid synthesis under hydrothermal conditions by Fischer-Tropsch-type reactions. Orig Life Evol Biosph 29:153-166

McCollom TM, Seewald JS (2001) A reassessment of the potential for reduction of dissolved CO_2 to hydrocarbons during serpentinization of olivine. Geochim Cosmochim Acta 65:3769-3778

McCollom TM, Seewald JS (2006) Carbon isotope composition of organic compounds produced by abiotic synthesis under hydrothermal conditions. Earth Planet Sci Lett 243:74-84

McCollom TM, Seewald JS (2007) Abiotic synthesis of organic compounds in deep-sea hydrothermal environments. ChemInform 38:382-401
Meinschein WG, Frondel C, Laur P, Mislow K (1966) Meteorites; optical activity in organic matter. Science 154:377-380
Mendeleev D (1877) L'origine du petrole. Revue Scientifique 8:409-416
Menez B, Pasini V, Brunelli D (2012) Life in the hydrated suboceanic mantle. Nature Geosci 5:133-137
Milkov AV (2004) Global estimates of hydrate-bound gas in marine sediments: how much is really out there? Earth-Science Rev 66:183-197
Moldowan JM, Dahl J, Huizinga BJ, Fago FJ, Hickey LJ, Peakman TM, Taylor DW (1994) The molecular fossil record of oleanane and Its relation to angiosperms. Science 265:768-771
Nagy B (1966) A study of the optical rotation of lipids extracted from soils, sediments and the Orgueil, carbonaceous meteorite. Proc Natl Acad Sci USA 56:389-398
Nagy G, Murphy TJ, Modzeleski VE, Rouser GE, Claus G, Hennesy DJ, Colombo U, Gazzarini F (1964) Optical activity in saponified organic matter isolated from the interior of the Orgueil meteorite. Nature 202:228-223
Ni H, Keppler H (2013) Carbon in silicate melts. Rev Mineral Geochem 75:251-287
Oakwood TS, Shriver DS, Fall HH, McAleer WJ, Wunz PR (1952) Optical activity of petroleum. Ind Eng Chem 44:2568-2570
O'Nions RK, Oxburgh ER (1988) Helium, volatile fluxes and the development of continental crust. Earth Planet Sci Lett 90:331-347
Parnell J (1988) Migration of biogenic hydrocarbons into granites: A review of hydrocarbons in British plutons: Mar Pet Geol 5:385-396
Pei F, Xu W, Yang D, Zhao Q, Liu X, and Hu Z (2007) Zircon U-Pb geochronology of basement metamorphic rocks in the Songliao Basin: Chinese Sci Bull 52:942-948
Peters KE, Walters CC, Moldowan JM (2005) The Biomarker Guide. Volume 1 Biomarkers and Isotopes in the Environment and Human History. Cambridge University Press, Cambridge
Pizzarello S, Cronin JR (2000) Non-racemic amino acids in the Murray and Murchison meteorites. Geochim Cosmochim Acta 64:329-338
Pizzarello S, Huang Y, Alexandre MR (2008) Molecular asymmetry in extraterrestrial chemistry: Insights from a pristine meteorite. Proc Natl Acad Sci USA 105:3700-3704
Ponnamperuma C, Pering K (1966) Possible abiogenic origin of some naturally occurring hydrocarbons. Nature 209:979-982
Prinzhofer A, Dos Santos Neto EV, Battani A (2010) Coupled use of carbon isotopes and noble gas isotopes in the Potiguar basin (Brazil): Fluids migration and mantle influence. Mar Pet Geol 27:1273-1284
Proskurowski G, Lilley MD, Seewald JS, Früh-Green GL, Olson EJ, Lupton JE, Sylva SP, Kelley DS (2008) Abiogenic hydrocarbon production at Lost City Hydrothermal Field. Science 319:604-607
Roedder E (ed) (1984) Fluid Inclusions. Reviews in Mineralogy, Volume 12. Mineralogical Society of America, Washington, DC
Roussel EG, Bonavita M-AC, Querellou J, Cragg BA, Webster G, Prieur D, Parkes RJ (2008) Extending the sub-sea-floor biosphere. Science 320:1046-1046
Safronov AF (2009) Vertical zoning of oil and gas formation: Historico-genetic aspects. Russian Geol Geophys 50:327-333
Salvi S, Williams-Jones AE (1997) Fischer-Tropsch synthesis of hydrocarbons during sub-solidus alteration of the Strange Lake peralkaline granite, Quebec/Labrador, Canada. Geochim Cosmochim Acta 61:83-99
Schrenk MO, Huber JA, Edwards KJ (2010) Microbial provinces in the subseafloor. Ann Rev Marine Sci 2:279-304
Schrenk MO, Brazelton WJ, Lang SQ (2013) Serpentinization, carbon, and deep life. Rev Mineral Geochem 75:575-606
Schutter SR (2003) Hydrocarbon occurrence and exploration in and around igneous rocks. Geol Soc London, Special Pub 214:7-33
Scott HP, Hemley RJ, Mao HK, Herschbach DR, Fried LE, Howard WM, Bastea S (2004) Generation of methane in the Earth's mantle: In situ high pressure–temperature measurements of carbonate reduction. Proc Natl Acad Sci USA 101:14023-14026
Seewald JS (2001) Aqueous geochemistry of low molecular weight hydrocarbons at elevated temperatures and pressures: constraints from mineral buffered laboratory experiments. Geochim Cosmochim Acta 65:1641-1664
Seifert WK, Moldowan JM (1978) Applications of steranes, terpanes and monoaromatics to the maturation, migration and source of crude oils. Geochim Cosmochim Acta 42:77-95
Selley RC (1985) Elements of Petroleum Geology. W.H. Freeman, San Francisco
Sephton MA (2002) Organic compounds in carbonaceous meteorites. Natural Product Reports 19:292-311
Sephton MA, Botta O (2005) Recognizing life in the Solar System: guidance from meteoritic organic matter. Int J Astrobiology 4:269-276

Sephton MA, Gilmour I (2001) Compound-specific isotope analysis of the organic constituents in carbonaceous chondrites. Mass Spectrometry Rev 20:111-120

Sephton MA, Pillinger CT, Gilmour I (2001) Normal alkanes in meteorites: molecular $\delta^{13}C$ values indicate an origin by terrestrial contamination. Precambrian Res 106:47-58

Sharma A, Cody GD, Hemley RJ (2009) *In situ* diamond-anvil cell observations of methanogenesis at high pressures and temperatures. Energ Fuel 23:5571-5579

Shcheka SS, Wiedenbeck M, Frost DJ, Keppler H (2006) Carbon solubility in mantle minerals. Earth Planet Sci Lett 245:730-742

Sherwood-Lollar B, Ballentine CJ (2009) Insights into deep carbon derived from noble gases. Nature Geosci 2:543-547

Sherwood-Lollar B, Lacrampe-Couloume G, Slater GF, Ward J, Moser DP, Gihring TM, Lin LH, Onstott TC (2006) Unravelling abiogenic and biogenic sources of methane in the Earth's deep subsurface. Chem Geol 226:328-339

Sherwood-Lollar B, Lacrampe-Couloume G, Voglesonger K, Onstott TC, Pratt LM, Slater GF (2008) Isotopic signatures of CH_4 and higher hydrocarbon gases from Precambrian Shield sites: A model for abiogenic polymerization of hydrocarbons. Geochim Cosmochim Acta 72:4778-4795

Sherwood-Lollar B, Westgate TD, Ward JA, Slater GF, Lacrampe-Couloume G (2002) Abiogenic formation of alkanes in the Earth's crust as a minor source for global hydrocarbon reservoirs. Nature 416:522-524

Silverman SR (1971) Influence of petroleum origin and transformation on its distribution and redistribution in sedimentary rocks. Proc Eight World Petroleum Cong. Applied Science Publishers, London, p 47-54

Simoneit BRT, Lein AY, Peresypkin VI, Osipov GA (2004) Composition and origin of hydrothermal petroleum and associated lipids in the sulfide deposits of the Rainbow field (Mid-Atlantic Ridge at 36°N). Geochim Cosmochim Acta 68:2275-2294

Tingle TN, Mathez EA, Hochella MF Jr (1991) Carbonaceous matter in peridotites and basalts studied by XPS, SALI, and LEED. Geochim Cosmochim Acta 55:1345-1352

Treibs A (1934) The occurrence of chlorophyll derivatives in an oil shale of the upper Triassic. Justus Liebigs Ann Chem 509:103-114

Ulmishek GF, Klemme HD (1991) Effective petroleum source rocks of the world: stratigraphic distribution and controlling depositional factors. AAPG Bull 75:1809-1851

Wang H, Frenklach M (1994) Calculations of rate coefficients for the chemically activated reactions of acetylene with vinylic and aromatic radicals. J Phys Chem 98:11465-11489

Wei Z, Zou Y-R, Cai Y, Tao W, Wang L, Guo J, Peng P (2012) Abiogenic gas: Should the carbon isotope order be reversed? J Petrol Sci Eng 84-85:29-32

Wenger LM, Davis CL, Isaksen GH (2002) Multiple controls on petroleum biodegradation and impact on oil quality. SPE Reservoir Evaluation and Engineering 5:375-383

Williams JA (1974) Characterization of oil types in the Williston Basin. Am Assoc Petrol Geol Bull 58:1243-1252

Wills C, Bada JL (2000) The Spark of Life. Perseus, Cambridge, Massachusetts

Winters JC, Williams JA (1969) Microbial alteration of crude oil in the reservoir Am Chem Soc, Division Petrol Chem, New York Meeting Preprints 14:22-31

Wöhler F (1828) Ueber künstliche Bildung des Harnstoffs. Ann Phys Chem 88:253-256

Wood BJ, Alison P, Frost DR (1996) Water and carbon in the Earth's mantle. Philos Trans R Soc London Ser A 354:1495-1511

Wood BJ, Bryndzia LT, Johnson KE (1990) Mantle oxidation state and its relationship to tectonic environment and fluid speciation. Science 248:337-345

Yuen G, Blair N, Des Marais DJ, Chang S (1984) Carbon isotope composition of low molecular weight hydrocarbons and monocarboxylic acids from Murchison meteorite. Nature 307:252-254

Reviews in Mineralogy & Geochemistry
Vol. 75 pp. 467-494, 2013

Laboratory Simulations of Abiotic Hydrocarbon Formation in Earth's Deep Subsurface

Thomas M. McCollom

Laboratory for Atmospheric and Space Physics
University of Colorado
Boulder. Colorado 80309, U.S.A.

mccollom@lasp.colorado.edu

INTRODUCTION

In recent years, methane and other light hydrocarbons with an apparently abiotic origin have been identified in an increasing number of geologic fluids on Earth. These compounds have been found in a variety of geologic settings, including seafloor hydrothermal systems, fracture networks in crystalline rocks from continental and oceanic crust, volcanic gases, and gas seeps from serpentinized rocks (e.g., Abrajano et al. 1990; Kelley 1996; Sherwood Lollar 2002, 2008; Fiebig et al. 2007, 2009; Proskurowski et al. 2008; Taran et al. 2010b). Understanding the origin of these compounds has significant implications for range of topics that includes the global carbon cycle, the distribution of life in the deep subsurface (Gold 1992), and the origin of life (Martin et al. 2008). There are even claims that abiotic sources are major contributors to global hydrocarbon reservoirs (Gold 1993; Glasby 2006; Kutcherov and Krayushkin 2010; Sephton and Hazen 2013). While most experts are highly skeptical of such broad claims, it seems possible that at least some petroleum and gas reservoirs could contain hydrocarbons with an abiotic origin.

Conceptually, there are two potential major sources of abiotic hydrocarbons to fluids in Earth's crust. First, abiotic hydrocarbons could migrate to the crust from deeper sources within Earth, through processes such as convective transport, grain boundary diffusion, or release of magmatic volatiles. Second, abiotic hydrocarbons could form *in situ* within the crust through reduction of inorganic carbon sources. Potential substrates for carbon reduction include CO_2 and CO in circulating fluids, and carbon-bearing solids such as carbonate minerals and graphite. In either case, the ultimate source of the inorganic carbon may be primordial (i.e., from the mantle) or recycled from Earth's surface.

This paper summarizes some of the recent laboratory experimental studies conducted to investigate potential pathways for the abiotic formation of organic compounds in subsurface geologic environments. Experimental studies of abiotic organic synthesis are far too numerous for comprehensive coverage in a brief chapter. Therefore, this overview focuses on the formation of methane (CH_4) and other light hydrocarbons, since these are the compounds that have most frequently been attributed to an abiotic origin in natural systems. In addition, only a selected subset of relevant studies is discussed, with the intent of providing a brief overview of current progress rather than an exhaustive review. The discussion intentionally takes a critical perspective; this is done not to disparage the results of any particular study (which in most cases have represented the cutting edge of research on the subject), but to emphasize evolving paradigms and identify directions for future research. Because the mantle and crust provide largely different environments for formation of abiotic organic compounds, they are discussed separately in the following sections.

1529-6466/13/0075-0015$00.00 DOI: 10.2138/rmg.2013.75.15

ABIOTIC HYDROCARBONS IN EARTH'S UPPER MANTLE

The chemical and physical environment of Earth's upper mantle

The stable form of carbon in the deep subsurface is dependent on factors that include temperature, pressure, and the local oxidation state of the system. Petrogaphic studies combined with thermodynamic models of fluids in the C-O-H system indicate that the shallower portions of the upper mantle favor oxidized forms of carbon, while greater depths increasingly favor reduced forms, including graphite, CH_4, and other hydrocarbons (Frost and McCammon 2008; Zhang and Duan 2009). Oxygen thermobarometry of mantle xenoliths indicate that oxygen fugacities in the shallower portions of the upper mantle are buffered to values near the fayalite-magnetite-quartz (FMQ) oxidation state reference benchmark (the oxygen fugacity, f_{O_2}, is in indicator of the relative oxidation state in geologic systems, with higher values indicating more oxidizing conditions and lower values indicative of more reducing conditions; Fig. 1a). At greater depths, however, the rocks become increasingly reducing, with f_{O_2} reaching levels equivalent to several log units below FMQ.

Equilibrium thermodynamic speciation for the C-O-H system at the prevailing oxidation state of mantle rocks indicate that conditions in the shallower parts of the mantle strongly favor CO_2 relative to other carbon species (Fig. 1). Consequently, fluids in equilibrium with rocks of the shallow upper mantle should contain predominantly CO_2, with little or no methane or other forms of reduced carbon. If bulk carbon contents are sufficiently high, the rocks would also contain graphite in equilibrium with the CO_2 (Holloway 1984). The oxidation state of the upper mantle would allow carbonate minerals to be present as well, but these minerals are generally unstable relative to silicate and oxide assemblages at mantle conditions, and in most circumstances would decompose at temperatures well below those present in the mantle (e.g., French 1971; Frost and McCammon 2008). At increasing depths, the relatively more reducing conditions are reflected in the speciation of carbon compounds, with CO_2 giving way to CH_4 and graphite as the predominant stable forms of carbon (Fig. 1). Methane in the fluid at depth is accompanied by molecular hydrogen (H_2) and smaller amounts of light hydrocarbons such as ethane (Fig. 2; e.g., Kenney et al. 2002; Zhang and Duan 2009; Spanu et al. 2011). For a typical subcontinental mantle geotherm, the transition from CO_2- to CH_4-dominated regimes occurs at temperatures between 900 and 1100 °C and pressures of 3.5 to 5 GPa, corresponding to depths of 125-140 km (Fig. 1). The suboceanic mantle is more reducing and has steeper geotherms (Pollack and Chapman 1977; Frost and McCammon 2008), so the transition to CH_4 would occur at somewhat shallower depths.

Experimental studies of hydrocarbons at mantle conditions

Experimental studies of C-O-H fluids demonstrate that carbon speciation is strongly dependent on oxidation state, and that conditions in the shallow upper mantle favor CO_2 as the predominant form of carbon relative to CH_4 and hydrocarbons (Jakobsson and Oskarsson 1990, 1994; Matveev et al. 1997). An example is given in Figure 2, which shows results of an experimental study performed by Matveev et al. (1997) to examine the affect of oxidation state on speciation of fluids in equilibrium with graphite at 1000 °C and 2.4 GPa, equivalent to conditions at depths of about 80 km in the subcontinental mantle (Pollack and Chapman 1977). At the strongly reducing conditions on the left side of the diagram, carbon was present in the experimental fluid predominantly as CH_4 with minor amounts of C_2H_6, accompanied by minor amounts of H_2. At more oxidizing conditions (i.e., towards the right on the diagram), CO_2 begins to displace CH_4 as the predominant carbon species. This transition occurred when the oxidation state present in the experiments was approximately equivalent to the wustite-magnetite (WM) buffer. Although the experiments were performed only to oxidation states slightly more oxidizing than the WM buffer, CO_2 already represented >95% of the carbon in the fluid. The distribution of species observed in the experiments agrees very closely with the

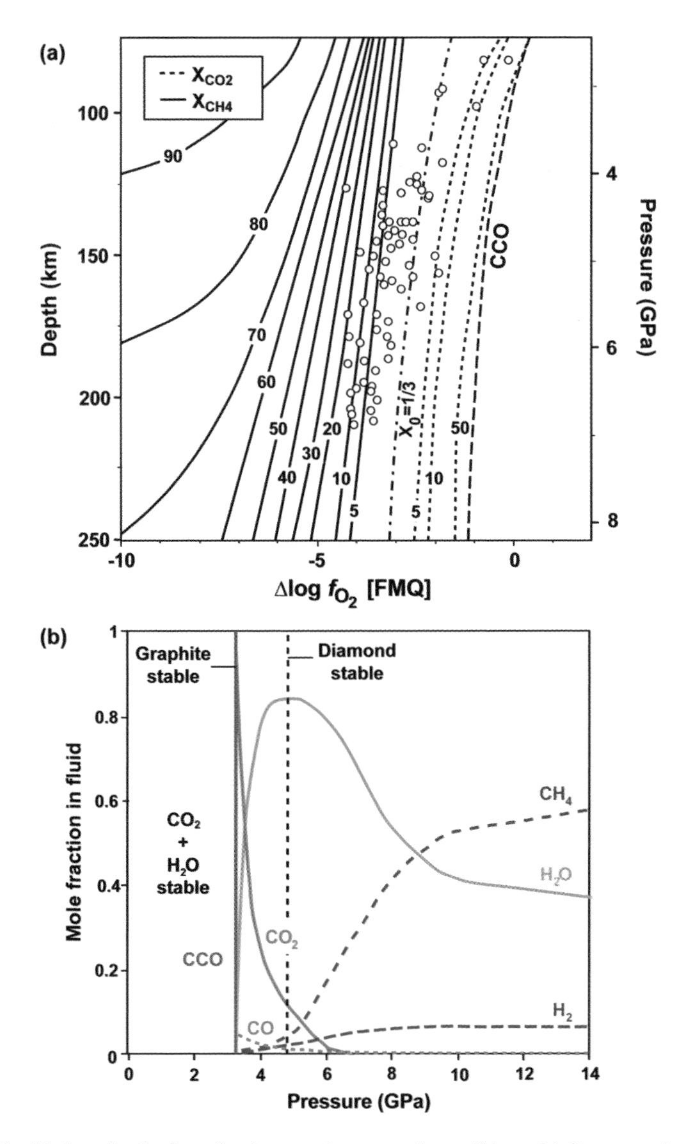

Figure 1. Equilibrium distribution of carbon species at mantle conditions. (a) Contours of mole fraction (X) for CO_2 (dashed lines) and CH_4 (solid lines) in equilibrium with graphite as a function of oxidation state relative to FMQ. Circles correspond to the pressure, temperature, and f_{O_2} values defined by study of mantle xenoliths from South Africa, Canada, and Russia. The line marked "X_O =1/3" defines conditions where O composes 33% of the elemental composition of the C-O-H fluid, and defines the point where CO_2 and CH_4 are in equal proportions. To the right of this line, CO_2 dominates the carbon species and CH_4 dominates to the left. The numbers labeling the lines are the mole percent of the major carbon component in the fluid (CH_4 or CO_2). The diagram is calculated for a mantle geothermal gradient of 45 °C/km. (b) C-O-H fluid speciation along an adiabat in the upper mantle defined by a potential temperature of 1200 °C and f_{O_2} values derived from study of mantle xenoliths. CCO refers to the reaction C(graphite) + O_2 = CO_2. In (a), graphite is unstable and the fluid is dominated by CO_2 to the right of the CCO line, while graphite is unstable to the left of the line in (b). [Diagram (a) modified permission from Elsevier after Zhang and Duan (2009), *Geochim Cosmochim Acta*, Fig. 7a, p. 2099; (b) used with permission of Annual Reviews from Frost and McCammon (2008), *Annu Rev Earth Planet Sci*, Fig. 6a, p. 406].

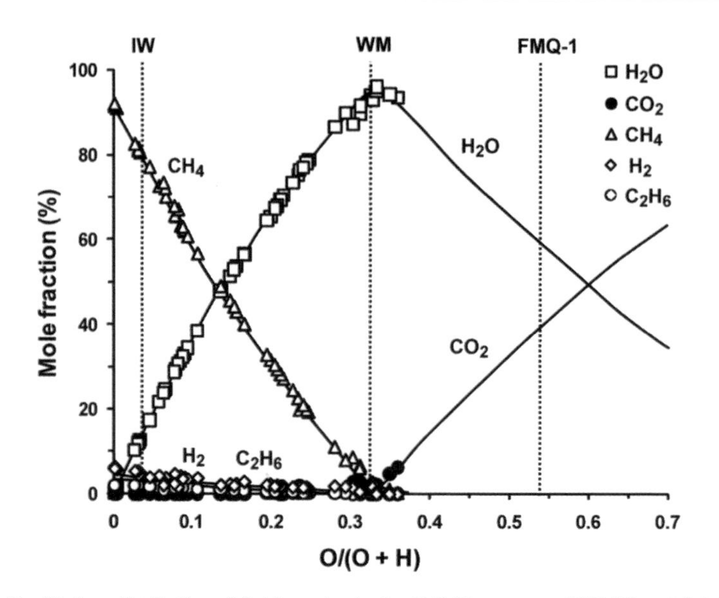

Figure 2. Equilibrium distribution of fluid species in the C-O-H system at 1000 °C and 2.4 GPa as a function of bulk fluid O/(O+H) ratio (a proxy of oxidation state). Symbols show results of laboratory experiments of Matveev et al. (1997), while lines represent equation of state model for the system from Zhang and Duan (2009). Also shown for reference are equivalent O/(O + H) values for the oxidation state buffers IW (iron-wüstite) and WM (wüstite-magnetite), and for an f_{O_2} one log unit below FMQ (FMQ-1). The temperature and pressure of the diagram and an oxidation state near FMQ-1 represent conditions at ~80 km depth in the mantle. [Diagram modified with permission from Elsevier after Zhang and Duan (2009), *Geochim Cosmochim Acta*, Fig. 3, p. 2096,]

predictions of the equilibrium distribution calculated using equation of state (EOS) parameters developed independent of the experiments (Fig. 2; Zhang and Duan 2009).

The experimental results of Matveev et al. (1997) as well as those of Jakobsson and Oskarsson (1990, 1994) demonstrate two key points relevant to the study of hydrocarbons in the mantle. First, the results confirm that oxidation states close to WM or below are required to favor stable formation of hydrocarbons relative to CO_2, although the exact point of transition will vary for temperatures and pressures that differ from the experiments. Since the oxidation state of the mantle, which is buffered by the stable mineral assemblage, is near FMQ at shallow depths and only approaches WM at depths below about 125 km, CH_4 and other light hydrocarbons can only be expected to be stable in the deeper parts of the mantle.

Second, even though the experiments were run for only a few days or less, it appears that equilibrium among the carbon species was achieved. This indicates that equilibration among carbon compounds is rapid at mantle temperatures, and suggests that equilibrium distributions of carbon compounds should be expected for mantle fluids. One caveat, however, is that diamond, which is thermodynamically stable relative to graphite at deeper mantle conditions (Fig. 1), may not be as rapidly reactive as the graphite employed in the experiments. Experimental studies performed at metamorphic conditions in the crust (up to 725 °C and 1 GPa) have indicated that disordered forms of graphite can persist metastably in contact with C–O–H fluids, at least on the timescales of the experiments (Ziegenbein and Johannes 1980; Pasteris and Chou 1998; Foustoukos 2012). Even if such metastable forms of graphite persist at the higher temperatures and longer residence times of the upper mantle, however, it is not likely to have a large impact on the speciation of carbon in fluids in equilibrium with the graphite, since the thermodynamic properties of ordered and disordered forms are probably very similar (see, for example, Fig. 6 of Pasteris and Chou 1998).

In other efforts to investigate formation of hydrocarbons in the mantle, several recent experimental studies have examined the production of CH_4 and other light hydrocarbons when calcite ($CaCO_3$) is exposed to mantle pressures and temperatures under reducing conditions (Kenney et al. 2002; Scott et al. 2004; Chen et al. 2008b; Sharma et al. 2009; Kutcherov et al. 2010). These experiments have been conducted at temperatures of 500-1500 °C, and pressures from 1 to 11 GPa. Ferrous iron-bearing solids including FeO and fayalite (Fe_2SiO_4), or native Fe, have been included among the reactants to provide the reducing conditions necessary to convert the inorganic carbon from calcite to hydrocarbons. Several of these experiments have been conducted using diamond anvil cells (DAC), which offer the distinct advantage that reactions can be monitored *in situ* using techniques such as Raman spectroscopy, X-ray diffraction, and optical microscopy (e.g., Scott et al. 2004).

During heating of the experiments, the calcite and other minerals decompose, and CH_4 is generated as the predominant carbon compound in the fluid. In some cases, minor amount of C_2-C_6 hydrocarbons have been reported as well (Kenney et al. 2002; Sharma et al. 2009; Kutcherov et al. 2010), but CO_2 is not observed. Few details of the solid reaction products of these experiments have been reported, but magnetite (Fe_3O_4) and $Ca(OH)_2$ (or CaO) appear to be the major products, with calcium ferrite possibly present in at least one set of experiments (Scott et al. 2004). Whether graphite was produced in any of these experiments is unclear. Using Raman spectrosropy, Sharma et al. (2009) also identified iron carbonyl among the products in some experiments, and suggested it may be an intermediate in carbon reduction reactions.

The formation of CH_4 in these experiments is consistent with previous experimental results and with thermodynamic expectations that it should be the stable carbon species in the fluid under reducing conditions at the elevated temperature and pressures of the experiments (Fig. 2; Jakobsson and Oskarsson 1990; Matveev et al. 1997; Zhang and Duan 2009). While none of the experiments provide sufficient information to determine the oxidation state of the system at experimental conditions, the use of FeO and Fe as reactants together with formation of magnetite as a reaction product suggests that most of the carbonate experiments resulted in oxidation states that were near the wüstite-magnetite buffer. The production of C_2H_6 and higher hydrocarbons in the carbonate decomposition experiments as well as in other experimental studies using other carbon sources (Fig. 2; Jakobsson and Oskarsson 1990; Matveev et al. 1997) is also consistent with thermodynamic considerations, which indicate that minor amounts of C_{2+} hydrocarbons should accompany CH_4 at equilibrium in C-O-H fluids at mantle conditions (e.g., Zhang and Duan 2009; Spanu et al. 2011). For instance, the reaction:

$$2CH_4 \rightarrow C_2H_6 + 2H_2 \tag{1}$$

demands that C_2H_6 be present at equilibrium at a finite level consistent with the law of mass action for the reaction:

$$\log K = \log a_{C_2H_6} + 2\log f_{H_2} - 2\log f_{CH_4} \tag{2}$$

where K is the equilibrium constant and f_X is the fugacity of species X. Thermodynamic constraints indicate that C_2H_6 should compose up to several percent of the carbons species under reducing conditions at mantle temperatures (Fig. 2). Although speciation calculations like those shown in Figure 2 have not yet considered hydrocarbons larger than C_2H_6 for mantle conditions, it is likely that these compounds should be present at equilibrium in diminishing amounts with increasing carbon number (Kenney et al. 2002).

Results of the carbonate decomposition experiments provide additional evidence for the rapidity of the reduction of inorganic carbon to CH_4 at mantle conditions. Methane production was observed in these experiments at reaction times lasting only minutes to hours and, although an evaluation of equilibrium status was not considered in the carbonate decomposition studies, it appears likely that equilibrium of the system was attained on the timescales of the experiments. Similar results are obtained when graphite is substituted for calcite as the carbon source (Sharma et al. 2009; Kutcherov et al. 2010), indicating that graphite can also equilibrate with reduced carbon species in the fluid at short timescales for mantle conditions.

The use of carbonate minerals in the experiments described above is probably best viewed as a means of introducing a source of C and O to the system rather than a literal indication that carbonate decomposition per se contributes to hydrocarbon formation in the mantle. Indeed, it is not immediately apparent where carbonate minerals would be present in the mantle under conditions that would be sufficiently reducing for this to occur. For typical mantle geotherms, carbonate decomposition occurs at pressures of ~3 GPa or less (Frost and McCammon 2008), and the oxidation state of the system favors their decomposition to CO_2 rather than CH_4 and other hydrocarbons (Fig. 1). Carbonates may penetrate to greater depths in deeply subducting slabs (e.g., Kerrick and Connolly 2001), but the oxidation states in subducting slabs tend to be even more oxidizing than those elsewhere in the mantle (Frost and McCammon 2008).

Other recent experiments have demonstrated that CH_4 exposed to mantle conditions will spontaneously polymerize to form C_{2+} hydrocarbons (Chen et al. 2008a; Kolesnikov et al. 2009). These experiments were performed in diamond anvil cells with *in situ* detection of hydrocarbon products by Raman spectroscopy. For example, Koleshnikov et al. (2009) observed formation of ethane, propane, and butane during heating of pure CH_4 to temperatures in the 1000-1500 K (727-1227 °C) range at pressures above 2 GPa, with ethane most abundant (Fig. 3). Other

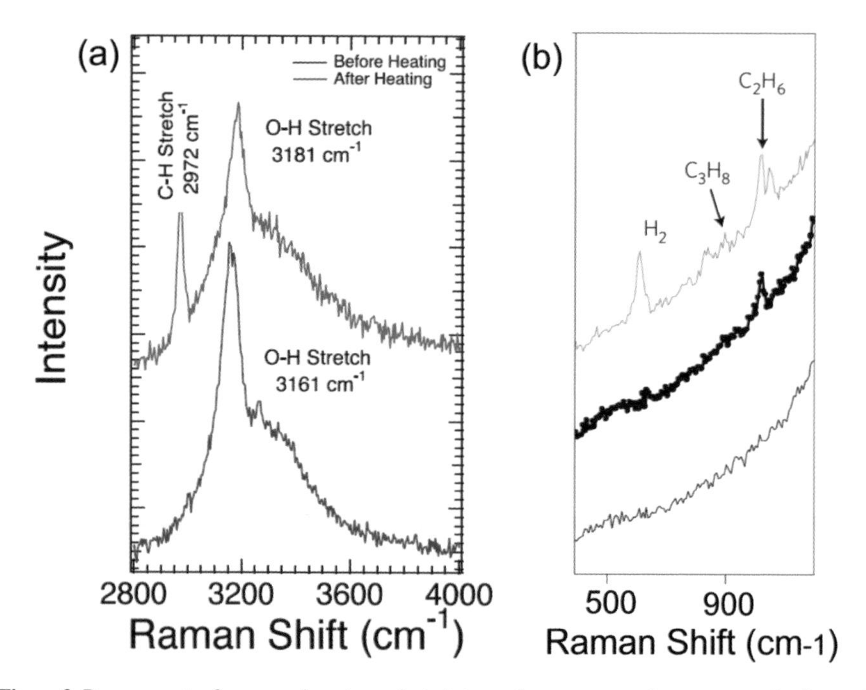

Figure 3. Raman spectra from experiments conducted at mantle pressures and temperatures in diamond anvil cells. (a) Spectrum obtained during heating of calcite, FeO, and water at 1,500 °C and 5.7 GPa (Scott et al. 2004). The peak at 2972 cm^{-1} corresponds to a C-H vibration stretching mode, and indicates formation of CH_4 during heating. (b) Spectrum obtained during heating of pure methane at 2 GPa at ~300 K (bottom line), ~900 K (middle) and ~1,500 K (top). [(b) is reprinted by permission from Macmillan Publishers Ltd: Nature Geosciences from Kolesnikov et al. (2009), *Nature Geosci*, Vol. 2, Fig. 2a, p. 567].

products included H_2 and graphite. Experiments performed by heating pure ethane at similar conditions produced CH_4, H_2, and graphite. As noted by Koleshnikov et al. (2009), at the lower end of the temperature range studied, the experiments were significantly more reducing than mantle conditions, but converge towards the oxidation state of mantle rocks at higher temperatures (see their Fig. 3a). Consequently, the experimental conditions are most directly applicable to the deeper parts of the upper mantle.

The results of Koleshnikov et al. (2009) are consistent with thermodynamic expectations that small amounts of higher hydrocarbons should coexist with CH_4 at mantle temperatures and pressures if conditions are sufficiently reducing. However, equilibrium states can only be approached if suitable reaction pathways exist that are not kinetically inhibited, and these experiments further demonstrate that reactions among carbon species proceed very rapidly under such conditions, even when the reactions involve formation of carbon-carbon bonds and reaction with graphite. Although the reaction mechanism was not determined, the experiments also suggest that formation and polymerization of methyl radicals might be one pathway for the formation of higher hydrocarbons in the deep subsurface.

Implications for mantle sources of hydrocarbons

The experimental studies summarized above demonstrate that reactions among carbon species proceed very rapidly at mantle temperatures and pressures. Furthermore, comparison of predicted equilibrium distributions of carbon species based on thermodynamic models agree closely with the relative abundances of compounds observed in laboratory experiments (e.g., Fig. 2). Together, these results indicate that the distribution of carbon compounds in the mantle under most circumstances will be controlled by chemical equilibrium. Since the equilibrium distribution of carbon species is strongly dependent on the oxidation state, the prevailing oxidation state of the mantle, which is buffered by reactions among minerals, will exert a dominant influence on which carbon compounds are present. Given the prevailing mantle oxidation state, it appears likely that CH_4 and hydrocarbons should exist in mantle rocks, but only in those rocks at depths greater than ~110-125 km (for a continental geotherm) where conditions are sufficiently reducing to allow these compounds to be stable (Fig. 1). For a suboceanic geotherm, sufficiently reducing conditions to favor the formation of CH_4 and hydrocarbons may occur at slightly shallower depths.

These considerations suggest that one of three sets of circumstances may be required for mantle sources to contribute to CH_4 and other hydrocarbons found in crustal fluids. First, hydrocarbons from deep within the mantle could be transported to near-surface environments under circumstances that prevent their re-equilibration at the relatively less reducing conditions of the shallow mantle. This transport might occur, for instance, during rapid migration of fluids in concert with migrating rocks from the deep mantle. The occurrence of diamond-bearing xenoliths in near-surface rocks that were erupted from depths of up to 150 km in the mantle (Shirey et al. 2013) suggests that conduits for relatively rapid transport may exist, although the distribution of such conduits appears to have been spatially and temporally limited during Earth's history. To date, there have been no experimental studies that have examined oxidation of methane and other reduced carbon compounds that would provide quantitative constraints on the rates of migration required to prevent re-equilibration. Second, mantle hydrocarbons could arise in shallower portions of the mantle that are anomalously more reducing than typical values (Fig. 1). For instance, deep subduction of organic-rich sediments might create pockets of rocks whose bulk composition is depleted in O relative to typical mantle, leading to a lower overall oxidation state within the rocks. A third possibility is that hydrocarbons might form and persist in disequilibrium with surrounding rocks in the shallow upper mantle. However, given the rapidity with which carbon compounds equilibrate at mantle temperatures in laboratory experiments, this seems like a rather remote possibility.

It has been suggested that the co-occurrence of CH_4 and other hydrocarbons with mantle-derived helium (i.e., gases with high $^3He/^4He$ ratios) in some hydrothermal fluids and natural gas reservoirs is an indication that the CH_4 has a mantle source (e.g., Welhan and Craig 1983; Gold 1993; see also Jenden et al. 1993; Sephton and Hazen 2013; and references therein). If so, the hydrocarbons would probably have to either come from deep within the mantle, or be present as trace constituents of fluids from the shallow mantle where CO_2 was the predominant carbon species. However, an alternative explanation is that the CH_4 and other hydrocarbons migrated out of the mantle as CO_2, and were then converted to hydrocarbons by reactions within the crust (see following section) (Jenden et al. 1993; McCollom 2008). For instance, this appears to be the source of methane-rich gases associated with high ^3He levels in serpentinized ultramafic rocks (Abrajano et al. 1990; Proskurowski et al 2008), as well as CH_4 and other hydrocarbons in volcanic gases (Fiebig et al. 2007, 2009).

ABIOTIC HYDROCARBON FORMATION IN CRUSTAL ENVIRONMENTS

Chemical and physical environments for hydrocarbon formation in the crust

The prevailing oxidation state of source regions of magmas in the upper mantle dictates that carbon speciation in pristine magmatic-derived fluids should be dominated by CO_2, with very little CH_4 present (e.g., Mathez 1984; Kelley 1996). This observation appears to be consistent, for example, with chemical analyses of magmatic volatiles trapped in vesicles within seafloor basalts, which are characterized by very high CO_2/CH_4 ratios (e.g., Pineau and Javoy 1983). At the same time, oxidizing conditions at Earth's surface ensure that CO_2 and bicarbonate are the predominant forms of dissolved carbon in seawater, in fracture-filling groundwater, and in shallow pore waters. As a consequence, the predominant inputs of carbon to Earth's crust from both above and below are in highly oxidized forms. Accordingly, any hydrocarbons found in fluids circulating in Earth's crust that do not derive from deep within in the mantle or from biologic sources must therefore be formed by non-biological reduction of inorganic carbon within the crust itself.

In general, two sets of conditions favor the reduction of inorganic carbon to hydrocarbons within the crust (McCollom and Seewald 2007; McCollom 2008). At the elevated temperatures and oxidation states that prevail in environments deep within the crust, CO_2 is thermodynamically stable relative to CH_4. However, decreasing temperatures increasingly favor the stability of CH_4 relative to CO_2. This trend is illustrated in Figure 4a, which shows that the log K for the reaction:

$$CO_2 + 4H_2 \leftrightarrow CH_4 + 2H_2O \tag{3}$$

becomes increasingly positive with decreasing temperature, indicating that lower temperatures favor the compounds on the right side of the reaction relative to those on the left side. As shown in Figure 4b, this means that CO_2 predominates relative to CH_4 at equilibrium for high temperatures (>~200 to 350 °C, depending on oxidation state), but CH_4 is the more thermodynamically stable compound at lower temperatures. Similar relationships can be shown for other hydrocarbons relative to CO_2. As a consequence, cooling of high-temperature fluids that contain dissolved CO_2 and H_2 will thermodynamically favor reduction of the CO_2 to CH_4 and other hydrocarbons (e.g., Shock 1990, 1992).

A second set of circumstances that can promote reduction of inorganic carbon within Earth's crust is the generation of reducing environments through fluid-rock interactions. These interactions are typically manifested by increasing abundances of H_2 as fluid-rock interactions proceed, shifting the equilibrium of reactions like the one written above towards the compounds on the right side. As a consequence, the predominant equilibrium carbon species can shift from CO_2 to CH_4 as fluids interact with rocks, even without a change in temperature.

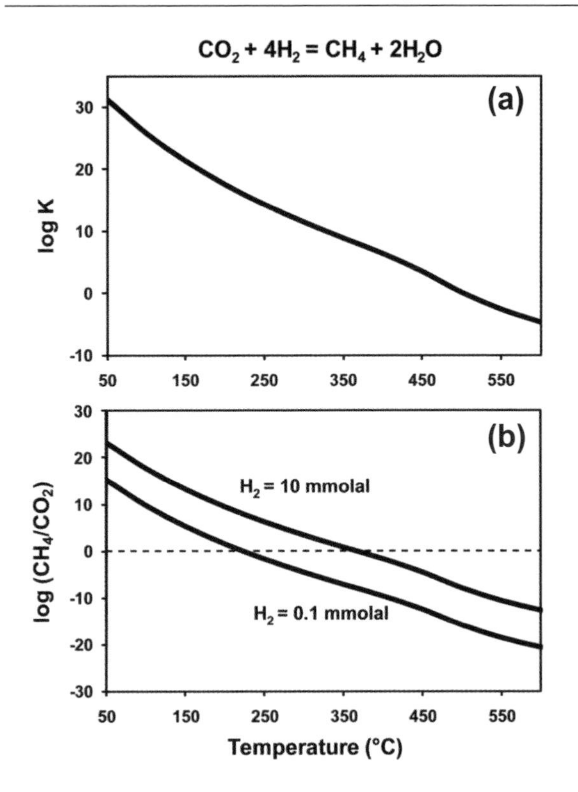

Figure 4. Thermodynamic relationships between dissolved CO_2 and CH_4 at elevated temperature and pressure. (a) Log *K* for reduction of CO_2 to CH_4. (b) Calculated equilibrium (CH_4/CO_2) ratios as a function of temperature at two values of H_2 concentration that bracket those found in reducing environments within the crust. Data shown are for a pressure of 50 MPa. Values for log *K* calculated using thermodynamic data from SUPCRT92 (Johnson et al. 1992) and Shock et al. (1989).

Because the interaction of aqueous fluids with ultramafic rocks is known to generate particularly large amounts of H_2, fluids circulating through these rocks have become the focus of many studies of abiotic hydrocarbon formation. For example, hydrothermal fluids circulating through ultramafic rocks below the seafloor have measured H_2 concentrations up to 15 mmol/kg, and evidence suggests that light hydrocarbons in these fluids have an abiotic origin (Charlou et al. 2002, 2010; Kelley et al. 2005; Proskurowski et al. 2008). Hydrous alteration of ultramafic rocks, which are composed predominantly of the minerals olivine and pyroxene, is known as serpentinization, owing to precipitation of the mineral serpentine as the primary alteration product (Schrenk et al. 2013). Serpentinization can be summarized by the general reaction:

$$\underset{\textit{Olivine } (Fo_{90})}{Mg_{1.8}Fe_{0.2}SiO_4} + aH_2O \rightarrow \underset{\textit{Serpentine}}{0.5(Mg,Fe)_3Si_2O_5(OH)_4} + \underset{\textit{Brucite}}{x(Mg,Fe)(OH)_2} + \underset{\textit{Magnetite}}{yFe_3O_4} + zH_2 \qquad (4)$$

The exact stoichiometry of this reaction, and thus the amount of H_2 generated, is dependent on a number of factors that affect partitioning of Fe among the reaction products, including temperature, rock composition, and water:rock ratio (e.g., Seyfried et al. 2007; McCollom and Bach 2009; Marcaillou et al. 2012). While ultramafic rocks have become a focal point for studies of abiotic hydrocarbon formation, serpentinization is by no means the only fluid-rock reaction that can produce sufficiently reducing conditions to favor carbon reduction, and many other rock types and reactions may be involved in H_2 production within Earth's crust (e.g., Charlou et al. 1996; Potter et al. 2004; Sherwood Lollar et al. 2006, 2007).

Although thermodynamic factors can favor reduction of inorganic carbon to hydrocarbons under circumstances like those outlined above, kinetic inhibitions can still prevent the reactions from occurring. In contrast to the experimental results obtained at mantle temperatures, which show rapid equilibration among carbon species, reactions involved in the reduction of inorganic carbon at temperatures and pressures relevant to environments in Earth's crust are susceptible to kinetic inhibitions (e.g., Seewald et al. 2006). Laboratory experiments provide a means to evaluate which conditions within the crust can allow carbon reduction to proceed. The remainder of this section describes some of the experimental studies performed in recent years to investigate potential reaction pathways for carbon reduction under crustal conditions. Since a detailed review of experimental studies of abiotic organic synthesis under geological conditions was recently published (McCollom and Seewald 2007), only the most salient results are summarized here, and readers interested in a more in depth discussion are referred to the earlier review.

Fischer-Tropsch-type synthesis

The most widely invoked pathway for the formation of hydrocarbons and other organic compounds in geologic environments is the Fischer-Tropsch synthesis. Accordingly, this process has also received the greatest attention in experimental studies. As originally described, Fischer-Tropsch synthesis refers to the surface-catalyzed reduction of CO by H_2 in gas mixtures. However, the term is often used in a broader context in the geological literature to refer to reduction of an inorganic carbon source to form organic compounds, regardless of the nature of the carbon source, the medium in which the reaction occurs, or the identity of the reductant. In many cases, dissolved CO_2 is inferred to be the primary carbon source for abiotic organic synthesis in geologic systems. We will follow the geologic convention here, and use the term Fischer-Tropsch-type (FTT) synthesis to refer in general to any surface-catalyzed reduction of an inorganic carbon source to organic matter.

During FTT synthesis, CO or CO_2 is reduced to organic compounds through a series of steps on the surface of a catalyst (Fig. 5). Typically, the primary products of the reaction are CH_4 and a homologous series of linear alkanes that show a regular decrease in abundance with increasing number of carbons (Fig. 6). However, the process also generates lesser amounts of other compounds, including alkenes, branched hydrocarbons, and oxygen-bearing compounds including alkanols and alkanoic acids (e.g., Anderson and White 1994; McCollom et al. 1999). If a source of nitrogen is present, the synthesis can also generate amino acids, amines, and other N-bearing compounds (Hayatsu et al. 1968; Yoshino et al. 1971).

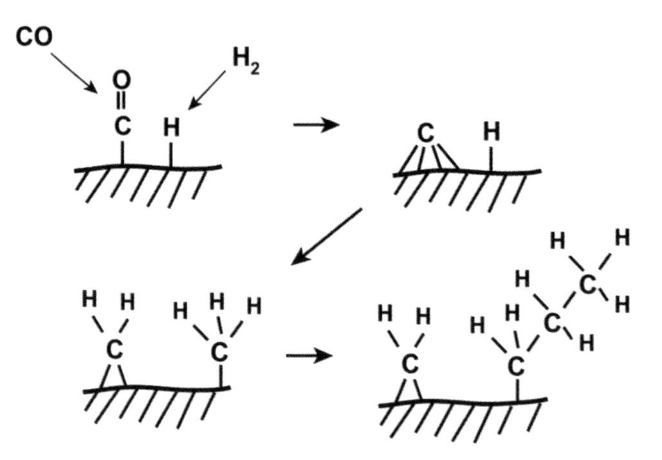

Figure 5. Generalized reaction mechanism for Fischer-Tropsch synthesis of hydrocarbons. The reaction is initiated with binding of CO to the catalyst surface to form a carbonyl unit (–CO), which then undergoes sequential reduction to surface-bound carbide (–C), methylene (–CH_2), and methyl (–CH_3) groups. Chain growth occurs as methylene groups polymerize to one another, and terminates when the growing chain combines with a methyl group or surface-bound H rather than another methylene.

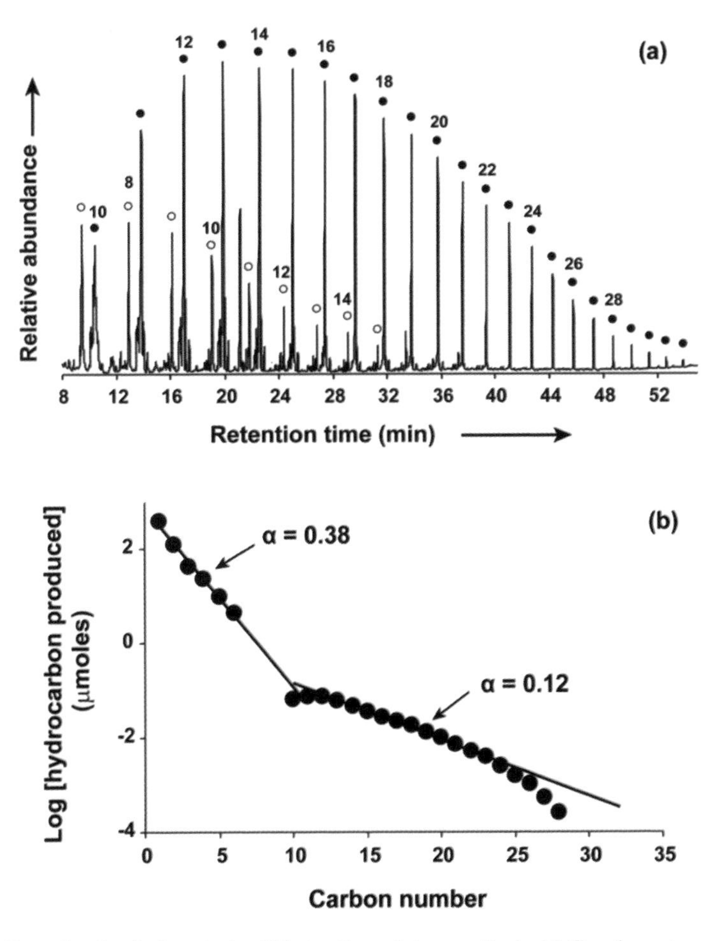

Figure 6. Example of typical products of Fischer-Tropsch-type synthesis. (a) Gas chromatogram of non-volatile products, showing predominance of linear saturate hydrocarbons (*n*-alkanes; filled circles) and alcohols (*n*-1-alkanols; open circles). (b) Relative abundance of saturated hydrocarbons (*n*-alkanes) as a function of carbon number, showing typical log-linear decrease. The slope of the decrease (α) is a gauge of the relative probabilities of continued growth of the carbon chain versus chain termination. The break in slope at carbon number around 10 is common for iron-based catalysts. Figure shows data from McCollom and Seewald (2006).

The Fischer-Tropsch reaction was originally developed as a means of converting coal-bed gases to petroleum, and over the years it has been the subject of hundreds of experimental studies directed at optimizing yields of hydrocarbons and other industrial products. Unfortunately, most of this vast literature has no clear relevance to the study of formation of carbon compounds within Earth because the reaction conditions are not directly comparable to geologic environments. For instance, industrial Fischer-Tropsch experiments are often performed in dry gas mixtures with no water present (except for the small amounts produced as a by-product of the reaction). In contrast, FTT synthesis in geologic systems is often inferred to take place in environments where liquid or supercritical water is the medium for the reactions. Industrial reactions also generally rely on purpose-designed synthetic catalysts that may or may not resemble mineral phases present in natural systems.

The challenge for experimental geochemists is to understand the extent to which FTT reactions can proceed at conditions that more closely resemble subsurface geologic environments. One key question is whether FTT type reactions can proceed in environments where reactants are dissolved in sub- or supercritical aqueous fluids. Another critical question is which naturally occurring minerals, if any, are effective in catalyzing the reaction, and under what circumstances. Some catalysts used in industrial Fischer-Tropsch synthesis are found in natural systems, but for industrial purposes their catalytic properties are typically enhanced in ways that may not occur in natural settings. For example, magnetite has long been employed as a catalyst for industrial Fischer-Tropsch studies, but it is usually pre-treated with a stream of H_2 and CO prior to use in these applications. This process generates pockets of highly reactive native Fe or Fe-carbides on the surface that appear to be active sites for catalysis (e.g., Dictor and Bell 1986; Satterfield et al. 1986). These kinds of sites would likely be destroyed very rapidly in natural systems, especially if H_2O is present. To address these issues, recent experimental studies concerning the potential contribution of FTT synthesis to hydrocarbon occurrences in geologic systems have largely focused on exploring the capacity for naturally occurring minerals to catalyze the reaction, and on evaluating the effectiveness of the reactions under hydrothermal conditions.

Several recent laboratory experiments indicate that FTT synthesis can indeed proceed readily under hydrothermal conditions in some circumstances (e.g., McCollom et al. 1999, 2010; McCollom and Seewald 2006). In these studies, CO or formic acid (HCOOH) and water were heated to temperatures of 175 or 250 °C and pressures ranging from steam-saturation to 25 MPa, with H_2 supplied by decomposition of formic acid ($HCOOH \rightarrow CO_2 + H_2$) or of native Fe included in the reaction vessel ($Fe + H_2O \rightarrow Fe_3O_4 + H_2$). Over periods of hours to days, several percent of the carbon is reduced to typical Fischer-Tropsch reaction products. These products include CH_4 and other light hydrocarbons, as well as long-chain *n*-alkanes, alkanols, and alkanoic acids (Fig. 6). The products exhibit a regular decrease in abundance with increasing carbon number that is characteristic of Fischer-Tropsch products.

While these experiments show that FTT synthesis is not necessarily inhibited by hydrothermal conditions, a couple of considerations may limit their direct applicability to natural systems. First, the synthesis in these experiments was probably catalyzed either by native Fe included in the reaction vessel, or by the walls of the steel tube used in some of the experiments. Thus, they may not represent catalysts present in natural environments. Second, it appears likely that the synthesis reactions took place either in the vapor headspace for reactions performed in fixed-volume tube reactors (McCollom et al. 1999) or in H_2-rich vapor bubbles formed on the surfaces of the solids in reactions performed at higher pressures (McCollom and Seewald 2006; McCollom et al. 2010). Consequently, while these experiments show that the presence of water-saturated vapors do not preclude efficient FTT synthesis, they do not indicate that the reactions can proceed for compounds dissolved in aqueous liquid. In addition, H_2 was also present in these experiments at very high levels (>200 mmol/kg) that are probably rarely approached in natural environments within Earth's crust.

The prospect that abiotic hydrocarbons might form during serpentinization of ultramafic rocks led to experimental investigation of the capacity for minerals found in serpentinites to catalyze carbon reduction reactions. The first study to focus on this possibility was that of Berndt et al. (1996), who monitored the production of light hydrocarbons during reaction of Fe-bearing olivine with an aqueous solution containing dissolved bicarbonate at elevated temperature and pressure (300 °C, 50 MPa). The experiment utilized a flexible-cell reaction apparatus, which allows reactions to proceed without a vapor phase present and also provides a means to monitor concentrations of compounds dissolved in the fluid as reactions proceed. Dissolved concentrations of H_2 and several light hydrocarbons (CH_4, C_2H_6, and C_3H_8) were observed to increase steadily during the experiment as the serpentinization reaction progressed, while the concentration of total dissolved CO_2 ($CO_{2,aq} + HCO_3^-$) declined (Fig. 7). The authors interpreted the small amounts of hydrocarbons generated over the course of the experiment to represent products of Fischer-Tropsch synthesis through reduction of dissolved CO_2. Magnetite formed in the experiments as a product of serpentinization (Eqn. 4) was suggested to be the catalyst. Although Berndt et al. (1996) also reported the presence of bi-lobed, carbonaceous particles among the solid reaction products that were interpreted to be high-molecular weight products of FTT synthesis, this claim has since been retracted (see Geology 1996, p. 671).

The groundbreaking results of Berndt et al. (1996) suggested that reduction of dissolved inorganic carbon to hydrocarbons could proceed readily with minerals common in hydrothermal systems serving as catalysts. Further investigation, however, showed that hydrocarbon formation at the conditions of their experiments was much more limited than initially thought. McCollom and Seewald (2001) performed an experiment under essentially identical conditions to those of Berndt et al. (1996), except that ^{13}C-labeled bicarbonate (99% $H^{13}CO_3^-$) was substituted as the inorganic carbon source in order to trace the origin of carbon in the hydrocarbon products. While the experiment yielded similar amounts of H_2 and C_1-C_3 hydrocarbons to those reported by Berndt et al. (1996), isotopic analysis of the hydrocarbon products indicated that only a small fraction of the CH_4 contained the ^{13}C label (2-15%), while none of the C_2H_6 or C_3H_8 was labeled. This result indicated that, except for a small fraction of the CH_4, the C_1-C_3 hydrocarbons generated in the experiments were not the product of reduction of dissolved CO_2, but were instead generated from thermal decomposition of other sources of reduced carbon already present among the reactants at the start of the experiment.

It is worth emphasizing that, at the H_2 concentrations attained in both of these experiments (up to 158 mmol/kg; Fig. 7), reduction of inorganic carbon to CH_4 and other hydrocarbons was strongly favored by thermodynamics, and essentially all of the dissolved CO_2 present should have been converted to CH_4 to attain equilibrium (Fig. 4). Yet, only a very small fraction (<<1%) of the available carbon was reduced to CH_4, even after nearly three months of heating at 300 °C. Thus, the results clearly demonstrated that reduction of dissolved inorganic carbon to light hydrocarbons is kinetically sluggish even at 300 °C. Furthermore, although magnetite

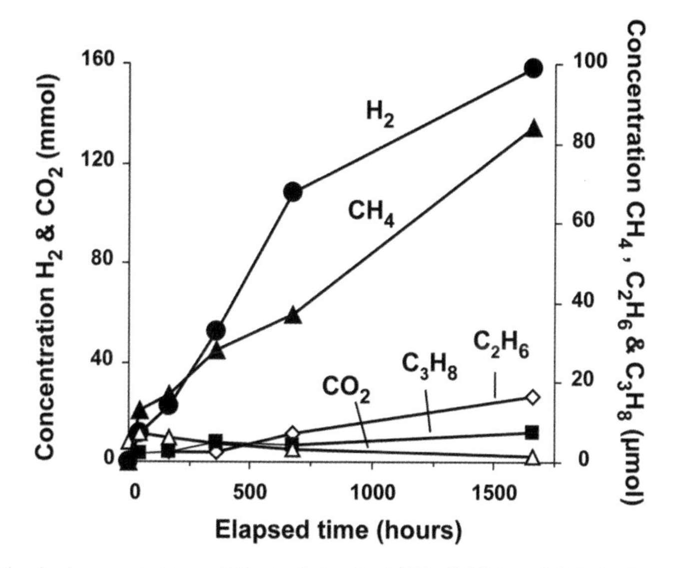

Figure 7. Dissolved concentrations of H_2, total dissolved CO_2 (ΣCO_2), and light hydrocarbons during reaction with serpentinized olivine in the experimental study of Berndt et al. (1996). Concentrations in mmol kg^{-1} (mmol) or mmol kg^{-1} (mmol).

was formed in abundance in both experiments as a product of serpentinization (Eqn. 3), the lack of significant hydrocarbon formation suggests that it is not a very effective catalyst for reduction of dissolved inorganic carbon in natural systems. Otherwise, a much larger fraction of the inorganic carbon present in the experiments would have been converted to CH_4 or other organic compounds.

In a series of follow-up experiments also using ^{13}C-labeled carbon sources, it was found that dissolved CO_2 (or HCO_3^-) was rapidly reduced by H_2 to HCOOH and CO under hydrothermal conditions (McCollom and Seewald 2003; Seewald et al. 2006). In contrast to CH_4, these compounds attained thermodynamic equilibrium within a few days, even at temperatures as low as 175 °C and in the absence of mineral catalysts. Methanol was also found to accumulate, but did not reach equilibrium proportions. As in the serpentinization experiments, ^{13}C-labeled CH_4 was produced in small amounts (mmolar concentrations) during these experiments, along with larger amounts of unlabeled CH_4 and other light hydrocarbons. Overall, the experimental results demonstrated that partial reduction of dissolved inorganic carbon proceeds rapidly and spontaneously to HCOOH and CO under hydrothermal conditions, but complete reduction to CH_4 proceeds only very slowly without catalysis.

Several other experimental studies have focused on evaluating the catalytic potential of individual minerals found in serpentinites, including NiFe-alloys, magnetite, chromite, and Ni-sulfide (pentlandite). Awaruite, a Ni-Fe alloy with compositions between Ni_2Fe and Ni_3Fe, occurs in many serpentinites as a by-product of the highly reducing conditions that can develop as a result of H_2 production during serpentinization (Frost 1985; Klein and Bach 2009). Horita and Berndt (1999) examined the reduction of dissolved CO_2 in the presence of NiFe-alloy, magnetite, water, and highly elevated H_2 concentrations (170-300 mmol/kg). Experiments were conducted at temperatures of 200-400 °C and 50 MPa. In contrast to the small amounts of CH_4 produced in the earlier experiments of Berndt et al. (1996), nearly complete conversion of CO_2 to CH_4 was observed in the 300 °C experiments in two weeks or less, and >40% conversion was observed after 3 months of reaction at 200 °C. Moreover, rates of conversion were found to increase when greater amounts of NiFe-alloy were included, and no CH_4 was generated in a control experiment in the absence of the alloy, demonstrating that the NiFe-alloy had catalyzed the production of CH_4. No C_2H_6 or C_3H_8 were found in the experiment, indicating that the catalysis exclusively promoted CH_4 synthesis. Lesser amounts of conversion of CO_2 to CH_4 were observed at 400 °C, which was attributed to passivation of the catalyst at higher temperature.

The experiments of Horita and Berndt (1999) provided the first documentation that the kinetic inhibitions to reduction of dissolved CO_2 to CH_4 could be effectively overcome by naturally occurring minerals, and that NiFe-alloy was a very effective catalyst in promoting the reaction. The experiments also showed that pure CH_4 could be produced by an abiotic process in hydrothermal environments, which was contrary to the widely held notion that only methanogenic microorganisms were capable of producing nearly pure CH_4 in geologic systems (e.g., Whiticar 1990). Although the alloy present in the experiments of Horita and Berndt (1999) was significantly enriched in Ni ($Ni_{50}Fe$ to $Ni_{10}Fe$) relative to awaruite found in natural serpentinites, there is no obvious reason to doubt that natural awaruite would have similar catalytic properties.

The catalytic potential of chromite and magnetite has been investigated in experiments by Foustoukos and Seyfried (2004). Two experiments were performed under hydrothermal conditions at 390 °C and 40 MPa, one containing chromite plus magnetite and the other including only magnetite, with both minerals synthesized from Fe and Cr oxides. A ^{13}C-labeled inorganic carbon source was used (99% $H^{13}CO_3^-$) to assess the origin of carbon in reaction products, and high levels of H_2 (>100 mmol/kg) provided conditions favorable for carbon reduction (Fig. 4). During heating for 44 to 120 days, $^{13}CH_4$ was observed to accumulate at concentrations up to 192 mmol/kg. Much smaller, but detectable, levels of ^{13}C-labeled C_2H_6 and

C_3H_8 were also observed. In each case, the labeled compounds represented only a small fraction of the total concentrations of C_1-C_3 hydrocarbons observed in the experiments.

A key finding of this study was that C_{2+} hydrocarbons could be produced from reduction of dissolved reactants, demonstrating that higher hydrocarbons as well as CH_4 could be generated without a vapor phase present in subsurface and hydrothermal environments. Although yields were low (<0.1% of added inorganic carbon), the concentrations attained were comparable to those of hydrocarbons in deep-sea hydrothermal systems thought to have an abiotic origin (e.g., Proskurowski et al. 2008; Charlou et al. 2010). The results show that it is possible for light hydrocarbons to be generated from reduction of inorganic carbon in subsurface environments even when there is no vapor phase present. Owing to the high levels of unlabeled hydrocarbons in the experiments, which apparently derived from traces of background reduced carbon included among the reactants, the use of labeled carbon source was a critical factor in the authors' ability to demonstrate that the hydrocarbons were derived from reduction of inorganic carbon.

Foustoukos and Seyfried (2004) inferred that the minerals catalyzed formation of the hydrocarbons and, based on the observation that yields of labeled hydrocarbons were higher in the experiment that contained chromite than in the experiment that contained only magnetite, the authors inferred that chromite was more effective than magnetite in promoting the hydrocarbon formation. However, subsequent studies have not observed any catalytic effect for the reduction of dissolved CO_2 by naturally occurring chromites (Lazar et al. 2012; Oze et al. 2012), and alternative explanations for the differences observed by Foustoukos and Seyfried (2004) are possible. In particular, the chromite-bearing experiment contained higher H_2 than the magnetite-only experiment (~220 mmol/kg vs. ~120 mmol/kg) and was also performed at a substantially lower pH (4.8 vs. 8.8, which changes the predominant carbon species from $CO_{2(aq)}$ to HCO_3^-), either of which may have affected the rate of reduction of inorganic carbon to hydrocarbons independent of any mineral catalysis. As a consequence, the capacity for chromite to catalyze the reduction of inorganic carbon to CH_4 and other light hydrocarbons under natural hydrothermal conditions remains uncertain.

Because the physical properties of water undergo substantial changes near the critical point (404 °C, 29 MPa for seawater salinity; Bischoff and Rosenbauer 1985), the near-critical conditions of the Foustoukos and Seyfried (2004) experiments might have had a role in allowing the reduction of carbon to occur. However, a more recent study by Ji et al. (2008) reported the reduction of dissolved $^{13}CO_2$ to C_1-C_5 hydrocarbons in experiments at 300 °C and 30 MPa with a cobalt-enriched magnetite catalyst. The authors reported isotopic analyses only for the C_3-C_5 hydrocarbons, but the results showed that at least 50% of carbon in linear alkanes was ^{13}C, while branched C_4 and C_5 alkanes contained none of the added ^{13}C. Yields of hydrocarbons in these experiments were an order of magnitude greater than those of Foustoukos and Seyfried (2004), although uncertainties in the abundance and isotopic composition of CO_2 and CH_4 preclude direct comparisons. In any case, the results of Ji et al. (2008) provide further evidence that synthesis of at least short-chain hydrocarbons is possible in the absence of a vapor phase.

The potential of magnetite to promote abiotic production of hydrocarbons under hydrothermal conditions was further investigated by Fu et al. (2007), who heated solutions of CO_2 and H_2 in the presence of magnetite at 400 °C and 50 kPa. In order to avoid background contributions of hydrocarbons from magnetite, the minerals were scrupulously treated prior to the experiments to reduce carbon contents, and control experiments without an added carbon source did not generate detectable levels of CH_4 or other light hydrocarbons. In contrast, when a carbon source was injected into the experiments (as CO_2 or HCOOH) together with H_2, concentrations of CH_4, C_2H_6, or C_3H_8 were found to increase over time. On the order of 0.2-0.3% of the inorganic carbon added to experiments was converted to hydrocarbons, comparable to the conversion amounts observed by Foustoukos and Seyfried (2004) at similar pressure and temperature. Based on a carbon imbalance in their experiments, Fu et al. (2007) also suggested that a large fraction of the inorganic carbon reactant was converted to additional unidentified organic products, although no supporting evidence was provided.

While the experiments of McCollom and Seewald (2001), Fu et al. (2007), and Foustoukos and Seyfried (2004) resulted in the synthesis of CH_4 and, in some case, other light hydrocarbons in the presence of magnetite, it must be pointed out that none of these studies provide definitive evidence that magnetite actually catalyzed the reactions. Using similar methods, experiments conducted by Seewald et al. (2006) with no minerals present generated small amounts of CH_4 comparable to those observed in magnetite-bearing experiments, and it seems possible that other hydrocarbons might be produced as well. Definitive studies showing that the presence of magnetite increases the production of hydrocarbons over that observed for the same conditions without magnetite have not yet been published. Additionally, in the experiments of Horita and Berndt (1999), the catalytic properties of NiFe-alloy was demonstrated by showing that CH_4 production increased with the amount of alloy present, but no similar set of experiments with variable amounts of magnetite have yet been published. Consequently, whether magnetite can catalyze carbon reduction in hydrothermal environments should be considered an unresolved question.

The potential for hydrocarbon synthesis in the presence of pentlandite (an Fe-Ni sulfide mineral) has also been explored by Fu et al. (2008), using similar conditions (400 °C, 50 MPa) and experimental methods to those of previous experiments with chromite and magnetite. The source of inorganic carbon in the experiment, and also of H_2, was ^{13}C-labeled HCOOH. During several weeks of reaction, μmolar concentrations of labeled $^{13}CH_4$, $^{13}C_2H_6$, and $^{13}C_3H_8$ accumulated in the fluid, along with much higher levels of unlabeled compounds. Overall conversion of inorganic carbon was <0.01%, and $^{13}CH_4$ yields were an order of magnitude lower than observed in the previous magnetite and chromite experiments (Foustoukos and Seyfried 2004). More recently, Lazar et al. (2012) reported that $^{13}CH_4$ generation was faster during experiments using natural komatiite and $H^{13}COOH$ as reactants at 300 °C than in parallel experiments with synthetic komatiite, and inferred that the increased rates were attributable to the presence of pentlandite in the natural sample. These studies suggest that Ni-bearing sulfides are another potential catalyst for reduction of inorganic carbon to hydrocarbons in subsurface hydrothermal environments.

While most of studies described above were performed at temperatures of 300 °C or above, several recent studies have explored the production of CH_4 during reaction of olivine with water at lower temperatures (Jones et al. 2010; Neubeck et al. 2011; Oze et al. 2012). In a series of experiments using methods similar to those of Berndt et al. (1996) and McCollom and Seewald (2001), Jones, Oze, and colleagues (Jones et al. 2010; Oze et al. 2012) reacted olivine and chromite with fluids containing variable amounts of HCO_3^- at 200 °C and 30 MPa. Steadily increasing levels of H_2 were observed during the experiments, resulting from serpentinization of olivine (Eqn. 4). Methane concentrations increased in parallel with the H_2, which was inferred by the authors to be an indication that the CH_4 was generated by FTT synthesis reactions catalyzed by magnetite. After reaction for up to 1321 hours, dissolved concentrations of 5-12 mmol/kg H_2 and 56-120 μmol/kg CH_4 were attained.

Although these authors inferred that the CH_4 observed in their experiments was the product of magnetitie-catalyzed reduction of dissolved inorganic carbon, results from previous experiments suggest that this inference must be viewed with caution. As described above, numerous experimental studies have demonstrated that minerals and other sources can produce background CH_4 during hydrothermal experiments in comparable amounts to the CH_4 concentration reported by Jones et al. (2010) and Oze et al. (2012). Furthermore, most of their experiments generated CH_4 even though no inorganic carbon source was included among the reactants, and dissolved inorganic carbon was below detection limits throughout the experiments. Methane generation actually decreased when an inorganic carbon source was added, opposite to expectations for carbon reduction. Since Jones et al. (2010) and Oze et al. (2012) did not assess background levels of CH_4 in their experiments, it is not possible to evaluate the actual source of the CH_4. However, based on results of other studies, it is likely that only a very small fraction of the CH_4 observed, if any, was derived from reduction of inorganic carbon during the experiments.

Neubeck et al. (2011) employed a totally different experimental approach to examine CH_4 generation from olivine at 30-70 °C. These investigators reacted olivine with a bicarbonate solution in partially filled glass vials, sealed with rubber stoppers and flushed with N_2. The headspace of the vials was monitored for production of H_2 and CH_4, and over nine months of reaction these compounds were found to accumulate at low nanomolar amounts, with higher yields observed with increasing temperature. Based on lower CH_4 production in mineral-free experiments and apparent absence of alternative carbon sources, Neubeck et al. (2011) inferred that the CH_4 observed in their experiments was derived from reduction of the dissolved HCO_3^-, with trace amounts of magnetite or chromite serving as catalysts.

If the results of Neubeck et al. (2011) are taken at face value, it would suggest a possible widespread source of CH_4 within the crust from low-temperature serpentinization of ultramafic rocks. However, as with results of Jones et al. (2010) and Oze et al. (2012), the claim of abiotic CH_4 formation in these experiments must be viewed with some caution. Neubeck et al. (2011) do not report any effort to assess background levels of CH_4 in their experiments, and the amounts of CH_4 produced in experiments that included no inorganic carbon source among the reactants were nearly identical to the levels generated in experiments performed with bicarbonate solutions. As a result, this study provides no evidence directly linking the CH_4 observed to reduction of bicarbonate, and previous results suggest that most, if not all, of the CH_4 observed may have been derived from reduced carbon sources among the reactants. Consequently, the potential for abiotic CH_4 production during low-temperature alteration of ultramafic rocks remains uncertain.

Isotopic fractionation during FTT synthesis. The carbon and hydrogen isotopic compositions of CH_4 and other light hydrocarbons have increasingly become key elements of efforts to identify hydrocarbons with an abiotic origin in natural systems (e.g., Sherwood Lollar et al. 2002, 2008; Fiebig et al. 2007; Proskurowski et al. 2008; Taran et al. 2010a). The potential for isotopes to be used as criteria for identification of abiotic hydrocarbons has led to renewed interest in recent years in experimental study of isotopic fractionation during FTT synthesis. The isotopic compositions of FTT reaction products have been reported for a number of laboratory experiments performed under a variety of reaction conditions, including CO/H_2 gas mixtures in flow-through reactors (Taran et al. 2007, 2010a; Shi and Jin 2011), gas mixtures in closed reaction vessels (Lancet and Anders 1970; Hu et al. 1998), and hydrothermal reactions (McCollom and Seewald 2006; Fu et al. 2007; McCollom et al. 2010). A variety of Fe-, Co-, and Ru-bearing catalysts have been used in these experiments.

The isotopic compositions of the light hydrocarbons generated in the various experiments exhibit both some similarities and some substantial differences. Nearly all of the hydrocarbons generated in the experiments are depleted in ^{13}C relative to the initial carbon source. However, the magnitude of the depletion varies considerably among the different experiments, and also varies substantially within individual experiments as a function of reaction times and for compounds with different number of carbon atoms. These trends are illustrated in Figure 8, which shows the carbon isotopic composition of a few selected examples of experimental reaction products. It should be noted that these examples represent only a portion of the full range of isotopic compositions that have been reported for experimental products. Within this selected set of results, however, ^{13}C depletions of the products relative to the carbon source range up to 33‰, while other results show no fractionation or even a slight enrichment in ^{13}C. Among the experiments, the isotopic compositions of hydrocarbons exhibit a wide variety of

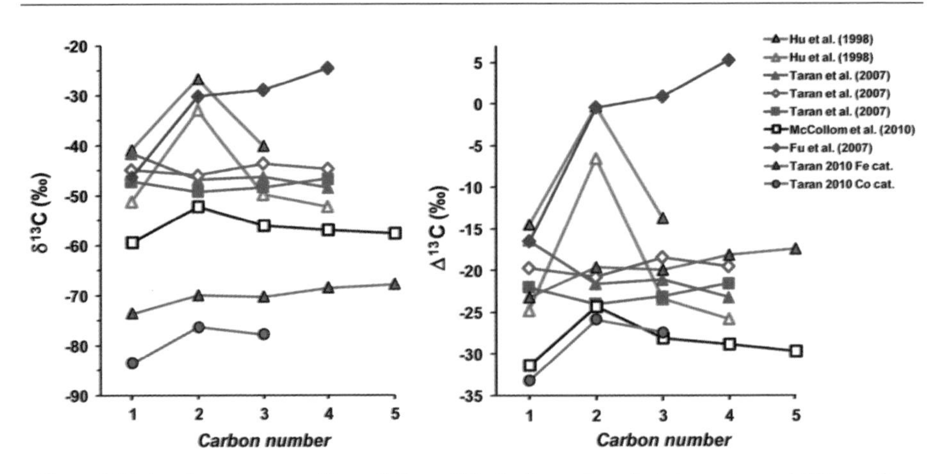

Figure 8. Carbon isotopic composition of CH_4 and C_2-C_5 alkanes for selected experimental FTT reaction products. Values are shown both as measured values (*left*) and relative to the composition of the initial carbon source (*right*) ($\Delta^{13}C = \delta^{13}C_{product} - \delta^{13}C_{source}$).

different trends as a function of carbon number. At present, however, no obvious explanation for the variation in trends among experiments has emerged, and there does not appear to be any consistent variations in trends with factors such as catalyst composition, reaction temperature, or closed- versus open-system reactions. Although not strictly a FTT reaction because no carbon-carbon bonds are formed, the reduction of dissolved CO_2 to CH_4 catalyzed by NiFe-alloy also results in a substantial depletion in ^{13}C, with depletions of 42-49‰ at 200 °C and 15-29‰ at 300 °C reported for the experiments of Horita and Berndt (1999).

Much less data is available on the hydrogen isotope composition of experimental FTT reaction products (Fu et al. 2007; Taran et al 2010b; McCollom et al. 2010). However, the limited data that are available for hydrogen isotopes show somewhat more consistent trends among experiments than the data for carbon isotopes. For the most part, these data show that CH_4 is depleted in ^2H by −35 to −80‰ relative to the initial H_2, and exhibit a regular trend of increasing ^2H abundance with increasing number of carbon atoms (Fig. 9). The data of Fu et al. (2007) deviate somewhat from this trend, but the CH_4 in this experiment may include some contribution from background sources.

In addition to the lack of consistency among isotopic trends observed among experimental products, the presently available experimental data do not show close agreement with isotopic trends observed for natural samples of light hydrocarbons that are thought to have an abiotic origin (Fig. 9). Although this might be construed to be an indication that the hydrocarbons in the natural samples do not really have an abiotic origin, a more likely explanation would appear to be that the experimental conditions employed to date do not accurately simulate the conditions of hydrocarbon formation in natural systems. At this point, however, it is not immediately apparent which aspects of the natural system are not adequately represented in the laboratory experiments.

While there are clear discrepancies among the broader datasets for experimental and natural systems, a consistent explanation may be emerging for at least a subset of those data. Sherwood Lollar et al. (2008) proposed a model to explain the isotopic composition of light hydrocarbons in deep fracture fluids from ancient Precambrian Shield settings that involves isotopic fractionation during formation of the first C–C bond that initiates chain growth, but subsequent additions of carbon atoms to the hydrocarbon backbone are non-fractionating.

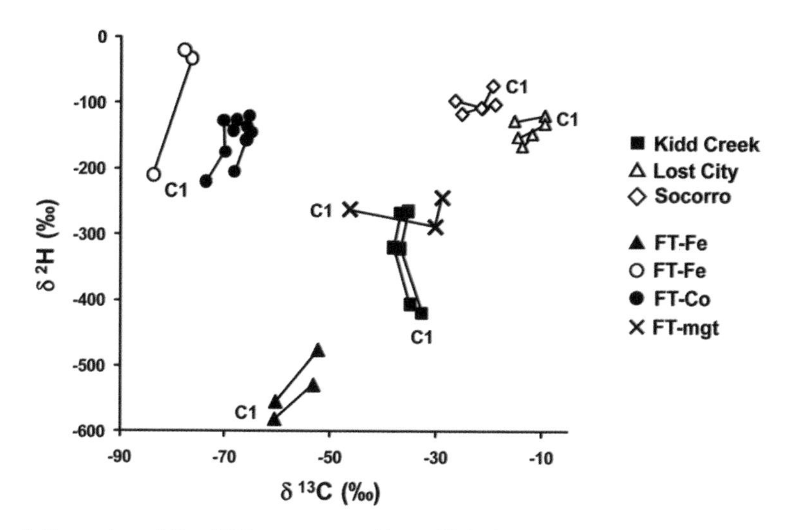

Figure 9. Comparison of C and H isotope compositions of light hydrocarbons produced during laboratory FTT synthesis experiments with compounds from natural systems thought to have an abiotic origin. For each connected series, the point labeled "C1" represents methane and each successive point represents an alkane with an additional carbon atom (from C_2 to C_5, depending on the study). Natural samples are from fracture fluids in ancient Precambrian Shield rocks (Kidd Creek; Sherwood Lollar et al. 2008), deep-sea hydrothermal vent fluids from the Lost City site, thought to be generated from serpentinization of ultramafic rocks (Proskurowski et al. 2008), and volcanic gases from Socorro Island, Mexico (Taran et al. 2010b). Data for FTT synthesis experiments are from Fu et al. (2007), Taran et al. (2010a) and McCollom et al. (2010). Catalysts used in the synthesis experiments are cobalt (Co), iron (Fe), or magnetite (mgt).

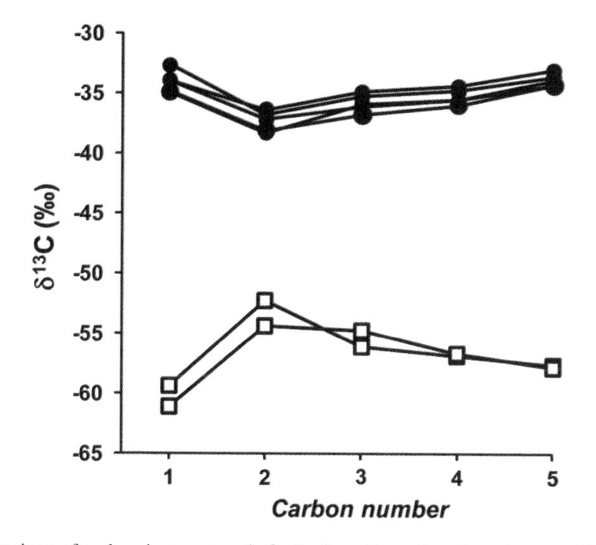

Figure 10. Comparison of carbon isotope trends for hydrocarbons thought to have an abiotic origin from Kidd Creek (circles; Sherwood Lollar et al. 2008) with those for hydrocarbons generated in FTT synthesis experiments performed under hydrothermal conditions (open squares; McCollom et al. 2010).

The outcome is an isotopic trend that shows a sharp fractionation between CH_4 and C_2H_6, while higher hydrocarbons converge towards the isotopic value of CH_4 (Fig. 10). McCollom et al. (2010) inferred a similar scenario to explain the carbon isotopic trend observed for C_1-C_5 hydrocarbons generated in their laboratory FTT synthesis experiments performed under hydrothermal conditions (Fig. 10). These two trends exhibit a difference in direction of isotopic fractionation between CH_4 and C_2H_6, but a similar convergence towards the isotopic composition of methane with increasing carbon number for C_{2+} compounds. The opposing directions of the initial step may be attributable to different reaction mechanisms or conditions effecting the fractionation during the chain initiation.

Implications for abiotic hydrocarbon formation by FTT synthesis within the crust. Several general conclusions can be drawn from the experiments described in the previous sections that are particularly relevant to evaluation of the production of abiotic hydrocarbons within Earth's crust. Several experiments demonstrate that pathways exist for the reduction of dissolved CO_2 (or HCO_3^-) to CH_4 and other hydrocarbons. However, except when NiFe-alloy was present, the amount of carbon converted to hydrocarbons was only a very small fraction (<<1%) of the inorganic carbon present in the experiments, even with very high temperatures and dissolved H_2 concentrations that in many cases are an order of magnitude higher than those observed in natural systems. In each of the experiments, thermodynamic constraints indicate that most or all of the carbon should have been converted to hydrocarbons to approach equilibrium, yet the reaction only proceeded to a very limited extent. While it is possible that the minerals present in these experiments (magnetite, chromite, pentlandite) promoted the reactions to some degree, none of these minerals could have been very effective in catalyzing the conversion of inorganic carbon to hydrocarbons, or much greater yields would have been attained. Thus, while industrial Fischer-Tropsch processes may conceptually invoke rapid synthesis of organic compounds, FTT reactions in natural systems may require significantly longer reaction times and produce a more limited range of compounds.

The same gradual reduction reaction rates observed in the experiments may prevail in natural systems as well. For instance, although CH_4 and other hydrocarbons with an apparent abiotic origin have been observed in high-temperature (>350 °C) ultramafic-hosted deep-sea hydrothermal systems (e.g., Charlou et al. 2002, 2010), the concentrations of CH_4 in these fluids are much lower than expected for thermodynamic equilibrium with the measured levels of dissolved CO_2 and H_2 in the fluids (McCollom and Seewald 2007; McCollom 2008). Consequently, while some reduction of inorganic carbon has apparently occurred as the fluids circulated through the hydrothermal system, the reactions remain far from equilibrium. Nevertheless, concentrations of CH_4 and other light hydrocarbons produced in the experiments are similar to those observed in fluids from the natural system.

The slow rates of reduction of dissolved inorganic carbon are in strong contrast to the rapid reaction rates and high hydrocarbon yields (typically >50%; e.g., Taran et al. 2010a) obtained in conventional gas-phase Fischer-Tropsch studies using transition metal catalysts. Rapid reaction rates and relatively high conversions (1-10%) have also been observed for FTT reactions in the presence of a water-saturated vapor phase (McCollom and Seewald 2006; McCollom et al. 2010), although these reactions were probably catalyzed by native Fe. This comparison suggests the possibility that the presence of a vapor phase may facilitate rapid reduction of inorganic carbon to hydrocarbons at crust temperatures, although it remains to be determined whether naturally occurring minerals can promote the reaction to the same degree as native transition metals. Vapor phase reactions might occur, for instance, during ascent of magmatic fluids to the surface. Alternatively, an H_2-rich vapor phase could develop in fluids circulating through the crust as a result of fluid-rock interactions that consume H_2O and produce H_2, such as serpentinization. With respect to the latter, it may be noteworthy the mineral assemblage magnetite + pentlandite + awaruite that is found in many serpentinites requires H_2 concentrations that are at, or very close to, the solubility limit of H_2 in water (Klein and Bach 2009), suggesting that an H_2-rich vapor may have exsolved locally during formation of this assemblage.

An exception to the above discussion, of course, is NiFe-alloy, which has been shown to effectively catalyze reduction of dissolved CO_2 to CH_4 (Horita and Berndt 1999). Where this mineral is present, rapid reduction of CO_2 and equilibration with CH_4 can be expected at temperatures at least as low as 200 °C. The formation of NiFe-alloys requires conditions that are more strongly reducing than are commonly found in Earth's crust, but such conditions are sometimes attained during serpentinization of ultramafic rocks (Frost 1985; Klein and Bach 2009). Consequently, fluids circulating through serpentinites could become enriched in abiotic CH_4 as a result of interaction with NiFe-alloy. However, since this process appears to exclusively catalyze formation of CH_4, additional processes would be required for formation of other light hydrocarbons in these environments. The high CH_4/C_{2+} ratios and ^{13}C-depleted isotopic signatures for CH_4 formed by NiFe-alloy catalysis are very similar to the characteristics of CH_4 produced during methanogenesis by autotrophic microorganisms (Whiticar 1990), indicating that it may be difficult to confidently distinguish between abitoic and biotic sources of CH_4 in many subsurface settings (Horita and Berndt 1999; Sherwood Lollar and McCollom 2006; Bradley and Summons 2010).

Alternative pathways for hydrocarbon formation in the crust

While most experimental studies have focused on FTT synthesis as a possible contributor of abiotic hydrocarbons to geologic systems, it is by no means the only pathway that could generate these compounds. A few of these alternative pathways are discussed briefly in the following sections. For most alternative pathways, however, there has been little or no experimental study of the reactions at conditions relevant to Earth's crust, and certainly none have been investigated to nearly the same extent as FTT synthesis.

Methane polymerization. The process of methane polymerization discussed above with respect to the mantle (Chen et al. 2008a; Kolesnikov et al. 2009) could also contribute to formation of C_{2+} hydrocarbons in the crust, provided a source of CH_4 is available. However, pure CH_4 might arise, for instance, through NiFe-alloy catalyzed reduction of CO_2 in serpentinites (Horita and Berndt 1999) or reaction of H_2-rich fluids with graphite in metamorphic rocks (Holloway 1984). To date, the potential for methane polymerization has apparently not been investigated experimentally at conditions relevant to the crust, although methane polymerization has been invoked as a possible explanation for isotopic trends in some deep-crustal fluids (Sherwood Lollar et al. 2008).

Polymerization of CH_4 to higher hydrocarbons can be represented by the general reaction:

$$nCH_4 \rightarrow C_nH_{n+2} + (n-1)H_2 \qquad (5)$$

As pointed out earlier, thermodynamic considerations demand that some finite amount of hydrocarbons must be present at equilibrium with CH_4 to satisfy the type of reactions represented by this equation, albeit the amount of hydrocarbons required at equilibrium may be vanishingly small under some conditions. According to the reaction, formation of hydrocarbons will be favored by lower levels of H_2, all other factors being equal.

Carbonate decomposition. Another process that could generate hydrocarbons in the crust is thermal decomposition of carbonate minerals. Carbonate minerals precipitate in a variety of geologic settings at relatively low temperatures, and these minerals will decompose when exposed to higher temperatures in metamorphic or hydrothermal environments. This decomposition can lead to the formation of reduced carbon compounds, particularly when the carbonates contain ferrous Fe that can serve as a reductant for the carbon (French 1971). For example, reduced carbon coexisting with magnetite in ancient metamorphosed rocks in the Isua Greenstone Belt in Greenland has been attributed to thermal decomposition of siderite (i.e.,

$FeCO_3 \rightarrow Fe_3O_4 + C$; Van Zuilen et al. 2002), and decomposition of Fe-bearing carbonates has also been suggested as source of reduced carbon in Martian meteorites (Zolotov and Shock 2000; McCollom 2003; Steele et al. 2012). It is possible that light hydrocarbons could be produced as a by-product of such reactions. In contrast to the experiments with calcite conducted at the elevated temperatures relevant to the mantle where thermodynamic equilibrium appears to have exerted a large influence on carbon speciation, thermal decomposition of carbonates at temperatures in the crust may allow the formation of metastable organic compounds.

Experimental investigations of the formation of hydrocarbons during thermal decomposition of carbonates at temperatures and pressures relevant to the crust have been limited. However, in one set of experiments, thermal degradation of siderite in the presence of water vapor at 300 °C was found to produce small amounts of organic products, predominantly alkylated and hydroxylated aromatic compounds (McCollom 2003). Only trace amounts of CH_4 and other light hydrocarbons were observed. The products of this experiment differed considerably from the typical products of Fischer-Tropsch synthesis, suggesting that an alternative reaction mechanism was responsible for the organic compounds generated. However, the actual process involved remains undetermined. The relatively small amounts of CH_4 generated during siderite decomposition at 300 °C compared with the larger amounts formed in carbonate decomposition experiments performed at higher temperatures and pressures (Kenney et al. 2002; Scott et al. 2004; Chen et al. 2008a; Kutcherov et al. 2010) appears to be related to the lower H/C and H/O ratios of the experimental charge rather than to the differences in reaction conditions. Whether thermal composition of siderite might produce greater amounts of hydrocarbons under other reaction conditions, and whether other carbonates might also generate organic compounds during decomposition at crust temperatures in reducing environments, are questions that remain to be explored with further experiments.

Organosulfur pathways. Abiotic synthesis reactions that proceed through organosulfur intermediates are another possible source of hydrocarbons in the crust. Scientific interest in the possibility that organosulfur compounds might contribute to abiotic organic synthesis was initially stimulated by the origin of life theories of Gunther Wächtershäuser and others. In these theories, reaction of carbon-bearing fluids with sulfide minerals in hydrothermal environments induced organic synthesis reactions that later evolved into primordial metabolic pathways (Wächtershäuser 1990, 1993). The first experimental study to test this theory was that of Heinen and Lauwers (1996), who heated solid FeS with H_2S, CO_2, and water in glass vials at temperatures of 25 to 90 °C, and observed formation of a homologous series of C_1-C_5 alkylthiols as major products. Reduction of carbon in this system presumably involved reaction of CO_2 with H_2 produced by the reaction: $FeS + H_2S \rightarrow FeS_2$ (pyrite) + H_2 (referred to by Wächtershäuser as "pyrite-pulled" reactions). The alkylthiols decrease in abundance with increasing number of carbons, and linear butane- and pentanethiol were more abundant than branched forms.

This initial set of experiments was followed up recently by Loison et al. (2010). Using similar reaction conditions to Heinen and Lauwers (1996), but replacing the water in the reactions with D_2O, these investigators showed that the alkylthiols incorporated D into their structures, thus proving they were formed during the reactions. Other reaction products identified by Loison et al. (2010) were polydeuterated C_1-C_4 carboxylic acids.

In a related study, Huber and Wächtershäuser (1997) reported formation of acetic acid during reaction of aqueous solutions of methanethiol (CH_3SH) with CO gas at 100 °C in the presence of metal sulfide minerals. The overall reaction can be expressed as:

$$CH_3SH + CO + H_2O \rightarrow CH_3COOH + H_2S \quad (6)$$

The key step in this reaction is the formation of a C–C bond, which the authors inferred took place by insertion of CO into the C–S bond on the surface of the sulfide mineral. The reaction was further investigated by Cody et al. (2000), who reacted nonanethiol ($C_9H_{19}SH$) with formic acid in the presence of iron sulfide at 250 °C and 50-200 MPa. Decomposition of formic acid resulted in a mixture of CO, CO_2, H_2, and H_2O in the reaction vessel at experimental conditions. Through a reaction analogous to Equation (6), decanoic acid ($C_9H_{19}COOH$) was found as a reaction product, although this compound was only a minor component of a broad spectrum of organic products dominated by dinonyldisulfide ($H_{19}C_9SSC_9H_{19}$). In a follow-up study testing the capacity of other sulfide minerals and native Ni to catalyze the reaction, C_{11}-C_{13} carboxylic acids were identified as reaction products in addition to decanoic acid (Cody et al. 2004). Although several metal sulfide minerals catalyzed the reaction, it was most strongly promoted by Ni- and Co-bearing sulfides.

Results of these experiments suggest the possibility that organic compounds might be generated in subsurface environments through the sequential formation of C–C bonds involving thiols or other organosulfur as reaction intermediates, with sulfide minerals serving as a catalyst. The products of these reactions have many similarities to the products of FTT synthesis, such as a preference for straight alkyl chains relative to branched forms, a regular decrease in abundance with increasing carbon number, and involvement of surface catalysis in the synthesis reaction. However, the apparent lack of alkane and alkene products similar to those that predominate during FTT synthesis (Fig. 6) suggests that a distinctly different reaction mechanism may be involved.

Based on a series of experiments with potential reaction intermediates, Loison et al. (2010) have proposed a scenario whereby a homologous series of alkylthiols and carboxylic acids with increasing carbon number would be generated through an iterative process involving formation and reduction of thioesters. In this scenario, CO is inserted into the C–S bond of an alkylthiol to form a thiocarboxylic acid with one additional carbon, which is then either converted to a carboxylic acid or reduced to an akylthiol, which can then undergo the same reaction cycle. McCollom and Seewald (2007) proposed a somewhat different iterative process, whereby CO is incorporated into a growing alkyl carbon chain through insertion into the C–S bond of the alkylthiol to form carboxylic acid, followed by reduction of the acid to an alcohol, which then is converted back to an alkylthiol that starts the sequence of steps over again.

Whatever the actual mechanism, the outcome would be formation of series of alkylthiols and carboxylic acids that could subsequently undergo reduction to form hydrocarbons or other organic matter. If this were to occur, the process could contribute to the light hydrocarbons observed in geologic fluids. Since the canonical wisdom is that most industrial catalysts for Fischer-Tropsch synthesis are "poisoned" by sulfur (though this remains to be rigorously tested for natural geologic materials), sulfide-catalyzed reactions could provide an alternative pathway for hydrocarbon formation in sulfur-rich subsurface environments.

Clay-catalyzed hydrocarbon synthesis. Clay minerals have catalytic properties for many types of chemical reactions, which have been exploited for many years by industry. Since clay minerals are widespread in geologic environments, including altered igneous rocks and sediments, it is possible that they could play a role in the abiotic formation of hydrocarbons. This possibility was investigated in experiments by Williams et al. (2005), who reacted aqueous methanol solutions with clay minerals at 300 °C and 100 MPa. The reactions produced dimethylether (a condensation product of two methanol molecules) as the principal product, but a number of other organic products were observed including CH_4, C_2-C_6 alkanes and alkenes, along with an assortment of alkylated cyclic aromatic compounds, such as alkylbenzenes, alkylphenols, alkylnaphthalenes, and alkylnaphthols. The predominance of alkylated aromatic compounds among the higher molecular weight reaction products distinguishes them from the linear saturated alkanes that are characteristic of FTT synthesis, suggesting either a different mechanism is involved or that alkanes were generated and then underwent secondary reactions to form aromatic compounds. The mechanism of the reaction remains to be determined, but the authors hypothesized that that the organic compounds may have formed within the interlayers of the clay structure, suggesting that clay minerals may provide unique microenvironments for organic synthesis.

SOME DIRECTIONS FOR FUTURE STUDIES

Hopefully, this review has provided a sense of the considerable insights that experimental studies have provided for the conditions that allow abiotic hydrocarbons to form in subsurface geologic environments. At the same time, it should be evident that there exist many significant gaps in understanding of these processes. Experimental studies demonstrate that carbon speciation will likely be controlled by thermodynamic considerations at mantle conditions, but kinetic inhibitions limit carbon reduction reactions at temperatures within the crust below about 400 °C. While experiments show that some minerals can allow these inhibitions to be overcome, the catalytic potential of most naturally occurring minerals remains uncertain. Further experimental studies are needed to evaluate whether minerals that occur in natural systems of interest can catalyze hydrocarbon synthesis, and under what conditions. Another important issue to resolve is the impact of vapor phase versus liquid aqueous phase conditions on catalysis and reaction rates.

It is also apparent that significant discrepancies exist between the isotopic compositions of hydrocarbons observed in experimental systems and those of apparently abiotic compounds in natural systems. While there is a tremendous potential to use isotope systematics to help determine whether hydrocarbons in geologic fluids have an abiotic origin, confidence in this criteria would be significantly improved if there was a firm experimental basis for interpretation of the isotopic signatures. At this time, however, it remains unclear what factors contribute to the wide variation observed in isotopic compositions of experimentally generated FTT reaction products, or which of the experimental products best represents the conditions of natural environment. A more systematic approach is needed to determine how variations in reaction conditions affect the isotopic composition of the products. Development of clumped isotope methods for hydrocarbons similar to those used for analyzing carbonate minerals (Eiler 2007) may also provide novel insights, although they would need to be calibrated with additional experimental studies. Resolution of these issues will provide a much more solid basis for evaluating sources of abiotic hydrocarbons in Earth's crust.

ACKNOWLEDGMENTS

The author's research in abiotic organic synthesis has been supported by the NSF Earth Sciences Division and the NASA Exobiology program. The author appreciates the support of the Hanse Wissenschaftskolleg, Delmenhorst, Germany during preparation of the manuscript.

REFERENCES

Abrajano TA, Sturchio NC, Kennedy BM, Lyon GL, Muehlenbachs K, Bohlke JK (1990) Geochemistry of reduced gas related to serpentinization of the Zambales ophiolite, Philippines. Appl Geochem 5:625-630

Anderson RR, White CM (1994) Analysis of Fischer-Tropsch by-product waters by gas chromatography. J High Resolut Chromatogr 17:245-250

Berndt ME, Allen DE, Seyfried WE (1996) Reduction of CO_2 during serpentinization of olivine at 300 °C and 500 bar. Geology 24:351-354

Bischoff JL, Rosenbauer RJ (1985) An empirical equation of state for hydrothermal seawater (3.2 percent NaCl). Am J Sci 285:725-763

Bradley AS, Summons RE (2010) Multiple origins of methane at the Lost City Hydrothermal Field. Earth Planet Sci Lett 297:34-41

Charlou JL, Donval JP, Fouquet Y, Jean-Baptiste P, Holm (2002) Geochemistry of high H_2 and CH_4 vent fluids issuing from ultramafic rocks at the Rainbow hydrothermal field (36° 14′N, MAR). Chem Geol 191:345-359

Charlou JL, Donval JP, Konn C, Ondréas H, Fouquet Y (2010) High production and fluxes of H_2 and CH_4 and evidence of abiotic hydrocarbons synthesis by serpentinization in ultramafic-hosted hydrothermal systems on the Mid-Atlantic Ridge. *In:* Diversity of Hydrothermal Systems on Slow Spreading Ocean Ridges. Rona PA, Devey C, Dyment J, Murton BJ (eds). Am Geophys Union, Washington, DC, p 265-296

Charlou JL, Fouquet Y, Donval JP, Auzende JM, Jean-Baptiste P, Stievenard M (1996) Mineral and gas chemistry of hydrothermal fluids on an ultrafast spreading ridge: East Pacific Rise, 17° to 19°S (Naudur cruise, 1993) phase separation processes controlled by volcanic and tectonic activity. J Geophys Res 101:15,899-15,919

Chen JY, Jin LJ, Dong JP, Zheng HF (2008a) *In situ* Raman spectroscopy study on dissociation of methane at high temperatures and at high pressures. Chin Phys Lett 25:780-782

Chen JY, Jin LJ, Dong JP, Zheng HF, Liu GY (2008b) Methane formation from $CaCO_3$ reduction catalyzed by high pressure. Chin Chem Lett 19:475-478

Cody GD, Boctor NZ, Brandes JA, Filley TR, Hazen RM, Yoder HS (2004) Assaying the catalytic potential of transition metal sulfides for abiotic carbon fixation. Geochim Cosmochim Acta 68:2185-2196

Cody GD, Boctor NZ, Filley TR, Hazen RM, Scott JH, Sharma A, Yoder HS (2000) Primordial carbonylated iron-sulfur compounds and the synthesis of pyruvate. Science 289:1337-1340, doi:10.1126/science.289.5483.1337

Dictor RA, Bell AT (1986) Fischer-Tropsch synthesis over reduced and unreduced iron oxide catalyst. J Catal 97:121-136

Eiler JM (2007) "Clumped Isotope" geochemistry – The study of naturally-occurring, multiply-substituted isotopologues. Earth Planet Sci Lett 262:309-327

Fiebig J, Woodland AB, D'Alessandro W, Püttmann W (2009) Excess methane in continental hydrothermal emissions is abiogenic. Geology 37:495-498, doi:10.1130/G25598A.1

Fiebig J, Woodland AB, Spangenberg J, Oschmann W (2007) Natural evidence for rapid abiogenic hydrothermal generation of CH_4. Geochim Cosmochim Acta 71:3028-3039

Foustoukos DI (2012) Metastable equilibrium in the C–H–O system: Graphite deposition in crustal fluids. Am Mineral 97:1373-1380

Foustoukos DI, Seyfried WE (2004) Hydrocarbons in hydrothermal vent fluids: The role of chromium-bearing catalysts. Science 304:1002-1005

French BM (1971) Stability relations of siderite ($FeCO_3$) in the system Fe-C-O. Am J Sci 271:37-78

Frost BR (1985) On the stability of sulfides, oxides, and native metals in serpentinite. J Petrol 26:31-63

Frost DJ, McCammon (2008) The redox state of Earth's mantle. Ann Rev Earth Planet Sci 36:398-420, doi:10.1029/annurev.earth.36.031207.124322

Fu Q, Foustoukos DI, Seyfried WE (2008) Mineral catalyzed organic synthesis in hydrothermal systems: An experimental study using time-of-flight secondary ion mass spectrometry. Geophys Res Lett doi:10.1029/2008GL033389

Fu Q, Sherwood Lollar B, Horita J, Lacrampe-Couloume G, Seyfried WE (2007) Abiotic formation of hydrocarbons under hydrothermal conditions: Constraints from chemical and isotope data. Geochim Cosmochim Acta 71:1982-1998

Glasby GP (2006) Abiogenic origin of hydrocarbons: An historical overview. Resource Geol 56:85-98

Gold T (1992) The deep, hot biosphere. Proc Natl Acad Sci USA 89:6045-6049, doi:10.1073/pnas.89.13.6045

Gold T (1993) The origin of methane in the crust of the Earth. U S Geol Surv Prof Paper 1570:57-80

Hayatsu R, Studier MH, Oda A, Fuse K, Anders E (1968) Origin of organic matter in early solar system – II. Nitrogen compounds. Geochim Cosmochim Acta 32:175-190

Heinen W, Lauwers AM (1996) Organic sulfur compounds resulting from the interaction of iron sulfide, hydrogen sulfide and carbon dioxide in an anaerobic aqueous environment. Orig Life Evol Biosphere 26:161-150

Holloway JR (1984) Graphite-CH_4-H_2O-CO_2 equilibria at low-grade metamorphic conditions. Geology 12:455-458

Horita J, Berndt ME (1999) Abiogenic methane formation and isotopic fractionation under hydrothermal conditions. Science 285:1055-1057

Hu G, Ouyang Z, Wang X, Wen Q (1998) Carbon isotopic fractionation in the process of Fischer-Tropsch reaction in primitive solar nebula. Sci China Ser D 41:202-207

Huber C, Wächtershäuser G (1997) Activated acetic acid by carbon fixation on (Fe,Ni)S under primordial conditions. Science 276:245-247

Jakobsson S, Oskarsson N (1990) Expperimental determination of fluid compositions in the system C–O–H at high P and T and low fO_2. Geochim Cosmochim Acta 54:355-362

Jakobsson S, Oskarsson N (1994) The system C–O in equilibrium with graphite at high pressure and temperature: An experimental study. Geochim Cosmochim Acta 58:9-17

Jenden PD, Hilton DR, Kaplan IR, Craig H (1993) Abiogenic hydrocarbons and mantle helium in oil and gas fields. *In:* The Future of Energy Gases. Howell DG (ed) US Geological Survey Professional Paper 1570:31-56

Ji F, Zhou H, Yang Q (2008) The abiotic formation of hydrocarbons from dissolved CO_2 under hydrothermal conditions with cobalt-bearing magnetite. Origins Life Evol Biosphere 38:117-125

Johnson JW, Oelkers EH, Helgeson HC (1992) SUPCRT92: A software package for calculating the standard molal thermodynamic properties of minerals, gases, aqueous species, and reactions from 1 to 5000 bar and 0 to 1000 °C. Comput Geosci 18:899-947

Jones LC, Rosenbauer RJ, Goldsmith JI, Oze C (2010) Carbonate controls of H_2 and CH_4 production in serpentinization systems at elevated P-Ts. Geophys Res Lett 37, doi:10.1029/2010GL043769

Kelley DS (1996) Methane-rich fluids in the oceanic crust. J Geophys Res 101:2943-2962

Kelley DS, Karston JA, Früh-Green GL, Yoerger DR, Shank TM, Butterfield DA, Hayes JM, Schrenk MO, Olson EJ, Proskurowski G, Jakuba M, Bradley A, Larson B, Ludwig K, Glickson D, Buckman K, Bradley AS, Brazelton WJ, Roe K, Elend MJ, Delacour A, Bernasconi SM, Lilley MD, Baross JA, Summons RE, Sylva SP (2005) A serpentinite-hosted ecosystem: The Lost City hydrothermal field. Science 307:1428-1434

Kenney JF, Kutcherov VA, Bendeliani NA, Alekseev VA (2002) The evolution of multicomponent systems at high pressures: VI. The thermodynamic stability of the hydrogen-carbon system: The genesis of hydrocarbons and the origin of petroleum. Proc Natl Acad Sci USA 99:10,976-10,981, doi:10.1073/pnas.172376899

Kerrick DM, Connolly JAD (2001) Metamorphic devolatilization of subducted marine sediments and the transport of volatiles into the Earth's mantle. Nature 411:293-296

Klein F, Bach W (2009) Fe-Ni-Co-O-S phase relations in peridotite-seawater interactions. J Petrol 50:37-59, doi:10.1093/petrology/egn071

Kolesnikov A, Kutcherov VG, Goncharov AF (2009) Methane-derived hydrocarbons produced under upper-mantle conditions. Nat Geosci 2:566-570, doi:10.1038/NGE0591

Kutcherov VG, Kolesnikov AY, Dyuzhheva TI, Kulikova LF, Nikolaev NN, Sazanova OA, Braghkin VV (2010) Synthesis of complex hydrocarbon systems at temperatures and pressures corresponding to the Earth's upper mantle conditions. Dokl Phys Chem 433:132-135

Kutcherov VG, Krayushkin VA (2010) Deep-seated abiogenic origin of petroleum: From geological assessment to physical theory. Rev Geophys, doi:10.1029/2008RG000270

Lancet MS, Anders EA (1970) Carbon isotope fractionation in the Fischer-Tropsch synthesis and in meteorites. Science 170:980-982

Lazar C, McCollom TM, Manning CE (2012) Abiogenic methanogensis during experimental komatiite serpentinization: Implications for the evolution of the early Precambrian atmosphere. Chem Geol 326-327:102-112

Loison A, Dubant S, Adam P, Albrecht P (2010) Elucidation of an iterative process of carbon-carbon bond formation of prebiotic significance. Astrobiology 10:973-988

Marcaillou C, Muñoz M, Vidal O, Parra T, Harfouche H (2012) Mineralogical evidence for H_2 degassing during serpentinization at 300 °C/300 bar. Earth Planet Sci Lett 308:281-290

Martin W, Baross J, Kelley D, Russell MJ (2008) Hydrothermal vents and the origin of life. Nat Rev Microbiol 6:805-814

Mathez EA (1984) Influence of degassing on oxidation states of basaltic magma. Nature 310:371-375

Matveev S, Ballhaus C, Fricke K, Truckenbrodt J, Ziegenbein D (1997) Volatiles in the Earth's mantle: I. Synthesis of CHO fluids at 1273 K and 2.4 GPa. Geochim Cosmochim Acta 61:3081-3088

McCollom TM (2003) Formation of meteorite hydrocarbons by thermal decomposition of siderite ($FeCO_3$). Geochim Cosmochim Acta 67:311-317

McCollom TM (2008) Observational, experimental, and theoretical constraints on carbon cycling in mid-ocean ridge hydrothermal systems. *In:* Modeling Hydrothermal Processes at Oceanic Spreading Centers: Magma to Microbe. Lowell RP, Seewald J, Perfit MR, Metaxas A (eds). Am Geophys Union, Washington, DC, p 193-213

McCollom TM, Ritter G, Simoneit BRT (1999) Lipid synthesis under hydrothermal conditions by Fischer-Tropsch-type reactions. Origins Life Evol Biosphere 29:153-166

McCollom TM, Seewald JS (2001) A reassessment of the potential for reduction of dissolved CO_2 to hydrocarbons during serpentinization of olivine. Geochim Cosmochim Acta 65:3769-3778

McCollom TM, Seewald JS (2006) Carbon isotope composition of organic compounds produced by abiotic synthesis under hydrothermal conditions. Earth Planet Sci Lett 243:74-84

McCollom TM, Seewald JS (2007) Abiotic synthesis of organic compounds in deep-sea hydrothermal environments. Chem Rev 107:382-401

McCollom TM, Sherwood Lollar B, Lacrampe-Couloume G, and Seewald JS (2010) The influence of carbon source on abiotic organic synthesis and carbon isotope fractionation under hydrothermal conditions. Geochim Cosmochim Acta 74:2717-2740

Neubeck A, Duc NT, Bastviken D, Crill P, Holm NG (2011) Formation of H_2 and CH_4 by weathering of olivine at temperatures between 30 and 70 °C. Geochem Trans 12:6, http://www.geochemicaltransactions.com/content/12/1/6

Nooner DW, Oro J (1979) Synthesis of fatty-acids by a closed system Fischer-Tropsch process. Hydrocarbon Synthesis 12:160-170

Oze C, Jones LC, Goldsmith JI, Rosenbauer RJ (2012) Differentiating biotic from abiotic methane genesis in hydrothermally active planetary surfaces. Proc Natl Acad Sci USA 109:9750-9754, doi:10.1073/pnas.1205223109

Pasteris JD, Chou I-M (1998) Fluid-deposited graphitic inclusions in quartz: Comparison between KTB (German Continental Deep-Drilling) core samples and artificially reequilibrated natural inclusions. Geochim Cosmochim Acta 62:109-122

Pineau F, Javoy M (1983) Carbon isotopes and concentrations in mid-ocean ridge basalts. Earth Planet Sci Lett 62:239-257

Pollack NH, Chapman DS (1977) On the regional variation of heat flow, geotherms, and lithospheric thickness. Tectonophysics 38:279-296

Potter J, Rankin AH, Treloar PJ (2004) Abiogenic Fischer-Tropsch synthesis of hydrocarbons in alkaline igneous rocks; fluid inclusion, textural and isotopic evidence from the Lovozero complex, N. W. Russia. Lithos 75:311-330

Proskurowski G, Lilley MD, Seewald JS, Früh-Green GL, Olson EJ, Lupton JE, Sylva SP, Kelley DS (2008) Abiogenic hydrocarbon production at Lost City Hydrothermal Field. Science 319:604-607

Satterfield CN, Hanlon RT, Tung SE, Zou Z, Papaefthymiou GC (1986) Initial behaviour of a reduced fused-magnetite catalyst in the Fischer-Tropsch synthesis. Ind Eng Chem Prod Res Dev 25:401-407.

Schrenk MO, Brazelton WJ, Lang SQ (2013) Serpentinization, carbon, and deep life. Rev Mineral Geochem 75:575-606

Scott HP, Hemley RJ, Mao H, Herschbach DR, Fried LE, Howard WM, Bastea S (2004) Generation of methane in the Earth's mantle: *In situ* high pressure-temperature measurements of carbonate reduction. Proc Natl Acad Sci USA 101:14,023-14,026, doi:10.1073/pnas.0405930101

Seewald JS, Zolotov MY, McCollom TM (2006) Experimental investigation of single carbon compounds under hydrothermal conditions. Geochim Cosmochim Acta 70:446-460

Sephton MA, Hazen RM (2013) On the origins of deep hydrocarbons. Rev Mineral Geochem 75:449-465

Seyfried WE, Foustoukos DI, Fu Q (2007) Redox evolution and mass transfer during serpentinization: An experimental and theoretical study at 200 °C, 500 bar with implications for ultramafic-hosted hydrothermal systems at Mid-Ocean Ridges. Geochim Cosmochim Acta 71:3872-3886

Sharma A, Cody GD, Hemley RJ (2009) *In situ* diamond-anvil cell observations of methanogenesis at high pressures and temperatures. Energy Fuels 23:5571-5579

Sherwood Lollar B, Lacrampe-Couloume G, Slater GF, Ward J, Moser DP, Gihring TM, Lin L, Onstott TC (2006) Unravelling abiogenic and biogenic sources of methane in the Earth's deep subsurface. Chem Geol 226:328-339

Sherwood Lollar B, Lacrampe-Couloume G, Voglesonger K, Onstott TC, Pratt LM, Slater GF (2008) Isotopic signatures of CH_4 and higher hydrocarbon gases from Precambrian Shield sites: A model for abiogenic polymerization of hydrocarbons. Geochim Cosmochim Acta 72:4778-4795

Sherwood Lollar B, McCollom TM (2006) Biosignatures and abiotic constraints on early life. Nature doi:10.1038/nature05499

Sherwood Lollar B, Voglesonger K, Lin L-H, Lacrampe-Couloume G, Telling J, Abrajano TA, Onstott TC, Pratt LM (2007) Hydrogeologic controls on episodic H_2 release from Precambrian fractured rocks – Energy for deep subsurface life on Earth and Mars. Astrobiology 7:971-986, doi:10.1089/ast.2006.0096

Sherwood Lollar B, Westgate TD, Ward JA, Slater GF, Lacrampe-Couloume G (2002) Abiogenic formation of alkanes in the Earth's crust as a minor source for global hydrocarbon reservoirs. Nature 416:522-524

Shi B, Jin C (2011) Inverse kinetic isotope effects and deuterium enrichment as a function of carbon number during formation of C-C bonds incobalt catalyzed Fischer-Tropsch synthesis. Appl Catal A 393:178-173

Shirey SB, Cartigny P, Frost DJ, Keshav S, Nestola F, Nimis P, Pearson DG, Sobolev NV, Walter MJ (2013) Diamonds and the geology of mantle carbon. Rev Mineral Geochem 75:355-421

Shock EL (1990) Geochemical constraints on the origin of organic compounds in hydrothermal systems. Origins Life Evol Biosphere 20:331-367

Shock EL (1992) Chemical environments of submarine hydrothermal systems. Origins Life Evol Biosphere 22:67-107

Shock EL, Helgeson HC, Sverjensky DA (1989) Calculation of the thermodynamic and transport properties of aqueous species at high pressures and temperatures: Standard partial molal properties of inorganic neutral species. Geochim Cosmochim Acta 53:2157-2183

Spanu L, Donadio D, Hohl D, Schwegler E, Galli G (2011) Stability of hydrocarbons at deep Earth pressures and temperatures. Proc Natl Acad Sci USA 108:6843-6845, doi:10.1073/pnas.1014804108

Steele A, McCubbin FM, Fries MD, Golden DC, Ming DW, Benning LG (2012) Graphite in the Martian meteorite Allan Hills 84001. Am Mineral 97:1256-1259

Taran YA, Kliger GA, Cienfuegos E, Shuykin AN (2010a) Carbon and hydrogen isotopic compositions of products of open-system catalytic hydrogenation of CO_2: Implications for abiogenic hydrocarbons in Earth's crust. Geochim Cosmochim Acta 71:4474-4487

Taran YA, Kliger GA, Sevastianov VS (2007) Carbon isotope effects in the open-system Fischer-Tropsch synthesis. Geochim Cosmochim Acta 71:4474-4487

Taran YA, Varley NR, Inguaggiato S, Cienfuegos E (2010b) Geochemistry of H_2- and CH_4-enriched hydrothermal fluids of Socorro Island, Revillagigedo Archipelago, Mexico. Evidence for serpentinization and abiogenic methane. Geofluids 10:542-555, doi:10.1111/j.1468-8123.2010.00314.x

van Zuilen MA, Lepland A, Arrhenius G (2002) Reassessing the evidence for the earliest traces of life. Nature 418:627-630

Wächtershäuser G (1990) Evolution of the first metabolic cycles. Proc Natl Acad Sci USA 87:200-204

Wächtershäuser G (1993) The cradle chemistry of life – on the origin of natural-products in a pyrite-pulled chemoautotrophic origin of life. Pure Appl Chem 65:1343-1348

Welhan JA, Craig H (1983) Methane, hydrogen and helium in hydrothermal fluids at 21 degrees N on the east Pacific Rise. *In:* Hydrothermal Processes at Seafloor Spreading Centres. Rona PA, Bostrom K, Laubier L, Smith KLJ (eds), Plenum, New York, p 391-409

Whiticar M (1990) A geochemical perspective of natural gas and atmospheric methane. Org Geochem 16:531-547

Williams LB, Canfield B, Vogelsonger KM, Holloway JR (2005) Organic molecules formed in a "primordial womb." Geology 33:913-916

Yoshino D, Hayatsu R Anders E (1971) Origin of organic matter in early solar system – III. Amino acids: catalytic synthesis. Geochim Cosmochim Acta 32:175-190

Zhang C, Duan Z (2009) A model for C–H–O fluid in the Earth's mantle. Geochim Cosmochim Acta 73:2089-2102

Ziegenbein D, Johannes W (1980) Graphite in C–H–O fluids: an unsuitable compound to buffer fluid composition at temperatures up to 700 °C. Neues Jahrbuch Mineral Monats 19:289–305

Zolotov MY, Shock EL (2000) An abitoic origin for hydrocarbons in the Allan Hills 84001 martian meteorite through cooling of magmatic and impact-generated gases. Meteorit Planet Sci 35:629-638

Reviews in Mineralogy & Geochemistry
Vol. 75 pp. 495-545, 2013

16

Hydrocarbon Behavior at Nanoscale Interfaces

David R. Cole[1,2], Salim Ok[1]

[1]School of Earth Sciences and [2]Department of Chemistry
The Ohio State University
Columbus, Ohio 43210, U.S.A.

cole.618@osu.edu ok.12@osu.edu

Alberto Striolo, Anh Phan

School of Chemical, Biological and Materials Engineering
The University of Okalhoma
Norman, Oklahoma 73019, U.S.A.

astriolo@ou.edu aphan@ou.edu

INTRODUCTION

Throughout Earth's crust and upper mantle, fluids play the dominant role in transporting and concentrating Earth's energy and mineral resources (Liebscher and Heinrich 2007). Furthermore, the flux of fluids, which act as both reaction media and reactants, strongly influences the genesis and evolution of many different kinds of rocks. Among many different types of fluids, those containing volatile carbon, hydrogen and oxygen (C-H-O) species tend to dominate in the lithosphere along with various electrolytes and silica. These fluids commonly contain methane as both a major constituent and an important energy source. Conventional natural gas deposits reside in sedimentary basins where fluid overpressure often results in brittle failure of the confining rocks. Industry exploration and exploitation of shale gas (e.g., the Marcellus, Utica, and Barnett formations) has refocused attention on understanding the fundamental behavior of volatile hydrocarbon—rock interactions. Recent observations of hydrocarbons emanating from non-sedimentary systems (abiogenic), such as mid-ocean ridge hydrothermal systems or occurring within some crystalline rock-dominated Precambrian shield environments have challenged the view that organic rich sediments provide the only significant source of crustal hydrocarbons (Potter and Konnerup-Madsen 2003; Sleep et al. 2004; Sherwood Lollar et al. 2006; McCollom 2013; Sephton and Hazen 2013). Geopressured-geothermal regimes contain C-H-O fluids with vast energy potential in the form of methane and hot water at high pressure. Even fluid inclusions from both metamorphic and igneous terrains record the presence of methane-bearing fluids reflecting reduced redox state conditions of formation.

The consequences of coupled reactive-transport processes common to most geological environments depend on the properties and reactivity of these crustal fluids over broad ranges of temperature, pressure and fluid composition. The relative strengths of complex molecular-scale interactions in geologic fluids, and the changes in those interactions with temperature, pressure, and fluid composition, are the fundamental basis for observed fluid properties. Complex intermolecular interactions of C-H-O-N-S fluids (H_2O, CO_2, CH_4, H_2, H_2S, N_2,) result in their unique thermophysical properties, including large deviations in the volumetric properties from ideality, vapor-liquid equilibria, and critical phenomena. Indeed, a key goal in geochemistry is to develop a comprehensive understanding of the thermophysical properties, structures, dynamics, and reactivity of complex geologic fluids and molecules (water and other C-H-O-N-S fluids, electrolytes, and organic-biological molecules) at multiple length scales

1529-6466/13/0075-00167$00.00 DOI: 10.2138/rmg.2013.75.16

(molecular to macroscopic) over wide ranges of temperature, pressure, and composition. This knowledge is foundational to advances in the understanding of other geochemical processes involving mineral-fluid interfaces and reactions. It is also becoming increasingly clear that organic molecules present as gas species, in aqueous and mixed-volatile fluids—ranging from simple hydrocarbons and carboxylic acids to branched and cyclic compounds, to proteins and humic substances—play major roles in controlling geochemical processes, not just at Earth's surface, but also deep within the crust. The origin of life may be partly attributable to the properties of such molecules in complex fluids under extreme conditions, as they appear to play an important role in mineral reactivity and templating of mineral precipitates.

Hydrocarbons (e.g., CH_4, C_2H_6, etc.), CO_2, and aqueous solutions can occupy the pores or fractures of numerous types of complex heterogeneous Earth materials present in the systems outlined above. This accessible porosity within the solids can span wide length scales (*d* as pore diameter or fracture aperture) including micro-, meso-, and macroporous regimes ($d < 2.0$ nm, $2.0 < d < 50$ nm, and $d > 50$ nm, respectively, as defined by IUPAC). Porous solid matrices include rock or soil systems that contain clays and other phyllosilicates, zeolites, coal, graphite, or other carbonaceous-rich units; and weathered or altered silicates (e.g., feldspar to clay; olivine to serpentine), oxides, and carbonates. Examples of micro- and mesoporous features in natural solids and synthetic engineered proxies for natural materials are given in Figure 1. A number of factors dictate how fluids, and with them reactants and products of intrapore transformations, migrate into and through these nano-environments, wet, and ultimately adsorb and react with the solid surfaces. Factors include the size, shape, distribution, and interconnectedness of confined geometries, the chemistry of the solid and the fluids, and their physical properties (Cole et al. 2004). The dynamic behavior of fluids and gases contained within solids is controlled by processes occurring at the interface between the various phases (e.g., water-water, water-solute, water-volatile, water-solid, solute-solid, volatile-solid, etc.), as well as the rates of supply and removal of mobile constituents.

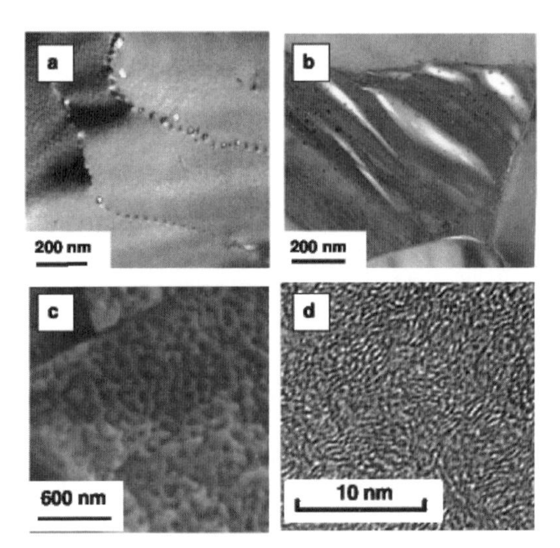

Figure 1. Electron microscopy images of micro- and mesoporous Earth and engineered materials: (a) pores along grain boundaries in weakly weathered basalt, (b) clay formation with large pores (white areas) at a grain boundary intersection in altered basalt, (c) controlled-pore glass, and (d) slit-like pores in carbon fiber monolith. [Used with kind permission of Springer Science+Business Media from Cole et al. (2009) *Neutron Applications in Earth, Energy and Environmental Sciences*, Fig. 1, p. 544.]

There is general agreement that the collective structure and properties of bulk fluids are altered by solid substrates, confinement between two mineral surfaces, or in narrow pores due to the interplay of the intrinsic length scales of the fluid and the length scale due to confinement (Gelb et al. 1999). However, compared with the effort expended to study bulk fluids, a fundamental understanding of the thermodynamic, structural and dynamic properties of volatile C-H-O fluids in nano-confined geometries, and their influence on the properties of the porous solid, is much less evolved, particularly for natural mineral substrates. Examples of experimental and computational efforts relevant to the behavior of Earth materials (defined as gases, solutions, and solids) include the study of CO_2 in thin pores (Belonoshko 1989), water structure and dynamics in clays (Skipper et al. 1995; Pitteloud et al. 2003; Wang et al. 2003; Skipper et al. 2006), ion adsorption into alumina mesoporous materials (Wang et al. 2002), and water within layered silicates at elevated pressure and temperature (Wang et al. 2004, 2005, 2006). These studies (and numerous others cited in Cole et al. 2006, 2009, 2010) demonstrate that a fluid can exhibit nano-confinement promoted phase transitions, including freezing, boiling, condensation, and immiscibility, which are intrinsic to the fluid-confining surface interactions (such as wetting and layering). Also crucial to the molecular behavior of fluids is the geometry of the pore, which can include simple planar walls (slits) such as encountered in clays and micas; cylinders, and spheres; and spheres linked with cylinders as observed in zeolites. Other factors that contribute to the modification of fluid properties include the randomness of the matrix and the connectivity of the pore network.

Given the complexity of natural C-O-H fluids and their roles in mediating surface interactions and reactivity with mineral phases, there can be no doubt that a quantitative understanding is needed of molecular-level fluid properties and fluid interactions with solids. A wide spectrum of analytical approaches can be brought to bear on Earth materials and engineered proxies, including, but certainly not limited to dynamic light scattering, IR, microscopy (e.g., electron; force), NMR, synchrotron-based X-rays, and neutron scattering and diffraction. When coupled with molecular simulation, this wide array of methods provides the means to which we can interrogate the structure and dynamics of fluids and their interactions with solids. Each of these methods provides a unique window into the properties and behavior of fluids and their reactivity. The inherent advantage of using engineered proxies for Earth materials is two-fold: (a) the mathematical rendering of details of the solid structure for simulation purposes is more straightforward when using synthetic materials and (b) interpretation of experimental results is less cumbersome.

The objective of this chapter is to provide fundamental, molecular- to microscopic-level descriptions of the sorptivity, structure, and dynamics of hydrocarbon (HC)-bearing fluids at mineral surfaces or within nanoporous matrices. The emphasis is on non-aqueous systems. Wherever possible we highlight results obtained from higher temperature-pressure sorption experiments, neutron scattering, NMR, and molecular-level modeling that have relevance to the deeper carbon cycle, although such studies are not common. We will not focus on the voluminous literature describing the behavior of hydrocarbons on activated carbon, carbon nanotubes, coal, synthetic gas storage materials such as metal-organic frameworks, and polymers. Rather we will emphasize key experimental and modeling results obtained on oxides and silicates, and synthetic engineered equivalents such a meso- and microporous silica and certain zeolites [the reader is referred to the book by Kärger et al. (2012) for more in-depth coverage of fluid behavior in zeolites]. We will assume as a first approximation that these engineered substrates act as reasonable proxies for the Earth materials. To best capture the behavior of hydrocarbons on mineral substrates or within nanoconfined volumes, we will address three key subtopics:

1. Adsorption-desorption behavior of methane and related HC fluids (and their mixtures) on a variety of substrates and in nanoporous matrices that yield microstructural insights,
2. Dynamical behavior of methane and related HC volatiles at mineral surfaces and within nanopores with and without surface H_2O present,
3. Molecular-level modeling results that provide important insights into the interfacial properties of these mineral-volatile systems, assist in the interpretation of experimental data and predict fluid behavior beyond the limits of current experimental capability.

Probing C-O-H behavior with neutron scattering and NMR

The properties of neutrons make them an ideal probe for comparing the properties of bulk hydrocarbons with those filling confined geometries (Pynn 2009). Neutrons can be scattered either coherently or incoherently, thus providing opportunity for various kinds of analysis of both structural and dynamic properties of confined liquids. Such analysis is possible due to the fact that the wavelengths of thermal and cold neutrons are comparable with intermolecular distances in condensed phases, while the neutron energy can be tailored to probe both high- (collective and single-particle vibrational) and low-frequency (single-particle diffusive) motions in the system. Importantly, the large incoherent scattering cross section of hydrogen compared to other elements allows obtaining scattering spectra dominated by the scattering from hydrogen-containing species (see recent review article by Neumann (2006)), whereas the X-ray scattering from such systems, which is virtually insensitive to hydrogen, would be dominated by the signal from the confining matrix. Last but not least, the large difference in the coherent and incoherent neutron scattering cross sections of hydrogen and deuterium allows selection of atoms to dominate the scattering signal by means of deuteration of the fragments of liquid molecules or the confining matrix.

Nuclear magnetic resonance (NMR) is a resonance concept between magnetization of nuclear spins and magnetic radio-frequency waves (Abragam 1961; Ernst 1987). NMR is employed in order to study local molecular properties of matter in detail regardless of the state of the system. One of the primary "fingerprints" of local structure probed by solid-state NMR is the chemical shielding, usually measured as the "chemical shift," which is the difference in NMR resonance frequency of the species of interest from that of a known reference compound. Isotropic chemical shifts are indicative of local environment, and are often tabulated with typical ranges given for specific types of local bonding environments. For example, ^{29}Si nuclei in silicon atoms bound through three bridging oxygen atoms to other silicon atoms and one oxygen that is part of a hydroxyl group are typically found in the range of −110 to −120 ppm of shift from the reference of a tetramethylsilane molecule in neat solution. However, when proposed chemical species are difficult to identify by shift alone, multiple-resonance NMR methods (Pantano et al. 2003) are used to probe and compare to quantum chemical calculations (Fry et al. 2006; Johnston et al. 2009) for determination of local ordering.

Many of the chemistries of interest in hydrocarbon systems will occur at surfaces of materials and "surface selective" NMR will aid in the understanding of adsorption and reactivity. Hydrocarbons whose under-confinement behaviors have been investigated by NMR include benzene, derivatives of benzene, diethyl ether, methane, and acetone (Stallmach et al. 2001; Krutyeva et al. 2007; Xu et al. 2007). These hydrocarbons and their derivatives were confined into various porous systems such as mesoporous and nanoporous MCM-41 (Stallmach et al. 2001; Xu et al. 2007), mesoporous Vycor glass (Dvoyaskin et al. 2007), trimethyl-silylized nanoporous silica gel (Fernandez et al. 2008), and silicate zeolite (Pampel et al. 2005). The first important issue regarding porous system surface characterization is the utilization of solid-state magic angle spinning (MAS) NMR. In MAS NMR, the effect of molecular dynamics on the NMR interactions is mimicked by fast sample spinning around an axis, including the "magic angle" of 54.7° with the external applied magnetic (B_0) field. Fast sample spinning such as MAS NMR experiments removes most anisotropic nuclear interactions leading to line broadening in NMR spectra of solids. Nowadays, typical rotation speeds are in the range of 10-70 kHz with various outer diameter rotors of 7 mm to 1.5 mm (Xu et al. 2007; Vogel 2010). Solid-state MAS NMR on porous systems include, for instance, quantifying accessible hydroxyl sites of porous surfaces (Fry et al. 2003; Pantano et al. 2003). The detection of specific species on the surfaces of materials is often accomplished with heteronuclear correlation methods such as cross-polarization (Fry et al. 2003, 2006; Tsomaia et al. 2003) or J-coupled NMR spectroscopy. Surface selective spin correlation methods can be used on both raw samples and these same samples during and after contact with fluids of interest to map out possible reactive structures as well as reaction products or sites depleted by interactions with the alkane species. NMR studies of the alkanes themselves, as well as any water or CO_2 incorporated with the alkanes, will focus on studies of chemical shift—primarily ^{13}C and ^{1}H—as well as spin relaxation and diffusion experiments using these same nuclides. Demonstrated efforts (Riehl and Koch 1972) have revealed the depth of information available from systematic measurements of longitudinal and transverse spin relaxation rates ($1/T_1$ and $1/T_2$, respectively) as well as diffusion coefficients (Dawson et al. 1970; Helbaek et al. 1996). In NMR experiments, upon the application of radio-frequency pulses to disturb the magnetization of nuclear spins, the equilibrium distribution state is reestablished by relaxation processes. The longitudinal relaxation time T_1 and transversal magnetization time T_2 are the magnetization components parallel and perpendicular to external magnetic field B_0, respectively. T_1 is an energy driven process, while T_2 is governed by entropy. Both T_1 and T_2 are important parameters to be measured since they are traditional ways to study molecular reorientations (Abragam 1961; Vogel 2010).

NON-AQUEOUS FLUID ADSORPTION BEHAVIOR: EXPERIMENTAL

Background on adsorption concepts and approaches

The interaction of short-length linear alkanes including methane, ethane, propane, and *n*-butane and their mixtures with high surface area solids has received considerable attention over many decades, driven largely by the separations and catalysis communities. Substrates that have received the most attention include various types of carbon (e.g., activated carbon, carbon black, carbon fibers, and coal; Jiang et al. 2005), pillared clays (Li et al. 2007), clathrates (Roman-Perez et al. 2010), and zeolites, both natural and synthetic (Denayer et al. 2008). More recently synthetic materials such as micro- and mesoporous silica and metal-organic frameworks have been used to adsorb selectively light hydrocarbons from various gas mixtures.

Precise determination of adsorption/desorption isotherms at geological conditions is key to identifying surface-fluid interactions. Sorption measurements (volumetric, gravimetric, and calorimetric) are important because they provide valuable insights into the interfacial interactions between the solid substrate and the fluid phase. The magnitude of adsorption, the shape of the adsorption isotherm, and the presence or absence of hysteresis and its magnitude reveal a great deal about the properties of the surface, adsorbate molecule configurations, and the interactions between both adsorbate molecules themselves and with the substrate (Rouquerol et al. 1999; Myers and Monson 2002). Volumetric or gravimetric-based isotherms provide the most direct measure of such interactions to which other structural data, such as small angle neutron scattering (SANS) and neutron reflectivity (NR), can be compared. In most cases, however, these types of measurements are conducted at low temperature and modest pressures. Various kinds of carbon-based porous materials, zeolites, and modified organic frameworks are widely studied for methane adsorption (Zhang et al. 1991; Menon and Komarneni 1998; Cavenati et al. 2004; Wu et al. 2009). Carbon-based materials are more efficient adsorbents compared to zeolite on a weight basis. The reverse trend is observed on a volume basis because of the high solid density of zeolites (Zhang et al. 1991). Few studies explored the effect of moisture on adsorption capacity relative to methane (Clarkson and Bustin 1996; Rodriguez et al. 1997). Adsorption of methane and wet hydrocarbon gases on natural clays is rarely studied due to lower adsorption capacity of natural minerals (Stoessell and Byrne 1982). Pires et al. (2008) studied the selective adsorption of CO_2, methane, and ethane on porous clay heterostructures at ambient temperature and demonstrated that alkanes adsorb proportionally to free volume of adsorbent but CO_2 shows an inverse trend. This anomalous behavior is attributed to clay composition as well as specific interactions of CO_2 with the surface. Cheng and Huang (2004) have reported a comparative study of adsorption of C_1-C_6 hydrocarbons in gas mixtures on a variety of clays and organic matter at lower pressures and temperatures up to 80 °C. This study confirms that despite the lower adsorption capacity for clay surfaces compared to coal substrates, the amounts adsorbed are significant (50-75% of the amount sorbed on coal). Similarly, few studies have been published on adsorption of methane and other hydrocarbon gases on natural silica (Wu et al. 1994). Although various studies discuss adsorption of CO_2 on coal substrates, zeolites, assorted metal oxides, and modified silica/clays, a limited number have tried to study CO_2 adsorption on natural minerals (e.g., Yong et al. 2002) particularly in the presence of water and aqueous carbonate species (e.g., Villalobos and Leckie 2000).

While adsorption-desorption phenomena have been the focus of many of these studies issues such as rates of sorption and transport behavior (diffusivity) have also been addressed (Schloemer and Krooss 2004; Kim and Dauskardt 2010). Techniques that have been used in these kinds of studies include but are not limited to volumetric and gravimetric sorption isotherm measurements, differential scanning and micro-calorimetry (DSC), NMR, FTIR, scattering (light, X-ray, and neutrons) and diffraction (X-ray, neutrons). For the most part these studies involve either high pressure at cryogenic to near-ambient temperature conditions or the converse, high temperature, but low pressures (few MPa). The use of extreme conditions of temperature and pressure is limited by the availability of novel high temperature-pressure sorption apparatus such as Rubotherm's magnetic suspension balance and appropriate high pressure-temperature scattering sample cells for *in situ* interrogation. Molecular dynamics, Monte Carlo, and *ab initio* methods are widely used to predict sorption and transport behavior (Combariza et al. 2011; Krishna and van Baten 2011) and/or help interpret experimental data (Lithoxoos et al. 2010). Collectively, results from these studies (and many more not formally cited) provide an important framework for gaining a fundamental understanding of the interfacial behavior of the light hydrocarbons interacting with synthetic "Earth" proxies, as well as natural mineral and rock matrices at conditions relevant to shallow crustal settings; ~200 °C and 100 MPa. However, to our knowledge, with the possible exception of coal, there has not been a systematic study of the effects of temperature, pressure, variable pore size, shape, roughness, and connectivity, and degree of surface hydrophobicity on light hydrocarbon-mineral interaction relevant to subsurface fine-grained sedimentary lithologies. Further, fundamental understanding of interfacial behavior involving interactions between hydrocarbons and aqueous films on minerals is very poor compared to studies that have explored H_2O wetting phenomena (e.g., Fenter 2002), and the role of confinement, surface charge, and electrochemical reactions on surface forces at mineral surfaces (e.g., Alcantar et al. 2003; Anzalone et al. 2006; Greene et al. 2009).

Behavior of hydrocarbons and related C-O-H fluids in the presence of complex solution chemistry (e.g., elevated electrolyte concentrations; silica) at elevated temperatures and pressures is obviously of more relevance to our understanding of the deep carbon cycle. A number of sorption studies have been performed on supercritical and near-critical fluids, mainly using volumetric techniques. Experimental adsorption isotherms obtained over wider ranges of pressure extending to compressed liquid or dense supercritical fluid revealed effects that were not present or could be neglected in low-density gas adsorption. Pronounced high-pressure depletion effects over a large region of fluid densities have been reported for argon, neon, krypton, nitrogen, and methane physisorbed to activated carbon (Malbrunot et al. 1992). Thommes et al. (1995) studied the sorption of supercritical SF_6 to mesoporous CPG-10 silica and found adsorption at low fluid density, but a strong decrease of the adsorbed amount of fluid in the vicinity of the critical point, and named the effect *critical depletion*. Rayendran et al. (2002) report the occurrence of critical depletion for N_2O sorption to silica gel. Several theoretical and simulation studies have been published on critical/high-density depletion phenomena (Maciolek et al. 1999; Brovchenko et al. 2004, 2005; Oleinikova et al. 2006; Brovchenko and Oleinikova 2008), with partly conflicting results. Part of the problem is that the common quantity measured in sorption experiments, excess adsorption, gives only the net sorption effect, but cannot provide a microscopic picture of the fluid-substrate interactions.

Observationally as one traverses from low to high density, the experimental adsorption isotherms for C-O-H fluids can exhibit a maximum at the density of bulk fluid approaching its critical value (Parcher and Strubinger 1989; Strubinger and Parcher 1989; Aranovich and Donohue 1998; Donohue and Aranovich 1999). At still higher densities, experimental excess adsorption isotherms may reach zero or even negative values. This behavior is independent of any additional volumetric effects, such as adsorbent swelling or deformation at elevated pressure. The decrease of the amount adsorbed with pressure may appear counterintuitive by implying a mechanical instability of the system. However, the quantity measured by all conventional methods does not represent the total amount of fluid present in the vicinity of solid surface but instead the excess adsorption (Gibbs surface excess; Fig. 2)—the difference between the actual amount of fluid contained inside a pore system and the hypothetical amount of fluid at bulk density filling the pore spaces, i.e. in absence of fluid-solid interactions.

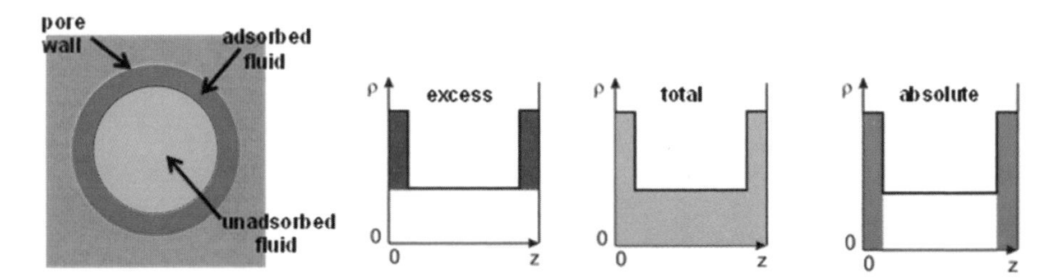

Figure 2. Schematic of a single pore (left panel) showing the adsorbed phase (red) and the unadsorbed fluid (blue). Also shown are the three types of sorption described in the text – excess, total and absolute. ρ refers to fluid density and z is the diameter of the pore. (G. Rother, pers. commun.)

C-O-H pore fluid densities

As noted above, excess adsorption, which can be measured without the knowledge of any microscopic properties of the adsorbed phase, is an important quantity used for thermodynamic analysis of many aspects of adsorption and is routinely used in modeling and control of technological processes (Sircar 1999). However, the properties of the adsorbed phase, including the average density of the pore-filling fluid, are essential in quantifying fluid-rock interactions in systems dominated by nano- to microscale pore features (Rother et al. 2007). Gruszkiewicz et al. (2012) reported results on propane (C_3H_8) and CO_2 obtained from a novel high temperature-high pressure vibrating tube densimeter (VTD) capable of measuring pore fluid density and total adsorption capacity in mesoporous solids. The densities were determined for propane at subcritical and supercritical temperatures (between 35 °C and 97 °C) and carbon dioxide at supercritical temperatures (between 32 °C and 50 °C) saturating hydrophobic mesoporous silica aerogel (0.2 g/cm^3, 90% porosity) synthesized inside Hastelloy U-tubes (Fig. 3). In this method the porous solid completely fills the tube, so that virtually no bulk fluid outside of the pore system is present in the measurement zone; i.e., the contact with the bulk fluid reservoir occurs outside of the vibrating cantilever. The mass of the pore fluid, proportional to its average density, is measured directly as the inertia of the cantilever containing the solid sample imbibed with fluid. Additionally, supercritical isotherms of excess adsorption for CO_2 and the same porous material were measured gravimetrically using a precise Rubotherm magnetically-coupled microbalance.

The densities of pore-filling propane measured at four subcritical temperatures (35, 70, 92, and 95 °C) and at 97.0 °C, about 0.3 °C above the critical temperature (T_c = 96.7 °C) are given in Figure 4 plotted against pressure. Also shown are the densities of bulk fluid calculated from the equation of state (Span and Wagner 1996; Lemmon et al. 2010). Pore fluid densities and

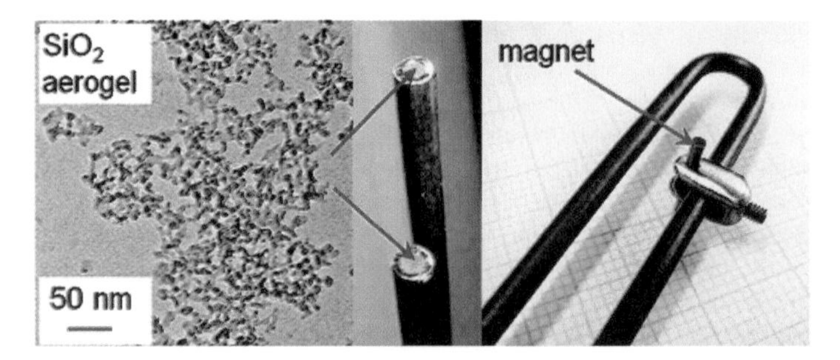

Figure 3. A transmission electron microscopy (TEM) image of the silica aerogel structure (silica strands are darker), the Hastelloy U-tube with the silica aerogel synthesized inside, and magnet clamp attached to the U-tube. [Used by permission of the American Chemical Society © 2012, from Gruszkiewicz et al. (2012) *Langmuir,* Vol. 28, Fig. 1, p. 5073.]

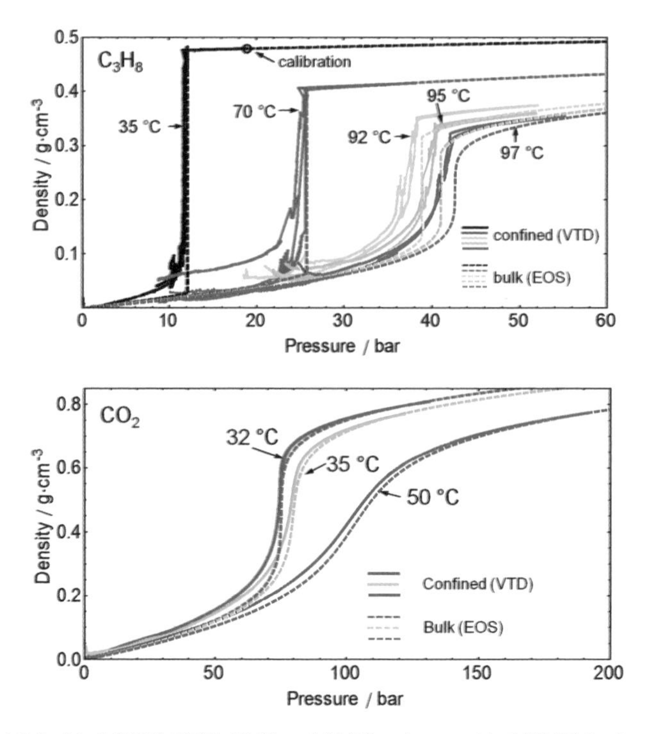

Figure 4. (*top*) Subcritical (35 °C, 70 °C, 92 °C, and 95 °C) and supercritical (97 °C) isotherms of confined fluid density for propane in silica aerogel. (*bottom*) Supercritical (32, 35 and 50 °C) isotherms of confined and bulk fluid density for carbon dioxide in silica aerogel. [Used by permission of the American Chemical Society © 2012, from Gruszkiewicz et al. (2012) *Langmuir,* Vol. 28, Fig. 4, p. 505 (*top panel*), Fig. 6, p. 5076 (*bottom panel*).]

total adsorption isotherms increased monotonically with increasing density of the bulk fluid, in contrast to excess adsorption isotherms, which reached a maximum and then decreased towards zero or negative values above the critical density of the bulk fluid (Fig. 4). Compression of the confined fluid significantly beyond the density of the bulk fluid at the same temperature was observed even at subcritical temperatures. The isotherms of confined fluid density and excess adsorption (not shown) contain complementary information. For instance, the maxima of excess adsorption occur below the critical density of the bulk fluid at the beginning of the plateau region in the total adsorption, marking the end of the transition of pore fluid to a denser, liquid-like pore phase. No measurable effect of pore confinement on the liquid-vapor critical point was found. The results for propane and carbon dioxide showed similarity in the sense of the principle of corresponding states. Good quantitative agreement was obtained between excess adsorption isotherms determined from VTD total adsorption results and those measured gravimetrically at the same temperature, confirming the validity of the vibrating tube measurements. The flatter initial slopes exhibited in the propane isotherms are indicative of relatively weak fluid-pore wall interactions. The steep increase in density at higher pressures can be indicative of pore condensation and/or stronger fluid-fluid interactions. The somewhat steeper initial slopes associated with lower pressures observed for the CO_2 are typical for a fluid that experiences a somewhat stronger fluid-pore wall interaction compared to propane.

To better compare these results and to emphasize the corresponding states' similarity between the fluids, the results are also shown in Figure 5 intermsofreduceddensities, $\square_r=\square/\square_c$, where $\square_c = 0.220$ g/cm^3 for C_3H_8 and $\square_c = 0.4676$ g/cm^3 for CO_2. The diagonal dashed straight line in Figure 5 represents the hypothetical condition where the confined fluid density is equal to the bulk fluid density; the deviations of the experimental isotherms from this line represent the excess density due to solid-fluid interactions. This figure demonstrates that the confined fluid densities, and consequently total adsorption isotherms, are non-decreasing functions of increasing bulk fluid density. Each of the subcritical isotherms features a plateau formed by a straight tie line extending between the densities of bulk vapor and liquid phases in equilibrium. The dotted curve in Figure 5 represents the vapor-liquid equilibrium envelope of bulk propane with the densities of the phases in equilibrium at each temperature and the bulk fluid critical point marked with symbols.

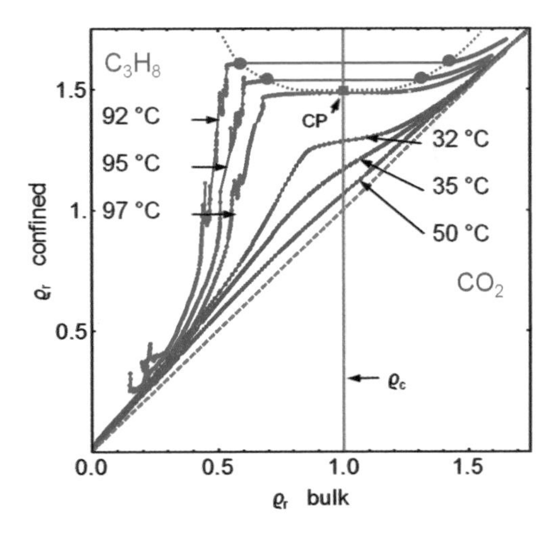

Figure 5. Total confined fluid reduced density (g/cm^3) isotherms for C_3H_8 and CO_2 plotted as a function of bulk fluid reduced density. [Used by permission of the American Chemical Society © 2012, from Gruszkiewicz et al. (2012) *Langmuir,* Vol. 28, Fig. 7, p. 5077.]

Hydrocarbon-interfacial microstructure

As noted previously, the sorption of gaseous subcritical fluids on solid substrates has been studied extensively (Schreiber et al. 2002; Sel et al. 2007), while only a few studies exist on the nanoscale structure and dynamics of interfacial fluids, and almost nothing is known about the interfacial properties of near-critical and supercritical fluids. In the context of hydrocarbons, this is an important *P-T* regime because these fluids will be present at supercritical conditions. A poorly constrained yet fundamentally important fluid behavior has been identified wherein at *P-T*-density conditions below the critical point, fluid volume and density increase as the critical point is approached, while above the critical density fluid volume remains essentially constant but density decreases—the so-called fluid depletion effect where negative values of excess adsorption are estimated (Malbrunot et al. 1992; Thommes et al. 1995; Rajendran et al. 2002). However, theoretical and simulation efforts to model these data give conflicting results

(Maciolek et al. 1998, 1999; Brovchenko et al. 2004, 2005; Oleinikova et al. 2006), which is due to the lack of experimental micro-structural characterization.

Many studies deal with the properties of fluids and fluid mixtures imbibed in the pores of engineered nanoporous materials. Porous silica (SiO_2) is frequently chosen because it can be synthesized with well-defined pore sizes in the range of less than 1 nm up to several tens of nm. The structural properties of confined liquids can be assessed using coherent scattering techniques, neutron diffraction (ND), and small-angle neutron scattering (SANS). The former allows one to measure the static structure factor, $S(Q)$, which can be then Fourier transformed to obtain the radial pair-distribution function, $g(r)$, that describes the distribution of the distances between the coherently scattering nuclei in the liquid. While ND measurements of liquids in confinement probe structural correlations not exceeding a few molecular diameters, SANS measurements provide coverage over much broader range in the real space (Radlinski 2006; Triolo and Agamalian 2009). This is because SANS involves measuring neutron intensities at very low values of the scattering vector, Q (i.e., at small angles).

SANS has been widely used in the study of fluid behavior in porous media and recently became the first technique capable to quantify the sorption properties of C-O-H fluids in porous media in terms of the mean density and volume of the adsorbed phase (Rother et al. 2007). In this study, the sorption properties of supercritical deuterated propane and CO_2 in silica aerogel with 96% porosity and pores ranging from 20-50 nm were investigated. SANS and neutron transmission data have been measured for fluid-saturated silica at different fluid densities and temperatures. The mean density ρ_3 and volume fraction ϕ_3 of the sorption phase were calculated from the SANS and neutron transmission data by application of a new model, which makes use of the three-phase model by Wu (1982) and a mass balance consideration of the pore fluid, which can be obtained from neutron transmission measurements or gravimetric sorption measurements. It was found that the fluid is adsorbed to the porous matrix at low fluid densities but depleted from the pore spaces at higher fluid densities (i.e., in the vicinity of the critical density and above). Figure 6 shows the evolution of the physical properties of the sorption phase, expressed in terms of ρ_3 as a function of temperature and (bulk) fluid density ρ_2. The bulk critical density of deuterated propane is $\rho_c \approx 0.27$ g/cm^3, and the critical temperature is $T_c \approx$ 91.0 °C. The fluid density in the adsorbed phase is up to about three times higher than ρ_2 in the low-pressure region, while it remains constant and below ρ_2 at and above the critical pressure.

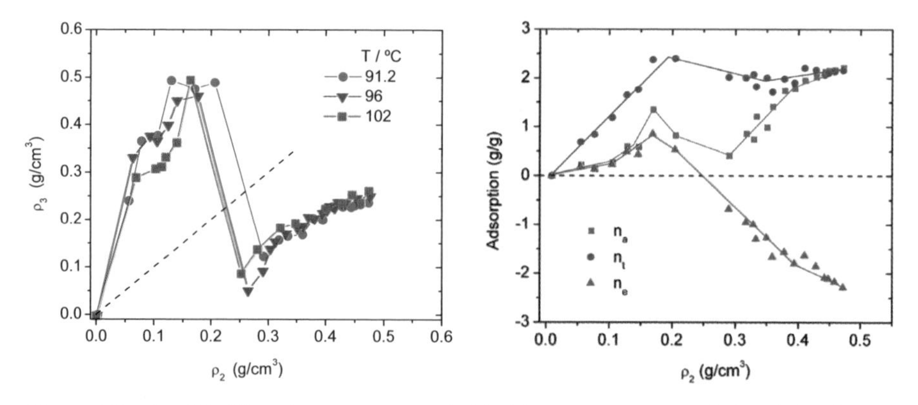

Figure 6. Results of SANS on deuterated propane inside silica mesoporous aerogel (0.1 g/cc) at three supercritical temperature from; (*left*) mean density ρ_3 of the sorption phase as a function of the density of the unadsorbed fluid; dashed line represents behavior for a fluid-porous solid system with no adsorption; (*right*) adsorption quantities calculated from the density and volume of the sorption phase. n_a = absolute adsorption; n_t = total adsorption; n_e = excess adsorption. [Used by permission of the American Chemical Society © 2007, from Rother et al. (2007) *J Phys Chem-C,* Vol. 111, Fig. 6, p. 15740 and Fig. 8, p. 15740.]

With the information on ρ_3 and ϕ_3 (not shown), calculation of the absolute adsorption, which is the relevant quantity for the application of the equation of adsorption and molecular modeling work, is possible without the introduction of further assumptions. The calculated values for the absolute adsorption (n_a), total adsorption (n_t) and excess adsorption (n_e) are given in Figure 6. The absolute adsorption is similar to the commonly measured excess adsorption only at low fluid densities but differ significantly at higher fluid densities. Cole et al. (2010) has been able to theoretically emulate the shape of this trend by using an integral equation approximation. From the sorption phase density and volume the excess sorption, total sorption and absolute sorption can be calculated. These neutron results have been compared to excess sorption data measured with gravimetric techniques and total sorption data measured with the vibrating tube densimeter (Rother et al 2012; Gruszkiewicz et al 2012). General agreement of the data has been found, verifying the validity of the neutron method.

In a related study, Kainourgiakis et al. (2010) studied the behavior of water, hexane, and hexane-water mixtures at ambient temperature imbibed in macroporous α-Al_2O_3 (34% porosity; ~180 nm pores; total pore volume of ~0.14 cm^3/g), using ultra-small angle neutron scattering (USANS). The intent was to quantify the multiphase pore-filling behavior of a wetting (water) and non-wetting fluid (hexane). They used a 7.3% H_2O/D_2O mixture that matched the scattering length of the alumina matrix, which allowed the study of the contributions to the scattering signal of the individual phases in water-water and water-hexane systems, and the hydrocarbon in hexane-air systems. In the case of the water-loaded samples (and to a certain extent also when water–hexane mixtures are used) the progressive hydration leads towards the formation of larger water-hydrocarbon clusters. These clusters can be considered as biphasic "aggregates," comprising continuous solid- and water-rich regions, which nevertheless can be characterized as homogeneous in terms of scattering behavior since the 7.3% H_2O/D_2O mixture used has the same scattering length density as alumina. This characterization is indeed confirmed by the practically zero intensity recorded when the pore space is fully occupied ($Vs_{water} = 1$) by the aqueous phase. On the other hand, the spectra of the samples impregnated with hexane exhibit, in practice, the same trend, while the slight variation of intensity is attributed to the reduced contrast achieved as increased quantities of the hydrocarbon are introduced in the pore network. Most interestingly, the autocorrelation function curves obtained from scattering for the cases of the partial filling (e.g., $Vs_{water} = 1/3$ or $2/3$ where Vs is defined as the fraction of total pore volume occupied by a certain fluid component) of the pore volume with only the aqueous phase and the complete saturation with an equivalent water–hexane mixture (i.e., $Vs_{water} = 1/3$ or $2/3$ and $Vs_{hexane} = 2/3$ or $1/3$) practically coincide despite the differentiation of the fluids occupying the pore volume and the significant variation therefore of the interfacial energies coupled in the respective systems. In practical terms this important observation provides direct experimental evidence that the spatial distribution of the fluid phases is related to their wetting/non-wetting relative behavior and is not affected significantly by the actual values of their particular interfacial properties. Simulations of this wetting process for different loadings are shown in Figure 7.

Zeolites are microporous aluminosilicate minerals that play an important role in many natural and industrial processes, including water purification through ion exchange, catalytic hydrocarbon cracking, and separation of pollutants from natural gas. They possess pore widths of typically a few tenths of a nanometer, making X-ray and neutron diffraction suitable tools for the study of these materials and guest molecules inside their pore systems. Neutron diffractometers exist in a variety of configurations for thermal and cold neutrons optimized for resolution or flux and interrogate length scales of up to 2 nm. The ND technique has been recently used by Mentzen to study the adsorption of hydrogen and benzene in MFI-type zeolites (Mentzen 2007). The positions of the guest molecules in the microporous host structure were determined through Rietveld analysis to define the binding sites. The sorption of several hydrocarbons, including heptane, in silicalite-1 zeolite was studied by Floquet et al. (2003,

Figure 7. 3D simulation images of (a) dry alumina structure; (b) spatial distribution of a wetting fluid for Vs = 1/3 and (c) Vs = 2/3. Each block is approximately 500 nm on a side. . [Used by permission of Elsevier © 2010, from Kainourgiakis et al. (2010) *Appl Surf Phys*, Vol. 256, Fig. 3, p. 5332.]

2007), who used ND. Their results indicate that heptane populates the straight channels of the silicalite pore network first, while the sinusoidal channels and intersections fill with heptane only above a heptane concentration of about 3.9 molecules per unit cell.

NON-AQUEOUS FLUID DYNAMICS AT INTERFACES: EXPERIMENTAL

Dynamical fluid behavior is controlled by processes occurring at the fluid-pore wall or fluid-fluid interface, as well as by the rates of supply and removal of mobile constituents. Key issues pertaining to these types of interactions that remain largely unresolved, particularly for conditions of elevated temperature and pressure, include the extent of possible hydrogen bonding, molecular translation and rotation, average times between molecular jumps, and self-diffusion, which involve specific interactions between neighboring molecules leading to preferred molecular orientation that are affected by reduced dimensionality and fluid-pore wall interactions (Cole et al. 2006, 2009). Of the techniques that are available, neutron scattering and nuclear magnetic resonance are probably the most heavily used to obtained dynamical information on hydrogen-bearing fluids, such as self-diffusion, translational, and rotational motion from hydrocarbon-matrix interactions. More specifically, Pulsed Field Gradient NMR (PFG-NMR) was the first "microscopic" method advanced for the measurement of diffusion in zeolites in the late 1970s. It allowed accessing molecular displacements down to 0.1-1 μm for times on the order of 1 ms and was thus instrumental in elucidating intracrystalline and intercrystalline transport phenomena for a wide variety of systems (Jobic and Theodorou 2007). Quasielastic neutron scattering (QENS) is a complementary method to NMR that tracks the diffusive motions that take place on the timescale of a pico- to nano-second, which corresponds to the energy scale from a fraction of μeV to several hundred μeV (Bee 2003). The more energetic (on the energy scale from several to several hundreds of meV) vibrational and librational modes are typically probed using dedicated neutron spectrometers with moderate energy resolution and reasonably high incident neutron energies. On the timescale of such spectrometers, rotational and translational motions are very slow and can be neglected. This type of measurement is known as inelastic neutron spectroscopy (INS). Compared to infrared spectroscopy, INS benefits from the absence of optical selection rules and the large incoherent scattering cross section of hydrogen.

QENS probe of hydrocarbons in nanopores

As noted above QENS is an excellent tool to probe the mobility of confined hydrogen-bearing fluids, the property affected the most by a confinement; a change by one to two orders of magnitude in the mobility of a confined liquid is common. QENS targets the signal from incoherently scattering nuclei such as H, leading to a description of self-diffusion. The signal measured by a spectrometer as a function on neutron energy transfer, E, is a Lorentzian with a half-width at half maximum (HWHM) $\Gamma = hDQ^2$, where Q is the momentum transfer to the particle in the scattering process and D is the diffusion coefficient (Cole et al. 2006). In general, QENS probes rotational and translational diffusive motions of molecules that result in the broadening of the elastic peak. In QENS measurements, the effects of faster vibrational and librational motions manifest themselves in the overall reduction of scattering intensities (Debye-Waller factor). It should be noted that knowledge of the resolution function and an extremely good energy resolution are of paramount importance in QENS. For this reason, time-of-flight and backscattering spectrometers built at cold neutron sources are frequently employed in this type of experiment.

Even though water has been by far the most extensively investigated medium in nano-confinement, other fluids in confined environments such as methane and other alkanes have attracted some attention. The majority of studies using QENS to interrogate hydrocarbon dynamics have focused on behavior in a variety of synthetic silicas and zeolites. For example, Benes et al. (2001) presented QENS results on the temperature dependence (200 to 250 K) of methane self-diffusion and molecular rotation in microporous silica with pores smaller than 1 nm. The self-diffusion coefficients of translational motion range from 1.1×10^{-8} m^2 s^{-1} at 200 K to 1.9×10^{-8} m^2 s^{-1} at 250 K with an estimated activation energy of 4 kJ mol^{-1}. The isotropic rotation diffusion constant is on the order of 10^{11} s^{-1}. Jobic (2000a, b) described the dynamics of complex hydrocarbon molecules in confinement (linear and branched alkanes, for hydrocarbon chains up to C-14, confined in ZSM-5 zeolite, Fig. 8a). In ZSM-5 zeolite there are two types of channels consisting of ten-membered oxygen rings. The straight elliptical channels (0.57-0.52 nm) are interconnected by near-circular channels (0.54 nm) in a zig-zag fashion, and there are four channel intersections per unit cell. Because of the relatively large size of the confined molecules, the diffusion could be observed within the time window of a backscattering spectrometer only at high temperatures. Branched alkanes were found to diffuse much more slowly than linear alkanes. Mitra and Mukhopadhyay (2004) reported on the residence times (τ), mean jump length (l), and translational motion (D) of propane in Na-Y zeolite (Na:Al = 1.7). The Na-Y zeolite structure is made up of a network of tetrahedrally connected pores (a-cages) of diameter ~1.18 nm. The pores are interconnected through windows of diameter ~0.8 nm. A schematic of Na-Y zeolite structure is shown in Figure 8b. They compared the experimental results with MD as shown in Table 1. Mamontov et al. (2005) explored the diffusion and relaxation dynamics of benzene (C_6H_6) in oriented 5-nm nanochannels of chrysotile [$Mg_3Si_2O_5(OH)_4$] asbestos fibers from 260 to 320 K (Fig. 9). The

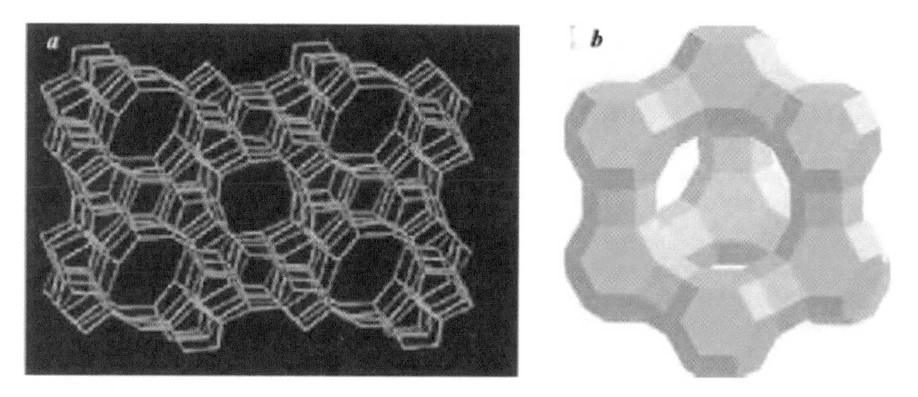

Figure 8. Schematic of cage pore structure in two commonly used zeolites (a) ZSM-5; channel diameter 0.52-0.57 nm and (b) Na-Y; channel diameter 0.8 nm. [Used by permission of Indian Academy of Sciences © 2003, from Mitra and Mukhopadhyay (2003) *Curr Sci*, Vol. 84, Fig. 3, p. 657.]

Table 1. Dynamical parameters for residence times (τ), jump length (l) and translational motion of propane adsorbed in Na-Y zeolite (Mitra and Mukhopadhyay 2004).

T (K)	τ (ps)	$(l^2)^{0.5}$ (nm)	D (× 10^{-5} cm^2 s^{-1}) QENS	D (× 10^{-5} cm^2 s^{-1}) MD
300	4.6 ± 0.5	0.25 ± 0.02	2.3 ± 0.3	3.6
324	4.3 ± 0.4	0.29 ± 0.02	3.2 ± 0.3	3.2
350	3.8 ± 0.3	0.30 ± 0.03	4.0 ± 0.4	3.0

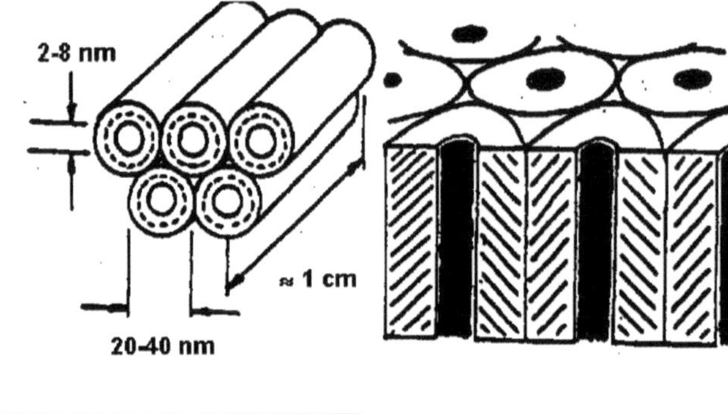

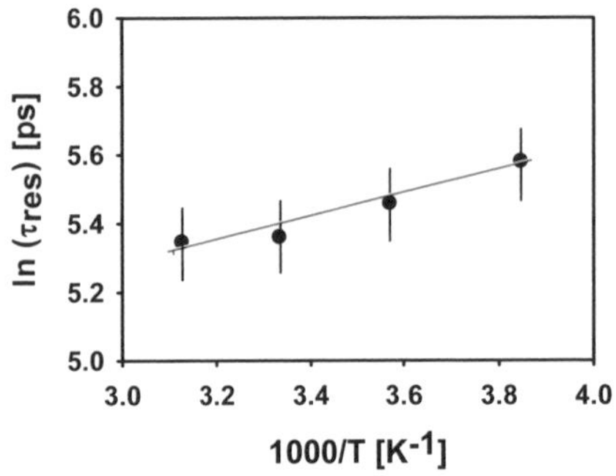

Figure 9. (*top*) Schematic picture of a bundle of chrysotile asbestos fibers. (*left*) temperature dependence of the average residence time (τ) between translational diffusion jumps of benzene and its fit with Arrhenius law. [Used by permission of American Physical Society © 2005, from Mamontov et al. (2005) *Phys Rev E*, Vol. 72, Fig. 1, p. 051502-1, and Fig. 6, p. 051502-5. *http://link.aps.org/doi/10.1103/PhysRevE.72.051502*]

macroscopic alignment of the nanochannels provided an opportunity to study the anisotropy of the dynamics of a confined fluid by means of collecting the data with the scattering vector either parallel or perpendicular to the fibers axes. The translational diffusive motion of benzene molecules was observed to be isotropic. Diffusivities were not strongly temperature dependent and ranged from 0.88 × 10^{-10} m^2 s^{-1} to 1.31 × 10^{-10} m^2 s^{-1}. Conversely, the residence times between translational jumps exhibited a weak temperature dependence (Fig. 9) and yielded a low activation energy of 2.8 kJ mol^{-1}.

A recent survey by Jobic and Theodorou (2007) provides an excellent overview of QENS studies of confined media in a number of zeolites and their synergy with molecular dynamic simulations. They compared the behavior of various alkanes in Al-Si-O-based zeolites with or without counterions such as Na (e.g., Si-only silicalite-1; Na-ZSM-5 [$Na_nAl_nSi_{96-n}O_{192}\cdot16H_2O$ ($0 < n < 27$)]: their unit cells contain 96 tetrahedral units with Si or Al as central atoms and oxygen as corner atoms. These structures (MFI; mordenite framework inverted) contain straight channels and zigzag channels, both with free apertures of about 0.55 nm in diameter. In general, self-diffusion coefficients decrease with increasing carbon number, are faster by a factor of 4-5 for zeolites without a counterion such as Na, and tend to be faster than comparable systems interrogated by pulse-field gradient NMR. In all probability, this discrepancy can be attributed to defects in the silicalite-1 structure. PFG-NMR, which measures displacements on the order of micrometers, is much more sensitive to such defects than QENS, which has an effective length scale of nanometers. Molecular simulations, which postulate a perfect crystalline structure, are closer to QENS than to PFG-NMR. For linear pores less than 1 nm in diameter there is a general tendency for the diffusion coefficients to decrease with increasing alkane chain length (Jobic et al 2010). This unexpected result (Fig. 10) can be explained by the fact *n*-butane keeps the same orientation in the 1D channels of the V^{4+}-metal organic framework (MIL-47) so that it loses less momentum than propane, which tumbles within the pore channel.

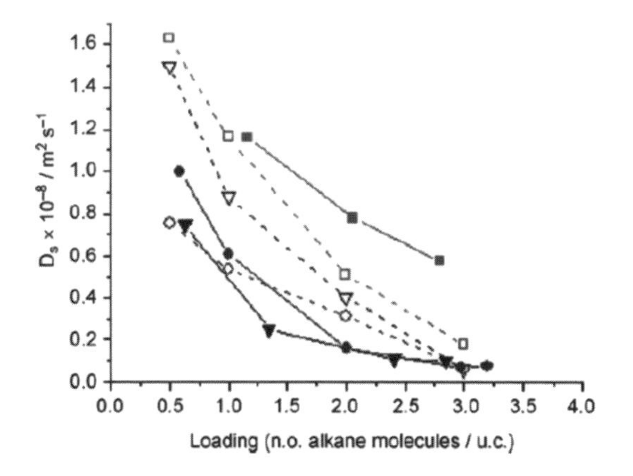

Figure 10. Self-diffusion coefficients (D_s) for ethane (square), propane (inverted triangle) and *n*-butane (circle) in V^{4+} MIL-47 metal organic framework as a function of the loading at 300 K. QENS (solid symbols), MD (open symbols). Note that the reported D_s values are orientationally averaged. [Used by permission of Wiley-VCH © 2010, from Jobic et al. (2010) *Chem Eur J*, Vol.16, Fig. 5, p. 10340.]

The properties of organic molecules in hydrated 2:1 clays have been studied by Skipper et al. (2006), who used QENS to show that methane interacts strongly with the clay, snugly fitting into the hexagonal ring sites on the clay surface, thus leading to a decrease of the diffusion coefficient by one order of magnitude as compared to bulk water-methane. Chathoth et al. (2010) presented results from a SNS-QENS study of the behavior of methane in a mesoporous carbon aerogel; a proxy for coal. The QENS portion of this study focused on the pressure and surface hydration effects on methane mobility. They observed a slowing of the motion for hydrated carbon pore walls compared to a "dry" carbon matrix. The pressure effect was non-linear with a subtle, yet measurable, maximum observed between ~3-4 MPa (maximum interrogated pressure was ~9 MPa). For the most part these various experiments were carried out at cryogenic to ambient temperatures and modest pressures, up to ~10 MPa.

NMR probes of hydrocarbons in nanopores

Diffusion considered as random motion of the elementary components of matter is among the most fundamental concepts in nature (Heitjans and Kärger 2005). This conceptual model is correct in the case of nanoporous systems as well. For this reason, studies of molecular diffusion in nanoporous materials attracted large interest. Among different techniques, pulsed field gradient (PFG) NMR has been proven to be very useful method to investigate systems with small molecules confined into the "subsurfaces" of nanoporous hosts (e.g., Meresi et al. 2001; Seland et al. 2001; Stallmach et al. 2000, 2001; Kärger et al. 2003; Krutyeva et al. 2007). PFG NMR diffusion measurements are based on pulse sequences forming a primary or stimulated spin echo of the magnetization of nuclei in resonance (Stallmach et al. 2000, 2001). In other words, the potential of PFG NMR depend on the amplitude and the rise and fall times of the field gradient pulses (Kärger et al. 2003). Applying appropriate pulsed magnetic field gradients of duration δ, intensity g, and observation time t during the defocusing and refocusing cycles of the NMR pulse sequence leads to the spin echo being sensitive to the translational motion of the molecules (Stallmach et al. 2000, 2001). Depending on the measuring conditions, PFG NMR experiments on molecular diffusion in cavities of nanoporous host systems result in valuable information on different aspects of mass transfer. The valuable information covers 1) molecular diffusion in the interior particles; 2) hindered transportation by outer surface of particles; and 3) long-range diffusion. Thus, PFG NMR may yield the probability distribution $P(x,t)$ function that, during time t, a randomly selected molecule of the sample is shifted over a distance x in the direction of applied field gradient (Kärger 2008). Such capabilities of PFG NMR provides an opportunity to investigate diffusion behavior of molecules confined to both mesoporous and nanoporous systems (Stallmach et al. 2000, 2001; Kärger et al. 2003, 2009), to explore surface permeability of nanoporous particles (Krutyeva et al. 2007), to study transport properties of supercritical fluids in confined geometry (Dvoyashkin et al. 2007), and to study diffusion of small molecules such as methane and carbon dioxide in carbon molecular sieve membranes (Mueller et al. 2012). However, there are some problems in performing NMR experiments of diffusion of mixtures within porous materials since they represent heterogeneous systems (Pampel et al. 2005; Fernandez et al. 2008). In such heterogeneous systems, the transverse proton magnetization is decayed due to a continuous dipolar interaction between the spins of interest, as well as differences in the internal magnetic fields. These effects, including the restricted mobility of adsorbed molecules, broaden the NMR signal and results in reduced resolution (Pampel et al. 2005; Fernandez et al. 2008). The combination of PFG and MAS techniques overcomes this problem (Nivarthi et al. 1994; Pampel et al. 2003; Gaede et al. 2004). MAS PFG NMR enabled, for example, studying complex formation in an acetone-alkane mixture confined to nanoporous host systems (Fernandez et al. 2008) and molecular diffusion in zeolites (Pampel et al. 2005). A more interesting MAS probe without FG was designed to investigate the evolution of adsorption on nanoporous solids (see Fig. 11 by Xu et al. 2007). The results obtained by this technique depicted clearly the viability of a new technique for performing *in situ* solid-state NMR investigations of adsorption processes into nanoporous host systems and possible subsequent reactions. In addition to such PFG NMR and MAS PFG NMR studies on hydrocarbon derivatives confined into various host systems, there are recent attempts for studying such heterogeneous systems at low temperatures and high pressures (Huo et al. 2009; Hoyt et al. 2011).

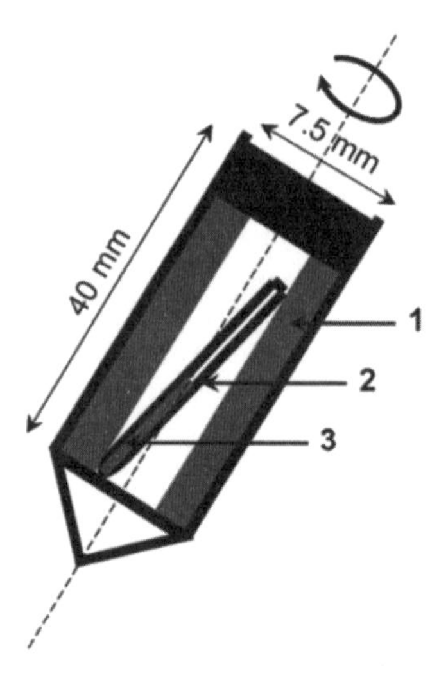

Figure 11. Schematic illustration of the setup for the *in situ* adsorption experiment inside the solid-state NMR rotor: 1) nanoporous material compacted on the walls of the rotor by prior spinning; 2) thin-walled glass capillary tube; 3) liquid to be adsorbed on the nanoporous material. [Used by permission of Wiley-VCH © 2007, from Xu et al. (2007) *Chem Phys Chem*, Vol. 8, Fig. 1, p. 1311.]

One more potential method in analyzing small molecules in confined geometry is low-field (LF) NMR. LF-NMR has been used not only for the fast determination of the water and oil components in several food samples (Aeberhardt et al. 2007; Straadt et al. 2008), but also was applied for studying protein aggregation (Indrawati et al. 2007), detecting heterogeneities in polymer networks (Saalwachter 2003), and water diffusivity in aggregated systems (Guichet et al. 2008). LF-NMR measures the response of ^{1}H protons after immersing the collection of nuclei into an external magnetic field. Then the protons inside the sample are polarized in the same direction with the external field. A radio-frequency field is applied to bring the protons to the perpendicular transverse plane. The magnetic signal processing in the transverse plane induces electric current, which is called the NMR signal or free induction decay (FID; Levitt 2001). The main advantage of LF-NMR experiments are based on time-domain relaxation measurements, such as transverse relaxation and longitudinal relaxation (no need to do Fourier transformed spectra analysis) and calibration of the instrument response or model-dependent analyses of the relaxation functions (Saalwachter 2003; Kantzas et al. 2005; Indrawati et al. 2007). The other benefits of LF-NMR technique are very short experimental time, very little if any sample preparation is necessary, and being non-destructive and non-invasive (Indrawati et al. 2007). In addition to relaxation measurements by LF-NMR, water diffusion coefficients in complex food products by low-field ^{1}H PFG-NMR were determined with standard errors lower than 0.5 % (Metais and Mariette 2003). Similar to this low-field, bench-top PFG NMR study on water diffusion in food products, Blümich et al. (2009) mentioned 2D relaxation exchange NMR of water in contact with nanoporous silica particles measured by low-field NMR. It seems that various NMR techniques ranging from low-field PFG NMR to high-pressure MAS NMR will contribute positively to the understanding of small molecule behaviors in confined geometry of subsurfaces.

Representative NMR studies

NMR studies of various hydrocarbons in nanoporous materials (mostly zeolites) are too voluminous to do justice to in this chapter, rather we will highlight some key results that illustrate the kinds of dynamical behavior hydrocarbons exhibit under nanoconfinement. Collectively these studies provide quantitative insights into hydrocarbon behavior as a function of particle size (i.e., transport length), pore diameter, pore geometry and pore intersection dimensions, cation size, ratio of Si:Al, hydrophobicity of the pore wall, load rate (i.e., amount of hydrocarbon), and hydrocarbon chain length.

Stallmach et al. (2001) demonstrated that PFG NMR can monitor anisotropic diffusion of slowly moving sorbate molecules arising from the inherent anisotropy of the MCM-41 nanopore system. As shown in Figure 12, there is significant deviation in non-exponentially decaying

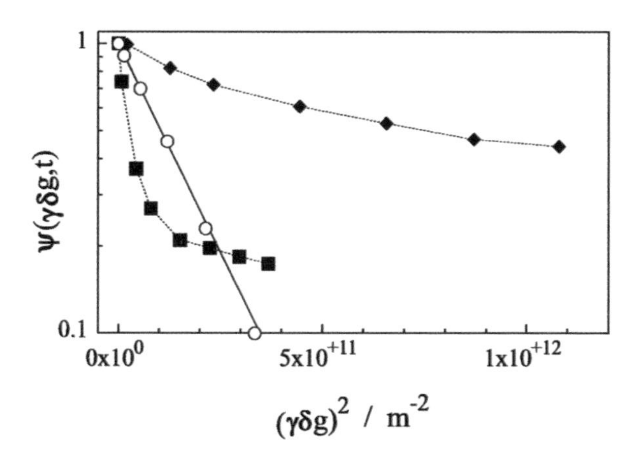

Figure 12. Examples for experimentally observed non-exponentially decaying PFG NMR spin echo intensities for benzene in two MCM-41 species (full symbols) compared with the exponential decay in the bulk liquid (open symbols) under the same conditions (T = 298 K, t = 3 ms). (♦) MCM-41 having particle size between 1 and 10 μm and pore radius 1.5 nm (■) Si-MCM-41 having particle size smaller than 1 μm and pore radius between 1.5 and 1.7 nm. [Used by permission of Elsevier © 2001, from Stallmach et al. (2001) *Microporous and Mesoporous Materials*, Vol. 44, Fig. 2, p. 750.]

PFG NMR spin echo intensities for benzene in two MCM-41 nanoporous hosts with respect to exponential decay in bulk benzene under the same experimental conditions. For instance, in sample MCM-41 the benzene exhibits slower decay than the bulk liquid, while in the case of Si-MCM-41 there is faster motion than the benzene in bulk.

Another host system of interest, zeolite NaX, has been studied by Kärger et al. (2003) with a mixture of *n*-butane and benzene. The zeolite hosts 0.8 molecules per supercage for *n*-butane and 2 molecules per supercage for benzene. The mobility of ethane in the nano-volume of zeolite NaX was also measured by PFG NMR. Figure 13 depicts temperature dependence of the coefficients of long-range diffusion of ethane upon confining into beds of zeolite NaX for two different loadings (Geier et al. 2002). The lower-temperature exponential dependence shifts to a much weaker dependence for higher temperatures. This change reflects the transition from Knudsen diffusion to bulk diffusion.

In a recent study by Dvoyashkin et al. (2007) using Vycor pore glass particles of about 500-μm size with mesoporous structure inside and 6-nm pore diameter, it was shown that a sufficiently large size of porous particles prevented extensive exchange between the bulk and the mesopores within the observation time $t = 3$ ms of the T_2 Hahn-echo pulse sequence (Figs. 14a,b). Both diffusivities D_b and D_p, in the bulk liquid and in the mesopores, respectively, increase at high temperatures following an Arrhenius law (the difference in absolute values attributed to the tortuosity of the porous space). Around 438 K an important deviation from the Arrhenius pattern in D_p was observed. However, at that temperature the diffusivity of the bulk liquid for the sample with Vycor did not exhibit noteworthy deviation from normal behavior.

One of the other NMR techniques employed in analyzing confined molecules behaviors was MAS PFG NMR. Fernandez et al. (2008) studied complex formation in acetone—*n*-alkanes (including hexane, heptane, and octane) mixtures by MAS PFG NMR diffusion measurements in two different specimens of trimethyl-silylized nanoporous silica gel. The silica gel nanoporous systems were synthesized from tetraethyl orthosilicate and had average pore sizes of 4 and 10 nm. MAS ^{1}H NMR was applied in order to resolve the signals of the acetone and the alkane constituent of interest. Upon resolving the signals of the two mixture components, comparing the CH_3 signals of acetone and the *n*-alkanes molecule yielded the acetone to *n*-alkane ratio. Selective diffusion measurements of acetone–*n*-alkane mixtures in both narrow (4 nm) and large

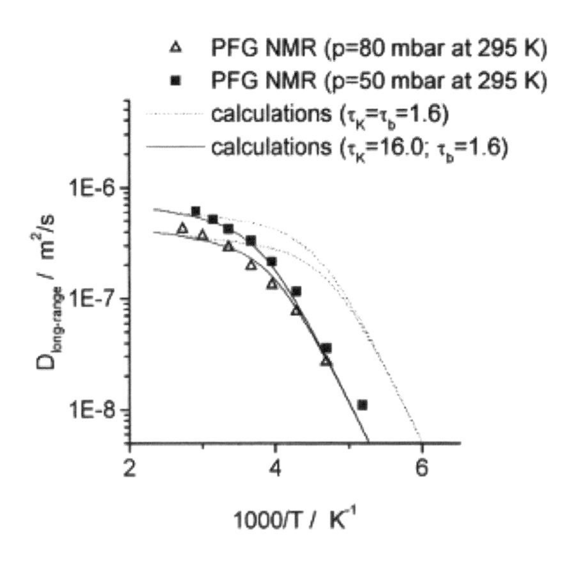

Figure 13. Temperature dependence of the coefficients of long-range diffusion of ethane measured by the PFG NMR method in beds of zeolite NaX for two different loadings (shown by the two different adsorbate pressures inside the sample tube at 295 K), comparison with the diffusitivities calculated for identical and different tortuosity factors $\tau_{K(b)}$ in the cases of Knudsen and bulk diffusion using a simple kinetic gas approach. [Used by permission of American Institute of Physics © 2002, from Geier et al. (2002) *J Chem Phys*, Vol. 117, Fig. 1, p. 1936.]

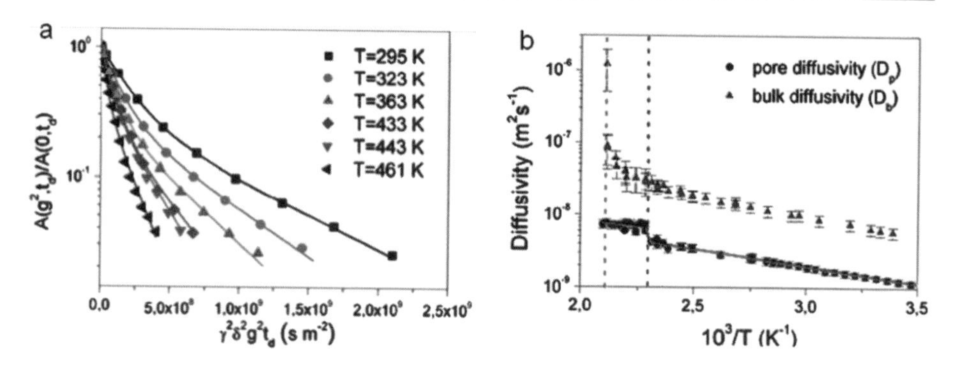

Figure 14. (a) Normalized spin-echo attenuation functions for *n*-pentane in Vycor porous glass obtained at different temperatures using PFG NMR. The solid lines show fits to the experimental data by two-exponential functions. (b) Arrhenius plot of the bulk and pore fluid diffusivities for *n*-pentane in Vycor porous glass. The solid line was calculated by assuming a transition to the supercritical state at $T = T_{cp}$. The vertical dashed lines show the positions of the bulk (left line) and pore (right line) critical points. [Used by permission of the American Chemical Society © 2007, from Dvoyashkin et al. (2007) *J Am Chem Soc*, Vol. 129, Fig. 1a, p. 10344-10345 (a), Fig. 1b, p. 10344-10345 (b).]

(10 nm) pore silica gels as a function of gradient amplitude at 298 K illustrated a notable deviation from a mono-exponential behavior observed in the free liquids. The deviation is stronger in the narrow-pore gel than in the large-pore one. This behavior was attributed to the deviation of the pore shapes (channel structure) from a cubic resulting in an orientation dependent anisotropic diffusion behavior. Additionally, examining the diffusion behavior of the two component mixtures in large and narrow pores indicated the following differences: 1) in large-pore gels the diffusivities increased by one or two orders of magnitude, 2) in large-pore gels the diffusion behavior difference of the *n*-alkanes and acetone disappeared, 3) in large-pore gels there is no longer any clear oscillation indication of the acetone diffusivities as a function of chain length of *n*-alkane molecules.

As mentioned in the introduction, Xu et al. (2007) developed a facile technique to handle MAS NMR for probing the adsorption of ^{13}C-labeled acetone on nanoporous MCM-41 and ZSM-5. MAS ^{13}C NMR spectra of acetone in MCM-41 pores acquired by spin rate of 3 kHz depicted that liquid acetone signals disappeared rapidly and were replaced by broader signals typical of acetone molecules in a less mobile environment (Fig. 15). Such a result indicates

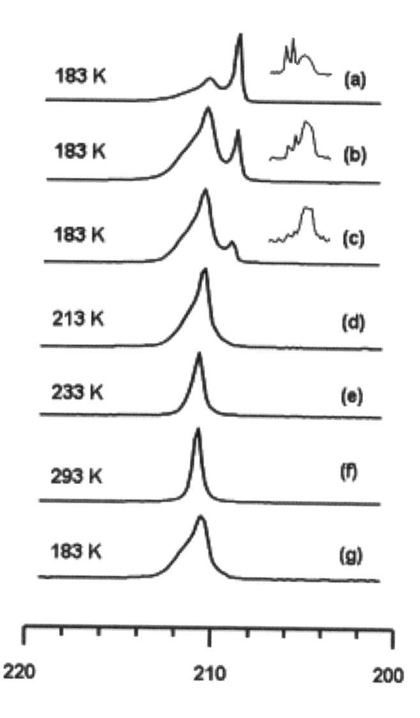

Figure 15. ^{13}C NMR spectra acquired as a function of time and temperature following adsorption of acetone on MCM-41, indicating the region of the spectrum for the (labeled) carbonyl carbon. In the spectra obtained at 183 K, the region corresponding to the (unlabeled) methyl carbon (between 27 and 32 ppm) is depicted as an inset. Each spectrum was acquired within 2 h. [Used by permission of Wiley-VCH © 2007, from Xu et al. (2007) *Chem Phys Chem*, Vol. 8, Fig. 2, p. 1311.]

that this technique may be provide a better understanding of adsorption processes into porous host systems.

Another study by Huo et al. (2009) employs MAS NMR on proton motion in confined geometry. Proton motion in HZSM-5, an acid catalyst widely used in petroleum industry, has been studied by low-temperature ^{1}H MAS NMR over the temperature range 150-295 K. The proton signal of the Brønsted acid sites (-OH) shifts to higher frequencies in the NMR spectra with decreasing temperature—a trend that is attributed to a gradual contraction of zeolite framework.

Another study reports a new high-pressure MAS NMR probe (Hoyt et al. 2011). Using this design, an internal pressure above 15 MPa was achieved without serious leakage problem during a time period of 72 h. The possible capability of this high-pressure MAS probe was demonstrated by carrying out *in situ* ^{13}C MAS NMR measurements of forsterite (Mg_2SiO_4) carbonation by mixture of supercritical CO_2 and H_2O at 15 MPa and 50 °C (Hoyt et al. 2011). The results indicate progressive carbonation reaction from solid phase reactants to solid-phase products—a result relevant to geological sequestration of carbon dioxide. This probe design will help scientists to investigate molecular interactions at the high temperatures and pressures expected in deep geologic reservoirs. Figure 16 summarizes *in situ* ^{13}C MAS NMR studies on Mg_2SiO_4 reacted with supercritical CO_2 and H_2O.

Quite often, NMR and QENS are used in concert with one another to quantify mobility of hydrocarbons in nanopores. For example, Jobic et al. (1995) determined the reasonably similar self-diffusivities of cyclohexane in microporous (pores < 2 nm) SiO_2 powder using both QENS and PFG-NMR. Since the time scale of the two methods are different; of the order of ms for PFG-NMR and ns for QENS, the mean-square displacements are of a different magnitude.

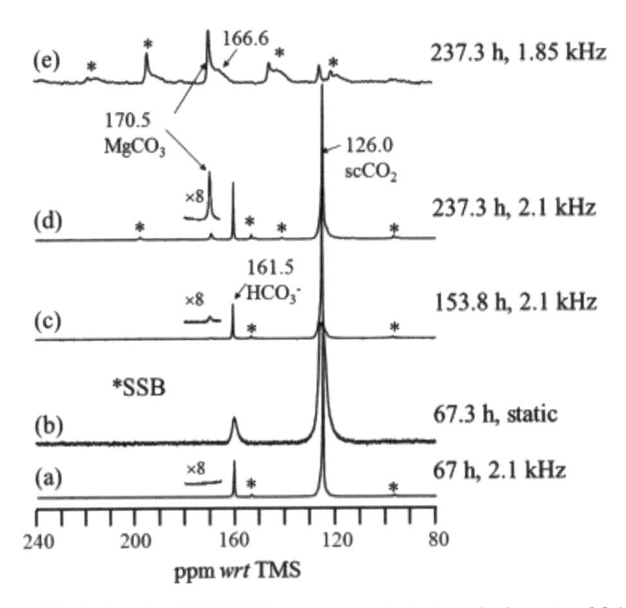

Figure 16. *In situ* ^{13}C single pulse MAS NMR spectra acquired at a spinning rate of 2.1 kHz on 0.27 g Mg_2SiO_4 + 0.1 g H_2O + 150 bar 14.3% ^{13}C enriched CO_2 + 1 g of extra H_2O separated from the forsterite powder at 50 °C for 67.3 h (including 17.3 h acquisition time) (a), 153.8 h (including 25 h second acquisition time period), (c) and 237.3 h (including 22 h third acquisition time period), respectively. (b) Static spectrum obtained immediately after (a). (e) MAS spectrum (d) and after the supercritical CO_2 was released. (*)s are spinning sidebands. The integrated peak area for the $MgCO_3$ normalized to per unit number of accumulation increase by 6.4 fold from (c) to (d). [Used by permission of Elsevier © 2011, from Hoyt et al, (2011) *J Magn Res*, Vol. 212, Fig. 6, p. 383.]

They amount to several μm in PFG-NMR and to a few nm in QENS. The agreement between the two methods, within experimental error, indicates that the two techniques are measuring the same process; i.e., long-range translational motion. This result means that there are no dramatic transport resistances with spacing above the nm scale. The presence of transport resistance would lead to a reduction of the NMR diffusivities, while the QENS results would remain essentially unaffected by them (Jobic et al. 1995). At low loading (0.02 and 0.04 g cyclohexane per g SiO_2), the diffusion coefficients appear to be insensitive to the concentration, whereas at high loading, (0.08 g cyclohexane per g SiO_2) lower diffusivities are measured. Jobic et al. (1995) attribute this decrease in mobility to mutual hindrance of the molecules, which evidently becomes effective only at high concentrations. Activation energies of 10.9 and 11.6 kJ mol^{-1} were determined for lower loadings and the higher loading, respectively.

Another good example of the power of using combined ^{2}H-NMR and QENS was described by Stepanov et al. (2003) in a study of the translational and rotational dynamics of *n*-hexane in ZSM-5 and 5A zeolites (the characteristics of these are described above). There are profound differences between the two systems. In ZSM-5, the molecule sits in the channel segments and the energy barrier between two sites is small. On the other hand, in 5A zeolite, the molecule spends a longer time in the α-cages before jumping to the next cage. The ^{2}H-NMR spectra point out the more confined adsorption geometry of the molecule in the ZSM-5 structure, in the form of anisotropic motions, whereas isotropically reorienting molecules are evidenced in the α-cages of 5A zeolite. This result is in agreement with the larger entropy variations measured in silicalite (the Al-free analog of ZSM-5; Millot et al. 1998) compared with 0.5 nm zeolite (Paoli et al. 2002). Finally, the long-range diffusion coefficient of *n*-hexane, derived from neutron scattering techniques, is more than 4 orders of magnitude larger in ZSM-5, at 300 K, compared with 0.5 nm zeolite. This difference illustrates the drastic effect of the pore size and shape on the diffusivity of molecules in microporous materials. The reader should consult the excellent review articles on NMR studies of liquids in confined geometry by Packer (2003), Buntkowsky et al. (2007), and Webber (2010). For recent reviews of the application of neutron scattering and MD to both natural and engineered materials and their interaction with fluids see Cole et al. (2006, 2009, 2010).

ATOMIC AND MOLECULAR-LEVEL SIMULATIONS

Properties of confined fluids: do they differ compared to the bulk?

We briefly discuss herein effects on (1) the diffusivity of confined gas/liquids; (2) fluid adsorption and vapor-liquid equilibria for the confined fluids; and (3) liquid-solid transitions under confinement. These observations will be used in the section on "*Selected simulations of alkanes within alumina and silica-based pores*," where we will discuss in a few more details the properties of confined alkanes. We provide a brief overview on simulation methods in the "*Simulation Details*" section.

Diffusion under confinement. The seminal works of Knudsen (1909), von Smoluchowski (1910), Pollard and Present (1948), and Mason et al. (1967) described confined fluid particles as hard spheres and took into consideration the momentum exchange between the fluid particles and the solid wall when the fluid-fluid interactions could be neglected, and later considered the effect of density on the mobility of the confined gases. More recently, theoretical developments have attempted to consider dispersive and longer-ranged fluid-fluid interactions (Guo et al. 2005, 2006). Building on the Chapman-Enskog kinetic theory approach (Davis 1992), Jepps et al. (2003) provided a significant improvement in our theoretical understanding of the diffusion of confined Lennard-Jones fluids by developing the "oscillator model" theory, which is exact at low confined fluid densities (i.e., low-pressure gases). The theory builds on the diffusive reflection model for fluid particles bouncing on the confining walls. To investigate the fluid diffusion within pores of nano-scale dimensions (e.g., "nanopores"), theoretical investigations have benefited from advances in molecular simulation techniques. For example, recent simulation results suggest that, for fluids confined in carbon nanotubes, the specular reflection might be more accurate than the diffusive reflection model invoked in the groundbreaking theories discussed above (Skoulidas et al. 2002; Sokhan et al. 2002; Bhatia et al. 2005; Striolo 2006).

Bhatia and Nicholson (2007) recently extended the theories above to include the effect of loading (i.e., increased pressure). They compared the theoretical results to molecular simulation data, achieving satisfactory, though not quantitative agreement. In narrow cylindrical pores they showed that the diffusivity of confined methane is constant as the density increases (the simulations show slight increase), while in wider pores (albeit of diameter comparable to that of confined molecules) the diffusivity decreases as the density increases. In even larger pores it has been reported that the diffusivity increases as the density of the confined fluid increases because of viscoelastic effects (Bhatia and Nicholson 2006; Nguyen et al. 2006). In the case of alkanes adsorbed within silica mesopores at low loadings, it was observed by molecular dynamics simulations that the diffusivity of the alkane chains increases at the loading increases because of screened fluid-pore interactions (Raghavan and Macelroy 1995). Recent interest has been devoted to enhancing the permeability of natural gas through rock formations (e.g., shale deposits), as well as in sequestering carbon dioxide in geological formations. It has also been proposed to employ carbon dioxide as a fluid to enhance the permeability of natural gas, achieving simultaneously carbon dioxide sequestration and natural gas extraction. Clearly, for these applications to succeed it is necessary to understand how each of the gases diffuses in sub-micron, often nanometer scale pores. We summarize a few contributions to highlight extreme confinement effects.

Zeolites are among the most widely used porous materials in industrial applications. We refer here to the zeolite ZK5, which has a framework consisting of two different cages (the largest cavity, known as the α-cage, has diameter ~1.16 nm; the other cavity, the γ-cage, has cross section 0.66 × 1.08 nm) interconnected via circular rings of diameter 0.39 nm composed by 8 oxygen atoms (Zorine et al. 2004; Baerlocher et al. 2007). The size of both cages as well as that of the connecting ring is comparable to the molecular dimensions of alkane molecules such as *n*-pentane. At most, the unit cell of ZK5 can contain 12 *n*-pentane molecules, 3 in each of the α-cages, and 1 in each of the γ-cages. The question of interest here is how fast *n*-pentane molecules can diffuse across the zeolite. Experimental data have been obtained by ^{13}C NMR techniques when the zeolite was filled with *n*-pentane (Magusin et al. 1999). The results are consistent with a "hopping" diffusion mechanism with molecules jumping between neighboring cages with a hopping rate of 1 to 10^3 jumps per second in the temperature range 247 to 317 K. The resultant self-diffusion coefficient was found to be of the order 10^{-18} to 10^{-15} m^2/s.

Because of the extremely long time that separates distinct hopping events traditional "brute force" molecular dynamics simulation techniques cannot yield satisfactory information regarding this phenomenon. Alternative approaches have been designed. Saengsawang et al. (2010) applied the "high-temperature configuration-space exploration (HTCE)" method of Schuring et al. (2007) to overcome the limitations of standard molecular dynamics. In short, the rare-event hopping event of one *n*-pentane molecule from one α-cage to another was investigated at extremely high temperature (i.e., 6000 K), which makes the event more likely to be observed, and the results were appropriately corrected to become applicable in the temperature range of experimental interest. The transition state theory was then applied to obtain self-diffusion coefficients from the relative probability of observing the *n*-pentane molecules either within the cages, or inside the connecting oxygen rings. The self-diffusion coefficients obtained show Arrhenius dependence as a function of temperature, in general agreement with experiments. In Figure 17 we report a schematic for the ZK5 zeolite, the two cages occupied by *n*-pentane to

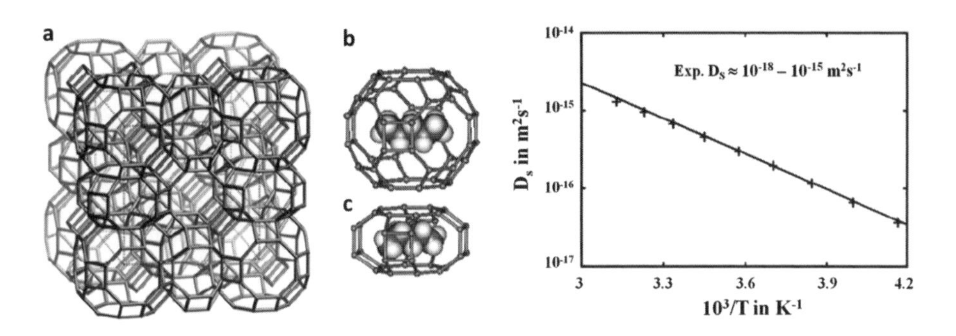

Figure 17. (*left*) Schematic representation of the ZK5 zeolite. Silicon or aluminum atoms constitute the vertexes of the structure in panel (a). The various cages and how they are interconnected are visible from the schematic. Panels (b) and (c) show one *n*-pentane molecule inside α- and γ-cage, respectively. (*right*) predicted self-diffusion coefficients as a function of temperature. The predictions are in relatively good agreement with experiments. [Used by permission of Elsevier © 2010, from Saengsawang et al, (2010) *Chem Phys*, Vol. 368, Fig. 1, p. 122 and Fig. 5, p. 125.]

illustrate the level of confinement considered, and the self-diffusion coefficients predicted as a function of *T*. Although some differences do exist between experimental and simulated results (in part due to the different *n*-pentane loadings considered), this example shows how advanced simulations can elucidate the mechanism of diffusion for fluid molecules within extremely narrow porous networks.

To prevent, and when necessary remediate, environmental and energy challenges, it has been proposed to store CO_2 in deep un-minable coal beds, with the possible additional advantage of extracting trapped natural gas in the process (Busch et al. 2004; White et al. 2005). For such an application to be practically realized it is important to understand, quantify, and predict several physical-chemical aspects, including the adsorption and diffusion of both CH_4 and CO_2 within typical coal bed porous networks. Molecular dynamics simulations are being used for such purposes (i.e., to identify the relative diffusion and adsorption propensity of the two gases in the carbon-based pores; Hu et al. 2010). Firouzi and Wilcox (2012) employed the dual control volume grand canonical molecular dynamics algorithm (Heffelfinger and Van Swol 1994; Cracknell et al. 1995; Ford and Glandt 1995) to investigate the flux of pure methane, pure CO_2, and their mixtures across representations of the carbon-based porous networks. The porous materials were obtained by implementing a geometric algorithm and the resultant models were characterized in terms of overall porosity and pore size distribution. Although some simplifications were necessary, the porous networks reproduced important experimental properties. Because of the large computational requirements, the fluid molecules were treated as spherical particles and all interactions were described by pair-wise Lennard-Jones potentials. One simulated system is reproduced in Figure 18, where an equimolar mixture of CH_4 (squares) and CO_2 (circles) are present in the two bulk reservoirs (left and right), and in the porous network (middle). The carbon atoms of the coal bed are not shown for clarity. The image represents pictorially the flux of both gases from the high-pressure region on the left (5 MPa) to the low pressure region on the right (2 MPa), and the presence of regions within the porous network that are not accessible to the fluid molecules, presumably because of the small pore size or because of lack of pore connectivity (the porosity of the material is 25% and the average pore size is 1.25 nm). Simulations were conducted as a function of the feed composition, the porosity of the substrate, and the difference in pressure between the two reservoirs. The results were quantified in terms of the permeability of the various gases. In the right panel of Figure 18 we reproduce simulation results for the permeability of the two gases in the mixture as a

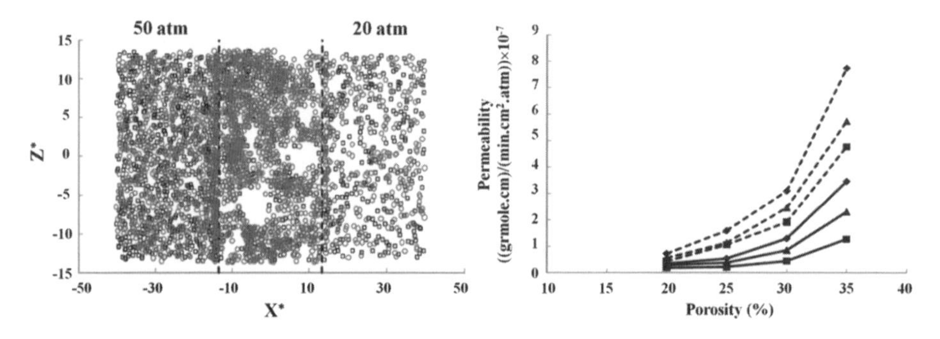

Figure 18. (*left*) Schematic representation of the system simulated by Firouzi and Wilcox to study the permeability of CH_4-CO_2 mixtures across model coal beds. Squares and circles are for CH_4 and CO_2 molecules, respectively. The carbon atoms in the porous network are not shown for clarity. (*right*) Permeability of CH_4 (continuous lines) and CO_2 (dashed lines) through carbon-based porous materials with average pore size 1.25 nm and varying porosity. Different symbols are for varying methane mole fraction in the feed: diamonds, triangles, and squares are for 0.75, 0.5, and 0.25 CH_4 mole fraction, respectively. [Used by permission of Elsevier © 2012, from Firouzi and Wilcox (2012) *Microporous and Mesoporous Materials*, Vol. 158, Fig. 7, p. 201 and Fig. 10, p. 201.]

function of total porosity. The average pore size in all cases was 1.25 nm. As expected, the permeability increases with porosity. Important is that no detectable permeability was reported at porosity lower than ~20% because of the lack of connectivity between the pores. Simulations were conducted at varying methane mole fractions in the feed. The interesting observation is that both CH_4 and CO_2 permeability increase as the CH_4 mole fraction in the feed increases. Fully understanding the molecular reasons for these observations will certainly enhance the possibility of sequestering CO_2 in coal beds.

The simulations discussed above contrast a number of simulations available in the literature in which the adsorption and the diffusion of various gases were considered in simple pores, varying, for example, the pore width. Restricting our analysis to the case of carbon dioxide and methane, the recent reports by Lim and coworkers appear to be important. For example, Lim et al. (2010) simulated the adsorption of carbon dioxide (simulated using two different models, one spherical, the other, more realistic, linear) in carbon slit pores of various widths. The simulations were conducted at various bulk pressures and at various temperatures. The simulated diffusions were found to be much larger compared to experimental measurements, and were interpreted invoking a modified Knudsen diffusion mechanism. The authors suggest that the predicted values could be the upper limit available, when membranes could be prepared with perfect carbon pores as long as pore-entrance and pore-exit effects can be neglected. Lim and Bhatia (2011) extended the study to methane, and found that in the temperature range 298-318 K, and pressure range from 0.001 to 8 MPa, the permeability of methane in pores of width 0.65-0.75 nm is controlled more by the adsorption of the gas within the pores than by the self-diffusion coefficient.

When simulation results in individual pores, such as those just summarized, are available for individual gases and their mixtures, and when the resistance for the various gases to enter/exit the various pores are known, it will be possible to predict the permeability across realistic materials by implementing pore network models such as the one presented by Seaton and coworkers for the diffusion in nanoporous carbons (Cai et al. 2008). The validation of such models could be done by comparing the predicted permeability to simulated ones, such as those presented by Firouzi and Wilcox (2012), provided that the same porous network is used in both approaches.

Regarding the diffusion of trace elements (e.g., heavy metal ions) confined within narrow pores, we point out the work of Ho et al. (2012). These authors employed brute force equilibrium molecular dynamics simulations to quantify the mobility of aqueous NaCl or CsCl within slit-shaped silica-based pores of size ~1 nm. These simulations face several challenges, including the long time required to achieve equilibrium, the uncertainty regarding the composition of the confined system, and the slow dynamics typical of aqueous solutions under confinement. Despite these problems, the enormous computational resources that have become available to researchers have allowed Ho et al. (2012) to estimate the mobility of the various ions as a function of the degree of protonation of the confining surfaces. In Figure 19 we report a schematic of the system simulated, together with predicted planar self-diffusion coefficients, and the density distributions across the pore width for the various confined molecules. The results have been interpreted based on the expectation that water molecules diffuse slowly near a silica-based substrate (Argyris et al. 2009). As the degree of protonation changes, the preferential distribution of the various electrolytes across the pore width changes. Those ions that are preferentially found near the pore center have higher mobility than those that are preferentially found near the solid substrate. Such observations, if verified experimentally, could be useful for predicting the environmental fate of heavy metal ions accidentally released during operations such as mining.

Adsorption and vapor-liquid equilibria under confinement. Gelb and Gubbins provided an extensive review on the extent to which confinement leads to significant deviations for fluid thermodynamics properties compared to bulk (Gelb et al. 1999). One such effect is on the vapor-liquid coexistence curve (Striolo et al. 2005). A Monte Carlo study by Singh et al. (2009) elucidated the phase behavior of methane, butane, and other short alkanes confined in narrow slit-shaped carbon and mica pores. The results showed that as the pore width decreases the critical temperature decreases. Representative results are shown in Figure 20 for butane in mica and in carbon-based pores. Confinement was found to have a strong effect on the densities of the coexisting phases, and it was also found that the vapor-liquid surface tension for the confined alkanes is significantly reduced compared to values obtained in the bulk, possibly a direct consequence of the lower critical temperature for confined fluids.

As in the case of diffusion, the simulation studies that are being conducted for adsorption and phase transitions of various fluids in porous materials will help the interpretation of experimental data, as well as better understanding the behavior of fluids in subsurface conditions. However, to be useful the simulations need to be conducted carefully. For example, it is often tempting to employ simplified representations for the confining material. Smooth surfaces and simple geometries (e.g., cylindrical and slit shaped) lead to enormous savings of computational resources. However, details on the porous wall structure can lead to significant changes in the simulated adsorption isotherms (used to obtain results such as those in Figure 20, but also to predict the permeability of various gases). The simulation community is striving to generate computer models for the solid adsorbent that are more and more realistic. The literature is vast and not summarized here for the sake of brevity. Instead, we highlight a contribution from Coasne et al. (2006), who simulated the adsorption of simple Lennard-Jones gases in two models of the silica-based porous material MCM-41. One model had atomically smooth pore walls, the other was rough at the atomic scale. The adsorption isotherms obtained showed large adsorption-desorption hysteresis loops in the case of the smooth pore, and more gradual adsorption curves in the case of the rough surface. Careful analysis suggested that the smooth model yields results in agreement with experiments at short length, but not at large scales, while the rough model reproduces the surface disorder at large length scales but does not at short length scales. When the structure of the confined fluid is considered at those relative pressures at which the pores are not completely filled, the simulation results are dramatically different, as shown in Figure 21. Within the smooth pore (left panel) a clear separation is observed between

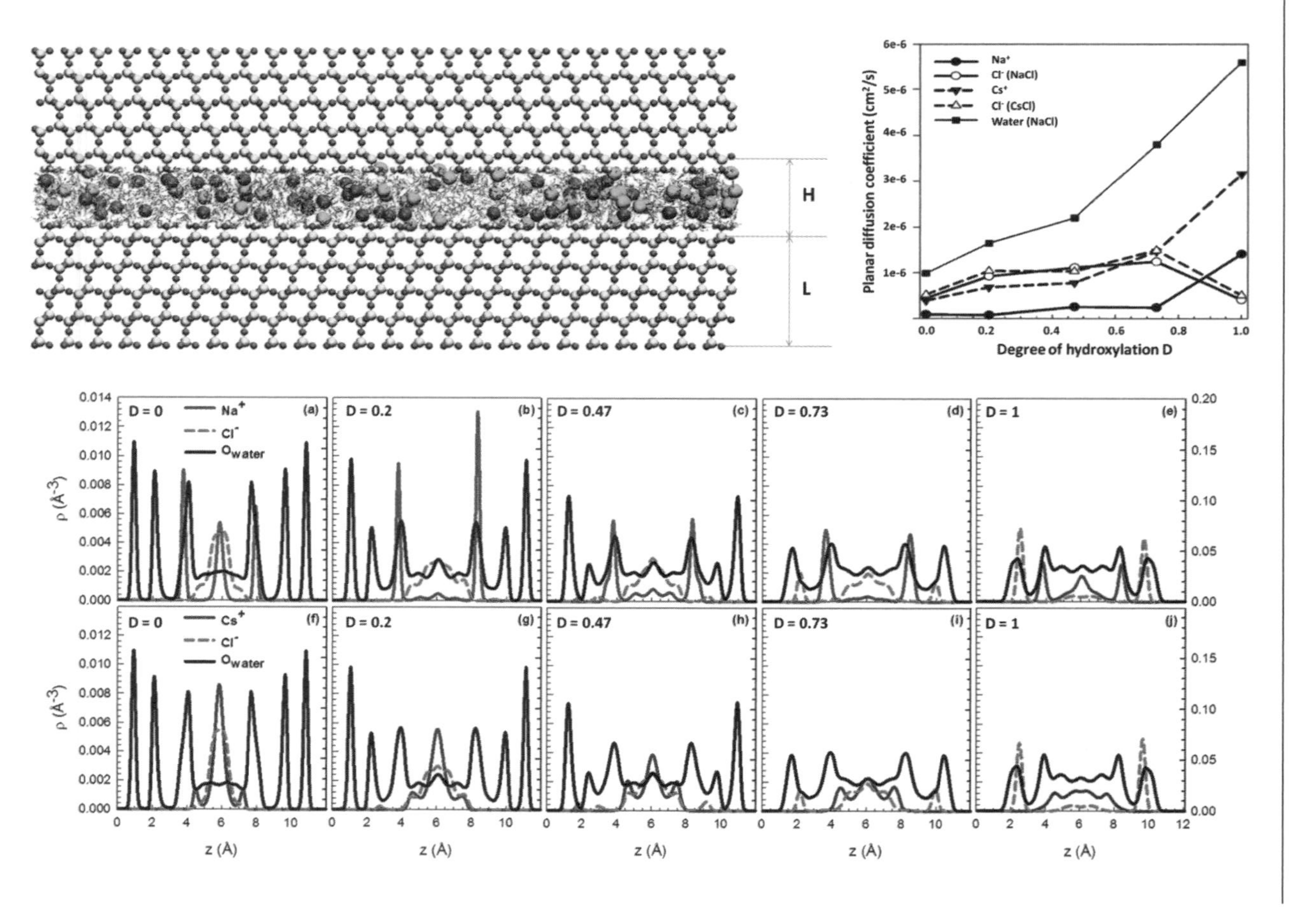

Figure 19. (*top left*) Schematic representation of the slit-shaped silica pores simulated by Ho et al. (2012). The pore width, H, is ~ 1 nm. (*top right*) Predicted in-plane self-diffusion coefficients for water and electrolytes within the pores as a function of the degree of protonation D. (*bottom*) Density distribution across the pore volume for oxygen atoms of water (dashed lines), Na^+ (purple), Cs^+ (blue), and Cl^- ions (dashed green lines) as a function of D. [Used by permission of the American Chemical Society © 2012, from Ho et al. (2012) *Langmuir*, Vol. 28, Fig. 1, p. 1257, Fig. 5, p. 1261, and Fig. 4, p. 1259.]

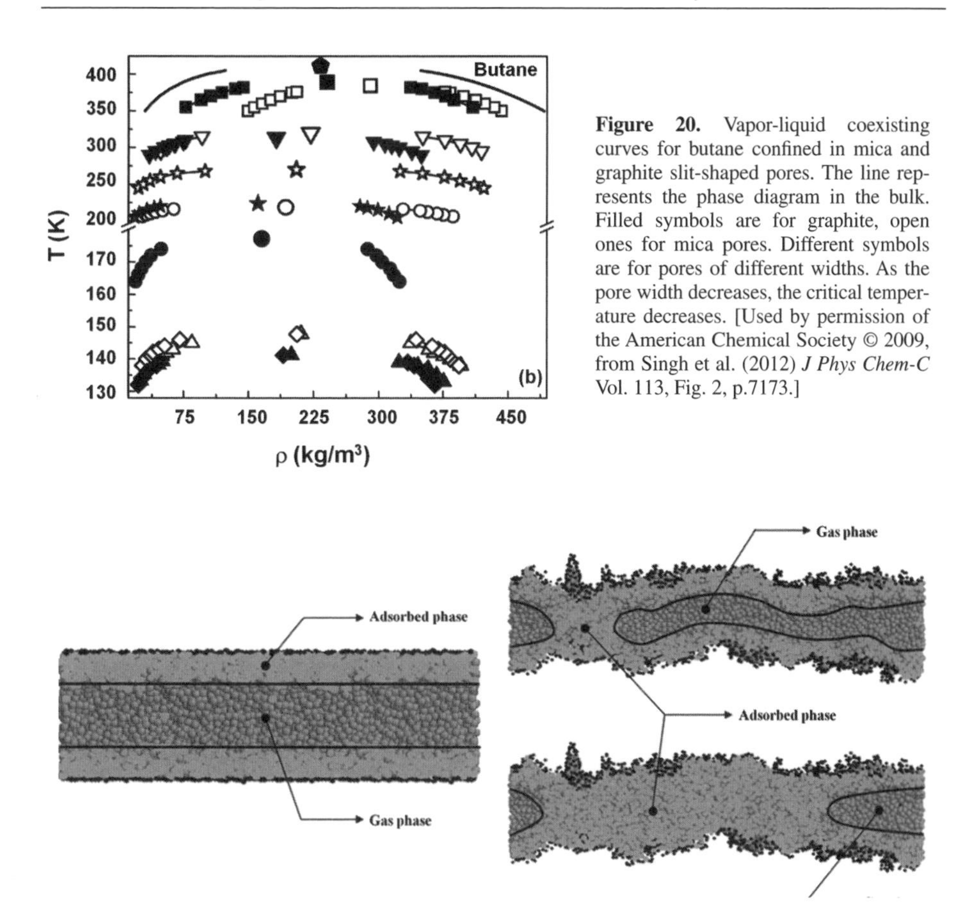

Figure 20. Vapor-liquid coexisting curves for butane confined in mica and graphite slit-shaped pores. The line represents the phase diagram in the bulk. Filled symbols are for graphite, open ones for mica pores. Different symbols are for pores of different widths. As the pore width decreases, the critical temperature decreases. [Used by permission of the American Chemical Society © 2009, from Singh et al. (2012) *J Phys Chem-C* Vol. 113, Fig. 2, p.7173.]

Figure 21. (*left*) Transverse section of a simulation snapshot illustrating Ar atoms adsorbed within a smooth model of the MCM-41 pore at 87K and relative pressure 0.69. The black lines divide the adsorbed fluid from the gaseous phase within the pore. (*right*) Transverse section of an atomically rough model pore partially filled with Ar at 87 K and relative pressures 0.44 (*top*) and 0.46 (*bottom*). Although the relative pressure is lower, the snapshots on the right suggest that the adsorbed phase fills more of the pore volume, and that the gaseous phase is trapped only within a few bubbles. [Used by permission of the American Chemical Society © 2006, from Coasne et al. (2006) *Langmuir*, Vol. 22, Fig. 7, p. 199 and Fig. 8, p. 200 .]

the adsorbed fluid and the confined gas at the center of the pore, while in the rough pore model (right panel) a few bubbles of gas are found trapped within the adsorbed fluid. It is very likely that the transport properties in this dual-phase system strongly depend on the structure of the confined fluid, and therefore the predictions obtained with the two models are expected to be very different [no transport properties were reported by Coasne et al. (2006)]. Because subsurface pores are likely to be very heterogeneous in size, shape, surface morphology, and even chemical composition, in our opinion the observations from Coasne et al. (2006) should be kept in mind when simulation data are compared to experiments.

It is very likely that the phase equilibrium within the pores will determine the diffusion of confined substances. Ho et al. (2012), for example, found that aqueous Na^+, Cs^+, and Cl^- ions partition at different distances from the solid surface within the pores depending on the protonation state of the surface, leading to significantly different mobility for the various ions (see Fig. 19). Analogous results obtained for organic systems could lead to strategies for enhancing the dislocation of various fluids trapped in subsurface formations. For example, the preferential adsorption of CO_2 compared to CH_4 near the carbon-based pores used in the simulations of Firouzi and Wilcox (2012) was invoked to explain the different permeability observed for the two gases, either pure or in binary mixtures (see Fig. 18). The preferential adsorption of CO_2 with respect to CH_4 in coal beds is the basis of the enhanced coal bed methane recovery (White et al. 2005). Unfortunately, this process is also responsible for the differential swelling of the coal (i.e., coal swells more when CO_2 is adsorbed than when CH_4 is adsorbed), which causes a reduced permeability of methane as CO_2 is injected. Brochard et al. (2012a) conducted a simulation study for the competitive adsorption of CO_2 and CH_4 in realistic models of carbon pores obtained with the reverse Monte Carlo method (Jain et al. 2006a,b; Nguyen et al. 2008). They considered temperatures and pressures representative for geological applications. They then applied a poromechanical model, developed independently by Brochard et al. (2012b), for predicting the material swelling as a function of the amount adsorbed. Representative results, shown in Figure 22, show that CO_2 adsorption is highly selective compared to CH_4 adsorption. Differential swelling is almost insensitive to the geological temperatures and pressures considered, while it appears to be proportional to the CO_2 mole fraction in the pores. Although swelling of coal beds is certainly dependent on additional factors, including the stress imposed in the rock formation by the surroundings, these results could aid interpretation and prediction of experimental observables.

As another example to illustrate preferential distributions of different fluids at contact with a solid substrate, in Figure 23 we compare preliminary simulation snapshots obtained for pure ethanol (left) and for an ethanol-water mixture (right) on a free-standing (0001) α-Al_2O_3 substrate at 298 K and 0.1 MPa. When ethanol is the only fluid present, the results suggest the formation of a dense adsorbed layer on the substrate. When a small amount of water is added to the system, water molecules dislocate the adsorbed ethanol, they form a dense hydration layer, and the ethanol molecules are segregated further from the surface. Because those molecules (either ethanol or water) found in proximity to the substrate tend to show low mobility, should results similar to those presented in Figure 23 be observed also within narrow pores it is likely that pure ethanol will have low permeability within alumina-based rock formations, and that the addition of small amounts of water could potentially allow for an increased ethanol mobility. It is also possible that if the pores are extremely narrow, water molecules would fill the pores and block ethanol molecules from permeating the rock formation. Clearly, careful verification of such predictions needs to be conducted. This brief discussion points out the need of carefully analyzing the composition of the subsurface fluids, together with details concerning the pore network (e.g., pore size, shape, entrance, and chemical composition) for predicting the permeability of different fluid systems.

Liquid-solid transitions under confinement. Using macroscopic thermodynamic arguments (equating the free energies of confined solid and liquid phases, Warnock et al. 1986), or calculating the chemical potential as a function of temperature for confined liquid and solid phases (Evans and Marconi 1987), it is possible to derive the Gibbs-Thomson equation to relate the freezing temperature for a confined fluid to the pore size. The latter relation is invoked in the experimental technique of thermoporometry, used to characterize porous materials (Eyraud et al. 1988). Unfortunately, the Gibbs-Thomson relationship fails as the pore size becomes comparable to the molecular diameter. Alba-Simionesco et al. (2006) provided a clear and concise review on the subject. Recent results show that the freezing temperature of the confined fluid can increase or decrease compared to bulk values, depending on whether the fluid-wall interactions are weaker or stronger than fluid-fluid interactions, respectively. Sometimes exotic phases, not observable for bulk systems, can be observed under confinement (e.g., the hexatic phase) (Radhakrishnan et al. 2000, 2002).

Combining experiments and simulations leads to important observations regarding the fluid-solid transition for confined systems. For example, Coasne et al. (2009) investigated, using

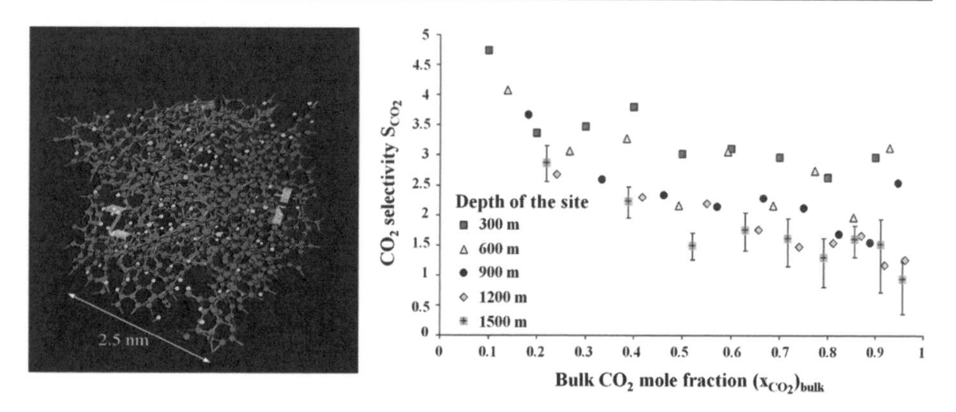

Figure 22. (*left*) Schematic representation of the model used to simulate a carbon adsorbent (i.e., coal bed). (*right*) Selectivity of CO_2 with respect to CH_4 adsorption as a function of bulk CO_2 mole fraction. Different symbols represent data obtained at different geological conditions (depths of the coal bed). Note that in most cases the selectivity is well above 1. [Used by permission of the American Chemical Society © 2012, from Brochard et al. (2012a) *Langmuir*, Vol. 22, Fig. 1, p. 2661 and Fig. 6, p. 2664.]

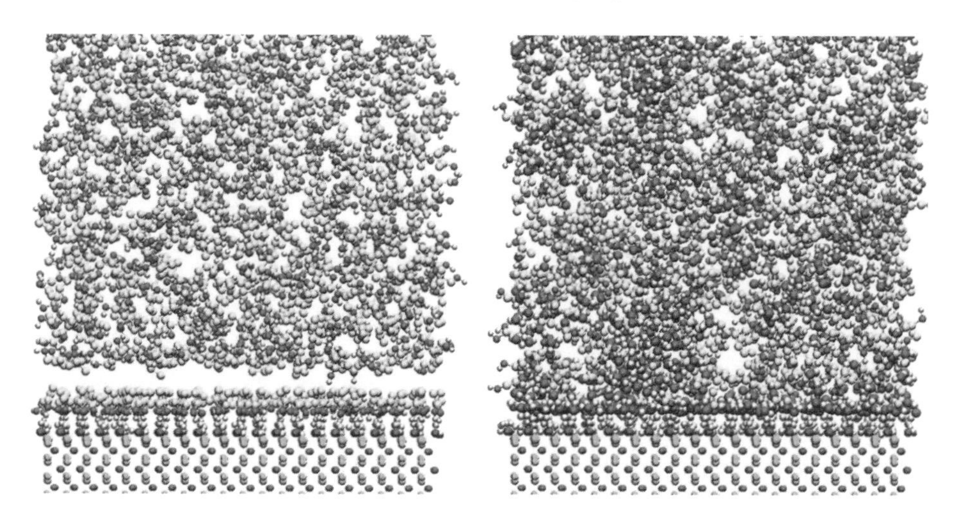

Figure 23. Preliminary simulation results for pure ethanol (*left panel*) and for a water-ethanol mixture containing 50% water molecules (*right panel*) at contact with a free-standing fully protonated (0001) α-Al_2O_3 surface. Ethanol molecules are shown in yellow (red and white are for the OH groups). Simulations performed at 298 K and 0.1 MPa. Note that when ethanol is at contact with the substrate, a dense layer forms near the surface. As water molecules are added to the system, these replace ethanol and form one dense hydration layer near the substrate. Because those fluid molecules near the substrate tend to show lower mobility than those far from it, these preliminary simulation results suggest that adding some water could enhance the permeability of ethanol through alumina-based rock formations.

parallel tempering Monte Carlo techniques, the freezing of pure methane and argon, as well as of their mixtures in a slit-shaped graphitic pore. Because the pore is strongly attractive to both fluids, the freezing temperature for both confined substances, as well as for their mixtures, was found to be larger than that in the bulk. The phase diagram for the confined mixture showed an azeotrope, similarly to that observed for the bulk mixture. Regarding azeotropic mixtures, Czwartos et al. (2005) reported dielectric spectroscopy experiments for CCl_4/C_6H_{12} mixtures confined in carbon fibers and grand canonical Monte Carlo simulations for Ar/CH_4 mixtures confined in carbon-slit pores. The results are summarized in Figure 24. Both experiments and simulations show that qualitatively the phase diagram for either mixture does not change upon confinement (note the azeotrope). However, because all fluids considered are strongly attracted to the confining pores, the coexistence lines are shifted to higher temperatures upon confinement. The other important observation is that the composition of the azeotrope is enriched in the component that is most strongly attracted to the carbon pores (C_6H_{12} in the case of the CCl_4/C_6H_{12} mixture, Ar in the case of the Ar/CH_4 mixture). A cautionary note should be made here, as the experiments could not assess the composition of the mixture under confinement. The authors speculate that the experimental mole fractions for the confined system under-estimate the mole fraction of C_6H_{12}, as this substance is more strongly attracted to the carbon surfaces.

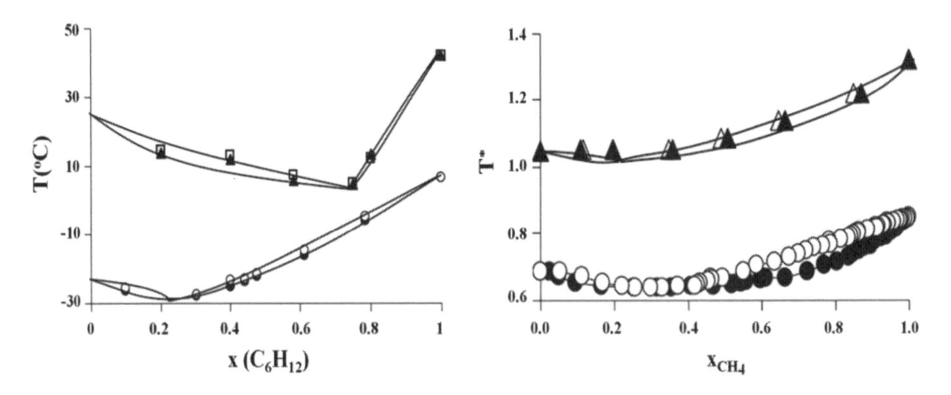

Figure 24. (*left*) Experimental phase diagram for CCl_4/C_6H_{12} mixtures (mole fraction) as measured by dielectric spectroscopy in activated carbon fibers. Circles are for the bulk mixtures, triangles and squares are for the mixtures at contact with the porous adsorbent. Because the experiments were conducted for fluid mixtures at contact with the porous material, the compositions reported are for the overall system. (*right*) Simulated phase diagram for Ar/CH_4 mixtures in graphitic slit pores. Circles are for the bulk mixtures, triangles for the mixtures under confinement. In both panels open and closed symbols are liquid and solid coexistence data, respectively; lines are guides to the eye. [Used by permission of Taylor and Frances © 2005, from J. Czwartos et al. (2005) *Molecular Physics*, Vol. 103, Fig. 3, p. 3107 and Fig. 9, p. 3110.]

Given the interest in carbon-based pores for the possible combined application of CO_2 sequestration and enhanced methane recovery, it is worth highlighting some results reported by Hung et al. (2005). These authors conducted Monte Carlo simulation studies for Lennard-Jones fluids confined within carbon nanotubes. The results showed that the tube diameter is very important in determining not only the solid-liquid transition for the confined fluid, but also the structure of the solid phase under confinement. In particular, when the tube diameter was ~9.6 times the diameter of one fluid molecule the simulations did not show a solid structure similar to the one observed in the bulk. Further, while the molecules in contact with the confining walls showed an increase in the freezing temperature, those close to the nanotube center showed a depression in the freezing temperature. The simulation results appeared to be in reasonable agreement with dielectric spectroscopy experiments. There is no doubt that freezing will have a strong effect on the mobility of the confined fluids. It is possible that some fluid, maybe present in small amounts within the porous network, by freezing effectively blocks some pores and reduces the rock permeability. These effects could be facilitated by the strong expansions (from high to ambient pressures) typically observed when natural gas and/or other geological fluids are extracted from geological formations.

Selected simulations of alkanes within alumina and silica-based pores

Simulations, in particular molecular dynamics (complemented by Monte Carlo and *ab initio* density functional theory, DFT), have become a routine tool to study the properties of bulk fluids, including those of water, alkanes, and their mixtures. Unfortunately, studying fluids at contact with a solid substrate is hindered by the uncertainties regarding the solid-fluid interaction potentials, and also by the fact that the relaxation time of the solid is typically much longer than that of the fluid. However, it is now recognized that simulation results can significantly improve the interpretation of experimental data, especially when direct observation is difficult, as in the case of confined fluids. Some examples have been given in the "*Properties of confined fluid*" section. More and more evidence suggests that semi-quantitative agreement can be obtained between simulation predictions (sometimes even obtained by implementing force fields not necessarily derived to reproduce experimental data for fluid-solid interfaces) and experimental observations. When some disagreement is observed, it could be used as a valid reason to improve the simulation models, but also to revisit the interpretation of the experimental data, which could lead to discoveries. We highlight below a few simulation studies for alkanes in contact with alumina and porous silica. Compared to the studies for simple fluids, which have been the basis for the general aspects discussed above, the studies for long alkanes are somewhat limited. As the importance of understanding fluid behavior and rock-fluid interactions in subsurface formations grows, so is the need to further the investigations summarized herein.

Alkanes near alumina. Alumina is frequently used as catalyst and catalyst support, and it also constitutes part of clay, a naturally occurring material. Understanding the behavior of fluids near to, or confined between, alumina surfaces is therefore of wide interest.

The Al-terminated α-Al_2O_3 surface shows significant relaxation of the solid atoms, as suggested by both theoretical (Konstadinidis et al. 1992; Manassidis and Gillan 1994; Streitz and Mintmire 1994) and experimental observations (Ahn and Rabalais 1997). It also readily reacts when exposed to water. The hydroxylated form of the α-Al_2O_3 surface shows limited relaxation (Wittbrodt et al. 1998). Whether or not the solid atoms near the interface relax compared to the positions they would occupy in the bulk of the material has important consequences in the structure predicted for interfacial fluids. For example, Li and Choi (2007) employed a number of techniques to study linear C-11 and C-200 alkanes on an alumina support. They first relaxed the Al-terminated alumina substrate using *ab initio* DFT, and then used molecular dynamics (MD) to study the structure of the liquid alkanes adsorbed onto the substrate, maintained rigidly. Representative results, shown in Figure 25, are interesting for a number of reasons: (1) the density profiles obtained for the short alkane are qualitatively, though not quantitatively, similar to those obtained for the long chains, which is consistent with surface force apparatus (SFA) experiments for alkanes of different lengths confined between mica surfaces; and (2) the relaxed alumina substrate promotes the formation of well-defined density layers formed by alkane segments (CH_3 groups) as the distance from the surface increases while the non-relaxed surface does not (these observations suggest that the relaxed alumina support is commensurate with the alkane molecules and promotes the formation of contact layers within which the alkane chains lay parallel to the solid surface).

Jin et al. (2000) obtained results to some extent analogous to those just discussed when they employed MD to study the structure of C-8, C-16, and C-32 alkanes confined between two parallel alumina surfaces. The alumina surfaces were hydroxylated, and the authors reported limited relaxation compared to bulk structures when DFT calculations were performed (Wittbrodt et al. 1998). Thus, a crystalline structure was assumed for the solid. The simulations were conducted at 300 K. Pronounced layering was observed in the atomic density profiles for up to 4-5 atomic layers (up to 2.0-2.5 nm) from each surface. The density profiles were qualitatively similar for all alkanes considered. This result (layering) appears to be consistent

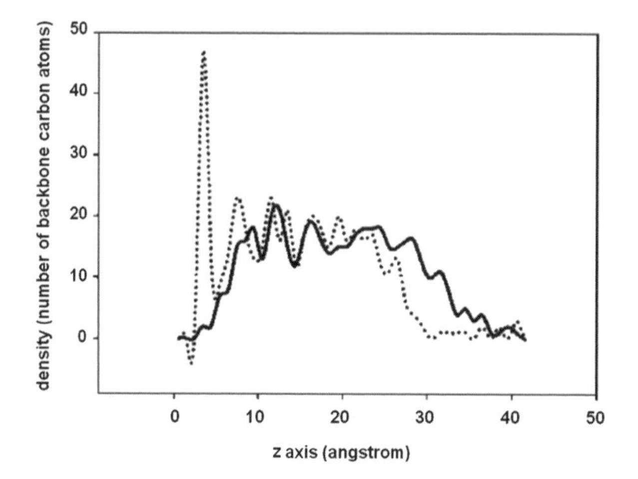

Figure 25. Density profiles for CH_2/CH_3 groups as a function of the distance from a flat alumina substrate. Results are shown before the substrate was relaxed (black line) and after the relaxation (dotted line). Relaxing the surface yields pronounced layering in the film. [Used by permission of the American Chemical Society © 2007, from Li and Choi (2007) *J Phys Chem-C*, Vol. 111, Fig. 4, p. 1751.]

with a number of simulation results for liquid straight-chain alkanes confined between two parallel surfaces (Bitsanis and Hadziioannou 1990; Vacatello et al. 1990; Balasubramanian et al. 1995, 1996; Dijkstra 1997; Gao et al. 1997a,b; Wang and Fichthorn 1998), structured or unstructured (Magda et al. 1985). Compared to results obtained by Li and Choi (2007), the simulations reported by Jin et al. (2000) were conducted for fluid molecules confined within two parallel surfaces, which in general promotes longer-ranged ordering than does one free-standing surface. This difference might be the reason why Li and Choi did not report ordering for alkanes adsorbed on a non-relaxed surfaces, whereas Jin et al. found pronounced ordering for any type of confining surfaces.

In addition, by analyzing the layer-by-layer intra-molecular radial distribution functions Jin et al. (2000) showed that the alkane chains adsorbed onto the alumina substrate have a tendency to adopt a trans conformation, which has also been found for alkanes adsorbed on gold (Balasubramanian et al. 1995). Finally, the dynamics of the adsorbed alkane chains were found to be significantly delayed compared to that of bulk alkanes, and infrequent exchanges were observed between alkanes belonging to different layers (the simulations were only conducted for at most 1 ns; current computational capabilities would allow researchers to reach up to 50-100 ns). This latter observation suggests the existence of a correlation between spatial distribution and dynamic properties. Namely, the alkane chains that are located near the surfaces show less mobility than those found away from the surfaces, suggesting that when a technology is designed to "detach" the alkanes from the pore surfaces, enhanced permeability could be achieved.

Along these lines, de Sainte Claire et al. (1997) conducted MD simulations for a system composed of a free-standing Al-terminated alumina surface, a thin layer of pre-adsorbed alkanes (butane, octane, and dodecane), and a small cluster of 30 water molecules. They found that the water molecules penetrate the pre-adsorbed thin film of alkane chains because they are attracted to the alumina substrate via long-range electrostatic interactions. During the penetration process, a competition was found between the displacement of the hydrocarbon chains, and the densification of the adsorbed thin film. Both phenomena were caused by the penetrating water molecules. The process of penetration was found to become more difficult as the alkane chain length increased and as the temperature decreased. In Figure 26 we reproduce the simulation results for the evolution of the probability of finding water molecules at various distances from the alumina substrate when the 30 water molecules were placed near the bare alumina (left) or near the alumina substrate covered by 2.6 monolayers of octane. In the case

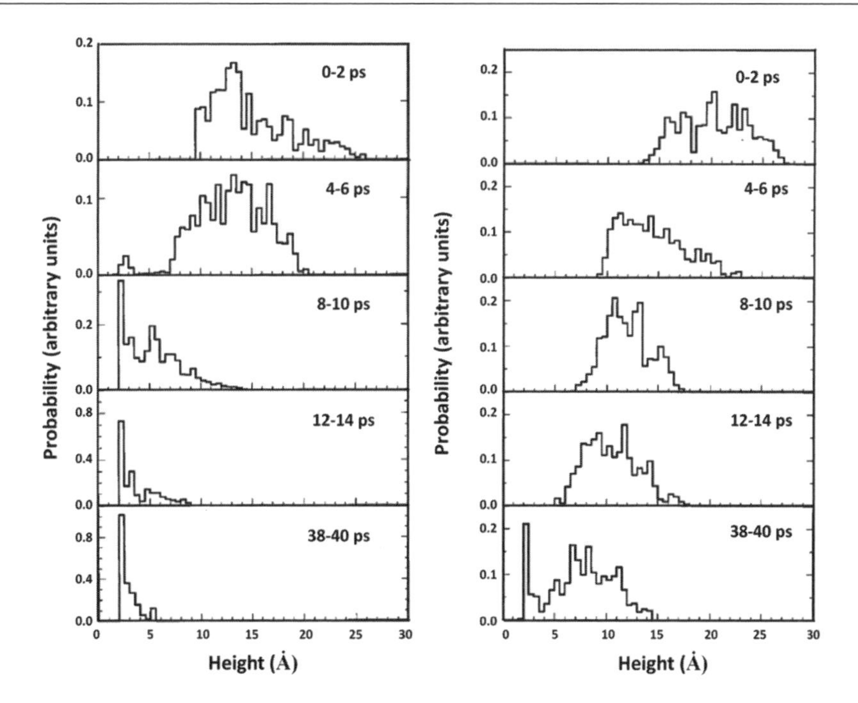

Figure 26. Time evolution of the density probability of observing water molecules at various vertical distances from the (0001) surface of α-Al_2O_3. In the left panel a cluster of 30 water molecules was placed initially at about 1.5 nm from the bare surface. On the right panel the cluster of 30 water molecules was placed on a surface on which 2.6 monolayers of octane were pre-adsorbed. The time intervals indicate the length of the MD simulations performed. Water molecules quickly adsorb and spread on the bare surface, yielding a monolayer (*left*). When the surface is covered by octane (*right*) water molecules manage to come in contact with the alumina substrate, although a perfect monolayer is not obtained within the 40 ps of the simulations presented. Simulations were performed at 300 K. [Used by permission of American Institute of Physics © 1997, from de Sainte Claire et al. (1997) *J Chem Phys*, Vol. 106, Fig. 6, p. 7339 and, Fig. 9, p. 7340.]

of water on bare alumina, water molecules readily form a monolayer on the substrate, which is in qualitative agreement with recent MD simulations (Argyris et al. 2011). When octane is present the water molecules still manage to come in contact with the substrate, although the process takes longer to complete and the distribution of water molecules after 40 ps spans the ~1.5 nm near the substrate. Note that in recent years computational capabilities have become highly evolved, and it would now be possible to perform simulations such as those of Figure 26 for several hundreds of ns.

Studies such as the one just summarized are important for providing interpretation to detailed experimental studies of alkane-mineral interfaces, such as, for example, the broadband sum frequency generation studies recently reported by Buchbinder et al. (2010) Other simulation studies along similar lines have investigated the deposition of waxes on hematite (San-Miguel and Rodger 2003), sometimes in the presence of other compounds (San-Miguel and Rodger 2010).

Regarding confined mixtures, it should be pointed out that smectites, present in sediments and soil, have been found to host a variety of hydrated organic molecules in their interlayer (Kaiser and Guggenberger 2000; Kennedy et al. 2002; Kawahigashi et al. 2006), and that large amounts of methane appear to be stored in clay minerals present in marine sediments (Hinrichs et al. 2006; Ertefai et al. 2010). These observations stimulated experimental (Cha et al. 1988;

Guggenheim and Koster van Groos 2003; Seo et al. 2009) and simulation (Sposito et al. 1999; Titiloye and Skipper 2000; Park and Sposito 2003; Cygan et al. 2004a; Zhang and Choi 2006; Zhou et al. 2011) studies for methane-water mixtures confined within various smectites. The simulation results suggest that under confinement one methane molecule is hydrated by ~12-13 water molecules and remains at contact with ~6 oxygen atoms on the clay surface. Because in bulk hydrates one methane molecule is hydrated by 21 waters (Koh et al. 2000), the simulations suggest that methane prefers to accumulate near the solid substrate when it is confined in smectites.

When MD simulations are conducted to study mixtures, one source of uncertainty is whether or not the system studied is at equilibrium. At equilibrium the chemical potential for each of the fluid molecules under confinement should be equal to that of the fluid molecules in a correspondent bulk mixture. Typically, grand canonical Monte Carlo simulations are conducted to ensure that the chemical potential of a fluid under confinement is equal to that of the fluid in the bulk (Nicholson and Parsonage 1982). Indeed, by implementing such algorithms it is possible to simulate adsorption isotherms for various fluids, pure or mixtures, into porous materials. However, for dense systems composed of long hydrocarbons, and maybe containing hydrogen-bonding fluids under confinement, the acceptance of addition/deletion Monte Carlo moves is likely to be very low, and advanced simulation schemes need to be implemented. To overcome this requirement, and taking advantage of the enormous computational resources that are becoming available, we have implemented MD procedures to attempt to investigate the equilibrium partition of different fluid molecules between bulk systems and nanometer-scale pores. In Figure 27 we show one simulation snapshot, obtained for a mixture of water and methane at contact with a narrow slit-shaped Al_2O_3 pore of width ~1 nm. The alumina substrate is fully hydroxylated. Although the simulations are still preliminary, and proper assessment of their reliability has yet to be completed (e.g., we need to ensure that similar results are obtained when simulations are conducted from different initial configurations), the results clearly show that the bulk behavior of the water-methane mixture differs substantially compared to the behavior of the mixture under confinement. Water appears to be strongly attracted to the slit pore, while methane prefers to stay in the bulk region. A few water molecules escape the pore volume, but remain close to the pore entrance, where they can form energetically advantageous hydrogen bonds with other water molecules, and also interact favorably with the solid atoms. Although most methane molecules remain outside of the pore in the bulk, some methane molecules are found trapped within the pore. Contrary to what has been suggested in the case of smectites, the simulation snapshots of Figure 27 (see expanded view in the bottom panel) suggest that methane molecules prefer to accumulate near the pore center, where they are completely surrounded by water. Understanding how the structure of hydrated methane changes as the pore width and the pore chemistry change is important for a number of practical applications that include hydraulic fracturing. It appears that simulations, especially when synergistically coupled with experimental characterization, have the potential of helping this important quest.

Alkanes in silica-based porous materials. Many studies have focused on uncovering the properties of alkanes confined within silica-based pores. In particular, large attention has been given to zeolites, which are alumino-silicates. The composition of the zeolite has important consequences on its chemical physical properties. For example, when the aluminum content is high the zeolite is expected to become highly polar; as a consequence, preferential adsorption of polar molecules (e.g., water) is expected. Under such circumstances even small amounts of water present in the system can affect the adsorption of other gases (Brandani and Ruthven 2004; Galhotra et al. 2009; see, as a only partially related example, the simulation snapshot of Fig. 27). For example, the FAU type X zeolite, with a Si to Al ratio of 1-1.5, has been used in the form of membranes for the separation of methanol and water from systems containing hydrogen and carbon dioxide, respectively, in a large temperature interval (Sandstrom et al. 2010).

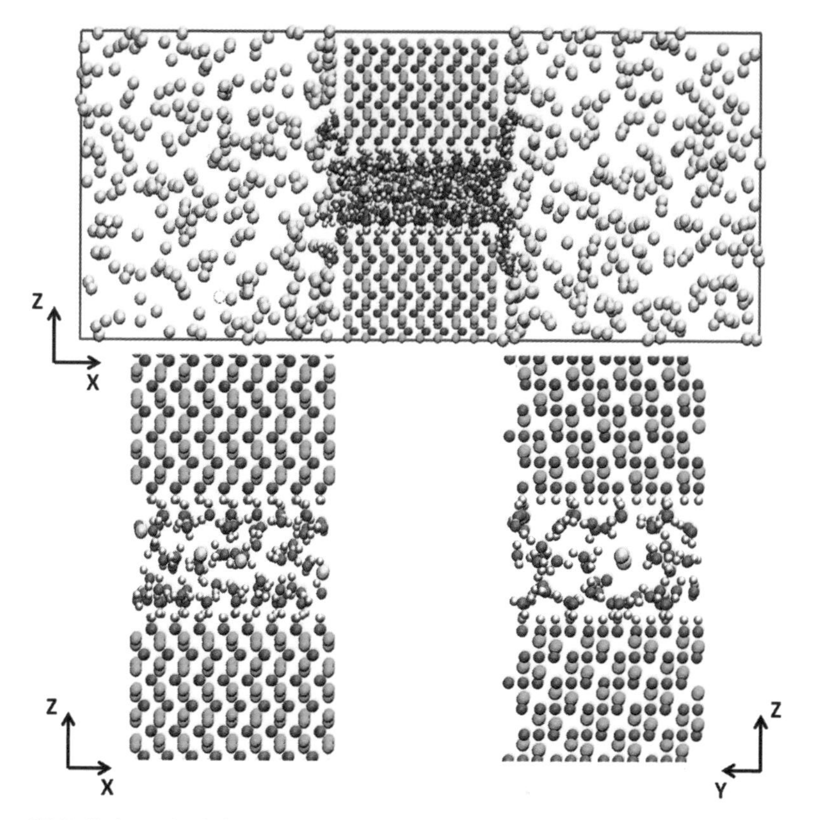

Figure 27. Preliminary simulation snapshot for a fluid system composed of methane (yellow spheres) and water molecules (red and white spheres represent oxygen and hydrogen atoms, respectively) partitioned between a bulk region (enriched in methane) and a slit-shaped pore of width ~1 nm. The pore is obtained by two facing surfaces of hydroxylated α-Al_2O_3. The bottom panels are enlargements of the simulation snapshot in the top panel, showing how the methane molecules trapped within the pore are preferentially found near the pore center, surrounded by water molecules. Simulations were performed at 300 K and 8-10 MPa for a mixture containing 50% methane molecules.

Due to a number of practical applications not limited to extraction of hydrocarbons from subsurface formations (e.g., catalysis), the study of alkanes, as well as of their mixtures confined in zeolites is growing rapidly. In Figure 28 we reproduce high-resolution scanning transmission electron microscopy images for the HY zeolite recently reported by Lu et al. (2012). In this study iridium is deposited on the zeolite for catalytic purposes, which is beyond the scopes of this chapter. The figure is important for remembering the high degree of confinement expected for fluids confined within zeolites (see, for example, Fig. 17 for a discussion on this topic). Because of space limitations we concentrate primarily on pioneering recent contributions.

Smit pioneered the use of simulations, including Monte Carlo and MD, to understand the behavior of molecules confined within zeolites for explaining observations in catalysis. He studied simultaneously the effects of alkane adsorption within the zeolite pores, their diffusion, and their catalytic conversion. For example, Smit and Maesen (1995) employed grand canonical Monte Carlo simulations to study the adsorption of butane, hexane, and heptane in silicalite. They demonstrated that unexpected, and at the time still unexplained, features in experimental adsorption isotherms could be explained by a phase transition experienced by confined hexane and heptane molecules that lead to a collective freezing. The phase transition could only be

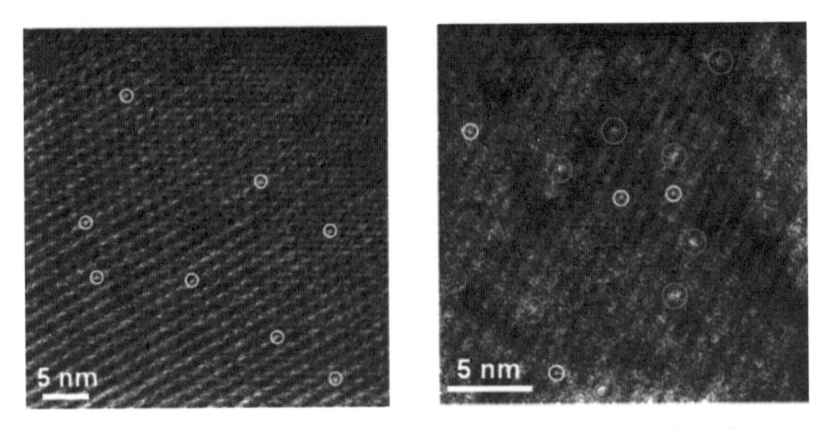

Figure 28. High resolution scanning transmission electron microscopy images of HY zeolite as prepared with $Ir(C_2H_4)$, (*left*), and after 1 h of flowing H_2 at 300 K (*right*). The treatment causes individual Ir atoms (circled in white) to coalesce into clusters, circled in red. [Used by permission of the American Chemical Society © 2012, from Lu et al. (2012) *J Am Chem Soc*, Vol. 134, Fig. 1, p. 5022.]

observed for hexane and heptane because their length is commensurate with the porous structure of silicalite. For shorter and longer chains, similar phase transitions could not occur because of thermodynamic considerations, and typical type-I adsorption isotherms were obtained, both experimentally and by simulations. More recently adsorption isotherms have been calculated using Monte Carlo simulations by Ndjaka et al. (2004) for several short alkanes (methane, ethane, and propane) in a few different zeolites. The various zeolites selected are characterized by a relatively wide range of pore sizes, and the results, obtained implementing the Lennard-Jones parameters proposed by Vlugt et al. (1999), yield good comparison to experimental observations.

The diffusivity of methane, ethane, and propane was assessed via MD simulations within mordenite, zeolite EU-1, and silicalite (Nowak et al. 1991). The analysis of the simulation trajectories showed that the mobility of methane on the three substrates is not isotropic but is instead determined by the porous structure. As the adsorbate size increases, because of hindered diffusion, the diffusion coefficient decreases. Reasonable agreement was observed between the predicted diffusivities and experimental pulse field gradient NMR data (Briscoe et al. 1988). This contribution demonstrates the advantage of molecular simulations compared to experiments, as this computational approach allows researchers to identify the preferential trajectories for each adsorbed species, thus allowing a complete understanding of the properties of the confined fluids and how such properties depend on the morphology, chemical composition, etc., of the confining pores. Of course, satisfactory comparison to experiments is needed to ratify the reliability of the physical picture obtained.

As the computational capabilities evolved, it became possible to study the diffusion of linear and branched alkanes on various zeolites, and even to calculate the activation energies by producing Arrhenius-type plots (Webb et al. 1999). The models implemented to describe the zeolites have also become more and more realistic. Webb et al. (1999) showed that the geometry of the pores, the presence of protrusions, and that of void spaces along the channel walls strongly influence the diffusion coefficient, as well as the activation energy for diffusion of linear alkanes. The effects become even more pronounced when the alkane chains are branched (the branched alkanes simulated in this contribution contained one methyl group linked at various positions along the otherwise linear chain). The calculations were performed at low loadings and for up to 40 ns. Skoulidas and Sholl (2002) extended these types of calculations, and investigated the effect of loading on the self-diffusion coefficient of small molecules (methane, helium, and argon) within silicalite. The results, in good experimental agreement, showed that the diffusion coefficient decreases as the loading increases. These authors also used the self-diffusion coefficients results to estimate transport diffusivities. The latter data have been instrumental for developing theoretical models to predict the permeability of zeolite membranes to various gases (Newsome and Sholl 2005).

Regarding studies on the effect of loading on permeability, it is worth remembering that Raghavan and Macelroy (1995) conducted equilibrium MD simulations for short alkanes adsorbed within models of silica gels at low loadings. They characterized the structure of the films adsorbed onto the solid substrate as a function of temperature. They found that the adsorbed hydrocarbon chains are characterized by a substantial broadening of the bond angle and dihedral angle distributions compared to the same molecules in the bulk. They also found that the diffusivity of the adsorbed chains increases as the loading increases because of screening of the fluid-pore interactions due to the pre-adsorbed molecules.

The simulation studies summarized here could be a starting point for addressing experimental observations of relevance for applications such as CO_2 sequestration. Key reactions and chemical mass transfer rates associated with progressive water-CO_2-rock interactions can be assessed by batch experiments such as those conducted by Gysi and Stefansson (2012). These authors conducted batch experiments at 40 °C for up to 260 days for four mineral assemblages: A1 – mixtures of Ca-Mg-Fe carbonates and altered basaltic glasses; A2 – Fe hydroxides and/or oxyhydroxides; A3 – mixtures of Ca-Mg-Fe clays and altered basaltic glasses; and A4 – mixtures of Ca-Fe clays and altered basaltic glasses. They maintained these samples at contact with water containing different amounts of initial dissolved CO_2 (ranging from 24 to 305 mmol/Kg). Then they monitored the concentration of the various species in the aqueous phase as a function of time. The results, summarized in Figure 29, suggest that increased aqueous CO_2 concentrations modify considerably the natural water-basalt reaction pathways during CO_2 mineralization. Equilibrium studies for the preferential adsorption of various compounds on the rock surfaces, such as those briefly described here, coupled with appropriate *ab initio* DFT studies could be useful for better understanding such phenomena at conditions of geological importance.

Simulation details

The interested reader is referred to specialized textbooks for a detailed discussion on simulation methods and algorithms (Nicholson and Parsonage 1982; Allen and Tildesley 1987; Frenkel and Smit 2002). This section is intended to summarize briefly some of the methods available to simulate systems of interest to subsurface applications, some of the models available to simulate alkanes, those that can be used to describe the solid substrate, and some of the experimental information that could validate the simulation results.

To study the properties of confined fluids from a theoretical/simulation perspective, a number of alternatives exist. The structure of confined fluids, as well as adsorption isotherms, can be calculated using classical density functional theory (Evans 1992) and integral equation theory (Henderson 1992) approaches. Both approaches require involved calculations for the minimization of the free energy, and often invoke approximations. An alternative, which is becoming widely accepted, is based on molecular simulations (Nicholson and Parsonage 1982). This latter approach is computationally expensive, and its reliability depends on the accuracy of the force fields implemented. It has the advantage of providing quantities that can be directly compared to experiments (e.g., self-diffusion coefficients, orientation of molecules near a substrate, radial distribution functions, vibrational frequencies, etc.). Very often molecular simulations are conducted within the MD formalism. However, it should be remembered that MD suffers from the difficulty of testing whether true equilibrium has been achieved. For example, the chromatography community has shown that MD simulations for confined systems

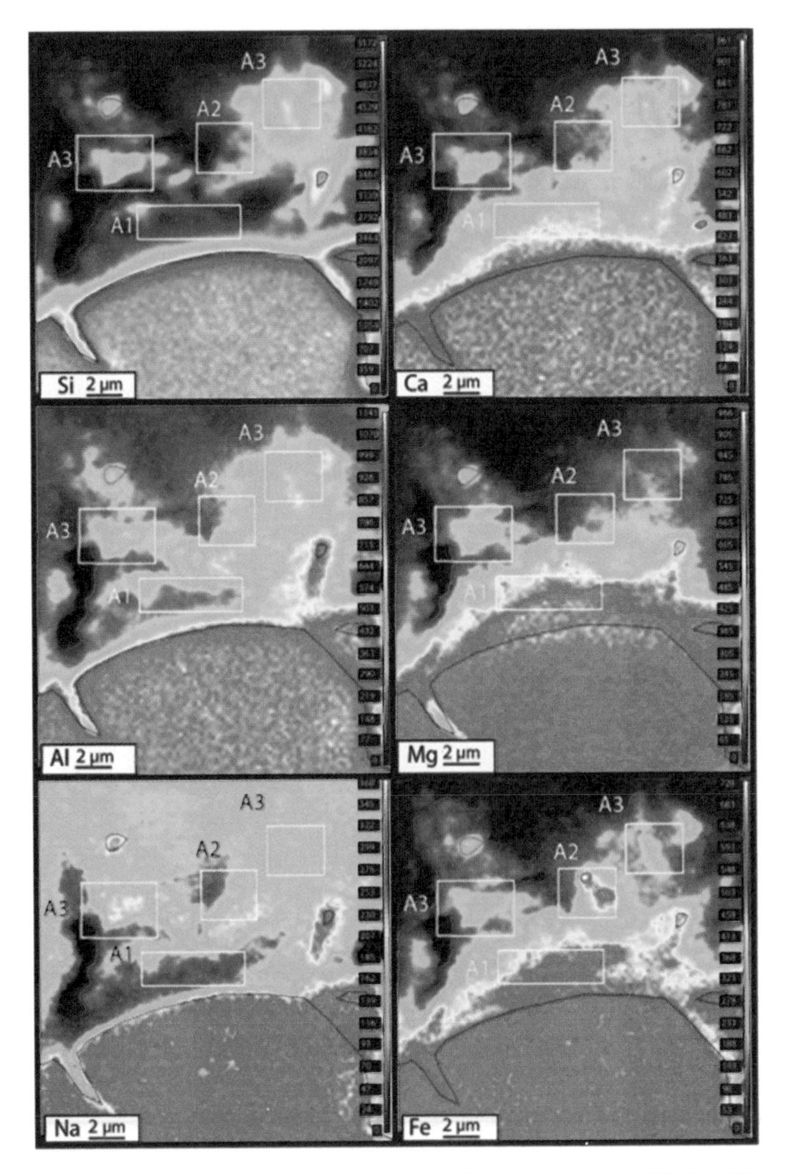

Figure 29. Elemental composition of mineral assemblages after a reaction time of 30 days. The samples were exposed to aqueous solutions with initial concentration of CO_2 < 50 mmol/Kg, and maintained at 40 °C. Compared to fresh basaltic glasses, A1 assemblages are depleted in Si, Al, and Na and enriched in Ca, Mg, and Fe; A2 assemblages are enriched in Fe ad depleted of all other elements; A3 and A4 assemblages are enriched in Si, Al, Na, Ca, Mg, and Fe, although sometimes the amount of Na and Mg was found to be moderate. [Used by permission of Elsevier Pub © 2012, from Gysi and Stefansson (2012) *Geochim Cosmochim Acta*, Vol. 81, Fig. 9, p. 140.]

containing alkanes (in their case grafted to a silica substrate) and water are strongly influenced by the initial configurations (Klatte and Beck 1996; Schure 1998). Specifically, it was found that for water it is difficult to permeate hydrocarbon chains that have previously collapsed onto a silica substrate. Monte Carlo techniques provide more reliable results, independent of the initial configuration (Zhang et al. 2005).

An alternative approach to molecular simulations, which could be used to study the equilibrium properties of fluids confined within porous materials, is based on equations of state. These approaches are attractive because their analytical form provides quick estimates for phase equilibria and mixture composition, although they rely on rather strong simplifications for fluid-fluid and fluid-pore interactions and they cannot provide density distributions within the pores, nor dynamical properties for the confined species. However, they could be used to estimate the composition of the confined fluids, which is necessary to initiate more involved molecular dynamics simulations. The development of equations of states for confined fluids is due to numerous contributions. Zhu et al. (1999) considered fluids in cylindrical mesopores (e.g., MCM-41). Schoen and Diestler (1998) applied the perturbation theory to describe fluids in slit-shaped pores. Giaya and Thompson (2002) applied the latter methodology successfully to describe water in cylindrical pores. Zarragoicoechea and Kuz (2002) extended the van der Waals equation of state to fluids in square pores. Derouane (2007) further extended the approach and obtained agreement with experimental data for the reduction in the critical temperature compared to bulk values for the confined fluid (Derouane 2007). Travalloni et al. (2010) applied the van der Waals approach to model simple fluids and their mixtures confined in porous solids. The model manages to reproduce several adsorption isotherm types, successfully correlates to experimental data for pure hydrocarbons adsorbed in pores, and reasonably well predicts binary mixtures adsorption without the need of binary interaction parameters. The mixtures considered included toluene and 1-propanol in DAY-13 zeolites, and methane and ethane in MCM-41.

Models for alkanes. One model available to describe alkanes has been developed by Smith and coworkers both at the explicit and at the united-atom levels (Smith and Yoon 1994). With subtle modifications, this model has been used to study, for example, interactions between polymer nanofibers (Buell et al. 2010), and it reliably reproduces structural and dynamical properties of bulk alkanes. Alternative force fields for alkanes include the widely applied united-atom version of the transferable potentials for phase equilibria (TraPPE-UA; Martin and Siepmann 1998, 1999; Chen et al. 2001; Stubbs et al. 2004). Bolton et al. (1999) employed *ab initio* DFT to parameterize the interactions between the alkane model of Smith and coworkers and Al-terminated alumina substrates. They then compared the results obtained when either the explicit or the united-atom models were used to simulate octane at contact with alumina. The results suggest that when the structure of interfacial octane is considered, united-atom and explicit models yield comparable results. However, rotational and translational dynamics are not satisfactory when the united atom model is employed, suggesting the need of including an explicit description of the hydrogen atoms within the hydrocarbon chain.

Significant work has also been done in predicting the stability of clathrates in the water-methane system. Considering the selection of appropriate force fields, the contribution from Alavi et al. (2007) should be highlighted. These authors employed MD simulations and implemented careful free-energy calculation techniques to establish the occupancy of methane in H clathrate hydrates at 300 K and 2 GPa. They compared the OPLS united atom Lennard-Jones potential, the Tse-Klein-McDonald five-site force field, and the Murad-Gubbins five-site potential to describe methane. The results show that the force field implemented strongly influences the predicted occupancy. The OPLS and the Tse-Klein-McDonald models yield occupancy of 5 methane molecules per cage, while the Murad-Gubbins model yields smaller occupancy. The differences in the results are due to the less attractive methane-methane potentials implemented by the Murad-Gubbins model.

Models for the solid substrates. Solid mineral substrates can be simulated implementing the popular CLAYFF force field (Cygan et al. 2004b). However, it is suggested to validate the force field by reproducing known experimental observations for the system of interest. When interested in studying carbon-based materials, the interested reader could take advantage of the vast literature developed to study adsorption in realistic models of activated carbons (Petersen et al. 2003; Palmer et al. 2010). However, a rational effort to investigate the effect of confinement on the properties of alkanes and of their mixtures requires, in our opinion, the employment of simplified models (e.g., the slit-shaped pore), especially when mixtures are considered. One microscopic model that is simple enough to allow researchers to obtain meaningful insights, yet complex enough to reproduce important experimental observations, has been recently proposed by Kowalczyk et al. (2010). This model has proven successful in simulating the adsorption of small hydrocarbons, as well as that of mixtures containing hydrocarbons and hydrogen. The results are satisfactory when compared to experimental data obtained in carbon molecular sieves. Additionally, the anomalous transport of small molecules (hydrogen vs. methane) in carbon molecular sieves can be explained by using this model. It was found that the size of the nanopore constrictions determines the anomalous diffusion, while the size of the cages does not influence the transport properties of the confined fluids.

Validation of simulation predictions. When possible, simulation data should be compared to desorption energies estimated from temperature-programmed desorption experiments (Slayton et al. 1995), chromatographic results, and batch techniques (Partyka and Douillard 1995; Askin and Inel 2001; Dabrowski et al. 2003; Diaz et al. 2004). Electron energy loss spectroscopy (Machida et al. 2002) and attenuated total reflection Fourier transform infrared spectroscopy (ATR-FTIR) (Dubowski et al. 2004; Almeida et al. 2008) can be used to study surface species on flat surfaces. Raman (Resini et al. 2005; Kim and Stair 2009) and transmission IR (Mawhinney and Yates 2001; Yeom et al. 2004) spectroscopies could provide information for hydrocarbons adsorbed on high-surface-area powders. One technique that is becoming very popular for studying structural details of molecules with any orientation with respect to an interface is the vibrational sum-frequency generation (SFG), which has been used previously to study molecules on oxide surfaces (Nanjundiah and Dhinojwala 2005; Hayes et al. 2009; Stokes et al. 2009). For instance, Sefler et al. (1995) studied liquid hexadecane at the silica interface, and found a preferential orientation parallel to the interface. Buchbinder et al. (2010) studied linear and cyclic alkanes and alkenes at the liquid-alumina interface. The results show that liquid alkanes, both linear and cyclic, prefer to lay parallel to the solid substrate. On the contrary, unsaturated olefins did not show consistent trends. As noted above, the diffusion constant of gases in pores can be measured experimentally by conducting macroscopic uptake rate measurements and microscopic ones by conducting pulsed field gradient nuclear magnetic resonance (PFG NMR) and quasi-elastic neutron scattering (Kärger and Ruthven 1991; June et al. 1992; Maginn et al. 1996; Reyes et al. 1997; Runnebaum and Maginn 1997). In our opinion microscopic observations (e.g., NMR and/or neutron scattering) are preferable to macroscopic ones, because they provide a direct comparison to the results predicted by simulations.

SUMMARY AND RECOMMENDATIONS

An atomistic to molecular-level understanding of how C-O-H fluids (e.g., water, CO_2, CH_4, higher hydrocarbons, etc.) interact with and participate in reactions with other solid Earth materials is central to the development of predictive models that aim to quantify a wide array of geochemical processes. The importance of the adsorption, microstructure and dynamics of important hydrocarbon and related fluids has been highlighted as well as the sensitivity of C-O-H fluid species to perturbations by a change in physical conditions or proximity to solute molecules and interfaces. Despite the large body of work that documents the nature of these processes and associated interactions with its local surroundings, it is premature to assume that we have a complete understanding of the mechanisms that give rise to the particular properties exhibited by hydrocarbon fluids. Understanding is more limited as one goes both above and below ambient conditions. For example, there is continuing discussion on the metastability of hydrocarbon species as a system experiences more extreme elevated temperature and pressure relevant to deep Earth environments. This is of particular interest, since we have seen that nanoporous confinement of C-O-H fluids at ambient conditions leads to structural and dynamical features that deviate markedly from bulk system behavior. In the context of natural systems, interrogation of fluids and fluid-solid interactions at elevated temperatures and pressures is an area requiring much more work, particularly for complex solutions containing geochemically relevant aqueous species interacting with relevant C-O-H fluids within the interfacial regime. We have tried to describe a series of prototypical interfacial and surface problems using a number of examples to stimulate the thinking of Earth scientists interested applying some of these outcomes to confined systems of mineralogical importance.

Our ability to predict the molecular-level properties of fluids and fluid-solid interactions relies heavily on the synergism between novel experiments, such as neutron scattering or nuclear magnetic resonance and molecular-based simulations. Tremendous progress has been made in closing the gap between experimental observations and predicted behavior based on simulations, owing to improvements in the experimental methodologies and instrumentation on the one hand, and the development of new potential models for C-O-H fluid species on the other. There has been an emergence of studies taking advantage of advanced computing power that can accommodate the demands of *ab initio* molecular dynamics. On the neutron instrumentation side, while much of the quasi-elastic work described above has been performed using instrumentation located at reactor based sources, the advent of second-generation spallation neutron sources like ISIS, new generation sources like the SNS at the Oak Ridge National Laboratory, and the low repetition rate second-target station at ISIS now offer significant opportunities for the study of interfacial and entrained liquids. Development of unique sample cells that allow interrogation of higher temperature-pressure conditions relevant to crust and mantle settings are needed that can allow interrogation of complex fluid-solid interactions. An improvement of the counting statistics by one to two orders of magnitude on many instruments, such as vibrational and time-of-flight spectrometers at SNS, will allow parametric studies of many hydrocarbon-based systems that otherwise would be prohibitively time consuming. The extended-Q SANS diffractometer and liquids reflectometer at SNS will offer very high intensity and unparalleled Q-range to extend the accessible length scale in real space, from 0.05 nm to 150 nm. The backscattering spectrometer will provide very high intensity and excellent energy resolution through unprecedented range of energy transfers, thereby allowing simultaneous studies of translational and rotational diffusion components in various systems. The vibrational spectrometer, with two orders of magnitude improvement in performance and the capability to perform simultaneous structural measurements, should present exciting opportunities as well, and support an new population of users interested in quantifying behavior of hydrocarbons and other fluid species relevant to Deep Earth.

ACKNOWLEDGMENTS

The authors are grateful to Robert Hazen and Dimitri Sverjensky for their thorough reviews and insightful comments. Support comes from the Sloan Foundation, Deep Carbon Observatory administered by the Geophysical Laboratory of the Carnegie Institution of Washington; and the Department of Energy, Office of Basic Energy Sciences, Geosciences Program. The authors wish to also thank David Tomasko at Ohio State, Gernot Rother, Mirek Gruszkiewicz, and Eugene Mamontov at Oak Ridge National Laboratory, and Karl Mueller at Pacific Northwest National Laboratory for their continued support of our experimental efforts.

REFERENCES

Abragam A (1961) The Principles of Nuclear Magnetism, Clarendon Press, Oxford

Aeberhardt K, Bui QD, Normand V (2007) Using low-field NMR to infer the physical properties of glassy oligosaccharide/water mixtures. Biomacromolecules 8:1038-1046

Ahn J, Rabalais JW (1997) Composition and structure of the Al_2O_3 {0001}-(1×1) surface. Surf Sci 388:121-131

Alavi S, Ripmeester JA, Klug DD (2007) Molecular dynamics study of the stability of methane structure H clathrate hydrates. J Chem Phys 126:124708-1 - 124708-6

Alba-Simionesco C, Coasne B, Dosseh G, Dudziak G, Gubbins KE, Radhakrishnan R, Sliwinska-Bartkowiak M (2006) Effects of confinement on freezing and melting. J Phys Condens Matter 18:R15-R68

Alcantar N, Israelachvili J, Boles J (2003) Forces and ionic transport between mica surfaces: Imlications for pressure solution. Geochim Cosmochim Acta 67:1289-1304

Allen MP, Tildesley D (1987) Computer Simulations of Liquids. Oxford University Press: New York

Almeida AR, Moulijn JA, Mul G (2008) *In situ* ATR-FTIR study on the selective photo-oxidation of cyclohexane over anatase TiO_2. J Phys Chem C 112:1552-1561

Anzalone A, Boles J, Greene G, Young K, Israelachvili J, Alcantar N (2006) Confined fluids and their role in pressure solution. Chem Geol 230:220-231

Aranovich G, Donohue M (1998) Analysis of adsorption isotherms: lattice theory predictions, classification of isotherms for gas–solid equilibria, and similarities in gas and liquid Adsorption behavior. J Colloid Interface Sci 200:273-290

Argyris D, Cole DR, Striolo A (2009) Dynamic behavior of interfacial water at the silica surface. J Phys Chem C 113:19591-19600

Argyris D, Ho TA, Cole DR, Striolo A (2011) Molecular dynamics studies of interfacial water at the alumina surface. J Phys Chem C 115:2038-2046

Askin A, Inel O (2001) Evaluation of the heat of adsorption of some *n*-alkanes on alumina and zeolite by inverse gas chromatography. Sep Sci Technol 36:381-397

Baerlocher C, McCusker LB, Olson DH (2007) Atlas of Zeolites Framework Types. Elsevier: Amsterdam

Balasubramanian S, Klein ML, Siepmann JI (1995) Monte-Carlo investigations of hexadecane films on a metal-substrate. J Chem Phys 103:3184-3195

Balasubramanian S, Klein ML, Siepmann JI (1996) Simulation studies of ultrathin films of linear and branched alkanes on a metal substrate. J Phys Chem 100:11960-11963

Bee M (2003) Localized and long-range diffusion in condensed matter: state of the art of QENS studies and future prospects. Chem Phys 292:121-141

Belonoshko AB (1989) The Thermodynamics of the aqueous carbon-dioxide fluid within thin pores. Geochim Cosmochim Acta 53:2581-2590

Benes NE, Jobic H, Verweij H (2001) Quasi-elastic neutron scattering study of the mobility of methane in microporous silica. Microporous Mesoporous Mater 43:147-152

Bhatia SK, Chen HB, Sholl DS (2005) Comparisons of diffusive and viscous contributions to transport coefficients of light gases in single-walled carbon nanotubes. Mol Simulation 31:643-649

Bhatia SK, Nicholson D (2006) Transport of simple fluids in nanopores: Theory and simulation. Aiche J 52:29-38

Bhatia SK, Nicholson D (2007) Anomalous transport in molecularly confined spaces. J Chem Phys 127:124701, doi: 10.1063/1.2768969

Bitsanis I, Hadziioannou G (1990) Molecular-dynamics simulations of the structure and dynamics of confined polymer melts. J Chem Phys 92:3827-3847

Blümich B, Casanova F, Appelt S (2009) NMR at low magnetic fields. Chem Phys Lett 477:231-240

Bolton K, Bosio SBM, Hase WL, Schneider WF, Hass KC (1999) Comparison of explicit and united atom models for alkane chains physisorbed on alpha-Al_2O_3 (0001). J Phys Chem B 103:3885-3895

Brandani F, Ruthven DM (2004) The effect of water on the adsorption of CO_2 and C_3H_8 on type X zeolites. Ind Eng Chem Res 43:8339-8344

Briscoe NA, Johnson DW, Shannon MD, Kokotailo GT, Mccusker LB (1988) The framework topology of zeolite Eu-1. Zeolites 8:74-76

Brochard L, Vandamme M, Pelenq RJM, Fen-Chong T (2012a) Adsorption-induced deformation of microporous materials: coal swelling induced by CO_2-CH_4 competitive adsorption. Langmuir 28:2659-2670

Brochard L, Vandamme M, Pellenq RJM (2012b) Poromechanics of microporous media. J Mech Phys Solids 60:606-622

Brovchenko I, Geiger A, Oleinikova A (2004) Water in nanopores: II. The liquid-vapour phase transition near hydrophobic surfaces. J Phys Condens Matter 16:S5345-S5370

Brovchenko I, Geiger A, Oleinikova A (2005) Surface critical behavior of fluids: Lennard-Jones fluid near a weakly attractive substrate. Eur Phys J B 44:345-358

Brovchenko I, Oleinikova A (2008) Complexity versus universality - fluid in the nearly solid surface. Chem Unserer Zeit 42:152-159

Buchbinder AM, Weitz E, Geiger FM (2010) Pentane, hexane, cyclopentane, cyclohexane, 1-hexene, 1-pentene, cis-2-pentene, cyclohexene, and cyclopentene at vapor/alpha-alumina and liquid/alpha-alumina interfaces studied by broadband sum frequency generation. J Phys Chem C 114:554-566

Buell S, Rutledge GC, Van Vliet KJ (2010) Predicting polymer nanofiber interactions via molecular simulations. Acs Appl Mater Inter 2:1164-1172

Buntkowsky G, Breitzke H, Adamczyk A, Roelofs F, Emmler T, Gedat E, Grünberg B, Xu Y, Limbach HH, Shenderovich I, Vyalikh A, Findenegg G (2007) Structural and dynamical properties of guest molecules confined in mesoporous silica materials revealed by NMR. Phys Chem Chem Phys 9:4843-4853

Busch A, Gensterblum Y, Krooss BM, Littke R (2004) Methane and carbon dioxide adsorption-diffusion experiments on coal: upscaling and modeling. Int J Coal Geol 60:151-168

Cai Q, Buts A, Seaton NA, Biggs MJ (2008) A pore network model for diffusion in nanoporous carbons: Validation by molecular dynamics simulation. Chem Eng Sci 63:3319-3327

Cavenati S, Grande CA, Rodrigues AE (2004) Adsorption equilibrium of methane, carbon dioxide, and nitrogen on zeolite ^{13}X at high pressures. J Chem Eng Data 49:1095-1101

Cha SB, Ouar H, Wildeman TR, Sloan ED (1988) A 3rd-surface effect on hydrate formation. J Phys Chem 92:6492-6494

Chathoth SM, Mamontov E, Melnichenko YB, Zamponi M (2010) Diffusion and adsorption of methane confined in nano-porous carbon aerogel: A combined quasi-elastic and small-angle neutron scattering study. Microporous Mesoporous Mater 132:148-153

Chen B, Potoff JJ, Siepmann JI (2001) Monte Carlo calculations for alcohols and their mixtures with alkanes. Transferable potentials for phase equilibria. 5. United-atom description of primary, secondary, and tertiary alcohols. J Phys Chem B 105:3093-3104

Cheng AL, Huang WL (2004) Selective adsorption of hydrocarbon gases on clays and organic matter. Org Geochem 35:413-423

Clarkson CR, Bustin RM (1996) Application of adsorption potential theory to coal/methane adsorption isotherms at elevated temperature and pressure; implications for reservoir characterization. Abstracts with Programs Geol Soc Am 28:42

Coasne B, Czwartos J, Sliwinska-Bartkowiak M, Gubbins KE (2009) Effect of pressure on the freezing of pure fluids and mixtures confined in nanopores. J Phys Chem B 113:13874-13881

Coasne B, Hung FR, Pellenq RJM, Siperstein FR, Gubbins KE (2006) Adsorption of simple gases in MCM-41 materials: The role of surface roughness. Langmuir 22:194-202

Cole DR, Chialvo AA, Rother G, Vlcek L, Cummings PT (2010) Supercritical fluid behavior at nanoscale interfaces: Implications for CO_2 sequestration in geologic formations. Philos Mag 90:2339-2363

Cole DR, Gruszkiewicz MS, Simonson JM, Chialvo AA, Melnichenko YB (2004) Influence of nanoscale porosity on fluid behavior. *In*: Water-Rock Interaction. Vol 1. Wanty R, Seal R (eds) p 735-739

Cole DR, Herwig K, Mamontov E, Larese LZ (2006) Neutron scattering and diffraction studies of fluids and fluid-solid interactions. Rev Mineral Geochem 63:313-362

Cole DR, Mamontov E, Rother G (2009) Structure and dynamics of fluids in microporous and mesoporous Earth and engineered materials. *In*: Neutron Applications in Earth, Energy, and Environmental Sciences. Liang L, Rinaldi R, Schober H (eds) Springer, p 547-570

Combariza AF, Sastre G, Corma A (2011) Molecular dynamics simulations of the diffusion of small chain hydrocarbons in 8-ring zeolites. J Phys Chem C 115:875-884

Cracknell RF, Nicholson D, Quirke N (1995) Direct molecular-dynamics simulation of flow down a chemical-potential gradient in a slit-shaped micropore. Phys Rev Lett 74:2463-2466

Cygan RT, Guggenheim S, Koster van Groos AF (2004a) Molecular models for the intercalation of methane hydrate complexes in montmorillonite clay. J Phys Chem B 108:15141-15149

Cygan RT, Liang JJ, Kalinichev AG (2004b) Molecular models of hydroxide, oxyhydroxide, and clay phases and the development of a general force field. J Phys Chem B 108:1255-1266

Czwartos J, Coasne B, Gubbins KE, Hung FR, Sliwinska-Bartkowiak M (2005) Freezing and melting of azeotropic mixtures confined in nanopores: experiment and molecular simulation. Molec Phys 103:3103-3113

Dabrowski A, Podkoscielny P, Bulow M (2003) Comparison of energy-distribution functions calculated for gas-solid and liquid-solid adsorption data. Colloid Surface A 212:109-114

Davis HT (1992) Kinetic theory of strongly inhomogeneous fluids. *In*: Fundamentals of Inhomogeneous Fluids. Henderson D (ed) Dekker: New York, p 551-592

Dawson R, Khoury F, Kobayash R (1970) Self-diffusion measurements in methane by pulsed nuclear magnetic resonance. AIChE J 16:725-729

Denayer JFA, Devriese LI, Couck S, Martens J, Singh R, Webley PA, Baron GV (2008) Cage and window effects in the adsorption of *n*-alkanes on chabazite and SAPO-34. J Phys Chem C 112:16593-16599

Derouane EG (2007) On the physical state of molecules in microporous solids. Microporous Mesoporous Mater 104:46-51

de Sainte Claire P, Hass KC, Schneider WF, Hase WL (1997) Simulations of hydrocarbon adsorption and subsequent water penetration on an aluminum oxide surface. J Chem Phys 106:7331-7342
Diaz E, Ordonez S, Vega A, Coca J (2004) Adsorption characterisation of different volatile organic compounds over alumina, zeolites and activated carbon using inverse gas chromatography. J Chromatogr A 1049:139-146
Dijkstra M (1997) Confined thin films of linear and branched alkanes. J Chem Phys 107:3277-3288
Donohue MD, Aranovich GL (1999) A new classification of isotherms for Gibbs adsorption of gases on solids. Fluid Phase Equilib 158:557-563
Dubowski Y, Vieceli J, Tobias DJ, Gomez A, Lin A, Nizkorodov SA, McIntire TM, Finlayson-Pitts BJ (2004) Interaction of gas-phase ozone at 296 K with unsaturated self-assembled monolayers: A new look at an old system. J Phys Chem A 108:10473-10485
Dvoyashkin M, Valiullin R, Kärger J, Einicke W-D, Glaser R (2007) Direct assessment of transport properties of supercritical fluids confined to nanopores. J Am Chem Soc 129:10344-10345
Ernst RR, Bodenhausen G, Wokaun A (1987) Principles of Nuclear Magnetic Resonance in One and Two Dimensions, Clarendon Press, Oxford
Ertefai TF, Heuer VB, Prieto-Mollar X, Vogt C, Sylva SP, Seewald J, Hinrichs KU (2010) The biogeochemistry of sorbed methane in marine sediments. Geochim Cosmochim Acta 74:6033-6048
Evans R (1992) Density functionals in the theory of nonuniform fluids. *In:* Fundamentals of Inhomogeneous Fluids. Henderson D (ed) Dekker, New York, p 85-176
Evans R, Marconi UMB (1987) Phase-equilibria and solvation forces for fluids confined between parallel walls. J Chem Phys 86:7138-7148
Eyraud C, Quinson JF, Brun M (1988) The role of thermoporometry in the study of porous solids. *In*: Characterization of Porous Solids I. Vol 39. Unger KK, Rouqerol J, Sing KSW, Kral H (eds) Studies in Surface Science and Catalysis, Elsevier, Amsterdam p 295-305
Fernandez M, Pampel A, Takahashi R, Sato S, Freude D, Kärger J (2008) Revealing complex formation in acetone-*n*-alkanes mixtures by MAS PFG NMR diffusion measurement in nanoporous hosts. Phys Chem Chem Phys 10:4165-4171
Firouzi M, Wilcox J (2012) Molecular modeling of carbon dioxide transport and storage in porous carbon-based materials. Microporous Mesoporous Mater 158:195-203
Floquet N, Coulomb JP, Bellat JP, Simon JM, Weber G, Andre G (2007) Heptane adsorption in silicalite-1: Neutron scattering investigation. J Phys Chem C 111:18182-18188
Floquet N, Coulomb JP, Weber G, Bertrand O, Bellatt JP (2003) Structural signatures of type IV isotherm steps: Sorption of trichloroethene, tetrachloroethene, and benzene in silicalite-I. J Phys Chem B 107:685-693
Ford DM, Glandt ED (1995) Molecular simulation study of the surface-barrier effect - dilute gas limit. J Phys Chem 99:11543-11549
Frenkel D, Smit B (2002) Understanding Molecular Simulations- From Algorithms to Applications. Academic Press: San Diego, CA
Fry RA, Kwon KD, Komarneni S, Kubicki JD, Mueller KT (2006) Solid-state NMR and computational chemistry study of mononucleotides adsorbed to alumina. Langmuir 22:9281-9286
Fry RA, Tsomaia N, Pantano CG, Mueller KT (2003) F-19 MAS NMR quantification of accessible hydroxyl sites on fiberglass surfaces. J Am Chem Soc 125:2378-2379
Gaede HC, Gawrisch K (2004) Multi-dimensional pulsed field gradient magic angle spinning NMR experiments on membranes. Magn Reson Chem 42:115-122
Galhotra P, Navea JG, Larsen SC, Grassian VH (2009) Carbon dioxide ($C^{16}O_2$ and $C^{18}O_2$) adsorption in zeolite Y materials: effect of cation, adsorbed water and particle size. Energy Environ Sci 2:401-409
Gao JP, Luedtke WD, Landman U (1997a) Origins of solvation forces in confined films. J Phys Chem B 101:4013-4023
Gao JP, Luedtke WD, Landman U (1997b) Structure and solvation forces in confined films: Linear and branched alkanes. J Chem Phys 106:4309-4318
Geier O, Vasenkov S, Kärger J (2002) PFG NMR study of long-range diffusion in beds of NaX zeolite: evidence for different apparent tortuosity factors in the Knudsen and bulk regimes. J Chem Phys 117:1935-1938
Gelb LD, Gubbins KE, Radhakrishnan R, Sliwinska-Bartkowiak M (1999) Phase separation in confined systems. Rep Prog Phys 62:1573-1659
Giaya A, Thompson RW (2002) Water confined in cylindrical micropores. J Chem Phys 117:3464-3475
Greene GW, Kristiansen K, Meyer EE, Boles J, Israelachvili J (2009) Role of electrochemical reactions in pressure solution. Geochim Cosmochim Acta 73:2862-2874
Gruszkiewicz MS, Rother G, Wesolowski DJ, Cole DR, Wallacher D (2012) Direct measurements of pore fluid density by vibrating tube densimetry. Langmuir 28:5070-5078
Guggenheim S, Koster van Groos AF (2003) New gas-hydrate phase: Synthesis and stability of clay-methane hydrate intercalate. Geology 31:653-656
Guichet X, Fleury M, Kohler E (2008) Effect of clay aggregation on water diffusivity using low field NMR. J Colloid Interface Sci 327:84-93

Guo ZL, Zhao TS, Shi Y (2005) Simple kinetic model for fluid flows in the nanometer scale. Phys Rev E 71:035301-035304
Guo ZL, Zhao TS, Xu C, Shi Y (2006) Simulation of fluid flows in the nanometer: kinetic approach and molecular dynamic simulation. Int J Comput Fluid D 20:361-367
Gysi AP, Stefansson A (2012) CO_2-water-basalt interaction. Low temperature experiments and implications for CO2 sequestration into basalts. Geochim Cosmochim Acta 81:129-152
Hayes PL, Chen EH, Achtyl JL, Geiger FM (2009) An optical voltmeter for studying cetyltrimethylammonium interacting with fused silica/aqueous interfaces at high ionic strength. J Phys Chem A 113:4269-4280
Heffelfinger GS, Vanswol F (1994) Diffusion in Lennard-Jones fluids using dual control-volume grand-canonical molecular-dynamics simulation (Dcv-Gcmd). J Chem Phys 100:7548-7552
Heitjans P, Kärger J (eds) (2005) Diffusion in Condensed Matter: Methods, Materials, Models. Springer, Berlin
Helbaek M, Hafskjold B, Dysthe DK, Sorland GH (1996) Self-diffusion coefficients of methane or ethane mixtures with hydrocarbons at high pressure by NMR. J Chem Eng Data 41:598-603
Henderson D (1992) Integral equation theories for inhomogeneous fluids. *In*: Fundamentals of Inhomogeneous Fluids. Henderson D (ed), Dekker: New York p 177-200
Hinrichs KU, Hayes JM, Bach W, Spivack AJ, Hmelo LR, Holm NG, Johnson CG, Sylva SP (2006) Biological formation of ethane and propane in the deep marine subsurface. Proc Natl Acad Sci USA 103:14684-14689
Ho TA, Argyris D, Cole DR, Striolo A (2012) Aqueous NaCl and CsCl solutions confined in crystalline slit-shaped silica nanopores of varying degree of protonation. Langmuir 28:1256-1266
Hoyt DW, Turcu RVF, Sears JA, Rosso KM, Burton SD, Felmy AR, Hu JZ (2011) High-pressure magic angle spinning nuclear magnetic resonance. J Magn Reson 212:378-385
Hu HX, Li XC, Fang ZM, Wei N, Li QS (2010) Small-molecule gas sorption and diffusion in coal: Molecular simulation. Energy 35:2939-2944
Hung FR, Gubbins KE, Radhakrishnan R, Szostak K, Beguin F, Dudziak G, Sliwinska-Bartkowiak M (2005) Freezing/melting of Lennard-Jones fluids in carbon nanotubes. Appl Phys Lett 86: 103110-1 - 103110-3
Huo H, Peng L, Grey CP (2009) Low temperature ^1H MAS NMR spectroscopy studies of proton motion in zeolite HZSM-5. J Phys Chem C 113:8211-8219
Indrawati L, Stroshine RL, Narsimhan G (2007) Low-field NMR: A tool for studying protein aggregation. J Sci Food Agric 87:2207-2216
Jain SK, Gubbins KE, Pellenq RJM, Pikunic JP (2006a) Molecular modeling and adsorption properties of porous carbons. Carbon 44:2445-2451
Jain SK, Pellenq RJM, Pikunic JP, Gubbins KE (2006b) Molecular modeling of porous carbons using the hybrid reverse Monte Carlo method. Langmuir 22:9942-9948
Jepps OG, Bhatia SK, Searles DJ (2003) Wall mediated transport in confined spaces: Exact theory for low density. Phys Rev Lett 91:1260102-1 - 126102-4
Jiang JW, Sandler SI, Schenk M, Smit B (2005) Adsorption and separation of linear and branched alkanes on carbon nanotube bundles from configurational-bias Monte Carlo simulation. Phys Rev B 72:045447-045457
Jin RY, Song KY, Hase WL (2000) Molecular dynamics simulations of the structures of alkane/hydroxylated alpha-Al_2O_3(0001) interfaces. J Phys Chem B 104:2692-2701
Jobic H (2000a) Diffusion of linear and branched alkanes in ZSM-5. A quasi-elastic neutron scattering study. J Mol Catal A Chem 158:135-142
Jobic H (2000b) Inelastic scattering of organic molecules in zeolites. Physica B 276:222-225
Jobic H, Bee M, Kärger J, Vartapetian RS, Balzer C, Julbe A (1995) Mobility of cyclohexane in a microporous silica sample - a quasi-elastic neutron-scattering and NMR pulsed-field gradient technique study. J Membr Sci 108:71-78
Jobic H, Rosenbach N, Ghoufi A, Kolokolov DI, Yot PG, Devic T, Serre C, Ferey G, Maurin G (2010) Unusual chain-length dependence of the diffusion of *n*-alkanes in the metal-organic framework MIL-47(V): The blowgun effect. Chem Eur J 16:10337-10341
Jobic H, Theodorou DN (2007) Quasi-elastic neutron scattering and molecular dynamics simulation as complementary techniques for studying diffusion in zeolites. Microporous Mesoporous Mater 102:21-50
Johnston JC, Iuliucci RJ, Facelli JC, Fitzgerald G, Mueller KT (2009) Intermolecular shielding contributions studied by modeling the C-13 chemical-shift tensors of organic single crystals with plane waves. J Chem Phys 131:144503-1 - 144503-11
June RL, Bell AT, Theodorou DN (1992) Molecular-dynamics studies of butane and hexane in silicalite. J Phys Chem 96:1051-1060
Kainourgiakis M, Steriotis T, Charalambopoulou G, Strobl M, Stubos A (2010) Determination of the spatial distribution of multiple fluid phases in porous media by ultra-small-angle neutron scattering. Appl Surf Sci 256:5329-5333
Kaiser K, Guggenberger G (2000) The role of DOM sorption to mineral surfaces in the preservation of organic matter in soils. Org Geochem 31:711-725

Kantzas A, Bryan JL, Mai A, Hum FM (2005) Applications of low field NMR techniques in the Ccharacterization of oil sand mining, extraction and upgrading processes. Can J Chem Eng 83:145-150
Kärger J, Caro J, Cool P, Coppens MO, Jones D, Kapteijn F, Rodriguez-Reinoso F, Stocker M, Theodorou D, Vansant EF, Weitkamp J (2009) Benefit of microscopic diffusion measurement for the characterization of nanoporous materials. Chem Eng Technol 32:1494-1511
Kärger J, Ruthven D (1991) Diffusion in Zeolites and Other Microporous Solids. Wiley, New York
Kärger J, Ruthven DM, Theodorou DN (2012) Diffusion in Nanoporous Materials. John Wiley & Sons, New York
Kärger J, Stallmach F, Vasenkov S (2003) Structure-mobility relations of molecular diffusion in nanoporous materials. Magn Reson Imaging 21:185-191
Kärger J (2008) Diffusion measurement by NMR techniques. *In*: Adsorption and Diffusion. Karge HG, Weitkamp J (ed) Springer, Berlin, p 85-133
Kawahigashi M, Kaiser K, Rodionov A, Guggenberger G (2006) Sorption of dissolved organic matter by mineral soils of the Siberian forest tundra. Global Change Biol 12:1868-1877
Kennedy MJ, Pevear DR, Hill RJ (2002) Mineral surface control of organic carbon in black shale. Science 295:657-660
Kim HS, Stair PC (2009) Resonance raman spectroscopic study of alumina-supported vanadium oxide catalysts with 220 and 287 nm excitation. J Phys Chem A 113:4346-4355
Kim TS, Dauskardt RH (2010) Molecular mobility under nanometer scale confinement. Nano Lett 10:1955-1959
Klatte SJ, Beck TL (1996) Microscopic simulation of solute transfer in reversed phase liquid chromatography. J Phys Chem 100:5931-5934
Knudsen M (1909) The laws of the molecular current and the internal friction current of gases by channels. Ann Phys Berlin 29:75-130
Koh CA, Wisbey RP, Wu XP, Westacott RE, Soper AK (2000) Water ordering around methane during hydrate formation. J Chem Phys 113:6390-6397
Konstadinidis K, Thakkar B, Chakraborty A, Potts LW, Tannenbaum R, Tirrell M, Evans JF (1992) Segment level chemistry and chain conformation in the reactive adsorption of poly(methyl methacrylate) on aluminum-oxide surfaces. Langmuir 8:1307-1317
Kowalczyk P, Gauden PA, Terzyk AP, Furmaniak S (2010) Microscopic model of carbonaceous nanoporous molecular sieves-anomalous transport in molecularly confined spaces. Phys Chem Chem Phys 12:11351-11361
Krishna R, van Baten JM (2011) A molecular dynamics investigation of the unusual concentration dependencies of Fick diffusivities in silica mesopores. Microporous Mesoporous Mater 138:228-234
Krutyeva M, Yang X, Vasenkov S, Kärger J (2007) Exploring the surface permeability of nanoporous particles by pulsed field gradient NMR. J Magn Reson 185:300-307
Lemmon EW, Huber ML, McLinden MO (2010) NIST Standard Reference Database 23: Reference Fluid Thermodynamic and Transport Properties-REFPROP, Version 9.0. National Institute of Standards and Technology, Standard Reference Data Program, Gaithersburg
Levitt MH (2001) Spin Dynamics Basics of Nuclear Magnetic Resonance. John Wiley and Sons, New York
Li CL, Choi P (2007) Molecular dynamics study of the adsorption behavior of normal alkanes on a relaxed α-Al_2O_3 (0001) surface. J Phys Chem C 111:1747-1753
Li WZ, Liu ZY, Che YL, Zhang D (2007) Molecular simulation of adsorption and separation of mixtures of short linear alkanes in pillared layered materials at ambient temperature. J Colloid Interface Sci 312:179-185
Liebscher A, Heinrich CA (eds) (2007) Fluid-Fluid Interactions. Reviews in Mineralogy and Geochemistry Vol. 65. Mineralogical Society of America, Chantilly VA
Lim YI, Bhatia SK (2011) Simulation of methane permeability in carbon slit pores. J Membrane Sci 369:319-328
Lim YI, Bhatia SK, Nguyen TX, Nicholson D (2010) Prediction of carbon dioxide permeability in carbon slit pores. J Membrane Sci 355:186-199
Lithoxoos GP, Labropoulos A, Peristeras LD, Kanellopoulos N, Samios J, Economou IG (2010) Adsorption of N_2, CH_4, CO and CO_2 gases in single walled carbon nanotubes: A combined experimental and Monte Carlo molecular simulation study. J Supercrit Fluid 55:510-523
Lu J, Aydin C, Browning ND, Gates BC (2012) Hydrogen activation and metal hydride formation trigger cluster formation from supported iridium complexes. J Am Chem Soc 134:5022-5025
Machida S, Hamaguchi K, Nagao M, Yasui F, Mukai K, Yamashita Y, Yoshinobu J, Kato HS, Okuyama H, Kawai M (2002) Electronic and vibrational states of cyclopentene on Si(100)(2×1). J Phys Chem B 106:1691-1696
Maciolek A, Ciach A, Evans R (1998) Critical depletion of fluids in pores: Competing bulk and surface fields. J Chem Phys 108:9765-9774

Maciolek A, Evans R, Wilding NB (1999) Effects of confinement on critical adsorption: Absence of critical depletion for fluids in slit pores. Phys Rev E 60:7105-7119
Magda JJ, Tirrell M, Davis HT (1985) Molecular-dynamics of narrow, liquid-filled pores. J Chem Phys 83:1888-1901
Maginn EJ, Bell AT, Theodorou DN (1996) Dynamics of long *n*-alkanes in silicalite: A hierarchical simulation approach. J Phys Chem 100:7155-7173
Magusin PCMM, Schuring D, van Oers EM, de Haan JW, van Santen RA (1999) *n*-Pentane hopping in zeolite ZK-5 studied with C-13 NMR. Magn Reson Chem 37:S108-S117
Malbrunot P, Vidal D, Vermesse J, Chahine R, Bose TK (1992) Adsorption measurements of argon, neon, krypton, nitrogen, and methane on activated carbon up to 650 MPa. Langmuir 8:577-580
Mamontov E, Kumzerov YA, Vakhrushev, SB (2005) Diffusion of benzene confined in oriented nanochannels of chrysotile asbestos fibers. Phys Rev E 72:051502-1 - 0515072-7
Manassidis I, Gillan MJ (1994) Structure and energetics of alumina surfaces calculated from first principles. J Am Ceram Soc 77:335-338
Martin MG, Siepmann JI (1998) Transferable potentials for phase equilibria. 1. United-atom description of *n*-alkanes. J Phys Chem B 102:2569-2577
Martin MG, Siepmann JI (1999) Novel configurational-bias Monte Carlo method for branched molecules. Transferable potentials for phase equilibria. 2. United-atom description of branched alkanes. J Phys Chem B 103:4508-4517
Mason EA, Malinaus AP, Evans RB (1967) Flow and diffusion of gases in porous media. J Chem Phys 46:3199-3216
Mawhinney DB, Yates JT (2001) FTIR study of the oxidation of amorphous carbon by ozone at 300 K - direct COOH formation. Carbon 39:1167-1173
McCollom TM (2013) Laboratory simulations of abiotic hydrocarbon formation in Earth's deep subsurface. Rev Mineral Geochem 75:467-494
Menon VC, Komarneni S (1998) Porous adsorbents for vehicular natural gas storage: A review. J Porous Mat 5:43-58
Mentzen BF (2007) Crystallographic determination of the positions of the monovalent H, Li, Na, K, Rb, and Tl cations in fully dehydrated MFI type zeolites. J Phys Chem C 111:18932-18941
Meresi G, Wang YZ, Cardoza J, Wen WY, Jones AA, Gosselin J, Azar D, Inglefield PT (2001) Pulse field gradient NMR study of diffusion of pentane in amorphous glassy perfluorodioxole. Macromolecules 34:4852-4856
Metais A, Mariette F (2003) Determination of water self-diffusion coefficient in complex food products by low filed ^{1}H PFG NMR: comparison between the standard spin-echo sequence and the T_1-wieghted spin-echo sequence. J Magn Reson 165:265-275
Millot B, Methivier A, Jobic H (1998) Adsorption of *n*-alkanes on silicalite crystals. a temperature-programmed desorption study. J Phys Chem B 102:3210-3215
Mitra S, Mukhopadhyay R (2003) Molecular dynamics using quasielastic neutron scattering. Current Sci 84:653-662
Mitra S, Mukhopadhyay R (2004) Quasi-elastic neutron scattering study of dynamics in condensed matter. Pramana J Phys 63:81-89
Mueller R, Kanungo R, Kiyono-Shimobe M, Koros WJ, Vasenkov S (2012) Diffusion of methane and carbon dioxide in carbon molecular sieve membranes by multinuclear Pulsed Field Gradient NMR. Langmuir 28:10296-10303
Myers AL, Monson PA (2002) Adsorption in porous materials at high pressure: Theory and experiment. Langmuir 18:10261-10273
Nanjundiah K, Dhinojwala A (2005) Confinement-induced ordering of alkanes between an elastomer and a solid surface. Phys Rev Lett 95: 154301-1 - 154301-4
Ndjaka JMB, Zwanenburg G, Smit B, Schenk M (2004) Molecular simulations of adsorption isotherms of small alkanes in FER-, TON-, MTW- and DON-type zeolites. Microporous Mesoporous Mater 68:37-43
Neumann DA (2006) Neutron scattering and hydrogenous materials. Mater Today 9:34-41
Newsome DA, Sholl DS (2005) Predictive assessment of surface resistances in zeolite membranes using atomically detailed models. J Phys Chem B 109:7237-7244
Nguyen TX, Bhatia SK, Jain SK, Gubbins KE (2006) Structure of saccharose-based carbon and transport of confined fluids: hybrid reverse Monte Carlo reconstruction and simulation studies. Molec Simulation 32:567-577
Nguyen TX, Cohaut N, Bae JS, Bhatia SK (2008) New method for atomistic modeling of the microstructure of activated carbons using hybrid reverse Monte Carlo simulation. Langmuir 24:7912-7922
Nicholson D, Parsonage NG (1982) Computer Simulation and the Statistical Mechanics of Adsorption. Academic Press, London
Nivarthi SSS, McCormick AV, Davis HT (1994) Diffusion anisotropy in molecular sieves: A Fourier transform PFG NMR study of methane in A1PO4-5. Chem Phys Lett 229:297-301

Nowak AK, Denouden CJJ, Pickett SD, Smit B, Cheetham AK, Post MFM, Thomas JM (1991) Mobility of adsorbed apecies in zeolites - methane, ethane, and propane diffusivities. J Phys Chem 95:848-854
Oleinikova A, Brovchenko I, Geiger A (2006) Behavior of a wetting phase near a solid boundary: vapor near a weakly attractive surface. Eur Phys J B 52:507-519
Packer KJ (2003) Magnetic resonance in porous media: forty years on. Magn Reson Imaging 21:163-168
Palmer JC, Llobet A, Yeon SH, Fischer JE, Shi Y, Gogotsi Y, Gubbins KE (2010) Modeling the structural evolution of carbide-derived carbons using quenched molecular dynamics. Carbon 48:1116-1123
Pampel A, Fernandez M, Freude D, Kärger J (2005) New options for measuring molecular diffusion in zeolites by MAS PFG NMR. Chem Phys Lett 407:53-57
Pampel A, Kärger J, Michel D. (2003) Lateral diffusion of a transmembrane peptide in lipid bilayers studied by pulsed field gradient NMR in combination with magic angle sample spinning. Chem Phys Lett 379:555-561
Pantano CG, Fry RA, Mueller KT (2003) Effect of boron-oxide on surface hydroxyl coverage of aluminoborosilicate glass fibers: a ^{19}F solid-state NMR Study. Phys Chem Glass 44:64-68
Paoli H, Methivier A, Jobic H, Krause C, Pfeifer H, Stallmach F, Kärger J (2002) Comparative QENS and PFG NMR diffusion studies of water in zeolite NaCaA. Microporous Mesoporous Mater 55:147-158
Parcher JF, Strubinger JR (1989) High-pressure adsorption of carbon-dioxide on super-critical-fluid chromatography adsorbents. J Chromatogr 479:251-259
Park SH, Sposito G (2003) Do montmorillonite surfaces promote methane hydrate formation? Monte Carlo and molecular dynamics simulations. J Phys Chem B 107:2281-2290
Partyka S, Douillard JM (1995) Nature of interactions between organic pure liquids and model rocks - a calorimetric investigation. J Petrol Sci Eng 13:95-102
Petersen T, Yarovsky I, Snook I, McCulloch DG, Opletal G (2003) Structural analysis of carbonaceous solids using an adapted reverse Monte Carlo algorithm. Carbon 41:2403-2411
Pires J, Bestilleiro M, Pinto M, Gil A (2008) Selective adsorption of carbon dioxide, methane and ethane by porous clays heterostructures. Sep Purif Technol 61:161-167
Pitteloud C, Powell DH, Gonzalez MA, Cuello GJ (2003) Neutron diffraction studies of ion coordination and interlayer water structure in smectite clays: lanthanide(III)-exchanged Wyoming montmorillonite. Colloid Surf A 217:129-136
Pollard WG, Present RD (1948) On gaseous self-diffusion in long capillary tubes. Phys Rev 73:762-774
Potter J, Konnerup-Madsen J (2003) A review of the occurrence and origin of abiogenic hydrocarbons in igneous rocks. Geol Soc Spec Pubs 214:151-173
Pynn R (2009) Neutron scattering – a non-destructive microscope for seeing inside matter. *In*: Neutron Applications in Earth, Energy, and Environmental Sciences. Liang L, Rinaldi R, Schober H (eds) Springer, Berlin, p 15-36
Radhakrishnan R, Gubbins KE, Sliwinska-Bartkowiak M (2000) Effect of the fluid-wall interaction on freezing of confined fluids: Toward the development of a global phase diagram. J Chem Phys 112:11048-11057
Radhakrishnan R, Gubbins KE, Sliwinska-Bartkowiak M (2002) Global phase diagrams for freezing in porous media. J Chem Phys 116:1147-1155
Radlinski AP (2006) Small-angle neutron scattering and the microstructure of rocks. Rev Mineral Geochem 63:363-397
Raghavan K, Macelroy JMD (1995) Molecular-dynamics simulations of adsorbed alkanes in silica micropores at low-to-moderate loadings. Molec Simulation 15:1-33
Rajendran A, Hocker T, Di Giovanni O, Mazzotti M (2002) Experimental observation of critical depletion: Nitrous oxide adsorption on silica gel. Langmuir 18:9726-9734
Resini C, Montanari T, Busca G, Jehng JM, Wachs IE (2005) Comparison of alcohol and alkane oxidative dehydrogenation reactions over supported vanadium oxide catalysts: *in situ* infrared, Raman and UV-vis spectroscopic studies of surface alkoxide intermediates and of their surface chemistry. Catal Today 99:105-114
Reyes SC, Sinfelt JH, DeMartin GJ, Ernst RH, Iglesia E (1997) Frequency modulation methods for diffusion and adsorption measurements in porous solids. J Phys Chem B 101:614-622
Riehl JW, Koch K (1972) NMR relaxation of adsorbed gases - methane on graphite. J Chem Phys 57:2199-2208
Rodriguez MA, Rubio J, Rubio F, Liso MJ, Oteo JL (1997) Application of inverse gas chromatography to the study of the surface properties of slates. Clay Clay Miner 45:670-680
Roman-Perez G, Moaied M, Soler JM, Yndurain F (2010) Stability, adsorption, and diffusion of CH_4, CO_2, and H_2 in clathrate hydrates. Phys Rev Lett 105:145901-145904
Rother G, Krukowski E, Wallacher D, Grimm N, Bodnar R, Cole DR (2012) Pore size effects on the sorption of supercritical carbon dioxide in mesoporous CPG-10 silica. J Phys Chem C 116(1):917-922
Rother G, Melnichenko YB, Cole DR, Frielinghaus H, Wignall GD (2007) Microstructural characterization of adsorption and depletion regimes of supercritical fluids in nanopores. J Phys Chem C 111:15736-15742
Rouquerol F, Rouquerol J, Sing K (1999) Adsorption by Powders and Porous Solids. Principles, Methodology and Applications. Academic Press, San Diego, CA

Runnebaum RC, Maginn EJ (1997) Molecular dynamics simulations of alkanes in the zeolite silicalite: Evidence for resonant diffusion effects. J Phys Chem B 101:6394-6408
Saalwachter K (2003) Detection of heterogeneities in dry and swollen polymer networks by proton low-field NMR spectroscopy. J Am Chem Soc 125:14684-14685
Saengsawang O, Schuring A, Remsungnen T, Hannongbua S, Newsome DA, Dammers AJ, Coppens MO, Fritzsche S (2010) Diffusion of *n*-pentane in the zeolite ZK5 studied by high-temperature configuration-space exploration. Chem Phys 368:121-125
Sandstrom L, Palomino M, Hedlund J (2010) High flux zeolite X membranes. J Membrane Sci 354:171-177
San-Miguel MA, Rodger PM (2003) Wax deposition onto Fe_2O_3 surfaces. Phys Chem Chem Phys 5:575-581
San-Miguel MA, Rodger PM (2010) Templates for wax deposition? Phys Chem Chem Phys 12:3887-3894
Schloemer S, Krooss BM (2004) Molecular transport of methane, ethane and nitrogen and the influence of diffusion on the chemical and isotopic composition of natural gas accumulations. Geofluids 4:81-108
Schoen M, Diestler DJ (1998) Analytical treatment of a simple fluid adsorbed in a slit-pore. J Chem Phys 109:5596-5606
Schreiber A, Bock H, Schoen M, Findenegg GH (2002) Effect of surface modification on the pore condensation of fluids: experimental results and density functional theory. Molec Phys 100:2097-2107
Schure MR (1998) Particle simulation methods in separation science. *In*: Advances in Chromatography. Vol 39. Brown PR (ed) Marcel Dekker, New York, p 435-441
Schuring A, Auerbach SM, Fritzsche S (2007) A simple method for sampling partition function ratios. Chem Phys Lett 450:164-169
Sefler GA, Du Q, Miranda PB, Shen YR (1995) Surface crystallization of liquid *n*-alkanes and alcohol monolayers studied by surface vibrational spectroscopy. Chem Phys Lett 235:347-354
Sel O, Brandt A, Wallacher D, Thommes M, Smarsly B (2007) Pore hierarchy in mesoporous silicas evidenced by *in situ* SANS during nitrogen physisorption. Langmuir 23:4724-4727
Seland JG, Ottaviani M, Hafskjold B (2001) A PFG-NMR study of restricted diffusion in heterogeneous polymer particles. J Colloid Interface Sci 239:168-177
Seo YJ, Seol J, Yeon SH, Koh DY, Cha MJ, Kang SP, Seo YT, Bahk JJ, Lee J, Lee H (2009) Structural, mineralogical, and rheological properties of methane hydrates in smectite clays. J Chem Eng Data 54:1284-1291
Sephton MA, Hazen RM (2013) On the origins of deep hydrocarbons. Rev Mineral Geochem 75:449-465
Sherwood Lollar B, Lacrampe-Couloume G, Slater GF, Ward J, Moser DP, Gihring TM, Lin LH, Onstott TC (2006) Unravelling abiogenic and biogenic sources of methane in the Earth's deep subsurface. Chem Geol 226:328-339
Singh SK, Sinha A, Deo G, Singh JK (2009) Vapor-liquid phase coexistence, critical properties, and surface tension of confined alkanes. J Phys Chem C 113:7170-7180
Sircar S (1999) Gibbsian surface excess for gas adsorption - revisited. Ind Eng Chem Res 38:3670-3682
Skipper NT, Lock PA, Titiloye JO, Swenson J, Mirza ZA, Howells WS, Fernandez-Alonso F (2006) The structure and dynamics of 2-dimensional fluids in swelling clays. Chem Geol 230:182-196
Skipper NT, Smalley MV, Williams GD, Soper AK, Thompson CH (1995) Direct measurement of the electric double-layer structure in hydrated lithium vermiculite clays by neutron-diffraction. J Phys Chem 99:14201-14204
Skoulidas AI, Ackerman DM, Johnson JK, Sholl DS (2002) Rapid transport of gases in carbon nanotubes. Phys Rev Lett 89:185901-1 - 185901-4
Skoulidas AI, Sholl DS (2002) Transport diffusivities of CH_4, CF_4, He, Ne, Ar, Xe, and SF_6 in silicalite from atomistic simulations. J Phys Chem B 106:5058-5067
Slayton RM, Aubuchon CM, Camis TL, Noble AR, Tro NJ (1995) Desorption-kinetics and adlayer sticking model of *n*-butane, *n*-hexane, and *n*-octane on Al_2O_3(0001). J Phys Chem 99:2151-2154
Sleep NH, Meibom A, Fridriksson T, Coleman RG, Bird DK (2004) H_2-rich fluids from serpentinization: Geochemical and biotic implications. Proc Natl Acad Sci USA 101:12818-12823
Smit B, Maesen TLM (1995) Commensurate freezing of alkanes in the channels of a zeolite. Nature 374:42-44
Smith GD, Yoon DY (1994) Equilibrium and dynamic properties of polymethylene melts from molecular-dynamics simulations. 1. *n*-tridecane. J Chem Phys 100:649-658
Sokhan VP, Nicholson D, Quirke N (2002) Fluid flow in nanopores: Accurate boundary conditions for carbon nanotubes. J Chem Phys 117:8531-8539
Span R, Wagner W (1996) A new equation of state for carbon dioxide covering the fluid region from the triple-point temperature to 1100 K at pressures up to 800 MPa. J Phys Chem Ref Data 25:1509-1596
Sposito G, Skipper NT, Sutton R, Park SH, Soper AK, Greathouse JA (1999) Surface geochemistry of the clay minerals. Proc Natl Acad Sci USA 96:3358-3364
Stallmach F, Graser A, Kärger J, Krause C, Jeschke M, Oberhagemann U, Spange S (2001) Pulsed field gradient NMR studies of diffusion in MCM-41 mesoporous solids. Microporous Mesoporous Mater 44:745-753
Stallmach F, Kärger J, Krause C, Jeschke M, Oberhagemann U (2000) Evidence of anisotropic self-diffusion of guest molecules in nanoporous materials of MCM-41 Type. J Am Chem Soc 122:9237-9242

Stepanov AG, Shegai TO, Luzgin MV, Jobic H (2003) Comparison of the dynamics of *n*-hexane in ZSM-5 and 5A zeolite structures. Eur Phys J E 12:57-61

Stoessell RK, Byrne PA (1982) Methane solubilities in clay slurries. Clay Clay Miner 30:67-72

Stokes GY, Chen EH, Walter SR, Geiger FM (2009) Two reactivity modes in the heterogeneous cyclohexene ozonolysis under tropospherically relevant ozone-rich and ozone-limited conditions. J Phys Chem A 113:8985-8993

Straadt IK, Thybo AK, Bertram HC (2008) NaCl-induced changes in structure and water mobility in potato tissue as determined by CLSM and LF-NMR. LWT-Food Sci Tech 41:1493-1500

Streitz FH, Mintmire JW (1994) Electrostatic-based model for alumina surfaces. Thin Solid Films 253:179-184

Striolo A (2006) The mechanism of water diffusion in narrow carbon nanotubes. Nano Lett 6:633-639

Striolo A, Gubbins KE, Gruszkiewicz MS, Cole DR, Simonson JM, Chialvo AA (2005) Effect of temperature on the adsorption of water in porous carbons. Langmuir 21:9457-9467

Strubinger JR, Parcher JF (1989) Surface Excess (Gibbs) Adsorption-isotherms of supercritical carbon-dioxide on octadecyl-bonded silica stationary phases. Anal Chem 61:951-955

Stubbs JM, Potoff JJ, Siepmann JI (2004) Transferable potentials for phase equilibria. 6. United-atom description for ethers, glycols, ketones, and aldehydes. J Phys Chem B 108:17596-17605

Thommes M, Findenegg GH, Schoen M (1995) Critical depletion of a pure fluid in controlled-pore glass - experimental results and grand-canonical ensemble Monte-Carlo simulation. Langmuir 11:2137-2142

Titiloye JO, Skipper NT (2000) Computer simulation of the structure and dynamics of methane in hydrated Na-smectite clay. Chem Phys Lett 329:23-28

Travalloni L, Castier M, Tavares FW, Sandler SI (2010) Thermodynamic modeling of confined fluids using an extension of the generalized van der Waals theory. Chem Eng Sci 65:3088-3099

Triolo R, Agamalian M (2009) Chapter 20: The combined ultra-small and small-angle neutron scattering (USANS/SANS) technique for Earth scieces. *In*: Neutron Applications in Earth, Energy, and Environmental Sciences. Liang L, Rinaldi R, Schober H (eds) Springer, Berlin p 571-594

Tsomaia N, Brantley SL, Hamilton JP, Pantano CG, Mueller KT (2003) NMR evidence for formation of octahedral and tetrahedral Al and repolymerization of the Si network during dissolution of aluminosilicate glass and crystal. Am Mineral 88:54-67

Vacatello M, Yoon DY, Laskowski BC (1990) Molecular arrangements and conformations of liquid normal-tridecane chains confined between 2 hard walls. J Chem Phys 93:779-786

Villalobos M, Leckie JO (2000) Carbonate adsorption on goethite under closed and open CO_2 conditions. Geochim Cosmoch Acta 64:3787-3802

Vlugt TJH, Krishna R, Smit B (1999) Molecular simulations of adsorption isotherms for linear and branched alkanes and their mixtures in silicalite. J Phys Chem B 103:1102-1118

Vogel M (2010) NMR studies on simple liquids in confinement. Eur Phys J 189:47-64

von Smoluchowski M (1910) Regarding the kinetic theory of transpiration and diffusion hyperdiffuse gases. Ann Phys Berlin 33:1559-1570

Wang JC, Fichthorn KA (1998) Effects of chain branching on the structure of interfacial films of decane isomers. J Chem Phys 108:1653-1663

Wang JW, Kalinichev AG, Kirkpatrick RJ (2004) Molecular modeling of water structure in nano-pores between brucite (001) surfaces. Geochim Cosmochim Acta 68:3351-3365

Wang JW, Kalinichev AG, Kirkpatrick RJ (2006) Effects of substrate structure and composition on the structure, dynamics, and energetics of water at mineral surfaces: A molecular dynamics modeling study. Geochim Cosmochim Acta 70:562-582

Wang JW, Kalinichev AG, Kirkpatrick RJ, Cygan RT (2005) Structure, energetics, and dynamics of water adsorbed on the muscovite (001) surface: A molecular dynamics simulation. J Phys Chem B 109:15893-15905

Wang YF, Bryan C, Xu HF, Gao HZ (2003) Nanogeochemistry: geochemical reactions and mass transfers in nanopores. Geology 31:387-390

Wang YF, Bryan C, Xu HF, Pohl P, Yang Y, Brinker CJ (2002) Interface chemistry of nanostructured materials: Ion adsorption on mesoporous alumina. J Colloid Interface Sci 254:23-30

Warnock J, Awschalom DD, Shafer MW (1986) Geometrical supercooling of liquids in porous-glass. Phys Rev Lett 57:1753-1756

Webb EB, Grest GS, Mondello M (1999) Intracrystalline diffusion of linear and branched alkanes in the zeolites TON, EUO, and MFI. J Phys Chem B 103:4949-4959

Webber JBW (2010) Studies of nano-structured liquids in confined geometries and at surfaces. Prog Nucl Magn Reson Spectrosc 56:78-93

White CM, Smith DH, Jones KL, Goodman AL, Jikich SA, LaCount RB, DuBose SB, Ozdemir E, Morsi BI, Schroeder KT (2005) Sequestration of carbon dioxide in coal with enhanced coalbed methane recovery - A review. Energ Fuel 19:659-724

Wittbrodt JM, Hase WL, Schlegel HB (1998) *Ab initio* study of the interaction of water with cluster models of the aluminum terminated (0001) alpha-aluminum oxide surface. J Phys Chem B 102:6539-6548

Wu H, Zhou W, Yildirim T (2009) Methane sorption in nanoporous metal-organic frameworks and first-order phase transition of confined methane. J Phys Chem C 113:3029-3035
Wu JG, Li SB, Li GQ, Li C, Xin Q (1994) FT-IR investigation of methane absorption on silica. Appl Surf Sci 81:37-41
Wu W (1982) Small-angle X-Ray study of particulate reinforced composites. Polymer 23:1907-1912
Xu M, Harris KDM, Thomas JM, Vaughan DEW. (2007) Probing the evolution of adsorption on nanoporous solids by *in situ* solid-state NMR spectroscopy. Chem Phys Chem 8:1311-1313
Yeom YH, Wen B, Sachtler WMH, Weitz E (2004) NOx reduction from diesel emissions over a nontransition metal zeolite catalyst: A mechanistic study using FTIR spectroscopy. J Phys Chem B 108:5386-5404
Yong Z, Mata V, Rodrigues AE (2002) Adsorption of carbon dioxide at high temperature - a review. Sep Purif Technol 26:195-205
Zarragoicoechea GJ, Kuz VA (2002) van der Waals equation of state for a fluid in a nanopore. Phys Rev E 65:021110-1 - 021110-4
Zhang JF, Choi SK (2006) Molecular dynamics simulation of methane in potassium montmorillonite clay hydrates. J Phys B At Mol Opt Phys 39:3839-3848
Zhang L, Sun L, Siepmann JI, Schure MR (2005) Molecular simulation study of the bonded-phase structure in reversed-phase liquid chromatography with neat aqueous solvent. J Chromatogr A 1079:127-135
Zhang SY, Talu O, Hayhurst DT (1991) High-pressure adsorption of methane in Nax, Mgx, Cax, Srx, and Bax. J Phys Chem 95:1722-1726
Zhou Q, Lu XC, Liu XD, Zhang LH, He HP, Zhu JX, Yuan P (2011) Hydration of methane intercalated in Na-smectites with distinct layer charge: Insights from molecular simulations. J Colloid Interface Sci 355:237-242
Zhu HY, Ni LA, Lu GQ (1999) A pore-size-dependent equation of state for multilayer adsorption in cylindrical mesopores. Langmuir 15:3632-3641
Zorine VE, Magusin PCM, van Santen RA (2004) Rotational motion of alkanes on zeolite ZK-5 studied from ^{1}H-^{13}C NMR cross-relaxation. J Phys Chem B 108:5600-5608

Reviews in Mineralogy & Geochemistry
Vol. 75 pp. 547-574, 2013

Nature and Extent of the Deep Biosphere

Frederick S. Colwell

College of Earth, Ocean, and Atmospheric Sciences
Oregon State University
Corvallis, Oregon 97331-5503, U.S.A.

rcolwell@coas.oregonstate.edu

Steven D'Hondt

Graduate School of Oceanography
University of Rhode Island
Narragansett, Rhode Island 02882, U.S.A.

dhondt@gso.uri.edu

INTRODUCTION

In the last three decades we have learned a great deal about microbes in subsurface environments. Once, these habitats were rarely examined, perhaps because so much of the life that we are concerned with exists at the surface and seems to pace its metabolic and evolutionary rhythms with the overt planetary, solar, and lunar cycles that dictate our own lives. And it certainly remains easier to identify with living beings that are in our midst, most obviously struggling with us or against us for survival over time scales that are easiest to track using diurnal, monthly or annual periods. Yet, research efforts are drawn again and again to the subsurface to consider life there. No doubt this has been due to our parochial interests in the resources that exist there (the water, minerals, and energy) that our society continues to require and that in some cases are created or modified by microbes. However, we also continue to be intrigued by the scientific curiosities that might only be solved by going underground and examining life where it does and does not exist.

But really, is life underground just a peculiarity of most life on the planet and only a recently discovered figment of life? Or is it actually a more prominent and fundamental, if unseen, theme for life on our planet? Our primary purpose in this chapter is to provide an incremental assembly of knowledge of subsurface life with the aim of moving us towards a more complete conceptual model of deep life on the planet. We aim to merge the consideration of the subseafloor and the continental subsurface because it is only through such a unified treatment that we can reach a comprehensive view of this underground life. We also provide some thoughts on a way forward with what we consider to be interesting new research areas, along with the methods by which they might be addressed as we seek new knowledge about life in this Stygian realm.

EARLY STUDIES AND COMPREHENSIVE REVIEWS

The earliest studies attempted to tease out the nature of subsurface life, albeit in qualitative ways. Edson Sunderland Bastin and his team, intrigued by the formation of sulfides in oil wells, examined oil and water pumped from wells in eastern Illinois (Bastin et al. 1926). Using classical methods of cultivating and detecting sulfate-reducing microbes, they established that

1529-6466/13/0075-0017$00.00 DOI: 10.2138/rmg.2013.75.17

cells were present in fluids produced from oil reservoirs even if their original provenance could not be ascertained. These land-based studies only preceded Claude ZoBell's first look into the seafloor by a few years. Even with short core lengths, ZoBell and Anderson discerned a trend towards increasing numbers of anaerobes relative to aerobes as the samples came from deeper and deeper in the top two meters of sediment (Zobell and Anderson 1936). Of course, much detail has been laid atop these early observations but this glimpse of the effect of plummeting redox still holds as a principle of subsurface studies.

A number of useful reviews of progress in the science of subsurface microbiology should be consulted to grasp the origin of this field and some of the important directions. Early work was described in a special issue of the journal Microbial Ecology (cf., Balkwill et al. 1988 and other papers in this volume) and an update to these findings was reported with a distinctive emphasis on continental habitats in the deep subsurface (Fredrickson and Onstott 1996). Shortly thereafter, two books similarly focused on terrestrial systems (Amy and Haldeman 1997; Fredrickson and Fletcher 2001). Perhaps prompted by the estimates presented by Whitman and colleagues (Whitman et al. 1998) and the first microbiological findings reported by scientists associated with the Ocean Drilling Program (ODP; cf., Whelan et al. 1986; Parkes et al. 1994) the last ten years has seen an upswing in papers published on the microbiology of subsurface of marine systems (D'Hondt et al. 2002b; Smith and D'Hondt 2006). Recent overview papers consider past research progress and directions for the future and begin to express the need to bind together our consideration of deep terrestrial and deep seafloor life in a more inclusive light (Fredrickson and Balkwill 2006; Onstott et al. 2009a; Schrenk et al. 2010, 2013; Edwards et al. 2012; Anderson et al. 2013; Meersman et al. 2013). These works were all paralleled by the five editions of the book Geomicrobiology (cf., Ehrlich 1990), now completed by the publication of H.L. Ehrlich's memoir (Ehrlich 2012).

WHERE WE ARE NOW – THE TERROIR OF SUBSURFACE LIFE

And what do we know now? In a broad sense we might start by examining the range of chemical and physical processes that sustain gradients in Earth and therefore may allow the establishment of microbial communities at depth. Regardless of where it exists, metabolically active life—as we know it—seems to require a gradient (Kappler et al. 2005). Living cells may transit places where there is no gradient and manage to survive if these cells can make the passage through harsh conditions. Cells that are locked in a permanent deep freeze (Vorobyova et al. 1997), although some manage activity under frigid conditions (Bakermans et al. 2003), or encased within materials such as amber (Cano and Borucki 1995) or in halite (Satterfield et al. 2005) may be unwitting travelers through a hostile geological medium that is thermodynamically static. However, if cells are shown to be active in any sense then they must be taking advantage of a thermodynamic disequilibrium within their geological setting (cf., Gaidos et al. 1999) even if detection of such disequilibria is beyond our current methods of measure and may only be computationally modeled.

Thermodynamic disequilibria are established in any environment by one or several chemical or physical regimes that determine the presence of some even furtive supply of electron donors and acceptors. Under such conditions microbes can make a living. And if conditions are ample in a large volume of earth then there may be enough active cells in such pockets to make a biogeochemical difference through their collective activities.

Numerous events or processes in and on Earth permit the conditions under which life has found or might find a way to be active (Fig. 1). Research has confirmed many of the processes depicted here; however, many of the processes are conjectural (e.g., that microbes respond to the long cycles of glacial compression and rebound). In addition to the type of event or process we might consider the time over which the respective phenomenon occurs or the spatial

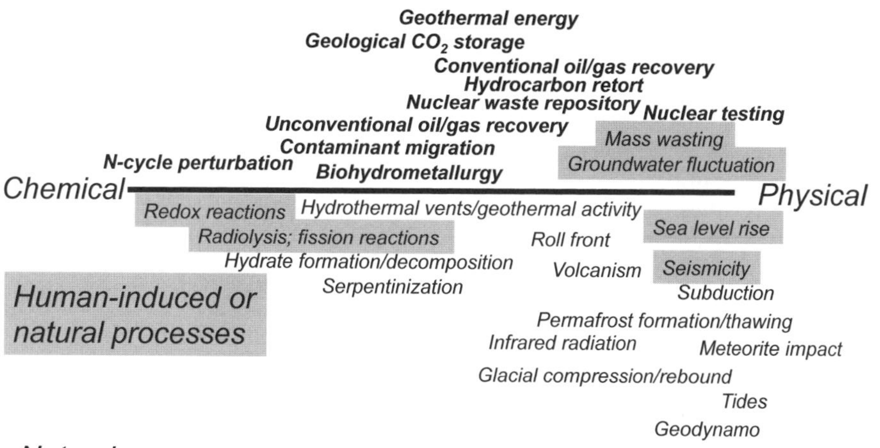

Figure 1. Subsurface phenomena that create chemical, physical, or mixed chemical-physical gradients within which microbial metabolic activity is known to occur or might occur. Various processes are plotted along an arbitrary axis that defines whether the process is primarily chemical or physical in origin. Processes are classified according to whether they are mainly natural associated with Earth (non-bold), mainly human-induced (bold), or best described as being either natural or human-induced (shaded).

extent of its influence as additional attributes pertinent to any microbes exploiting a subsurface niche. For example, the acute effect of a meteorite impact or nuclear test would be followed by a chronic stage of alteration of the surrounding medium that plays out over millennia as the system stabilizes. By comparison, processes governed by the diurnal sighing of tidal action or (on a much longer period) the compression and rebound associated with advance and retreat of continental glaciers are related in a temporally cyclic manner and lack a concussive onset. The spatial boundaries over which these phenomena occur range from submicron perturbations that any of these processes could impose on individual cells to continental- or global-scale processes or events such as sea level rise or seismicity.

Natural processes that are reasonably well studied and that have been demonstrated to stimulate life include hydrothermal vent systems, geothermal activity or volcanism, and serpentinized environments. Noteworthy here is that many of these sites have been sampled at the surface, as windows to the subsurface; however, deep coring rarely occurs either because of the expense of seagoing expeditions or because the fragility and aesthetic value of sites like the thermal features of Yellowstone National Park is too much to risk. It is also exceedingly difficult to collect competent cores of representative quality from deep within porous and fractured rocks of volcanic provenance and where thermal fluids circulate, because the same properties that make these fluidically active habitats also lead to the likelihood that drilling fluids will contaminate the interiors of these cores. For a number of the natural processes we can only speculate that there are microbes in the subsurface that are managing to use to their benefit the chemical or physical energy inherent to the system. These processes would include roll front development for various ore bodies (e.g., uranium), plate subduction, hydrate formation and decomposition. Receiving attention recently, some in deep subsurface environments, are processes that include permafrost thawing (Mackelprang et al. 2011), infrared radiation (Beatty et al. 2005), seismic activity (Hirose et al. 2011), and asteroid or comet impact (Cockell et al. 2012).

And now, in the Anthropocene, many human-induced processes intentionally or unintentionally have manipulated the subsurface to favor microbial activity and survival. Probably the most prominent example is the remediation of contaminants in the subsurface with resulting changes in microbial activities. With thirty years of experimentation, scientists and engineers have encouraged bioremediation of numerous human-introduced chemicals by introducing to the subsurface nutrients, electron acceptors, and electron donors and by altering the direction and rates of groundwater flow (Hazen 1997). Numerous refinements of these processes include organic destruction, metal redox changes, and precipitation in mineral form, and even so-called natural attenuation in which the contaminants are monitored as they disappear in the presence of naturally occurring microbes shown to adapt and degrade the waste. Other purposeful geomicrobial processes include microbial enhanced oil recovery, which aims to alter the mobility of hydrocarbons in porous media, and biohydrometallurgy, whereby oxygenated acidic fluids circulated through crushed metal ore bodies stimulate microbes that oxidize the iron and sulfide minerals and thereby leach metals from the decomposing rocks.

Human-induced processes that are conducted irrespective of whether they alter subsurface microbial activities almost certainly do alter such activities. Little research has been performed to determine how underground microbes respond to our engineering of repositories for nuclear or carbon-based wastes, hydrocarbon retort, nuclear weapons testing, unconventional oil and gas recovery, or geothermal energy exploitation. In these cases, the array of geotechnical phenomena that overwhelmingly relate to our extraction of energy from Earth or our storage of waste products of energy production also quite appropriately relate to the ways in which microbial processes can derive energy in the Earth. By engineering Earth, humans have imposed redox stratification, fluid movement, fracturing, seismicity, and groundwater fluctuation some of which are mirrored by natural phenomena; these changes most likely reform subsurface deserts into oases for microbial life. Whether life pre-existed in these deep settings may not ever be discovered. However, it has almost certainly become richer in these locations due to our efforts.

THE TOOLS THAT WE NEED

Over thirty years of concerted, effort scientists have examined different locations of the planet's subsurface and have adapted and acquired a range of tools for conducting this research (Table 1). As for many system-science disciplines today, a step back to examine the approaches used for sample collection, characterization, and description reveals impressive progress.

Because access to the subsurface is a necessity in order to study it, scientists should consider how they can obtain deep samples. Once sample depths are beyond the reach of simple push cores or augers, coring while using circulating drilling fluids is required. Although this process is expensive and dirty it cannot be avoided if certain environments are to be explored. Coring remains a mainstay of subsurface sampling and numerous reviews summarize the approaches that should be used (cf., Kieft et al. 2007).

When sea- or land-based coring is used, researchers encounter the dual problems of low sample quantity and high cost per quantity of sample material recovered. To avoid drilling, many new sampling efforts have turned to new means of getting underground or maintaining a lasting sampling opportunity in the form of observatories or long-term sampling devices. Many researchers have taken advantage of deep mines that reach into Earth for minerals or to find secure geological strata that serve as repositories (Onstott et al. 1997; Pedersen 1999; Satterfield et al. 2005; Edwards et al. 2006; Rastogi et al. 2009). This approach means that some of our surveys of deep-Earth microbes are canted toward formations that are rich in economic minerals or are proximal to such formations, or else are abnormally quiescent (i.e., locations likely to be undisturbed as required for waste repositories). Elsewhere, sampling devices are placed into drill holes or squeezed into casing and allowed to incubate in place in order to

Table 1. Methods that have advanced or will advance our understanding and communication of subsurface microbiological processes. Shown are example studies from the mid-1980s to present including both subsurface and surface investigations. References are provided in the text.

Method	Enabling technologies	Some key studies
Sample collection	Drill ships, mine access, observatories, CORKS, SCIMPIs	IODP cruises, South African gold mines, long-term ecological research sites
Field analysis and manipulation	Stable isotope probing, "mark and recapture", biogeophysics	Methanotrophy, denitrification, biodegradation, transport
Molecular science instrumentation	Extraction of nucleic acids, lipids or proteins; amplification, sequencing, comparison of nucleic acids; mass spectrometry of lipids and proteins; flow cytometry; single-cell techniques	-omics (e.g., genomics, transcriptomics, proteomics) studies of environmental microbial communities in soils, anoxic methane-rich sediments, thawing permafrost, ultra-low pH mine drainage
Cultivation	New bioreactor designs, high-throughput, miniaturization	anammox; SAR11 clade and *Pelagibacter;* Iron Mountain, CA
Imaging	Fluorescent *in situ* hybridization, NanoSIMS; CT scanning, synchrotrons	Anaerobic methane oxidation, anammox
Computational simulation science	Bioinformatics, reaction-path modeling, thermodynamic modeling	Yellowstone hot spring mat communities, U mill tailings remediation
Data science	Internet	Census of Marine Life, the Tetherless World Constellation
Visualization, engagement	Internet, GoogleEarth	International Census of Marine Microbes

sample indigenous microbes. So-called CORKs or SCIMPIs (Davis et al. 1992; Moran et al. 2006) are used in the seafloor and parallel the multi-level samplers (Smith et al. 1991; Lehman et al. 2004), *in situ* flow cells (Nielsen et al. 2006), and flow through *in situ* reactors (FTISR) (Lehman 2007) used in the continental subsurface. Hybrids of these technologies and ingenious smaller systems that are deployed using clever means by which to sample microbes and chemistry include passive gas samplers (Spalding and Watson 2006), U-tube systems (Freifeld et al. 2005), and the ever-changing osmosampler (Orcutt et al. 2010). And, with pressure-coring tools becoming more common, there are opportunities to collect deep cores, return them to the surface while maintaining *in situ* pressure, and then transfer these samples into analysis systems without decompression. Still, only limited work has been done with such equipment (Parkes et al. 2009); however, along with other methods of restricting pressure (Bowles et al. 2011), the chance to examine microbial activities as they may actually occur *in situ* has been expanded.

Despite our ability to obtain samples over extended periods, it is important to note the bias associated with such samples. This concern was first considered in the late 1980s with observations of altered microbial communities following the coring of holes (Hirsch and Rades-Rohlkohl 1988). Lehman summarized the limitations of samplers incubated *in situ* with a guarded appraisal of the technologies (Lehman 2007). It is important to bear these cautions in mind when analyzing the chemistry and microbiology of subsurface samples so obtained. Direct and rapid characterization of freshly acquired core and porewaters appears to be the best means by which to investigate native microbes in rocks.

In some cases, direct field methods of analysis or manipulation of microbial communities can provide essential data to understand the fundamental ecology of the subsurface or to determine the outcome of engineered processes in the subsurface. The new discipline of biogeophysics, an expansion from several geophysical methods of examining geological strata, keys on microbially-induced alterations of geological materials and how these alterations influence underground electrical signals in order to image the presence or activity of microbial communities (Allen et al. 2007; Revil et al. 2010). At this stage of development, these remote sensing strategies are mainly able to detect profound changes in such properties as electrical conductivity or specific conductance and are used mostly in places where purposeful stimulation of the microbiological characteristics of formations has occurred.

Microbiologists and hydrologists have also experimented with a miniaturized version of the macro-ecologist's "mark and recapture" experiment. In such investigations, microbes from the environment are collected and then cultivated in the presence of a ^{13}C-labelled substrate. The result is a population that is uniquely tagged with the heavy isotope and that can then be released in a well-field and hopefully collected down gradient to determine the degree to which microbes may be transported in the aquifer (Holben and Ostrom 2000; DeFlaun et al. 2001). This method is similar to stable isotope probing that uses a heavy isotope label to determine the most active communities in a sample based on their ability to take up the label and deposit the label in their DNA (Radajewski et al. 2003). Push-pull tests are another ingenious method of examining *in situ* activities and, if done with discretion, can provide useful information about the real metabolic capabilities of microbes in Earth (Haggerty et al. 1998; Pombo et al. 2005; Urmann et al. 2005).

Scientists also now possess a range of tools that can be used for biological characterization once samples have been recovered from the subsurface and are returned to the lab. Significant advances have been made in the ability to characterize the molecular capabilities of cells (i.e., genomics, transcriptome, proteome, metabolome, lipidome) that allow many of the same diagnostic tools as used in modern medicine. The Richmond Mine site has provided a wealth of information related to the function of highly constrained microbial communities of limited diversity. In a classic application of metagenomic information, the researchers investigating this site gained enough genomic data of the simple communities present that they were able to devise appropriate cultivation methods of uncultured members of the microbial assemblage (Tyson et al. 2005). The same communities have assisted the understanding of ecological divergence of similar but slightly distinct microbes according to the genomic and proteomic patterns detected in these cells (Denef et al. 2010). These relatively simple systems are teaching us how best to transfer this skill to environments that are intrinsically more biologically complex or where a limited number of samples may be available.

It is acknowledged that few microbes can be nurtured using traditional cultivation-based approaches. However, startling progress has been made in culturing pelagic marine microorganisms by taking advantage of high-throughput robotic systems to handle samples, parsing them into numerous novel media formulations, and then recognizing that populations of these cells simply do not achieve high density (Connon and Giovannoni 2002; Stevenson et al. 2004). Along with the aforementioned metagenomic approach applied in the Richmond Mine, the approach used for pelagic microbes could be applied to microbes from Earth's subsurface. Matching these cultivation approaches with single-cell manipulation techniques (Stepanauskas and Sieracki 2007) may tease geologically inclined microbes into culture and also help to minimize the influence that PCR-based amplification has on our view of microbial diversity. When these approaches are combined with imaging techniques that hinge on fluorescence microscopy (Amann et al. 2001), synchrotron-based characterization (Holman et al. 1998), interferometry (Davis and Luttge 2005), atomic force microscopy (Warren et al. 2001), and different electron microscopy techniques, we gain new resolution of the relationships between these microbes and the minerals on which they depend.

To complete our ability to comprehend life underground, there are new opportunities in the sciences that are not directly associated with the field and lab. Collectively, computational and simulation sciences have made advances that enable rapid and detailed modeling of porous media, where microbes participate in the alteration and dissolution of mineral species. Reactive transport models (Steefel et al. 2005) are now merged with *in-silico* models and bioinformatics approaches (King et al. 2009) to yield simulations that explain or predict the active microbial taxa in a given subsurface setting under a given set of environmental conditions (Scheibe et al. 2009; Li et al. 2010; Zhuang et al. 2011). Thermodynamic modeling has been pacing these studies and also can provide essential insight into what microbial processes are likely to be active and when they are active (Spear et al. 2005) and, when combined with kinetic models, how rapidly the activities may occur (Jin and Bethke 2005).

Finally, advances in data science and in visualization are poised to help scientists who study the subsurface with their task of communicating the results of their findings. Web science and its new ways of using the internet (Fox and Hendler 2011) will allow a binding together of disparate disciplines such that the data acquired by scientists in different fields can begin to sketch relationships between the living and non-living in the subsurface, as is happening in other fields. We can now take advantage of work done to explore interwoven features of the oceans (Amaral-Zettler et al. 2010; Tittensor et al. 2010), or of the cosmos (Szalay and Gray 2001) and proteins (Askenazi et al. 2011), to do the same for places inside Earth. This visualization capability should lead us to the point of greater interaction with the public where engagement through websites and museums, and possibly problem-solving through crowdsourcing will draw our knowledge and questions about the subsurface into the vernacular of non-scientists.

THERE'S NO PLACE LIKE HOME

As already noted, the earliest intensive investigations of underground life often targeted locations of known resource prospects for fossil energy or metals (e.g., Taylorsville Basin, South African Gold mines), sites where groundwater or soils were contaminated (e.g., numerous U.S. Department of Energy [DOE] and other sites), planned nuclear waste repositories (e.g., Äspö, Yucca Mountain, Waste Isolation Pilot Plant), or simply where it was convenient to collect subsurface material (e.g., ODP sites). This pragmatic approach has surely biased our understanding of subsurface life. Many of the more recent sites that have been investigated were selected in order to test hypotheses grounded in prior knowledge of life in the subsurface and what appear to be the limits of the biosphere. The *Early studies and comprehensive reviews* section identifies papers (especially, Fredrickson and Balkwill 2006; Onstott et al. 2009a; Schrenk et al. 2010, 2013; Edwards et al. 2012) that highlight different locations where subsurface research has been conducted. We will not reiterate the reports of these papers but rather point to some selected studies that have helped us to understand the range of underground locations already investigated. Within the next few years, the newly initiated Census of Deep Life is expected to provide a catalog of subsurface life and the disparate geological settings where it has been detected.

The DOE is responsible for waste released into a range of subsurface environments over 40 years following World War II (Riley et al. 1992). Despite this very practical concern, the DOE Office of Science supported research paths to understand the basic properties of life underground. Initially, DOE-funded scientists were unfettered with the need to examine actual waste sites; rather, they "cut their teeth" at pristine locations where the limits of subterranean life could be explored. This freedom led to reports of microbial life in deep and shallow coastal plain sediments and rocks on the southeastern coastal plain of the U.S. (Fredrickson et al. 1991), in Oyster, Virginia (Zhang et al. 1997), and in the Taylorsville Triassic Basin (Onstott et al. 1998); thick sedimentary zones in arid regions (McKinley et al. 1997) and in thick flood deposits (Brockman et al. 1992; Kieft et al. 1998) in southwestern Washington state; unsaturated and saturated fractured basalts of the Snake River Plain Aquifer (Colwell and Lehman 1997; Lehman et al. 2004) and the Columbia River Basalt Group (Stevens et al. 1993; Stevens and McKinley 1995); hot methane-charged fractured sandstones in the Piceance Basin (Colwell et al. 1997); volcanic tuffs in the Great Basin (Amy et al. 1992; Russell et al. 1994); and ancient marine sediments into which volcanic dikes impinged in New Mexico (Fredrickson et al. 1997; Krumholz et al. 1997).

Studies of deep seafloor locations for the presence of microbes have now reached well beyond the original skin of sediment examined by the first marine microbiologists. The ODP and later the Integrated Ocean Drilling Program (IODP), using the drilling platforms JOIDES *Resolution* and the *Chikyu*, catalyzed the first seagoing coring expeditions aimed at answering questions about the deep marine biosphere. Microbiologists participated in earlier drilling legs and continue to be opportunistically involved in expeditions.

Through targeted drilling and coring studies of different subseafloor environments, we are slowly accumulating information about subsurface communities in a broader range of environments. The ODP Leg 201 was the first dedicated microbiology drilling leg. Leg 201 scientists used procedures to control contamination (Smith et al. 2000) that helped to reach out to a new scientific community. This sampling cruise occurred off the west coast of South America and explored deep sediment in near shore and pelagic sites (D'Hondt et al. 2004; Inagaki et al. 2006). More recent dedicated microbiology investigations sponsored by the IODP have explored the eastern flank of the Juan de Fuca Ridge, a hydrologically active basalt aquifer (Fisher et al. 2011); the South Pacific Gyre, in which sediment and basalt underlie waters of extremely low productivity (D'Hondt et al. 2011); a hydrothermal field in the Okinawa Trough (Takai et al. 2011); and basalt and sediment of North Pond, a sediment-filled basin off the main axis of the Mid-Atlantic Ridge (Expedition 336 Scientists 2012).

One of the curiosities of subsurface microbiology studies is that only rarely do these subsurface environments lack measureable life. Teams of scientists are typically able to tease evidence of life out of most subsurface rock or sediment that is within a temperature regime that embraces the known limits of life and that has enough connected pore space. Even environments contaminated with high levels of radioactive elements contain microbes (Fredrickson et al. 2004). Exceptions include the dry and thick unsaturated zone in the Eastern Snake River Plain (Colwell et al. 1992) and some sections of the deep, massive sandstones of the Piceance Basin (Colwell et al. 1997). It is possible that where life appears to be missing from moderate temperature regimes at moderate depths, the problem is either detection limit (with life present but not detectable) or inability to survey large enough samples due to the limited amount of material collected by coring.

Of course, our inquiries of the subsurface for microbes remain inadequate to survey the life there. Much of the field research has an exploratory element. We rarely acquire true replicate samples because drilling identical holes in the same location and sampling the same depths is both difficult and expensive. Furthermore, innumerable subsurface environments have thus far been ignored or simply been too difficult or expensive to reach. More sampling along lengthy, confined horizontal flow paths, similar to past studies on the southeastern coastal plain of the U.S. (Murphy et al. 1992), would provide excellent data about microbes limited by geological constraints. Notably static environments (Vreeland et al. 2000) will offer insight into how long cells can last and possibly the adaptations that they require in order to last on timescales of thousands to millions of years (Lomstein et al. 2012; Røy et al. 2012). Arctic and Antarctic deep-Earth environments remain under-sampled, though we are gradually accumulating examples (Mikucki and Priscu 2007; D'Elia et al. 2008; Onstott et al. 2009b; Pham et al. 2009; Colwell et al. 2011). Deep samples from thick vadose zones are lacking in general as are studies that examine microbes present near faults in seismically active areas.

IS DIVERSITY THE SPICE OF SUBSURFACE LIFE?

Biologists, other scientists, and even non-scientists are justifiably attracted to the incredible diversity of life at Earth's surface and the essential aspect of surface life's diversity to the health of ecosystems is well recognized (Daily and Matson 2008; Rockstrom et al. 2009; Stein and Nicol 2011). Certainly, the amazing diversity of life at the surface is almost harrowing to those who must reach into the subsurface to get samples where biomass is typically low. We have devised ways of keeping our precious deep samples isolated from surface samples and also ways of determining whether surface contamination has occurred (Lehman et al. 1995; Masui et al. 2008). By relying on new methods of molecular characterization and accumulating enough information from the relatively rare sampling events, the nature of life's diversity underground is becoming clearer.

Bacteria and archaea are the common targets of investigations that aim to study subsurface diversity. Eukarya are rarely targeted. Subsurface studies often show that bacteria are more abundant in some subsurface environments than archaea (Schippers et al. 2005; Lin et al. 2006; Rastogi et al. 2009; Briggs et al. 2012), but others show that archaea predominate (Biddle et al. 2006) and some indicate more equal representation of the two domains (Pham et al. 2009). Microbiologists are now familiar with subsurface habitats like the sulfate-methane transition zone in the subseafloor, where the supply of sulfate from seawater and methane from deeper sediments dictate that both bacteria and archaea will be present. But this environment is now well defined and frequently sampled. While bacterial (vs. archaeal) abundance seems to be a common theme in the subsurface, more work is required to determine the relative abundance of these groups.

Before we can describe the patterns of abundance of these two domains in the subsurface, some technical hurdles associated with the methods of analysis must be overcome. Extraction of DNA from archaeal cells in deep sediments is difficult (Lipp et al. 2008) and, relative to bacteria, archaea typically have poorer representation in the gene databases that are required for construction of the amplification primers. Polymerase chain reaction (PCR) based amplification depends on these primers and some version of PCR is often used to assess microbial diversity. As usual, a robust approach using multiple methods of analysis (e.g., fluorescence *in situ* hybridization, DNA sequencing, and intact polar lipid characterization) is the best way to present a detailed description of an environmental microbial community. And new methods like single-cell sorting followed by whole genome amplification can help to explain where partiality associated with typical primer-based amplification has occurred.

Clone library-based investigations of the 16S rRNA and other genes obtained from DNA extracted from a number of samples find evidence of many taxa in the subsurface (cf., Biddle et al. (2006) and Inagaki et al. (2006), as well as examples as reported in Fredrickson and Balkwill (2006)). It is not unusual to detect "new" microbes in these surveys based on the presence of unique genes in the libraries. The results of diversity studies can be influenced by the manner of sampling and the types of sample used. That attached subsurface communities are different than free-living subsurface communities has been understood for some time (Hazen et al. 1991) but surveys of large volumes of subsurface space (e.g., pumping and filtering or concentrating aquifer samples for free-living microbial cells) may yield different findings than surveys based on a few grams of solid material. Examinations of subsurface microbial diversity rarely seem to address the question of how much of the subsurface was queried; however, adhering to guidelines for how diversity is reported relative to the amount of material and the type material sampled (i.e., water, solids, or both) would help to develop our view of subsurface variations in communities (cf., Lehman et al. 2001; Lehman 2007).

That microbial diversity in the subsurface is typically lower than in surface systems is not necessarily surprising. Unique niches occur in the subsurface though certainly not as many as in surface systems. It seems that microbes that survive in the subsurface often need to overcome certain barriers that may not be unheard of at the surface (e.g., pressure, temperature, confinement) but may be more severe and persistent in the subsurface. Thus, species that can last at depth have been winnowed by the chemical and physical realities of their habitat once they arrive in the subsurface. Investigations of a low-pH system in the Richmond Mine at Iron Mountain in California attest to the stable presence of five dominant microbes making up most of the community (Tyson et al. 2004). Microbial communities obtained from deeply occurring fractures in the Mponeng Mine in South Africa provide another such example. There, a single suflate-reducing microbe represents over 99.9% of the community (Lin et al. 2006); microbial ecosystems where diversity is so low are not commonly reported. Given the conditions of that environment and the genomic characteristics of this solo microbe, it seems to have the functional attributes needed in order to survive there, and indeed appears to be well distributed in the deep environment of the Witwatersrand Basin (Chivian et al. 2008). While both sites mentioned here are mine environments that are subject to unnatural forces during their creation, it seems plausible that similar low diversity communities can be found in utterly pristine locations of the subsurface.

Relatively low biomass and relatively low diversity may be the norm in much of the subsurface; however, some locations may not be so biologically depleted. Recent studies in the ocean crust at the East Pacific Rise reveal notable diversity (Santelli et al. 2008). These samples were not from the subsurface but might contain microbes representative of deeper crustal materials. While not as diverse as such well-studied surface systems as farm soils, these basaltic communities are considerably more complex than pelagic marine microbial communities (Santelli et al. 2008). Similar places that exhibit much milder geothermal gradients and probably milder flux through the system may foster simpler communities (Edwards et al. 2011). More open subsurface environments like those crustal basalts that benefit from a porous and fractured architecture and fluid circulation driven by geothermal processes (Delaney et al. 1998; Edwards et al. 2005) could be as abundant in the subsurface as the distribution of large igneous provinces around the planet (Saunders 2005) and volumetrically may be a significant source of the planet's underground diversity.

The search for eukarya was often a part of early investigations of aquifers (Sinclair and Ghiorse 1989; Sinclair et al. 1993; Novarino et al. 1997); however, many surveys of life underground may not even look for eukarya. The obvious spatial constraints in porous media prevent the occurrence of larger cells or multicellular organisms. However, we now have an example of a nematode that lives in deep fractures, apparently managing to subsist through grazing on microbes (Borgonie et al. 2011), and evidence of Collembola that exist in deep limestone caves (Jordana et al. 2012). These studies promote the idea that any deep system with cavities large enough (several microns?) such as the aforementioned crustal systems or fractured rocks may also be open enough to support eukarya if adequate unicellular biomass can be generated to provide food for the higher organisms. However, it should be noted that sterols, a key structural component of the membranes of eukarya, are difficult to make anaerobically and this may limit the extent to which these cells might penetrate anoxic zones.

Early considerations of the presence of viruses in the subsurface focused largely on shallow systems with considerable contact with surface environments (Gerba and Bales 1990; Matthess 1990). Certainly, in the subsurface viruses confront several factors such as low host biomass (Wiggins 1985), generally patchy, disconnected microbial communities (Brockman and Murray

1997), and limited fluid exchange between communities. These factors could minimize the effectiveness by which phages can infect bacteria or archaea. Some subsurface environments might be more likely to contain viruses (i.e., where microbes are abundant) and sand columns have been used to examine their distribution (Yates et al. 1997).

In recent years, studies in different environmental settings with new methods have revealed huge numbers of viruses (Anderson et al. 2013). These small packages of genetic information are now believed to provide significant paths for moving molecular information among microbial hosts and represent a massive means of turnover for living biomass (cf., Suttle 2005; Anderson et al. 2011a, 2013). Accordingly, there have been more studies of viruses in the subsurface. A metagenomic study conducted on marine sediments found highly diverse, largely unrecognized phage populations and identified marine sediments as a massive "reservoir of sequence space" (Breitbart et al. 2004). Finding more temperate than lytic phage suggests that an important infection strategy for subsurface viruses might be incorporation into the host genome rather than destroying the host. Shallow sediments worldwide contain large numbers of phage and collectively these have profound impact on biogeochemical cycles, at least as pelagic microbes are buried in the uppermost sediment layers (Danovaro et al. 2008). Microbial cell death due to phage in benthic systems of deep waters may significantly haze the viable microbes that are buried and increase the amount of organic detritus in the uppermost centimeters and alter the disposition of buried organic matter.

Recent identification of a "microbial immune system," the so-called clustered regularly interspaced short palindromic repeat or CRISPRs, found within host genomes offers a new means of detecting phage populations and connecting them to their respective hosts (Banfield and Young 2009). The CRISPR approach was applied to hydrothermal vent samples and the results indicate that large numbers of microbial hosts are infected with viruses and that the hosts represent a diverse range of microbes (Anderson et al. 2011b). These vent fluids speak to the processes and populations in the subsurface and naturally lead to considering how phage may play a role in the transfer of genetic information in subsurface environments other than vents. That phages dictate the genetic diversity and evolution of microbial communities in vent systems of the shallow subsurface (Anderson et al. 2011a) suggests that microbial communities in other subsurface locations where fluids are actively moving may benefit from the enhanced genetic fitness and functional capacities that are conferred by prophages and that may assist subsurface survival.

BIOMASS OF SUBSURFACE LIFE

The pioneering study of global biomass by Whitman and colleagues (Whitman et al. 1998) proposed that subsurface bacteria and archaea comprise 35 to 47% of Earth's total biomass, nearly equal to plants in their total carbon content. Microbes in subseafloor sediment comprised nearly 1/3 of their global biomass estimate. Microbes in terrestrial subsurface sediment comprised between 1/50 and 1/5 of their global biomass estimate. Their study was a great starting point for estimates of global microbial biomass. However, it was based on the relatively sparse data that were available in the mid-1990s. Estimates of subsurface biomass are changing as more data become available.

Cell abundance data for terrestrial subsurface sediment have not significantly improved since 1998. However, data for subseafloor sediment have improved greatly. Subsequent studies have generally yielded lower estimates than Whitman et al. (1998) for subseafloor sedimentary biomass (Parkes et al. 2000; Lipp et al. 2008; Kallmeyer et al. 2012). The most recent studies show that cell concentrations in the broad expanses of open-ocean sediment beneath the Pacific gyres are orders of magnitude lower than earlier counts from subseafloor sediment, which were largely limited to organic-rich sediment that underlies oceanic upwelling zones (D'Hondt et al.

2009; Kallmeyer et al. 2012). Consequently, total microbial abundance varies between sites by five orders of magnitude (Kallmeyer et al. 2012). This variation strongly co-varies with mean sedimentation rate and distance from shore. Based on these correlations, total cell abundance in subseafloor sediment is ~3 × 10^{29} cells, corresponding to ~4 petagram C and ~0.6% of Earth's total biomass (Kallmeyer et al. 2012).

Most estimates of subseafloor sedimentary biomass are based on visual cell counts, which do not include spores (the fluorescent dyes used for cell counts generally do not penetrate spores). However, a recent study of dipicolinic acid and muramic acid concentrations indicate that bacterial endospores are approximately as abundant as counted cells in deep subseafloor sediment of the Peru Margin (Lomstein et al. 2012; dipicolinic acid is limited to endospores and muramic acid is much more abundant in endospores than in vegetative cells). Because endospore abundance is not yet known for other subsurface habitats (or even for most sediment of the world ocean), this will be an intriguing avenue of research in the near future.

A further complication for global estimates of subsurface biomass is that the biomass resident in large subsurface habitats is not yet known. For example, biomass in the vast volume of fractured igneous basement in continents and oceans cannot yet be quantified, because the data do not yet exist.

Discussions of subsurface biomass to date have relied on counts of stained cells (e.g., (Thierstein and Störrlein 1991; Parkes et al. 1994, 2000; D'Hondt et al. 2004, 2009) or abundance of intact biomarkers (Lipp et al. 2008). These are powerful techniques that census overlapping, but non-identical subsets of a microbial community. Counts of cells stained with non-specific DNA-binding compounds (e.g., acridine orange or SYBR-Green) include intact vegetative bacteria and intact archaea, but do not include bacterial endospores. Molecular probes (fluorescence *in situ* hybridization [FISH] probes) specific to RNA in bacteria, archaea or more narrowly defined phylogenetic groups have also been used, albeit much less frequently (e.g., Mauclaire et al. 2004; Schippers et al. 2005). Intact biomarker assays have focused on archaeal biomarkers and consequently estimate subsurface archaeal biomass, not total subsurface biomass.

The results of these techniques (cell counts and biomarker assays) beg discussion of the distinction between "intact" and living cells. It is not yet certain how long cell membranes, their included nucleic acids, or their diagnostic phospholipids remain intact after cell death in deep subsurface environments. This said, RNA-based FISH counts are generally interpreted to suggest that a large fraction of counted subseafloor sedimentary cells are living or at least recently alive; RNA is widely recognized to degrade far more readily than DNA and RNA-based FISH counts constitute several percent to several tens of percent of DNA-based counts in subsurface environments (Mauclaire et al. 2004; Schippers et al. 2005). Compelling independent evidence that the majority of counted subseafloor cells in individual samples are alive was recently provided by experiments with isotopically-labeled organic substrates and sediment from hundreds of meters beneath the seafloor in the Japan Sea (Morono and al. 2011); in these experiments, as many as 76% of the counted cells assimilated the isotope-labeled substrates.

PHYSIOLOGICAL PROCESSES OF SUBSURFACE LIFE

Subsurface microorganisms include both heterotrophs (which consume organic matter) and lithoautotrophs (which consume inorganic compounds). Electron donors in subsurface environments include buried organic matter, reduced chemicals (such as reduced iron and reduced sulfur), and reduced compounds created by water-rock interactions; examples include H_2 from radioactive splitting of water (Pedersen 1997; Lin et al. 2006) and H_2 and CH_4 from serpentinization reactions (Kelley et al. 2005; Nealson et al. 2005). All of these electron donors occur in a broad range of subsurface environments. For example, organic matter that was photosynthesized in the overlying ocean is the principal electron donor for microbes in subseafloor sediment (D'Hondt et al. 2004) and also circulates in dissolved form with seawater through oceanic basalt. The primary electron donors in subseafloor basaltic aquifers include reduced chemicals in mineral phases (e.g., Bach and Edwards 2003), which also commonly occur in both terrestrial and marine sediment. Hydrogen produced by natural radioactive splitting of water appears to sustain microbial life in deep continental aquifers (Lin et al. 2006) and may also be a significant electron donor in very organic-poor marine sediment (Blair et al. 2007).

Rates of subsurface microbial respiration have been most commonly quantified for subsurface sedimentary communities. Calculations based on concentration profiles of dissolved electron acceptors and products of microbial respiration indicate that the subsurface microbes of both terrestrial sedimentary aquifers (Chapelle and Lovley 1990; Phelps et al. 1994) and subseafloor sediment (D'Hondt et al. 2002a, 2004; Røy et al. 2012) respire orders of magnitude more slowly than microbes in the surface world (Onstott et al. 1999; Price and Sowers 2004).

Given the extraordinarily low rates of microbial respiration in many subsurface environments, subsurface microbes are generally assumed to reproduce very slowly, if at all. D'Hondt et al. (2002a) speculated that most subseafloor sedimentary microbes are either inactive (dormant) or adapted for extraordinarily low metabolic activity. Price and Sowers (2004) suggested that subsurface sedimentary microbes exhibit survival metabolism (sufficient to repair macromolecular damage but insufficient to sustain growth or motility). Whether they are actually growing or merely repairing macromolecular damage, amino acid racemization ratios indicate that subseafloor sedimentary biomass turns over very slowly, on timescales of hundreds to thousands of years (Lomstein et al. 2012). We do not yet know whether the microbes of these subsurface environments reproduce at these slow rates of biomass turnover or live without dividing for millions to tens of millions of years.

These extraordinarily slow rates of respiration and biomass turnover beg consideration of the factor(s) that control(s) rates of microbial activity in subsurface ecosystems. Where electron acceptors are present, areal or volumetric rates of subsurface microbial activities (e.g., activity in a square-meter sediment column or in a cubic meter of sediment or rock) are broadly related to electron donor availability. For example, areal rates of microbial respiration are orders of magnitude higher in subseafloor sediment rich in organic matter (D'Hondt et al. 2004) than in subseafloor sediment where organic matter is extremely dilute (D'Hondt et al. 2009). However, where electron acceptors are vanishingly rare, electron donors can build to extraordinarily high concentrations. For example, in some fractures intersected by deep South African gold mines, dissolved hydrogen from water radiolysis is present in millimolar concentrations but electron acceptors are scarce, indicating that microbial activity in those fractures is far too low to keep up with very low rates of hydrogen production on timescales of tens to hundreds of millions of years (Lin et al. 2006).

Such examples suggest that, in a broad sense, subsurface rates of bulk microbial activities are controlled by energy availability. However, the situation is much more problematic at closer inspection. For example, why don't sedimentary microbial communities oxidize all available organic matter within the first few centimeters of the seafloor? In other words, how does buried organic matter survive microbial activity to sustain slow rates of activity for millions to hundreds of millions of years? Why don't rapidly respiring cells outcompete the slowly respiring cells by oxidizing all available organic matter over a much shorter interval of geologic time?

The situation is also perplexing on a per-cell basis. For example, aerobic microbial communities of subseafloor sediment in the North Pacific Gyre exhibit per-cell rates of microbial activity (Røy et al. 2012) that are not vastly different from per-cell rates in anaerobic communities of subseafloor sediment in the Peru Margin and the equatorial Pacific Ocean

(D'Hondt et al. 2002a; 2004) although areal rates of activity differ by orders of magnitude between the oxic gyre sediment and the anoxic Peru Margin sediments. In both environments, mean per-cell rates of respiration are orders of magnitude lower than per-cell rates in surface sediment or laboratory cultures. What are the limits to survival that allow microbial hunger artists to eke out a living at such extraordinarily slow rates in both environments?

Finally, the extent to which subsurface organisms are (i) microbial zombies, incapable of being revived to a normal state, or (ii) capable of metabolism, growth, and reproduction at rates typical of the surface world is not yet known for many subsurface ecosystems. Isolation of many microbial strains from deep subsurface environments (e.g., (Balkwill 1989; Takai et al. 2001; D'Hondt et al. 2004; Batzke et al. 2007) has demonstrated that at least some deep subsurface microbes can emerge into the surface world, grow, and multiply. However, these few hundreds of laboratory isolates may not represent the majority of subsurface microbes. A recent study by Morono et al. (2011) sheds light on this issue. In short, Morono and colleagues demonstrated that many microbes from sediment hundreds of meters beneath the seafloor take up measurable quantities of isotopically-labeled substrates. In doing so, they effectively showed that many deeply-buried organisms maintain the potential to metabolize and grow, regardless of what they are doing deep beneath the seafloor (Jørgensen 2011). This result effectively demonstrates that the metabolic potential of long-buried microbes can be activated at much higher rates when they emerge into a moderate environment.

WHERE AND WHEN DOES LIFE IN THE SUBSURFACE REALLY MATTER TO US?

It is fair to ask when and where deep life matters to the life and processes at Earth's surface. Can we identify ecosystem services that are provided by life underground? The question might be considered for both naturally occurring subsurface microbes and those that are a part of a human-engineered process. A number of engineered systems that utilize or involve microbes and their activities are the result of stimulation of subsurface life. Microbes underground are responsible for numerous variations on the general theme of bioremediation. Where wastes have been carelessly released into aquifers, we now depend upon microbial communities to decontaminate these freshwater resources. The processes can take decades to be complete; however, it is usually far more economical to track these *in situ* reactors over time as they eliminate contaminants than it is to dig the waste out of the ground. Perhaps this situation is similar to how we depend upon subsurface microbes and the biogeochemical processes that they carry out to purify tainted water that enters the subsurface prior to our use of the water when it is collected down gradient. The mingled biological, chemical, and physical processes that are inherent to deep Earth can eliminate the human pathogens that are simply not able to survive.

Similar processes have been conceived for conducting *in situ* mining or biohydrometallurgy, where low-grade ores may be attacked by well-understood microbial processes under controlled conditions to extract the metals within (Das et al. 2011). This approach is conducted worldwide in managed "heap leach" operations, where the rubblized rock is piled onto a large impermeable pad, irrigation networks trickle the "lixiviant" fluid through the system, and the metal-rich liquid that results is collected (Rawlings 2002). This biologically driven process is responsible for the recovery of most of the world's copper (Rawlings 2002), as well as uranium and gold, and is contemplated for manganese extraction (Das et al. 2011).

Microbes and their astounding metabolic activities have been considered also for processes that would convert hydrocarbons deep in Earth into products that can more readily be extracted. Studies of anaerobic modification of hydrocarbons are relatively new as many aliphatic and aromatic structures were long considered to be inert (Heider et al. 1998). A better understanding of the distribution of microbes in hydrocarbon-rich geological formations and the constraints under which they survive and modify the organic matter therein (Head et al. 2003) has also led to considering ways by which microbes might alter hydrocarbons in place where oxygen is absent. Oxygen-free reactions, including hydroxylation, methylation, fumarate addition, and reverse methanogenesis (anaerobic methane oxidation), allow microbes metabolic access to complex hydrocarbons and broaden our view of how organic matter can be converted in the subsurface (Heider 2007).

Many subsurface environments are used not for their resources but rather for their remoteness, stability, or controllability. Such subsurface settings are ideal repositories for nuclear waste, carbon dioxide, or as artificial reservoirs for natural gas. In each case, microbial activities may play a role in the security of the materials deposited therein. The microbiology of nuclear waste storage locations has been investigated to determine the degree to which biological activity may alter the waste in a range of geological environments designated as candidate underground repositories (Stroes-Gascoyne and West 1997; Pedersen 1999; Pitonzo et al. 1999; Jolley et al. 2003; Horn et al. 2004; Nazina et al. 2004). Locations close to the waste canisters shortly after enclosure may create conditions that are outside the range of microbial survival due to high-temperature or high-radiation fields of the newly deposited waste. But at some distance away from the waste, and over time as the radioactivity decays, these extreme conditions will moderate and microbes may recolonize the geologic niches.

Deep-Earth storage of carbon dioxide as a means to remove it from the atmosphere has received considerable attention. Most studies have focused on the physical or chemical controls on carbon dioxide stability in the subsurface (Benson and Cook 2005); however, microbes can survive in many environments suitable for CO_2 storage and for these settings we must also consider biogeochemical aspects of stability. To date, few studies have examined microbial communities where CO_2 would be secured or the conditions to which these cells would be exposed. These investigations make it clear that some microbes—likely as spores—can survive in the presence of supercritical CO_2 (Mitchell et al. 2008, 2009; Dupraz et al. 2009). Native communities in several geological habitats may resist the solvent properties of the supercritical CO_2 or survive proximal to the highest concentrations of the solvent (Morozova et al. 2010). Geochemical modeling suggests that subsurface microbes in some environments where CO_2 could be disposed (e.g., basalts) might be able to alter the disposition of the carbon (Onstott 2004). Assurance of the stability of deeply sequestered CO_2 is important and so there should be an effort to understand the biogeochemistry where life can survive.

Even though these microbial reactions occur only on the dimensions of single microbial cells or microcolonies or minerals or dissolved compounds, the large size of the biomass, its ability to permeate living space, and the relentless nature of this metabolism mean that the effects can translate to scales of hundreds of kilometers over millennia. An example that displays the cumulative effect of pervasive, sustained microbial activity is the accumulation of biogenic methane in continental shelf sediments where conditions are met for methanogenesis and capture of the methane in the form of hydrates (Hazen et al. 2013). The release of methane from this "large, dynamic microbially-mediated gas hydrate capacitor" (Dickens 2003) is one explanation for how massive quantities of isotopically-light carbon were injected into the Earth system at the Paleocene-Eocene thermal maximum and possibly at other times in Earth's history. Computational modeling of how microbially-generated methane accumulates in sediments as hydrates, free-gas, or dissolved gas; how it is oxidized by microbes under normal conditions of leakage; and how it may escape from sediment and enter the overlying water or atmosphere and act as a greenhouse gas have matched the observed $\delta^{13}C$ excursions in the sediment records (Dickens 2003; Gu et al. 2011). Although Earth-system models indicate that the current phase of planetary warming is unlikely to cause large-scale release of methane present as hydrates (Archer 2007), modeling efforts that focus on high-latitude sediment suggest that more immediate release of methane from hydrates is possible (Reagan and Moridis 2009). It seems that sediment and deep permafrost containing accumulations of biologically-produced methane are perhaps especially precarious (Westbrook et al. 2009; Ruppel 2011). That methane plumes can transition from the seafloor, through the water column, and then to the atmosphere is notable (Solomon et al. 2009). Polar field sites may be excellent places to observe how subsurface biota and their processes respond to the surface system (and how humans are changing it) and may be responsible for accelerating (i.e., by making methane) or quenching (i.e., by consuming the methane) the changes that are underway on the planet's skin.

Another example of microbes in their native state that may contribute significantly to processes of concern at the surface are those present in deep aquifers covered by the oceans and contained within large igneous provinces or proximal to spreading centers or seamounts (Schrenk et al. 2010; 2013). By virtue of their activity, these cells likely play an important role in planetary elemental cycling. These regions of considerable fluid movement are driven by advection of seawater into the crustal materials at the seafloor and by thermal convection cells generated when geothermally-heated waters circulate through the porous geological structure (Delaney et al. 1998; Edwards et al. 2005). Life in the crust consists of microbes that form complete ecosystems with lithoauthotrophy and heterotrophy present (cf., Cowen et al. 2003; Santelli et al. 2008; Mason et al. 2010; Smith et al. 2011). It has been estimated that ca. 1×10^{12} g C/yr of primary biomass may accumulate based on the volume of accessible crustal material; the amount of water cycling through these sponge-like materials; and the iron-, sulfur- and hydrogen-based metabolisms upon which these ecosystems rely (Bach and Edwards 2003). Thus, crustal communities mediate the flux of crucial elements from the mantle to the overlying water, where chemical energy is converted into microbial cells (Menez et al. 2012). The surface (seafloor) exposures of these deep aquifers are windows through which the geochemical and microbiological fluxes may shine into the overlying water. These surface windows exhibit diverse and complex accumulations of life based on interaction of the released fluids and the microbes with the seawater into which they emerge (Bernardino et al. 2012; Thurber et al. 2012). And because of their distance from us, we have not yet completely seen deeply into these windows where there may be complex ecosystems projecting into the crustal materials.

Also of some importance is the concept that the subsurface was once a refuge for life when the surface was too harsh to allow survival (Stevens 1997). Early in the planet's history, perhaps after life started, but still when surface conditions were austere, the subsurface might have been relatively stable, perhaps even much as it is today. Bolide impacts might have routinely sterilized the surface proximal to the impact; however, at some distance the resultant fractures and fluid movement (Cockell et al. 2012) might have provided conditions that would enhance survival. The same might have been true in some regions that sustained ice cover. Here, at some depth (as is the case in present high-latitude locations), the balance between low surface temperatures and high subsurface temperatures would offer thermally optimal conditions for long-term stability of microbial communities. It is sobering to think that billions of years from now, as the Sun sears the surface of Earth, life may make its final stand in the refugial depths of the planet.

PROJECTIONS AND PRIORITIES FOR FUTURE STUDIES

The future of subsurface microbiology research is rich with opportunities to understand the peculiarities of this environment and how these characteristics define it as an important component of the biosphere. The priorities for future studies can be divided into topics that deal with how the subsurface is sampled and envisioned and also into topics that relate to traits of subsurface microbes or the ecology, diversity, biomass, activity, and constraints of microbial communities.

Imagining how we might sample and visualize deep life

Numerous research tools are available to those who would study microbes in the subsurface; however, new tools would enable even more information to be gleaned from our deep investigations. New experiments could help to address the bias that is expected to occur associated with sampling. Although there are few examples of such experiments (cf., Hirsch and Rades-Rohlkohl 1988), we can expect that the mere act of coring, pumping, or excavating in an underground environment may stimulate microbes in an otherwise quiescent setting. The creation of open boreholes that serve as vertical conduits for fluids where none existed before, or of hydrologic gradients associated with the strenuous pumping of aquifers, are certain stimuli to microflora that are used to stasis. Crustal systems are notoriously difficult geological environments to sample because of their inherent porosity and the tendency for fluid imbibition that can carry drilling materials into the rocks.

Equally delicate conditions are those required for studies that would manipulate the subsurface and determine the biogeochemical responses. Numerous examples reveal how microbes respond to purposeful changes in their environment in shallow or accessible systems derived from bioremediation research, but fewer studies have been conducted in deep or hard to access habitats and experimental alterations in such places are more complicated. Observatories such as CORKS and SCIMPIs in the subseafloor have yielded exciting results, but even these devices are not always straightforward to deploy. As an example, settings where methane concentration, pressures, and temperatures are conducive to gas hydrate formation are not yet amenable to the observatory approach because hydrate formation prevents easy recovery of the samplers. Elsewhere, in methane-rich sediments with increased leakage of the gas, where fractures are important in distributing microbes, where shale gas is extracted, or in formations designated as CO_2 repositories, we can envision that sampling systems such as CORKS, osmosamplers, or FTISR will be essential for puzzling out the implications of microbial activity.

Advances are pending in the ways that we see or imagine the subsurface. From a computational perspective, new modeling approaches will strengthen the relationships of scientists dwelling on the disciplinary boundaries and drive new studies that can be approached using experimental work or field collections. The coupling of genome-scale models to reactive transport models has been accomplished in soils systems (cf., O'Donnell et al. 2007) and even in boutique subsurface settings like the Rifle uranium mill tailing cleanup site in western Colorado (Scheibe et al. 2009). The Rifle research sets a splendid example for other researchers, who must explain the biogeochemistry of their target environment underground. We eagerly anticipate the application of the same visualization and engagement tools that have been used for developing visual observatories in astronomy (Szalay and Gray 2001), mapping human proteins (Askenazi et al. 2011), detecting new evolutionary relationships in a spatiotemporal context (Kidd and Liu 2008; Sidlauskas et al. 2009), seeing the biogeographical distribution of large species (Kidd 2010), and observing how disease propagates through time (Janies et al. 2007). By using data science and visualization approaches, we can move towards conceptual models of subsurface life that will help us to realize principles of that life comparable to how progress has been made in Earth system models, simulations of deep astronomical time, and human behavior (Wright and Wang 2011). Perhaps just beyond the computational models of subsurface life will be the physical models, the holograms and GeoWalls, that allow us to see in our museums or auditoriums the inside of a living Earth.

Unexplored adaptations of subsurface microbes

Because the subsurface presents such extreme conditions for microbial life compared to life at the surface where most of our studies of life occur, it is a challenge to understand how processes that we recognize as essential for life, or even routine for life, can occur in the subsurface. We have already noted the records for long-term survival that life underground appears to sustain; and subsurface cells may depend substantially on the formation of inert endospores (Lomstein et al. 2012). However, if cells remain marginally vegetative in some sense their protracted temporal survival evokes new questions:

- Do such vegetative cells grow at all or do they exercise some sort of contact inhibition when they lie in such intimate space with minerals?
- Can these living cells evolve—and if so, at what rate—without undergoing the cell division that normally seems so essential to introducing genetic change to a population? Might new evolutionary mechanisms be discovered such as CRISPRs (Banfield and Young 2009) or the apparent importance of viruses (Anderson et al. 2011a, 2013) in these deep dwelling cells?
- Are there specific adaptations (e.g., efficient nucleic acid repair) associated with survival proximal to minerals that may be undergoing radioactive decay (Arrage et al. 1993)?
- For cells that enter into extended dormancy, how do they muster the bare metabolic activity to repair damaged (i.e., oxidized or racemized) molecules that are associated with survival in aqueous media?
- Have these cells found new ways to resist high temperatures, as would appear to be true for some that live under the dual stress of high temperatures and pressures (Takai et al. 2008)?
- As some may be fixed in place for millennia, have these cells devised structures such as "nanowires" or pili to explore neighboring pores and fractures for the thermodynamic disequilibria that are essential for gaining energy (Nielsen et al. 2010)?

Unstudied physiologies and genotypes for the subsurface

The demands of life in Earth's deep reaches appear to invoke as-yet unstudied strategies by which microbes achieve the necessary energy. These strategies may not be new to life, but certainly may be new to science simply because they are discreet compared to the physiological capabilities that blare in surface systems that have been well investigated. What evocative new approaches to survival are awaiting discovery?

- What is the relative importance of "latent" redox systems in subsurface environments where microbes cling to life (Valentine 2011)? An example is the radiolytic splitting of molecules to generate transient reactive species. Such metastable oxidants (e.g., peroxides, oxidized Mn, Fe, or S species) could allow incremental metabolic activity in systems that are otherwise deprived of oxidants (Chivian et al. 2008; D'Hondt et al. 2009). Another example is dehalorespiration, whereby chlorinated organics that are buried in the sediments may serve as oxidants (Futagami et al. 2009). The means by which these halogenated species (of human origin) serve as electron acceptors have been understood for years (Lee et al. 1998); however, we have not fully explored the process by which low concentrations of naturally occurring versions of these organics may be accessed by microbial communities in seafloor settings. Is the anoxic production of oxygen at the expense of methane and nitrite (Ettwig et al. 2010) common in subsurface systems given the correct chemistry and might this be another reasonable oxidant source for cells that do not require much? Collectively, are these just metabolic eddies and mere curios or bona fide survival strategies by which microbes can and do survive the subsurface in broad measure?
- Are there strategies that would appear to be thermodynamically insurmountable that have been solved by subsurface microbes? These capabilities might be analogues to the still incompletely understood anaerobic oxidation of methane, where life

manages to exist at the extreme edge of thermodynamic probability. For example, methanogens appear to exist, albeit not in high numbers, in sediment where methane levels are high enough to make additional methane production exceedingly difficult. Is it possible that hydrates may serve as a sink for additional methane production for proximal methanogens? Could questions like these be solved by cultivation studies that creatively determined how to cultivate seawater microbes that resisted normal laboratory media (Connon and Giovannoni 2002)?

- Does the subsurface have a "rare" biosphere just as was found in the surface oceans (Sogin et al. 2006)? If there are numerous rare taxa in samples from the subsurface then what does this say about functional resilience of the subsurface? Do keystone taxa exist in the subsurface; that is, microbes that are hallmarks of the subsurface and therefore playing some fundamental roles underground? The Census of Deep Life, currently underway and a part of the Deep Carbon Observatory, may inform us.

Subsurface coupling of the living and the non-living

We now understand new ways that abiotic and biotic systems of our planet are inextricably linked to each other and to human systems (Liu et al. 2007; Watkins and Freeman 2008; Stafford 2010). The subsurface is no different. As we learn more about the ways in which life survives underground, the activities and identities of these cells, questions arise related to how deep Earth changes life and, in reciprocity, is changed as a result of the life therein. What are the various connections between large-scale Earth processes and subsurface microbes that require thermodynamic disequilibria to conserve energy for metabolic activity, however slim that activity might be (Fig. 1)?

- Does the fluid movement associated with tidal forces (Tolstoy et al. 2002) influence subsurface microbial communities? And to include another large-scale phenomenon, if earthquakes influence tidal activity (Glasby and Kasahara 2001), how then are these seismic events tied to microbial community activity?
- We understand that hydrogen may be key to microbial survival in the subsurface (Morita 2000; Sleep et al. 2004). Does hydrogen production from seismic activity represent yet another way that microbes in the subsurface can be kept alive by Earth movements (Hirose et al. 2011)?
- Besides hydrogen, what other microbial provisions might be generated by seismic energy release? Can new space in the subsurface in the form of fractures and porosity, as well as access to oxidants from newly cracked surfaces, be supplied to deep microbes that are otherwise so limited in this regard?
- Are continental margins, where tectonics dictate the adjustment of plates and promote fluid movement (Torres et al. 2002; Wood et al. 2002) by opening fractures and by changing the stability of gas hydrates within the sediments, also places where blooms of subsurface microbial activity can be expected?
- How do annual planetary cycles determine what may occur in the subsurface? For example, does the seasonal accumulation and then melting of snow, which loads and unloads Earth's surface in places like northeastern Japan and causes deformation at the land surface (Heki 2001), also simulate a subterranean bellows opening and closing on an annual cycle that microbes might take advantage of?
- Do global events with longer cyclic periods, such as the current change in climate at Earth's surface, cause changes realized by life underground? In high latitudes, where warming processes appear to have provoked the incipient thaw of permafrost and degeneration of methane hydrates (cf., Shakhova et al. 2010; Ruppel 2011; Walter

Anthony et al. 2012), will changes eventually translate into the subsurface sediments and prompt microbial activity as a result of renewed fluid movement in long-frozen materials? How will the microbes in these systems, as they become more active as they can in more surficial Arctic settings (Mackelprang et al. 2011), impose themselves on the fluxes of greenhouse gases that we are so concerned with?

SUMMARY

As investigators of life underground, we anticipate the chance to learn more about the unseen world of small life deep in Earth, but only if we continue to engage in essential collaborations with all those who possess complementary knowledge of the planet's history and systems. Soon, we hope, there will be more complete synthetic models of how the living and non-living aspects of Earth's uppermost layers function and integrate with one another. These models will guide our search for new deep life, its niches, capabilities, and adaptations. We look forward to new explanations for how the subsurface biosphere is sustained and how its expression matters to life at the surface.

ACKNOWLEDGMENTS

We thank the Deep Carbon Observatory funded by the Alfred P. Sloan Foundation and the Department of Energy, Office of Science, Office of Biological and Environmental Research, Subsurface Biogeochemical Research Program (Award Number DE-SC0001533) for support. We greatly appreciate the editorial support provided by Sachi and Tomi Nakama.

REFERENCES

Allen JP, Atekwana EA, Atekwana EA, Duris JW, Werkema DD, Rossbach S (2007) The microbial community structure in petroleum-contaminated sediments corresponds to geophysical signatures. Appl Environ Microbiol 73:2860-2870

Amann R, Fuchs BM, Behrens S (2001) The identification of microorganisms by fluorescence *in situ* hybridisation. Curr Opin Biotech 12:231-236

Amaral-Zettler L, Artigas LF, Baross J, Bharathi PAL, Boetius A, Chandramohan D, Herndl G, Kogure K, Neal P, Pedros-Alio C, Ramette A, Schouten S, Stal L, Thessen A, de Leeuw J, Sogin M (2010) A global census of marine microbes. *In*: Life in the World's Oceans: Diversity, Distribution and Abundance. McIntyre AD (ed) Wiley-Blackwell, Oxford, p 223–245

Amy P, Haldeman D, Ringelberg D, Hall D, Russell C (1992) Comparison of identification systems for classification of bacteria isolated from water and endolithic habitats within the deep subsurface. Appl Environ Microbiol 58:3367-3373

Amy PS, Haldeman DL (eds) (1997) The Microbiology of the Terrestrial Deep Subsurface. CRC Press, Boca Raton Florida

Anderson RE, Brazelton WJ, Baross J (2011a) Is the genetic landscape of the deep subsurface biosphere affected by viruses? Front Microbiol 2:doi: 10.3389/fmicb.2011.00219

Anderson RE, Brazelton WJ, Baross JA (2011b) Using CRISPRs as a metagenomic tool to identify microbial hosts of a diffuse flow hydrothermal vent viral assemblage. FEMS Microbiol Lett 77:120-133

Anderson RE, Brazelton WJ, Baross JA (2013) The deep viriosphere: assessing the viral impact on microbial community dynamics in the deep subsurface. Rev Mineral Geochem 75:649-675

Archer D (2007) Methane hydrate stability and anthropogenic climate change. Biogeosciences 4:521-544

Arrage AA, Phelps TJ, Benoit RE, White DC (1993) Survival of subsurface microorganisms exposed to UV radiation and hydrogen peroxide. Appl Environ Microbiol 59:3545-3550

Askenazi M, Webber JT, Marto JA (2011) mzServer: web-based programmatic access for mass spectrometry data analysis. Molecular Cell Proteomics 10, doi: 10.1074/mcp.M110.003988

Bach W, Edwards K (2003) Iron and sulfide oxidation within the basaltic ocean crust: implications for chemolithoautotrophic microbial biomass production. Geochim Cosmochim Acta 67:3871-3887

Bakermans C, Tsapin AI, Souza-Egipsy V, Gilichinsky DA, Nealson KH (2003) Reproduction and metabolism at -10 °C of bacteria isolated from Siberian permafrost. Environ Microbiol. 5:321-326

Balkwill D (1989) Numbers, diversity, and morphological characteristics of aerobic, chemoheterotrophic bacteria in deep subsurface sediments from a site in South Carolina. Geomicrobiol J 7:33-52

Balkwill DL, Leach FR, Wilson JT, McNabb JF, White DC (1988) Equivalence of microbial biomass measures based on membrane lipid and cell wall components, adenosine triphosphate, and direct counts in subsurface aquifer sediments. Microb Ecol 16:73-84

Banfield JF, Young M (2009) Variety - the splice of life - in microbial communities. Science 326:1198-1199

Bastin ES, Greer FE, Merritt CA, Moulton G (1926) Bacteria in oil field waters. Science 63:21-24

Batzke A, Engelen B, Sass H, Cypionka H (2007) Phylogenetic and physiological diversity of cultured deep-biosphere bacteria from Equatorial Pacific Ocean and Peru Margin sediments. Geomicrobiol J 24:261-273

Beatty JT, Overmann J, Lince MT, Manske AK, Lang AS, Blankenship RE, Van Dover CL, Martinson TA, Plumley FG (2005) An obligately photosynthetic bacterial anaerobe from a deep-sea hydrothermal vent. Proc Natl Acad Sci USA 102:9306-9310

Benson SM, Cook P (2005) Chapter 5: Underground geological storage. *In*: IPCC Special Report on Carbon Dioxide Capture and Storage, Intergovernmental Panel on Climate Change, Interlachen, Switzerland, p 5-1 to 5-134

Bernardino AF, Levin LA, Thurber AR, Smith CR (2012) Comparative composition, diversity and trophic ecology of sediment macrofauna at vents, seeps and organic falls. PLoS One 7, doi: 10.31371/journal.pone.0033515

Biddle JF, Lipp JS, Lever MA, Lloyd KG, Sorensen KB, Anderson R, Fredricks HF, Elvert M, Kelly TJ, Schrag DP, Sogin ML, Brenchley JE, Teske A, House CH, Hinrichs K-U (2006) Heterotrophic Archaea dominate sedimentary subsurface ecosystems off Peru. Proc Natl Acad Sci USA 103:3846-3851

Blair CC, D'Hondt S, Spivack AJ, Kingsley RH (2007) Potential of radiolytic hydrogen for microbial respiration in subseafloor sediments. Astrobiology 7:951-970

Borgonie G, Garcıa-Moyano A, Litthauer D, Bert W, Bester A, van Heerden E, Moller C, Erasmus M, Onstott TC (2011) Nematoda from the terrestrial deep subsurface of South Africa. Nature 474:79-82

Bowles MW, Samarkin VA, Joye SB (2011) Improved measurement of microbial activity in deep-sea sediments at *in situ* pressure and methane concentration. Limnol Oceanogr Meth 9:499-506

Breitbart M, Felts B, Kelley S, Mahaffy JM, Nulton J, Salamon P, Rohwer F (2004) Diversity and population structure of a near-shore marine-sediment viral community. Proc Royal Soc London Ser B Biol Sci 271:565-574

Briggs BR, Inagaki F, Morono Y, Futagami T, Huguet C, Rosell-Mele A, Lorenson TD, Colwell FS (2012) Bacterial dominance in subseafloor sediments characterized by methane hydrates. FEMS Microbiol Ecol 81:88-98

Brockman FJ, Kieft TL, Fredrickson JK, Bjornstad BN, Li SW, Spangenburg W, Long PE (1992) Microbiology of vadose zone paleosols in south-central Washington State. Microb Ecol 23:279-301

Brockman FJ, Murray CJ (1997) Microbiological heterogeneity in the terrestrial subsurface and approaches for its description. *In*: The Microbiology of the Terrestrial Deep Subsurface. Amy PS, Haldeman DL (eds) CRC Press, Boca Raton Florida, p 75-102

Cano RJ, Borucki MK (1995) Revival and identification of bacterial spores in 25- to 40-million- year-old Dominican amber. Science 268:1060-1064

Chapelle F, Lovley D (1990) Rates of microbial metabolism in deep coastal plain aquifers. Appl Environ Microbiol 56:1865-1874

Chivian D, Brodie EL, Alm EJ, Culley DE, Dehal PS, DeSantis TZ, Gihring TM, Lapidus A, Lin L-H, Lowry SR, Moser DP, Richardson PM, Southam G, Wanger G, Pratt LM, Andersen GL, Hazen TC, Brockman FJ, Arkin AP, Onstott TC (2008) Environmental genomics reveals a single-species ecosystem deep within earth. Science 322:275-278

Cockell CS, Voytek MA, Gronstal AL, Finster K, Kirshtein JD, Howard K, Reitner J, Gohn GS, Sanford WE, J.W. HJ, Kallmeyer J, Kelly L, Powars DS (2012) Impact disruption and recovery of the deep subsurface biosphere. Astrobiology 12:231-246

Colwell F, G. Stormberg, T. Phelps, S. Birnbaum, J. McKinley, S. Rawson, C. Veverka, S. Goodwin, P. Long, B. Russell, T. Garland, D. Thompson, Skinner P, Grover S (1992) Innovative techniques for collection of saturated and unsaturated subsurface basalts and sediments for microbiological characterization. J Microbiol Meth 15:279-292

Colwell F, Schwartz A, Briggs B (2011) Microbial community distribution in sediments from the Mount Elbert Gas Hydrate Stratigraphic Test Well, Alaska North Slope. Mar Petrol Geol 28:404-410

Colwell FS, Lehman RM (1997) Carbon source utilization profiles for microbial communities from hydrologically distinct zones in a basalt aquifer. Microb Ecol 33:240-251

Colwell FS, Onstott TC, Delwiche ME, Chandler D, Fredrickson JK, Yao Q-J, McKinley JP, Boone DR, Griffiths R, Phelps TJ, Ringelberg D, White DC, LaFreniere L, Balkwill D, Lehman RM, Konisky J, Long PE (1997) Microorganisms from deep, high temperature sandstones: constraints on microbial colonization. FEMS Microbiol Rev 20:425-435

Connon SA, Giovannoni SJ (2002) High-throughput methods for culturing microorganisms in very-low-nutrient media yield diverse new marine isolates. Appl Environ Microbiol 68:3878-3885

Cowen JP, Giovannoni SJ, Kenig F, Johnson HP, Butterfield D, Rappe MS, Hutnak M, Lam P (2003) Fluids from aging ocean crust that support microbial life. Science 299:120-123

D'Hondt S, Inagaki F, Alvarez Zarikian CA, Expedition 329 Scientists (2011) Proceedings of the Integrated Ocean Drilling Program Management International, Inc. doi: 10.2204/iodp.proc.329.2011

D'Hondt S, Jorgensen BB, Miller DJ, Batzke A, Blake R, Cragg BA, Cypionka H, Dickens GR, Ferdelman T, Hinrichs K-U, Holm NG, Mitterer R, Spivack A, Wang G, Bekins B, Engelen B, Ford K, Gettemy G, Rutherford SD, Sass H, Skilbeck CG, Aiello IW, Guerin G, House CH, Inagaki F, Meister P, Naehr T, Niitsuma S, Parkes RJ, Schippers A, Smith DC, Teske A, Wiegel J, Padilla CN, Acosta JLS (2004) Distributions of microbial activities in deep subseafloor sediments. Science 306:2216-2221

D'Hondt S, Rutherford S, Spivak AJ (2002a) Metabolic activity of subsurface life in deep-sea sediments. Science 295:2067-2070

D'Hondt S, Smith DC, Spivak AJ (2002b) Exploration of the marine subsurface biosphere. JOIDES J 28:51-54

D'Hondt S, Spivack AJ, Pockalny R, Ferdelman TG, Fischer JP, Kallmeyer J, Abrams LJ, Smith DC, Graham D, Hasiuk F, Schrum H, A.M. S (2009) Subseafloor sedimentary life in the South Pacific Gyre. Proc Natl Acad Sci USA 106:11651-11656

Daily GC, Matson PA (2008) Ecosystem services: from theory to implementation. Proc Natl Acad Sci USA 105:9455-9456

Danovaro R, Dell'Anno A, Corinaldesi C, Magagnini M, Noble R, Tamburini C, Weinbauer M (2008) Major viral impact on the functioning of benthic deep-sea ecosystems. Nature 454:1084-1087

Das AP, Sukla LB, Pradhan N, Nayak S (2011) Manganese biomining: a review. Bioresour Technol 102:7381-7387

Davis EE, Becker K, Pettigrew T, Carson B, MacDonald BR (1992) CORK: a hydrological seal and downhole observatory for deep-ocean boreholes. Proc Ocean Drill Program Initial Rep 139:43-53

Davis KJ, Luttge A (2005) Quantifying the relationship between microbial attachment and mineral surface dynamics using vertical scanning interferometry (VSI). Am J Sci 305:727-751

DeFlaun MF, Fuller ME, Zhang P, Johnson WP, Mailloux BJ, Holben WE, Kovacik WP, Balkwill DL, Onstott TC (2001) Comparison of methods for monitoring bacterial transport in the subsurface. J Microbiol Meth 47:219-231

Delaney JR, Kelley DS, Lilley MD, Butterfield DA, Baross JA, Wilcock WSD, Embley RW, Summit M (1998) The quantum event of oceanic crustal accretion: impacts of diking at mid-ocean ridges. Science 281:222-230

D'Elia T, Veerapaneni R, Rogers SO (2008) Isolation of microbes from Lake Vostok accretion ice. Appl Environ Microbiol 74:4962-4965

Denef VJ, Kalnejais LH, Mueller RS, Wilmes P, Baker BJ, Thomas BC, VerBerkmoes NC, Hettich RL, Banfield JF (2010) Proteogenomic basis for ecological divergence of closely related bacteria in natural acidophilic microbial communities. Proc Natl Acad Sci USA 107:2383-2390

Dickens GR (2003) Rethinking the global carbon cycle with a large, dynamic and microbially mediated gas hydrate capacitor. Earth Planet Sci Lett 213:169-183

Dupraz S, Parmentier M, Ménez B, Guyot F (2009) Experimental and numerical modeling of bacterially induced pH increase and calcite precipitation in saline aquifers. Chem Geol 265:44-53

Edwards KJ, Bach W, McCollom TM (2005) Geomicrobiology in oceanography: microbe - mineral interactions at and below the seafloor. Trends Microbiol 13:449-456

Edwards KT, Becker K, Colwell F (2012) The deep, dark energy biosphere: intraterrestrial life on earth. Annu Rev Earth Planet Sci 40:551-568

Edwards KT, Glazer BT, Rouxel OJ, Bach W, Emerson D, Davis RE, Toner BM, Chan CS, Tebo BM, Staudigel H, Moyer CL (2011) Ultra-diffuse hydrothermal venting supports Fe-oxidizing bacteria and massive umber deposition at 5000 m off Hawaii. ISME J 5:1748-1758

Edwards RA, Rodriguez-Brito B, Wegley L, Haynes M, Breitbart M, Peterson DM, Saar MO, Alexander S, Alexander EC, Rohwer F (2006) Using pyrosequencing to shed light on deep mine microbial ecology under extreme hydrogeologic conditions. BMC Genomics 7:57

Ehrlich HL (1990) Geomicrobiology. Marcel Dekker, New York

Ehrlich HL (2012) Reminiscences from a career in geomicrobiology. Ann Rev Earth Planet Sci 40:1-21

Ettwig KF, Butler MK, Le Paslier D, Pelletier E, Mangenot S, Kuypers MMM, Schreiber F, Dutilh BE, Zedelius J, de Beer D, Gloerich J, Wessels HJCT, van Alen T, Luesken F, Wu ML, van de Pas-Schoonen KT, Op den Camp HJM, Janssen-Megens EM, Francoijs K-J, Stunnenberg H, Weissenbach J, Jetten MSM, Strous M (2010) Nitrite-driven anaerobic methane oxidation by oxygenic bacteria. Nature 464:543-548

Expedition 336 Scientists (2012) Mid-Atlantic Ridge microbiology: initiation of long-term coupled microbiological, geochemical, and hydrological experimentation within the seafloor at North Pond, western flank of the Mid-Atlantic Ridge. IODP Preliminary Report 336. doi: 10.2204/iodp.pr.336.2012

Fisher AT, Tsuji T, Petronotis K, Expedition 327 Scientists (2011) Proceedings of the Integrated Ocean Drilling Program Management International, Inc. doi: 10.2204/iodp.proc.327.2011
Fox P, Hendler J (2011) Changing the equation on scientific data visualization. Science 331:705-708
Fredrickson JK, Balkwill D, Zachara J, Li S, Brockman F, Simmons M (1991) Physiological diversity and distributions of heterotrophic bacteria in deep Cretaceous sediments of the Atlantic coastal plain. Appl Environ Microbiol 57:402-411
Fredrickson JK, Balkwill DL (2006) Geomicrobial processes and biodiversity in the deep terrestrial subsurface. Geomicrobiol J 23:345-356
Fredrickson JK, Fletcher M (eds) (2001) Subsurface Microbiology and Biogeochemistry. Wiley-Liss, New York
Fredrickson JK, McKinley JP, Bjornstad BN, Long PE, Ringelberg DB, White DC, Suflita JM, Krumholz L, Colwell FS, Lehman RM, Phelps TJ (1997) Pore-size constraints on the activity and survival of subsurface bacteria in a Late Cretaceous shale-sandstone sequence, northwestern, New Mexico. Geomicrobiol J 14:183-202
Fredrickson JK, Onstott TC (1996) Microbes deep inside the Earth. Sci Am 275:68-73
Fredrickson JK, Zachara JM, Balkwill DL, Kennedy D, Li SMW, Kostandarithes HM, Daly MJ, Romine MF, Brockman FJ (2004) Geomicrobiology of high-level nuclear waste-contaminated vadose sediments at the Hanford Site, Washington State. Appl Environ Microbiol 70:4230-4241
Freifeld BM, Trautz RC, Kharaka YK, Phelps TJ, Myer LR, Hovorka SD, Collins DJ (2005) The U-tube: a novel system for acquiring borehole fluid samples from a deep geologic CO_2 sequestration experiment. J Geophys Res-Solid Earth 110, doi: 10.1029/2005JB003735
Futagami T, Morono Y, Terada T, Kaksonen AH, Inagaki F (2009) Dehalogenation activities and distribution of reductive dehalogenase homologous genes in marine subsurface sediments. Appl Environ Microbiol 75:6905-6909
Gaidos EJ, Nealson KH, Kirschvink JL (1999) Life in ice-covered oceans. Science 284:1631-1632
Gerba CP, Bales RC (1990) Virus transport in the subsurface. First International Symposium on Microbiology of the Deep Subsurface, WSRC Information Services Section Publications Group, p 7-23 to 7-31
Glasby GP, Kasahara J (2001) Influence of tidal effects on the periodicity of earthquake activity in diverse geological settings with particular emphasis on submarine hydrothermal systems. Earth-Sci Rev 52:261-297
Gu G, Dickens GR, Bhatnagar G, Colwell FS, Hirasaki GJ, Chapman WG (2011) Abundant Early Palaeogene marine gas hydrates despite warm deep-ocean temperatures. Nature Geosci 4: 848-851
Haggerty R, Schroth MH, Istok JD (1998) Simplified method of "push-pull" test data analysis for determining *in situ* reaction rate coefficients. Ground Water 36:314-324
Hazen RM, Downs RT, Jones AP, Kah L (2013) Carbon mineralogy and crystal chemistry. Rev Mineral Geochem 75:7-46
Hazen TC (1997) Bioremediation. *In*: The Microbiology of the Terrestrial Deep Subsurface. Amy PS, Haldeman DL (eds) CRC Press, Boca Raton Florida, p 247-266
Hazen TC, Jimenez L, Lopez de Victoria G (1991) Comparison of bacteria from deep subsurface sediment and adjacent groundwater. Microb Ecol 22:293-304
Head IM, Jones DM, Larter SR (2003) Biological activity in the deep subsurface and the origin of heavy oil. Nature 426:344-352
Heider J (2007) Adding handles to unhandy substrates: anaerobic hydrocarbon activation mechanisms. Curr Opin Chem Biol 11:188-194
Heider J, Spormann AM, Beller HR, Widdel F (1998) Anaerobic bacterial metabolism of hydrocarbons. FEMS Microbiol Rev 22:459-473
Heki K (2001) Seasonal modulation of interseismic strain buildup in northeastern Japan driven by snow loads. Science 293:89-92
Hirose T, Kawagucci S, Suzuki K (2011) Mechanoradical H_2 generation during simulated faulting: implications for an earthquake-driven subsurface biosphere. Geophys Res Lett 38, doi: 10.1029/2011GL048850
Hirsch P, Rades-Rohlkohl E (1988) Some special problems in the determination of viable counts of groundwater microorganisms. Microb Ecol 16:99-113
Holben WE, Ostrom PH (2000) Monitoring bacterial transport by stable isotope enrichment of cells. Appl Environ Microbiol 66:4935-4939
Holman H-YN, Perry DL, Hunter-Cevera JC (1998) Surface-enhanced infrared absorption-reflectance (SEIRA) microspectroscopy for bacteria localization on geologic material surfaces. J Microbiol Meth 34:59-71
Horn JM, Masterson BA, Rivera A, Miranda A, Davis MA, Martin S (2004) Bacterial growth dynamics, limiting factors, and community diversity in a proposed geological nuclear waste repository environment. Geomicrobiol J 21:273-286
Inagaki F, Nunoura T, Nakagawa S, Teske A, Lever M, Lauer A, Suzuki M, Takai K, Delwiche M, Colwell FS, Nealson KH, Horikoshi K, D'Hondt S, Jørgensen BB (2006) Biogeographical distribution and diversity of microbes in methane hydrate-bearing deep marine sediments on the Pacific Ocean Margin. Proc Natl Acad Sci USA 103:2815-2820

Janies D, Hill AW, Guralnick R, Habib F, Waltari E, Wheeler WC (2007) Genomic analysis and geographic visualization of the spread of avian influenza (H5N1). Syst Biol 56:321-329

Jin Q, Bethke CM (2005) Predicting the rate of microbial respiration in geochemical environments. Geochim Cosmochim Acta 69:1133-1143

Jolley DM, Ehrhorn TF, Horn J (2003) Microbial impacts to the near-field environment geochemistry: a model for estimating microbial communities in repository drifts at Yucca Mountain. J Contam Hydrol 62-3:553-575

Jordana R, Baquero E, Reboleira S, Sendra A (2012) Reviews of the genera *Schaefferia* Absolon, 1900, *Deuteraphorura* Absolon, 1901, *Plutomurus* Yosii, 1956 and the *Anurida* Laboulbène, 1865 species group without eyes, with the description of four new species of cave springtails (Collembola) from Krubera-Voronya cave, Arabika Massif, Abkhazia. Terrest Arth Rev 5:35-85

Jørgensen BB (2011) Deep subseafloor microbial cells on physiological standby. Proc Natl Acad Sci USA 108:18193-18194

Kallmeyer J, Pockalny R, Adhikari RR, Smith DC, D'Hondt S (2012) Global distribution of microbial abundance and biomass in subseafloor sediment. Proc Natl Acad Sci USA doi: 10.1073/pnas.1203849109

Kappler A, Emerson D, Edwards KJ, Amend JP, Gralnick JA, Grathwohl P, Hoehler TM, Straub KL (2005) Microbial activity in biogeochemical gradients - new aspects of research. Geobiol 3:229-233

Kelley DS, Karson JA, Früh-Green GL, Yoerger DA, Butterfield DA, Hayes J, Shank T, Schrenk MO, Olson EJ, Proskurowski G, Jakuba M, Bradley A, Larson B, Ludwig K, Glickson D, Buckman K, Bradley AS, Brazelton WJ, Roe K, Elend MJ, Delacour A, Bernasconi SM, Lilley MD, Baross JA, Summons RE, Sylva SP (2005) A serpentinite-hosted submarine ecosystem: the Lost City Hydrothermal Field. Science 307:1428-1434

Kidd DM (2010) Point of view: geophylogenies and the map of life. Syst Biol 59:741-752

Kidd DM, Liu X (2008) GEOPHYLOBUILDER 1.0: an ARCGIS extension for creating 'geophylogenies'. Molec Ecol Resour 8:88-91

Kieft TL, Murphy EM, Haldeman DL, Amy PS, Bjornstad BN, McDonald EV, Ringelberg DB, White DC, Stair J, Griffiths RP, Gsell TC, Holben WE, Boone DR (1998) Microbial transport, survival, and succession in a sequence of buried sediments. Microb Ecol 36:336-348

Kieft TL, Phelps TJ, Fredrickson JK (2007) Drilling, coring, and sampling subsurface environments. *In*: Manual of Environmental Microbiology. Vol 3. Hurst CJ, Crawford RL, Garland JL, Lipson DA, Mills AL, Stetzenbach LD (eds) ASM Press, Washington, DC, p 799-817

King EL, Tuncay K, Ortoleva P, Meile C (2009) In silico *Geobacter sulfurreducens* metabolism and its representation in reactive transport models. Appl Environ Microbiol 75:83–92

Krumholz LR, McKinley JP, Ulrich FA, Suflita JM (1997) Confined subsurface microbial communities in Cretaceous rock. Nature 386:64-66

Lee MD, Odom JM, Buchanan RJ (1998) New perspectives on microbial dehalogenation of chlorinated solvents: insights from the field. Ann Rev Microbiol 52:423-452

Lehman RM (2007) Understanding of aquifer microbiology is tightly linked to sampling approaches. Geomicrobiol J 24:331-341

Lehman RM, Colwell FS, Bala GA (2001) Attached and unattached microbial communities in a simulated basalt aquifer under fracture- and porous-flow conditions. Appl Environ Microbiol 67:2799-2809

Lehman RM, Colwell FS, Ringelberg D, White DC (1995) Combined microbial community-level analyses for quality assurance of terrestrial subsurface cores. J Microbiol Meth 22:263-281

Lehman RM, O'Connell SP, Banta A, Fredrickson JK, Reysenbach A-L, Kieft TL, Colwell FS (2004) Microbiological comparison of core and groundwater samples collected from a fractured basalt aquifer with that of dialysis chambers incubated *in situ*. Geomicrobiology J 21:169-182

Li L, Steefel CI, Kowalsky MB, Englert A, Hubbard SS (2010) Effects of physical and geochemical heterogeneities on mineral transformation and biomass accumulation during biostimulation experiments at Rifle, Colorado. J Contam Hydrol 112:45-63

Lin L-H, Wang P-L, Rumble D, Lippmann-Pipke J, Boice E, Pratt LM, Lollar BS, Brodie EL, Hazen TC, Andersen GL, DeSantis TZ, Moser DP, Kershaw D, Onstott TC (2006) Long-term sustainability of a high-energy, low-diversity crustal biome. Science 314:479-482

Lipp JS, Morono Y, Inagaki F, Hinrichs K-U (2008) Significant contribution of Archaea to extant biomass in marine subsurface sediments. Nature 454:991-994

Liu J, Dietz T, Carpenter SR, Alberti M, Folke C, Moran E, Pell AN, Deadman P, Kratz T, Lubchenco J, Ostrom E, Ouyang Z, Provencher W, Redman CL, Schneider SH, Taylor WW (2007) Complexity of coupled human and natural systems. Science 317:1513-1516

Lomstein BA, Langerhuus AT, D'Hondt S, Jørgensen BB, Spivack AJ (2012) Endospore abundance, microbial growth and necromass turnover in deep sub-seafloor sediment. Nature 484:101-104

Mackelprang R, Waldrop MP, DeAngelis KM, David MM, Chavarria KL, Blazewicz SJ, Rubin EM, Jansson JK (2011) Metagenomic analysis of a permafrost microbial community reveals a rapid response to thaw. Nature 480:368-371

Mason OU, Nakagawa T, Rosner M, Van Nostrand JD, Zhou J, Maruyama A, Fisk MR, Giovannoni SJ (2010) First investigation of the microbiology of the deepest layer of ocean crust. PLoS ONE 5, doi: 15310.11371/journal.pone.0015399

Masui N, Morono Y, Inagaki F (2008) Microbiological assessment of circulation mud fluids during the first operation of riser drilling by the deep-earth research vessel Chikyu. Geomicrobiol J 25:274-282

Matthess G (1990) Hydrogeological controls of bacterial and virus migration in subsurface environments. First International Symposium on Microbiology of the Deep Subsurface, WSRC Information Services Section Publications Group, p 7-33 to 7-46

Mauclaire L, Zepp K, Meister P, McKenzie J (2004) Direct *in situ* detection of cells in deep-sea sediment cores from the Peru Margin (ODP Leg 201, Site 1229). Geobiology 2:217-223

McKinley JP, Stevens TO, Fredrickson JK, Zachara JM, Colwell FS, Wagnon KB, Smith SC, Rawson SA, Bjornstad BN (1997) Biogeochemistry of anaerobic lacustrine and paleosol sediments within an aerobic unconfined aquifer. Geomicrobiol J 14:23-39

Meersman F, Daniel I, Bartlett DH, Winter R, Hazael R, McMillain PF (2013) High-pressure biochemistry and biophysics. Rev Mineral Geochem 75:607-648

Menez B, Pasini V, Brnelli D (2012) Life in the hydrated suboceanic mantle. Nat Geosci 5:133-137

Mikucki JA, Priscu JC (2007) Bacterial diversity associated with Blood Falls, a subglacial outflow from the Taylor Glacier, Antarctica. Appl Environ Microbiol 73:4029-4039

Mitchell AC, Phillips AJ, Hamilton MA, Gerlach R, Hollis WK, Kaszuba JP, Cunningham AB (2008) Resilience of planktonic and biofilm cultures to supercritical CO_2. J Supercrit Fluids 47:318-325

Mitchell AC, Phillips AJ, Hiebert R, Gerlach R, Spangler LH, Cunningham A (2009) Biofilm enhanced geologic sequestration of supercritical CO_2. Int J Greenhouse Gas Control 3:90-99

Moran K, Farrington S, Massion E, Paull C, Stephen R, Trehu A, Ussler III W (2006) SCIMPI: A New Seafloor Observatory System. OCEANS 2006, doi: 10.1109/OCEANS.2006.307103

Morita RY (2000) Is H_2 the universal energy source for long-term survival? Microb Ecol 38:307-320

Morono Y, Takeshi T, Nishizawa M, Ito M, Hillion F, Takahata N, Sano Y, Inagaki F (2011) Carbon and nitrogen assimilation in deep subseafloor microbial cells. Proc Natl Acad Sci USA 108:18295-18300

Morozova D, Wandrey M, Alawi M, Zimmer M, Vieth A, Zettlitzer M, Wurdemann H (2010) Monitoring of the microbial community composition in saline aquifers during CO_2 storage by fluorescence *in situ* hybridisation. Int J Greenhouse Gas Control 4:981-989

Murphy E, Schramke J, Fredrickson J, Bledsoe H, Francis A, Sklarew D, Linehan J (1992) The influence of microbial activity and sedimentary organic carbon on the isotope geochemistry of the Middendorf Aquifer. Water Resour Res 28:723-740

Nazina TN, Kosareva IM, Petrunyaka VV, Savushkina MK, Kudriavtsev EG, Lebedev VA, Ahunov VD, Revenko YA, Khafizov RR, Osipov GA, Belyaev SS, Ivanov MV (2004) Microbiology of formation waters from the deep repository of liquid radioactive wastes Severnyi. FEMS Microbiology Ecology 49:97-107

Nealson KH, Inagaki F, Takai K (2005) Hydrogen-driven subsurface lithoautotrophic microbial ecosystems (SLiMEs): do they exist and why should we care? Trends Microbiol 13:405-410

Nielsen LP, Risgaard-Petersen N, Fossing H, Christensen PB, Sayama M (2010) Electric currents couple spatially separated biogeochemical processes in marine sediment. Nature 463:1071-1074

Nielsen ME, Fisk MR, Istok JD, Pedersen K (2006) Microbial nitrate respiration of lactate at *in situ* conditions in ground water from a granitic aquifer situated 450 m underground. Geobiology 4:43-52

Novarino G, Warren A, Butler H, Lambourne G, Boxshall A, Bateman J, Kinner NE, Harvey RW, Mosse RA, Teltsch B (1997) Protistan communities in aquifers: a review. FEMS Microbiol Rev 20:261-275

O'Donnell AG, Young IM, Rushton SP, Shirley MD, Crawford JW (2007) Visualization, modelling and prediction in soil microbiology. Nat Rev Microbiol 5:689-699

Onstott TC (2004) Impact of CO_2 injections on deep subsurface microbial ecosystems and potential ramifications for the surface biosphere. *In*: The CO_2 Capture and Storage Project (CCP) Thomas DC, Benson SM (eds) Lawrence Berkeley National Laboratory, Berkeley, CA, p 1207-1239

Onstott TC, Colwell FS, Kieft TL, Murdoch L, Phelps TJ (2009a) New horizons for deep subsurface microbiology. Microbe 4:499-505

Onstott TC, McGown DJ, Bakermans C, Ruskeeniemi T, Ahonen L, Telling J, Soffientino B, Pfiffner SM, Sherwood Lollar B, Frape S, Stotler R, Johnson EJ, Vishnivetskaya TA, Rothmel R, Pratt LM (2009b) Microbial communities in subpermafrost saline fracture water at the Lupin Au Mine, Nunavut, Canada. Microb Ecol 58:786-807

Onstott TC, Phelps TJ, Colwell FS, Ringelberg D, White DC, Boone DR, McKinley JP, Stevens TO, Long PE, Balkwill DL, Griffin WT, Kieft T (1998) Observations pertaining to the origin and ecology of microorganisms recovered from the deep subsurface of Taylorsville Basin, Virginia. Geomicrobiol J 15:353-385

Onstott TC, Phelps TJ, Kieft T, Colwell FS, Balkwill DL, Fredrickson JK, Brockman F (1999) A global perspective on the microbial abundance and activity in the deep subsurface. *In*: Enigmatic Microorganisms and Life in Extreme Environments, Seckbach J (ed) Kluwer Academic Publishers, Netherlands, p 487-500

Onstott TC, Tobin K, Dong H, DeFlaun M, Fredrickson J, Bailey T, Brockman F, Kieft T, Peacock A, White DC, Balkwill D, Phelps TJ, Boone DR (1997) The deep gold mines of South Africa: windows into the subsurface biosphere. Proc. SPIE, International Society for Optical Engineering 3111:344–357

Orcutt BN, Wheat CG, Edwards KJ (2010) Subseafloor ocean crust microbial observatories: development of FLOCS (FLow-through Osmo Colonization System) and evaluation of borehole construction materials. Geomicrobiol J 27:143-157

Parkes RJ, Cragg BA, Bale SJ, Getliff JM, Goodman K, Rochelle PA, Fry JC, Weightman AJ, Harvey SM (1994) Deep bacterial biosphere in Pacific Ocean sediments. Nature 371:410-413

Parkes RJ, Cragg BA, Wellsbury P (2000) Recent studies on bacterial populations and processes in subseafloor sediments: a review. Hydrogeol J 8:11-28

Parkes RJ, Sellek G, Webster G, Martin D, Anders E, Weightman AJ, Sass H (2009) Culturable prokaryotic diversity of deep, gas hydrate sediments: first use of a continuous high-pressure, anaerobic, enrichment and isolation system for subseafloor sediments (DeepIsoBUG). Environ Microbiol 11:3140-3153

Pedersen K (1997) Microbial life in deep granitic rock. FEMS Microbiol Rev 20:399-414

Pedersen K (1999) Subterranean microorganisms and radioactive waste disposal in Sweden. Eng Geol 52:163-176

Pham VD, Hnatow LL, Zhang S, Fallon RD, Jackson SC, Tomb J-F, DeLong EF, Keeler SJ (2009) Characterizing microbial diversity in production water from an Alaskan mesothermic petroleum reservoir with two independent methods. Environ Microbiol 11:176-187

Phelps TJ, Murphy EM, Pfiffner SM, White DC (1994) Comparison between geochemical and biological estimates of subsurface microbial activities. Microb Ecol 28:335-349

Pitonzo BJ, Amy PS, Rudin M (1999) Effect of gamma radiation on native endolithic microorganisms from a radioactive waste deposit site. Radiation Res 152:64-70

Pombo SA, Kleikemper J, Schroth MH, Zeyer J (2005) Field-scale isotopic labeling of phospholipid fatty acids from acetate-degrading sulfate-reducing bacteria. FEMS Microbiol Ecol 51:197-207

Price PB, Sowers T (2004) Temperature dependence of metabolic rates for microbial growth, maintenance, and survival. Proc Natl Acad Sci USA 101:4631-4636

Radajewski S, McDonald IR, Murrell JC (2003) Stable-isotope probing of nucleic acids: a window to the function of uncultured microorganisms. Curr Opin Biotechnol 14:296-302

Rastogi G, Stetler LD, Peyton BM, Sani RK (2009) Molecular analysis of prokaryotic diversity in the deep subsurface of the former Homestake gold mine, South Dakota, USA. J Microbiol 47:371-384

Rawlings DE (2002) Heavy metal mining using microbes. Annu Rev Microbiol 56:65-91

Reagan MT, Moridis GJ (2009) Large-scale simulation of methane hydrate dissociation along the West Spitsbergen Margin. Geophys Res Lett 36, doi: 10.1029/2009gl041332

Revil A, Mendonça CA, Atekwana EA, Kulessa B, Hubbard SS, Bohlen KJ (2010) Understanding biogeobatteries: where geophysics meets microbiology. J Geophys Res 115, doi: 10.1029/2009JG001065

Riley RG, Zachara JM, Wobber FJ (1992) Chemical Contaminants on DOE Lands and Selection of Contaminant Mixtures for Subsurface Science Research. US Department of Energy, Office of Energy Research, Subsurface Science Program, Washington, DC

Rockstrom J, Steffen W, Noone K, Persson A, Chapin FS, Lambin EF, Lenton TM, Scheffer M, Folke C, Schellnhuber HJ, Nykvist B, de Wit CA, Hughes T, van der Leeuw S, Rodhe H, Sorlin S, Snyder PK, Costanza R, Svedin U, Falkenmark M, Karlberg L, Corell RW, Fabry VJ, Hansen J, Walker B, Liverman D, Richardson K, Crutzen P, Foley JA (2009) A safe operating space for humanity. Nature 461:472-475

Røy H, Kallmeyer J, Adhikari RR, Pockalny R, Jorgensen BB, D'Hondt S (2012) Aerobic microbial respiration in 86-million-year-old deep-sea red clay. Science 336:922-925

Ruppel CD (2011) Methane hydrates and contemporary climate change. Nature Education Knowledge 3(10): 29

Russell CE, Jacobson R, Haldeman DL, Amy PS (1994) Heterogeneity of deep subsurface microoranisms and correlations to hydrogeological and geochemical parameters. Geomicrobiol J 12:37-51

Santelli CM, Orcutt BN, Banning E, Bach W, Moyer CL, Sogin ML, Staudigel H, Edwards KJ (2008) Abundance and diversity of microbial life in ocean crust. Nature 453:653-656

Satterfield CL, Lowenstein TK, Vreeland RH, Rosenzweig WD, Powers DW (2005) New evidence for 250 Ma age of halotolerant bacterium from a Permian salt crystal. Geology 33:265-268

Saunders AD (2005) Large igneous provinces: origin and environmental consequences. Elements 1:259-263

Scheibe TD, Mahadevan R, Fang Y, Garg S, Long PE, Lovley DR (2009) Coupling a genome-scale metabolic model with a reactive transport model to describe *in situ* uranium bioremediation. Microb Biotechnol 2:274-286

Schippers A, Neretin LN, Kallmeyer J, Ferdelman TG, Cragg BA, John Parkes R, Jorgensen BB (2005) Prokaryotic cells of the deep sub-seafloor biosphere identified as living bacteria. Nature 433:861-864

Schrenk MO, Brazelton WJ, Lang SQ (2013) Serpentinization, carbon, and deep life. Rev Mineral Geochem 75:575-606

Schrenk MO, Huber JA, Edwards KJ (2010) Microbial provinces in the subseafloor. Ann Rev Mar Sci 2:279-304

Shakhova N, Semiletov I, Salyuk A, Yusupov V, Kosmach D, Gustafsson O (2010) Extensive methane venting to the atmosphere from sediments of the East Siberian Arctic Shelf. Science 327:1246-1250

Sidlauskas B, Ganapathy G, Hazkani-Covo E, Jenkins KP, Lapp H, McCall LW, Price S, Scherle R, Spaeth PA, Kidd DM (2009) Linking the big: the continuing promise of evolutionary synthesis. Evolution, doi: 10.1111/j.1558-5646.2009.00892.x

Sinclair J, Kampbell D, Cook M, Wilson J (1993) Protozoa in subsurface sediments from sites contaminated with aviation gasoline or jet fuel. Appl Environ Microbiol 59:467-472

Sinclair JL, Ghiorse WC (1989) Distribution of aerobic bacteria, protozoa, algae, and fungi in deep subsurface sediments. Geomicrobiol J 7:15-31

Sleep NH, Meibom A, Fridriksson T, Coleman RG, Bird DK (2004) H_2-rich fluids from serpentinization: geochemical and biotic implications. Proc Natl Acad Sci USA 101:12818-12823

Smith A, Popa R, Fisk MR, Nielsen M, Wheat CG, Jannasch HW, Fisher AT, Becker K, Sievert M, Flores G (2011) *In situ* enrichment of ocean crust microbes on igneous minerals and glasses using an osmotic flow-through device. Geochem Geophys Geosyst 12, doi: 10.1029/2010GC003424

Smith DC, D'Hondt S (2006) Exploration of life in deep subseafloor sediments. Oceanography 19:58-70

Smith DC, Spivak AJ, Fisk MR, Haveman SA, Staudigel H (2000) Tracer-based estimates of drilling-induced microbial contamination of deep sea crust. Geomicrobiol J 17:207-219

Smith RL, Harvey RW, LeBlanc DR (1991) Importance of closely spaced vertical sampling in delineating chemical and microbiological gradients in groundwater studies. J Contam Hydrol 7:285-300

Sogin ML, Morrison HG, Huber JA, Welch DM, Huse SM, Neal PR, Arrieta JM, Herndl GJ (2006) Microbial diversity in the deep sea and the underexplored "rare biosphere". Proc Natl Acad Sci USA 103:12115-12120

Solomon EA, Kastner M, MacDonald IR, Leifer I (2009) Considerable methane fluxes to the atmosphere from hydrocarbon seeps in the Gulf of Mexico. Nature Geosci 2, doi: 10.1038/ngeo574

Spalding BP, Watson DB (2006) Measurement of dissolved H_2, O_2, and CO_2 in groundwater using passive samplers for gas chromatographic analyses. Environ Sci Technol 40:7861-7867

Spear JR, Walker JJ, McCollom TM, Pace NR (2005) Hydrogen and bioenergetics in the Yellowstone geothermal ecosystem. Proc Natl Acad Sci USA 102:2555-2560

Stafford SG (2010) Environmental science at the tipping point. BioScience 60:94-95

Steefel CI, DePaolo DJ, Lichtner PC (2005) Reactive transport modeling: an essential tool and a new research approach for the Earth sciences. Earth Planet Sci Lett 240:539-558

Stein LY, Nicol GW (2011) Grand challenges in terrestrial microbiology. Front Microbiol 2, doi: 10.3389/fmicb.2011.00006

Stepanauskas R, Sieracki ME (2007) Matching phylogeny and metabolism in the uncultured marine bacteria, one cell at a time. Proc Natl Acad Sci USA 104:9052-9057

Stevens TO (1997) Subsurface microbiology and the evolution of the biosphere. *In*: The Microbiology of the Terrestrial Deep Subsurface. Amy PS, Haldeman DL (eds) CRC Press, Boca Raton Florida, p 205-223

Stevens TO, McKinley JP (1995) Lithoautotrophic microbial ecosystems in deep basalt aquifers. Science 270:450-454

Stevens TO, McKinley JP, Fredrickson JK (1993) Bacteria associated with deep, alkaline, anaerobic groundwaters in southeast Washington. Microb Ecol 25:35-50

Stevenson BS, Eichorst SA, Wertz JT, Schmidt TM, Breznak JA (2004) New strategies for cultivation and detection of previously uncultured microbes. Appl Environ Microbiol 70:4748-4755

Stroes-Gascoyne S, West JM (1997) Microbial studies in the Canadian nuclear fuel waste management program. FEMS Microbiol Rev 20:573-590

Suttle CA (2005) Viruses in the sea. Nature 437:356-361

Szalay A, Gray J (2001) The world-wide telescope. Science 293:2037-2040

Takai K, Moser DP, Onstott TC, Spoelstra N, Pfiffner SM, Dohnalkova A, Fredrickson JK (2001) *Alkaliphilus transvaalensis* gen. nov., sp. nov., an extremely alkaliphilic bacterium isolated from a deep South African gold mine. Int J Syst Evol Microbiol 51:1245-1256

Takai K, Mottl MJ, Nielsen SH, Expedition 331 Scientists (2011) Proceedings of the Integrated Ocean Drilling Program Management International, Inc. doi: 10.2204/iodp.proc.331.2011

Takai K, Nakamura K, Toki T, Tsunogai U, Miyazaki M, Miyazaki J, Hirayama H, Nakagawa S, Nunoura T, Horikoshi K (2008) Cell proliferation at 122 °C and isotopically heavy CH_4 production by a hyperthermophilic methanogen under high-pressure cultivation. Proc Natl Acad Sci USA 105:10949-10954

Thierstein HR, Störrlein U (1991) Living bacteria in Antarctic sediments from Leg 119. *In*: Scientific Results, Ocean Drilling Program, Leg 119. Barron J, Larsen B (eds) Ocean Drilling Program, College Station, TX, p 687-692

Thurber AR, Levin LA, Orphan VJ, Marlow JJ (2012) Archaea in metazoan diets: implications for food webs and biogeochemical cycling. ISME J 6:602-612

Tittensor DP, Mora C, Jetz W, Lotze HK, Ricard D, Vanden Berghe E, Worm B (2010) Global patterns and predictors of marine biodiversity across taxa. Nature 466:1098-1101

Tolstoy M, Vernon FL, Orcutt JA, Wyatt FK (2002) Breathing of the seafloor: tidal correlations of seismicity at Axial volcano. Geology 30:503-506

Torres ME, McManus J, Hammond DE, de Angelis MA, Heeschen KU, Colbert SL, Tryon MD, Brown KM, Suess E (2002) Fluid and chemical fluxes in and out of sediments hosting methane hydrate deposits on Hydrate Ridge, OR, I: hydrological provinces. Earth Planet Sci Lett 201:525-540

Tyson GW, Chapman J, Hugenholtz P, Allen EE, Ram RJ, Richardson PM, Solovyev VV, Rubin EM, Rokhsar DS, Banfield JF (2004) Community structure and metabolism through reconstruction of microbial genomes from the environment. Nature 428:37-43

Tyson GW, Lo I, Baker BJ, Allen EE, Hugenholtz P, Banfield JF (2005) Genome-Directed isolation of the key nitrogen fixer *Leptospirillum ferrodiazotrophum* sp. nov. from an acidophilic microbial community. Appl Environ Microbiol 71:6319-6324

Urmann K, Gonzalez-Gil G, Schroth MH, Hofer M, Zeyer J (2005) New field method: gas push-pull test for the *in situ* quantification of microbial activities in the vadose zone. Environ Sci Technol 39:304-310

Valentine DL (2011) Emerging topics in marine methane biogeochemistry. Annu Rev Mar Sci 3:147-171

Vorobyova E, Soina V, Gorlenko M, Minkovskaya N, Zalinova N, Mamukelashvili A, Gilichinsky D, Rivkina E, Vishnivetskaya T (1997) The deep cold biosphere: facts and hypotheses. FEMS Microbiol Rev 20:277-290

Vreeland RH, Rosenzweig WD, Powers DW (2000) Isolation of a 250 million-year-old halotolerant bacterium from a primary salt crystal. Nature 407:897-900

Walter Anthony KM, Anthony P, Grosse G, Chanton J (2012) Geologic methane seeps along boundaries of Arctic permafrost thaw and melting glaciers. Nature Geosci 5:419-426

Warren LA, Maurice PA, Parmar N, Ferris FG (2001) Microbially mediated calcium carbonate precipitation: implications for interpreting calcite precipitation and for solid-phase capture of inorganic contaminants. Geomicrobiol J 18:93-115

Watkins NW, Freeman MP (2008) GEOSCIENCE: natural complexity. Science 320:323-324

Westbrook GA, Thatcher KE, Rohling EJ, Piotrowski AM, Palike H, Osborne AH, Nisbet EG, Minshull TA, Lanoiselle M, Huhnerbach V, Green D, Fisher RE, Crocker AJ, Chabert A, Bolton C, Beszczynska-Moller A, Berndt C, Aquilina A (2009) Escape of methane gas from the seabed along the West Spitsbergen continental margin. Geophys Res Lett 36, doi: 10.1029/2009GL039191

Whelan JK, Oremland R, Tarafa M, Smith R, Howarth R, Lee C (1986) Evidence for sulfate-reducing and methane-producing microorganisms in sediments from Sites 618, 619, 622. *In*: Scientific Results, Ocean Drilling Program, Leg 119. Bouma AH, Coleman JM, Meyer AW (eds) Ocean Drilling Program, College Station, TX, doi: 10.2973/dsdp.proc.96.147.1986

Whitman WB, Coleman DC, Wiebe WJ (1998) Prokaryotes: the unseen majority. Proc Natl Acad Sci USA 95:6578-6583

Wiggins B, Alexander, M. (1985) Minimum bacterial density for bacteriophage replication: implications for significance of bacteriophages in natural ecosystems. Appl Environ Microbiol 49:19-23

Wood WT, Gettrust JF, Chapman NR, Spence GD, Hyndman RD (2002) Decreased stability of methane hydrates in marine sediments owing to phase-boundary roughness. Nature 420:656-660

Wright DJ, Wang S (2011) The emergence of spatial cyberinfrastructure. Proc Natl Acad Sci USA 108:5488-5491

Yates M, Thompson S, Jury W (1997) Sorption of viruses during flow through saturated sand columns. Environ Sci Technol 31:548-555

Zhang CL, Lehman RM, Pfiffner SM, Scarborough SP, Palumbo AV, Phelps TJ, Beauchamp JJ, Colwell FS (1997) Spatial and temporal variations of microbial properties at different scales in shallow subsurface sediments. Appl Biochem Biotechnol 63-5:797-808

Zhuang K, Izallalen M, Mouser P, Richter H, Risso C, Mahadevan R, Lovley DR (2011) Genome-scale dynamic modeling of the competition between *Rhodoferax* and *Geobacter* in anoxic subsurface environments. ISME J 5:305-316

Zobell CE, Anderson DQ (1936) Vertical distribution of bacteria in marine sediments. Bull Am Assoc Petrol Geol 20:258-269

Reviews in Mineralogy & Geochemistry
Vol. 75 pp. 575-606, 2013

18

Serpentinization, Carbon, and Deep Life

Matthew O. Schrenk, William J. Brazelton

Department of Biology
East Carolina University
Greenville, North Carolina 27858-2502, U.S.A.

schrenkm@ecu.edu brazeltonw@ecu.edu

Susan Q. Lang

Department of Earth Sciences
ETH Zürich
8092 Zürich, Switzerland

susan.lang@erdw.ethz.ch

INTRODUCTION

The aqueous alteration of ultramafic rocks through serpentinization liberates mantle carbon and reducing power. Serpentinization occurs in numerous settings on present day Earth, including subduction zones, mid-ocean ridges, and ophiolites and has extended far into Earth's history, potentially contributing to the origins and early evolution of life. Serpentinization can provide the energy and raw materials to support chemosynthetic microbial communities that may penetrate deep into Earth's subsurface. Microorganisms may also influence the composition and quantity of carbon-bearing compounds in the deep subsurface. However, conditions created by serpentinization challenge the known limits of microbial physiology in terms of extreme pH, access to electron acceptors, and availability of nutrients. Furthermore, the downward transport of surface carbon and subsequent mixing with calcium-rich fluids at high pH contributes to the precipitation and immobilization of carbonate minerals. The following chapter will explore the physiological challenges presented by the serpentinite environment, data from studies of serpentinite-hosted microbial ecosystems, and areas in need of further investigation.

THE PROCESS OF SERPENTINIZATION

Physical and chemical consequences of serpentinization

Serpentinization is an alteration process of low-silica ultramafic rocks, characteristic of the lower oceanic crust and upper mantle. These rocks are rich in the minerals olivine and pyroxene. Water-rock reactions result in the oxidation of ferrous iron from olivine and pyroxene, resulting in the precipitation of ferric iron in magnetite (Fe_3O_4) and other minerals, and in the release of diatomic hydrogen (H_2). At low temperatures (< ~150 °C) the reaction results in extremely high pH, commonly above 10. The combination of H_2 and CO_2 or CO under highly reducing conditions leads to formation of methane and other hydrocarbons through Fischer-Tropsch Type (FTT) synthesis (McCollom and Seewald 2001; Charlou et al. 2002; Proskurowski et al. 2008; McCollom 2013). Serpentinization also results in volume changes in the altered materials, making serpentinites less dense than their parent materials and facilitating uplift due to volume expansion. These reactions are highly exothermic, and may contribute to hydrothermal fluid circulation through the fractured materials (Lowell and Rona 2002; Allen and Seyfried 2004).

1529-6466/13/0075-0018$00.00 DOI: 10.2138/rmg.2013.75.18

Furthermore, the Ca^{2+} ions liberated from the water-rock reactions react with carbonate ions at high pH to induce calcium carbonate precipitation (Barnes et al. 1978; Neal and Stanger 1983; Fritz et al. 1992; Palandri and Reed 2004; Kelley et al. 2005). Carbonates can exist as fracture infillings in the host rock or can manifest as travertines or chimneys upon exiting the subsurface (Fig. 1).

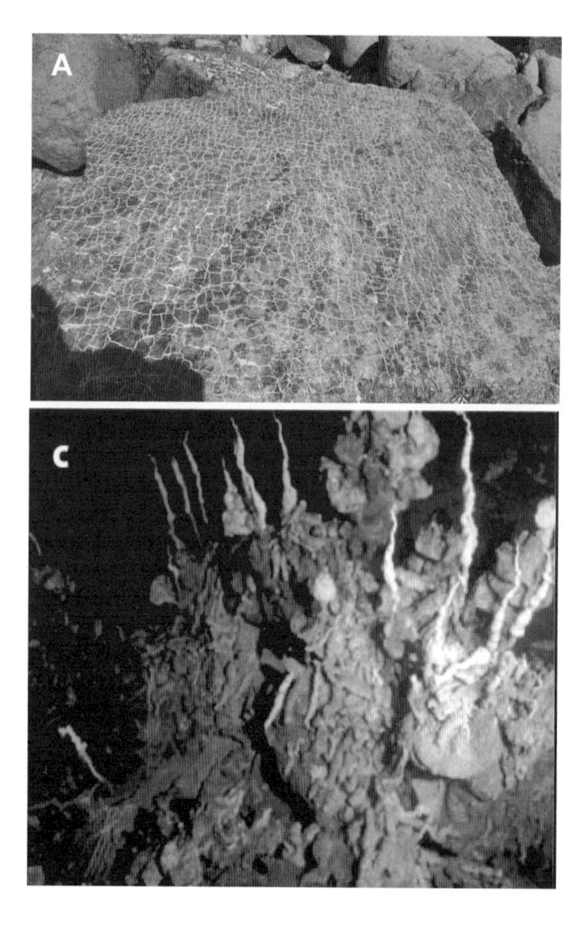

Figure 1. Examples of serpentinization and calcium carbonate mineralization. Fractured serpentinite rock (A) from the Tablelands Ophiolite, Newfoundland, Canada, showing the complex mosaic of fractures and infilling associated with pervasive serpentinization. Carbonate chimneys formed in the marine serpentinite environments at the (B) Lost City Field and (C) Mariana Forearc. (B) reprinted with permission of AAAS from Kelley et al. (2005). (C) reprinted with permission of Annual Reviews from Fryer (2012).

Types of serpentinizing habitats

Serpentinization occurs in a diverse range of locations on Earth, particularly where tectonic processes lead to the uplift and exposure of mantle materials. Additionally, serpentinization may occur in the deepest habitable portions of Earth's subsurface, ultimately constrained by the depths of fluid circulation.

The characterization of several of the systems described in more detail below is in the early stages (Tables 1 and 2). In some cases it is not yet possible to differentiate between locations where serpentinization is an active process from those locations where fluids pass through reacted serpentinites, leading to the dissolution of the rock.

Identifying serpentinization habitats. The large carbonate deposits that commonly form when alkaline waters exit from peridotites are often the most readily recognized feature of active serpentinization. The first descriptions of springs affected by serpentinization are from continental locations, some of which are associated with large travertine deposits (Barnes et al. 1978). Detection of active serpentinization in oceanic settings is more challenging. In some instances, such as the Rainbow vent field, a hydrothermal system with primarily basalt-hosted characteristics (high-temperature metal-rich fluids) may also have characteristics indicative of reaction with ultramafic rocks, i.e., unexpectedly high hydrogen and methane concentrations (Table 1; Charlou et al. 2002). But the discovery of the low-temperature, alkaline Lost City system came about much in the same way as many of those found in land-based systems: through the serendipitous observation of large carbonate deposits (Kelley et al. 2001). Detecting the presence of a Lost City-type system is challenging, as fluids do not reach the temperatures of basalt-hosted systems and low metal content in the fluids result in only minor particle precipitation. Therefore, the typical methods to detect active venting of marine basalt-hosted hydrothermal systems, such as water column temperature and optical backscatter anomalies are of limited usefulness.

Serpentinizing ophiolites. Remnants of seafloor serpentinization are present along continental margins in ophiolite sequences. Continental springs seep alkaline, volatile-rich fluids from ultramafic rocks. The ophiolites themselves originate from a number of different tectonic events including subduction, extension, and plume-related events (Dilek and Furnes 2011). The distinctive plant flora associated with serpentinites has been the subject of study for several decades, although the microbiology of these systems is less well understood. The serpentinization of ophiolites can continue for hundreds of millions of years after emplacement. Seep fluid compositions are generally lacking in dissolved ions as the original source water is meteoric, although in some cases dissolved ions such as sulfate may be present from the contributions of paleo-seawater, or the dissolution of mineral salts (Table 2). In some cases accretion of marine sediments and organic matter can contribute to the carbon and nutrient budgets of these environments (Hosgormez et al. 2008).

The Coast Range and Josephine ophiolites in the western United States were amongst the first serpentinizing ophiolites to be intensively studied (Barnes et al. 1967; Barnes and O'Neil 1971). Multiple Ca-OH type springs and seeps emanate from rocks that are Jurassic-Cretaceous in age at temperatures less than ~50 °C.

The Samail ophiolite in the Sultanate of Oman is one of the most extensive known continental ophiolites, comprised of oceanic crust and upper mantle (ultramafic) lithologies that are >350 km long, ~40 km wide and have an average thickness of ~5 km (Neal and Stanger 1983; Nicolas et al. 2000). Stable isotope and ^{14}C dating of the carbonate veins indicate the majority of veins were formed at low (<60 °C) temperatures in the recent (<50,000 years) past, during interaction with meteoric groundwater (Barnes et al. 1978; Nicolas et al. 2000; Kelemen and Matter 2008; Kelemen et al. 2011). Alkaline water and gases emanating from the peridotites have high concentrations of methane, hydrogen, and short-chain hydrocarbons, and

Table 1. Geochemical characteristics of marine serpentinite springs

			Central Indian Ridge	Mariana Forearc	
		Seawater[a]	Kairei 25°19′S[b]	Conical Seamount 19° 32.5′ N ODP Site 780[c]	South Chamorro Seamount 13° 47′ N ODP Site 1200[d]
Sample Type	**Units**		vent fluids	pore fluids	pore fluids
Water Depth	**meters**		~2,400	3,083	2,960
Sediment Depth	**mbsf**			130	71
Temperature	**°C**	2			
pH		7.8		12.5	12.5
$H_{2(aq)}$	**mM**	0.0004	8		
δD, H_2	**(‰ VSMOW)**				
$H_2S_{(aq)}$	**mM**	0			
$CH_{4(aq)}$	**mM**	0.0003	0.5	2	2
$\delta^{13}C$, CH_4	**(‰ VPDB)**				
δD, CH_4	**(‰ VSMOW)**				
ΣCO_2	**mM**	2.3			
formate	**μM**			0 to ~2250	
acetate	**μM**			0 to ~210	
NO_3^-	**mM**				
SO_4^{2-}	**mM**	28.2		46	28
$O_{2(aq)}$	**mM**	0.1			
Cl	**mM**	546		260	510
$SiO_{2(aq)}$	**mM**	<0.2			

[a]Charlou et al. 2002
[b]Keir 2010
[c]Haggerty and Fisher 1992; Mottl et al. 2003
[d]Mottl et al. 2003
[e]Kelley et al. 2001, 2005; Proskurowski et al. 2006; Lang et al. 2010
[f]Charlou et al. 1998, 2002; Douville et al. 2002
[g]Charlou et al. 1998, 2002, 2010; Douville et al. 2002; Proskurowski et al. 2006
[h]Charlou et al. 2010
[i]Melchert et al. 2008; Schmidt et al. 2011

mean residence times on the order of years to decades (Fritz et al. 1992). In recent years, this ophiolite has been studied as a potential site for the sequestration of atmospheric carbon dioxide through geo-engineering strategies aimed at mineral carbonation (Kelemen and Matter 2008).

The Zambales Ophiolite in the Philippines is one of the first locations where stable isotope geochemistry was applied to determine the source of volatile gases (Abrajano et al. 1988, 1990). The authors of these studies determined that the hydration of peridotites through serpentinization contributes to the gas compositional signatures at the site (Abrajano et al. 1990).

The Tekirova Ophiolites of Turkey have emitted significant quantities of H_2 and CH_4 gas emissions for millennia, contributing an estimated 150 to 190 tons of CH_4 per year (Etiope et al. 2011). Gases at the Chimaera seep at this site have been aflame for centuries and can be attributable to the "eternal flame" mentioned in ancient Greek literature. The methane in the fluids is generated largely through serpentinization and to a lesser extent the thermogenic degradation of complex organic matter (Hosgormez et al. 2008).

The Bay of Islands Ophiolite in Newfoundland, Canada was emplaced nearly 500 Ma ago. The site contains pH >12 springs enriched in H_2 (~1 mg L^{-1}) and CH_4 (~0.3 mg L^{-1}) that

Mid-Atlantic Ridge						
Lost City 30° N[e]	**Rainbow 36°14′ N[f]**	**Logatchev 1 14°45′ N[g]**	**Logatchev 2 14°43′ N[h]**	**Ashadze 1 12°58′ N[h]**	**Ashadze 2 12°59′ N[h]**	**Nibelungen 8°18′ S[i]**
vent fluids 700 to 800	vent fluids 2300	vent fluids 3000	vent fluids 2700	vent fluids 4088	vent fluids 3263	vent fluids 3,000
40 to 91	365	347 to 352	320	355	296	>192 to 372
9 to ~11	2.8	3.3 to 3.9	4.2	3.1	4.1	2.9
<1 to 15	16	12		8 to 19	26	11.4
−689 to −605		−372		−343 to −333	−270	
	1.2	0.5 to 0.8	1.9	1		0.035 to 1.1
1 to 2	2.5	2.1	1.2	0.5 to 1.2	0.8	1.4
−13.6 to −9.5	−15.8	−13.6 to −6	−6.1	−14.1 to −12.3	−8.7	
−141 to −99		−109				
0.0001 to 0.0026	5 to 16	10.1	6.2	3.7		
36 to 158						
1 to 35						
1 to 4	~0	~0	0	0	0	
	0	0				
~550	750	515	127	614	326	
	6.9	8.2	11.5	6.6	7.3	12.7 to 13.7

emanate from largely barren, heavily serpentinized terrain (Szponar et al. 2012). The extremely long time period of serpentinization is speculated to be punctuated and linked to geophysical changes in the environment, such as glaciation/deglaciation events.

Ultramafic rocks of the Gruppo di Voltri (Liguria, Italy) host highly reducing, pH 10.5-12 springs forming carbonate deposits (Cipolli et al. 2004). At Cabeço de Vide (Portugal), deep wells access aquifers in the serpentinite subsurface (Marques et al. 2008)

Oceanic core complexes and mid-ocean ridges. Ocean core complexes are sections of deep oceanic lithosphere exhumed to the seafloor by detachment faults formed along the flanks of slow to intermediate spreading ridges. The extensive faulting and fracturing associated with this uplift as well as the latent heat of the rock promotes fluid circulation that can result in serpentinization. Ultramafic rocks may constitute up to 20% of slow spreading mid-ocean ridges; therefore the process of serpentinization represents a significant, yet understudied phenomenon (Früh-Green et al. 2004).

Since its discovery in 2000, the Lost City Hydrothermal Field (LCHF), near the Mid-Atlantic Ridge, has been one of the most intensively studied sites of active serpentinization. At

Table 2. Geochemical characteristics of continental serpentinite springs

Ophiolite		Semail	Coast Range Cazadero, CA	Coast Range Del Puerto	Bay of Islands Tablelands
Country	**Units**	**Oman[a]**	**USA[b]**	**USA[c]**	**Canada[d]**
Temperature	°C	25 to 35.7	20	17.8 to 24.2	
pH		11.1 to 12.1	11.54	8.6	11.8 to 12.3
Eh	mV	–630 to 165			–609 to 121
Fluid Chemistry					
ΣCO_2	mg/L as HCO_3^-	0	0	466 to 639	1.1 to 27.25
Na^+	mg/L	110 to 603	19	5.4 to 9.6	
K^+	mg/L	3.6 to 27.8	1.1	0.31 to 0.6	
Ca^{2+}	mg/L	55.2 to 120	40	3.5 to 8.1	
Mg^{2+}	mg/L	<0.1 to 0.2	0.3	110 to 150	0.06 to 7.57
Cl^-	mg/L	140 to 858	63	4.8 to 9.5	166 to 479
HS^-	mg/L				
NO_3^-	mg/L	0.18 to 31.2	0.1		
SO_4^{2-}	mg/L	0.19 to 34.1	0.4	10 to 16	
SiO_2	mg/L	<0.1 to 0.2	0.4	5.6 to 13	
Gas Chemistry		***(gas bubbles)***			***(dissolved gas)***
H_2	vol%	0 to 99			0.03 to 0.6 mM
$\delta D, H_2$	(‰ VSMOW)	–733 to –697			
CH_4	vol%	0 to 4.3			0 to 23.7 μM
$\delta^{13}C, CH_4$	(‰ VPDB)	–14.7 to –12.0			–28.5 to –15.9
$\delta D, CH_4$	(‰ VSMOW)	–251 to –210			
Ethane	vol%	0.005 to 0.0078			0 to 1.3 μM
Propane	vol%	0.001 to 0.0018			0 to 1.1 μM
iso-Butane	vol%	0.001 to 0.002			0 to 0.3 μM
n-Butane	vol%	0.0001 to 0.0012			0 to 0.3 μM
Total hydrocarbons	vol%	0.0008 to 0.0114			0 to 3.0 μM
CO	%	0 to 0.00001			

[a]Barnes et al. 1978; Neal and Stanger 1983; Bath et al. 1987; Fritz et al. 1992
[b]Barnes et al. 1978
[c]Blank et al. 2009
[d]Szponar et al. 2012
[e]Barnes et al. 1978
[f]Marques et al. 2008
[g]Cipolli et al. 2004
[h]Abrajando et al. 1988, 1990
[i]Etiopeet al. 2011; Hosgormez et al. 2008

Lost City, the Atlantis Massif has been uplifted ~5,000 m relative to the surrounding terrain. Tall calcium carbonate chimneys at the LCHF rise up to 60 m from the seafloor and serve as conduits for hot, highly reducing, high-pH hydrothermal fluids (Table 1; Fig. 1b). Venting fluids at the LCHF range from 40 to 91 °C, and are rich in hydrogen and methane (Kelley et al. 2001, 2005). Hydrothermal activity has been sustained at the LCHF for at least 30,000 years and potentially >100,000 years, as evidenced by ^{14}C and U/Th dating of the extensive calcium carbonate deposits that abound in the vent field (Früh-Green et al. 2003; Ludwig et al. 2011).

Attempts to access the serpentinite subsurface of the Atlantis Massif through the Integrated Ocean Drilling Program at Hole 1309D during Expeditions 304 and 305 sampled igneous rocks for microbiological analysis from up to 1,391 m below the seafloor (Mason et al. 2010). The

New Caledonia[e]	Yugoslavia[e]	Cabeço de Vide Portugal[f]	Liguria Gruppo di Voltri Italy[g]	Zambales Los Fuegos Eternos Phillippines[h]	Tekirova Chimaera Turkey[i]
23 to 34	29	17.1 to 19.8	10.5 to 23		
9.2 to 10.8	11.75	10.7 to 11.1	9.95 to 11.86		
		−177 to −39	−525 to −388		
20 to 56.2	0	1.14 to 2.82	0.53 to 160	<0.1 to 0.03 (vol%)	0.01 to 0.18 (vol%)
7.7 to 26.1	35	37.0 to 55.9	3.9 to 84		
1.4 to 3.3	1.5	4.15 to 5.22	0.51 to 10.8		
9 to 23	29	5.1 to 22.5	0.6 to 61.9		
2.3 to 5.9	7	0.16 to 0.3	0.001 to 15.4		
8.5 to 30	20		8.96 to 97.4		
			0.06 to 1.81		
	0.05	5.57 to 7.60	<0.01 to 1.51		
0.75 to 5.8		0.96 to 12.60	0.1 to 25.3		
0.4 to 3.7	1.9	5.5 to 6.1	0.09 to 22.3		
			(dissolved gas)	***(gas bubbles)***	***(gas bubbles)***
				8.4 to 45.6	7.5 to 11.3
				−599 to −581	
			0.6 to 867 μM	13 to 55.3	65.24 to 93.22
				−7.5 to −6.1	−12.5 to −7.9
				−137 to −118	−129 to −97
				0.04 to 0.15	0.17 to 0.43
					0.09 to 0.13
					0.027 to 0.031
					0.05 to 0.07
					~0.57

study found a highly heterogeneous petrology, consisting largely of gabbros, but with regions of serpentinized peridotites.

In addition to the LCHF, slow- and ultraslow-spreading mid-ocean ridges support a number of well-studied high-temperature (>300 °C) hydrothermal systems influenced by serpentinization. In these systems, the flow pathway for magmatically driven hydrothermal circulation passes through ultramafic rocks, facilitating water-rock reaction, imparting a strong volatile signature upon the hydrothermal fluids. Compared to the LCHF, fluids from these systems have higher temperatures, higher metal concentrations, and are more acidic, but, in contrast to purely basalt-hosted systems, have highly elevated hydrogen and methane concentrations (Table 1). The Rainbow (36° 14′ N) and Logatchev (14° 45′ N) vent fields along the Mid-Atlantic Ridge are the best studied of these systems. Fluids from both fields exit large sulfide deposits at temperatures >300 °C and contain elevated concentrations of hydrogen, methane, and C_2-C_5 hydrocarbons (Charlou et al. 2002). Over the past decade, several more high-temperature vent fields hosted in peridotites have been discovered along the Mid-Atlantic Ridge, including a second field close to Logatchev, named Logatchev-2 (14° 43′ N) and, further south, the Ashadze I and II (12° 58′ N) and Nibelungen vent fields (8° 18′ S) (Melchert et al. 2008; Charlou et al. 2010; Schmidt et al. 2011).

Similar settings occur on the Central Indian Ridge and the Southwest Indian Ridge in the Indian Ocean. Kairei Field at the Central Indian Ridge contains features of both ultramafic- and basalt-hosted hydrothermal fields, including high H_2 and Si concentrations and a low CH_4/H_2 ratio (Takai et al. 2004). These characteristics can be explained by the serpentinization of troctolites (olivine-rich gabbros) and subsequent reaction with basalt.

Recent discoveries may expand the known breadth of hydrothermal systems that are influenced by serpentinization. Studies of the ultraslow spreading Gakkel Ridge in the Arctic Ocean indicate extensive hydrothermal activity hosted in ultramafic rocks at high latitudes (Edmonds et al. 2003). Additionally, the deepest hydrothermal vents known to date, up to 5,000 m below the sea surface, have been discovered near the Mid-Cayman Rise in the Caribbean Sea (German et al. 2010). Vent plumes along the Mid-Cayman Rise are consistent with the presence of both high-temperature "black smoker type" venting as well as moderate-temperature, serpentinization-influenced vent fields. Ongoing investigations are serving to better define the characteristics of these systems and their associated biological communities.

A potential shallow sea analog to the LCHF may exist in the Bay of Prony, New Caledonia, where brucite-rich, carbonate towers vent 22-43 °C, freshwater, pH 11 fluids (Cox et al. 1982; Launay and Fontes 1985). This site, which was visited by a French research cruise in 2011, may represent an interesting transition between marine and terrestrial serpentinite hydrothermal settings.

Subduction zones and mud volcanoes. A second important location of serpentinization in the marine environment is associated with the hydration of oceanic crust at subduction zones. At these sites, fluids from the subducting plate can hydrate the mantle of the overriding plate and cause serpentinization. Density changes associated with serpentinization of the mantle lead to diapirs of the hydrated materials that in some cases reach the seafloor, forming "mud volcanoes". Fluid compositions change with distance from the trench, indicating variability in the conditions of serpentinization (Mottl et al. 2003). The best studied of these systems is the Mariana Forearc, where the Pacific Plate is subducted beneath the Philippine Plate (Fryer 2012). Mud volcanoes vent high pH fluids (up to 12.5) rich in hydrogen, methane, formate, and acetate (Haggerty and Fisher 1992; Mottl et al. 2003). High ^{13}C contents of methane and relatively low C_1/C_2 ratios in pore fluids point to a mantle source for these carbon species. A second serpentinization-influenced seep site has been identified in the Southern Mariana Forearc (Ohara et al. 2011), and similar mud volcanoes exist in regions associated with subduction such as in the Caribbean Sea.

Precambrian shields. Ultramafic rocks found in Precambrian continental shield environments in the Fennoscandian Shield, South Africa, and Canada are important additional sites of serpentinization, potentially contributing to global hydrogen and hydrocarbon budgets (Sherwood Lollar et al. 1993). As fluids circulate through these systems, serpentinization of ultramafic rocks leads to hydrogen generation and potentially to abiogenic synthesis of organic matter (McCollom 2013; Sephton and Hazen 2013).

Serpentinization on other planets. Evidence of past serpentinization has been detected on the surface of Mars and has been important for helping to model planetary habitability and define the source of atmospheric gases (Ehlmann et al. 2010). Serpentinization may even drive hydrothermal circulation in small planetary bodies of the outer solar system and contribute to the energy budget of Europa's subsurface ocean (Vance et al. 2007).

BIOLOGICAL CONSEQUENCES OF SERPENTINIZATION

Although the picture of the global serpentinite biosphere is still emerging, themes can be gleaned from both geochemical and microbiological studies of serpentinite-hosted microbial ecosystems. The impact of serpentinization upon fluid chemistry imposes a unique set of conditions upon the organisms and biological activities operative in such ecosystems (Tables 1 and 2, Fig. 2). Serpentinizing environments are typically rich in electron donors such as H_2 and CH_4, and in some cases, short-chain hydrocarbons and formate. The availability of terminal electron acceptors is frequently limited, however, particularly in continental systems. Moreover, the microbial communities in serpentinizing ecosystems face physiological challenges in terms of high pH and low concentrations of dissolved inorganic carbon.

Common biological observations in the most extreme alkaline fluids at many of the sites include low cell abundances (typically less than 10^5 cells/ml, and as low as 10^2 cells/mL) and low taxonomic diversity (Schrenk et al. 2004; Tiago et al. 2004; Brazelton et al. 2010). Greater biomass (but not necessarily greater diversity) can be seen in surface-attached habitats, where serpentinization fluids mix with ambient water; the densely-populated biofilms of Lost City carbonate chimneys are a dramatic example. In many of these systems the cycling of hydrogen, methane, sulfur, and fermentative processes appear to be important metabolic activities.

Metabolic strategies in serpentinite-hosted ecosystems

An important consideration for the ability of life to persist in extreme environments is the balance between the energy necessary for biosynthesis and activity, and that required for repair and maintenance (Hoehler 2007). Developing a framework for the productivity of serpentinite-hosted subsurface environments requires a greater understanding of microbial physiology in such ecosystems and the energetic costs associated with coping strategies. Therefore, an important step toward developing quantitative models of habitability in these ecosystems is to inventory the taxonomies and metabolic capabilities of the native microorganisms. A review of our as yet limited knowledge on this topic follows.

Hydrogen cycling. The most obvious electron donor in serpentinizing habitats is the copious quantity of hydrogen produced through serpentinization reactions (Tables 1 and 2). The oxidation of this hydrogen under aerobic or anaerobic conditions can provide abundant metabolic energy to local communities (McCollom 2007, 2013). The ubiquity of this energy source in serpentinizing environments is reflected in the microbiology of these systems through genomic and cultivation-based studies (Tables 3 and 4; Takai et al. 2004; Perner et al. 2010; Brazelton et al. 2012).

Hydrogen stimulates the metabolic activity of Lost City chimney biofilms, as evidenced by microcosm experiments (Brazelton et al. 2011). Work by Perner et al. (2010) at the Logatchev vent field (14° 4′ N) showed molecular and geochemical evidence for carbon fixation linked to hydrogen oxidation. The diversity of hydrogenase genes (associated with hydrogen oxidation) increased with increasing hydrogen concentrations (Perner et al. 2007, 2010).

Spring fluids from the Tablelands Complex within the Bay of Islands Ophiolite (Newfoundland, Canada) have been sampled and used in comparative community (meta-) genomic analyses. Metagenomic data contained a high proportion of sequences related to Betaproteobacteria within the order Burkholderiales who appear to oxidize hydrogen as a source of energy and assimilate carbon using the Calvin-Benson-Bassham (CBB) pathway (Brazelton et al. 2012). The metagenomic data also identified Clostridiales-like organisms that

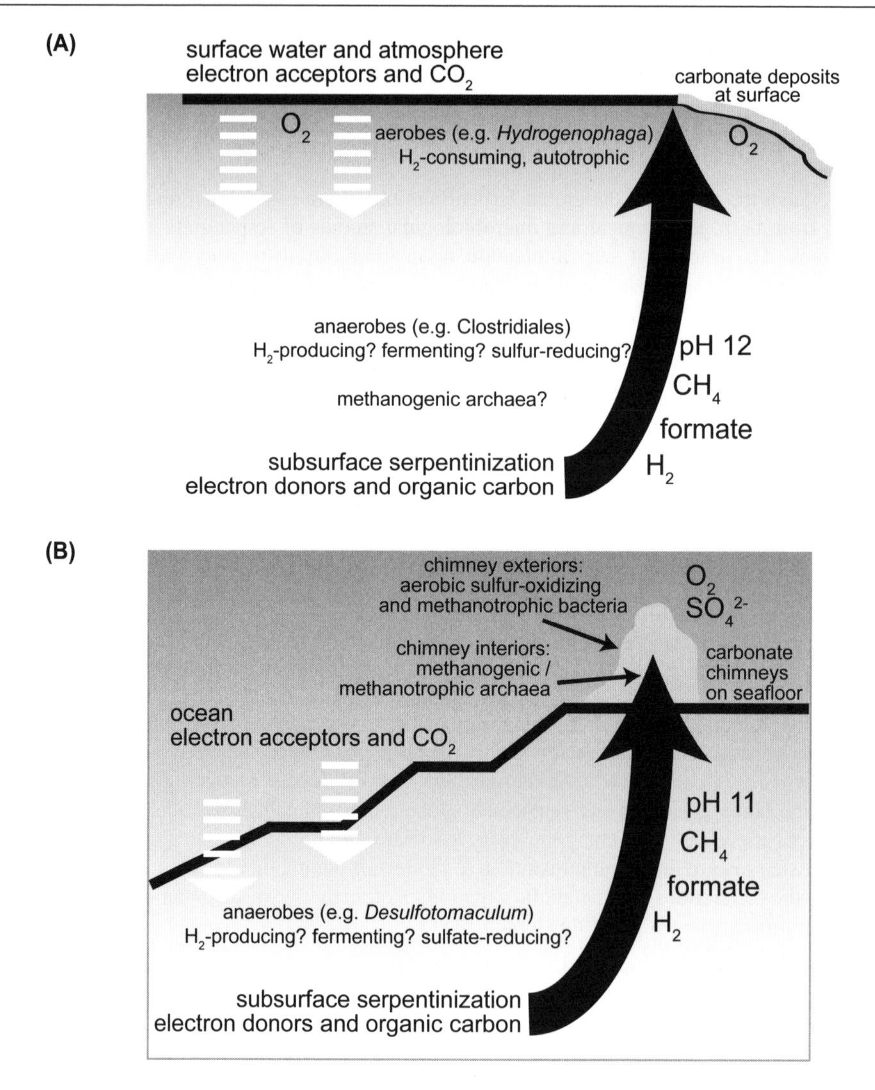

Figure 2. Schematic of putative biogeochemical processes in continental (A) and marine (B) serpentinization settings. Question marks indicate sources of uncertainty due to lack of data and should be considered speculative. Each schematic represents an idealized setting and reflects reality to varying degrees for each actual natural environment. This diagram is most representative of low-temperature, actively serpentinizing systems such as the Lost City hydrothermal field. Many serpentinite-hosted marine systems (e.g. Rainbow and Logatchev) feature sulfide chimneys rather than carbonate chimney deposits and have lower pH fluids compared to this diagram.

possess genes for hydrogen production typically associated with fermentation and are presumed to inhabit anoxic regions of the subsurface underlying the springs. Therefore, these serpentinite springs harbor the genetic potential for both microbial oxidation and production of hydrogen.

Recent studies of a subsurface Outokumpu borehole in Finland have also reported phylotypes within the Betaproteobacteria and Clostridiales in waters up to 1,400 m below the land surface (Itävaara et al. 2011). Elevated cell abundances are associated with a layer containing serpentinized peridotites, although the pH of the bulk borehole water is ~9.

Table 3. Archaea characteristic of serpentinite habitats

	Habitat	Types of Analysis	Reference
Crenarchaeota			
Desulfurococcales	Kairei Field	16S rDNA	Takai et al. 2004
	Rainbow	16S rDNA	Roussel et al. 2011
Euryarchaeota			
Archaeoglobales	Kairei Field	16S rDNA	Takai et al. 2004
	Rainbow	*dsrAB*	Nercessian et al. 2005
		16S rDNA	Roussel et al. 2011
Thermococcales	Kairei Field	cultivation, 16S rDNA	Takai et al. 2004
	Lost City	16S rDNA	Brazelton et al. 2006
	Rainbow	16S rDNA	Roussel et al. 2011
Methanococcales	Kairei Field	cultivation, 16S rDNA	Takai et al. 2004
	Logatchev	16S rDNA	Perner et al. 2007, 2010
	Rainbow	*mcrA* gene	Nercessian et al. 2005
		16S rDNA, *mcrA*	Roussel et al. 2011
Methanobacteriales	Coast Range	16S rDNA, *mcrA*	Blank et al. 2009
	South Chamorro	16S rDNA	Curtis and Moyer 2005
Methanopyrales	Kairei Field	16S rDNA	Takai et al. 2004
	Rainbow	*mcrA* gene	Nercessian et al. 2005
		16S rDNA, *mcrA*	Roussel et al. 2011
Methanosarcinales (non-ANME)	Kairei Field	16S rDNA	Takai et al. 2004
	Lost City	16S rDNA	Schrenk et al. 2004
	Rainbow	*mcrA* gene	Nercessian et al. 2005
	South Chamorro	16S rDNA	Curtis and Moyer 2005
ANME-2	Logatchev	16S rDNA	Perner et al. 2007
	Rainbow	16S rDNA	Roussel et al. 2011
ANME-1	Lost City	16S rDNA	Brazelton et al. 2006, 2010

The detection of both Betaproteobacteria (typically order Burkholderiales) and Clostridia (typically order Clostridiales) appears to be a common theme in serpentinite-hosted ecosystems. As mentioned above, both taxa have been identified in high-pH serpentinite springs in Canada and Finland, and they are also present (but not dominant) in Lost City carbonate chimneys (Brazelton et al. 2010). These taxa also dominate the microbial community structure of a 3-km borehole in South Africa (Moser et al. 2005). Fluids from this borehole are moderately basic (pH ~9) and enriched in H_2 (up to 3.7 mM), but serpentinization does not appear to be active at this site. Therefore, further work should investigate whether the presence of particular phylotypes of Betaproteobacteria and Clostridia are diagnostic of serpentinization-driven ecosystems and whether these phylotypes are ubiquitous in other high pH, H_2-rich environments irrespective of serpentinizing activity.

As explained above, metagenomic evidence indicates that the Betaproteobacteria harbor genes for hydrogen oxidation (Brazelton et al. 2012), a partial explanation for their ubiquity in these environments. It is unknown whether they also harbor adaptations for survival at high pH. While the Betaproteobacteria seem most likely to inhabit transition zones where hydrogen and oxygen are both available, the Clostridia seem to be more abundant in the deepest, most

Table 4. Bacteria characteristic of serpentinite habitats

	Habitat	Types of Analysis	Reference
Aquificae			
Aquificales	Kairei Field	cultivation, 16S rDNA	Takai et al. 2004
	Logatchev	16S rDNA	Perner et al. 2007
Bacteroidetes			
Chimaereicella	Cabeço de Vide	cultivation	Tiago et al. 2006
	Coast Range	16S rDNA, *mcrA*	Blank et al. 2009
Flavobacteria	Lost City	16S rDNA	Brazelton et al. 2006, 2010
	Tablelands	metagenomics	Brazelton et al. 2012
Sphingobacteria	Lost City	16S rDNA	Brazelton et al. 2010
	Outokumpo	16S rDNA	Itävaara et al. 2011
Betaproteobacteria			
Hydrogenophaga	Tablelands	metagenomics	Brazelton et al. 2012
	Outokumpo	16S rDNA	Itävaara et al. 2011
Gammaproteobacteria			
Methylococcales	Lost City	16S rDNA	Brazelton et al. 2006, 2010
	Rainbow	*pmoA*	Nercessian et al. 2005
	Rainbow	*pmoA*	Roussel et al. 2011
Thiomicrospira	Lost City	16S rDNA, metagenomics	Brazelton et al. 2006, 2010ab
Marinobacter alkaliphila	South Chamorro	cultivation	Takai et al. 2005
Chromatiales	Rainbow	*pmoA*	Roussel et al. 2011
	Tablelands	metagenomics	Brazelton et al. 2012
Deltaproteobacteria			
Desulfovibrionales	Lost City (rare)	16S rDNA	Brazelton et al. 2010
	Outokumpo	*dsrB*	Itävaara et al. 2011
Desulfobacterales	Logatchev	16S rDNA	Perner et al. 2007
	Lost City (rare)	16S rDNA	Brazelton et al. 2010
	Outokumpo	*dsrB*	Itävaara et al. 2011
	Rainbow	*dsrAB*	Nercessian et al. 2005
	South Chamorro	lipid biomarker	Mottl et al. 2003
Other Deltaproteobacteria	Semail	cultivation	Bath et al. 1987
Epsilonproteobacteria			
Campylobacterales	Kairei Field	16S rDNA	Takai et al. 2004
	Logatchev	16S rDNA	Perner et al. 2007, 2010
	Lost City	16S rDNA	Brazelton et al. 2006, 2010
Sulfurovum-like	Kairei Field	16S rDNA	Takai et al. 2004
	Logatchev	16S rDNA	Perner et al. 2007, 2010
	Lost City	16S rDNA	Brazelton et al. 2006, 2010
Other Epsilonproteobacteria	Kairei Field	16S rDNA	Takai et al. 2004
	Logatchev	16S rDNA	Perner et al. 2007, 2010
Thermodesulfobacteria			
Thermodesulfobacterium	Rainbow	*dsrAB*	Nercessian et al. 2005

(Table 4 is continued on facing page)

Table 4 (cont.).

	Habitat	Types of Analysis	Reference
Actinobacteria			
Actinomycetales	Cabeço de Vide	cultivation	Tiago, et al. 2004
Misc. Actinobacteria	Logatchev	16S rDNA	Perner et al. 2007
	Lost City	16S rDNA	Brazelton et al. 2006, 2010
	Outokumpo	16S rDNA	Itävaara et al. 2011
	Tablelands	metagenomics	Brazelton et al. 2012
Firmicutes			
Bacillales	Cabeço de Vide	cultivation	Tiago et al. 2004
	Tablelands	metagenomics	Brazelton et al. 2012
	Semail	cultivation	Bath et al. 1987
Desulfotomaculum	Lost City	16S rDNA	Brazelton et al. 2006
		dsrAB	Gerasimchuk et al. 2010
	Outokumpo	16S rDNA, *dsrB*	Itävaara et al. 2011
	Tablelands	metagenomics	Brazelton et al. 2012
Other Clostridia	Cabeço de Vide	cultivation	Tiago et al. 2004
	Outokumpo	16S rDNA, *dsrB*	Itävaara et al. 2011
	Tablelands	metagenomics	Brazelton et al. 2012
	Semail	cultivation	Bath et al. 1987
Erysipleotrichi	Outokumpo	16S rDNA	Itävaara et al. 2011
	Tablelands	metagenomics	Brazelton et al. 2012

anoxic portions of serpentinizing ecosystems (Fig. 2; Brazelton et al. 2010, 2012; Itävaara et al. 2011). The Clostridia in these environments are expected to be capable of fermentation that can involve the use of a single organic compound as both oxidant and reductant. This mode-of-growth may allow them to overcome the lack of exogenous oxidants in the most extreme zones of these environments. However, other strains of alkaliphilic Clostridia isolated from soda lakes are capable of autotrophic or mixotrophic growth (Sorokin et al. 2008). Deciphering the carbon assimilation strategies of subsurface Clostridiales will be of critical importance in studying the biogeochemistry of the serpentinizing subsurface.

Sulfur cycling. The presence of sulfate in continental and marine settings, when combined with the hydrogen that is a by-product of serpentinization reactions, presents the ideal conditions for microbial sulfate reduction. In freshwater systems dissolved ions are frequently scarce, and sulfate may not be present in appreciable concentrations (Table 2). In marine settings, deep seawater contains high concentrations of sulfate that would be available to microbial communities in the mixing zones with hydrothermal fluids. In some instances such as the Lost City field, where fluids do not reach high enough temperatures to induce the precipitation of anhydrite ($CaSO_4$), sulfate can also be present in the end-member hydrothermal fluid (Kelley et al. 2005).

One of the earliest published studies on the microbiology of serpentines examined the alkaline springs of Oman using culture-dependent approaches and documented the presence of sulfate-reducing bacteria (Bath et al. 1987). Recent cultivation-independent studies, however, are notable for their lack of evidence for typical sulfate-reducing bacteria such as Deltaproteobacteria (Brazelton et al. 2006, 2010). Some of the ubiquitous Clostridia discussed above (in particular the *Desulfotomaculum* group) are suspected to be capable of sulfate reduction (Moser et al. 2005; Lin et al. 2006; Chivian et al. 2008). Taxonomy, however, is not a reliable predictor of sulfate-reduction capability, so additional genetic and physiological experiments are required to assess the role of Clostridia in the sulfur cycle of serpentinite springs.

Nevertheless, several lines of evidence point to sulfate reduction as an active process at the Lost City hydrothermal field. The continued study of the LCHF towers and the Atlantis Massif has led to improved resolution of the distribution of microbial communities and their relationship to the geochemistry at the site (Fig. 2). The concentrations of hydrogen, sulfate, and sulfide vary widely across the field and indicate that sulfate reduction actively influences fluid compositions (Proskurowski et al. 2008; Lang et al. 2012). Bacterial sulfate reduction was also detected at 5 to 8 °C in mat samples from the carbonate towers of Lost City (higher temperatures were not tested), and taxonomic and functional genes (*dsrAB*) related to *Desulfotomaculum* were present (Dulov et al. 2005; Gerasimchuk et al. 2010). Work by Delacour et al. (2008a) also showed geochemical evidence for sulfate reduction in the Atlantis Massif below the LCHF.

The reduced, dissolved sulfide that results from sulfate reduction can also contribute to the electron donor budget and can support microbial carbon fixation. Metagenomic analyses of hydrothermal chimneys from Lost City found high abundances of genes related to the autotrophic sulfur-oxidizing Gammaproteobacterium *Thiomicrospira crunogena* (Brazelton and Baross 2010). These organisms are presumed to be abundant at oxic-anoxic interfaces within the chimney walls. Fluids venting from the chimneys appear to have a different microbial community composition than the chimneys themselves, suggesting that the carbonate-hosted biofilms may not accurately represent conditions in the subsurface (Brazelton et al. 2006).

Methane-cycling. The presence of hydrogen and reducing conditions make hydrogenotrophic methanogenesis a thermodynamically favorable process in many serpentinization systems. The occurrence of biological methane production may be limited, however, by the availability of dissolved inorganic carbon at high pH, as discussed in more detail below. Abiogenic methane is also an abundant source of electrons in many of these systems, and the potential for both aerobic and anaerobic methanotrophy is reflected in the occurrence of *pmoA* genes (Mason et al. 2010) and ANME (Anaerobic Methanotrophic archaea) phylotypes and associated *mcrA* genes (Mottl et al. 2003; Kelley et al. 2005), respectively.

Methane is frequently present in fluids passing through serpentinizing ophiolites, even when hydrogen concentrations are below detection limits (Table 2). Genes related to known methanogens were detected in the Del Puerto Ophiolite in the California Coast Range (Blank et al. 2009), although the abundance of the putative methanogens is unknown. Sequences related to methanogens were present but extremely rare in the metagenomic dataset from the Bay of Islands Ophiolite, Newfoundland, but no genes diagnostic of methanogenesis were identified (Brazelton et al. 2012). It is unclear whether the low abundance of these sequences reflects a small contribution from surface soil methanogens or instead suggests the presence of a more inaccessible, methanogen-rich deep subsurface habitat. Surprisingly, this metagenomic dataset also lacks any *pmo*A genes or other sequences expected to represent methanotrophic bacteria (Brazelton et al. 2012). This initial data from the Bay of Islands Ophiolite thus implies that methane cycling in continental serpentinites may be limited by as yet unknown factors.

Studies of the actively venting carbonate towers at Lost City have revealed low-diversity microbial communities dominated by a single archaeal phylotype (based upon 16S ribosomal RNA gene sequences), termed Lost City Methanosarcinales (LCMS; Schrenk et al. 2004). LCMS have been found in numerous samples of different chimneys associated with active venting at the Lost City (Brazelton et al. 2006) and are marginally related to the ANME taxa commonly found at gas hydrates and associated with anaerobic methane oxidation. However, further studies have shown extensive functional and physiological diversification within the LCMS biofilms that is not reflected by the rRNA data (Fig. 3). Imaging of the biofilms revealed

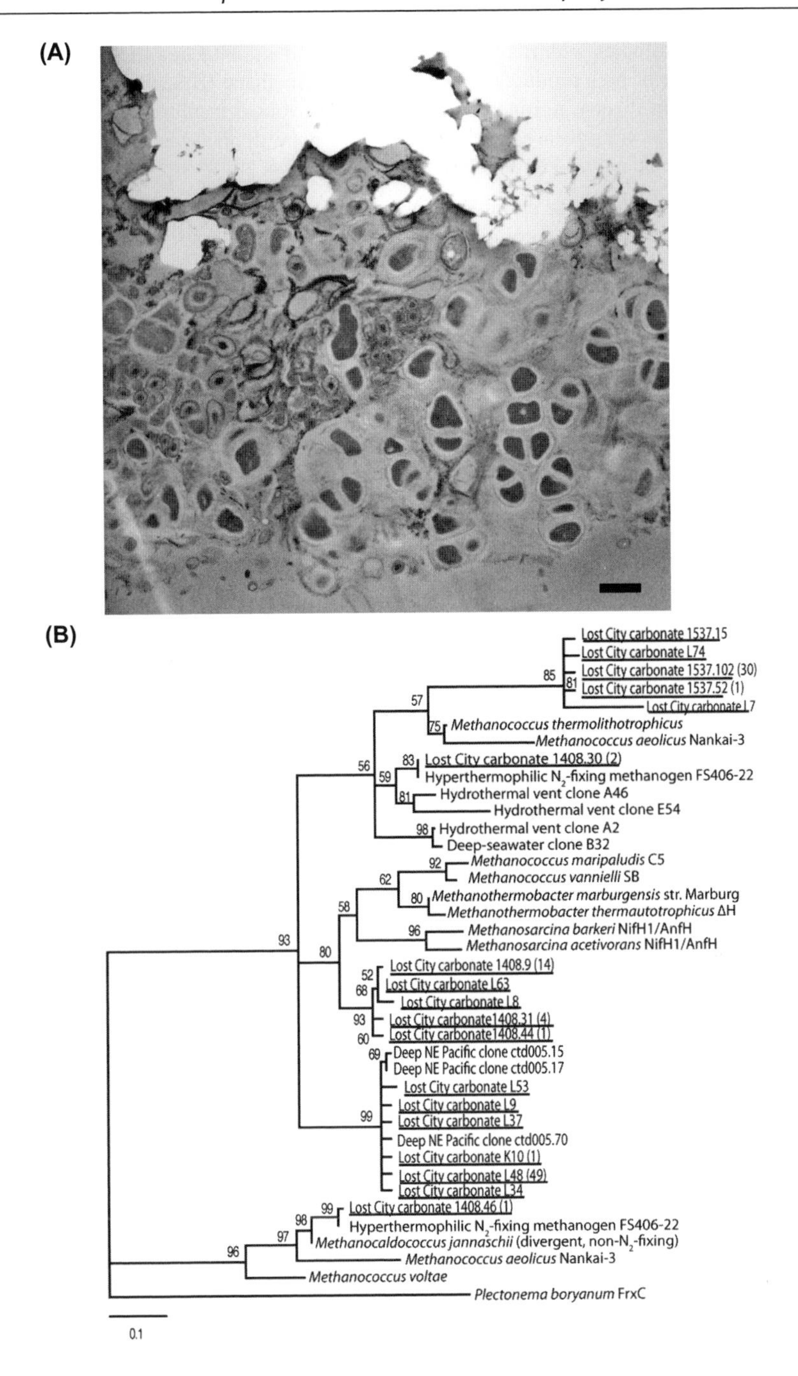

Figure 3. Physiological and phylogenetic diversification of "single species" archaeal biofilms from the LCHF hydrothermal chimneys. Panel A is a TEM thin section through a biofilm of LCMS showing the morphological diversity of cells within the carbonate chimney. Scale bar is 1 µm. Panel B shows the phylogenetic diversity of nitrogenase (*nif*H) genes from the same biofilms. The single species biofilms harbored at least 18 different clusters of methanogen-related *nif*H genes likely involved in nitrogen fixation. [Figures are reproduced with permission of the American Society for Microbiology from Brazelton et al. (2011).]

multiple cell morphologies within a single cluster of carbonate-attached LCMS cells, and the biofilms are capable of both production and oxidation of methane (Brazelton et al. 2011). The biofilms also contain diverse nitrogenase genes (*nifH*; involved in nitrogen fixation) and the highest percentage of transposase genes (involved in gene duplication and transfer) of any environmental sample analyzed to date (Brazelton and Baross 2009; Brazelton et al. 2011).

As the Lost City carbonate chimneys age and are no longer exposed to actively venting fluids, the microbial community composition shifts from LCMS-dominated populations, to a distinct group of ANME-1 archaea (Brazelton et al. 2010). Interestingly, little evidence exists for the presence of archaea in the subsurface rocks obtained from drilling the Atlantis Massif (Mason et al. 2010). Renewed efforts to study the root zone beneath the LCHF in a highly systematic manner are planned over the next several years.

Methanogenic and methanotrophic archaea have been observed at other marine serpentinizing settings as well (Table 3). Methanogenic archaea within the order Methanococcales have been noted at some sites at Logatchev (Perner et al. 2007, 2010), and high abundances of methanogenic archaea related to the Methanococcales have been reported from the Central Indian Ridge and the Southwest Indian Ridge (Takai et al. 2004). This latter report detected the presence of *Methanopyrus kandleri* strain 116, a hyperthermophilic microorganism isolated from another environment, and grown in culture at temperatures as high as 122 °C (the current upper temperature limit for life) under elevated hydrostatic pressure (Takai et al. 2008). Finally, microbiological studies of IODP Site 1200 core samples at South Chamorro Seamount have shown a high abundance of archaea that, based on the increasing concentrations of carbonate and bisulfide in ascending fluids, appear to be anaerobically oxidizing ascending CH_4 while reducing sulfate (Mottl et al. 2003).

Putative aerobic methanotrophic bacteria (e.g. *Methylococcales*) are widespread in Lost City chimneys and fluids, and they have also been detected at the Rainbow hydrothermal field (Table 4; Nercessian et al. 2005; Roussel et al. 2011). Presumably, these bacteria occupy zones near the surface where both methane and oxygen are available. Methane and oxygen-rich zones are also found in continental serpentinite settings (such as the Bay of Islands Ophiolite, discussed above) that seem to lack methanotrophic bacteria, though, so additional factors may favor the growth of these methanotrophic bacteria in marine settings.

Heterotrophy/fermentation. Serpentinite settings include many sources of organic carbon; indeed, these environments are unusual in that organic carbon is typically much more biologically available than inorganic carbon (discussed in detail below). Processes associated with serpentinization can lead to the abiogenic generation of numerous short-chain hydrocarbons in addition to methane (Proskurowski et al. 2008). Mantle-derived carbon exists in the ultramafic rocks within fluid inclusions and along grain boundaries (Kelley and Früh-Green 1999). Additionally, downwelling fluids can mix with the serpentinizing fluids and contribute photosynthetically-derived organic carbon (Abrajano et al. 1990). For example, seawater-derived organics were observed in drill cores from beneath the Atlantis Massif (Delacour et al. 2008b). A third source of organic carbon could include the biomass and metabolic byproducts produced by autotrophs in the deep sea, seafloor, and continental subsurface. Also worth consideration is a recent study of a granitic deep subsurface habitat suggests that viral lysis of autotrophic microorganisms supplies dissolved organic matter that can feed heterotrophic microbial communities and limit overall population sizes (Pedersen 2012)

These sources of organic compounds may support the growth of heterotrophic and fermentative archaea and bacteria. Alkaliphilic heterotrophs were observed using culture-dependent approaches in the alkaline springs of Oman (Bath et al. 1987). A number of heterotrophic alkaliphiles have also been isolated from high pH wells in Portugal (Tiago et al. 2004, 2005, 2006). At Cabeço de Vide in Portugal, alkaliphilic heterotrophic bacilli and

Actinobacteria are common in aerobic enrichment cultures (Table 4; Tiago et al. 2004). They are speculated to use organic materials near more oxidizing mixing zones within the ophiolite. In at least one case a novel alkaliphilic heterotroph within the Gammaproteobacteria that was capable of growth up to pH 12.4 was isolated from South Chamorro Seamount and other locations of the Mariana forearc (Takai et al. 2005).

Elevated concentrations of potentially fermentable substrates (hydrocarbons and organic acids) have been observed in many locations (Table 1 and 2; Haggerty and Fisher 1992; Proskurowski et al. 2008; Lang et al. 2010; Itävaara et al. 2011), but *in situ* fermentation activity has yet to be confirmed. Analysis of microbial populations from the igneous rocks obtained from drilling the Atlantis Massif revealed functional genes involved in hydrocarbon degradation (Mason et al. 2010). Potentially fermentative Clostridia have been detected in many serpentinite settings (discussed above), and it is possible that fermentation is an advantageous metabolic strategy in environments where oxidants are scarce.

Challenges of high pH

One of the primary challenges for life in high pH environments is the maintenance of a proton motive force across the cytoplasmic membrane. Proton gradients sustain ATP synthesis through oxidative phosphorylation or photophosphorylation in a large fraction of characterized species (Fig. 4). Typically, the proteins involved in the main pH homeostasis mechanisms of these extremophiles are constitutively expressed, so that these microorganisms are prepared for sudden shifts to the extreme end of the pH range (Krulwich et al. 2011). Alternatively, organisms can rely upon metabolic processes such as fermentation to obtain ATP via substrate level phosphorylation. However, these processes generate less energy per mole of reactant than oxidative processes, and ionic gradients are still required for molecular transport across the cytoplasmic membrane. At high pH, the calculated proton-motive force (PMF) may drop to almost zero (Fig. 4; Krulwich 1995). Microorganisms found in alkaline soda lakes rely upon ionic gradients involving Na^+ or K^+ to substitute for protons and to generate ATP (Krulwich 1995). It is unknown whether organisms in low ionic strength solutions, such as serpentinizing ophiolites, use alternative ions in their membrane transport mechanisms.

High pH also provides problems in terms of the stability of ribonucleic acid (RNA), a critical molecule in the transcription, translation, and regulation of genes. It is well known that RNA is unstable in alkaline solutions, and high pH is commonly avoided in laboratory procedures aimed at RNA isolation. Hydroxyl groups attack phosphate groups, thereby disrupting the polymeric backbone of the RNA molecule through transesterfication reactions (Li and Breaker 1999). The chemical impacts of ultrabasic conditions upon RNA structure may be an important reason why cytoplasmic pH, even within alkaliphiles, is typically orders of magnitude more neutral than external pH (Krulwich 1995). However, more information about molecular strategies enhancing RNA stability still remains to be explored, and in fact these may be attractive targets for biotechnological applications.

Limitations to carbon fixation

Serpentinization supplies copious energy to the subsurface environment in terms of reducing power. When mixed with oxidants from surface and subsurface sources, these systems provide a strong thermodynamic drive in terms of chemical disequilibria that can be harnessed by autotrophic microbial populations (McCollom 2007). One of the primary challenges in these systems, however, is the limited availability of inorganic carbon. Under high pH conditions, the predominant form of dissolved inorganic carbon (DIC) is carbonate rather than bicarbonate. During water-rock reactions, concentrations of calcium ions become elevated in the fluids; when combined with the high pH conditions that are also a result of serpentinization reactions, inorganic carbon is rapidly precipitated as calcite and aragonite (Barnes et al. 1978; Neal and Stanger 1983; Fritz et al. 1992; Palandri and Reed 2004; Kelley et al. 2005). As a result,

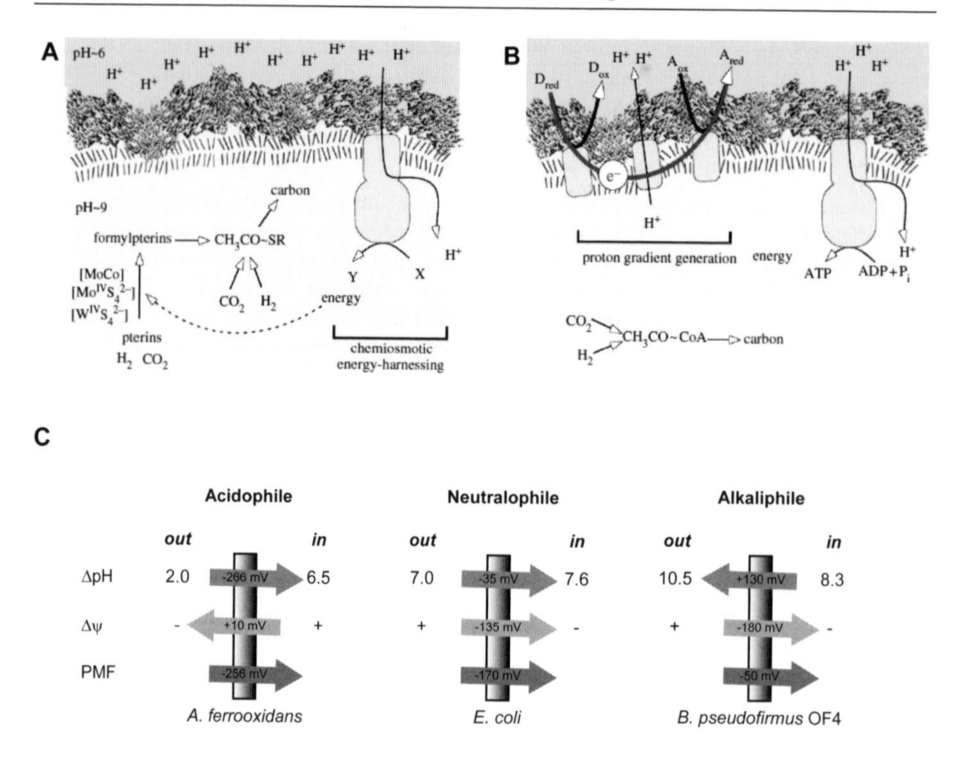

Figure 4. Diagram of ionic gradients in serpentinizing ecosystems and their relationship to biochemistry. The theorized role that chemical gradients between alkaline hydrothermal fluids and an acidic ocean may have played in the development of the proton motive force (PMF) in prebiotic settings (A) and comparison to the modern day PMF (B) across biological membranes. (C) depicts the pH and electrochemical gradients across the cytoplasmic membrane of microorganisms inhabiting different pH ranges and how they contribute to the PMF. $\Delta\psi$ is the transmembrane electrical potential, ΔpH is the transmembrane pH gradient, R and F are the gas and Faraday constants, respectively. These electrochemical gradients are harnessed by modern biochemistry to make ATP, to facilitate transport, and for mechanical processes such as motility. Alkaliphiles grow optimally at pH > 9 and most known alkaliphiles maintain a cytoplasmic pH < 10. [(A) and (B) are reproduced with permission of the Royal Society from Martin and Russell (2007). (C) was adapted with permission of Nature Publishing Group from Krulwich et al. (2011).]

extremely low concentrations of DIC are typical in high pH fluids from marine and terrestrial systems (Table 1 and 2).

At the alkaline marine Lost City field, the lack of any radiocarbon in methane indicates that the precursor carbon source is mantle derived, implying that any modern seawater DIC is precipitated prior to the abiogenic formation of methane (Proskurowski et al. 2008). Consequently, biological communities may be inorganic carbon-limited (Bradley et al. 2009b). Microbial communities in these systems may rely on carbon sources other than DIC. The majority of serpentinite-hosted ecosystems have high concentrations of CH_4 that may provide a carbon source if microorganisms have the ability to oxidize it and assimilate it into biomass. Thermodynamic studies indicate that in addition to CH_4 and *n*-alkanes, the formation of CO and organic acids as metastable intermediaries can be thermodynamically favorable under the reducing, high H_2 conditions present in many serpentinizing environments (Shock 1992; Shock and Schulte 1998; McCollom and Seewald 2001, 2003; Seewald et al. 2006). Both CO and formate have been formed abiogenically in controlled laboratory studies at temperatures >175 °C (McCollom and Seewald 2001, 2003; Seewald et al. 2006). At the high temperature, acidic Rainbow field, CO is present at detectable concentrations of ~5 μmol/kg (Charlou et al. 2002). Fluids from both Lost City and the Marianas Forearc have elevated concentrations of the organic acids formate and acetate (Haggerty and Fisher 1992; Lang et al. 2010). There are several indications that formate at Lost City is indeed formed abiogenically in the subsurface, although the presence of acetate is more likely to be due to the degradation of biomass (Lang et al. 2010).

When high-pH, calcium-rich fluids exit the subsurface and are again exposed to inorganic carbon either in the form of seawater DIC (marine systems) or atmospheric CO_2 (terrestrial systems), solid carbonate precipitates. Microorganisms could mobilize solid carbonate minerals by dissolution, perhaps by the localized secretion of organic acids. Some organisms living at the oxic/anoxic interface may also be able to access the inorganic carbon before it precipitates. A recent study used the ^{14}C content of biomass in the chimneys at the Lost City field to distinguish between seawater-derived and mantle-derived carbon sources (Lang et al. 2012). In some locations, the carbon source assimilated by the chimney communities was >50% mantle-derived.

Functional and metagenomic studies have revealed a number of genes associated with carbon fixation in these systems. These include the genes encoding the RuBisCO enzyme of the Calvin Benson Bassham cycle and genes associated with the reverse tricarboxylic acid (rTCA) cycle. The metagenomic data from the Lost City field included putative homologs of Type I RuBisCO (Brazelton and Baross 2010), which is the form typically expressed when carbon dioxide is limiting (Dobrinski et al. 2005; Berg 2011). Genes for carbonic anhydrase and carboxysomes, which may aid in concentration of carbon dioxide within the cell, were also found (Brazelton and Baross 2010). As with the methane metabolisms, to date no studies have demonstrated the *in situ* expression of carbon fixation genes in serpentinite-hosted ecosystems.

Recently, metagenomic analyses of terrestrial serpentinites have also detected Type I RuBisCO in pH 12 waters from the Tablelands Ophiolite in Canada. The sequences appear to be associated with facultatively anaerobic Betaproteobacteria that were also found to harbor genes required for oxidation of H_2 and carbon monoxide (CO). Utilization of CO is an intriguing possible strategy in environments that experience low to moderate levels of serpentinization, where high pH conditions can limit CO_2, and H_2 may be limiting due to its rapid consumption by many organisms. Therefore, some organisms may occupy a niche where CO, even at low concentrations, could provide reducing power as well as a carbon source, thereby obviating any need for either H_2 or CO_2 (Brazelton et al. 2012).

In ultramafic-hosted high-temperature seafloor vents, Perner et al. (2010) recently demonstrated the connection between H_2 availability and carbon fixation rates. The study found carbon fixation genes important in both the CBB and reverse tricarboxylic acid (rTCA) cycles. These results were mirrored by a much higher diversity of NiFe hydrogenase genes involved in H_2 uptake in an ultramafic hydrothermal vent site, compared to a basaltic site. The authors also demonstrated that addition of H_2 stimulated carbon fixation rates under anoxic conditions.

Sources of nutrients

To date, the concentrations of macro-nutrients essential to microbial communities such as nitrate, ammonia, and phosphate are unknown in the majority of these environments. Available N-sources may be low, and the propensity for nitrogen fixation has been noted in serpentinite-hosted ecosystems. Multiple varieties of *nifH* genes have been found in chimney materials from the LCHF (Fig. 3b; Brazelton et al. 2011). Their closest relatives are found in methanogenic archaea, such as the predominant LCMS phylotypes found in the chimneys. Studies of fluids ascending through South Chamorro Seamount sediments also showed a small increase in ammonia concentrations that could reflect microbial nitrogen fixation (Mottl et al. 2003).

Phosphorus (P) is critical to the synthesis of new biomass and the functioning of cells. The mineral brucite, often formed through the mixing of high pH serpentinite springs with carbonate-bearing waters, is effective at scavenging P from solution. Holm and colleagues pointed out the importance of Na^+ in solubilizing P from brucite and its analogies to a primitive phosphate pump (Holm and Baltscheffsky 2011). Additionally, under phosphorus-limited conditions, microorganisms have been known to substitute sulfur for phosphorus in their membrane lipids (Van Mooy et al. 2006). A similar phosphorus conservation strategy has been proposed for microbial populations inhabiting carbonate chimneys at the LCHF, where glyocsyl head groups replace phosphatadyl head groups in bacterial membrane lipids (Bradley et al. 2009a).

Microbe-mineral interactions

In addition to direct impacts upon carbon flux through their growth and metabolism, microbes can impact carbon flow in serpentinite habitats through their interaction with solid phases. The dense biofilm communities of the LCHF can create favorable niches for themselves within the carbonate chimneys that may serve to buffer the organisms against the effects of oxygen and high pH. The cells and their polymeric matrix can serve as nucleation sites for carbonate precipitation (Fig. 5; Blank et al. 2009). Microorganisms can provide the conditions required for precipitation of carbonates: elevated dissolved inorganic carbon (respiration), and nucleation sites from extracellular polymeric substances (EPS), or degradation of EPS resulting in the release of cations. However, microbial activities may also inhibit the precipitation of carbonates, by cation capture by EPS, consumption of DIC, and acidification (sulfide oxidation; Blank et al. 2009).

Microbial communities in direct contact with ultramafic rocks potentially utilize solid electron acceptors. Some of the byproducts of serpentinization are the production of magnetite and other Fe (III)-bearing minerals. Menez et al. (2012) reported the association of organic matter, potentially the by-product of biology, with ferric minerals in the ocean crust. In this setting, perhaps the minerals play a role in the sustenance of subsurface microbial communities. The ability of microbial activities to contribute to the release of solid carbon phases from mantle rocks also has not been explored. Whether microbes play an active or passive role in the process of serpentinization remains to be investigated.

Serpentinization and the origins of life

Prior to the differentiation of the lithosphere on early Earth, exposed ultramafic rocks were probably more prevalent than they are today (Sleep et al. 2004). Throughout the Archean Eon, the crust was likely to have been more mafic than the modern crust and hence more likely to support serpentinization (Arndt 1983; Nisbet and Fowler 1983; Nna-Mvondo and Martinez-Frias 2007). Evidence consistent with serpentinization and its metamorphic products occurs in exposures of the approximately 3.8 Ga Isua supercrustal belt of western Greenland (Friend et al. 2002; Sleep et al. 2011). Hydrogen and oxygen isotopic properties of serpentine minerals in these rocks suggest that they were formed by reaction of ultramafic rocks with seawater. Serpentinization is also evident in >2.5 Ga Archean komatiites of the Kuhmo greenstone belt in Finland (Blais and Auvray 1990). In general, the meager geological evidence that is available for the Hadean and Archean Eons is consistent with serpentinization being more widespread during that time than it is today (Sleep et al. 2011).

Sites of active serpentinization are attractive venues for origin of life scenarios for several reasons outlined below, all of which are consequences of the highly reducing and high pH fluids generated by exothermic serpentinization-associated reactions. Uncertainty of the early atmosphere's redox state has challenged the formation of a consensus by the origin of life community regarding the most thermodynamically favorable prebiotic chemical pathways. Serpentinization-driven systems, however, provide highly reducing local environments with high H_2 concentrations where prebiotic organic synthesis is clearly favored regardless of

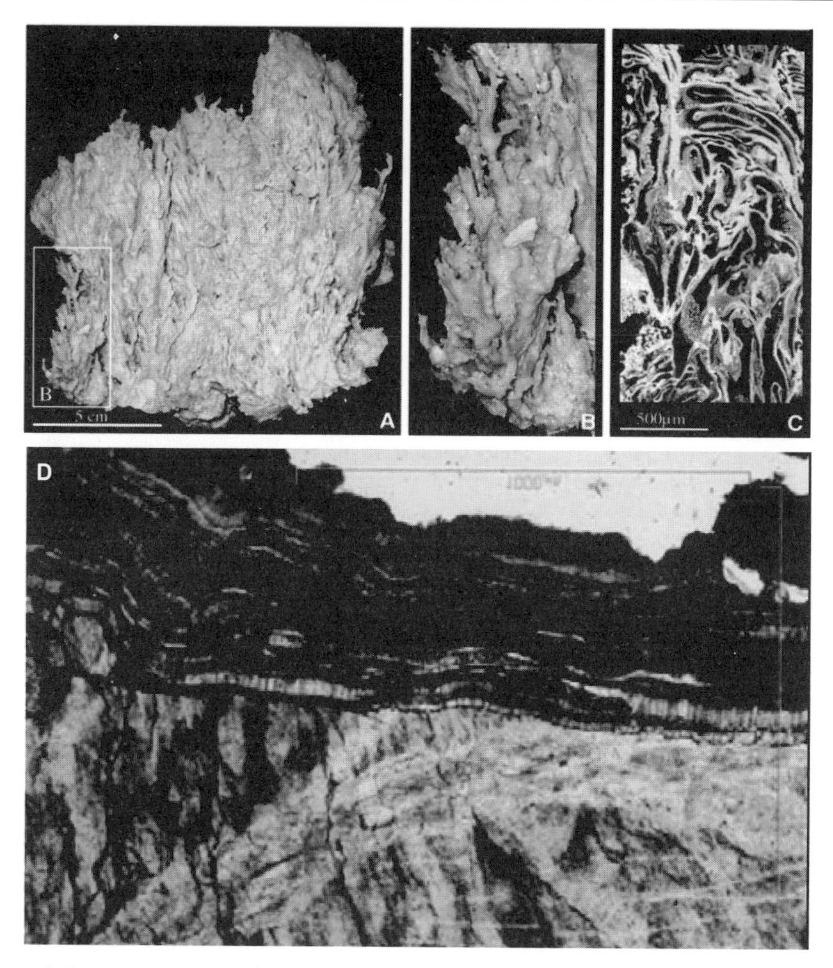

Figure 5. Examples of mineral nucleation associated with microbiological processes in serpentinite ecosystems. Panels A through C are photomicrographs of carbonate hydrothermal chimney from the Lost City Field, near the Mid Atlantic Ridge, where pore spaces within the chimneys take the shape of microbial biofilms pervasive along the exterior of the structures. Panel D shows a lithified microbial biofilm (top) coating a serpentinite rock in the Coast Range Ophiolite of California, USA. [A-C are reproduced with permissionof AAAS from Kelley et al. (2005). D is reproduced with permission of Elsevier from Blank et al. (2009).]

atmospheric conditions because abiotic organic synthesis can be observed in such environments even today (Proskurowski et al. 2008). Earlier work has suggested that H_2 may play a critical role in the sustenance and long-term survival of microbial communities (Morita 2000). If this is true, it may be especially important in the discontinuous and ancient niches presented in many serpentinite rocks.

In addition to promoting abiotic synthesis of organics, H_2-rich environments are ideal locations for early metabolic evolution because hydrogen transfer is at the heart of almost every biochemical reduction or oxidation reaction in modern metabolic pathways (Nealson et al. 2005). Thus it is parsimonious to suppose that the first metabolic pathways were fueled by geochemically derived H_2. Interestingly, enzymes that catalyze H_2 oxidation or production (i.e., hydrogenases) contain iron and/or nickel at their catalytic sites. Both iron and nickel are generally enriched in serpentinites, but the exact mineral phases involved in organic synthesis reactions associated with serpentinization have yet to be identified (Foustoukos and Seyfried 2004; Sleep et al. 2004; McCollom and Seewald 2007).

Most enzymes involved in modern biological carbon fixation pathways also have minerals as essential components of their catalytic sites. The reductive acetyl-CoA (or Wood-Ljungdahl) pathway is utilized for both carbon fixation and ATP generation by methanogenic archaea and acetogenic bacteria, making it the only carbon fixation pathway shared by both archaea and anaerobic bacteria (Berg et al. 2010). Remarkably, the iron/nickel minerals found at the active sites of Wood-Ljungdahl enzymes can catalyze at least some of the steps in this pathway on their own without any organic components (Huber and Wächterhäuser 1997; Cody et al. 2000; Cody 2004). The simplicity (Fuchs and Stupperich 1985; Berg et al. 2010) and phylogeny (Pereto et al. 1999) of proteins involved in the Wood-Ljungdahl pathway are also consistent with their ancient origin. Almost all organisms that utilize this pathway today are fueled by H_2, and the exceptions almost certainly represent later evolutionary innovations (Bapteste et al. 2005). The rare congruence among these geological, chemical, and biological data supports the emerging view that the earliest biochemical pathways were driven by H_2 and evolved as mimicry of pre-existing geochemical reactions (Cody and Scott 2007) that would have been favored in serpentinizing environments.

The greatest appeal of hydrothermal environments, in general, as key sites in the origin of life is the presence of diverse catalytic mineral surfaces in geological, physical, and chemical gradients that are formed as a result of the dynamic mixing associated with hydrothermal circulation (Baross and Hoffman 1985; Martin et al. 2008). In the presence of strong gradients, chemical reactants are more likely to be far from equilibrium with respect to each other, and therefore the thermodynamic favorability of them reacting to generate new products is greatly improved (Shock and Schulte 1998). Serpentinite-hosted hydrothermal systems, in particular, feature characteristic gradients in temperature, redox, geochemistry, porosity, and pH that have their own advantages for prebiotic and early evolution scenarios. In the detailed model proposed by Martin and Russell (2007), the pH gradient between serpentinization-derived fluids and ambient seawater causes protons to leak out of iron sulfide compartments that they predict would form on ancient chimney deposits. The iron sulfide "bubbles" are considered to be the precursors to the modern lipid membrane (Russell et al. 1994; Russell and Hall 1997), and their leakage of protons would have resulted in a chemiosmotic potential (Fig. 4a; Lane and Martin 2010) that could have been harnessed by the first enzymes to catalyze H_2-fueled carbon fixation. Regardless of whether these particular details are exact descriptions of how the origin of life actually occurred, it is clear that chemical potential gradients involving hydrogen species were important aspects of early metabolic processes, just as they are critical in all organisms today (Sleep et al. 2011).

Although a discussion of possible prebiotic pathways for the synthesis of specific biomolecules is outside the scope of this review, it should be noted that most proposals for the evolution of genetic information systems (e.g. "RNA world" theories) are compatible with a H_2-rich setting. The building blocks of any genetic information system require an energy source and favorable thermodynamic conditions for organic synthesis, and serpentinite-hosted systems clearly fit these criteria, as we have described above. The H_2-rich chimneys of serpentinization-driven hydrothermal systems, for example, contain abundant micro-compartments for concentration of reactants and feature temperature ranges similar to those in the polymerase chain reaction (PCR), ideal for nucleic acid synthesis (Kelley et al. 2005; Baaske et al. 2007).

The mineralogy of serpentinites may have also provided advantages for prebiotic chemistry. Phosphate availability on early Earth could have been a severe limitation to the origin of nucleic acids, but Nisbet and Sleep (2001) have noted that the "RNA world" could have existed within pores in serpentinites that are rich in the phosphate-containing mineral hydroxyapatite. As mentioned earlier, the mineral brucite, commonly associated with active marine serpentinization, is an effective scavenger of both boron and phosphorous. Phosphates, including pyrophosphate, can accumulate in brucite over millions of years on the seafloor, and then become desorbed when exposed to high Na^+ concentrations and the high pH conditions within serpentinites. It has been shown that pentoses, such as ribose, that make up the building blocks of RNA can be stabilized by boron (Ricardo et al. 2004), so high pH fluids circulating through serpentinites that contain boron-enriched brucite may have supported RNA synthesis (Holm et al. 2006). As the geochemical reaction of serpentinization consumes water, discontinuous fracture networks or surface-exposed serpentinites could have provided the additional benefit of hydration-dehydration reactions and concentration by evaporation that may have been necessary for polymer synthesis.

WHERE DOES THE ABIOTIC CARBON CYCLE END AND BIOGEOCHEMISTRY BEGIN?

Studies of serpentinizing environments to date have shown that these ecosystems host low-abundance, low-diversity microbial communities. However, these habitats coincide with environments where abiotic carbon transformations are taking place. In addition to basic ecological questions about the relative balance of autotrophy and heterotrophy in various niches within serpentinite habitats, it is also intriguing to consider the boundary between living and non-living. Where do biogenic processes end, and where does abiogenic organic geochemistry become the predominant process? What does this imply for the magnitude and extent of the global subsurface biosphere (Schrenk et al. 2010)? What role does abiogenic organic chemistry play in the flux of carbon from the deep Earth into the surface biosphere?

Abiogenesis in thermodynamic and experimental studies

The reducing conditions and high hydrogen concentrations arising from the serpentinization reactions can make the abiogenic synthesis of organic carbon molecules thermodynamically favorable. Biologically relevant compounds such as methane, hydrocarbons, carboxylic acids, alcohols, and amino acids are thermodynamically favored over inorganic constituents when buffered at the appropriate temperatures, redox conditions, and H_2 fugacities, or when reducing hydrothermal fluids mix with oxic seawater (Shock 1990; Amend and Shock 1998; Shock and Schulte 1998; McCollom 2013). Methane is typically the most thermodynamically stable organic compound, but its formation may be kinetically inhibited, leading to the formation of other organic compounds.

Numerous experimental studies have focused on the abiogenic formation of methane and n-alkanes, with synthesis pathways attributed to Fischer-Tropsch-type and/or Sabatier-type reactions (for in depth recent reviews, see McCollom and Seewald 2007; Proskurowski 2010; McCollom 2013). One of the earliest experiments demonstrated elevated concentrations of methane, ethane, and propane when an aqueous solution was reacted with olivine at high temperatures and pressures (Berndt et al. 1996), although a later study used ^{13}C-labed bicarbonate under similar conditions to demonstrate that most of these compounds were in fact generated from the thermal decomposition of organic matter present in the reaction vessel or the catalysts (McCollom and Seewald 2001). Nonetheless, these experiments spurred a large number of variations over the past decade, focusing on the importance of the starting carbon source (e.g. bicarbonate, formate, oxalic acid, CO), the type of mineral catalyst (e.g. NiFe-alloy, magnetite, chromite, olivine, hematite), as well as the roles of temperature and pressure (Horita and Berndt 1999; McCollom et al. 1999; Rushdi and Simoneit 2001; McCollom and Seewald 2003; Foustoukos and Seyfried 2004; Rushdi and Simoneit 2004). These experiments have repeatedly demonstrated the abiogenic synthesis of methane, branched and straight-chained alkanes up to C-27, alkenes, alkenones, formate, and long-chain alcohols.

A largely independent series of experiments has focused on the synthesis and stability of amino acids, peptides, and proteins. Multiple laboratory studies have demonstrated that amino acids can by synthesized from inorganic constituents under aqueous conditions designed to simulate hydrothermal (although not necessarily serpentinite) environments (Óro et al. 1959; Lowe et al. 1963; Wolman et al. 1971; Kamaluddin and Egami 1979; Hennet et al. 1992; Yanagawa and Kobayashi 1992; Marshall 1994; Islam et al. 2001; Aubrey et al. 2009). In these experiments, glycine is frequently the amino acid synthesized in the highest yield. While many organic compounds are thermodynamically stable, and able to be synthesized abiogenically in laboratory conditions, identifying their presence in the environment is often complicated by the presence of biologically produced compounds.

Distinguishing biotic from abiotic processes

A grand challenge in the study of microbial ecosystems near the limits of habitability is developing criteria to accurately discriminate between biological and abiological processes. In serpentinizing ecosystems some of the classical discriminants for life, e.g. organic carbon compounds, are also potentially produced by abiotic processes (Sephton and Hazen 2013). The microbial contribution to the net flux of methane from serpentinizing environments is one of the most intriguing questions in this field. Although serpentinization of ultramafic rock is associated with only a small fraction of total hydrothermal circulation (<10%), it can supply up to ~75% of the abiogenic methane from mid-ocean ridges (Cannat et al. 2010; Keir 2010). In many cases isotopic and geochemical evidence indicates that abiogenic FTT and/or Sabatier reactions lead to the production of CH_4 and, occasionally, higher hydrocarbons. In other cases, the thermogenic alteration of organic matter contributes to methane production (Hosgormez et al. 2008). Due to the overlap of biogenic and abiogenic processes, and complex physiological adaptations of microbial populations (e.g. under carbon limitation), serpentinite-hosted ecosystems present a challenge to deciphering biogenic from abiogenic methane sources (Bradley and Summons 2010). For these reasons, serpentinites are being explored as astrobiological analogs to aid in understanding potential sources of methane on Mars (Mumma et al. 2009).

The most thorough studies of the provenance of organic compounds are associated with the study of methane and *n*-alkanes. Early studies relied upon the isotopic ratios of carbon ($^{13}C/^{12}C$) and hydrogen ($^{2}H/^{1}H$) in methane to distinguish the sources (Fig. 6; Abrajano et al. 1988). Stable isotope evidence alone can be misleading however. The fractionation factors associated with the abiogenic formation of CH_4 through FTT-synthesis may be as large as those associated with biological processes (McCollom and Seewald 2006). Additionally, the $\delta^{13}C$ value of methane from most serpentinization environments does not reflect this full fractionation factor, possibly due to carbon limitation (Proskurowski et al. 2008). In these cases, ^{14}C data may be better suited to differentiate between distinct carbon sources and can potentially constrain abiotic versus biotic origins (Lang 2012). Further studies of hydrocarbon distribution patterns, coupled with isotope systematics have provided additional means to separate the various sources of small organic molecules (Sherwood Lollar et al. 2006). These interpretations are complicated by the fact that many of these compounds have both biogenic and abiogenic origins, and can be influenced by contributions from the thermogenic degradation of sedimentary organic matter (Hosgormez et al. 2008; Bradley and Summons 2010; Szponar et al. 2012).

It is less clear when and where microbial communities contribute to methane production and consumption. Isotopic and genetic evidence from these systems provide strong clues that the process is occurring but would be strengthened by a better understanding of the organisms catalyzing the process and their physiologies. As discussed above, genes found in methanogenic taxa have been documented in a number of environments, including Del Puerto Ophiolite and at the LCHF (Table 3; Kelley et al. 2005; Blank et al. 2009). The expression of these genes has not yet been reported, however. To our knowledge, there have not been any reports of laboratory cultivation of pure methanogenic isolates from high-pH serpentinite habitats.

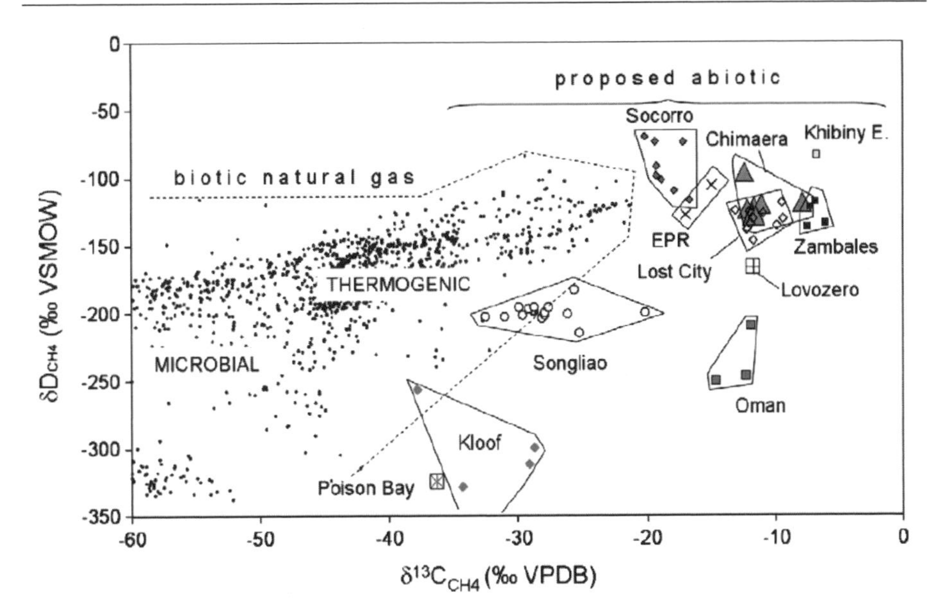

Figure 6. Plot showing the carbon ($\delta^{13}C$) and hydrogen (δD) isotopic ratios of methane from diverse environments. Methane from serpentinizing environments tends to be enriched in ^{13}C compared to locations where it is derived predominantly from biological methanogenesis or the thermogenic degradation of organic matter. Reproduced with permission of Elsevier from Etiope et al. (2011).

However, stable isotope microcosm experiments have demonstrated the capability of biomass from the LCHF towers to both produce and consume methane (Brazelton et al. 2011). Estimates of methanogenic production rates can be used to approximate the maximum contributions of biotic methane (Bradley and Summons 2010). Quantifying the impacts of methane cycling relative to abiogenic processes is important to quantifying carbon flux through these systems.

The isotope signatures of biomass and biomarkers such as membrane lipids can provide clues to the sources of carbon for deep life in serpentinites, and the biochemical pathways used to sustain cell growth. Diether lipids at the LCHF display an extraordinary enrichment in ^{13}C that has been attributed to extreme carbon limitation in the chimney ecosystem (Bradley et al. 2009b). Studies of organic compounds in the Atlantis Massif associated with IODP Hole 1309D demonstrate that the majority of the organic matter associated with serpentinites at that location are seawater-derived (Delacour et al. 2008b). In many cases, differentiating between mantle, thermogenic, and biological processes remains equivocal. Problematically, high contributions of biologically derived compounds may swamp small amounts of abiogenically derived organic compounds that are large enough to represent the next steps of pre-biotic synthesis. A thorough carbon budget of these systems, in addition to an improved understanding of the microbiological processes, is necessary to complete the picture.

Linking abiotic and biological processes

A missing link between the abiotic synthesis of organics and the microbial isolates obtained to date are that none of the organisms in culture have been demonstrated to utilize the small organic compounds (e.g. ethane, propane, formate) that can result from serpentinization Radiocarbon analysis of organic carbon and biomass in the LCHF chimneys, for example, is indicative of biological utilization of mantle-derived carbon (Lang et al. 2012). However, it has not yet been determined whether this organic carbon results from direct assimilation of abiogenic organic matter or if a microbial community first oxidizes these reduced compounds to a more accessible form. Many of the microorganisms that appear to inhabit the deepest, most anoxic portions of serpentinite ecosystems are closely related to the bacterial order Clostridiales, a group known to include fermentative organisms (Brazelton et al. 2006, 2012; Itävaara et al. 2011), but further work is required to establish the metabolic strategies of the Clostridiales-like organisms in serpentinites. If these organisms utilize organic carbon derived from serpentinization associated abiotic reactions, then fermentation in these systems could be considered, somewhat non-intuitively, to be a kind of primary production as it would be the generation of new biomass from non-biological carbon and energy. In completing the picture of carbon flow in the serpentinite subsurface, it is important to consider the metabolic products of such processes and their influence upon fluid chemistry.

COMMON THEMES AND UNCHARTED TERRITORY

One of the most interesting features of serpentinites is that they allow access to observe a set of variables that may constrain the limits of life on Earth. While there is substantial energy in terms of electron donors to support microbes in these systems, electron acceptors are typically limiting. They have discontinuous fluid circulation pathways that may both benefit and trap subsurface life. Furthermore, because these rocks originate in Earth's deep interior, they may extend beyond the thermal and pressure limits of habitability. Clearly, there is a balance between energy production and energy demand that needs to be incorporated into models and tested empirically. These studies should be coordinated with physiological studies of microorganisms from serpentinite habitats to determine whether they host unique adaptations to cope with the environmental stresses of the high pH environment. Additionally, it is important to study dormancy and survival in these populations. Many of the species recovered from deep subsurface habitats are related to those known to produce spores. Spore formation could be an important dispersal and survival strategy in serpentinites and other deep subsurface environments.

It is critical to document the extent of the deep biosphere to include in global compilations of microbial biomass and to constrain their activities and contributions to subsurface biogeochemistry. Quantification of the rock-hosted subsurface biosphere is in its nascent stages and is completely unaccounted for in compilations of global microbial abundances. Important facets of this problem include developing strategies to decipher abiogenic and biogenic sources of methane and other organic molecules. Additionally, the magnitude of microbial contributions to biogeochemical cycles needs further investigation in serpentinizing ecosystems, linking genomic and geochemical approaches. Finally, it is imperative to better understand the ability of microorganisms to interact with solid phases in terms of either mobilizing deep carbon or inducing carbonate precipitation. This becomes particularly important as researchers are looking to serpentinites as a site of carbon sequestration (Kelemen and Matter 2008).

Serpentinization reactions can lead to the abiotic synthesis of small organic molecules, as has been shown through experiment, observation, and theory. Ultramafic rocks were likely more prevalent early in Earth's history during the origin and evolution of the biosphere. Some have speculated that water-rock reactions played a role in the origins of life associated with deep-sea hydrothermal vents. Furthermore, serpentinites have been documented on the surface of Mars and serpentinization may be operative elsewhere in our solar system. Microorganisms in serpentinite settings may host relicts of these ancient microbe-mineral processes. Although delineating biotic from abiotic processes is challenging at the edge of the biosphere, it is critical that we are not deterred as it may help us to define the limits of our biosphere, and ultimately the transition between prebiotic Earth and life.

ACKNOWLEDGMENTS

We gratefully acknowledge funding support from the Deep Carbon Observatory (Alfred P. Sloan Foundation) and the NASA Astrobiology Institute (CAN-5) through the Carnegie Institution for Science. We are extremely appreciative to our colleagues studying various facets of serpentinization worldwide, especially Katrina Twing (ECU), Igor Tiago (U. of Coimbra) and Gretchen Früh-Green (ETH-Zurich) for helpful discussions and insight.

REFERENCES

Abrajano TA, Sturchio NC, Bohlke JK, Lyon GL, Pordea RJ, Stevens CM (1988) Methane-hydrogen gas seeps, Zambales ophiolite, Philippines: Deep or shallow origin? Chem Geol 71(3):211-222

Abrajano TA, Sturchio NC, Kennedy BM, Muelenbachs K, Lyon GL, Bohlke JK (1990) Geochemistry of reduced gas related to serpentinization of the Zambales Ophiolite, Philippines. Appl Geochem 5:625-630

Allen DE, Seyfried Jr. WE (2004) Serpentinization and heat generation: Constraints from Lost City and Rainbow hydrothermal systems. Geochim Cosmochim Acta 68(6):1347-1354

Amend JP, Shock E (1998) Energetics of amino acid synthesis in hydrothermal ecosystems. Science 281:1659-1662

Arndt NT (1983) Role of a thin, komatiite-rich oceanic crust in the Archean plate-tectonic process. Geology 11:372-375

Aubrey AD, Cleaves HJ, Bada JL (2009) The role of submarine hydrothermal systems in the synthesis of amino acids. Origins Life Evol Biosphere 39:91-108

Baaske P, Weinert FM, Duhr S, Lemke KH, Russell MJ, Braun D (2007) Extreme accumulation of nucleotides in simulated hydrothermal pore systems. Proc Natl Acad Sci USA 104(22):9346-9351

Bapteste E, Brochier C, Boucher Y (2005) Higher-level classification of the Archaea: evolution of methanogenesis and methanogens. Archaea 1:353-363

Barnes I, LaMarche VC Jr, Himmelberg G (1967) Geochemical evidence of present day serpentinization. Science 156:830-832

Barnes I, O'Neil JR (1971) Calcium-magnesium carbonate solid solutions from Holocene conglomerate cements and travertines in the Coast Range of California. Geochim Cosmochim Acta 35:699-718

Barnes I, O'Neil JR, Trescases JJ (1978) Present day serpentinization in New Caledonia, Oman and Yugoslavia. Geochim Cosmochim Acta 42:144-145

Baross JA, Hoffman SE (1985) Submarine hydrothermal vents and associated gradient environments as sites for the origin and evolution of life. Origins Life 15:327-345

Bath AH, Christofi N, Philp JC, Cave MR, McKinley IG, Berner U (1987) Trace element and microbiological studies of alkaline groundwaters in Oman, Arabian Gulf: A natural analog for cement pore-waters. FLPU Report 87-2, British Geological Survey

Berg IA (2011) Ecological aspects of the distribution of different autotrophic CO_2 fixation pathways. Appl Environ Microbiol 77(6):1925-1936, doi: 10.1128/AEM.02473-10

Berg IA, Kockelkorn D, Ramos-Vera WH, Say RF, Zarzycki J, Hügler M, Alber BE, Fuchs G (2010) Autotrophic carbon fixation in archaea. Nature Rev Microbiol 8:447-460

Berndt ME, Allen DE, Seyfried WE (1996) Reduction of CO_2 during serpentinization of olivine at 300 °C and 500 bar. Geology 24(4):351-354

Blais S, Auvray B (1990) Serpentinization in the Archean komatiitic rocks of the Kuhmo greenstone belt, eastern Finland. Can Mineral 28:55-66

Blank JG, Green SJ, Blake D, Valley JW, Kita NT, Treiman A, Dobson PF (2009) An alkaline spring system within the Del Puerto Ophiolite (California, USA): A Mars analog site. Planet Space Sci 57:533-540

Bradley AS, Fredricks H, Hinrichs K-U, Summons RE (2009a) Structural diversity of diether lipids in carbonate chimneys at the Lost City Hydrothermal Field. Org Geochem 40:1169-1178

Bradley AS, Hayes JM, Summons RE (2009b) Extraordinary ^{13}C enrichment of diether lipids at the Lost City Hydrothermal Field indicates a carbon-limited ecosystem. Geochim Cosmochim Acta 73:102-118

Bradley AS, Summons RE (2010) Multiple origins of methane at the Lost City Hydrothermal Field. Earth Planet Sci Lett 297:34-41

Brazelton WJ, Baross JA (2009) Abundant transposases encoded by the metagenome of a hydrothermal chimney biofilm. ISME J 3:1420-1424

Brazelton WJ, Baross JA (2010) Metagenomic comparison of two *Thiomicrospira* lineages inhabiting contrasting deep-sea hydrothermal environments. PLoS One 5(10):e13530, doi: 10.1371/journal.pone.0013530

Brazelton WJ, Ludwig KA, Sogin ML, Andreishcheva EN, Kelley DS, Shen C-C, Edwards RL, Baross JA (2010) Archaea and bacteria with surprising microdiversity show shifts in dominance over 1,000-year time scales in hydrothermal chimneys. Proc Natl Acad Sci USA 107(4):1612-1617

Brazelton WJ, Mehta MP, Kelley DS, Baross JA (2011) Physiological differentiation within a single-species biofilm fueled by serpentinization. mBio 4(2), doi: 10.1128/mBio.00127-11
Brazelton WJ, Nelson B, Schrenk MO (2012) Metagenomic evidence for H_2 oxidation and H_2 production by serpentinite-hosted subsurface microbial communities. Frontiers in Microbiology 2, doi: 10.3389/fmicb.2011.00268
Brazelton WJ, Schrenk MO, Kelley DS, Baross JA (2006) Methane and sulfur metabolizing microbial communities dominate in the Lost City Hydrothermal Field ecosystem. Appl Environ Microbiol 72(9):6257-6270
Cannat M, Fontaine F, Escartin J (2010) Serpentinization and associated hydrogen and methane fluxes at slow-spreading ridges. *In:* Diversity of Hydrothermal Systems on Slow Spreading Ocean Ridges. Rona PA, Devey CW, Dyment J, Murton BJ (eds), Washington, D. C. American Geophysical Union, p 241-264
Charlou JL, Donval JP, Fouquet Y, Jean-Baptiste P, Holm N (2002) Geochemistry of high H_2 and CH_4 vent fluids issuing from ultramafic rocks at the Rainbow hydrothermal field (36°14′N, MAR). Chem Geol 191:345-359
Charlou JL, Donval JP, Konn C, Ondréas H, Fouquet Y (2010) High production and fluxes of H_2 and CH_4 and evidence of abiotic hydrocarbon synthesis by serpentinization in ultramafic-hosted hydrothermal systems on the Mid-Atlantic Ridge. *In:* Diversity of Hydrothermal Systems on Slow Spreading Ocean Ridges. Rona PA, Devey CW, Dyment J, Murton BJ (eds), Washington, D. C. American Geophysical Union, p 265-296
Chivian D, Brodie EL, Alm EJ, Culley DE, Dehal PS, DeSantis TZ, Gihring TM, Lapidus A, Lin L-H, Lowry SR, Moser DP, Richardson PM, Southam G, Wanger G, Pratt LM, Andersen GL, Hazen TC, Brockman FJ, Arkin AP, Onstott TC (2008) Environmental genomics reveals a single-species ecosystem deep within Earth. Science 322:275-278
Cipolli F, Gambardella B, Marini L, Ottonello G, Zuccolini MV (2004) Geochemistry of high-pH waters from serpentinites of the Gruppo di Voltri (Genova, Italy) and reaction path modeling of CO_2 sequestration in serpentinite aquifers. Appl Geochem 19:787-802
Cody GD (2004) Transition metal sulfides and the origins of metabolism. Annu Rev Earth Planet Sci 32:569-599
Cody GD, Boctor NZ, Filley TR, Hazen RM, Scott JH, Sharma A, Yoder Jr. HS (2000) Primordial carbonylated iron-sulfur compounds and the synthesis of pyruvate. Science 289:1337-1340
Cody GD, Scott JH (2007) The roots of metabolism. *In:* Planets and Life: The Emerging Science of Astrobiology. Sullivan III WT, Baross J (eds) Cambridge, UK Cambridge University Press, p 174-186
Cox ME, Launay J, Paris JP (1982) Geochemistry of low temperature geothermal systems in New Caledonia. *In*: Proc Pacific Geothermal Conf 1982, University of Auckland, New Zealand, p 453-459
Delacour A, Früh-Green GL, Bernasconi SM, Kelley DS (2008a) Sulfur in peridotites and gabbros at Lost City (30°N, MAR): Implications for hydrothermal alteration and microbial activity during serpentinization. Geochim Cosmochim Acta 72:5090-5110
Delacour A, Früh-Green GL, Bernasconi SM, Schaeffer P, Kelley DS (2008b) Carbon geochemistry of serpentines in the Lost City Hydrothermal System (30°N, MAR). Geochim Cosmochim Acta 72:3681-3702
Dilek Y, Furnes H (2011) Ophiolite genesis and global tectonics: Geochemical and tectonic fingerprinting of ancient oceanic lithosphere. Geol Soc Am Bull 123:387-411
Dobrinski KP, Longo DL, Scott KM (2005) The carbon-concentrating mechanism of the hydrothermal vent chemolithoautotroph *Thiomicrospira crunogena*. J Bacteriol 187(16):5761-5766, doi: 10.1128/JB.187.16.5761–5766.2005
Douville E, Charlou JL, Oelkers EH, Bienvenu P, Colon CFJ, Donval JP, Fouquet Y, Prieur D, Appriou P (2002) The rainbow vent fluids (36 degrees 14′ N, MAR): the influence of ultramafic rocks and phase separation on trace metal content in Mid-Atlantic Ridge hydrothermal fluids. Chem Geol 184:37-48
Dulov LE, Lein AY, Dubinina GA, Pimenov NV (2005) Microbial processes at the Lost City Vent Field, Mid-Atlantic. Microbiology 74(1):111-118
Edmonds HN, Michael PJ, Baker ET, Connelly DP, Snow JE, Langmuir CH, Dick HJB, Mühe R, German CR, Graham DW (2003) Discovery of abundant hydrothermal venting on the ultraslow-spreading Gakkel ridge in the Arctic Ocean. Nature 421:252-256
Ehlmann BL, Mustard JF, Murchie SL (2010) Geologic setting of serpentine deposits on Mars. Geophys Res Lett 37:L06201, doi: 10.1029/2010GL042596
Etiope G, Schoell M, Hosgörmez H (2011) Abiotic methane flux from the Chimaera seep and Tekirova ophiolites (Turkey): Understanding gas exhalation from low temperature serpentinization and implications for Mars. Earth Planet Sci Lett 310:96-104
Foustoukos DI, Seyfried Jr. WE (2004) Hydrocarbons in hydrothermal vent fluids: The role of Chromium-bearing catalysts. Science 304:1002-1005

Friend CRL, Bennett VC, Nutman AP (2002) Abyssal peridotites >3800 Ma from southern West Greenland: field relationships, petrography, geochronology, whole-rock and mineral chemistry of dunite and harzburgite inclusions in the Itsaq Gneiss Complex. Contrib Mineral Petrol 143:71-92, doi: 10.1007/s00410-001-0332-7

Fritz P, Clark ID, Fontes JC, Whiticar MJ, Faber E (1992) Deuterium and ^{13}C evidence for low temperature production of hydrogen and methane in a highly alkaline groundwater environment in Oman. *In:* Proceedings of the 7th International Symposium on Water-Rock Interaction. Kharaka Y, Maest AS (eds), Rotterdam Balkama, p 792-796

Früh-Green GL, Connolly JAD, Plas A, Kelley DS, Grobéty B (2004) Serpentinization of oceanic peridotites: Implications for geochemical cycles and biological activity. *In:* The Subseafloor Biosphere at Mid-Ocean Ridges, Geophysical Monograph Series. WD Wilcock, DeLong EF, Kelley DS, Baross JA, Cary SC (eds) Washington, DC, American Geophysical Union, p 119-136

Früh-Green GL, Kelley DS, Bernasconi SM, Karson JA, Ludwig KA, Butterfield DA, Boschi C, Proskurowski G (2003) 30,000 years of hydrothermal activity at the Lost City Vent Field. Science 301:495-498

Fryer P (2012) Serpentinite mud volcanism: observations, processes, and implications. Ann Rev Mar Sci 4:345-373

Fuchs G, Stupperich E (1985) Evolution of autotrophic CO_2 fixation. *In:* Evolution of the Procaryotes. Schleifer KH, Stackebrandt E (eds), London, UK Academic Press, p 235-251

Gerasimchuk AL, Shatalov AA, Novikov AL, Butorova OP, Pimenov NV, Lein AY, Yanenko AS, Karnachuk OV (2010) The search for sulfate reducing bacteria in mat samples from the Lost City Hydrothermal Field by molecular cloning. Microbiology 79(1):96-105

German CR, Bowen A, Coleman ML, Honig DL, Huber JA, Jakuba MV, Kinsey JC, Kurz MD, Leroy S, McDermott JM, Lépinay BMd, Nakamura K, Seewald JS, Smith JL, Sylva SP, Dover CLV, Whitcomb LL, Yoerger DR (2010) Diverse styles of submarine venting on the ultraslow spreading Mid-Cayman Rise. Proc Natl Acad Sci USA 10(32):14020-14025

Haggerty JA, Fisher JB (1992) Short-chain organic acids in interstitial waters from Mariana and Bonin forearc serpentinites: Leg 1251. *In:* Proceedings of the Ocean Drilling Program, Scientific Results. Fryer P, Coleman P, Pearce JA, Stokking LB (eds) College Station, Texas Ocean Drilling Program, p 387-395, doi: 10.2973/odp.proc.sr.125.125.1992

Hennet RJC, Holm NG, Engel MH (1992) Abiotic synthesis of amino acids under hydrothermal conditions and the origin of life - a perpetual phenomenon. Naturwissenschaften 79:361-365

Hoehler TM (2007) An energy balance concept for habitability. Astrobiology 7(6):824-838

Holm NG, Baltscheffsky H (2011) Links between hydrothermal environments, pyrophosphate, Na^+, and early evolution. Origins Life Evol Biosphere, doi: 10.1007/s11084-011-9235-4

Holm NG, Dumont M, Ivarsson M, Konn C (2006) Alkaline fluid circulation in ultramafic rocks and formation of nucleotide constituents: a hypothesis. Geochem Trans 7(7), doi: 10.1186/1467-4866-7-7

Horita J, Berndt ME (1999) Abiogenic methane formation and isotopic fractionation under hydrothermal conditions. Science 285:1055-1057

Hosgormez H, Etiope G, Yalçin MN (2008) New evidence for a mixed inorganic and organic origin of the Olympic Chimaera fire (Turkey): a large onshore seepage of abiogenic gas. Geofluids 8:263-273

Huber C, Wächterhäuser G (1997) Activated acetic acid by carbon fixation on (Fe, Ni)S under primordial conditions. Science 276:245-247

Islam MN, Kaneko T, Kobayashi K (2001) Determination of amino acids formed in a supercritical water flow reactor simulating submarine hydrothermal systems. Anal Sci 17:1631-1634

Itävaara M, Nyyssönen M, Kapanen A, Nousiainen A, Ahonen L, Kukkonen I (2011) Characterization of bacterial diversity to a depth of 1500 m in the Outokumpu deep borehole, Fennoscandian Shield. FEMS Microbiol Ecol 77:295-309

Kamaluddin HY, Egami F (1979) Formation of molecules of biological interest from formaldehyde and hydroxylamine in a modified sea medium. J Biochem Tokyo 85:1503-1508

Keir RS (2010) A note on the fluxes of abiogenic methane and hydrogen from mid-ocean ridges. Geophys Res Lett 37: L24609, doi: 10.1029/2010GL045362

Kelemen PB, Matter J (2008) In situ carbonation of peridotite for CO_2 storage. Proc Natl Acad Sci USA 105(45):17295-17300

Kelemen PB, Matter J, Streit EE, Rudge JF, Curry WB, Blusztajn J (2011) Rates and mechanisms of mineral carbonation in peridotite: natural processes and recipes for enhanced, in situ CO_2 capture and storage. Annu Rev Earth Planet Sci 39:545-576, doi: 10.1146/annurev-earth-092010-152509

Kelley DS, Früh-Green GL (1999) Abiogenic methane in deep-seated mid-ocean ridge environments: insights from stable isotope analyses. J Geophys Res 104:10439-10460

Kelley DS, Karson JA, Blackman DK, Früh-Green GL, Butterfield DA, Lilley MD, EJ Olson, Schrenk MO, Roe KK, Lebon GT, Rivizzigno P, AT3-60 Shipboard Party (2001) An off-axis hydrothermal vent field near the Mid-Atlantic Ridge at 30°N. Nature 412:145-149

Kelley DS, Karson JA, Früh-Green GL, Yoerger DR, Shank TM, Butterfield DA, Hayes JM, Schrenk MO, Olson EJ, Proskurowski G, Jakuba M, Bradley A, Larson B, Ludwig K, Glickson D, Buckman K, Bradley AS, Brazelton WJ, Roe K, Elend MJ, Delacour A, Bernasconi SM, Lilley MD, Baross JA, Summons RE, Sylva SP (2005) A serpentinite-hosted ecosystem: The Lost City Hydrothermal Field. Science 307:1428-1434

Krulwich TA (1995) Alkaliphiles: "basic" molecular problems of pH tolerance and bioenergetics. Mol Microbiol 15(3):403-410

Krulwich TA, Sachs G, Padan E (2011) Molecular aspects of bacterial pH sensing and homeostasis Nature Rev Microbiol 9:330-343

Lane N, Martin W (2010) The energetics of genome complexity. Nature 467:929-934

Lang SQ, Butterfield DA, Schulte M, Kelley DS, Lilley MD (2010) Elevated concentrations of formate, acetate and dissolved organic carbon found at the Lost City hydrothermal field. Geochim Cosmochim Acta 74:941-952

Lang SQ, Früh-Green GL, Bernasconi SM, Lilley MD, Proskurowski G, Méhay S, Butterfield DA (2012) Microbial utilization of abiogenic carbon and hydrogen in a serpentinite-hosted system. Geochim Cosmochim Acta 92:82-99

Launay J, Fontes J-C (1985) Les sources thermales de Prony (Nouvelle--Caledonie) et leurs precipites chimiques. Exemple de formation de brucite primaire. Geologie de la France 1:83-100

Li Y, Breaker RR (1999) Kinetics of RNA degradation by specific base catalysis of transesterification involving the 2′-hydroxyl group. J Am Chem Soc 121(23):5364-5372, doi: 10.1021/ja990592p

Lin L-H, Wang P-L, Rumble D, Lippmann-Pipke J, Boice E, Pratt LM, Lollar BS, Brodie EL, Hazen TC, Andersen GL, DeSantis TZ, Moser DP, Kershaw D, Onstott TC (2006) Long-term sustainability of a high-energy, low-diversity crustal biome. Science 314:479-482

Lowe CU, Markham R, Rees MW (1963) Synthesis of complex organic compounds from stable precursors-formation of amino acids, amino acid polymers, fatty acids and purines from ammonium cyanide. Nature 199:219-220

Lowell RP, Rona PA (2002) Seafloor hydrothermal systems driven by the serpentinization of peridotite. Geophys Res Lett 29(26):1-5

Ludwig KA, Shen C-C, Kelley DS, Cheng H, Edwards RL (2011) U-Th systematics and ^{230}Th ages of carbonate chimneys at the Lost City Hydrothermal Field. Geochim Cosmochim Acta 75:1869-1888

Marques JM, Carreira PM, Carvalho MR, Matias MJ, Goff FE, Basto MJ, Graça RC, Aires-Barros L, Rocha L (2008) Origins of high pH mineral waters from ultramafic rocks, Central Portugal. Appl Geochem 23(12):3278-3289

Marshall WL (1994) Hydrothermal synthesis of amino acids. Geochim Cosmochim Acta 58:2099-2106

Martin W, Baross J, Kelley D, Russell MJ (2008) Hydrothermal vents and the origin of life. Nature Rev Microbiol 6(11):805-814

Martin W, Russell MJ (2007) On the origin of biochemistry at an alkaline hydrothermal vent. Philos Trans Royal Soc London B 367:1887-1925

Mason OU, Nakagawa T, Rosner M, Nostrand JDV, Zhou J, Maruyama A, Fisk MR, Giovannoni SJ (2010) First investigation of the microbiology of the deepest layer of ocean crust. PloS One(11):e15399

McCollom TM (2007) Geochemical constraints on sources of metabolic energy for chemolithoautotrophy in ultramafic-hosted deep-sea hydrothermal systems. Astrobiology 7(6):933-950

McCollom TM (2013) Laboratory simulations of abiotic hydrocarbon formation in Earth's deep subsurface. Rev Mineral Geochem 75:467-494

McCollom TM, Ritter G, Simoneit BRT (1999) Lipid synthesis under hydrothermal conditions by Fischer-Tropsch-type reactions. Origins Life Evol Biosphere 29:153-166

McCollom TM, Seewald JS (2001) A reassessment of the potential for reduction of dissolved CO_2 to hydrocarbons during serpentinization of olivine. Geochim Cosmochim Acta 65(21):3769-3778

McCollom TM, Seewald JS (2003) Experimental constraints on the hydrothermal reactivity of organic acids and acid anions: I. Formic acid and formate. Geochim Cosmochim Acta 67:3625-3644

McCollom TM, Seewald JS (2006) Carbon isotope composition of organic compounds produced by abiotic synthesis under hydrothermal conditions. Earth Planet Sci Lett 243(1-2):74-84

McCollom TM, Seewald JS (2007) Abiotic synthesis of organic compounds in deep-sea hydrothermal environments. Chem Rev 107:382-401

Melchert B, Devey CW, German CR, Lackschewitz KS, Seifert R, Walter M, Mertens C, Yoerger DR (2008) First evidence for high-temperature off-axis venting of deep crustal/mantle heat: The Nibelungen hydrothermal field, southern Mid-Atlantic Ridge. Earth Planet Sci Lett 275:61-69

Ménez B, Pasini V, Brunelli D (2012) Life in the hydrated suboceanic mantle. Nature Geoscience 5:133-137

Morita RY (2000) Is H_2 the universal energy source for long-term survival? Microbiol Ecol 38:307-320

Moser DP, Gihring TM, Brockman FJ, Fredrickson JK, Balkwill DL, Dollhopf ME, Lollar BS, Pratt LM, Boice E, Southam G, Wanger G, Baker BJ, Pfiffner SM, Lin L-H, Onstott TC (2005) *Desulfotomoaculum* and *Methanobacterium* spp. dominate a 4- to 5-kilometer-deep fault. Appl Environ Microbiol 71(12):8773-8783

Mottl MJ, Komor SC, Fryer P, Moyer CL (2003) Deep-slab fluids fuel extremophilic Archaea on a Mariana forearc serpentinite mud volcano: Ocean Drilling Program Leg 195. Geochem Geophys Geosyst 4(11):2003GC000588

Mumma MJ, Villanueva GL, Novak RE, Hewagama T, Bonev BP, DiSanti MA, Mandell AM, Smith MD (2009) Strong release of Methane on Mars in northern summer 2003. Science 323:1041-1045

Neal C, Stanger G (1983) Hydrogen generation form mantle source rocks in Oman. Earth Planet Sci Lett 66:315-320

Nealson KH, Inagaki F, Takai K (2005) Hydrogen-driven subsurface lithoautotrophic microbial ecosystems (SLiMEs): do they exist and why should we care? Trends Microbiol 13(9):405-410

Nercessian O, Bienvenu N, Moriera D, Prieur D, Jeanthon C (2005) Diversity of functional genes of methanogens, methanotrophs and sulfate reducers in deep-sea hydrothermal environments. Environ Microbiol 7(1):118-132

Nicolas A, Boudier F, Ildefonse B, Ball E (2000) Accretion of Oman and United Arab Emirates ophiolite - Discussion of a new structural map. Mar Geophys Res 21(3-4):147-180, doi: 10.1023/A:1026769727917

Nisbet EG, Fowler CMR (1983) Model for Archean plate tectonics. Geology 11:376-379

Nisbet EG, Sleep NH (2001) The habitat and nature of early life. Nature 409:1083-1091

Nna-Mvondo D, Martinez-Frias J (2007) Komatiites: from Earth's geological settings to planetary and astrobiological contexts Earth, moon, and planets 100(3-4):157-179, doi: 10.1007/s11038-007-9135-9

Ohara Y, Reagan MK, Fujikura K, Watanabe H, Michibayashi K, Ishii T, Stern RJ, Pujana I, Martinez F, Girard G, Ribeiro J, Brounce M, Komori N, Kino M (2011) A serpentinite-hosted ecosystem in the Southern Mariana Forearc. Proc Natl Acad Sci USA 109(8):2831-2835, doi: 10.1073/pnas.1112005109

Óro J, Kimball A, Fritz R, Master F (1959) Amino acid synthesis from formaldehyde and hydroxylamine. Arch Biochem Biophys 85:115-130

Palandri JL, Reed MH (2004) Geochemical models of metasomatism in ultramafic systems: Serpentinization, rodingitization, and sea floor carbonate chimney precipitation. Geochim Cosmochim Acta 68(5):1115-1133

Pedersen K (2012) Subterranean microbial populations metabolize hydrogen and acetate under *in situ* conditions in granitic groundwater at 450 m depth in the Äspö Hard Rock Laboratory, Sweden. FEMS Microbiol Ecol 81:217-229

Pereto JG, Velasco AM, Becerra A, Lazcano A (1999) Comparative biochemistry of CO_2 fixation and the evolution of autotrophy. Int Microbiol 2:3-10

Perner M, Kuever J, Seifert R, Pape T, Koschinsky A, Schmidt K, Strauss H, Imhoff JF (2007) The influence of ultramafic rocks on microbial communities at the Logatchev hydrothermal field, located at 15°N on the Mid-Atlantic Ridge. FEMS Microbiol Ecol 61:97-109

Perner M, Petersen JM, Zielinski F, Gennerich H-H, Seifert R (2010) Geochemical constraints on the diversity and activity of H_2-oxidizing microorganisms in diffuse hydrothermal fluids from a basalt- and an ultramafic-hosted vent. FEMS Microbiol Ecol 74:55-71

Proskurowski G (2010) Abiogenic Hydrocarbon Production at the Geosphere-Biosphere Interface via Serpentinization Reactions. *In:* Handbook of Hydrocarbon and Lipid Microbiology. KN Timmis (ed) Berlin Heidelberg Springer, p 215-231

Proskurowski G, Lilley MD, Seewald JS, Früh-Green GL, Olson EJ, Lupton JE, Sylva SP, Kelley DS (2008) Abiogenic hydrocarbon production at Lost City hydrothermal field. Science 319:604-607

Ricardo A, Carrigan MA, Olcott AN, Benner SA (2004) Borate minerals stabilize ribose. Science 303:196

Roussel EG, Konn C, Charlou J-L, Donval J-P, Fouquet Y, Querellou J, Prieur D, Bonavita M-AC (2011) Comparison of microbial communities associated with three Atlantic ultramafic hydrothermal systems. FEMS Microbiol Ecol 77:647-665

Rushdi AI, Simoneit BRT (2001) Lipid formation by aqueous Fischer-Tropsch-type synthesis over a temperature range of 100 to 400 degrees C. Origins Life Evol Biospheres 31:103-118

Rushdi AI, Simoneit BRT (2004) Condensation reactions and formation of amides, esters, and nitriles under hydrothermal conditions. Astrobiology 4:211-224

Russell MJ, Daniel RM, Hall AJ, Sherringham JA (1994) A hydrothermally precipitated catalytic iron sulphide membrane as a first step toward life. J Mol Evol 39(3):231-243

Russell MJ, Hall AJ (1997) The emergence of life from iron monosulphide bubbles at a submarine hydrothermal redox and pH front. J Geol Soc London 154:377-402

Schmidt K, Garbe-Schönberg D, Koschinsky A, Strauss H, Jost CL, Klevenz V, Königer P (2011) Fluid elemental and stable isotope composition of the Nibelungen hydrothermal field (8°18′S, Mid-Atlantic Ridge): Constraints on fluid-rock interaction in heterogeneous lithosphere. Chem Geol 280(1):1-18

Schrenk MO, Bolton SA, Kelley DS, Baross JA (2004) Low archaeal diversity linked to subseafloor geochemical processes at the Lost City Hydrothermal Field, Mid-Atlantic Ridge. Environ Microbiol 6(10):1086-1095

Schrenk MO, Huber JA, Edwards KJ (2010) Microbial provinces in the subseafloor. Annu Rev Mar Sci 2:279-304

Seewald JS, Zolotov M, McCollom TM (2006) Experimental investigation of carbon speciation under hydrothermal conditions. Geochim Cosmochim Acta 70:446-460

Sephton MA, Hazen RM (2013) On the origins of deep hydrocarbons. Rev Mineral Geochem 75:449-465

Sherwood Lollar B, Frape SK, Weise SM, Fritz P, Macko SA, Welhan JA (1993) Abiogenic methanogenesis in crystalline rocks. Geochim Cosmochim Acta 57:5087-5097

Sherwood Lollar B, Lacrampe-Couloume G, Slater GF, Ward J, Moser DP, Gihring TM, Lin L-H, Onstott TC (2006) Unravelling abiogenic and biogenic sources of methane in the Earth's deep subsurface. Chem Geol 226:328-339

Shock EL (1990) Geochemical constraints on the origin of organic compounds in hydrothermal systems. Origins Life Evol Biosphere 20:331-367

Shock EL (1992) Stability of peptides in high temperature aqueous solutions. Geochim Cosmochim Acta 56:3481-3492

Shock EL, Schulte MD (1998) Organic synthesis during fluid mixing in hydrothermal systems. J Geophys Res 103(No. E12):28513-28527

Sleep NH, Bird DK, Pope EC (2011) Serpentinite and the dawn of life. Philos Trans Royal Soc London B 366:2857-2869

Sleep NH, Meibom A, Fridriksson T, Coleman RG, Bird DK (2004) H_2-rich fluids from serpentinization: Geochemical and biotic implications. Proc Nat Acad Sci USA 101(35):12818-12823

Sorokin DY, Tourova TP, Mußmann M, Muyzer G (2008) *Dethiobacter alkaliphilus* gen. nov. sp. nov., and *Desulfurivibrio alkaliphilus* gen. nov. sp. nov.: two novel representatives of reductive sulfur cycle from soda lakes. Extremophiles 12:431-439

Szponar N, Brazelton WJ, Schrenk MO, Bower DM, Steele A, Morrill P (2012) Geochemistry of a continental site of serpentinization in the Tablelands Ophiolite, Gros Morne National Park: a Mars analogue. Icarus, doi: 10.1016/j.icarus.2012.07.004

Takai K, Gamo T, Tsunogai U, Nakyama N, Hirayama H, Nealson KH, Horikoshi K (2004) Geochemical and microbiological evidence for a hydrogen-based, hyperthermophilic subsurface lithoautotrophic microbial ecosystem (HyperSLiME) beneath an active deep-sea hydrothermal field. Extremophiles 8:269-282

Takai K, Moyer CL, Miyazaki M, Nogi Y, Hirayama H, Nealson KH, Horikoshi K (2005) *Marinobacter alkaliphilus* sp. nov., a novel alkaliphilic bacterium isolated from subseafloor alkaline serpentine mud from Ocean Drilling Program Site 1200 at South Chamorro Seamount, Mariana Forearc. Extremophiles 9:17-27

Takai K, Nakamura K, Toki T, Tsunogai U, Miyazaki M, Miyazaki J, Hirayama H, Nakagawa S, Nunoura T, Horikoshi K (2008) Cell proliferation at 122 °C and isotopically heavy CH_4 production by a hyperthermophilic methanogen under high-pressure cultivation. Proc Natl Acad Sci USA 105(31):10949-10954

Tiago I, Chung AP, Veríssimo A (2004) Bacterial diversity in a nonsaline alkaline environment: heterotrophic aerobic populations. Appl Environ Microbiol 70(12):7378-7387

Tiago I, Mendes V, Pires C, Morais PV, Veríssimo A (2006) *Chimaereicella alkaliphila* gen. nov., sp. nov., a Gram-negative alkaliphilic bacterium isolated from a nonsaline alkaline groundwater. Syst Appl Microbiol 29:100-108

Tiago I, Pires C, Mendes V, Morais PV, Costa Md, Veríssimo A (2005) *Microcella putealis* gen. nov., sp. nov., a Gram-positive alkaliphilic bacterium isolated from a nonsaline alkaline groundwater. Syst Appl Microbiol 28:479-487

Van Mooy BAS, Rocap G, Fredricks HF, Evans CT, Devol AH (2006) Sulfolipids dramatically decrease phosphorus demand by picocyanobacteria in oligotrophic marine environments. Proc Natl Acad Sci USA 103(23):8607-8612

Vance S, Harnmeijer J, Kimura J, Hussmann H, Demartin B, Brown JM (2007) Hydrothermal systems in small ocean planets. Astrobiology 7(6):987-1005

Wolman Y, Miller SL, Ibanez J, Oro J (1971) Formaldehyde and ammonia as precursors to prebiotic amino acids. Science 174:1039-1041

Yanagawa H, Kobayashi K (1992) An experimental approach to chemical evolution in submarine hydrothermal systems. Origins Life Evol Biosphere 22:147-159

Reviews in Mineralogy & Geochemistry
Vol. 75 pp. 607-648, 2013

High-Pressure Biochemistry and Biophysics

Filip Meersman

Department of Chemistry, University College London
20 Gordon Street, LondonWC1H 0AJ, United Kingdom

Rousselot-Expertise Centre, R&D Laboratory
Meulestedekaai 81, 9000 Gent, Belgium

f.meersman@ucl.ac.uk

Isabelle Daniel

Laboratoire de Sciences de la Terre, Université Claude Bernard Lyon 1
Bâtiment Géode, 6 rue Raphael Dubois, F-69622 Villeurbanne cedex, France

Douglas H. Bartlett

Scripps Institution of Oceanography, University of California, San Diego
9500 Gilman Drive, La Jolla California, 92093, U.S.A.

Roland Winter

Physical Chemistry I - Biophysical Chemistry, TU Dortmund University
10 Otto- Hahn Str. 6, D-44227 Dortmund, Germany

Rachael Hazael, Paul F. McMillan

Department of Chemistry, University College London
20 Gordon Street, LondonWC1H 0AJ, United Kingdom

INTRODUCTION

By the end of the 19th century British and French oceanographic expeditions had shown that life exists in the deepest ocean trenches. Since then, microorganisms have been found to thrive in diverse environments characterized by a wide range of pressure-temperature-composition (*P-T-X*) conditions (Rothschild and Mancinelli 2001). The range of physicochemical conditions under which microbial life has been observed has continued to expand with greater access to extreme environments and greatly improved tools for sampling and assessing the diversity and physiology of microbial communities. This exploration now includes examination of subseafloor and continental subsurface settings—key goals of the Deep Life Directorate within the Deep Carbon Observatory (DCO) Program. Bacterial metabolic activity has been described at temperatures as low as −40 °C (Rivkina et al. 2000; Price and Sowers 2004; Panikov and Sizova 2007; Collins et al. 2010) and a methanogen has been cultured at 122 °C under hydrostatic pressure (Takai et al. 2008). Moreover, bacteria can withstand ionizing radiation levels up to 30,000 grays (Rainey et al. 2005), and can grow over a pH range between 0 and 12.5 (Takai et al. 2001; Sharma et al. 2012), at salinities up to 5.2 M NaCl (Kamekura 1998), and at hydrostatic pressures up to 130 MPa (Yayanos 1986). Bacterial survival has also been demonstrated up into the GPa range (Sharma et al. 2002; Vanlint et al. 2011).

Because of the difficulty to access deep pressure-affected environments compared to most other extreme environments, less is known about deep-sea and deep-continental microbial com-

1529-6466/13/0075-0019$00.00 DOI: 10.2138/rmg.2013.75.19

munities and their physiological adaptation to high hydrostatic pressure, even though high-pressure environments are more voluminous in nature than other extreme environments. Our current knowledge about life at high pressure currently derives from studies of deep-sea microorganisms that possess adaptations for growth at pressures roughly in the 10-130 MPa range (Bartlett 2002; Lauro and Bartlett 2008; Oger and Jebbar 2010). Such pressures are far below those typically used to assess the survival of microbes or to interrogate biophysically their isolated macromolecular systems. Nevertheless, the growth and reproduction adaptations at even these modest pressures provide valuable information on physiological properties, complex quaternary assemblages, and enzyme architectures necessary to understand the adaptation of life in a pressurized world. The technologies associated with growing deep-sea microbes in pressurized vessels are well described (Jannasch et al. 1996; Prieur and Marteinsson 1998; Yayanos 2001; Kato 2006), although additional technological developments continue to be made (Hiraki et al. 2012). For this reason, high-pressure biology has become an important topic, ranging from physiological studies of deep-sea organisms under *in situ* conditions, to the nature and function of extremophile organisms inhabiting the rocky subsurface. Complementing and underpinning the biological investigations are studies of the high-pressure physics and chemistry of the macromolecules essential for life that are important to a broad range of disciplines including food science, biomedicine, and nanotechnology.

In this chapter we first review effects of high pressure on lipid membranes, proteins, and nucleic acids, which are the principal macromolecules of cells. The pressures necessary to initiate protein unfolding and lipid phase transitions are generally higher than the maximum pressures observed for growth of organisms. We also discuss the intermolecular interactions that are relatively pressure sensitive and how biophysical studies on model systems can provide molecular explanations for biological observations such as the increased content of unsaturated lipids in deep-sea organisms. Also discussed are the recent applications of molecular methods to better understand pressure effects at the genetic level.

PROTEINS AND POLYPEPTIDES

Structures of proteins and polypeptides

Proteins are the chief macromolecules of the cell. They catalyze small molecule transformations, they allow cells to move around and to do work, and they maintain internal cell rigidity. Furthermore, they control the genes that determine the cell constitution and function, transport molecules across membranes, direct the synthesis of themselves and other macromolecules, and protect other macromolecules against denaturing conditions (Lodish et al. 1995). It is difficult to consider life processes as currently understood without proteins.

From the chemical viewpoint proteins are linear, heterogeneous polymers assembled from 20 different amino acid residues linked by covalent peptide bonds into the polypeptide chain. However, their most surprising characteristic is the fact that each polypeptide chain folds into a unique three-dimensional structure that is defined by the amino acid sequence. This feature makes proteins stand out among all macromolecules, biological and synthetic. The acquisition of the three-dimensional structure is necessary for a protein to be biologically functional. Some proteins, however, only adopt a tertiary structure upon interaction with their target molecule, whereas others can maintain functionality only through interaction with other proteins or macromolecules (Wright et al. 1999).

As the nascent polypeptide chain comes off the ribosome it will start to fold. This process involves the adoption of well-defined secondary structure elements such as the α-helix and β-sheet structures. The assembly of these structures in three-dimensional space results in the tertiary structure that is generally referred to as the native state. Some proteins may form complexes with themselves or with other proteins. Such assemblies represent the quaternary structure of proteins. For instance, hemoglobin is a heterotetramer made up of two α- and two β-subunits with polypeptide chains that have a conformation resembling that of myoglobin, and the trans-membrane proteins associated with transport of ions and molecular species across cell walls typically form pentameric or heptameric complexes that define hydrophilic regions inside the central channel and a hydrophobic exterior containing cavities or "pockets" in contact with the lipid bilayer.

The driving forces for protein folding are the non-covalent interactions (hydrophobic effect, hydrogen bonding, and other electrostatic interactions) between the amino acids and their interaction with the surrounding aqueous milieu. Non-covalent interactions, with typical energies of 4-40 kJ mol^{-1}, are weak compared to covalent bonds (300-400 kJ mol^{-1}). As a result, proteins are only marginally stable and can easily break down into a less-ordered state, the so-called unfolded state. However, this state is highly unstable under physiological conditions and the protein readily reassumes its native state. It has always been assumed that the unfolded state is a random coil structure in which no side chain-side chain interactions occur. However, a large body of evidence now suggests that the unfolded state is, in fact, a heterogeneous ensemble of varying compactness and often contains large amounts of residual structure (Shortle 1996; Smith et al. 1996; Klein-Seetharaman et al. 2002).

Protein stability is defined as the difference in free energy, ΔG_{stab}, between the native and the unfolded state under physiological conditions (Creighton 1990; Pace et al. 1991). However, here ΔG will refer to the free energy change of unfolding, which equals $-\Delta G_{stab}$. Typical values of ΔG are in the range 20 to 40 kJ mol^{-1}. The reason for this low stability lies in the fact that proteins require sufficient conformational flexibility for transport across membranes, natural turnover, binding of substrates, and processes like allostery and signal transduction (Daniel et al. 1996).

Thermodynamic considerations: volume *versus* compressibility arguments

As described above, proteins exist with distinct structures and conformational states determined by the P and T conditions as well as the chemical (X) environments. For a reversible, two-state folding/unfolding process between N (Native) ~ U (Unfolded) states, the pressure (P) and temperature (T) dependence of ΔG, the difference in Gibbs free energy between U and N, is given by

$$d(\Delta G) = -\Delta S dT + \Delta V dP \tag{1}$$

where ΔS is the difference in entropy and ΔV is the volume change between the native and unfolded states. At constant T, the derivative of ΔG with respect to P is given by ΔV as summarized by the principle of Le Châtelier, which states that a pressure increase will shift a given equilibrium to the side that occupies the smallest volume. Integration of Equation (1) leads to:

$$\Delta G(P,T) = \Delta G^{o} - \Delta S^{o}(T - T_{o}) + \Delta C_{p}[(T - T_{o}) - T\ln(T / T_{o})] + \Delta V^{o}(P - P_{o}) - \frac{\Delta\beta}{2}(P - P_{o})^{2} + \Delta\alpha(T - T_{o})(P - P_{o}) \tag{2}$$

where ΔG^o, ΔV^o and ΔS^o refer to the reference conditions, usually taken to be $P_o = 0.1$ MPa and $T_o = 298$ K. The second order terms $\Delta\alpha$, $\Delta\beta$ and ΔC_P are proportional to differences in thermal expansion, compressibility, and heat capacity between the unfolded and the native state of the protein, respectively. These parameters are assumed to be P- and T-independent, and are defined as follows:

$$\begin{aligned} \Delta\alpha &= \left(\partial\Delta V / \partial T\right)_P = -\left(\partial\Delta S / \partial P\right)_T \\ \Delta\beta &= -\left(\partial\Delta V / \partial P\right)_T \\ \Delta C_P &= T\left(\partial\Delta S / \partial T\right)_P \end{aligned} \tag{3}$$

Equation (2) originates as a Taylor expansion of $\Delta G(P,T)$, with a cut-off after the second-order terms. The precise meaning and measurement of the α, β and C_P parameters are developed and discussed elsewhere (Chalikian 2003; Meersman et al. 2006). At constant T, Equation (2) can be rewritten as:

$$\Delta G(P) = \Delta G^{\circ} + \Delta V^{\circ}(P - P_{\circ}) - \frac{\Delta\beta}{2}(P - P_{\circ})^2 \tag{4}$$

The last term on the right reflects the P dependence of ΔV, which, at high pressures, can no longer be predicted from ΔV° alone. The compressibility factor $\Delta\beta$ is related to the isothermal compressibility β_T [$\beta_T = -V^{-1}(\partial V/\partial P)_T$], which is the second derivative of ΔV with respect to pressure, *via* $\Delta\beta = V\,\Delta\beta_T$. The isothermal compressibility of a system is of particular interest because its difference between the native and unfolded states reflects the pressure dependence of ΔV, and therefore influences the relative response of the two protein conformations to densified conditions. In addition, there exists a relationship developed *via* statistical mechanics between the isothermal compressibility and volume fluctuations within the system (Heremans and Smeller 1998):

$$\left\langle \delta V^2 \right\rangle = \mathrm{k_B}\, T\, V \beta_T \tag{5}$$

Here k_B is the Boltzmann constant, T is the absolute temperature, and V is the intrinsic volume of the system. In this case, the system volume has to be correlated with the partial molar volume of the protein. Hence the isothermal compressibility not only provides insight into the effect of pressure on protein structure, but also into the dynamics of the native protein in terms of volume fluctuations. It must be emphasized, however, that protein volume fluctuations and macromolecular flexibility parameters are not strictly identified with each other.

The protein volume paradox

One early question that arose during efforts to understand protein folding was related to the nature of the forces that drive a polypeptide chain to adopt a collapsed, globular conformation, but with a high degree of functional specificity. The dominant force was suggested to be the hydrophobic effect that results in clustering of non-polar residues to minimize their interaction with solvent water. The hydrophobic effect has been modeled by the transfer of non-polar compounds, such as pentane, from non-aqueous to aqueous media. This process is highly disfavored both entropically and energetically and it is accompanied by a large increase in heat capacity, a characteristic that is typically observed during the thermal unfolding of proteins. Moreover, there is a close resemblance between the temperature dependence of protein folding events and the temperature dependence of the free energy for the transfer of non-polar compounds from water into non-polar media. Thus the liquid hydrocarbon model has been quite successful in explaining the energetic properties of thermal unfolding.

Based on such studies the volume change upon unfolding of proteins is predicted to have a large negative absolute value at ambient conditions. However, it is also predicted that ΔV for this transfer should become positive with increasing pressure. In contrast, at 0.1 MPa, depending on the temperature of unfolding, amongst other factors, the sign of ΔV can become positive or negative, and it may depend on the nature of the observed transition, e.g., native-to-molten globule or native-to-unfolded processes (Chalikian 2003). However, at high pressure protein unfolding is invariably accompanied by small and negative volume changes, typically on the order of -10 to -100 mL mol^{-1} (Royer 2002). This apparent contradiction is termed the "*protein volume paradox*" and it was first recognized by Kauzmann, who stated that "*the liquid hydrocarbon model fails almost completely when one attempts to extend it to the effects of pressure on protein unfolding*" (Kauzmann 1987).

What is the molecular interpretation of this apparently anomalous volume change? The partial molar volume of a protein i in solution, V_i, is defined as the change in volume of the solution as a small amount of solute is added, divided by the total number of moles of added solute while keeping the amount of the other components constant. For an ideal solution, V_i would be the difference between the solution volume and the original solvent volume. However, due to hydration effects dissolution of a protein will also affect the solvent volume. Therefore, V_i can be expressed as the sum of both an intrinsic term and a hydration term:

$$V_i = V_{atom} + V_{cavities} + \Delta V_{hydration} \qquad (6)$$

where V_{atom} is the sum of the van der Waals volumes of the constituent atoms, $V_{cavities}$ is the volume of the cavities that originate from imperfect packing in the native conformation, and $\Delta V_{hydration}$ is the volume change resulting from the interaction of the protein with the solvent (Heremans and Smeller 1998). Upon protein unfolding, the van der Waals volumes will not change, so the volume change accompanying the unfolding can be written as:

$$\Delta V = \Delta V_{cavities} + \Delta\Delta V_{hydration} \qquad (7)$$

Evidence for the role of cavities in the folded structure comes from mutagenesis experiments, where the creation of new cavities as a result of amino acid mutations results in larger negative volume changes upon unfolding compared to the native protein (Torrent et al. 1999). Contributions to $\Delta\Delta V_{hydration}$ arise from changes in hydration of hydrophobic and hydrophilic groups and from the hydration of cavities previously devoid of water. Note that the hydrophobic contribution is probably very small and its sign is often unclear. In principle, the largest contribution to $\Delta\Delta V_{hydration}$ would arise from the exposure to or burial from water of charged groups due to electrostriction effects: the formation of an ion in solution results in a strong attraction of the dipoles of nearby water molecules by the Coulombic field of the ion. The overall volumetric properties of proteins, however, seem to be largely if not primarily determined by their internal solvent-excluded void volumes and the tendency of these volumes to expand with increasing temperature. Differential hydration, on the other hand, appears to contribute less to the volume change of unfolding (Rouget et al. 2011; Royer and Winter 2011).

Mechanistic aspects of pressure-induced protein unfolding

Pioneering observations of pressure effects on the behavior of proteins were made independently by Percy W. Bridgman (1914) and Keizo Suzuki (1960). Both found that, contrary to expectation, the rates of the pressure-induced unfolding *increase* as the temperature is *reduced*, implying that the process is characterized by a negative activation enthalpy. Such negative activation energies have also been observed in the urea-induced unfolding of proteins. To explain his observations Suzuki proposed the following mechanism:

$$\mathcal{P} + n\mathrm{H_2O} \leftrightarrow \mathcal{P}(\mathrm{H_2O})_n \rightarrow \mathcal{P}_\mathrm{U} \qquad (8)$$

where $\mathcal{P}$ is the native protein, $\mathcal{P}(H_2O)_n$ is the hydrated protein and $\mathcal{P}_U$ is the unfolded protein. This model suggests that the application of pressure results in the penetration of water molecules into the protein interior in a strongly exothermic step that results in unfolding. Several lines of evidence in support of this model have now been obtained. For instance, it was shown that lysozyme remains globular at high pressure, although its hydrodynamic volume has increased by 60-80% and fluorescence probes have undergone a blue shift, indicative of an increased polarity of their environment (Silva and Weber 1993). In another study the distance dependence of chromophore-solvent interactions in cytochrome c was determined and it was found that, as a lower estimate, the solvent had to be within ~4.5 Å of the chromophore in order to cause a blue shift in the fluorescence spectrum (Lesch et al. 2004). This distance is much smaller than the radius of cytochrome c, suggesting that water indeed had to penetrate the protein to explain the blue shift. Others have used neutron and X-ray scattering techniques to determine the radius of gyration of proteins under pressure (Paliwal et al. 2004; Panick et al. 1998). For example, the R_g of staphylococcal nuclease (Snase) increased from 16.3 Å at 0.1 MPa to 34.7 Å at 310 MPa, an expansion that is comparable to the increase in R_g resulting from urea-induced unfolding ($R_g \approx 33$ Å at 8 M urea), but is still much less than in the case of heat-induced unfolding ($R_g \approx 65$ Å) (Paliwal et al. 2004). The latter value approximates the value expected for a random coil. The data also indicated that the protein remained globular at 310 MPa, even though an increase in the R_g by a factor of two corresponds to an eight-fold increase in volume. Further characterization of the unfolded states of Snase and several other proteins indicates a persistence of at least some native secondary structure at high pressure (Zhang et al. 1995; Panick et al. 1998; Meersman et al. 2002; Paliwal et al. 2004). Thus the pressure-unfolded state can be considered to be a swollen, hydrated globular structure with a partially unfolded conformation, as illustrated for Snase (Fig. 1). The persistence of secondary structure is interesting, as it indicates that the penetration of water molecules into the protein does not cause further unfolding through, for example, competition of protein-protein hydrogen bonds for protein-water hydrogen bonds.

In recent years computer simulations, for example, using pairs of methane molecules in water as a simple model for the hydrophobic effect, have provided further microscopic details of the pressure-unfolding mechanism that support the empirical model (Payne et al. 1997; Hummer et al. 1998; Ghosh et al. 2001, 2002). The potential of mean force for a pair of methane molecules in contact with each other ($r \approx 0.39$ nm) is destabilized relative to the a pair of molecules separated by solvent ($r \approx 0.79$ nm) as the pressure increases (Fig. 2), implying a weakening of the hydrophobic contact. The latter can be rationalized by the supposition that, as pressure increases, the average number of water molecules surrounding another water molecule increases and the average binding energy of the water molecules decreases (Sciortino et al. 1991). Thus, as a result of a reduction in the tetrahedral symmetry of the hydrogen bond network, the relative cost of inserting water molecules into an unfavorable non-polar environment is also reduced. Using a water-soluble polymer as a model system, the increased level of hydration of both hydrophobic and polar moieties under pressure could also be demonstrated experimentally (Meersman et al. 2005). However, one should keep in mind that the above description of water under pressure is based on current levels of simulations for bulk water (Sciortino et al. 1991). In addition to influencing the structure of water by reducing its tetrahedral framework, pressure is also a necessary requirement for keeping water within the protein. It is well known from hydrogen exchange experiments, for example, that water molecules can penetrate into and escape from the protein interior on picosecond to millisecond timescales. A molecular dynamics (MD) simulation demonstrated that water molecules inserted between a hydrophobic pair of amino acids only remained there at high pressure, whereas at 0.1 MPa the original hydrophobic contact was restored within the simulation time (Paliwal et al. 2004).

We can now return to the *volume paradox* presented above. Given the fact that pressure-induced unfolding corresponds to the penetration of water into the protein core rather than to the exposure of the core residues to the solvent, as is usually the case in heat-induced unfolding, any estimation of ΔV on the basis of a random coil-like unfolded state will overestimate the hydrophobic hydration. Moreover, although the compressibility change $\Delta\beta$ is often assumed to be zero, Prehoda et al. (1998) showed that $\Delta\beta$ is significantly different from this value in the case of ribonuclease A and found that $\Delta V = -21$ mL mol^{-1} compared with -59 mL mol^{-1} when $\Delta\beta$ was assumed to be zero. $\Delta\beta$, however, was found to be quite small in the case of Snase (Seemann et al. 2001). Thus a proper understanding of the structure of the pressure-unfolded state, an improved estimate of the contribution of cavities to the volume change, and the pressure dependence of the volume change could provide a solution to Kauzmann's apparent volume paradox.

Pressure effects on multimeric proteins and aggregates

So far we have considered the effect of pressure on monomeric proteins, which generally become unfolded between 400-800 MPa. Moderate pressures of 100-300 MPa are also known to dissociate protein oligomers into their monomers (Silva and Weber 1993). The latter can maintain their native conformation or may denature in this process. These pressure limits have

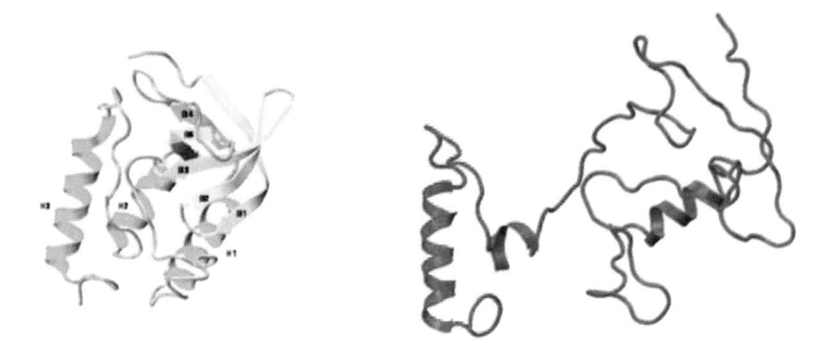

Figure 1. Structure of native and unfolded staphylococcal nuclease at 0.1 (*left*) and 800 MPa (*right*), respectively. These drawings have been obtained from molecular dynamics simulations. [Redrawn after Fig. 10 from Paliwal et al. 2004.]

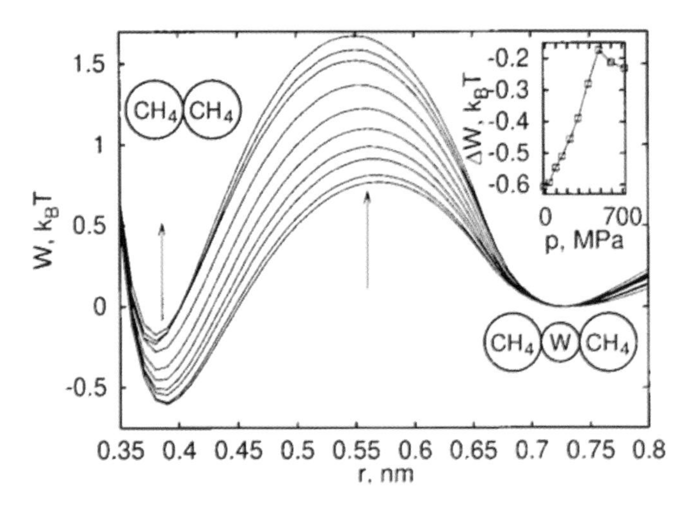

Figure 2. Potential of mean force (*W*) for methane association at various pressures. The arrows indicate the changes with increasing pressure. Note that the minimum of methane pair separated by a water molecule "W" remains largely unaffected by a pressure increase. The inset shows the difference in free energy between the contact pair and the solvent separated pair. [Used by permission of the US National Academy of Sciences, from Hummer et al. (1998), *Proceedings of the National Academy of Sciences USA*, Vol. 95, Fig. 2, p. 1553.]

been incorporated into discussions of the likely maximal pressures for survival of organisms and also technological applications of pressure-induced sterilization procedures. Of particular interest is the pressure-induced depolymerization of larger protein assemblies, such as cytoskeletal proteins, which have been shown to result in morphological changes in both eukaryotic and bacterial cells (Wilson et al. 2001; Molina-Höppner et al. 2003; Ishii et al. 2004). Note that these changes occur at low pressures (~50 MPa) and that, in some cases, the original cell morphology is restored after the pressure is returned to ambient. In case of irreversible depolymerization of the cytoskeleton, however, this will impair cell growth and viability.

Pressure effects on protein energy landscapes

Energy landscapes reflect cooperative structural relaxation processes, from protein folding to glass transitions, and they describe the energy of interaction between atoms or molecules as their relative positions are rearranged in order to achieve the overall ground state or metastable equilibrium structures and conformations. When dealing with proteins, one should consider free energy rather than potential energy landscapes, as the conformational entropy of the polypeptide chain plays a major role in determining the relative stability of the different states. The process of protein folding involves a change in free energy when moving from the unfolded ensemble to folded (native) ensemble. However, due to the dynamical behavior of proteins one can also explore changes in volume and energy within a single ensemble, e.g., the native state, and depict this variation within a single ensemble in terms of a free energy landscape. Thus, an apparent single well (a local energy minimum) on the overall folding landscape contains many other local minima (Fig. 3; Fenimore et al. 2004). We address the influence of pressure on these two aspects of free energy landscapes.

Protein folding free energy landscapes. In general, the pressure dependence of a reaction rate k is given by:

$$\left(\frac{\partial \ln k}{\partial P}\right)_T = -\frac{\Delta V^{\#}}{RT} \tag{9}$$

where R is the ideal gas constant and $\Delta V^{\#}$ is the activation volume. Any reaction that is accompanied by a negative $\Delta V^{\#}$, i.e., if the transition state has a smaller volume than the product, will be accelerated by pressure and *vice versa*. Pressure is a useful variable to investigate reaction

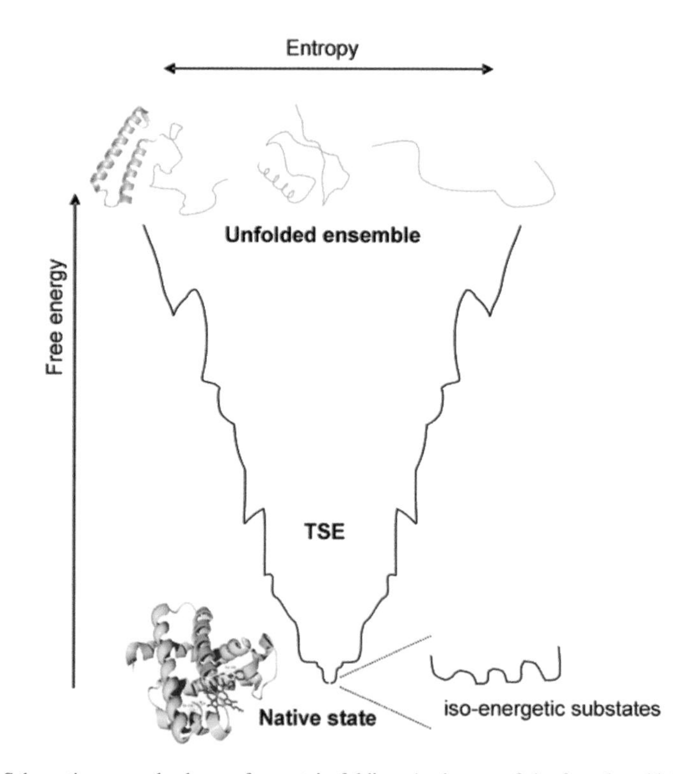

Figure 3. Schematic energy landscape for protein folding. At the top of the funnel, at high free energy, one can find the various conformations that make up the unfolded ensemble. At the bottom of the funnel the global energy minimum represents the native state, here illustrated for myoglobin. Even in the global minimum one can discern various quasi-energetic substates represented by different wells (enlarged on the right-hand side of the funnel).

mechanisms as studies of model systems have shown that pressure (i.e., density) effects often determine the mechanism, whereas temperature mainly changes the frequency of the motions.

The effects of pressure on folding and unfolding rates have been investigated by pressure-jump and high-pressure stopped-flow experiments for a number of proteins and in most cases pressure is found to decrease the folding rate and to increase the unfolding rate (Panick et al. 1998; Pappenberger et al. 2000; Jacob et al. 2002; Brun et al. 2006; Korzhnev et al. 2006). Using an off-lattice minimalist model, Hillson et al. (1999) demonstrated that, depending on the nature of the atomic interactions involved in the transition state, pressure may lower or increase the transition state free energy. Thus the folding rate can, in principle, increase or decrease with pressure, although only decreases in folding rates have been observed so far with an increase in pressure. The latter phenomenon is due to the fact that, as pressure increases, the diffusion of the polypeptide chain as it adopts its final structure, characterized by the reconfigurational diffusion coefficient, becomes slower, and this effect in practice dominates any pressure-induced lowering of the transition state energy. Because the reconfigurational diffusion coefficient is a function of the fold of the native protein and the roughness of the energy landscape, one can conclude that the free energy landscape is rougher at pressures different from ambient. An important consequence of this conclusion is that metastable states may reside in their local minima for longer times, thereby enabling their characterization. In this respect it is also of interest to note that the pressure-unfolded states of several proteins have been suggested to resemble intermediates in the folding process (Zhang et al. 1995; Meersman et al. 2002). For example, in the case of ribonuclease A hydrogen-deuterium exchange protection factors and the secondary structure of the pressure-unfolded state are similar to those found for a previously characterized early folding intermediate (Zhang et al. 1995), suggesting that high-pressure studies may provide important information on such intermediate conformations.

In order to go from the unfolded to the folded state, the polypeptide chain has to cross a free energy barrier, which corresponds to the transition state. The properties of this transition state are rather elusive given its transient nature; structural information has been obtained mainly through mutational (ϕ-value) analysis, and computational methods (Vendruscolo et al. 2005). The transition state has been found to be a rather heterogeneous ensemble of conformations, whose major, defining feature is an overall native-like topology. One important question concerns the role of water in the folding mechanism and whether the rate-limiting step involves desolvation (Rhee et al. 2004). This question can be addressed by determining the hydration properties of the transition state ensemble (TSE). Pressure studies can provide information on the TSE by measuring the activation volumes of folding ($\Delta V_f^{\#}$) and unfolding ($\Delta V_u^{\#}$). In the case of Snase, for instance, the respective activation volumes are +56 and −8 mL mol^{-1}, indicating that the TSE is closer to the native than to the unfolded state on the reaction coordinate and that it is largely dehydrated (Fig. 4; Brun et al. 2006). This is the case for most proteins studied so far, a finding that seems to differ from the conclusions of most computational studies and ϕ-value analyses (Brun et al. 2006). However, it has been shown for Snase that although the wild-type protein has a highly dehydrated transition state, some of its mutants containing ionizable residues have a hydrated TSE, i.e., the absolute value of $\Delta V_f^{\#} < \Delta V_u^{\#}$ (Fig. 4; Brun et al. 2006). Taken together these data suggest that the degree of hydration of the TSE depends on the properties of the particular protein as well as on the experimental conditions. Moreover, a recent simulation study comparing implicit and explicit solvation models showed that, although both models are qualitatively in agreement with each other, the explicit model does indicate that the TSE is more hydrated (Rhee et al. 2004). Part of the apparent contradiction mentioned above can therefore be related to the fact that most simulations deal with water implicitly and that ϕ-value analysis is also interpreted in terms of an implicit role of the solvent. As a consequence, high-pressure methods may provide the best, if not the only, experimental approach to characterize the TSE in terms of hydration.

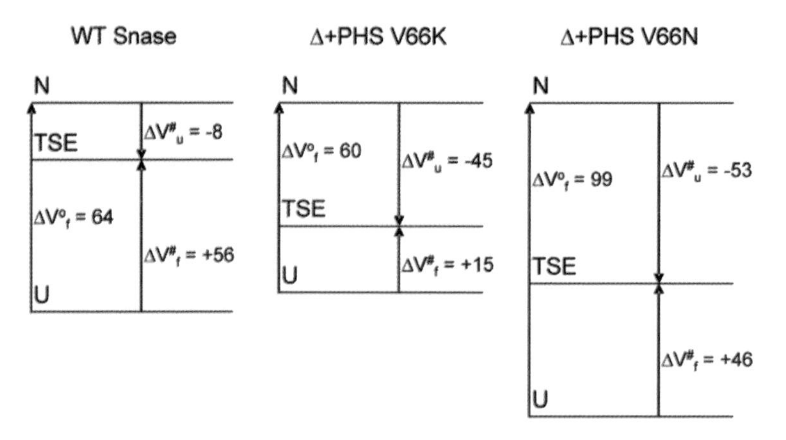

Figure 4. Schematic representation of the folding-unfolding reaction of wild-type (WT) staphylococcal nuclease (Snase) and two of its mutants (Δ+PHS V66K and Δ+PHS V66N). The folded state is abbreviated by F, U is the unfolded state, T represents the transition state ensemble, and ΔV^{o}_{f} is the overall volume change upon folding. [Used with permission of the American Chemical Society from Brun et al. (2006).]

Protein dynamics: accessing conformational substates. Proteins are dynamic molecules, a characteristic that enables them to perform functions such as ligand or substrate binding and alteration and release. The fluctuations that underlie this dynamic behavior cause the protein to adopt numerous conformations, which are commonly referred to as conformational substates (Frauenfelder et al. 1990; Fenimore et al. 2004). Thus the native state of a protein is actually an ensemble of nearly isoenergetic substates (Fig. 3), which may perform different functions. Experiments have shown that within these substates one can also identify statistical substates, which perform the same function, but with different rates. Pressure is a useful tool to explore the conformational substates in an energy landscape as it can shift the population from one substate to another on the basis of the volumetric properties of the respective substates. In addition, pressure can also change the reaction rate k with which a given substate performs its function, as its value depends on the activation volume ($\Delta V^{\#}$; Eqn. 9), which may be different for different substates, as well as on the properties of the solvent (e.g., viscosity). This type of experiment can lead to new insights into the dynamics and reactions of proteins, such as the binding mechanism of carbon monoxide and oxygen to myoglobin (Frauenfelder et al. 1990; Fenimore et al. 2004). In a recent example, pressure modulation in combination with FTIR spectroscopy was applied to reveal equilibria between spectroscopically resolved substates of the lipidated signaling protein N-Ras. The conformational dynamics of N-Ras in its different nucleotide binding states in the absence and presence of a model membrane were probed by pressure perturbation. It was shown that not only nucleotide binding, but also the presence of the membrane has a drastic effect on the conformational dynamics and selection of conformational substates of the protein. Moreover, a previously unknown substate that appears upon membrane binding was observed using this pressure perturbation approach (Kapoor et al. 2012a,b).

From free energy landscapes to *P-T* phase diagrams

Life on Earth can thrive in environments characterized by a wide range of pressures and temperatures. In order to understand this ability on the molecular level it is necessary to consider pressure effects on proteins in particular, and living systems in general, over a wide temperature range. A plot of the transition midpoint for pressure unfolding versus temperature yields an elliptical phase diagram (Fig. 5), which, interestingly, is also found when plotting

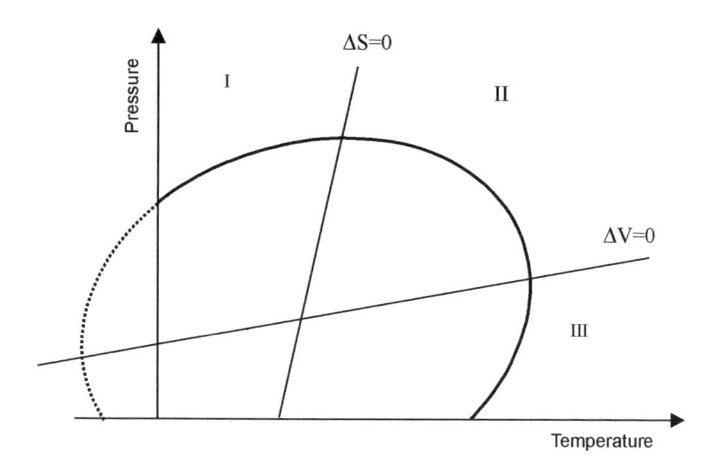

Figure 5. Schematic representation of a pressure-temperature stability diagram for proteins. Inside the ellipse the protein adopts its native conformation, outside the ellipse it is unfolded. See main text for full description of regions I to III.

inactivation rates of microorganisms (Hashizume et al. 1995; Yayanos 1998) or the phase separation behavior of water-soluble polymers (Meersman et al. 2006). On the basis of the contours of the phase diagram, with its re-entrant behavior at low temperature, cold unfolding of proteins was predicted. It is worth noting that, at least at elevated pressures, cold and pressure unfolding are thermodynamically (Fig. 5) and mechanistically similar (Meersman et al. 2002). This close relationship between the effects of pressure and cold may explain why the cellular responses to these variables are similar. Pressure is often used in cold unfolding experiments because pressures of ~200 MPa reduce the freezing point of water by ~20 °C, thus enabling experiments at low temperature in the liquid state.

The phase diagram can be described by Equation (2), often referred to as the Hawley equation. The main advantage of this equation over purely empirical equations lies in the fact that all parameters can be given a physical interpretation. However, using this equation one can only obtain the differences in compressibility, thermal expansion, and heat capacity between the unfolded and the native state. Therefore, other techniques are required to determine β_T, C_P and α_P. These thermodynamic quantities are of particular interest as they can be related to fluctuations in volume, energy, and a cross-correlation of volume and energy, respectively (Heremans and Smeller 1998). Such fluctuations underlie the dynamic behavior of proteins in aqueous conditions (Chalikian 2003).

At the phase boundary, the Gibbs free energy change for unfolding, $\Delta G = G_D - G_N$, is zero. Within the elliptic contour the protein is in the native conformation ($\Delta G > 0$), outside the contour the protein is unfolded ($\Delta G < 0$). At the highest pressure (P_{max}), where the native state is stable, the slope of the tangent on the ellipse is zero. At the highest temperature (T_{max}) the slope is infinite. At these points ΔS and ΔV, respectively, are equal to zero and these parameters can be represented by a straight line in P-T space (Fig. 5). It can be seen that these lines divide the $\Delta G = 0$ contour into three regions. In the first region (I), where ΔS and ΔV are both negative, an increase in temperature will stabilize the protein against pressure unfolding. It can be derived from the van't Hoff equation that in this region the enthalpy change, ΔH, will be negative. In the second region (II) ΔS is positive and ΔV is negative. Here increasing temperature lowers the unfolding pressure, and vice versa. In the third region (III) ΔH, ΔV, and ΔS are all positive. One of the interesting features of the phase diagram is that for a number of proteins dT_m/dP is positive at low pressures and high temperatures, suggesting that pressure increases the stability of the protein towards thermal unfolding. This effect may provide the molecular basis for cells and their constituents to survive at combinations of high temperature and high pressure. As a result it is possible to refold a thermally unfolded protein (at 0.1 MPa), at temperatures just above the unfolding temperature, by increasing the pressure. The fact that in this low pressure-high temperature region the unfolding is associated with a positive volume change has been attributed to the difference in the thermal expansion of the folded and unfolded states (Seemann et al. 2001). This can be seen from the pressure-temperature dependence of the volume change ΔV:

$$\Delta V(P,T) = \Delta V^{\circ} + \Delta\alpha(T - T_{o}) - \Delta\beta(P - P_{o}) \tag{10}$$

where the second term $\Delta\alpha(T - T_0)$ represents the temperature dependence of the volume. The volume change ΔV is found to have a strong temperature dependence, with ΔV becoming less negative as the temperature increases (Seemann et al. 2001). However, in the case of ribonuclease A the changes in ΔV with temperature have been shown to depend on other experimental conditions (Yamaguchi et al. 1995).

The signs of ΔV and ΔS provide a thermodynamic basis for the mechanistic and conformational differences between the pressure and heat unfolding of proteins, and rationalize the similarities between the pressure and cold unfolding (Meersman et al. 2002). The slope of the equilibrium line in the diagram (Fig. 5) is given by:

$$\frac{dT}{dP} = \frac{\Delta V^{\circ} - \Delta\beta(P - P_{o}) + \Delta\alpha(T - T_{o})}{\Delta S^{\circ} - \Delta\alpha(P - P_{o}) + \Delta C_{P}\left((T - T_{o})/T_{o}\right)} \tag{11}$$

Note that, if $\Delta\beta$, $\Delta\alpha$, and ΔC_P are zero then this equation is reduced to the classical Clapeyron equation ($dT_m/dP = T_m\Delta V/\Delta H$), which describes the behavior under pressure of, for instance, lipids. This demonstrates clearly the importance of the second-order terms in the elliptic nature of the phase diagrams of proteins. Higher-order terms, describing the temperature and pressure dependence of $\Delta\beta$, $\Delta\alpha$, and ΔC_P, have been ignored in Equation (2), but for at least one protein, ribonuclease A, a pressure dependence of ΔC_P has been reported (Yamaguchi et al. 1995). Inclusion of such higher-order terms distorts the diagram, but does not change completely its overall elliptical appearance.

In practice, the stability of a protein will depend strongly on solution conditions such as pH, the presence of chemical denaturants, or co-solutes. Zipp and Kauzmann (1973) studied the phase diagram of myoglobin over a wide pH range. They observed that the shape of the diagram changed at extreme pH values, where the difference between the cold and heat unfolding temperatures becomes smaller and the unfolding pressure is lowered. Likewise, the presence of co-solutes such as urea and salts also affect the position and shape of the phase diagram, although the outcome depends strongly on their kosmotropic or chaotropic nature (Fig. 6; Herberhold et al. 2004). The ad-

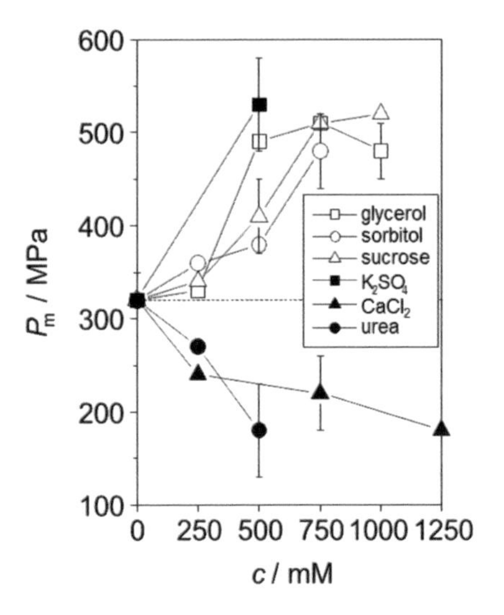

Figure 6. Effect of co-solutes and their concentration (c) dependence on the unfolding pressure (P_m) of staphylococcal nuclease (Snase). [Used by permission of the American Chemical Society, from Herberhold et al. (2004), *Biochemistry*, Vol. 43, Fig. 3, p. 3338.]

dition of co-solutes or co-solvents can have a large effect on the volume change, and Scharnagl et al. (2005) have given a comprehensive thermodynamic description of the effect of co-solutes and co-solvents on the stability of the protein.

Other factors, such as macromolecular crowding as occurs in organisms, have been virtually unexplored with respect to pressure stability. Moreover, food scientists, investigating the inactivation of microorganisms in foods, are well aware of the fact that the pressure sensitivity of vegetative bacteria depends on the composition of the food matrix. For instance, when plotting the decimal reduction time D in the P-T plane, the inactivation of *Escherichia coli* in carrot juice follows a linear pattern, whereas in HEPES buffer a typical elliptical outline can be observed (Van Opstal et al. 2005). This example clearly shows that pressure sensitivity strongly depends on the nature of the experimental medium. Hence, one should be careful when extrapolating data from *in vitro* (buffer) systems to real life systems, as the latter involve a large number of unknown factors that we cannot yet fully understand or model.

Kinetic aspects of the phase diagram

In many instances, the rate of unfolding as a function of pressure and temperature is studied yielding a P-T-k diagram, where k is the rate constant of inactivation or unfolding. A mathematical analysis of the isokineticity curves yields the activation parameters for the unfolding. The change of the free energy of activation as a function of pressure and temperature is typically expressed by:

$$d\Delta G^{\#} = \Delta V^{\#} dP - \Delta S^{\#} dT \tag{12}$$

where $\Delta G^{\#} = -RT \ln k$, being the difference in free energy between the transition state and the native state. We do note, however, that this expression is developed for equilibrium thermodynamics conditions, and other formulations and approaches may prove to be significant in the future.

According to the transition state theory the activation volume is defined as:

$$\Delta V^{\#} = -RT \frac{\partial \ln k}{\partial P} \tag{13}$$

Similar to the P-T phase diagram, the P-T-k diagram can be divided in three regions based on the signs of $\Delta V^{\#}$, $\Delta H^{\#}$ and $\Delta S^{\#}$. This similarity can easily be understood from the thermodynamic background of the kinetic theory of the transition state. An important aspect that has to be taken into account is the irreversibility of the protein unfolding, which is generally due to protein aggregation at high temperatures. Such a phenomenon can be represented by the following mechanism (Heremans and Smeller 1997):

$$N \underset{k_2}{\overset{k_1}{\rightleftharpoons}} D \xrightarrow{k_3} I \tag{14}$$

where N and D are the native and reversibly unfolded protein, and I is the irreversibly unfolded protein. From the viewpoint of the phase diagram two conditions are worth considering. First, if $k_3 << k_1,k_2$, then there is a fast exchange between N and D, while the transformation of D into I is slow. Under this condition the apparent rate constant, k_{obs}, can be defined as:

$$k_{obs} = \left(\frac{k_1}{k_2}\right) k_3 = Kk_3 \tag{15}$$

Here K is the equilibrium constant for the N to D transition. Thus, the overall rate of the reaction is mainly determined by the formation of I. Secondly, when $k_3 >> k_1,k_2$, then all the reversibly unfolded molecules will be incorporated into an intermolecular aggregation network before they can refold. In this case the unfolding of N into D is the rate-limiting step.

The temperature dependence of the rate constants is given by the Arrhenius equation:

$$k_3 = A \exp\left(-\frac{\Delta G^{\#}}{RT}\right) \tag{16}$$

where A is a pre-exponential factor. From this equation it is clear that the first condition ($k_3 << k_1,k_2$) is most probable at low temperature and high pressure, whereas the second condition ($k_3 >> k_1,k_2$) will likely take place at high temperature. It explains why protein aggregation is often observed during thermal unfolding experiments and not during pressure experiments.

Relevance of biophysical studies on proteins to deep carbon

Most studies presented herein investigate the effects of pressure on proteins obtained from mesophilic microorganisms or even multicellular organisms. The conclusions derived from these studies are likely to be general; i.e., they can be extended to proteins found in extremophiles. Little is known about the pressure stability of proteins from piezophiles, but research on proteins from thermophiles and psychrophiles suggests that these proteins shift their thermal stability by increasing the number of stabilizing (non-covalent) interactions (Daniel et al. 1996). The pressure range in which proteins unfold tends to be higher than that in which cell survival is compromised. What could be more critical in the context of pressure effects on cellular growth and viability is the maintenance of protein-protein, protein-nucleic acid, and protein-lipid interactions (vide infra) that are more pressure sensitive. Likewise, relatively low pressures could suffice to influence protein dynamics, and hence protein activity, although few studies have addressed this issue. Low pressures will also influence reaction rates, thereby affecting important cellular processes responsible, for instance, for turnover of cellular constituents and catabolic reactions. Pressure effects could be particularly relevant in view of the long generation times observed in subseafloor microorganisms (Jørgensen and Boetius 2007).

All these observations are further complicated when changing the nature of the matrix in which protein stability is studied from water to a more complex one that resembles the intracellular environment. Here, effects such as molecular crowding may have a crucial influence on protein stability and dynamics. Moreover, the *P-T* diagram indicates a close relationship between pressure and temperature effects, and could explain why proteomics studies on cells that have been exposed to pressure stress have revealed so far that the proteins for which the expression becomes upregulated are similar to those induced by heat or cold (Hörmann et al. 2006). It remains a question, however, whether or not these mainly chaperone (heat shock) proteins play the same role in coping with pressure as with heat.

NUCLEIC ACIDS

Deoxyribonucleic acid (DNA) and ribosomal ribonucleic acid (rRNA) represent the hereditary blueprint of every cell. Other RNA's, such as messenger and transfer RNA, are involved in the translation of the genetic information into proteins. Whereas DNA adopts essentially one structure, the iconic double helix of Watson and Crick, the various RNA molecules display a much greater conformational variability.

The thermal stability of the DNA double helix has been well characterized structurally and biochemically and it is well known that the two complementary DNA strands dissociate into single-stranded coils by heating. The midpoint of the melting transition T_m (at atmospheric pressure) depends on the base pair composition and the sequence of the DNA, as well as on the salt concentration, indicating that the stability of DNA is intimately related to its hydration (Dubins et al. 2001; Rayan et al. 2005). After his observation of an elliptical *P-T* phase diagram for proteins, Hawley (1971) investigated the pressure-temperature stability of nucleic acids to assess whether a similar diagram could be developed for DNA (Hawley et al. 1974). He found that

dT_m/dP is linear and positive (up to 600 MPa) with slightly steeper slopes at higher salt concentration. Because the melting enthalpy ΔH_m is positive at atmospheric pressure (0.1 MPa), one can derive from the Clapeyron equation that ΔV is positive, indicating that pressure stabilizes the helix conformation of *Clostridium perfringens* DNA (Hawley et al. 1974). This conformation is not unexpected as base stacking and hydrogen bonds are stabilized by high pressure. The pressure insensitivity of the DNA double helix was recently confirmed by high-pressure crystallography and NMR experiments (Girard et al. 2007; Wilton et al. 2008). Both studies indicate that the double helix undergoes only a minor distortion under pressure (Fig. 7), although the details of the distortion were not the same in the crystalline versus solution state. By plotting the slope of the coexistence lines as a function of T_m, Hawley observed that the slope changes sign at $T_m \approx 59$ °C, implying that below this temperature ΔV would be negative and thus pressure would destabilize the DNA double helix. Note that 59 °C is well below the T_m of natural chromosomes under physiological conditions. Indeed, work on synthetic DNA or RNA duplexes revealed that pressure destabilizes the double-stranded conformation at T_m values below approximately 50 °C (Dubins et al. 2001); in other words dT_m/dP becomes negative at high salt concentrations and low temperatures. For instance, the midpoint for the pressure-induced melting of the poly(dA)poly(rU) DNA/RNA duplex is 60 MPa at 25 °C. Such a change in sign of dT_m/dP also has been observed for water soluble synthetic polymers, depending on the nature and concentration of the added salt (Kunugi et al. 1999). The effect of salts has been attributed to changes in water structure and can be related to the Hofmeister series (Cacace et al. 1997; Zhang et al. 2006). This again underscores the importance of hydration in the processes considered in this chapter. Dubins et al. (2001) also calculated the phase diagram in an extended *P-T* region. The calculated phase diagram reveals that as pressure is further increased dT_m/dP changes sign, and above 600 MPa dT_m/dP is close to zero (regardless of T_m at 0.1 MPa). Unfortunately, these authors did not explore this pressure range experimentally in order to verify the correctness of their prediction. Also, Hawley et al. (1974) did not observe any transition in *C. perfringens* DNA under pressures in excess of 900 MPa. In contrast to DNA, computer simulations of a RNA hairpin show that it does unfold under pressure (Garcia et al. 2007). Presumably this conformation change is due to the fact that RNA, similar to proteins, has a tertiary structure held together by non-covalent interactions. The temperature dependence of the unfolding pressure again follows an elliptic outline.

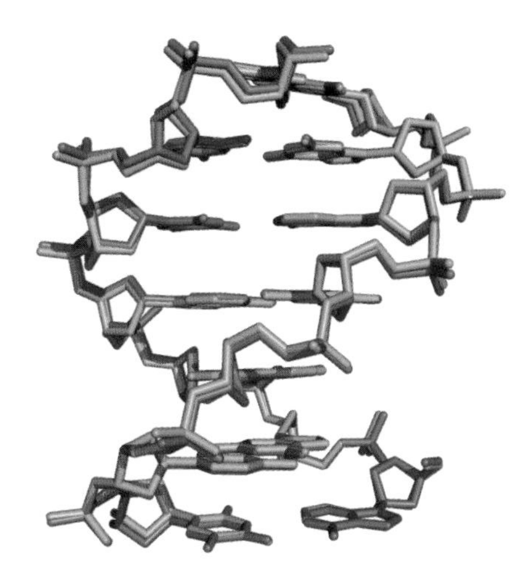

Figure 7. Structure of B-DNA at ambient pressure (green) and 200 MPa (red) as determined by NMR spectroscopy. This figure was made with PyMol using PDB codes 2VAH (low-pressure structure) and 2VAI (high-pressure structure).

When comparing the heat- and pressure-induced helix-to-coil transition it was found that the cooperative length, being the number of base pairs that melt as one unit, of the pressure-induced transition is two-fold greater than the one for the heat transition. This difference suggests that these processes are mechanistically different. Moreover, on the basis of the hypochromicity of several infrared bands and a comparison of thermodynamic variables (α_P, β_T, ΔV), it was concluded that the pressure-induced single-strand DNA is more structured than the heat-induced coil, due to the greater amount of stacking at high pressure (Rayan et al. 2005). These findings are reminiscent of the differences in high pressure vs. high temperature behavior of proteins.

Several cellular processes involving nucleic acids, such as replication, transcription and recombination, depend on the correct recognition of protein and DNA or RNA binding partners. X-ray crystallography of protein-DNA complexes can identify the non-covalent interactions involved in the complex. Electrostatic and hydrophobic interactions are the primary forces involved, and these will be destabilized by pressure (see above). Therefore, pressure can provide information on the stoichiometry and thermodynamic parameters of the association. A typical dissociation constant for *Bam*HI-DNA complex is 4.6 ± 0.4 nM at 50 MPa vs. 0.7 ± 0.1 nM at 0.1 MPa, demonstrating a clear destabilization at high pressure under the test conditions used. Molecular dynamics studies show that pressure forces water into the protein-DNA complex and that it is sequestered at the intermolecular interface, similar to the effect of pressure on protein oligomers. Moreover, as most DNA-interacting proteins are oligomers, high hydrostatic pressure studies can also reveal information on the effect of DNA binding on their stability. Depending on the protein involved, DNA has been found both to stabilize and to destabilize protein oligomers (Silva et al. 2002). For instance, the tetrameric LacI repressor protein is stabilized by the inducer, but destabilized by DNA (Royer et al. 1990). In contrast, the dimeric LexA repressor, involved in the regulation of the transcription of the SOS system in *E. coli*, is stabilized upon DNA binding (Mohana-Borges et al. 2000). Pressure effects have also been studied on RNA-RNA interactions (e.g., GAAA tetraloop-receptor motif; Downey et al. 2007). Here the volume change associated with the pressure-induced dissociation (−5 to −9 mL mol^{-1}) was smaller than those typically observed for protein unfolding or protein-DNA dissociation. In addition, the effect of the co-solutes sucrose and glycerol was found to be opposite of what is seen in the case of proteins.

In a recent study of the binding of a highly conserved protein involved in DNA repair in prokaryotes, RecA, to single stranded DNA (ssDNA) was investigated (Merrin et al. 2011). As expected, pressures of 70-130 MPa were sufficient to disrupt the RecA-ssDNA interaction. When comparing the RecA of a mesophile with that of a thermophilic organism it was observed that the increased thermal stability correlated with increased pressure stability. Moreover, a pressure-temperature plot for the dissociation also displays the same contour as that for protein unfolding. This similarity supports the conclusion that the effect of pressure on protein-DNA interactions is mainly due to changes in structure and/or hydration at the interface, rather than being caused by changes in DNA structure (Wilton et al. 2008). More importantly, the pressure range in which this important interaction is disrupted corresponds to the upper pressure limit for microbial growth in nature. Taken together the above suggests that the disruption of protein-protein and protein-nucleic acid interactions is potentially key to the loss of pressure survivability in microorganisms.

LIPIDS AND CELL MEMBRANES

Lamellar lipid bilayer phases

Lyotropic lipid mesophases are formed by amphiphilic molecules, mostly phospholipids, in the presence of water. They exhibit a rich structural polymorphism, depending on their molecular structure, hydration level, pH, ionic strength, temperature, and pressure. The basic structural element of biological membranes consists of a lamellar phospholipid bilayer matrix (Fig. 8). Even though most lipids possess two acyl-chains and one hyphrophilic headgroup, the composition of the chains and the headgroup can vary significantly in cellular membranes. Also, the lipid composition is very different in different cell types of the same organism, or even in different organelles of the same cell. Not only is the entire cell membrane very complex, containing a large variety of different lipid molecules and a large body (ca. 50%) of proteins performing versatile biochemical functions, but also the simplest lipid bilayer consisting of only one or two kinds of lipid molecules already exhibits a very complex phase behavior. Lipid

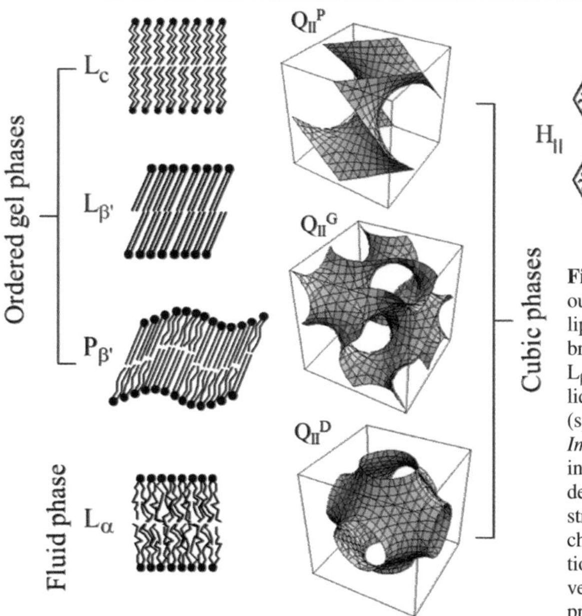

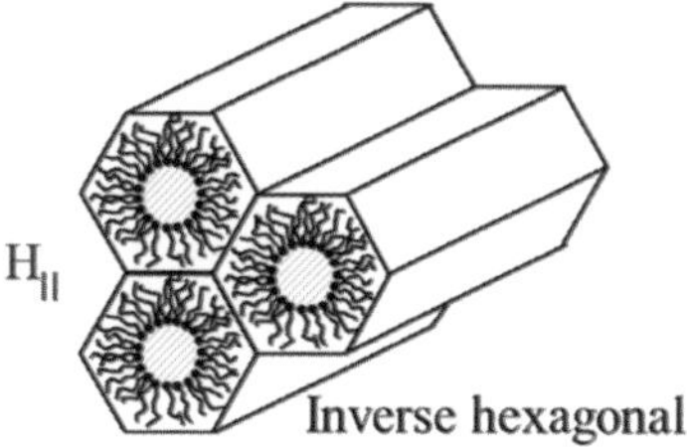

Figure 8. Schematic drawing of various lamellar and nonlamellar lyotropic lipid mesophases adopted by membrane lipids: L_c, lamellar crystalline; $L_{\beta'}$, $P_{\beta'}$, lamellar gel; L_α, lamellar liquid-crystalline (fluid-like); Q_{II}^G (space group *Ia3d*); Q_{II}^P (space group *Im3m*); Q_{II}^D (space group *Pn3m*); H_{II}, inverse hexagonal. Numerous factors determine the particular mesophase structure, e.g., the type of lipids, lipid chain length and degree of unsaturation, headgroup area and charge, solvent properties, pH, temperature, and pressure.

bilayers display various phase transitions, including a chain melting transition. In excess water, saturated phospholipids often exhibit two thermotropic lamellar phase transitions, a gel-to-gel ($L_{\beta'}$-$P_{\beta'}$) pretransition and a gel-to-liquid-crystalline ($P_{\beta'}$-L_α) main (chain melting) transition at a higher temperature (Fig. 8). Phosphatidylcholines display a tilt angle of about 30°, while phosphatidylethanolamines do not. In the fluid-like L_α phase, the acyl-chains of the lipid bilayers are conformationally disordered ("melted"), whereas in the gel phases the chains are more extended and ordered. The lipids in the $L_{\beta'}$ phase are arranged on a two-dimensional triangular lattice in the membrane phase. This phase is also called solid-ordered (s_o) phase. Besides neutral or zwitterionic lipids, negatively-charged lipids are also present in the cell membranes. The melting temperature of negatively-charged lipid membranes generally increases when neutralizing the charges by proteins or divalent ions. In addition to these thermotropic phase transitions, a range of pressure-induced phase transformations has also been observed (Winter et al. 1989, 2000, 2004; Landwehr et al. 1994a,b; Hammouda et al. 1997; Czeslik et al. 1998; Winter 2001).

Because the average end-to-end distance of disordered hydrocarbon chains in the L_α-phase is smaller than that of ordered (all-trans) chains, the bilayer becomes thinner during melting at the $P_{\beta'}/L_\alpha$-transition, even though the partial lipid volume increases. This behavior is demonstrated in Figure 9, which shows the temperature dependence of the specific partial lipid volume V_L of DMPC* in water (Böttner et al. 1994). The change of V_L near 14 °C corresponds to a small

* Abbreviations: DMPC 1,2-dimyristoyl-sn-glycero-3-phosphatidylcholine (di-C14:0); DMPS 1,2-dimyristoyl-sn-glycero-3-phosphatidylserin (di-C14:0); DPPC 1,2-dipalmitoyl-sn-glycero-3-phosphatidylcholine (di-C16:0); DPPE 1,2-dipalmitoyl-sn-glycero-3-phosphatidylethanolamine (di-C16:0); DOPC 1,2-dioleoyl-sn-glycero-3-phosphatidylcholine (di-C18:1,cis); DOPE 1,2-dioleoyl-sn-glycero-3-phosphatidylethanolamine (di-C18:1,cis); POPC 1-palmitoyl-2-oleoyl-sn-glycero-3-phosphatidylcholine (C16:0,C18:1,cis); DLPC **1,2-dilauroyl-sn-glycero-3-phosphocholine**.

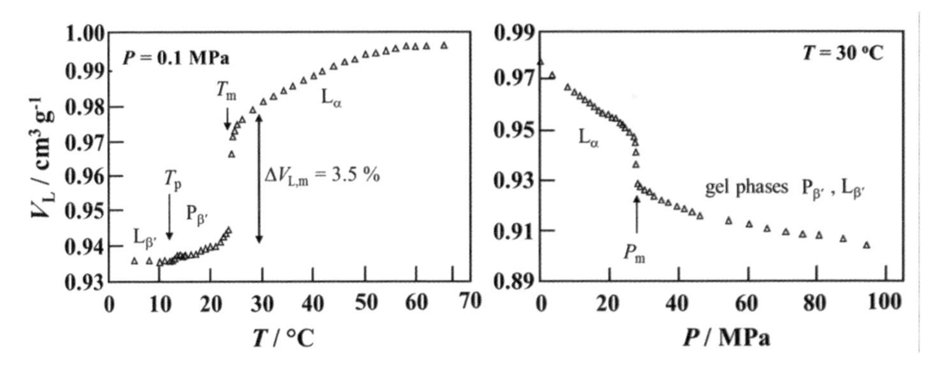

Figure 9. Effect of temperature (left) and pressure (at $T = 30$ °C) (right) on the partial lipid volume V_L of DMPC bilayers as obtained from densimetric measurements. The gel-to-gel ($L_{\beta'}$ to $P_{\beta'}$,) and gel-to-fluid (L_α) lamellar phase transitions appear at temperatures T_p and T_m, respectively.

volume change in course of the $L_{\beta'}$-to-$P_{\beta'}$ transition. The main transition at $T_m = 23.9$ °C for this phospholipid is accompanied by a pronounced 3% change in volume, which is mainly due to changes of the chain cross-sectional area, because the chain disorder increases drastically at the transition. The compression of the bilayer as a whole is anisotropic, lateral shrinking being accompanied by a increase in thickness due to a straightening of the acyl-chains. Figure 9 exhibits the pressure dependence of V_L at a temperature above T_m; e.g., 30 °C. Increasing pressure triggers the phase transformation from the L_α to the gel phase, as can be seen from the rather abrupt decrease of the lipid volume at 27 MPa. The volume change, ΔV_m, at the main transition decreases slightly with increasing temperature and pressure along the main transition line.

Biological lipid membranes can also melt. Typically, such melting transitions are found about 10 °C below body or growth temperatures. It seems that biological membranes adapt their lipid compositions such that the temperature distance to the melting transition is maintained. The same may hold true for adaptation to high-pressure conditions. Hence it is likely that such behavior serves a purpose in the biological cell. In lipid bilayers the fluctuations in enthalpy, volume, and area are higher close to the melting transition. High enthalpy fluctuations lead to high heat capacity, high volume fluctuations lead to high volume compressibility, and high area fluctuations lead to a high area compressibility. In turn, area fluctuations lead to fluctuations in curvature and bending elasticity. These properties may be required for optimal physiological function.

A common slope of ~0.22 °C MPa^{-1} has been observed for the gel-fluid phase boundary of saturated phosphatidylcholines as shown in Figure 10 (Winter 2001; Winter et al. 2000, 2004). Assuming the validity of the Clapeyron relation describing first-order phase transitions for this quasi-one-component lipid system, $dT_m/dP = T_m\Delta V_m(T_m, P)/\Delta H_m(T_m, P)$, the positive slope can be explained by an endothermic enthalpy change, ΔH_m, and a partial molar volume increase, ΔV_m, for the gel-to-fluid transition, which have indeed been determined in direct thermodynamic measurements (Seeman et al. 2003; Janosch et al. 2004; Krivanek et al. 2008). The transition enthalpy at atmospheric pressure is about 36 kJ mol^{-1}, for DPPC at ambient pressure and decreases slightly with pressure; $(d\Delta H_m/dP) = -0.034$ kJ mol^{-1} MPa^{-1} (Potekhin et al. 2008). As $d\Delta H_m/dP = -T_m(d\Delta V_m/dT)_P + \Delta C_{P,m}(dT_m/dP)$, the drop of enthalpy change with pressure evidences a significant difference in the coefficients of thermal expansion of the two phases. Similarly, ΔV_m decreases linearly with increasing pressure (from 22.9 cm^3 mol^{-1} at 0.1 MPa to ~13 cm^3 mol^{-1} at 200 MPa, i.e., $d\Delta V_m/dP = -0.0493$ cm^3 mol^{-1} MPa^{-1}; Potekhin et al. 2008). According to $d\Delta V_m/dP = (d\Delta V_m/dP)_T + (d\Delta V_m/dT)_p(dT_m/dP)$, this decrease is due to the significant difference in the lipid compressibility coefficients in the fluid and gel phase,

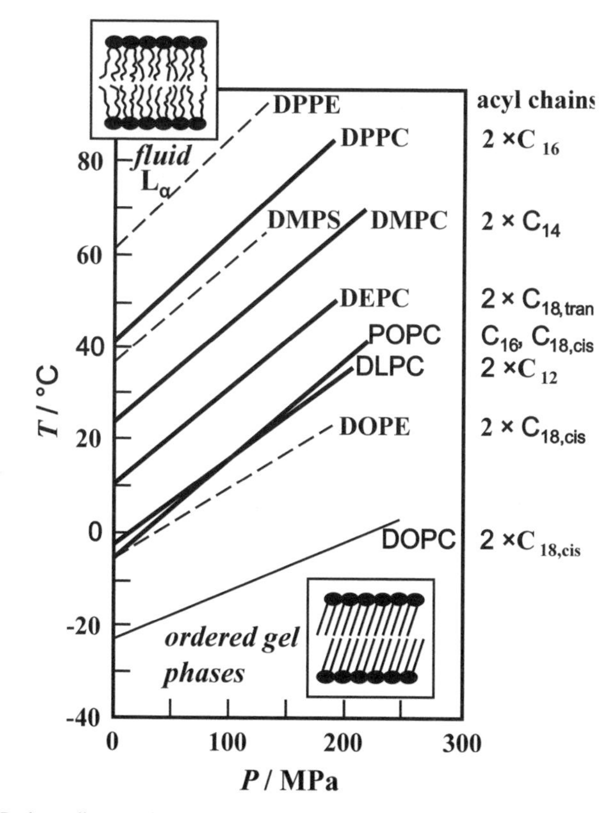

Figure 10. *T-P* phase diagram for the main (chain-melting) transition of different phospholipid bilayer systems. The fluid-like (liquid-crystalline) L_α-phase is observed in the low-pressure, high-temperature region of the phase diagram; the ordered gel phase regions appear at low temperatures and high pressures, respectively. The gel-to-fluid transition lines of the phosphatidylcholines are drawn as solid lines, those of the phospholipids with different headgroups as dashed lines. The lengths and degree of unsaturation of the acyl-chains of the various phospholipids are denoted on the right-hand side of the figure. See footnote on page 623 for abbreviations.

respectively. The transition half-width ($\Delta T_{m,1/2}$), which can be estimated as the ratio of the calorimetric peak area $\Delta H_{m,\mathrm{cal}}$ to its amplitude $C_{P,max}$, can be determined from the van't Hoff enthalpy change by using $\Delta H_{m,\mathrm{vH}} = 4\mathrm{R}T_m^2 C_{P,max}/\Delta H_{m,\mathrm{cal}} = 4\mathrm{R}T_m^2/\Delta T_{m,1/2}$. The transition half-width does not change with pressure, and the average number of lipid molecules ($N = \Delta H_{m,\mathrm{cal}}/\Delta H_{m,\mathrm{vH}}$) comprising the coooperative unit N of the transition grows slightly with the increase of pressure and temperature.

Similar transition slopes have been determined for the mono-*cis*-unsaturated lipid POPC, the phosphatidylserine DMPS, and the phosphatidylethanolamine DPPE. Only the slopes of the di-*cis*-unsaturated lipids DOPC and DOPE have been found to be markedly smaller. The two *cis*-double bonds of DOPC and DOPE lead to very low transition temperatures and slopes, as they impose kinks in the linear conformations of the lipid acyl-chains, thus creating significant free volume fluctuations in the bilayer so that the ordering effect of high pressure is reduced. Hence, in order to remain in a physiologically relevant, fluid-like state at high pressures, more of such *cis*-unsaturated lipids are incorporated into cellular membranes of deep-sea organisms, another example of homeoviscous adaptation (Yayanos 1986; Behan et al. 1992). For example, the ratio of unsaturated to saturated fatty acids of the piezophilic deep-sea bacterium CNPT3 is linearly dependent on the hydrostatic pressure at which they were cultivated (DeLong et al. 1985). The unsaturated to saturated ratio increases from 1.9 at ambient pressure up to about 3 at 69 MPa at 2 °C.

As seen in Figure 10, pressure generally increases the order of membranes, thus mimicking the effect of cooling. But we note that applying high pressure can lead to the formation of additional ordered phases, which are not observed under ambient pressure conditions, such as a partially interdigitated high pressure gel phase, $L_{\beta i}$, found for phospholipid bilayers with acyl-chain lengths $\approx C_{16}$ (Landwehr et al. 1994a, b; Hammouda et al. 1997). To illustrate this phase variety, the results of a detailed small-angle X-ray and neutron scattering and FTIR spectroscopy study of the *P-T* phase diagram of DPPC in excess water are shown in Figure 11. At much higher pressures as shown here, even further ordered gel phases appear, differing in the tilt angle of the acyl-chains and the level of hydration in the headgroup area. Even at pressures where the bulk water freezes, the lamellar structure of the membrane is preserved (Czeslik et al. 1998).

Lipid mixtures, cholesterol, and peptides

To increase the level of complexity, *T-P* phase diagrams of binary mixtures of saturated phospholipids have been determined as well (Winter et al. 1999a,b,c, 2000, 2004; Winter 2001). They are typically characterized by lamellar gel phases at low temperatures, a lamellar fluid phase at high temperatures, and an intermediate fluid-gel coexistence region (Fig. 12). The narrow fluid-gel coexistence region in the DMPC(di-C_{14})-DPPC(di-C_{16}) system indicates a nearly ideal mixing behavior of the two components (isomorphous system). In comparison, the coexistence region in the DMPC(di-C_{14})-DSPC(di-C_{18}) system is broader and reveals pronounced deviations from ideality. As seen in Figure 12, with increasing pressure the gel-fluid coexistence region of the binary lipid systems is shifted toward higher temperatures. A shift of about 0.22 C MPa^{-1} is observed, similar to the slope of the gel-fluid transition line of the pure lipid components (Landwehr et al. 1994a, b; Winter et al. 2000, 2004; Winter 2001).

Membranes also contain other (macro)molecules, such as cholesterol and peptides. Cholesterol (Chol) thickens liquid-crystalline bilayers and increases the packing density of the lipid acyl-chains in a way that has been referred to as "condensing effect" (Jorgensen et al. 1995; Seemann et al. 2003; Krivanek et al. 2008). An increase in pressure up to the 100 MPa range is much less effective in suppressing water permeability than cholesterol embedded in fluid DPPC bilayers at high concentration levels. It has been shown that sterols can efficiently regulate the structure, motional freedom, and hydrophobicity of lipid membranes, so that they can withstand even drastic changes in environmental conditions, such as in external pressure and temperature.

Membrane proteins can constitute about 30% of the entire protein content of a cell and act as various anchors, enzymes, or transporters on, within, and traversing the lipid environment. Membrane lipids and proteins influence each other directly as a result of their biochemical nature and in reaction to environmental changes. Pressure studies on this interaction, however, are still scarce. One example is the channel peptide gramicidin D (GD) on the structure and phase behavior of phospholipid bilayers (Zein et al. 2000; Eisenblatter et al. 2005). Gramicidin is polymorphic, being able to adopt a range of structures with different topologies. Common forms are the dimeric single-stranded right-handed $\beta^{6.3}$-helix with a length of 24 Å, and the antiparallel double-stranded $\beta^{5.6}$-helix, being approximately 32 Å long. For comparison, the hydrophobic fluid bilayer thickness is about 30 Å for DPPC bilayers, and the hydrophobic thickness of the gel phases is larger by 4-5 Å. Depending on the gramicidin concentration, significant changes of the lipid bilayer structure and phase behavior were observed. These changes include disappearance of certain gel phases formed by the pure DPPC system, and the formation of broad two-phase coexistence regions at higher gramicidin concentrations (Fig. 11b). Likewise, the lipid environment influences peptide conformation. Depending on the phase state and lipid acyl-chain length, gramicidin adopts at least two different types of

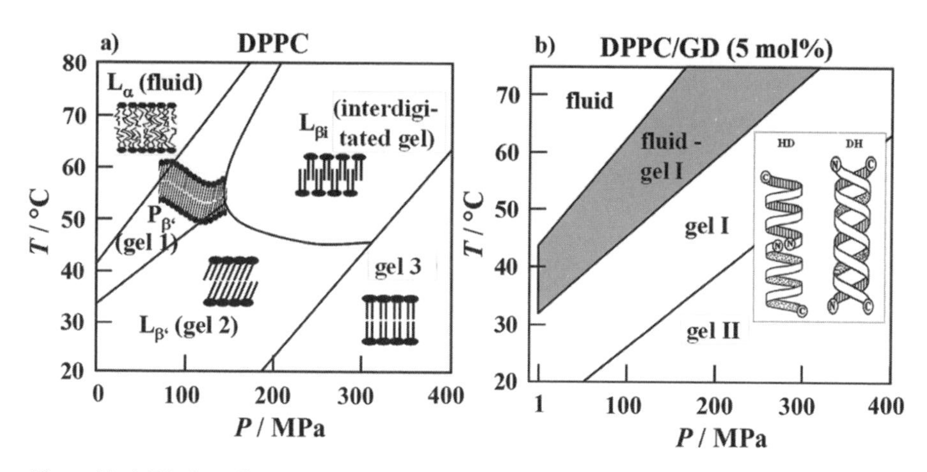

Figure 11. a) *T-P* phase diagram of DPPC bilayers in excess water. Besides the Gel 1 ($P_{\beta'}$), Gel 2 ($L_{\beta'}$) and Gel 3 phase, an additional crystalline gel phase (L_c) can be induced in the low-temperature regime after prolonged cooling, which is not shown here. b) Phase diagram of DPPC-gramicidin D (GD) (5 mol%) in excess water as obtained from diffraction and spectroscopic data. The inset shows a schematic view of the helical dimer (HD) and double helix (DH) conformation of GD.

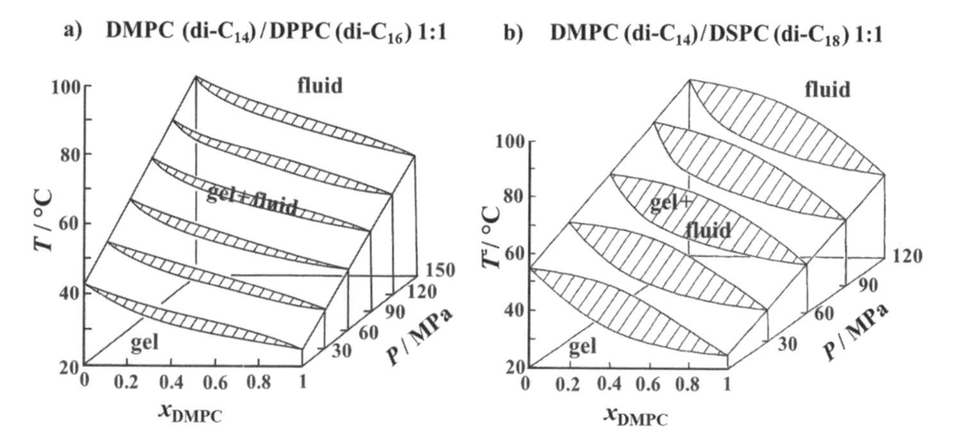

Figure 12. *T-P-X* phase diagram of equimolar DMPC/DPPC (di-C_{14}/di-C_{16}) and DMPC/DSPC (di-C_{14}/di-C_{18}) multi-lamellar vesicles in excess water. x is the weight fraction of lipid.

quaternary structures in the bilayer environment, a double helical pore (DH) and a helical dimer channel (HD; inset Fig. 11b). When the bilayer thickness changes at the gel-to-fluid main phase transition of DPPC, the conformational equilibrium of the peptide also changes (Zein et al. 2000). Hence, not only the lipid bilayer structure and *T-P*-dependent phase behavior drastically depends on the polypeptide concentration, but also the peptide conformation (and hence function) can be significantly influenced by the lipid environment. No pressure-induced unfolding of the polypeptide is observed up to 1.0 GPa. For large integral and peripheral proteins, however, pressure-induced changes in the physical state of the membrane may lead to a weakening of protein-lipid interactions as well as to protein dissociation.

Studies were also carried out on the phase behavior of cholesterol containing ternary lipid mixtures, generally containing an unsaturated lipid like a phosphatidylcholine and a saturated lipid-like sphingomyelin (SM) or DPPC. Such lipid systems are supposed to mimic distinct liquid-ordered lipid regions, called "rafts," which also seem to be present in cell membranes and are thought to be important for cellular functions such as signal transduction and the sorting and transport of lipids and proteins (Munro 2003; Janosch et al. 2004; Nicolini et al. 2005, 2006; Jeworrek et al. 2008; Weise et al. 2009; Kapoor et al. 2012a,b). Lipid domain formation can be influenced by temperature, pH, calcium ions, protein adsorption, and may be expected to change upon pressurization as well. Recently, we determined the liquid-disordered/liquid-ordered (l_d/l_o) phase coexistence region of canonical model raft mixtures such as POPC/SM/Chol (1:1:1), which extends over a rather wide temperature range. An overall fluid phase without domains is only reached at temperatures above ~50 °C (Nicolini et al. 2006). Upon pressurization at ambient temperatures (20-40 °C), an overall (liquid- and solid-) ordered state is reached at pressures of about 100-200 MPa. A similar behavior has been observed for the model raft mixture DOPC/DPPC/Chol (1:2:1; Fig. 13; Kapoor et al. 2012a, b).

Interestingly, in this pressure range of ~200 MPa, cessation of membrane protein function in natural membrane environments has been observed for a variety of systems (De Smedt et al. 1979; Chong et al. 1985; Kato et al. 2002; Ulmer et al. 2002; Powalska et al. 2007; Linke et al. 2008; Periasamy et al. 2009), which might be related to the membrane matrix reaching a physiologically unacceptable overall ordered state at these pressures. Moreover, many bacteria have been shown to completely lose their biologically relevant activity at these pressures.

Nonlamellar lipid phases

For a series of lipid molecules, nonlamellar lyotropic phases, such as inverse bicontinuous cubic (Q_{II}) or hexagonal (H_{II}) phases, are observed as thermodynamically stable phases or as long-lived metastable phases (Seddon et al. 1993; Winter et al. 2000, 2004; Winter 2001;

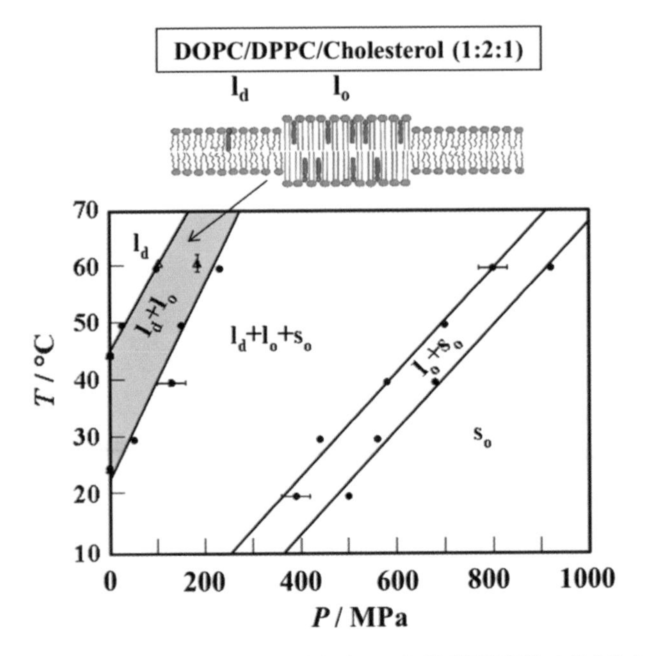

Figure 13. *T-P* phase diagram of the ternary lipid mixture DOPC/DPPC/Chol (1:2:1) in excess water as obtained from FTIR spectroscopy (•) and small-angle X-ray scattering (Δ) data. The l_d+l_o two-phase coexistence region is marked in grey and depicted schematically in the adjacent drawing. The liquid disordered phase is represented by l_d, whereas l_o and s_o are the liquid and solid ordered phases, respectively.

Fig. 8). Lipids, which can adopt a hexagonal phase, are present at substantial levels in biological membranes, usually with at least 30 mol% of the total lipid content. Fundamental metazoan cell processes, such as endo- and exocytosis, fat digestion, membrane budding, and fusion, involve a rearrangement of biological membranes, where such nonlamellar highly curved lipid structures are probably involved, but probably also static cubic structures (cubic membranes) occur in biological cells. The cubic lipid phases are mostly bicontinuous unilamellar lipid bilayer phases with periodic three-dimensional order.

In recent years, the temperature- and pressure-dependent structure and phase behavior of series of phospholipid systems, including phospholipid/fatty acid mixtures (e.g., DLPC/LA, DMPC/MA, DPPC/PA) and monoaclyglycerides (MO, ME), exhibiting nonlamellar phases have been studied (Erbes et al. 1996; Templer et al. 1998; Winter et al. 1999a,b,c, 2000, 2004). Contrary to DOPC which shows a lamellar L_β-to-L_α transition (Fig. 10), the corresponding lipid DOPE with ethanolamine as (smaller) headgroup exhibits an additional phase transition from the lamellar L_α to the nonlamellar, inverse hexagonal H_{II} phase at high temperatures (Fig. 14). As pressure forces a closer packing of the lipid chains, which results in a decreased number of *gauche* bonds and kinks in the chains, both transition temperatures, of the L_β-L_α and the L_α-H_{II} transition, increase with increasing pressure. The L_α-H_{II} transition observed in DOPE/water and also in egg-PE/water (egg-PE is a natural mixture of different phosphatidylethanolamines) is the most pressure-sensitive lyotropic lipid phase transition found to date ($dT/dP \approx 0.40$ °C·MPa^{-1}). The reason why this transition has such a strong pressure dependence is the strong P dependence of the chain length and volume of its *cis*-unsaturated chains. Generally, at sufficiently high pressures, hexagonal and cubic lipid mesophases give way to lamellar structures as they exhibit smaller partial lipid volumes. Interestingly, in these systems inverse cubic phases Q_{II}^D and Q_{II}^P can be induced in the region of the L_α-H_{II} transition by subjecting the sample to extensive temperature or pressure cycles across the phase transition. It has been shown that for conditions,

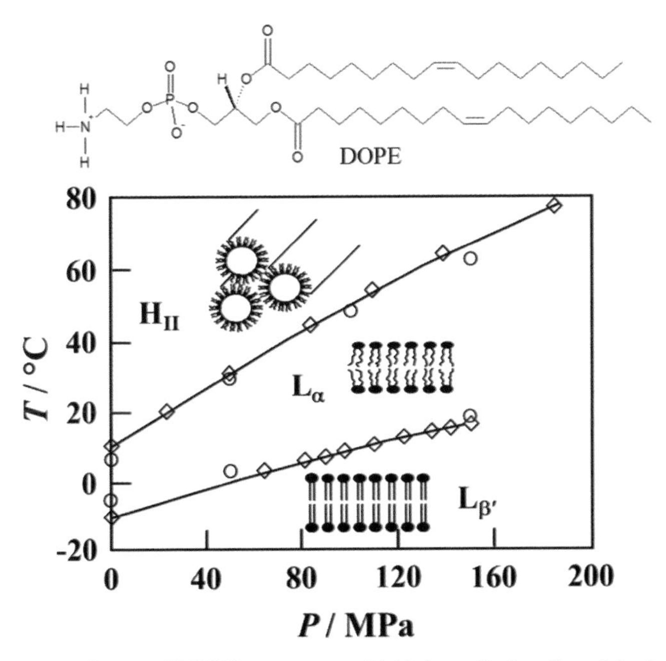

Figure 14. *T-P* phase diagram of DOPE in excess water. Lipid phases: L_β, lamellar gel; L_α, lamellar liquid-crystalline (fluid-like), and H_{II} inverse hexagonal.

which favor a spontaneous curvature of a lipid monolayer that is not too high, the topology of an inverse bicontinuous cubic phase (Fig. 8) can have a similar or even lower free energy than the lamellar or inverse hexagonal phase, as the cubic phases are characterized by a low curvature free energy and do not suffer the extreme chain packing stress predominant in the H_{II}-phase.

Biological and reconstituted membranes

It has generally been observed that at sufficiently high pressures of several hundred MPa, membrane protein function ceases, and integral and peripheral proteins may even become detached from the membrane when its bilayer is sufficiently ordered by pressure, and depolymerization of cytoskeletal proteins may be involved as well. In a detailed study, the influence of hydrostatic pressure on the activity of Na^+, K^+-ATPase enriched in the plasma membrane from rabbit kidney outer medulla was studied using a kinetic assay that couples ATP hydrolysis to NADH oxidation. The data shown in Figure 15 reveal that the activity, k, of Na^+, K^+-ATPase is inhibited by pressures below 200 MPa. The plot of ln$k$ vs. P revealed an apparent activation volume of the pressure-induced inhibition reaction which amounts to $\Delta V^{\#} = 47$ mL mol^{-1}. At higher pressures, exceeding 200 MPa, the enzyme is inactivated irreversibly in agreement with literature data (De Smedt et al. 1979; Chong et al. 1985). Kato et al. (2002) suggested that the activity of the enzyme shows at least three step changes induced by pressure: at pressures below and around 100 MPa, a decrease in the fluidity of the lipid bilayer and a reversible conformational change in the transmembrane protein is induced, leading to functional disorder of the membrane associated ATPase activity. Pressures of 100-200 MPa cause a reversible phase transition and the dissociation or conformational changes in the protein subunits, and pressures higher that 220 MPa irreversibly destroy the membrane structure due to protein unfolding and interface separation. To be able to explore the effect of the lipid matrix on the enzyme activity, the Na^+, K^+-ATPase was also reconstituted into various lipid bilayer systems of different chain length, configuration, phase state and heterogeneity including model raft mixtures. In the low-pressure region, around 10 MPa, a significant increase of the activity was observed for the enzyme reconstituted into DMPC and DOPC bilayers. It was found that the enzyme activity decreases upon further compression, reaching zero activity around 200

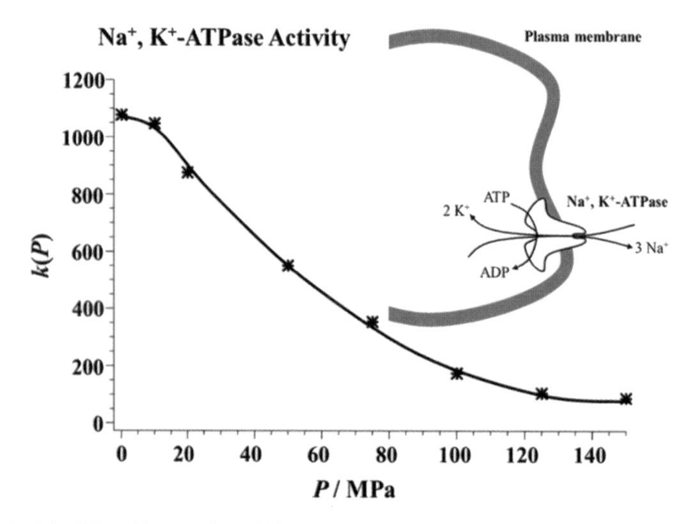

Figure 15. Activity k (in arbitrary units) of Na^+, K^+-ATPase — as measured using an enzymatic assay — at selected pressures and $T = 37$ °C. The free energy of hydrolysis of one ATP molecule is converted to uphill transport by actively transporting 3 Na^+ ions out of and 2 K^+ into the cell.

MPa for all reconstituted systems measured, similar to the natural system. A similar behavior has been found for the chloroplast ATP-synthase (Souza et al. 2004).

The effect of pressure was also determined for the HorA activity of the bacterium *Lactobacillus plantarum* (Ulmer et al. 2002), an ATP-dependent multi-drug-resistance transporter of the ABC family. Changes were determined in the membrane composition of *L. plantarum* induced by different growth temperatures and their effect on the pressure inactivation, and a temperature-pressure phase diagram was constructed for the *L. plantarum* membranes that could be correlated with the respective kinetics of high-pressure inactivation. Upon pressure-induced transitions to rigid (e.g., gel-like) membrane structures at pressures around 50-150 MPa for temperatures between 20 and 37 °C, fast inactivation of HorA was observed.

The effect of pressure and the influence of the lipid matrix on lipid-protein interactions was also studied for the multidrug resistance protein LmrA, which was expressed in the bacterium *Lactococcus lactis* and functionally reconstituted in different model membrane systems (Periasamy et al. 2009). The membrane systems were composed of DMPC, DOPC, DMPC+10 mol% Chol, and the model raft mixture DOPC:DPPC:Chol (1:2:1). Teichert (2008) showed that a sharp pressure-induced fluid-to-gel phase transition without the possibility for lipid sorting, such as in DMPC bilayers, has a drastic inhibitory effect on the LmrA activity. As inferred from the experiments performed so far, inactivation of membrane protein function upon entering a rigid gel-like (solid-ordered) membranous state seems to be a rather common phenomenon. Otherwise, an overall fluid-like membrane phase over the whole pressure range covered, with suitable hydrophobic matching, such as for DOPC, prevents the membrane protein from total high-pressure inactivation even up to 200 MPa. Also the systems exhibiting thicker membranes with higher lipid order parameters, such as DMPC/10 mol% Chol and the model raft mixture, show remarkable pressure stabilities. The results also revealed that an efficient packing with optimal lipid adjustment to prevent (also pressure-induced) hydrophobic mismatch might be a particular prerequisite for the homodimer formation, and hence function of LmrA.

Recently, high-pressure-induced dimer dissociation of membrane proteins *in vivo* was studied using the ToxR inner membrane-spanning transcription factor present in some piezosensitive and piezophilic bacteria. Analyses of ToxR derived from the mesophilic bacterium *Vibrio cholerae* were carried out by introducing protein variants in *Escherichia coli* reporter strains carrying *a ToxR activatable reporter gene* fusion. Dimerization ceased at 20 to 50 MPa, depending on the nature of the transmembrane segment rather than as a result of changes in the pressure-induced lipid bilayer environment (Linke et al. 2008). Similar results were also obtained for ToxR derived from the piezophilic deep-sea bacterium *Photobacterium profundum* strain SS9 in both *E. coli* and SS9 backgrounds (Linke et al. 2009).

Relevance of lipid biophysics for deep carbon

The biophysical results discussed above demonstrate that organisms are able to modulate the physical state of their membranes in response to pressure and temparature changes in the external environment by regulating the fractions of the various lipids in a cell membrane differing in chain length, chain unsaturation, or headgroup structure ("homeoviscous adaption"). Moreover, they have further means to regulate membrane fluidity, such as by changing the membrane concentration of their sterols and by a lateral redistribution of their various lipid components and domains. In fact, several studies have demonstrated that membranes are significantly more fluid in barophilic and/or psychrophilic species, which is principally a consequence of an increase in the unsaturated to saturated lipid ratio. It needs to be emphasized that, similar to the case for proteins, the effects of cold temperatures and pressure are the same, whereas adaptation to high temperature conditions induces a reduction of the degree of unsaturated lipids.

Archaea also contain a significant fraction of tetraether lipids in their membranes (Hanford and Peeples 2002). The pressure behavior of these lipids, in which the aliphatic chains are connected to the glycerol backbone via ether bonds instead of ester linkages, remains to be elucidated. Further research into the effects of pressure on protein-lipid interactions and their consequences for membrane protein activity is required.

HIGH-PRESSURE MICROBIOLOGY AND BIOCHEMICAL CYCLES

The oceans have an average depth of 3,800 m at a pressure of 38 MPa. Pressure increases with depth at a rate of ~10 MPa km^{-1} in the oceanic water column and in general about 20-30 MPa km^{-1} below sediments and hard-rocks. Deep-sea microbiology has its origins in the pioneering expeditions in the 19th century by the French scientists Certes and Regnard, who collected microbes from depths as great as 8,200 m. Certes actually performed some high-pressure experiments on bacteria and both scientists concluded that the deep-sea microbes were present *in situ* in a state of suspended animation (reviewed by Deming and Baross 2000). The inspiring discovery of microbial populations preferentially active under deep-sea high-pressure conditions was reported much later by ZoBell, his student Morita and others during the Danish *Galathea* "Round the World" expedition of 1950-1952 (ZoBell and Morita 1959; Bartlett et al. 2008 and refs therein). ZoBell first introduced the concept of "piezophily" (termed "barophily" in those early days) by comparing cell-density estimates of various physiological groups of microbes incubated at atmospheric pressure and elevated pressure on mud samples from 5.8 km off the coast of Bermuda. Even after the observational reports of ZoBell, the issue of whether piezophiles actually existed was still debated until 1979 when Yayanos and his team succeeded in obtaining and maintaining a pure culture of the piezophile that later came to be identified as *Psychromonas sp.* CNPT-3 and then a few years later the obligate piezophile *Colwellia sp.* MT-41 (Yayanos et al. 1979, 1981). Yayanos and colleagues then correlated the rates of piezophilic growth with capture depth and explored several facets of piezophile lipid and protein adaptation to high pressure (Delong and Yayanos 1985, 1986, 1987). Along the way Yayanos suggested the formal name change for these organisms and their behavior from "barophile" to "piezophile," since the Greek term *piezo* is that associated for pressure (Yayanos 1995). For life forms existing under high hydrostatic pressure conditions Jannash and Taylor (1984) defined the deep ocean as those organisms living in oceanic water below 1000 m, i.e., under pressures higher than 10 MPa. That concept was later extended to all high-pressure environments (Fig. 16), and now extends to environments ranging from the cold and oligotroph deep ocean, to hydrothermal vents rich in nutrients along spreading oceanic ridges, and subseafloor and subcontinental environments with highly variable sources of energy and thermal gradients. Considering the very large extent of the high-pressure biotopes identified to date, the deep biosphere could therefore represent a largely "unseen majority" of life on Earth, perhaps constituting up to 30% of the total living biomass, or even more (Whitman et al. 1998; Parkes et al. 2000). Although the upper limit of such projections have been questioned recently (Kallmeyer et al. 2012; Colwell and D'Hondt 2013), the subseafloor remains potentially the largest ecosystem on Earth. These organisms constitute an important and active cycling reservoir of carbon resources on Earth (D'Hondt et al. 2002; Jørgensen and D'Hondt 2006; Jørgensen and Boetius 2007). They are also the most difficult to access due to the technological challenges of drilling and working beneath hundreds to thousand of meters of seawater (see Schrenk et al. 2010 for a detailed review).

Who's down there?

Within the water column of ocean basins the groups of microbial species (more specifically operational taxonomic units, OTUs) present at depth are distinct from those above them, and conversely are more related to those OTUs from other deep-sea water masses, including across ocean basins (Eloe et al. 2011a). Examples of highly abundant deep-living clades of marine

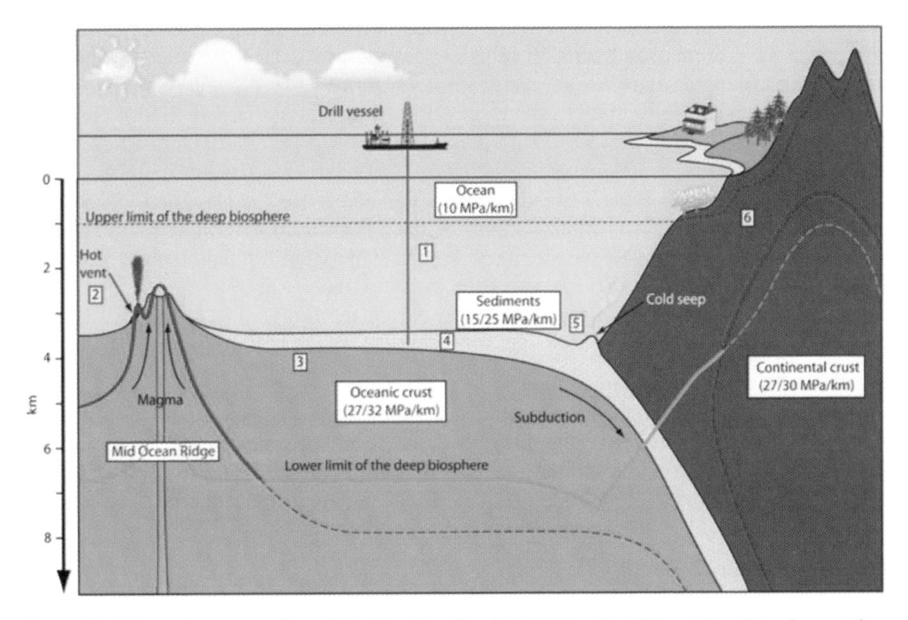

Figure 16. Schematic cross-section of Earth's subsurface highlighting the different deep biosphere settings (not to scale). 1: deep sea; 2: deep-sea hydrothermal vents; 3: oceanic crust; 4: sedimentary subseafloor; 5: deep-sea cold seep; 6: continental crust. The red and blue lines represent the current temperature and pressure limits for life, respectively. Solid lines highlight the parameter, which currently limits the depth of the deep biosphere. The upper dashed red line symbolizes the 10 MPa arbitrarily defined upper limit of the deep biosphere. [Used with permission of Elsevier, from Oger and Jebbar (2010), *Research in Microbiology*, Vol. 161, Fig. 1, p. 800.]

microbes include the order Rhizobiales within the Alphaproteobacteria, SAR324 within the Deltaproteobacteria, SAR406 within the Marine Group A, the phylum Thaumarchaea, the Group II Euryarchaeota and basidiomycete fungi. Extending from the deep-sea pelagic environment into the benthos microbial communities are highly stratified along narrow bands of redox gradients (Nealson 1997). Relatively few microbiological studies have yet been performed on sediments present at great depth, regardless of the combination of water depth or sediment depth. Consortia of methane-oxidizing archaea and sulfate-reducing bacteria related to those first reported in shallower cold-seep environments by others (Boetius 2000) have been recorded at depths up to 7.4 km in the Japan Trench (Kato et al. 2008). Deep-sea hydrothermal vent environments have long been proposed to provide a view into the deep subsurface biosphere (Deming and Baross 1993; Colwell and D'Hondt 2013; Schrenk et al. 2013). These systems harbor diverse microbial communities, existing over broad ranges of temperature and pH, and responsible for extensive cycling of sulfur, hydrogen, and methane. The microbes range from aerobic mesophiles to anaerobic hyperthermophiles (Jaeschke et al. 2012). The archaeal domain dominates at the highest temperatures and lowest pHs still supporting life and include both Euryarchaeota and Crenarchaeota. Within the bacteria Aquificales and members of the Epsilonproteobacteria are often dominant. A useful examination of the genes and processes present within these communities has been reported by Xie et al. (2011).

Cultured piezophilic isolates from deep biosphere locations provide useful resources for investigations into the adaptations responsible for life at high pressure. Some of these studies are described below, but much awaits the more detailed biophysical studies described in the preceding sections of this chapter. To date all such isolates come from the deep sea.

The psychrophilic and psychrotolerant piezophilic isolates obtained to date come from deep-sea seawater, animal, and surficial sediment samples. Most of these microbes belong to the five bacterial genera *Colwellia*, *Moritella*, *Photobacterium*, *Psychromonas*, and *Shewanella* within the Gammaproteobacteria (Kato et al. 2008; Lauro and Bartlett 2008). Additional microbes displaying modest degrees of high-pressure adaptation are represented by an anaerobic sulfate-reducing member of the genus *Desulfovibrio* and one Gram-positive member of genus *Carnobacterium* (e.g., see Table 7 in Orcutt et al. 2011 and Table 2 in Oger and Jebbar 2010). More recently low-nutrient cultivation strategies have been used to isolate a slow-growing, low-biomass producing obligate piezophile within the subphylum Alphaproteobacteria and the genus *Roseobacter* (Eloe et al. 2011b). This latter isolate is likely to be much more representative than others obtained to date of the bulk of deep-ocean heterotrophic life forms existing under exceedingly low concentrations of organic carbon.

At the other end of the temperature scale adaptation to elevated pressure is also evident. One dramatic example is strain CH1 isolated from hydrothermal vent smoker material collected at a depth of 4,100 m on the Mid-Atlantic ridge (Zeng et al. 2009). Strain CH1 grows optimally at 98 °C and 52 MPa and belongs to the genus *Pyrococcus*, within the Euryarchaeota lineage of the archaea domain. It is the first obligately piezophilic and hyperthermophilic microorganism known so far.

Genomic attributes at depth

Just as the communities of microbes present at depth are distinct from the microbial consortia above them, so are their genes (Colwell and D'Hondt 2013). The relatively few metagenomics studies of deep-sea microbial populations completed thus far have indicated much about the selective pressures and metabolic pathways present in the pelagic portion of the deep, dark biosphere (Eloe et al. 2011a). They have come from the North Pacific Gyre (4000 m depth), the Mediterranean (3000 m depth), and the Puerto Rico Trench (6000 m depth). The results indicate that the genome sizes of deep-sea microbes are substantially larger than those of their surface-water counterparts. They typically possess larger intergenic distances, expanded regulatory and signal transduction capacities, and diverse transport and metabolic pathways. Examples of expanded transcriptional regulation include alternative sigma factors such as the RpoE sigma factor that has been shown to play a role in growth at low temperature and high pressure. Examples of the expanded signal transduction capabilities are the PAS domain-containing proteins that function as internal sensors of redox potential. The transporters (especially within the Puerto Rico Trench) include many associated with heavy metal resistance. A few highlights of the expanded metabolic capabilities are the overabundance of aerobic carbon monoxide (CO) oxidation, as well as oxidative carbohydrate metabolic components for butanoate, glyoxylate, and dicarboxylate metabolism. In addition to considering the over-represented genes, one category of genes is dramatically under represented in the deep-sea genomes. Gene products associated with light-driven processes, including photosynthesis, rhodopsin photoproteins, and photorepair of DNA damage are largely absent from dark deep-oceanic environments.

The most thoroughly studied genome of a deep-sea bacterium is that belonging to the moderate piezophile *P. profundum* species strain SS9. SS9 is a deep-sea Gammaproteobacterium growing over a wide range of pressures (0.1-90 MPa, pressure optimum of 28 MPa) and temperatures (2-20 °C). Its ability to grow as colonies at atmospheric pressure has enabled the development of a limited set of genetic tools for complementation analysis, in-frame deletion construction, reporter gene usage and transposon mutagenesis. The *P. profundum* SS9 genome consists of two chromosomes of 4.1 and 2.2 Mbp in size along with a 80 kbp plasmid. The *P. profundum* SS9 genome encodes 15 rRNA operons, which is the highest number known for any bacterial species. The high number of rRNA operons is thought to enable *P. profundum* SS9 to rapidly adapt to changing environmental conditions, perhaps reflecting a feast and famine existence associated with sporadic and variable nutrient fluxes into its bathyal environment.

One of the best studied adaptations of deep-sea piezophiles such as strain SS9 is their need to counteract the compression effect of high pressure on membrane physical structure by producing high levels of unsaturated fatty acids. Many piezophiles produce not just high levels of monounsaturated fatty acids but also omega-3 polyunsaturated fatty acids (PUFAs) using a biosynthetic process related to that of polyketide antibiotics. The importance of unsaturated fatty acids to growth at high pressure has been demonstrated, although the relative importance of mono- and poly-unsaturated fatty acid varies among species. Genetic and fatty acid supplementation experiments have demonstrated the critical role of the monounsaturated fatty acids, and the dispensable role of the PUFA eicosapentaenoic acid (EPA), in high-pressure growth of *P. profundum* SS9 (Allen et al. 1999; Allen and Bartlett 2000). However, in the case of the piezophile *Shewanella violacea* DSS12, genetic and phospholipid feeding experiments have clearly demonstrated the requirement for EPA in the high-pressure growth and cell division of this species (Kawamoto et al. 2011). The basis for the difference in fatty acid needs at high pressure between the two species is unknown but could relate to differences in physiology, fermentation versus respiration, and resulting lipid-protein interactions.

Insight into additional nonessential genes important for the growth of SS9 at depth was obtained following transposon mutagenesis and screening for cells with growth defects at either low temperature or elevated pressure. Many of the genes influencing low-temperature and high-pressure growth were involved in signal transduction and adaptation. Genes for ribosome assembly and function were found to also be important for both low-temperature and high-pressure growth. The largest fraction of loci specific to cold sensitivity were involved in the biosynthesis of extracellular polysaccharide. The largest fraction of loci associated with pressure sensitivity were involved in chromosomal structure and function.

The connection between pressure and chromosome function, specifically DNA replication, was further studied. Transposon insertions into the genes encoding DiaA, a positive regulator, and SeqA, a negative regulator, of the initiation of DNA replication displayed opposite phenotypes. *diaA* mutants were pressure sensitive and *seqA* mutants were high-pressure growth enhanced. These SS9 genes were found to restore DNA replication synchrony in *E. coli* strains lacking homologous gene function. In addition, overproduction of the SS9 SeqA protein in SS9 converted this strain into a piezosensitive species. These results indicate that the activation of DNA replication in SS9 is hypersensitive to the influence of pressure and more specifically that the ratio of the activities of DiaA and SeqA effectively tune the pressure-growth characteristics of the cells.

Genetic investigations in SS9 have also uncovered another complex system that displays adaptation to elevated pressure. Flagellar motility is one of the most pressure-sensitive cellular processes in mesophilic bacteria. The SS9 genome contains two flagellar gene clusters: a polar flagellum gene cluster (PF) and a putative lateral flagellum gene cluster (LF). Mutants bearing in-frame deletions of the PF flagellin or motor protein genes are defective in motility under all conditions. However, deletion mutants in the LF flagellin or motor protein genes are defective only under conditions of high pressure and high viscosity, conditions that also induce LF gene expression. Direct swimming velocity measurements obtained using a high-pressure microscopic chamber (*http://bartlettlab.ucsd.edu/Motility_at_HP.html*) indicated that elevated pressure strongly represses the motility of the mesophile *E. coli*, turning off all motility by 50 MPa, and produces a gradual reduction in swimming speed for the piezotolerant *P. profundum* strain 3TCK, which was capable of some movement up to 120 MPa, while strain SS9 actually increased swimming velocity up to 30 MPa, and maintained motility up to a maximum pressure of 150 MPa, well above the known upper pressure limit for life. These results indicate the evolution of pressure-optimized motility systems in the piezophile *P. profundum* SS9, a feature that presumably extends to all motile deep-sea microbes. The mechanisms responsible for this piezo-adaptation are unknown.

Metabolism: organic matter, energy and nutrients

Microbial populations and their metabolic activities rely on substrate diversity and availability. Understanding such metabolic cycles is central to understanding carbon cycling by the deep biosphere. Although it has been sometimes asserted that the deep subseafloor microbial cells could be mostly dormant or even dead, both field and experimental studies (Price and Sowers 2004; Morono et al. 2011) have now shown that deep life is able to proceed, although in extreme slow motion, with a mean metabolic rate four orders of magnitude slower than at the surface (D'Hondt et al. 2002; Jørgensen and D'Hondt 2006), leading to generation times of subseafloor communities that range between only a few hours to thousands of years under nutrient- and/or energy limited conditions (Fig. 17).

The proliferation of subsurface life requires the availability of organic matter that derives in such dark environments from the deposition of terrigeneous sediments along the margins, in some cases from primary productivity in overlying surface waters or at hydrothermal vents, and importantly from recycling the microbial necromass over timescales of hundreds to thousands of years (Lomstein et al. 2012).

Other studies show that microorganisms operate extremely efficient catabolic systems and they may not necessarily inhabit only the most apparently favorable environments. It has been suggested that as little as −4.5 kJ mol^{-1} of free energy supply could support bacterial growth, and that bacterial metabolism can proceed near equilibrium in syntrophic associations (Jackson and McInerney 2002). In the deep-sea water column that is mostly oxic, aerobic respiration will dominate. Below the seafloor, oxygen becomes rapidly depleted within the sediments and other electron terminal acceptors, including nitrate and sulfate are utilized by facultative and obligate microorganisms for metabolism. Within the oxic oligotrophic sediments, the activity is generally

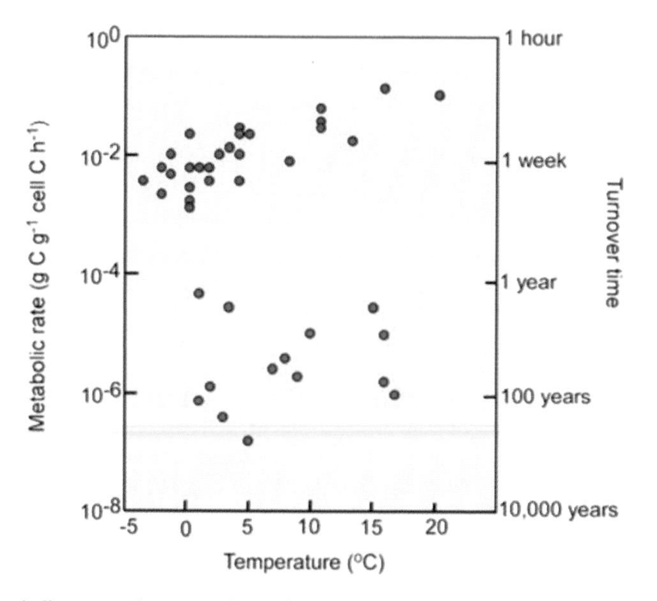

Figure 17. Metabolic rates and turnover times of natural communities of microorganisms. Blue indicates nutrient-rich environments. Red indicates nutrient starved environments such as subsurface sediments. Left axis shows metabolized organic carbon per cell carbon per unit time. Right axis shows the corresponding turnover time of cell carbon, approximately corresponding to minimum potential generation times. [Used with permission of the US National Academy of Sciences, from Jørgensen (2011), *Proceedings of the National Academy of Sciences USA*, Vol. 108, Fig. 1, p. 18193.]

low but it may increase by several orders of magnitude depending on the lithology (Picard and Ferdelman 2011). Under the deeper anoxic conditions, the terminal electron acceptors are used in sequential series, according to the free energy yield of the redox reactions, starting from nitrate reduction and denitrification, to dissimilatory Mn(IV) and Fe(III) reduction, to sulfate reduction and finally to methanogenesis, which is of primary importance to the DCO Deep Life mission (Fang and Bazylinski 2008). This activity is even lower but is stimulated at solid-fluid interfaces (D'Hondt et al. 2002; Parkes et al. 2005).

Metabolic activity of model piezosensitive microorganisms has been investigated under high-pressure conditions, including the yeast *Saccharomyces cerevisiae* for which the physiological response to high pressure is known as the best characterized eukaryotic cell (Fernandes 2005; Abe 2007). Although alcoholic fermentation was predicted to stop at 50 MPa due to inactivation of the enzyme phosphofructokinase as the cytoplasm becomes too acid, experiments showed that the fermentation continues to 87 ± 7 MPa. At 10 MPa, both the rate and the yield of ethanol production are enhanced, showing that pressure-enhanced catabolism might not be specific to piezophiles (Picard et al. 2007). Dissimilatory Se(IV) and Fe(III) reduction by the model bacterium *Shewanella oneidensis* MR-1 has also been investigated as a function of hydrostatic pressure. The catabolic activity of the piezosensitive *S. oneidensis* extends well beyond its anabolic limits (Picard et al. 2011, 2012), suggesting that piezosensitive strains could potentially ensure their maintenance in most of the deep subsurface environments at moderate pressures of 40-50 MPa.

ACQUISITION OF RESISTANCE TO GIGAPASCAL PRESSURES

Exploring extreme pressure limits for life

Although this topic falls out with any census of life relevant to the carbon cycle on Earth, it is important to examine and understand the ultimate limits for survival and adaptation of organisms to the most extreme conditions of high *P*, *T*, and chemical environments. Those studies are then relevant to the existence and origin of life on Earth as well as elsewhere in the universe, as well as for practical applications including food technology and bio-nanomaterials fabrication. Based on the results of biochemical/biophysical research combined with the growing body of information on the pressure limits of the integrity of cell membranes, proteins, and intracellular apparatus, it was thought until relatively recently that most organisms could not survive beyond approximately ~120 MPa (Zeng et al. 2009). Industrial processing units for Pascalization treatments typically operate at between 200-300 MPa for flow systems, or up to an upper limit around 500-700 MPa for batch conditions, depending on the organisms and biochemical conditions targeted. In this section we will discuss recent work on the evolution of organisms resistant to pressures that are far beyond the pressure limits currently experienced by organisms on Earth.

In 2002, researchers from the Geophysical Laboratory in Washington DC reported a remarkable result that samples of *E. coli* and *S. oneidensis* showed signs of metabolic activity at pressures extending into the GPa range (1.4-1.7 GPa), mainly based on *in situ* Raman spectroscopic investigations of product/reactant ratios (Sharma et al. 2002). However, that result received immediate criticism, questioning both the results and their interpretations (Yayanos 2002). One of the criteria that Sharma et al. (2002) used for survivability was formate oxidation. Formate oxidation is a metabolic reaction that is vital for bacteria and it is also quite easy to detect using spectroscopy. The main problem, however, is that this reaction can also occur even if the cells are not viable; i.e., if they are inactivated by death or dormancy. Studies have shown that this reaction can occur in purified enzyme solutions; therefore, it is not considered by the microbiological community as a satisfactory indicator of the presence of a living cell.

More recently a directed evolution study of *E. coli* confirmed that microbes can in fact adapt and survive to at least 2 GPa (Vanlint et al. 2011; Fig. 18). That work subjected a strain of *E. coli* to progressively higher pressures extending into the GPa range. Following decompression, the survivors were recovered and allowed to form colonies and their pressure resistance was examined. Using this technique, Vanlint et al. (2011) could identify clones that had a pressure resistance extending up to at least 2 GPa, greatly exceeding that of the parent strain that became extinct above 700 MPa.

Kish et al. (2012) recently published results suggesting that high-pressure tolerance is due to mechanical properties of the cell, including cell envelope structure and intracellular salts. This study is the first to investigate the role of these intracellular salts in both Gram-negative *E. coli* MG1655, *Chromohalobacter salexigens* and Gram-positive (*Deinococcus radiodurans* R1) and archaea (*Halobacterium salinarum* NRC-1) bacterial strains. The strains were subjected to pressures up to 400 MPa and the authors concluded that even without directed evolution bacterial strains can acquire piezo-resistance from adaptations to other environmental factors.

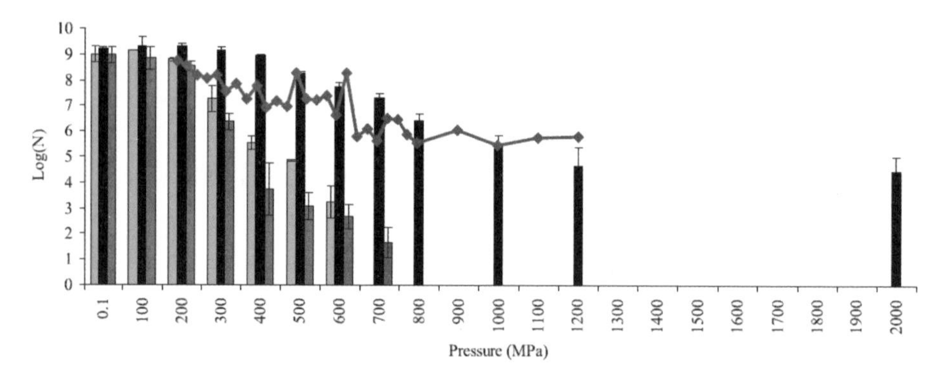

Figure 18. Directed evolution of *E. coli* K-12 MG1655 toward high pressure (ambient temperature). [Used with permission of American Society for Microbiology from Vanlint et al. (2011), *mBio*, Vol. 2, Fig. 1, p. 2.]

Acquisition of gigapascal pressure resistance by higher organisms

Such studies of extreme pressure resistance have not been limited to unicellular organisms, and the survival of several dehydrated biological systems to pressures of 7.5 GPa has been investigated. These mainly plant- or animal-related organisms are subjected to high pressure in a fully dehydrated state, in which they are metabolically inactive. They are then observed to recover relatively well after their exposure to the extreme hyperbaric conditions, following decompression to ambient *P* and rehydration over periods of time extending up to a few days. Table 1 summarizes such studies on pressure resistance of multicellular systems carried out to date. A similar observation has been made for temperature survival in the case of *Milnesium tardigradum*, a small invertebrate animal that survives temperatures as low as −273 °C and as high as +151 °C in its dehydrated state. The crux of these survival phenomena is the lack of water in the system, and in this sense they are completely different from those reported on microorganisms. Although these studies have been interpreted in terms of pressure-induced selection, it is difficult to see how such a process could occur on a dehydrated and latent system. Overall, this work emphasizes the role of water in the disruption of hydrated biological structures at high pressure as outlined in the first part of this chapter.

Table 1. Resistance of higher biological systems to GPa pressures.

Species	Form	Duration of pressure exposure (hr)	Survival and ability to develop	Refs.
Milnesium tardigradum	whole cryptobiotic	3	100%	[1]
		6	100%	
		12	25%	
		24	0%	
Artemia	dried eggs	< 48	80-90%	[2]
Ptychomitrium	spores	< 48	80-90%	[3]
		< 72	80%	
		144	32%	
Venturiella sinensis	spores	< 24	100%	[4]
		72	70-90%	
		144	ca. 4%	
Trifolium lepens L.	seeds	1	Germination of stems (leaves)	[5]
		< 24	Germination of roots only	

References: [1] Ono et al. (2008); [2] Minami et al. (2010), Ono et al. (2010); [3] Nishihira et al. (2010); [4] Ono et al. (2009); [5] Ono et al. (2012)

Resistance to extreme shock pressures

Other remarkable survival stories concern the exposure of bacterial organisms to shock conditions. It is well known that impacts from comets and meteorites, including ejecta from other planets as well as non-planetary bodies, have influenced the chemical evolution of Earth, and investigations or speculations about these extra-solar bodies or remnants of early solar system components have instigated theories of "panspermia" and possible origins of life elsewhere in the universe with subsequent transportation to early Earth (Melosh 1988). Such theories propose that primitive life forms extending to bacteria can survive in extreme environments, such as those of interplanetary space, for a sufficient time for these life forms to become trapped in debris (Fajardo-Cavazos et al. 2009), and then to be transported intact to Earth *via* a planetary impact, meteor bombardment, or cometary interaction (Willis et al. 2006). Earth then provides a hospitable habitat for those life forms to evolve and develop. Such organism survival studies following shock impact have been examined experimentally.

Experiments on organic matter at hypervelocities have been conducted using light gas guns on various broths, spores, and bacterial organisms from 1-8 GPa. These hypervelocity impacts are integral to several hypotheses arising from origin-of-life questions. Experiments have now tested several strains of bacteria, including *Rhodococcus erythropolis*, *Bacillis subtilis*, *E. coli, Enterococcus faecalis*, and the eukaryote *Zygosaccharomyces bailii* at shock pressures (Burchell et al. 2001, 2004; Hazell et al. 2010, 2009). In some experiments the bacteria are subjected to shock pressures of up to 78 GPa and survive (Burchell et al. 2004). While the survival rates vary between runs, these cells, whether in a broth or as a spore, show a resistance to the pressures they have experienced. In addition to surviving these extreme shock pressures, the organisms appear to exhibit subsequent growth and continued existence post pressure shock. However, to date no truly systematic studies have been carried out, especially controlling the temperatures achieved during the shock experiments.

CONCLUSIONS

The early expeditions of the late 19th century revolutionized our awareness of life in the deep sea, and Bridgman's developments in high-pressure experimentation opened up the possibility of exploring molecules and life at high pressure. As we come close to the centenary of Bridgman's pioneering observations on the pressure-induced unfolding of proteins, obligate piezophiles have been isolated and the physical chemistry underlying pressure effects on biomacromolecules have been largely elucidated, although intermolecular (e.g., protein-lipid) interactions remain to be investigated in greater detail. Yet, the existence of diverse organisms in extreme environments, some of which may display generation times of years to decades, highlights the many challenges that remain in understanding life in high-pressure environments. Physiological adaptation of microorganisms to high-pressure environments may also broaden our understanding of the adaptation of organisms to other extreme conditions of pH, salinity, and low temperatures (Kish et al. 2012), because organisms from extreme environments are often exposed to multiple stressors: high pressure, low or high temperatures, low nutrient concentrations, and more. Thus extremophiles may reveal elements of cellular evolution and ultimately provide greater insight on the origins of life (Daniel et al. 2006). Studies of life at extreme conditions also hold the potential for technological advances; for example, by the discovery of new molecules that hold the potential to cure human diseases (Wilson and Brimble 2009; Lutz and Falkowski 2012). The use of advanced technologies, such as neutron scattering, proteomics, and genomics, will enable us to probe even more complex systems, and directed evolution in the lab will allow us to interrogate the cellular response to high pressure. There is a bright future for high-pressure biophysics and microbiology that will lead to an understanding of the evolution and adaptation of cells and their macromolecules to high pressure. Moreover, given the spatial magnitude of pressure-affected environments, it is possible that a significant portion of the global organic carbon produced on Earth is mediated by pressure-affected microbial communities.

ACKNOWLEDGMENTS

Work in PFMs group at UCL is supported by funding from the Leverhulme Trust (UK) and the Deep Life directorate of the Deep Carbon Observatory of the Sloan Foundation (USA) through a block grant to East Carolina State University. Work in DHB's laboratory is supported by funding from the National Science Foundation, the National Aeronautics and Space Administration, the Prince Albert II Foundation, the Avatar Global Foundation, and the Deep Life directorate of the Deep Carbon Observatory of the Sloan Foundation (USA) through a block grant to East Carolina State University. RW gratefully acknowledges financial support from the DFG. FM thanks Drs Paliwal, Asthagiri, and Paulaitis for providing data to reproduce Figure 1, and Professor K. Van Hecke for assitance in preparing Figure 7.

REFERENCES

Abe F (2007) Induction of DANITIR yeast cell wall mannoprotein genes in response to high hydrostatic pressure and low temperature. FEBS Lett 581:4993-4998, doi: 10.1016/j.febslet.2007.09.039

Allen EE, Bartlett DH (2000) FabF is required for Piezoregulation of cis-vaccenic acid levels and piezophilic growth of the deep-sea bacterium *Photobacterium profundum* strain SS9. J Bacteriol 182:1264-1271, doi: 10.1128/JB.182.5.1264-1271.2000

Allen EE, Facciotti D, Bartlett DH (1999) Monounsaturated but not polyunsaturated fatty acids are required for growth at high pressure and low temperature in the deep-sea bacterium *Photobacterium profundum* strain SS9. Appl Environ Microbiol 65:1710-1720

Bartlett DH (2002) Pressure effects on in vivo microbial processes. Biochim Biophys Acta 1595:367-381, doi: 10.1016/S0167-4838(01)00357-0

Bartlett DH, Ferguson G, Valle G (2008) Adaptations of the psychrotolerant piezophile Photobacterium profundum strain SS9. *In*: High-pressure Microbiology. Michiels C, Bartlett D, Aertsen A (ed) ASM Press, Washington DC p 319-337

Behan MK, Macdonald AG, Jones GR, Cossins AR (1992) Homeoviscous adaptation under pressure: the pressure dependence of membrane order in brain myelin membranes of deep-sea fish. Biochim Biophys Acta 1103:317-323, doi: 10.1016/0005-2736(92)90102-R

Boetius A, Ravenschlag K, Schubert C, Rickert D, Widdel F, Gleseke A, Amann R, Jorgensen B, Wilde U, Pfannkuche O (2000) A marine microbial consortium apparently mediating anaerobic oxidation of methane, Nature Lett 407:623-626, doi: 10.1038/35036572

Böttner M, Ceh D, Jacobs U, Winter R (1994) High pressure volumetric measurements on phospholipid bilayers. Z Phys Chem 184:205-218, doi: 10.1524/zpch.1994.184.Part_1_2.205

Bridgman PW (1914) The coagulation of albumen with pressure. J Biol Chem 19:511-512

Brun L, Isom DG, Velu P, Garcia-Moreno B, Royer CA (2006) Hydration of the folding transition state ensemble of a protein. Biochem 45:3473-3480, doi: 10.1021/bi052638z

Burchell MJ, Mann J, Bunch AW, Brand PFB (2001) Survivability of bacteria in hypervelocity impact. Icarus 154:545-547, doi:10.1006/icar.2001.6738

Burchell MJ, Mann JR, Bunch AW (2004) Survival of bacteria and spores under extreme shock pressures. Mon Not R Astron Soc 352:1273-1278, doi: 10.1111/j.1365-2966.2004.08015.x

Cacace MG, Landau EM, Ramsden JJ (1997) The Hofmeister series: salt and solvent effects on interfacial phenomena. Quart Rev Biophys 30:241-277, doi: 10.1017/S0033583597003363

Chalikian TV (2003) Volumetric properties of proteins. Annu Rev Biophys Biomol Struct 32:207-235, doi: 10.1146/annurev.biophys.32.110601.141709

Chong PL, Fortes PA, Jameson DM (1985) Mechanisms of inhibition of (Na,K)-ATPase by hydrostatic pressure studied with fluorescent probes. J Biol Chem 260: 14484-14490

Collins RE, Rocap G, Deming JW (2010) Persistence of bacterial and Archaeal communities in sea ice through an Arctic winter. Environ Microbiol 12:1828-1841, doi: 10.1111/j.1462-2920.2010.02179.x

Colwell FS, D'Hondt S (2013) Nature and extent of the deep biosphere. Rev Mineral Geochem 75:547-574

Creighton TE (1990) Protein folding. Biochem J 270:1-16

CzeslikC,ReisO,WinterR,RappG(1998)Effectofhighpressureonthestructureofdipalmitoylphosphatidylcholine bilayer membranes: a synchrotron-X-ray diffraction and FT-IR spectroscopy study using the diamond anvil technique. Chem Phys Lipids 91:135-144, doi: 10.1016/S0009-3084(97)00104-7

Daniel I, Oger P and Winter R (2006) Origins of life and biochemistry under high pressure conditions. Chem Soc Rev 35:858-875, doi: 10.1039/b517766a

Daniel RM, Dines M, Petach H (1996) The denaturation and degradation of stable enzymes at high temperatures. Biochem J 317: 1-11

DeLong EF, Yayanos AA (1985) Adaptation of the membrane lipids of a deep-sea bacterium to changes in hydrostatic pressure. Science 228:1101-1103

DeLong EF, Yayanos AA (1986) Biochemical function and ecological significance of novel bacterial lipids in deep-sea procaryotes. Appl Environ Microbiol 51:730-737

DeLong EF, Yayanos AA (1987) Properties of the glucose transport system in some deep-sea bacteria. Appl Environ Microbiol 53:527-532

Deming JW, Baross JA (1993) Deep-sea smokers: windows to a subsurface biosphere? Geochim Cosmochim Acta 57:3219-3230

Deming JW, Baross JA (2000) Survival, dormancy and non-culturable cells in deep-sea environments. *In*:Non-Culturable Microorganisms in the Environment (ed) ASM Press, Washington, DC, p 147-197

De Smedt H, Borghgraef R, Ceuterick F, Heremans K (1979) Pressure effects on lipid-protein interactions in ($Na^+ + K^+$)-ATPase. Biochim Biophys Acta 556:479-489, doi: 10.1016/0005-2736(79)90135-4

D'Hondt S, Rutherford S and Spivack A (2002) Metabolic activity of subsurface life in deep sea sediments. Science 295:2067-2070, doi: 10.1126/science.1064878

Downey CD, Crisman RL, Randolph TW, Pardi A (2007) Influence of hydrostatic pressure and cosolutes on RNA tertiary structure. J Am Chem Soc 129:9290-9291, doi: 10.1021/ja072179k

Dubins DN, Lee A, Macgregor RB, Chalikian TV (2001) On the stability of double stranded nucleic acids. J Am Chem Soc 123:9254-9259, doi: 10.1021/ja004309u

Eisenblatter J, Winter R (2005) Pressure effects on the structure and phase behavior of phospholipid, polypeptide bilayers: A synchrotron small-angle x-ray scattering and 2H-NMR spectroscopy study on DPPC,Gramicidin lipid bilayers. Z Phys Chem 219:1321-1345

Eloe E, Shulse CN, Fadrosh DW, Williamson SJ, Allen EE, Bartlett DH (2011a). Compositional differences in particle-associated and free-living microbial assemblages from an extreme deep-ocean environment. Environ Microbiol Rep 3:449-458, doi: 10.1111/j.1758-2229.2010.00223.x

Eloe E, Malfatti F, GutlerreJ, Hardy K, Schmidt W, Pogllano K, Azam F, Bartlett DH (2011b) Isolation and characterisation of a psychropiezophilic alphaproteobacterium. Appl Environ Microbiol 77:8145-8153, doi: 10.1128/AEM.05204-11

Erbes J, Winter R, Rapp G (1996) Rate of phase transformations between mesophases of the 1:2 lecithin/fatty acid mixtures DMPC/MA and DPPC/PA - a time-resolved synchrotron X-ray diffraction study. Ber Bunsenges Phys Chem 100:1713-1722, doi: 10.1002/bbpc.19961001008

Fajardo-Cavazos P, Langenhorst F, Melosh HJ, Nicholson WL (2009) Bacterial spores in granite survive hypervelocity launch by spallation: Implications for lithopanspermia. Astrobiology 9:647-57, doi: 10.1089/ast.2008.0326

Fang J, Bazylinski D (2008) Deep sea geomicrobiology. *In*. High Pressure Microbiology. Michiels C, Bartlett D, Aertsen A (eds), ASM Press, Washington, DC, p 237-263

Fenimore PW, Frauenfelder H, McMahon BH, Young RD (2004) Bulk-solvent and hydration-shell fluctuations, similar to α and β fluctuations in glasses, control protein motions and functions. Proc Natl Acad Sci USA 101:15469-15472, doi: 10.1073/pnas.0607168103

Fernandes PMB (2005) How does yeast respond to pressure? Brazilian J Med Biol Res 38:1239-1245, doi: 10.1590/S0100-879X2005000800012

Frauenfelder H, Alberding NA, Ansari A, Braunstein D, Cowen BR, Hong MK, Iben IET, Johnson JB, Luck S (1990) Proteins and pressure. J Phys Chem 94:1024-1037, doi: 10.1021/j100366a002

Garcia AE, Paschek D (2007) Simulation of the pressure and temperature folding/unfolding equilibrium of a small RNA hairpin. J Am Chem Soc 130:815-817, doi: 10.1021/ja074191i

Ghosh T, Garcia AE, Garde S (2001) Molecular dynamics simulations of pressure effects on hydrophobic interactions. J Am Chem Soc 123:10997-11003, doi: 10.1021/ja010446v

Ghosh T, Garcia AE, Garde S (2002) Enthalpy and entropy contributions to the pressure dependence of hydrophobic interactions. J Chem Phys 116:2480, doi: 10.1063/1.1431582

Girard E, Prangé T, Dhaussy A-C, Migianu-Griffoni E, Lecouvey M, Chervin J-C, Mezouar M, Kahn R, Fourme R (2007) Adaptation of the base-paired double-helix molecular architecture to extreme pressure. Nucleic Acids Res 35:4800-4808, doi: 10.1093/nar/gkm511

Hammouda B, Worcester D (1997) Interdigitated hydrocarbon chains in C20 and C22 phosphatidylcholines induced by hydrostatic pressure. Physica B Condens Matter 241-243:1175-1177

Hanford MJ, Peeples TL (2002) Archeal tetraether lipids. unique structures and applications. Appl Biochem Biotechnol 97:45-62, doi: 10.1385/ABAB:97:1:45

Hashizume C, Kimura K, Hayashi R (1995) Kinetic analysis of yeast inactivation by high pressure treatment at low temperatures. Biosci Biotechnol Comm 59:1455-1458

Hawley SA (1971) Reversible pressure-temperature denaturation of chymotrypsinogen. Biochem 10:2436-2442, doi: 10.1021/bi00789a002

Hawley SA, Macleod RM (1974) Pressure-temperature stability of DNA in neutral salt solutions. Biopolymers 13:1417-1426, doi: 10.1002/bip.1974.360130712

Hazell PJ, Beveridge C, Groves K, Stennett C, Elert M, Furnish MD, Anderson WW, Proud WG, Butler WT (2009) Shock compression and recovery of microorganism-loaded broths and an emulsion. Am Inst Phys Conf Proc 2009:1395-1398

Hazell PJ, Beveridge C, Groves K, Appleby-Thomas G (2010) The shock compression of microorganism-loaded broths and emulsions: Experiments and simulations. Int J Impact Eng 37:433-440, doi: 10.1016/j.ijimpeng.2009.08.007

Herberhold H, Royer CA, Winter R (2004) Effects of chaotropic and kosmotropic cosolvents on the pressure-Induced unfolding and denaturation of proteins: An FT-IR study on staphylococcal nuclease. Biochem 43:3336-3345, doi: 10.1021/bi036106z

Heremans K, Smeller L (1997) Pressure versus temperature behavior of proteins. Eur J Solid State Inorg Chem 34:745-758

Heremans K, Smeller L (1998) Protein structure and dynamics at high pressure. Biochim Biophys Acta 1386:353-370, doi: 10.1016/S0167-4838(98)00102-2

Hillson N, Onuchic JN, Garcia AE (1999) Pressure-induced protein-folding/unfolding kinetics. Proc Natl Acad Sci USA 96:14848-14853, doi:10.1073/pnas.96.26.14848

Hiraki T, Sekiguchi T, Kato C, Hatada Y, Maruyama T, Abe F, Konishi M (2012) New type of pressurized cultivation method providing oxygen for piezotolerant yeast. J Biosci Bioeng 113:220-223, doi: 10.1016/j.jbiosc.2011.09.017

Hörmann S, Scheyhing C, Behr J, Pavlovic M, Ehrmann M, Vogel RF (2006) Comparative proteome approach to characterize the high-pressure stress response of Lactobacillus sanfranciscensis DSM 20451. Proteomics 6:1878-1885, doi: 10.1002/pmic.200402086

Hummer G, Garde S, Garcia AE, Paulaitis ME, Pratt LR (1998) The pressure dependence of hydrophobic interactions is consistent with the observed pressure denaturation of proteins. Proc Natl Acad Sci USA 95:1552-1555

Ishii A, Sato T, Wachi M, Nagai K, Kato C (2004) Effects of high hydrostatic pressure on bacterial cytoskeleton FtsZ polymers *in vivo* and *in vitro*. Microbiol 150:1965-1972, doi: 10.1099/mic.0.26962-0

Jackson B, McInerney M (2002) Anaerobic microbial metabolism can proceed close to thermodynamic limits. Nature Lett 415:454-456, doi: 10.1038/415454a

Jacob MH, Saudan C, Holtermann G, Martin A, Perl D, Merbach AE, Schmid FX (2002) Water contributes actively to the rapid crossing of a protein unfolding barrier. J Mol Biol 318:837-845, doi: 10.1016/S0022-2836(02)00165-1

Jaeschke A, Jorgensen SL, Bernasconi SM, Pedersen RB, Thorseth IH, Fruh-Green GL (2012) Microbial diversity of Loki's Castle black smokers at the Arctic Mid-Ocean Ridge. Geobiology 10:548-561, doi: 10.1111/gbi.12009

Jannasch HW, Taylor CD (1984) Deep sea microbiology. Annu Rev Microbiol 38:487-541

Jannasch HW, Wirsen CO, Doherty KW (1996) A pressurized chemostat for the study of marine barophilic and oligotrophic bacteria. Appl Environ Microbiol 62:1593-1596

Janosch S, Nicolini C, Ludolph B, Peters C, Volkert M, Hazlet TL, Gratton E, Waldmann H, Winter R (2004) Partitioning of dual-lipidated peptides into membrane microdomains lipid sorting vs peptide aggregation. J Am Chem Soc 126:7496-7503, doi: 10.1021/ja049922i

Jeworrek C, Puahse M, Winter R (2008) X-ray kinematography of phase transformations of three-component lipid mixtures: A time-resolved Synchrotron x-ray scattering study using the pressure-jump relaxation technique. Langmuir 24:11851-11859, doi: 10.1021/la801947v

Jorgensen K, Mouritsen OG (1995) Phase separation dynamics and lateral organization of two-component lipid membranes. Biophys J 69: 942-954

Jørgensen BB, D'Hondt S (2006) Ecology - A starving majority deep beneath the seafloor. Science 314:932-934, doi: 10.1126/science.1133796

Jørgensen B, Boetius A (2007) Feast and famine- microbial life in the deep sea bed. Nature Rev 5:770-781, doi: 10.1038/nrmicro1745

Jørgensen B (2011) Deep subseafloor microbial cells on physiological standby. Proc Natl Acad Sci USA 108:18193-18194, doi: 10.1073/pnas.1115421108

Kallmeyer J, Pockalny R, Adhikari RR, Smith DC, D'Hondt S (2012) Global distribution of microbial abundance and biomass in subseafloor sediment. Proc Natl Acad Sci USA 109:16213-16216, doi: 10.1073/pnas.1203849109

Kamekura M (1998) Diversity of extremely halophilic bacteria. Extremophiles 2:289-295, doi: 10.1007/s007920050071

Kapoor S, Triola G, Vetter IR, Erlkamp M, Waldmann H, Winter R (2012a) Revealing conformational substates of lipidated N-Ras protein by pressure modulation. Proc Natl Acad Sci USA 109:460-465, doi: 10.1073/pnas.1110553109

Kapoor S, Weise K, Erlkamp M, Triola G, Waldmann H and Winter R (2012b) The role of G-domain orientation and nucleotide state on the Ras isoform-specific membrane interaction. Eur Biophys J 41:801-813, doi: 10.1007/s00249-012-0841-5

Kato M, Hayashi R, Tsuda T, Taniguchi K (2002) High pressure-induced changes of biological membrane. Eur J Biochem 269:110-118, doi: 10.1046/j.0014-2956.2002.02621.x

Kato C (2006) Handling of piezophilic microorganisms. Methods Microbiol 35:733-741, doi: 10.1016/S0580-9517(08)70034-5

Kato C, Nogi Y, Arakawa S (2008) Isolation, cultivation and diversity of deep sea piezophiles. *In*: High Pressure Microbiology. Michiels C, Bartlett D, Aertsen A (eds), ASM Press, Washington, DC, p 203-217

Kauzmann W (1987) Thermodynamics of unfolding. Nature 325:763-764, doi: 10.1038/325763a0

Kawamoto J, Sato T, Nakasone K, Kato C, Mihara H, Esaki N, Lurihara T (2011) Favourable effects of eicosapentaenoic acid on the late step of the cell division in a piezophilic bacterium, Shewanella violacea DSS12, at high-hydrostatic pressures. Environ Microbiol 13:2293-2298, doi: 10.1111/j.1462-2920.2011.02487.x

Kish A, Griffin PL, Rogers KL, Fogel ML, Hemley RJ, Steele A (2012) High pressure tolerance in Halobacterium salinarum NRC-1 and other non-piezophilicprokaryotes. Extremophiles 16:355-361, doi: 10.1007/s00792-011-0418-8

Klein-Seetharaman J, Oikawa M, Grimshaw SB, Wirmer J, Duchardt E, Ueda T, Imoto T, Smith LJ, Dobson CM, Schwalbe H (2002) Long-range interactions within a nonnative protein. Science 295:1719-1722, doi: 10.1126/science.1067680

Korzhnev DM, Bezsonova I, Evanics F, Taulier N, Zhou Z, Bai Y, Chalikian TV, Prosser RS, Kay LE (2006) Probing the transition state ensemble of a protein folding reaction by pressure-dependent NMR relaxation dispersion. J Am Chem Soc 128:5262-5269, doi: 10.1021/ja0601540

Krivanek R, Okoro L, Winter R (2008) Effect of cholesterol and ergosterol on the compressibility and volume fluctuations of phospholipid-sterol bilayers in the critical point region: A molecular acoustic and calorimetric study. Biophys J 94:3538-3548, doi: 10.1529/biophysj.107.122549

Kunugi S, Yamazaki Y, Takano K, Tanaka N, Akashi M (1999) Effects of ionic additives and ionic comonomers on the temperature and pressure responsive behavior of thermoresponsive polymers in aqueous solutions. Langmuir 15:4056-4061, doi: 10.1021/la981184m

Landwehr A, Winter R (1994a) The T,x,p-phase diagram of binary phospholipid mixtures. Ber Bunsenges Phys Chem 98:1585-1589

Landwehr A, Winter R (1994b) High-pressure differential thermal analysis of lamellar to lamellar and lamellar to non-lamellar lipid phase transitions. Ber Bunsenges Phys Chem 98:214-218

Lauro F, Bartlett D (2008) Prokaryotic lifestyles in deep sea habitats. Extremophiles 12:15-25, doi: 10.1007/s00792-006-0059-5

Lesch H, Schlichter J, Friedrich J, Vanderkooi JM (2004) Molecular probes: What is the range of their interaction with the environment? Biophys J 86:467-472, doi: 0006-3495/04/01/467/06

Linke K, Periasamy N, Ehrmann M, Winter R, Vogel RF (2008) Influence of high pressure on the dimerization of ToxR, a protein involved in bacterial signal transduction. Appl Environ Microbiol 74:7821-7823, doi: 10.1128/AEM.02028-08

Linke K, Periasamy N; Eloe EA, Ehrmann M, Winter R, Bartlett DH, Vogel RF (2009) Influence of membrane organization on the dimerization ability of ToxR from Photobacterium profundum under high hydrostatic pressure. High Press Res 29:431-442, doi: 10.1080/08957950903129114

Lodish H, Berk A, Zipursky L, Matsudaira P, Baltimore D, Darnell J (1995) Molecular Cell Biology (3rd Ed). Scientific American Books, New York

Lomstein B, Langerhuus A, D'Hondt S Jorgensen B, Spivack A (2012) Endospore abundance, microbial growth and mecromass turnover in deep sub-seafloor sediment. Nature Lett 484:101-104, doi: 10.1038/nature10905

Lutz RA, Falkowski PG (2012) A dive to Challenger Deep. Science 336:301-302, doi: 10.1126/science.1222641

Meersman F, Smeller L, Heremans K (2002) Comparative Fourier transform infrared spectroscopy study of cold-, pressure-, and heat-induced unfolding and aggregation of myoglobin. Biophys J 82:2634-2644, doi: 10.1016/S0006-3495(02)75605-1

Meersman F, Wang J, Wu Y, Heremans K (2005) Pressure effect on the hydration properties of poly(N-isopropylacrylamide) in aqueous solution studied by FTIR spectroscopy. Macromolecules 38:8923-8928, doi: 10.1021/ma051582d

Meersman F, Smeller L, Heremans K (2006) Protein stability and dynamics in the pressure/temperature plane. Biochim Biophys Acta 1764:346-354, doi: 10.1016/j.bbapap.2005.11.019

Melosh HJ (1988) The rocky road to panspermia. Nature 332:687-688, doi: 10.1038/332687a0

Merrin J, Kumar P, Libchaber A (2011) Effects of pressure and temperature on the binding of RecA protein to single-stranded DNA. Proc Natl Acad Sci USA 108:19913-19918, doi: 10.1073/pnas.1112646108

Minami K, Ono F, Mori Y, Takarabe K, Saigusa M, Matsushima Y, Saini N L Yamashita M (2010) Strong environmental tolerance of Artemia under very high pressure. J Phys Conf Ser 215:012164, doi: 10.1088/1742-6596/215/1/012164

Mohana-Borges R, Pacheco ABF, Sousa FJR, Foguel D, Almeida DF, Silva JL (2000) LexA repressor forms stable dimers in solution. J Biol Chem 275:4718-4712, doi: 10.1074/jbc.275.7.4708

Molina-Höppner A, Sato T, Kato C, Gänzle MG, Vogel RF (2003) Effects of pressure on cell morphology and cell division of lactic acid bacteria. Extremophiles 7:511-516, doi:10.1007/s00792-003-0349-0

Morono Y, Terada T, Nishizawa, Ito M, Hillion F, Takahata N, Sano S, Inagaki F (2011) Carbon and nitrogen assimilation in deep subseafloor microbial cells. Proc Natl Acad Sci USA 108:18295-18300, doi: 10.1073/pnas.1107763108

Munro S (2003) Lipid rafts: elusive or iillusive? Cell 115:377-388, doi: 10.1016/S0092-8674(03)00882-1

Nealson K H (1997) Sediment bacteria: who's there, what are they doing, and what's new? Annu Rev Earth Planet Sci 25:403-434, doi: 10.1146/annurev.earth.25.1.403

Nicolini C, Baranski J, Schlummer S, Palomo J, Lumbierres-Burgues M, Kahms M, Kuhlmann J, Sanchez S, Gratton E, Waldmann H, Winter R (2005) Visualizing association of N-Ras in lipid microdomains: Influence of domain structure and interfacial adsorption. J Am Chem Soc 128:192-201, doi: 10.1021/ja055779x

Nicolini C, Kraineva J, Khurana M, Periasamy N, Funari SS, Winter R (2006) Temperature and pressure effects on structural and conformational properties of POPC/SM/cholesterol model raft mixtures, FT-IR, SAXS, DSC, PPC and Laurdan fluorescence spectroscopy study. Biochim Biophys Acta 1758:248-254, doi: 10.1016/j.bbamem.2006.01.019

Nishihira N, Shindo A, Saigusa M, Ono F, Matsushima Y, Mori Y, Takarabe K, Saini NL, Yamashita M (2010) Preserving life of moss Ptychomitrium under very high pressure. J Phys Chem Solids 71:1123-1126, doi: 10.1016/j.jpcs.2010.03.018

Oger P, Jebbar M (2010) The many ways of coping with pressure. Res Microbiol 161:799-809, doi: 10.1016/j.resmic.2010.09.017

Ono F, Saigusa M, Uozumi T, Matsushima Y, Ikeda H, Saini NL, Yamashita M (2008) Effect of high hydrostatic pressure on to life of the tiny animal tardigrade. J Phys Chem Solids 69:2297-2300, doi: 10.1016/j.jpcs.2008.04.019

Ono F, Mori Y, Takarabe K, Nishihira N, Shindo, Saigusa M, Matsushima Y, Saini NL, Yamashita M (2009) Strong environmental tolerance of moss Venturiellaunder very high pressure. J Phys Conf Ser 215, 012165, doi: 10.1088/1742/6596/215/1/012165

Ono F, Minami K, Saigusa M, Matsushima Y, Mori Y, Takarabe K, Saini NL, Yamashita M (2010) Life of Artemia under very high pressure. J Phys Chem Solids 71, doi: 10.1016/j.jpcs.2010.03.019

Ono F, Mori Y, Sougawa M, Takarabe K, Hada Y, Nishihira N, Motose H, Saigusa M, Matsushima Y, Yamazaki D, Ita E, Saini NL (2012) Effect of very high pressure on life of plants and animals. J Phys Conf Ser 377:012053, doi: 10.1088/1742-6596/377/1/012053

Orcutt B, Sylvan J, Knab N, Edwards K (2011) Microbial ecology of the dark ocean above, at, and below the seafloor. Microbiol Mol Biol Rev 75:361-422, doi: 10.1128/MMBR.00039-10

Pace CN, Heinemann U, Hahn U, Saenger W (1991) Ribonuclease T1: Structure, function, and stability. Angew Chem Int Ed (English) 30:343-360

Paliwal A, Asthagiri D, Bossev DP, Paulaitis ME (2004) Pressure denaturation of staphylococcal nuclease studied by neutron small-angle scattering and molecular simulation. Biophys J 87:3479-3492, doi: 10.1529/biophysj.104.050526

Panick G, Malessa R, Winter R, Rapp G, Frye KJ, Royer CA (1998) Structural characterization of the pressure-denatured state and unfolding/refolding kinetics of staphylococcal nuclease by synchrotron small-angle X-ray scattering and Fourier-transform infrared spectroscopy. J Mol Biol 275:389-402, doi: 10.1006/jmbi.1997.1454

Panikov NS, Sizova MV (2007) Growth kinetics of microorganisms isolated from Alaskan soil and permafrost in solid media frozen down to −35 degrees C. FEMS Microbiol Ecol 59:500-512, doi: 10.1111/j.1574-6941.2006.00210.x

Pappenberger G, Saudan C, Becker M, Merbach AE, Kiefhaber T (2000) Denaturant-induced movement of the transition state of protein folding revealed by high-pressure stopped-flow measurements. Proc Natl Acad Sci USA 97:17-22, doi: 10.1073/pnas.97.1.17

Parkes RJ, Cragg B, Wellsbury P (2000) Recent studies on bacterial populations and processes in subseafloor sediments: A review. Hydrogeology J 8:11-28, doi: 10.1007/PL00010971

Parkes RJ, Webster G, Cragg BA, Weightman A, Newberry C, Ferdelman T, Kallmeyer J, Jorgenson B, Aiello I, Fry J (2005) Deep subb-seafloor prokaryotes stimulated at interfaces over geological time. Nature Lett 436:390-394, doi: 10.1038/nature03796

Payne VA, Matubayasi N, Murphy LR, Levy RM (1997) Monte Carlo study of the effect of pressure on hydrophobic association. J Phys Chem B 101:2054-2060, doi: 10.1021/jp962977p

Periasamy N, Teichert H, Weise K, Vogel RF, Winter R (2009) Effects of temperature and pressure on the lateral organization of model membranes with functionally reconstituted multidrug transporter LmrA. Biochim Biophys Acta 1788:390-341, doi: 10.1016/j.bbamem.2008.09.017

Picard A, Daniel I, Montagnac G, Oger P (2007) *In situ* monitoring by quantitative Raman spectroscopy of alcoholic fermentation by Saccharomyces cerevisiae under high pressure. Extremophiles 11:445-452, doi: 10.1007/s00792-006-0054-x

Picard A, Daniel I, Testemale D, Kieffer I, Bleuet P, Cardon H, Oger P (2011) Monitoring microbial redox trannsformations of metal and metalloid elements under high pressure using in situ X-ray absorption spectroscopy. Geobiol 9:196-204, doi: 10.1111/j.1472-4669.2010.00270.x

Picard A, Ferdelman TG (2011) Linking microbial heterotrophic activity and sediment lithology in oxic, oligotrophic sub-seafloor sediments of the North Atlantic Ocean. Front Microbiol 2:263, doi: 10.3389/fmicb.2011.00263

Picard A, Testemale D, Hazemann J-L, Daniel I (2012) The influence of high hydrostatic pressure on bacterial dissimilatory iron reduction. Geochim Cosmochim Acta 88:120-129, doi: 10.1016/j.gca.2012.04.030

Potekhin SA, Senin AA, Abdurakhmanov NN, Khusainova RS (2008) High pressure effect on the main transition from the ripple gel P phase to the liquid crystal phase in dipalmitoylphosphatidylcholine. Microcalorimetric study. Biochim Biophys Acta 1778:2588-2593, doi: 10.1016/j.bbamem.2008.08.001

Powalska E, Janosch S, Kinne-Saffran E, Kinne RKH, Fontes CFL, Mignaco JA, Winter R (2007) Fluorescence spectroscopic studies of pressure effects on Na+,K+-ATPase reconstituted into phospholipid bilayers and model raft mixtures. Biochem 46:1672-1683, doi: 10.1021/bi062235e

Prehoda KE, Mooberry ES, Markley JL (1998) Pressure denaturation of proteins: Evaluation of compressibility effects. Biochem 37:5784-5790, doi: 10.1021/bi980384u

Price PB, Sowers T (2004) Temperature dependence of metabolic rates for microbial growth, maintenance and survival. Proc Natl Acad Sci USA 101:4631-4636, doi: 10.1073/pnas.0400522101

Prieur D, Marteinsson VT (1998) Prokaryotes living under elevated hydrostatic pressure. *In*. Advances in Biochemical Engineering Biotechnology; Biotechnology of extremophiles. G Antranikian (ed), Springer-Verlag, Berlin, p 23-35

Rainey FA, Ray K, Ferreira M, Gatz BZ, Nobre MF, Bagaley D, Rash BA, Park MJ, Earl AM, Shank NC, Small AM, Henk MC, Battista JR, Kämpfer P, da Costa MS (2005) Extensive diversity of ionizing-radiation-resistant bacteria recovered from Sonoran Desert soil and description of nine new species of the genus Deinococcus obtained from a single soil sample. Appl Environ Microbiol 71:5228-5235, doi: 10.1128/AEM.71.9.5225-5235.2005

Rayan G, Macgregor RB (2005) Comparison of the heat- and pressure-induced helix coil transition of two DNA copolymers. J Phys Chem B 109:15558-15565, doi: 10.1021/jp050899c

Rhee YM, Sorin E, Jayachandran G, Lindahl E, Pande V (2004) Simulations of the role of water in the protein-folding mechanism. Proc Natl Acad Sci USA 101:6456-6461, doi: 10.1073/pnas.0307898101

Rivkina EM, Friedmann EI, McKay CP, Gilichinsky DA (2000) Metabolic activity of permafrost bacteria below the freezing point. Appl Environ Microbiol 66:3230-3233, doi: 10.1128/AEM.66.8.3230-3233.2000

Rothschild LJ, Mancinelli RL (2001) Life in extreme environments. Nature 409:1092-1101

Rouget J-B, Aksel T, Roche J, Saldana J-L, Garcia AE, Barrick D, Royer CA (2011) Size and sequence and the volume change of protein folding. J Am Chem Soc 133:6020-6027, doi: 10.1021/ja200228w

Royer CA, Chakerian AE, Matthews KS (1990) Macromolecular binding equilibria in the lac repressor system: studies using high-pressure fluorescence spectroscopy. Biochem 29:4959-4966

Royer CA (2002) Revisiting volume changes in pressure-induced protein unfolding. Biochim Biophys Acta 1595:201-209, doi: 10.1016/S0167-4838(01)00344-2

Royer CA, Winter R (2011) Protein hydration and volumetric properties. Curr Opin Coll Interface Sci 16:568-571, doi: 10.1016/j.cocis.2011.04.008

Scharnagl C, Reif M, Friedrich J (2005) Stability of proteins: Temperature, pressure and the role of the solvent. Biochim Biophys Acta 1749:187-213, doi: 10.1016/j.bbapap.2005.03.002

Schrenk MO, Huber JA, Edwards KJ (2010) Microbial provinces in the subseafloor. Annu Rev Mar Sci 2:279-304, doi: 10.1146/annurev-marine-120308-081000

Schrenk MO, Brazelton WJ, Lang SQ (2013) Serpentinization, carbon, and deep life. Rev Mineral Geochem 75:575-606

Sciortino F, Geiger A, Stanley HE (1991) Effect of defects on molecular mobility in liquid water. Nature 354:218-221, doi: 10.1038/354218a0

Seddon JM, Templer RH (1993) Cubic phases of self-assembled amphiphilic aggregates. Philos Trans R Soc A 344:377-401, doi: 10.1098/rsta.1993.0096

Seemann H, Winter R (2003) Volumetric properties, compressibilities and volume fluctuations in phospholipid-cholesterol bilayers. Z Phys Chem 217:831-846, doi: 10.1524/zpch.217.7.831.20388

Seemann H, Winter R, Royer CA (2001) Volume, expansivity and isothermal compressibility changes associated with temperature and pressure unfolding of staphylococcal nuclease. J Mol Biol 307:1091-1102, doi: 10.1006/jmbi.2001.4517

Sharma A, Scott J, Coldy G, Fogel M, Hazen RM, Hemley RJ, Huntress W (2002) Microbial activity at GigaPascal pressures. Science 295:1514-1516, doi: 10.1126/science.1068018

Sharma A, Kawarabayasi Y, Satyanarayana T (2012) Acidophilic bacteria and archaea: acid stable biocatalysts and their potential applications. Extremophiles 16:1-19, doi: 10.1007/s00792-011-0402-3

Shortle D (1996) The denatured state (the other half of the folding equation) and its role in protein stability. FASEB J 10:27-34

Silva JL, Weber G (1993) Pressure stability of proteins. Annu Rev Phys Chem 44:89-113, doi: 10.1146/annurev.pc.44.100193.000513

Silva JL, Oliveira AC, Gomes AMO, Lima LMTR, Mohana-Borges R, Pacheco ABF, Foguel D (2002) Pressure induces folding intermediates that are crucial for protein DNA recognition and virus assembly. Biochim Biophys Acta 1595:250-265, doi: 10.1016/S0167-4838(01)00348-X

Smith LJ, Fiebig KM, Schwalbe H, Dobson CM (1996) The concept of a random coil: Residual structure in peptides and denatured proteins. Fold Des 1:R95-R106

Souza MO, Creczynski-Pasa TnB, Scofano HM, Graber P, Mignaco JA (2004) High hydrostatic pressure perturbs the interactions between CF0F1 subunits and induces a dual effect on activity. Int J Biochem Cell Biol 36:920-930, doi: 10.1016/j.biocel.2003.10.011

Suzuki K (1960) Studies on the kinetics of protein unfolding under high pressure. Rev Phys Chem Japan 29:49-56

Takai K, Moser DP, Onstott TC, Spoelstra N, Pfiffner SM, Dohnalkova A, Fredrickson JK (2001) *Alkaliphilus transvaalensis* gen. nov. sp nov., an extremely alkaliphilic bacterium isolated from a deep South African gold mine. Int J Syst Evol Microbiol 51:1245-1256

Takai K, Nakamura K, Toki T, Tsunogai U, Miyazaki M, Miyazaki J, Hirayama H, Nakagawa S, Nunoura T, Horikoshi K (2008) Cell proliferation at 122 degrees C and isotopically heavy CH(4) production by a hyperthermophilic methanogen under high pressure cultivation. Proc Natl Acad Sci USA 105:10949-10954, doi: 10.1073/pnas.0712334105

Teichert H, Periasamy N, Winter R, Vogel RF (2009) Influence of membrane lipid composition on the activity of functionally reconstituted LmrA under high hydrostatic pressure. High Press Res 29:344-357, doi: 10.1080/08957950902941030

Templer RH, Seddon JM, Duesing PM, Winter R, Erbes J (1998) Modeling the phase behavior of the inverse hexagonal and inverse bicontinuous cubic phases in 2:1 fatty acid/phosphatidylcholine mixtures. J Phys Chem B 102:7262-7271

Torrent J, Connelly JP, Coll MG, Ribo M, Lange R, Vilanova M (1999) Pressure versus heat-induced unfolding of ribonuclease A: The case of hydrophobic interactions within a chain-folding initiation site. Biochem 38:15952-15961, doi: 10.1021/bi991460b

Ulmer HM, Herberhold H, Fahsel S, Gänzle MG, Winter R, Vogel RF (2002) Effects of pressure-induced membrane phase transitions on inactivation of HorA, an ATP-dependent multidrug resistance transporter, in Lactobacillus plantarum. Appl Environ Microbiol 68:1088-1095, doi: 10.1128/AEM.68.3.1088-1095.2002

Vanlint D, Mitchell R, Bailey E, Meersman F, McMillan P F, Aertsen A, Michiels C (2011) Rapid acquisition of Gigapascal high pressure resistance by Escherichia coli. Mbio 2:e00130-10, doi: 10.1128/mBio.00130-10

Van Opstal I, Vanmuysen SCM, Wuytack EY, Masschalck B, Michiels CW (2005) Inactivation of Escherichia coli by high hydrostatic pressure at different temperatures in buffer and carrot juice. Int J Food Microbiol 98:179-191, doi: 10.1016/j.ijfoodmicro.2004.05.022

Vendruscolo M, Dobson CM (2005) Towards complete descriptions of the free energy landscapes of proteins. Philos Trans R Soc A 363:433-452, doi: 10.1098/rsta.2004.1501

Weise K, Triola G, Brunsveld L, Waldmann H, Winter R (2009) Influence of the lipidation motif on the partitioning and association of N-Ras in model membrane subdomains. J Am Chem Soc 131:1557-1564, doi: 10.1021/ja808691r

Whitman W, Coleman D, Wiebe W (1998) Prokaryotes: The unseen majority. Proc Natl Acad Sci USA 95:6578-6583, doi: 10.1073/pnas.95.12.6578

Willis MJ, Ahrens TJ, Bertani LE, Nash CZ (2006) Bugbuster, survivability of living bacteria upon shock compression. Earth Planet Sci Lett 247:185-196, doi: 10.1016/j.epsl.2006.03.054

Wilson RG Jr, Trogadis JE, Zimmerman S, Zimmerman AM (2001) Hydrostatic pressure induced changes in the cytoarchitecture of pheochromocytoma (PC-12) cells. Cell Biol Int 25:649-666, doi: 10.1006/cbir.2000.0692

Wilson ZE, Brimble MA (2009) Molecules derived from the extremes of life. Nat Prod Rep 26:44-71, doi: 10.1039/B800164M

Wilton DJ, Ghosh M, Chary KVA, Akasaka K, Williamson MP (2008) Structural change in a B-DNA helix with hydrostatic pressure. Nucleic Acids Res 36:4032-4037, doi: 10.1093/nar/gkn350

Winter R, Pilgrim WC (1989) A SANS study of high pressure phase transitions in model biomembranes. Ber Bunsenges Phys Chem 93:708-717

Winter R, Erbes J, Templer RH, Seddon JM, Syrykh A, Warrender NA, Rapp G (1999a) Inverse bicontinuous cubic phases in fatty acid/phosphatidylcholine mixtures: the effects of pressure and lipid composition. Phys Chem Chem Phys 1:887-893, doi: 10.1039/A808950G

Winter R, Gabke A, Czeslik C, Pfeifer P (1999b) Power-law fluctuations in phase-separated lipid membranes. Phys Rev E 60:7354-7359, doi: 10.1103/PhysRevE.60.7354

Winter R, Jonas J (1999c) High Pressure Molecular Science. Kluwer Academic Publisher. NATO Science Series E358

Winter R, Czeslik C (2000) Pressure effects on the structure of lyotropic lipid mesophases and model biomembrane systems. Z Kristallogr 215:454-474, doi: 10.1524/zkri.2000.215.8.454

Winter R (2001) Effects of hydrostatic pressure on lipid and surfactant phases. Curr Opin Coll Interface Sci 6:303-312, doi: 10.1016/S1359-0294(01)00092-9

Winter R, Kohling R (2004) Static and time-resolved synchrotron small-angle x-ray scattering studies of lyotropic lipid mesophases, model biomembranes and proteins in solution. J Phys: Cond Matt 16:S327, doi: 10.1088/0953-8984/16/5/002

Wright PE, Dyson HJ (1999) Intrinsically unstructured proteins: re-assessing the protein structure-function paradigm. J Mol Biol 293:323-331, doi: 10.1006/jmbi.1999.3110

Xie W, Wang FP, Guo L, Chen ZL, Sievert SM, Meng J, Huang G, Li Y, Yan Q, Wu S, Wang X, Chen S, He G, Xiao X, Xu A (2011) Comparative metagenomics of microbial communities inhabiting deep-sea hydrothermal vent chimneys with contrasting chemistries. ISME J 5:414-426, doi: 10.1038/ismej.2010.144

Yamaguchi T, Yamada H, Akasaka K (1995) Thermodynamics of unfolding of ribonuclease A under high pressure. A study by proton NMR. J Mol Biol 250:689-694, doi: 10.1006/jmbi.1995.0408

Yayanos AA, Dietz A, Van Boxtel R (1979) Isolation of a deep sea barophilic bacterium and some of its growth characteristics. Science 205:808-810, doi: 10.1126/science.205.4408.808

Yayanos AA, Dietz A, Van Boxtel R (1981) Obligately barophilic bacterium from the Mariana Trench. Proc Natl Acad Sci USA 78:5212-5215, doi: 10.1073/pnas.78.8.5212

Yayanos AA (1986) Evolutional and ecological implications of the properties of deep-sea barophilic bacteria. Proc Natl Acad Sci USA 83:9542-9546, doi: 10.1073/pnas.83.24.9542

Yayanos AA (1995) Microbiology to 10,500 meters in the deep sea. Annu Rev Microbiol 49777-805

Yayanos AA (1998) Empirical and theoretical aspects of life at high pressure in the deep sea. *In*: Extremophiles-Microbial life in extreme environments Horikoshi K, Grant WD (ed), Wiley-Liss, New York, p 47-92.

Yayanos AA (2001) Deep-sea piezophilic bacteria. Methods Microbiol 30:615-637

Yayanos AA (2002) Are cells viable at GigaPascal pressures. Science 297:295, doi: 10.1126/science.297.5580.295a
Zein M, Winter R (2000) Effect of temperature, pressure and lipid acyl chain length on the structure and phase behavior of phospholipid-gramicidin bilayers. Phys Chem Chem Phys 2:4545-4551, doi: 10.1039/B003565N
Zeng X, Birrien J-L, Fouquet Y, Cherkashov G, Jebbar M, Querellou, Oger P, Cambon-Bonavita M-A, Xiao X, Prieur D (2009) Pyrococcus CH1, an obligate piezophilic hyperthermophile: extending the upper pressure-temperature limits for life. ISME J 3:873-876, doi: 10.1038/ismej.2009.21
Zhang J, Peng X, Jonas A, Jonas J (1995) NMR study of the cold, heat, and pressure unfolding of ribonuclease A. Biochem 34:8631-8641, doi: 10.1021/bi00027a012
Zhang Y, Cremer PS (2006) Interactions between macromolecules and ions: the Hofmeister series. Curr Opin Chem Biol 10:658-663, doi: 10.1016/j.cbpa.2006.09.020
Zipp A, Kauzmann W (1973) Pressure denaturation of metmyoglobin. Biochem 12:4217-4228, doi: 10.1021/bi00745a028
Zobell C, Morita R (1959) Deep sea bacteria. Galathea Report, Copenhagen 1:139-154

Reviews in Mineralogy & Geochemistry
Vol. 75 pp. 649-675, 2013

The Deep Viriosphere: Assessing the Viral Impact on Microbial Community Dynamics in the Deep Subsurface

Rika E. Anderson[1], William J. Brazelton[1,2], John A. Baross[1]

[1]*School of Oceanography and Astrobiology Program*
University of Washington
Seattle, Washington 98195, U.S.A.

[2]*Department of Biology*
East Carolina University,
Greenville, North Carolina 27858, U.S.A.

jbaross@u.washington.edu

INTRODUCTION

All regions of Earth's biosphere that we have studied—the waters of Earth's oceans, the soil beneath our feet, and even the air we breathe—teem with viruses. Viral particles are among the smallest biological entities on the planet, with the average viral particle measuring about 100 nm in length: a size so small that five thousand viruses, lined end to end, would fit across the thickness of a human fingernail. What they lack in size, though, they compensate with sheer abundance. If we were to line up all the viruses in the ocean, they would stretch across the diameter of the Milky Way galaxy one hundred times (Suttle 2007). Those viruses are responsible for up to 10^{23} infections per second in the oceans (Suttle 2007). With each new infection, viruses can have a profound impact on their hosts: they can alter the structure of a microbial population, break up cellular biomass into its constituent organic matter, or introduce new genes into their hosts. Through this activity, viruses play a role in top-down as well as bottom-up processes, and can potentially alter the course of evolution.

The importance of viruses in the surface oceans is now well recognized, and research is increasingly dedicated to improving our understanding of their role in important marine processes. The viral role in the deep subsurface, however, is rarely considered. Deep within the crust and sediment below the ocean, viruses may play a profound role in altering biogeochemical cycles, structuring microbial diversity, and manipulating genetic content. Yet many questions remain unanswered: Are certain species or strains in the deep subsurface more susceptible to viral infection than others? What role do viruses play in driving natural selection and evolution in the deep biosphere? Is it more common for viruses to persist as protein-bound virion particles, or do they more commonly incorporate their genomes into that of their hosts? What impact do viruses have on their hosts while incorporated as stable symbionts? Can viruses provide their hosts with the keys to survival in the extreme environments of our planet?

The following chapter seeks to address these topics by exploring the nature of the relationship between viruses and their microbial hosts across a range of environments within the deep subsurface biosphere. We begin with a review of viral diversity by briefly describing the diversity of viral morphologies, nucleic acid types, and genetic content, and provide an overview of viral life cycles. We briefly discuss what is known of the viral impact on microbial biogeochemistry, microbial population structure and diversity, and on genetic content and expression patterns. We then apply these concepts to the deep subsurface, a region where unique

1529-6466/13/0075-0020$00.00 DOI: 10.2138/rmg.2013.75.20

attributes such as low nutrient and energy levels, enclosed pore spaces, and fluid flux may combine to produce an environment in which viruses play a significant role in manipulating the genetic landscape of deep subsurface microbial communities. We discuss completed and ongoing work that seeks to address some of these issues. Finally, we ask whether these viruses may have been involved in the origin of life in the subsurface. Viruses of the deep may play an important role in altering the evolutionary trajectory of their microbial hosts, and in doing so they complicate the concepts of parasitism and symbiosis in the microbial world, both now and in life's deep past. Ultimately, it is possible that the smallest biological entities on the planet have their most profound influence in its deepest realms, both now and in Earth's early history.

DIVERSITY IN THE VIRAL WORLD

Viruses infect all three domains of life and in doing so they adopt a wide variety of morphologies, lifestyle strategies, and genetic materials. These differences in viral types can have important implications for the nature of the virus-host relationship, and for the ways in which viruses can manipulate microbial community structure and evolution. By understanding the types of viruses that predominate in a given system, we can predict the nature of their impact on the host community. Here, we provide a brief overview of different viral types and life cycles, and then describe what types of viruses we might expect to predominate in the deep subsurface, given the environmental conditions.

Viral life cycles

Viruses infecting archaea and bacteria assume two different lifestyle strategies, each with significant implications for the viral relationship with the host and for the nature of virus-host co-evolution. Here, we provide a very simplified overview of viral life cycles; these are illustrated schematically in Figure 1. In the *lytic cycle*, viral particles attach to the outside of the host and inject their genetic material into the host cytoplasm. This genetic material then mounts a takeover of cell machinery for immediate synthesis of viral particles, which accumulate within the cell until it bursts, or lyses, releasing the viral particles into the surrounding medium, ready to infect a new host (Fig. 1A). Viruses employing the *lysogenic cycle*, in contrast, incorporate their genome into the host genome upon infection. Incorporated viral genomes are known as "prophage" or "proviruses," and can lie latent within the cellular genome for many generations. A glossary (Table 1) defines these terms and others used throughout the manuscript. Cells can maintain one or several prophage,

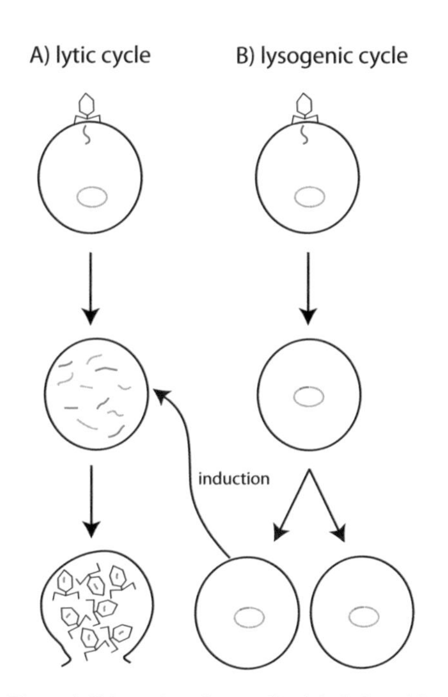

Figure 1. Schematic and generalized depiction of the lytic and lysogenic cycles of phage. A) Lytic cycle, in which a virus lands on the cellular membrane, injects genetic material into the cytoplasm, resulting in viral takeover of cellular machinery. New viral capsids are synthesized, packaged with viral genomes, and lyse the cell. B) Lysogenic cycle, in which a virus lands on the cellular membrane and injects its genetic material into the cytoplasm, and then integrates into the host genome. This "prophage" lies latent for several generations, then enters the lytic cycle in response to an induction event. See text for details.

Table 1. Glossary of terms used throughout the manuscript.

Term	Description
16S small subunit ribosomal RNA	A polynucleotide of about 1500 base pairs in length that makes up part of the small subunit of the prokaryotic ribosome; the highly conserved 16S ribosomal RNA sequence is used to determine evolutionary relationships between organisms.
Archaea	One of the three domains of life (in addition to Bacteria and Eukarya). Also single-celled, archaea are often found living in extreme or energy-limited environments, though they are also found in temperate environments such as the open ocean. Distinguishable from bacteria through genetic differences (including 16S rRNA sequence), a lack of peptidoglycan in the cell wall, differences in lipid composition, and many other biochemical differences.
Bacteria	One of the three domains of life (in addition to Archaea and Eukarya). These single-celled organisms have been found in nearly every environment investigated on the planet, with tremendously high diversity of morphology and metabolism.
Capsid	The protein coat that encapsulates viral genetic material.
Cytoplasm	Cellular contents within the cellular membrane.
DNA polymerase	An enzyme that catalyzes the polymerization of deoxyribonucleotides into a single strand. The sequence is used in some cases by viral ecologists to distinguish between viral types.
Lysogeny/lysogenic cycle	Viral life cycle in which the viral genetic material is integrated into the genome of the host.
Metagenome/ metagenomics	Method used by microbial ecologists to characterize a microbial community by collecting and sequencing genetic material directly from the environment, yielding thousands of sequencing reads currently ranging from ~50-1000 base pairs in length. Target organisms can be archaea, bacteria, viruses, or even single-celled eukaryotes.
Myoviridae, Podoviridae, Siphoviridae	Three families within the *Caudovirales* order, which are categorized according to morphology. All *Caudovirales* are characterized by a head-tail morphology and are extremely common in our oceans.
Phage	Short for bacteriophage ("phage" is Greek for "to devour.") Viruses that infect the bacteria; sometimes applied to archaeal viruses as well.
Phenotype	The physical manifestation of an organism's genetic material (or genotype).
Prokaryote	Organisms that lack membrane-bound organelles (generally including the archaea and the bacteria).
Prophage	Term for viral genetic material that is in the lysogenic state, integrated in a host chromosome.
Recombination	The process by which genetic material from two separate organisms are brought together.
Syntrophy	A relationship between two organisms in which they combine metabolic capabilities to derive energy from a net reaction that neither could metabolize independently.
Transduction	Virally-mediated horizontal gene transfer.
Virus	A parasitic element with a DNA or RNA genome that relies on a host to replicate, but has an extracellular state in which the genetic material is contained within a protein and/or lipid coat. Infects all three domains of life.

which can sometimes provide immunity from superinfection by other viruses. Viral genes can be expressed while integrated into the host genome, and can thereby influence the cellular phenotype. Generally, these viruses are induced, or triggered to enter the lytic cycle, in response to an environmental stimulus. At this point, the viral genome removes itself from the host genome and takes over the cellular machinery to create new viral particles, which then lyse the cell to begin the infection cycle anew (Fig. 1B). We believe it to be likely that the lysogenic cycle predominates among viruses in the deep biosphere, for reasons we will discuss below.

Viral sizes and morphologies

Viruses can range in size from 20 nm to well over 800 nm, and adopt myriad shapes, genome sizes, and replication strategies. Most viruses possess genomes ranging between a few to ~100 kilobases (kb), but recently the giant amoeba-infecting Mimivirus was discovered to possess a genome of 1,185 kb, and the virus structure itself is larger than some of the smallest cells (La Scola et al. 2003; Raoult et al. 2004). On the other end of the spectrum, the tiny Sputnik virus possesses a genome of only 18 kb, and parasitizes not a cell, but the Mimivirus itself (La Scola et al. 2008). RNA viruses are often among the smallest of the viruses, with some RNA viruses possessing genomes of only about 2 kb. Giant viruses continue to be discovered in various biomes of the globe (Fischer et al. 2010), and much remains to be learned about their lifestyles, replication mechanisms, and their evolutionary and ecological impacts on their hosts.

Viruses of the archaea and bacteria, our focus here, are represented by a wide variety of morphologies, including filamentous, icosahedral, and head-tail viruses. Many of the archaeal viruses possess particularly unusual shapes that have only recently been discovered. Examples of typical archaeal and bacterial virus morphologies are depicted in Figure 2. The most commonly observed phages (bacterial viruses) in the oceans are the head-tail viruses (Suttle 2005), all of which have linear double-stranded DNA genomes. Among the dsDNA viruses, morphology can give an indication of lifestyle and host range. In the marine realm the most abundant viruses are from the *Podoviridae* family, which have short, non-contractile tails and tend to infect only a narrow range of hosts, usually only particular strains within a species (Suttle 2005). In contrast, the members of the *Myoviridae* family, with contractile tails, and the *Siphoviridae*, with long non-contractile tails, tend to have a broader host range. Consequently, environments dominated by *Myoviridae* or *Siphoviridae* are more likely to be sites of interspecies viral infections.

However, viruses are not limited to the use of double-stranded DNA. Viruses also use single-stranded DNA (ssDNA) as their genetic material, and ssDNA viruses are increasingly found to be important members of the marine viral community. A recent study found that *Microviridae,* a family of ssDNA viruses, is one of the most common viral types in marine waters (Angly et al. 2006). Viruses also use RNA as their genetic material, and can be double- or single-stranded with plus or minus sense RNA strands. RNA viruses have been found to be important constituents of the marine ecosystem (Culley et al. 2003, 2006), infecting members across the trophic levels, from bacteria to whales. Retroviruses are one type of RNA virus that use an enzyme called reverse transcriptase to produce DNA from their RNA genomes, and then integrate this DNA into the genome of the host. Retroviruses also occur in both double-stranded and single-stranded forms. Interestingly, while retroviruses are common in eukaryotes, none have yet been found to naturally infect either the archaea or the bacteria.

While much is known about the morphologies and nucleic acid types of bacteriophages, very little is known about archaeal viruses. The few archaeal viruses isolated thus far have morphologies vastly different from those seen in bacterial viruses (Pina et al. 2011; Prangishvili et al. 2006) (Fig. 2). Some archaeal viruses possess a never-before-seen ability to change their morphology outside of the host, extruding tails on each end of an initially lemon-shaped viral capsid after release from the host (Häring et al. 2005). The unusual viral shapes encountered among the archaeal viruses are occasionally accompanied by unique release mechanisms from

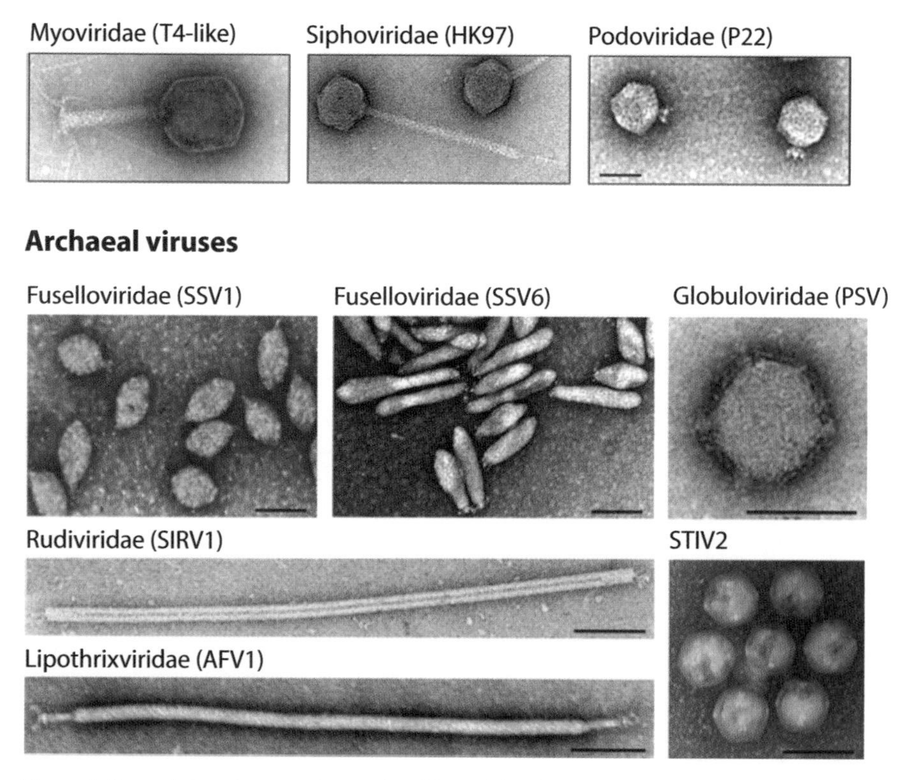

Figure 2. Transmission electron micrographs of bacterial and archaeal viruses. Scale bars for bacterial viruses are 50 nm; scale bars for archaeal viruses are 100 nm. Reprinted with permission from American Society of Microbiology from Krupovic et al. (2011).

the host cell, such as the formation of pyramid-like structures in archaeal membranes that serve as virus outlet sites (Bize et al. 2009; Brumfield et al. 2009). Most archaeal viruses studied to date are double-stranded DNA viruses, with only a single ssDNA archaeal virus discovered thus far (Pietilä et al. 2009). However, these results almost certainly reflect the nature of the detection techniques that have been utilized thus far and not the true diversity of archaeal viruses in natural environments. Considering the similarities between the eukaryotic and archaeal transcription apparatus, the discovery of archaeal RNA viruses and retroviruses seems imminent and may have great potential for yielding important insights into viral evolution. In the deep subsurface biosphere, where archaea constitute a larger proportion of the community than in surface oceans (Biddle et al. 2006), archaeal viruses may dominate, and further study may reveal as yet unknown morphologies or life strategies. Finally, a recent metagenomics study in an acidic, high-temperature lake in Lassen Volcanic Park, USA, uncovered a viral genome sequence suggesting recombination between an RNA and a DNA virus (Diemer and Stedman 2012). While the host of this particular virus was most likely eukaryotic, this study points to the possibility of such recombination events, which may occur between bacterial or archaeal RNA and DNA viruses as well.

Genetic diversity

An important question in viral ecology is the degree to which viral types are restricted to a given environment, or whether there is movement across biomes. In this sense viruses represent a further test of the null hypothesis of microbial biogeography: "Everything is everywhere, but the environment selects" (Baas Becking 1934; O'Malley 2007). One of the great challenges in assessing viral diversity and biogeography is the lack of a universal "barcoding" gene, analogous to the 16S small ribosomal subunit among the archaea and bacteria, which might be used to compare across all groups. Therefore, other techniques are used to assess virus biogeography. Steward et al. (2000) compared the relative genome sizes of viruses using pulsed-field gel electrophoresis, and found that certain genome size classes are found in many different marine environments. Similarly, Breitbart et al. (2004) investigated the environmental distribution of the T7 phage DNA polymerase gene, and found that the same sequences were found in a wide variety of diverse biomes, indicating a ubiquity of similar viruses across diverse environmental types. In this scenario, viral diversity is high locally, but viral types are distributed globally. The observation of globally-distributed viral types implies extensive movement among biomes and potential infection of (and sharing genes between) a wide array of hosts (Breitbart and Rohwer 2005).

Other studies present a contrasting picture of viral biogeography. For example, genomic analysis of a thermophilic virus of *Sulfolobus* revealed that viruses and their hosts tend to be spatially restricted in hot springs (Held and Whitaker 2009). Metagenomic analysis of viruses in stromatolites and thrombolites found a similar geographic restriction (Desnues et al. 2008), and metagenomic characterization of viruses in soil found distinctions between viral assemblages in soil samples and those in marine or fecal samples (Fierer et al. 2007). On a larger scale, metagenomic studies have revealed that while certain types of viruses, such as the myoviruses, were ubiquitous across sample sites, others, such as podoviruses and siphoviruses, had more site-specific distributions (Williamson et al. 2008a). Thus, an opposing paradigm suggests that distinct groups of viruses are tied closely to specific hosts, resulting in spatial restriction (Thurber 2009). While further study will provide greater insight into this story, it seems that some viral types are globally distributed, while others are much more spatially restricted. Furthermore, spatial distribution is likely to be determined by host specificity, but this relationship is mostly unexplored. Future work aimed at distinguishing between widely distributed viral types and more locally restricted (and presumably more host-specific) viral types may give insight into which viral types are most likely to facilitate gene flow between biomes.

In this context, the viruses of the deep subsurface represent an interesting case. It might be expected that viruses in the deep subsurface, on the one hand, should have reduced mobility as a result of being restricted within a sediment or rock matrix, and therefore have limited and patchy geographic distribution. This limited range may be particularly the case in sedimented regions away from the ridge axis. On the other hand, fluid flux within the subsurface in regions closer to the ridge axis, as well as allochthonous input from above, might facilitate movement of hosts and therefore of viruses from one locality to the next (Anderson et al. 2011a). It seems entirely possible that some viral types are restricted to particular regions of the subsurface, while others are more ubiquitous across the deep biosphere, perhaps in biogeographic correlation with their hosts. Further study will be required to resolve these questions.

VIRAL IMPACTS ON HOST ECOLOGY AND EVOLUTION

Viruses are a peculiarly potent force in that they can impact host community structure through both bottom-up and top-down control, and they can influence host genetic content through horizontal gene transfer and lysogenic conversion. Here, we provide a brief overview of what is known thus far of the viral impact on host microbial communities, with the aim of better understanding the ecological and evolutionary dynamics of the deep subsurface habitat.

Bottom-up effects: the biogeochemical impact

Through lysis of microbial hosts, viruses convert biomass to dissolved and particulate organic matter (Proctor and Fuhrman 1990). Estimates show that viral lysis removes approximately 20-40% of prokaryotic biomass in the ocean daily, though quantifying mortality rates due to viral lysis is difficult (Suttle 2007). This rapid turnover has tremendous biogeochemical implications, as viral lysis converts organic matter from biomass into the pool of dissolved organic matter (DOM), redirecting it from higher trophic levels and effectively short-circuiting the microbial loop. This phenomenon has been dubbed the "viral shunt," and has the effect of stimulating bacterial production by providing a source of DOM and thus stimulating respiration (Suttle 2007). Moreover, the "viral shunt" is thought to stimulate the ocean's biological pump by accelerating sinking rates of lysed cells or transforming bacterial biomass into dissolved organic matter, though it is unclear what percentage of this lysed material is recalcitrant or labile (Jiao et al. 2010). This is depicted schematically in Figure 3A.

The impact of the viral shunt on the deep biosphere naturally depends on virus-to-cell ratios, which impact the rate of infection. As this ratio varies according to depth and location, it is difficult to calculate the net impact of viruses on prokaryotic mortality in the deep subsurface. Danovaro et al. (2008) showed that viruses become the predominant source of prokaryotic mortality as depth increases in the sediments; in continental margin sediments off of Chile, it was estimated that viruses were responsible for mortality of 38-144% of bacterial net production (Middelboe et al. 2006). In mud volcanoes, viruses account for up to 33% of cells killed daily, and also contributed an estimated 49 mg carbon per square meter per day—a substantial contribution to the total carbon budget (Corinaldesi et al. 2011). Thus it can be expected that viruses in the deep biosphere will have a significant, if poorly constrained, impact on microbial mortality and, by extension, biogeochemical cycles. The extent of viral impact will also necessarily depend upon the predominant life cycle of viruses in the subsurface: if lysis predominates, the virus to cell ratio will be the most important factor in determining the importance of viruses in microbial mortality and trophic cycling; whereas if lysogeny predominates, the viral impact on mortality will also be dependent on the frequency and pattern of induction events within each environment.

Top-down effects: altering population structure

As predators, viruses also control population structure from the top-down; in the deep subsurface, where other predators such as grazers are likely to be absent, viruses may constitute the sole inducer of cell mortality, aside from natural decay. The question that then arises is how the dynamics of viral host range, lifestyle and infection frequency can alter the structure of host microbial communities.

One of the most influential ideas related to viral control of population structure is the notion of "kill the winner" (Thingstad and Lignell 1997). Several authors have observed that viral infection rates are dependent upon cell density and growth rate (e.g., Middelboe 2000); as most viruses have a fairly limited host range, this dependence implies that if a particular microbial group becomes dominant in a population, those cells are most susceptible to viral infection as a result of their increased density. This concept is depicted schematically in Figure 3A. Consequently, viruses may act as a homogenizing agent on the diversity of microbial communities, effectively maintaining high species evenness. Studies have shown that viruses are instrumental in the termination of certain types of plankton blooms (Bratbak et al. 1993). Moreover, Rodriguez-Valera et al. (2009) found that regions with the greatest variability within a given species' genome were regions coding for surface receptors, which are potential phage-recognition targets. They argue that viruses maintain high diversity in a system through

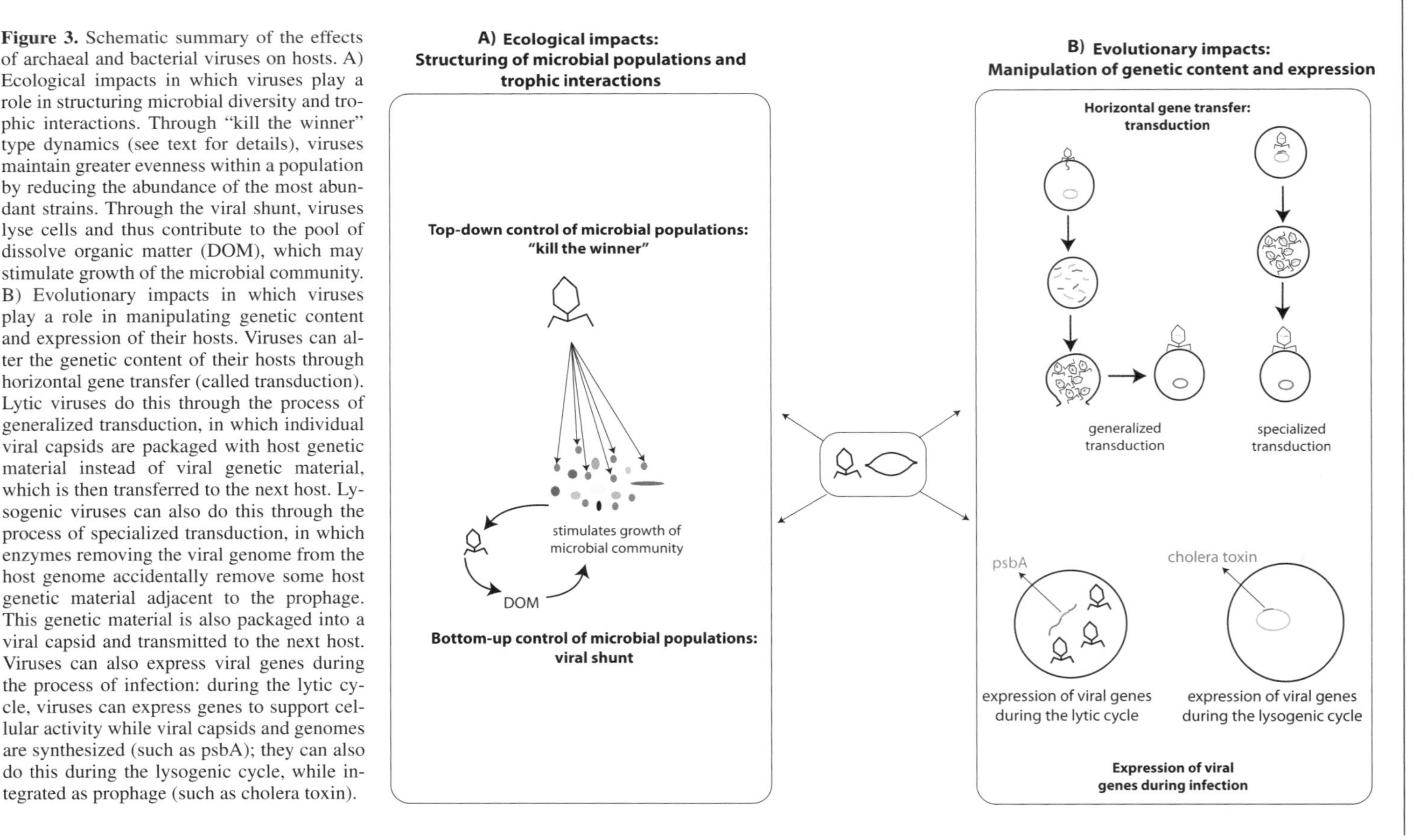

Figure 3. Schematic summary of the effects of archaeal and bacterial viruses on hosts. A) Ecological impacts in which viruses play a role in structuring microbial diversity and trophic interactions. Through "kill the winner" type dynamics (see text for details), viruses maintain greater evenness within a population by reducing the abundance of the most abundant strains. Through the viral shunt, viruses lyse cells and thus contribute to the pool of dissolve organic matter (DOM), which may stimulate growth of the microbial community. B) Evolutionary impacts in which viruses play a role in manipulating genetic content and expression of their hosts. Viruses can alter the genetic content of their hosts through horizontal gene transfer (called transduction). Lytic viruses do this through the process of generalized transduction, in which individual viral capsids are packaged with host genetic material instead of viral genetic material, which is then transferred to the next host. Lysogenic viruses can also do this through the process of specialized transduction, in which enzymes removing the viral genome from the host genome accidentally remove some host genetic material adjacent to the prophage. This genetic material is also packaged into a viral capsid and transmitted to the next host. Viruses can also express viral genes during the process of infection: during the lytic cycle, viruses can express genes to support cellular activity while viral capsids and genomes are synthesized (such as psbA); they can also do this during the lysogenic cycle, while integrated as prophage (such as cholera toxin).

kill-the-winner-like purges of ecotypes carrying the same surface receptors, which they coined the "constant-diversity" (CD) model. These viral purges contrast with the natural selection purges in the theory of "periodic selection," in which occasional changes in environmental conditions drastically reduce diversity by eliminating all groups not adapted to those conditions (Cohan 2002). Thus, viruses can contribute to ecosystem stability by maintaining high levels of diversity, even though they are agents of mortality.

Should viruses be a potent force in the deep subsurface, these impacts on population structure should not be discounted, and the impact is likely to vary depending on the nature of the environment. In stagnant sediments with little fluid flux, for example, environmental conditions may be fairly stable, and thus the CD model posited by Rodriguez-Valera et al. may be a primary mechanism for maintaining diversity among strains in subsurface communities. However, in more dynamic environments, such as hydrothermal vent systems, community diversity may be structured through a synergistic combination of periodic selective sweeps through environmental change as well as CD dynamics through viral predation.

Viral manipulation of genetic content and expression

We have reviewed several processes by which viruses influence the evolution of biological communities throughout the globe: viruses are agents of cell mortality and nutrient recycling, they stimulate co-evolution with their hosts through the virus-host arms race, and they play a hand in the structuring of communities and therefore in the generation of new ecotypes and species. Additionally, viruses are known to manipulate genetic content and expression through horizontal gene transfer and lysogenic conversion. Through these mechanisms, viruses may facilitate adaptation to specific niches within a given ecosystem and thereby exert profound impacts on the evolution of their hosts.

Transduction. Viruses facilitate horizontal gene transfer through the process of transduction, which occurs when a virus introduces foreign genetic material into a host during the course of infection. This transfer can occur during the course of lytic infection in a process known as *generalized transduction*, depicted schematically in Figure 3B. During the lytic cycle, a virus degrades host DNA and synthesizes viral particles. In this process, host DNA can be accidentally incorporated into a new virus capsid. The resulting *transducing particles* can then infect a new host, introducing genetic material from a previous host into a new one, which can then recombine into the genome of the new host. In this process, almost any region of genetic material may be transferred from the donor cell to the recipient.

In the process of *specialized transduction*, lysogenic viruses excise their genomes incorrectly, incorporating a small region of adjacent host genetic material into the viral genome. This mechanism is shown schematically in Figure 3B. Combined, generalized and specialized transduction can have a significant impact on the genetic content of viral hosts: one study estimated that up to 10^{14} transduction events occur per year in Tampa Bay Estuary alone (Jiang and Paul 1998).

A more recent discovery may increase the estimated rates for transduction even further. Gene transfer agents, or GTAs, are viral-like transducing particles, most likely defective phages, which seem to have been usurped by the host to facilitate the process of horizontal gene transfer. While most have been found in Alphaproteobacteria such as *Rhodobacter* (Lang and Beatty 2000) or *Brachyspira* (Matson et al. 2005), GTAs have also been found in methanogens and other groups (Stanton 2007) and may be widespread throughout the archaeal and bacterial domains. A recent study suggested that GTA transduction rates may be over one million times higher than previously reported viral transduction rates in the marine environment (McDaniel et al. 2010). Because of their small size (Matson et al. 2005) GTA particles should be well-represented in "viral" metagenomes, but positive identifications of GTAs are difficult due to the scarcity of sequenced GTAs thus far (Kristensen et al. 2009). Nevertheless, GTAs may constitute a crucial source of genetic innovation in all biomes of the planet, including the deep subsurface. Isolation of strains encoding GTAs may be necessary to enable metagenomic identifications and to increase our knowledge of the scope of their impact.

Expression of genes during the course of infection. Viruses can also carry genes that are expressed during the course of infection. These genes often serve to improve host fitness, which in turn improves virus fitness while the virus is dependent on the host. Some of these genes are expressed by the virus during the lytic cycle, presumably as a means to support host machinery while viral genomes and capsids are replicating within the cell. The most well-known examples of this effect are photosynthesis genes expressed by cyanophage infecting *Prochlorococcus* and *Synechococcus* (Mann et al. 1993). Genes encoding the photosystem core reaction center D1 are expressed during the course of infection in *Prochlorococcus* phage, and it was proposed that these genes improve phage fitness by supporting host photosynthesis during infection (Lindell et al. 2005).

Lysogenic viruses can also manipulate host genetic content while integrated as prophage in the cell. As prophage, lysogenized viruses depend on the host for survival over a longer term, and thus benefit from improving host fitness while integrated in the host genome. In some lysogenic viruses, selection has favored the maintenance of genes known as "fitness factors"—genes that are encoded and expressed by prophage that alter host phenotype and enhance fitness. One of the most well-known examples of this process is the production of cholera toxin by a filamentous bacteriophage integrated into the genomes of virulent *Vibrio cholerae* strains (Waldor and Mekalanos 1996). Studies have shown that infection by a prophage can drastically alter the phenotypic range for a given species (Vidgen et al. 2006). In some cases, phage can also alter host phenotype by suppressing certain metabolic capabilities; it has been suggested that these phage can act to slow host metabolism to shut down unnecessary pathways and conserve energy in environments with low nutrient or energy resources (Paul 2008)—conditions that are expected to be quite common in many deep subsurface ecosystems.

Thus, in addition to influencing host evolution through top-down or bottom-up control of microbial communities, viruses can directly manipulate host genetic content in multiple ways: through generalized or specialized transduction, or through expression of genes either during the lytic cycle or as prophage. Next, we highlight how the unique attributes of the deep subsurface biosphere may create an environment ripe for viral manipulation of host genetic content.

VIRAL MANIPULATION OF THE DEEP SUBSURFACE BIOSPHERE

As we have reviewed above, the viral impact on the microbial ecology of the deep subsurface is potentially profound but largely unexplored. The deep biosphere is not a homogenous biome (Colwell and D'Hondt 2013; Schrenk et al. 2013), and as such, the potential role of viruses is likely to vary depending on the region, particularly with distance from ridge axes, as fluid flux declines and sedimentation increases. Here we examine the potential roles of viruses in these regions, with a focus on the marine subsurface.

Hydrologically active regions of the subsurface

Much of the ocean crust experiences fluid flux to a certain degree; it is estimated that at least 60% of the ocean crust is hydrologically active (Edwards et al. 2011). The volume of fluid fluxing through the crust is at its highest near active hydrothermal systems at mid-ocean ridges, but does not immediately dissipate. The degree of fluid flux varies depending on the sediment cover, as sediments tend to restrict fluid flow (Edwards et al. 2005). This dynamic situation is illustrated schematically in Figure 4. Most fluid flux occurs through connected channels in ocean crust, such as around breccia zones and around pillow basalts or flow boundaries

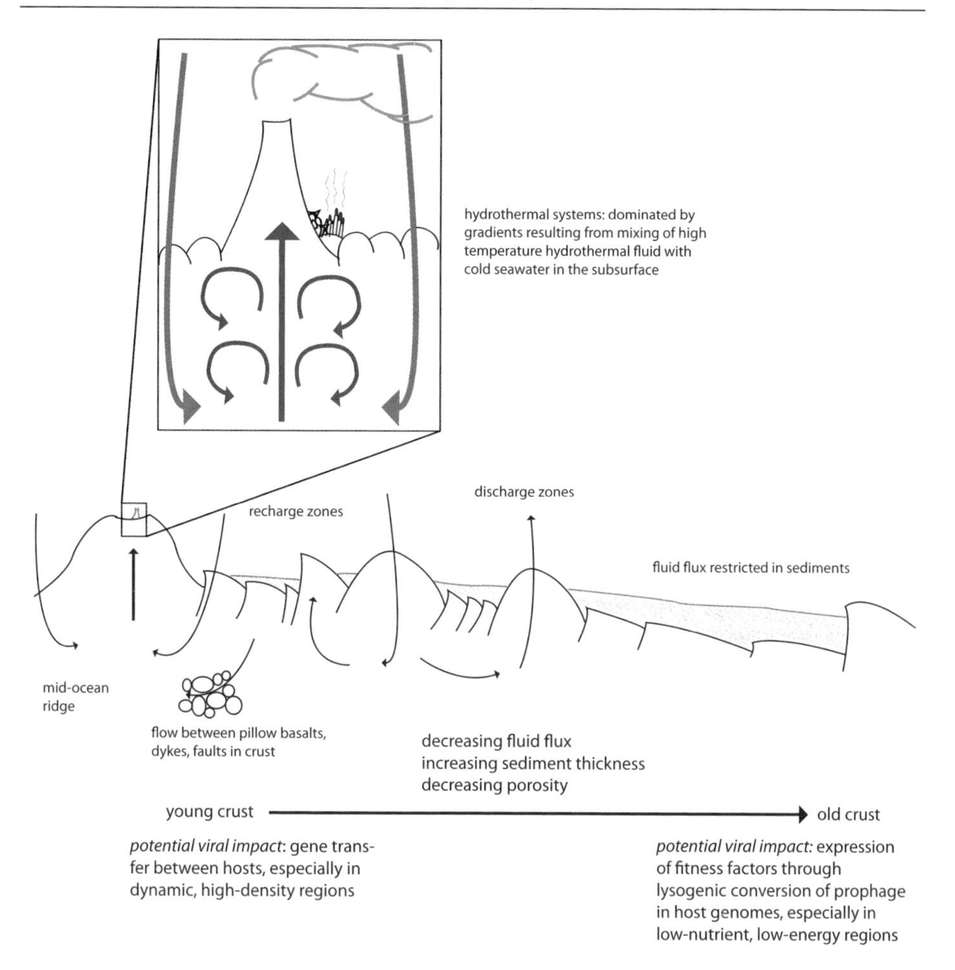

Figure 4. Schematic depicting fluid flux and porosity in the marine subsurface. Arrows depict fluid flux across the seawater/subsurface interface, and throughout the subsurface. Inset shows detail of fluid flux and mixing of seawater and high-temperature hydrothermal fluid in mid-ocean ridge hydrothermal systems, where high-temperature hydrothermal fluid (red) rises from high-temperature water rock reactions deeper in the subsurface and mixes with colder seawater (blue). Crust ages as it moves away from the mid-ocean spreading ridge.

(Fisher and Becker 2000). Seawater flows through seamounts, ridge flanks and recharge zones away from the ridge axis, with residence times ranging from days to years, depending on the location (Johnson and Pruis 2003). Thus a substantial portion of the ocean subsurface biosphere is exposed to dynamic fluid flux.

Connectivity of habitats within the subsurface via fluid flux from one region to the next has the potential to bring members of separate microbial communities, including archaea, bacteria, and their viruses, into contact with each other. This connectivity, in turn, could facilitate gene flow between each of these communities. Cells could exchange genes through conjugation, whereby DNA exchange occurs by direct cell-to-cell contact. However, conjugation requires two cells to be in close proximity and physiologically equipped for this process to occur. Cells can also acquire novel genetic material through transformation, whereby free DNA from lysed cells is taken up into a competent cell. The availability of free DNA varies depending on DNA degradation rates, which tend to be lower in marine sediments (Lorenz and Wackernagel 1994). It is also possible that DNA is stabilized in surface-attached biofilm communities, which are discussed below. In general, though, viruses that have the ability to travel between hosts and to protect genetic material within a protein capsid may represent one of the most important vectors of gene transfer in these environments.

It should be noted that while much of the focus of this review is on regions of the marine subsurface, viruses have been observed to be present in terrestrial subsurface systems as well. While not much is known about viruses in the terrestrial subsurface, observations have been made of viruses in deep granitic groundwater at abundances ranging from 10^5-10^7 per milliliter (Kyle et al. 2008). These viruses seem to be somewhat diverse morphologically, though the viruses isolated thus far have a fairly narrow host range (Eydal et al. 2009). Viruses are thus a potentially important ecological and evolutionary force in the terrestrial subsurface as well, yet much remains unknown about their role.

Hydrothermal vent systems. Hydrothermal vent systems are found at mid-ocean ridge spreading centers or seamounts, and are driven by high-temperature water-rock reactions that occur when seawater comes into close contact with a magma chamber. These basalt-hosted systems are characterized by high-temperature, low-pH fluids that are enriched in reduced compounds, transition metals, sulfide, CO_2, helium, methane, and hydrogen, and are depleted in magnesium (Von Damm 1990; Schrenk et al. 2013). Upon reaching the ocean-crust interface, these fluids precipitate sulfide minerals to create sulfide chimneys that play host to complex microbial communities, which in turn form the trophic basis of macrofaunal communities hosting worms, mussels, crabs, and shrimp. One of the defining characteristics of these systems is the dominance of chemical, physical, and mineralogical gradients that shape the structure of the microbial communities inhabiting these systems (Baross and Hoffman 1985; Schrenk et al. 2003). As seawater mixes with hydrothermal fluid, this results in temperatures that range from 2° to 400°C, acidities that range from 2 to 8, and chemical composition that spans the spectrum of reduced hydrothermal fluid to oxidized seawater. As a result, vents play host to a wide range of archaeal and bacterial species: hyperthermophiles, thermophiles, and psychrophiles, including both heterotrophs and autotrophs.

Few studies have been conducted on viruses inhabiting these systems. Ortmann and Suttle (2005) found that the abundance of viral-like particles in diffuse flow fluids was approximately 10^6 per milliliter, about ten times that of cells, a ratio that is also typical of surface seawater. Williamson et al. (2008b) found that a higher abundance of viral-like particles were induced from hydrothermal vent microbial communities exposed to a mutagen compared to those from the upper water column, suggesting that lysogeny is a more predominant lifestyle at vents than in the upper water column. Metagenomics has revealed that the marine vent viral assemblage has the potential to infect a wide variety of bacterial and archaeal hosts from a range of thermal regimes, reflecting the gradient-dominated nature of the environment (Anderson et al. 2011b). Together, these studies suggest that many different archaeal and bacterial groups may have prophage integrated into their genomes that potentially introduce novel genetic material or express fitness factors. If this is the case, then genomic analyses of subsurface archaea and bacteria that contain prophage should be a fairly efficient, though clearly biased, approach for exploring the diversity of subsurface viruses. Ideally, such analyses would be coupled with a metagenomic census of free viral particles (e.g., Anderson et al. 2011b) in order to compare lysogenic and lytic viruses. Clearly, much more research remains to be done to better understand the nature of viral roles in the subsurface.

Deeply buried sediments

Regions of the deep subsurface with more restricted fluid flux, particularly in regions with high sedimentation such as on continental margins, present a drastically different set of conditions for microbial inhabitants. Within the sediment matrix, viral mobility may be reduced, resulting in a potentially lower host contact rate. This reduced contact rate would be the case especially if cell abundances are low in deeply buried sediments, such as in the sediments of the South Pacific Gyre (D'Hondt et al. 2009). On the other hand, viruses that form small but hardy particles may be less affected by restrictions on fluid flux than other mechanisms of genetic exchange, which would accentuate the importance of viruses in the ecology and evolution of these isolated communities. The challenges faced by microbial communities inhabiting these regions include limitations on nutrient and energy levels, particularly in deeply buried sediments, where organic carbon and potential oxidizing agents are scarce (Jørgensen and D'Hondt 2006), and metabolic rates have been shown to be extremely low (Røy et al. 2012). The low activity and long doubling times of cells in these regions likely provides further resistance to viral infection, which is largely dependent on the density and activity of the host (Fuhrman 2009). As with cellular abundances, viral abundance decreases with depth in the sediments. Middelboe et al. (2011) quantified viral abundance with depth on the eastern margin of the Porcupine Seabight and observed approximately 10^8 VLPs/cm^{-3} at 4 meters below the seafloor (mbsf), to about 10^6 VLPs/cm^{-3} at 96 mbsf. However, another study by Engelhardt et al. (2012) found that the virus-to-cell ratio increased with depth in the sediments, potentially indicating continued viral production at these depths in sediments, rather than long-term viral preservation, as had been suggested previously.

Lysogeny is expected to be a common viral lifestyle in the deep subsurface, resulting from selection for a viral lifestyle that limits the necessity for finding hosts in a sparse soil matrix and in harsh environmental conditions. Work by Engelhardt et al. (2011) has demonstrated that nearly half of the bacterial isolates tested from a deep-sea sediment core harbored prophage, and a subsequent study found that all deeply buried isolates of a common deep-sea species, *Rhizobium radiobacter,* were lysogenic (Engelhardt et al. 2012). The ubiquity of lysogeny could have interesting implications for cellular survival in these energy-limited systems, where archaea and bacteria are likely to be under strong selection pressure to harness alternate forms of energy when they are available, and to minimize energy use when energy sources are limiting. Previous studies have shown that certain lysogenic phage can actively repress host metabolic genes, and therefore repress wasteful host metabolic processes when conditions are not favorable (Paul 2008), a trait that would be particularly useful in the energy-limited subsurface. Further work in deeply buried sediments will reveal what genes are expressed by these lysogenized phages, perhaps revealing that the relationship between virus and host transcends the parasitic, becoming instead a mutualistic symbiosis.

Viral impacts on surface-attached communities

Regardless of whether a deep subsurface habitat is hydrologically active or not, most of the inhabitants probably live as biofilms (i.e., communities attached to some sort of hard surface, which could be hard rock, vent chimney deposit, or sediment). Generally, biofilms have much higher cell density than the surrounding medium, so there is potential for biofilms to be hotspots of viral activity. Viruses are known to accumulate in biofilms growing in drinking water systems (Skraber et al. 2005), and viral lysis is frequent during biofilm development in *Staphylococcus aureus* (Resch et al. 2005). Metagenomic sequencing of biofilms from an acid mine drainage site recovered complete viral genomes and extensive evidence that bacterial genomes are continually influenced by viral infection (Andersson and Banfield 2008). Viral genes are highly expressed in *Pseudomonas aeruginosa* biofilms (Whiteley et al. 2001), and viral-mediated cell death is a normal component of biofilm development (Webb et al. 2003). Beyond these basic detections of viruses and viral genes in biofilm habitats, however, surprisingly little is known about the molecular mechanisms and ecological impacts of virus-biofilm interactions, especially in the subsurface.

The polysaccharide-rich extracellular matrix of biofilms is probably a barrier against infection, but it is clear that viruses can penetrate the barrier, in some cases via enzymatic digestion (Weinbauer 2004). Another complication is the recent finding that a human virus can generate its own biofilm-like matrix (Thoulouze and Alcover 2011). The prevalence of viruses encased within extracellular matrices is entirely unexplored in subsurface ecosystems, and it is likely that such surface-attached viral populations can evade detection and depress counts of viral-like particles in fluid samples. Therefore, interpretations of viral abundance and activity data from subsurface fluid samples must consider how well the fluid samples represent the rocks and sediments that provide habitat for most bacteria, archaea, and viruses in the subsurface.

In addition to high cell density, many biofilm communities also have high genetic and phenotypic diversity, resulting in complex interactions among many species on microscopic spatial scales (Stoodley et al. 2002). One potential consequence is that viruses with high host specificity may have greater difficulty finding their host in a tightly-packed, diverse biofilm, resulting in a large total number of viruses, each capable of infecting only a tiny proportion of the diverse biofilm community. This scenario is one possible explanation of the "infectivity paradox": the observation that many habitats have high viral abundance but low infectivity (Weinbauer 2004). Preliminary data (Filippini et al. 2006) suggest that biofilms exemplify the infectivity paradox, but no such studies have been conducted in the deep subsurface.

Many studies have demonstrated the importance of lateral gene transfer in biofilm communities (Molin and Tolker-Nielsen 2003), and in some cases, viruses have been identified as the agents of transfer (Whiteley et al. 2001; Webb et al. 2003). It is clear that in biofilms, gene transfer is not a rare curiosity but a fundamental aspect of biofilm formation and development, notably as a mechanism for a phenomenon known as "phenotype switching." In *Staphylococcus epidermidis* biofilms, for example, genomic insertion of a mobile genetic element results in a stable population of variant cells unable to produce the biofilm matrix (Ziebuhr et al. 1999). The effect is reversible because the inserted DNA is frequently excised, restoring biofilm production. Other species also exhibit reversible phenotype switching associated with biofilm formation, and viruses have been implicated in at least one case (Webb et al. 2004). The evolutionary dynamics of such processes have not been explored experimentally, but one simulation study predicted that the coexistence of multiple phenotypes in a biofilm community can be promoted by continual gene transfer. If two phenotypes are linked, as in a syntrophic partnership, the fitness of each member is dependent on the fitness of the other. Therefore, natural selection of such cells living in a dense community could result in complex inter-species relationships that are dependent on (potentially viral-mediated) transfer of genetic content. In summary, we expect future research to reveal that biofilms in subsurface habitats exemplify the concept described above that viral activity in the subsurface is likely to have complex and varied evolutionary consequences that extend beyond just cell mortality.

Tools for analysis: viral metagenomics in the deep subsurface

Given the current state of knowledge about viruses in the deep subsurface, how can we gain further insight into the role they play in manipulating geochemical cycles, altering diversity, and influencing the course of evolution in their hosts? One method by which we can probe the viral world is through metagenomics, in which a sample of community DNA is extracted and sequenced directly from the environment. While metagenomics in the microbial realm has traditionally focused on asking "who is there?" and "what are they doing?", viral metagenomics presents a unique set of challenges. Viruses are generally separated from the microbial fraction through size fractionation, which may exclude large viral particles or include small cells, so contamination is an issue of concern. Moreover, one of the primary challenges facing viral metagenomics is the large proportion of unknown sequences. The average percentage of viral metagenomic sequences with no match to existing databases ranges from about 60 to over 90%, depending on the read length (i.e., Breitbart et al. 2002; Angly et al. 2006; Desnues et al. 2008; Anderson et al. 2011b; Rosario and Breitbart 2011).

The vast number of sequences with no match to existing databases presents a challenge to viral ecologists seeking to understand who viruses infect, what impacts they have on their hosts, and what types of genes they encode and transfer.

One goal of viral ecology in any environment is the identification of which archaeal or bacterial groups play host to those viruses. This information is key to understanding how viruses may impact a given microbial community. If only certain groups are most susceptible to viral attack, this selectivity may have further implications for microbial population structure or biogeochemistry. Some information about potential hosts can be gleaned by identification of known viral groups: *Rudiviridae* and *Fuselloviridae*, for example, are only known to infect the archaea. As we have noted, classification of viral metagenomic sequences is tremendously challenging, and even if it is successful, only limited information is gained because many families of viruses infect wide ranges of hosts.

One method that has been used to identify potential hosts of a viral assemblage is to identify clustered regularly interspaced palindromic repeats (CRISPRs), an immune system used by archaea and bacteria to combat invasive genetic material, including viruses and plasmids (Barrangou et al. 2007; Brouns et al. 2008; Sorek et al. 2008; van der Oost et al. 2009; Horvath et al. 2010; Labrie et al. 2010; Marraffini and Sontheimer 2010). CRISPRs are found on bacterial and archaeal genomes and are structured as a series of short repeats of about 20-50 bp in length, interspersed by slightly longer sequences, called "spacers," that are 25-75 bp in length. These spacers are synthesized to match a short region of foreign DNA, such as a virus or a plasmid that has invaded the cell. This short spacer region is then inserted into the CRISPR locus. Subsequently, if foreign DNA containing a region with a close match to a CRISPR sequence invades the cell in the future, an immune response is mobilized via CRISPR-associated, or *Cas*, genes that work in conjunction with small RNAs derived from the CRISPR spacers to bind and cleave invading DNA (Garneau et al. 2010; Jore et al. 2011).

CRISPRs have great potential for the study of viral ecology and viral-host coevolution because each locus can be treated as a library of previous viral infections for a given strain. Previous studies have used CRISPRs as an ecological tool to track the impacts of viruses over time and space within and between microbial populations (Tyson and Banfield 2008; Held and Whitaker 2009). We have previously used CRISPRs as a tool to identify potential hosts of the viral assemblage of hydrothermal vent diffuse flow fluids by constructing a CRISPR spacer database to query viral metagenomes and identify potential hosts based on matches of metagenomic reads to known CRISPR sequences in archaeal and bacterial genomes (Anderson et al. 2011b). This analysis indicated that viruses in the hot subsurface near vent systems have the potential to infect several different taxa of archaea and bacteria from a wide range of thermal regimes. Expanding this analysis to viral assemblages from other regions of the subsurface would indicate whether viral assemblages throughout the deep subsurface are similarly broad in host range, or whether only certain groups tend to be susceptible to viral infection.

Another outstanding question regarding viral roles in the subsurface is the degree to which viruses mediate horizontal gene transfer, or manipulate archaeal and bacterial genomes through incorporation as prophage. One way to address this question is to examine viral and cellular metagenomes for sequences potentially associated with mobile elements, such as those that encode transposases, integrases, and recombinases. The presence of abundant genes that encode such enzymes provides one piece of evidence that the organisms in a particular community actively exchange genes. Table 2 shows the percentage of reads in a set of viral metagenomes that had a match to a transposase, recombinase, or integrase domain (maximum e-value: 10^{-5}). Each of the viral metagenomes listed in the table was sequenced with pyrosequencing technology, and each of the reads for these metagenomes ranged between approximately 90-300 base pairs. The hydrothermal vent viral metagenome contained one of

Table 2. Percent of reads in a set of viral metagenomes with a match to a transposase, recombinase, or integrase protein domain. tblastn searches were conducted with a set of Pfam seed sequences matching transposase, recombinase, and integrase domains as the query. Hits were designated as a "match" if they had an e-value below 10^{-5}.

Metagenome	Number of reads	Number of hits	% of reads with a hit	Reference
Antarctic Lake summer	30515	66	0.22	Lopez-Bueno et al. 2009
Hydrothermal vent	231246	227	0.098	Anderson et al. 2011b
Arctic Ocean	688590	605	0.088	Angly et al. 2006
Bay of British Columbia	138347	81	0.059	Angly et al. 2006
Antarctic Lake spring	31691	13	0.041	Lopez-Bueno et al. 2009
Gulf of Mexico	263908	106	0.040	Angly et al. 2006
Microbialites	621110	246	0.040	Desnues et al. 2008
Coral	36354	14	0.039	Dinsdale et al. 2008
Sargasso Sea	399343	17	0.0043	Angly et al. 2006

the highest proportions of reads matching a potential component of a mobile element, perhaps implying that viruses in the vent viral assemblage actively transduce genes or integrate their genomes into that of their hosts. This trend is consistent with the hypothesis that lysogeny is a prevalent lifestyle in hydrothermal systems and that viruses actively transduce genetic material from one host to the next. Interestingly, the authors of the paper describing the Antarctic Lake virome noted that there was a distinct shift from ssDNA viruses in the spring to dsDNA viruses in the summer (Lopez-Bueno et al. 2009). They suggest that under cover of the ice, metabolic rates are restricted, and thus the only actively replicating phages may be lytic ssDNA viruses with small genomes. In contrast, in the summer, larger dsDNA phages (which presumably were integrated as prophage in the winter and spring) were able to enter the lytic phase and therefore would have been collected in the viral size fraction sampled for the metagenome. This transition is reflected in the relative abundance of mobile elements in each of the metagenomes: the summer fraction, in which the lysogenic phage have entered the lytic phase and are therefore captured in the sample, contains a higher abundance of sequences potentially associated with mobile elements. It would therefore be quite interesting to examine a viral metagenome from deeply buried sediments, where we also hypothesize lysogeny to be common, and calculate the relative abundance of mobile elements. Induction of the sample prior to sampling would cause lysogenized viruses to enter the lytic cycle, so that they can be sampled in the viral size fraction (Anderson et al. 2011a).

Similarly, an analysis of viral metagenomes from a range of different environments indicates that DNA ligases are also particularly abundant in the viral assemblage from diffuse flow hydrothermal fluids (Anderson et al. 2011a), with over ten times the percentage of reads encoding a DNA ligase compared to the metagenome with the next highest abundance. As ligases are known to play a role in repairing double-stranded breaks, it is possible that these ligases play a role in facilitating horizontal gene transfer in the subsurface, though further research will be required to support or disprove this hypothesis.

If genes are in fact being transferred by viruses in these environments, a subsequent question to ask is the extent of gene diversity within the viral gene pool: does the viral assemblage contain a high diversity of protein-encoding genes that can be potentially transferred, or is the pool of genes relatively small? One way to address this question is to construct rarefaction curves comparing the relative gene diversity of viral metagenomes. Figure 5 depicts rarefaction curves of protein-encoding genes in viral metagenomes from several different environments. ORFs derived from reads in these metagenomes identified by FragGeneScan in the MG-RAST pipeline (Meyer et al. 2008) were clustered to 40% similarity using UCLUST (Edgar 2010). Rarefaction curves were generated with mothur (Schloss et al. 2009). The slopes of these curves indicate that most of the marine viral assemblages have fairly high diversity compared to the Antarctic Lake metagenomes and two of the microbialite metagenomes, perhaps reflecting the diversity of the hosts they infect in each of these environments. Further work will elucidate the functions of the genes encoded by these viral assemblages, and comparative metagenomics between viral and cellular fractions may provide insight into which types of genes are enriched in the viral fraction compared to the cellular fraction, perhaps indicating which genes are selected to be maintained in the viral gene pool.

In summary, metagenomics holds great potential for illuminating the virus-host relationship, and further sequencing of both viral and cellular metagenomes from regions of the deep subsurface will contribute much to our understanding of which organisms are most susceptible to viral infection, whether the lytic or lysogenic lifestyle is more common in the subsurface, and the nature of the role viruses play in facilitating horizontal gene transfer in the deep subsurface. The isolation of virus-host systems from organisms in pure culture would also provide further insights into the nature of the host-virus relationship in the subsurface. Virus-host systems could uncover new morphologies or escape mechanisms, particularly among the archaeal viruses, or new relationships between the virus and the host. Experiments with virus-host systems could reveal new insights into the nature of viral manipulation of the host by identifying which genes are expressed, upregulated or downregulated when prophage are integrated into a host genome.

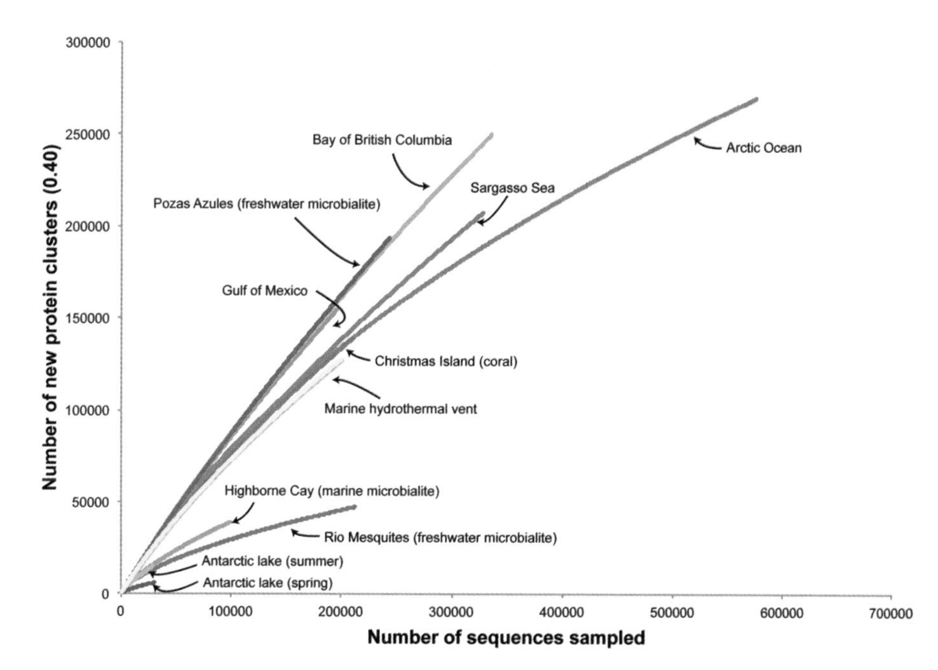

Figure 5. Rarefaction curves of clusters of translated open reading frames from viral metagenomic sequences from around the globe. ORFs were clustered to 0.40 similarity using UCLUST (Edgar 2010). Rarefaction curves were generated in mothur (Schloss et al. 2009).

VENTS, VIRUSES AND THE ORIGIN OF LIFE

The evidence appears to be clear that viruses can have a substantial impact on the evolution of their hosts, and have most likely been doing so for billions of years. But for how long has this mutual evolutionary relationship persisted? When and how did viruses originate? The question is particularly relevant here because the deep subsurface, and hydrothermal systems in particular, are often considered the most ancient continuously inhabited ecosystems on the planet (Reysenbach and Shock 2002), and indeed, are often thought to have been an important setting for the origin of life on Earth (e.g., Baross and Hoffman 1985; see Hazen 2005).

Hydrothermal vents and the deep subsurface: key settings in the origin of life

On the Hadean Earth, about 4 billion years ago, hydrothermal vent systems would have been present in perhaps even greater abundance than they are today. Residual heat of formation would have resulted in a volcanically and seismically more active planet, with longer mid-ocean ridges and more plate tectonic activity (Hargraves 1986), resulting in a higher incidence of water-rock reactions at the bottom of the ocean. Both basalt-hosted and peridotite-hosted hydrothermal systems are likely to have been present in the Hadean Earth. Metal-sulfide minerals in basalt-hosted systems, including pyrite, have been implicated as key catalysts in several important prebiotic reactions in which CO or CO_2 is fixed into simple organic compounds (Wächtershäuser 1988a, 1988b, 1990; Cody 2004). Peridotite-hosted systems, formed off-axis and powered by serpentinization through the interaction of seawater with peridotite, are characterized by lower-temperature fluids with high pH and form large calcium carbonate structures (Kelley et al. 2001; Schrenk et al. 2013).These vents may have also acted as a source for key organic compounds generated in the process of serpentinization, including formate, acetate, methane, organic sulfur compounds, and larger hydrocarbons (Heinen and Lauwers 1996; Proskurowski et al. 2008; Lang et al. 2010). Moreover, the calcium carbonate porous structures formed in these systems may have acted as a concentrating mechanism for early prebiotic compounds (Baaske et al. 2007).

One of the most appealing aspects of hydrothermal vents as a setting for the origin of life is the formation of geological, physical, and chemical gradients in these systems (Baross and Hoffman 1985). These gradients provide a wide range of environmental conditions within a relatively small physical space with fluid flow between them, facilitating the occurrence of multiple chemical processes across a multiplicity of environmental conditions in parallel. These gradients extend beyond hydrothermal vent fields themselves to other regions of the deep subsurface. The minerals catalyzing reactions in one region, such as at basalt-hosted hydrothermal vents, would have differed from those in other regions of the subsurface, such as at peridotite-hosted systems or in sedimented regions. As mentioned above, much of the ocean crust is linked by fluid flux, which moves at different flow rates and volumes depending on the depth and degree of porosity in the crust. Thus, compounds synthesized in one region of the ocean crust, whether at a hydrothermal system or more distal to a mid-ocean ridge, could be transferred from one region of the subsurface to the next.

In this sense, the deep subsurface may have acted as a natural laboratory for the origin of life, in which multiple "experiments" could have been carried out in tandem. Later, the products of these natural experiments could have been combined to form an autocatalytic network. Several studies have suggested that the chemiosmotic gradients at vent sites, combined with enclosed pore spaces and organic syntheses, could have resulted in the first autocatalytic networks (Koonin and Martin 2005; Martin and Russell, 2007; Martin et al. 2008; Lane et al. 2010). Martin, Russell and others describe a model in which a chemiosmotic potential is generated across the membrane of an iron-sulfide bubble, which they presume would form in an anoxic Hadean ocean. The potential could then have been harnessed to yield a protometabolism based on the reduction of CO_2 by H_2.

The question that arises is how the first self-replicating entities formed and evolved in these settings. Koonin and Martin (2005) suggest that self-replicating networks could have formed within the walls of these iron-sulfide compartments. In their scenario, each component of the network consisted of a selfish RNA molecule encoding one or a few proteins, with the original selection pressures favoring rapid self-replication. The authors refer to these replicating entities as "virus-like RNA molecules," which Koonin then elaborated upon in a later publication detailing the "Virus World" (Koonin et al. 2006). In this model, the authors describe a scenario in which viruses emerged early from the various replicating entities and networks that formed part of the RNA world. These scenarios suggest that viruses may have played a primary role at the earliest stages of life's evolution.

The viral role in the origin of life

Viruses have not always been considered to be primordial elements. Historically, three theories were put forward regarding the origin of viruses: first, that viruses were originally parasitic cells that evolved into a viral-like form (the "reduction hypothesis"); second, that viruses were rogue genetic elements from cells that developed a protein capsid to survive in an extracellular state (the "escape hypothesis"); and third, that viruses originated in parallel with cells (the "virus first hypothesis") (Prangishvili et al. 2006). The last hypothesis, however, has been gaining favor as scientists have found that viruses infecting different domains of life share certain "hallmark genes" that are missing from cellular genomes, perhaps pointing to an early origin that predates the divergence of the three domains of life (Koonin et al. 2006). Others have suggested that DNA as a genetic material first arose in a virus, which later spread to the cellular world (Forterre 2006).

The tremendous diversity of viruses, though, greatly complicates an elucidation of their origin. Viruses encompass one portion of a spectrum of mobile genetic elements, which range in size and complexity from simple introns and transposons, to GTAs, to RNA viruses, viroids, and satellite viruses, to dsDNA viruses and the giant Mimivirus described above. These elements may not share a common origin, yet in many cases, many virus-like elements share genes that are not found in the cellular world (Koonin et al. 2006), or share structural features in their protein capsids (Bamford et al. 2005). Many of these shared attributes transcend domains, leading many to consider viruses to be ancient.

The attribute that all viruses share is their dependence on a host for the purposes of replication: in a word, these are parasites. Consideration of the role of parasites in the origin of life is not a new concept. In an RNA-protein world, or even a pre-RNA world, parasites could have undermined replication networks, as they could take resources from these replication networks (or "hosts," in a sense) without benefiting them, and thus destroy the cycle (Maynard-Smith and Szathmáry 1999). In these networks, elements are linked such that each element replicates another element in the cycle. Parasites emerge when a mutant of one of the elements is preferentially replicated, but does not replicate another element in the cycle (Fig. 6A). It has been suggested that containing replication cycles within a compartment, or at least confining them to a surface, may circumvent this problem by placing the selective pressure not on the individual elements within a replication cycle, but on the cycle as a whole (Maynard-Smith and Szathmáry 1999). This compartmentalization effectively provides a basis for heredity and competition between individuals, which are required for natural selection to occur. Moreover, spatial structuring of the environment may reduce the spread of parasites from one hypercycle to the next (Boerlijst and Hogeweg 1991). However, spatial structuring may prevent "sharing" of new functions through horizontal gene transmission (Poole 2009).

Yet as we have discussed here, parasites can at times improve the fitness of the host they depend upon. Just as modern viruses can express fitness factors to boost the fitness of their host, the same may have applied in life's early evolution: for example, if a selfish element were to

A) Parasitic elements in replication networks

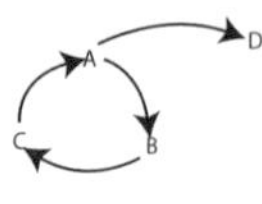

B) Parasitic elements contribute to fitness of host network

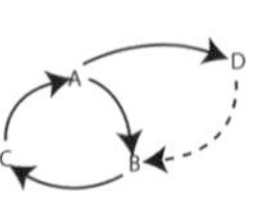

C) Parasitic elements connect networks: expansion of functionality

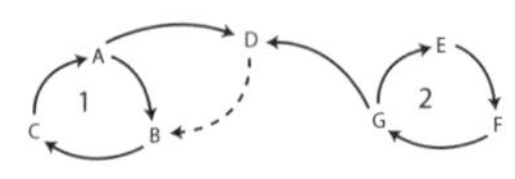

D) Spatial structuring of networks in the subsurface enables selection and competition; fluid flux enables mixing of important prebiotic engredients

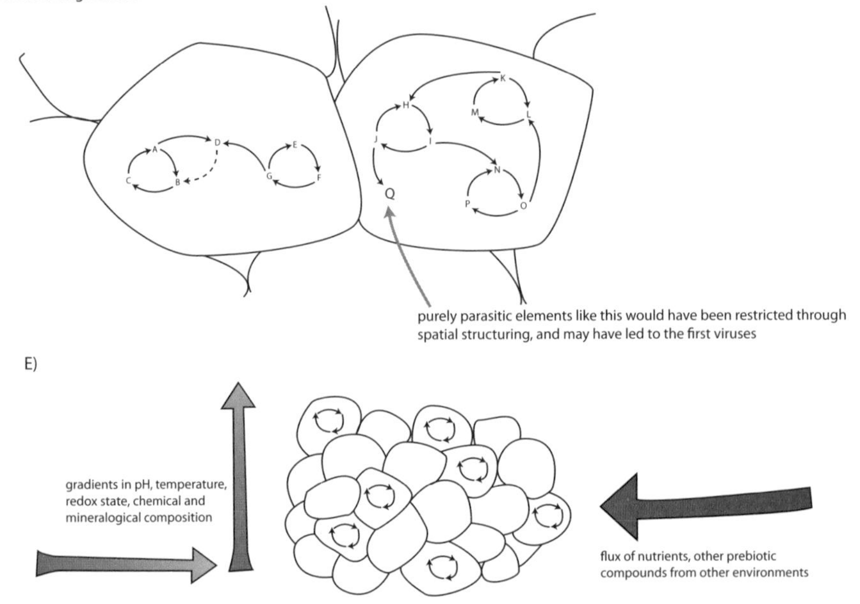

Figure 6. Role of parasitic elements in early replication cycles. A) Cooperative replication network. Element D is a parasite to the network because it uses network resources, but it does not contribute to network fitness. B) Selection may favor elements that are able to contribute to the fitness of the host network. Here, element D contributes to the stability of element B, thus improving network fitness. This feedback improves the fitness of D as well. C) If element D is also replicated by element G in another network, this replication could link the two networks, thereby increasing network functionality. D) Spatial structuring through restriction to mineral surfaces, or enclosure in pore spaces, could restrict the degree to which parasitic elements (like element Q, considered "parasitic" here because it does not contribute to the parent network or any other networks) spread between networks. E) Scenario in which replicating networks operate in the subsurface, with diverse replicating networks defined by gradients in environmental conditions, and fed by an influx of prebiotic compounds through fluid flux in the subsurface.

contribute toward the overall fitness of a given replication cycle, such as through stabilizing another element in the cycle, this change would improve the fitness of the whole cycle and therefore the fitness of the element as well (Fig. 6B). Indeed, this process may have been the means by which new functions were added to replication networks. Just as modern viruses can contribute novel genetic material through transduction or expression of fitness-boosting genes, ancient parasites may have increased the functionality of the networks they were part of by linking together disparate networks. Rather than (or in addition to) presenting a problem for early replication networks, early viruses may have provided a means by which to increase their functionality. Viral-like particles or selfish elements may have acted as a means to share genes between networks, and ultimately may have allowed early genomes to expand (Fig. 6C).

In this sense the meaning of words like "parasite" or "mutualist" become blurred, as selection at this level may favor varying degrees of parasitism or mutualism; in the RNA world, selection operated at the level of both the individual elements and at the level of whole networks.

This scenario is consistent with previously published ideas about the origin of life that may or may not have explicitly specified roles for viruses. For example, the idea that all life today evolved from primitive cells in an early biofilm-like community evolving "through prolific genetic exchange with other 'precells' in the community, perhaps involving structures resembling transposable genetic elements and viral-like particles" (Baross and Hoffman 1985) was inspired by the initial discovery of prolific subsurface life evident at hydrothermal vents. Woese (1998, 2002) has developed in detail the idea of a communal ancestor in which horizontal gene transfer is the primary driver of evolution and viral-like elements could very well have been one of the mediators of this gene transfer.

However, as with modern viruses, these parasitic elements likely would have had a wide host range, in which some could spread rapidly to other networks, whereas others were more restricted. Modern viruses also exhibit a range of virulence, in which some were almost entirely parasitic whereas others are almost entirely mutualistic, and we envision prebiotic selfish elements to have had similar characteristics. In this sense, indiscriminate horizontal gene transfer in the communal ancestor may have been disruptive by facilitating the spread of the more virulent, entirely deleterious parasites (Poole 2009). It is also unclear how the communal ancestor would have evolved "as a unit" (Woese 1998) without competition or selection with other units. We therefore envision a scenario in which spatial structuring of the environment facilitates selection at the level of an entire network, rather than on individual elements. This structuring would also serve to restrict the movement of wide-ranging, deleterious parasites. Iron-sulfur bubbles or pores in hydrothermal systems could have acted as a structuring mechanism prior to the emergence of lipid membranes (Russell and Hall 1997; Koonin and Martin 2005; Martin and Russell 2007) (Fig. 6D).

On a larger scale in the prebiotic world, there would have been extensive fluid flux in the subsurface, and this transport could have served as a conduit for nutrients or products of prebiotic reactions from other environments. Subsurface flow would have connected these networks and facilitated some degree of gene sharing between them, as well as provided them with important prebiotic precursors (Fig. 6E). Gradients in temperature, pH, chemical and mineralogical composition through the subsurface or in hydrothermal structures would have generated diversity in these replicating networks, creating variation in the population and facilitating selection among them. In this sense, the environment may have fostered the earliest stages of natural selection, and these early viral-like or "selfish" elements may have been important in facilitating gene transfer between these networks, allowing them to grow and change.

Regardless of the degree to which horizontal gene transfer occurred, subpopulations within the ancestral community became more resistant to genetic exchange with other subpopulations over time, which may have arisen as a defense against parasitic genetic elements: the first viruses. The crossing of this "Darwinian threshold" from one ancestral community to many independent cells marked the origin of speciation and the emergence of the life forms we know today (Woese 2002), and it is possible that viruses or viral-like elements were intimately involved in this critical stage in the evolution of life.

CONCLUSION

There is clearly much work that remains to be done to understand the nature of the viral impact on biogeochemical cycling, microbial community structure, and evolution in the deep subsurface. Yet the few available details provide tantalizing hints that the role of viruses in the deep subsurface could be profound on many levels. Viral infection may significantly impact biogeochemical cycling in the subsurface through lysis of cellular biomass, releasing nutrients and compounds that would otherwise be entrained in biomass. Viruses are also known to alter the structure of the microbial communities they infect, potentially increasing overall diversity through lysis of cells that become most abundant in a given region. Through the process of lysogeny and transduction, viruses may manipulate the genomes and expression of the hosts they infect throughout the subsurface, effectively resulting in a mutualistic, symbiotic relationship between host and virus that transcends traditional notions of viruses as parasites. Indeed, this role of virus as parasite, as mutualist, and as a sharer of information through gene transfer may be a fundamental underpinning of life in the deep subsurface that extends back in time to the dawn of life itself. Finally, the hot subsurface environments associated with hydrothermal systems harbor many of the most deeply rooted microorganisms on the universal phylogenetic tree of life; most are hyperthermophilic archaea. Very little is known about the viruses that are associated with these microorganisms. Given their antiquity, including their primordial setting, it is possible they harbor viruses and virus-like particles that could lead to a better understanding of the origin of viruses and their role in the early evolution of life.

REFERENCES

Anderson RE, Brazelton WJ, Baross JA (2011a) Is the genetic landscape of the deep subsurface biosphere affected by viruses? Front Extreme Microbiol 2:00219 doi: 10.339/fmicb.2011.00219

Anderson RE, Brazelton WJ, Baross JA (2011b) Using CRISPRs as a metagenomic tool to identify microbial hosts of a diffuse flow hydrothermal vent viral assemblage. FEMS Microbiol Ecol 77:120-133, doi: 10.1111/j.1574-6941.2011.01090.x/full

Andersson A, Banfield J (2008) Virus population dynamics and acquired virus resistance in natural microbial communities. Science 320:1047-1050, doi: 10.1126/science.1157358

Angly FE, Felts B, Breitbart M, Salamon P, Edwards RA, Carlson C, Chan AM, Haynes M, Kelley S, Liu H, Mahaffy JM, Mueller JE, Nulton J, Olson R, Parsons R, Rayhawk S, Suttle CA, Rohwer F (2006) The marine viromes of four oceanic regions. PLoS Biol 4:e368, doi: 10.1371/journal.pbio.0040368

Baas Becking LGM (1934) Geobiologie of Inleiding tot de Milieukunde. Van Stockum & Zoon, The Hague

Baaske P, Weinert FM, Duhr S, Lemke KH, Russell MJ, Braun D (2007) Extreme accumulation of nucleotides in simulated hydrothermal pore systems. Proc Natl Acad Sci USA 104:9346-9351, doi: 10.1073/pnas.0609592104

Bamford DH, Grimes JM, Stuart DI (2005) What does structure tell us about virus evolution? Curr Opin Struc Biol 15:655-663, doi: 10.1016/j.sbi.2005.10.012

Baross JA, Hoffman SE (1985) Submarine hydrothermal vents and associated gradient environments as sites for the origin and evolution of life. Origins Life Evol B 15:327–345

Barrangou R, Fremaux C, Deveau H, Richards M, Boyaval P, Moineau S, Romero DA, Horvath P (2007) CRISPR provides acquired resistance against viruses in prokaryotes. Science 315:1709-1712, doi: 10.1126/science.1138140

Biddle JF, Lipp JS, Lever MA, Lloyd KG, Sørensen KB, Anderson R, Fredricks HF, Elvert M, Kelly TJ, Schrag DP, Sogin ML, Brenchley JE, Teske A, House CH, Hinrichs K-U (2006) Heterotrophic Archaea dominate sedimentary subsurface ecosystems off Peru. Proc Natl Acad Sci USA 103:3846-3851, doi: 10.1073/pnas.0600035103

Bize A, Karlsson EA, Ekefjärd K, Quax TEF, Pina M, Prevost M-C, Forterre P, Tenaillon O, Bernander R, Prangishvili D (2009) A unique virus release mechanism in the Archaea. Proc Natl Acad Sci USA 106:11306-11311, doi: 10.1073/pnas.0901238106

Boerlijst MC, Hogeweg P (1991) Spiral wave structure in pre-biotic evolution: Hypercycles stable against parasites. Physica D 48:17-28, doi: 10.1016/0167-2789(91)90049-F

Bratbak G, Egge J, Heldal M (1993) Viral mortality of the marine alga *Emiliania huxleyi* (Haptophyceae) and termination of algal blooms. Mar Ecol-Prog Ser 93:39-48, doi: 10.3354/meps093039

Breitbart M, Miyake JH, Rohwer F (2004) Global distribution of nearly identical phage-encoded DNA sequences. FEMS Microbiol Lett 236:249-256, doi: 10.1111/j.1574-6968.2004.tb09654.x

Breitbart M, Rohwer F (2005) Here a virus, there a virus, everywhere the same virus? Trends Microbiol 13:278-284. doi: 16/j.tim.2005.04.003

Breitbart M, Salamon P, Andresen B, Mahaffy JM, Segall AM, Mead D, Azam F, Rohwer F (2002) Genomic analysis of uncultured marine viral communities. Proc Natl Acad Sci USA 99:14250-14255, doi: 10.1073/pnas.202488399

Brouns SJJ, Jore MM, Lundgren M, Westra ER, Slijkhuis RJ. H, Snijders AP. L, Dickman MJ, Makarova KS, Koonin EV, van der Oost J (2008) Small CRISPR RNAs guide antiviral defense in prokaryotes. Science 321:960-964, doi: 10.1126/science.1159689

Brumfield SK, Ortmann AC, Ruigrok V, Suci P, Douglas T, Young MJ (2009) Particle assembly and ultrastructural features associated with replication of the lytic archaeal virus Sulfolobus turreted icosahedral virus. J Virol 83:5964-5970, doi: 10.1128/JVI.02668-08

Cody GD (2004) Transition metal sulfides and the origins of metabolism. Annu Rev Earth Planet Sci 32:569-599, doi: 10.1146/annurev.earth.32.101802.120225

Cohan FM (2002) What are bacterial species? Annu Rev Microbiol 56:457-487, doi: 10.1146/annurev.micro.56.012302.160634

Colwell FS, D'Hondt S (2013) Nature and extent of the deep biosphere. Rev Mineral Geochem 75:547-574

Corinaldesi C, Dell'anno A, Danovaro R (2011) Viral infections stimulate the metabolism and shape prokaryotic assemblages in submarine mud volcanoes. ISME J 6:1250-1259, doi: 10.1038/ismej.2011.185

Culley AI, Lang AS, Suttle CA (2003) High diversity of unknown picorna-like viruses in the sea. Nature 424:1054-1057, doi: 10.1038/nature01886

Culley AI, Lang AS, Suttle CA (2006) Metagenomic analysis of coastal RNA virus communities. Science 312:1795-1798, doi: 10.1126/science.1127404

D'Hondt S, Spivack AJ, Pockalny R, Ferdelman TG, Fischer JP, Kallmeyer J, Abrams LJ, Smith DC, Graham D, Hasiuk F, Schrum H, Stancin AM (2009) Subseafloor sedimentary life in the South Pacific Gyre. Proc Natl Acad Sci USA 106:11651-11656, doi: 10.1073/pnas.0811793106

Danovaro R, Dell'Anno A, Corinaldesi C, Magagnini M, Noble R, Tamburini C, Weinbauer M (2008) Major viral impact on the functioning of benthic deep-sea ecosystems. Nature 454:1084-1087, doi: 10.1038/nature07268

Desnues C, Rodriguez-Brito B, Rayhawk S, Kelley S, Tran T, Haynes M, Liu H, Furlan M, Wegley L, Chau B, others, Ruan Y, Hall D, Angly FE, Edwards RA, Li L, Thurber RV, Reid RP, Siefert J, Souza V, Valentine DL, Swan BK, Breitbart M, Rohwer F (2008) Biodiversity and biogeography of phages in modern stromatolites and thrombolites. Nature 452:340–343, doi: 10.1038/nature06735

Diemer GS, Stedman KM (2012) A novel virus genome discovered in an extreme environment suggests recombination between unrelated groups of RNA and DNA viruses. Biol Direct 7:13, doi: 10.1186/1745-6150-7-13

Dinsdale EA, Edwards RA, Hall D, Angly F, Breitbart M, Brulc JM, Furlan M, Desnues C, Haynes M, Li L, McDaniel L, Moran M, Nelson K, Nilsson C, Olson R, Paul J, Brito BR, Ruan Y, Swan BK, Stevens R, Valentine DL, Thurber RV, Wegley L, White BA, Rohwer F (2008) Functional metagenomic profiling of nine biomes. Nature 452:629-632, doi: 10.1038/nature06810

Edgar RC (2010) Search and clustering orders of magnitude faster than BLAST. Bioinformatics 26:2460-1, doi: 10.1093/bioinformatics/btq461

Edwards KJ, Bach W, McCollom TM (2005) Geomicrobiology in oceanography: microbe-mineral interactions at and below the seafloor. Trends Microbiol 13:449–456

Edwards KJ, Wheat CG, Sylvan JB (2011) Under the sea: microbial life in volcanic oceanic crust. Nature Rev Microbiol 9:703-712, doi: 10.1038/nrmicro2647

Engelhardt T, Sahlberg M, Cypionka H, Engelen B (2011) Induction of prophages from deep-subseafloor bacteria. Environ Microbiol 3:459-465, doi: 10.1111/j.1758-2229.2010.00232.x

Engelhardt T, Sahlberg M, Cypionka H, Engelen B (2012) Biogeography of Rhizobium radiobacter and distribution of associated temperate phages in deep subseafloor sediments. ISME J, advance online publication, doi: 10.1038/ismej.2012.92

Eydal HSC, Jägevall S, Hermsson M, Pedersen K (2009) Bacteriophage lytic to *Desulfovibrio aespoeensis* isolated from deep groundwater. ISME J 3:1139-1147, doi: 10.1038/ismej.2009.66

Fierer N, Breitbart M, Nulton J, Salamon P, Lozupone C, Jones R, Robeson M, Edwards RA, Felts B, Rayhawk S, Knight R, Rohwer F, Jackson RB (2007) Metagenomic and small-subunit rRNA analyses reveal the genetic diversity of bacteria, archaea, fungi, and viruses in soil. Appl Environ Microb 73:7059-7066, doi: 10.1128/AEM.00358-07

Filippini M, Buesing N, Bettarel Y, Sime-Ngando T, Gessner MO (2006) Infection paradox: high abundance but low impact of freshwater benthic viruses. Appl Environ Microbiol 72:4893-4898, doi: 10.1128/AEM.00319-06

Fischer MG, Allen MJ, Wilson WH, Suttle CA (2010) Giant virus with a remarkable complement of genes infects marine zooplankton. Proc Natl Acad Sci USA 107:19508

Fisher A, Becker K (2000) Channelized fluid flow in oceanic crust reconciles heat-flow and permeability data. Nature 403:71-74, doi: 10.1038/47463

Forterre P (2006) The origin of viruses and their possible roles in major evolutionary transitions. Virus Res 117:5-16, doi: 10.1016/j.virusres.2006.01.010

Fuhrman JA (2009) Microbial community structure and its functional implications. Nature 459:193-199, doi: 10.1038/nature08058

Garneau JE, Dupuis MÈ, Villion M, Romero DA, Barrangou R, Boyaval P, Fremaux C, Horvath P, Magadán AH, Moineau S (2010) The CRISPR/Cas bacterial immune system cleaves bacteriophage and plasmid DNA. Nature 468:67-71

Hargraves RB (1986) Faster spreading or greater ridge length in the Archean? Geology 14:750, doi: 10.1130/0091-7613(1986)14<750:FSOGRL>2.0.CO;2

Häring M, Vestergaard G, Rachel R, Chen L, Garrett RA, Prangishvili D (2005) Virology: independent virus development outside a host. Nature 436:1101-1102, doi: 10.1038/4361101a

Hazen RM (2005) Genesis: The Scientific Quest for Life's Origins. Joseph Henry Press, Washington, DC

Heinen W, Lauwers AM (1996) Organic sulfur compounds resulting from the interaction of iron sulfide, hydrogen sulfide and carbon dioxide in an anaerobic aqueous environment. Origins Life Evol Biosph 26:131-150, doi: 10.1007/BF01809852

Held NL, Whitaker RJ (2009) Viral biogeography revealed by signatures in Sulfolobus islandicus genomes. Environ Microbiol 11:457-466, doi: 10.1111/j.1462-2920.2008.01784.x

Horvath P, Barrangou R, Hovarth P (2010) CRISPR/Cas, the Immune System of Bacteria and Archaea. Science 327:167-170, doi: 10.1126/science.1179555

Jiang SC, Paul JH (1998) Gene transfer by transduction in the marine environment. Appl Environ Microbiol 64:2780-2787

Jiao N, Herndl GJ, Hansell DA, Benner R, Kattner G, Wilhelm SW, Kirchman DL, Weinbauer MG, Luo T, Chen F, Azam F (2010) Microbial production of recalcitrant dissolved organic matter: long-term carbon storage in the global ocean. Nature Rev Microbiol 8:593-599, doi: 10.1038/nrmicro2386

Johnson HP, Pruis MJ (2003) Fluxes of fluid and heat from the oceanic crustal reservoir. Earth Planet Sci Lett 216:565-574, doi: 10.1016/S0012-821X(03)00545-4

Jore MM, Lundgren M, van Duijn E, Bultema JB, Westra ER, Waghmare SP, Wiedenheft B, Pul Ü, Wurm R, Wagner R, Beijer MR, Barendregt A, Zhou K, Snijders APL, Dickman MJ, Doudna JA, Boekema EJ, Heck AJR, van der Oost J, Brouns SJJ, Pul Ü, Wurm R, Wagner R, Beijer MR, Barendregt A, Zhou K, Snigders APL, Dickman MG, Doudna JA, Boekma EJ, Heck AJR, van der Oost J, Brouns SJJ (2011) Structural basis for CRISPR RNA-guided DNA recognition by Cascade. Nature Struct Mol Biol 18:529-536, doi: 10.1038/nsmb.2019

Jørgensen BB, D'Hondt S (2006) Ecology. A starving majority deep beneath the seafloor. Science 314:932-934, doi: 10.1126/science.1133796

Kelley DS, Karson JA, Blackman DK, Früh-Green GL, Butterfield DA, Lilley MD, Olson EJ, Schrenk MO, Roe KK, Lebon GT, Rivizzigno P (2001) An off-axis hydrothermal vent field near the Mid-Atlantic Ridge at 30 degrees N. Nature 412:145-149, doi: 10.1038/35084000

Koonin EV, Martin W (2005) On the origin of genomes and cells within inorganic compartments. Trends Genet 21:647-654, doi: 10.1016/j.tig.2005.09.006

Koonin EV, Senkevich TG, Dolja VV (2006) The ancient Virus World and evolution of cells. Biol Direct 1:29, doi: 10.1186/1745-6150-1-29

Kristensen DM, Mushegian AR, Dolja VV, Koonin EV (2009) New dimensions of the virus world discovered through metagenomics. Trends Microbiol 18:11-19

Krupovic M, Prangishvili D, Hendrix RW, Bamford DH (2011) Genomics of bacterial and archaeal viruses: Dynamics within the prokaryotic virosphere. Microbiol Mol Biol Rev 75:610-635, doi: 10.1128/MMBR.00011-11

Kyle JE, Eydal HSC, Ferris FG, Pedersen K (2008) Viruses in granitic groundwater from 69 to 450m depth of the Äspö hard rock laboratory, Sweden. ISME J 2:571-574, doi: 10.1038/ismej.2008.18

La Scola B, Audic S, Robert C, Jungang L, de Lamballerie X, Drancourt M, Birtles R, Claverie J-M, Raoult D (2003) A giant virus in amoebae. Science 299:2033, doi: 10.1126/science.1081867

La Scola B, Desnues C, Pagnier I, Robert C, Barrassi L, Fournous G, Merchat M, Suzan-Monti M, Forterre P, Koonin E, Raoult D (2008) The virophage as a unique parasite of the giant mimivirus. Nature 455:100-104, doi: 10.1038/nature07218

Labrie SJ, Samson JE, Moineau S (2010) Bacteriophage resistance mechanisms. Nature Rev Microbiol 8:317-327, doi: 10.1038/nrmicro2315

Lane N, Allen JF, Martin W (2010) How did LUCA make a living? Chemiosmosis in the origin of life. BioEssays 32:271-280, doi: 10.1002/bies.200900131

Lang AS, Beatty JT (2000) Genetic analysis of a bacterial genetic exchange element: The gene transfer agent of *Rhodobacter capsulatus*. Proc Natl Acad Sci USA 97:859-864, doi: 10.1073/pnas.97.2.859

Lang SQ, Butterfield DA, Schulte M, Kelley DS, Lilley MD (2010) Elevated concentrations of formate, acetate and dissolved organic carbon found at the Lost City hydrothermal field. Geocim Cosmochim Act 74:941-952, doi: 10.1016/j.gca.2009.10.045

Li W, Godzik A (2006) cd-hit: a fast program for clustering and comparing large sets of protein or nucleotide sequences. Bioinformatics 22:1658-1659, doi: 10.1093/bioinformatics/btl158

Lindell D, Jaffe JD, Johnson ZI, Church GM, Chisholm SW (2005) Photosynthesis genes in marine viruses yield proteins during host infection. Nature 438:86-89, http://dx.doi.org/10.1038/nature04111

Lopez-Bueno A, Tamames J, Velazquez D, Moya A, Quesada A, Alcami A (2009) High diversity of the viral community from an Antarctic lake. Science 326:858

Lorenz MG, Wackernagel W (1994) Bacterial gene transfer by natural genetic transformation in the environment. Microbiol Mol Biol Rev 58:563-602

Mann NH, Cook A, Millard A, Bailey S, Clokie M (1993) Bacterial photosynthesis genes in a virus. Environ Microbiol 59:3736–3743

Marraffini LA, Sontheimer EJ (2010) Self versus non-self discrimination during CRISPR RNA-directed immunity. Nature 463:568-571, doi: 10.1038/nature08703

Martin W, Baross JA, Kelley D, Russell MJ (2008) Hydrothermal vents and the origin of life. Nature Rev Microbiol 6:805-814, doi: 10.1038/nrmicro1991

Martin W, Russell MJ (2007) On the origin of biochemistry at an alkaline hydrothermal vent. Philos Trans R Soc London Ser B 362:1887-1925, doi: 10.1098/rstb.2006.1881

Matson EG, Thompson MG, Humphrey SB, Zuerner RL, Stanton TB (2005) Identification of genes of VSH-1, a prophage-like gene transfer agent of *Brachyspira hyodysenteriae*. J Bacteriol 187:5885-5892, doi: 10.1128/JB.187.17.5885-5892.2005

Maynard-Smith J, Szathmáry E (1999) The Origins of Life: From the Birth of Life to the Origin of Language. Oxford University Press, New York

McDaniel LD, Young E, Delaney J, Ruhnau F, Ritchie KB, Paul JH (2010) High frequency of horizontal gene transfer in the oceans. Science 330:50

Meyer F, Paarmann D, D'Souza M, Olson R, Glass EM, Kubal M, Paczian T, Rodriguez A, Stevens R, Wilke A, Wilkening J, Edwards RA (2008) The metagenomics RAST server - a public resource for the automatic phylogenetic and functional analysis of metagenomes. BMC Bioinformatics 9:386, doi: 10.1186/1471-2105-9-386

Middelboe M (2000) Bacterial growth rate and marine virus-host dynamics. Microbial Ecol 40:114-124, doi: 10.1007/s002480000050

Middelboe M, Glud RN, Wenzhöfer F, Oguri K, Kitazato H (2006) Spatial distribution and activity of viruses in the deep-sea sediments of Sagami Bay, Japan. Deep-Sea Res Part 1 53:1-13, doi: 16/j.dsr.2005.09.008

Molin S, Tolker-Nielsen T (2003) Gene transfer occurs with enhanced efficiency in biofilms and induces enhanced stabilisation of the biofilm structure. Curr Opin Biotech 14:255-261, doi: 10.1016/S0958-1669(03)00036-3

O'Malley MA (2007) The nineteenth century roots of "everything is everywhere". Nature Rev Microbiol 5:647-651, doi:10.1038/nrmicro1711

Ortmann AC, Suttle CA (2005) High abundances of viruses in a deep-sea hydrothermal vent system indicates viral mediated microbial mortality. Deep-Sea Res Part 1 52:1515–1527

Paul JH (2008) Prophages in marine bacteria: dangerous molecular time bombs or the key to survival in the seas? ISME J 2:579-589

Pietilä MK, Roine E, Paulin L, Kalkkinen N, Bamford DH (2009) An ssDNA virus infecting archaea: a new lineage of viruses with a membrane envelope. Mol Microbiol 72:307-319, doi:10.1111/j.1365-2958.2009.06642.x

Pina M, Bize A, Forterre P, Prangishvili D (2011) The archaeoviruses. FEMS Microbiol Rev 35:1035-1054, doi: 10.1111/j.1574-6976.2011.00280.x

Poole AM (2009) Horizontal gene transfer and the earliest stages of the evolution of life. Res Microbiol 160:473-480, doi: 10.1016/j.resmic.2009.07.009

Prangishvili D, Forterre P, Garrett RA (2006) Viruses of the Archaea: a unifying view. Nature Rev Microbiol 4:837-848

Proskurowski G, Lilley MD, Seewald JS, Früh-Green GL, Olson EJ, Lupton JE, Sylva SP, Kelley DS (2008) Abiogenic hydrocarbon production at lost city hydrothermal field. Science 319:604-607, doi: 10.1126/science.1151194

Raoult D, Audic S, Robert C, Abergel C, Renesto P, Ogata H, La Scola B, Suzan M, Claverie J-M (2004) The 1.2-megabase genome sequence of Mimivirus. Science 306:1344-1350, doi: 10.1126/science.1101485

Resch A, Fehrenbacher B, Eisele K, Schaller M, Götz F (2005) Phage release from biofilm and planktonic *Staphylococcus aureus* cells. FEMS Microbiol Lett 252:89-96, doi: 10.1016/j.femsle.2005.08.048

Reysenbach A-L, Shock E (2002) Merging genomes with geochemistry in hydrothermal systems. Science 296:1077-1082, doi: 10.1126/science.1072483

Rodriguez-Valera F, Martin-Cuadrado AB, Rodriguez-Brito B, Pasic L, Thingstad TF, Rohwer F, Mira A (2009) Explaining microbial population genomics through phage predation. Nature Rev Microbiol 7:828-836, doi: 10.1038/nrmicro2235

Rosario K, Breitbart M (2011) Exploring the viral world through metagenomics. Curr Opin Virol 1:289-297, doi: 10.1016/j.coviro.2011.06.004

Røy H, Kallmeyer J, Adhikari RR, Pockalny R, Jørgensen BB, D'Hondt S (2012) Aerobic microbial respiration in 86-milion-year-old deep-sea red clay. Science 336:922-925, doi: 10.1126/science.1219424

Russell MJ, Hall AJ (1997) The emergence of life from iron monosulphide bubbles at a submarine hydrothermal redox and pH front. J Geol Soc London 154:377-402, doi: 10.1144/gsjgs.154.3.0377

Schloss PD, Westcott SL, Ryabin T, Hall JR, Hartmann M, Hollister EB, Lesniewski RA, Oakley BB, Parks DH, Robinson CJ, Sahl JW, Stres B, Thallinger GG, Van Horn DJ, Weber CF (2009) Introducing mothur: open-source, platform-independent, community-supported software for describing and comparing microbial communities. Appl Environ Microbiol 75:7537-7541, doi: 10.1128/AEM.01541-09

Schrenk MO, Kelley DS, Delaney JR, Baross JA (2003) Incidence and diversity of microorganisms within the walls of an active deep-sea sulfide chimney. Appl Environ Microbiol 69:3580-3592, doi: 10.1128/AEM.69.6.3580-3592.2003

Schrenk MO, Brazelton WJ, Lang SQ (2013) Serpentinization, carbon, and deep life. Rev Mineral Geochem 75:575-606

Skraber S, Schiven J, Gantzer C, de Roda Husman AM (2005) Pathogenic viruses in drinking-water biofilms: A public health risk? Biofilms 2:105-117

Sorek R, Kunin V, Hugenholtz P (2008) CRISPR--a widespread system that provides acquired resistance against phages in bacteria and archaea. Nature Rev Microbiol 6:181-186, doi: 10.1038/nrmicro1793

Stanton TB (2007) Prophage-like gene transfer agents–Novel mechanisms of gene exchange for *Methanococcus*, *Desulfovibrio*, *Brachyspira*, and *Rhodobacter* species. Anaerobe 13:43-49

Steward GF, Montiel JL, Azam F Genome size distributions indicate variability and similarities among marine viral assemblages from diverse environments. Linmol Oceanogr 45:1697-1706

Stoodley P, Sauer K, Davies DG, Costerton JW (2002) Biofilms as complex differentiated communities. Annu Rev Microbiol 56:187-209, doi: 10.1146/annurev.micro.56.012302.160705

Suttle CA (2005) Viruses in the sea. Nature 437:356–361, doi: 10.1038/nature04160

Suttle CA (2007) Marine viruses--major players in the global ecosystem. Nature Rev Microbiol 5:801-812, doi: 10.1038/nrmicro1750

Thingstad T, Lignell R (1997) Theoretical models for the control of bacterial growth rate, abundance, diversity and carbon demand. Aquat Microb Ecol 13:19-27, doi: 10.3354/ame013019

Thoulouze M-I, Alcover A (2011) Can viruses form biofilms? Trends Microbiol 19:257-262, doi: 10.1016/j.tim.2011.03.002

Thurber RV (2009) Current insights into phage biodiversity and biogeography. Curr Opin Microbiol 12:582-587, doi: 10.1016/j.mib.2009.08.008

Tyson GW, Banfield JF (2008) Rapidly evolving CRISPRs implicated in acquired resistance of microorganisms to viruses. Environ Microbiol 10:200-207, doi: 10.1111/j.1462-2920.2007.01444.x

van der Oost J, Jore MM, Westra ER, Lundgren M, Brouns SJJ (2009) CRISPR-based adaptive and heritable immunity in prokaryotes. Trends Biochem Sci 34:401-407, doi: 10.1016/j.tibs.2009.05.002

Vidgen M, Carson J, Higgins M, Owens L (2006) Changes to the phenotypic profile of *Vibrio harveyi* when infected with the *Vibrio harveyi* myovirus-like (VHML) bacteriophage. J Appl Microbiol 100:481-487, doi: 10.1111/j.1365-2672.2005.02829.x

Von Damm KL (1990) Seafloor hydrothermal activity: Black smoker chemistry and chimneys. Annu Rev Earth Planet Sci 18:173-204, doi: 10.1146/annurev.ea.18.050190.001133

Wächtershäuser G (1988a) Pyrite Formation, the first energy source for life: a hypothesis. Syst Appl Microbiol 10:207-210, doi: 10.1016/S0723-2020(88)80001-8

Wächtershäuser G (1988b) Before enzymes and templates: theory of surface metabolism. Microbiol Rev 52:452-484

Wächtershäuser G (1990) Evolution of the first metabolic cycles. Proc Natl Acad Sci USA 87:200-204, doi: 10.1073/pnas.87.1.200

Waldor MK, Mekalanos JJ (1996) Lysogenic conversion by a filamentous phage encoding cholera toxin. Science 272:1910-1914, doi: 10.1126/science.272.5270.1910

Webb JS, Givskov M, Kjelleberg S (2003) Bacterial biofilms: prokaryotic adventures in multicellularity. Curr Opin Microbiol 6:578-585, doi: 10.1016/j.mib.2003.10.014

Webb JS, Lau M, Kjelleberg S (2004) Bacteriophage and phenotypic variation in Pseudomonas aeruginosa biofilm development. J Bacteriol 186:8066-8073, doi: 10.1128/JB.186.23.8066-8073.2004

Weinbauer MG (2004) Ecology of prokaryotic viruses. FEMS Microbiol Rev 28:127-181, doi: 10.1016/j.femsre.2003.08.001

Whiteley M, Bangera MG, Bumgarner RE, Parsek MR, Teitzel GM, Lory S, Greenberg EP (2001) Gene expression in *Pseudomonas aeruginosa* biofilms. Nature 413:860-864, doi: 10.1038/35101627.

Williamson SJ, Cary SC, Williamson KE, Helton RR, Bench SR, Winget D, Wommack KE (2008a) Lysogenic virus–host interactions predominate at deep-sea diffuse-flow hydrothermal vents. ISME J 2:1112–1121

Williamson SJ, Rusch DB, Yooseph S, Halpern AL, Heidelberg KB, Glass JI, Andrews-Pfannkoch C, Fadrosh D, Miller CS, Sutton G, Frazier M, Venter JC (2008b) The Sorcerer II Global Ocean Sampling Expedition: metagenomic characterization of viruses within aquatic microbial samples. PloS One 3:e1456, doi: 10.1371/journal.pone.0001456

Woese C (1998) The universal ancestor. Proc Natl Acad Sci USA 95:6854-6859, doi: 10.1073/pnas.95.12.6854

Woese CR (2002) On the evolution of cells. Proc Natl Acad Sci USA 99:8742-8747, doi: 10.1073/pnas.132266999

Ziebuhr W, Krimmer V, Rachid S, Lossner I, Gotz F, Hacker J (1999) A novel mechanism of phase variation of virulence in *Staphylococcus epidermidis*: evidence for control of the polysaccharide intercellular adhesin synthesis by alternating insertion and excision of the insertion sequence element IS256. Mol Microbiol 32:345-356, doi: 10.1046/j.1365-2958.1999.01353.x

INDEX

Note: Page numbers followed by "f" and "t" indicate figures and tables.

1529-6466/13/0060-0ind$00.00 DOI: 10.2138/rmg.2013.75.ind